Encyclopedia of VIROLOGY

THIRD EDITION

EDITORS-IN-CHIEF

Dr BRIAN W J MAHY
and
Dr MARC H V VAN REGENMORTEL

Amsterdam • Boston • Heidelberg • London • New York • Oxford
Paris • San Diego • San Francisco • Singapore • Sydney • Tokyo
Academic Press is an imprint of Elsevier

ELSEVIER

ACADEMIC PRESS

Academic Press is an imprint of Elsevier
Linacre House, Jordan Hill, Oxford, OX2 8DP, UK
525 B Street, Suite 1900, San Diego, CA 92101-4495, USA

Copyright © 2008 Elsevier Inc. All rights reserved

The following articles are US government works in the public domain and are not subject to copyright:
Bovine Viral Diarrhea Virus, Coxsackieviruses, Prions of Yeast and Fungi, Human Respiratory Syncytial Virus, Fish Rhabdoviruses, Varicella-Zoster Virus: General Features, Viruses and Bioterrorism, Bean Common Mosaic Virus and Bean Common Mosaic Necrosis Virus, Metaviruses, Crimean-Congo Hemorrhagic Fever Virus and Other Nairoviruses, AIDS: Global Epidemiology, Papaya Ringspot Virus, Transcriptional Regulation in Bacteriophage.

Nepovirus, Canadian Crown Copyright 2008

No part of this publication may be reproduced, stored in a retrieval system or transmitted in any form or by any means electronic, mechanical, photocopying, recording or otherwise without the prior written permission of the publisher

Permissions may be sought directly from Elsevier's Science & Technology Rights Department in Oxford, UK: phone (+44) (0) 1865 843830; fax (+44) (0) 1865 853333; email: permissions@elsevier.com. Alternatively you can submit your request online by visiting the Elsevier web site at (http://elsevier.com/locate/permission), and selecting *Obtaining permission to use Elsevier material*

Notice
No responsibility is assumed by the publisher for any injury and/or damage to persons or property as a matter of products liability, negligence or otherwise, or from any use or operation of any methods, products, instructions or ideas contained in the material herein. Because of rapid advances in the medical sciences, in particular, independent verification of diagnoses and drug dosages should be made

British Library Cataloguing in Publication Data
A catalogue record for this book is available from the British Library

Library of Congress Catalog Number: 200892260

ISBN: 978-0-12-373935-3

For information on all Elsevier publications
visit our website at books.elsevier.com

PRINTED AND BOUND IN SLOVENIA
08 09 10 11   10 9 8 7 6 5 4 3 2 1

Working together to grow
libraries in developing countries

www.elsevier.com | www.bookaid.org | www.sabre.org

ELSEVIER   BOOK AID International   Sabre Foundation

Encyclopedia of VIROLOGY
THIRD EDITION

EDITORS-IN-CHIEF

Brian W J Mahy MA PhD ScD DSc
Senior Scientific Advisor,
Division of Emerging Infections and Surveillance Services,
Centers for Disease Control and Prevention,
Atlanta GA, USA

Marc H V Van Regenmortel PhD
Emeritus Director at the CNRS,
French National Center for Scientific Research,
Biotechnology School of the University of Strasbourg,
Illkirch, France

ASSOCIATE EDITORS

Dennis H Bamford, Ph.D.
Department of Biological and Environmental Sciences
and Institute of Biotechnology, Biocenter 2,
P.O. Box 56 (Viikinkaari 5),
00014 University of Helsinki,
Finland

Charles Calisher, B.S., M.S., Ph.D.
Arthropod-borne and Infectious Diseases Laboratory
Department of Microbiology, Immunology and Pathology
College of Veterinary Medicine and Biomedical Sciences
Colorado State University
Fort Collins
CO 80523
USA

Andrew J Davison, M.A., Ph.D.
MRC Virology Unit
Institute of Virology
University of Glasgow
Church Street
Glasgow G11 5JR
UK

Claude Fauquet
ILTAB/Donald Danforth Plant Science Center
975 North Warson Road
St. Louis, MO 63132

Said Ghabrial, B.S., M.S., Ph.D.
Plant Pathology Department
University of Kentucky
201F Plant Science Building
1405 Veterans Drive
Lexington
KY 4050546-0312
USA

Eric Hunter, B.Sc., Ph.D.
Department of Pathology and Laboratory Medicine, and
Emory Vaccine Center
Emory University
954 Gatewood Road NE
Atlanta Georgia 30329
USA

Robert A. Lamb, Ph.D., Sc.D.
Department of Biochemistry,
Molecular Biology and Cell Biology
Howard Hughes Medical Institute
Northwestern University
2205 Tech Dr.
Evanston
IL 60208-3500
USA

Olivier Le Gall
IPV, UMR GDPP, IBVM,
INRA Bordeaux-Aquitaine, BP 81,
F-33883 Villenave d'Ornon Cedex
FRANCE

Vincent Racaniello, Ph.D.
Department of Microbiology
Columbia University
New York, NY 10032
USA

David A. Theilmann, Ph.D., B.Sc., M.Sc
Pacific Agri-Food Research Centre
Agriculture and Agri-Food Canada
Box 5000, 4200 Highway 97
Summerland
BC V0H 1Z0
Canada

H. Josef Vetten, Ph.D.
Julius Kuehn Institute, Federal Research Centre for
Cultivated Plants (JKI)
Messeweg 11-12
38104 Braunschweig
Germany

Peter J Walker, B.Sc., Ph.D.
CSIRO Livestock Industries
Australian Animal Health Laboratory (AAHL)
Private Bag 24
Geelong
VIC 3220
Australia

PREFACE

This third edition of the *Encyclopedia of Virology* is being published nine years after the second edition, a period which has seen enormous growth both in our understanding of virology and in our recognition of the viruses themselves, many of which were unknown when the second edition was prepared. Considering viruses affecting human hosts alone, the worldwide epidemic of severe acute respiratory syndrome (SARS), caused by a previously unknown coronavirus, led to the discovery of other human coronaviruses such as HKU1 and NL63. As many as seven chapters are devoted to the AIDS epidemic and to human immunodeficiency viruses. In addition, the development of new molecular technologies led to the discovery of viruses with no obvious disease associations, such as torque-teno virus (one of the most ubiquitous viruses in the human population), human bocavirus, human metapneumovirus, and three new human polyomaviruses.

Other new developments of importance to human virology have included the introduction of a virulent strain of West Nile virus from Israel to North America in 1999. Since that time the virus has become established in mosquito, bird and horse populations throughout the USA, the Caribbean and Mexico as well as the southern regions of Canada.

As in the two previous editions, we have tried to include information about all known species of virus infecting bacteria, fungi, invertebrates, plants and vertebrates, as well as descriptions of related topics in virology such as antiviral drug development, cell- and antibody-mediated immunity, vaccine development, electron microscopy and molecular methods for virus characterization and identification. Many chapters are devoted to the considerable economic importance of virus diseases of cereals, legumes, vegetable crops, fruit trees and ornamentals, and new approaches to control these diseases are reviewed.

General issues such as the origin, evolution and phylogeny of viruses are also discussed as well as the history of the different groups of viruses.

To cover all these subjects and new developments, we have had to increase the size of the Encyclopedia from three to five volumes.

Throughout this work we have relied upon the 8th Report of the International Committee on Taxonomy of Viruses published in 2005, which lists more than 6000 viruses classified into some 2000 virus species distributed among more than 390 different genera and families. In recent years the criteria for placing viruses in different taxa have shifted away from traditional serological methods and increasingly rely upon molecular techniques, particularly the nucleotide sequence of the virus genome. This has changed many of the previous groupings of viruses, and is reflected in this third edition.

Needless to say, a work of this magnitude has involved many expert scientists, who have given generously of their time to bring it to fruition. We extend our grateful thanks to all contributors and associate editors for their excellent and timely contributions.

Brian W J Mahy
Marc H V van Regenmortel

HOW TO USE THE ENCYCLOPEDIA

Structure of the Encyclopedia

The major topics discussed in detail in the text are presented in alphabetical order (see the Alphabetical Contents list which appears in all five volumes).

Finding Specific Information

Information on specific viruses, virus diseases and other matters can be located by consulting the General Index at the end of Volume 5.

Taxonomic Groups of Viruses

For locating detailed information on the major taxonomic groups of viruses, namely virus genera, families and orders, the Taxonomic Index in Volume 5 (page...) should be consulted.

Further Reading sections

The articles do not feature bibliographic citations within the body of the article text itself. The articles are intended to be a first introduction to the topic, or a 'refresher', readable from beginning to end without referring the reader outside of the encyclopedia itself. Bibliographic references to external literature are grouped at the end of each article in a Further Reading section, containing review articles, 'seminal' primary articles and book chapters. These point users to the next level of information for any given topic.

Cross referencing between articles

The "See also" section at the end of each article directs the reader to other entries on related topics. For example. The entry *Lassa, Junin, Machupo and Guanarito Viruses* includes the following cross-references:

See also: Lymphocytic Choriomeningitis Virus: General Features.

CONTRIBUTORS

S T Abedon
The Ohio State University, Mansfield, OH, USA

G P Accotto
Istituto di Virologia Vegetale CNR, Torino, Italy

H-W Ackermann
Laval University, Quebec, QC, Canada

G Adam
Universität Hamburg, Hamburg, Germany

M J Adams
Rothamsted Research, Harpenden, UK

C Adams
University of Duisburg–Essen, Essen, Germany

E Adderson
St. Jude Children's Research Hospital, Memphis, TN, USA

S Adhya
National Institutes of Health, Bethesda, MD, USA

C L Afonso
Southeast Poultry Research Laboratory, Athens, GA, USA

P Ahlquist
University of Wisconsin – Madison, Madison, WI, USA

G M Air
University of Oklahoma Health Sciences Center, Oklahoma City, OK, USA

D J Alcendor
Johns Hopkins School of Medicine, Baltimore, MD, USA

J W Almond
sanofi pasteur, Lyon, France

I Amin
National Institute for Biotechnology and Genetic Engineering, Faisalabad, Pakistan

J Angel
Pontificia Universidad Javeriana, Bogota, Republic of Colombia

C Apetrei
Tulane National Primate Research Center, Covington, LA, USA

B M Arif
Great Lakes Forestry Centre, Sault Ste. Marie, ON, Canada

H Attoui
Faculté de Médecine de Marseilles, Etablissement Français Du Sang, Marseilles, France

H Attoui
Université de la Méditerranée, Marseille, France

H Attoui
Institute for Animal Health, Pirbright, UK

L Aurelian
University of Maryland School of Medicine, Baltimore, MD, USA

L A Babiuk
University of Alberta, Edmonton, AB, Canada

S Babiuk
National Centre for Foreign Animal Disease, Winnipeg, MB, Canada

A G Bader
The Scripps Research Institute, La Jolla, CA, USA

S C Baker
Loyola University of Chicago, Maywood, IL, USA

T S Baker
University of California, San Diego, La Jolla, CA, USA

J K H Bamford
University of Jyväskylä, Jyväskylä, Finland

Y Bao
National Institutes of Health, Bethesda, MD, USA

M Bar-Joseph
The Volcani Center, Bet Dagan, Israel

H Barker
Scottish Crop Research Institute, Dundee, UK

A D T Barrett
University of Texas Medical Branch, Galveston, TX, USA

J W Barrett
The University of Western Ontario, London, ON, Canada

T Barrett
Institute for Animal Health, Pirbright, UK

R Bartenschlager
University of Heidelberg, Heidelberg, Germany

N W Bartlett
Imperial College London, London, UK

S Basak
University of California, San Diego, CA, USA

C F Basler
Mount Sinai School of Medicine, New York, NY, USA

T Basta
Institut Pasteur, Paris, France

D Baxby
University of Liverpool, Liverpool, UK

P Beard
Imperial College London, London, UK

M N Becker
University of Florida, Gainesville, FL, USA

J J Becnel
Agriculture Research Service, Gainesville, FL, USA

K L Beemon
Johns Hopkins University, Baltimore, MD, USA

E D Belay
Centers for Disease Control and Prevention, Atlanta, GA, USA

M Benkő
Veterinary Medical Research Institute, Hungarian Academy of Sciences, Budapest, Hungary

M Bennett
University of Liverpool, Liverpool, UK

M Bergoin
Université Montpellier II, Montpellier, France

H U Bernard
University of California, Irvine, Irvine, CA, USA

K I Berns
University of Florida College of Medicine, Gainesville, FL, USA

P Biagini
Etablissement Français du Sang Alpes-Méditerranée, Marseilles, France

P D Bieniasz
Aaron Diamond AIDS Research Center, The Rockefeller University, New York, NY, USA

Y Bigot
University of Tours, Tours, France

C Billinis
University of Thessaly, Karditsa, Greece

R F Bishop
Murdoch Childrens Research Institute Royal Children's Hospital, Melbourne, VIC, Australia

B A Blacklaws
University of Cambridge, Cambridge, UK

C D Blair
Colorado State University, Fort Collins, CO, USA

S Blanc
INRA–CIRAD–AgroM, Montpellier, France

R Blawid
Institute of Plant Diseases and Plant Protection, Hannover, Germany

G W Blissard
Boyce Thompson Institute at Cornell University, Ithaca, NY, USA

S Blomqvist
National Public Health Institute (KTL), Helsinki, Finland

J F Bol
Leiden University, Leiden, The Netherlands

J-R Bonami
CNRS, Montpellier, France

L Bos
Wageningen University and Research Centre (WUR), Wageningen, The Netherlands

H R Bose Jr.
University of Texas at Austin, Austin, TX, USA

H Bourhy
Institut Pasteur, Paris, France

P R Bowser
Cornell University, Ithaca, NY, USA

D B Boyle
CSIRO Livestock Industries, Geelong, VIC, Australia

C Bragard
Université Catholique de Louvain, Leuven, Belgium

J N Bragg
University of California, Berkeley, Berkeley, CA, USA

R W Briddon
National Institute for Biotechnology and Genetic Engineering, Faisalabad, Pakistan

M A Brinton
Georgia State University, Atlanta, GA, USA

P Britton
Institute for Animal Health, Compton, UK

J K Brown
The University of Arizona, Tucson, AZ, USA

K S Brown
University of Manitoba, Winnipeg, MB, Canada

J Bruenn
State University of New York, Buffalo, NY, USA

C P D Brussaard
Royal Netherlands Institute for Sea Research, Texel, The Netherlands

J J Bugert
Wales College of Medicine, Heath Park, Cardiff, UK

J J Bujarski
Northern Illinois University, DeKalb, IL, USA and Polish Academy of Sciences, Poznan, Poland

R M Buller
Saint Louis University School of Medicine, St. Louis, MO, USA

J P Burand
University of Massachusetts at Amherst, Amherst, MA, USA

J Burgyan
Agricultural Biotechnology Center, Godollo, Hungary

F J Burt
University of the Free State, Bloemfontein, South Africa

S J Butcher
University of Helsinki, Helsinki, Finland

J S Butel
Baylor College of Medicine, Houston, TX, USA

M I Butler
University of Otago, Dunedin, New Zealand

S Bühler
University of Heidelberg, Heidelberg, Germany

P Caciagli
Istituto di Virologia Vegetale – CNR, Turin, Italy

C H Calisher
Colorado State University, Fort Collins, CO, USA

T Candresse
UMR GDPP, Centre INRA de Bordeaux, Villenave d'Ornon, France

A J Cann
University of Leicester, Leicester, UK

C Caranta
INRA, Montfavet, France

G Carlile
CSIRO Livestock Industries, Geelong, VIC, Australia

J P Carr
University of Cambridge, Cambridge, UK

R Carrion, Jr.
Southwest Foundation for Biomedical Research, San Antonio, TX, USA

J W Casey
Cornell University, Ithaca, NY, USA

R N Casey
Cornell University, Ithaca, NY, USA

S Casjens
University of Utah School of Medicine, Salt Lake City, UT, USA

R Cattaneo
Mayo Clinic College of Medicine, Rochester, MN, USA

D Cavanagh
Institute for Animal Health, Compton, UK

A Chahroudi
University of Pennsylvania School of Medicine, Philadelphia, PA, USA

S Chakraborty
Jawaharlal Nehru University, New Delhi, India

T J Chambers
Saint Louis University School of Medicine, St. Louis, MO, USA

Y Chang
University of Pittsburgh Cancer Institute, Pittsburgh, PA, USA

J T Chang
Baylor College of Medicine, Houston, TX, USA

D Chapman
Institute for Animal Health, Pirbright, UK

D Chattopadhyay
University of Calcutta, Kolkata, India

M Chen
University of Arizona, Tucson, AZ, USA

J E Cherwa
University of Arizona, Tucson, AZ, USA

V G Chinchar
University of Mississippi Medical Center, Jackson, MS, USA

A V Chintakuntlawar
University of Oklahoma Health Sciences Center, Oklahoma City, OK, USA

W Chiu
Baylor College of Medicine, Houston, TX, USA

J Chodosh
University of Oklahoma Health Sciences Center, Oklahoma City, OK, USA

I-R Choi
International Rice Research Institute, Los Baños, The Philippines

P D Christian
National Institute of Biological Standards and Control, South Mimms, UK

M G Ciufolini
Istituto Superiore di Sanità, Rome, Italy

P Clarke
University of Colorado Health Sciences, Denver, CO, USA

J-M Claverie
Université de la Méditerranée, Marseille, France

J R Clayton
Johns Hopkins University Schools of Public Health and Medicine, Baltimore, MD, USA

R J Clem
Kansas State University, Manhattan, KS, USA

C J Clements
The Macfarlane Burnet Institute for Medical Research and Public Health Ltd., Melbourne, VIC, Australia

L L Coffey,
Institut Pasteur, Paris, France

J I Cohen
National Institutes of Health, Bethesda, MD, USA

J Collinge
University College London, London, UK

P L Collins
National Institute of Allergy and Infectious Diseases, Bethesda, MD, USA

A Collins
University of Wisconsin School of Medicine and Public Health, Madison, WI, USA

D Contamine
Université Versailles St-Quentin, CNRS, Versailles, France

K M Coombs
University of Manitoba, Winnipeg, MB, Canada

J A Cowley
CSIRO Livestock Industries, Brisbane, QLD, Australia

J K Craigo
University of Pittsburgh School of Medicine, Pittsburgh, PA, USA

M St. J Crane
CSIRO Livestock Industries, Geelong, VIC, Australia

J E Crowe, Jr.
Vanderbilt University Medical Center, Nashville, TN, USA

H Czosnek
The Hebrew University of Jerusalem, Rehovot, Israel

T Dalmay
University of East Anglia, Norwich, UK

B H Dannevig
National Veterinary Institute, Oslo, Norway

C J D'Arcy
University of Illinois at Urbana-Champaign, Urbana, IL, USA

A J Davison
MRC Virology Unit, Glasgow, UK

W O Dawson
University of Florida, Lake Alfred, FL, USA

L A Day
The Public Health Research Institute, Newark, NJ, USA

J C de la Torre
The Scripps Research Institute, La Jolla, CA, USA

X de Lamballerie
Faculté de Médecine de Marseille, Marseilles, France

M de Vega
Universidad Autónoma, Madrid, Spain

P Delfosse
Centre de Recherche Public-Gabriel Lippmann, Belvaux, Luxembourg

B Delmas
INRA, Jouy-en-Josas, France

M Deng
University of California, Berkeley, CA, USA

J DeRisi
University of California, San Francisco, San Francisco, CA, USA

C Desbiez
Institut National de la Recherche Agronomique (INRA), Station de Pathologie Végétale, Montfavet, France

R C Desrosiers
New England Primate Research Center, Southborough, MA, USA

A K Dhar
Advanced BioNutrition Corp, Columbia, MD, USA

R G Dietzgen
The University of Queensland, St. Lucia, QLD, Australia

S P Dinesh-Kumar
Yale University, New Haven, CT, USA

L K Dixon
Institute for Animal Health, Pirbright, UK

C Dogimont
INRA, Montfavet, France

A Domanska
University of Helsinki, Helsinki, Finland

L L Domier
USDA–ARS, Urbana, IL, USA

L L Domier
USDA-ARS, Urbana-Champaign, IL, USA

A Dotzauer
University of Bremen, Bremen, Germany

T W Dreher
Oregon State University, Corvallis, OR, USA

S Dreschers
University of Duisburg–Essen, Essen, Germany

R L Duda
University of Pittsburgh, Pittsburgh, PA, USA

J P Dudley
The University of Texas at Austin, Austin, TX, USA

W P Duprex
The Queen's University of Belfast, Belfast, UK

R E Dutch
University of Kentucky, Lexington, KY, USA

B M Dutia
University of Edinburgh, Edinburgh, UK

M L Dyall-Smith
The University of Melbourne, Parkville, VIC, Australia

J East
University of Texas Medical Branch – Galveston, Galveston, TX, USA

A J Easton
University of Warwick, Coventry, UK

K C Eastwell
Washington State University – IAREC, Prosser, WA, USA

B T Eaton
Australian Animal Health Laboratory, Geelong, VIC, Australia

H Edskes
National Institutes of Health, Bethesda, MD, USA

B Ehlers
Robert Koch-Institut, Berlin, Germany

R M Elliott
University of St. Andrews, St. Andrews, UK

A Engel
National Institutes of Health, Bethesda, MD, USA and
D Kryndushkin
National Institutes of Health, Bethesda, MD, USA

J Engelmann
INRES, University of Bonn, Bonn, Germany

L Enjuanes
CNB, CSIC, Madrid, Spain

A Ensser
Virologisches Institut, Universitätsklinikum, Erlangen, Germany

M Erlandson
Agriculture & Agri-Food Canada, Saskatoon, SK, Canada

K J Ertel
University of California, Irvine, CA, USA

R Esteban
Instituto de Microbiología Bioquímica CSIC/University de Salamanca, Salamanca, Spain

R Esteban
Instituto de Microbiología Bioquímica CSIC/University of Salamanca, Salamanca, Spain

J L Van Etten
University of Nebraska–Lincoln, Lincoln, NE, USA

D J Evans
University of Warwick, Coventry, UK

Ø Evensen
Norwegian School of Veterinary Science, Oslo, Norway

D Falzarano
University of Manitoba, Winnipeg, MB, Canada

B A Fane
University of Arizona, Tucson, AZ, USA

R-X. Fang
Chinese Academy of Sciences, Beijing, People's Republic of China

D Fargette
IRD, Montpellier, France

A Fath-Goodin
University of Kentucky, Lexington, KY, USA

C M Fauquet
Danforth Plant Science Center, St. Louis, MO, USA

B A Federici
University of California, Riverside, CA, USA

H Feldmann
National Microbiology Laboratory, Public Health Agency of Canada, Winnipeg, MB, Canada

H Feldmann
Public Health Agency of Canada, Winnipeg, MB, Canada

F Fenner
Australian National University, Canberra, ACT, Australia

S A Ferreira
University of Hawaii at Manoa, Honolulu, HI, USA

H J Field
University of Cambridge, Cambridge, UK

K Fischer
University of California, San Francisco, San Francisco, CA, USA

J A Fishman
Massachusetts General Hospital, Boston, MA, USA

B Fleckenstein
University of Erlangen – Nürnberg, Erlangen, Germany

R Flores
Instituto de Biología Molecular y Celular de Plantas (UPV-CSIC), Valencia, Spain

T R Flotte
University of Florida College of Medicine, Gainesville, FL, USA

P Forterre
Institut Pasteur, Paris, France

M A Franco
Pontificia Universidad Javeriana, Bogota, Republic of Colombia

T K Frey
Georgia State University, Atlanta, GA, USA

M Fuchs
Cornell University, Geneva, NY, USA

S Fuentes
International Potato Center (CIP), Lima, Peru

T Fujimura
Instituto de Microbiología Bioquímica CSIC/University of Salamanca, Salamanca, Spain

R S Fujinami
University of Utah School of Medicine, Salt Lake City, UT, USA

T Fukuhara
Tokyo University of Agriculture and Technology, Fuchu, Japan

D Gallitelli
Università degli Studi and Istituto di Virologia Vegetale del CNR, Bari, Italy

F García-Arenal
Universidad Politécnica de Madrid, Madrid, Spain

J A García
Centro Nacional de Biotecnología (CNB), CSIC, Madrid, Spain

R A Garrett
Copenhagen University, Copenhagen, Denmark

S Gaumer
Université Versailles St-Quentin, CNRS, Versailles, France

R J Geijskes
Queensland University of Technology, Brisbane, QLD, Australia

T W Geisbert
National Emerging Infectious Diseases Laboratories, Boston, MA, USA

E Gellermann
Hannover Medical School, Hannover, Germany

A Gessain
Pasteur Institute, CNRS URA 3015, Paris, France

S A Ghabrial
University of Kentucky, Lexington, KY, USA

W Gibson
Johns Hopkins University School of Medicine, Baltimore, MD, USA

M Glasa
Slovak Academy of Sciences, Bratislava, Slovakia

Y Gleba
Icon Genetics GmbH, Weinbergweg, Germany

U A Gompels
University of London, London, UK

D Gonsalves
USDA, Pacific Basin Agricultural Research Center, Hilo, HI, USA

M M Goodin
University of Kentucky, Lexington, KY, USA

T J D Goodwin
University of Otago, Dunedin, New Zealand

A E Gorbalenya
Leiden University Medical Center, Leiden, The Netherlands

E A Gould
University of Reading, Reading, UK

A Grakoui
Emory University School of Medicine, Atlanta, GA, USA

M-A Grandbastien
INRA, Versailles, France

R Grassmann
University of Erlangen – Nürnberg, Erlangen, Germany

M Gravell
National Institutes of Health, Bethesda, MD, USA

M V Graves
University of Massachusetts–Lowell, Lowell, MA, USA

K Y Green
National Institutes of Health, Bethesda, MD, USA

H B Greenberg
Stanford University School of Medicine and Veterans Affairs Palo Alto Health Care System, Palo Alto, CA, USA

B M Greenberg
Johns Hopkins School of Medicine, Baltimore, MD, USA

I Greiser-Wilke
School of Veterinary Medicine, Hanover, Germany

D E Griffin
Johns Hopkins Bloomberg School of Public Health, Baltimore, MD, USA

T S Gritsun
University of Reading, Reading, UK

R J de Groot
Utrecht University, Utrecht, The Netherlands

A J Gubala
CSIRO Livestock Industries, Geelong, VIC, Australia

D J Gubler
John A. Burns School of Medicine, Honolulu, HI, USA

A-L Haenni
Institut Jacques Monod, Paris, France

D Haig
Nottingham University, Nottingham, UK

F J Haines
Oxford Brookes University, Oxford, UK

J Hamacher
INRES, University of Bonn, Bonn, Germany

J Hammond
USDA-ARS, Beltsville, MD, USA

R M Harding
Queensland University of Technology, Brisbane, QLD, Australia

J M Hardwick
Johns Hopkins University Schools of Public Health and Medicine, Baltimore, MD, USA

D Hariri
INRA – Département Santé des Plantes et Environnement, Versailles, France

B Harrach
Veterinary Medical Research Institute, Budapest, Hungary

P A Harries
Samuel Roberts Noble Foundation, Inc., Ardmore, OK, USA

L E Harrington
University of Alabama at Birmingham, Birmingham, AL, USA

T J Harrison
University College London, London, UK

T Hatziioannou
Aaron Diamond AIDS Research Center, The Rockefeller University, New York, NY, USA

J Hay
The State University of New York, Buffalo, NY, USA

G S Hayward
Johns Hopkins School of Medicine, Baltimore, MD, USA

E Hébrard
IRD, Montpellier, France

R W Hendrix
University of Pittsburgh, Pittsburgh, PA, USA

L E Hensley
USAMRIID, Fort Detrick, MD, USA

M de las Heras
University of Glasgow Veterinary School, Glasgow, UK

S Hertzler
University of Illinois at Chicago, Chicago, IL, USA

F van Heuverswyn
University of Montpellier 1, Montpellier, France

J Hilliard
Georgia State University, Atlanta, GA, USA

B I Hillman
Rutgers University, New Brunswick, NJ, USA

S Hilton
University of Warwick, Warwick, UK

D M Hinton
National Institutes of Health, Bethesda, MD, USA

A Hinz
UMR 5233 UJF-EMBL-CNRS, Grenoble, France

A E Hoet
The Ohio State University, Columbus, OH, USA

S A Hogenhout
The John Innes Centre, Norwich, UK

T Hohn
Basel university, Institute of Botany, Basel, Switzerland

J S Hong
Seoul Women's University, Seoul, South Korea

M C Horzinek
Utrecht University, Utrecht, The Netherlands

T Hovi
National Public Health Institute (KTL), Helsinki, Finland

A M Huger
Institute for Biological Control, Darmstadt, Germany

L E Hughes
University of St. Andrews, St. Andrews, UK

R Hull
John Innes Centre, Colney, UK

E Hunter
Emory University Vaccine Center, Atlanta, GA, USA

A D Hyatt
Australian Animal Health Laboratory, Geelong, VIC, Australia

T Hyypiä
University of Turku, Turku, Finland

T Iwanami
National Institute of Fruit Tree Science, Tsukuba, Japan

A O Jackson
University of California, Berkeley, CA, USA

P Jardine
University of Minnesota, Minneapolis, MN, USA

J A Jehle
DLR Rheinpfalz, Neustadt, Germany

A R Jilbert
Institute of Medical and Veterinary Science, Adelaide, SA, Australia

P John
Indian Agricultural Research Institute, New Delhi, India

J E Johnson
The Scripps Research Institute, La Jolla, CA, USA

R T Johnson
Johns Hopkins School of Medicine, Baltimore, MD, USA

W E Johnson
New England Primate Research Center, Southborough, MA, USA

S L Johnston
Imperial College London, London, UK

A T Jones
Scottish Crop Research Institute, Dundee, UK

R Jordan
USDA-ARS, Beltsville, MD, USA

Y Kapustin
National Institutes of Health, Bethesda, MD, USA

P Karayiannis
Imperial College London, London, UK

P Kazmierczak
University of California, Davis, CA, USA

K M Keene
Colorado State University, Fort Collins, CO, USA

C Kerlan
Institut National de la Recherche Agronomique (INRA), Le Rheu, France

K Khalili
Temple University School of Medicine, Philadelphia, PA, USA

P H Kilmarx
Centers for Disease Control and Prevention, Atlanta, GA, USA

L A King
Oxford Brookes University, Oxford, UK

P D Kirkland
Elizabeth Macarthur Agricultural Institute, Menangle, NSW, Australia

C D Kirkwood
Murdoch Childrens Research Institute Royal Children's Hospital, Melbourne, VIC, Australia

R P Kitching
Canadian Food Inspection Agency, Winnipeg, MB, Canada

P J Klasse
Cornell University, New York, NY, USA

N R Klatt
University of Pennsylvania School of Medicine, Philadelphia, PA, USA

R G Kleespies
Institute for Biological Control, Darmstadt, Germany

D F Klessig
Boyce Thompson Institute for Plant Research, Ithaca, NY, USA

W B Klimstra
Louisiana State University Health Sciences Center at Shreveport, Shreveport, LA, USA

V Klimyuk
Icon Genetics GmbH, Weinbergweg, Germany

N Knowles
Institute for Animal Health, Pirbright, UK

R Koenig
Biologische Bundesanstalt für Land- und Forstwirtschaft, Brunswick, Germany

R Koenig
Institut für Pflanzenvirologie, Mikrobiologie und biologische Sicherheit, Brunswick, Germany

G Konaté
INERA, Ouagadougou, Burkina Faso

C N Kotton
Massachusetts General Hospital, Boston, MA, USA

L D Kramer
Wadsworth Center, New York State Department of Health, Albany, NY, USA

P J Krell
University of Guelph, Guelph, ON, Canada

J Kreuze
International Potato Center (CIP), Lima, Peru

M J Kuehnert
Centers for Disease Control and Prevention, Atlanta, GA, USA

R J Kuhn
Purdue University, West Lafayette, IN, USA

G Kurath
Western Fisheries Research Center, Seattle, WA, USA

I Kusters
sanofi pasteur, Lyon, France

I V Kuzmin
Centers for Disease Control and Prevention, Atlanta, GA, USA

M E Laird
New England Primate Research Center, Southborough, MA, USA

R A Lamb
Howard Hughes Medical Institute at Northwestern University, Evanston, IL, USA

P F Lambert
University of Wisconsin School of Medicine and Public Health, Madison, WI, USA

A S Lang
Memorial University of Newfoundland, St. John's, NL, Canada

H D Lapierre
INRA – Département Santé des Plantes et Environnement, Versailles, France

G Lawrence
The Children's Hospital at Westmead, Westmead, NSW, Australia and
University of Sydney, Westmead, NSW, Australia

H Lecoq
Institut National de la Recherche Agronomique (INRA), Station de Pathologie Végétale, Montfavet, France

B Y Lee
Seoul Women's University, Seoul, South Korea

E J Lefkowitz
University of Alabama at Birmingham, Birmingham, AL, USA

J P Legg
International Institute of Tropical Agriculture, Dar es Salaam, Tanzania,
UK and
Natural Resources Institute, Chatham Maritime, UK

P Leinikki
National Public Health Institute, Helsinki, Finland

J Lenard
University of Medicine and Dentistry of New Jersey (UMDNJ), Piscataway, NJ, USA

J C Leong
University of Hawaii at Manoa, Honolulu, HI, USA

K N Leppard
University of Warwick, Coventry, UK

A Lescoute
Université Louis Pasteur, Strasbourg, France

D-E Lesemann
Biologische Bundesanstalt für Land- und Forstwirtschaft, Brunswick, Germany

J-H Leu
National Taiwan University, Taipei, Republic of China

H L Levin
National Institutes of Health, Bethesda, MD, USA

D J Lewandowski
The Ohio State University, Columbus, OH, USA

H-S Lim
University of California, Berkeley, Berkeley, CA, USA

M D A Lindsay
Western Australian Department of Health, Mount Claremont, WA, Australia

R Ling
University of Warwick, Coventry, UK

M L Linial
Fred Hutchinson Cancer Research Center, Seattle, WA, USA

D C Liotta
Emory University, Atlanta, GA, USA

W Ian Lipkin
Columbia University, New York, NY, USA

H L Lipton
University of Illinois at Chicago, Chicago, IL, USA

A S Liss
University of Texas at Austin, Austin, TX, USA

J J López-Moya
Instituto de Biología Molecular de Barcelona (IBMB), CSIC, Barcelona, Spain

G Loebenstein
Agricultural Research Organization, Bet Dagan, Israel

C-F Lo
National Taiwan University, Taipei, Republic of China

S A Lommel
North Carolina State University, Raleigh, NC, USA

G P Lomonossoff
John Innes Centre, Norwich, UK

M Luo
University of Alabama at Birmingham, Birmingham, AL, USA

S A MacFarlane
Scottish Crop Research Institute, Dundee, UK

J S Mackenzie
Curtin University of Technology, Shenton Park, WA, Australia

R Mahieux
Pasteur Institute, CNRS URA 3015, Paris, France

B W J Mahy
Centers for Disease Control and Prevention, Atlanta, GA, USA

E Maiss
Institute of Plant Diseases and Plant Protection, Hannover, Germany

E O Major
National Institutes of Health, Bethesda, MD, USA

V G Malathi
Indian Agricultural Research Institute, New Delhi, India

A Mankertz
Robert Koch-Institut, Berlin, Germany

S Mansoor
National Institute for Biotechnology and Genetic Engineering, Faisalabad, Pakistan

A A Marfin
Centers for Disease Control and Prevention, Atlanta, GA, USA

S Marillonnet
Icon Genetics GmbH, Weinbergweg, Germany

G P Martelli
Università degli Studi and Istituto di Virologia vegetale CNR, Bari, Italy

M Marthas
University of California, Davis, Davis, CA, USA

D P Martin
University of Cape Town, Cape Town, South Africa

P A Marx
Tulane University, Covington, LA, USA

W S Mason
Fox Chase Cancer Center, Philadelphia, PA, USA

T D Mastro
Centers for Disease Control and Prevention, Atlanta, GA, USA

A A McBride
National Institutes of Health, Bethesda, MD, USA

L McCann
National Institutes of Health, Bethesda, MD, USA

M McChesney
University of California, Davis, Davis, CA, USA

J B McCormick
University of Texas, School of Public Health, Brownsville, TX, USA

G McFadden
University of Florida, Gainesville, FL, USA

G McFadden
The University of Western Ontario, London, ON, Canada

D B McGavern
The Scripps Research Institute, La Jolla, CA, USA

A L McNees
Baylor College of Medicine, Houston, TX, USA

M Meier
Tallinn University of Technology, Tallinn, Estonia

P S Mellor
Institute for Animal Health, Woking, UK

X J Meng
Virginia Polytechnic Institute and State University, Blacksburg, VA, USA

A A Mercer
University of Otago, Dunedin, New Zealand

P P C Mertens
Institute for Animal Health, Woking, UK

T C Mettenleiter
Friedrich-Loeffler-Institut, Greifswald-Insel Riems, Germany

H Meyer
Bundeswehr Institute of Microbiology, Munich, Germany

R F Meyer
Centers for Disease Control and Prevention, Atlanta, GA, USA

P de Micco
Etablissement Français du Sang Alpes-Méditerranée, Marseilles, France

B R Miller
Centers for Disease Control and Prevention (CDC), Fort Collins, CO, USA

C J Miller
University of California, Davis, Davis, CA, USA

R G Milne
Istituto di Virologia Vegetale CNR, Torino, Italy

P D Minor
NIBSC, Potters Bar, UK

S Mjaaland
Norwegian School of Veterinary Science, Oslo, Norway

E S Mocarski
Emory University School of Medicine, Atlanta, GA, USA

E S Mocarski, Jr.
Emory University School of Medicine, Emory, GA, USA

V Moennig
School of Veterinary Medicine, Hanover, Germany

P Moffett
Boyce Thompson Institute for Plant Research, Ithaca, NY, USA

T P Monath
Kleiner Perkins Caufield and Byers, Menlo Park, CA, USA

R C Montelaro
University of Pittsburgh School of Medicine, Pittsburgh, PA, USA

P S Moore
University of Pittsburgh Cancer Institute, Pittsburgh, PA, USA

F J Morales
International Center for Tropical Agriculture, Cali, Colombia

H Moriyama
Tokyo University of Agriculture and Technology, Fuchu, Japan

T J Morris
University of Nebraska, Lincoln, NE, USA

S A Morse
Centers for Disease Control and Prevention, Atlanta, GA, USA

L Moser
University of Wisconsin – Madison, Madison, WI, USA

B Moury
INRA – Station de Pathologie Végétale, Montfavet, France

J W Moyer
North Carolina State University, Raleigh, NC, USA

R W Moyer
University of Florida, Gainesville, FL, USA

E Muller
CIRAD/UMR BGPI, Montpellier, France

F A Murphy
University of Texas Medical Branch, Galveston, TX, USA

A Müllbacher
Australian National University, Canberra, ACT, Australia

K Nagasaki
Fisheries Research Agency, Hiroshima, Japan

T Nakayashiki
National Institutes of Health, Bethesda, MD, USA

A A Nash
University of Edinburgh, Edinburgh, UK

N Nathanson
University of Pennsylvania, Philadelphia, PA, USA

C K Navaratnarajah
Purdue University, West Lafayette, IN, USA

M S Nawaz-ul-Rehman
Danforth Plant Science Center, St. Louis, MO, USA

J C Neil
University of Glasgow, Glasgow, UK

R S Nelson
Samuel Roberts Noble Foundation, Inc., Ardmore, OK, USA

P Nettleton
Moredun Research Institute, Edinburgh, UK

A W Neuman
Emory University, Atlanta, GA, USA

A R Neurath
Virotech, New York, NY, USA

M L Nibert
Harvard Medical School, Boston, MA, USA

L Nicoletti
Istituto Superiore di Sanità, Rome, Italy

N Noah
London School of Hygiene and Tropical Medicine, London, UK

D L Nuss
University of Maryland Biotechnology Institute, Rockville, MD, USA

M S Oberste
Centers for Disease Control and Prevention, Atlanta, GA, USA

W A O'Brien
University of Texas Medical Branch – Galveston, Galveston, TX, USA

D J O'Callaghan
Louisiana State University Health Sciences Center, Shreveport, LA, USA

W F Ochoa
University of California, San Diego, La Jolla, CA, USA

M R Odom
University of Alabama at Birmingham, Birmingham, AL, USA

M M van Oers
Wageningen University, Wageningen, The Netherlands

M B A Oldstone
The Scripps Research Institute, La Jolla, CA, USA

G Olinger
USAMRIID, Fort Detrick, MD, USA

K E Olson
Colorado State University, Fort Collins, CO, USA

A Olspert
Tallinn University of Technology, Tallinn, Estonia

G Orth
Institut Pasteur, Paris, France

J E Osorio
University of Wisconsin, Madison, WI, USA

N Osterrieder
Cornell University, Ithaca, NY, USA

S A Overman
University of Missouri – Kansas City, Kansas City, MO, USA

R A Owens
Beltsville Agricultural Research Center, Beltsville, MD, USA

M S Padmanabhan
Yale University, New Haven, CT, USA

S Paessler
University of Texas Medical Branch, Galveston, TX, USA

P Palese
Mount Sinai School of Medicine, New York, NY, USA

M A Pallansch
Centers for Disease Control and Prevention, Atlanta, GA, USA

M Palmarini
University of Glasgow Veterinary School, Glasgow, UK

P Palukaitis
Scottish Crop Research Institute, Invergowrie, Dundee, UK

I Pandrea
Tulane National Primate Research Center, Covington, LA, USA

O Papadopoulos
Aristotle University, Thessaloniki, Greece

H R Pappu
Washington State University, Pullman, WA, USA

S Parker
Saint Louis University School of Medicine, St. Louis, MO, USA

C R Parrish
Cornell University, Ithaca, NY, USA

R F Pass
University of Alabama School of Medicine, Birmingham, AL, USA

J L Patterson
Southwest Foundation for Biomedical Research, San Antonio, TX, USA

T A Paul
Cornell University, Ithaca, NY, USA

A E Peaston
The Jackson Laboratory, Bar Harbor, ME, USA

M Peeters
University of Montpellier 1, Montpellier, France

J S M Peiris
The University of Hong Kong, Hong Kong, People's Republic of China

P J Peters
Centers for Disease Control and Prevention, Atlanta, GA, USA

M Pfeffer
Bundeswehr Institute of Microbiology, Munich, Germany

H Pfister
University of Köln, Cologne, Germany

O Planz
Federal Research Institute for Animal Health, Tuebingen, Gemany

L L M Poon
The University of Hong Kong, Hong Kong, People's Republic of China

M M Poranen
University of Helsinki, Helsinki, Finland

K Porter
The University of Melbourne, Parkville, VIC, Australia

A Portner
St. Jude Children's Research Hospital, Memphis, TN, USA

R D Possee
NERC Institute of Virology and Environmental Microbiology, Oxford, UK

R T M Poulter
University of Otago, Dunedin, New Zealand

A M Powers
Centers for Disease Control and Prevention, Fort Collins, CO, USA

D Prangishvili
Institut Pasteur, Paris, France

C M Preston
Medical Research Council Virology Unit, Glasgow, UK

S L Quackenbush
Colorado State University, Fort Collins, CO, USA

F Qu
University of Nebraska, Lincoln, NE, USA

B C Ramirez
CNRS, Paris, France

A Rapose
University of Texas Medical Branch – Galveston, Galveston, TX, USA

D V R Reddy
Hyderabad, India

A J Redwood
The University of Western Australia, Crawley, WA, Australia

M Regner
Australian National University, Canberra, ACT, Australia

W K Reisen
University of California, Davis, CA, USA

T Renault
IFREMER, La Tremblade, France

P A Revill
Victorian Infectious Diseases Reference Laboratory, Melbourne, VIC, Australia

A Rezaian
University of Adelaide, Adelaide, SA, Australia

J F Ridpath
USDA, Ames, IA, USA

B K Rima
The Queen's University of Belfast, Belfast, UK

E Rimstad
Norwegian School of Veterinary Science, Oslo, Norway

F J Rixon
MRC Virology Unit, Glasgow, UK

Y-T Ro
Konkuk University, Seoul, South Korea

C M Robinson
University of Oklahoma Health Sciences Center, Oklahoma City, OK, USA

G F Rohrmann
Oregon State University, Corvallis, OR, USA

M Roivainen
National Public Health Institute (KTL), Helsinki, Finland

L Roux
University of Geneva Medical School, Geneva, Switzerland

J Rovnak
Colorado State University, Fort Collins, CO, USA

D J Rowlands
University of Leeds, Leeds, UK

P Roy
London School of Hygiene and Tropical Medicine, London, UK

L Rubino
Istituto di Virologia Vegetale del CNR, Bari, Italy

R W H Ruigrok
CNRS, Grenoble, France

C E Rupprecht
Centers for Disease Control and Prevention, Atlanta, GA, USA

R J Russell
University of St. Andrews, St. Andrews, UK

B E Russ
The University of Melbourne, Parkville, VIC, Australia

W T Ruyechan
The State University of New York, Buffalo, NY, USA

E Ryabov
University of Warwick, Warwick, UK

M D Ryan
University of St. Andrews, St. Andrews, UK

E P Rybicki
University of Cape Town, Cape Town, South Africa

K D Ryman
Louisiana State University Health Sciences Center at Shreveport, Shreveport, LA, USA

K D Ryman
Louisiana State University Health Sciences Center, Shreveport, LA, USA

K H Ryu
Seoul Women's University, Seoul, South Korea

M Safak
Temple University School of Medicine, Philadelphia, PA, USA

M Salas
Universidad Autónoma, Madrid, Spain

S K Samal
University of Maryland, College Park, MD, USA

J T Sample
The Pennsylvania State University College of Medicine, Hershey, PA, USA

C E Sample
The Pennsylvania State University College of Medicine, Hershey, PA, USA

R M Sandri-Goldin
University of California, Irvine, Irvine, CA, USA

H Sanfaçon
Pacific Agri-Food Research Centre, Summerland, BC, Canada

R Sanjuán
Instituto de Biología Molecular y Cellular de Plantas, CSIC-UPV, Valencia, Spain

N Santi
Norwegian School of Veterinary Science, Oslo, Norway

C Sarmiento
Tallinn University of Technology, Tallinn, Estonia

T Sasaya
National Agricultural Research Center, Ibaraki, Japan

Q J Sattentau
University of Oxford, Oxford, UK

C Savolainen-Kopra
National Public Health Institute (KTL), Helsinki, Finland

B Schaffhausen
Tufts University School of Medicine, Boston, MA, USA

K Scheets
Oklahoma State University, Stillwater, OK, USA

M J Schmitt
University of the Saarland, Saarbrücken, Germany

A Schneemann
The Scripps Research Institute, La Jolla, CA, USA

G Schoehn
CNRS, Grenoble, France

J E Schoelz
University of Missouri, Columbia, MO, USA

L B Schonberger
Centers for Disease Control and Prevention, Atlanta, GA, USA

U Schubert
Klinikum der Universität Erlangen-Nürnberg, Erlangen, Germany

D A Schultz
Johns Hopkins University School of Medicine, Baltimore, MD, USA

S Schultz-Cherry
University of Wisconsin – Madison, Madison, WI, USA

T F Schulz
Hannover Medical School, Hannover, Germany

P D Scotti
Waiatarua, New Zealand

B L Semler
University of California, Irvine, CA, USA

J M Sharp
Veterinary Laboratories Agency, Penicuik, UK

M L Shaw
Mount Sinai School of Medicine, New York, NY, USA

G R Shellam
The University of Western Australia,
Crawley, WA, Australia

D N Shepherd
University of Cape Town, Cape Town, South Africa

N C Sheppard
University of Oxford, Oxford, UK

F Shewmaker
National Institutes of Health, Bethesda, MD, USA

P A Signoret
Montpellier SupAgro, Montpellier, France

A Silaghi
University of Manitoba, Winnipeg, MB, Canada

G Silvestri
University of Pennsylvania, Philadelphia, PA, USA

T L Sit
North Carolina State University, Raleigh, NC, USA

N Sittidilokratna
Centex Shrimp and Center for Genetic Engineering and Biotechnology, Bangkok, Thailand

M A Skinner
Imperial College London, London, UK

D W Smith
PathWest Laboratory Medicine WA, Nedlands, WA, Australia

G L Smith
Imperial College London, London, UK

L M Smith
The University of Western Australia,
Crawley, WA, Australia

E J Snijder
Leiden University Medical Center, Leiden, The Netherlands

M Sova
University of Texas Medical Branch – Galveston, Galveston, TX, USA

J A Speir
The Scripps Research Institute, La Jolla, CA, USA

T E Spencer
Texas A&M University, College Station, TX, USA

P Sreenivasulu
Sri Venkateswara University, Tirupati, India

J Stanley
John Innes Centre, Colney, UK

K M Stedman
Portland State University, Portland, OR, USA

D Stephan
Institute of Plant Diseases and Plant Protection, Hannover, Germany

C C M M Stijger
Wageningen University and Research Centre, Naaldwijk, The Netherlands

L Stitz
Federal Research Institute for Animal Health, Tuebingen, Gemany

P G Stockley
University of Leeds, Leeds, UK

M R Strand
University of Georgia, Athens, GA, USA

M J Studdert
The University of Melbourne, Parkville, VIC, Australia

C A Suttle
University of British Columbia, Vancouver, BC, Canada

N Suzuki
Okayama University, Okayama, Japan

J Y Suzuki
USDA, Pacific Basin Agricultural Research Center, Hilo, HI, USA

R Swanepoel
National Institute for Communicable Diseases, Sandringham, South Africa

S J Symes
The University of Melbourne, Parkville, VIC, Australia

G Szittya
Agricultural Biotechnology Center, Godollo, Hungary

M Taliansky
Scottish Crop Research Institute, Dundee, UK

P Tattersall
Yale University Medical School, New Haven, CT, USA

T Tatusova
National Institutes of Health, Bethesda, MD, USA

S Tavantzis
University of Maine, Orono, ME, USA

J M Taylor
Fox Chase Cancer Center, Philadelphia, PA, USA

D A Theilmann
Agriculture and Agri-Food Canada, Summerland, BC, Canada

F C Thomas Allnutt
National Science Foundation, Arlington, VA, USA

G J Thomas Jr.
University of Missouri – Kansas City, Kansas City, MO, USA

J E Thomas
Department of Primary Industries and Fisheries, Indooroopilly, QLD, Australia

H C Thomas
Imperial College London, London, UK

A N Thorburn
The University of Melbourne, Parkville, VIC, Australia

P Tijssen
Université du Québec, Laval, QC, Canada

S A Tolin
Virginia Polytechnic Institute and State University, Blacksburg, VA, USA

L Torrance
Scottish Crop Research Institute, Invergowrie, UK

S Trapp
Cornell University, Ithaca, NY, USA

S Tripathi
USDA, Pacific Basin Agricultural Research Center, Hilo, HI, USA

E Truve
Tallinn University of Technology, Tallinn, Estonia

J-M Tsai
National Taiwan University, Taipei, Republic of China

M Tsompana
North Carolina State University, Raleigh, NC, USA

R Tuma
University of Helsinki, Helsinki, Finland

A S Turnell
The University of Birmingham, Birmingham, UK

K L Tyler
University of Colorado Health Sciences, Denver, CO, USA

A Uchiyama
Cornell University, Ithaca, NY, USA

C Upton
University of Victoria, Victoria, BC, Canada

A Urisman
University of California, San Francisco, San Francisco, CA, USA

J K Uyemoto
University of California, Davis, CA, USA

A Vaheri
University of Helsinki, Helsinki, Finland

R Vainionpää
University of Turku, Turku, Finland

A M Vaira
Istituto di Virologia Vegetale, CNR, Turin, Italy

N K Van Alfen
University of California, Davis, CA, USA

R A A Van der Vlugt
Wageningen University and Research Centre, Wageningen, The Netherlands

M H V Van Regenmortel
CNRS, Illkirch, France

P A Venter
The Scripps Research Institute, La Jolla, CA, USA

J Verchot-Lubicz
Oklahoma State University, Stillwater, OK, USA

R A Vere Hodge
Vere Hodge Antivirals Ltd., Reigate, UK

H J Vetten
Federal Research Centre for Agriculture and Forestry (BBA), Brunswick, Germany

L P Villarreal
University of California, Irvine, Irvine, CA, USA

J M Vlak
Wageningen University, Wageningen, The Netherlands

P K Vogt
The Scripps Research Institute, La Jolla, CA, USA

L E Volkman
University of California, Berkeley, Berkeley, CA, USA

J Votteler
Klinikum der Universität Erlangen-Nürnberg, Erlangen, Germany

D F Voytas
Iowa State University, Ames, IA, USA

J D F Wadsworth
University College London, London, UK

E K Wagner
University of California, Irvine, Irvine, CA, USA

P J Walker
CSIRO Australian Animal Health Laboratory, Geelong, VIC, Australia

A L Wang
University of California, San Francisco, CA, USA

X Wang
University of Wisconsin – Madison, Madison, WI, USA

C C Wang
University of California, San Francisco, CA, USA

L-F Wang
Australian Animal Health Laboratory, Geelong, VIC, Australia

R Warrier
Purdue University, West Lafayette, IN, USA

S C Weaver
University of Texas Medical Branch, Galveston, TX, USA

B A Webb
University of Kentucky, Lexington, KY, USA

F Weber
University of Freiburg, Freiburg, Germany

R P Weir
Berrimah Research Farm, Darwin, NT, Australia

R A Weisberg
National Institutes of Health, Bethesda, MD, USA

W Weissenhorn
UMR 5233 UJF-EMBL-CNRS, Grenoble, France

R M Welsh
University of Massachusetts Medical School, Worcester, MA, USA

J T West
University of Oklahoma Health Sciences Center, Oklahoma City, OK, USA

E Westhof
Université Louis Pasteur, Strasbourg, France

S P J Whelan
Harvard Medical School, Boston, MA, USA

R L White
Texas A&M University, College Station, TX, USA

C A Whitehouse
United States Army Medical Research Institute of Infectious Diseases, Frederick, MD, USA

R B Wickner
National Institutes of Health, Bethesda, MD, USA

R G Will
Western General Hospital, Edinburgh, UK

T Williams
Instituto de Ecología A.C., Xalapa, Mexico

K Willoughby
Moredun Research Institute, Edinburgh, UK

S Winter
Deutsche Sammlung für Mikroorganismen und Zellkulturen, Brunswick, Germany

J Winton
Western Fisheries Research Center, Seattle, WA, USA

J K Yamamoto
University of Florida, Gainesville, FL, USA

M Yoshida
University of Tokyo, Chiba, Japan

N Yoshikawa
Iwate University, Ueda, Japan

L S Young
University of Birmingham, Birmingham, UK

R F Young, III
Texas A&M University, College Station, TX, USA

T M Yuill
University of Wisconsin, Madison, WI, USA

A J Zajac
University of Alabama at Birmingham, Birmingham, AL, USA

S K Zavriev
Shemyakin and Ovchinnikov Institute of Bioorganic Chemistry, Russian Academy of Sciences, Moscow, Russia

J Ziebuhr
The Queen's University of Belfast, Belfast, UK

E I Zuniga
The Scripps Research Institute, La Jolla, CA, USA

CONTENTS

Editors-in-Chief	v
Associate Editors	vii
Preface	ix
How to Use the Encyclopedia	xi
Contributors	xiii

VOLUME 1

A

Adenoviruses: General Features	B Harrach	1
Adenoviruses: Malignant Transformation and Oncology	A S Turnell	9
Adenoviruses: Molecular Biology	K N Leppard	17
Adenoviruses: Pathogenesis	M Benkő	24
African Cassava Mosaic Disease	J P Legg	30
African Horse Sickness Viruses	P S Mellor and P P C Mertens	37
African Swine Fever Virus	L K Dixon and D Chapman	43
AIDS: Disease Manifestation	A Rapose, J East, M Sova and W A O'Brien	51
AIDS: Global Epidemiology	P J Peters, P H Kilmarx and T D Mastro	58
AIDS: Vaccine Development	N C Sheppard and Q J Sattentau	69
Akabane Virus	P S Mellor and P D Kirkland	76
Alfalfa Mosaic Virus	J F Bol	81
Algal Viruses	K Nagasaki and C P D Brussaard	87
Allexivirus	S K Zavriev	96
Alphacryptovirus and *Betacryptovirus*	R Blawid, D Stephan and E Maiss	98
Anellovirus	P Biagini and P de Micco	104
Animal Rhabdoviruses	H Bourhy, A J Gubala, R P Weir and D B Boyle	111
Antigen Presentation	E I Zuniga, D B McGavern and M B A Oldstone	121
Antigenic Variation	G M Air and J T West	127
Antigenicity and Immunogenicity of Viral Proteins	M H V Van Regenmortel	137

Antiviral Agents	H J Field and R A Vere Hodge	142
Apoptosis and Virus Infection	J R Clayton and J M Hardwick	154
Aquareoviruses	M St J Crane and G Carlile	163
Arboviruses	B R Miller	170
Arteriviruses	M A Brinton and E J Snijder	176
Ascoviruses	B A Federici and Y Bigot	186
Assembly of Viruses: Enveloped Particles	C K Navaratnarajah, R Warrier and R J Kuhn	193
Assembly of Viruses: Nonenveloped Particles	M Luo	200
Astroviruses	L Moser and S Schultz-Cherry	204

B

Baculoviruses: Molecular Biology of Granuloviruses	S Hilton	211
Baculoviruses: Molecular Biology of Mosquito Baculoviruses	J J Becnel and C L Afonso	219
Baculoviruses: Molecular Biology of Sawfly Baculoviruses	B M Arif	225
Baculoviruses: Apoptosis Inhibitors	R J Clem	231
Baculoviruses: Expression Vector	F J Haines, R D Possee and L A King	237
Baculoviruses: General Features	P J Krell	247
Baculoviruses: Molecular Biology of Nucleopolyhedroviruses	D A Theilmann and G W Blissard	254
Baculoviruses: Pathogenesis	L E Volkman	265
Banana Bunchy Top Virus	J E Thomas	272
Barley Yellow Dwarf Viruses	L L Domier	279
Barnaviruses	P A Revill	286
Bean Common Mosaic Virus and Bean Common Mosaic Necrosis Virus	R Jordan and J Hammond	288
Bean Golden Mosaic Virus	F J Morales	295
Beet Curly Top Virus	J Stanley	301
Benyvirus	R Koenig	308
Beta ssDNA Satellites	R W Briddon and S Mansoor	314
Birnaviruses	B Delmas	321
Bluetongue Viruses	P Roy	328
Border Disease Virus	P Nettleton and K Willoughby	335
Bornaviruses	L Stitz, O Planz and W Ian Lipkin	341
Bovine and Feline Immunodeficiency Viruses	J K Yamamoto	347
Bovine Ephemeral Fever Virus	P J Walker	354
Bovine Herpesviruses	M J Studdert	362
Bovine Spongiform Encephalopathy	R G Will	368
Bovine Viral Diarrhea Virus	J F Ridpath	374
Brome Mosaic Virus	X Wang and P Ahlquist	381
Bromoviruses	J J Bujarski	386

Bunyaviruses: General Features *R M Elliott*	390
Bunyaviruses: Unassigned *C H Calisher*	399

C

Cacao Swollen Shoot Virus *E Muller*	403
Caliciviruses *M J Studdert and S J Symes*	410
Capillovirus, Foveavirus, Trichovirus, Vitivirus *N Yoshikawa*	419
Capripoxviruses *R P Kitching*	427
Capsid Assembly: Bacterial Virus Structure and Assembly *S Casjens*	432
Cardioviruses *C Billinis and O Papadopoulos*	440
Carlavirus *K H Ryu and B Y Lee*	448
Carmovirus *F Qu and T J Morris*	453
Caulimoviruses: General Features *J E Schoelz*	457
Caulimoviruses: Molecular Biology *T Hohn*	464
Central Nervous System Viral Diseases *R T Johnson and B M Greenberg*	469
Cereal Viruses: Maize/Corn *P A Signoret*	475
Cereal Viruses: Rice *F Morales*	482
Cereal Viruses: Wheat and Barley *H D Lapierre and D Hariri*	490
Chandipura Virus *S Basak and D Chattopadhyay*	497
Chrysoviruses *S A Ghabrial*	503
Circoviruses *A Mankertz*	513
Citrus Tristeza Virus *M Bar-Joseph and W O Dawson*	520
Classical Swine Fever Virus *V Moennig and I Greiser-Wilke*	525
Coltiviruses *H Attoui and X de Lamballerie*	533
Common Cold Viruses *S Dreschers and C Adams*	541
Coronaviruses: General Features *D Cavanagh and P Britton*	549
Coronaviruses: Molecular Biology *S C Baker*	554
Cotton Leaf Curl Disease *S Mansoor, I Amin and R W Briddon*	563
Cowpea Mosaic Virus *G P Lomonossoff*	569
Cowpox Virus *M Bennett, G L Smith and D Baxby*	574
Coxsackieviruses *M S Oberste and M A Pallansch*	580
Crenarchaeal Viruses: Morphotypes and Genomes *D Prangishvili, T Basta and R A Garrett*	587
Crimean–Congo Hemorrhagic Fever Virus and Other Nairoviruses *C A Whitehouse*	596
Cryo-Electron Microscopy *W Chiu, J T Chang and F J Rixon*	603
Cucumber Mosaic Virus *F García-Arenal and P Palukaitis*	614
Cytokines and Chemokines *D E Griffin*	620
Cytomegaloviruses: Murine and Other Nonprimate Cytomegaloviruses *A J Redwood, L M Smith and G R Shellam*	624
Cytomegaloviruses: Simian Cytomegaloviruses *D J Alcendor and G S Hayward*	634

VOLUME 2

D

Defective-Interfering Viruses	L Roux	1
Dengue Viruses	D J Gubler	5
Diagnostic Techniques: Microarrays	K Fischer, A Urisman and J DeRisi	14
Diagnostic Techniques: Plant Viruses	R Koenig, D-E Lesemann, G Adam and S Winter	18
Diagnostic Techniques: Serological and Molecular Approaches	R Vainionpää and P Leinikki	29
Dicistroviruses	P D Christian and P D Scotti	37
Disease Surveillance	N Noah	44
DNA Vaccines	S Babiuk and L A Babiuk	51

E

Ebolavirus	K S Brown, A Silaghi and H Feldmann	57
Echoviruses	T Hyypiä	65
Ecology of Viruses Infecting Bacteria	S T Abedon	71
Electron Microscopy of Viruses	G Schoehn and R W H Ruigrok	78
Emerging and Reemerging Virus Diseases of Plants	G P Martelli and D Gallitelli	86
Emerging and Reemerging Virus Diseases of Vertebrates	B W J Mahy	93
Emerging Geminiviruses	C M Fauquet and M S Nawaz-ul-Rehman	97
Endogenous Retroviruses	W E Johnson	105
Endornavirus	T Fukuhara and H Moriyama	109
Enteric Viruses	R F Bishop and C D Kirkwood	116
Enteroviruses of Animals	L E Hughes and M D Ryan	123
Enteroviruses: Human Enteroviruses Numbered 68 and Beyond	T Hovi, S Blomqvist, C Savolainen-Kopra and M Roivainen	130
Entomopoxviruses	M N Becker and R W Moyer	136
Epidemiology of Human and Animal Viral Diseases	F A Murphy	140
Epstein–Barr Virus: General Features	L S Young	148
Epstein–Barr Virus: Molecular Biology	J T Sample and C E Sample	157
Equine Infectious Anemia Virus	J K Craigo and R C Montelaro	167
Evolution of Viruses	L P Villarreal	174

F

Feline Leukemia and Sarcoma Viruses	J C Neil	185
Filamentous ssDNA Bacterial Viruses	S A Overman and G J Thomas Jr.	190
Filoviruses	G Olinger, T W Geisbert and L E Hensley	198
Fish and Amphibian Herpesviruses	A J Davison	205
Fish Retroviruses	T A Paul, R N Casey, P R Bowser, J W Casey, J Rovnak and S L Quackenbush	212

Fish Rhabdoviruses *G Kurath and J Winton*	221
Fish Viruses *J C Leong*	227
Flaviviruses of Veterinary Importance *R Swanepoel and F J Burt*	234
Flaviviruses: General Features *T J Chambers*	241
Flexiviruses *M J Adams*	253
Foamy Viruses *M L Linial*	259
Foot and Mouth Disease Viruses *D J Rowlands*	265
Fowlpox Virus and Other Avipoxviruses *M A Skinner*	274
Fungal Viruses *S A Ghabrial and N Suzuki*	284
Furovirus *R Koenig*	291
Fuselloviruses of Archaea *K M Stedman*	296

G

Gene Therapy: Use of Viruses as Vectors *K I Berns and T R Flotte*	301
Genome Packaging in Bacterial Viruses *P Jardine*	306
Giardiaviruses *A L Wang and C C Wang*	312

H

Hantaviruses *A Vaheri*	317
Henipaviruses *B T Eaton and L-F Wang*	321
Hepadnaviruses of Birds *A R Jilbert and W S Mason*	327
Hepadnaviruses: General Features *T J Harrison*	335
Hepatitis A Virus *A Dotzauer*	343
Hepatitis B Virus: General Features *P Karayiannis and H C Thomas*	350
Hepatitis B Virus: Molecular Biology *T J Harrison*	360
Hepatitis C Virus *R Bartenschlager and S Bühler*	367
Hepatitis Delta Virus *J M Taylor*	375
Hepatitis E Virus *X J Meng*	377
Herpes Simplex Viruses: General Features *L Aurelian*	383
Herpes Simplex Viruses: Molecular Biology *E K Wagner and R M Sandri-Goldin*	397
Herpesviruses of Birds *S Trapp and N Osterrieder*	405
Herpesviruses of Horses *D J O'Callaghan and N Osterrieder*	411
Herpesviruses: Discovery *B Ehlers*	420
Herpesviruses: General Features *A J Davison*	430
Herpesviruses: Latency *C M Preston*	436
History of Virology: Bacteriophages *H-W Ackermann*	442
History of Virology: Plant Viruses *R Hull*	450
History of Virology: Vertebrate Viruses *F J Fenner*	455

Hordeivirus J N Bragg, H-S Lim and A O Jackson	459
Host Resistance to Retroviruses T Hatziioannou and P D Bieniasz	467
Human Cytomegalovirus: General Features E S Mocarski Jr. and R F Pass	474
Human Cytomegalovirus: Molecular Biology W Gibson	485
Human Eye Infections J Chodosh, A V Chintakuntlawar and C M Robinson	491
Human Herpesviruses 6 and 7 U A Gompels	498
Human Immunodeficiency Viruses: Antiretroviral Agents A W Neuman and D C Liotta	505
Human Immunodeficiency Viruses: Molecular Biology J Votteler and U Schubert	517
Human Immunodeficiency Viruses: Origin F van Heuverswyn and M Peeters	525
Human Immunodeficiency Viruses: Pathogenesis N R Klatt, A Chahroudi and G Silvestri	534
Human Respiratory Syncytial Virus P L Collins	542
Human Respiratory Viruses J E Crowe Jr.	551
Human T-Cell Leukemia Viruses: General Features M Yoshida	558
Human T-Cell Leukemia Viruses: Human Disease R Mahieux and A Gessain	564
Hypovirulence N K Van Alfen and P Kazmierczak	574
Hypoviruses D L Nuss	580

VOLUME 3

I

Icosahedral dsDNA Bacterial Viruses with an Internal Membrane J K H Bamford and S J Butcher	1
Icosahedral Enveloped dsRNA Bacterial Viruses R Tuma	6
Icosahedral ssDNA Bacterial Viruses B A Fane, M Chen, J E Cherwa and A Uchiyama	13
Icosahedral ssRNA Bacterial Viruses P G Stockley	21
Icosahedral Tailed dsDNA Bacterial Viruses R L Duda	30
Idaeovirus A T Jones and H Barker	37
Iflavirus M M van Oers	42
Ilarvirus K C Eastwell	46
Immune Response to Viruses: Antibody-Mediated Immunity A R Neurath	56
Immune Response to Viruses: Cell-Mediated Immunity A J Zajac and L E Harrington	70
Immunopathology M B A Oldstone and R S Fujinami	78
Infectious Pancreatic Necrosis Virus Ø Evensen and N Santi	83
Infectious Salmon Anemia Virus B H Dannevig, S Mjaaland and E Rimstad	89
Influenza R A Lamb	95
Innate Immunity: Defeating C F Basler	104
Innate Immunity: Introduction F Weber	111
Inoviruses L A Day	117
Insect Pest Control by Viruses M Erlandson	125
Insect Reoviruses P P C Mertens and H Attoui	133
Insect Viruses: Nonoccluded J P Burand	144

Interfering RNAs *K E Olson, K M Keene and C D Blair*	148
Iridoviruses of Vertebrates *A D Hyatt and V G Chinchar*	155
Iridoviruses of Invertebrates *T Williams and A D Hyatt*	161
Iridoviruses: General Features *V G Chinchar and A D Hyatt*	167

J

Jaagsiekte Sheep Retrovirus *J M Sharp, M de las Heras, T E Spencer and M Palmarini*	175
Japanese Encephalitis Virus *A D T Barrett*	182

K

Kaposi's Sarcoma-Associated Herpesvirus: General Features *Y Chang and P S Moore*	189
Kaposi's Sarcoma-Associated Herpesvirus: Molecular Biology *E Gellermann and T F Schulz*	195

L

Lassa, Junin, Machupo and Guanarito Viruses *J B McCormick*	203
Legume Viruses *L Bos*	212
Leishmaniaviruses *R Carrion Jr, Y-T Ro and J L Patterson*	220
Leporipoviruses and Suipoxviruses *G McFadden*	225
Luteoviruses *L L Domier and C J D'Arcy*	231
Lymphocytic Choriomeningitis Virus: General Features *R M Welsh*	238
Lymphocytic Choriomeningitis Virus: Molecular Biology *J C de la Torre*	243
Lysis of the Host by Bacteriophage *R F Young III and R L White*	248

M

Machlomovirus *K Scheets*	259
Maize Streak Virus *D P Martin, D N Shepherd and E P Rybicki*	263
Marburg Virus *D Falzarano and H Feldmann*	272
Marnaviruses *A S Lang and C A Suttle*	280
Measles Virus *R Cattaneo and M McChesney*	285
Membrane Fusion *A Hinz and W Weissenhorn*	292
Metaviruses *H L Levin*	301
Mimivirus *J-M Claverie*	311
Molluscum Contagiosum Virus *J J Bugert*	319
Mononegavirales *A J Easton and R Ling*	324
Mouse Mammary Tumor Virus *J P Dudley*	334
Mousepox and Rabbitpox Viruses *M Regner, F Fenner and A Müllbacher*	342
Movement of Viruses in Plants *P A Harries and R S Nelson*	348

Mumps Virus	*B K Rima and W P Duprex*	356
Mungbean Yellow Mosaic Viruses	*V G Malathi and P John*	364
Murine Gammaherpesvirus 68	*A A Nash and B M Dutia*	372
Mycoreoviruses	*B I Hillman*	378

N

Nanoviruses	*H J Vetten*	385
Narnaviruses	*R Esteban and T Fujimura*	392
Nature of Viruses	*M H V Van Regenmortel*	398
Necrovirus	*L Rubino and G P Martelli*	403
Nepovirus	*H Sanfaçon*	405
Neutralization of Infectivity	*P J Klasse*	413
Nidovirales	*L Enjuanes, A E Gorbalenya, R J de Groot, J A Cowley, J Ziebuhr and E J Snijder*	419
Nodaviruses	*P A Venter and A Schneemann*	430
Noroviruses and Sapoviruses	*K Y Green*	438

O

Ophiovirus	*A M Vaira and R G Milne*	447
Orbiviruses	*P P C Mertens, H Attoui and P S Mellor*	454
Organ Transplantation, Risks	*C N Kotton, M J Kuehnert and J A Fishman*	466
Origin of Viruses	*P Forterre*	472
Orthobunyaviruses	*C H Calisher*	479
Orthomyxoviruses: Molecular Biology	*M L Shaw and P Palese*	483
Orthomyxoviruses: Structure of Antigens	*R J Russell*	489
Oryctes Rhinoceros Virus	*J M Vlak, A M Huger, J A Jehle and R G Kleespies*	495
Ourmiavirus	*G P Accotto and R G Milne*	500

VOLUME 4

P

Papaya Ringspot Virus	*D Gonsalves, J Y Suzuki, S Tripathi and S A Ferreira*	1
Papillomaviruses: General Features of Human Viruses	*G Orth*	8
Papillomaviruses: Molecular Biology of Human Viruses	*P F Lambert and A Collins*	18
Papillomaviruses of Animals	*A A McBride*	26
Papillomaviruses: General Features	*H U Bernard*	34
Paramyxoviruses of Animals	*S K Samal*	40
Parainfluenza Viruses of Humans	*E Adderson and A Portner*	47
Paramyxoviruses	*R E Dutch*	52
Parapoxviruses	*D Haig and A A Mercer*	57

Partitiviruses of Fungi *S Tavantzis*	63
Partitiviruses: General Features *S A Ghabrial, W F Ochoa, T S Baker and M L Nibert*	68
Parvoviruses of Arthropods *M Bergoin and P Tijssen*	76
Parvoviruses of Vertebrates *C R Parrish*	85
Parvoviruses: General Features *P Tattersall*	90
Pecluvirus *D V R Reddy, C Bragard, P Sreenivasulu and P Delfosse*	97
Pepino Mosaic Virus *R A A Van der Vlugt and C C M M Stijger*	103
Persistent and Latent Viral Infection *E S Mocarski and A Grakoui*	108
Phycodnaviruses *J L Van Etten and M V Graves*	116
Phylogeny of Viruses *A E Gorbalenya*	125
Picornaviruses: Molecular Biology *B L Semler and K J Ertel*	129
Plant Antiviral Defense: Gene Silencing Pathway *G Szittya, T Dalmay and J Burgyan*	141
Plant Reoviruses *R J Geijskes and R M Harding*	149
Plant Resistance to Viruses: Engineered Resistance *M Fuchs*	156
Plant Resistance to Viruses: Geminiviruses *J K Brown*	164
Plant Resistance to Viruses: Natural Resistance Associated with Dominant Genes *P Moffett and D F Klessig*	170
Plant Resistance to Viruses: Natural Resistance Associated with Recessive Genes *C Caranta and C Dogimont*	177
Plant Rhabdoviruses *A O Jackson, R G Dietzgen, R-X Fang, M M Goodin, S A Hogenhout, M Deng and J N Bragg*	187
Plant Virus Diseases: Economic Aspects *G Loebenstein*	197
Plant Virus Diseases: Fruit Trees and Grapevine *G P Martelli and J K Uyemoto*	201
Plant Virus Diseases: Ornamental Plants *J Engelmann and J Hamacher*	207
Plant Virus Vectors (Gene Expression Systems) *Y Gleba, S Marillonnet and V Klimyuk*	229
Plum Pox Virus *M Glasa and T Candresse*	238
Poliomyelitis *P D Minor*	243
Polydnaviruses: Abrogation of Invertebrate Immune Systems *M R Strand*	250
Polydnaviruses: General Features *A Fath-Goodin and B A Webb*	257
Polyomaviruses of Humans *M Safak and K Khalili*	262
Polyomaviruses of Mice *B Schaffhausen*	271
Polyomaviruses *M Gravell and E O Major*	277
Pomovirus *L Torrance*	283
Potato Virus Y *C Kerlan and B Moury*	288
Potato Viruses *C Kerlan*	302
Potexvirus *K H Ryu and J S Hong*	310
Potyviruses *J J López-Moya and J A García*	314
Poxviruses *G L Smith, P Beard and M A Skinner*	325
Prions of Vertebrates *J D F Wadsworth and J Collinge*	331
Prions of Yeast and Fungi *R B Wickner, H Edskes, T Nakayashiki, F Shewmaker, L McCann, A Engel and D Kryndushkin*	338

Pseudorabies Virus *T C Mettenleiter*	342
Pseudoviruses *D F Voytas*	352

Q

Quasispecies *R Sanjuán*	359

R

Rabies Virus *I V Kuzmin and C E Rupprecht*	367
Recombination *J J Bujarski*	374
Reoviruses: General Features *P Clarke and K L Tyler*	382
Reoviruses: Molecular Biology *K M Coombs*	390
Replication of Bacterial Viruses *M Salas and M de Vega*	399
Replication of Viruses *A J Cann*	406
Reticuloendotheliosis Viruses *A S Liss and H R Bose Jr.*	412
Retrotransposons of Fungi *T J D Goodwin, M I Butler and R T M Poulter*	419
Retrotransposons of Plants *M-A Grandbastien*	428
Retrotransposons of Vertebrates *A E Peaston*	436
Retroviral Oncogenes *P K Vogt and A G Bader*	445
Retroviruses of Insects *G F Rohrmann*	451
Retroviruses of Birds *K L Beemon*	455
Retroviruses: General Features *E Hunter*	459
Rhinoviruses *N W Bartlett and S L Johnston*	467
Ribozymes *E Westhof and A Lescoute*	475
Rice Tungro Disease *R Hull*	481
Rice Yellow Mottle Virus *E Hébrard and D Fargette*	485
Rift Valley Fever and Other Phleboviruses *L Nicoletti and M G Ciufolini*	490
Rinderpest and Distemper Viruses *T Barrett*	497
Rotaviruses *J Angel, M A Franco and H B Greenberg*	507
Rubella Virus *T K Frey*	514

S

Sadwavirus *T Iwanami*	523
Satellite Nucleic Acids and Viruses *P Palukaitis, A Rezaian and F García-Arenal*	526
Seadornaviruses *H Attoui and P P C Mertens*	535
Sequiviruses *I-R Choi*	546
Severe Acute Respiratory Syndrome (SARS) *J S M Peiris and L L M Poon*	552
Shellfish Viruses *T Renault*	560
Shrimp Viruses *J-R Bonami*	567

Sigma Rhabdoviruses *D Contamine and S Gaumer*	576
Simian Alphaherpesviruses *J Hilliard*	581
Simian Gammaherpesviruses *A Ensser*	585
Simian Immunodeficiency Virus: Animal Models of Disease *C J Miller and M Marthas*	594
Simian Immunodeficiency Virus: General Features *M E Laird and R C Desrosiers*	603
Simian Immunodeficiency Virus: Natural Infection *I Pandrea, G Silvestri and C Apetrei*	611
Simian Retrovirus D *P A Marx*	623
Simian Virus 40 *A L McNees and J S Butel*	630
Smallpox and Monkeypox Viruses *S Parker, D A Schultz, H Meyer and R M Buller*	639
Sobemovirus *M Meier, A Olspert, C Sarmiento and E Truve*	644
St. Louis Encephalitis *W K Reisen*	652
Sweetpotato Viruses *J Kreuze and S Fuentes*	659

VOLUME 5

T

Taura Syndrome Virus *A K Dhar and F C T Allnutt*	1
Taxonomy, Classification and Nomenclature of Viruses *C M Fauquet*	9
Tenuivirus *B C Ramirez*	24
Tetraviruses *J A Speir and J E Johnson*	27
Theiler's Virus *H L Lipton, S Hertzler and N Knowles*	37
Tick-Borne Encephalitis Viruses *T S Gritsun and E A Gould*	45
Tobacco Mosaic Virus *M H V Van Regenmortel*	54
Tobacco Viruses *S A Tolin*	60
Tobamovirus *D J Lewandowski*	68
Tobravirus *S A MacFarlane*	72
Togaviruses Causing Encephalitis *S Paessler and M Pfeffer*	76
Togaviruses Causing Rash and Fever *D W Smith, J S Mackenzie and M D A Lindsay*	83
Togaviruses Not Associated with Human Disease *L L Coffey,*	91
Togaviruses: Alphaviruses *A M Powers*	96
Togaviruses: Equine Encephalitic Viruses *D E Griffin*	101
Togaviruses: General Features *S C Weaver, W B Klimstra and K D Ryman*	107
Togaviruses: Molecular Biology *K D Ryman, W B Klimstra and S C Weaver*	116
Tomato Leaf Curl Viruses from India *S Chakraborty*	124
Tomato Spotted Wilt Virus *H R Pappu*	133
Tomato Yellow Leaf Curl Virus *H Czosnek*	138
Tombusviruses *S A Lommel and T L Sit*	145
Torovirus *A E Hoet and M C Horzinek*	151
Tospovirus *M Tsompana and J W Moyer*	157
Totiviruses *S A Ghabrial*	163

Transcriptional Regulation in Bacteriophage	R A Weisberg, D M Hinton and S Adhya	174
Transmissible Spongiform Encephalopathies	E D Belay and L B Schonberger	186
Tumor Viruses: Human	R Grassmann, B Fleckenstein and H Pfister	193
Tymoviruses	A-L Haenni and T W Dreher	199

U

Umbravirus	M Taliansky and E Ryabov	209
Ustilago Maydis Viruses	J Bruenn	214

V

Vaccine Production in Plants	E P Rybicki	221
Vaccine Safety	C J Clements and G Lawrence	226
Vaccine Strategies	I Kusters and J W Almond	235
Vaccinia Virus	G L Smith	243
Varicella-Zoster Virus: General Features	J I Cohen	250
Varicella-Zoster Virus: Molecular Biology	W T Ruyechan and J Hay	256
Varicosavirus	T Sasaya	263
Vector Transmission of Animal Viruses	W K Reisen	268
Vector Transmission of Plant Viruses	S Blanc	274
Vegetable Viruses	P Caciagli	282
Vesicular Stomatitis Virus	S P J Whelan	291
Viral Killer Toxins	M J Schmitt	299
Viral Membranes	J Lenard	308
Viral Pathogenesis	N Nathanson	314
Viral Receptors	D J Evans	319
Viral Suppressors of Gene Silencing	J Verchot-Lubicz and J P Carr	325
Viroids	R Flores and R A Owens	332
Virus Classification by Pairwise Sequence Comparison (PASC)	Y Bao, Y Kapustin and T Tatusova	342
Virus Databases	E J Lefkowitz, M R Odom and C Upton	348
Virus Entry to Bacterial Cells	M M Poranen and A Domanska	365
Virus Evolution: Bacterial Viruses	R W Hendrix	370
Virus-Induced Gene Silencing (VIGS)	M S Padmanabhan and S P Dinesh-Kumar	375
Virus Particle Structure: Nonenveloped Viruses	J A Speir and J E Johnson	380
Virus Particle Structure: Principles	J E Johnson and J A Speir	393
Virus Species	M H V van Regenmortel	401
Viruses and Bioterrorism	R F Meyer and S A Morse	406
Viruses Infecting Euryarchaea	K Porter, B E Russ, A N Thorburn and M L Dyall-Smith	411
Visna-Maedi Viruses	B A Blacklaws	423

W

Watermelon Mosaic Virus and Zucchini Yellow Mosaic Virus *H Lecoq and C Desbiez* 433

West Nile Virus *L D Kramer* 440

White Spot Syndrome Virus *J-H Leu, J-M Tsai and C-F Lo* 450

Y

Yatapoxviruses *J W Barrett and G McFadden* 461

Yeast L-A Virus *R B Wickner, T Fujimura and R Esteban* 465

Yellow Fever Virus *A A Marfin and T P Monath* 469

Yellow Head Virus *P J Walker and N Sittidilokratna* 476

Z

Zoonoses *J E Osorio and T M Yuill* 485

Taxonomic Index 497

Subject Index 499

Papaya Ringspot Virus

D Gonsalves, J Y Suzuki, and S Tripathi, USDA, Pacific Basin Agricultural Research Center, Hilo, HI, USA
S A Ferreira, University of Hawaii at Manoa, Honolulu, HI, USA

Published by Elsevier Ltd.

Introduction

The term papaya ringspot virus (PRSV) was first used in the 1940s to describe a viral disease of papaya. The name was used primarily to describe the ringspots that appeared on fruits from infected plants. Early investigations showed that the virus was transmitted by several species of aphids in a nonpersistent manner. That is, the aphid vector could acquire the virus in a short period of time while feeding on infected plants and likewise transmit the virus in a span of few seconds to less than a minute during subsequent feeding. In the same decade researchers from India and other places like Puerto Rico reported the occurrence of an aphid-transmitted disease of papaya; based on the symptoms on the leaves, it was identified as papaya mosaic virus. Work in the 1980s showed that the aphid-transmitted papaya mosaic virus and PRSV were really the same, and the name of PRSV was adopted. PRSV is a member of the family *Potyviridae*, a large and arguably the most economically important group of plant viruses. Today, the term papaya mosaic virus is reserved for a virus that is not aphid transmitted, belongs to the family *Potexviridae*, and causes the papaya mosaic disease which is seldom observed and not important commercially.

The systemic host range of PRSV is confined to plants in the families Caricaceae and Cucurbitaceae, with the primary economically important host being papaya and a range of cucurbits such as squash, watermelon, and melons. It does cause local lesions on plants of the family Chenopodiaceae such as *Chenopodium quinoa* and *C. amaranticolor*. The disease on cucurbits was, early on, referred to as being caused by watermelon mosaic virus-1 (WMV-1). Later serological and molecular characterization showed that PRSV and WMV-1 are virtually identical. Based on their close relationship, a single name was adopted to unify both viruses into one group. The name PRSV was chosen due to its being named before WMV-1. To clarify host range, 'P' (PRSV-P) or 'P type' is used to designate virus infecting papaya and cucurbits, while 'W' (PRSV-W) or 'W type' refers to virus infecting cucurbits only. The virus symptoms on cucurbits are identical to those on Caricaceae. Leaves of infected plants show severe mosaic, and chlorosis, are deformed, and often exhibit shoestring-type symptoms. The fruits are also often deformed and bumpy. In papaya, PRSV infection is characterized by mosaic and chlorosis symptoms on leaves, water-soaked streaks on the petiole, and deformation of leaves that can result in shoestring-like symptoms that resemble mite damage (**Figure 1**). The virus can cause deformation and ringspot symptoms on the fruit, hence the name PRSV. Commercial PRSV-resistant transgenic papaya expressing the coat protein (*CP*) gene of the virus has been used to control PRSV P in Hawaii, as will be discussed later.

General Properties of PRSV

The virus particles are flexuous rods about 760–800 nm $\times$ 12 nm with single RNA of about 10 326 b in length. Virus particles consist of 94.5% protein and 5.5% nucleic acid by weight. It has a single coat protein (CP) of about 36 kDa. Analysis of purified virus preparations that are stored show that the CP degrades to smaller proteins of *c.* 31–34 and 26–27 kDa proteins, possibly due to proteolytic degradation. The density of the virion in purified preparations is 1.32 g cm^{-3} in CsCl.

PRSV should not be confused with another potyvirus, papaya leaf distortion mosaic virus (PLDMV), which occurs in Okinawa and other parts of Asia, such as Taiwan. This virus causes very similar symptoms as PRSV on papaya and cucurbits but is serologically unrelated and its CP shares only 55–59% similarity to that of PRSV.

PRSV Genome

A genetic map of PRSV genome with polyprotein processing sites and products is presented in **Figure 2** and their possible functions in **Table 1**. Much of the knowledge on the genome of PRSV has been obtained from extensive work done by the laboratory of Dr. Shyi-Dong Yeh of National Chung-Hsing University in Taiwan. The genomic RNA of PRSV is 10 326 nt in length excluding the poly(A) tract and contains one large open reading frame that encodes a polyprotein of 3344 amino acids starting at nucleotide position 86 and ending at position 10 120. A VPg protein is linked to the 5′ end of the RNA while a poly(A) tract is at the 3′ end. The polyprotein is cleaved into proteins designated (name (size in M_r)): P1 (63K), helper component (HC-Pro, 52K), P3 (46K), cylindrical inclusion protein (CI, 72K), nuclear inclusion protein a (NIa, 48K), nuclear inclusion protein b (NIb, 59K), coat protein (CP, 35K), as well as two other proteins 6K1 (6K) and 6K2 (6K). The cleaved proteins are arranged on the genome starting from the 5′ in order as: P1–(HC-Pro)–P3–6K1–CI–6K2–NIa–NIb–CP (**Figure 2**).

Figure 1 Symptoms of PRSV on papaya.

The cleavage proteins mentioned above have been identified by immunoprecipitation and dynamic precursor studies with PRSV-p as well as by extensive studies on proteolytic processing of polyproteins from other potyviruses. Three virus-encoded proteinases are responsible for at least seven cleavages: the P1 protein from N-terminus of the polyprotein autocatalytically liberates its own C-terminus, the HC-Pro also cleaves its own C-terminus, and NIa is responsible for *cis*- and *trans*-proteolytic processing to generate the CI, 6K, NIa, NIb, and CP proteins. NIa has also been shown to contain an internal cleavage site for delimitation of the genome-linked protein (VPg) and the proteinase (Pro) domains. Thus, the genomic organization and processing of the polyprotein of PRSV is similar to those of other potyviruses.

A rather interesting feature of the PRSV is that sequence analysis predicts two potential cleavage sites at the N-terminus of the CP. One of the sites (VFHQ/SKNF) predicts a CP of 33K and a NIb of 537 amino acids about 20 amino acids larger than those of other potyviruses. The second predicted cleavage site (VYHE/SRGTD) generates a CP of 35K and an NIa of 517 amino acids. There is no firm evidence to suggest that only one cleavage site is used. If both sites are used in polyprotein processing, one would expect heterogeneous products. This may explain why the analysis of purified CP preparations that are stored frequently shows the major ~36K form in addition to smaller CPs that are 2–5K smaller.

Sequence Diversity and Evolution

Knowledge of the sequence diversity among isolates of a virus has great implications in developing an effective virus disease management program and in understanding the origin and biology of the virus. Recently, numerous PRSV P and W sequences from virus isolated from different parts of the world have been reported in the sequence database. Amino acid and nucleotide sequence divergence among PRSV isolates differ by as much as 14%. These differences, interestingly, are considerably less than that found among isolates of other potyviruses, such as yam mosaic virus (YMV), that differ by as much as 28%. Although initial data from the USA and Australia suggested that there was little variation among PRSV isolates within these countries, more recent data from India and Mexico have suggested that the sequence

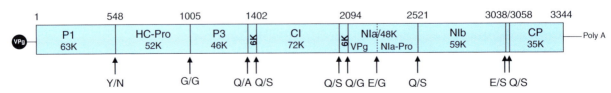

Figure 2 Genome map of PRSV. Vertical arrows indicate the proteolytic cleavage sites.

Table 1 PRSV proteins and their possible functions

Viral protein	Size (M_r)	Functions
P1	63K	Proteinase
		Cell-to-cell movement
HC-Pro	52K	Vector transmission
		Proteinase
		Pathogenicity
		Suppressor of RNA silencing
		Cell-to-cell movement
P3	46K	Unknown, but possible role in replication
6K1	6K	Unknown, but possible role in replication
CI	72K	Genome replication (RNA helicase)
		Membrane attachment
		Nucleic acid-stimulated ATPase activity
		Cell-to-cell movement
6K2	6K	Unknown, but possible roles in:
		• Replication
		• Regulation; inhibition of NIa nuclear translocation
NIaVPg	21K	Genome replication (primer for initiation of RNA synthesis)
NIaPro	27K	Major proteinase
NIb	59K	Genome replication (RNA-dependent RNA polymerase, RdRp)
CP	35K	RNA encapsidation
		Vector transmission
		Pathogenicity
		Cell-to-cell movement

variation between PRSV isolates in other countries may be greater than previously recognized. Heterogeneity in *CP* length ranging from 840 to 870 nt has been noted. The observed size differences in *CP* sequence occurred in multiples of three, preserving the reading frame between genes of different genomes and resulting in CPs of between 280 amino acids (Indian P isolate KA2) and 290 amino acids (VNW-38 from central Vietnam). Interestingly, the CP-coding region of all isolates from Thailand were 286 amino acids in length, while those from India and Vietnam demonstrated considerable heterogeneity in CP length, at 280–286 and 285–290 amino acids, respectively. The first 50 or so amino acids of the N-terminal region of the PRSV *CP* gene were found to be highly variable and all differences in CP length were confined to this region. The differences in this region that did exist consisted of conservative amino acid substitutions. The majority of the size differences occurred in one of two hypervariable regions and most were due to differences in the number of EK repeats.

A phylogenetic study based on *CP* sequences from 93 isolates of type P and W PRSV from different geographic locations was done by generating a phylogenetic tree using the neighbor-joining method. In the phylogenetic tree, sequences from one Sri Lankan isolate and two Indian isolates formed a sister cluster to the rest of the sequences. The other isolates formed two major lineages: I included all isolates from the Americas, Puerto Rico, Australia, and a few from South Asia; and II included isolates from Southeast Asia and the Western Pacific. Lineage I was the major of the two, containing three clusters of Brazilian isolates, two Indian isolates, and Australian, Mexican and US isolates. Within lineage I, the Brazilian and Mexican isolates were more diverse than the US and Australian isolates. Lineage II included all of the isolates from the Southeast Asia and Western Pacific, including China, Indonesia, Thailand, Vietnam, Taiwan, Japan, and the Philippines. However, the subclustering of isolates did not correlate well with their geographic origins; rather, they appeared to be a single mixed population with some well-defined subpopulations. These observations suggest that considerable movement of PRSV isolates has occurred among the Southeast Asian countries. Thai isolates of the P type diverged together, whereas PRSV W type diverged with other Southeast Asian isolates. Both P and W isolates of PRSV from Vietnam were intermingled with other Asian isolates. Sequence analysis showed that all Vietnamese isolates (except the P type from the southern part of the country) diverged from a common branch with P isolates from Japan and Taiwan while PRSV isolates from South Vietnam were diverged compared to those from the Philippines and seemed closely related to several W types from Thailand.

An interesting feature of PRSV is the origin of types P and W. As noted above, a number of viral diseases on cucurbits were historically associated with WMV-1 and not with any diseases of papaya. Did type P originate from W, or was it the reverse, or did they evolve independently? Evidences from various sources indicate that PRSV is primarily a pathogen of cucurbits, and that PRSV P originated from PRSV W. Work in Australia suggests that the recent outbreak of PRSV P came from the population of PRSV W already present in Australia. This suggestion is also supported by the diversity in cucurbit-infecting potyviruses that are phylogenetically related to PRSV.

Infectious Transcripts of Recombinant PRSV Help to Reveal Potential Determinants of Several Biological Characteristics

Determinants for Host Range Specificity

Studies utilizing the technique of producing infectious transcripts from recombinant viruses followed by bioassays of the transcripts have demonstrated that sequences of the PRSV genome responsible for determining papaya and cucurbit host specificity are not in the region of the *CP* gene. However, nucleotides 6509–7700 encoding the NIa gene and parts of the NIb gene were critical for papaya infection. Amino acids of the NIb gene of this region (nucleotides 7644–7700) between PRSV-P and PRSV-W type are identical, whereas sequence comparison of nucleotides 6509–7643 of four type P and two type W viruses showed that two amino acids at positions 2309 (K → D) and 2487 (I → V) of PRSV are significantly different between papaya-infecting type P and non-papaya-infecting type W. Further point mutational studies in these sites indicated that these two amino acids located in the NIa proteinase are responsible for conferring the ability to infect papaya.

Determinants for Local Lesion Formation on *Chenopodium*

As noted earlier, PRSV causes local lesions on *C. amaranticolor* and *C. quinoa*. The severe strain PRSV HA strain from Hawaii causes local lesions on *C. quinoa* but a mild nitrous acid mutant of it, PRSV HA 5-1, does not. Recombinant infectious viruses were generated by exchanging genome parts between PRSV HA and PRSV HA 5-1. The study revealed that the pathogenicity-related region is present between nucleotide positions 950 and 3261 of the PRSV HA genome and mutations in the *P1* and *HC-Pro* genes resulted in the attenuation of PRSV HA symptoms and the loss of ability to produce local lesions on *C. quinoa*. The *HC-Pro* gene of PRSV is the major determinant factor for local lesion formation.

Determinants on Severity of Symptoms, Suppression of Gene Silencing, Infection of Transgenic Papaya

Virus–host interaction studies based on recombinant analyses between severe and mild strains of PRSV indicated that the *HC-Pro* gene plays an important role in viral pathogenicity and virulence and acts as a suppressor of the gene-silencing defense mechanism in the papaya host plant. In addition, the comparative reaction of recombinant PRSV with chimeric *CP* gene sequences showed that heterologous sequences and their position in the *CP* gene influences their pathogenicity on PRSV-resistant transgenic papaya.

An interesting phenomenon of strain-specific cross-protection was observed in papaya and horn melon provided by the mild strain HA 5-1 of PRSV. The PRSV mild mutant HA 5-1 provided 90–100% protection against the severe parental strain PRSV HA in greenhouse and field conditions. However, the degree of protection provided in horn melon by HA 5-1 against a PRSV type-W strain from Taiwan was only 20–30%. Studies on strain-specific cross-protection phenomenon indicated that the recombinant HA 5-1 carrying both the heterologous *CP* and the 3′ untranslated region (UTR) of the PRSV W from Taiwan significantly enhanced the protection against the Taiwan strain in cucurbits. However, chimeric HA 5-1 virus carrying either heterologous *CP* or heterologous 3′ UTR showed reduced effectiveness of protection against PRSV HA in papaya when compared to protection by the native mild HA 5-1.

Similar to other potyviruses, PRSV is also transmitted by insect vector aphids (*Myzus persicae* and *Aphis gossypii*) in a nonpersistent manner. Detailed studies with other potyviruses show that HC-Pro/virions interaction is essential for aphid transmission of potyviruses. Although not empirically tested, it would seem likely that aphid transmission of PRSV would similarly be governed by HC-Pro/virion interactions.

Pathogen-Derived Resistance for Controlling PRSV: The Hawaii Case

The control or management of PRSV has been approached through practices such as quarantine, eradication, avoidance by planting papaya in areas isolated from the virus, continual rogueing of infected plants, use of tolerant lines to reduce damage caused by PRSV, cross-protection through the use of mild virus strains, and resistance using the approach of 'pathogen-derived resistance'. The efforts to control PRSV in Hawaii are described because it involves all of the above practices. Ultimately, the most successful has been the 'pathogen-derived resistance' approach.

The state of Hawaii consists of eight main islands that are in rather close proximity to each other with the shortest and farthest distance between islands being 7 mile between Maui and Kahoolawe and 70 mile between Kauai and Oahu. Travels between the islands are prevalent with the exception of Kahoolawe, which is not inhabited, and Niihau, which is privately owned. PRSV was first detected in the 1940s on Oahu where Hawaii's papaya industry was located at that time. Efforts to control the virus on Oahu largely consisted of state officials and farmers continually monitoring for infected plants and rogueing them, especially in areas where the virus was not prevalent. However, by the late 1950s, PRSV was causing extensive damage, which caused the papaya industry to relocate to Puna on the island of Hawaii.

The relocation of the industry to Puna was timely and effective because Puna had an abundance of land that was suitable to grow Kapoho, a cultivar of excellent quality that adapted to the volcanic soil base there, allowing excellent drainage, had high rainfall and yet lots of sunshine, and the land there could be bought or leased at reasonable prices. By the 1970s, the Kapoho papaya grown in Puna accounted for 95% of the state's papaya production, making papaya the second most important fruit crop behind pineapple.

Despite strict quarantine on movement of papaya seedlings between islands, PRSV was discovered in the town of Hilo which was only about 18 miles away from the center of the papaya-growing area of Puna. However,

PRSV was indeed discovered in Puna in May 1992 (**Figure 3(a)**) and the Hawaiian papaya industry would be forever changed. By 1995, a third of the papaya growing area was completely infected and much of the rest of Puna had widespread infection (**Figure 3(b)**). By 1998, the production of papaya in Puna had dropped to 27 million pounds of papaya from 52 million pounds in 1992 when PRSV was discovered in Puna. In retrospect, the efforts of quarantine, monitoring and rogueing of infected plants in Hilo, and suppression efforts of PRSV in Puna all played key roles in helping Hawaii's papaya industry, because it gave researchers time to develop control measures for PRSV.

Research to develop tolerant varieties and cross-protection measures were started in the 1970s. Since resistance to PRSV has not been identified in *Carica papaya*, researchers have used tolerant germplasm in an attempt to develop papaya cultivars with acceptable PRSV tolerance and horticultural characteristics. However, tolerance to PRSV is apparently governed by a family of genes that is inherited quantitatively, which makes it technically difficult to develop cultivars of acceptable horticultural quality. Furthermore, the tolerant lines do become infected with PRSV, although fruit production continues still at a lower level. Indeed, in Thailand, the Philippines, and Taiwan, a number of tolerant lines have been developed and are used. However, Hawaii grows the small 'solo'-type papaya and efforts to introduce tolerance into acceptable 'solo' papaya cultivars have not been successful.

Efforts to use cross-protection were similarly started in the late 1970s to control PRSV in Hawaii. Cross-protection can be defined as the use of a mild strain of virus to infect plants that are subsequently protected against economic damage caused by a severe strain of the same virus. This practice has been used successfully for many years to minimize damage by citrus tristeza virus in Brazil, for example. In the early 1980s, a mild strain of PRSV (described above as PRSV HA 5-1) was developed through nitrous acid treatment of a severe strain, PRSV HA isolated from Oahu island. This mild strain was tested in Hawaii on Oahu island and showed good protection against damage by severe strains but produced symptoms that were very obvious on certain cultivars, such as Sunrise, especially in the winter months. This prominent symptom induction on certain cultivars and the logistics of mild strain buildup and inoculation of plants, among others, were factors that caused it not to be consistently used on the island of Oahu. There was no justification to use it on the island of Hawaii because PRSV was not yet found in Puna during the 1980s. Interestingly, the mild strain was used extensively for several years in Taiwan, but it did not afford sufficient protection against the severe strains from Taiwan and thus its use was abandoned after several years.

Transgenic Resistance

In the mid-1980s, an exciting development on tobacco mosaic virus (TMV) provided a rationale that resistance to plant viruses could be developed by expressing the viral *CP* gene in a transgenic plant. This approach was called CP-mediated protection, and, at about the same time, a report introduced the concept of 'parasite-derived resistance'. The report on transgenic resistance to TMV set off a flurry of work in many laboratories to determine if this approach could be used for developing resistance to other plant viruses. Likewise, work was initiated in 1985 to use this approach for developing PRSV-resistant transgenic papaya for Hawaii.

Key requirements for successful development and commercialization of transgenic virus-resistant plants are the isolation and engineering of the gene of interest, vectors for mobilization into and expression of the gene in the host, transformation and subsequent regeneration of the host cells into plants, effective and timely screening of transformants, testing of transformants, and the ability to deregulate and commercialize the product.

The *CP* gene of the mild strain of PRSV was chosen as the 'resistance' gene because it had been recently cloned and it was of the PRSV P type. The gene was engineered into a wide host range vector that could replicate in *Escherichia coli* as well as in *Agrobacterium tumefaciens*, the bacterium used for one of the most widely used methods of plant transformation. The commercial cultivars Kapoho, Sunrise, and Sunset were chosen for transformation. Initially, transformation of papaya was attempted using the *Agrobacterium*–leaf piece approach, where leaf pieces would be infected with *Agrobacterium* harboring the *CP* gene and transformed cells would be regenerated via organogenesis into transgenic plants. The latter is the direct regeneration of cells from an organ such as the leaf. This approach did not work due to our failure to develop plants from leaf pieces. A shift to the transformation of somatic embryos via the biolistic (often referred to as the gene gun) approach resulted in obtaining of about a dozen transgenic papaya lines, four of which expressed the *CP* gene. In 1991, tests of the R_0 lines identified a transgenic Sunset that expressed the *CP* gene of PRSV HA 5-1, and showed resistance to PRSV from Hawaii. A field trial of R_0 plants was started in April 1992 on as the island of Oahu, and a month later PRSV was discovered at Puna in May 1992, as discussed above.

The Oahu field trial showed that R_0 plants of line 55-1 were resistant, and line 55-1 was further developed to obtain the cultivar 'SunUp' which is line 55-1 that has the *CP* gene in a homozygous state, and 'Rainbow' which is an F_1 hybrid of SunUp and the nontransgenic 'Kapoho'. SunUp is red-fleshed and Rainbow is yellow-fleshed. In 1995, SunUp and Rainbow were tested in a subsequent field trial in Puna and showed excellent resistance (**Figure 3(c)**). Due to its

Figure 3 (a) Healthy Puna papaya in 1992. (b) Severely infected papaya orchards in Puna in 1994. (c) Field trial of transgenic papaya. PRSV-infected nontransgenic papaya on left and PRSV-resistant transgenic papaya on right. (d) Commercial planting of transgenic papaya one year after releasing seeds of PRSV-resistant transgenic papaya. (e) Transgenic papaya commonly sold in supermarkets. (f) Risk of growing nontransgenic papaya still exists in 2005. Foreground is PRSV-infected nontransgenic papaya that are cut, and background shows healthy PRSV-resistant transgenic papaya.

yellow flesh and good shipping qualities, Rainbow was especially preferred by the growers. Line 55-1 was deregulated by the US government and commercialized in May 1998. The deregulation also applied to plants that were derived from line 55-1. The timely commercialization of the transgenic papaya in 1998 was crucial since PRSV had decreased papaya production in Puna by 50% that year compared to 1992 production levels. The transgenic Rainbow papaya was quickly adopted by growers and recovery of papaya production in Hawaii was underway

(**Figure 3(d)**). The transgenic papaya is sold throughout Hawaii (**Figure 3(e)**) and the mainland USA, and to Canada where it was deregulated in 2003. However, several challenges remain: coexistence, exportation of nontransgenic papaya to Japan, deregulation of transgenic papaya in Japan, and the adoption of transgenic papaya in other countries that suffer from PRSV.

Hawaii still needs to grow nontransgenic papaya to satisfy the lucrative Japanese market as well as for production of organic papaya, for example. Interestingly, the islands of Kauai and Molokai do not have PRSV but grow only limited acreage of papaya. This situation illustrates the point that many factors influence decisions on the localities and crops that are grown. In Hawaii, Puna is the best place to grow papaya for the reasons mentioned above; there is a lot of land, farmers there are intuned to growing the crop, the region receives plenty of water, sunshine, and has a well-drained lava-based 'soil' structure, and there are high-quality cultivars adapted to the local growing conditions. The disadvantage is PRSV, but that disadvantage was overcome through the introduction of the PRSV-resistant Rainbow papaya that has good commercial attributes plus virus resistance. Puna accounts for 90% of Hawaii's papaya and, as of 2005, Rainbow represents 66% of the papaya grown in Puna. Growing nontransgenic papaya can be risky because PRSV is still around (**Figure 3(f)**), but judicious use of isolation from virus sources and constant rogueing can provide a means of raising nontransgenic papaya. However, a major market is Japan, which still has not deregulated the transgenic papaya. To maintain the lucrative Japanese market, Hawaii has to continue the exportation of nontransgenic papaya to Japan. The immediate solution is to concurrently grow nontransgenic and transgenic papaya, and subsequently to deregulate the transgenic papaya so it can be freely shipped into Japan. What approaches are being taken?

Currently, Japan accepts nontransgenic papaya but it needs to be free of 'contamination' by transgenic papaya. The Hawaii Department of Agriculture (HDOA) and the Japan Ministry of Agriculture, Forestry and Fisheries (MAFF) have agreed on an 'identity preservation protocol' (IPP), in which nontransgenic papaya can even be grown in close proximity (coexistence) to transgenic papaya in Puna, for example, and still be shipped to Japan. The protocol involves a series of monitoring and checkpoints in Hawaii that allow direct marketing of the papaya without delay while samples of the shipment are spot-checked by MAFF officials in Japan. The process has worked very well and allowed Hawaii to maintain its market share in Japan. This represents a practical case of 'coexistence' of transgenic and nontransgenic papaya and fruitful collaboration between governments (Hawaii and Japan) that provides mutual benefits to all parties.

The ideal situation would be, however, to freely export nontransgenic and transgenic papaya to Japan. To this end, efforts are underway to deregulate the transgenic papaya in Japan by obtaining approval from Japanese governmental agencies such as MAFF, the Ministry of Health, Labor, and Welfare (MHLW), and the Ministry of the Environment (MOE). MAFF has provisionally approved Hawaii's transgenic Rainbow and SunUp papayas, and efforts to present the final documentation to all three agencies are nearing completion. Deregulation of the transgenic papaya in Japan not only would expand the Hawaiian transgenic market but would also be a good case study for evaluating the effectiveness of commercialization and marketing of fresh transgenic products, since the transgenic papaya in Japan would be labeled, and subsequently consumers there would be given an opportunity to make a personal choice between a transgenic and nontransgenic product. The implications of this opportunity are obvious given the current 'controversial' climate over genetically modified organisms (GMOs) in the world. It is indeed rare that the previously little known PRSV could perhaps provide an example that would help us to resolve such controversies.

Summary Remarks

PRSV has been thoroughly characterized and is a typical member of the family *Potyviridae*, arguably the largest and economically most important plant virus group. The complete genome sequence has been elucidated and infectious transcripts have provided a means to determine the genetic determinants of some important biological functions such as host range and virulence. Furthermore, pathogen-derived resistance has been used to control PRSV in Hawaii through the use of virus-resistant transgenic papaya. In the US, only three virus-resistant transgenic crops have been commercialized: squash, papaya, and potato. The transgenic virus-resistant papaya provides a potential means to test the global acceptance of GMOs while presenting a plausible approach to control a disease affecting papaya worldwide.

See also: Plum Pox Virus; Potato Virus Y; Plant Resistance to Viruses: Engineered Resistance; Watermelon Mosaic Virus and Zucchini Yellow Mosaic Virus; Potyviruses.

Further Reading

Bateson M, Henderson J, Chaleeprom W, Gibbs A, and Dale J (1994) Papaya ringspot potyvirus: Isolate variability and origin of PRSV type P (Australia). *Journal of General Virology* 75: 3547–3553.

Fuchs M and Gonsalves D (2007) Safety of virus-resistant transgenic plants two decades after their introductions: Lessons from realistic field risk assessments studies. *Annual Review of Phytopathology* 45: 173–202.

Gonsalves D (1998) Control of papaya ringspot virus in papaya: A case study. *Annual Review of Phytopathology* 36: 415–437.

Gonsalves D, Gonsalves C, Ferreira S, et al. (2004) Transgenic virus resistant papaya: From hope to reality for controlling of papaya ringspot virus in Hawaii. APSnet Feature, American Phytopathological Society, Aug.–Sept. http://www.apsnet.org/online/feature/ringspot/ (accessed January 2008).

Gonsalves D, Vegas A, Prasartsee V, Drew R, Suzuki JY, and Tripathi S (2006) Developing papaya to control papaya ringspot virus by transgenic resistance, intergeneric hybridization, and tolerance breeding. Plant Breeding Reviews 26: 35–78.

Yeh SD and Gonsalves D (1994) Practices and perspective of control of papaya ringspot virus by cross protection. In: Harris KF (ed.) Advances in Disease Vector Research, pp. 237–257. New York: Springer.

Yeh SD, Jan FJ, Chiang CH, et al. (1992) Complete nucleotide sequence and genetic organization of papaya ringspot virus RNA. Journal of General Virology 73: 2531–2541.

Papillomaviruses: General Features of Human Viruses

G Orth, Institut Pasteur, Paris, France

© 2008 Elsevier Ltd. All rights reserved.

History

The recognition of different clinical types of warts and the suspicion of the venereal transmission of anogenital warts date from Greco-Roman times. Transmission of skin warts to humans by cell-free wart extracts demonstrated their viral etiology in 1907. Rabbit skin warts and derived carcinomas induced by the cottontail rabbit (Shope) papillomavirus (PV) provided the first model of viral carcinogenesis in mammals in the 1930s. The role of PVs in human cancer remained unexplored for a long time, mainly because of the lack of tissue culture systems allowing their propagation and study. It was long assumed that all types of human warts were due to a single virus, the human wart virus, until advances in DNA technology allowed the demonstration of the plurality of the human papillomavirus (HPV) in the late 1970s. It was proposed that the diversity of HPV-associated lesions resulted from distinct biological properties of the different HPV types. More than 100 HPV genotypes have been characterized so far.

The specific association of HPV type 5 (HPV-5) with the skin cancers of epidermodysplasia verruciformis (EV), a rare genetic disease, provided the first molecular evidence for the role of HPVs in human cancer. In the mid-1970s, koilocytes in cervicovaginal smears were recognized as HPV-infected cells, and their frequent association with features of cervical dysplasia suggested a role of HPV in cervical carcinogenesis. Specific HPV types (HPV-16 and HPV-18) were isolated from cervical cancers in the early 1980s. A wealth of virological, epidemiological, and molecular data now support the conclusion that specific HPV genotypes are etiologically related to over 95% of cervical carcinomas, the second most common cancer in women worldwide. For a long time, the prevention of cervical cancer relied only on the Papanicolaou (Pap) test, which involved the cytological detection of intraepithelial cancer precursors. Virological tests allowing the detection of the potentially oncogenic HPV genotypes now represent a means to improve the sensitivity of the screening of women at risk. A vaccine aimed at preventing infection with HPV-16 and HPV-18 is now available, paving the way for primary prevention of cervical cancer.

Virion Structure and Genome Organization

HPVs are characterized by a small (52–55 nm diameter), nonenveloped, icosahedral capsid composed of 72 pentameric capsomeres (**Figure 1**). Their genome is a double-stranded, covalently closed, circular DNA molecule of 7500–8000 bp. The viral DNA is associated with cellular histones to form a chromatin-like structure. The genetic information is located on a single DNA strand and consists of at least eight open reading frames (ORFs). The viral genome is divided into three regions (**Figure 2**): (1) a noncoding, long regulatory region (LRR) or upstream regulatory region (URR), which contains transcription and replication regulatory elements; (2) the early (E) region, which encodes proteins (E6 and E7) interacting with cellular proteins regulating the cell cycle, proteins (E1 and E2) involved in the replication and transcriptional regulation of the viral genome, a protein (E4) expressed late in the viral life cycle, and a small hydrophobic protein with growth-promoting properties, which is encoded by an ORF present in most HPV types either between E2 and L2 (E5) or overlapping E6 (E8); (3) the late (L) region, which encodes the capsid proteins, the major L1 protein constituting the capsomeres, and the minor L2 protein involved in viral DNA transport and encapsidation.

Taxonomy and Classification

The International Committee on the Taxonomy of Viruses recognized the family *Papillomaviridae* in 2005. More than 100 HPV types have been described so far, based on the

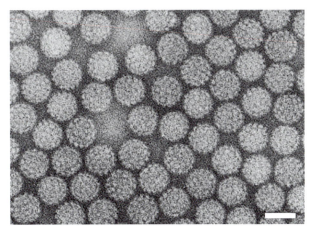

Figure 1 HPV-1 particles purified from plantar warts. Scale = 50 nm. Courtesy of Dr. O. Croissant.

isolation of their complete genome and comparison of their nucleotide sequences. Two PV genomes are recognized as distinct types (genotypes) if they share less than 90% identical nucleotides in the L1 ORF, the most conserved genomic region. Types were designated by a number (1–106 so far), following the chronological order of their characterization. The PV taxonomy adopted recently is based on the phylogenetic relationships of the L1 ORF of animal and HPVs (**Figure 3**). Higher-order assemblages are considered as genera (identified by Greek alphabetical prefixes) and lower-order assemblages as species (each species identified by the number of the best known type). Genera share less than 60% nucleotide identity in the L1 ORF, species within a genus share 60–70%, and types within a species share 71–89%.

HPV types are distributed in five of the 16 PV genera (**Figure 3**). The genus *Alphapapillomavirus* comprises types differing by their genital, oral, or cutaneous tropism and by their pathogenicity (low-risk types associated with benign proliferations and high-risk types associated with invasive carcinomas). The genera *Gammapapillomavirus*, *Mupapillomavirus*, and *Nupapillomavirus* comprise types associated with cutaneous warts, whereas the types associated with EV belong to the genus *Betapapillomavirus* (**Figure 3** and **Table 1**). The genera differ by the presence of an E5 (*Alphapapillomavirus*) or E8 (*Gammapapillomavirus*, *Mupapillomavirus*) ORF or the absence of an E5/E8 ORF (*Betapapillomavirus*). Species comprise related PV types that often share similar biological and pathogenic properties. A large number of presumed novel HPV types, most of them belonging to the genera *Betapapillomavirus* and *Gammapapillomavirus*, have been partially characterized by sequencing short amplicons obtained with degenerate or consensus L1 primers. The number of HPV types is likely to exceed 200.

Within types, DNA sequence identities in the range of 90–98% correspond to subtypes, and isolates with more than 98% sequence identity represent variants. Whereas subtypes have only been described for a few HPV types

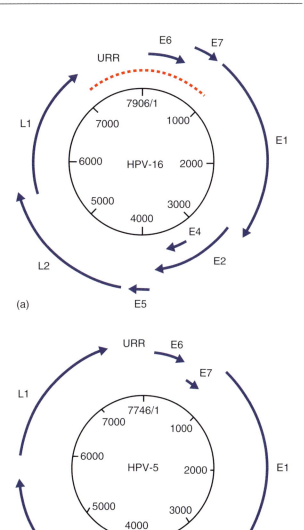

Figure 2 Genetic maps of (a) HPV-16 and (b) HPV-5. Note the lack of an E5/E8 ORF in HPV-5. The minimal HPV-16 genome fragment retained in carcinoma cells after integration into the host genome is indicated in red.

(HPV-5, HPV-20, HPV-34, HPV-44, and HPV-68), variants have been identified for each HPV type, especially for the most clinically relevant genital types and HPV-5. Phylogenetic grouping of the HPV-16 L1 DNA variants identified in different geographic regions and ethnic groups has allowed recognition of five lineages (African-1, African-2, European, Asian, and African-American). A similar clustering of variants has also been reported for other genital types. The intratype variability does not usually exceed 2% in the coding region and 5% in the LRR. Some studies suggest that HPV-16 variants differ by their biological properties and oncogenic potential. More studies are needed to understand the impact of genetic intratypic variability on the persistence of infections with

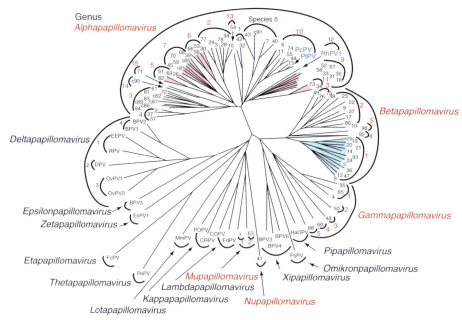

Figure 3 Phylogenetic tree of human and animal PVs based on the L1 ORF sequences of 118 PV types. Genera and species (the latter numbered according to the groups listed in **Table 1**) comprising HPV types are indicated in red. The species containing high-risk types associated with anogenital cancers are shaded in pink. The species containing HPV types associated with EV skin carcinomas is shaded in light blue. Types infecting nonhuman primates are indicated in blue. Reproduced from de Villiers EM, Fauquet C, Broker TR, Bernard HU and zur Hausen H (2004) Classification of Papillomaviruses. *Virology* 324: 20, with permission from Elsevier.

carcinogenic HPV types and the risk of malignant progression of the associated lesions. It should be stressed that HPV variants are stable entities. The observed HPV intratypic variability likely reflects the very slow co-evolution of these viruses with their host.

Host Range, Tissue Tropism, and Transmission

HPVs are highly host and tissue specific. Humans are not susceptible to infection by animal PVs, and HPVs do not infect other species. However, HPV capsids agglutinate mouse erythrocytes, and oncogenic genital HPVs are capable of transforming established rodent cell lines, which points to a ubiquitous cellular receptor. Tumors developed after grafting HPV-16-transformed cells to syngeneic mice provide models to test immunotherapeutic approaches. Transgenic mice for the HPV-16 or HPV-18 E6 and E7 genes develop benign or malignant cutaneous tumors.

HPVs infect keratinocytes of keratinizing or nonkeratinizing, pluristratified squamous epithelia. Some HPV types infect the epidermis of the skin, while others infect the epithelial component of the anogenital and oral mucous membranes. A major target for oncogenic genital HPVs is the squamous epithelium resulting from glandular metaplasia, at the squamocolumnar junction of the uterine cervix. Almost all cervical intraepithelial precancerous lesions and invasive carcinomas originate in this so-called transformation zone. The reserve cells located beneath the endocervical single-layered, columnar epithelium are likely to develop into either squamous metaplastic cells or columnar (glandular) cells, and this may account for the development of HPV-associated squamous cell carcinoma (SCC) or adenocarcinoma and their precursor lesions. Esophageal, tracheobronchial, nasal, conjunctival, and urinary epithelia are possible sites of infection with the known HPVs.

HPVs are very stable viruses, which are released from the desquamating superficial, differentiated keratinocytes of the infected epithelia. Infection of basal cells proceeds through minor abrasions of the epithelium. Transmission of cutaneous types occurs through direct contact with an infected tissue or contaminated objects and surfaces. Genital HPVs are generally transmitted by sexual contact. Transmission of HPV-6 and HPV-11 by passage through an infected birth canal is at the origin of laryngeal papillomatosis of young children.

Virus Replication

In vivo, the life cycle of low-risk and high-risk HPVs is closely linked to the biology of the keratinocyte. Keratinocytes are responsible for the renewal, cohesion, and barrier function of pluristratifying epithelia. HPVs most likely target the slow-cycling, self-renewing stem cells located in the basal layer of the epithelium and in the bulge of hair follicles, as well as their transient amplifying cell progeny.

Candidate cell-surface receptors are heparan sulfate proteoglycans and α6 integrin. Virus entry may result in a latent, asymptomatic infection, which is a still poorly understood phenomenon. The maintenance of viral DNA in stem cells, as autonomous replicating episomes, would only require the E1 and E2 proteins. Infection may also lead to the development of a productive lesion within 1–3 months. Depending on the tissue and the HPV type, these lesions are known as warts, papillomas, or low-grade anogenital intraepithelial lesions (**Table 1**). Lesions develop as a consequence of the proliferation of infected basal and parabasal cells maintaining episomal viral genomes at low copy number. This involves the expression of E1, E2, E5/E8, E6, and E7 genes. The E5/E8, E6, and E7 proteins likely release stem cells from their tight micro-environmental control and, through their interaction with host cell proteins, activate the host cellular DNA synthesis machinery and inhibit apoptosis in response to unscheduled DNA synthesis. This strategy allows the productive phase of the life cycle to proceed in the growth-arrested, terminally differentiating keratinocytes. Vegetative viral DNA replication, transcription of the E4, L1, and L2 genes from a differentiation-dependent promoter, synthesis of the late proteins, and viral particle assembly take place as the differentiating cells migrate toward the most superficial layers. Virion-containing squames are released into the environment (**Figures 4** and **5**). HPV replication provokes a cytopathic effect, such as the accumulation of cytoplasmic

Table 1 Main HPV genotypes and associated diseases

Genus and species[a]	Group[a]	Types[b]	Tropism	Diseases
Alphapapillomavirus				
Human papillomavirus 32	1	32	Oral	Focal epithelial hyperplasia
		42	Genital	Low-grade lesions
Human papillomavirus 10	2	3, 10, 28, 29, 77	Cutaneous	Flat and common warts
Human papillomavirus 61	3	61, 72, 81, 83, 84	Genital	Low-grade lesions
Human papillomavirus 2	4	2, 27, 57	Cutaneous	Common warts
Human papillomavirus 26	5	26, **51**, 69, **82**	Genital	Low- and high-grade lesions, invasive carcinomas
Human papillomavirus 53	6	30, 53, **56**, **66**	Genital	Low- and high-grade lesions, invasive carcinomas
Human papillomavirus 18	7	**18, 39, 45, 59, 68**, 70	Anogenital	Low- and high-grade lesions, invasive carcinomas
Human papillomavirus 7	8	7	Cutaneous	Warts found in butchers
		40, 43	Genital	Low-grade lesions
Human papillomavirus 16	9	**16, 31, 33, 35, 52, 58**, 67	Anogenital	Low- and high-grade lesions, invasive carcinomas
Human papillomavirus 6	10	6, 11	Anogenital	Warts, cervical low-grade lesions, Buschke–Löwenstein tumors
			Laryngeal	Recurrent papillomatosis
		13	Oral	Focal epithelial hyperplasia
Human papillomavirus 34	11	34, **73**	Genital	Low- and high-grade lesions, invasive carcinomas
Human papillomavirus 54	13	54	Genital	Low-grade lesions
Human papillomavirus 71	15	71	Genital	Low-grade lesions
Betapapillomavirus				
Human papillomavirus 5	1	**5, 8**, 12, **14**, 19, **20**, 21, 24, 25, 36, **47**, 93	Cutaneous	EV
Human papillomavirus 9	2	9, 15, **17**, 22, 23, 37, 38, 80	Cutaneous	EV
Human papillomavirus 49	3	49, 75, 76	Cutaneous	EV
Human papillomavirus c and 92	4	92	Cutaneous	Premalignant/malignant lesions?
Gammapapillomavirus				
Human papillomavirus 4	1	4, 65	Cutaneous	Warts
Human papillomavirus 50	2	50	Cutaneous	Warts
Human papillomavirus 48	3	48	Cutaneous	Warts
Human papillomavirus 60	4	60	Cutaneous	Warts
Mupapillomavirus				
Human papillomavirus 1	1	1	Cutaneous	Plantar warts
Human papillomavirus 63	2	63	Cutaneous	Warts
Nupapillomavirus				
Human papillomavirus 41	1	41	Cutaneous	Warts

[a]Genera and species are defined in the *Eighth Report of the International Committee on Taxonomy of Viruses*. Groups of phylogenetically related types corresponding to species were numbered earlier according to a different scheme, which is utilized in **Figures 3** and **6**.
[b]Genotypes found in cancers are indicated in bold type.

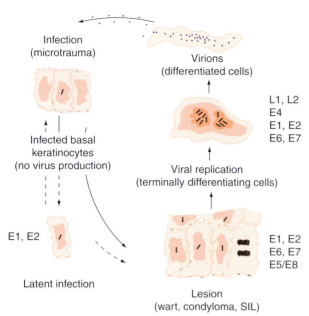

Figure 4 Life cycle of HPVs in keratinocytes. Viral genomes are retained as low-copy number episomes in the nuclei of latently infected cells and in basal and parabasal cells of benign lesions. Vegetative viral DNA replication occurs in more superficial, terminally differentiating cells. ORFs expressed are indicated.

inclusions (containing the viral E4 protein) pathognomonic of HPV-1-induced skin warts, or the perinuclear cavity, binucleation, and heavily stained nuclei that characterize the koilocytes observed in anogenital lesions. In low-grade anogenital lesions, such as cervical intraepithelial neoplasia grade 1 (CIN1), basaloid cells and mitotic figures occupy the lower third of the epithelium, and koilocytosis is frequently marked. In high-grade lesions, neoplastic basaloid cells and mitotic figures occupy the lower two-thirds of the epithelium in CIN grade 2 (CIN2) or the full thickness of the epithelium in CIN grade 3 (CIN3). Koilocytes are often still identified in CIN2 but little, if any, virus production occurs in CIN3. Infection of endocervical glandular cells does not lead to the production of virions.

Routine propagation of HPVs *in vitro* has not yet been achieved. Until recently, the production of infectious viral particles has relied on cumbersome methods involving xenografts or organotypic raft cultures. Oncogenic genital HPV types are capable of immortalizing human keratinocytes in culture, and cell lines (e.g., HeLa or Caski cells) have been derived from cervical cancers. The viral genome is integrated into the cellular genome, which precludes virus multiplication in such cell lines. Cell lines harboring viral episomes have been isolated from cervical intraepithelial lesions or obtained by transfecting cloned genomes of genital HPVs into primary keratinocyte cultures. The late viral functions are expressed and viral particles are produced when cells are grown on collagen rafts at an air–liquid interface, in conditions allowing terminal differentiation of keratinocytes (organotypic cultures).

High-yield methods have been developed recently for producing HPV pseudoviruses carrying a reporter gene or for obtaining infectious HPV particles without any requirement for epithelial cell differentiation (co-transfection of codon-optimized L1 and L2 expression plasmids and full-length HPV genomic DNA). Pseudoviruses and infectious virions represent useful tools for studying the first steps of infection and performing neutralization assays.

Evolution

It is acknowledged that PVs have co-evolved with their hosts, with estimated nucleotide change rates of 1% per 100 000 to 1 000 000 years. All known HPV types are assumed to have already infected the anatomically modern humans who emerged in East Africa about 200 000 years ago and spread out of Africa some 60 000–50 000 years ago. The time required for the emergence of species, types, and subtypes can be inferred from a co-evolutionary model on the basis of the phylogenetic relationships between alphapapillomaviruses of monkeys (rhesus macaque (RhPV)), apes (common chimpanzee (PtPV) and pigmy chimpanzee (PcPV)), and humans (HPVs) (**Figure 3**). Specifically, the divergence between

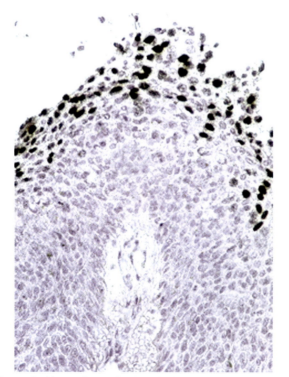

Figure 5 Vegetative replication of HPV-16 DNA in the nuclei of superficial cells of a high-grade cervical intraepithelial neoplasia, as detected by *in situ* hybridization. Courtesy of Dr. O. Croissant.

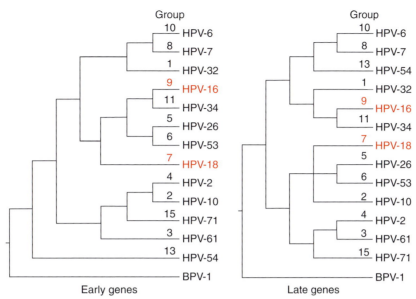

Figure 6 Phylogenetic trees of human alphapapillomaviruses based on the nucleotide sequences of early (E6, E7, E1, E2) and late (L2, L1 genes). The species corresponding to the groups shown are listed in **Table 1**. HPV-16 (group 9) and HPV-18 (group 7) show different groupings in the early and late phylogenies. From Narechania A, Chen Z, DeSalle R, and Burk RD (2005) Phylogenetic incongruence among oncogenic genital alpha human papillomaviruses. *Journal of Virology* 79: 15503–15510, with permission from the American Society for Microbiology.

Old World monkeys and the common ancestor of apes and humans, between the chimpanzee and human lineages, and between common and pigmy chimpanzees occurred approximately 25, 6, and 0.9 million years ago, respectively. Thus, emergence of a PV species would take 10–25 million years, with types diverging within a few million years and subtypes within less than 1 million years.

The mechanisms that have led to the extreme diversity among HPVs and their distinct tissue tropisms are little understood. It is likely that both selection and random genetic drift as well as founder effects were involved, as in human evolutionary history. The extent to which interspecies transmission and recombination events contributed to diversification remains an open question. Comparison of phylogenies of early and late genes of oncogenic genital HPV types disclosed a phylogenetic incongruence, which suggests that their evolution may not have been monophyletic across all genes (**Figure 6**). This may have involved an ancient recombination event in the lineage derived from a common ancestral oncogenic PV or an asymmetric genome convergence driven by selective constraints on particular genes.

Serological Relationships and Variability

Virions can be purified from lesions for some skin HPV types only, and early studies led to the identification of type-specific antigens and group-specific (type-common) antigens, using antibodies against intact or disrupted viral particles. HPV capsids are composed of 360 L1 molecules (72 capsomeres) and 12–30 L2 molecules displaying their N-terminal region on the capsid surface. When produced in various eukaryotic expression systems, L1 proteins self-assemble into virus-like particles (VLPs), which are morphologically similar to virions. When co-expressed with L1, L2 is incorporated stably into the capsids. Monoclonal antibodies raised against L1 VLPs have allowed the identification of both conformational, usually type-specific epitopes and surface or cryptic linear epitopes that show various levels of cross-reactivity when analyzed by an ELISA test using intact or denatured VLPs. Distinct serotypes have been identified only for HPV-5. Only the closely related types HPV-6 and HPV-11, or HPV-18 and HPV-45, exhibited detectable cross-reactions when the serological relatedness of genital HPV types was assessed using a hemagglutination inhibition test.

Neutralizing antibodies can be demonstrated by a xenograft infectivity assay, a cell culture focus assay (HPV/bovine PV-1 pseudovirions), assays based on HPV pseudoviruses carrying a reporter gene, or antibody displacement tests. Type-specific neutralizing antibodies develop in response to natural HPV infections and VLP immunization. They recognize conformational epitopes mapped to surface-exposed, hypervariable loops of the L1 protein. HPV-16 variants belong to a single serotype. VLPs of closely related types (HPV-6/HPV-11, HPV-16/HPV-31, HPV-31/HPV-33, and HPV-18/HPV-45) share one or more cross-neutralization epitopes that are less immunogenic than type-specific epitopes. Broadly cross-neutralizing epitopes shared by cutaneous and mucosal human and animal PVs have been found at the N-terminus of the L2 protein.

Epidemiology

Epidemiologic studies on HPV infections have been hindered by the great multiplicity and phylogenetic diversity of HPV types and by methodological problems. The Southern blot hybridization technique (without prior amplification of viral sequences), the first gold standard for detecting and identifying HPVs, has a low sensitivity and is labor intensive. The US FDA-approved Hybrid Capture 2 test, based on liquid hybridization of HPV DNA sequences with a mixture of 13 high-risk HPV RNA probes and the chemiluminescent detection of hybrids, does not distinguish individual types. This test is useful as an adjunct to Pap smear for the management of equivocal cervical cytology and screening for cervical cancer. HPV detection and genotyping methods currently used for research purposes rely on amplification of viral DNA sequences (polymerase chain reaction (PCR)-based tests) in fresh specimens (cervical scrapes, skin swabs, plucked hairs, or biopsy specimens) or archival smears, and biopsies. These sensitive tests allow screening for the whole spectrum of HPV infections, from asymptomatic to overt infections. A broad spectrum of genital or cutaneous HPV types can be detected with degenerate or consensus primers targeting a conserved region (usually the L1 ORF). HPV amplicons are detected by hybridization with generic probes or cocktails of probes and identified by dot blot hybridization or reverse blot assays, using type-specific oligonucleotide probes. Sequencing PCR products allows identification of types and variants or putative novel HPV types. Viral loads can be determined by type-specific, real-time PCR. Seroepidemiological studies are based on ELISA using L1 VLPs and detect past or current infections.

Genital and cutaneous HPV infections are widespread and have a worldwide distribution. An impressive diversity of known human beta- and gammapapillomaviruses or related putative novel types (over 100) was identified in the normal skin of healthy adults. Skin HPVs cause highly prevalent, asymptomatic infections likely to be acquired very early in infancy, which may suggest a commensal nature. High detection rates of DNA sequences of HPV-5 (associated with EV carcinomas) and of anti-HPV-5 antibodies have been reported in patients with psoriasis and other conditions involving extensive keratinocyte proliferation. Over 40 HPV types of the genus *Alphapapillomavirus* infect the anogenital tract. Genital HPV infections are very common in sexually active young women (about one-third of normal women in their teens and early twenties) and men. The 3-year cumulative incidence of HPV infection among young women has been found to exceed 40% in several studies. The concurrent or sequential detection of more than one type is a common finding. Prevalence of cervical infections decreases with increasing age, with, sometimes, a second peak after age 55. HPV-16 is one of the more widespread types. Most infections are asymptomatic and transient. Most women appear to have cleared their infection after 1 or 2 years, with a median duration of HPV DNA detectability of approximately 1 year. Cytological signs of infection (koilocytes) are observed at three- to tenfold lower rates than HPV DNA detection. Most cytological abnormalities correspond to low-grade lesions and most regress spontaneously within a year and represent transient productive infections. Only a minority of women exposed to genital HPVs will show persistent infection (approximately 10% being still HPV DNA positive after 5 years) and have a substantial risk to develop a high-grade cervical lesion and, possibly, an invasive carcinoma when infected with high-risk types. High-grade CIN may develop from persistent low-grade lesions or arise without any history of mild cytological atypia. CIN2 lesions are virologically more heterogeneous than CIN3, and have a greater chance to regress and a lower risk to progress to invasive cancer. As a safety measure, all lesions diagnosed as CIN2 or worse have to be treated. Type-specific differences are observed in the natural history of HPV infections in terms of prevalence, regional distribution, association with cytological abnormalities, duration, and risk for the development of a high-grade lesion. HPV-16 persists longer than most other types and has a higher prevalence, and persistent HPV-16 infections are more likely to cause cervical precancerous and malignant disease.

Sexual behavior (multiple lifetime or recent sex partners) is the major risk factor for genital HPV infection. Immune factors play an important role in the natural history of HPV infections. For instance, human immunodeficiency virus (HIV)-infected women display two- to fourfold higher rates of cervical HPV infection and homosexual male AIDS patients have a high prevalence (about 50%) of anal HPV infection. Genetic factors are predominantly involved in two HPV-associated diseases, oral focal epithelial hyperplasia (Heck's disease) and EV.

Pathogenicity

Each HPV type has a specific pathogenicity. A number of HPV types induce benign, self-limiting proliferations of the skin or the mucous membranes (**Table 1**), which usually regress spontaneously or after treatment. Each clinical type of skin wart described is preferentially associated with specific HPV types. HPV-6 and HPV-11 cause anogenital warts. Both types are associated with the rare anogenital Buschke-Löwenstein tumors (verrucous carcinomas) and single cases of carcinomas of the respiratory tract, but are almost never detected in cervical carcinomas. Other HPV types are etiologically associated with the development of invasive carcinomas (**Table 1**). In 2005, the International Agency for Research on Cancer recognized HPV-16, HPV-18, HPV-31, HPV-33, HPV-35, HPV-39, HPV-45, HPV-51, HPV-52, HPV-56, HPV-58, HPV-59, and HPV-66 as carcinogenic to humans and HPV-5 and HPV-8 as carcinogenic to patients with EV.

Contrasting with the many betapapillomaviruses associated with EV benign lesions (EV-specific HPVs), only a few (HPV-5 and, occasionally, HPV-8, HPV-14, HPV-17, HPV-20, and HPV-47) are found in EV premalignant lesions and SCCs of the skin. Viral genomes are maintained as high copy number episomes and E6 and E7 transcripts are easily detected. Invalidating mutations in both alleles of either the *EVER1/TMC6* or the *EVER2/TMC8* gene, encoding transmembrane proteins of unknown function, usually cause the abnormal susceptibility of EV patients to infection by betapapillomaviruses. DNA sequences of a large variety of cutaneous HPV types are detected in low amounts (usually much less than one copy of viral genome per cell) in 25–55% of cutaneous basal cell carcinomas and SCCs in the general population, and in an even higher proportion of SCC (up to 90%) in immunosuppressed patients. Similar detection rates are found for premalignant skin lesions (actinic keratoses and Bowen's disease), and also for healthy skin. The role of HPVs in skin carcinogenesis in non-EV patients remains a matter of debate.

It is now firmly established that persistent infection by specific, sexually transmitted high-risk alphapapillomaviruses is causally associated with the development of over 95% of invasive carcinomas of the uterine cervix and their immediate intraepithelial precursors. About 85% of cervical cancers are SCCs and most of the other cases are adenocarcinomas. This casual relationship is based on virological, epidemiological, and molecular evidence. It is first supported by the consistency of HPV DNA detection in invasive cancers and the specificity of the types detected. Of the 40 HPV types associated with low-grade CIN, 16 are found in cancers. The seven most frequently detected types are, by decreasing order of frequency, HPV-16 (about 55%), HPV-18 (about 15%), HPV-45, HPV-31, HPV-33, HPV-52, and HPV-58, and account for about 87.5% of invasive cancers, worldwide. Other types are found at frequencies lower than 2% (HPV-35, HPV-39, HPV-51, HPV-56, HPV-59, HPV-66, HPV-68, HPV-73, and HPV-82). Some geographical variations have been observed, with a higher prevalence of cervical cancer associated with HPV-45 in Africa or HPV-52 and HPV-58 in China and Japan. The same HPV types are found in CIN3, the immediate precursor of SCC. While HPV-16 predominates in SCCs, HPV-16 and HPV-18 are almost as frequent in adenocarcinomas.

A strong association between the detection of high-risk HPV DNA and high-grade CIN or invasive cancer was demonstrated in case-control studies. The odds ratio values, that is, the ratio of the odds of being infected among cases and controls, ranged from 45 to 430 for 13 HPV types. Cohort studies showed that women positive at baseline for high-risk types (especially HPV-16) have a higher risk of developing CIN3 and invasive cancer and that this risk is closely linked to persistent infections. Infection thus precedes the development of precancer and cancer, a prerequisite for establishing causality.

Experimental studies support the role of high-risk HPV types in the whole process of cervical carcinogenesis. Integration of the viral genome into the cellular genome is observed in the majority of cancers (**Figure 7**), and in high-grade lesions. Disruption of the E1/E2 region inactivates the negative regulation exerted by the E2 protein on transcription of the E6/E7 region, which results in constitutive expression of E6 and E7. These multifunctional oncoproteins complex and inactivate cellular proteins, such as p53 protein (E6) and pRB protein (E7), involved in negative regulation of the cell cycle and the response to DNA-damaging agents. This leads to cell proliferation and genetic and chromosomal instability. CIN3 may evolve into invasive SCC after years or decades. This involves the accumulation of genetic and epigenetic events targeting cellular oncogenes or tumor-suppressor genes.

HPV infection is a necessary, but not sufficient, cause of cervical cancer. Long-term use of hormonal contraception, high parity, tobacco smoking, and coinfection with HIV increase the risk of cancer in HPV-infected women. Other probable cofactors are coinfection with herpes simplex virus type 2 or *Chlamydia trachomatis* and dietary factors. Viral factors, such as HPV variants, viral load, and viral integration, are likely to play a part. Host genetic factors remain little understood. The positive or negative

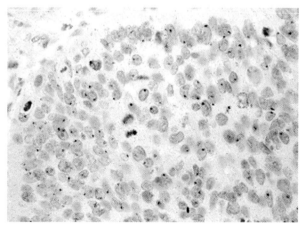

Figure 7 Integrated HPV-16 DNA sequences in an invasive cervical squamous cell carcinoma, as detected in nuclei (dots) by *in situ* hybridization. Courtesy of Dr. O. Croissant.

associations of CIN and/or invasive carcinoma with certain human leukocyte antigen (HLA) class II alleles or haplotypes suggest an immunogenetic control.

High-risk genital HPV types, mostly HPV-16, are also associated with other anogenital invasive cancers and their precursors: the great majority of vaginal and anal carcinomas and about 50% of vulvar and penile carcinomas (basaloid and warty tumors). HPV-16 DNA has been detected in approximately 20% of oropharyngeal carcinomas and in a limited number of carcinomas of the oral cavity.

Clinical Features of Infection

Skin warts (papillomas) occur predominantly in children over 5 years and in young adults. At least two-thirds of cases regress spontaneously within 2 years. Palmo-plantar myrmecia warts are deep and painful. Common warts are elevated, hyperkeratotic and located mainly on the hands, fingers, and knees. Flat warts are usually multiple, slightly raised, and are localized mainly on the hands and face. EV is a lifelong genodermatosis characterized by disseminated flat, wart-like lesions and macules. In about half of the patients, some lesions start converting to premalignant lesions and invasive SCCs, usually on light-exposed areas, about 20 years after the onset of the disease.

Genital HPV infections represent the most commonly diagnosed viral sexually transmitted diseases. Genital warts and Bowenoid papules are found mostly in young adults, on the external genitalia and in the anal region. Anogenital warts are soft, exophytic proliferations. Bowenoid papules, also recognized as vulvar, penile, and anal intraepithelial neoplasia (VIN, PIN, and AIN), are flat or somewhat elevated, often pigmented, usually multiple, and display a rather benign clinical course in spite of their usual association with HPV-16 and histological features of intraepithelial neoplasia. This is in contrast to solitary lesions of genital Bowen's disease (high-grade intraepithelial neoplasia) affecting individuals over 50 years. Anal lesions with a variety of morphological appearance and features of intraepithelial neoplasia are found at the anorectal junction in HIV-infected males. Infection of the cervix results mostly in flat, acetowhitening areas corresponding to low-grade CIN. Cytological evidence of HPV infection is found in cervicovaginal smears of 2–3% of women in mass-screening programs. Infection with genital HPVs are often multiple, and associated diseases can affect the entire anogenital region.

Oral focal epithelial hyperplasia is observed in children and adults. The disease is highly prevalent in American Indians and Eskimos but rarely observed in Caucasians. It is characterized by multiple papules, which never undergo malignant transformation. Laryngeal papillomatosis is a rare recurrent disease that can occur at any age. The most frequent site affected is the vocal cords. Trachea, lung, oral, and nasal cavities can also be involved. Respiratory papillomatosis may be life-threatening by obstructing airways.

Pathology and Histopathology

Benign lesions are characterized by variable degrees of hyperplasia of the various epithelial layers and hypertrophy of dermal papillae (papillomatosis). Some features may reflect distinct biological properties of the associated HPV types. The extent of papillomatosis is prominent for HPV-1, HPV-2, HPV-4, HPV-6, HPV-7, or HPV-11, much less pronounced for HPV-3 or HPV-10, and often absent for EV-HPVs. The cytopathic effect associated with viral replication in differentiating cells is recognized as koilocytosis for anogenital alphapapillomaviruses, and presents specific features (number, size, shape, and structure of cytoplasmic inclusions or extent of perinuclear vacuolization) for cutaneous lesions induced by alpha-, beta-, gamma-, mu-, and nupapillomaviruses.

Anogenital lesions with features of squamous intraepithelial neoplasia (CIN, VIN, PIN, AIN) may show different degrees of severity. The grading of the lesions depends on the proportion of epithelium occupied by basal-type cells. Basaloid cells restricted to the lower third of the epithelium, usually showing mild atypia, and a well-preserved maturation define low-grade (grade 1) intraepithelial neoplasia. In high-grade lesions, neoplastic basaloid cells occupy up to two-thirds of the epithelium (grade 2) or the whole epithelium (grade 3). The disorganized architecture of the epithelium includes abnormal mitotic figures and nuclear atypia. For cervical lesions, CIN1 corresponds to mild dysplasia, CIN2 to moderate dysplasia, and CIN3 to severe dysplasia and carcinoma *in situ*, according to the dysplasia-carcinoma *in situ* classification. The Bethesda System of nomenclature for the evaluation of cervicovaginal Pap smears has introduced low-grade squamous intraepithelial lesions (SIL), encompassing HPV infection (koilocytes) and CIN1, and high-grade SIL, encompassing CIN2 and CIN3. Equivocal atypical squamous cells (ASC) are labeled ASC-US (undetermined significance) or ASC-H (cannot exclude a high-grade lesion). Features specific to Bowen's disease, especially individual cell keratinization, are observed in Bowenoid papules and EV lesions undergoing malignant conversion.

Immune Response

Strong immune responses are not usually generated because HPVs most often induce asymptomatic infections and because productive infection does not lead to host-cell lysis. A major problem remains the availability of well-standardized, sensitive, and specific assays.

Humoral immune responses upon natural infections or immunization are evaluated using VLP-based ELISA tests or pseudovirus-based neutralization assays. IgG and IgA antibody responses to HPV infections are weak and not always detected (50–75% of women with HPV-6/HPV-11- or HPV-16-associated disease). Seroconversion is a slow process. IgG antibody levels are generally stable during several years. Men show lower seropositivity rates when compared to women at similar risk. Although predominantly type-specific, the VLP-based ELISA test may also recognize non-neutralizing, cross-reactive, linear epitopes disclosed by ill-assembled VLPs. The neutralization assay may be substantially more type-specific. Neutralizing antibodies are found in about 75% of women with prevalent HPV-16 or HPV-18 infections. Such antibodies are likely to prevent repeat infections. Antibodies to native E6 or E7 oncoproteins are rarely detected in patients with precancerous cervical lesions but are found in about 50% of patients with late-stage invasive cervical carcinomas. They constitute virus type- and cancer-specific markers, with no prognostic value.

HPV infections and HPV-associated lesions usually clear spontaneously. Cell-mediated immunity (CMI) is thought to play a major role in this outcome. A high incidence of cutaneous and anogenital HPV-associated diseases is observed among patients with acquired or iatrogenic CMI deficiencies. Infiltrates of CD4+ and CD8+ T-lymphocytes and macrophages are observed in regressing flat warts, genital warts, and CIN. This suggests that cytotoxic T-lymphocyte (CTL) responses and delayed-type hypersensitivity reactions are important in the regression of HPV-induced lesions. The E1 and E2 proteins (required for viral genome episomal maintenance and replication) and E6 and E7 proteins (upregulated during tumor progression) are likely targets of the adaptive CMI responses. Several HLA-restricted CTL and T-helper (Th) cell epitopes have been identified in HPV-16 early proteins. Memory CD4+ Th cell responses to HPV-16 E2 and E6 have been detected in approximately half of healthy women but found to be absent or impaired in patients with high-grade CIN or invasive cancers. CTL responses against E6 and E7 have been found occasionally in patients with cervical HPV-16 associated infection or disease but no obvious relationship has been found so far with regression or progression of the disease. There is some evidence for defective natural killer cell function in patients with HPV-16 associated disease.

Genital high-risk HPV types, especially HPV-16, have evolved mechanisms to evade innate and adaptive immunity. E6 and E7 oncoproteins interact with cellular proteins involved in the synthesis of, or response to, type 1-interferons. Unlike dermal dendritic cells, Langerhans cells, the resident epithelial antigen-presenting cells, are not activated by uptake of HPV VLPs. An altered expression of HLA class I antigens (downregulation) and class II antigens (upregulation) is observed in a substantial proportion of CIN and cervical carcinomas. Tumor cells may thus evade immune surveillance by altering their interaction with CD8+ CTLs or turn into nonprofessional antigen-presenting cells, which may result in T-cell anergy.

Patients suffering from certain primary immunodeficiency diseases are prone to chronic cutaneous and/or genital HPV disease: (1) patients with the WHIM syndrome (mutations in the gene encoding the CXCR4 receptor) and (2) late after hemopoietic stem-cell transplantation, patients with severe combined immune deficiency caused by common γc cytokine receptor subunit or Janus kinase-3 deficiency. Inactivation of *EVER* genes may result in the development of EV lesions by complementing for the lack of an E5/E8 ORF in EV-specific HPVs. EV could thus represent a primary deficiency of intrinsic immunity against certain HPV types.

Prevention and Control

Education on prevention of sexually transmitted diseases should include appropriate information on HPV infections, their role in cervical cancer, and the available prevention strategies, which makes invasive cervical cancer an almost entirely preventable disease.

There are no specific anti-HPV therapeutic agents (targeting E6, E7, E1, or E2). Treatment modalities are mostly aimed at destroying or removing lesions. Conventional treatments involve topical applications of caustic agents, physical destruction, surgical removal, or chemotherapeutic agents. Immune response modifiers could be of value (interferon or imiquimod, an agonist of Toll-like receptor 7). Therapeutic vaccines aimed at generating specific CTL responses to eradicate premalignant lesions, such as CIN2/3, have targeted the HPV E6 and E7 oncoproteins. Phase I/II trials have been performed or are in progress to test vaccines based on recombinant viruses, naked DNA, fusion proteins, peptides, or pulsed dendritic cells. Contrasting with preclinical studies in murine models, most published trials showed little efficacy in patients, but some promising approaches are currently being tested.

The implementation of population-based cervical cancer screening programs, based on the Pap test, proved to be a most successful strategy for dramatically decreasing cervical cancer incidence and mortality. To increase the sensitivity of the detection of pre-invasive lesions and cervical cancer, several US organizations recommend the use of an FDA-approved HPV DNA test for high-risk HPV types in conjunction with cytology (conventional Pap test or liquid-based cytology), as a screening option for women aged 30 years and older. Future screening strategies may combine HPV testing and a cytological triage of women found HPV positive.

Two prophylactic subunit vaccines have recently been developed, the tetravalent (Merck) and bivalent (GlaxoSmithKline, GSK) vaccines, which contain HPV-16, HPV-18, HPV-6 and HPV-11 L1 VLPs and HPV-16 and HPV-18 L1 VLPs, respectively. Clinical trials have established their safety and consistent immunogenicity in young adult women and their high efficacy in preventing persistent infections, premalignant lesions, and genital warts caused by the vaccine HPV types. The tetravalent vaccine was licensed by the US FDA in 2006, as a vaccine for girls and women 9–26 years of age (prior to exposure to the risk of HPV infection) to prevent cervical cancer, genital warts, high-grade cervical, vulvar and vaginal lesions, and CIN1 caused by HPV-6, HPV-11, HPV-16, and HPV-18. Approval of the bivalent vaccine is expected in 2007. Both vaccines should theoretically protect against ~70% of cervical cancers, 60–70% of CIN2/3 and 25–30% of CIN1, and the tetravalent vaccine also against over 90% of genital warts. A general agreement is the need for regular screening of vaccinated women according to current guidelines, to check for the long-term efficacy of the vaccine and detect lesions associated with non-vaccine HPV types. Several issues remain to be addressed: duration of protection (need for boosters), long-term safety, degree of cross-protection against non-vaccine HPV types, and efficacy in men. A major concern is the availability of the vaccine in developing countries, where over 80% of worldwide cervical cancers occur and screening programs are not available. Incorporating additional high-risk HPV types or, for instance, taking advantage of the broadly cross-neutralizing epitopes borne by the L2 protein, should provide strategies for second generation vaccines.

See also: Papillomaviruses: Molecular Biology of Human Viruses; Papillomaviruses of Animals; Tumor Viruses: Human; Viral Pathogenesis.

Further Reading

Bosch FX, Lorincz A, Munoz N, Meijer CJ, and Shah KV (2002) The causal relation between human papillomavirus and cervical cancer. *Journal of Clinical Pathology* 55: 244–265.

Bosch FX, Schiffman M, and Solomon D (eds.) (2003) Future directions in epidemiologic and preventive research on human papillomaviruses and cancer. *Journal of the National Cancer Institute Monographs* 31: 1–130.

Campo MS (ed.) (2006) *Papillomavirus Research: From Natural History to Vaccines and Beyond*. Wymondham, UK: Caister Academic Press.

de Villiers EM, Fauquet C, Broker TR, Bernard HU, and zur Hausen H (2004) Classification of papillomaviruses. *Virology* 324: 17–27.

Frazer IH, Thomas R, Zhou J, et al. (1999) Potential strategies utilised by papillomaviruses to evade host immunity. *Immunological Reviews* 168: 131–141.

Harper DM, Franco EL, Wheeler CM, et al. (2006) Sustained efficacy up to 4.5 years of a bivalent L1 virus-like particle vaccine against human papillomavirus types 16 and 18: Follow-up from a randomised control trial. *Lancet* 367: 1247–1255.

Howley PM and Lowy DR (2007) Papillomaviruses. In: Knipe DM, Howley PM, Griffin DE, et al. (eds.) *Fields Virology*, 5th edn., pp. 2229–2354. Philadelphia, PA: Wolters Kluwer/Lippincott Williams and Wilkins.

IARC (2006) *IARC Monographs on the evaluation of carcinogenic risks to humans, Vol. 90: Human Papillomaviruses*. Lyon: International Agency for Research on Cancer Press.

Münger K, Baldwin A, Edwards KM, et al. (2004) Mechanisms of human papillomavirus-induced oncogenesis. *Journal of Virology* 78: 11451–11460.

Munoz N, Bosch FX, de Sanjosé S, et al. (2003) Epidemiologic classification of human papillomavirus types associated with cervical cancer. *New England Journal of Medicine* 348: 518–527.

Narechania A, Chen Z, DeSalle R, and Burk RD (2005) Phylogenetic incongruence among oncogenic genital alpha human papillomaviruses. *Journal of Virology* 79: 15503–15510.

Orth G (2006) Genetics of epidermodysplasia verruciformis: Insights into host defense against papillomaviruses. *Seminars in Immunology* 18: 362–374.

Orth G and Jablonska S (eds.) (1997) Papillomaviruses: Part two. *Clinics in Dermatology* 15: 303–470.

Villa LL, Costa RLR, Andrade RP, et al. (2006) High sustained efficacy of a prophylactic quadrivalent human papillomavirus types 6/11/16/18 virus-like particle vaccine through 5 years of follow-up. *British Journal of Cancer* 95: 1459–1466.

zur Hausen H (2002) Papillomaviruses and cancer: From basic studies to clinical application. *Nature Reviews Cancer* 2: 342–350.

Papillomaviruses: Molecular Biology of Human Viruses

P F Lambert and A Collins, University of Wisconsin School of Medicine and Public Health, Madison, WI, USA

© 2008 Published by Elsevier Ltd.

Introduction

Papillomaviruses (PVs) are nonenveloped, double-stranded DNA tumor viruses that infect mammalian and avian hosts. There are more than 100 genotypes of human papillomaviruses (HPVs) that can either infect mucosal or cutaneous epithelia. PVs all share the common property of inducing benign proliferative lesions within the epithelial tissues they infect. In these lesions, the virus is replicated in a manner that is exquisitely tied to the differentiation program of the epithelial host cell. Within the mucosal genotypes, HPVs are further subcategorized as low- or high-risk genotypes, reflective of the latter's additional association with frank cancer. High-risk HPVs are best known as the etiologic agent of virtually all cervical cancer, other anogenital cancers, as well as a subset of head and neck cancers. The high-risk HPV most commonly found in these cancers is HPV-16. In this article, the nature of the papillomaviral life cycle as discovered through the study of this high-risk HPV is described. We also compare and contrast its life cycle to that of other PVs, where possible.

Genome

PVs contain an ~8000 bp, circular, double-stranded DNA genome. The encapsidated genome is thought to exist in a chromatinized state within the virus particle; however, there is one report claiming an absence of cellular histones in virus particles. In the infected host cell, the viral genome is delivered to the nucleus where it remains in an extrachromosomal state throughout the viral life cycle. Integration of the viral DNA into the host chromosome is not a normal part of the viral life cycle but can occur in some PV-associated cancers.

The viral genome is organized into three regions: the long control region (LCR), which contains many *cis*-elements involved in the regulation of both DNA replication and transcription; as well as the early (E) and late (L) regions that encode subsets of genes so denoted by their onset of expression in the context of the viral life cycle (**Figure 1**). Transcription of the viral genome occurs unidirectionally and is directed from multiple promoters that are located within the LCR and the 5' end of the E region. Multiple splicing and polyadenylation signals are used depending on the stage of the viral life cycle, leading to a variety of transcripts with differential coding capacity. The viral genome codes for eight viral genes. The early gene products, which are coded from the E region, contribute to the transcription, replication, and maintenance of the genome as a stable, nuclear plasmid within the poorly differentiated compartment of the infected tissue, as well as its amplification and encapsidation within the more terminally differentiated cellular compartment to make progeny virus. A subset of the early genes (E5, E6, and E7) possesses transforming properties in tissue culture, and tumorigenic properties in experimental animal models. The late structural genes encode for the major capsid protein, L1, and the minor capsid protein, L2. The individual role of each viral gene in the life cycle is described in this article.

Pathogenesis

PVs cause warts: benign, self-regressing lesions of the skin, oral cavity, and anogenital tract. The mucosotropic HPVs are the leading viral cause of sexually transmitted disease in women, with over 50% of sexually active women having been infected in their lifetime. A subset of these mucosotropic HPVs, the so-called 'high-risk' genotypes, are the cause of virtually all cervical cancer, with approximately 500 000 new cases diagnosed yearly, and approximately half as many deaths due to this cancer, making it the second leading cause of death by cancer among women worldwide. Precursor lesions to cervical cancer, called cervical intraepithelial neoplasia (CIN) grades 1–3 (with grade 3 having the most dysplastic characteristics), can be detected by the Pap smear, a cytological screening of cervical scrapes. Pap smears have led to a threefold reduction in the incidence of cervical cancer in those countries employing it as a routine preventative screening technique.

Recently approved prophylactic vaccines that prevent infection by two of the most common high-risk HPVs, HPV-16 and HPV-18, are predicted to lead to an 80% reduction of cervical cancer worldwide beginning in 2040 if the vaccine is universally administered to the world population of women. Socioeconomic conditions may limit the utility of these vaccines.

Other less frequently observed anogenital cancers, of the anus, penis, vulva, and vagina, are also caused by these high-risk HPVs. On average, there is a latency of over a decade between infection and onset of these anogenital cancers. In these cancers, one can find retention of the HPV genome, often in the integrated state, with a subset of viral oncogenes, specifically E6 and E7, upregulated in their expression. In mouse models, the high-risk HPV E6 and E7 oncogenes have been demonstrated to predispose mice to cervical cancer.

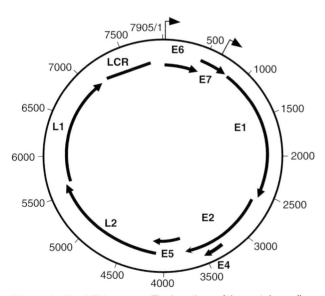

Figure 1 The HPV genome. The locations of the protein-coding sequences of early (E) and late (L) genes, and those of the long control region (LCR) and transcriptional promoters (arrows), are shown.

The same high-risk HPVs, particularly HPV-16, are etiologically associated with approximately 25% of head and neck squamous cell carcinoma (HNSCC), particularly of the oropharynx where the etiological correlation is 50% or greater. Again, the continued expression of E6 and E7 is found in HPV-positive HNSCC, and, in a mouse model, the HPV-16 E6 and E7 oncogenes have been demonstrated to predispose mice to HNSCC.

Cutaneous HPVs are also associated with a subset of skin cancers. This has been demonstrated most clearly for a subset of cutaneous HPVs that cause epidermodysplasia verruciformis (EV), a skin lesion that appears to arise preferentially in certain families, arguing a genetic predisposition. Patients with EV are highly prone to develop skin cancers. In mouse models, EV-associated HPVs have been experimentally demonstrated to predispose animals to skin cancer.

Life Cycle

With the exception of the so-called fibropapillomaviruses (e.g., bovine papillomaviruses (BPVs) and other PVs that infect ungulate hosts), most PVs are specifically epitheliotropic. Their life cycle occurs within stratified squamous epithelia, which are composed of basal and suprabasal compartments (**Figure 2**). The basal cells within the three-dimensional architecture of such epithelial tissues comprise a single layer sheet of poorly differentiated, proliferative cells that are physically attached to the underlying basement membrane that separates the epithelium from the underlying stroma. The suprabasal compartment comprises the multiple layers of cells positioned above the basal compartment that are quiescent and, as they become more superficial in their position, progressively more differentiated. When basal cells undergo cell division, one daughter cell loses contact with the basement membrane, elevating it to the suprabasal compartment where it exits the cell cycle and begins a program of terminal differentiation. As these cells move upward, they become progressively more terminally differentiated and lose nuclear membrane integrity, leading to the production of terminally differentiated squames, which eventually slough off into the environment. PVs simultaneously exploit and disrupt these normal cellular programs of differentiation and quiescence for their life cycle and the production of progeny virions.

Fibropapillomaviruses differ from other PVs in that they can also infect the underlying dermal fibroblasts, leading to a hyperplasia of the cells. Lesions induced by fibropapillomaviruses can be quite large due to this underlying fibroplasia. The life cycle of these fibropapillomaviruses, however, still relies directly upon the infection of the epithelial cell, where the production of progeny virus specifically occurs.

Entry

It is presumed, though it has never been directly demonstrated, that PVs initiate infection of the host epithelia via entry into cells within the basal compartment. The specific mechanism(s) PVs use to gain access to, bind, and enter basal cells is not clear. The integrin $\alpha 6:\beta 4$ heterodimer has been implicated as a cellular receptor for the virus, allowing for entry into cells upon its expression. However, in the absence of this integrin, some PVs retain the ability to infect cells. Other factors such as heparin sulfate proteoglycans are thought to play a role in cellular binding/entry by certain HPV genotypes at least in some cell types. Both the major (L1) and minor (L2) capsid proteins are required for viral particle infectivity, and both have been postulated to bind to cellular receptors. PVs have been argued to enter cells via clathrin-dependent or caveolin-dependent pathways and to require endosomal acidification. Once inside, L2 is thought to bind to β-actin and this is thought to facilitate movement of the virus particle through the cytoplasm to the nucleus. Alternatively, it has been argued that L2 binds to the microtubule motor dynein, facilitating intracytoplasmic transport. How the encapsidated DNA actually enters the nucleus is currently unknown; however, once there it appears to associate with subnuclear domains called promyelocytic leukemia protein oncogenic domains (PODs).

Establishment of the Nonproductive Infectious State

Once an HPV particle enters the host cell and translocates its DNA to the nucleus, it must rely primarily on cellular machinery to replicate its genome since HPVs only encode one component of the machinery required for initiating DNA replication, E1, a DNA helicase. It is partially for this reason that it is presumed that HPVs initiate their viral life cycle by infecting cells within the basal compartment, as these are the only cells within

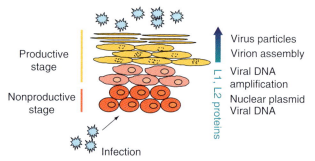

Figure 2 The HPV life cycle. A schematic cross section of stratified squamous epithelium is shown with basal cells at the bottom and terminally differentiated squames at the top. The various processes that arise within the nonproductive and productive stages of the viral life cycle are indicated at relevant positions on the periphery.

the epithelia that are normally proliferating and have active cellular replication machinery. In these infected basal cells, the HPV genome becomes established as a low-copy-number nuclear plasmid. The 'early' promoter (designated P_{97} in the case of HPV-16) directs expression of the early genes, some of which are required for the replication of the viral genome, with transcripts being observed as early as 4 h post infection. This early promoter appears to stay active throughout the viral life cycle, whereas the 'late' promoter (designated P_{742} in HPV-16) is only active in cells upon their differentiation. Thus within the basal compartment, early but not late viral genes that encode the structural proteins are expressed, and therefore no progeny virus is produced. Consequently, the infectious state found in the basal cells is referred to as the 'nonproductive' stage of the infectious life cycle.

Maintenance of the Nonproductive Infectious State

A hallmark of HPV infections is their long-term persistence over many years. In the case of high-risk HPVs, this persistence is a prerequisite for the development of cancers associated with these viruses. For the virus to persist long term, it must maintain its genome within the basal compartment over multiple cell divisions, allowing for the continual regeneration of daughter cells that can provide a reservoir of proliferation-competent cells harboring the viral genome. Several early genes have been implicated in the establishment, as well as maintenance of viral genomes (see below). The virus must also affect an amplification of cells within the basal compartment that harbor the viral genome, as seen in warts where one finds a localized expansion of cells within the infected epithelia.

One possible explanation for both persistence and localized expansion of infected cells is that the virus infects epithelial stem cells. From these stem cells are derived the transiently amplifying cell population that populates the local area with HPV-positive cells. This is an attractive hypothesis; however, stem cells are infrequent. Therefore, the necessity to infect stem cells would indicate that productive infection is an inefficient process. An alternative explanation is that HPVs infect the transiently amplifying nonstem cells and extend their lifespan, thereby leading both to viral persistence and an expansion of infected cells at the original site of infection. Perhaps reflective of this alternative explanation is the observation that high-risk viruses have the unique capability among PVs to immortalize epithelial cells. This immortalization potential of high-risk HPVs, in turn, reflects the unique properties of their oncogenes E6 and E7. Whether the high-risk HPVs are unique in their capacity to establish persistent infections without the need to infect stem cells, as might be predicted by their immortalization potential, remains unclear.

Productive Stage

Normally when a basal cell divides, it produces two daughter cells, one of which separates physically from the basement membrane, transits to the suprabasal compartment, and begins a program for terminal differentiation that includes exiting the cell cycle and halting new rounds of DNA synthesis. When a daughter cell that harbors HPV genomes transits to the suprabasal compartment, it begins the so-called 'productive' stage of the HPV life cycle in which progeny viruses are made. In this stage of the viral life cycle, the HPV genome induces a delay and perturbation of the terminal differentiation program, which, importantly, includes the retention of an ability of these suprabasal cells to support DNA synthesis. This DNA synthesis-competent state is critical to allow for the amplification of the viral genome that is necessary for the production of progeny virus. Production of progeny virus within the suprabasal compartment also relies upon the selective expression of the late viral genes that encode the viral capsid proteins. These steps are spatially and temporally regulated. Specifically, induction of suprabasal DNA synthesis and a delay in early steps of epithelial cell differentiation both occur early in the productive stage and within the lower strata. Late gene expression, viral DNA amplification, and progeny virus production arise in the upper strata and are associated with further perturbations in terminal differentiation. Thus, specific life cycle events correspond intimately with different alterations in the terminal differentiation process within certain regions of the infected epithelium.

The multistage mode of replication of viral DNA genomes exemplifies the role that differentiation plays in the viral life cycle. HPV genomes harbored extrachromosomally within the basal compartment are replicated on average once per cell cycle and are stably maintained within this proliferating cell compartment at a low copy number. In contrast, the viral DNA is amplified to a much higher copy number in the upper layers of the epithelia strata. This difference in replication state in basal versus suprabasal cells correlates with a switch in the mode of replication from theta to, seemingly, rolling-circle DNA replication, and is induced by cell differentiation.

Functions of Individual Viral Genes

The Role of E1

PVs rely heavily on cellular machinery to replicate their DNA. However, PVs do encode one gene that is enzymatically involved in viral DNA replication and is required for plasmid replication, E1. E1 binds the viral origin of replication (ori) and possesses DNA helicase activity that permits the unwinding of the viral genomic template for replication. *In vivo* viral replication is mediated by E1, as well as the viral transactivator, E2.

E2 is able to bind specific DNA sequences, E2 binding sites (E2BSs), as well as the E1 protein; both of these E2 properties help drive E1-mediated plasmid replication. The current model proposes a multistep, interdependent process for viral replication in which E1 homodimers and E2 homodimers cooperatively bind to ori, then undergo conformational changes that allow E1 to assemble into a helicase-competent hexamer. Many cellular factors contribute to E1-dependent viral replication, including DNA polymerase α, chaperone proteins, histone H1, and replication factor A. In some cases, E1 has been shown to recruit these factors directly.

While E1 is required for the establishment of papillomaviral DNA replication, E1 is dispensable for plasmid maintenance of the BPV genome. In bovine papillomavirus type 1 (BPV-1)-transformed mouse C127 cells, in which that genome was already established as a nuclear plasmid, E1 was not required for stable maintenance of the viral plasmid. An E1-independent form of replication could also be detected in *Saccharomyces cerevisiae* and the *cis*-requirements for this replication were mapped. The ability of PVs to employ an E1-independent replication pathway may be a property of or reflect various modes of replication that the virus utilizes during different stages of its life cycle. In contrast, or in addition, the replication mechanism may reflect the ability of the virus to take advantage of a different host cell environment.

The Role of E2

In addition to a role in viral DNA replication, E2 also regulates viral transcription and the maintenance of the extrachromosomal viral genome. The ability of E2 to regulate transcription is mediated by its binding to E2BSs within the papillomaviral genome. In the PV genome there are multiple E2BSs that have different affinities for E2. Their relative locations are thought to determine the effects of E2 on transcription. If E2 binds an E2BS that is near to, but not overlapping, promoter elements, E2 activates transcription from that promoter. However, if the E2BS overlaps promoter elements, then it can repress transcription by sterically hindering binding of cellular factors necessary for transcription from that promoter.

In the context of the high-risk HPV genomes, E2's ability to suppress transcription from the viral promoter that drives expression of the viral early genes is thought to reflect one reason why integration of the viral genome into the host chromosome, and the consequent disruption of E2, contributes to increased expression of the viral oncogenes E6 and E7. However, these integration events are likely to have pleiotropic effects, as it has been shown that integration results in increased stability of the E6/E7 mRNAs due to the loss of an mRNA instability element located in the 3′ end of the early region. The ability of E2 to repress transcription of the viral oncogenes has been exploited to turn off expression of E6 and E7 in cervical cancer cells lines and thereby demonstrate the requirements for constitutive expression of E6 and E7 in these cancer-derived cell lines. Interestingly, E2 only appears to repress transcription of viral genomes when they are integrated into the host genome. This may reflect the fact that E2BSs in extrachromosomal HPV genomes can be methylated, and this methylation is known to prevent E2 binding. It is likely that epigenetic events also modulate the expression of integrated copies of the viral genome in cervical cancers by altering the chromatin state of the viral genome. In addition to the full-length E2 gene product, E2-mediated transcriptional regulation can also be mediated by N-terminally truncated forms of E2, E8^E2, and E2TR, which are synthesized from alternative transcripts. These E2 proteins can act as dominant negative factors to suppress E2-mediated transcriptional activation either by competing themselves for binding to the E2BSs or through heterodimerization with the full-length E2 protein.

PV genomes are maintained as nuclear plasmids throughout the viral life cycle. Therefore, there is a need for viral plasmids to be segregated efficiently to daughter cells during cell division. For BPV-1, the full-length E2 protein has been shown to mediate this process and is thought to reflect E2's ability to associate with host chromosomes during mitosis. E2 can bind to the chromosomal attachment protein, bromodomain 4 (Brd4), and, in yeast, the ability of BPV-1 E2 to mediate plasmid maintenance relies upon the co-expression of mammalian Brd4. These data suggest that Brd4 contributes to BPV-1 E2's plasmid maintenance function. Further study is required to ascertain whether this proposed mechanism is relevant to HPVs. Of concern is the observation that mutant HPV-31 genomes that contain amino acid substitutions in the E2 protein predicted to disrupt Brd4 binding, based upon prior studies in BPV-1 E2, retain an ability to replicate stably as a nuclear plasmid, though less efficiently than wild-type HPV-31 genomes.

The Role of E1^E4

As with the other PV proteins, numerous functions are ascribed to the E1^E4 gene product. The small (10–20 kDa) regulatory viral protein is generated from a spliced mRNA that fuses the E1 and E4 translational open reading frames (see **Figure 1**). E1^E4 is poorly expressed in undifferentiated monolayer cultures containing HPV, as well as within basal cells of raft cultures. However, E1^E4 expression is increased substantially in the upper strata of papillomas. Within raft cultures the greatest E1^E4 expression is seen in the more superficial layers of epithelium that also contain amplified viral genomes. Owing to colocalization of E1^E4 and amplified viral genomes, it was proposed that E1^E4 functions in the productive stage of the viral life cycle, specifically viral DNA amplification. Indeed, when severe truncation mutations of E1^E4 were made in the context of the entire cottontail rabbit papillomavirus (CRPV), HPV-16, or HPV-31 genomes, defects in viral DNA amplification were observed, indicating that E1^E4 contributes to the amplification of viral genomes. However, effects in viral genome establishment were also seen with some HPV-16 E1^E4 mutants. How E1^E4 mediates viral genome establishment is not clear.

E1^E4 contains three major motifs postulated to contribute to the role of E1^E4 in the viral life cycle. A leucine-rich motif in the N-terminus is required for association with keratins and perhaps to contribute to viral DNA amplification. A proline-rich region in the central section, containing threonine residues, is required for induction of cell cycle arrest at the G2/M phase and sequestering of cyclinB/cdk1 complexes in the cytoplasm. The C-terminus contains a mucosal homology domain, and regulates the ability of E1^E4 to oligomerize, bind the DEAD-box RNA helicases, and induce collapse of keratin filamentous networks. In addition, the association of E1^E4 with cornified cell envelopes (CCEs) and PODs has not been mapped, but both interactions are postulated to contribute to the role of E1^E4 in the PV life cycle.

The association of E1^E4 with both CCEs and keratin filaments is thought to aid egress of the virus. PVs are not lytic; rather, they assemble their progeny virions within the granular and cornified layers of the epithelia. The progeny virions are thought to be released in the context of the natural process of epithelial sloughing. These E1^E4 interactions are proposed to aid the disruption of cellular structure and thus the ability of progeny virions to be released into the environment.

The associations of E1^E4 with cyclinB/cdk1, PODs, and the E1^E4-associated DEAD-box RNA helicase are each postulated to mediate a role for E1^E4 in viral replication or amplification. Overexpression of E1^E4 leads to a cell cycle profile consistent with cells that are arrested at G2/M. It is thought that E1^E4 mediates G2 arrest via decreasing the soluble nature of the cyclin complex and preventing it from translocating to the nucleus. It is hypothesized that these cells retain an ability to support viral DNA amplification. In HPV-1-induced warts, E1^E4 associates with promyelocytic leukemia protein (PML) and the *in vitro* overexpression of E1^E4 induces relocalization of PML to E1^E4 nuclear inclusion bodies. These data suggest that PODs, with E1^E4, may play a role in the life cycle of HPVs.

The Role of E5

E5 is a small, dimeric, hydrophobic membrane protein that localizes to the Golgi membrane and endoplasmic reticulum. Initially, most studies to understand the function and contribution of E5 to transformation, replication, or oncogenesis by PVs were performed within the context of BPV. In the case of BPV-1, E5 is the major transforming protein and can induce immortalization and transformation of rodent fibroblasts and keratinocytes. In contrast, E6 and E7 are the major transforming proteins of HPV. Though, in the context of the HPV genome, E5 is not absolutely required for the induction of DNA synthesis or immortalization, it contributes quantitatively to both of these processes. Furthermore, E5 can independently transform rodent keratinocytes, induce anchorage-independent growth, and override cellular growth arrest signals to promote proliferation. Therefore, E5 appears to play an important yet secondary role in transformation and oncogenesis induced by HPVs, whereas it plays a primary role in BPV-induced transformation.

The ability of E5 to transform cells is attributed to its ability to activate growth factor receptors and to inhibit the 16 kDa vacuolar ATPase (vATPase). BPV-1 E5 can induce the activation of platelet-derived growth factor receptor β (PDGFRβ) in a ligand-independent fashion. This activation occurs through direct binding of BPV-1 E5 to PDGFRβ, which induces receptor oligomerization and activation. In contrast, HPV E5 increases the signaling capacity of the epidermal growth factor receptor (EGFR). This difference in target growth receptors is presumably mediated by tissue tropism specificity. Fibroblasts, which BPVs infect along with keratinocytes, contain abundant amounts of PDGFR. In keratinocytes, which are the normal host cells of HPVs, EGFR is a key growth factor receptor. EGFR phosphorylation, a marker of its activation, is increased in the presence of E5, at least when it is overexpressed. Expression of EGF in concert with E5 leads to greater proliferation of keratinocytes. However, when HPV E5 was studied in the context of the complete viral genome, HPV E5 did not induce a detectable increase in either levels or phosphorylation of EGFR. E5 contributes with E6 and E7 to immortalize keratinocytes and contributes quantitatively to the proliferative capacity of HPV and the ability of HPV to override growth arrest signals. Therefore, while *in vitro* and heterologous system studies predicted that HPV E5 contributes to the viral life cycle through upregulation of EGFR signaling, EGFR-independent pathways may actually be responsible for the proliferative contribution of E5 to the HPV life cycle. Alternatively, E5 may mediate its effects via upregulation of EGFR but the threshold level modified by E5 may not be distinguishable with current technical methodologies.

While E5 does not contribute to the nonproductive stage of the HPV-16 or HPV-31 life cycle, it does contribute quantitatively to the productive stage of both HPV-16 and HPV-31. In the absence of a full-length E5 gene, HPV genomes are defective in their ability to override the normal differentiation program. Normally, HPV-16 and HPV-31 can induce DNA synthesis and proliferation markers in cells induced to undergo differentiation, which are normally quiescent. However, in the absence of E5, DNA synthesis and proliferation marker expression was quantitatively less than the induction of these markers by wild-type HPV genomes. E5 was also shown to be required for efficient viral DNA amplification (HPV-16 and HPV-31), late viral gene expression, and re-entry into the cell cycle (HPV-31). These data illustrate that E5 contributes quantitatively to many aspects of the productive stage of the HPV life cycle. Further support for a role of E5 in productive infection lies in *in vivo* studies that examine papilloma formation by E5 alone or E5 mutant genomes. In the context of CRPV, E5 mutant genomes are less effective at inducing papillomas than the wild-type CRPV genome. In addition, HPV-16 E5 transgenic mice develop spontaneous papillomas and hyperplasia. Thus, E5 contributes to proliferation both in the context of animal models and organotypic raft culture viral life cycle studies.

The Role of E6

E6, an approximately 150 amino acid residue protein, is encoded by the first translational open reading frame (ORF) within the HPV genome and is one of the major HPV oncoproteins. High-risk HPV E6 is able to transform rodent fibroblasts and immortalize keratinocytes, as well as other epithelial cells, in conjunction with E7, the other major HPV oncogene. In addition, *in vivo* E6 can inhibit epithelial differentiation, induce hyperplasia, suppress DNA damage-induced cell cycle arrest, and contribute to carcinogenic promotion and progression. All of these activities of E6 are thought to contribute to its oncogenicity. E6 does not possess any intrinsic enzymatic activity; rather, its oncogenic capabilities are mediated through its ability to act as a scaffold and regulate protein–protein interactions. Within E6 there are two motifs that are generally thought to mediate such interactions: two zinc finger domains and an α-helix-binding domain. High-risk E6 also contains a C-terminal PSD-95/Dlg/ZO-1 (PDZ) domain as well. Some of the protein interactions that have been mapped to these E6 domains include: p53, E6-associated protein (E6AP), E6-binding protein (E6BP), c-myc, p300/CBP, paxillin, PDZ proteins, interferon regulatory factor 3, and Bcl-2 homologous antagonist/killer (Bak).

The most studied of these interactions is the E6–p53 interaction. p53 is a transcription factor and tumor suppressor that is activated upon aberrant DNA replication, cellular stress, or cellular damage signals. As a response to cellular damage signals, the tumor suppressor p53 is upregulated and can induce cell cycle arrest or apoptosis through its transcriptional activity. p53 is a major regulator of DNA synthesis inhibition through cell cycle arrest and induction of cell death through apoptosis, which is presumably the reason that p53 is one of the most commonly mutated proteins in human cancer. Through the association of E6 with E6AP, a ubiquitin ligase, E6 is able to bind and induce the degradation of p53, though recent studies argue that other factors might also mediate E6's ability to induce p53 degradation. The ability of E6 to deregulate p53 correlates, at least in part, with its ability to transform cells in tissue culture and induce tumors in mice.

Clearly, the E6–p53 interaction is important for the oncogenic capabilities of E6. However, *in vivo* studies of the high-risk HPV-16 E6 using HPV transgenic mouse models indicate that an E6 mutant that retains the ability to inactivate p53 is partially defective for oncogenic phenotypes observed in mice transgenic for wild-type HPV-16 E6, including its abilities to induce skin tumors and cervical cancer. Thus, p53 inactivation is not sufficient to account for E6's oncogenic properties. The E6 mutant studied, competent to inactivate p53, is defective in its interaction with PDZ proteins. The ability of E6 to bind PDZ proteins also correlates with its ability to transform baby rat kidney cells. Only E6 proteins encoded by high-risk HPVs can bind PDZ proteins; E6 proteins encoded by low-risk HPVs do not contain the four-residue C-terminal, PDZ interaction domain. PDZ proteins are named for the protein family members that were initially identified to possess copies of a motif of approximately 80–100 residues, termed PDZ domains. These proteins are scaffolding proteins that generally interact with the C-terminus of their protein partners, which possess a charged four-residue motif, the PDZ interaction domain. The evolutionarily conserved PDZ proteins are required for development, cell adhesion, cellular growth and differentiation, and cell cycle processes. The E6 interaction with PDZ proteins is reported to lead to degradation of the PDZ proteins. Data indicate that the E6–PDZ interaction allows for the dysregulation of normal cellular growth controls and E6-induced hyperproliferation, transformation, and carcinogenesis. Therefore, while the E6–p53 interaction is required for E6-induced transformation and proliferation, the E6–PDZ interaction also contributes to E6 oncogenicity in a fashion that is distinct from p53.

E6 contributes to the viral life cycle. In HPV-31, as well as HPV-16, E6 is required for viral episome maintenance. In transient replication assays, HPV-31 E6-null genomes are able to replicate, indicating that E6 is not required for the initial establishment of HPV genomes. However, these E6-null genomes are not able to maintain the viral episome in long-term assays. HPV-31 genomes that are specifically defective in their ability to interact with PDZ proteins are still able to exhibit properties of intact nonproductive and productive stages, yet are slightly defective at each step of the life cycle. In contrast, there are clear defects in the viral life cycle when E6–p53 interactions are disrupted. Although the ability to bind p53 is not required for the establishment of viral episomes, the ability of viral episomes to be maintained correlates with the ability of E6 to induce degradation of p53. How the degradation of p53 contributes to maintenance of the viral episome is not clear. E6-mediated p53 degradation may be needed for viral maintenance to protect against induction of cellular damage signals induced during viral replication, especially as a consequence of the dysregulation of the tumor-suppressor protein, retinoblastoma (pRb), by E7.

The Role of E7

E7 is a 98-residue protein composed of two N-terminal conserved regions and a zinc finger domain in its C-terminus. The C-terminus regulates E7 multimerization. The N-terminus is partially homologous to two other virally coded oncoproteins, adenovirus (Ad) E1A and simian virus 40 T antigen (SV40TAg). E7 contains a pocket protein-binding motif, LXCXE, that confers an ability for E7 to bind pRb, as well as the two other 'pocket protein' family members, p107 and p130. The LXCXE motif is required for the ability of E7 to transform rodent fibroblasts and, in conjunction with E6, to immortalize human fibroblasts and keratinocytes efficiently. In addition, the LXCXE motif is required for E7 repression of pRb-induced senescence of cells harboring HPV genomes; and, in mice, the ability of E7 to bind pRb is essential for E7's induction of epithelial hyperplasia. This conserved region within high-risk E7s, and their interaction with the pocket proteins, correlate with their oncogenic properties. However, E7 is a multifunctional protein that is reported to bind not only the pocket proteins but also as many as 100 other cellular factors. Other potentially relevant partners include: the pocket proteins (pRb, p107, and p130), the cyclin-dependent kinase inhibitors p21 and p27, CK2 (formerly casein kinase II), and histone deacetylase (HDAC).

The pocket proteins are cell cycle regulators that modulate cell cycle progression and DNA synthesis. pRb is a tumor suppressor and is part of one of the most frequently mutated regulatory pathways in human cancers. p107 and p130 are closely related to pRb and possess similar biochemical activities. Together the pocket proteins regulate overlapping and distinct parts of the cell cycle. Both low- and high-risk E7 are able to bind the pocket proteins, although the relative binding preference for the pocket proteins appears to differ between low- and high-risk E7 proteins. In addition, high-risk HPV E7s can induce degradation of the pocket proteins. Two other DNA tumor viruses, Ad and SV40, also bind and inactivate the pocket proteins and affect the cell cycle.

The most studied E7–pocket protein interaction is the E7–pRb interaction. Normally, pRb is hypophosphorylated early in the cell cycle. In its hypophosphorylated state, pRb binds the transcription factor E2F/DP complex. The E2F/DP complex is a transcriptional activation complex that drives the expression of S phase genes. In early G1 phase of the cell cycle, hypophosphorylated pRb binds to E2F/DP and consequently inactivates the transcriptional complex. In late G1, pRb is phosphorylated by G1-specific cyclin/cdk complexes composed of cyclin D and cyclin E, and this leads to release of the E2F/DP complex from pRb. The free E2F/DP complex then becomes transcriptionally active and S-phase promoting genes are transcribed. In the context of HPV infection, E7 expression negates the requirement for pRb phosphorylation to activate the E2F/DP transcriptional complex. Specifically, E7 binding preferentially to hypophosphorylated pRb leads to the release of E2F/DP complexes from pRb and the consequent activation of the E2F complex. Therefore, E7 inactivation of pRb bestows on HPV the ability to override the pRb inhibition of the cell cycle.

In the context of the entire genome, the overall mechanistic requirements for E7 differ between genotypes. In HPV-16, E7 is required for the productive stage of the viral life cycle. Specifically, an HPV-16 genome carrying a null mutation of E7, while able to become established stably as a nuclear plasmid in human keratinocytes, is unable to support many aspects of the life cycle that arise in the differentiating compartment of stratified epithelium, including the inhibition of differentiation, reprogramming of suprabasal cells to support DNA synthesis, and viral DNA amplification. A similar requirement for E7 specifically in the productive stage of the life cycle was reported for HPV-18, and subsequent studies with HPV-16 demonstrated the importance of the E7–pocket protein interactions in mediating its role in the life cycle. An HPV-16 mutant genome carrying a mutation within the E7 ORF that disrupts the ability of E7 to bind pRb and the other pocket proteins displayed all of the defects in the productive life cycle that were observed with the E7-null HPV-16 genome. In contrast, a mutant genome in which E7 is disrupted in its ability to induce the degradation of the pocket proteins retained a partial ability to induce suprabasal DNA synthesis and viral DNA amplification but was completely defective in inhibiting cellular differentiation. Thus, there are differential requirements for E7's binding versus degradation of pRb and the other pocket proteins in E7's contribution to the various stages of the viral life cycle.

In contrast to what was observed with HPV-16 and HPV-18, the intermediate-risk HPV-31 genome appears to require E7 during the nonproductive stage of the viral life cycle as well as during the productive stage. E7-null HPV-31 genomes are unable to be maintained stably as nuclear plasmids in cultures of poorly differentiated human keratinocytes, whereas E7-null HPV-16 and HPV-18 genomes do stably replicate as nuclear plasmids. This difference apparently reflects genotype-specific differences.

The Role of L1 and L2

L1 and L2 are late viral genes encoding the major and minor capsid proteins, respectively. The PV capsid contains 72 pentameric capsomeres arranged on a $T=7$ icosahedral lattice and is approximately 55 nm in diameter, as determined by cryoelectron microscopy. The majority of the viral capsid is composed of the major capsid protein, L1, with L2 composing a minor fraction. Expression of L1 alone allows for the formation of pseudovirions or virus-like particles, VLPs, which are visually indistinguishable from native virions. If L2 is expressed in conjunction with L1, it is also incorporated into VLPs, but L2 is not required for capsid formation. L1 and L2 are specifically expressed in the most superficial layers of stratified keratinocytes, where progeny virus production arises.

Although L2 is not required for capsid formation, it does possess activities that likely contribute to a role in the viral life cycle. Some of these have been described above, under the section on virus entry, and include binding to cell surface receptors, actin, and PML, all of which have been implicated in early steps of virus infection. In addition, L2 is thought to be required for viral DNA encapsidation, although this requirement is obviated in some pseudovirion assembly systems. In life cycle studies carried out with HPV-31, L2-null genomes were able to support the nonproductive stage and most aspects of the productive viral life cycle in transfected human keratinocytes, with the exception that the level of encapsidated DNA in the progeny virions was reduced significantly and these progeny viruses were further reduced in their infectivity. Therefore, L2 seems to have a role in the viral life cycle in at least two distinct steps, encapsidation and infectivity. Whether the role of L2 in infectivity is at the level of receptor recognition or viral trafficking or both is something that should be determined in the future.

See also: Papillomaviruses: General Features; Papillomaviruses: General Features of Human Viruses; Papillomaviruses of Animals.

Further Reading

Bernard HU (2002) Gene expression of genital human papillomaviruses and potential antiviral approaches. *Antiviral Therapy* 7: 219–237.

Collins A, Nakahara T, and Lambert PF (2005) Contribution of pocket protein interactions in mediating HPV16 E7's role in the papillomavirus life cycle. *Journal of Virology* 79: 14769–14780.

Demeret C, Desaintes C, Yaniv M, and Thierry F (1997) Different mechanisms contribute to the E2 mediated transcriptional repression of human papillomavirus type 18 viral oncogenes. *Journal of Virology* 71: 9343–9349.

Evander M, Frazer IH, Payne E, Qi YM, Hengst K, and McMillan NA (1997) Identification of the α6 integrin as a candidate receptor for papillomaviruses. *Journal of Virology* 71: 2449–2456.

Lambert PF, Ozbun MA, Collins A, Holmgren S, Lee D, and Nakahara T (2005) Using an immortalized cell line to study the HPV life cycle in organotypic "raft" cultures. *Methods in Molecular Medicine* 119: 141–155.

Longworth MS and Laimins LA (2004) Pathogenesis of human papillomaviruses in differentiating epithelia. *Microbiology and Molecular Biology Reviews* 68: 362–372.

Munger K, Baldwin A, Edwards KM, *et al.* (2004) Mechanisms of human papillomavirus-induced oncogenesis. *Journal of Virology* 78: 11451–11460.

Nakahara T, Doorbar J, and Lambert PF (2005) HPV16 E1^E4 contributes to multiple facets of the papillomavirus life cycle. *Journal of Virology* 79: 13150–13165.

Ozbun MA (2002) Human papillomavirus type 31b infection of human keratinocytes and the onset of early transcription. *Journal of Virology* 76: 11291–11300.

Pyeon D, Lambert PF, and Ahlquist P (2005) Production of infectious human papillomavirus independently of viral replication and epithelial cell differentiation. *Proceedings of the National Academy of Sciences, USA* 102: 9311–9316.

Sanders CM and Stenlund A (2001) Mechanism and requirements for bovine papillomavirus, type 1, E1 initiator complex assembly promoted by the E2 transcription factor bound to distal sites. *Journal of Biological Chemistry* 276: 23689–23699.

Schiller JT and Lowy DR (2006) Prospects for cervical cancer prevention by human papillomavirus vaccination. *Cancer Research* 66: 10229–10232.

Skiadopoulos MH and McBride AA (1998) Bovine papillomavirus type 1 genomes and the E2 transactivator protein are closely associated with mitotic chromatin. *Journal of Virology* 72: 2079–2088.

zur Hausen H (2002) Papillomaviruses and cancer: From basic studies to clinical application. *Nature Reviews. Cancer* 2: 342–350.

Papillomaviruses of Animals

A A McBride, National Institutes of Health, Bethesda, MD, USA

© 2008 Elsevier Ltd. All rights reserved.

History

Warts or papillomas have been recognized in animals for centuries. The first description of transmissible animal papillomas was in the ninth century in *Al-Kheyl wal Beytareh*, a book of horse medicine by Ibn Akhi Hazam, the stablemaster for the Caliph of Baghdad. By the late 1800s, there were several examples of experimental transmission of warts in animals such as dogs and horses and in 1907 Ciuffo demonstrated that human warts could be transmitted by sterile filtrates, indicating a viral etiology. Cottontail rabbit or Shope papillomavirus (CRPV) was isolated in 1933 by Richard Shope and this was the first papillomavirus (PV) to be studied in detail.

Papillomas are normally benign, but there are several examples of animal warts undergoing malignant transformation to carcinomas. This was one of the earliest indications that viruses could be involved in the development of cancer and it helped initiate the field of viral oncology. This potential for malignant progression was later important in the recognition that human papillomaviruses (HPVs) were associated with cervical cancer. Relative to other viruses, PV research was hampered because of difficulties in propagating the virus; papillomaviruses (PVs) are species specific and epitheliotrophic and require a stratified and differentiated epithelium for productive infection. In the 1980s, research progressed more rapidly with the advent of molecular cloning because the small viral genomes could be readily propagated, and genetically modified, in a bacterial plasmid. Bovine papillomavirus type 1 (BPV-1) became a very popular prototype because of its unusual ability to transform and maintain its genome in rodent cells in culture. This allowed extensive analysis of the transformation, replication, and transcriptional properties of BPV-1. By the 1990s, several systems had been developed that allowed limited propagation of PVs in culture in artificial skin equivalents or epithelial xenografts in laboratory animals.

Much PV research has centered on the functions of the viral proteins. These small viruses encode only a handful of proteins and most function by interacting with a plethora of cellular proteins. Because viruses target pivotal regulatory functions in the cell, this research has led not only to a detailed understanding of the PV life cycle but has provided great insight into the function of key cellular proteins.

Taxonomy, Classification, and Evolutionary Relationships

Initially, PVs were grouped with polyomaviruses in the family *Papovaviridae*. However, it was later recognized that these viruses are not closely related and the International Committee on the Taxonomy of Viruses (ICTV) now recognizes the PVs as a new family, the *Papillomaviridae*. The *Papillomaviridae* have been further classified into genera and species based upon phylogenetic clustering of viral capsid protein L1 gene sequences (see **Figure 1**). Different genera have less than 60% and species have 60–70% nucleotide sequence identity. Individual PV types have 71–89% nucleotide identity. Differences of 2–10% identity define a subtype and less than 2%, a variant. Each genus is identified by a Greek letter, for example, *Alphapapillomavirus*. Each PV type is designated according to the host species and sometimes the site of infection, PV for papillomavirus, and finally, if several PVs have been isolated from the same species, a number. Examples are ROPV (rabbit oral PV) and BPV-1.

PVs have co-evolved with their hosts over millions of years and all types have most likely existed since the speciation of their host. The slow evolution of PVs is probably due to their use of the high-fidelity host DNA polymerase to replicate their genomes. It has been estimated that the most variable parts of PV genomes change at a rate of 0.25% per 10 000–20 000 years and there is no evidence for recombination between different viral types.

Host Range and Tissue Tropism

PVs are widespread. There are hundreds of different human types and PVs have been isolated from a diverse range of animals. To date, most animal PVs have been isolated from mammals, but viruses have also been identified in birds (see **Table 1**). PVs have been isolated from rodents, such as Syrian hamsters, the European harvest mouse, and the African multimammate rat. Unfortunately, no virus has yet been found that can infect laboratory mice. Multiple viral types have been found in those species that have been examined carefully (e.g., primate, cow, and rabbit) and many subclinical infections have been identified in primates (human and nonhuman). Presumably, more extensive investigation of each host would reveal a number of PV types that might rival the hundreds of PVs found in humans. Therefore, it would not be unreasonable to predict that there might exist over 100 000 different PV types.

All PVs are epitheliotrophic; each virus is species specific and infects and replicates in either cutaneous or mucosal epithelium, often at particular sites in the host. The reason for this tropism is not well understood, but it does not seem to depend on cell surface receptors for the virus and is thought to be due to very precise interactions between viral and host proteins that are required for a successful, productive infection. An exception to this is the fibropapillomaviruses. In addition to the productive infection of the epithelium, these viruses also nonproductively infect the underlying dermal fibroblasts, resulting in a fibroma. Probably because of this, these viruses have a less restricted host range and can nonproductively infect related host species and can transform and nonvegetatively replicate in cells from other hosts. For example, BPV-1 can cause equine sarcoids and can transform rodent cells.

Transmission, Clinical Features, Pathology, and Pathogenicity

PV-associated disease ranges from clinically inapparent infections, through a variety of benign warts, to malignant carcinoma. These differences in pathology are due to different viral types, different epithelial host cells, and the immune response of the host. The complete viral life cycle requires a stratified, differentiating epithelium

28 Papillomaviruses of Animals

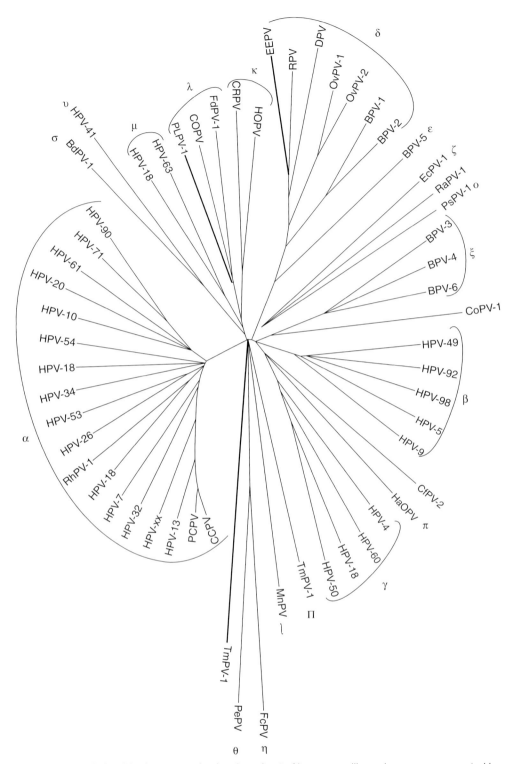

Figure 1 The evolutionary relationships between animal and a subset of human papillomaviruses as represented by a phylogenetic tree. Adapted from Van Doorslaer K, Rector A, Vos P, and Van Ranst M (2006) Genetic characterization of the *Capra hircus* papillomavirus: A novel close-to-root artiodactyl papillomavirus. *Virus Research* 118: 164–169, with permission from Elsevier.

(see **Figure 2**). In normal skin, only the basal cells can divide; after each division, one daughter cell remains in the basal layer and the other is pushed upward to begin the differentiation process. The latter cells withdraw from the cell cycle and begin to synthesize proteins that provide strength and barrier function to the epithelium.

To initiate infection, PVs usually require direct contact with the appropriate epithelia and must access the

Table 1 Notable animal papillomaviruses (by genus)

Alphapapillomavirus: mucosal and cutaneous lesions in order Primates
    PCPV-1: Pygmy chimpanzee (*Pan paniscus*) papillomavirus
    RhPV-1: Rhesus monkey (*Macaca mulatta*) papillomavirus type 1
Deltapapillomavirus: fibropapillomas in superorder Ungulata
    BPV-1, BPV-2: Bovine (*Bos taurus*) papillomavirus types 1 and 2
    DPV: Deer (*Cervus*) papillomavirus
    EEPV: European elk (*Alces alces*) papillomavirus
    OvPV-1, OvPV-2: Ovine (*Ovis aries*) papillomavirus types 1 and 2
    RPV: Reindeer (*Rangifer tarandus*) papillomavirus
Epsilonpapillomavirus: cutaneous lesions in family Bovidae
    BPV-5: Bovine (*Bos taurus*) papillomavirus type 5
Zetapapillomavirus: cutaneous lesions in order Perissodactyla
    EcPV-1: Equus caballus (horse) papillomavirus type 1
Etapapillomavirus: cutaneous lesions in class Aves (Passeriformes)
    FcPV: Fringilla coelebs (chaffinch) papillomavirus
Thetapapillomavirus: cutaneous lesions in class Aves (Psittaciformes)
    PePV: Psittacus erithacus timneh (African grey parrot) papillomavirus
Iotapapillomavirus: cutaneous lesions in order Rodentia
    MnPV: Mastomys natalensis (multimammate rat) papillomavirus
Kappapapillomavirus: cutaneous and mucosal lesions in order Lagomorpha
    CRPV: Cottontail rabbit (*Sylvilagus*) papillomavirus
    ROPV: Rabbit (*Oryctolagus cuniculus*) oral papillomavirus
Lambdapapillomavirus: cutaneous and mucosal lesions in order Carnivora
    COPV: Canine (*Canis familiaris*) oral papillomavirus
    FdPV: Felis domesticus (cat) papillomavirus
Xipapillomavirus: cutaneous and mucosal lesions in family Bovidae
    BPV-3, BPV-4, BPV-6: Bovine (*Bos taurus*) papillomavirus types 3, 4, and 6
Omikronpapillomavirus
    PsPV: Phocoena spinipinnis (porpoise) papillomavirus
Pipapillomavirus
    HaOPV: Hamster (*Mesocricetus auratus*) oral papillomavirus
Unassigned
    BPV-7: Bovine (*Bos taurus*) papillomavirus type 7
    CCPV-1: Common chimpanzee (*Pan troglodytes*) papillomavirus type 1
    CfPV-2, CfPV-3: Canine (*Canis familiaris*) papillomavirus types 2 and 3
    CgPV-1, CgPV-2: Colobus monkey (*Colobus guereza*) papillomavirus types 1 and 2
    ChPV-1: Capra hircus (goat) papillomavirus type 1
    EdPV: Erethizon dorsatum (porcupine) papillomavirus
    MfPV: Macaca fasicularis (long-tailed macaque) papillomavirus
    MmPV: Micromys minutus (European harvest mouse) papillomavirus
    PlPV-1: Procyon lotor (raccoon) papillomavirus type 1
    RaPV-1: Rousettus aegyptiacus (Egyptian fruit bat) papillomavirus type 1
    TmPV-1: Trichechus manatus latirostris (manatee) papillomavirus type 1
    TtPV-1, TtPV-2: Tursiops truncates (bottlenose dolphin) papillomavirus types 1 and 2
    TvPV: Trichosurus vulpecula (brushtall possum) papillomavirus

dividing cells in the basal layer of the epithelium through micro-abrasions or wounds. Viral DNA is established and maintained in the nuclei of dividing basal cells as an extrachromosomal replicating element. The infected basal cells serve as a reservoir of infected cells in the continual, progressive vertical differentiation that occurs in the maturation of the epidermis. Viral genome amplification, capsid protein synthesis, and particle production are restricted to the overlying, terminally differentiated cells. PVs disrupt the normal differentiation process, primarily because they must maintain cells in an S-phase-like state so that DNA replication enzymes are available to amplify the viral genome in differentiated cells. The infectious process is associated with a proliferation of the epidermal layers and in acanthosis (hyperplasia of the spinous layer), parakeratosis (persistence of the nuclei into the stratum corneum), hyperkeratosis (thickened cornified layer), and papillomatosis (undulating epithelium). Koilocytes (large, round, vacuolated cells with pyknotic nuclei) appear in the stratum spinosum and granulosum and abnormal keratohyalin granules are produced.

This process can result in a broad spectrum of papilloma-induced morphologies. For example, BPV-1-induced fibropapillomas in cattle can be sessile or pedunculate and lobate, fungiform, or verrucate. Other BPV types cause flat or filiform teat papillomas or alimentary papillomas. CRPV-induced papillomas occur as dark, highly keratinized masses. They range in size from 0.5 to 1 cm in

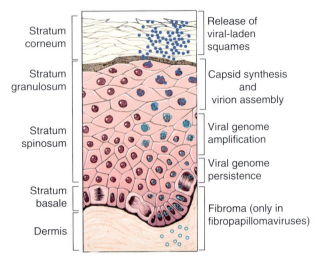

Figure 2 Life cycle of papillomaviruses. The diagram shows a model of stratified epithelium. The lowermost basal layer provides the germinal cells necessary for regeneration of the epidermis. In the stratum spinosum, desmosomes can be observed between cells giving them a characteristic spiny appearance. The cells of the stratum granulosum accumulate dense basophilic keratohyalin granules that contain lipids, which help form a waterproof barrier function. The outermost layer consists of dead cells filled with mature keratin. The stages of the viral replicative life cycle that occur in each cell layer are indicated. Fibropapillomaviruses also infect fibroblasts in the dermis, resulting in a fibroma.

diameter and can reach several centimeters in height, resulting in cutaneous 'horns'. In fact, it is thought that these 'horns' inspired the legend of the jackalope, a mythical creature that is a cross between a jackrabbit and an antelope. At the other end of the spectrum, healthy appearing skin from many animals has been found to harbor PVs.

Replication Cycle

All PVs have similar life cycles, closely linked to keratinocyte differentiation, although there are variations in the timing of each stage of infection with respect to differentiation that are probably linked to the host cell, the viral type, and the immune response. PVs infect the basal cells and probably use a common cell surface receptor. $\alpha 6\beta 4$ integrin and heparan sulfate proteoglycans are candidate receptors, but there is evidence to suggest that other molecules may be important. In the basal cells, the viral genome is maintained in a low-copy, extrachromosomal state and is replicated along with cellular DNA in a cell cycle-dependent manner. PV infections are usually long-lived and persistent and these cells must provide a reservoir of infected cells for the overlying virus producing tissue. As the infected cells differentiate and migrate upward to the stratum spinosum, there are changes in viral gene expression and vegetative viral DNA replication begins. Expression of the viral capsid proteins, L1 and L2, is first detected in cells of the stratum spinosum and virus-specific cytopathic effects are most pronounced in the stratum granulosum. Virions are assembled in the upper differentiated layers of the papilloma and are found throughout the nuclei, frequently organized into paracrystalline arrays, in cells which are destined to be sloughed from the epidermis. Viral transcription, translation, and replication are regulated through both positive and negative cellular processes that change during terminal differentiation. In fibropapillomas, there is also a proliferation of infected fibroblasts in the underlying dermis. Although the viral genome is maintained in the infected fibroblasts, there is no late gene expression or virion production.

Genome Organization and Expression

All PVs have a double-stranded, circular DNA genome of 7–8 kbp. There is a region of approximately 1 kbp that is called the long control region (LCR), upstream regulatory region (URR), or noncoding region (NCR). This region contains transcriptional enhancers and promoters, the DNA replication origin, and sequences required for genome maintenance (MME; minimal maintenance element). The viral promoters are regulated by cellular factors and the viral E2 proteins. In most viruses, the genes are organized in the order LCR–E6–E7–E2/E4–(E5)–SIR–L2–L1, where SIR represents the short intergenic region (see **Figure 3**). The coding region is divided into the early and late regions. The early region is expressed in the lower, more undifferentiated layers of a papilloma and proteins expressed from this region are designated E1 through E8 (see below). The capsid antigens, L1 and L2, are encoded by the late region and are expressed in the more superficial, differentiated cells of a papilloma. One exception is the E4 protein, which is encoded by the early region but is expressed abundantly in the upper layers of a wart. All viral RNA species are transcribed from one strand and are extensively processed to give rise to alternatively spliced mRNA species. The early polyadenylation site is located between the early and late regions while the late transcripts use a second site at the end of the late region.

Each viral protein is relatively well conserved from one virus to another. However, proteins from different viral types interact with distinct cellular proteins as well as those common to many viral types, and this likely explains their highly selective species and tissue tropism. One exception to the common genomic organization is the E5 gene, which is not present in all viruses. Another is genus *Xipapillomavirus*, which contains certain bovine papillomaviruses (BPV-3, BPV-4, BPV-6), that do not have an E6 gene and instead have a different gene in this position that has been designated E8. E8 is a membrane

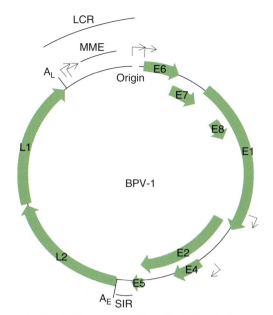

Figure 3 The BPV-1 genome. E and L designate the early and late open reading frames, respectively; promoters are indicated by arrows. LCR, long control region; SIR, short intragenic region; A_E and A_L, early and late polyadenylation sites, respectively; MME, minichromosome maintenance element.

protein with properties analogous to the E5 proteins of other PVs but there is no evidence that they are evolutionarily related. PePV, the African grey parrot PV, also has two genes at the beginning of the early region that have little similarity with the classical E6 and E7 genes and have instead been designated E8 and E9.

Functions of Viral Proteins

The functions of the individual viral proteins encoded by animal viruses are somewhat analogous to their HPV counterparts. The E5 protein of BPV-1 is the primary transforming protein and promotes proliferation of infected cells. It is a membrane protein that can transform fibroblasts by inducing constitutive, ligand-independent activation of the β-type platelet-derived growth factor receptor (PDGF-Rβ), and by interfering with Golgi acidification. BPV-4 E5 also transforms fibroblasts and disrupts gap junction-mediated intercellular communication. Many PV E5 proteins downregulate surface expression of major histocompatibility complex (MHC) class I molecules, which helps the virus to evade the host immune system.

The E1 protein is the primary replication protein. It is an ATP-dependent helicase that specifically binds and unwinds the viral DNA replication origin to allow access of the cellular replication proteins. It binds cooperatively to the origin in concert with the E2 protein. The E1 protein then converts to a double hexamer that encircles the DNA and unwinds the origin. The structure and function of the E1 protein is analogous to that of the polyomavirus SV40 T antigen. Much of our understanding about PV replication originates from key studies in BPV-1.

The E2 gene encodes multiple proteins that are the result of expression from multiple promoters and alternative RNA splicing. In BPV-1, there are three E2 proteins (E2TA, E2TR, and E8/E2). They share sequence-specific DNA binding and dimerization activities and the longest protein, E2TA, also contains a transcriptional activating domain. E2TA stimulates viral promoters by binding to multiple 12 bp palindromic sequences in E2-specific enhancers within the LCR. This activity is modulated by the E8/E2 and E2TR proteins which antagonize the functions of E2TA. The E2TA protein also represses transcription from viral promoters when the E2 binding sites overlap essential promoter elements. The E2 protein has an additional role in maintaining and partitioning the genome in the dividing basal cells. The E2 protein is a sequence-specific binding protein that binds to multiple E2 binding sites in the viral genome (in the MME) and tethers them to the cellular mitotic chromosomes through the transactivation activation domain. This has been best characterized for BPV-1, and the cellular bromodomain protein, Brd4, which binds to acetylated histones on chromatin, is an important cellular protein in this tethering complex. E2 also has a role at late stages of infection because expression of BPV-1 E2 proteins is greatly increased in cells that are vegetatively amplifying viral DNA.

Restricting high levels of viral DNA replication and antigens to the more superficial layers of an epithelium and releasing virus by the natural process of desquamation might be important for immune evasion by the virus. However, this means that the virus needs to replicate its DNA in cells that have normally withdrawn from the cell cycle and are undergoing differentiation. One of the main functions of the E6 and E7 proteins seems to be to sustain cells in an S-phase-like state so that the virus has available the enzymes to replicate its own DNA. This aberrant state induces cell cycle sensors that would normally arrest cells and perhaps cause them to undergo apoptosis. The molecular mechanism by which the E6 and E7 proteins of the 'high-risk' HPVs avert this process, by inactivating the retinoblastoma (pRb) and p53 proteins, is well understood. However, it is not yet clear how many of the animal viruses and 'low-risk' HPVs fulfill this function. The BPV-1 E6 protein disrupts the actin cytoskeleton and binds a number of cellular proteins, including ERC-55, the focal adhesion protein paxillin, the E3 ubiquitin ligase E6AP, and the clathrin adaptor complex AP-1. BPV-1 E7 lacks the binding motif that mediates direct binding of the HPV E7 proteins to pRb. However, both BPV and HPV E7 proteins bind a cellular protein, p600, a unique pRb- and calmodulin-binding protein. In the nucleus, p600 and pRB, seem to act as a chromatin scaffold and, in the cytoplasm, p600 forms a meshwork structure with clathrin. Viral-induced hyperplasia is believed to be induced by the viral early gene products, and results from both increased division of the basal cells and delayed maturation of the committed keratinocytes of the spinous layer (acanthosis). An unintended result of this cell cycle deregulation in some viruses is the immortalization of the infected cells and the continual division of cells that have sustained DNA damage, which can lead to a malignant phenotype.

The E4 protein, usually expressed from a spliced mRNA as E1^E4, is expressed at very high levels in the productively infected cells of a wart. It can interfere with the cell cycle, which may be important for deregulation of cell division and differentiation. E4 can also disrupt keratin filaments and induce abnormalities in the cornified envelope. These functions may be important for egress of the virus from the outer layers of a wart.

The L1 protein is the major capsid protein. The minor capsid protein, L2, is important late in infection for packaging the viral genome into the capsid. L2 is also important early in infection to transport the viral DNA to the nucleus and to establish a permissive site within the nucleus to initiate viral transcription and replication.

Virion Structure and Properties

PV virions form a nonenveloped, icosahedral structure of 55–60 nm diameter (see **Figure 4**). The capsid is composed of the major capsid protein, L1, and the minor capsid protein, L2, which form 72 pentameric capsomeres with an icosahedral symmetry of $T = 7$. The viral genome is packed in a nucleohistone complex. The L1 protein can self-assemble into virus-like particles (VLPs); VLPs present the conformational epitopes required for generating high-titer neutralizing antibodies and are the basis of very successful vaccines in humans and animals.

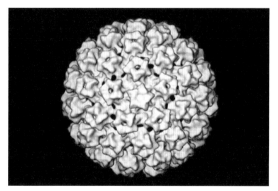

Figure 4 BPV-1 capsid. Reproduced from Trus BL, Roden RB, Greenstone HL, Vrhel M, Schiller JT, and Booy FP (1997) Novel structural features of bovine papillomavirus capsid revealed by a three-dimensional reconstruction to 9 Å resolution. *Nature Structural Biology* 4: 413–420.

Virus Propagation

The complete life cycle of PVs requires a stratified, differentiated epithelium because vegetative viral DNA replication and late gene expression can only take place in differentiated keratinocytes. To reproduce this in the laboratory, several xenograft techniques have been developed. Small pieces of epithelial tissue infected with virus and implanted in the renal capsule of an immunocompromised mouse will produce viral particles. Skin from various species can also be grafted onto immunocompromised mice and either infected with virus or transfected with viral DNA. The infected xenograft will form a papilloma-like lesion that will produce virion particles. Artificial skin equivalents (organotypic rafts) can also be established in tissue culture from keratinocytes and fibroblasts of various species. These rafts support the viral life cycle and produce viral particles. Relatively large quantities of infectious virions can also be purified from cells that have been co-transfected with the viral genome along with expression vectors for the L1 and L2 capsid proteins.

Notable Animal Papillomaviruses

PVs have been isolated and characterized from a multitude of animals including cattle, sheep, deer, horses, rabbits, dogs, mice, birds, and nonhuman primates. Many are listed in **Table 1** and a few of the better-studied viruses are described in detail below.

BPV-1. BPV-1 causes fibropapillomas on the cutaneous epithelium of cattle. It is readily transmitted among herd animals through direct contact of abraded skin. BPV-1 belongs to an unusual class of PVs that cause fibropapillomas in ungulates. These viruses have a broader host range than most PVs; BPV-1 can naturally infect horses, giving rise to sarcoids, can cause tumors in hamsters, and can morphologically transform mouse fibroblasts in culture. The infection is nonproductive in each of these cases.

In 1980, Lowy and co-workers demonstrated that cloned BPV-1 DNA could morphologically transform mouse cells in culture. The viral DNA replicated as an extrachromosomal element within these cells. Because of this, BPV-1 became the molecular prototype of the PVs and viral functions responsible for transformation, DNA replication, and transcriptional regulation were first characterized for BPV-1.

BPV-4. BPV-4 causes benign papillomas of the alimentary tract in cattle. In certain regions of Scotland, when cattle graze on bracken, these papillomas progress at a high rate to malignant carcinoma. Bracken grown in these regions contains at least one identified co-carcinogen, a flavenoid called quercetin, which promotes malignant progression of the papillomas.

CRPV. Cottontail rabbit papillomavirus (CRPV) naturally infects the cutaneous epithelium of wild cottontail rabbits and is also able to infect jackrabbits and snowshoe rabbits. In contrast, experimental infection of domestic rabbits results in nonproductive papillomas that support normal early viral gene expression and genome replication, but are unable to support late gene expression and virus particle production. CRPV-induced papillomas can either persist or regress, depending on host genetic factors, and persistent papillomas can progress to carcinomas in both wild and domestic rabbits. CRPV DNA can induce papillomas on scarified rabbit skin and this has allowed a genetic assessment of which viral functions are required to induce papillomas. The CRPV model has also been used for the development of preventive and therapeutic PV vaccines.

COPV. Canine oral PV (COPV) induces warts on the oral mucosa of dogs. Infection is normally followed by spontaneous immune-mediated regression. COPV is a good model for mucosal PV infection and has been very useful in studying the immune response and the development of PV vaccines.

RhPV-1. Rhesus papillomavirus type 1 (RhPV-1) is a sexually transmitted PV associated with genital disease that was first isolated from a penile squamous cell carcinoma. RhPV-associated disease progression closely resembles that seen in human genital HPV infections and RhPV-1 is phylogenetically closely related to the 'high-risk' HPVs that are associated with cervical cancer. RhPV-1 will likely become an important model to study human genital PV infections.

Detection and Diagnosis

Serological tests for PV infection have been unreliable and detection usually relies on testing for viral DNA. Specific viral types can be detected using either hybrid capture or highly sensitive molecular techniques such as nested polymerase chain reaction (PCR) with the use of degenerate primers.

Immune Response, Prevention, and Control

PV infections can be prolonged and persistent, but usually regress spontaneously. This immune-mediated regression is effected by T-cells while reinfection is prevented by humoral immunity. This phenomenon was noted in 1898 by M'Fadyean and Hobday who concluded after experiments with canine oral papillomas that, "the animal is left in a measure protected against an infection of the same kind." This spontaneous and simultaneous regression of papillomas by systemic immunity has been noted in many animals.

Much of our understanding of the immunology of PV infection comes from studies of animal PVs. Humoral immunity to subsequent infections is due to neutralizing antibodies directed against the capsid antigens. This immunity is type specific and can be bypassed by using viral DNA to induce papillomas in rabbit skin. Effective animal vaccines were produced that consist of crude wart extract, and these have been quite effective in cattle and dogs. The use of highly purified VLPs of COPV, CRPV, BPV, and *Equus caballus* papillomavirus (EcPV) as effective prophylactic vaccines laid the groundwork for the recently licensed HPV vaccines.

Cellular immunity is crucial for regression of papillomas. Dense infiltrates of T-lymphocytes can be observed in regressing warts in many animal species and immunosuppression can result in severe papillomatosis. The early, noncapsid proteins are important antigens for cell-mediated immunity and these might prove to be effective therapeutic vaccines for existing infections. Vaccination with CRPV early viral gene products has been shown to clear papillomas in rabbits. However, one complication is that PVs encode several functions that enable them to evade the immune system; they are able to inhibit interferon-dependent innate immunity and disrupt viral antigen presentation, which might also interfere with therapeutic vaccination.

See also: Papillomaviruses: General Features; Papillomaviruses: General Features of Human Viruses; Papillomaviruses: Molecular Biology of Human Viruses; Simian Virus 40.

Further Reading

Campo MS (2002) Animal models of papillomavirus pathogenesis. *Virus Research* 89: 249–261.

de Villiers EM, Fauquet C, Broker TR, Bernard HU, and zur Hausen H (2004) Classification of papillomaviruses. *Virology* 324: 17–27.

Nicholls PK and Stanley MA (2000) The immunology of animal papillomaviruses. *Veterinary Immunology and Immunopathology* 73: 101–127.

Peh WL, Middleton K, Christensen N, *et al.* (2002) Life cycle heterogeneity in animal models of human papillomavirus-associated disease. *Journal of Virology* 76: 10401–10416.

Trus BL, Roden RB, Greenstone HL, Vrhel M, Schiller JT, and Booy FP (1997) Novel structural features of bovine papillomavirus capsid revealed by a three-dimensional reconstruction to 9 Å resolution. *Nature Structural Biology* 4: 413–420.

Van Doorslaer K, Rector A, Vos P, and Van Ranst M (2006) Genetic characterization of the *Capra hircus* papillomavirus: A novel close-to-root artiodactyl papillomavirus. *Virus Research* 118: 164–169.

Papillomaviruses: General Features

H U Bernard, University of California, Irvine, Irvine, CA, USA

© 2008 Elsevier Ltd. All rights reserved.

Introduction

Papillomaviruses (PVs) are small DNA viruses that infect mucosal and cutaneous epithelia (skin). More than 100 PV types have been isolated from humans, and one or some few PV types have been found in virtually every carefully studied mammal and bird. PVs are strictly host species specific – human papillomaviruses (HPVs) cannot infect any other mammals, and no animal PV infects humans. Historically, PV research began with certain mammalian PV types (cottontail rabbit papillomavirus and bovine papillomavirus), as the large lesions caused by these viruses were sources for substantial virus preparations. Today's refined techniques led to the isolation and analysis of numerous PV types from a variety of human lesions, notably carcinomas of the cervix uteri (HPV-16), genital (HPV-6), and common warts (HPV-2). PVs cause benign and malignant neoplasia as they express oncoproteins with pleiotropic functions resulting from interactions with numerous cellular proteins. These oncoproteins induce continuing cell divisions in peripheral epithelial layers, while in the absence of PV oncoproteins, cell divisions are restricted to cells of the basal layer of epithelia. The consequence of this deregulation is a localized growth of the epithelium, leading to a papilloma (or wart), which gave this virus family its name. Within the broad spectrum of PV research, most efforts were directed to understand HPV-mediated carcinogenesis. These molecular studies have an increasing impact on medical practice in the form of DNA diagnosis and prophylactic vaccination.

Papillomavirus Particles and Genomes

PVs have icosahedral, nonenveloped particles (capsids or virions) with a diameter of approximately 55 nm. The particle is composed of 72 capsomers, and each of these capsomers consists of five identical L1 proteins, which are encoded by the PV genome. The virions also contain the minor capsid protein L2, which, however, is not part of the capsomers and is not required to generate a complete virion.

The PV particle contains the circular, double-stranded DNA genome in the form of chromatin. Many important PV types have genome sizes very close to 7900 bp; other viral genomes have sizes of a few hundred basepairs below or above this number. Most PV types encode eight open reading frames (ORFs): E1, E2, E4, E5, E6, E7, L1, and L2 (Figure 1). These ORFs are considered genes, as each of them is sufficient to encode a protein, although differential splicing also leads to alternative uses of some ORFs. The exact role and regulation of some alternative splices are still insufficiently understood. PV mRNAs are normally polycistronic (i.e., contain several ORFs), a dramatic deviation from the monocistronic mRNAs typical for eukaryotes and their viruses. The efficient translation of ORFs that are located 3′ of other ORFs is still little understood. Some few PVs lack the E5 or E6 genes. The L1 and E6 genes are separated by a genomic segment constituting about 10% of the PV genome, which contains many *cis*-responsive elements required for viral transcription and replication. This segment is called the long control region (LCR).

The Most Commonly Studied Papillomaviruses

More than 200 PV types exist in humans, and half of these have been isolated and formally described. The number of additional PV types in mammals is probably unlimited, although so far only a few dozen have been described. The tremendous diversity of PVs is not such a formidable barrier to understand PV biology as one might fear, as most research was based on only nine HPVs and three PVs from other mammals. Human papillomavirus 1 (HPV-1) induces plantar (foot-sole) warts and HPV-2 common (hand or face) warts. HPV-5 and HPV-8 are associated with epidermodysplasia verruciformis, a skin neoplasia linked to a genetic risk factor. HPV-6 and HPV-11 cause genital and laryngeal warts. HPV-16, HPV-18, and HPV-31 are the most prominent types causally linked to anogenital and some head and neck carcinomas. Bovine papillomavirus 1 (BPV-1) causes fibropapillomas and BPV-4 mucosal lesions in cattle, and the cottontail rabbit papillomavirus (CRPV) cutaneous lesions in rabbits. Owing to its medical importance and high prevalence, HPV-16 has probably been the object of more research than all remaining types together. The information in this article dealing with PV proteins and gene regulation is therefore biased in favor of HPV-16. Many – but clearly not all – functions of this virus are shared with other PVs.

Taxonomy and Evolution

PVs and polyomaviruses share two properties, namely small circular double-stranded DNA genomes and nonenveloped isosahedral capsids. As a consequence, they had once been

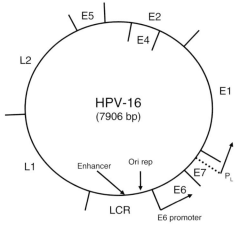

Figure 1 Genome organization of human papillomavirus type 16 (HPV-16). The genome consists of double-stranded DNA with a size of 7906 bp. LCR: Long control region, containing numerous cis-responsive elements for the viral transcription and replication, including the epithelial specific enhancer and the replication origin (ori rep). P_L: late promoter; E6 and E7: the two principal oncogenes; E1: replication initiation protein; E2: transcription factor, support of replication, segregation of viral DNA; E4: virion assembly and release; E5: minor oncogene; L2: minor capsid protein, intracellular transport of viral DNA after infection; L1: major capsid protein.

lumped into one virus family, *Papovaviridae*. When it became clear that they have different genome organizations and no genomic similarity, they were split into the two families *Papillomaviridae* and *Polyomaviridae*.

The PV literature uses the term 'type' to identify unrelated PV isolates, and the terms 'subtype' and 'serotype' are inappropriate. There are presently more than 100 formally described HPV types. New PV types have been found in all carefully examined mammal species, but so far in only two birds, and none in other animal taxa. Only one or some very few PV types have been detected in each animal host, and it is unclear whether the high number of HPV types has resulted from the more intense study of humans, or whether humans had been a more 'fertile' environment for PV evolution than other mammals.

The taxonomy of PVs was built on the comparison of genomes rather than on amplifiable viruses or on serology, since PVs do not multiply efficiently in cell culture, and since no consistent serology could be established based on natural infections. The genome-based taxonomy of PVs was founded in the 1980s with DNA hybridization data, and was later shifted to nucleotide sequence comparisons. By definition, a PV type is unique when the nucleotide sequence of its L1 gene differs by at least 10% from the L1 sequence of any other PV type. As a consequence, PV types are genotypes and not 'serotypes', although a retrospective study would likely reveal that most PV types are serologically distinct. PV types are designated by one or two capital letters identifying their host, and numbered according to the sequence of isolation, for example, HPV-16 for human papillomavirus type 16.

In order to classify PV types following traditional taxonomic terminology, the term 'type' has been placed in the taxonomic hierarchy below the term 'species'. In other words, phylogenetically closely related PV types with similar or identical biological and pathogenic properties have become lumped into PV species. While this procedure was justified to arrange PVs taxonomically in a manner similar to other virus families, it has created species with potentially confusing numeric designations that may not become generally used in the literature. In order to give an example, HPV types 16, 31, 33, 35, 52, 58, and 67 form the species *Human papillomavirus 16*. At the taxonomic level between species and the family (*Papillomaviridae*), PV species form genera that are designated by Greek prefixes (*Alpha-* through *Pi-papillomavirus*). Most medically important HPV types belong to genus *Alphapapillomavirus*.

Repeated isolates of the same PV type differ by up to 2% of their genomic sequence, and are termed variants of the original prototype isolate. Evidence suggests that this diversity evolved over a period of several hundred thousand years in linkage to the evolution of *Homo sapiens*. The term 'subtype' is used for rare PV isolates that show diversity in between the type and the variant level. PVs do not rapidly diversify like many RNA viruses, and do not form quasi-species.

Host Range

PVs are host species-specific, with the exception of a few PVs that were found in several different domestic hoofed animals, possibly transmitted by close contact between these animals. It has not been carefully studied whether host specificity is based on the lack of efficient contact or, more likely, on molecular incompatibility between PVs and heterologous hosts.

Within a particular host, notably humans, PVs often have a preferential target tissue. For example, HPV-1 is typically found in flat warts of the foot sole, HPV-2 in common warts of the skin elsewhere, HPV-6 in mucosal but also cutaneous lesions of the genitals, and HPV-16 mostly in anogenital and head and neck mucosal epithelia. This may be based either on molecular restrictions to replication in other types of epithelia, or on the lack of symptoms of established viral infections in inappropriate target epithelia. An example for the latter scenario is the fact that HPV-16 and some related types can cause cancer in women, but infect men normally asymptomatically. This possibility is also supported by the observation of HPV types that cause 'epidermodysplasia verruciformis' in patients with a genetic risk factor, while these same types can be detected in the skin of many individuals of the general human population in the absence of any symptoms.

While nearly all PV types are only found in epithelia, certain PV types replicate in cattle (like BPV-1) and other hoofed animals; also in mesenchymal tissue (fibropapillomaviruses).

Functions of Papillomavirus Proteins

The PV proteins that participate in viral replication, transcription, and transformation are identified by the letter E (for early), and those found in the capsid by the letter L (for late), each followed by a number according to its size in the first genome characterized (E1 and L1 being the two largest proteins). In the following paragraphs, these proteins will be discussed in the order of their numbers.

The 649-amino-acid residue E1 protein controls the duplication of the PV genome. It initiates DNA replication at the single viral replication origin about 100 bp 5′ of the E6 promoter. By convention, PV genomes are numbered in such a way that the E1 binding site spans the genome position 1. The E1 binding site is only loosely conserved as a 20 bp A+T-rich segment, and is therefore also only poorly recognized by the E1 protein. This failure is compensated by the E2 protein, which has binding sites close to the E1 target. E2 binding sites are much more sequence-specific than those for E1, and by forming a complex with E1, E2 is crucial for binding the replication origin.

E1 binds initially as a dimer and then undergoes polymerizations during the subsequent steps of replication initiation. Central to E1 function is an ATP-dependent helicase, and E1 hexamers melt and unwind the double-stranded DNA of the replication origin and translocate DNA single strands. Beyond these basic E1 functions, PV DNA duplication depends on the host replication machinery. E1 recruits cellular replication proteins to the PV origin of replication, including DNA polymerase α, replication protein A, and topoisomerase I. There is also evidence that E1 is a target for cellular signals that couple PV replication to the cell cycle.

The E2 protein has a size of 365-amino-acid residues and can be divided into three domains, an N-terminal transcription activation domain, a central hinge region, and a C-terminal DNA-binding domain. E2 proteins have at least three functions: stimulation of replication, modulation of transcription, and attachment of PV genomes to cellular chromosomes during mitosis.

All three functions require the DNA-binding domain. E2 proteins form dimers, which bind the palindromic sequence ACCGNNNNCGGT or slightly degenerated targets. The genomes of HPV-16 (and those of many related medically important HPV types) have four E2 binding sites, two of them positioned close to the E6 promoter, one 5′ of the E1 binding site, and one in the center of the LCR. The two E2 binding sites at the E6 promoter overlap with binding sites for TFIID and the Sp1 factor, and E2 represses the promoter by displacing these activators. The E2 binding site 5′ to the E1 binding sites, and one of the two E2 sites at the E6 promoter, lead to cooperation between E1 and E2 in origin function, as described above. The function of the E2 binding site in the center of the LCR is not well understood.

The E2 protein was for the first time studied in BPV-1 and described as an activator of transcription. Occupation of E2 binding sites in the LCR of BPV-1 strongly stimulates the E6 promoter, as these sites form an E2-dependent enhancer. This E2 function depends on the N-terminal transcription activation domain. The N-terminal domain of the HPV-16 E2 protein is homologous to the BPV-1 E2 domain, and the HPV-16 protein functions as a transcriptional activator in the context of BPV-1 genomes. Strangely, however, no enhancer activation function could ever be detected in HPV-16 genomes, and, consequentially, E2 is known only as a repressor of this virus. Many PVs also express shortened transcripts of E2 that solely encode the C-terminal DNA-binding domain and function as repressors by competing with full-size E2 proteins for binding to E2 cis-responsive elements.

The E4 ORF is positioned within the E2 gene, overlapping with the E2 hinge. Its 95-amino-acid residues are translated in a different reading frame from E2. A splice (E1^E4) fuses some N-terminal amino acids of the E1 ORF to E4. Some E4 transcripts terminate downstream of the early genes, while others extend into the late genes and are spliced to L2 or L1. E4 is sometimes considered a late protein due to the resulting co-expression with L2 and L1 and the fact that E4 is highly expressed in differentiating cells. E4 is known to support viral genome amplification in the productive part of the PV life cycle, capsid protein expression, virion assembly, and virion release, although the detailed mechanisms behind these E4 functions are still poorly understood. They include interactions with and reorganization of the cytokeratin network, reorganization of nuclear ND10 domains, and effects on cell cycle regulators.

E5 proteins are small (83-amino-acid residues in HPV-16 and only 44 residues in BPV-1) and do not show inter-type sequence similarities. They have functional homologies, however, as they are highly hydrophobic and thereby localized to the cell membrane and the endoplasmatic reticulum. Cell culture experiments defined E5 proteins as transforming proteins, although they are not expressed in many anogenital carcinomas due to interruption of the viral genome. They are pleiotropic and share at least three functions: (1) association with the 16 kDa subunit of the vacuolar H^+-ATPase, influencing the half-life of tyrosine kinase receptors; (2) direct interactions with tyrosine kinase receptors including the platelet-derived growth factor and epidermal growth factor receptors, and (3) downregulation of major histocompatibility complex class I molecules.

E6 and E7 are the two principal transforming proteins of the oncogenic HPV types as judged by cell culture transformation, continued expression in carcinomas, reversion of the transformed phenotype of cancer-derived cell lines after annihilation of E6/E7 expression, and complex pleiotropic molecular functions. They are encoded in the 5′ segment of polycistronic mRNAs transcribed from the E6 promoter, although many transcripts lack the ability to translate E6 due to an internal (E6*) splice.

The E6 protein of HPV-16 has a size of 149-amino-acid residues. The position of eight cysteine residues suggests the formation of two zinc fingers, which are larger than zinc fingers reported from cellular proteins. Although one would expect that these would help to induce a strong three-dimensional conformation, there is no information yet about the structure of E6. E6 shares this problem with the E7 protein, which has one single zinc finger with similar sequence properties. The lack of information about the structure of E6 and E7 proteins has remained a major obstacle in efforts to design drugs that would specifically interfere with these important oncoproteins.

More than 20 different cellular targets of E6 proteins have been described, and the contributions of these diverse interactions to the viral life cycle and carcinogenic processes are still much debated. It should be noted that PV oncoprotein functions evolved to support the latent and productive HPV life cycle, that is, the creation of a molecular environment favorable to virus replication. In spite of the medical importance of HPV-induced carcinogenesis, this pathological outcome should be considered an aberration and a fortuitous byproduct of the viral biology.

The most prominent target of E6 is the cell-cycle regulator and tumor suppressor p53. E6 and p53 form a trimeric complex including a protein called E6AP. E6AP is a ubiquitin ligase and the trimeric association results in p53 degradation. As a consequence, p53 is lost as inducer of the cdk inhibitor p21CIP, and the cell cycle of the infected cell is set free to undergo G1/S transition, creating an environment favorable to viral DNA replication, and establishing a prerequisite for oncogenic transformation. p53 is also known to be an inducer of apoptosis, and p53 elimination by E6 protects against this mechanism, which would otherwise eliminate infected cells and terminate PV replication. Yet other functions of E6 include the modulation of transcription by affecting the cofactor CBP/p300, effects on the immune response by interactions with the interferon regulatory factor-3, and alterations of cell shape and signaling by reaction with hDlg in a manner similar to a pathway that is induced by mutations during carcinogenesis of the colon.

The E7 oncoprotein of HPV-16 has a size of only 99 amino acid residues. E7 binds the retinoblastoma (RB) cell cycle regulator and tumor suppressor. In its normal function, RB represses the transcription factor E2F, which, in the absence of RB, induces genes required for G1/S transition of the cell cycle. The normal RB and E2F interactions are controlled by cyclin-dependent kinases. When E7 binds RB, E2F is released from the RB-E2F complex in an uncontrolled manner. Elimination of p53 and RB cell cycle control is a fundamental event in carcinogenesis, as many cancers whose etiology does not depend on viruses carry p53 and RB mutations. E7 is pleiotropic, and its functions include an affinity with the centromere. This leads to chromosomal abnormalities and genomic instability, resulting in a 'mutator phenotype' and promoting establishment of the malignant phenotype.

The 531-amino-acid residue L1 protein is the principal building block of the virion by forming 72 pentamers, which arrange into structurally complete icosahedral capsids in the absence of viral DNA or any other viral protein. This property has led to the production of HPV-6, HPV-11, HPV-16, and HPV-18 particles in heterologous expression systems as prophylactic anti-HPV vaccines. The L1 protein is central to the first step of PV infections by binding to proteoglycans and integrins at the surface of epithelial target cells. The infection does not select specific target cells, and the epithelial specificity of PV infections is established by the transcriptional environment in epithelial cells.

The L2 protein is part of the viral particle, where it forms a complex with the PV DNA. After uptake into endosomes, L2 accompanies the viral DNA by contact to the microtubule network via the motor protein dynein to the nucleus and subsequently to the subnuclear promyelocytic leukemia protein bodies, suggesting that it may be involved in the intracytoplasmic and nuclear transport of the PV DNA.

Gene Expression

The expression of PV genes, that is, ultimately the availability of each viral protein in the infected cell, is influenced by numerous mechanisms, such as frequency of transcription, recognition of transcription termination sites, differential splicing, mRNA stability, translation efficiency, and protein stability. There are many similarities of these mechanisms among different PVs, notably among the medically important HPV types, but strict generalizations are not possible. An example of exceptions is the transcription factor E2, which is in BPV-1 an enhancer-activating factor, but a negative regulator in HPV-16.

Transcription of all eight PV genes starts a few nucleotides 5′ of the ATG of E6 (E6 promoter), and continues unidirectionally around the whole viral genome. The E6 promoter has a TATA box that binds the general transcription factors including TFIID, and a promoter element, bound by Sp1. Among the numerous mRNAs generated from the E6 promoter are those that encode E2. E2 has two binding sites at the E6 promoter, and E2 proteins bound to these sites displace Sp1 and TFIID. Increasing mRNA levels derived from the E6 promoter lead to increased E2 expression and decreased use of this promoter, a perfect negative feedback loop, one of several adaptions of PV genomes to an inefficient (latent) rather than fulminant infection. HPV-16 genomes often recombine with chromosomal DNA in carcinomas such that the E2 gene is disconnected from the E6 promoter, releasing this repression mechanism.

The E6 promoter of HPVs has a low activity and requires activation by an enhancer centered 300 bp upstream of the promoter. This enhancer is only active in epithelial cells and determines the epithelial specificity of PVs. It does not bind E2, but depends on a variety of transcription factors including AP1, NFI, and glucocorticoid and progesterone receptors. Interestingly, the stimulation of PV oncogene expression by progesterone correlates with epidemiological data pointing to a high risk of cancer progression in women with multiple pregnancies and long-term anti-ovulant usage. A second promoter (p_L), relevant for the expression of L1 and L2, is positioned within the E7 gene. Activation of the TATA-less p_L is reminiscent of an early-late switch and still little understood.

PV transcription is strongly regulated by epigenetic mechanisms, that is, conformational changes of the PV chromatin, specifically of two nucleosomes that bind the HPV-16 enhancer and promoter. Histone acetylation and deacetylation can permit or restrict transcription. A variety of mechanisms can influence these parameters, notably the antagonistic factors CDP and AP1, which couple PV transcription to epithelial differentiation, and DNA methylation, which may determine whether an infection takes a latent or productive course.

Transmission and Epidemiology

PV infections are only stable when the pathogen reaches basal layers of an epithelium, for example, in wounds. PV infections result from physical contact between healthy and infected epithelia that peripherally release PV particles (e.g., during desquamation of the skin). As sexual intercourse leads to physical contact between genital epithelia, infections by many HPV types are considered sexually transmitted diseases. Epidemiological data support this mechanistic concept and document a rapid increase of genital PV infections in male and female individuals after commencement of sexual activity. Epidemiological evidence also supports the view that the risk to develop cervical carcinomas increases with young age at the start of sexual activity and with the number of sexual partners per lifetime. Additional risk factors are long-term use of anti-ovulants, multiparity, and tobacco smoking. Epidemiology provided a foundation for the concept of 'high-risk' and 'low-risk' HPV types. Both groups of HPVs are regularly found in exfoliated cells from anogenital sites, but only high-risk HPVs are frequent in cancer. 18 of the roughly 40 HPV types found in anogenital mucosas are considered high-risk types (e.g., HPV-16, HPV-18, and HPV-31), while HPV-6 and HPV-11, the cause of genital warts, are low-risk types, as they are rare in malignancies.

PVs are frequently transferred from mother to child during birth. So far, however, there is no evidence for efficient establishment of infections with high-risk HPV types toward anogenital carcinogenesis by this infection route. It is very possible, however, that the relatively high fraction of juvenile patients affected by laryngeal papillomatosis have acquired their HPV-6 and HPV-11 infections from their mothers.

Pathogenicity

PVs are well established as the cause of a variety of cutaneous and mucosal neoplastic lesions. As discussed above, the molecular pathology of such lesions is based on changes of epithelial homeostasis affected by the viral oncoproteins. In spite of the detailed knowledge of many molecular mechanisms, many aspects of PV pathogenicity are not as obvious as it is often assumed.

An example is the fact that many PV types have only been found in latent infections, and even those viruses typically associated with benign and malignant neoplasia can regularly be detected in healthy tissue. It is not well understood whether every PV infection leads to neoplasia, unless suppressed by an immune response, or whether the viral biology includes latent stages, maintenance of PV DNA without changes of the infected cell. This is exemplified by the fact that the same HPV types that can induce carcinogenesis of the cervix infect the penis without easily detectable symptoms. While male individuals function efficiently as transient hosts, they are rarely affected by disease. Yet another example is the fact that HPV-associated genital carcinogenesis in women is rare in the vagina, but typically arises from the transformation zone of the cervix, a tissue where adjacent endocervical columnar epithelia and ectocervical squamous epithelia change in a process called squamous metaplasia. As HPV tumors at yet other sites (anal carcinomas and laryngeal papillomas) are also associated with boundaries between different epithelia, one may speculate that yet poorly understood differentiation processes may synergize with PV molecular biology in order to induce neoplastic changes. Lastly, it is well confirmed that progression of precancerous lesions to malignant carcinomas is not just the result of PV-encoded mechanisms, but requires the accumulation of additional genomic changes, which are documented in extensive databases of chromosomal aberrations detected in cervical carcinomas. The molecular identity of these mutations is still a matter of research.

Diagnosis

HPV infections are diagnosed by the detection of a cutaneous or mucosal neoplasia, that is, a wart. Wart-like lesions, condylomata acuminata, can also occur at the cervix under the influence of the low-risk HPV-6 and HPV-11. High-risk HPVs induce flat condylomas which can be diagnosed by visual inspection through a colposcope. The traditional diagnosis of cervical PV infections is the Papanicolaou test ('pap test'), a technique developed in the 1950s and predating all knowledge about PVs. The Pap test aims to detect and classify dysplasia by microscopic observation of stained cervical exfoliated cells. Diagnostic criteria include a perinuclear halo (believed to result from accumulation of E4 protein) and nuclear enlargement (a consequence of polyploidies). In recent years, Pap tests have become complemented by DNA diagnostic detection of PVs.

Treatment

The treatment of many benign cutaneous or mucosal PV lesions will typically be a 'wait and see' approach. For those cases where treatment is necessary, surgical procedures can be based on excision by knife, laser, cryotherapy, or caustic substances. The application of concentrated solutions of salicylate is a traditional treatment of common warts, and podophyllin has been used in the treatment of genital warts. Antiviral drugs targeting PVs include interferons and imiquimode. Cervical precancerous lesions are surgically removed by excision of a cone-shaped wedge from the cervix or loop excision of the transformation zone. For information about treatment of malignancies, appropriate handbooks about gynecological oncology, etc., should be consulted.

Immunology and Vaccination

Anti-PV immune responses can be directed against the viral capsid, that is, targeting the L1 protein, or against virally infected cells, targeting any of the six early proteins.

Particles of HPV-16 and HPV-18 consisting only of L1 protein have been developed as vaccines by Merck and GlaxoSmithKline, the product of the former company also including HPV-6 and HPV-11 capsids. The vaccines became available in 2006/07, and have been cleared by the Food and Drug Administration of the United States for prophylactic vaccination of women aged 9–26. Vaccination during extensive clinical studies efficiently protected PV-uninfected women against *de novo* PV infections over periods exceeding 5 years. The vaccinations led to the stimulation of humoral immune responses by more than an order of magnitude beyond levels found in natural infections. The success of this approach is apparently based on anti-PV immune globulin concentrations in cervical mucus that suffice to neutralize PV particles before they inject the viral DNA into target cells.

Induction of immunity against the early PV proteins could form the basis of therapeutic vaccination, a potentially splendid strategy for anticancer immune therapy. Unfortunately, while scientists are well aware of this possibility, no major success has yet been achieved, and evidence points to numerous major obstacles. Investigations of naturally infected individuals generally showed weak immune responses against early PV proteins, and those humoral or cellular responses that were detectable correlated poorly with pathology or detection of PV DNA. A reason for this fact may be that PV early proteins are poor antigens, as suggested by comparison in animal systems with various other antigens. In addition, many PV proteins are only expressed at low concentrations, notably the oncoprotein E6 present in all cervical carcinomas, and rarely enter the circulation, as they are shed from the epithelium with the infected cell population.

See also: Papillomaviruses: General Features of Human Viruses; Papillomaviruses: Molecular Biology of Human Viruses; Papillomaviruses of Animals; Polyomaviruses.

Further Reading

Bernard HU (2002) Gene expression of genital human papillomaviruses and potential antiviral approaches. *Antiviral Therapy* 7: 219–237.

Campo MS (ed.) (2006) *Papillomavirus Research: From Natural History to Vaccines and Beyond.* Wymondham, UK: Caister Academic Press.

Davy C and Doorbar J (eds.) (2005) *Human Papillomaviruses, Methods and Protocols.* Totowa, NJ: Humana Press.

de Villiers EM, Fauquet C, Broker TR, Bernard HU, and zur Hausen H (2004) Classification of papillomaviruses. *Virology* 324: 17–27.

Lowy DR and Schiller JT (2006) Prophylactic human papillomavirus vaccines. *Journal of Clinical Investigation* 116: 1167–1173.

Mantovani F and Banks L (2001) The human papillomavirus E6 protein and its contribution to malignant progression. *Oncogene* 20: 7874–7887.

Munger K, Basile JR, Duensing S, et al. (2001) Biological activities and molecular targets of the human papillomavirus E7 oncoprotein. *Oncogene* 20: 7888–7898.

Munoz N, Bosch FX, de Sanjosé S, et al. (2003) Epidemiological classification of human papillomavirus types associated with cervical cancer. *New England Journal of Medicine* 348: 518–527.

zur Hausen H and de Villiers EM (1994) Human papillomaviruses. *Annual Review of Microbiology* 48: 427–447.

Paramyxoviruses of Animals

S K Samal, University of Maryland, College Park, MD, USA

© 2008 Elsevier Ltd. All rights reserved.

Glossary

Emerging virus A virus that has never before been recognized.
Phenotype The collective structural and biological properties of a cell or an organism.
Reverse genetics A technique whereby infectious virus is produced entirely from complementary DNA.
Syncytia Formation of fused or multinucleated cells.
Viremia The presence of a virus in the blood.
Zoonotic diseases Diseases that can be transmitted from animals to humans.

Introduction

The family *Paramyxoviridae* contains a large number of viruses of animals (**Table 1**), including a number of major animal pathogens (such as Newcastle disease virus (NDV), canine distemper virus, and rinderpest virus), zoonotic pathogens (such as Hendra and Nipah viruses), and a number of somewhat obscure viruses whose natural histories are poorly understood. New paramyxoviruses are being isolated on an ongoing basis from a wide variety of animals. For example, new paramyxoviruses have emerged that are pathogenic for marine mammals such as seals, dolphins, and porpoises (e.g., cetacean morbillivirus). Other paramyxoviruses that have been identified from various sources during the last few decades, such as Salem virus, Mossman virus, J-virus, and Beilong virus, are not associated with known diseases and are poorly understood. The recently identified Hendra and Nipah viruses came to light when they crossed species barriers and infected humans, causing severe, often fatal, zoonotic diseases. There are many animal paramyxoviruses, but only a few effective vaccines are currently available. Previously, genetic manipulation of paramyxoviruses was not possible because the genome is not infectious alone and RNA recombination is essentially nonexistent. This posed an impediment to the molecular and biological characterization of these viruses. However, in the last decade, methods of producing virus entirely from cDNA clones (reverse genetics) have been developed and have allowed manipulation of the genome of paramyxoviruses. This has greatly improved our understanding of the functions of each gene in replication and pathogenesis of these viruses. Another important aspect of this new technology is that vaccines can now be designed for some of the animal paramyxoviruses for which either vaccines are not currently available or the available vaccines are not satisfactory.

Taxonomy and Classification

Paramyxoviruses (some of which are sometimes also called parainfluenza viruses) belong to the family *Paramyxoviridae* of the order *Mononegavirales*. The order contains four families of enveloped viruses possessing linear, nonsegmented, negative-sense, single-stranded RNA genomes. The family *Paramyxoviridae* is further divided into two subfamilies: *Paramyxovirinae* and *Pneumovirinae* (**Table 1**). The two subfamilies differ in several features, most notably: (1) differences in nucleocapsid diameter (18 nm in *Paramyxovirinae* and 13–14 nm in *Pneumovirinae*); (2) possession of six to seven transcriptional units in *Paramyxovirinae*, and eight to ten transcriptional units in *Pneumovirinae*; (3) presence of an additional nucleocapsid-associated protein (M2-1) and an RNA regulatory protein (M2-2) in *Pneumovirinae*; (4) structural differences in the attachment protein; and (5) lack of RNA editing of the P mRNA in *Pneumovirinae*. The subfamily *Paramyxovirinae* comprises five genera, *Rubulavirus*, *Avulavirus*, *Respirovirus*, *Henipavirus*, and *Morbillivirus*, as well as a number of unclassified viruses that might become the basis of one or more additional future genera, in *Paramyxovirinae* (**Table 1**). The division of this subfamily into five genera and the unclassified group is based on: (1) amino acid sequence relationship between the corresponding proteins; (2) the number of transcriptional units; (3) RNA editing products of the P gene; and (4) the presence of neuraminidase and hemagglutinin activities in the attachment protein. The subfamily *Pneumovirinae* contains two genera: *Pneumovirus* and *Metapneumovirus*. These two genera differ by (1) presence of two additional genes, NS1 and NS2, in pneumovirus; (2) the pneumovirus gene order SH–G–F–M2, as opposed to metapneumovirus gene order F–M2–SH–G; and (3) amino acid sequence relationship between the corresponding proteins.

Host Range and Virus Propagation

Animal paramyxoviruses have been isolated from many different vertebrate animal hosts including mice, rats, bats, dogs, dolphins, seals, birds, cattle, pigs, horses, reptiles, tree shrews, and monkeys. In general, paramyxoviruses

Table 1 The genera and species of animal paramyxoviruses

Subfamily	Genus	Animal virus	Animal host	Disease
Paramyxovirinae	Rubulavirus	Parainfluenza virus 5 (formerly simian virus 5)	Dogs, pigs, monkeys	Respiratory disease
		Simian virus 41	Monkeys	Respiratory disease
		Porcine rubulavirus (La-Piedad-Michoacan-Mexico virus)	Pigs	Encephalitis, reproductive failure, corneal opacity
		Mapuera virus	Bats	Unknown
		Menangle virus (tentative species in the genus)	Pigs, bats	Reproductive failure
		Tioman virus (tentative species in the genus)	Bats	Unknown
	Avulavirus	Newcastle disease virus (avian paramyxovirus 1)	Domestic and wild fowl	Respiratory and neurological disease
		Avian paramyxoviruses 2–9	Domestic and wild fowl	Respiratory disease
	Respirovirus	Bovine parainfluenza virus 3	Cattle, sheep, and other mammals	Respiratory disease
		Sendai virus (murine para influenza virus 1)	Mice, rats, and rabbits	Respiratory disease
		Simian virus 10	Monkeys	Respiratory disease
	Henipavirus	Hendra virus	Bats, horses, humans	Severe respiratory disease
		Nipah virus	Bats, pigs, humans	Encephalitis
	Morbillivirus	Canine distemper virus	Carnivora species	Severe generalized and central nervous system disease
		Cetacean morbillivirus	Dolphins and porpoises	Severe respiratory and generalized disease
		Peste des petits ruminants virus	Sheep and goats	Severe generalized disease
		Phocine distemper virus	Seal	Severe generalized and central nervous system disease
		Rinderpest virus	Cattle, wild ruminants	Severe generalized disease
	Unclassified	Nariva virus		Unknown
		J-virus		Unknown
		Mossman virus		Unknown
		Tupia paramyxovirus	Tree shrews	Unknown
		Salem virus	Horses	Unknown
		Fer de lance virus	Snakes	Fatal disease
		Beilong virus	Rodents (?)	Unknown
Pneumovirinae	Pneumovirus	Bovine respiratory syncytial virus	Cattle	Respiratory disease
		Pneumonia virus of mice	Mice	Respiratory disease
	Metapneumovirus	Avian metapneumovirus	Turkeys, chickens	Severe respiratory disease in turkeys
				Swollen head syndrome in chickens

are restricted in host range. However, in recent years, some animal paramyxoviruses have been found to cross species barriers and infect other animal species and humans. In some cases, the animal viruses are highly virulent in the new host, as exemplified by Nipah and Hendra viruses, and pose a major public health concern. Interestingly, fruit bats in the genus *Pteropus* have been implicated as a reservoir of a number of new and emerging zoonotic animal paramyxoviruses. Other paramyxoviruses, such as NDV and bovine parainfluenza virus 3, can experimentally infect a variety of non-natural hosts, including rodents and monkeys, but typically are highly attenuated in these hosts. Many different primary and established cell cultures are used to grow animal paramyxoviruses. Some viruses do not readily grow in cell culture (e.g., avian metapneumovirus) and require adaptation by several passages in the cell cultures. Cell cultures derived from homologous species are generally used for cultivation of morbilliviruses and pneumoviruses. However, a number of paramyxoviruses grow well in cells of different host origin. For example, avian metapneumoviruses grow well in monkey kidney (Vero) cells, and bovine parainfluenza virus-3 grows well in monkey kidney (LLC-MK2 and Vero) cells and in baby hamster kidney (BHK_{21}) cells. Avian paramyxoviruses grow well in embryonated chicken eggs or cells derived from avian species. Some paramyxoviruses require the addition of protease, such as trypsin, α-chymotrypsin, or allantoic fluid (as a source of secreted protease), to the medium for growth in cell culture. This is necessary for cleavage activation of the viral fusion F protein (see below). Characteristic cytopathic effects of paramyxoviruses include the formation of syncytia (multinucleated giant cells) and eosinophilic cytoplasmic inclusion bodies.

Properties of Virion

The virions are 150–350 nm in diameter, pleomorphic, but usually spherical in shape. They consist of a nucleocapsid surrounded by a lipid envelope. Virion M_r is around 500×10^6. Virion buoyant density in sucrose is 1.18–1.20 g cm^{-3}. Some viruses (particularly of *Pneumovirinae*) are also produced in long filamentous form. Virions are highly sensitive to dehydration, heat, detergents, lipid solvents, formaldehyde, and oxidizing agents. Virus stability varies from stable (NDV) to very labile (rinderpest, canine distemper, bovine respiratory syncytial virus, and avian metapneumovirus). The schematic of a typical paramyxovirus is shown in **Figure 1**.

Genome

The genome consists of a single segment of negative-sense RNA (i.e., complementary to mRNA) that is 13–19 kbp in length and contains six to ten genes encoding up to 12 different proteins. The genome contains neither a 5' cap nor a 3' end poly(A) tail. At the 3' and 5' ends of the genome are short extragenic (noncoding) regions known as the 'leader' and 'trailer' region, respectively. The length of the leader is approximately 50 nt, whereas the length of the trailer is 23–161 nt. The leader region (*Pneumovirinae*) or the leader region and adjacent upstream end of the adjacent N gene (*Paramyxovirinae*) contains a single genomic promoter that is involved in the synthesis of the mRNAs as well as a complete positive-sense replicative intermediate called the antigenome. Generally, the first 10–12 nt of the leader and trailer are complementary, reflecting a conservation of promoter sequences present at the end of the genome and antigenome. At the beginning and end of each gene are conserved transcriptional control signals involved in initiation and termination/polyadenylation of the mRNAs. These conserved sequences are known as 'gene-start' and 'gene-end' sequences. The genes are separated by short intergenic regions that are not copied into mRNA. There is one exception in bovine respiratory syncytial virus where the L gene-start sequence is located upstream of the gene-end sequence of the upstream M2 gene, resulting in overlapping genes. The intergenic region is a conserved trinucleotide for respiroviruses, morbilliviruses, and henipaviruses, but is variable in length for all other paramyxoviruses. Thus, this might be a potential signal in some viruses, but not in others. The gene map of a representative member of each genus is shown in **Figure 2**. The nucleotide lengths of the genomes of members of subfamily *Paramyxovirinae* are even multiples of six, which is required for efficient RNA replication and is known as the 'rule of six'. However, the rule of six does not apply to the members of the subfamily *Pneumovirinae*. The genome size of a number of animal paramyxoviruses has been determined: 15 384 nt for Sendai virus; 15 456 nt for bovine parainfluenza virus 3; 15 246 nt for simian virus 5; 15 450 nt for simian virus 41; 15 186 nt for NDV; 15 882 nt for rinderpest virus; 15 948 nt for peste des petits ruminants virus; 15 690 nt for canine distemper virus; 15 702 nt for cetacean morbillivirus; 18 234 nt for Hendra virus; 18 246 nt for Nipah virus; 15 140 nt for bovine respiratory syncytial virus; 14 886 nt for pneumonia virus of mice; 15 522 nt for Tioman virus; 16 236 nt for avian paramyxovirus type 6; 13 373 nt for avian metapneumovirus type A; and 14 150 nt for avian metapneumovirus type C.

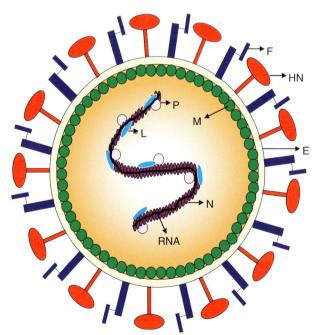

Figure 1 Schematic diagram of a paramyxovirus. N, nucleocapsid protein; P, phosphoprotein; L, large polymerase protein; M, matrix protein; F, fusion protein; HN, hemagglutinin-neuraminidase protein; E, envelope.

Proteins

All paramyxoviruses contain two glycosylated surface envelope proteins, a fusion protein (F) and an attachment protein (G or H or HN). The F protein mediates viral penetration by inducing fusion between the viral envelope and the host cell plasma membrane. In paramyxoviruses,

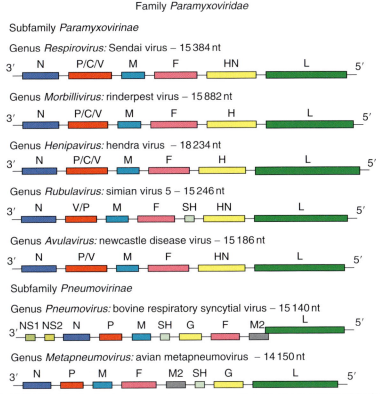

Figure 2 Map of genomic RNA (3' to 5') of animal paramyxoviruses representing the seven genera of the family *Paramyxoviridae*. Each box represents a separate gene; multiple distinct open reading frames (ORFs) within a single gene are indicated by slashes. For the P gene, the product encoded by the unedited mRNA is given first. In bovine respiratory syncytial virus of genus *Pneumovirus*, there is a transcriptional overlap at M2 and L genes.

the fusion event occurs at neutral pH. The F protein is synthesized as an inactive precursor (F_0), which is activated following cleavage by cellular protease(s) to generate two disulfide-linked F_1 and F_2 subunits. Some paramyxoviruses have multiple basic residues (arg and lys) at the cleavage site and thus are readily cleaved by furin-like proteases found intracellularly in most tissue types. Other paramyxoviruses have few, or only one, basic residues at the cleavage site and thus are cleaved extracellularly by a trypsin-like protease secreted in the respiratory and intestinal tracts, which limits virus replication. Hence, the number of arg and lys residues at the F protein cleavage site is a major determinant of paramyxovirus virulence. However, other viral proteins also contribute to the virulence of paramyxoviruses. The attachment protein binds to cell surface receptor and facilitates viral penetration. The attachment proteins of rubulaviruses, avulaviruses, and respiroviruses are designated HN since they possesses both hemagglutination activity, which is due to binding to sialic acid, and neuraminidase activity, which cleaves sialic acid on the cell surface and facilitates release. The attachment protein of morbilliviruses is designated H because it possesses hemagglutination activity, which is due to binding to the signaling lymphocyte activation molecule (SLAM) receptor, but not neuraminidase activity. Attachment proteins that generally lack hemagglutination and neuraminidase activities are designated G (glycoprotein). These occur in members of genera *Henipavirus*, *Pneumovirus*, and *Metapneumovirus*. The envelopes of the genera *Rubulavirus*, *Pneumovirus*, and *Metapneumovirus* contain a third integral membrane protein called small hydrophobic (SH) protein. The SH protein might play a role in cell fusion or in morphogenesis and also has been reported to interfere with cytokine-mediated intracellular signaling. The viral matrix protein (M) forms the inner layer of the virus envelope and plays an important role in virus assembly.

Inside the virion envelope lies the helical nucleocapsid. The genome and antigenome of paramyxoviruses are never found as free RNA either intracellularly or in the virion, but rather are tightly associated with the viral nucleocapsid protein (N) in the form of a ribonucleoprotein core. Virion nucleocapsids contain two other proteins, the phosphoprotein (P) and the large protein (L) that together constitute the viral RNA-dependent RNA polymerase complex. Virion nucleocapsids of the subfamily *Pneumovirinae* contain an additional protein (M2-1), which is a transcription elongation factor. A nonabundant protein (M2-2) is also produced from the second open reading frame (ORF) of the M2 gene and is involved in the balance between genome replication and transcription. The RNA within the nucleocapsid is resistant to nucleases.

Members of subfamily *Paramyxovirinae* encode multiple proteins from the P gene, due in part to a mechanism called 'RNA editing'. The P gene contains an editing site at which nontemplated G residues are added into the P mRNA by stuttering during transcription. The inclusion of additional G residues has the potential to shift the reading frame to access alternate frames, thus creating one or more chimeric proteins in which N-terminal domain is encoded by the P ORF upstream of the editing site and the C-terminal domain is encoded by the alternative ORF downstream of the editing site. In almost all members of the subfamily *Paramyxovirinae*, two of the major products of the P gene are the P and V proteins. For the respiroviruses, morbilliviruses, avulaviruses, and henipaviruses, the unedited mRNA of the P gene produces the P protein. Addition of one G nucleotide at the editing site produces an mRNA that encodes the V protein. In rubulaviruses, the unedited P mRNA encodes the V protein and addition of two G nucleotides produces the P mRNA. The V proteins of respiroviruses and morbilliviruses are nonstructural; whereas, the V proteins of rubulaviruses and avulaviruses are structural components of the virions. The respiroviruses, henipaviruses, and morbilliviruses also encode a third major protein from the P gene, namely the C protein. The C protein is synthesized from a +1 reading frame that overlaps the P and V reading frames. The V and C accessory proteins play important roles counteracting host cell antiviral defense mechanisms, especially the interferon system, and have been reported to be involved in other activities such as RNA synthesis and virion morphogenesis. *Pneumovirinae* lacks RNA editing. The members of genus *Pneumovirus* produce two additional nonstructural proteins (NS1 and NS2) from separate, promoter-proximal genes, which play a role in counteracting host cell antiviral defense mechanisms.

Replication and Virus Assembly

Paramyxovirus gene expression and RNA replication occur in the cytoplasm of infected cells, and progeny virions bud from the plasma membrane. Various cell surface molecules serve as receptors. Respiroviruses, rubulaviruses, and avulaviruses utilize sialic acid residues on various cellular glycoproteins (e.g., glycophorin) and gangliosides as receptors. Morbilliviruses utilize SLAM (also known as CD150) as a receptor. Infection by respiratory syncytial virus *in vitro* involves glycosaminoglycans, and Hendra and Nipah viruses use ephrin-B2 for infection of human cells. The F protein mediates fusion of the viral envelope and the plasma membrane of the host cell. As a result of the fusion, the viral nucleocapsid is released into the cytoplasm. Once in the cytoplasm, the nucleocapsid initiates transcription. The viral polymerase enters at the promoter located at the 3′-end of the genome. This promoter serves the dual function of mRNA and antigenome synthesis. Transcription is linear, sequential, and involves a stop-start mechanism guided by the gene-start and gene-end signals. As polymerase molecules progress along the genome, there is some dissociation at each gene junction, leading to a gradient of mRNA abundance that decreases according to distance from the 3′ end of the genome. The viral mRNAs are 5′-capped by the viral polymerase and contain a 3′ poly(A) tail that is produced by stuttering on the gene-end sequence. The intracellular accumulation of viral nucleocapsid-associated proteins results in the initiation of RNA replication. During RNA replication, the gene-start and gene-end signals are ignored and an exact complementary copy of the genome (antigenome) is synthesized. RNA synthesis is tightly linked to encapsidation of the progeny molecule. A promoter located at the 3′ end of the antigenome is used to synthesize genome.

The viral M protein plays a major role in mediating association of the nucleocapsids with patches in the plasma membrane where the viral envelope proteins have accumulated. It is thought that the M protein assembles the virion by forming a bridge between the cytoplasmic tails of envelope proteins and the nucleocapsids. Both the final assembly and budding of the virus occur at the plasma membrane of infected cells.

Reverse Genetics

Reverse genetics refers to the generation of subviral particles or complete infectious virus entirely by expression of cloned cDNAs. This provides a method for introducing desired changes into the viral genome. A number of animal paramyxoviruses have been recovered from cDNAs using reverse genetics, including simian virus 5, NDV, bovine parainfluenza virus 3, Sendai virus, canine distemper virus, rinderpest virus, bovine respiratory syncytial virus, and avian metapneumovirus. The basic method involves transfecting cultured cells with plasmids encoding the viral N, P, and L proteins, as well as the viral antigenome, all under the control of the T7 promoter. The positive-sense antigenome typically is expressed rather than the negative-sense genome to avoid hybridization with the positive-sense mRNAs, but virus has also been recovered (less efficiently) by expressing the genome. The bacteriophage T7 RNA polymerase is provided either by infection with a recombinant vaccinia virus expressing T7 RNA polymerase or by transfecting into cell lines that constitutively express T7 RNA polymerase. The recovery of bovine respiratory syncytial virus requires expression of an additional plasmid encoding the transcription elongation factor M2-1. Intracellular synthesis of the viral N, P, and L proteins and antigenome RNA results in the assembly of a biologically viral nucleocapsid that launches an infection leading to production of infectious virus. It is now feasible to genetically engineer attenuated viruses for use as live virus vaccines for several animal paramyxoviruses for which effective vaccines are not currently available. Perhaps even more exciting is the potential to use animal paramyxoviruses as vaccine vectors to design multivalent vaccines or to use more stable vectors to express antigens from less stable pathogens. At present, several animal paramyxoviruses, such as Sendai virus, NDV, and bovine parainfluenza virus 3, are being evaluated as vaccine vectors for other animal pathogens and also for use in humans as host-range-restricted vectors expressing antigens of human pathogens.

Genetic and Serologic Relationships

The relationships among paramyxoviruses can be deduced from nucleotide and amino acid sequence relatedness and serological analysis. Paramyxoviruses show very little amino acid sequence conservation among members of different genera. The sequence relatedness varies greatly within a genus, some members showing higher levels of sequence relatedness than others. The overall sequence conservation of paramyxovirus structural proteins in descending order seems to be L>M>F>N>H/HN/G>P. The L protein has five short regions of high homology near the center of the protein, which are not only conserved among paramyxoviruses, but are also conserved among all nonsegmented negative-strand RNA viruses. The C-terminal, domain of the V protein, is also conserved among all paramyxoviruses and contains seven invariant cysteine residues. Some animal paramyxoviruses show high levels of relatedness by sequence and serology with human paramyxoviruses. This implies that they have close evolutionary relationships and may have arisen by crossing species boundaries. Examples of pairs of related animal and human viruses include bovine and human parainfluenza virus 3, bovine and human respiratory syncytial virus, Sendai virus and human parainfluenza virus 1, simian virus 5 and human parainfluenza virus 2, and avian and human metapneumoviruses. In addition, all viruses within the genus *Morbillivirus* are related by sequence and serology. Rinderpest virus is more closely related to measles virus than to peste des petits ruminants virus and canine distemper virus. It is thought that the rinderpest virus is the archetype from which the other members of the genus morbillivirus have probably evolved, a process that involved crossing species boundaries.

Epidemiology

Some paramyxoviruses, such as NDV, canine distemper virus, bovine parainfluenza virus 3, and Sendai virus, have a worldwide distribution. Peste des petits ruminants virus is widespread in all countries lying between the Sahara and the Equator, in the Middle East, and in Southeast Asia. Avian metapneumovirus subtypes A, B, and D are present in Europe, but only subtype C is prevalent in the US. Nipah and Hendra viruses have emerged as new pathogens in Malaysia and Australia, respectively. Outbreaks of Menangle virus infection have been reported only in Australia.

The diseases caused by animal paramyxoviruses depend in part on their tissue tropism: as described below, some remain restricted to the respiratory tract and cause disease at that site, whereas others can disseminate by viremia to other tissues and cause disease that depends on the site of viral replication and pathogenesis. Immunity against viruses whose pathogenesis involves viremia tends to be relatively strong and long-lived, likely reflecting the long life of the serum antibody response. For example, rinderpest virus, which was once present on most continents, has been eradicated from Europe, America, and most of Asia. It remains enzootic only in parts of Asia and Africa. In contrast, immunity against viruses that remain localized in the superficial epithelium of the respiratory tract, such as bovine parainfluenza virus 3 and respiratory syncytial virus, is less effective and long-lived, and reinfection is common.

Transmission and Pathogenesis

Paramyxoviruses such as NDV and the morbillliviruses are highly infectious. The respiratory tract is the primary portal of entry for most paramyxoviruses and, for many, is the major site of viral replication; a few paramyxoviruses also infect via the enteric tract. Infection occurs by several different routes, including aerosols (NDV, bovine respiratory syncytial virus, avian metapneumovirus) and contaminated feed and water (Newcastle disease, canine distemper, and rinderpest viruses). Transmission of paramyxoviruses from fruit bats to animals is thought to occur by the fecal–oral route. In some viruses, the replication is confined to the respiratory mucosal surface (bovine parainfluenza virus 3, bovine respiratory syncytial virus, avian metapneumovirus), while in others, the initial replication on the respiratory tract is followed by systemic spread. Virulent strains of NDV initially infect the upper respiratory tract and then spread via the blood in the spleen and kidney, producing a secondary viremia. This leads to infection of other target organs, such as lung, intestine, and central nervous systems. In morbilliviruses, after initial replication in the respiratory tract, the virus multiplies further in regional lymph nodes, then enters the bloodstream, carried within lymphocytes, to produce primary viremia that spreads the virus to reticuloendothelial systems. Viruses produced from these sites are carried by lymphocytes to produce secondary viremia, which leads to infection of target tissues, such as lung, intestine, and central nervous systems.

Diseases

Paramyxoviruses are responsible for a wide variety of diseases in animals. Many paramyxoviruses primarily cause respiratory disease (bovine parainfluenza virus 3, bovine respiratory syncytial virus, avian metapneumovirus), while others cause serious systemic disease (rinderpest, virulent strains of Newcastle disease, canine distemper). Many diseases caused by animal paramyxoviruses also have a neurological component (canine distemper, Newcastle disease, Nipah virus) or a reproductive disease component (parainfluenza virus 5 in pigs and Menangle virus). Interestingly, the type of disease caused by Newcastle disease virus can vary, depending on the strain of the virus. Some strains cause only respiratory tract disease, some cause generalized hemorrhagic lesion, while others cause neurological disease. Most Newcastle disease virus strains replicate in the respiratory tract, while some predominantly replicate in the intestinal tract. Certain members of the genus *Morbillivirus*, canine distemper virus, phocine distemper virus, and cetacean viruses, cause high levels of central nervous system (CNS) diseases in their natural hosts, but CNS diseases are not associated with other members of genus *Morbillivirus*, such as rinderpest and peste des petits ruminant viruses. The severity of clinical disease also varies among animal paramyxoviruses. Some viruses cause asymptomatic or mild respiratory disease (bovine parainfluenza virus 3, simian virus 5, avian paramyxovirus types 2–9), while other viruses can cause severe disease leading to 90–100% mortality in susceptible hosts (rinderpest, Newcastle disease virus, canine distemper virus). There is also extreme variation in the pathogenicity of strains of some paramyxoviruses. For example, Newcastle disease virus strains range from avirulent to highly virulent (causing 100% mortality in chickens).

Immune Response

Paramyxoviruses induce both local and systemic antibody-mediated and cell-mediated immunity. Secretory IgA and cytotoxic T-lymphocytes play major roles in resolving infection and protecting against reinfection, but are somewhat short-lived, especially following a primary infection. Serum antibodies can also contribute to resolving infection and usually provide durable protection against reinfection. As already noted, serum antibodies are particularly effective against viruses whose pathogenesis involves viremia. Local immune factors play a greater role against viruses that remain localized in the respiratory tract. The envelope glycoproteins, H/HN/G and F, are the major neutralization and protective antigens of paramyxoviruses, although all of the viral proteins have the potential to contain epitopes for cellular immune responses. In some viruses (e.g., bovine parainfluenza virus 3), HN protein is the major protective antigen, while in other viruses (e.g., Newcastle disease virus), F protein is the major protective antigen. Most or all paramyxoviruses have evolved mechanisms that suppress the synthesis of interferon and the establishment of an interferon-mediated antiviral state.

Prevention and Control

Vaccination is a very effective means of controlling paramyxovirus infections. Both live-attenuated and inactivated vaccines have been developed for major animal paramyxovirus pathogens. Live-attenuated vaccines typically are more effective than the inactivated vaccines. Currently, effective live-attenuated vaccines are available for rinderpest, canine distemper, and Newcastle disease. However, satisfactory live-attenuated or inactivated vaccines are not available for diseases caused by bovine respiratory syncytial virus, Nipah virus, Hendra virus, and avian metapneumoviruses. Although the live-attenuated vaccines for rinderpest, canine distemper, and Newcastle diseases are generally very effective, there have been concerns about their potential safety and reversion to virulence. Furthermore, these vaccines cannot serologically distinguish vaccinated animals from naturally infected animals. Therefore, new and highly effective animal paramyxovirus vaccines are being engineered using reverse genetics techniques.

Future Perspectives

Some of the animal paramyxoviruses cause devastating diseases of animals, while others appear to be nonpathogenic. Many of the animal paramyxoviruses lack an effective vaccine. Development of reverse genetics systems has not only improved our understanding of the biology of these viruses, but has also provided methods for engineering effective vaccines. It is now possible to adjust the attenuation phenotype of a vaccine, introduce genetic markers into the vaccine viruses for differentiation between vaccine and wild-type strains, and to engineer thermostable vaccines for use in developing countries. The next steps will be to test these vaccines using a large number of animals, and to have them commercially available for vaccination purposes. Another advantage of reverse genetics is the use of animal paramyxoviruses as vectors to express foreign genes. This makes possible the use of one animal paramyxovirus vaccine to protect from multiple animal diseases. Since recombination involving members of *Paramyxoviridae* is essentially nonexistent, they will be particularly valuable as vectors to express the antigens of recombination-prone viruses such as coronaviruses. Some animal paramyxovirus-based vectors can be useful for the development of vaccines against emerging human infections such as H5N1 avian influenza, severe acute respiratory syndrome (SARS), and those caused by Ebola, Marburg, Nipah, and Hendra viruses. Since animal paramyxoviruses can be chosen that are serologically unrelated to common human pathogens, the general human population is susceptible to immunization with animal paramyxovirus-vectored vaccines. Reverse genetics systems are currently available for many but not all animal paramyxoviruses. Therefore, there is a great need to develop reverse genetics systems for the remaining animal paramyxoviruses. Furthermore, it is necessary to develop reverse genetics systems of local paramyxovirus strains for development of effective vaccines against the prevailing virus strains. In addition to vaccine development, it is also important to understand the pathogenesis and determinants of virus virulence and the mechanisms of interspecies transmission of the viruses. Due to the availability of reverse genetics systems for these viruses, we are confident that the next decade will bring a significant improvement in our understanding of their biology and we will witness development of better and safer vaccines against animal diseases.

See also: Measles Virus; Mumps Virus; Parainfluenza Viruses of Humans; Viral Pathogenesis; Human Respiratory Syncytial Virus; Rinderpest and Distemper Viruses.

Further Reading

Conzelmann KK (2004) Reverse genetics of mononegavirales. *Current Topics in Microbiology and Immunology* 283: 1–41.

Easton AJ, Domachowske JB, and Rosenberg HF (2004) Animal pneumoviruses: Molecular genetics and pathogenesis. *Clinical Microbiology Reviews* 17: 390–412.

Kurath G, Batts WN, Ahme W, and Winton JR (2004) Complete genome sequence of fer-de-lance virus reveals a novel gene in reptilian paramyxoviruses. *Journal of Virology* 78: 2045–2056.

Lamb RA, Collins PL, Kolakofsky D, et al. (2005) *Paramyxoviridae.* In: Fauquet CM, Mayo MA, Maniloff J, Desselberger U, and Ball LA (eds.) *Virus Taxonomy: Eighth Report of the International Committee on Taxonomy of Viruses*, pp. 655–668. San Diego, CA: Elsevier Academic Press.

Lamb RA and Parks GD (2006) *Paramyxoviridae*: The viruses and their replication. In: Knipe DM, Howley PM, Griffin DE, et al. (eds.) *Fields Virology*, 5th edn., pp. 1449–1496. Philadelphia: Lippincott Williams and Wilkins.

Li Z, Yu M, Zhang H, et al. (2006) Beilong virus, a novel paramyxovirus with the largest genome of non-segmented negative-stranded RNA viruses. *Virology* 346: 219–228.

Wang LF, Harcourt BH, Yu M, et al. (2001) Molecular biology of Hendra and Nipah viruses. *Microbes and Infection* 3: 279–287.

Parainfluenza Viruses of Humans

E Adderson and A Portner, St. Jude Children's Research Hospital, Memphis, TN, USA

© 2008 Elsevier Ltd. All rights reserved.

Introduction

The human parainfluenza viruses (hPIVs) are an important cause of respiratory disease in infants and children. Four types were discovered between 1956 and 1960. hPIV-1, hPIV-2, and hPIV-3 were first isolated from infants and children with lower respiratory tract (LRT) disease and subsequently shown to be a major cause of croup (type 1) and pneumonia and bronchiolitis (type 3). hPIV-4 was initially isolated from young adults and has been associated with mild upper respiratory tract disease of children and adults. Other viruses antigenically and structurally related to the human paramyxoviruses have been isolated from animals. Sendai virus, a natural pathogen of mice and not of humans, was the first PIV isolated and is antigenically related to human PIV-1. Simian virus (SV5) now PIV-5, recovered from primary monkey kidney cells, causes croup in dogs and is related to human type 2, and bovine shipping fever virus is a subtype of type 3.

Taxonomy and Classification

The PIVs belong to two genera, human parainfluenza virus types 1 and 3 to *Respirovirus* and human parainfluenza virus types 2, 4a, and 4b to *Rubulavirus*, of the subfamily *Paramyxovirinae* in the family *Paramyxoviridae*. Some other species found in the *Rubulavirus* are mumps virus, which causes disease in humans and in the genus *Respirovirus* Sendai virus in mice and Bovine parainfluenza type 3. The family *Paramyxoviridae* belongs to the order *Mononegavirales*, the distinctive feature of which is a negative-stranded RNA genome and a similar strategy of replication, suggesting that all negative-stranded viruses may have evolved from an archetypal virus.

Virion Structure, Genome Organization, and Protein Composition

The hPIVs are roughly spherical, lipoprotein enveloped particles 150–250 nm in diameter with an internal helical nucleocapsid containing the negative-sense single-stranded RNA genome. Projecting from the surface of the virion are the hemagglutinin–neuraminidase (HN) and fusion (F) glycoproteins. These glycoproteins are anchored to the plasma membrane of the infected cell or the virion envelope by a hydrophobic transmembrane region. In paramyxoviruses the transmembrane domain of HN is located near the N-terminus of the molecule and F near the C-terminus. Extending into the cytoplasm of the infected cell or inside the virion membrane is a short hydrophilic tail region, which plays a role in viral assembly through its interaction with the matrix (M) protein which lines the inner surface of the plasma membrane and itself interacts with the nucleocapsid. Inside the lipid bilayer of the viral particle is the RNA nucleocapsid which houses the nonsegmented negative-stranded RNA genome. The genomes of hPIV-1, hPIV-2, and hPIV-3 are approximately 15 000 nt and all genes except L have been sequenced for hPIV-4a and -4b and are certain to fall in this range once the sequencing is complete. The RNA genome serves as a template for transcription of mRNAs specifying six virion structural proteins linked in tandem in the order of 3′-nucleoprotein (NP)–polymerase-associated protein (P/V)–matrix protein (M)–fusion protein (F)–hemagglutinin-neuraminidase (HN)–large protein (L)-5′. The P gene alone is unique in its capacity to express, in addition to the P protein, various other proteins by utilizing internal initiation codons in the same or different reading frames or by RNA editing of the P gene mRNA through insertion of nontemplated G residues. Besides P, these other proteins have been designated V, C, and D, depending on the PIV. The helical nucleocapsid is 18 nm in diameter, contains approximately 2000 molecules of NP bound to the RNA genome, about 200 P and 20 L molecules. The P protein, which forms a polymerase complex with L, is essential for the enzymatic processes of viral RNA transcription and replication including the 3′ addition of poly(A), modification of the 5′ end, and nucleotide polymerization of viral transcripts. The approximate molecular weights in daltons of the PIV proteins as exemplified by PIV-3 are: NP, 58 000; P, 68 000; M, 40 000; F, 63 000; HN, 72 000; and L, 256 000. A cartoon and an electron micrograph of a naturally occurring PIV are shown in **Figure 1**.

Attachment Protein

The process of infection is initiated by the action of the HN glycoprotein, which binds the virion to sialic acid-containing glycoprotein or glycolipid receptors on the host cell surface. The same process is responsible for the hemagglutination of avian and mammalian erythrocytes. HN also causes the enzymatic (neuraminidase) cleavage of sialic acid residues from the carbohydrate moiety of glycoproteins and glycolipids, which functionally serves to prevent the self-aggregation of virus during release and likely aids in the spread of virus from infected cells. Besides attachment, HN provides an unknown function that is either essential or enhances the fusion activity of the F protein. It is proposed that HN directly interacts with F, possibly altering its conformation, thereby stimulating the fusion activity of F.

The morphology of HN based on studies of PIVs such as Sendai virus is envisioned as an N-terminal stalk region

Membrane proteins
- HN: receptor-binding, neuraminidase, fusion promotion
- F: membrane fusion
- M: assembly

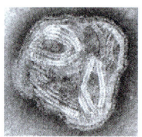

Nucleocapsid
- NP: encapsidate viral RNA
- P: forms polymerase complex with L
- L: RNA synthesis, capping, polyadenylation

Figure 1 Parainfluenza virus structure.

of approximately 130 amino acids anchoring a large glycosylated hydrophilic globular head region to the viral envelope. A small uncharged hydrophobic peptide located near the N-terminus spans the viral envelope, and a small hydrophilic domain is internal to the membrane. The globular head contains the active site for virus attachment, neuraminidase activity, and antigenic determinants that induce neutralizing antibodies for hPIVs, as well as the other members of the *Respirovirus* and *Rubulavirus* such as Sendai virus. HN exits on the surface of the virion as disulfide-linked homodimers or tetramers.

Solution of the three-dimensional structure of the HN protein of NDV and more recently of hPIV-3 and SV5 has resolved many of the previous issues concerning the structure and functions of HN. A significant advance was the determination that the HN sialic acid-binding site and the neuraminidase active site were the same and for NDV that conformational change of the site switches HN activity from sialic acid binding (attachment) to sialic acid hydrolysis (neuraminidase activity). However, similar conformational changes may not occur for hPIV-3 and SV5 HN.

Fusion Protein

Following virus attachment, F, the other surface glycoprotein, mediates the fusion of the virion and host cell-surface membranes, which allows the nucleocapsid to be deposited in the cell cytoplasm where gene expression begins. F expressed on the surface of infected cells also mediates fusion, allowing the extension of infection to uninfected cells.

All F proteins are synthesized as inactive precursors (FO) which are post-translationally cleaved by a host cell trypsin-like protease to form the biologically active molecule. For the paramyxoviruses in general, the cleavage site is located about 100 amino acids from the N-terminus of FO. The cleavage site is characterized by a short span of basic amino acids on the amino side of the site and a longer stretch of about 30 hydrophobic residues on the carboxyl side. The number and location of basic amino acids on the amino side of the cleavage site varies with individual PIVs. hPIV-1 and Sendai virus have one basic amino acid immediately adjacent to the cleavage site (Arg), whereas PIV-2 and PIV-3 have two (Arg–Lys) and PIV-5 has five (all Arg). The motif of paired basic amino acids at the cleavage site increases the efficiency of cleavage, host and tissue tropism, and pathogenicity, as the dibasic motif is recognized by a ubiquitous protease, whereas the enzymes that cleave at a monobasic site are found in a limited number of tissues. Cleavage of FO results in two disulfide-linked fragments; the larger one, F1, forms the new hydrophobic N-terminus, which causes membrane fusion. The smaller product, F2, is the original approximately 100 N-terminal residues of FO.

Recent solutions of the crystal structure of the F in its prefusion and postfusion conformations reveals dramatic alterations in the F protein structure during membrane fusion and offers insight into the mechanism of F activation.

M Protein

The M protein lines the inner surface of the viral envelope and is thought to play a role in virus maturation by interacting with the envelope glycoproteins and the nucleocapsid. During infection M associates with the inner leaflet of the plasma membrane where it orchestrates the release (budding) of progeny virus by interacting with specific sites on the cytoplasmic tail of the viral glycoproteins and the nucleocapsid and in transport of viral components to the budding site.

Transcription, Translation, and Replication

The negative-strand strategy of viral replication involves the synthesis of unique mRNA species of each paramyxovirus gene during infection. After introduction of the infecting nucleocapsid into the host cell, transcription is the first step in gene expression. Transcriptional regulation of mRNA abundance is determined by the gene order; the closer the gene is to the 3′ end of the genome, the more efficient is the transcription. Thus, the abundance of paramyxovirus proteins is determined mainly by the polarity of the genome. Once protein synthesis is underway, replication of the genome begins, which provides additional templates for transcription and replication.

Development of 'reverse genetics' systems for SV and hPIV-3 as well as other members of the paramyxovirus family, in which the RNA genomes of these viruses are expressed from a DNA template, has provided tools and facilitated our understanding of PIV protein structures and function, virus assembly, regulation of RNA transcription and replication, and viral pathogenesis.

Geographic and Seasonal Distribution

All of the hPIV types 1–4 have a wide geographic distribution. PIV-1–PIV-3 have been identified in most areas where facilities are available for the study of childhood respiratory tract diseases. PIV-4 has been isolated in fewer areas but this is likely to be due to the difficulty of isolation and it is probably widely distributed.

Host Range and Viral Propagation

The four hPIV types were originally isolated from humans: they cause disease in humans and humans are the primary host. Other laboratory animals can serve as experimental hosts; hamsters, guinea pigs, and ferrets can be infected with hPIV-1, hPIV-2, and hPIV-3 but these infections are usually asymptomatic. hPIV-3 can also infect cotton rats, rhesus, and patas monkeys and chimpanzees, but these animals are not good model systems for studying disease caused by PIVs.

Embryonated hen's eggs can support the growth of some strains of hPIV-1, hPIV-2, and hPIV-3, but are much less sensitive than monkey kidney cells for primary isolation. Sendai virus, a murine subtype of PIV-1, is an exception in that it grows exceedingly well in eggs.

All four hPIV types grow well in primary monkey or human kidney cells, which are also used in the isolation of virus from clinical samples. LLC-MK2, a rhesus monkey kidney cell line, offers an efficient system for the isolation of PIVs and an experimental tissue culture system. PIV-2 and PIV-1 require trypsin in the medium to cleave the F glycoprotein for cell growth, but not PIV-3 strains. Viral infection of tissue culture can be detected by hemadsorption with chicken and guinea pig erythrocytes and the cytopathic effects produced by the viruses.

Genetics and Evolution

The entire genomic sequences of hPIV-1, hPIV-2, and hPIV-3 have been completed. The genomes of other members of the genera *Respirovirus* and *Rubulavirus* (Sendai virus, bPIV-3, and PIV-5) have been sequenced as well. In general, the genome organization is remarkably similar, but differences exist in sequence, intergenic regions, and nonstructural proteins expressed from the P gene. Sequence and immunological analyses suggest that human hPIV-1 and Sendai virus are closely related type 1 PIVs, as is PIV-3 of human and bovine origin. hPIV-1 and -3 are more closely related to each other than to hPIV-2, PIV-5, mumps virus and NDV, which in turn show a closer evolutionary relationship to each other. Evolutionary divergence of hPIV-3 and hPIV-1 is greatest for the P protein (a phenomenon of paramyxovirus P proteins in general), less for the HN and F glycoproteins, and least for M, NP, and L. Similarly, hPIV-1 and Sendai virus, and human and bovine PIV-3, show the greatest divergence in the P protein and least for NP.

Serologic Relationships and Variability

The human paramyxoviruses can be divided into two antigenic groups, comprised of PIV-1/PIV-3 and PIV-2/PIV-4/mumps virus. Human PIVs share certain antigenic determinants, but structural and antigenic variation occurs between serotypes and, to a lesser degree, within strains belonging to the same type. Antigenic diversity is not generally progressive, but may contribute to PIVs' propensity to cause recurrent infection. Antibody to F and HN correlates with neutralizing antibody and protective immunity.

Epidemiology and Transmission

The hPIVs are an important cause of respiratory tract infections (**Table 1**). hPIV-3 infections are endemic, whereas infections caused by other PIVs occur in outbreaks. Infections are common before 2 years, especially with hPIV-3, and almost universal by 5 years. hPIVs are highly communicable by respiratory droplets and by contact with surfaces contaminated by respiratory secretions. The incubation period is 2–6 days. Virus is typically shed for 4–21 days.

Pathogenicity

hPIVs bind to specific receptors on upper respiratory tract epithelial cells and may subsequently spread to the LRT. Optimal fusion efficiency requires the coordinated action of both F and HN. The virus and host factors contributing to disease pathogenesis are not completely understood. Both viral (F and HN structure) and host features (genetic susceptibility, host cell proteases, immunocompetence) are likely to influence the severity of infection.

Table 1 Proportion of viral respiratory tract infections caused by PIV

	Proportion (%)	Most common type
Rhinitis (common cold)	20	
Pharyngitis	20	
Laryngotracheitis (croup)	50–70	PIV-1
Bronchiolitis	20	PIV-3
Pneumonia	10	PIV-1, PIV-3
Otitis media (middle ear)	10	

Clinical Features

hPIVs cause between 15% and 65% of viral respiratory infections, most notably laryngotracheitis (croup) and bronchiolitis (**Table 1**). Although infections are generally mild, PIVs are responsible for over 10% of pediatric hospitalizations for respiratory infection. Life-threatening LRT infection occurs in patients with cellular immunodeficiencies, particularly those with severe combined immunodeficiency and hematopoietic stem cell and solid organ transplants. Rare cases of parotiditis, myocarditis, aseptic meningitis, and encephalitis have been reported.

The gold standard for diagnosis of hPIV infections is culture on primary monkey kidney, human embryonic kidney, or human lung carcinoma cells, which typically requires incubation for 4–7 days. Detection of epithelial cell-associated viral antigens in nasopharyngeal or bronchoalveolar washings is rapid and has a sensitivity and specificity of >80% compared to culture. Polymerase chain reaction (PCR) is rapid and more sensitive, but is not widely available. Detection of specific IgM and IgG in paired acute and convalescent sera has limited utility in the diagnosis of acute infections.

Pathology

hPIV has direct cytopathic effects, causing ciliary damage, epithelial necrosis, and recruitment of a mononuclear inflammatory response. Recent studies suggest that host inflammatory responses, rather than the direct effects of viral replication, are most accountable for the signs and symptoms of PIV infection. Alterations in HN receptor-binding or -activity influence host inflammatory responses independent of viral replication or the ability to infect epithelial cells. PIV infection may trigger long-term bronchial hyperreactivity in genetically susceptible persons.

Immune Response

Viral shedding is more prolonged and disease severity increased in persons with T-cell immunodeficiencies, implying cellular immunity is important in the control of acute hPIV infection. Protection from reinfection is associated with the development of serum and secretory antibody against F and HN. The duration of protection after acute illness is relatively short however, especially in infants, and protection from infection by heterotypic stains is incomplete.

Prevention and Control

Most hPIV infections require supportive care only. Children with moderate-to-severe croup benefit from inhaled vasoconstrictors or inhaled or systemic corticosteroids, which reduce airway edema. Anecdotal reports describe the successful use of ribavirin in immunocompromised patients. There are currently no licensed vaccines to prevent hPIV. Early formalin-inactivated vaccines were poorly immunogenic and had the potential to enhance pulmonary inflammatory responses to infection with wild-type virus.

Future Perspectives

Novel antiviral therapies being developed for PIV infections include selective inhibitors of HN, fusion, and IMP dehydrogenase, and siRNA against phosphoprotein mRNA. Vaccines currently in early clinical trials include live attenuated human PIV-3, bovine PIV-3, human-bovine chimeras, and bovine PIV or Sendai virus expressing heterologous human PIV F and HN proteins. Besides vaccines, antiviral drugs may be designed that curtail viral replication or prevent viral spread. Knowledge of the three-dimensional structure of HN and F, especially those regions involved in viral attachment, penetration, and release, will be important in drug design.

See also: Paramyxoviruses of Animals; Human Respiratory Viruses; Viral Membranes.

Further Reading

Karron RA and Collins PL (2006) Parainfluenza viruses. In: Knipe DM, Howley PM, Griffin DE, et al. (eds.) *Fields Virology,* 5th edn., pp. 1497–1526. Philadelphia, PA: Lippincott Williams and Wilkins.

Lamb RA and Parks GD (2006) *Paramyxoviridae*: The viruses and their replication. In: Knipe DM, Howley PM, Griffin DE, et al. (eds.) *Fields Virology,* 5th edn., pp. 1449–1496. Philadelphia, PA: Lippincott Williams and Wilkins.

Lamb RA, Paterson RG, and Jardetzky TS (2006) Paramyxovirus membrane fusion: Lessons from the F and HN atomic structures. *Virology* 344(1): 30–37.

Morrison T and Portner A (1991) Structure, function and intracellular processing of the glycoproteins of the *Paramyxoviridae*. In: Kingsbury DW (ed.) *The Paramyxoviruses*, pp. 347–382. New York: Plenum.

Subbarao K (2003) Parainfluenza viruses. In: Long SS, Pickering LK, and Prober CG (eds.) *Principles and Practice of Pediatric Infectious Diseases,* 2nd edn., p. 1131. Philadelphia, PA: Churchill Livingstone.

Takimoto T and Portner A (2004) Molecular mechanism of paramyxovirus budding. *Virus Research* 106: 133–145.

Paramyxoviruses

R E Dutch, University of Kentucky, Lexington, KY, USA

© 2008 Elsevier Ltd. All rights reserved.

Glossary

Antigenome A complementary copy of the entire viral genome.
RNA-dependent RNA polymerase An enzyme that synthesizes new strands of RNA from an RNA template.
Zoonotic virus Virus that can be transmitted between animals and humans.

Introduction

The family *Paramyxoviridae* includes numerous important viral pathogens for both animals and humans. For humans, measles virus remains a significant cause of mortality, particularly in children, causing approximately 500 000 deaths annually. Human respiratory syncytial virus (HRSV), human metapneumovirus (HMPV), and the human parainfluenza viruses (HPIVs) are important causative agents of respiratory disease, especially in the very young. Paramyxoviruses such as Newcastle disease virus and rinderpest virus are potent animal pathogens with significant economic consequences. Finally, two recently identified paramyxoviruses, Hendra virus and Nipah virus, infect multiple species, and are thus the first zoonotic members of the family. The lethality of Hendra and Nipah viruses in humans and the lack of effective antiviral treatment led to their classification as biosafety level 4 (BSL-4) level pathogens. Thus, members of this important viral family infect a wide variety of species, and many of the members cause potentially lethal disease.

Taxonomy and Classification

The family *Paramyxoviridae* contains two subfamilies: the *Paramyxovirinae* and the *Pneumovirinae*. Differentiation between subfamilies is based on morphological characterization, genomic differences, and variability in function of viral proteins. *Paramyxovirinae* genomes typically encode either six or seven genes, with RNA editing of the P gene leading to additional proteins. Five genera are present in the *Paramyxovirinae*: *Rubulavirus*, which includes mumps virus; *Avulavirus*, which contains the avian Newcastle disease virus; *Respirovirus*, whose members include Sendai virus and hPIV 1 and 3; *Morbillivirus*, including measles virus; and the recently created *Henipavirus* genus, which is comprised of the Hendra and Nipah viruses. Viruses in the subfamily *Pneumovirinae* encode eight to ten genes and are morphologically distinct from those in the *Paramyxovirinae*. Two genera are present in the *Pneumovirinae*: *Pneumovirus*, which includes HRSV; and *Metapneumovirus*, which contains the recently discovered human respiratory pathogen, HMPV. In addition, a number of unclassified paramyxoviruses exist, including J-virus, Fer de lance virus, and Beilong virus.

Virion Structure and Genome Organization

Paramyxoviruses are enveloped particles with the lipid envelope derived from the plasma membrane of the infected cells which produced the progeny virus (**Figure 1**). Viral particles range in size from 150 to 350 nm in diameter. While the majority of viruses are spherical, varying shapes including filamentous forms have been observed. The major viral glycoproteins, the fusion (F) protein, and the attachment (HN, H, or G) protein, extend as spike-like projections from the virion envelope, and thus are positioned to mediate attachment and entry. Contained within the envelope is the viral nucleocapsid core, 15 000–19 000 nt of negative-sense RNA wrapped with the viral nucleocapsid (N) protein. Attached to this are the L (large-polymerase) and P (phosphoprotein) proteins. The viral RNA, N, L, and P proteins are referred to as the nucleocapsid core. The nucleocapsid is helical, with the nucleocapsids of the subfamily *Paramyxovirinae* having an average diameter of 18 nm, and those of the subfamily *Pneumovirinae* having a 13–14 nm diameter.

The *Paramyxoviridae* genome is a single segment of negative-sense RNA. This genome contains between six and ten genes, depending on the virus, but the presence of overlapping reading frames in the P gene results in the production of up to 12 proteins. Noncoding regions are present at the 3′ end (the approximately 50 nt 'leader') and the 5′ end (the 23–161 nt 'trailer'); they are essential for the transcription and replication of the genome. The gene order for the subfamily *Paramyxovirinae* is N–P/C/V–M–F–HN(H)–L, with the exception of the genus *Rubulavirus*, which contains an additional gene encoding the SH protein between the F and HN genes. The subfamily *Pneumovirinae* contains additional genes, with the gene order NS1–NS2–N–P–M–SH–G–F–M2–L for the genus *Pneumovirus* and N–P–M–F–M2–SH–G–L for the genus *Metapneumovirus*. Intergenic regions not present in the transcribed mRNA are present between each set of genes. The one exception is the overlapping M2 and L genes of the genus *Pneumovirus*, as the gene start for L is present within the M2 gene. Sequences at the beginning ('gene start') and end ('gene end') are critical for the initiation and termination/polyadenylation of each gene (**Figure 2**).

Viral Proteins

Attachment (HN, H, or G) Protein

The paramyxovirus attachment proteins perform the critical function of primary attachment of the virus to the target cell via interactions with cell surface receptors. Paramyxovirus attachment proteins are termed HN, H, or G depending on their ability to bind and cleave sialic acid. Paramyxovirus HN proteins, such as those from Newcastle disease virus and mumps virus, both bind cell surface sialic acid moieties during attachment and cleave sialic acid molecules from the carbohydrates of other viral glycoproteins and from the surface of the infected cell during assembly and budding. Viruses with HN attachment proteins utilize sialic acid on cell surface molecules as the receptor for virus binding. Paramyxovirus H proteins, such as those from the *Morbillivirus* measles, can promote sialic acid binding, but lack sialic-acid cleaving neuraminidase activity. The primary receptor for the measles attachment protein is not sialic acid moities, however. Instead, measles H protein binds the cell surface proteins CD46 and SLAM, providing an explanation for the restriction of measles infection to higher primates, which contain these molecules. Finally, paramyxovirus attachment proteins which neither bind nor cleave sialic acid are termed G, for glycoprotein. Viruses with G attachment proteins include the Hendra and Nipah viruses in the *Paramyxovirinae* and all members of the *Pneumovirinae*. The Hendra and Nipah G proteins utilize Ephrin B2 (and Ephrin B3) as the viral receptor. *Paramyxovirinae* attachment proteins also play a second important role in entry, as the triggering of the membrane fusion activity of the fusion protein generally requires the presence of the homotypic attachment protein for members of this subfamily. *Pneumovirinae* G proteins vary greatly from

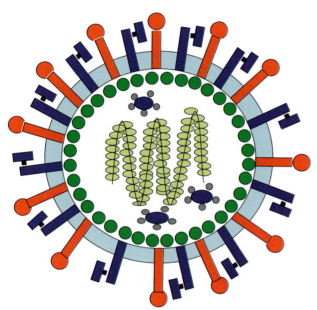

Figure 1 Schematic diagram of a paramyxovirus. The lipid bilayer (envelope) is shown in light blue. The fusion (blue) and attachment (red) glycoproteins extend outward from the bilayer. Lining the interior of the envelope is the matrix protein (green). The RNA genome is coated with nucleocapsid protein (yellow). The RNA-dependent RNA polymerase composed of the large polymerase protein (blue) and phosphoprotein (gray) is also packed in the interior, and interacts with the ribonuclecapsid particle.

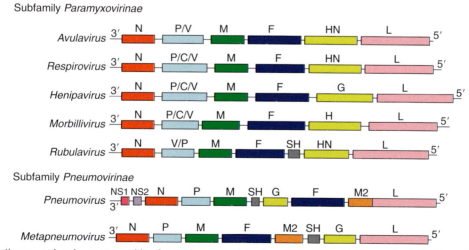

Figure 2 Genetic maps showing gene position for each genus of the *Paramyxoviridae*. Genes are indicated by colored boxes, with gene name designated above the box. For genes encoding more than one open reading frame, the name of each is shown above, separated by a '/'. For the *Pneumovirus* genus, the M2 and L genes overlap.

those of the *Paramyxovirinae*, as they are considerably smaller in size and are not essential for fusion protein function. In fact, recombinant HRSV and HMPV lacking their respective G proteins are both competent for replication, suggesting that the F protein can facilitate attachment to the target cell.

Fusion (F) Protein

Paramyxovirus F proteins are critical for viral entry, as they promote fusion of the viral membrane with the target cell membrane, enabling release of the nucleocapsid core into the cytoplasm of the target cell. F proteins can also promote fusion of infected cells with neighboring cells, leading to the formation of multinucleated giant cells, termed syncytia. Paramyxovirus F proteins are functional trimers which undergo both N-linked glycosylation and post-translational proteolytic processing. Proteolytic processing of the precursor form F_0 to the disulfide-linked heterodimer F_1+F_2 is necessary for the fusogenic activity of all F proteins, but the mechanisms of cleavage differ. F proteins from the majority of paramyxoviruses are cleaved by the host enzyme furin at a position following a series of multibasic amino acids. Furin cleavage occurs in the *trans*-Golgi network, resulting in cleaved, activated F protein on the plasma membrane and on the surface of budded virions. A small number of F proteins, including those from Sendai virus, hPIV1, and HMPV, have a single basic residue at the proteolytic cleavage site and are processed by extracellular proteases following viral release. This mechanism of proteolytic processing restricts viral propagation to tissue containing the required protease. Finally, Hendra and Nipah virus F proteins are proteolytically processed after a single basic residue by the cellular protease cathepsin L, in a mechanism that requires endocytosis from and subsequent recycling to the plasma membrane. The majority of F proteins require their homotypic attachment protein for fusion activity, and it is hypothesized that receptor binding by the attachment protein triggers the F protein to undergo a series of conformational changes which lead to fusion. Some members of the subfamily *Paramyxovirinae* (SV5 F and measles F) and of the subfamily *Pneumovirinae* (HRSV F and HMPV F) can promote fusion in the absence of the attachment protein. The mechanism for triggering of fusion for these F proteins is less clear, though HMPV F protein-promoted fusion can be triggered by low pH. Following the triggering event, a series of conformational changes leads to formation of a six-helix bundle between conserved heptad repeat regions, an event that is intimately tied to the membrane fusion process. Recent structural analysis of the prefusion and postfusion conformations of paramyxovirus F proteins has shed considerable light on this process, but many details of the mechanism remain unclear. Interestingly, paramyxovirus F proteins contain many conserved elements seen in fusion proteins from other viruses, including the HIV env protein, the Ebola virus glycoprotein, and the influenza virus hemagglutinin, suggesting a common ancestry for these important viral proteins.

Matrix (M) Protein

The M protein of *Paramyxoviridae* is the most abundant protein in the virion, and plays a central role in virus assembly. M protein interacts with multiple components of the budding virus, and these interactions are likely critical for M protein function. M is a peripheral membrane protein, and evidence suggests that it underlies the viral membrane. M also interacts with nucleocapsids and with the cytoplasmic tails of the attachment and fusion proteins. In addition, the M protein self-associates. These key interactions make the M protein the central player in orchestrating viral budding. Interestingly, the inactivation of M protein may contribute to persistent paramyxovirus infections, such as subacute sclerosing panencephalitis, a persistent, fatal measles virus infection of the brain.

Nucleocapsid (N) Protein

The paramyxovirus N protein plays a number of critical roles in the viral life cycle. The viral RNA genome is encapsidated by N proteins, protecting the genome from degradation. This tight N protein–RNA association is maintained during transcription and replication. Paramyxovirus N proteins do not contain previously identified RNA-binding motifs, but N protein RNA-binding activity has been localized to the conserved N-terminus of the protein. The N protein also facilitates interactions with the P and L proteins that are critical for transcription and replication. It is hypothesized that the concentration of free N protein in the cell is important for relative rates of transcription and replication. Finally, N protein interactions with the viral M protein likely assist in proper viral assembly.

Polymerase (L) Protein and Phosphoprotein (P)

Paramyxovirus RNA-dependent RNA polymerase activity, needed for transcription and replication, is promoted by a complex between the viral L and P proteins. The L protein, the largest of the paramyxovirus proteins at approximately 2200 amino acids, contains a number of small domains homologous to those found in other viral RNA-dependent RNA polymerases, indicating that L likely contains the catalytic activity for RNA polymerization. Experimental evidence suggests that the L protein also promotes polyadenylation of viral mRNA, and L has been implicated in phosphorylation of the viral N and P proteins. The L protein is present in only approximately

50 copies per virion, making it the least abundant of the paramyxovirus structural proteins. This low abundance is at least partly due to the position of the L gene in the genome, as it is the last to be transcribed, and thus has the lowest amount of mRNA produced. The active P–L complex contains five to ten P proteins per L protein. While the L protein likely contains the catalytic domains required for polymerization and polyadenylation, the P protein plays the critical role of tethering L to the N protein – RNA template.

For members of the subfamily *Paramyxovirinae*, multiple protein products are produced from the P gene through a fascinating mechanism known as RNA editing. In this process, the L–P RNA-dependent RNA polymerase adds, at a specific location known as the editing site, one, two, or four G nucleotides not encoded in the P gene. This addition of nontemplated nucleotides results in production of additional mRNAs and protein products, including the V, W, D, and I proteins. While many of the functions of these proteins remain unclear, V has been shown to interact with a cellular protein DDB1 that in turn forms a protein complex, a viral degradation complex (VDC) that mediates proteasomal destruction of the transcription factor, STAT 1. STAT 1 is a critical factor in establishing the interferon-induced antiviral state. Viruses from the genera *Morbillivirus, Respirovirus,* and *Henipavirus* also contain two overlapping reading frames in the P gene region. The viral C protein, also shown to play roles in countering the antiviral response, is produced from this alternative reading frame.

Additional Proteins

Several proteins are produced only by a subset of the *Paramyxoviridae*. SH (small hydrophobic) is a small transmembrane domain containing protein produced by members of the genera *Rubulavirus, Pneumovirus,* and *Metapneumovirus*. The SV5 and mumps SH proteins have been demonstrated to block apoptosis (programmed cell death) during infection, and are suggested to interfere with tumor necrosis factor alpha (TNFα) signaling. The SH proteins of the pneumoviruses and metapneumoviruses differ considerably from each other and from those of the rubulaviruses, and the role of these proteins is currently unclear.

Members of the subfamily *Pneumovirinae* also contain an M2 gene, which encodes two proteins, M2-1 and M2-2. M2-1 is a transcription elongation factor which, recent evidence suggests, interacts with the viral P protein. The M2-2 protein has been implicated in controlling the balance between transcription and replication. Finally, members of the genus *Pneumovirus* contain two additional genes at the 3′ end of the genome which encode the NS1 and NS2 proteins. Both NS1 and NS2 have also been implicated in counteracting antiviral defense within infected cells.

Replication

Paramyxovirus replication (**Figure 3**) begins with viral attachment to the plasma membrane of the target cell, a process mediated by the attachment protein through interactions with sialic acid moieties or other specific receptors. Subsequent fusion of the viral envelope with the target cell plasma membrane is promoted by the viral F protein, and results in delivery of the viral ribonucleocapsid core into the cytoplasm. Both transcription and replication occur exclusively in the cellular cytoplasm. As the RNA genome is negative sense (anti-coding sense), transcription of the genome to generate mRNA is then performed. The RNA-dependent RNA polymerase, composed of the L and P proteins, binds to a promoter present at the 3′ end of

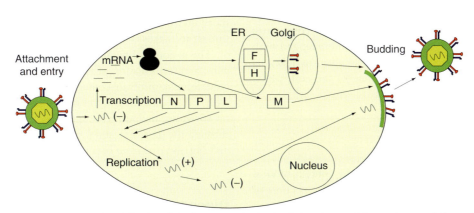

Figure 3 Life cycle of a paramxyovirus infection. Attachment and entry leads to release of the nucleocapsid (wavy line) into the cytoplasm. Transcription then occurs, generating mRNA which is translated to viral proteins by the host translation machinery. The F (fusion) and H (attachment) proteins are co-translationally transported into the endoplasmic reticulum (ER), followed by transport through the Golgi to the plasma membrane. Newly synthesized N (nucleocapsid), P (phosphoprotein), and L (large polymerase) proteins allow for replication of the genome, first to antigenome (+), then to genome (−) forms. M (matrix) protein assists in viral assembly, when the newly synthesized ribonucleoprotein core is incorporated into an enveloped particle containing the viral glycoproteins.

the genome, with a second internal promoter needed for efficient initiation of transcription for members of the subfamily *Paramyxovirinae*. The viral RNA polymerase begins transcription at this 3' end, synthesizing a leader RNA and then sequentially transcribing the genes in the 3'-to-5' direction. After termination of transcription of each gene, the polymerase moves on to reinitiation of transcription at the next gene, always moving in a 3'-to-5' direction. However, reinitiation of transcription is not completely efficient, and this process therefore results in a sequential decrease in the mRNA levels from genes at the 3' end to those at the 5' end, with the final L gene having the lowest level of mRNA synthesized. Newly synthesized viral mRNAs are then translated to viral protein using the host translation machinery. The viral glycoproteins are co-translationally transported in the endoplasmic reticulum, where they are folded, modified by addition of N-linked carbohydrates, and transported through the secretory pathway to the plasma membrane. The same viral RNA polymerase which carries out transcription of mRNA is also responsible for viral replication, and the 3' end promoter is also critical for this process. Many important questions remain concerning the switch from transcription (generating separate mRNAs) to replication (which ignores gene start and end regions, thus generating a complete copy of the genome). However, evidence does suggest that the accumulation of viral proteins, in particular N and M2-2, is critical for the switch to replication. Replication results first in an antigenome, a complementary copy of the entire viral genome. The newly synthesized antigenome can then be used as a template for synthesis of new negative-sense genomes. Replication of the viral genome is closely linked to encapsidation, where the newly created viral ribonucleocapsid cores are packaged into new virion particles. This assembly process is thought to occur at the plasma membrane for all members of the *Paramyxoviridae*. The M protein is a key player in the proper assembly of new progeny virus, as it interacts with the cytoplasmic tail of the glycoproteins, with the lipid bilayer of the plasma membrane, and with the ribonucleocapsid cores. The budding virus takes its envelope from the plasma membrane of the cell, which now contains viral glycoproteins.

Reverse Genetics

Analysis of specific mutations within the context of viral infection allows for a detailed molecular dissection of events in the viral life cycle. An important breakthrough which facilitated this type of detailed molecular analysis for the study of paramxyoviruses was the development of reverse genetics systems which allow synthesis of new viral progeny with specifically introduced mutations. This process was technically challenging as the paramyxovirus genome is a negative-sense RNA, and thus cannot be utilized by cellular machinery for direct translation of proteins. Reverse genetics methods allow for synthesis of viral mRNA, proteins, and eventual viral progeny by introduction of a plasmid containing the viral genome along with plasmids expressing the viral N, P, and L proteins, with a T-7 promoter present on all four plasmids. Expression of the T-7 polymerase leads to synthesis of the viral N, P, and L proteins along with synthesis of antigenome RNA. These elements then drive viral mRNA and protein synthesis, and eventual assembly of new viral particles containing the desired mutations. Experiments utilizing recombinant viruses generated using these techniques have led to significant new understanding of this important viral family. In addition, reverse genetics techniques are now being utilized for creation of new potential vaccines.

Host Range and Transmission

Paramyxoviruses infect a wide range of vertebrates, but to date no infection of invertebrates by members of this family has been described. A large number of paramyxoviruses are human pathogens, including measles virus, mumps virus, HRSV, and HMPV. The family also includes viruses which infect a variety of animal species. Newcastle disease virus and a series of avian paramyxoviruses infect both domestic and wild fowl, while avian metapneumovirus targets both chickens and turkeys. A wide range of animals, including dogs, raccoons, and skunks, are infected by canine distemper virus. Bats are infected by a number of paramyxoviruses, including Menangle virus and Tioman virus, and are likely the major animal reservoir for both Hendra and Nipah virus. Rinderpest virus infects cattle, as does bovine respiratory syncytial virus. Even sea creatures and reptiles are not immune from paramyxovirus infection. Phocine distemper virus causes severe disease in seals, while Cetacean morbillivirus infects both dolphins and porpoises. Fatal disease in snakes is associated with Fer de lance virus.

The majority of paramyxoviruses are respiratory pathogens, and transmission of virus generally occurs via virus-containing aerosols or contact with surfaces contaminated with virus-containing secretions. A small number of paramyxoviruses, including Newcastle disease virus, can infect through the gastrointestinal tract. Transmission of Hendra and Nipah virus from bats appears to involve contact with contaminated feces.

Disease, Prevention, and Treatment

Paramyxoviruses most commonly cause respiratory disease, but a number of members of the family also cause systemic infections, and some have been shown to cause encephalitis. Measles virus, which is still responsible for approximately 500 000 deaths per year, causes a systemic infection with fever, rash, and respiratory symptoms, and measles-related complications include pneumonia and encephalitis. HRSV is the leading cause of bronchiolitis and pneumonia in infants. HMPV and the HPIVs are also major pathogens causing severe respiratory disease in infants and young children. Nipah virus was first identified in 1999, and has already been the cause of a number of outbreaks in Southeast Asia, with mortality rates as high as 75% and evidence of human-to-human transmission. Nipah is associated with encephalitis resulting in fever, brain inflammation, and seizures, but respiratory symptoms have also been observed. Animal paramyxoviruses can also cause localized respiratory infection or systemic infections, depending on the animal and virus.

Vaccines have been effective at preventing diseases caused by a number of paramyxoviruses. The first vaccine for measles became available in 1963, and routine vaccination with MMR vaccine (measles-mumps-rubella) has decreased incidence of these infections in the developed world to extremely low levels. A number of vaccines for animal paramyxoviruses are routinely utilized, including vaccines for canine distemper, rinderpest, and Newcastle disease virus. However, development of successful vaccines for a number of important paramyxoviruses, including human and bovine respiratory syncytial viruses, has not been successful to date, and this remains a significant challenge for scientists in the field. In addition, few antiviral agents are approved for treatment of infection by these viruses. Development of effective antiviral therapies will likely be aided by our growing understanding of the molecular details of paramyxovirus infection.

See also: Henipaviruses; Human Respiratory Syncytial Virus; Measles Virus; Mumps Virus; Paramyxoviruses of Animals; Parainfluenza Viruses of Humans; Rinderpest and Distemper Viruses.

Further Reading

Eaton BT, Broder CC, Middleton D, and Wang L-F (2006) Hendra and Nipah viruses: Different and dangerous. *Nature Reviews Microbiology* 4(1): 23–35.

Lamb RA and Parks GD (2006) *Paramyxoviridae:* The viruses and their replication. In: Knipe DM, Howley PM, Griffin DE, et al. (eds.) *Fields Virology,* 5th edn., pp. 1449–1496. Philadelphia, PA: Lippincott Williams and Wilkins.

Loughlin GM and Moscona A (2006) The cell biology of acute childhood respiratory disease: Therapeutic implications. *Pediatric Clinics of North America* 53(5): 929–959.

Pekosz A, He B, and Lamb RA (1999) Reverse genetics of negative-strand RNA viruses: Closing the circle. *Proceedings of the National Academy of Sciences, USA* 96: 8804–8806.

van den Hoogen BG, de Jong JC, Groen J, et al. (2001) A newly discovered human pneumovirus isolated from young children with respiratory tract disease. *Nature Medicine* 7(6): 719–724.

Parapoxviruses

D Haig, Nottingham University, Nottingham, UK
A A Mercer, University of Otago, Dunedin, New Zealand

© 2008 Elsevier Ltd. All rights reserved.

Introduction

Parapoxviruses (PPVs) are epitheliotropic viruses found worldwide. The individual viruses generally exhibit a narrow host range and infect via scarified or damaged skin and give rise to pustular lesions of the skin and occasionally the buccal mucosa. These lesions are associated with low mortality and high morbidity. In addition to a narrow host range, most of the PPVs can also infect humans.

There are numerous historical references to diseases of domesticated animals such as sheep and cattle that we would now suspect to be the result of infection by PPVs. These references include Jenner's 'spurious' cowpox which is likely to have been caused by the PPV, pseudocowpox virus (PCPV). In the latter part of last century, reports appeared in the scientific literature which recognized the distinct identities of the diseases caused by members of this genus. Following an extensive study of contagious pustular dermatitis of sheep, Aynaud produced a report in 1923 which included the observation that the disease could be transmitted by a 'filterable' agent. The isolation of each of the viruses in cell culture was reported in the period from 1957 to 1963. Detailed reports of the transmission of each disease to humans appeared in 1933 (orf virus (ORFV)), 1963 (PCPV), and 1967 (bovine papular stomatitis virus (BPSV)). The first molecular analyses of PPV genomes appeared in 1979 with publication of restriction endonuclease cleavage site maps and reports of G + C contents. These were followed, in 1989, by the first description of the DNA sequence of a region of a PPV genome and in 2004 with full genome sequence for two strains of ORFV and one of BPSV.

Taxonomy and Classification

The genus *Parapoxvirus* belongs to the subfamily *Chordopoxvirinae* of the family *Poxviridae*. The type species of the genus is *Orf virus*, and the other species recognized as members are *Bovine papular stomatitis virus*, *Pseudocowpox virus*, *Squirrel parapoxvirus*, and a recently identified member, *Parapoxvirus of red deer in New Zealand*. Synonyms for the viruses include contagious pustular dermatitis virus and contagious ecthyma virus for ORFV and milker's nodule virus and paravaccinia virus for PCPV. Tentative species of this genus are Auzduk disease virus (camel contagious ecthyma virus), chamois contagious ecthyma virus, and sealpox virus. A virus that infects red squirrels and induces a pustular skin disease was thought to be a PPV, but recent DNA sequence data suggest that this is not the case.

The three original members of the genus were classified as separate species on the basis of the host animal and/or the pathology of the disease. Likewise, the observation of a parapox-like virus in red deer first suggested that this might represent another species. These separations have been supported by later studies which employed DNA/DNA hybridization, restriction endonuclease profiling, sequence data, or serology.

Host Range, Epidemiology, and Virus Propagation

Natural infection by ORFV has been reported in domestic, bighorn, and thinhorn sheep, domestic and Rocky Mountain goat, chamois, Himalayan thar, musk-ox, reindeer, steenbook, and humans. Experimental inoculations have shown that monkeys are susceptible to ORFV but a wide range of other animals including mouse, rabbit, dog, cat, and domesticated chicken are resistant. BPSV and PCPV both establish infection in cattle and humans but all other species tested, including sheep, are resistant. The PPV of red deer induces only very mild lesions on sheep and has not been tested in other species. The PPV of seals has been reported in a range of seals and sea lions. PPVs do not produce lesions on the chorioallantoic membrane of the developing chick embryo.

The PPVs of cattle and sheep are found throughout the world, essentially wherever their host animal occurs. The viruses are maintained in populations by a combination of chronic infection, frequent reinfection, and the environmentally resistant nature of the viruses.

ORFV shed in scab material can remain infective under dry conditions for lengthy periods (at least 4 months and possibly years) and infection of naive animals by virus persisting in heavily contaminated areas such as barns, yards, and sheep camps is likely to play a major role in maintaining the disease. One study has shown that if the scab material is ground up so as to release the virus, then exposure to field conditions quickly results in inactivation of ORFV. Several studies have shown that a productive infection can be established in animals which have recovered from a previous infection. Such reinfections result in lesions that are smaller and resolve more quickly than primary infections. This short-lived immunity is likely to contribute to the persistence of the disease.

In the case of PCPV, it is apparent that infection can be spread within dairy herds by contamination of milking machinery and milkers' hands. The introduction of procedures which reduce damage to teats and improve general hygiene at milking time can control the spread of the disease.

The most widely used cell culture systems have been primary ovine or bovine cells derived from sources such as testis, skin biopsy and embryonic kidney, lung, and muscle. There have also been reports of ORFV isolates adapted to growth in established cell lines. Yields of infectious ORFV from cell culture tend to be 10- to 100-fold lower than those achieved with vaccinia virus.

Serologic Relationships

PPVs show extensive antigenic cross-reactivity, although monoclonal antibodies have been able to distinguish each of the species. There are also antigens shared with other poxvirus genera but there is no cross-protection between PPVs and either orthopoxviruses or capripoxviruses.

Clinical Features

PPVs cause proliferative lesions that are confined to the skin and oral mucosa with no evidence of systemic spread (**Figure 1**). Infection is initiated in abrasions and generally proceeds through an afebrile, self-limiting lesion that resolves within 3–9 weeks without leaving a scar. ORFV lesions are most generally seen around the mouth and nares; hence, the infection is commonly referred to as scabby mouth or sore mouth. Lesions are also observed on other parts of the body, for example, the coronet, udder, or vulva. Following experimental inoculation of scarified skin, lesions progress through erythema, papule, vesicle, pustule, and scab before resolving. Large, proliferative, tumor-like lesions have been observed. It is likely that these are a result of an immune impairment of the host animal.

Figure 1 Orf in *Ovis aries*. Note the pustular lesions around the mouth and nares.

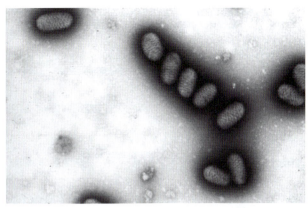

Figure 2 Electron micrograph of a cluster of negatively stained orf virus (ORFV) particles.

Lesions around the mouth can interfere with feeding or suckling and especially in young animals result in failure to thrive. Teat lesions can have similar effects through the inhibition of suckling. Lesions on growing deer antler can affect antler growth and severely affect marketability of the product.

It is probable that all PPVs are able to infect humans, although a human case of the PPV of red deer has not been reported. Transmission to humans occurs readily although there is little evidence of human to human transmission. Progression of the lesions is essentially as seen in sheep and cattle such that the infection is benign and confined to pustular lesions on the skin at the points of infection. More severe progressive disease can occur in immune-compromised individuals. Severe reactions have also been recorded in otherwise normal individuals in cases of burns and in cases of atopic dermatitis. Erythema multiformae reactions in the form of rashes on the backs of the hands and on the legs and ankles are common.

Properties of the Virion

PPV particles are ovoid in shape and measure 220–300 nm × 147–170 nm (**Figure 2**). In these characteristics they resemble other poxviruses except that PPVs are a little smaller than most other *Chordopoxvirinae*, which are also more commonly described as brick-shaped rather than ovoid. A distinctive feature of PPV virions is their 'ball-of-yarn' appearance when negatively stained specimens are viewed by electron microscopy. This results from a single 10–20 nm wide thread arranged as spiral coil around the particle. This unique morphology has been the basis for confirming a suspected PPV infection. However, this feature has recently been observed in poxviruses that are not members of the genus *Parapoxvirus* and DNA analysis is required to provide definitive identification.

Analysis of thin sections of virions has revealed a lipoprotein bilayer surrounding a biconcave core and two associated lateral bodies. Some particles have an external membranous structure. It seems likely that this is equivalent to the Golgi-derived membrane which forms the outer layer of the extracellular enveloped form of vaccinia virus.

Physical Properties

PPVs are resistant to desiccation and, in a dried state within scab material, the viruses retain infectivity for at least 4 months and possibly years. Under laboratory conditions, infectivity may be maintained over many years. UV light, γ-irradiation, or heating at 56 °C for 1 h will inactivate PPVs.

PPV Genomes

The PPV genome is a single, linear, double-stranded (ds) DNA molecule of 130–140 kbp with inverted terminal repeats of about 3.5 kbp in the case of ORFV (**Figure 3**). The ends of the genome are cross-linked. The G+C content of the genome is high (average of 64%). Restriction endonuclease cleavage site maps have been produced for each of the PPVs except that infecting red deer. These revealed some variability between isolates of the same species but conserved patterns, consistent with the classification of the genus, were apparent.

Full genome sequences have been obtained for three strains of tissue-culture adapted ORFV (OV-IA82 and NZ2 isolated from sheep and OV-SA00 isolated from a goat) and a tissue-culture-adapted strain of BPSV (BV-AR02). These encode an estimated 133 BPSV or 132 ORFV genes, of which 129 are collinear in the viral genomes and 88 conserved in all *Chordopoxvirinae* studied to date. The majority of these are essential genes involved in virus replication, packaging, and export. There are

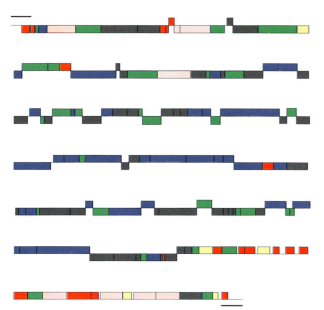

Figure 3 Orf virus (ORFV) genetic map. Boxes above the line represent genes transcribed rightward and those below the line genes transcribed leftward. Each line except the last corresponds to 20 kbp. The boxes are colored to indicate predicted functions. Those involved in virion structure or assembly are black, DNA replication or RNA transcription blue, host response modulation yellow, and ankyrin-repeat proteins are indicated with horizontal red lines. Genes of unknown function are further subdivided by their presence in most poxviruses (green), in parapoxviruses only (red), or in parapoxviruses and molluscipoxviruses only (red diagonal stripes).

fewer nucleotide metabolism genes compared with other poxviruses. Seventeen BPSV and eighteen ORFV open reading frames (ORFs) lack amino acid similarity to other poxvirus or cellular proteins. The PPV genomes have some similar features (including three orthologous genes) to that of molluscum contagiosum virus – a skin-tropic poxvirus pathogen of humans. Putative and known pathogenesis-related genes are concentrated in the terminal regions of the genome. In ORFV there are ∼25 such genes with unknown structure or function. In one study, the early transcribed genes of ORFV were mapped to identify candidate pathogenesis genes. cDNAs for 38 genes were identified, seven of these unrelated to any poxvirus or cellular gene in public databases.

The identification of transcriptional start points of PPV early genes highlighted adjacent sequences which are very similar to sequences shown to act as transcriptional promoters in vaccinia virus. These observations have been supported by data indicating that A+T-rich transcriptional control sequences characterized in vaccinia virus function in very similar ways in PPVs. Analysis of expression of ORFV antigens expressed from vaccinia virus recombinants carrying large multigene fragments of ORFV DNA suggests that ORFV late genes are also faithfully transcribed from their own promoters by vaccinia virus.

Passage of ORFV in cell culture results in genomic rearrangements. A detailed study of one such rearranged isolate showed that 19.3 kbp from the right end of the genome had been duplicated and replaced 6.6 kbp at the left end, resulting in a net deletion of 3.3 kbp at the left end. This recombination occurred between nonhomologous sequences and caused the deletion of three genes. These alterations to the genome attenuated the growth of the virus in sheep skin. Similar rearrangements have been observed in other isolates.

Properties of Viral Proteins

SDS-polyacrylamide gel electrophoresis analyses of ORFV and PCPV virions have detected 30–40 polypeptides ranging in size from 10 to 220 kDa. Controlled degradation of the virions into core and surface fractions indicated the presence of 10–13 surface polypeptides. Prominent among these were a polypeptide of 39 kDa (ORFV) or 42–45 kDa (PCPV), which was suggested to be the subunit of the virion surface tubule protein, and a 15 kDa polypeptide. The 39 kDa ORFV polypeptide appears to be a dominant antigen and several independently derived monoclonal antibodies are directed against it.

The genes encoding two major structural proteins of ORFV have been identified. One of these is a 42 kDa protein with strong amino acid sequence similarity to the vaccinia virus protein encoded by gene F13L. This vaccinia virus polypeptide is the major, nonglycosylated, 37 kDa protein specific for the extracellular enveloped form of the virus. Cloning and expression of the ORFV gene has shown that both antibody and T-cell responses are directed against this protein during infection by ORFV. A second ORFV protein, p10k, is a homolog of the vaccinia virus 14 kDa fusion protein encoded by gene A27L. This vaccinia virus protein has a surface location and is associated with intracellular mature virus but is required for the formation of extracellular enveloped virus. The ORFV protein may be involved in the expression of the characteristic basket-weave pattern on the virions. The F1L product of ORFV is a major heparin-binding protein and may be involved in virus binding to cells to initiate the infection cycle.

Pathogenesis-related proteins are discussed in a later section.

Replication

As with all poxviruses, the replication of PPVs occurs in the cytoplasm of infected cells. Studies in cell culture have shown very similar patterns in the replication of ORFV and PCPV. DNA replication begins 4–8 h post infection (h.p.i.) and reaches a plateau between 25 and

30 h.p.i. The first viral or viral-induced polypeptides appear from 10 h.p.i. Viral particles appear 16–24 h.p.i. and continue to be produced until at least 40 h.p.i. Viral replication is accompanied by inhibition of host DNA and protein synthesis. Some ORFV genes have been shown to be transcribed when infection occurs in the presence of an inhibitor of DNA synthesis and therefore fit the definition of early poxvirus genes. The transition between early and late replication events occurs about 8–10 h.p.i. The presence in ORFV of genes with homology to vaccinia virus intermediate genes encoding late gene transactivators suggests that expression of ORFV genes follows a regulated cascade (early–intermediate–late) similar to that reported for vaccinia virus.

Pathogenesis and the Host Immune Response to Infection

The histological sequence of events in the skin of sheep after ORFV infection is similar in primary and reinfection lesions, in spite of differences in the magnitude of the lesions and the time taken to resolve. Antibodies to ORFV envelope proteins have been used to detect ORFV antigen in epidermal keratinocytes, particularly those regenerating the damaged skin. Basal keratinocytes at the root of hair follicles can also contain virus. Some infected cells show evidence of a ballooning-like degeneration. There is no evidence that ORFV infects other, nonepithelial cell-types *in vivo*. ORFV lesions often exhibit epidermal downgrowths (rete formation) into the dermis. This is particularly marked in primary lesions. Another characteristic feature is extensive capillary dilation and proliferation.

ORFV lesions contain a dense accumulation of immune and inflammatory cells underneath and adjacent to virus-infected cells. These include neutrophils, lymphocytes (T and B cells), and dendritic cells that stain intensely with major histocompatibility complex class II antigens. This dense network of dendritic cells is characteristic of orf lesions in sheep. The function of these cells is not known. The accumulating cells increase and decrease in number in parallel with the presence of virus in epidermal cells. The histology of human ORFV lesions is generally similar to that described in sheep. A comparison of ORFV and PCPV lesions in humans has not revealed any histopathological differences.

Immune Response to Infection

PPVs, in common with other poxviruses, stimulate a vigorous immune and inflammatory response in their hosts, and have evolved to replicate in the presence of this response. In sheep experimentally infected with ORFV, studies in the skin and lymph draining into (afferent lymph) and out of (efferent lymph) local lymph nodes have demonstrated that activated CD4+ (helper) and CD8+ (cytotoxic) T cells, B cells, and antibodies are generated as part of the sheep-acquired immune response to infection. The cytokines generated in lymph in response to virus reinfection are typical of type 1 antiviral cell-mediated immune responses and include interleukin (IL)-1β, IL-2, tumor necrosis factor (TNF)-α, granulocyte-macrophage colony-stimulating factor (GM-CSF), interferon (IFN)-α, and IFN-γ. Studies of ORFV reinfection in sheep depleted of specific lymphocyte subsets or treated with the immunosuppressant drug cyclosporin-A indicated that at least CD4+ T cells and interferons are important components of the host-protective-response against infection. These studies also indicated that the cutaneous damage sustained during ORFV infection is due in large part to the virus rather than host immune-mediated.

Sheep infected with ORFV mount detectable antibody responses to a small number of viral antigens but there is considerable individual qualitative and quantitative variation in the response. There is a lack of neutralizing antibody. There is no apparent correlation between antibody titers and severity of viral lesions and passive transfer of antibody does not confer protection against virus challenge.

Immunomodulation by ORFV

In general, PPV infection is mild and localized with more severe disease only occurring in stressed or otherwise immune-impaired individuals. An intriguing feature of PPVs is the ability to repeatedly infect animals despite an apparently typical antiviral immune response to infection. This may in part be a result of the action of viral-encoded immune modulators that are a feature of large DNA viruses. There are now several examples of these that have been identified within the terminal regions of the genome of ORFV.

Viral Vascular Endothelial Growth Factor

The first of these to be reported was a homolog of mammalian vascular endothelial growth factor (VEGF). The viral VEGF is expressed early, shows 16–27% amino acid sequence identity to mammalian VEGFs (VEGF-A,-B, -C,-D) and has been classified as VEGF-E. It is unique among the VEGF family in that it interacts with VEGF receptor 2 (VEGFR-2) but not with either VEGFR-1 or VEGFR-3. VEGFs are specific mitogens for endothelial cells and regulate normal and pathological angiogenesis, including the vascularization of solid tumors. Extensive capillary proliferation and dilation is a feature of ORFV lesions and the viral VEGF plays a major role in generating these features. It may also indirectly enhance proliferation of the epithelium, the target tissue for virus replication, as epidermal proliferation was reduced in lesions with a VEGF gene knock-out virus compared to wild-type virus lesions. A report that human VEGF can inhibit the functional maturation of dendritic cells hints at a possible immune modulating role for the viral VEGF. A VEGF-like gene has not been reported in any other poxvirus.

Viral Interferon Resistance Protein

ORFV encodes an interferon resistance protein (OVIFNR) produced by the 020 gene early in infection, which inhibits the interferon-induced shutdown of cellular protein translation (via PKR) by binding to viral dsRNA, thus preventing activation of the protein kinase PKR. 020 is a homolog of the vaccinia virus E3L gene (31% predicted amino acid sequence identity) that has the same function.

Viral IL-10

ORFV IL-10 (vIL-10) is the product of an early viral gene and the predicted protein is very similar to ovine IL-10 over the C-terminal two-thirds of the molecule while differing substantially at the N-terminus (80% amino acid sequence identity overall). The vIL-10 exhibited similar anti-inflammatory and immunostimulatory activities to ovine IL-10. Both cytokines suppressed TNF-α production from macrophages and suppressed IFN-γ production from peripheral blood cells. Both vIL-10 and ovine IL-10 stimulated mast cell proliferation provided co-stimulatory cytokines were present (IL-3 or IL-4). vIL-10 is a virulence protein as recombinant virus lacking the vIL-10 gene is attenuated compared to wild-type or vIL-10 gene-reconstituted virus in sheep infection experiments. The ORFV IL-10 may act to suppress elements of the antiviral immune response, thereby delaying virus clearance. IL-10-like genes have been reported in two other poxvirus genera (*Capripoxvirus* and *Yatapoxvirus*) although these proteins show markedly less sequence similarity to mammalian IL-10 than does ORFV IL-10.

Viral GM-CSF Inhibitory Factor

The protein product of ORFV ORF 117 encodes an early virus protein with novel function. Viral GM-CSF inhibitory Factor (GIF) binds to and inhibits the biological activity of the cytokines GM-CSF and IL-2. Among other properties, GM-CSF regulates the recruitment, differentiation, and activation of macrophages, neutrophils, and dendritic cells. Activated T cells are an important source of GM-CSF during immune responses to pathogens. IL-2 is produced predominantly by T cells and stimulates the expansion and activation of T cells and natural killer cells among other cell types. GIF has probably evolved to perform its dual function from an ancestral type-1 cytokine receptor, as it shares important structure–function features of this receptor family while exhibiting an otherwise divergent primary amino acid sequence. A GIF protein is also expressed in other PPVs, and is conserved in ORFV strains though there is only ∼40% amino acid identity between these and BPSV GIF. The role of GIF in ORFV pathogenesis is currently not known.

Viral Chemokine-Binding Protein

ORFV encodes a chemokine-binding protein, the product of ORF 112, which is expressed early in the virus life cycle. It is structurally and functionally related to the type II CC chemokine-binding proteins of the *Orthopoxvirus* and *Leporipoxvirus* genera. Viral chemokine-binding protein (vCBP) binds CC chemokines, including monocyte-chemotactic protein-1, macrophage inflammatory protein-1α and RANTES (regulated upon activation, normal T-cell-expressed and secreted). These regulate T cell and monocyte/macrophage recruitment to sites of infection. In addition, and uniquely among poxvirus CBPs, vCBP binds to lymphotactin, a C-chemokine that recruits T and B cells, and neutrophils. vCBP may well function to inhibit key aspects of an antiviral immune response.

PPVs as Vectors and Immunomodulators

ORFV could prove useful as a vector for delivering microbial antigens to the immune system. Proof of concept for this has been obtained with protective immunity generated to pseudorabies virus and Borna disease virus infection of rodents. The restricted host ranges of the PPVs and lack of systemic spread even in immuno-compromised animals make them good viral vector candidates. Furthermore, inactivated ORFV particles have been shown to exhibit nonspecific immuno-modulatory effects that enhance immunity to a variety of pathogens in several species. This is thought to be mediated by IFN and a type 1 immune response that is downregulated at later stages by IL-10, among other cytokines. This feature of ORFV continues to be exploited commercially.

Prevention and Control

ORFV vaccines have been available for many years and are widely used to protect lambs against the debilitating effects of natural infection. These vaccines consist of live and essentially nonattenuated virus that is applied to a scratch on the skin of a leg. The ensuing infection does not interfere with feeding and provides significant protection against infection for some months. However, the scab derived from vaccination lesions is likely to contaminate the environment and contribute to the perpetuation of the disease. New vaccines that induce protection but do not shed infectious virus are highly desirable. This might be achieved by deleting genes encoding viral virulence determinants or by delivering the protective antigens of the virus in an appropriate way.

Humans may be infected by PPVs following contact with animal lesions or scab. The viruses in scab associated with animal products such as wool or farm equipment can remain infectious for lengthy periods. Care should be taken to avoid contact between any skin wound and potentially contaminated material.

Future Perspectives

Recent studies with ORFV have raised the possibility of developing a vaccine able to protect animals against infection by this virus without generating significant amounts of infectious virus. Such a vaccine would also be likely to reduce the frequency of human infections.

The use of ORFV and perhaps other PPVs as vaccine vectors is proving a useful and robust strategy. The generation of ORFV recombinants and other PPVs lacking pathogenesis-related genes will continue to advance our understanding of the pathogenesis of PPV diseases. An understanding of PPV pathogenesis and the viral proteins involved, coupled to recombinant virus generation, should lead to improved control strategies for this important group of viruses.

See also: Capripoxviruses; Poxviruses; Vaccinia Virus.

Further Reading

Delhon G, Tulman ER, Afonso CL, et al. (2004) Genomes of the parapoxviruses orf virus and bovine papular stomatitis virus. *Journal of Virology* 78: 168–177.

Haig DM (2006) Orf virus infection and host immunity. *Current Opinion in Infectious Diseases* 19: 127–132.

Haig DM and McInnes CJ (2002) Immunity and counter-immunity during infection with the parapoxvirus orf virus. *Virus Research* 88: 3–16.

Mercer AA, Ueda N, Friederichs SM, et al. (2006) Comparative analysis of genome sequences of three isolates of orf virus reveals unexpected sequence variation. *Virus Research* 116: 146–158.

Partitiviruses of Fungi

S Tavantzis, University of Maine, Orono, ME, USA

© 2008 Elsevier Ltd. All rights reserved.

Introduction

The vast majority of mycoviruses (viruses infecting fungi) are 'cryptoviruses', that is, they are not usually associated with an overt pathology of their fungal hosts. So, it is not surprising that the first mycovirus paper, by M. Hollings published in *Nature* (London) in 1962, was about virus-like particles (VLPs) thought to be the cause of a serious dieback disease of the cultivated mushroom *Agaricus bisporus*. Following Hollings' groundbreaking paper, progress in mycovirology was slow, as in addition to their cryptic nature, fungal viruses can be transmitted only through intracellular routes. Most of the characterized mycoviruses possess double-stranded RNA (dsRNA) genomes but this high proportion might be biased since most of the published data refer to fungal screenings for the presence of dsRNA followed by transmission electron microscopy or virus characterization studies.

Mycoviruses are classified into nine families (*Barnaviridae, Chrysoviridae, Hypoviridae, Metaviridae, Narnaviridae, Partitiviridae, Pseudoviridae, Reoviridae, Totiviridae*), and one genus (*Rhizidiovirus*) not classified into a specific family. Viruses belonging to the family *Partitiviridae* usually cause cryptic infections in fungi and plants, and are classified into three genera, *Partitivirus, Alphacryptovirus,* and *Betacryptovirus.* The name *Partitiviridae* originates from the Latin *partitus*, which means 'divided' and refers to the fact that the genome of these viruses is bipartite (separately encapsidated) and consists of two monocistronic dsRNAs that are similar in size.

Taxonomy and Classification

This section focuses on the genus *Partitivirus*, which consists of viruses that infect filamentous fungi. The genera *Alphacryptovirus* and *Betacryptovirus* include viruses that are found in plants. The recognized species of the genus *Partitivirus* are listed in **Table 1** whereas tentative members are shown in **Table 2**. The sequence accession numbers refer to the RNA-dependent RNA polymerase (RdRp) genes of the respective partitiviruses.

Table 1 Virus members in the genus *Partitivirus*

Virus	Abbreviation	Accession number
Agaricus bisporus virus 4	AbV-4	
Aspergillus ochraceous virus	AoV	
Atkinsonella hypoxylon virus	AhV	L39125
Discula destructiva virus 1	DdV-1	AF316992
Discula destructiva virus 2	DdV-2	AY033436
Fusarium poae virus	FpV-1	AF047013
Fusarium solani virus 1	FsV-1	
Gaeumannomyces graminis virus 019/6-A	GgV-019/6-A	
Gaeumannomyces graminis virus T1-A	GgV-T1-A	
Gremmeniella abietina RNA virus MS1	GarV-MS1	AY08993
Helicobasidium mompa virus	HmV	
Heterobasidion annosum virus	HaV	
Penicillium stoloniferum virus S	PsVS	AY156521
Rhizoctonia solani virus 717	RhsV-717	AF133250

Table 2 Tentative members of the genus *Partitivirus*

Virus	Abbreviation	Accession number
Ceratocystis polonica partitivirus	CpPV	
Ceratocystis resinifera partitivirus	CrPV	
Diplocarpon rosae virus	DrV	
Gremmeniella abietina RNA virus MS2	GaRV-MS2	AY615211
Helicobasidium mompa partitivirus	HmPV	
Ophiostoma partitivirus 1	OPV-1	AM087202
Penicillium stoloniferum virus F	PsV-F	AY758336
Phialophora radicicola virus 2-2-A	PrV-2-2-A	
Pleurotus ostreatus virus	PoV	AY533038
Rosellinia necatrix virus 1-W8	RnV-1-W8	AB113347

Virion Properties

The virion M_r estimates range from 6×10^6 to 9×10^6 and the $S_{20,w}$ values (in Svedberg units) range from 101 to 145, whereas the $S_{20,w}$ values of particles lacking nucleic acid range from 66 to 100S. Virion buoyant density in CsCl is $1.34–1.37 \text{ g cm}^{-3}$ for particles with nucleic acid and $1.29–1.30 \text{ g cm}^{-3}$ for particles devoid of nucleic acid. Purified virion preparations contain, in addition to mature virions and empty virions, sedimenting and density components that are thought to be replicative intermediates containing single-stranded RNA (ssRNA) and particles with both ssRNA and dsRNA.

Virion Structure and Composition

Viruses of the genus *Partitivirus* have isometric particles with icosahedral symmetry ranging from 30 to 35 nm in diameter (**Figure 1**). Negatively stained virions that are devoid of nucleic acid have capsids with dark centers as they are penetrated by stain (**Figure 1**). The virion capsid is not enveloped and consists of 12 capsomers and 120 capsid protein (CP) subunits with a molecular mass ranging from 57 to 76 kDa. Virion-associated RNA polymerase

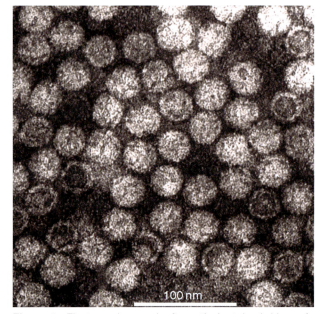

Figure 1 Electron micrograph of negatively stained virions of Rhizoctonia solani virus 717 (RhsV 717), a representative species of the genus *Partitivirus*. Adapted from Tavantzis SM and Bandy BP (1988) Properties of a mycovirus from *Rhizoctonia solani* and its virion-associated RNA polymerase. *Journal of General Virology* 69: 1465–1477, with permission from Society for General Microbiology.

activity is attributed to RdRp with sizes ranging from 77 to 86 kDa in fungal partitiviruses sequenced to date.

Antigenic Properties

Antisera to purified virus preparations appear to contain antibodies to the virion RdRp polypeptide. This argument is supported by two lines of evidence: (1) antiserum to the rhizoctonia solani virus 717 (RhsV 717) reacted to both virion proteins CP and RdRp and (2) antiserum to the aspergillus ochraceous virus (AoV) reacted to the AoV CP and RdRp as well as the RdRp of the 'slow' component (PsV-S) of penicillium stoloniferum virus complex. Unlike the respective CPs, the RdRp polypeptides of AoV and PsV-S share a high degree of similarity, and this is congruent with the phylogenetic analysis of CPs and RdRps of members of the genus *Partitivirus* which shows a higher rate of evolutionary changes occurring in the *CP* genes (**Figure 4(b)**) as compared to the rate in the *RdRp* genes (**Figure 4(a)**).

Genome Organization

Virions contain two separately encapsidated linear dsRNAs, dsRNA1 and dsRNA2, that are 1.5–2.4 kbp in size. The dsRNA components of a particular virus are of similar size (differing by 20–200 bp). Additional dsRNAs (satellite or defective) may be present in purified fungal partitivirus particles. The larger segment, dsRNA1, codes for an RdRp, whereas the smaller segment, dsRNA2, codes for a CP (**Table 3**). The 5′ ends of the coding strands of dsRNA1 and dsRNA2 are often highly conserved, and usually contain inverted repeats that may form stable stem–loop structures (**Figure 2**).

It has been hypothesized that partitiviruses may have evolved from totiviruses by dividing their genomes between two dsRNA segments. Like the totiviruses, the partitivirus genome consists of two genes (CP and RdRp) but, unlike the totiviruses, these genes are located on separate dsRNAs (bipartite genomes). In contrast to partitiviruses, many totiviruses express their RdRp as a CP–RdRp fusion protein. Totiviruses, such as the helminthosporium victoriae 190S virus (Hv190SV), that express RdRp as a separate nonfused protein might be evolutionary intermediates between ancestral totiviruses and partitiviruses.

Genome Expression and Virus Multiplication

As described above, the bisegmented genome of fungal partitiviruses consists of two dsRNA segments of similar

Table 3 Genome organization of the RhsV 717 partitivirus

Genome	ORF	Frame	Nucleotide coordinates	Amino acids	Mol. mass (kDa)	Putative function	Kozak sequence	Inverted repeats[a]
dsRNA1 (2363 bases)	1	2	86–2275	730	85.8	RdRp[b]	502–510	736–747
dsRNA2 (2206 bases)	1	1	79–2130	683	76.4	CP	None	None

[a]Inverted repeats longer than 12 bases.
[b]The Rhs 717 partitivirus RdRp (dsRNA 1, ORF 1) contains the dsRNA RdRp motifs according to Bruenn (1993). The residue coordinates of these motifs are as follows: Motif I, 242-LVTGT-246; Motif II, 315-FLKSFPTMM-323; Motif III, 363-ARKPECCIMYG-373; Motif IV, 397-IDWSGYDQRL-406; Motif V, 479-GVPSGMLLTQFLDSFGNLY-LII-500; Motif VI, 519-FIMGDDNSIF-528; Motif VII, 569-IETLSYRC-576; Motif VIII, 584-DVEK-587.
Reproduced from Strauss EE, Lakshman DK, and Tavantzis SM (2000) Molecular characterization of the genome of a partitivirus from the basidiomycete *Rhizoctonia solani. Journal of General Virology* 81: 549–555, with permission from Society for General Microbiology.

```
dsRNA1   - GUAGUCUUUUAGUAUCGAUCCCUCGACUCUCGACCGCACUAAAUCUCAUC  - 50
            ::::::::::::::::::::::::::::::      :::::::::::::::
dsRNA2   - G-AGUCUUUUAGUAUCGAUCCCUCGACU-----CGCACUAAAUCUCAUU  - 43

dsRNA1   - GUUAUACGAACGAACUCUCUUCAAUCAACACACAAUGCUCUACAACUUC  - 100
            :::  ::::::::  ::::::  :  ::   ::   ::::  ::
dsRNA2   - GUUUUUAAGAACCAAACUCUCAACUCUCGCAACCUGAUGCCUUCGCCAAAG - 93
```

Figure 2 Homology at the 5′ ends of the coding strands of the two genomic segments of the RhsV 717. If a 6-base gap is allowed in dsRNA2, the two dsRNAs are highly conserved. The identity drops to 65% around the region that includes the translation initiation sites. AUG codons are shown in bold letters and the CAA motifs (possibly involved in translational enhancement) are double-underlined. Identical positions are shown with colons beneath the sequence. The underlined bases are complementary and are capable of forming stable stem–loop structures. Adapted from Strauss EE, Lakshman DK, and Tavantzis SM (2000) Molecular characterization of the genome of a partitivirus from the basidiomycete *Rhizoctonia solani. Journal of General Virology* 81: 549–555, with permission from Society for General Microbiology.

size. The larger segment, dsRNA1, encodes an RdRp, whereas the smaller component, dsRNA2, encodes a CP. Northern blot analysis of purified virions as well as total RNA from partitivirus-infected mycelial tissue revealed no subgenomic RNA species in the form of discrete RNA bands with sizes smaller than that of full-length dsRNA1 or dsRNA2. This is consistent with analysis of the coding potential of these genomic RNAs, showing that the long open reading frames encoding the RdRp and CP polypeptides cover the entire length of dsRNA1 and dsRNA2, respectively.

Mycoviruses, including members of the genus *Partitivirus*, have been difficult to transfect using an inoculum consisting of purified virions. Thus, methods of synchronous viral infection and multiplication have not been available for studying partitivirus gene expression or virion replication *in vivo*. The recent transfection of the rosellinia necatrix virus W8 (RnV-W8) into protoplasts of its fungal host is a positive development with regard to unveiling the life cycle of fungal partitiviruses. Current knowledge of partitivirus gene expression, dsRNA replication, and virus multiplication is based on studies of RdRp activity in purified virus fractions, virus-related RNA content in infected mycelia, and *in vitro* translation studies of denatured dsRNA1 or dsRNA2 or full-length cDNA clones representing dsRNA1 or dsRNA2.

Populations of virions include particles of different $S_{20,w}$ values depending on their RNA content. Thus, purified virus preparations consist of mature virions (L1 or L2) containing a single molecule of dsRNA, virus particles with a single molecule of ssRNA corresponding to the plus-strand of the respective dsRNAs (M1, M2), a heterogeneous subpopulation of heavy (H) particles representing various replication stages and containing particles with a dsRNA genomic component and ssRNA tails or one molecule each of dsRNA and ssRNA; particles with two molecules of dsRNA (P) one of which is a product ('P') of replication (**Figure 3**). Data from *in vitro* RdRp activity studies in purified preparations of PsV-S and RhsV 717 are congruent with the replication scheme presented in **Figure 3**. However, experimental evidence for a number of the steps depicted in this replication scheme is currently lacking, and its verification awaits *in vivo* studies involving transfection of host mycelia with purified virions (or cDNAs representing dsRNAs 1 and 2) resulting in synchronous virus replication.

Phylogenetic Relationships among Fungal Partitiviruses

Phylogenetic analyses of the predicted amino acid sequences of fungal partitivirus RdRps and CPs (**Figures 4(a)** and **4(b)**, respectively) show that these viruses are closely related. Both analyses suggest that fungal partitiviruses form two clusters. The RdRp clusters have 100% bootstrap support (**Figure 4(a)**), whereas the CP clusters are also well supported (**Figure 4(b)**). These results are in agreement with

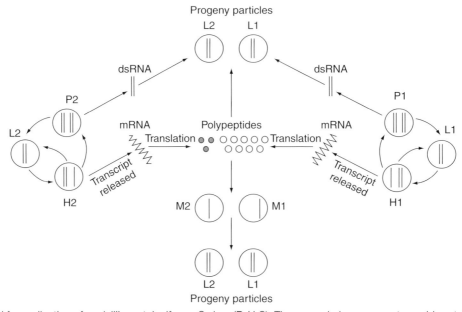

Figure 3 Model for replication of penicillium stoloniferum S virus (PsV-S). The open circles represent capsid protein subunits and the closed circles represent RNA polymerase subunits. Solid lines represent RNA strands whereas wavy lines represent mRNA. Adapted from Seventh Report of the International Committee on Taxonomy of Viruses, http://www.virustaxonomyonline.com/virtax/lpext.dll/vtax/agp-0013/dr06/dr06-fg/dr06-fig-0002, with permission from Elsevier.

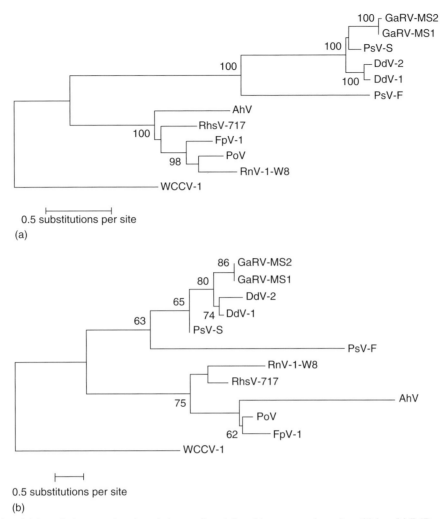

Figure 4 A neighbor-joining phylogram showing phylogenetic relationships among fungal partitivirus (a) RdRp and (b) CP amino acid sequences. The JTT model of amino acid evolution was applied in phylogenetic reconstruction using the software MEGA3.1. The tree was outgroup-rooted with the respective proteins of white clover cryptic virus 1 (WCCV-1), a plant partitivirus. Numbers above branches represent bootstrap support values and bar represents amino acid substitutions per site. See **Tables 1** and **2** for abbreviations of fungal partitivirus names.

recently reported phylogenetic analyses showing that the evolutionary rate of CPs is higher than that of the RdRps, and are in congruence with data suggesting a cross-serological relationship between the RdRps of AoV and PsV-S.

Transmission of Partitiviruses

Like most mycoviruses, partitiviruses may be transmitted by anastomosis between hyphae of genetically related fungal genotypes. Moreover, the ineffectiveness of hyphal tip isolation to eliminate partitivirus particles or dsRNA in conjunction with the detection of partitiviruses in apical hyphal sections by electron microscopy suggest that these viruses are transmitted vertically by sexual spores (basidiospores or rarely ascospores) as well as asexual spores (conidiospores) derived from virus-infected fruit bodies. The proportion of spores containing partitivirus dsRNAs varies widely even within the same species and apparently depends on the combination of fungal host and virus genotypes. For example, in different isolates of the basidiomycete *Heterobasidion annosum*, 3–55% of germinated conidia, and 10–84% of germinated basidiospores, contained partitivirus dsRNAs.

Partitivirus Pathogenesis

Partitiviruses occur in all parts of an infected mycelial mat, but virion concentrations are higher in the older hyphal sectors than in hyphal tips. This adaptation to multiply mainly in areas of the fungal mycelium that are less involved in fungal growth might be one of the reasons that partitiviruses, although they replicate to relatively high levels, are usually associated with a symptomless infection of their respective fungal hosts. Recently, the latency of a partitivirus infection was demonstrated by direct experimental evidence in the case of ascomycete *Rosellinia necatrix* and the RnV-1-W8 partitivirus. Protoplasts of *R. necatrix* were transfected with purified particles of RnV-1-W8. The resulting mycelium contained transmissible RnV-1-V8 virions but showed no observable symptoms associated with the presence and replication of this virus. In contrast, in the *H. annosum* study (see previous section), germination frequency of basidiospores was significantly reduced ($P < 0.05$) by the presence of partitivirus dsRNA in the parental fruit bodies. Use of isogenic fungal lines may be needed to verify these results.

See also: Fungal Viruses; Chrysoviruses; Partitiviruses: General Features; Totiviruses.

Further Reading

Bruenn JA (1993) A closely related group of RNA-dependent RNA polymerases from double-stranded RNA viruses. *Nucleic Acids Research* 21: 5667–5669.

Ghabrial SA (1998) Origin, adaptation and evolutionary pathways of fungal viruses. *Virus Genes* 16: 119–131.

Ghabrial SA, Buck KW, Hillman BI, and Milne RG (2005) Partitiviridae. In: Fauquet CM, Mayo MA, Maniloff J, Desselberger U, and Ball LA (eds.) *Virus Taxonomy: Eighth Report of the International Committee on Taxonomy of Viruses*, pp. 581–590. San Diego, CA: Elsevier Academic Press.

ICTVdB Management (2006) 00.049.0.01. Partitivirus. In: Büchen-Osmond C (ed.) *ICTVdB – The Universal Virus Database, version 4*, New York: Columbia University.

Kim JW, Choi EY, and Kim YT (2006) Intergeneric relationship between the aspergillus ochraceous virus F and the penicillium stoloniferum virus S. *Virus Research* 120: 212–215.

Strauss EE, Lakshman DK, and Tavantzis SM (2000) Molecular characterization of the genome of a partitivirus from the basidiomycete *Rhizoctonia solani*. *Journal of General Virology* 81: 549–555.

Tavantzis SM and Bandy BP (1988) Properties of a mycovirus from *Rhizoctonia solani* and its virion-associated RNA polymerase. *Journal of General Virology* 69: 1465–1477.

Partitiviruses: General Features

S A Ghabrial, University of Kentucky, Lexington, KY, USA
W F Ochoa and T S Baker, University of California, San Diego, La Jolla, CA, USA
M L Nibert, Harvard Medical School, Boston, MA, USA

© 2008 Elsevier Ltd. All rights reserved.

Glossary

Cryo-transmission electron microscopy Transmission electron microscopy of unstained, unfixed, frozen-hydrated (vitrified) specimens, preserved as close as possible to their native state.

Hyphal anastomosis The union of a hypha with another resulting in cytoplasmic exchange.

Icosahedral symmetry Arrangement of 60 identical objects or asymmetric units adopted by many isometric (spherical) viruses, with a combination of two-, three-, and fivefold rotations equivalently relating the units about a point in space.

Mycoviruses Viruses that infect and multiply in fungi.

'$T = 2$' symmetry The so-called 'forbidden' triangulation symmetry, in which 120 chemically identical protein monomers, form 60 identical, asymmetric dimers arranged with icosahedral symmetry. Monomers in a '$T = 2$' structure occupy two distinct, nonequivalent positions and therefore differ from the 'allowed' symmetries where monomers are either equivalently or quasi-equivalently related.

Virus capsid Generally a protective shell, composed of multiple copies each of one or more distinct protein subunits, that encapsidate the viral genome.

Introduction

In the early 1960s, interest in the antiviral activities associated with cultural filtrates of *Penicillium* spp. led to the discovery of double-stranded RNA (dsRNA) isometric viruses in these and other filamentous fungi. Fungal viruses, or mycoviruses, are now known to be of common occurrence. The isometric dsRNA viruses isolated from *Penicillium* spp. were among the first to be molecularly characterized and were shown to have segmented genomes. Those with bipartite genomes are currently classified in the genus *Partitivirus* in the family *Partitiviridae*. Interestingly, plant viruses (also called cryptoviruses) with bipartite dsRNA genomes and very similar properties to the fungal partitiviruses were discovered in the late 1970s and are presently classified in the genera *Alphacryptovirus* and *Betacryptovirus* in the family *Partitiviridae*. The fungal partitiviruses and plant partitiviruses are discussed elsewhere in this encylopedia. The goals of this article are to examine the similarities and differences between these two groups of viruses and to discuss future perspectives of partitivirus research.

Biological Properties

Fungal and plant partitiviruses are both associated with latent infections of their respective hosts. There are no known natural vectors for any partitivirus. Fungal partitiviruses are transmitted intracellularly during cell division and sporogenesis (vertical transmission) as well as following hyphal anastomosis, that is, cell fusion, between compatible fungal strains (horizontal transmission). In some ascomycetes (e.g., *Gaeumannomyces graminis*), virus is usually eliminated during ascospore formation. Experimental transmission of fungal partitiviruses has been reported by transfecting fungal protoplasts with purified virions. The plant cryptoviruses, on the other hand, are not horizontally transmitted by grafting or other mechanical means, but are vertically transmitted by ovule and/or pollen to the seed embryo. Thus, whereas sexual reproduction of the host is required for the survival of plant partitiviruses, it is detrimental to the continued existence of the fungal partitiviruses that infect some ascomycetes. Transmission of fungal partitiviruses through asexual spores, however, can be highly efficient, with 90–100% of single conidial isolates having received the virus. In summary, transmission of fungal partitiviruses by asexual spores and plant partitiviruses by seed provide the primary or only means for disseminating these viruses.

Both fungal and plant partitiviruses are generally associated with symptomless infections of their hosts. While cryptoviruses are present in very low concentrations in plants (e.g., 200 μg of virions per kg of tissue for white clover cryptic virus), fungal partitiviruses can accumulate to very high concentrations (at least 1 mg of virions per g of mycelial tissue for penicillium stoloniferum virus F (PsV-F)). Mixed infections of fungal or plants hosts with two distinct partitiviruses are not rare. PsV-S and PsV-F represent one example in which two partitiviruses infect the same fungus, *Penicillium stoloniferum*. Interestingly, a significant increase in cryptovirus concentration has been observed in mixed infections with unrelated viruses belonging to other plant virus families.

Virion Properties

The buoyant densities of virions of members of the partitivirus family range from 1.34 to 1.39 gm cm^{-3}, and the sedimentation coefficients of these virions range from 101S to 145S (S_{20w} in Svedberg units). Generally, each virion contains only one of the two genomic dsRNA segments. However, with some viruses like the fungal partitivirus PsV-S, purified preparations can contain other distinctly sedimenting forms that include empty particles and replication intermediates (see the next section).

All fungal and plant partitiviruses examined to date have been shown to possess virion-associated RNA-dependent RNA polymerase (RdRp) activity, which catalyzes the synthesis of single-stranded RNA (ssRNA) copies of the positive strand of each of the genomic dsRNA molecules. The *in vitro* transcription reaction occurs by a semi-conservative mechanism, whereby the released ssRNA represents the displaced positive strand of the parental dsRNA molecule and the newly synthesized positive strand is retained as part of the duplex.

Genome Organization and Replication

Virions of members of the partitivirus family contain two unrelated segments of dsRNA, in the size range of 1.4–2.3 kbp, one encoding the capsid protein (CP) and the other encoding the RdRp. The two segments are usually of similar size and are encapsidated separately, that is, each particle generally contains only one dsRNA segment. The genomes of at least 16 members of the genus *Partitivirus* have recently been completely sequenced (**Table 1**). In contrast, the complete genome sequences of only three alphacryptoviruses and no betacryptoviruses have yet been determined (**Table 1**). The genomic structure of Atkinsonella hypoxylon virus (AhV-1), the type species of the genus *Partitivirus*, comprising segment 1 (2180 bp, encoding the RdRp) and segment 2 (2135 bp, encoding the CP), is schematically represented in **Figure 1**.

The presence of one or more additional dsRNA segments is common among members of the family *Partitiviridae*. For example, in addition to the two genomic segments, preparations of AhV-1 contain a third dsRNA segment of 1790 bp. With the exception of the termini, this third segment is also unrelated to the other two. The absence of any long open reading frame (ORF) on either strand of segment 3 of AhV-1 suggests that it is a satellite segment, not required for replication. Satellite dsRNAs are often associated with infections by members of the family *Partitviridae* (**Table 1**).

Limited information on how fungal viruses in the genus *Partitivirus* replicate their dsRNAs is derived from *in vitro* studies of virion-associated RdRp and the isolation from naturally infected mycelium of particles that represent various stages in the replication cycle. The RdRp is believed to function as both a transcriptase and a replicase. The transcriptase activity within an assembled virion catalyzes the synthesis of progeny positive-strand

Table 1 List of viruses in the family *Partitiviridae* with sequenced genomic dsRNAs

Virus[a]	Abbreviation	dsRNA segment no. (size in bp; encoded protein, size in kDa)	GenBank accession no.
Genus: *Partitivirus*			
Atkinsonella hypoxylon virus *	AhV	1 (2180; RdRp, 78)	L39125
		2 (2135; CP, 74)	L39126
		3 (1790; satellite)	L39127
Ceratocystis resinifera virus	CrV	1 (2207; RdRp, 77)	AY603052
		2 (2305; CP, 73)	AY603051
Discula destructiva virus 1 *	DdV1	1 (1787; RdRp, 62)	NC_002797
		2 (1585; CP, 48)	NC_002800
		3 (1181; satellite)	NC_002801
		4 (308; satellite)	NC_002802
Discula destructiva virus 2 *	DdV2	1 (1781; RdRp, 62)	NC_003710
		2 (1611; CP, 50)	NC_003711
Fusarium poae virus 1 *	FpV-1	1 (2203; RdRp, 78)	NC_003884
		2 (2185; CP, 70)	NC_003883
Fusarium solani virus 1 *	FsV-1	1 (1645; RdRp, 60)	D55668
		2 (1445; CP, 44)	D55669
Gremmeniella abietina virus MS1 *	GaV-MS1	1 (1782; RdRp, 61)	NC_004018
		2 (1586; CP, 47)	NC_004019
		3 (1186; satellite)	NC_004020
Helicobasidium mompa virus *	HmV	V1–1 (2247; RdRp, 83)	AB110979
		V1–2 (1776; RdRp, 63)	AB110980
		V-70 (1928; RdRp, 70)	AB025903
Heterobasidion annosum virus *	HaV	1 (2325; RdRp, 87)	AF473549
Ophiostoma partitivirus 1	OPV-1	1 (1744; RdRp, 63)	AM087202
		2 (1567; CP, 46)	AM087203
Oyster mushroom virus	OMV	1 (2038; RdRp, 70)	AY308801
Penicillium stoloniferum virus F *	PsV-F	1 (1677; RdRp, 62)	NC_007221
		2 (1500; CP, 47)	NC_007222
		3 (677; satellite)	NC_007223
Penicillium stoloniferum virus S *	PsV-S	1 (1754; RdRp, 62)	NC_005976
		2 (1582; CP, 47)	NC_005977
Pleurotus ostreatus virus	PoV	1 (2296; RdRp, 82)	NC_006961
		2 (2223; CP, 71)	NC_006960
Rhizoctonia solani virus 717 *	RhsV-717	1 (2363; RdRp, 86)	NC_003801
		2 (2206; CP, 76)	NC_003802
Rosellinia necatrix virus 1	RnV-1	1 (2299; RdRp, 84)	NC_007537
		2 (2279; CP, 77)	NC_007538
Genus: *Alphacryptovirus*			
Beet cryptic virus 3 *	BCV-3	2 (1607; RdRp, 55)	S63913
Vicia cryptic virus *	VCV	1 (2012; RdRp, 73)	NC_007241
		2 (1779; CP, 54)	NC_007242
White clover cryptic virus 1 *	WCCV-1	1 (1955; RdRp, 73)	NC_006275
		2 (1708; CP, 54)	NC_006276
Unclassified viruses in the family *Partitiviridae*			
Cherry chlorotic rusty spot associated partitivirus	CCRSAPV	1 (2021; RdRp, 73)	NC_006442
		2 (1841; CP, 55)	NC_006443
Fragaria chiloensis cryptic virus	FCCV	1 (1743; RdRp, 56)	DQ093961
Pyrus pyrifolia cryptic virus	PpV	1 (1592; RdRp, 55)	AB012616
Raphanus sativus cryptic virus 1 (or Radish yellow edge virus*)	RasV-1 (RYEV)	1 (1866; RdRp, 67)	NC_008191
		2 (1791; CP, 56)	NC_008190
Raphanus sativus cryptic virus 2	RasV-2	1 (1717; RdRp, 55)	DQ218036
		2 (1521; unknown)	DQ218037
		3 (1485; unknown)	DQ218038

[a]An asterisk next to the virus name indicates it is presently recognized by ICTV as a member or a tentative member in the family *Partitiviridae*. Family members or tentative members that have not been sequenced to date are not included.

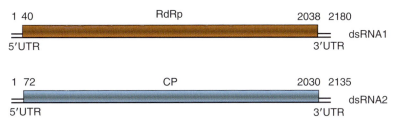

Figure 1 Genome organization of atkinsonella hypxylon virus (AhV), the type species of the genus *Partitivirus*. dsRNA1 contains the RdRp ORF (nt positions 40–2038) and dsRNA2 codes for the CP ORF (nt positions of 72–2030). The RdRp and CP ORFs are represented by rectangular boxes. Reproduced from Ghabrial SA, Buck KW, Hillman BI, and Milne RG (2005) *Partitiviridae*. In: Fauquet CM, Mayo MA, Maniloff J, Desselberger U, and Ball LA (eds.) *Virus Taxonomy: Eighth Report of the International Committee on Taxonomy of Viruses*, pp. 581–590. San Diego, CA: Elsevier Academic Press, with permission from Elsevier.

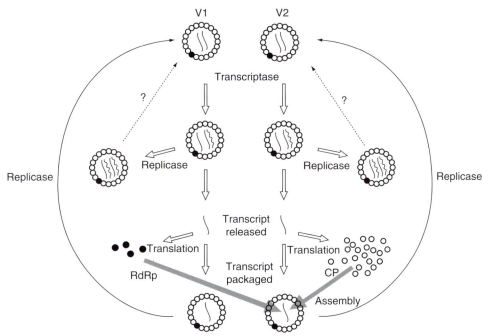

Figure 2 Model for replication of penicillium stoloniferum virus S (PsV-S). The open circles represent capsid protein (CP) subunits and the closed circles represent RNA-dependent RNA polymerase (RdRp) subunits. Solid lines represent parental RNA strands whereas wavy lines represent newly synthesized progeny RNA strands. Reproduced from Ghabrial SA and Hillman BI (1999) Partitiviruses-fungal (*Partitiviridae*). In: Granoff A and Webster RG (eds.) *Encyclopedia of Virology*, 2nd edn., pp. 1477–1151. San Diego: Academic Press, with permission from Elsevier.

RNA from the parental dsRNA template, accompanied by the displacement of the parental positive strand and its release from the virion. This can presumably occur through repeated rounds, giving rise to multiple positive-strand copies. Each released positive strand can then be consecutively or alternatively (1) used as a template for protein (CP or RdRp) translation by the host machinery or (2) packaged by CP and RdRp into an assembling progeny virion. The replicase activity within this assembling virion then catalyzes the synthesis of negative-strand RNA on the positive-strand template, reconstituting a genomic dsRNA segment. The replication of plant partitiviruses is presumed to mimic that described for the fungal viruses.

Partially purified virion preparations are known to include a small proportion of particles that contain only one ssRNA molecule corresponding to the genomic positive strand. These may be particles in which the replicase reaction is defective or has not yet occurred. In addition, partially purified virion preparations are known to contain a relatively large proportion of a heterogeneous population of particles more dense than the mature virions. These dense particles may represent various stages in the replication cycle including particles containing the individual genomic dsRNAs with ssRNA tails of varying lengths, particles with one molecule of dsRNA and one molecule of its ssRNA transcript and particles with two molecules of dsRNA (**Figure 2**).

Structures of Partitivirus Virions

The isometric dsRNA mycoviruses are classified within the families *Totiviridae*, *Chrysoviridae*, and *Partitiviridae*. The capsid structures of representative members of each of the first two families have been determined, at least one at near atomic resolution using X-ray crystallography, and the others at low to moderate resolutions (∼1.5–2.5 nm) using cryo-transmission electron microscopy (cryo-TEM) combined with three-dimensional (3D) image reconstruction. We have recently initiated systematic cryo-TEM and image reconstruction studies of three viruses, PsV-S, PsV-F, and FpV-1, within the genus *Partitivirus*. Based on phylogenetic analysis of partitivirus CPs or RdRps, PsV-S, and PsV-F form sister clades in the same cluster, whereas FpV-1 is placed in a separate and distinct cluster within this genus (see section on 'Taxonomic and phylogenetic considerations'). These viruses are structurally distinguished primarily on the basis of significant differences in the sizes of the respective CPs and percent nucleotide and amino acid sequence identity of their RdRps and CPs. Structures of partitviruses from the genera *Alphacrypto-* and *Betacryptovirus*, all of which are plant viruses, have yet to be reported.

Preliminary studies of PsV-S, PsV-F, and FpV-1 strongly suggest that all members of the genus *Partitivirus* share a number of common features. The structure of PsV-S is used to highlight these features, and significant differences are identified below.

PsV-S represents one of the simplest of the dsRNA viruses. Each virion comprises of one of the two genome segments (S1, 1754 bp and S2, 1582 bp), one or two copies of the 62 kDa RdRp, and 120 copies of the 47 kDa CP. Virions exhibit an overall spherical morphology in negative stain or in vitrified solution (**Figure 3**). Reports of virion diameters based on measurements from stained specimens likely underestimate the true virion size, owing to distortions and other potential artifacts associated with the imaging of negatively stained specimens. Based on cryo-TEM images, the maximum diameter of fully hydrated PsV-S virions is 35 nm. 3D reconstruction of the PsV-S structure at ∼21 Å resolution demonstrates that the capsid is a contiguous shell (average thickness ∼2 nm) of subunits arranged with icosahedral symmetry (**Figure 4**(a)). The external surface of PsV-S displays 60 prominent protrusions that rise approximately 3 nm above an otherwise relatively featureless, spherical shell. Each protrusion and a portion of the underlying shell, constitutes one asymmetric unit (1/60th) of the capsid. The dimeric morphology of each protrusion is

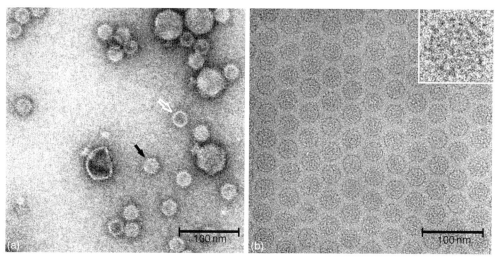

Figure 3 Electron micrographs of penicillium stoloniferum virus S (PsV-S). Samples were negatively stained in 2% uranyl acetate (a) or prepared unstained and vitrified (b). Micrographs were recorded on a CCD detector in an FEI Polara transmission electron microscope operated at 200 keV with samples at (a) room or (b) liquid nitrogen temperatures. In (a) bacteriophage P22 (five largest particles) was mixed with PsV-S to serve as a calibration reference. Heavy metal stain surrounds virions (e.g., black arrow) and contrasts their surfaces against the background carbon support film. Stain penetrates into the interior of 'empty' capsids (e.g., white arrow), resulting in particle images in which only a thin, annular shell of stain-excluding material (capsid) is seen. In (b) the unstained PsV-S sample was vitrified in liquid ethane. Here particles appear dark (higher density) against a lighter background of surrounding water (lower density). The inset shows a three-times enlarged view of an individual particle, in which several knobby surface features are clearly visible.

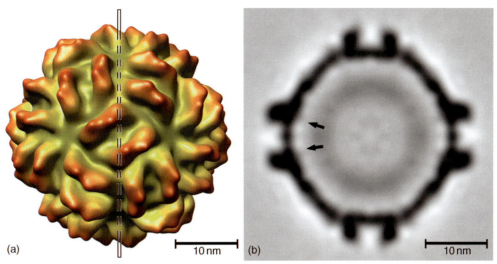

Figure 4 Three-dimensional (3D) structure of PsV-S. (a) Shaded, surface representation of PsV-S 3D reconstruction viewed along an icosahedral twofold axes. The 3D map is color-coded to emphasize the radial extent of different features (yellows and greens highlight features closest to the particle center, and oranges and reds those farthest from the center). A total of 60 prominent protrusions extend radially outward from the capsid surface. Each protrusion exhibits an approximate dyad symmetry, which is consistent with the expectation that the partitivirus capsid consists of 120 capsid protein monomers, organized as 60 asymmetric dimers in a so-called '$T=2$' lattice. (b) Density projection image of a central, planar section through the PsV-S 3D reconstruction (from region marked by dashed box in (a)). Darker shades of gray correspond to higher electron densities in the map section and lighter shades represent low-density features such as water outside as well as inside the particles. The capsid shell appears darkest because it contains a closely packed, highly ordered (icosahedral) arrangement of capsid subunits. The genomic dsRNA on the inside appears at lower density, in part because the RNA is not as densely packed and in part because the RNA adopts a less ordered arrangement. The protrusions seen in (a) appear as large 'bumps' in the central section view that decorate the outside of a contiguous, ~2 nm thick, shell. Arrows point to faint density features that appear to form contacts between the inner surface of the protein capsid and the underlying RNA. These contacts occur close to the fivefold axes of the icosahedral shell.

consistent with an asymmetric unit being composed of two CP monomers. Hence, there are 120 protein subunits arranged as 60 asymmetric dimers packed in a so-called '$T=2$' ('forbidden') lattice, which requires that the monomers do not interact equivalently or quasiequivalently as often occurs, especially in smaller, simpler virus capsids (e.g., with triangulation symmetries such as $T=1, 3, 4$).

An additional, symmetric ball of weak density within the capsid is attributed to the genomic dsRNA, and does not appear to adopt any regular structure at the limited resolution of this initial 3D reconstruction (**Figure 4(b)**). Lack of detectable genome organization may, in part, be attributed to averaging effects that occur when images of both PsV-S particle types are combined to produce the reconstructed 3D density map. Potential interactions between the genome and the inner wall of the capsid are suggested by faint lines of density observed in thin, planar sections through the density map (arrows; **Figure 4(b)**).

PsV-F and PsV-S co-infect and are co-purified from the fungal host, *Penicillium stoloniferum*. The coat protein of PsV-F (420 aa; 47 kDa) is very similar in size to that of PsV-S, but they only share 17% sequence identity. PsV-F virions are ~37 nm in diameter and the gross morphology and $T=2$ organization of the capsid is remarkably similar to that of PsV-S (not shown). However, the 60 protrusions in PsV-F are narrower, they extend further above the shell (~6.5 nm), and their long axes are rotationally aligned ~18° in a more anti-clockwise orientation compared to those in PsV-S.

Fusarium poae virus (FpV-1) also exhibits a '$T=2$' arrangement of its more massive coat protein subunits (637 aa; 70 kDa) (not shown). These subunits assemble to form ~42 nm diameter virions, significantly larger than both PsV-S and PsV-F. The average thickness of the FpV-1 shell varies considerably and is much less uniform than the shells of the smaller partitiviruses. The dimeric protrusions in FpV-1 extend ~3 nm above the capsid shell and have a wide base that appears to form more extensive interactions or connections to the underlying shell compared with the other two partitiviruses. Central sections through a low-resolution density map of FpV-1 exhibit three concentric layers of weak density, with a 3 nm spacing consistent with a relative close packing of nucleic acid as found in other dsRNA virus capsids.

Current studies are aimed at obtaining higher-resolution 3D reconstructions for each of the fungal partitiviruses mentioned above. In the absence of any crystallographic structure determinations, higher-resolution cryo-TEM reconstructions should enable more detailed comparisons to be made, and potentially provide a means to locate the

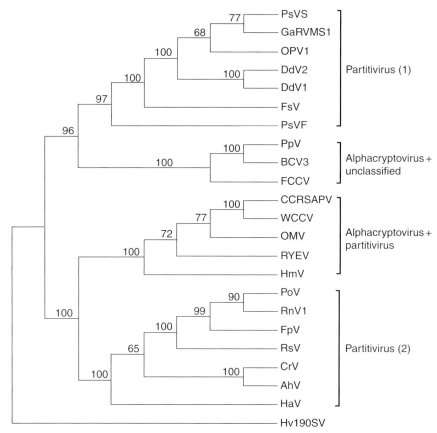

Figure 5 Phylogenetic analysis of the RdRp conserved motifs and flanking sequences derived from aligned, deduced amino acid sequences of members of the family *Partitiviridae* using the program CLUSTAL X. Motifs 3 through 8 and the sequences between the motifs, as previously designated by Jiang and Ghabrial (*Journal of General Virology* 85, 2111–2121, 2004) were used. See **Table 1** for virus name abbreviations. Bootstrap numbers out of 1000 replicates are indicated at the nodes. The tree was rooted with the RdRp of Helminthosporium victoriae 190S virus (Hv190SV), a member of the genus *Totivirus* in the family *Totiviridae* (GenBank accession no. NC_003607), which was included as an outgroup.

RdRps that are presumed to be fixed inside the capsid shell and possibly associated with a specific recognition site on each of the separately encapsidated genomic dsRNA segments. Such studies will also allow evolutionary links to the '$T=2$' cores of other well-studied dsRNA viruses that infect both fungal and nonfungal hosts to be explored. Structural studies expanded to include members of the genera *Alphacrypto-* and *Betacryptovirus* also have the potential to systematically characterize the similarities as well as differences in the life cycles of fungal and plant partitiviruses.

Taxonomic and Phylogenetic Considerations

Recent phylogenetic analyses based on amino acid sequences of RdRps (conserved motifs) of members of the family *Partitiviridae* led to the identification of four clusters, two of which are large and comprise only members of the genus *Partitivirus* (**Figure 5**). One large cluster with strong bootstrap support includes the partitiviruses DdV1, DdV2, FsV, GaV-MS1, OPV, and PsV-S (see **Table 1** for virus abbreviations). The CPs of these viruses share significant amino acid sequence identities (34–62%), but less than the RdRps (55–70% identity), suggesting that the CPs have evolved at a faster rate. The second partitivirus RdRp cluster consists of AhV, CrV, FpV, PoV, RhsV, and RnV1. These two large clusters were proposed to comprise two subgroups (subgroups 1 and 2) of the genus *Partitivirus* (**Figure 5**). Interestingly, the CPs of these two subgroups differ significantly in size with average sizes of 47 and 74 kDa for subgroups 1 and 2, respectively. Of the remaining two RdRp clusters (**Figure 5**), one consists of BCV (genus *Alphacryptovirus*) and two other, unclassified plant cryptoviruses (FCCV and PpV). The fourth cluster consists of a mixture of plant viruses (WCCV and RYEV, genus *Alphacryptovirus*) and fungal viruses (OMV and HmV, genus *Partitivirus*), together with CCRSAV, which may be either a fungal or a plant virus, as yet to be determined. This raises the interesting possibility of horizontal transfer of members of the family *Partitiviridae* between fungi and plants. This is a reasonable possibility because some of the viruses in these clusters have fungal hosts that are pathogenic to plants. Based on our current knowledge of partitiviruses, the taxonomy of the family *Partitiviridae* will probably need to be reconsidered. As additional molecular and structural information from a wider range of plant and fungal partitiviruses is gathered, the need for new or revised taxonomic classifications may become even more apparent.

Future Perspectives

Development of Infectivity Assays for Fungal Partitiviruses

The recent success of using purified RnV-1 virions to infect virus-free isolates of the fungal host *Rosellinia necatrix* is very promising since there is considerable need to advance our knowledge of the biology of fungal partitiviruses (e.g., host-range, virus–host interaction, and molecular basis of latent infection). The natural host range of partitiviruses is restricted to the same or closely related vegetative compatibility groups that allow lateral transmission. At present, there are no known experimental host ranges for fungal partitiviruses because suitable infectivity assays have yet to be developed. With the success of the *Rosellinia necatrix*-RnV-1 infectivity assays, it is expected that future research would explore the potential experimental host range of RnV-1 including other fungal species in the genus *Rosellinia* and related genera. It is also anticipated that comparable infectivity assays using purified virions would be developed for other fungal partitiviruses. Because of the possibility alluded to earlier, of horizontal transfer of partitiviruses between fungi and plants, it is interesting to test the infectivity, using fungal protoplasts, of plant partitiviruses such as the alphacryptoviruses WCCV-1 and RYEV, which are known to be more closely related to certain fungal partitiviruses (e.g., OMV, HaV, and HmV) than to other plant partitiviruses. Alternatively, it would likewise be interesting to explore the infectivity of fungal partitiviruses such as OMV, HaV, and HmV to appropriate plant protoplasts.

Fungal partitiviruses are known to be associated with latent infections of their natural hosts. But, conceivably, some partitiviruses may induce phenotypic changes and/or virulence attenuation in one or more of their experimental host fungi. If true, this would provide excellent opportunities for exploiting fungal partitiviruses for biological control and for basic studies on host–pathogen interactions. Furthermore, transfection assays using fungal protoplasts electroporated with full-length, *in vitro* transcripts of cloned cDNA to viral dsRNAs should also be applicable to partitiviruses. This would allow for the development of partitivirus-based vectors for expressing heterologous proteins in fungi. Because many partitiviruses are known to support the replication and encapsidation of satellite dsRNA, this property would facilitate the construction of recombinant vectors. Genes of interest could be inserted into the satellite molecule between the conserved termini that are presumed to be required for replication and encapsidation. Considering the high level of partitivirus accumulation in their fungal hosts, the use of partitiviral vectors provides an attractive and cost-effective means for the overproduction of valuable proteins in filamentous fungi.

Reconsideration of the Taxonomy of the Family *Partitiviridae*

There is an urgent need to characterize at the molecular level, a broad range of plant partitiviruses that should include representatives of the genus *Betacryptovirus*. Taxonomic considerations of partitiviruses may benefit greatly from elucidating the capsid structure of representative members of each of the four clusters delineated by phylogenetic analysis. It may, however, require concerted efforts to generate purified virions of alphacryptoviruses and betacryptoviruses in quantities needed for structural studies since they generally occur at very low concentrations in their hosts.

See also: *Alphacryptovirus* and *Betacryptovirus*; Fungal Viruses; Partitiviruses of Fungi.

Further Reading

Antoniw JF (2002) *Alphacryptovirus (Partitiviridae)*. In: Tidona CA and Darai G (eds.) *The Springer Index of Viruses*, pp. 676–679. New York: Springer.

Antoniw JF (2002) *Betacryptovirus (Partitiviridae)*. In: Tidona CA and Darai G (eds.) *The Springer Index of Viruses*, pp. 680–681. New York: Springer.

Crawford LJ, Osman TAM, Booy FP, et al. (2006) Molecular characterization of a partitivirus from ophiostoma himal-ulmi. *Virus Genes* 33: 33–39.

Ghabrial SA (2001) Fungal viruses. In: Maloy O and Murray T (eds.) *Encyclopedia of Plant Pathology*, pp. 478–483. New York: Wiley.

Ghabrial SA (2002) *Partitivirus (Partitiviridae)*. In: Tidona CA and Darai G (eds.) *The Springer Index of Viruses*, pp. 685–688. New York: Springer.

Ghabrial SA and Hillman BI (1999) Partitiviruses-fungal (*Partitiviridae*). In: Granoff A and Webster RG (eds.) *Encyclopedia of Virology*, 2nd edn., pp. 1477–1151. San Diego: Academic Press.

Ghabrial SA, Buck KW, Hillman BI, and Milne RG (2005) *Partitiviridae*. In: Fauquet CM, Mayo MA, Maniloff J, Desselberger U, and Ball LA (eds.) *Virus Taxonomy: Eighth Report of the International Committee on Taxonomy of Viruses*, pp. 581–590. San Diego, CA: Elsevier Academic Press.

Sasaki A, Kanematsu S, Onoue M, Oyama Y, and Yoshida K (2006) Infection of *Rosellinia necatrix* with purified viral particles of a member of *Partitiviridae* (RnPV1-W8). *Archives of Virology* 151: 697–707.

Parvoviruses of Arthropods

M Bergoin, Université Montpellier II, Montpellier, France
P Tijssen, Université du Québec, Laval, QC, Canada

© 2008 Elsevier Ltd. All rights reserved.

Glossary

Densonucleosis virus Original name of densovirus.
Densovirus Insect parvovirus.

History

The discovery of invertebrate parvoviruses goes back to the early 1960s when an epizootic decimated a mass rearing of the greater wax moth, *Galleria mellonella*, larvae reared as fishing bait on the French side of Lake of Geneva. Electron microscopic examination of thin sections from diseased larvae revealed hypertrophied nuclei containing an electron-dense viroplasm with thousands of tiny viral particles. After purification, these particles were observed as small isometric, nonenveloped virions, 22–25 nm in diameter, and their biochemical analysis showed that they contained single-stranded DNA. The virus causing the disease was designated as densonucleosis virus, which later was simplified to the name densovirus (DNV). Diseases presenting the same type of histopathological features, and caused by the same type of viruses, were later described not only in Lepidoptera but also in other insect orders and in the Decapoda phylum of Arthropoda with the discovery of DNV infections in crabs and shrimps. The cloning and sequencing of several DNV genomes during the last two decades, while confirming their relatedness to the vertebrate parvoviruses, revealed an extraordinary diversity, reflecting very likely the evolutionary diversity of their hosts. Presently, parvoviruses of arthropods are recognized as *bona fide* members of the family *Parvoviridae* in which they form the subfamily *Densovirinae*.

Classification of DNVs

The family *Parvoviridae* musters all viruses that contain a linear, single-stranded DNA genome and has been subdivided into two subfamilies. Vertebrate parvoviruses are classified into the subfamily *Parvovirinae*, whereas the *Densovirinae* groups all parvoviruses from arthropods.

Dedicated to the memory of Alan Simpson, who, under the supervision of Prof. Michael Rossmann at Purdue University, solved the first densovirus structure by X-ray crystallography. Alan, 38, sadly drowned after getting into difficulties while on a solo kayaking trip in Scotland on March 2007.

Consequently, arthropod parvoviruses are usually named DNVs. Vertebrate and arthropod parvoviruses share hallmark features such as a $T=1$ capsid structure, usually possess a structural protein with a phospholipase A2 (PLA2) motif, and contain similar-sized genomes of 4–6 kb that replicate by self-priming and hairpin transfer. However, there is little sequence identity, except for a few enzyme motifs, between the viruses of these subfamilies.

Another isometric, single-stranded DNA virus, bombyx mori densovirus type 2 (BmDNV-2), has been isolated from the silkworm. Despite its morphological similarity with DNVs, the genome of BmDNV-2 is much larger (about 12.5 kb), segmented, and contains, in contrast to parvoviruses, its own DNA polymerase motif. Because of these differences, the ICTV has rejected its inclusion in the *Densovirinae* and the naming of this virus as a DNV is unfortunate.

The grouping of the DNVs into genera is based on a limited number of viruses and becomes strained with an expanding number of characterized viruses and will probably undergo a continuous revision. So far, molecular characterization of DNVs allowed their classification into at least four genera: *Densovirus*, *Iteravirus*, *Pefudensovirus*, and *Brevidensovirus*. Many DNVs remain unclassified since only roughly half of the *c.* 30 reported DNV isolates are sufficiently characterized with respect to their molecular biology. On the other hand, many more DNVs remain to be isolated from the million-odd insect species.

DNV genera are currently distinguished in the ICTV classification according to the genome organization by the following criteria: (1) monosense versus ambisense; (2) presence or absence of inverted terminal repeats (ITRs); (3) split or single open reading frame (ORF) coding for the capsid proteins. Sequence identities may be very low despite overall similarity in genome organization and expression strategies. The members of the genus *Densovirus* have an ambisense genome, organized with cassettes for VP (structural proteins) and NS (nonstructural or Rep proteins) on the 5′ halves of the complementary strands and with long ITRs (*c.* 550 nt). Examples are the junonia coenia densovirus (JcDNV) and galleria mellonella densovirus (GmDNV). The viruses in the genus *Iteravirus* have a monosense 5 kb genome with *c.* 230 nt ITRs (hence the name of the genus) and are so far only found in the order Lepidoptera. Examples are the casphalia extranea densovirus (CeDNV) and bombyx mori densovirus type 1 (BmDNV-1). The members of the genus *Brevidensovirus* have a 4 kb genome with a monosense organization and terminal hairpins but lacking ITRs.

The members of this genus are found in Diptera, such as *Aedes aegypti* and *Aedes albopictus*, but at least some DNVs from shrimps, such as infectious hypodermal and hematopoietic necrosis virus (IHHNV), are closely related. The VPs of the members of both genera *Iteravirus* and *Brevidensovirus* are coded from a single ORF.

Previously, all ambisense DNVs isolated from different insect orders were classified into the single genus *Densovirus* making this genus quite heterogeneous. Presently, we distinguish the DNV members into three subgroups (A, B, and C; see **Table 1**) with increased uniformity. Subgroup A contains the prototypical DNV genus members such as GmDNV and JcDNV. Although the viruses in this genus have high sequence identities, they have different ecological niches. The ORF for VP of subgroup B members is split and requires splicing of the ORFs for expression of the largest VP. Moreover, their genomes are shorter (5.5 kb), and their ITRs have an I- instead of a Y-shape. Although subgroup B members share a common genome organization, that is distinct from subgroup A members, they have very low sequence identities (10–30%) among each other and with the subgroup A members. One of them, periplaneta fuliginosa densovirus (PfDNV), was reclassified in the genus *Pefudensovirus* in the latest ICTV report, but not enough is known about the expression strategies at the molecular level of other members of this subgroup for co- or separate classification. The ITRs of PfDNV, acheta domesticus densovirus (AdDNV), and blatella germanica densovirus (BgDNV) of this subgroup vary greatly in size (125–215 nt) and these viruses have been isolated from Orthoptera, Hemiptera, and Dictyoptera but not from Lepidoptera as the prototypical *Densovirus* members. So far, there is a single member in subgroup C, culex pipiens densovirus (CpDNV), from the mosquito *Culex pipiens*, that uses, in addition to two promoters on the strand encoding NS polypeptides, a complex splicing in the NS genes.

Table 1 Names of viruses, their abbreviations, hosts and accession numbers

Ambisense densoviruses			
Subgroup A densoviruses (genus *Densovirus*, 6 kb genome with 0.55 kb ITRs, single ORF for VP)			
Junonia coenia densovirus (JcDNV)	*Junonia coenia*	Lepidoptera	NC_004284
Galleria mellonella densovirus (GmDNV)	*Galleria mellonella*	Lepidoptera	NC_004286
Diatraea saccharalis densovirus (DsDNV)	*Diatraea saccharalis*	Lepidoptera	NC_001899
Helicoverpa armigera densovirus (HaDNV)	*Helicoverpa armigera*	Lepidoptera	Sequenced, not submitted
Mythimna loreyi densovirus (MlDNV)	*Mythimna loreyi*	Lepidoptera	NC_005341
Pseudoplusia includens densovirus (PiDNV)	*Pseudoplusia includens*	Lepidoptera	Sequenced, not submitted
Subgroup B densoviruses (5.5 kb genome with 0.2 kb ITRs, split ORFs for VP; PfDNV is classified in a new genus, *Pefudensovirus*, while others remain to be classified)			
Acheta domesticus densovirus (AdDNV)	*Acheta domesticus*	Orthoptera	Dictyoptera NC_005041
Blattella germanica densovirus (BgDNV)	*Blattella germanica*	Dictyoptera	
Myzus persicae densovirus (MpDNV)	*Myzus persicae*	Hemiptera	NC_005040
Planococcus citri densovirus (PcDNV)	*Planococcus citri*	Hemiptera	NC_004289
Periplaneta fuliginosa densovirus (PfDNV)	*Periplaneta fuliginosa*	Dictyoptera	AB028936
Subgroup C densoviruses (6 kb genome with 0.3 kb ITRs, split ORFs for VP and NS)			
Culex pipiens densovirus (CpDNV)	*Culex pipiens*	Diptera	Sequenced, not submitted
Monosense densoviruses			
Genus *Iteravirus* (5 kb genome, 0.25 kb ITRs)			
Casphalia extranea densovirus (CeDNV)	*Casphalia extranea*	Lepidoptera	AF375296
Bombyx mori densovirus (BmDNV-1)	*Bombyx mori*	Lepidoptera	AY033435
Dendrolimus punctatus (DpDNV-1)	*Dendrolimus punctatus*	Lepidoptera	NC_006555
Genus *Brevidensovirus* (4 kb genome, no ITRs but terminal hairpins, no phospholipase A2 motif in VP)			
Aedes aegypti densovirus (AaeDNV)	*Aedes aegypti*	Diptera	
Aedes albopictus densovirus (AalDNV-1)	C6/36 cell line	Diptera	NC_004285
Aedes albopictus densovirus (AalDNV-2)	C6/36 cell line	Diptera	AY095351
Penaeus stylirostris densovirus (PstDNV)	*Penaeus stylirostris*	Decapoda	AF273215
Unassigned (6.3 kb, no ITRs but hairpins, no phospholipase A2 motif in VP)			
Penaeus monodon densovirus (PmoDNV)	*Penaeus monodon*	Decapoda	DQ002873
Noncharacterized densoviruses			
Agraulis vanillae densovirus (AvDNV)	*Agraulis vanillae*	Lepidoptera	
Euxoa auxilliaris densovirus (EaDNV)	*Euxoa auxilliaris*	Lepidoptera	
Lymantria dispar densovirus (LdiDNV)	LPC Ld-D52 cell line	Lepidoptera	
Pieris rapae densovirus (PrDNV)	*Pieris rapae*	Lepidoptera	
Sibine fusca densovirus (SfDNV)	*Sibine fusca*	Lepidoptera	
Haemagogus equinus densovirus (HeDNV)	GML-HE-12 cell lines	Diptera	
Toxorhynchites amboinensis densovirus (TaDNV)	TRA-171 cell lines	Diptera	
Simulium vittatum densovirus (SvDNV)	*Simulium vittatum*	Diptera	
Leucorrhinia dubia densovirus (LduDNV)	*Leucorrhinia dubia*	Odonata	

Access number for the most recently complete DNA sequences are indicated.

Biophysical and Structural Features of Virions

All the capsids of DNVs are built on the same $T=1$ model with 60 structural proteins but the relative number of constituent polypeptides involved in virus assembly varies from one genus to another. The virions of the members of the genus *Densovirus* contain four clearly distinct polypeptide bands upon sodium dodecyl sulfate polyacrylamide gel electrophoresis (SDS-PAGE), ranging in size between 45 and 110 kDa and with varying relative amounts. These structural proteins were shown to be N-extended isoforms. As exemplified by BmDNV-1 and CeDNV, iteravirus particles contain five structural proteins, which are also N-extended isoforms, with molecular masses between 48 and 76 kDa. Both the capsids of the members of the genera *Iteravirus* and *Densovirus* possess, like virtually all vertebrate parvoviruses, phospholipase A2 activity. In contrast, the viruses of the third genus, *Brevidensovirus*, do not display phospholipase A2 activity and have only two structural proteins (N-extended isoforms) of about 40 kDa. The brevidensoviruses have all been isolated from mosquitoes and shrimps.

The virions of the various DNVs share with those of the vertebrate parvoviruses a high buoyant density in CsCl of c. 1.40–1.44 g cm^{-3}, linked to a high nucleic acid content (about 20–30%), and sedimentation coefficients of about 110S. GmDNV, JcDNV, and BmDNV-1 particles were shown to contain polyamines (spermine, spermidine, and putrescine) that could partially neutralize the viral DNA, whereas so far no reports of polyamines in vertebrate parvoviruses have been published.

The near-atomic structure of the particle of GmDNV was solved by X-ray crystallography and revealed some striking differences with the structure of vertebrate parvoviruses (**Figures 1(a)** and **1(b)**). The β-barrel of the GmDNV VP4 capsid protein must be rotated and radially translated to superimpose it on the β-barrel of the capsid protein of canine parvovirus (CPV). Another difference with the vertebrate parvoviruses was the absence of loop 4 in GmDNV capsid protein, resulting in a β-annulus-type structure instead of a spike around the threefold axes. The βA strand folds back to its own subunit in vertebrate parvovirus structures thus far solved. In contrast, in GmDNV, the βA strand is linearly extended across the twofold axes, allowing it to hydrogen bond with the βB strand of the neighboring subunit ('swapping domain'). Overall, the outside of the GmDNV capsid is much smoother than that of vertebrate parvoviruses, perhaps as a result of a different evolutionary pressure (**Figure 1(a)**).

The three-dimensional (3-D) structure of aedes albopictus densovirus type 2 (AalDNV-2), a member of the genus *Brevidensovirus*, has been solved to a resolution of

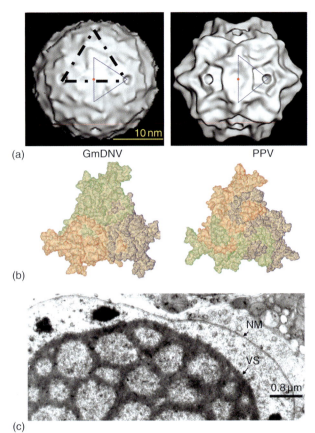

Figure 1 Comparative surface topology (a) and capsid protein interactions (b) between GmDNV and PPV (Porcine parvovirus) as representatives of subfamilies *Densovirinae* and *Parvovirinae*. The large triangle (broken line) connecting three fivefold axes represents the size of a trimer (surface representation in (b)) and the small triangle the approximate size occupied by a single structural protein. The most conserved sequences/structures are on the inside face of the trimers, whereas the outside faces, as shown here by the GmDNV and PPV trimers, differ considerably in structure sequence and protein interactions. (c) Thin section through the nucleus (NM, nuclear membrane) of a heavily infected *Sibine fusca* midgut cell. The viral particles assemble from a reticulated electron-dense virogenic stroma (VS). Photo J. L. Duthoit.

about 16 Å by cryoelectron microscopy. The sequence identity of the structural proteins of this virus with those of GmDNV is less than 20% which is translated into some distinctive structural features. For example, AalDNV-2 particles display density ridges around the threefold axes and prominent protusions at the fivefold axes that are both absent in the GmDNV particle. Nevertheless, the twofold proximal depression as well as the inner surfaces and large cavities under the fivefold vertices are conserved. Similarly, inner surfaces were most conserved among vertebrate parvoviruses.

Finally, X-ray diffraction data recently collected will allow solving of both the IHHNV shrimp parvovirus and the BmDNV-1 atomic structure to about 3 Å in the near future. Interestingly, the IHHNV capsid seems to lack both the A and G β-strands that are involved, in the case of GmDNV, in inter-subunit interactions.

Host Range and Pathology

As illustrated in **Figure 1(c)**, the hypertrophy of infected nuclei where thousands of viral particles are produced from electron-dense virogenic stroma is the striking histopathological feature in all DNV infections.

Most of the DNVs isolated so far are known to be highly pathogenic and fatal for their natural host and to be responsible for epizootics in natural populations of noxious or commercially reared insects. Examples of hosts are phytophagous members of the Lepidoptera or the Orthoptera, scavengers belonging to the Dictyoptera, and vectors like aphids and mosquitoes. Unfortunately, they also appeared to be highly pathogenic in mass-reared arthropods, such as the silkworm *Bombyx mori* or shrimps, causing serious economic losses.

The host range of the different DNVs varies considerably. Some, like GmDNV, are monospecific, whereas others, such as the closely related mythimna loreyi densovirus (MlDNV) or JcDNV, infect a large range of different hosts. Similarly, PfDNV infects many different members of the genus *Periplaneta*, whereas AdDNV infects readily the European cricket but not the American cricket. BmDNV-1 infects certain strains of *B. mori* whereas other strains are resistant while BmDNV-2 shows the opposite tropism. The mosquito DNVs of the genus *Brevidensovirus* seem all to have a wide host range.

The symptoms of viral infection vary from host to host. Some DNVs, such as GmDNV and JcDNV, infect all tissues except the midgut. Others, such as BmDNV-1, SfDNV, and CeDNV, infect only the midgut, whereas PfDNV infects the hindgut and shows a hypertrophied fat body. Not surprisingly, external signs of infection vary greatly. Larvae infected with a member of the DNV genus become lethargic and anorexic and become progressively whitish and paralyzed, followed by a slow melanization. Moulting and metamorphosis is also inhibited. Infection of the susceptible silkworms with BmDNV-1 also leads to flaccidity and is usually fatal within a week. Infection of slug caterpillar pests of the oil palm, *Casphalia extranea* and *Sibine fusca* (Lepidoptera *Limacodidae*), with their respective DNVs induces intensive proliferation and progressive thickening and opacity of the midgut wall, leading to tumor-like lesions. The *Casphalia* larvae become anorexic and their color changes from green to yellowish brown after infection. Infection of the cockroach, *Periplaneta fuliginosa*, with PfDNV leads to paralysis of the hind legs and their movements become uncoordinated. Females are particularly affected by PfDNV; the abdomen is swollen due to the hypertrophied fat body, which is milky white in contrast to the brownish tissues in uninfected specimens. The most striking feature of the infection of European crickets with AdDNV is their decreasing ability to jump and they end up lying motionless on their back. The most obvious pathological change was seen in the gut. Infected females always had less food in the gut, and especially striking was the completely empty digestive ceca. They usually die within 10 days. Infection of *Aedes aegypti* larvae with DNV led to paralysis and their bodies were distorted and curved. They usually hung near the water surface.

Potential Use of DNVs for Biological Control of Insect Pests

Despite their high virulence for their natural hosts, their usually limited host range, and their lack of pathogenicity for vertebrates, DNVs have so far not received the attention they deserve for their potential in controlling economically and medically important insects. Indeed, attempts to use SfDNV and CeDNV to control outbreaks of their host, *C. extranea* and *S. fusca*, two major pests of oil palm industrial plantations in Côte d'Ivoire and Columbia, respectively, were very successful. Mortalities above 90% were recorded 2 weeks after a single airplane spraying over hundreds of hectares with a suspension containing 5–100 heavily infected larval homogenate per hectare. Similarly, a commercial formulation (Viroden) of AeDNV was used for the control of *A. aegypti* larvae in different areas of the former Soviet Union and a PfDNV formulation is presently produced in China for the control of cockroaches. The homologies found in some specific regions of the NS and VP genes between DNV and vertebrate parvovirus genomes raised a concern about safety for the use of DNVs as biopesticides. However, a number of data indicate that DNVs can be safely used. These include (1) the complete lack of pathological symptoms in mice and rabbits inoculated with milligrams of different types of infectious DNVs used to raise antisera; (2) the lack of pathogenicity and normal development of suckling mice following intracranian inoculation of GmDNV, AaeDNV, and AalDNV, and (3) the inability of GmDNV and MlDNV genomes to replicate in vertebrate cells. These data are certainly good arguments for the safety toward animals and humans of DNVs, but their use as pesticides will require more studies before approval by safety agencies.

Genome Organization and Expression Strategies

The overall structure and organization of coding sequences of members in the different genera of DNVs is depicted in **Figure 2**. DNVs have a small, compact

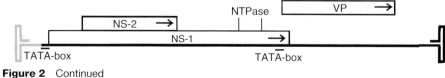

Figure 2 Continued

genome with limited coding capacities but well-developed complex strategies of alternative splicing and leaky scanning mechanisms for regulation of the expression of sets of overlapping genes in a timely and coordinated quantitative fashion. Leaky scanning expression strategy is widely used in invertebrate parvoviruses; it dictates the generation of N-terminal extended isoforms of the structural proteins from the same reading frame or of alternative nonstructural proteins from different reading frames. Leaky scanning of DNV mRNAs has not yet been studied in detail but is remarkable in that it often results in the coding of four to five isoforms of their structural proteins with scanning often exceeding 1000 nt for the most downstream initiation AUG. It is not clear whether this process includes ribosome shunting (discontinuous ribosomal scanning).

Another widely used expression strategy of DNVs to express their NS proteins NS-1 and NS-2 is the presence of two overlapping ORFs which have their initiation codons separated by a few nucleotides (usually four). These initiation codons, of which the 5′-proximal AUG is usually in a less favorable context, are thus completely covered by the footprint of about 13–15 nt of the small ribosomal subunit.

The VP cassette can be recognized by PLA2 motif for all parvoviruses except for those belonging to the genera *Amdovirus* and *Brevidensovirus*. This motif consists of a GPG calcium-binding site followed after about 18 amino acids by the HDxxY catalytic site motif (**Figure 3**).

Although DNVs from other genera have two NS proteins, NS-1 and NS-2, most ambisense DNVs have a third NS protein (NS-3). Sequence identity of NS-3 with proteins from non-parvoviruses (# AY293731 from ClGV granulosis virus and S78547 from BmDNV-2) suggests a horizontal transmission of this gene early in evolution of the ambisense DNVs. So far, the knowledge of expression strategies of DNVs is limited to those with an ambisense genome.

DNVs with an Ambisense Genome

Subgroup A

The subgroup A DNV members have 6 kb genomes with long ITRs of *c.* 550 nt representing about 20% of the viral genome (**Figure 2**). The terminal ~130 nt of the ITRs can be folded into typical Y-shaped hairpins occurring in two orientations 'flip' and its reverse complement 'flop', as described for vertebrate parvoviruses of the genus *Dependovirus*. Transcript mapping revealed that the rest of the ITRs contain the TATA-boxes at the border of the unique sequence and upstream promoter sequences for the NS and VP gene cassettes, respectively. These genomes have about 80–90% sequence identity and their proteins show strong serological cross-reactivities. Their structural proteins are N-terminally extended isoforms of 45, 53, 58, and 89 kDa synthesized by a leaky scanning mechanism from an unspliced 2.6 kb transcript. Their nonstructural proteins are synthesized from two species of mRNAs. An unspliced 2.5 kb transcript is translated from its 5′-most proximal initiation codon into NS-3, a 20 kDa protein, whereas the precise excision of NS-3 coding sequence from the 2.5 kb transcript gives rise to a spliced 1.8 kb mRNA which is translated into NS-1 (60 kDa) and NS-2 (30 kDa) by alternative initiation at the tandem initiation codons of NS-1/-2 cassette. In this subgroup, ORFs of the NS and VP genes are separated by *c.* 30 nt only. As a consequence, the NS and VP transcripts of GmDNV, JcDNV, and MlDNV have a terminal antisense sequence of about 60 nt.

Subgroup B

So far all ambisense DNVs of subgroup B have shorter genomes than those of subgroup A, about 5.5 kb, and use splicing and leaky scanning to generate the structural proteins. The ITRs are about 125–215 nt long, and occupy only about 7% of the total genome length. Their folding by self-complementarity generates I-shaped hairpin structures. As for subgroup A, the 3′ extremities of ITRs of subgroup B DNVs contain the TATA-boxes and their upstream promoter elements for both the NS and VP genes. The precise mapping of AdDNV NS transcripts revealed a great similarity with the transcription pattern of subgroup A genomes with an unspliced 2.5 kb and a spliced 1.8 kb transcript for NS-3 and NS-1–NS-2 translation, respectively. In contrast, expression of VP gene combines a complex splicing/frameshift strategy with leaky scanning

Figure 2 Schematic representation of the structure and expression strategies of DNVs. Two main groups can be distinguished, the ambisense DNVs with thus far three subgroups, and the monosense DNVs with thus far two genera, the iteraviruses and the brevidensoviruses. The genomes are represented with their 5′ and 3′ hairpin termini, whereas transcripts from left to right (NS coding strand) are depicted above the genome and transcripts from right to left (VP coding strand) below the genome. The positions of splicing, as so far known, is shown by a V (thin line) in or below the transcripts. Members of subgroups A and B produce two NS transcripts, one unspliced that yields NS-3 and one that undergoes splicing to remove NS-3 and allows translation of both NS-1 and NS-2 by leaky scanning. Whereas VPs of subgroup A viruses are produced by leaky scanning from an unspliced mRNA, the VPs of subgroup B viruses are coded by two ORFs that need to be connected in frame. For subgroup C, the position of NS ORFs is shown for both the primary transcripts as well as the mature spliced mRNAs. The two NS transcripts are not generated by splicing, as for subgroups A and B, but by using two distinct promoters. Although the large VP ORF for this virus is preceded by a small ORF, VPs are only generated from the large ORF by leaky scanning. Transcript mapping is still lacking for the iteraviruses but genome analysis indicates that NS and VP use different promoters and that no splicing is used. Brevidensoviruses use essentially the same strategy.

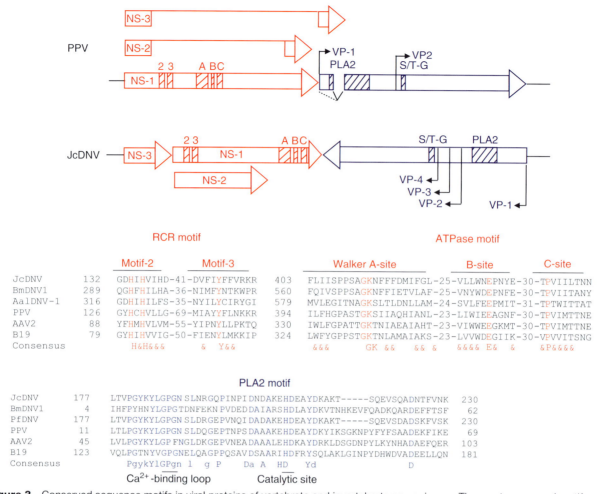

Figure 3 Conserved sequence motifs in viral proteins of vertebrate and invertebrate parvoviruses. The most-conserved motifs between members of the subfamilies *Parvovirinae* and *Densovirinae* are the rolling-circle replication (RCR) and ATPase (helicase) motifs of NS-1 and phospholipase A2 (PLA2) motifs of VP1. Consensus sequences are indicated below the motifs and represent bulky, hydrophobic amino acids (M, V, L, F, Y, V, W).

to generate four isoforms of the structural proteins. Surprisingly, splicing does not remove an initiation codon or put the two ORFs in frame, its most common objective in parvoviruses, but removes the sequences required for frameshifting. The VP and NS ORFs are separated by 3 nt for BgDNV, 15 nt for panonychus citri densovirus (PcDNV), 24 nt for AdDNV, and 30 nt for PfDNV. The overlap of NS and VP AdDNV transcripts generates a 32 antisense sequence.

Subgroup C

With a 6 kbp genome and c. 400 nt ITRs, able by folding to form an asymmetric T-shaped structure, the CpDNV is the only representative of this subgroup. The analysis of the unique sequence of CpDNV genome (unpublished) revealed several striking features in the strand encoding NS proteins. These features include both the organization of coding sequences and transcription modalities of the genome, and clearly differentiate CpDNV genome from those of subgroup A and B members. Indeed, instead of being in the same frame and separated from NS-1 coding sequence by a single stop codon, as in the other subgroups, NS-3 is in another frame and separated from NS-1–NS-2 tandem sequences by a 256 nt gap. Furthermore, the NS-1–NS-2 coding sequences are not continuous but split into four ORFs: two upstream encompassing the N-terminal and two downstream encompassing the C-terminal moieties, respectively (see **Figure 2**). Finally, expression of NS genes is not regulated by a single promoter as in subgroups A and B but by two. One promoter with a TATA-box located at the limit of the left ITR generates a 2.3 kb transcript that is used for NS-3 translation, whereas a second promoter located in the gap between NS-3 and NS-1 5′-most ORF drives expression of NS-1 and NS-2 proteins through a 1.7 kb transcript. A consequence of this organization is that by removing a short intron (53 nt) the N- and C-terminal sequences of both NS-1 and NS-2 are spliced in frame. It is not yet clear whether this small intron serves a particular role or that removal of corresponding intron sequence in the virus genome would be without consequence. The three VP proteins of CpDNV capsid (85, 64, and 57 kDa) are generated by leaky scanning from an unspliced 2.2 kb transcript. As for members of subgroups A and B, the overlap of NS and VP transcripts generates a 14 nt antisense sequence. Taken together, the properties of CpDNV genome justify, in our opinion, the creation of an additional genus for DNVs with an ambisense genome.

DNVs with a Monosense Genome

Members of two genera of DNVs, *Iteravirus* and *Brevidensovirus*, share a genomic organization similar to that of vertebrate parvoviruses by having their sequences coding for NS and VP proteins located on the same strand. They possess only two overlapping NS ORFs encoding NS-1 and NS-2 (NS-3 ORF is lacking) and a single VP ORF located in the 5' half and 3' half of the genome, respectively (**Figure 2**). In addition to their host range, iteraviruses differ from brevidensoviruses by their genome size and structure of the genome termini.

Genus Iteravirus

This genus is presently represented by only three viruses – BmDNV-1, CeDNV, and dendrolimus punctatus densovirus (DpDNV) – all isolated from Lepidoptera. Owing to its deleterious impact on silkworm production, BmDNV-1 was discovered early and has been studied in detail, particularly in Japan. Successful genetic selection of silkworm strains resistant to this virus has been achieved. The structure of the 5 kb genome is highly conserved among the three iteravirus representatives. They possess ITRs *c.* 230 nt in length, the first 159 of which are able to fold into a typical J-shaped terminal hairpin occurring in two flip/flop orientations.

The expression of iteraviruses has not yet been studied by transcript mapping but two putative TATA-boxes, one upstream from NS-1 ORF the other overlapping its 3' extremity, and polyadenylation signals have been identified. Furthermore, the ORFs in BmDNV, CeDNV, and DpDNV are in identical positions and share 80–90% homologies, suggesting that they represent actual genes. The putative initiation codons for both CeDNV, DpDNV, and BmDNV NS-1 and NS-2 are all in a favorable context. As previously shown for BmDNV-1 by peptide mapping, and more recently for CeDNV by expression in a baculovirus system, the five structural proteins of iteravirus capsid are five N-terminal extended isoforms of VP generated by leaky scanning.

Genus Brevidensovirus

With the shortest (4 kb) genomes among parvoviruses, the lack of ITRS, and a natural host range encompassing several species of medically important mosquitoes (*Aedes* and *Anopheles* species) and of economically important shrimp species (*Paeneus stylirostris*), brevidensoviruses constitute an original group of invertebrate parvoviruses. Recently, noninfectious IHHNV-related sequences were reported in the genome of *Penaeus monodon*. The fact that larval stages of most mosquitoes are aquatic and that many species can afford to develop in salty waters support the hypothesis of a common ancestor for mosquito and shrimp DNVs. The aedes aegypti densovirus (AaeDNV) and aedes albopictus densovirus type 1 (AalDNV-1) from cell line C6/36 are the prototypes. AalDNV strains have been isolated from several laboratories around the world handling C6/36 or other mosquito cell lines.

All the genomes of brevidensoviruses so far sequenced have terminal hairpins able to fold into perfect T-shaped structures of *c.* 60 nt, but with different sequences at 5' and 3' extremities. Their unique sequences are very divergent from the other DNVs, and like members of the genus *Amdovirus* (Aleutian disease virus) of vertebrate parvoviruses their VP ORF does not contain the PLA2 motif. This may be one of the reasons why the small (*c.* 1.2 kb) VP ORF codes for only two structural proteins, VP-1 and VP-2, of *c.* 40 kDa each, that is, a size much smaller than those of other parvoviruses.

Two putative promoter TATA-boxes at map units 7 and 61 are conserved among these viruses. Northern blot analysis of AeDNV transcripts revealed two species of apparently unspliced mRNAs of 3.5 and 1.2 kb, encoding NS and VP proteins, respectively. These sizes indicate that the VP and NS transcripts co-terminate in the vicinity of the polyadenylation signal near the 3' end of the genome, a property shared by most vertebrate parvovirus transcripts, except those of erythroviruses. Furthermore, primer extension and mutational analyses of the structural gene promoter of AaeDNV revealed that the TATA-sequence at map unit 61 is apparently not involved in gene expression but a CAGT initiator (Inr) sequence 60 nt upstream of this sequence and an upstream TATA-box at map unit 60 were found to be critical for efficient gene expression. The same CAGT box is also present upstream of P61 TATA box of AalDNV-1 genome and has been shown to be important for many arthropod and mammalian promoters.

Unclassified DNVs

In addition to a number of DNV isolates that have not been further characterized (**Table 1**), the genome of the hepatopancreatic parvovirus (renamed penaeus monodon densovirus, PmDNV) of the black tiger prawn *P. monodon* was recently cloned and sequenced. The virus packages (−)-polarity single-stranded DNA with a length of 6321 nt and terminal hairpins of 222 (5′) and 215 nt (3′), respectively. The genome has a monosense organization with three major predicted coding regions (left, mid-, and right ORF). Sequence identities indicated that these ORFs code for NS-1 (mid-ORF), NS-2 (left ORF), and VP (right ORF). Surprisingly, the coding capacity of the right ORF is approximately 92 kDa, whereas the two capsid proteins are only about 57 and 54 kDa. N-terminal protein sequencing suggested a proteolytic cleavage but it is not clear whether this cleavage is maturation related or that the N-terminal 35 kDa protein serves a nonstructural function.

Phylogenetic Comparison of DNVs

Several motifs in the proteins of DNVs having essential enzymatic functions can serve to establish phenetic relationships among these viruses and their vertebrate counterparts. In particular, the rolling-circle replication (RCR) and ATPase (Walker) motifs in NS-1 present through the whole family *Parvoviridae* and the PLA2 motif found in the N-terminal sequence of most of the VP1 structural polypeptide (**Figure 3**) are useful for these comparisons. On the other hand, only some members of subgroup A of ambisense DNVs contain an NS-3 gene. This gene is, however, found in other viruses such as the granuloviruses. Obviously, the choice of NS-3 for phylogenetic purposes may show a closer relationship of such viruses to granulosis viruses than to other NS-3-less DNVs. Analysis of relationships using the NS-1 motifs revealed three distinct virus subgroups (**Figure 4**) corresponding to ambisense DNVs, brevidensoviruses, and iteraviruses. Differences can still be substantial within each group, for example, sequence identities among the ambisense DNVs often are less than 20%.

Conclusions

The genomes of DNVs, while fitting the main criteria of those of vertebrate parvoviruses, have revealed a remarkable diversity in both their noncoding terminal palindromic sequences, gene organization, and transcription modalities. It is particularly noteworthy that viruses with a similar genome organization that have evolved in different insect orders have strikingly different genome sequences. It is far from clear whether the lack of sequence identities led to essential different expression strategies and classification according to overall genome organization.

The forthcoming data on genome sequence and organization of several insects from different orders could provide keys for a better understanding of the evolution and adaptation of DNVs to such a variety of hosts.

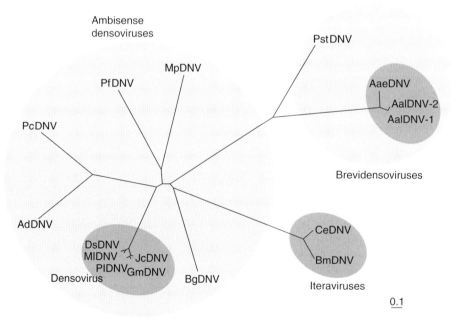

Figure 4 The RCR and ATPase motifs of the NS-1 (**Figure 3**) were used to construct an unrooted tree. The scale bar represents 10 mutations per 100 sequence positions. The length of the branch thus represents the phylogenetic distance between the various DNVs. Three distinct DNV groups are recognized, one ambisense virus group and two monosense virus groups. The tree was generated by distance matrix analysis (Phylip program package PROTDIST), using Dayhoff PAM 001 scoring matrix followed by FITC, applying the global search option.

See also: Insect Pest Control by Viruses; Parvoviruses: General Features.

Further Reading

Bergoin M and Tijssen P (2000) Molecular biology of *Densovirinae*. In: Faisst S and Rommelaere J (eds.) *Parvoviruses: From Molecular Biology to Pathology and Therapeutic Uses*, pp. 12–32. Basel: Karger.

Simpson AA, Chipman PR, Baker TS, *et al.* (1998) The structure of an insect parvovirus (galleria mellonella densovirus) at 3.7 Å resolution. *Structure* 15: 1355–1367.

Tattersal P, Bergoin M, Bloom E, *et al.* (2005) Family *Parvoviridae*. In: Fauquet CM, Mayo MA, Maniloff J, Desselberger U, and Ball LA (eds.) *Virus Taxonomy: Eighth Report of the International Committee on Taxonomy of Viruses*, pp. 353–369. San Diego, CA: Elsevier Academic Press.

Tijssen P, Bando H, Li Y, *et al.* (2006) Evolution of densoviruses. In: Kerr JR, Cotmore SF, Bloom ME, Linden RM, and Parrish CR (eds.) *Parvoviruses*, pp. 55–68. London: Hodder Arnold.

Tijssen P and Bergoin M (1995) Densonucleosis viruses constitute an increasingly diversified subfamily among the parvoviruses. *Seminars in Virology* 6: 347–355.

Tijssen P, Szelei J, and Zadori Z (2006) Phospholipase A2 domains in structural proteins of parvoviruses. In: Kerr JR, Cotmore SF, Bloom ME, Linden RM, and Parrish CR (eds.) *Parvoviruses*, pp. 96–105. London: Hodder Arnold.

Parvoviruses of Vertebrates

C R Parrish, Cornell University, Ithaca, NY, USA

© 2008 Published by Elsevier Ltd.

Glossary

Inverted terminal repeats Sequences at the termini of the viral genomes that form imperfect palindromes, which are required for DNA replication.

Panleukopenia Reduction in the numbers of white blood cells of all classes.

History

Parvoviruses are the cause of a variety of diseases that range from severe and even lethal to subclinical and in some cases cryptic diseases, and so their discovery was delayed by the relative lack of specific clinical signs and in some cases by the difficulties of culturing the agents. Diseases of cats caused by the feline panleukopenia virus (FPV) were first recognized in the early part of the twentieth century, and a similar virus was recognized as the cause of enteritis in mink in the late 1940s. Other diseases caused by parvoviruses were recognized in the 1950s, and those included the Kilham rat virus isolated from rats, and the H1 virus isolated from tissues grown in rats. The adeno-associated viruses (AAVs) were identified in the early 1960s as contaminants of adenovirus isolates, and those were soon shown to be dependent on the adenovirus as a helper for their replication. FPV was isolated in the early 1960s in tissue culture once it was recognized that its replication depended on dividing cells and that the virus itself was only poorly cytopathic, both common properties of parvoviruses. Many different viruses were identified over the next decades, including the minute virus of mice (MVM) that was isolated in the 1960s originally as a contaminant of cultured mouse cells, and which was subsequently found to be related to a number of different parvoviruses that are widespread in rodents. The porcine parvovirus was identified in the 1970s as the cause of porcine fetal infections, and the human B19 parvovirus was also identified in the 1970s as the cause of the erythema contagiosum (fifth disease) in children and anemia in people with underlying problems with erythropoiesis. The canine parvovirus (CPV) emerged in the late 1970s as the cause of new diseases of dogs; it was closely related to the FPV, and it seemed likely that CPV was a mutant of the feline virus with mutations that gave it an expanded host range. The list of parvoviruses and their diseases has continued to grow over the years due to the discovery of viruses in vertebrates as well as invertebrates and crustacea, and those have included viruses of rabbits, horses, cows, and other species which are not clearly the cause of specific diseases. DNA analysis seeking cryptic virus sequences has also identified a number of new parvoviruses in humans and other animals. Those human viruses are mostly not directly associated with any disease, and the recently described human bocavirus appears to be associated with respiratory disease in neonates who are infected with other agents. It appears likely that there are parvoviruses of most animals, but that most are the cause of only very mild disease or subclinical infections.

The biology and distribution of the viruses vary widely, and each virus has its own specific properties and natural history. Among the human parvoviruses related to B19 in the genus *Erythrovirus*, three genetically distinct lineages have been identified which have different temporal or geographic distributions. The B19 viruses have a global distribution and are found to infect people of all ages. The type 2 viral sequences are primarily found in the tissues of people born before c. 1973, suggesting that they were infecting people prior to that year, but that they are not currently circulating. The type 3 viruses have been mostly found in people in Africa or people who migrated to other countries from Africa.

Taxonomy and Classification

All of the viruses fall into the family *Parvoviridae*, which includes viruses with linear single-stranded DNA (ssDNA) genomes, and small nonenveloped capsids around 25 nm in diameter. The family is divided into two subfamilies, *Parvovirinae* and *Densovirinae*, to host respectively vertebrate and invertebrate viruses. The invertebrate viruses are not considered in this article. The subfamily *Parvovirinae* includes the genera *Parvovirus*, *Erythrovirus*, *Dependovirus*, *Amdovirus*, and *Bocavirus*. The viruses are classified based on their structure, serological relationships, and their genome organization and sequence relationships.

Geographical and Seasonal Distribution

The viruses are relatively resistant in the environment and are shed in various secretions depending on their specific pathogenesis and tissue tropism. Some infect the intestinal epithelial cells and are shed in the feces (CPV and FPV), while others infect cells in the kidneys and are shed in the urine (many rodent parvoviruses), or replicate in a limited number of tissues in the body and are likely spread by respiratory routes (human B19 parvovirus). This results in a variety of different means of spread and the epidemiology of the viruses can differ significantly. Some viruses require close contact between animals to get efficient transmission, as appears to be the case with the Aleutian mink disease virus (AMDV) which has been controlled in many cases by serological testing and culling infected animals. Other viruses can spread widely and rapidly with no direct contact between hosts. While the tracking of viruses and their spread can be difficult, there is clear evidence for the rapid and extensive spread of some parvoviruses. For example, the CPV spread around the world during 1978 after it first emerged, and variants of the original CPV infected animal strain also spread globally within a year of their first emerging. Global distribution of viral genotypes of B19 has also been observed. Although there is no strict seasonality to the infections by most parvoviruses, the fact that many of the viruses infect fetal or neonatal animals means that the distribution of the infection and disease may be seasonal for animals that breed at particular times of the year.

Host Ranges and Virus Propagation

Most of the parvoviruses appear to have relatively narrow natural host ranges, although they can sometimes be shown to infect cells from a broader range of hosts when those are inoculated in tissue culture. DNA replication of the autonomous parvoviruses occurs during S phase in mitotically active cells, and hence the tissue tropism of the infection depends on the age of the animal, which will affect the populations of dividing cells that are available. This also leads to age-related diseases, with many viruses showing a severe infection of fetuses or neonates, but only a mild or subclinical infection of older animals. Within the host some of the viruses may have a narrow tissue tropism, as is seen in the case of the human B19 virus which is largely restricted to the erythroid precursor cells in the bone marrow in humans after birth. The mechanisms that control host range or tissue tropisms of the viruses can vary depending on the virus, but have been associated with receptor binding by the capsids, with intracellular blocks to replication that are dependent on the function of the viral NS2 protein, or to the regulation of specific gene expression or splicing in different differentiated host cells.

Viral Capsids, Their Activities, and Cell Infection

The viral capsid is nonenveloped and around 25 nm in diameter (**Figure 1**). The capsid is assembled from two or three versions of a single protein. The main capsid protein is designated VP2 or VP3 (depending on the virus), and it contains the major structural motif that allows assembly to occur and makes up ∼90% of the protein in the capsid. Longer versions of the capsid protein are also included and required for infection, most likely because they contain a phospholipase A2 enzyme activity, as well as basic sequences that control the nuclear transport of the capsids and capsid proteins during infection and replication. The phospholipase A2 enzymatic activity can vary significantly depending on the virus being examined, but that appears to be required for successful cell infection, most likely by modifying the endosomal membrane to allow viral release into the cytoplasm. The capsid is assembled from a total of 60 copies of the combined VP1 and -2 (and -3 where that is present) proteins. The N-terminus of the largest (VP1) protein is sequestered or buried within the capsid when they are produced within the cell, and this can become exposed to the outside during the process of infection by an unknown mechanism. The capsid packages the viral ssDNA genome, and in some viruses capsids package mainly the minus strand, while others can package either the plus or minus DNA strands with similar efficiency.

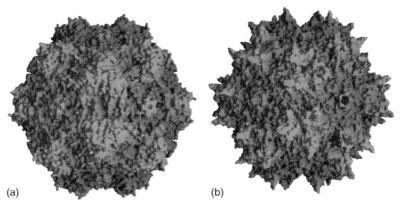

Figure 1 A surface view of (a) the CPV and (b) the AAV-2 virus capsid as derived from the atomic structure of the viruses. The diameter of the particles is ~25 nm, and they are assembled from 60 copies of a combination of the VP2 protein (CPV) or VP3 protein (AAV) (about 55 copies) and the slightly longer VP1 protein (about five copies). The binding sites of the cell receptor and many neutralizing antibodies are located on the raised regions found around the threefold axes of icosahedral symmetry. A pore that penetrates through the capsid is seen at the fivefold axes of symmetry.

The capsids can bind a variety of different receptors on cells leading to infection. Receptors range from sugars such as specific sialic acids (MVM) and heparan sulfate proteoglycan (many serotypes of AAV), to cell surface glycoprotein receptors of a variety of different types. Receptor or co-receptor proteins identified so far include the basic fibroblast growth factor receptor (AAV type 2), αV β2 integrins (AAV type 2), and transferrin receptor (canine and feline parvoviruses).

Cell Infection and Viral Replication

Although all of the parvoviruses appear to enter and infect cells through receptor-mediated endocytic processes, the specific pathways and mechanisms involved are still being defined, and it appears that many viruses are able to use multiple endosomal routes of infection. Uptake often involves clathrin-mediated pathways, but for many viruses other pathways may also be used. While signaling by the receptor or receptor-associated molecules may occur and in some cases enhances the infection, it is not clear that it is required for infection by all viruses. The viral particles traffic within the endosomal system of the cell, and in many cases remain in endosomes for long periods. The particle in an intact or mostly intact form enters the cytoplasm in a process that is controlled or assisted by the PLA2 activity of the VP1 unique region. Within the cytoplasm, the capsid may be able to associate with microtubular-based molecular motors such as dynein, and become localized in the pericentreolar region of the cell near the microtubular organizing center. Exactly how the virus or the viral DNA enters the cell nucleus is not clear. The small parvovirus particle could likely pass through the nuclear pore intact, and some evidence suggests that basic sequences in the N-terminal region of the VP1 protein are functional nuclear localization motifs. However, other studies suggest that the capsid may also be able to directly affect the nuclear envelope, perhaps allowing the virus to enter the nucleus for replication. It is also possible that both mechanisms can apply under different circumstances. In either case, the capsid and the associated ssDNA enters the nucleus. How the DNA is removed from the capsid is not known, but possibilities are that the capsid comes apart in the nucleus, or that the DNA is extracted from the capsid, possibly by the activity of the host cell DNA polymerase initiating replication on the 3′ end of the DNA.

The Viral Genome

The viruses of vertebrates have many of their genomic features in common, and although there is wide divergence between the sequences of many of the viruses, shared sequences in a conserved region of the nonstructural protein suggest that all the viruses share a single common ancestor and have many properties in common. The genomes are all between ~4500 and ~5200 bases in length, and each contains two large open reading frames, as well as short sequences near the 3′ and 5′ end of the genome, and either inverted terminal repeats or terminal palindromes (**Figure 2**). The terminal sequences consist of imperfect palindromes, in some cases with specific mismatched sequences that are involved in the control of DNA replication. In the standard genome orientation, the left-hand open reading frame encodes the nonstructural proteins (called NS1 (and -2) or Rep) and the right-hand open reading frame encodes the structural protein genes (in AAVs termed the Cap genes) (VP1 and -2, or VP1–3, depending on the virus). Genomes may contain

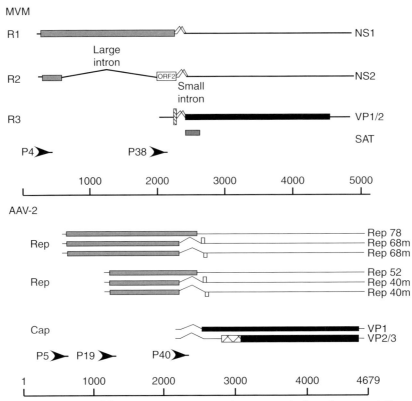

Figure 2 Diagrammatic representations of the genome structures and gene organization of representatives of the autonomous parvoviruses (MVM) and the AAVs (AAV type 2). The nonstructural 1 and 2 (NS1 and -2) and replication (Rep) proteins of the viruses are encoded by the left half of the genome, while the viral proteins 1 and 2 (VP1 and VP2), and, in the case of the AAV, the capsid (Cap) genes, are encoded in the right half of the genome.

between one and three functional promoters for RNA transcription. Whereas often there is only a single promoter that is found near the left of the genome, additional promoters may be used to express the capsid proteins (in many viruses, including the viruses related to MVM) or smaller versions of the nonstructural proteins (as seen in AAV). Each gene comes in multiple versions, through the use of alternative transcription start sites, splicing to give different transcripts, and by internal initiation of translation. In some viruses, there are short additional open reading frames besides the NS and capsid proteins, and in many viruses a short open reading frame (called SAT) that overlaps the beginning of the major capsid protein has been identified, while an 11 kDa protein is encoded in the sequence of the B19 parvovirus after the capsid protein gene open reading frame.

The larger NS1 or Rep are multifunctional proteins that serve various functions involved in viral replication. The protein serves as a DNA helicase, site-specific nickase, and it becomes covalently attached to the 5′ end of the ss- and double-stranded (ds)-forms of the viral DNA. The activities of the proteins are regulated by site-specific phosphorylation. The functions of the NS2 protein are not completely understood, but it interacts with the Crm1 protein and appears to regulate the transport and assembly of capsids within the cells of some hosts. As mentioned above, the VP or Cap proteins spontaneously assemble into capsids when expressed in cells.

Evolution and Variation

As mentioned, the parvoviruses appear to be related through one or a few common ancestors, and phylogenetic analysis shows that the viruses can be divided into several major clades or related groups. The timeline of the evolution of the viruses is unclear as the molecular clock of sequence variation over long time periods is not known. When CPV and B19 sequences with known dates were analyzed, the rates of evolution proved to be surprisingly high, with around 10^{-4}–10^{-5} substitutions per nucleotide per year, similar to the rates observed for some RNA viruses. Intrahost variation of the viral sequences has been seen during persistent infections by MVM in immunodeficient mice, and by B19 in humans. The viral DNA is replicated by host cell DNA polymerases, and while the true fidelity of the replication is not known, it appears that it must be considerably lower than the fidelity seen for the replication of the host genome. This probably results from the mode of DNA replication, the single-stranded nature of the viral genome, and the rapid turnover of the viruses in nature.

Pathogenesis

The parvoviruses cause a wide variety of diseases, ranging from subclinical or inapparent to severe and lethal diseases. When infected alone, the AAVs do not complete their replication or cause any disease in animals. The tissue tropisms of the autonomous parvoviruses require the presence of mitotically active cells, and so there are many more tissues available for virus replication in a fetus or neonatal animal, and fewer in older animals. For some viruses such as porcine parvovirus, the disease seen is primarily in fetuses that are infected *in utero*, and after birth infections cause only a mild disease. More severe disease associated with PPV is seen in a co-infection with the porcine circovirus type 2, where infection by the parvovirus apparently causes immune stimulation that enhances the replication and pathogenicity of the circovirus. It appears that the human bocavirus may cause only a mild infection by itself, but it is associated with more severe disease in cases of co-infection with other viruses. The CPV and FPV infect the dividing cells in various lymphoid organs which are most active in younger animals, while in the intestine they also infect the rapidly proliferating cells in the crypts of Leiberkuhn, and therefore cause both a lymphopenia and lymphoid depletion, and also diarrhea resulting from loss of epithelial cells in the gut and subsequent loss of osmoregulation. The human B19 has a tropism for the rapidly dividing erythroid precursor cells in the bone marrow, and can cause a loss of erythrocyte production. Since erythrocytes generally have a long life span, the loss of production only results in anemia where there is an underlying condition that causes loss of erythrocytes, or chronic loss of production. However, the large amounts of viral antigen produced in the circulation result in a transient antibody–antigen immune complex disease which is seen as erythema infectiosum (fifth disease) in children, or as arthritis in adults presumably due to deposition of immune complexes in the joints.

Some viruses cause a more chronic disease, the best understood being AMDV. Infection of mink generally results in a chronic persistent infection, where the virus is only partly neutralized by the antibody responses that develop, and so there is a continuing production of viral antigen–antibody complexes. The disease is more prominent in the background of the Aleutian coat color phenotype minks, which have a condition similar to human Chédiak–Higashi syndrome; they are therefore deficient in the ability of removing the immune complexes from circulation and show increased levels of circulating immune complexes and a more rapid course of disease.

Immune Responses, Diagnosis, Prevention, and Control

The immune responses to parvoviruses develop rapidly after infection, and in most cases the virus is largely or completely controlled. Where they have been tested, antibodies are quite protective and also are effective at controlling most virus infections, and those would work in conjunction with the cell-mediated responses. Maternal immunity is effective at protecting neonatal animals from infection in many cases. Most of the parvoviruses do not have obvious mechanisms for manipulating or evading the host immune responses, and how the viruses that establish persistent infections do so in face of a strong antibody response is not known. In the case of Aleutian mink disease, the virus is not effectively neutralized by the antibodies that develop and so it continues to infect and spread even in the face of strong immune responses.

Vaccines are effective against most of the parvoviruses (with the exception of AMDV where disease can be enhanced by vaccination). Both modified live and inactivated vaccines have been developed, and both can be effective at protecting against the disease as long as they can generate an effective immune response. Experimental subunit vaccines have been tested and shown to be effective at protecting animals against disease, and in one case the vaccine contained peptides derived from the sequence of the N-terminus of the VP2 protein, and the antibodies generated both neutralized the virus and protected the animals. Vaccination in most cases is interfered with by preexisting maternal immunity, and hence when vaccinating young animals it is necessary to vaccinate when the maternal immunity has waned to such a low level that the immune responses to the vaccine can be developed.

Diagnostic approaches vary depending on the purpose of the test. Antibodies are long-lived, and serological tests are generally sensitive and will show the existence of prior infections, but those may not be a reflection of the current infection status of the animal. Polymerase chain reaction (PCR) is frequently used to detect the presence of virus, but it will also detect the presence of persistent viral DNA left over from a prior viral infection which may not be actively replicating at the time of testing. Virus isolation can be used for some viruses, but others grow poorly in tissue culture (e.g., AMDV) or grow only in specific differentiated cells that are difficult to maintain in routine culture (B19). The wild-type viruses are often poorly cytopathic in culture; hence, other methods may be required to detect the presence of any viruses isolated.

Control is difficult for many parvoviruses, and in some cases the low incidence of clinical disease makes vaccination unnecessary. Some viruses such as the canine and feline parvoviruses appear to be widespread and to infect animals readily, and those are controlled only with difficulty. In those cases, quarantine may be used to prevent infection of young susceptible animals with wild-type virus before the animals are vaccinated. Vaccines to PPV are given to the sow before breeding to prevent infection of the fetuses. Parvoviruses that infect rodents in experimental colonies are difficult to control due to their being persistently shed from infected animals. In this case, control involves cesarean derivation of uninfected animals and establishment of a clean colony.

See also: Parvoviruses: General Features; Vaccine Strategies.

Further Reading

Brown KE and Young NS (1997) The simian parvoviruses. *Reviews in Medical Virology* 7: 211–218.

Farr GA, Zhang LG, and Tattersall P (2005) Parvoviral virions deploy a capsid-tethered lipolytic enzyme to breach the endosomal membrane during cell entry. *Proceedings of the National Academy of Sciences, USA* 102: 17148–17153.

Girod A, Wobus CE, Zadori Z, et al. (2002) The VP1 capsid protein of adeno-associated virus type 2 is carrying a phospholipase A2 domain required for virus infectivity. *Journal of General Virology* 83: 973–978.

Kaufmann B, Simpson AA, and Rossmann MG (2004) The structure of human parvovirus B19. *Proceedings of the National Academy of Sciences, USA* 101: 11628–11633.

Kerr JR, Cotmore SF, Bloom ME, Linden RM, and Parrish CR (2006) *Parvoviruses.* London: Hodder Arnold.

Young NS and Brown KE (2004) *Parvovirus B19.* New England Journal of Medicine 350: 586–597.

Parvoviruses: General Features

P Tattersall, Yale University Medical School, New Haven, CT, USA

© 2008 Elsevier Ltd. All rights reserved.

The Family *Parvoviridae*: An Overview

The family *Parvoviridae* comprises small, isometric, nonenveloped viruses that contain single-stranded DNA genomes that are linear, a set of properties unique in the known biosphere. The parvovirus virion contains a single genomic molecule 4–6 kbp long, which terminates in short palindromic sequences that can fold back on themselves to create duplex hairpin telomeres. These terminal hairpins are different from one another, both in sequence and predicted structure, in heterotelomeric genomes, while in homotelomeric genomes they form part of a terminal repeat (TR) that can be inverted or direct. These terminal hairpins are essential for the unique rolling-hairpin strategy of parvovirus DNA replication, and are therefore an invariant hallmark of the family.

The bulk of the parvovirus genome lies between these hairpins, and comprises two gene cassettes, with transcripts from one half of the genome, by convention the left-hand side, programming synthesis of the DNA replication initiator protein(s), while the right-hand encodes an overlapping set of capsid polypeptides. All parvovirus initiator proteins incorporate two separate enzymatic cores. The N-terminal of these is a site-specific single-strand nuclease domain, comprising active-site motifs common to all rolling-circle initiator proteins, which recognize and nick the replication origin(s). Carboxy-terminal to this nickase lies a helicase domain, which belongs to the superfamily III (SF3) group of viral 3′-to-5′ helicases. Parvovirus genomes replicated using a unidirectional, strand-displacement mechanism called rolling-hairpin replication (RHR), which appears to be an evolutionary adaptation of the ancient rolling-circle replication mechanism used by bacteriophages with circular single-stranded DNA genomes. The parvovirus genome contains two viral origins of DNA replication, one embedded in each hairpin telomere. During RHR, these hairpins can unfold to be copied, then refold to allow continuous amplification of the linear template. Viral replication exhibits two distinct phases, the first of which is an amplification phase, during which high-molecular-weight duplex replicative form (RF) DNA intermediates are generated and subsequently processed back down to monomeric duplexes by the action of the viral nickase. Later in the infectious process, a genome displacement phase predominates, in which individual progeny single strands are displaced from duplex forms and encapsidated.

To initiate the first phase of RHR, the 3′ nucleotide of incoming virion DNA pairs with an internal base to create a DNA primer, which allows a host polymerase to initiate synthesis of a complementary DNA strand. This generates a monomer-length, duplex intermediate in which the two strands, designated plus or minus with respect to transcription, are covalently linked, or hairpinned, by a single copy of the original viral telomere. For homotelomeric parvoviruses, these hairpinned structures create replication origins that can be acted upon by NS1 or Rep proteins, in a process called terminal resolution, which converts them into duplex palindromic telomeres. For heterotelomeric viruses, however, the left-end telomere in its hairpin form can be refractory to nicking by NS1, and replication proceeds through an obligatory dimer RF. For both types of parvovirus, further DNA amplification proceeds by unidirectional strand displacement, where the initiator protein also serves as the 3′-to-5′ replicative helicase, while all other replication proteins are commandeered from the host cell. Replication continues to the end of the genome where the hairpin is displaced and copied to create a palindromic, or extended form, telomere. These terminal palindromes can be melted and rearranged in hairpins such that the 3′ end can act as a primer for further replication, using the newly synthesized product as a template. Where the template ends in a hairpinned telomere, this is copied by the replication fork, switching the direction of synthesis back toward the initiating telomere, creating a dimer. Thus the RHR process creates a series of palindromic duplex dimeric and tetrameric concatemers, in which the unit-length genomes are fused in left-end:left-end and right-end:right-end combinations.

Site-specific single-strand nicks are subsequently introduced, by the viral initiator protein, into the duplex telomeric origins embedded in these concatemers, allowing successive rounds of replication to be initiated, or progeny single strands to be excised and ultimately displaced. Nicking involves a trans-esterification reaction that generates a base-paired 3′ nucleotide and leaves the initiator nuclease covalently attached, by a phosphotyrosine bond, to the 5′ nucleotide at the nick. Displaced progeny single strands re-enter the replication pool unless sequestered by the packaging machinery and encapsidated into empty particles. Thus packaging appears to be driven by ongoing viral DNA synthesis, and is entirely dependent upon the availability of preformed capsids. The efficiency of excision, displacement, and packaging of each strand sense depends upon the efficiency of the replication origin producing it. Thus it follows that all homotelomeric parvoviruses generate equal numbers of each strand as packaging precursors, and therefore produce virion populations containing equal numbers of plus and minus strands, each packaged in a separate particle. The ratio of plus to minus strands packaged by the heterotelomeric parvoviruses depends upon the relative efficiencies of the two different DNA replication origins embedded in their disparate telomeres. In all cases where this has been determined, the origin at the 5′ end of the transcription template strand, by convention the right-hand end of the minus strand, is the predominant origin, and drives the displacement and packaging of the minus strand at ratios between 10:1 and 100:1 over that of the plus strand. Thus, all of the heterotelomeric parvoviruses examined to date are effectively negative-strand DNA viruses.

Independent of the sense of strand they contain, parvovirus particles are physically very stable and are antigenically and structurally quite simple. A combination of protein chemistry, X-ray crystallography and cryo-electron microscopy, has established that the virion is an icosahedral structure exhibiting $T=1$ symmetry, constructed from two to four species of structural protein. These polypeptides are encoded as a nested sequence set in the right half of the genome, and the capsid shell comprises 60 copies of the common, usually C-terminal, 60–70 kDa region of the polypeptide set. Each polypeptide contains an eight-membered β-barrel, or jelly-roll, fold, found in most viral capsid proteins. The surface of the particle is formed by several of the loops between these β-strands, such that differences in length and primary amino acid sequence of these loops gives rise to the marked differences in topology observed between members of separate genera, as shown in **Figure 1**. Virions are resistant to inactivation by organic solvents, thus lack essential lipids, and there is no evidence that any of the capsid polypeptides are glycosylated, although they may be modified post-translationally by phosphorylation. The N-terminal region of the largest structural polypeptide, VP1, is an extension of the common structural region, and, in most parvoviruses, contains a functional phospholipase A_2 (PLA_2) enzymatic core whose activity is essential for escape from endosomal compartment into the cytosol, early in viral entry.

Only two antigenic sites have been identified on the virion surface, as defined by mutations that allow escape from neutralization by monoclonal antibodies. Since this relatively simple antigenic structure is very stable, serotype has been a useful adjunct to sequence-based phylogenetic analysis for taxonomic classification, particularly for the parvoviruses of vertebrates. Parvoviruses encode several well-recognized functional protein domains that might serve as linearly descended evolutionary 'tags'. Foremost among these is the relatively contiguous set of well-defined functional subdomains, called Walker boxes, within the SF3 helicase domain of the replication initiator protein of all known parvoviruses. Interestingly, SF3 helicases have also been identified within genes encoded by DNA viruses as diverse as members of the families *Poxviridae, Baculoviridae, Papillomaviridae, Polyomaviridae*, and *Circoviridae*, as well as in the genomes of small RNA viruses such as members of the families *Picornaviridae* and *Comoviridae*. However, this class of helicase has not been found encoded in cellular genomes, and their phylogenetic branch within the AAA + ATPase superfamily diverged from the rest of the tree of life before the separation of the archaea, bacteria, and eukarya. This suggests that the SF3 class of helicases might have originally evolved in primitive replicons that are only represented by viruses in the present biosphere, and that the SF3 helicase

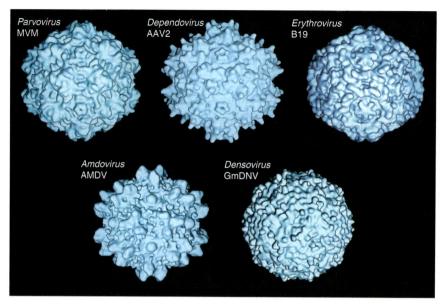

Figure 1 Molecular topography of representative virions from the family *Parvoviridae*. Low-resolution surface maps of minute virus of mice (MVM), adeno-associated virus-2 (AAV2), B19, mature aleutian mink disease virus (AMDV), and galleria mellonella densovirus (GmDNV), at 13 Å resolution. The surface map image of AMDV was generated from a pseudoatomic model built into cryo-EM density, while the others were generated directly from atomic coordinates, as described by Agbandje-McKenna and Chapman in 2006, and were graciously provided by Lakshmanan Govindasamy and Mavis Agbandje-McKenna.

domain evident in present-day parvoviruses might represent an uninterrupted vertical link to the first common ancestor of the entire family *Parvoviridae*.

The family *Parvoviridae* is divided into two subfamilies, the subfamily *Parvovirinae*, whose members infect vertebrate hosts, and the subfamily *Densovirinae*, whose members infect arthropods. Using DNA sequence-based phylogenetic analysis, members of the subfamily *Parvovirinae* have been divided into five genera, and those of the subfamily *Densovirinae* into four genera, as listed in **Table 1**. As listed in **Tables 2** and **3**, these nine genera contain 34 species accepted as such by the International Committee for the Taxonomy of Viruses, with 25 species tentatively assigned to individual genera and a further nine virus isolates currently unassigned within the subfamily *Densovirinae*.

Genera in the Subfamily *Parvovirinae*

Genus *Parvovirus*

Viruses belonging to species within the genus *Parvovirus* have heterotelomeric genomes ∼5 kbp in length. The packaged strands are predominantly negative-sense with respect to their monosense transcription strategy, although parvovirus LuIII virus packages both strands, in separate particles, in an approximately equimolar ratio. The terminal hairpin at the 3′-end of the negative strand, by convention the left-hand end of the genetic map, is 115–121 nucleotides (nt) long, whereas their right-hand hairpins

Table 1 The taxonomic structure of the family *Parvoviridae*

Subfamily	Genus
Parvovirinae	Parvovirus
	Dependovirus
	Erythrovirus
	Amdovirus
	Bocavirus
Densovirinae	Densovirus
	Pefudensovirus
	Iteravirus
	Brevidensovirus

vary between 200 and 248 nt in length. As shown in **Figure 2**, there are two transcriptional promoters, at map units ∼4 and ∼40, from the left-hand end, and transcripts co-terminate at a single polyadenylation site near the right-hand end of the genome. Many viruses belonging to this genus encode a second, smaller, nonstructural protein, NS2, in addition to the major DNA replication initiator protein NS1. NS2 is involved in several aspects of viral replication and capsid assembly, but is dispensable in some cell types. Infecting virions lack accessory proteins, chromatin or even a duplex transcription template, and therefore remain silent within their host cell nucleus until the cellular synthetic machinery manufactures a complementary DNA strand, creating the first transcription template. Typically, this occurs when the host cell enters S-phase, of its own volition, and it is followed rapidly by expression of viral transcripts driven from the left-hand promoter. These viruses can persist both in nondividing cells in

Table 2   Subfamily *Parvovirinae*

Genus *Parvovirus*
Species
    Minute virus of mice – type species   (MVM)
    Feline panleukopenia virus   (FPV)
    H-1 parvovirus   (H-1PV)
    Kilham rat virus   (KRV)
    LuIII virus   (LuIIIV)
    Mouse parvovirus 1   (MPV-1)
    Porcine parvovirus   (PPV)
    HB parvovirus[a]   (HBPV)
    Lapine parvovirus[a]   (LPV)
    RT parvovirus[a]   (RTPV)
    Tumor X virus[a]   (TXV)
Tentative species[b]
    Hamster parvovirus   (HaPV)
    Rat minute virus 1   (RMV-1)
    Rat parvovirus 1   (RPV-1)
Genus *Dependovirus*
Species
    Adeno-associated virus-1   (AAV1)
    Adeno-associated virus-2 – type species   (AAV2)
    Adeno-associated virus-3   (AAV3)
    Adeno-associated virus-4   (AAV4)
    Adeno-associated virus-5   (AAV5)
    Avian adeno-associated virus   (AAAV)
    Bovine adeno-associated virus   (BAAV)
    Duck parvovirus   (BDPV)
    Goose parvovirus   (GPV)
    Canine adeno-associated virus[a]   (CAAV)
    Equine adeno-associated virus[a]   (EAAV)
    Ovine adeno-associated virus[a]   (OAAV)
Tentative species[b]
    Adeno-associated virus-7   (AAV7)
    Adeno-associated virus-8   (AAV8)
    Adeno-associated virus-9   (AAV9)
    Adeno-associated virus-10   (AAV10)
    Adeno-associated virus-11   (AAV11)
    Adeno-associated virus-12   (AAV12)
    Serpentine adeno-associated virus   (SAAV)
    Caprine adeno-associated virus   (Go.1 AAV)
    Bovine parvovirus type 2[c]   (BPV2)
Genus *Erythrovirus*
Species
    Human parvovirus B19 – type species[d]   (B19V-Au)
Tentative species[b]
    Chipmunk parvovirus   (ChpPV)
    Pig-tailed macaque parvovirus   (PmPV)
    Rhesus macaque parvovirus   (RmPV)
    Simian parvovirus   (SPV)
    Bovine parvovirus type 3[c]   (BPV3)
Genus *Amdovirus*
Species
    Aleutian mink disease virus – type species   (ADV-G)
Genus *Bocavirus*
Species
    Bovine parvovirus 1 – type species   (BPV1)
    Canine minute virus   (CnMV)
Tentative species[b]
    Human bocavirus[c]   (HBoV)

[a]Formally accepted by ICTV, but no representative genomes have been sequenced.
[b]Representative genome sequenced, but not formally accepted as a species by ICTV.

culture and within the intact host animal, and, in the latter case, frequently re-emerge following immunosuppression. Despite convincing serologic and PCR evidence for long-term persistence by many parvoviruses, at present essentially nothing is known of the mechanisms underlying this type of latency. Members of many species within this genus, particularly those whose natural host are rodents, have been found to be markedly oncolytic in transformed human cells, both in culture and in xenotransplanted tumor models.

Genus *Dependovirus*

The genus *Dependovirus*, as the name implies, originally comprised only the helper-dependent adeno-associated viruses. However, phylogenetic analysis now places the autonomously replicating viruses from the *Goose parvovirus* and *Duck parvovirus* species solidly within this genus. Except for members of these two avian parvovirus species, all other viruses belonging to this genus that have been isolated to date are dependent upon helper adenoviruses or herpes viruses for efficient replication. Most primate adeno-associated virus (AAV) serotypes isolated to date appear to be simian viruses. On the other hand, sero-epidemiologic evidence clearly indicates that AAV2 and AAV3 are human viruses, while AAV5, which is most closely related to caprine AAV, has been isolated from humans only once.

So far, all members of this genus have been found to be homotelomeric, and their virions contain equivalent numbers of positive or negative DNA strands, between 4.7 and 5.1 kbp in size. Typically, the AAV genome has TRs of ~145 nt, the first ~125 nt of which form a palindromic hairpin sequence, while members of the autonomously replicating avian parvovirus species have ITRs that are much larger, between 444 and 457 nt. As shown in **Figure 2**, three distinct transcriptional strategies have been elucidated for members of this genus. Two subtypes, represented by AAV2 and AAV5, have three transcriptional promoters (P5, P19, and P40), but differ in the positions of their functional polyadenylation sites. For both of these subtypes there is one polyadenylation site at the right-hand end of the genome, but for the AAV5-like viruses another site, located in the middle of the genome, is also used. The GPV-like viruses have a polyadenylation site arrangement like that of AAV5, but differ from the other two subtypes by not having a functional middle, P19, promoter.

[c]Genome sequenced or partially sequenced by PCR-based virus discovery techniques, but virus not yet isolated physically or biologically.
[d]Three major genotypes, represented by B19V-Au, B19-LaLi, and B19-V9. Species names are italicized, abbreviations for individual viruses are given in parentheses.

Table 3 Subfamily *Densovirinae*

Genus *Densovirus*
Species
   Junonia coenia densovirus – type species
      Junonia coenia densovirus   (JcDNV)
   Galleria mellonella densovirus
      Galleria mellonella densovirus   (GmDNV)
Tentative species[a]
   Diatraea saccharalis densovirus   (DsDNV)
   Mythimna loreyi densovirus   (MlDNV)
   Toxorhynchites splendens densovirus   (TsDNV)
   Pseudoplusia includens densovirus   (PiDNV)

Genus *Pefudensovirus*
Species
   Periplaneta fuliginosa densovirus – type species
      Periplaneta fuliginosa densovirus   (PfDNV)

Genus *Iteravirus*
Species
   Bombyx mori densovirus – type species
      Bombyx mori densovirus   (BmDNV)
Tentative species[a]
   Casphalia extranea densovirus   (CeDNV)
   Sibine fusca densovirus   (SfDNV)

Genus *Brevidensovirus*
Species
   Aedes aegypti densovirus – type species
      Aedes aegypti densovirus   (AaeDNV)
   Aedes albopictus densovirus
      Aedes albopictus densovirus   (AalDNV)
Tentative species[a]
   Penaeus stylirostris densovirus[b]   (PstDNV)
   Aedes pseudoscutellaris densovirus   (ApDNV)
   Simulium vittatum densovirus   (SvDNV)

[a]Representative genome sequenced, but not formally accepted as a species by ICTV.
[b]This is also called infectious hypodermal and hematopoietic necrosis virus (IHHNV) of penaeid shrimps.
Species names are italicized, acronyms for individual viruses are given in parentheses.

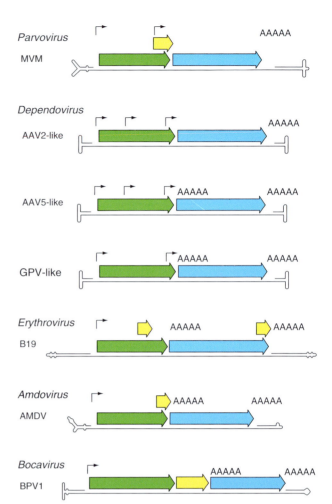

Figure 2 Genetic strategies of representative viruses from genera in the subfamily *Parvovirinae*. Genomes from viruses belonging to the type species of each genus, and an additional two subtypes from the genus *Dependovirus*, are denoted as a single line terminating in hairpin structures. The hairpins are drawn to represent their predicted structures, and are scaled about 20× with respect to the rest of the genome. Open reading frames are represented by arrowed boxes, colored green for the major SF3 domain-containing replication initiator protein, blue for the structural proteins of the capsid, and yellow for the ancillary nonstructural proteins. Transcriptional promoters are indicated by solid arrows and polyadenylation sites by the AAAAA sequence block.

Under certain conditions, such as the treatment of host cells with mutagens or hydroxyurea, AAV DNA replication can be detected in the absence of helper viruses. However, infections without helper virus generally result in a persistent latent infection. Three modes of persistence have been demonstrated for AAV. First, AAV2 and some other serotypes can integrate their genomes site-specifically into a 4 kbp locus on human chromosome 19q13-qter, designated AAVS1. This occurs by a replication-dependent mechanism, requiring low-level expression of Rep, and depends upon sequences in AAVS1 that can function as an AAV DNA replication origin. The second mode, which is independent of Rep gene expression, is primarily seen with recombinant AAV (rAAV) genomes used in gene transduction scenarios, and proceeds through nonspecific integration. Typically, a transgene, flanked by AAV ITRs and packaged in an AAV capsid, is used to infect target tissues in a host animal. Integration of rAAVs occurs at multiple positions throughout the host genome, with a bias toward actively transcribed loci, and is enhanced by the presence of double-strand breaks in host DNA. A third form of persistence, again exploited in rAAV-based gene transduction strategies, results from the establishment of monomeric, and later concatemeric, circular duplex episomes, following vector delivery at high copy number to postmitotic tissues, such as liver or skeletal muscle. Similar episomes have also been detected *in vivo* in latently infected human

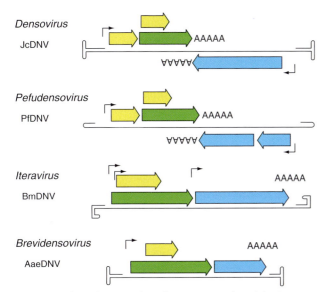

Figure 3 Genetic strategies of a representative of the type species from genera in the subfamily *Densovirinae*. Genomes from viruses belonging to the type species of each genus are diagrammed as in the legend to **Figure 2**.

tissues, although it is not known whether they are created by cellular repair mechanisms or by annealing of complementary strands. Recent studies have shown that the tissue most efficiently targeted by rAAV vectors depends upon the serotype of AAV providing the capsid gene, although a consensus hierarchy of target preference across all of the serotypes continues to evolve.

Genus *Erythrovirus*

Until recently, the only known human pathogenic member of the family *Parvoviridae* was human parvovirus B19 (B19V). This virus is responsible for transient aplastic crisis in patients with a variety of hemolytic anemias, and is the causative agent of *Erythema Infectiosum*, a widespread childhood rash-like disease also referred to as 'fifth disease'. Intrauterine transmission of B19V from an infected mother to her fetus can result in *hydrops fetalis*, particularly in the second trimester of pregnancy. Recently it has become apparent that there are at least three distinct genotypes of B19V, circulating in different human subpopulations. While these are all serotypically equivalent, and thus are considered members of the same species, retrospective PCR analysis of archival tissues has shown that they have predominated in the human population during different periods in the recent past. Additional erythroviruses have been identified in other primate species, which share B19V's specificity for erythroid progenitor cells and cause a similar disease in their hosts.

The erythrovirus genome is homotelomeric, with inverted terminal repeats (ITRs) of 383 nt, and populations of mature B19V virions contain equivalent numbers of positive- and negative-sense DNA strands, each ~5.5 kbp in size. As shown in **Figure 2**, viral transcription is driven by a single promoter at map unit 6 and terminates at alternative polyadenylation sites, one near the middle of the genome, the other at the right-hand end, which generates transcripts encoding NS1 and VP1, respectively. The B19V genomes contain two additional small open reading frames (ORFs), which are accessed by alternatively splicing and translation initiation. One of these encodes an 11 kDa protein containing three proline-rich regions containing consensus Src homology 3 domains, while the other encodes a 7.5 kDa polypeptide of unknown function.

Genus *Amdovirus*

The genus *Amdovirus* contains a single species, *Aleutian mink disease virus*. Mature Aleutian mink disease virus (AMDV) virions contain a predominantly negative-strand, heterotelomeric genome of 4748 nt whose sequence is markedly divergent from other members of the subfamily *Parvovirinae*. As shown in **Figure 2**, the ADMV transcription strategy resembles that of the erythroviruses, in having a single promoter and two alternative polyadenylation sites. The major distinguishing feature of AMDV is its VP1 N-terminus, which is quite short and completely lacks a phospholipase 2A enzymatic core. As seen in **Figure 1**, AMDV virion structure is also somewhat distinctive. Permissive replication has only been achieved in cell culture for the ADV-G isolate, which replicates in Crandell feline kidney cells and is relatively apathogenic in mink. Several highly pathogenic strains of AMDV exist and differ in their disease potential depending upon the age and genetic background of the infected host. Thus infected mink kits develop interstitial pneumonia involving direct attack on type II pneumocytes, while adult mink develop an autoimmune disease characterized by massive hypergammaglobulinemia and fatal glomerulonephritis. As the virus' name suggests, the latter condition is exacerbated in mink carrying the Aleutian coat color allele, which is tightly linked to an antigen presentation disorder in this genetic background. While mostly studied in mink, serologic evidence suggests that this virus also circulates naturally in several other species within the superfamily *Musteloidea*.

Genus *Bocavirus*

The bocaviruses package heterotelomeric genomes ~5.5 kbp in length, and the bovine virus, bovine parvovirus 1 (BPV1), has been shown to package negative-sense and positive-sense DNA strands at a 10:1 ratio. The bocavirus transcription strategy resembles that of the erythroviruses and AMDV in having a single promoter at the left-hand end, and two alternative polyadenylation sites. The bocaviruses differ, however, from all other members of the subfamily *Parvovirinae* in encoding a 22.5 kDa nuclear phosphoprotein, NP1, that is distinct from any other parvovirus-encoded polypeptide. As shown in **Figure 2**, the ORF encoding NP1 sits immediately downstream of that encoding NS1, and immediately upstream of the middle polyadenylation site. NP1, whose function is currently unknown, is abundantly expressed from transcripts that have been spliced to remove the NS1 ORF, and can be polyadenylated at either site. Human bocavirus (HBoV) was discovered in 2005 and subsequent PCR-based surveys have indicated that it is a common human respiratory pathogen, mostly of infants and young children. Although sero-epidemiologic assays remain to be developed, studies so far indicate that HBoV is likely the cause of a substantial proportion of undiagnosed respiratory tract disease, and appears to be distributed worldwide.

Genera in the Subfamily *Densovirinae*

Genus *Densovirus*

The homotelomeric densovirus genome is typically about 6 kbp in length, with long ITRs. The genome contains a >500 nt long ITR, the first ~100 nt of which are predicted to be able to fold as a T-shaped structure. These viruses, along with the pefudensoviruses described below, represent a paradigm shift for the members of the family *Parvoviridae*, since they exhibit ambisense organization. Although they appear to maintain the division of the genome into separate nonstructural and structural gene cassettes typical of all other family members, these cassettes are inverted with respect to one another, as shown in **Figure 3**. Populations of virions encapsidate equal numbers of positive and negative strands. The positive strand, by convention the strand encoding the nonstructural proteins, contains three ORFs, which are predicted to encode three NS proteins, and which are accessed by alternative splicing of mRNAs transcribed rightward from a promoter just within the ITR at the left-hand end. The negative, or complementary, strand expresses four structural proteins by alternative translation initiation, from an mRNA transcribed leftward from the homologous promoter just inside the right-hand ITR.

Genus *Pefudensovirus*

Like the members of the genus *Densovirus*, viruses belonging to the genus *Pefudensovirus* exhibit an ambisense organization. The homotelomeric genome is ~5.5 kbp in length, with a rightward promoter at 3 map units and a leftward promoter at 97 map units, located within the opposing 201 nt ITRs. As can be seen in **Figure 3**, the VP gene is located in the 5′ half of the complementary strand, and is split into a small upstream ORF and a large downstream ORF, which appear to be spliced in order to code for the largest VP, expression of which also appears to require frameshifting after splicing. Unlike other members of the family *Parvoviridae*, in which the PLA2 domain is expressed in the N-terminus of the largest structural protein, in pefudensoviruses this motif is centered 60–70 amino acids from the C-terminus of the ORF predicted to encode the small VP protein. The ORFs predicted to encode the three pefudensovirus nonstructural proteins are organized in the same way as for those of the genus *Densovirus*, and are of similar sizes.

Genus *Iteravirus*

The homotelomeric iteravirus genome is ~5 kbp in length, and populations of virions encapsidate equal numbers of plus and minus strands, in separate particles. The monosense genome, diagrammed in **Figure 3**, has an ITR of 230 nt, the first 159 of which are predicted to fold in a 'J-shaped' hairpin structure. There are two ORFs for nonstructural proteins and one ORF encoding the structural proteins, located on the same strand. A predicted transcriptional promoter resides upstream of each nonstructural ORF, such that each is expressed from its own transcript. A further putative promoter, lying downstream of the nonstructural genes, is predicted to drive production of the structural gene transcript, from which the synthesis of four or five overlapping structural proteins is proposed to occur by a leaky scanning translational initiation process.

Genus *Brevidensovirus*

The monosense genomes of members of the genus *Brevidensovirus* are the smallest within the family, being about 4 kbp in length. This is the only genus within the subfamily *Densovirinae* that have heterotelomeric genomes, and their virions encapsidate predominantly (85%) negative-sense strands. The left-hand end of a representative of the type species, *Aedes aegypti densovirus*, comprises a palindromic sequence of 146 nt while the right-hand end is a palindromic sequence of 164 nt. While different in sequence, both terminal sequences are predicted to fold into a T-shaped structure. As shown in **Figure 3**, the genome contains two transcriptional promoters, at map units 7 and 60. Two overlapping ORFs encoding the nonstructural proteins occupy more than half of the genome, the balance being occupied by a single ORF encoding the structural polypeptides. The genome does not contain any translated sequence recognizable as a PLA2 domain. Viruses belonging to this genus are evolutionarily extremely remote from all other members of the family *Parvoviridae*, and infect arthropod species as diverse as mosquito and shrimp. The latter include viruses that cause hypodermal and hematopoietic necrosis, infecting all multiple organs of ectodermal and mesodermal origin of the shrimp, but not the midgut. These viruses quite closely resemble the mosquito viruses in both genome size and organization. Recently, however, a number of parvoviruses have been isolated from shrimp that infect hepatopancreatic epithelial cells. While these cluster phylogenetically with the brevidensoviruses when their common sequences are compared, they are more than 50% larger, and differ from them substantially in genome organization. Namely, their two nonstructural ORFs are arranged in tandem, and their structural protein genes are also substantially longer, although they are still devoid of a recognizable PLA2 domain. Thus, in contrast to the founding members of this genus, these shrimp viruses have the largest genomes in the family *Parvoviridae*, and it is not yet clear whether they represent a new genus separate from the genus *Brevidensovirus*.

Acknowledgments

The author would like to thank Susan Cotmore, David Pintel, and Peter Tijssen for much helpful discussion, and Lakshmanan Govindasamy and Mavis Agbandje-McKenna for providing the images for **Figure 1**. The author was supported by PHS grants AI26109 and CA29303 from the National Institutes of Health.

See also: Parvoviruses of Vertebrates.

Further Reading

Agbandje-McKenna M and Chapman M (2006) Correlating structure with function in the viral capsid. In: Kerr J, Cotmore SF, Bloom ME, Linden RM, and Parrish CR (eds.) *The Parvoviruses*, ch. 10, pp. 125–139. London: Hodder Arnold.

Carter BJ, Burstein H, and Peluso RW (2004) Adeno-associated virus and AAV vectors for gene delivery. In: Templeton NS (ed.) *Gene and Cell Therapy: Therapeutic Mechanisms and Strategies*, ch. 5, pp. 71–102. New York: Dekker.

Cotmore SF and Tattersall P (2006) Genome structure and organization. In: Kerr JR, Cotmore SF, Bloom ME, Linden RM, and Parrish CR (eds.) *The Parvoviruses*, ch. 7, pp. 73–94. London: Hodder Arnold.

Cotmore SF and Tattersall P (2006) Parvoviruses. In: DePamphilis M (ed.) *DNA Replication and Human Disease*, ch. 29, pp. 593–608. Cold Spring Harbor, NY: Cold Spring Harbor Laboratory Press.

Qiu J, Yoto Y, Tullis G, and Pintel DJ (2006) Parvovirus RNA processing strategies. In: Kerr J, Cotmore SF, Bloom ME, Linden RM, and Parrish CR (eds.) *The Parvoviruses*, ch. 18, pp. 253–273. London: Hodder Arnold.

Tattersall P (2006) The evolution of parvoviral taxonomy. In: Kerr J, Cotmore SF, Bloom ME, Linden RM, and Parrish CR (eds.) *The Parvoviruses*, ch. 1, pp. 5–14. London: Hodder Arnold.

Tattersall P, Bergoin M, Bloom ME, *et al.* (2005) *Parvoviridae*. In: Fauquet CM, Mayo MA, Maniloff J, Desselberger U, and Ball LA (eds.) *Virus Taxonomy: Eighth Report of the International Committee on Taxonomy of Viruses*, pp. 353–369. San Diego, CA: Elsevier Academic Press.

Tijssen P, Bando H, Li Y, *et al.* (2006) Evolution of densoviruses. In: Kerr J, Cotmore SF, Bloom ME, Linden RM, and Parrish CR (eds.) *The Parvoviruses*, ch. 5, pp. 55–68. London: Hodder Arnold.

Wu Z, Asokan A, and Samulski RJ (2006) Adeno-associated virus serotypes: Vector toolkit for human gene therapy. *Molecular Therapy* 14: 316–327.

Young NS and Brown KE (2004) Parvovirus B19. *New England Journal of Medicine* 350: 586–597.

Pecluvirus

D V R Reddy, Hyderabad, India
C Bragard, Université Catholique de Louvain, Leuven, Belgium
P Sreenivasulu, Sri Venkateswara University, Tirupati, India
P Delfosse, Centre de Recherche Public-Gabriel Lippmann, Belvaux, Luxembourg

© 2008 Elsevier Ltd. All rights reserved.

Glossary

Coiled-coil motif A protein structure in which two to six α-helices of polypeptides are coiled together like the strands of a rope.
Hetero-encapsidation Partial or full coating of the genome of one virus with the coat protein of a differing virus. Also termed transcapsidation or heterologous encapsidation.
Leaky scanning mechanism Mechanism by which the ribosomes fail to initiate translation at the first AUG start codon, and scan downstream for the next AUG codon.
Post-transcriptional gene silencing Mechanism for sequence-specific RNA degradation in plants.
t-RNA-like structure Structure mimicking a t-RNA.
Virus-like particles Consist of the structural proteins of a virus. These particles resemble virions meaning that they are not infectious.

History

Pecluviruses, responsible for the 'clump' disease in peanut (=groundnut, *Arachis hypogaea*), have been reported from West Africa and the Indian subcontinent and contribute globally to annual losses estimated to exceed 38 million US dollars. Pecluviruses also cause economic losses also to other dicotyledonous crops such as pigeonpea, chilli, cowpea, and monocotyledonous crops such as wheat, barley, sorghum, pearl millet, foxtail millet, maize, and sugarcane.

Taxonomy and Classification

The genus *Pecluvirus* comprise two species: isolates from West Africa were grouped under the species name *Peanut clump virus* (African peanut clump virus), and those that occur in India were grouped under *Indian peanut clump virus*. Once the molecular features of the genomes of Indian peanut clump virus (IPCV) and peanut clump virus (PCV) were reported, the ICTV in 1997 assigned them to the newly established genus *Pecluvirus* (siglum from peanut clump). The two viruses differ in host range, antigenic properties, and genomic sequences (see below). Furthermore, the viruses can be distinguished from their geographical location.

Geographic and Seasonal Distribution

IPCV occurs in India in the states of Andhra Pradesh, Gujarat, Punjab, Rajasthan, and Tamil Nadu, and in Pakistan in the provinces of Sindh and Punjab. In Africa, PCV has been reported from Benin, Burkina Faso (Saria, Kamboinsé, Bobo Dioulasso, Niangoloko), Chad, Congo, Côte d'Ivoire, Niger (Sadoré, Maradi), Mali (Segou, Koutiala, Bamako), Gabon, Senegal (Bambey, Cap-Vert, Thies, Sine, Saloum, Pout, Mbour, Kirene), and Sudan. IPCV occurs mainly in the rainy season (July–November) in peanut crops. In post-rainy season (December–March), crops escape the disease. In West Africa, the majority of the peanut crops are grown during the rainy season.

Biological Properties

Host Range and Symptoms

PCV infects economically important crops such as peanut, sorghum, sugarcane, maize, pearl millet, finger millet, and cowpea, whereas IPCV infects peanut, cowpea, pigeon pea, chilli, wheat, barley, pearl millet, finger millet, sorghum, and maize. Both the viruses also infect several monocotyledonous weeds (*Cynodon dactylon, Cyperus rotundus*) that can play a vital role in the survival and dissemination.

Affected plants are conspicuous in the field as a result of their dark green appearance, stunting, and occurrence in patches. IPCV occurs at a high incidence in Rajasthan, India, where peanuts are grown on over 250 000 ha in rotation with irrigated winter wheat and barley and with rain-fed pearl millet during summer. Early-infected plants will not yield and even late-infected crops showed reduction in crop yields up to 60%.

The cereal hosts act as reservoirs of inoculum. The Hyderabad isolate of IPCV (IPCV-H) in young wheat plants up to 3 weeks old induced symptoms similar to rosette caused by soil-borne wheat mosaic virus (SBWMV). Diseased plants are stunted with poorly developed root system. Grain yield losses up to 58% were recorded. Wheat CV RR-21 infected with Durgapura isolate (IPCV-D) showed reduced growth without any overt symptoms on leaves. IPCV-infected barley plants were stunted and bushy, with chlorotic or necrotic leaves, and the majority of these plants died. Those plants that reached maturity produced poorly developed spikes. IPCV-H could also infect finger millet, foxtail, or Italian millet, pearl millet, and sorghum plants. In maize, IPCV-H is responsible for aerial biomass losses up to 33% and grain loss up to 36%.

PCV caused red mottle and chlorotic streaks or stripes on sugarcane with incidence up to 50%. The symptoms varied considerably depending on the cultivar. Yield reductions up to 6% were recorded and the sugar yield was reduced by 14%. In Niger, PCV caused yield losses to sorghum grain up to 62%, to sorghum straw up to 45%, and to pearl millet straw up to 15%, whereas no effect on millet grain yield was observed.

Various of PCV and IPCV were readily detected in the cells of roots, stems, and leaves of systemically infected hosts. PCV particles in wheat cells were found in the cytoplasm, near the nucleus or along the plasmalemma, and arranged in angled-layer aggregates. IPCV and PCV have wide experimental host ranges which include both dicotyledonous and monocotyledonous plants. *Nicotiana clevelandii* × *Nicotiana glutinosa* hybrid and *Phaseolus vulgaris* (Top Crop) are suitable for propagation and as assay/diagnostic hosts, respectively, for IPCV. The isolates of IPCV collected from clump-diseased peanut crops from different locations differed slightly in their host ranges. *Canavalia ensiformis* and *N. clevelandii* × *N. glutinosa* hybrid were found to be useful for distinguishing the IPCV isolates. *Chenopodium amaranticolor, Nicotiana benthamiana, N. glutinosa, P. vulgaris*, and *Triticum aestivum* are of diagnostic value for PCV. The symptoms induced in *C. amaranticolor* by various PCV isolates collected from Senegal, Burkina Faso, and Niger were shown to differ markedly. *Nicotiana benthamiana* and *P. vulgaris* are the propagation species of PCV.

Transmission

Pecluviruses can be transmitted by sap and through peanut seed (PCV up to 6% and IPCV up to 24%). IPCV is transmissible through the seed of pearl millet, finger millet, foxtail millet, wheat, and maize generally at rates of <2%. IPCV has been shown to be transmitted by the plasmodiophorid, *Polymyxa graminis. Sorghum arundinaceum* (bait plant) is a suitable host for testing vector transmission of PCV to roots. Convincing evidence for PCV transmission by *Polymyxa* is still lacking. The thick-walled resting spores of the vector probably carry the virus and contribute to its survival. *P. graminis* from tropical and subtropical regions differed in ribotype and temperature requirements. For North American and European *P. graminis* f. sp. *temperata*, optimum temperature for survival is between 15 and 20 °C as opposed to the narrow optimum temperature range (close to 30 °C) for the Indian *P. graminis* f. sp. *tropicalis*. *Polymyxa graminis* in the tropics has a wide host range, including monocotyledonous and dicotyledonous plants, as opposed to the narrow host range (largely restricted to monocotyledons) for *P. graminis* f. sp. *temperata*. The plasmodiophorid needs a cereal host for completing its life cycle.

Serological Diagnosis

PCV is highly immunogenic and its polyclonal antibodies did not react with IPCV, barley stripe mosaic (BSMV), tobacco mosaic (TMV), beet necrotic yellow vein, potato mop top (PMTV), and SBWMV viruses. Wide serological diversity exists among isolates of PCV and IPCV. Therefore serological tests have limitations to detect more than one isolate from a single antibody source. Antisera to different isolates of IPCV facilitated the grouping of isolates into three serotypes, IPCV-H, IPCV-D, and IPCV-Ludhiana (IPCV-L). All IPCV serotypes are serologically distinct from PCV isolates, and vice versa. Utilizing four different formats of enzyme-linked immunosorbent assay (ELISA) and a panel of monoclonal antibodies raised against a PCV isolate, a number of PCV isolates were grouped into five serotypes. Interestingly, one of the monoclonal antibodies reacted with IPCV-D in triple antibody sandwich ELISA.

Molecular Diagnosis

To detect pecluviruses in disease surveys, to eliminate virus-contaminated sources in quarantine, and to devise strategies for disease management, it is essential to utilize diagnostic tools that are highly sensitive and broadly specific. Nucleic acid probes (both radioactive and nonradioactive) for the conserved regions were found to be ideal for pecluvirus diagnosis. Initially probes derived from the 742 nt at the 3′ end of IPCV-RNA1 were used. Subsequently, probes targeting the P14 gene and a probe (CGAGCCATAGAGCACGGTTGTGGG) derived from the conserved 3′ terminal ends of both RNA1 and RNA2 of IPCV facilitated highly sensitive detection of a range of IPCV and PCV isolates. Of the various methods tested, reverse transcription-polymerase chain reaction (RT-PCR) was found to be the most suitable one.

Molecular Properties

Virus Particles

ICPCV virions are nonenveloped rigid rods, 24 nm in diameter, with two predominant lengths of *c*. 250 and *c*. 180 nm. PCV particles contain two predominant rigid rods measuring *c*. 245 and *c*. 190 nm and 21 nm in diameter with a clear axial canal. They contain a single coat protein (CP) of 23 kDa.

Genome

The pecluvirus genome is bipartite. PCV and IPCV virions contain positive-sense, single-stranded RNAs (4% by weight) RNA1 is *c*. 5900 nt long and RNA2 is *c*. 4500 nt long. The RNAs of PCV showed little sequence identity with those of IPCV, with the exception for the 3′ terminal 273 nt, which are conserved between the two RNAs (**Figure 1**).

Coding Sequences

RNA1 encodes two proteins involved in viral RNA replication (P131 and P191). P191 is a C-terminally extended form of P131 produced by translational readthrough of the UGA termination codon (P131) (**Table 1**). Additionally RNA1 codes for P15, a suppressor of post-transcriptional gene silencing (PTGS). The P15 open reading frame (ORF) is downstream of the P191 ORF and separated from it by a noncoding region of about 60 nt. The amino acid sequence of P191 contains methyltransferase, helicase, and RNA-dependent RNA polymerase domains. P15 on RNA1 is a cysteine-rich protein (CRP), translated from a relatively abundant sub-genomic RNA1 (**Figure 1**). P15 resembles CRPs of BSMV, poa semilatent virus, and SBWMV. These proteins have been suggested to act as regulatory factor during virus replication as well as for long-distance movement and contribute to virulence factor. RNA1 is able to replicate in the absence of RNA2 in protoplasts of tobacco BY-2 cells. However, both RNA1 and RNA2 are needed for infection. Experiments using enhanced 5′ green fluorescent protein and 5′-bromouridine 5′-triphosphate labels have suggested that PCV replication complexes co-localize with endoplasmic reticulum green fluorescent bodies accumulating around the nucleus during infection. P15 does not act directly at sites of viral replication but intervenes indirectly to control viral accumulation levels. It acts as a suppressor of PTGS, though it shares no sequence similarities with previously described anti-PTGS molecules encoded by other viruses. The P15 possesses four C-terminal proximal heptad

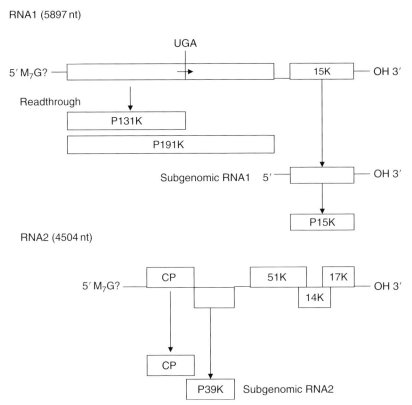

Figure 1 Illustrates the organization of the pecluvirus genome.

repeats that can generate a coiled-coil interaction and is targeted to peroxisomes via a C-terminal SKL motif. Such a motif is conserved among pecluviruses from both Africa and India. It has been demonstrated that a coiled-coil motif is necessary for the anti-PTGS activity of P15, but the peroxisomal localization signal is not, although it is required for efficient intercellular movement of the virus.

RNA2 is relatively more complex and encodes five polypeptides (**Figure 1**). The 5′-proximal ORF (23K) encodes the CP. The following ORF (ORF2) encodes P39, a putative vector transmission factor, which is expressed by a leaky scanning mechanism *in vitro*. ORF2 starts 1 nt upstream of the first residue of UGA stop codon of the CP cistron. The remaining three ORFs encode P51, P14, and P17 and form a triple gene block (TGB) (**Table 1**). TGB plays a role in virus movement. P51 is translated from a relatively abundant subgenomic RNA2, but subgenomic RNAs responsible for the synthesis of P14 and P17 have not yet been detected. However, analogy with other TGB-containing viruses suggests that both proteins are probably translated from a low-abundance subgenomic RNA with a 5′ terminus upstream of the P14 gene.

Noncoding Sequences

The 5′ and 3′ noncoding regions (NCRs) of RNA1 are about 130 and 300 nt in length, respectively (**Figure 1**).

Those of the RNA2 are more diverse, between c. 390 and c. 500 nt in length. There is no distinct 5′ sequence feature common to all pecluviruses. RNA1 and RNA2 have similar 5′ NCRs except for six to seven nucleotides and these sequences are shared among pecluvirus species. The 3′ NCRs are c. 300 nt in length, and c. 100 terminal nucleotides are identical among pecluvirus RNAs sequenced so far. Such sequence similarity has enabled the development of a hybridization probe corresponding to the 3′-terminal 700 nt of IPCV-H. This probe detected all the currently known IPCV serotypes as well as an isolate of PCV.

The 3′ NCR of pecluviruses, as in the case of furoviruses, forms a t-RNA-like structure (TLS) that is capable of high-efficiency valylation and aids in the replication of both PCV RNAs. The internal NCRs in RNA1 are present between P191 and P15 (60 nt) and in RNA2 between P39 and P51 (145 nt).

Sequence Comparisons

The RNA1-encoded polypeptides of PCV and IPCV share identities ranging from 75% (P15) to 95% (readthrough part of P191) and show significant similarities with furoviruses (e.g., 56% identity with polymerase of SBWMV). The proteins encoded by RNA2 are 39% (P39) to 89% (P14) identical between species. The CPs are

Table 1 Pecluvirus open reading frames, polypeptides, and their functions

Genomic RNA	ORF	M_r of polypeptide (kDa)		Function
RNA1	P131	131	}	Methyltransferase, helicase, replicase
	P191	191		
RNA2	P15	15		Suppressor of PTGS
	P23	23		Coat protein
	P39	39		Putative vector transmission factor
	P51	51	} Triple gene block	
	P14	14	}	Virus movement
	P17	17	}	

c. 60% identical and also have significant similarity (c. 30% identity) with the CP of BSMV (genus *Hordeivirus*). The TGB proteins resemble those of PMTV (genus *Pomovirus*).

The nucleotide sequence of IPCV-H RNA1, is similar to that of PCV and the polypeptides encoded by this are 60–95% identical. Comparison of the P15 gene shows a close relationship between IPCV isolates and a relatively high diversity among PCV isolates (**Figure 2**).

The five RNA2-encoded ORFs of IPCV-L are between 32% and 93% identical to those encoded by PCV RNA2. The partial nucleotide sequences of RNA2 of IPCV-H and -D showed that the polypeptides encoded by the two 5′-proximal ORFs (CP and P39) are similar to those of the IPCV-L serotype. A conserved motif 'F-E-X_6-W' is present near the CP C-terminus of all three IPCV serotypes and PCV, as in the CPs of other rod-shaped viruses (TMV and tobacco rattle virus).

The full-length sequence comparison between RNA2 of four isolates of PCV and two isolates of IPCV have revealed a high degree of variability in size (between 58% and 79%). Amino acid sequence alignments of each of the five ORFs of RNA2 showed that ORF4, encoding P14 of TGB, is highly conserved (90–98% identical), whereas the P39 encoded by ORF2 is less conserved (25–60% identical). The CP of eight isolates showed amino acid sequence identities between 37% and 89%. Phylogenetic comparisons, based on complete RNA2 sequences, showed that the eight isolates could be grouped into two distinct clusters with no geographical distinction between PCV and IPCV isolates. Phylogenetic tree topologies for individual ORFs revealed an overall similarity with that obtained from complete RNA2 sequences, but the relative positions of individual isolates varied within each cluster. Further, such studies indicate that there is substantial divergence among the RNA2's of pecluviruses and suggest that different polypeptides have evolved differently, possibly due to different selection pressures.

Several PCV isolates propagated in *N. benthamiana* contain an RNA2 shorter than that of the type isolate. Partial characterization of two such isolates revealed that their RNA2's have undergone deletions in ORF2. The impact of deletions in ORF2, implicated in vector transmission, is to be established.

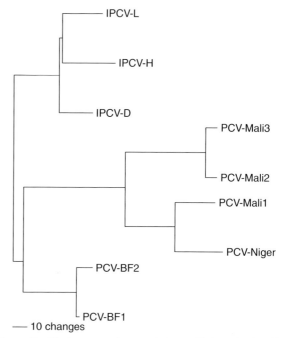

Figure 2 Maximum parsimony phylogenetic tree obtained by a heuristic search from 352 characters of the P15 ORF of IPCV and PCV. Multiple alignments of nucleotide sequences of the P15 ORF from different sources of IPCV (D, H, L serotypes) and PCV isolates from Mali, Burkina Faso (BF), and Niger were obtained using CLUSTAL W 2.08b with the suggested default settings. Phylogenetic analyses were performed using PAUP 4.0 beta 1 (Sinauer Associates, Inc., Sunderland, MA) (Dieryck and Bragard, unpublished data).

Assembly of Virus Particles

The origin of assembly sequence (OAS) positions has been identified in the ssRNA genomes of several rod-shaped viruses. By testing the ability of different RNA1 and RNA2 deletion mutants to be encapsidated *in vivo*, the RNA1 and RNA2 sequences required for assembly into PCV virions have been established. A putative OAS was mapped in the 5′-proximal part of the P15 gene of RNA2. Nevertheless, the nonencapsidation of subgenomic RNA that encodes P15 raises questions about the mechanism underlying the encapsidation process. Two sequence positions that could drive encapsidation of RNA2 have been identified. One is in the 5′-proximal CP gene and the other in the P14 gene near the RNA2 3′ terminus. No obvious sequence similarities between different assembly initiation sequences have been noted. The initiation of PCV assembly, like that of TMV, probably involves interaction of CP with a relatively short sequence which is presented by an RNA secondary structure in a special configuration.

Interestingly, the possible localization of an OAS in the CP gene of IPCV was realized as a result of formation of virus-like particles (VLPs) in *Escherichia coli* and *N. benthamiana*, confirming the results observed for PCV.

The assembly of VLPs in either bacteria or plants has been proposed as a way to protect and accumulate specific mRNA, as a means to study the molecular assembly of the capsid and for the production of oral vaccines. In genetically transformed *N. benthamiana* or *E. coli* cells, the cloned IPCV-H CP gene is expressed and assembled into VLPs. The monomer VLP size approximately corresponds to the one expected, according to the length of the encapsidated CP gene transcript RNA. Using immunocapture RT-PCR(IC-RT-PCR), such VLPs have been demonstrated to contain RNA encoding IPCV-H CP. When transgenic *N. benthamiana* expressing IPCV-H CP gene inoculated with the serologically distinct IPCV-L serotype, accumulated virus particles that contained both types of CP, the possibility of hetero-encapsidation in transgenic plants was suggested.

Epidemiology and Management

Introduction

The control of peanut clump depends upon the accurate and sensitive detection of IPCV and PCV in plants, seeds, and soil. Peanut clump disease is largely restricted to sandy soils and sandy loams. Detection is needed to identify infested fields and implement appropriate management strategies; second, to eliminate seed lots infected by the viruses and to assess the resistance of peanut, pearl millet, sorghum, and sugarcane breeding lines.

Life Cycle

Dicotyledonous hosts restrict the multiplication of the plasmodiophorid vector and hence are considered as fortuitous hosts that may not contribute to perpetuation of virus inoculum. Indeed, either virus-infected peanut roots or seed could transmit or establish the disease. Monocotyledonous plants such as maize, pearl millet, and sorghum are 'preferred' hosts of the vector and contribute to the build-up of vector inoculum potential in the soil. Seed of millets, maize, and wheat and rhizomatous grasses (e.g., *Cynodon dactylon*) are likely to contribute to the disease establishment in new areas by supporting the multiplication of both the virus and plasmodiophorid vector.

The role of rainfall and temperature in the dynamics of infection by IPCV-H and its vector were analyzed on various crops. Wheat followed by barley showed the highest virus incidence, although *P. graminis* was rarely observed in the roots of wheat and was not detected in those of barley. The roots of maize, pearl millet, and sorghum plants, colonized by *P. graminis*, showed the presence of the virus. Peanut is a systemic host for the virus but no *P. graminis* was found in its roots. High rainfall soon after summer months resulted in high incidences of the disease. Weekly rainfall of 14 mm is sufficient for the vector to initiate infection. Temperatures (27–30 °C) prevailing during the rainy season were found to be conducive to virus transmission. At temperatures below 23 °C, infection did not occur and the development of the plasmodiophorid was delayed. This appears to be the major reason for the absence of clump disease on crops raised during the post-rainy season in India.

Cultural Practices

Continuous cropping with fortuitous hosts such as peanut, cowpea, or pigeonpea is likely to reduce the plasmodiophorid population in soil. Seed-borne inoculum from dicotyledonous hosts does not aid in disease establishment. However, seed-borne inoculum from cereal hosts and rhizomatous grasses can contribute to disease establishment.

Initial experiments showed that application of soil biocides (e.g., dibromochloropropane, DD, fumigant nematicides that also have fungicidal action) and soil solarization were effective in reducing disease incidence. Nonetheless, they are not economical to adopt and additionally are hazardous to use.

The following cultural practices can reduce disease incidence of IPCV:

1. early planting of groundnut before the onset of the monsoon under judicious irrigation;
2. trap cropping with pearl millet, that is, sowing a pearl millet crop at a high density and then ploughing the entire crop, 2 weeks after germination, into the soil and then planting with peanut;
3. avoiding rotation of peanut with such highly susceptible crops as sorghum, wheat, and maize; and
4. maintain continuous cropping with dicotyledonous crops (peanut, pigeonpea, cowpea, and marigold) for at least three growing seasons.

The above-recommended measures are ecofriendly and economical and are practicable even under marginal farming conditions.

Host Plant Resistance

No resistance to IPCV was found in over 9000 cultivated *Arachis* germplasm lines. Resistance to IPCV was identified in a wild *Arachis* sp., but it is yet to be incorporated into cultivated peanut. Transgenic peanut lines carrying virus genes (CP and replicase) are currently being evaluated. However, they are unlikely to be available in the near future for cultivation.

Future Perspectives

The suspected role of P39 in vector transmission of the virus needs to be confirmed probably by mutational analysis to exploit this ORF in transgenic research. The presence of subgenomic RNAs encoding P14 and P17 in infected tissues needs to be verified. The cultural control measures tested for IPCV are worth exploiting in West Africa to minimize the impact of PCV.

The most economical way to control pecluviruses is by developing virus-resistant cultivars. The organizations involved must seek approvals from appropriate licensing agencies so that the transgenic plants will become available for cultivation.

See also: Furovirus.

Further Reading

Bragard C, Doucet D, Dieryck B, and Delfosse P (2006) Detection of pecluviruses. In: Rao GP, Kumar PL, and Holguin-Peña RJ (eds.) *Characterization, Diagnosis & Management of Plant Viruses: Vegetable and Pulse Crops*, 1st edn., vol. 3, pp. 125–140. Houston: Studium Press.

Bragard C, Duncan GH, Wesley SV, Naidu RA, and Mayo MA (2000) Virus-like particles assemble in plants and bacteria expressing the coat protein gene of Indian peanut clump virus. *Journal of General Virology* 84: 267–272.

Delfosse P, Reddy AS, Thirumala Devi K, *et al.* (2002) Dynamics of *Polymyxa graminis* and Indian peanut clump virus (IPCV) infection on various monocotyledonous crops and groundnut during the rainy season. *Plant Pathology* 51: 546–560.

Dunoyer P, Pfeffer S, Fritsch C, Hemmer O, Voinnet O, and Richards KE (2002) Identification, subcellular localization and some properties of a cystein-rich suppressor of gene silencing encoded by peanut clump virus. *Plant Journal* 19: 555–567.

Fritsch C and Dollet M (2000) Genus *Pecluvirus*. In: Van Regenmortel MHV, Fauquet CM, Bishop DHL, *et al.* (eds.) *Virus Taxonomy: Seventh Report of the International Committee on Taxonomy of Viruses*, pp. 913–917. San Diego: Academic Press.

Hemmer O, Dunoyer P, Richards K, and Fritsch C (2003) Mapping of viral RNA sequences required for assembly of peanut clump virus particles. *Journal of General Virology* 84: 2585–2594.

Herzog E, Guilley H, Manohar SK, *et al.* (1994) Complete sequence of peanut clump virus RNA-1 and relationships with other fungus transmitted rod shaped viruses. *Journal of General Virology* 75: 3147–3155.

Herzog E, Hemmer O, Hauser S, Meyer G, Bouzoubaa S, and Fritsch C (1998) Identification of genes involved in replication and movement of peanut clump virus. *Virology* 248: 312–322.

Legrève A, Delfosse P, and Maraite H (2002) Phylogenetic analysis of *Polymyxa* species based on nuclear 5.8S and internal transcribed spacers ribosomal DNA sequences. *Mycological Research* 106: 138–147.

Legrève A, Delfosse P, Vanpee B, Goffin A, and Maraite H (1998) Differences in temperature requirements between *Polymyxa* sp. of Indian origin and *Polymyxa graminis* and *Polymyxa betae* from temperate areas. *European Journal of Plant Pathology* 104: 195–205.

Manohar SK, Dollet M, Dubern J, and Gargani D (1995) Studies on variability of peanut clump virus: Symptomatology and serology. *Journal of Phytopathology* 143: 233–238.

Miller JS, Wesley SV, Naidu RA, Reddy DVR, and Mayo MA (1996) The nucleotide sequence of RNA-1 of Indian peanut clump furovirus. *Archives of Virology* 141: 2301–2312.

Naidu RA, Miller JS, Mayo MS, Wesley SV, and Reddy AS (2000) The nucleotide sequence of Indian peanut clump virus RNA 2: Sequence comparisons among pecluviruses. *Archives of Virology* 145: 1857–1866.

Reddy AS, Hobbs HA, Delfosse P, Murthy AK, and Reddy DVR (1998) Seed transmission of Indian peanut clump virus (IPCV) in peanut and millets. *Plant Disease* 82: 343–346.

Reddy DVR, Mayo MA, and Delfosse P (1999) Pecluviruses. In: Granoff A and Webster RG (eds.) *Encyclopedia of Virology*, 2nd edn., vol. 2, pp. 1196–1200. New York: Academic Press.

Pepino Mosaic Virus

R A A Van der Vlugt, Wageningen University and Research Centre, Wageningen, The Netherlands
C C M M Stijger, Wageningen University and Research Centre, Naaldwijk, The Netherlands

© 2008 Elsevier Ltd. All rights reserved.

Glossary

IEM (immunosorbent electron microscopy) Identification of a virus by visualization in the electron microscope of the reaction of the virus coat with a specific antiserum.

RFLP (restriction fragment length polymorphism) A technique whereby DNA fragments of approximately the same length but with different sequences can be differentiated by the patterns of fragments that are generated by digestion with restriction enzymes.

RT-PCR (reverse transcriptase polymerase chain reaction) A PCR reaction whereby first a cDNA copy of the viral RNA has to be generated by reverse transcriptase (RT) before PCR-amplification is possible.

Virus symptomatology Typical local or systemic symptoms on natural host plants or experimental test plants that can be used to characterize and differentiate plant viruses and plant virus strains.

Introduction

Pepino mosaic virus (PepMV) was first found in 1974 in field samples of pepino plants (*Solanum muricatum*) located near some potato fields, collected in the Canete valley in coastal Peru. These plants showed some yellow mosaic in young leaves. Electron microscope (EM) investigations showed the presence of typical filamentous potexvirus particles of approximately 500 nm, which did not react with antisera against potato virus X (PVX). Serologically, it appeared most closely related to narcissus mosaic virus (NaMV) but the host ranges of both viruses differ considerably. It was concluded that pepino mosaic virus was a new and distinct potexvirus.

After its initial description in 1980 the virus was not again reported and was not considered to be of any agricultural significance until it was again found in 1999 in commercial protected tomato crops (*Lycopersicon esculentum*) in the UK and the Netherlands. Since then the virus has spread very rapidly through commercial tomato crops worldwide and was reported from Belgium, Canada, Chile, China, France, Germany, Italy, Finland, Morocco, Norway, Peru, Poland, Portugal, Spain, the Netherlands, UK, Ukraine, and the USA.

Host Range and Symptomatology

The host range of PepMV is quite narrow and seems mainly restricted to the Solanaceae of which many species are infected systemically. Its main hosts appear to be *Lycopersicon* sp. and *Solanum* sp. The originally described pepino strain was found to infect wild and commercial tuber-bearing *S. tuberosum* sp., mostly with a symptomless systemic infection or with mild mosaic symptoms, but in two local Peruvian *S. tuberosum* cultivars and *S. stoloniferum* PI 230557 it caused a severe systemic necrosis. The virus was shown to be transmitted through the tubers.

Its best-known natural hosts are pepino (*S. muricatum*) and cultivated tomato (*L. esculentum*) but surveys showed infection with PepMV of several related wild *Lycopersicon* sp. These include *L. peruvianum*, *L. parviflorum*, *L. chilense*, *L. chmielewskii*, and *L. pimpinellifolium*. Infections in these wild species were generally symptomless.

Several weed hosts were found to be hosts for PepMV. The original virus description reports weed hosts in Peru, including *Datura stramonium*, *Nicandra physaloides*, and *Physalis peruvianum*. More recent studies on material from mainland Spain and the Canary Islands showed that a large number of weeds species were latently infected (i.e., showing no symptoms).

The initially described PepMV caused a distinct yellow mosaic on young leaves of pepino and most infected plants also showed dark green enations on the lower surface of some leaves. In *Lycopersicon* sp., it caused a symptomless systemic infection as shown by back-inoculation on a sensitive indicator plants such as *Nicotiana glutinosa*. Only *Datura metel*, *D. stramonium*, and a number of *Nicotiana* sp. showed distinct symptoms upon systemic infection. The virus failed to infect 13 plant species in six other families (see **Table 1**).

In contrast to the original pepino strain, the virus found in 1999 in commercial tomato crops in the UK and the Netherlands did cause distinct symptoms in tomato. Symptoms may depend on the tomato cultivar, the age of the plant when first infected, and growing conditions, and included yellow leaf spots, yellow-green mosaic, mottle on the older leaves and/or slight curling in the top leaves (nettlehead symptom), or a grayish appearance of the top of the plant. Fruit symptoms range from alteration in color, mild sometimes concentric yellow/orange mottling, and uneven ripening to netting and cracking and shape distortion. Basically, the tomato strain causes much more pronounced plant symptoms: more severe leaf chlorosis, bright yellow mosaic, leaf distortions ('bubbling'), green striation on the stem and sepals. Some isolates are reported to cause more or less pronounced leaf and stem necrosis. Because of fruit symptoms, fruit quality can be affected which may result in economic damage. Symptom expression is reported to depend on environmental conditions with low temperatures and low light conditions favoring more pronounced symptoms. Symptoms are also reported to become less pronounced when conditions change. There are indications that the different virus strains may induce different symptoms.

Inoculation studies on sets of indicator plants clearly showed a number of differences between the original Pepino strain and tomato isolates (see **Table 1**). Especially, *D. metel*, *D. stramonium*, and *N. glutinosa* can be used to distinguish between strains.

Virion Properties

PepMV has filamentous particles typical of potexviruses with a normal length of 510 nm. Particles are comprised of a single capsid protein (CP) of approximately 26 kDa. Ultrathin section of infected leaf material may contain inclusions consisting of arrays of filamentous virus-like particles. The virus particles are less clearly cross-banded than those of PVX.

PepMV is fairly stable. In endpoint dilution studies, sap from infected *N. glutinosa* was always infectious at

Table 1 Host range and symptoms of different pepino mosaic virus strains and isolates

	Pepino isolate	Dutch tomato type isolate	Italian tom isolate	Spanish tom isolate (Sp13)
Capsicum annuum	−/(vc,vn)	−/−	−/−	−/−
Chenopodium amaranticolor	−/−	−/−	+,cl/−	−/−
C. quinoa	−/−	−/−	+, cl/−	−/−
Cucumis sativus	−/−	−/−	−/−	−/−
Cucurbita pepo			−/−	
Datura stramonium	(nl)/nl	−/c,m	+, cs/+, m	
D. metel	nl/l	−/m		
Lycopersicon esculentum	−/l	−/(m)	+, cs/+, m	−/−
Nicotiana benthamiana	−/cl	−/c	+, cs/+, m	?/m, c,ld
N. bigelovii	nr/m	−/l		
N. clevelandii			+, cl−	?/m, c,ld
N. debneyi	−/l	−/l		
N. glutinosa	−/vc,cr,m	−/−	−/−	−/−
N. hesperis-67A	nl,n/d,nl,n	nl,n/d,nl,n		
N. megalosiphon			+, cs, n/+, m.n	
N. occidentalis-P1	cl,nl/c,d,n,nl	cl,nl/c,d,n,nl		
N. rustica	nl/l	−/m	+, cl−	−/−
N. tabacum 'WB'	−/(vc)	−/−	+, l/−	−/−
Petunia hybrida	−/−	−/−	−/−	
Phaseolus vulgaris	−/−	−/−		−/−
Physalis floridana	−/(cl,m,nl)	−/−		
Solanum melongena	−/l	−/m	+, cs/+, m	
S. muricatum	−/(m)	−/(m)	+, l/ +, l	
S. nigrum			+, cs/ +, m	
S. tuberosum	−/−	−/−	−/−	
Vicia faba	−/−	−/−		

The pepino isolate is the original Peruvian pepino isolate. The Dutch tomato type isolate is the isolate first described from the Netherlands. The Italian tomato isolate is the deviant 'chenopodium' isolate originally isolated from tomato plants. The Spanish tomato isolate is a typical Spanish tomato isolate.
Abbreviations: cl, chlorotic lesions; m, mosaic; n, necrosis; l, latent infection; c, chlorosis; cr, chlorotic rings; d, dwarfing; ld, leaf deformation; mo, mottle; nl, necrotic lesions; vc, veinal chlorosis; vn, veinal necrosis; ( ), symptoms observed occasionally.

dilutions of 10^{-4}, occasionally at 10^{-5} but never at 10^{-6}. Sap lost most of its infectivity after 10 min at 65 °C and was no longer infectious at 70 °C. Sap stored at 20 °C still shows some infectivity after 3 months while leaves of *N. glutinosa* desiccated over silica gel were still infectious after 6 months. Under practical conditions in greenhouses and fields, the virus may easily survive for several weeks in plant debris and on surfaces that have come in contact with virus-infected leaves or fruits.

A polyclonal antiserum raised against the original pepino isolate showed no reaction with PVX and potato aucuba mosaic virus (PAMV), the only two other potexviruses known to infect tomato. Both in immunosorbent electron microscopy (IEM) and in ELISA, it clearly reacted with different tomato isolates. Comparisons of antisera raised against the pepino and tomato strains showed differences in heterologous titers between them.

Genome Organization and Expression

The genome of PepMV resembles that of typical potexviruses. Its positive single-stranded RNA is 6410 nucleotides (nt) long, capped at the 5′ end, polyadenylated at the 3′ end and contains 5′- and 3′-nontranslated regions. It encodes five putative partly overlapping open reading frames (ORFs 1–5, see **Figure 1**).

The 5′-nontranslated region (5′-NTR) of the virus is 85 nt longs and like all other potexviruses it starts with the 5′-GAAAA pentanucleotide.

ORF1 (nt 86–4406) encodes a 164 kDa protein of 1439 amino acids (aa). It contains a putative methyltransferase domain (aa 59–224) specific for the supergroup of 'Sindbis-like' viruses, a NTPase/helicase domain (aa 708–934) with the NTP-binding motifs GCGGSGKS and VVIFDD and an RNA-dependent RNA polymerase (RdRp) domain (aa 1217–1374) characterized by the SGEGPTFDANT-X22-GDD motif.

The stop codon of the first ORF is followed by a short intergenic region (IR1) of 25 nt and a set of three partially overlapping ORFs typically known as the triple gene block (TGB).

ORF2 (nt 4432–5136) encodes the first TGB protein (TGBp1), a 234 aa protein of 26 kDa. This protein contains a typical NTPase/helicase motif (aa 26–233) characterized by seven conserved motifs, two of which may be involved in NTP binding. TGBp1 belongs to the superfamily I of RNA helicases.

Figure 1 Schematic organization of the RNA genome of pepino mosaic virus with its five open reading frames (ORFs). M7G = 5′ cap, AAAAA$_n$ = poly(A) tail.

ORF3 (nt 5117–5488) overlaps 19 nucleotides with the 3′-end of ORF2 and extends 148 nt past the start codon of ORF4. It encodes a small 14 kDa protein of TGBp2 which contains a potexvirus-specific consensus motif: PxxGDxxHxL/FPxGGxYxDGTKxxxY.

ORF4 (nt 5340–5594) encodes the third TGB protein (TGBp3), a 85 aa protein of 9 kDa. This protein is the most variable among potexviruses. It contains a CxV/IxxxG consensus motif among potexvirus TGBp3 proteins.

The second intergenic region (IR2) of 38 nt (nt 5595–5632) precedes ORF5 (nt 5633–6346) which encodes the 238 aa coat protein (CP) of 25 kDa. This CP contains the amphipathic core sequence KFAAFDFFDGVT. A similar sequence is also found in the CP of other potexviruses and might be responsible for binding of virus RNA to the CP through hydrophobic interactions.

The 64 nt long 3′-NTR (nt 6347–6410) precedes the poly(A) tail and contains the hexamer 5′-ACUUAA sequence which is also present in the 3′-NTR of all potexviruses sequenced so far. This motif is proposed to be a *cis*-acting element involved in the positive and negative viral RNA synthesis. The 5′-AAUAAA polyadenylation signal terminates the RNA genome, whereby the AAA portion forms the first A residues of the poly(A) tail.

Virus Isolates and Strains

The tomato isolate of PepMV described in 1999 clearly differed in a number of characteristics from the original pepino isolate from Peru. The differences are most pronounced in the reactions of both isolates on tomato (*L. esculentum*) in which the pepino isolate is symptomless and the tomato isolate causes distinct symptoms. Also on a number of test plants, the two isolates can be distinguished, most clearly on *N. glutinosa* and *D. stramonium* (see **Table 1**).

Since 1999 a large number of different PepMV isolates have been described. These were either isolated from commercial tomato crops in different countries or from wild *Lycopersicon* sp. Most show the typical characteristics of the tomato strain but some deviant isolates were observed. A number of isolates show more severe symptoms (including leaf and stem necrosis) in tomato or wild *Lycopersicon* sp.

Sequence data of most PepMV isolates are available from DNA fragments generated by RT-PCR reactions using degenerate primer sets based either on conserved potexvirus motifs or on PepMV specific sequences. Sequence comparisons of RT-PCR DNA fragments obtained from the RdRp region of the pepino isolate and a number of tomato isolates from different geographical origins show approximately over 99% sequence identity between the tomato isolates and only around 95% between all tomato isolates and the pepino isolate. Other studies confirmed the genetic distance between the original pepino isolate and the tomato isolates. Based on the available molecular and biological data, the tomato and pepino isolates are now considered two distinct strains of PepMV.

Comparison of full-length nucleotide sequences and deduced amino acid sequences of four tomato isolates from France and Spain confirmed the very high levels of sequence identity (over 99%) within the tomato isolates. Full-length sequence data were also obtained from an isolate collected from a wild *L. peruvianum* during a survey in Peru. This isolate induced only very mild mosaic and leaf distortions in *L. peruvianum* and was symptomless in *L. esculentum* upon mechanical inoculation. This isolate shows slightly lower levels of sequence identity with the isolates from the tomato strain (95.2–98.1%).

Interestingly, comparisons of sequence data obtained from RT-PCR fragments from many different tomato isolates identified a number of aberrant isolates. One Italian isolate obtained in 2000 from symptomatic tomato plants in Sardinia is divergent from the tomato and pepino strain in a 295 nt PCR-product derived from the methyltransferase motif in ORF1. Additionally, this isolate was described to infect *Chenopodium quinoa* and *C. amaranticolor*, both of which are not infected by either the tomato or pepino strain, and it also shows a distinctive single-stranded conformation polymorphism (SSCP) pattern. It is not clear whether this chenopodium-infecting isolate truly constitutes a new PepMV-strain.

In the USA the full sequences of two isolates collected in 2000 were obtained directly from dried symptomatic tomato leaves. The sequence of the 3′-end of a third isolate was determined directly from infected fresh tomato fruits. Direct nucleotide sequence comparisons showed only 79–82% overall identity between European tomato isolates from France and Spain and US isolates and 86% overall nucleotide identity between the RNA genomes of both US isolates. The 3′-end of the genome of the third US isolate showed between 92.9% (TGBp3) and 100% (CP) nt sequence identity to the two European tomato strain sequences.

An restriction fragment length polymorphism (RFLP) analysis of PCR products of the RdRp region obtained from 102 tomato samples collected at various locations in Spain had already identified three distinct virus types: the typical tomato and pepino isolates and a clearly distinct third type. The cloned PCR product of isolate 3253, representative of the third type, had 80.5% identity with a tomato strain isolate and 79.3% with the previously mentioned Peruvian LP2001 isolate. Phylogenetic analysis showed it had only 60–69% identity to sequences of other potexviruses and it was most closely related but distinct from both the tomato and pepino PepMV strains. Interestingly, US isolates showed 98% identity to the sequence of Sp3253 indicating a very close relationship between these isolates.

Currently available biological data (i.e., host and indicator plant symptoms), serological relationships, and multiple nucleotide sequence alignments suggest at least three clusters of relationship between the PepMV isolates investigated so far:

1. The tomato (*L. esculentum*) type strain isolates, all highly similar.
2. The pepino (*S. muricatum*) type strain isolates, originally from pepino and wild *Lycopersicon* sp.
3. Isolates US1, US2, and Sp3253.

The status of the Italian *Chenopodium* isolate remains unclear by lack of sufficient biological and molecular data.

Virus Epidemiology

PepMV clearly originates from South America. Studies showed that the virus was widespread in *Lycopersicon* sp., in Peru, even in isolated wild populations, suggesting that Peru might be its center of origin. Interestingly, most infected *Lycopersicon* sp. showed no distinct symptoms. Several studies have confirmed that the original pepino isolate is characterized by the absence of, or only very mild symptoms on commercial tomato crops. In contrast, the strain now known as the tomato strain of PepMV causes more distinct symptoms in tomato. These symptoms were first noticed in commercial tomato crops in 1999 in the UK and the Netherlands and soon after that in many other countries worldwide.

Studies employing either strain-specific primers or comparisons of viral gene sequences revealed that the tomato strain is prevalent in all countries where the virus has been found. All tomato isolates found are highly similar and cannot be classified according to geographic origin, or symptoms on naturally or experimentally infected plants. These data indicate a fairly recent introduction and expansion of the tomato strain into commercial tomato crops, as the most likely cause for the virus epidemic.

Detailed population studies from Spain on material collected between 1998 and 2004 confirmed the prevalence and homogeneity of the tomato strain isolates as well as their presence since 1998. However, there were clear indications for several independent introductions of this virus strain, both on the Canary Islands and the Spanish mainland. Interestingly in addition to the tomato strain, the original pepino strain and the US2 strain were also found, but always in mixed infections with the tomato strain. Both strains have been shown to be present in Spain since 2000.

Although the first known occurrence of PepMV in North America can be traced back to 2000, the virus seems to differ from that in the EU. In addition to US3, an isolate highly homologous to the tomato strain, two distinct strains were found; US1 and US2. Both strains were identified from pooled symptomatic leaf material collected in 2000 in Arizona. No reports are known about the presence of the pepino strain in North America.

The data above suggest that different strains of PepMV, at different occasions, have spread from South America, most likely from different host plants, either pepino, wild *Lycopersicon* sp. or other as yet unidentified hosts, resulting in different introductions of the virus in different parts of the world. A number of virus characteristics are likely to have contributed to this:

- The original pepino strain is (nearly) symptomless in commercial tomato.
- Proper diagnostic tests for the virus only became available after the recognition of the tomato strain in commercial tomato crops.
- As a potexvirus, PepMV is easily mechanically transmissible and remains infectious for several weeks in plant debris or on contaminated surfaces.
- Tomato fruits harvested from infected plants contain high concentrations of virus.

PepMV introductions in commercial tomato crops may thus have remained largely unnoticed until the tomato strain manifested itself. At this moment no data are available on the possible occurrence of this strain in pepino or wild *Lycopersicon* sp. in South America and the origin of this strain remains unknown. It has been suggested that the tomato isolates show an increased fitness and virulence in commercial tomato. This might explain the fast worldwide spread of the virus.

Several studies report mixed infections of the tomato strains with either the pepino strain or US2 strain or of the pepino and US2 strains and in Spain recombinants between the tomato and US2 strain have been identified. It is not known to what extent these mixed infections and recombination events have contributed or will contribute to the spread of the virus.

Virus Damage and Control

Initially, PepMV infections in commercial tomato crops were reported as relatively insignificant with no apparent plant or fruit symptoms and no or very limited yield reduction. Studies from the UK, however, reported significant effects on fruit quality. Smaller-sized fruits with different grades of uneven ripening and discoloration and occasionally mis-shaped fruits, led to production unsuited for the fresh UK market.

Studies indicate that symptoms induction by PepMV is highly dependent on environmental conditions. In Spain, PepMV-infected tomato plants showed only symptoms from autumn through spring and symptoms disappeared with higher temperatures in late spring. Similar effects of high light conditions and high temperatures have been reported from other countries. However, symptoms of PepMV-infected plants can be highly variable, ranging from very mild leaf symptoms to severe leaf and stem necrosis and fruit symptoms. Mild, chlorotic, or necrotic PepMV isolates have been reported and it seems that the capacity to induce different symptoms is a property of each isolate. Mixed infections of different strains of PepMV occur and are likely to be relatively common. The precise effects of the mixed infections on the symptomatology of the virus are unknown, as are the occurrences of these mixed infections. In addition, it should be noted that it is not always clear what the possible contribution of other plant pathogens in the expression of symptoms may be.

PepMV is a mechanically transmitted virus. The most important transmission routes are through contaminated tools, clothes, and surfaces. The virus is relatively stable at room temperature and can survive and stay infectious for several weeks in plant debris and on contaminated surfaces. Fruits from infected plants can contain high concentrations of virus and do not necessarily show symptoms. Long-distance transmission of the virus is likely to occur. Implementation of strict hygiene protocols during the growing season and thorough cleaning of greenhouses at the end of the growing season can effectively control the introduction and spread of the virus.

There are conflicting reports on the possibility of seed transmission. Sensitive techniques like real-time RT-PCR have detected the virus in seed lots harvested from infected plants, but transmission to progeny plants has not been conclusively demonstrated. In Europe, the EC directive 2004/200/EC prohibits the trade of seeds collected from PepMV-infected plants or not officially tested for the absence of PepMV. This effectively eliminates any possible chance of the transmission of the virus through seed lots.

Effective control of the virus in commercial tomato crop should come from the use of virus-free seeds and planting material, strict hygiene measures, and a constant monitoring for possible infections.

See also: Potexvirus.

Further Reading

Jones RAC, Koenig R, and Lesemann D-E (1980) Pepino mosaic virus, a new potexvirus from pepino (*Solanum muricatum*). *Annals of Applied Biology* 94: 61.

Pagán I, Córdoba-Sellés M, Martinéz-Priego L, et al. (2006) Genetic structure of the population of *Pepino mosaic virus* infecting tomato crops in Spain. *Phytopathology* 96: 274.

Van der Vlugt RAA, Cuperus C, Vink J, et al. (2002) Identification and characterization of *Pepino mosaic potexvirus* in tomato. *EPPO Bulletin* 32: 503.

Persistent and Latent Viral Infection

E S Mocarski and A Grakoui, Emory University School of Medicine, Atlanta, GA, USA

© 2008 Elsevier Ltd. All rights reserved.

Concepts

By nature of severity and ready association of symptoms with infection, acute viral diseases have played a dominant role in developing concepts of viral pathogenesis. Acute viral diseases were the first to succumb to vaccines that were based largely on the knowledge that initial exposure to a pathogen leaves the host with immune memory, reducing levels of secondary infection and disease. As a result, vaccination now controls a variety of acute viral diseases. Chronic viral infection underlies a wide variety of other medically important diseases, that either follow directly from primary infection or that require months, years, or decades to develop. Chronic viral infections are often widespread or even universal within a host species; however, disease rarely occurs in more than a small fraction of the infected population. Also, acute as well as chronic disease can follow infection with a single pathogen. For chronic disease, proof of etiology may only become evident once vaccination has successfully controlled the acute disease. Thus, measles vaccination provides protection from acute measles as well as the chronic disease, subacute sclerosing panencephalitis (SSPE). Vaccination against hepatitis B virus

(HBV) prevents acute and chronic hepatitis as well as hepatocellular carcinoma arising from chronic infection. Similarly, vaccination against varicella-zoster virus (VZV) prevents chickenpox (varicella) as well as shingles (zoster) and the neuropathic pain that often follows disease recurrence. Most recently, a human papillomavirus (HPV) vaccine specifically targeting chronic disease states, condyloma and cervical carcinoma, has been developed. Other pathogens associated with significant disease, such as human immunodeficiency virus (HIV), hepatitis C virus (HCV), and a number of herpesviruses (Table 1), have not yet surrendered to vaccination. Antiviral therapies have succeeded in controlling some of these; however, the chronic nature of the underlying infection complicates antiviral strategies and favors selection of drug-resistant progeny within the chronically infected host. The list of chronic diseases associated with long-term viral infection has grown, with experimental models suggesting that chronic virus infection contributes to certain cancers, as well as to chronic diseases such as diabetes and atherosclerosis. Many of these chronic diseases remain uncontrolled, and a link to viral (or microbial) pathogens might open avenues to therapy as well as vaccination.

Acute, Persistent, and Latent Infection

Chronic viral infection is distinguished from acute infection by timing and completeness of clearance. Initial

Table 1 Persistent and latent human viral infection and disease

Virus	Genome maintenance	Cell and tissue tropism	Chronic disease syndrome
Flavivirus			
HCV	Persistent infection	Hepatocyte	Chronic hepatitis[b]
			Hepatocellular carcinoma
Retrovirus			
HIV-1 and -2	Integrated provirus-persistent infection	CD4$^+$ T lymphocyte Mono/Macs/DCs	AIDS[b]
HTLV-1 and -2	Integrated provirus-persistent infection	T lymphocyte	Leukemia
Hepadnavirus			
HBV	Persistent infection	Hepatocyte	Chronic hepatitis[a,b] and hepatocellular carcinoma[a]
Herpesvirus			
HSV-1 and -2	Latent episome	Sensory neuron	Recurrent vesicular lesions[b]
VZV	Latent episome	Sensory ganglion	Zoster (shingles)[a,b]
EBV	Latent episome	B lymphocyte	Lymphoproliferative disease, nasopharyngeal carcinoma, African Burkitt's lymphoma and Hodgkin's lymphoma
CMV	Latent episome	Myeloid progenitor	Congenital disease, opportunistic disease[b]
HHV-6A,-6B, and -7	Latent episome	CD4$^+$ T Lymphocyte	Opportunistic disease
KSHV (HHV-8)	Latent episome	B lymphocyte (?)	Kaposi's sarcoma
Adenovirus			
Adenoviruses	Persistent infection	Adenoid lymphocyte and other sites	(None)
Papovavirus			
JC	Latent episome	Ductal epithelial, lymphoid, astrocyte	Progressive multifocal leukencephalopathy[b]
BK	Latent episome	Ductal epithelial	(None)
HPV	Latent episome	Basal epithelial	Warts, condyloma[a], and cervical cancer[a]
Poxvirus			
Molluscum contagiousum	Persistent infection	Basal epithelial	Skin lesion
Parvovirus			
AAV	Integrated-persistent infection	Epithelial cell and lymphocyte	(None)
B19	Persistent infection	Erythroid progenitor	Erythroblast crisis and persistent anemia
Paramyxovirus			
Measles	Persistent infection	CNS neuron	Subacute sclerosing panencephalitis and inclusion body encephalitis
Prion	Autocatalytic protein	CNS neuron	Transmissible spongiform encephalopathies

[a]Controlled by vaccination.
[b]Antiviral.

exposure to any virus results in primary infection that may be resolved in any of the three general infection patterns (**Figure 1**). One common pattern for viruses is 'acute infection', which may be accompanied by acute disease (**Figure 2**), and is typically controlled by the host adaptive immune response such that the pathogen is completely eliminated by the host response to infection. Immunological memory that initiates following primary infection provides protection from a subsequent, secondary infection. Immunological memory is the reason why vaccines substitute for primary infection and provide long-lived immunity. Exceptions where immunity is not sufficient to protect from reinfection (**Figure 2**), such as the well-known example of annual influenza virus epidemics, may be due to genetic changes in the virus (so-called antigenic 'drift' and 'shift') that allows this RNA virus to escape from existing immunity. Due to the nature of vaccines, immunity from vaccination often wanes with time, such as has been observed with measles, mumps, and, most recently, VZV. A booster vaccination can raise the levels of immunity. Continuous replication, or 'persistent infection' following primary infection (**Figure 1**), follows from incomplete immune clearance, sometimes occurring because infection occurs at an immune-privileged host site, and may continue without disease or in a chronic or progressive disease pattern (**Figure 2**). A biological reservoir of quiescent virus is called 'latent infection' (**Figure 1**). Latent infection is associated with maintenance of viral genome without active viral replication, and occurs in a cellular lineage specific for the particular virus. Latent infection is accompanied by host immune control that completely suppresses continuous replication; however, this surveillance does not prevent latent infection or the sporadic reactivation and recurrent virus replication that is a hallmark of latency (**Figure 1**). Such patterns may be associated with recurrent, chronic, or opportunistic disease patterns (**Figure 2**). Persistence and latency/recurrence patterns sustain pathogens within individuals and improve transmission within populations. The propensity of a particular virus to initiate chronic infection in either of the two general patterns (persistent or latent infection) depends on the type of virus as well as host cell tropism and immune response determinants. The replication scheme and particular cellular niche of the virus are important. A wide range of immunomodulatory viral functions have been identified in latent viruses, and these functions contribute to escape from host immunity. Immunological memory to latent viruses may be dysregulated as has been shown for persistent viruses leaving the host response ineffective in preventing recurrent infection.

Lifelong latent infection depends upon the participation of host cells that are not susceptible to productive replication but that form a biologic reservoir for reactivation, and reactivation can be dictated by changes in status or differentiation state of host cells. It is easy to imagine that latent infection may be established as soon as a virus reaches the appropriate cell lineage, and that this occurs while active replication continues at other body sites. This view of latency focuses on cells rather than the intact host, and follows from the fact that host cell populations may be permissive or nonpermissive. While a reservoir of latently infected cells may certainly be established concurrent with primary infection, the events of acute infection may leave many cells that are abortively infected or that are transiently nonpermissive and do not survive to become a biological reservoir. The term 'latency' best describes the period when actively replicating virus is no longer present anywhere in the host. Latency is a property of some virus groups (**Table 1**) and may be determined through the way the viral genome is maintained in host cells. The nature of genome maintenance allows precise distinction between latent and persistent infection. Finally, it is sometimes difficult

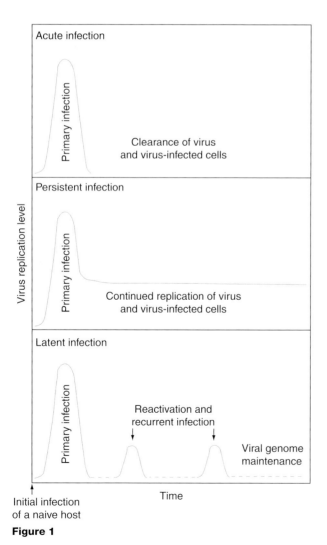

Figure 1

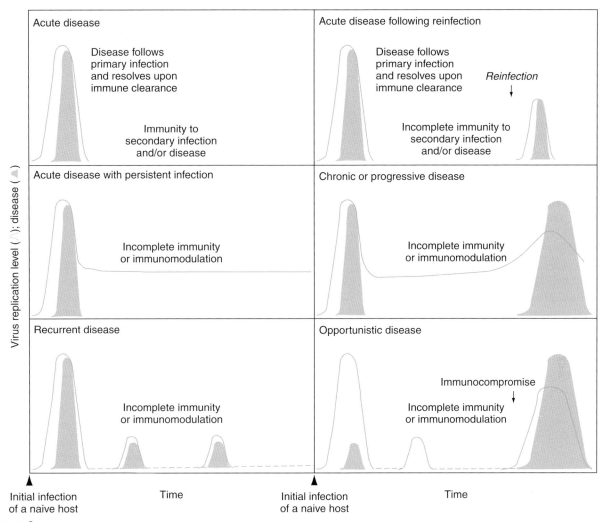

Figure 2

to distinguish abortive infection from latent infection. Abortive infection leaves either the complete viral genome or genome fragments deposited in host cells but these do not constitute a biological reservoir for reactivation, although abortive infection of nonpermissive cells may underlie pathogenesis such as the contribution of some DNA viruses to cancer.

Pathogenesis of Chronic Infection

Pathogenesis refers to the complex biological process of infection in the host animal and is influenced by the levels and distribution of virus during primary infection, just as characterized for acute viral infections. The diversity of viruses is immense; however, every virus in Nature infects only a defined set of susceptible host animals, which represents its species range. Although exceptions exist, human chronic viral pathogens tend to be species-restricted and exhibit evidence of a long co-evolution with the host species. This co-existence and co-evolution as the host species evolve enables these pathogens to reach a balance (or *détente*) with host-clearance mechanisms (cell-intrinsic, innate immune, and adaptive immune responses). The pathogenesis of chronic virus infection depends on the interplay of genetic and environmental determinants that are specific to a particular virus type as well as to the host animal species. Further, there may also be susceptibility determinants that vary between individuals within a particular species. Reactivation and recurrent viral replication are important components of pathogenesis of latent virus infections. Long-term persistent replication or latency may have a dramatic impact on the physiology of cells, altering normal cellular function. Thus, chronic viral damage underlies a variety of known conditions, including immunosuppression, organ dysfunction, and certain cancers, and has been suggested to play a role in many others based on experimental models.

Every member of some animal virus groups, including retroviruses, papovaviruses, adenoviruses, hepadnaviruses, and herpesviruses, relies on persistent and/or latent infection patterns. Some members of other virus groups rely only on persistent infection, sometimes in a host-dependent fashion, including flaviviruses, iridoviruses, and a range of others. In other cases, properties of the host dictate outcome, as with the regulatory mechanisms first characterized in lysogenic bacteriophages, or the persistent infection patterns that are intrinsic to plant and insect viruses. Disease pathogenesis (**Figure 2**) may follow various patterns. (1) Persistent infection may drive disease, such as with HIV-associated acquired immune deficiency syndrome (AIDS) or HCV-associated chronic liver damage and hepatocellular carcinoma. (2) Latent viral infection may alter host cell behavior such as with HPV-associated cervical cancer or Epstein–Barr virus (EBV) lymphoproliferative disease. (3) Reactivation of latent infection may lead to disease such as in herpes simplex virus (HSV)-recurrent cold sores or cytomegalovirus (CMV)-associated opportunistic infections in immunocompromised hosts. With various acute or chronic viral pathogens, reinfection may also contribute to disease patterns; however, existing immunity generally reduces the likelihood of significant disease. Common principles of chronic viral pathogenesis as well as in the diseases that arise from chronic infection have become more widely appreciated.

Immune Control

Immune control of chronic infections relies on cell-intrinsic, innate, and adaptive immunity in mammalian hosts, just the same as acute infections, although the biology of chronic infection relies on pathogen functions that can deflect or modulate host clearance. Cell-intrinsic and innate immunity reduce or delay acute replication levels, and are sometimes even sufficient to prevent disease through well-known mediators such as interferons and natural killer (NK) lymphocytes. The innate immune system is also responsible for recognition and processing of antigens to prime the adaptive immune response to infectious agents. Immune control of infection is influenced by the way in which a pathogen interacts with antigen-presenting cells (APCs): dendritic cells (DCs) and macrophages (Macs). This may occur through direct infection and antigen presentation, or indirectly through pathogen phagocytosis and cross-presentation to lymphocytes in secondary lymphoid tissues to dictate the ultimate breadth and effectiveness of the adaptive immune response.

All host cells sense and mount intrinsic responses to intracellular pathogens such as viruses, triggering cellular signaling cascades, programmed cell death, induction of interferon responses, and repression of viral genome transcription or replication. DCs and Macs, classically considered part of the reticuloendothelial system, produce an abundance of interferons, other cytokines, and a range of physiologically active mediators such as prostaglandins, vasoregulators, and hormones that may all be considered part of the innate response to infection. Cytokines and interferons initially drive the activation of NK lymphocytes, cytotoxic cells that recognize infected cells through reduced major histocompatibility complex (MHC) class I antigen levels, and increased MHC-like stress protein expression at the cell surface. This process is independent of the particular pathogen, and leads to the production of important cytokines that regulate the behavior of other leukocytes. DCs and Macs are considered crucial to innate responses because they acquire virus in the periphery and migrate to secondary lymphoid organs (lymph nodes, tonsils, and spleen) where an adaptive immune response unfolds. As the major producers of interferons and cytokines during viral infection, various subsets of DCs (and likely Macs) carry out the range of pattern recognition, antigen presentation, and co-stimulation activities with lymphocytes to guide both innate and adaptive immune responses. The antigen-specific adaptive immune response plays a crucial role in control of primary infection, as well as in establishing a reservoir of primed lymphocytes that constitute a memory response to secondary infection and prevents disease upon re-exposure to pathogens. Antibodies, produced by B lymphocytes following antigen-driven differentiation into plasma cells with the help of CD4+ T lymphocytes, form one critical arm of the adaptive response. Antibodies bind to and neutralize viruses, trigger complement-dependent lysis of virus-infected cells, and facilitate phagocytosis. Passive antibodies, or vaccines that only induce humoral immunity, are sufficient to control various acute viral pathogens, including influenza virus, paramyxoviruses, rubella virus, picornaviruses, rotaviruses, and vector-borne viruses. The B-lymphocyte response is central to control and clearance and occurs via neutralization of viral infectivity. Adaptive cellular immunity carried out by CD8+ T lymphocytes functioning as cytotoxic cells, often in collaboration with CD4+ cells, form the other critical arm. Control of persistent and latent viral infection, which is highly cell associated, relies on cytotoxic T-cell as well as antibody clearance mechanisms. In these viruses, cellular immunity is typically critical. CD4+ and CD8+ T lymphocytes also produce cytokines that contribute to suppressing replication and may induce immunopathology. There is sufficient evidence to say that both arms of the adaptive immune response facilitate long-lived immunity to reinfection, sometimes dependent on transmission route or virus type, so generalities may not apply to particular viral pathogens. For example, there is a strong interest in antibody-based immunity in control of HIV infection.

Immune Regulation and Immunopathology

The immune response is regulated, distinguishing self from non-self, while responding to infectious agents. Immune response processes may contribute to disease, by the direct recognition of viral antigen or by breaking the natural tolerance to self. Both antigen-specific and nonspecific damage to tissues as well as collateral damage to uninfected tissues may underlie disease. The damage to liver from chronic HBV (or chronic HCV) replication as well as the stromal keratitis due to recurrent HSV infection is triggered by viral infection but proceeds as a result of active antiviral immunity driving immunopathology. Although antibodies and lymphocytes are important effectors of immune control, dysregulation of the immune response during chronic infection may miscue processes that are normally beneficial to the host. Chronic infection can confound one important property of the adaptive immune response and cause a breakdown in the ability of the immune response to discriminate self from nonself. Thus, a balance of immunologic memory and the regulation of immunity, currently believed to be the job of regulatory CD4+ T lymphocytes as well as cell surface proteins that regulate memory T-lymphocytes, can be achieved. The capacity of persistent and latent viruses to deflect or avoid host immune clearance through immunomodulatory functions likely also contributes to infection and disease patterns. Immunologic memory associated with B or T lymphocytes contributes to secondary (anemnestic) responses upon subsequent exposures to the same or closely related viruses. Incompleteness in the immune response opens the way to reactivation as well as reinfection. Chronic viruses (papillomaviruses, retroviruses, herpesviruses, adenoviruses) rely on reactivation, but often benefit from transmission patterns that include reinfection with closely related strains. Immunopathology may be promoted by chronic infection but is by no means a requisite of chronic infection; this depends on the type of virus. Even acute infections that are typically cleared by adaptive immune effector mechanisms may trigger immunopathology upon re-exposure and this may manifest as chronic disease.

Immunomodulation

Like the immune response that the host mounts to control chronic infection, viruses carry an arsenal of defensive functions that undermine the effectiveness of immune clearance. Some of these functions go beyond defense and actively subvert or even exploit intrinsic and innate responses to infection. Although the main challenge to vaccination is often viewed as antigenic variation, such as characterizes HIV, or the number of viral strains in circulation, such as characterizes HCV, viral functions that deflect, subvert, or exploit host responses themselves may undermine vaccine strategies to control chronic infections. When viewing the large number of immunomodulatory functions encoded by large DNA viruses (poxviruses and herpesviruses), or the more restricted numbers encoded by smaller viruses (retroviruses and flaviviruses), the overwhelming impression is that viruses have found diverse ways to deal with a set of common challenges for chronic infection (**Table 2**). In many cases, the pathways that viruses target are critical to clearance, so neither the pathogen nor the host ultimately wins. The standoff that ensues is what characterizes chronic infection. The presence of immunomodulatory functions in a particular virus reveals the importance of specific immune pathways in control of virus infection. Remarkably, viruses with small genomes that encode fewer than 20 genes (retroviruses, papovaviruses, hepadnaviruses, flaviviruses, paramyxoviruses) have succeeded as well as viruses encoding greater than 20 genes (adenoviruses, poxviruses, herpesviruses), suggesting that there are many successful ways to reach this balance and achieve persistent or latent infection. Here, we will discuss a subset of these processes to provide a perspective on this growing field.

Certain cell-intrinsic and innate responses to infection are used throughout invertebrate and vertebrate hosts to detect and eliminate pathogens. Cell stress resulting from sensing an intruder and initiating pattern recognition and signaling cascades are initially important. Intrinsic stress leading to programmed cell death (apoptosis, autophagy, and other forms) is an important and ancient mechanism to get rid of a host of intracellular pathogens. Extrinsic apoptosis due to signaling via tumor necrosis factor (TNF) family ligands is a development of vertebrates to expand the ways that apoptosis can be induced. The induction of apoptosis stops viruses before they have a chance to replicate and disseminate. Virus-encoded cell death suppressors are common (**Table 2**) in persistent and latent virus infection, likely delaying cell death.

The sensing of virus infection by cellular receptors that detect pattern recognition, such as Toll-like receptors (TLRs), are important to induce interferons and subversion of TLR signaling is common in persistent and latent viruses. Various components of TLR signaling pathways, including IRF-3, IRF-7, RIG-I, and NF-κB, may be targeted.

A variety of cellular enzymes may block steps in viral replication. These need to be overcome by viral or cellular functions in order for replication to proceed, so these have been found to play roles in establishing latency. The impact of histone deactylases (HDACs) on herpesviruses is one well-established example, and these likely also play into papillomavirus and adenovirus infection. APOBEC, a cytoplasmic cytidine deaminase, is a potent inhibitor

Table 2 Immunomodulation in chronic viral infection

Virus	Cell-intrinsic	Innate	Adaptive
Flavivirus			
HCV	TLR, RIGI, IRF3, APOP	IFN, NK, APC	VAR, CD8,
Retrovirus			
HIV-1 and -2	TLR, APOP, AUTO, APOBEC	IFN, NK, PKR, CYTK, APC	VAR, MHC-I, CD4
Hepadnavirus			
HBV	APOP	IFN, APC	
Herpesvirus			
HSV-1 and -2	TLR, IRF3, APOP, AUTO, HDAC	IFN, NK, PKR, APC	MHC-I, MHC-II, Ab, C', CD8, CD4
VZV	TLR, IRF3, APOP, HDAC	IFN, NK, APC	MHC-I, MHC-II, Ab, CD8, CD4
EBV	TLR, IRF3, IRF7, APOP, AUTO, HDAC	IFN, NK, CYTK, APC	MHC-I, MHC-II, Ab, CD8, CD4
CMV	TLR, IRF3, APOP, AUTO, HDAC	IFN, NK, PKR, CYTK, APC	MHC-I, MHC-II, Ab, CD8, CD4
HHV-6A, -6B, and -7	TLR, IRF3, APOP, HDAC	IFN, NK, PKR, CYTK	MHC-I, Ab, CD8
KSHV (HHV-8)	TLR, IRF3, IRF7, APOP, HDAC	IFN, NK, PKR, CYTK	MHC-I, MHC-II, Ab, C', CD8, CD4
Adenovirus			
Adenoviruses	TLR, IRF3, APOP, AUTO, HDAC	IFN, NK, PKR, CYTK	MHC-I, CD8, CD4
Papillomavirus			
HPV	APOP, HDAC	IFN, NK, CYTK, APC	MHC-I
Poxvirus			
Molluscum contagiousum	APOP	IFN, NK, PKR, CYTK	MHC-I

Cell-intrinsic: TLR, Toll-like receptor; RIG-I, retinoic acid-inducible gene-I; IRF3/IRF7, interferon-regulated factor-3 and -7; APOP, apoptosis; AUTO, autophagy; APOBEC, family of cytosine deaminases; HDAC, histone deacetylases. Innate: IFN, interferon; PKR, protein kinase R; CYTK, cytokines; NK, natural killer lymphocytes; APC, antigen-presenting cell. Adaptive: MHC-I, major histocompatability class I; MHC-II, major histocompatability class II; Ab, antibody; C', complement; CD8, CD8+ T lymphocytes; CD4, CD4+ T lymphocytes; VAR, variation in genome sequence.

of HIV and restricts viral replication by introducing mutations into the viral genome.

Persistent and latent infections are controlled over the course of the first few days by innate clearance, making subversion of innate clearance mechanisms an important area. Inflammatory cytokines, the most ancient of which are in the interleukin-1 (IL-1) family, but joined by a range of pro-inflammatory cytokines (IL-6, TNF family, interferons), are induced by viral infection with a range of specific as well as nonspecific inhibitors of these pathways.

In addition to the impact on cytokines, NK cells effectively reduce levels of virus-infected cells within the first few days post infection, and some viruses, particularly in the herpesvirus family, have a number of NK subversion functions.

A variety of these subversion pathways can be imagined to be active in infected APCs. For example, some proteins produced by replicating EBV directly inhibit the antigen-presentation pathway by specifically preventing their own presentation on the cell surface, thereby influencing how well the adaptive immune response to the virus is primed. However, many immune-evasion strategies that have been studied in detail have been found to be effective only in somatic cells and are more important in reducing the potency of adaptive immune clearance mechanisms that are induced.

Persistent and latent viruses have found myriad ways to subvert the effectiveness of host immune effector functions, ranging from deflecting antibody and complement to reducing MHC-I and MHC-II levels to make infected cells less recognizable by T lymphocytes. Many viruses make cytokine/cytokine receptor and chemokine/chemokine receptor mimics that likely subvert both innate and adaptive phases of the immune response by altering cell differentiation or migration.

Chronic Disease Manifestations

Latent and persistent viral infections contribute to chronic disease manifestations in a variety of ways including causing (1) congenital infection with long-term sequelae, (2) transformation of infected cells into malignant cells, (3) virus-mediated organ damage, (4) immune-mediated organ damage. Viruses that are able to infect a newborn during delivery or cross the placenta during pregnancy pose both acute and chronic disease risks. Such pathogens may initiate spontaneous abortion or cause progressive diseases in newborns. The herpesvirus, HSV, causes a rare but fatal (if untreated) systemic infection with neurological damage in newborns exposed at delivery. Cytomegalovirus (CMV) is a common congenital viral infection worldwide that may cause severe neurological disease, but more often causes subtle, progressive hearing, eyesight, and learning disabilities. This virus causes congenital infection with chronic progressive disease consequences. Neonates born to mothers acutely infected with CMV acquire the infection through transplacental transmission. Although the mother's infection is often asymptomatic, the infant with severe congenital CMV disease (also called symptomatic infection) may have microencephaly (an abnormally small head), hepatosplenomegaly (an enlarged spleen and liver), a bruise-like rash, retinitis, and progressive central nervous system complications including deafness, blindness, and psychomotor retardation. More than half of those infants with symptomatic CMV infections at birth will have neurologic complications later in life. Immunocompromised patients are susceptible to reactivation of a wide variety of latent viruses, with CMV most frequently observed. Other herpesviruses, adenoviruses, and papovaviruses may also cause chronic disease manifestations.

In some viruses, proliferative diseases and malignancies may result directly during latent infection, without any requirement for reactivation or replication. Viruses such as EBV and HPV directly transform cells in which they reside latently. Thus, EBV specifically targets naive B lymphocytes, converts them to a memory phenotype as if they had encountered antigen, and immortalizes these cells such that they remain latent hosts for life. EBV has several distinct viral latency 'programs' characterized by differential viral protein expression and associated chronic lymphoproliferative diseases as well as malignancies including endemic Burkitt's lymphoma, some varieties of Hodgkin's lymphoma, NK/T-cell lymphomas, and nasopharyngeal lymphoma. HPV is a second example of a virus causing both transformation of infected cells into malignant cells and virus-mediated organ damage depending on the viral genotype. For example, HPVs 16 and 18 are considered high-risk genotypes with propensity to lead to malignant transformation and are implicated in the development of cervical cancer. The E6 and E7 proteins from these high-risk genotypes are oncogenic, whereas the E6 and E7 proteins from lower-risk genotypes are not. Some HPV types have virtually no oncogenic potential but cause chronic damage to the skin in the form of plantar warts (HPV 1) or common warts (HPV 2). Similarly, HSV can be a chronic problem as a cause of recurrent aphthous stomatitis or painful genital mucosal ulcerations. Although the chronic damage by low-risk HPV or mucosal ulcerations of HSV may not be life threatening, other viruses can directly and indirectly cause grave chronic illnesses. Although the mechanism is unclear, parvovirus B19 has been associated with cessation of red blood cell, white blood cell, and platelet production by the bone marrow. This aplastic anemia can be fatal. Certainly HIV, known to rapidly deplete the host of its important CD4+ T lymphocytes, progresses to life-endangering AIDS when untreated. Even when the virus itself is not implicated in causing cytopathic damage to host organ systems, the immune response to the viral infection may instead wreak havoc on critical tissue. For example, HBV and HCV, while not strictly cytopathic themselves, lead to persistent inflammation in the liver with the host immune response largely responsible for the eventual development of cirrhosis and end-stage liver disease. HCV-related end-stage liver disease is now the single leading indication for liver transplantation in the United States illustrating the impact of persistent infection on public health.

Many viruses are controlled adequately in the immunocompetent host only to cause significant disease in immunocompromised hosts, and in this instance are referred to as opportunistic infections. There are many reasons why a host may be immunocompromised including HIV disease in which there is a progressive loss of CD4+ T-cell function, iatrogenic immunosuppression to maintain organ viability following transplantation, or chemotherapy to suppress cancerous cell growth. HIV patients with very low CD4+ T-cell counts, rendering them severely immunocompromised with AIDS, often manifest with multiple opportunistic infections. For example, whereas HSV usually causes self-limited disease processes in most immunocompetent hosts, in immunosuppressed hosts, HSV can cause an AIDS-related encephalitis. Similarly, HHV 8 has been implicated in the development of Kaposi's sarcoma in AIDS patients. CMV is an example of a virus that is usually well controlled by immunocompetent hosts but presents life-threatening infection in immunosuppressed transplant patients. HPV can similarly lead to development of verruca and condyloma recalcitrant to treatment in the immunosuppressed host with more frequent development of HPV-related carcinoma; and the poxvirus responsible for molluscum contagiosum generally causes skin lesions only in children, whose immune systems are still developing, and adults who are immunocompromised. Finally, hosts co-infected with HIV and HCV demonstrate more rapid development of HCV-related liver disease than immunocompetent patients.

In summary, the interplay between viruses and the infected host determines whether a virus is acutely resolved, becomes latent or persistent, and causes minimal or significant disease manifestations. Many challenges lie ahead to determine strategies sufficient to alter viral pathogenesis, augment host immunity, and change the course of virus-induced disease manifestations. These strategies may ultimately be in the form of prophylactic or therapeutic vaccination or may capitalize on host immune response modulation.

Acknowledgments

The authors acknowledge support from the Cancer Research Institute Investigator Award (A. Grakoui), Woodruff Health Sciences Fund (A. Grakoui and E. S. Mocarski), the Georgia Cancer Coalition (E. S. Mocarski), and the USPHS (A. Grakoui and E. S. Mocarski).

See also: Apoptosis and Virus Infection.

Further Reading

Biron C and Sen G (2006) Innate immune responses to viral infection. In: Knipe DM, Howley PM Griffin DE, *et al.* (eds.) *Fields Virology,* 5th edn., pp. 249–278. Philadelphia: Lippincott Williams and Wilkins.

Braciale T, Hahn YS, and Burton D (2006) Adaptive immune responses to viral infection. In: Knipe DM, Howley PM, Griffin DE, *et al.* (eds.) *Fields Virology,* 5th edn., 279–325. Philadelphia: Lippincott Williams and Wilkins.

Virgin S (2006) Pathogenesis of viral infection. In: Knipe DM, Howley PM, Griffin DE, *et al.* (eds.) *Fields Virology,* 5th edn., pp. 2701–2772. Philadelphia: Lippincott Williams and Wilkins.

Phycodnaviruses

J L Van Etten, University of Nebraska–Lincoln, Lincoln, NE, USA
M V Graves, University of Massachusetts–Lowell, Lowell, MA, USA

© 2008 Elsevier Ltd. All rights reserved.

History

Since the early 1970s, viruses or virus-like particles (VLPs) have been been reported in at least 44 taxa of eukaryotic algae, which include members in 10 of the 14 classes of algae. However, most of the early reports described single accounts of microscopic observations. The VLPs were not characterized because they were difficult to obtain in reasonable quantities. Several factors contributed to the low virus concentrations: (1) often only a few algal cells contained particles; (2) usually the cells only contained particles at one stage of the algal life cycle; (3) cells containing particles tended not to lyse; (4) in most cases the particles were not infectious; and (5) some hosts could not be cultured easily.

However, this situation began to change with the discovery of a family of large double-stranded DNA (dsDNA)-containing viruses that infect and replicate in certain strains of unicellular, eukaryotic, exsymbiotic, chlorella-like green algae. The first such 'chlorella viruses' were discovered in 1978 in chlorella symbiotic with *Paramecium bursaria* and in 1981 in chlorella symbiotic with the green coelenterate *Hydra viridis*. The algae from *P. bursaria* can be grown free of the paramecium in culture, and these cultured, naturally endosymbiotic *Chlorella* strains (NC64A and Pbi or their equivalents) serve as hosts for many similar viruses. The lytic chlorella viruses can be produced in large quantities and assayed by plaque formation using standard bacteriophage techniques. Recently, a plaque-forming virus that infects chlorella symbiotic with the heliozoon *Acanthocystis turfacea* was described. This virus does not infect *Chlorella* NC64A or *Chlorella* Pbi. The prototype chlorella virus is PBCV-1, which stands for *Paramecium bursaria* chlorella virus. The genomes (313–370 kbp) of several of the chloroviruses have either been sequenced or are in the process of being sequenced.

Large polyhedral, dsDNA-containing viruses that infect certain marine algae are also under active investigation. These include viruses that infect filamentous brown algae, *Ectocarpus* sp. (EsV viruses) and *Feldmannia* sp. (FsV viruses) (**Table 1**), and viruses that infect *Emiliania huxleyi* (EhV viruses). The genomes of some of these viruses have also been sequenced recently. Although all of these algal viruses arose from a common ancestor, they can have different lifestyles. For example, EsV and FsV viruses have a lysogenic phase in their life cycle and are only expressed as virus particles in sporangial cells of their host. In contrast, the chlorella viruses and EhV viruses are lytic.

The first algal viruses to be discovered were large dsDNA viruses; consequently, it was assumed for several years that algae were only infected by large dsDNA viruses. However, this scenario is changing rapidly. A positive-sense 9.1 kbp single-stranded RNA (ssRNA) virus has been discovered that infects a toxic bloom-forming alga, *Heterosigma akashiwo* (called HaRNAV) that is related to the picorna-like virus superfamily. A dsRNA, reo-like virus that infects a microalga, *Micromonas pusilla*, has been reported and finally a virus (CsNIV) with an unusual genome structure that infects diatoms in the genus *Chaetoceros* has been described. The CsNIV genome consists of a single molecule of covalently closed circular single-stranded DNA (ssDNA) (6005 nucleotides) as well as a segment of linear ssDNA (997 nucleotides). These recently discovered algal viruses are described in other articles in this encyclopedia.

Table 1 Taxonomy and general characteristics of some phycodnaviruses

Genus[a]	Type species[a]	Known host range[a]	Source	Particle diameter (nm)	Genome size (kbp) and conformation	Latent period (h)	Burst size
Chlorovirus	Paramecium bursaria chlorella virus 1 (PBCV-1)	Chlorella NC64A Chlorella Pbi Chlorella SAG 3.83	FW	190	313–370 Closed linear dsDNA, hairpin termini	6–8	200–350[b]
Coccolithovirus	Emiliania huxleyi virus 86 (EhV-86)	Emiliania huxleyi	MW	160–200	407–415 Circular	4–6	400–1000
Phaeovirus	Ectocarpus siliculousus virus 1 (EsV-1)	Phaeophyceae Ectocarpus siliculousus Ectocarpus fasciculatus Feldmannia simplex Feldmannia irregularis Feldmannia species Hincksia hincksiae Myriotrichia clavaeformis Pilayella littoralis	MW	150–170	170–340 Open linear, single stranded regions	ND	>1×10^6
Prasinovirus	Micromonas pusilla virus SP1 (MpV-SP1)	Micromonas pusilla	MW	115	200	7–14	1800–4100
Prymnesiovirus	Chrysochomulina brevifilum virus PW1 (CbV-PW1)	Haptophyceae (aka Prymnesiophyceae) Chrysochomulina brevifilum Chrysochomulina strobilus Chrysochomulina globosa	MW	145–170	510	ND	800–1000
Raphidovirus	Heterosigma akashiwo virus 01 (HaV01)	Heterosigma akashiwo	MW	202	294	30–33	770

[a]Data abstracted from http://www.ncbi.nlm.nih.gov/ICTVdb/Ictv/index.htm.
[b]PFU/cell.
FW, freshwater; MW, marine/coastal water; ND, not determined.
Reproduced from Dunigan DD, Fitzgerald LA, and Van Etten JL (2006) Phycodnaviruses: A peek at genetic diversity. *Virus Research* 117: 119–132, with permission from Elsevier.

Taxonomy and Classification

Members and prospective members of the family *Phycodnaviridae* constitute a genetically diverse, but morphologically similar group of viruses with eukaryotic algal hosts from both fresh and marine waters. Accumulating genetic evidence indicates that the phycodnaviruses together with the poxviruses, iridoviruses, asfarviruses, and the 1.2 Mbp mimivirus have a common evolutionary ancestor, perhaps, arising at the point of eukaryogenesis, variously reported to be 2.0–2.7 billion years ago. All of these viruses share nine gene products and 33 more gene products are present in members of at least two of these five viral families. Collectively, these viruses are referred to as nucleocytoplasmic large DNA viruses (NCLDV).

Phycodnaviruses are large (mean diameter 160 ± 60 nm) icosahedrons, which encapsidate 160–560 kbp dsDNA genomes. Where known, the viruses have an internal-membrane that is required for infection. Phylogenetic analyses of their δ-DNA polymerases indicate that they are more closely related to each other than to other dsDNA viruses and that they form a monophyletic group, consistent with a common ancestor. However, the phycodnaviruses fall into six clades which correlate with their hosts and each has been given genus status. Often the genera can be distinguished by additional properties, for example, lytic versus lysogenic lifestyles or linear versus circular genomes. Members of the genus *Chlorovirus* infect freshwater algae, whereas, members of the other five genera (*Coccolithovirus*, *Phaeovirus*, *Prasinovirus*, *Prymnesiovirus*, and *Raphidovirus*) infect marine algae.

Structure and Composition

Chlorella virus particles are large (molecular weight $\sim 1 \times 10^9$ Da) and complex. The PBCV-1 virion contains more than 100 different virus-encoded proteins. The PBCV-1 54 kDa major capsid protein is a glycoprotein and comprises $\sim 40\%$ of the total protein. The major capsid protein consists of two eight-stranded, antiparallel β-barrel, jelly-roll domains related by pseudo sixfold rotation. This structure resembles the major coat proteins from some other dsDNA viruses that infect all three domains of life including bacteriophage PRD1, human adenoviruses, and a virus STIV infecting the Archaea, *Sulfolobus solfataricus*. This finding led to the suggestion that these three viruses may also have a common evolutionary ancestor with the NCLDVs, even though there is no significant amino acid sequence similarity among their major capsid proteins.

Cryoelectron microscopy and three-dimensional image reconstruction of the PBCV-1 virion (**Figure 1**) indicate that the outer capsid is icosahedral and covers a lipid

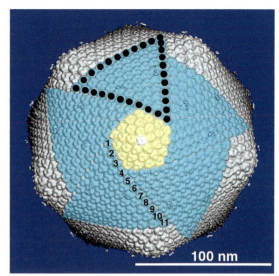

Figure 1 Three-dimensional image reconstruction of chlorella virus PBCV-1 from cryoelectron micrographs. The virion capsid consists of 12 pentasymmetrons and 20 trisymmetrons. Five trisymmetrons are highlighted in the reconstruction (blue) and a single pentasymmetron is colored yellow. A pentavalent capsomer (white) lies at the center of each pentasymmetron. Each pentasymmetron consists of one pentamer plus 30 trimers. Eleven capsomers form the edge of each trisymmetron (black dots) and therefore each trisymmetron has 66 trimers. Reprinted, with permission, from Van Etten JL (2003) Unusual life style of giant chlorella viruses. *Annual Review of Genetics* 37: 153–195, ©2003 by Annual Reviews.

bilayered membrane. The membrane is required for infection because the virus loses infectivity after exposure to organic solvents. The outer diameter of the virus capsid ranges from 1650 Å along the two- and threefold axes to 1900 Å along the fivefold axis. The capsid shell consists of 1680 doughnut-shaped trimeric capsomers plus 12 pentameric capsomers at each icosahedral vertex. The trimeric capsomers are arranged into 20 triangular facets (trisymmetrons, each containing 66 trimers) and 12 pentagonal facets (pentasymmetrons, each containing 30 trimers and one pentamer at the icosahedral vertices). Assuming all the trimeric capsomers are identical, the outer capsid of the virus contains 5040 copies of the major capsid protein. The virus has a triangulation number of 169. However, PBCV-1 is not the largest phycodnavirus; phaeocystis pouchetti virus (PpV01) has an icosahedral capsid with a triangulation number of 219.

Structural proteins of many viruses, such as herpesviruses, poxviruses, and paramyxoviruses, as well as the chlorella viruses, are glycosylated. Typically, viral proteins are glycosylated by host-encoded glycosyltransferases located in the endoplasmic reticulum (ER) and Golgi and then transported to a host membrane. Nascent viruses acquire the glycoprotein(s) and only become infectious by budding through the membrane, usually as they are released from the cell. Consequently, the glycan portion of virus glycoproteins is host specific.

However, glycosylation of PBCV-1 major capsid protein differs from this paradigm. Accumulating evidence indicates that PBCV-1 encodes most, if not all, of the enzymes involved in constructing the complex oligosaccharides attached to its major capsid protein and that the process occurs independently of the ER and Golgi. Furthermore, five of six putative PBCV-1-encoded glycosyltransferases are predicted to be located in the cytoplasm. PBCV-1 also encodes several additional proteins involved in post-translational modification that may alter virus structural proteins. These include a prolyl-4-hydroxylase and several protein kinases and a phosphatase.

Genomes

The 331 kbp PBCV-1 genome is linear and nonpermuted. The genome termini consist of 35 nucleotide long, incompletely base-paired, covalently closed hairpin loops that exist in one of two forms (flip and flop). Each hairpin loop is followed by an identical 2.2 kbp inverted repeat sequence; the remainder of the genome consists primarily of single-copy DNA. The PBCV-1 genome has ~695 open reading frames (ORFs) that have 65 or more codons, of which ~366 are probably protein encoding. The putative protein-encoding genes are evenly distributed on both strands and intergenic space is minimal, 275 ORFs are separated by less than 100 nucleotides (**Figure 2**). One exception is a 1788 bp sequence near the middle of the genome. This sequence has a polycistronic gene containing 11 tRNAs (**Figure 2**).

Approximately 40% of the 366 PBCV-1 gene products resemble proteins in the databases, many of which have not previously been associated with viruses (some are listed in **Figure 3**). Eighty-four ORFs have paralogs within PBCV-1, forming 26 groups. The size of these groups ranges from two to six members.

Some PBCV-1 genes are interrupted by introns; a gene encoding a transcription factor-like protein contains a self-splicing type I intron, whereas the δ-DNA polymerase gene contains a spliceosomal-processed type of intron. In addition, one of the PBCV-1 tRNA genes is predicted to have an intron.

One unusual feature of PBCV-1 DNA, as well as the other chlorella virus DNAs, is that they contain methylated bases. Chlorella virus genomes contain 5-methylcytosine in amounts varying from 0.1% to 48% of the total cytosines. Many viral DNAs also contain N^6-methyladenine with concentrations up to 37% of the total adenines. This led to the discovery that many chlorella viruses encode multiple DNA methyltransferases, as well as site-specific (restriction) endonucleases.

The 407 kbp circular dsDNA genome of one of the EhV viruses is predicted to have 472 protein-encoding genes. Only 66 (14%) of these 472 gene products match a sequence in GenBank. The EhV virus encodes several unexpected genes never found in a virus before, including four gene products involved in sphingolipid biosynthesis and two additional gene products which encode desaturases.

The structure of the 335 kbp EsV genome is unknown. Although several experiments suggest that the genome is circular, DNA sequencing indicates that it has defined ends with inverted repeats. The virus is predicted to contain 231 protein-encoding genes, 48% of which resemble various proteins in the public databases. About 12% of the EsV genome consists of tandem repeats and portions of the genome have ssDNA regions. Collectively, PBCV-1, EsV, and EhV viruses have in excess of 1000 unique ORFs. A total of 123 putative ORFs from these three viruses is organized into metabolic domains (**Figure 3**). Interestingly, these three viruses only have 14 gene products in common. Not surprisingly, several of these common ORFs are involved in DNA replication, such as the δ-DNA polymerase, large and small subunits of ribonucleotide reductase, proliferating cell nuclear antigen (PCNA), superfamily II and III helicases, and the newly recognized archaeo-eukaryotic primases. Another common ORF among the three phycodnaviruses is the major capsid protein.

Virus Replication

PBCV-1 attaches rapidly, specifically, and irreversibly to the external surface of cell walls, but not to protoplasts, of host *Chlorella* NC64A. Attachment always occurs at a virus vertex, followed by degradation of the wall at the attachment point. Following wall degradation, the internal membrane of the virus probably fuses with the host membrane resulting in entry of the virus DNA and virion-associated proteins into the cell, leaving an empty capsid on the surface. Several observations suggest that the infecting DNA, plus associated proteins, is rapidly transported to the host cell nucleus and the host transcription machinery is reprogrammed to transcribe virus RNAs. This process occurs rapidly because early PBCV-1 transcripts can be detected within 5–10 min post infection (p.i.). PBCV-1 translation occurs on cytoplasmic ribosomes and early PBCV-1-encoded proteins can be detected within 10 min p.i.

PBCV-1 DNA synthesis and late virus transcription begins 60–90 min p.i. Ultrastructural studies of PBCV-1-infected chlorella suggest that the nuclear membrane remains intact, at least during early stages of virus replication. At approximately 2–3 h p.i., assembly of virus

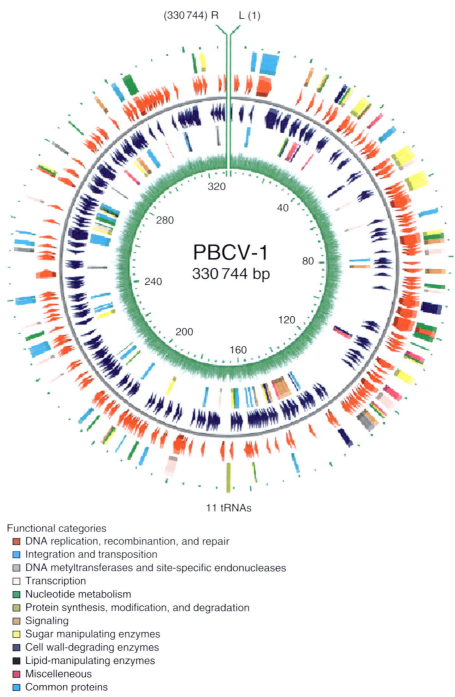

Figure 2 Map of the chlorella virus PBCV-1 genome visualized as a circle using Circular Genome Viewer (Paul Stothard, Genome Canada). However, the genome is a linear molecule and the ends are depicted at the top of the figure as green lines (L and R represent the left and right ends of the genome, respectively). The red and blue arrows represent the 366 protein-encoding genes; red arrows depict genes transcribed in the rightwards direction and blue arrows genes transcribed in the leftwards direction. The two rings that flank the protein-encoding genes show the predicted functions of the proteins, color-coded by function (see insert in the figure). The location of the polycistronic gene encoding the 11 tRNAs is indicated in the outermost ring (i.e., this gene is transcribed in the rightwards direction). The innermost ring (in green) represents the A + T content determined using a 25 bp window. Note that the A + T content is fairly constant over the genome (60% A + T).

capsids begins in localized regions in the cytoplasm, called virus assembly centers, which become prominent at 3–4 h p.i.. By 5 h p.i., the cytoplasm is filled with infectious progeny virus particles (∼1000 particles/cell) and by 6–8 h p.i. localized lysis of the host cell releases progeny. Of the progeny released, 25–50% of the particles are infectious. Mechanical disruption of cells releases infectious virus 30–50 min prior to cell lysis, indicating the

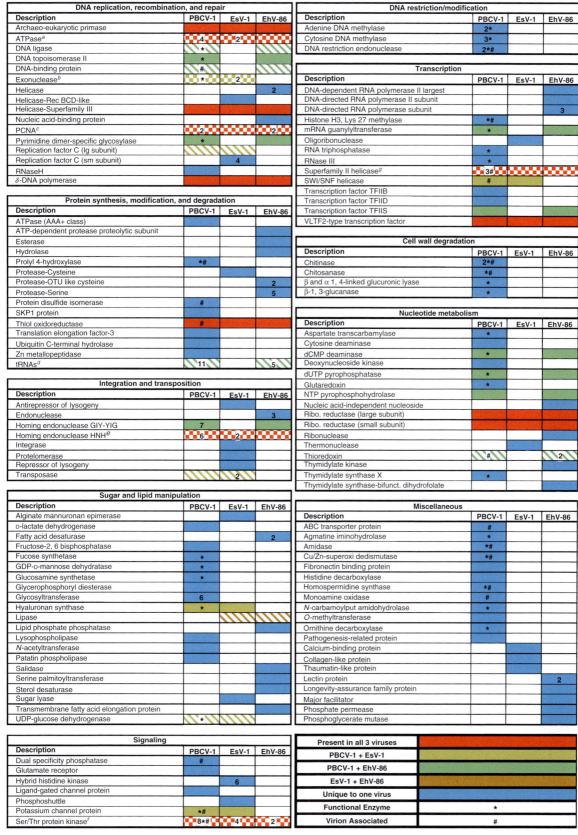

Figure 3 Continued

virus does not acquire its glycoprotein capsid by budding through the host plasma membrane as it is released from the cell. Other chlorella viruses have longer replication cycles than PBCV-1. For example, virus NY-2A requires approximately 18 h for replication and consequently forms smaller plaques.

Virus EsV initiates its life cycle by infecting free-swimming, wallless gametes of its host. Virus particles enter the cell by fusion with the host plasma membrane and release a nucleoprotein core particle into the cytoplasm, leaving remnants of the capsid on the surface. The viral core moves to the nucleus within 5 min p.i. One important feature that distinguishes the EsV life cycle from the other phycodnaviruses is that the viral DNA is integrated into the host genome and is transmitted mitotically to all cells of the developing alga. The viral genome remains latent in vegetative cells until it is expressed in the algal reproductive cells, the sporangia or gametangia. Massive viral DNA replication occurs in the nuclei of these reproductive cells, followed by nuclear breakdown and viral assembly that continues until the cell becomes densely packed with virus particles. Virus release is stimulated by the same factors that induce discharge of gametes from the host, that is, changes in temperature, light, and water composition. This synchronization facilitates interaction of viruses with their susceptible host cells.

In contrast to the chlorella and EsV viruses, very little is known about the EhV life cycle, including how it infects its host. One property that clearly distinguishes EhV from the other phycodnaviruses is that it encodes six RNA polymerase subunits; in contrast, neither PBCV-1 nor EsV encodes a recognizable RNA polymerase component. Thus like the poxviruses, EhV may carry out its entire life cycle in the cytoplasm of its host. The latent period for EhV is 4–6 h with a burst size of 400–1000 particles per cell.

Virus Transcription

Detailed studies on transcription are lacking for all of the algal viruses and only a few general statements can be made about PBCV-1 transcription:

1. PBCV-1 infection rapidly inhibits host RNA synthesis.
2. Viral transcription is programmed and early transcripts appear within 5–10 min p.i. Late viral transcription first begins about 60–90 min p.i.
3. Some early viral transcripts are synthesized in the absence of *de novo* protein synthesis. As expected, the synthesis of later transcripts requires translation of early virus genes.
4. Early and late genes are dispersed in the PBCV-1 genome.
5. PBCV-1 ORFs are tightly packed on both DNA strands and the coding regions of some of the genes overlap. The largest distance between PBCV-1 ORFs is a 1788-nucleotide stretch in the middle of the genome. This sequence has a polycistronic gene containing 11 tRNAs (**Figure 2**).
6. Consensus promoter regions for early and late genes have not been identified, although the 50 nucleotides

---

Figure 3 Selected ORFs in the PBCV-1, EsV, and EhV genomes are arranged by their metabolic domains. If a genome encodes a putative protein more than once, a number in the box indicates the number of genes of this type per genome. Color-coding is indicated on the figure and is used to depict the relationship between viruses. Red indicates proteins that are encoded by all three viruses; yellow indicates proteins that are encoded by PBCV-1 and EsV, but not EhV; green indicates proteins that are encoded by PBCV-1 and EhV, but not EsV; orange indicates proteins that are encoded by EsV and EhV, but not PBCV-1; and blue indicates there are no shared homologs. Solid colored boxes indicate that the putative proteins are homologs. A diagonally stripped box indicates that the putative proteins are nonhomologous, and a checkered box indicates that the putative proteins are a mix of homologous, nonhomologous, or unique ORFs. In this case, a footnote has been added to clarify the specific differences; in parentheses the ORF has been defined by the gene number and any ORF beginning with an 'A' is from the PBCV-1 genome, an 'EsV' is from the EsV-1 genome, and an 'EhV' is from the EhV-86 genome. Proteins known to be functional are indicated with a star (*) and proteins known to be associated with the virion are indicated with a pound sign (#). Superscript 'a' indicates ATPase – One homolog between all three viruses (A392R, EsV-26, and EhV072), one homolog between PBCV-1 and EsV (A565R and EsV-171), and two PBCV-1 ATPases which have no homologs in EsV or EhV (A561L and A554/556/557L). Superscript 'b' indicates Exonuclease – One homolog between PBCV-1 and EsV (A166R and EsV-64) and one unique to EsV (EsV-126). Supercript 'c' indicates PCNA – One homolog between all three viruses (A193L, EsV-132, and EhV020), one homolog between PBCV-1 and EhV (A574L and EhV020). However, EhV-440, another EhV-encoded PCNA, has no homologs in PBCV-1 or EsV. Superscript 'd' indicates tRNAs – PBCV-1 has 11 tRNA genes, encoding AA Leu (2), Ile, Asn (2), Lys (3), Tyr, Arg, and Val. EhV has five tRNA genes, encoding AA Leu, Ile, Gln, Asn, and Arg. Supercript 'e' indicates Homing endonuclease HNH – One homolog between two viruses (A422R and EhV087). PBCV-1 (A87R) and EsV (EsV-119) are homologous. Four other HNH endonucleases are unique to PBCV-1 (A267L, A354R, A478R, and A490L) and one is unique to EsV (EsV-1–16). Superscript 'f' indicates Ser/Thr protein kinase – Four PBCV-1, two EsV-1, and one EhV-86 S/T kinases ORFs are homologous (A248R, A277L, A282L, A289L, EsV-82, EsV-111, and EhV451). The remaining S/T kinase ORFs from the three genomes (A34R, A278L, A614L, A617R, EsV-104, EV-156, and EhV-402) are unique. Suerscript 'g' indicates that superfamily II helicase, PBCV-1 (A153R), and EsV (EsV-66) are homologous. The two additional PBCV-1-encoded helicases (A241R and A363R) and EhV104 are unique. Reproduced from Dunigan DD, Fitzgerald LA, and Van Etten JL (2006) Phycodnaviruses: A peek at genetic diversity. *Virus Research* 117: 119–132, with permission from Elsevier.

preceding the ATG start codon of most functional PBCV-1 genes are at least 70% A+T.

7. Transcription of some PBCV-1 genes appears to be complex. For example, some gene transcripts exist as multiple bands and these patterns change between early and late times in the virus life cycle.

Additional *Chlorella* Viruses

Several hundred plaque-forming chloroviruses have been characterized to various degrees. They infect either *Chlorella* NC64A cells (NC64A viruses), an endosymbiont of *P. bursaria* isolated from North America, or *Chlorella* Pbi cells (Pbi viruses) that are endosymbiotic with a paramecium isolated in Europe. Like PBCV-1, each of these viruses contain many structural proteins, a large (>300 kbp) dsDNA genome, and they are chloroform sensitive. The DNAs of some of these viruses hybridize strongly with PBCV-1 DNA, while others hybridize poorly.

Three additional chlorella virus genomes have been sequenced recently and others are nearing completion. The largest, the 370 kbp genome of virus NY-2A, contains ~400 protein-encoding genes. Most common genes are colinear in viruses PBCV-1 and NY-2A, which infect the same host chlorella. However, almost no colinearity exists between common genes in Pbi virus MT325 and those in PBCV-1, NY-2A, and AR158 suggesting plasticity in the chlorella virus genomes. Additionally, the G + C contents of the three NC64A viruses range from 40% to 41% whereas the G + C contents of the Pbi viruses are approximately 45%. These last two observations suggest that these two virus groups have been separated for considerable evolutionary time. Viruses morphologically similar to the NC64A and Pbi viruses have also been isolated from chlorella symbiotic in the coelenterate *Hydra viridis* and very recently from the heliozoon *Acanthocystis turfacea*. The symbiotic hydra chlorella have not been cultured and so very little is known about these viruses. However, the *A. turfacea* viruses can be isolated by plaque formation.

Other Algal Viruses

Field isolates representing at least six genera of filamentous brown algae contain virus particles that are morphologically similar to EsV and FsV. Virus expression is variable; virions are rarely observed in vegetative cells but often are common in unilocular sporangia (FsV) or both unilocular and plurilocular sporangia (EsV). EsV viruses only infect the free swimming, zoospore stage of *Ectacarpus* sp. All natural isolates of *Feldmania* sp. are infected with virus and so infection studies cannot be conducted.

Viruses that infect *Emiliania huxleyi*, *Micromonas pusilla*, and *Chrysochromulina brevfilum* have been isolated from many marine environments. These viruses can be distinguished by DNA restriction patterns.

Phycodnavirus Genes Encode Some Interesting and Unexpected Proteins

Many chlorella virus-encoded enzymes are either the smallest or among the smallest proteins of their class. In addition, homologous genes in the chloroviruses can differ in nucleotide sequence by as much as 50%, which translates into amino acid differences of 30–40%. Therefore, comparative protein sequence analyses can identify conserved amino acids in proteins as well as regions that tolerate amino acid changes. The small sizes and the finding that many chlorella virus-encoded proteins are 'user friendly' have resulted in the biochemical and structural characterization of several PBCV-1 enzymes. Examples include: (1) The smallest eukaryotic ATP-dependent DNA ligase, which is the subject of intensive mechanistic and structural studies. (2) The smallest type II DNA topoisomerase. The virus enzyme cleaves dsDNAs about 50 times faster than the human type II DNA topoisomerase; consequently, the virus enzyme is being used as a model enzyme to study the mechanism of topoisomerase II DNA cleavage. (3) An RNA guanylyltransferase that was the first enzyme of its type to have its crystal structure resolved. (4) A small prolyl-4-hydroxylase that converts Pro-containing peptides into hydroxyl-Pro-containing peptides in a sequence-specific fashion. (5) The smallest protein (94 amino acids) to form a functional potassium ion channel. These minimalist enzymes may represent evolutionary precursors of contemporary proteins, but it is also possible that they are products of evolutionary optimization during viral evolution.

The chloroviruses are also unusual because they encode enzymes involved in sugar metabolism. Three PBCV-1 encoded enzymes, glutamine:fructose-6-phosphate amindotransferase, UDP-glucose dehydrogenase, and hyaluronan synthase, are involved in the synthesis of hyaluronan, a linear polysaccharide composed of alternating β-1,4-glucuronic acid and β-1,3-N-acetylglucosamine residues. All three genes are transcribed early in PBCV-1 infection and hyaluronan accumulates on the external surface of the infected cells.

Two PBCV-1-encoded enzymes, GDP-D-mannose dehydratase and fucose synthase, comprise a three-step pathway that converts GDP-D-mannose to GDP-L-fucose. The function of this putative pathway is unknown. However, fucose, a rare sugar, is present in the glycans attached to the major capsid protein.

PBCV-1 encodes four enzymes involved in polyamine biosynthesis: ornithine decarboxylase (ODC), homospermidine synthase, agmatine iminohydrolase, and N-carbamoyl-putrescine amidohydrolase. ODC catalyzes the decarboxylation of ornithine to putrescine, which is the first and the rate-limiting enzymatic step in polyamine biosynthesis. Not only is the PBCV-1-encoded ODC the smallest known ODC, the PBCV-1 enzyme is also interesting because it decarboxylates arginine more efficiently than ornithine.

The genome sequences of other phycodnaviruses have revealed several interesting and unexpected gene products. However, except for an aquaglyceroporin encoded by chlorella virus MT325, none of these products have been expressed and tested for enzyme function.

Ecology

Eukaryotic algae are important components of both freshwater and marine environments; however, the significance of viruses in these systems is only beginning to be appreciated. The chlorella viruses are ubiquitous in freshwater collected throughout the world and titers as high as 100 000 infectious particles per mililiter have been reported in native waters. Typically, the titer is 1–100 infectious particles per milliliter. The titers are seasonal with the highest titers in the spring. It is not known whether chlorella viruses replicate exclusively in algae symbiotic with paramecia or if the viruses have another host(s). In fact, it is not known if paramecium chlorellae exist free of their hosts in natural environments. However, the chlorellae are protected from virus infection when they are in a symbiotic relationship with the paramecium.

The concept that viruses might have a major impact on the marine environment began about 15 years ago with the discovery that seawater contains as many as 10^7 VLPs per milliliter. This huge population consists of both bacterial and algal viruses and is important because phytoplankton, consisting of cyanobacteria and eukaryotic microalgae, fix 50–60% of the CO_2 on Earth. At any one time, 20–40% of these photosynthetic organisms are infected with a virus. Consequently, these viruses contribute to microbial composition and diversity, as well as, nutrient cycling in aqueous environments. Thus viruses, including the phycodnaviruses, have a major impact on global carbon/nitrogen cycles that is only beginning to be appreciated by scientists, including those who model such cycles.

Algal viruses are also believed to play major roles in the termination of marine algal blooms and so there are active research efforts to understand the natural history of these algal/virus systems. For example, the coccolithophorid *Emiliania huxleyi* is a unicellular alga found throughout the world; the alga can form immense coastal and mid-oceanic blooms at temperate latitudes that cover 10 000 km^2 or more. One of the primary mechanisms for terminating *E. huxleyi* blooms is lysis by the EhV viruses described above.

The filamentous brown algae, *Ectocarpus* sp. and *Feldmania* sp. isolated from around the world, are infected with lysogenic EsV and FsV viruses, respectively. Lysogeny is consistent with the observation by early investigators that VLPs appear infrequently in eukaryotic algae and only at certain stages of algal development. The apparent lack of infectivity by many of the previously observed VLPs in eukaryotic algae is also consistent with a lysogenic lifestyle. The VLPs might either infect the host and resume a lysogenic relationship or be excluded by preexisting lysogenic viruses.

Perspectives

Sequence analyses of three phycodnaviruses suggest that this family may have more sequence diversity than any other virus family. These three viruses only have 14 homologous genes, which means that there are in excess of 1000 unique ORFs in just these three viruses. Despite the large genetic diversity in these three sequenced phycodnaviruses, phylogenetic analyses of δ-DNA polymerases and DNA primases indicate that the phycodnaviruses group into a monophyletic clade within the NCLDVs. A recent analysis using eight concatenated core NCLDV genes also indicates that the phycodnaviruses cluster together and are members of the NCLDV 'superfamily'. However, it is obvious that the identification and characterization of phycodnaviruses is in its infancy. Metagenomic studies, such as DNA sequences from the Sargasso Sea samples, indicate that translation products of many unknown sequences are more similar to PBCV-1 proteins than the next known phycodnavirus.

See also: African Swine Fever Virus; Algal Viruses; Iridoviruses of Invertebrates; Marnaviruses; Poxviruses.

Further Reading

Brussaard CPD (2004) Viral control of phytoplankton populations – A review. *Journal of Eukaryotic Microbiology* 51: 125–138.

Dunigan DD, Fitzgerald LA, and Van Etten JL (2006) Phycodnaviruses: A peek at genetic diversity. *Virus Research* 117: 119–132.

Iyer ML, Balaji S, Koonin EV, and Aravind L (2006) Evolutionary genomics of nucleocytoplasmic large DNA viruses. *Virus Research* 117: 156–184.

Muller DG, Kapp M, and Knippers R (1998) Viruses in marine brown algae. *Advances in Virus Research* 50: 49–67.

Van Etten JL (2003) Unusual life style of giant chlorella viruses. *Annual Review of Genetics* 37: 153–195.

Van Etten JL, Graves MV, Muller DG, Boland W, and Delaroque N (2002) Phycodnaviridae – large DNA algal viruses. *Archives of Virology* 147: 1479–1516.

Wilson WH, Schroeder DC, Allen MJ, et al. (2005) Complete genome sequence and lytic phase transcription profile of a Coccolithovirus. *Science* 309: 1090–1092.

Yamada T, Onimatsu H, and Van Etten JL (2006) Chlorella viruses. *Advances in Virus Research* 66: 293–336.

Relevant Websites

http://www.giantvirus.org – Giantvirus.org.
http://greengene.uml.edu – Greengene at UML.

Phylogeny of Viruses

A E Gorbalenya, Leiden University Medical Center, Leiden, The Netherlands

© 2008 Elsevier Ltd. All rights reserved.

Introduction: Evolution, Phylogeny, and Viruses

Biological species, including viruses, change through generations and over time in the process known as evolution. These changes are first fixed in the genome of successful individuals that give rise to genetic lineages. Due to either limited fidelity of the replication apparatus copying the genome or physico-chemical activity of the environment, nucleotides may be changed, inserted, or deleted. Genomes of other origin may also be a source of innovation for a genome through the use of specially evolved mechanisms of genetic exchange (recombination). Accepted changes, known as mutations, may be neutral, advantageous, or deleterious, and depending on the population size and environment, the mutant lineage may proliferate or go extinct. Overall, advantageous mutations and large population size increase the chances for a lineage to succeed. The lineage fit is constantly re-assessed in the ever-changing environment and lineages that, due to mutation, became a success in the past could be unfit in the new environment. Due to the growing number of mutations accumulating in the genomes, lineages diverge over time, although occasionally, due to stochastic reasons or under similar selection pressure, they may converge.

The relationship between biological lineages related by common descent is called phylogeny; the same term also embodies the methodology of reconstructing these relationships. Phylogeny deals with past events and, therefore, it is reconstructed by quantification of differences accumulated between lineages. Due to the lack of fossils and (relatively) high mutation rate, viruses were not considered to provide a recoverable part of phylogeny until the advent of molecular data proved otherwise. Comparison of nucleotide and amino acid sequences, and, occasionally, other quantitative characteristics such as distances between three-dimensional structures of biopolymers, have been used to reconstruct virus phylogeny. Results of phylogenetic analysis are commonly depicted in the form of a tree that may be used as a synonym for phylogeny. For instance, all-inclusive phylogeny of cellular species is depicted as the Tree of Life (ToL).

With few exceptions, virus phylogeny follows the theory and practice developed for phylogeny of cellular life forms. For inferring phylogeny, differences between the sequences of species members, assumed to be of a discernable common origin, are analyzed. If species in all lineages evolve at a uniform constant rate, like clock ticks, their evolution conforms to a molecular clock model. The utility of this model in relation to viruses may be very limited. Rather, related virus lineages may evolve at different and fluctuating rates and some sites may mutate repeatedly with each new mutation erasing a record about the prior change. As a result, the accumulation of inter-species differences may progress nonlinearly with the time elapsed. At present, our understanding of these parameters of virus evolution is poor and this limits our ability to assess the fit between a reconstructed phylogeny and the true phylogeny, with the latter practically remaining unknown for most virus isolates. This gap in our knowledge does not eliminate the conceptual strength of phylogentic analysis for reconstructing the relationships between biological species.

The ultimate goal of virus phylogeny is reconstructing the relationships between 'all' virus isolates and species. In contrast to cellular species, which form three compact domains (kingdoms) and whose origin is traced back to a common ancestor in the ToL, major virus classes may combine species that have originated from different ancestors. Thus, reconstructing the comprehensive virus phylogeny requires comparisons that involve genomes of virus and cellular origins. This formidable task remains largely 'work in progress'. In fact, most efforts in virus phylogeny are invested in reconstructing the relationships at the micro, rather than grand, scale and they focus on well-sampled lineages that have practical (e.g., medical) relevance. Phylogeny itself or in combination with other data may provide a deep insight into virus evolution and diverse aspects of virus life cycles, including virus interactions with their hosts.

Our knowledge about contemporary virus diversity has been steadily advancing with new viruses being constantly described by systematic efforts as well as occasional discoveries. These developments indicate that only a small part of virus diversity has so far been unraveled and has become available for phylogenetic studies. It is also likely that many more lineages existed in the past; some of these lineages are likely to have ancestral relationships with contemporary lineages.

Tree Definitions

Species share similarity that varies depending on the rate of evolution and time of divergence. The entire process of generating contemporary species diversity from a common ancestor is believed to proceed through a chain of intermediate ancestors specific for different subsets of the analyzed species. The relationship between the common ancestor, intermediate ancestors, and contemporary species may be likened to the relationship between, respectively, root, internal nodes, and terminal nodes (leaves) of a tree, an abstraction that is widely used for the visualization of this relationship. Trees are also part of graph theory, a branch of mathematics, whose apparatus is used in phylogeny. Formally and due to a strong link between phylogeny and taxonomy, leaves may be called operational taxonomy units (OTUs) and internal nodes and roots, since they have not been directly observed, are known as hypothetical taxonomy units (HTUs). Nodes are connected by branches or edges.

The tree may be characterized by topology, length of branches, shape, and the position of the root. The topology is determined by relative position of internal and terminal nodes; it defines branching events leading to contemporary species diversity. If two or more trees obtained for different data sets feature a common topology, these trees are called congruent. The branch length of a tree may define either the amount of change fixed or the time passed between two nodes connected in a tree, and is known as 'additive' or 'ultrametric', respectively. The tree shape may be linked to particulars of evolutionary process and reflect changes in the population size and diversity due to genetic drift and natural selection. The position of the root at the tree defines the direction of evolution. Species that descend from an internal node in a rooted tree form a cluster and the node is called most recent common ancestor (MRCA) of the cluster that thus has a monophyletic origin. The branch lengths and the root position may be left undefined for a tree that is then called 'cladogram' and 'unrooted tree', respectively.

Phylogenetic Analysis

Multiple alignment of polynucleotide or amino acid sequences representing analyzed species and maximized for similarity is traditionally used as input for phylogenetic analysis. The quality of alignment is among the most significant factors affecting the quality of phylogenetic inference. Due to the redundancy of the genetic code, changes in polynucleotide sequences are accumulated at a higher rate than those in amino acid sequences. In viruses, including RNA viruses, this difference is not counterbalanced by other constraints linked to dinucleotide frequency and RNA secondary (tertiary) structure. Because of these differences, phylogeny of closely related species is commonly inferred using polynucleotide sequences, while protein sequences, preserving better phylogenetic signal, may be used to infer phylogeny of distantly related species.

Differences between species, as calculated from alignment, may be quantified as either pairwise distances forming a distance matrix or position-specific substitution columns (discrete characters of states of alignment), the latter preserving the knowledge about location of differences. The respective methods dealing with these quantitative characteristics are known as distance and discrete (character state). The distance methods are praised for their speed and are considered a technique of choice for analysis of large data sets. They are often designed to converge on a unique phylogeny, with none others being even considered. The unweighted pair group method with arithmetic means (UPGMA) in which a constantly recalculated distance matrix is used to define the hierarchy of similarities through systematic and stepwise merging of most similar pairs at a time was the first technique introduced for clustering. The neighbor-joining (NJ) method uses a more sophisticated algorithm of clustering that minimizes branch lengths, and is the most popular among distance methods. Although different trees may be compared in how they fit a distance matrix, it is distance character-based methods that are routinely used to assess numerous alternative phylogenies in search for the best one in a computationally very intensive process. Due to the calculation time involved, assessing all possible phylogenies is found to be impractical for data sets including more than 10 sequences; for larger data sets different heuristic approximations are used that may not guarantee a recovered phylogeny to be the best overall. There are two major criteria for selecting the best phylogeny using character-state based information through either maximum parsimony (MP) or maximum likelihood (ML). In MP analysis, a phylogeny with a minimal number of substitutions separating analyzed species is sought. The ML analysis offers a statistical framework for comparing the likelihood of fitting different trees into the data in search for one with the best fit. The latter approach is mathematically robust and its statistical power may also be used in combination with other techniques of tree generation. Most recently, a Bayesian variant of the ML approach has gained popularity. It utilizes prior knowledge about the evolutionary process in combination with repeated sampling from subsequently derived hypotheses.

After a tree is chosen, it is common to assign support for internal nodes through assessing nodes' persistence in trees related to the chosen tree. One particular technique, called bootstrap analysis, in which trees are generated for numerous randomly modified derivatives of the original data set, is most frequently used. Each internal node in the original tree is characterized by a so-called bootstrap value that is equal to the number of nodes appearing in all tested trees. Although the relationship between bootstrap and statistical values is not linear, nodes with very high bootstrap values are considered to be reliable. If species evolve according to a molecular clock model, the root position in a tree could directly be calculated from the observed inter-species differences as a midpoint of cumulative inter-species differences. Alternatively, the root position may be assigned to a tree from knowledge about analyzed species that was gained independently from phylogenetic analysis. Commonly, this knowledge comes in the form of a single or more species which are assumed (or known) to have emerged before the 'birth' of the analyzed cluster. These early diverged species are collectively defined as 'outgroup', while the analyzed species may be called 'in-group'. Also, a tree may be generated unrooted, a common practice in phylogenetic analysis of viruses for which the applicability of the molecular clock model remains largely untested and reliable outgroups may not be routinely available. In the unrooted tree, grouping of species in separate clusters may be apparent, although these clusters may not be treated as monophyletic as long as the direction of evolution has not been defined. These challenges are addressed by the development of new approaches that infer rooted trees without artificially restricting species evolution to a constant rate (known as relaxed molecular clock models).

Virus phylogeny can be inferred using genomes or distinct genes and each of these approaches, standard in phylogenomics, may be considered as complementary. Under the first approach, genome-wide alignments are used for analysis. Due to complexities of the evolutionary process that may be region specific, reliable genome-wide alignments can routinely be built only for relatively closely related viruses whose analysis, however, may be further complicated by recombination events (see below). Using the second approach, genes with no evidence for recombination may be merged (concatenated) in a single data set that may be used to produce a superior phylogenetic signal compared to those generated for distinct genes or entire genomes. For viruses with small genomes or for a diverse set of viruses, it is common practice to use a single gene to infer virus phylogeny. Although the results produced may be the best models describing evolutionary history of a group of viruses, the validity of this gene-based approach for the genome-wide extrapolation remains a point of debate.

When the tree is reconstructed for (part of) genomes, an underlying assumption is that the analyzed data set has a uniform phylogeny. This condition may be violated due to homologous recombination between (closely) related viruses. In phylogenetic terms, recombination may be revealed through incongruency of trees built for a genome region, where recombination occurs, and other regions. Trees may also become incongruent due to various technical reasons related to the size and diversity of a virus data set and deviations of the evolutionary process among lineages. These characteristics complicate interpretation of the congruency test, which is widely used in different programs to identify recombination in viruses.

Applications of Phylogeny in Virology

Phylogenetic analysis is used in a wide range of studies to address both applied and fundamental issues of virus research, including epidemiology, diagnostics, forensic studies, phylogeography, origin, evolution, and taxonomy of viruses. First question to be answered during an outbreak of a virus epidemic concern the virus identity and origin. Answers to these questions form the basis for implementing immediate practical measures and prospective planning enabling specific and rapid virus detection and epidemic containment, which may include the use and development of antiviral drugs and vaccines. Among different analyses performed for virus identification at the early stage of a virus epidemic, the phylogenetic characterization is used for determining the relationship of a newly identified virus with all other previously characterized and sequenced viruses.

Results of this analysis may be sufficient to provide answers to the questions posed, as regularly happens with closely monitored viruses that include most human viruses of high social impact, for example, influenza, human immunodeficiency virus (HIV), hepatitis C virus (HCV), poliovirus, and others. For these viruses, there exist large databases of previously characterized isolates and strains that comprehensively cover the natural diversity. Should a newly identified virus belong to one of these species, chances are that it has evolved from a previously characterized isolate or a close variant and this immediately becomes evident in the clustering of these viruses in the phylogenetic tree. Combining the results of gene-specific and genome-wide phylogenetic analysis allows one to determine whether recombination contributed to the isolate origin. For instance, recombination was found to be extremely uncommon in the evolution of HCV, but not for poliovirus lineages that recombine promiscuously, also with closely related human coxsackie A viruses, both of which belong to human enteroviruses.

When an emerging infection is caused by a new never-before-detected virus, the phylogenetic analysis is instrumental for classification of this virus and in the case of a zoonotic infection, for determining the dynamic of virus introduction into the (human) population and initiating the search for the natural virus reservoir. This was the case with many emerging infections including those caused by most recently introduced Nipah virus, a paramyxovirus, and SARS coronavirus (SARS-CoV). With the latter virus, poor sampling of the coronavirus diversity in the SARS-CoV lineage at the time, some uncertainty over the relationship between phylogeny and taxonomy of coronaviruses, and the complexity of phylogenetic analysis of a virus data set including isolated distant lineages led to considerable controversy over the exact evolutionary position of SARS-CoV among coronaviruses. Since then, the matter has largely been resolved but this experience illustrates some challenges in inferring virus phylogeny.

The search for a zoonotic reservoir of an emerging virus may involve a significant and time-consuming effort that requires numerous phylogenetic analyses of ever-expanding sampling of the virus diversity generated in pursuit of the goal. In this quest, phylogenetic analysis canalizes the effort and provides crucial information for reconstructing parameters of major evolutionary events that promoted the virus origin and spread. For instance, intertwining HIV and simian immunodeficiency virus (SIV) lineages in the primate lentivirus tree led to postulation that the existing diversity of HIV in the human population originated from several ancestral viruses independently introduced from primates over a number of years. Similar phylogenetic reasoning was used to trace the origin of a local HIV outbreak to a common source of HIV introduction through dental practice (known as 'HIV dentist' case). These are typical examples illustrating the utility of phylogenetic analysis for epidemiological and forensic studies.

Geographic distribution of places of virus isolation is another important characteristic relative to which virus phylogeny may be evaluated. This field of study belongs to phylogeography. The evolution of human JC polyomavirus provides an example of confinement of circulation of virus clusters to geographically isolated areas, represented by three continents. Recent identification of West Nile virus in the USA illustrates geographical expansion of an Old World virus into the New World. Analysis of phylogenies of field isolates of rabies virus of the family *Rhabdoviridae* sampled from different animals over Europe led to the recognition that interspecies virus expansion is occuring faster when compared to geographical expansion.

Phylogenies also reveal information about the relative strength of the virus–host association over time. In some virus families (e.g., the *Coronaviridae*) host-jumping events may be relatively frequent, including the emergence of at least two human viruses, dead-end SARS-CoV and successfully circulating human coronavirus OC43 (HCoV-OC43). At the other end of the spectrum one finds the family *Herpesviridae*. Extensive phylogenetic analysis of herpesviruses and their hosts showed a remarkable congruency of topologies of trees indicating that this virus family may have emerged some 400 million years ago and that herpesviruses cospeciate with their hosts.

Phylogenetic analysis becomes increasingly important in virus classification (taxonomy) and relies on complex multicharacter rules applied to separate virus families by respective 'study groups'. For viruses united in high-rank taxa above the genus level, phylogenetic clustering for most conserved replicative genes is commonly observed and used in the decision making process. For instance, human hepatitis E virus, originally classified as a calicivirus using largely virion properties, was eventually expelled from the family due to poor fit of genome characteristics, including results of phylogenetic analysis. Phylogenetic considerations also played an important role in forming recently established families, for example, the *Marnaviridae* and *Dicistroviridae*. In contrast, phylogenetic analysis has been of relatively little use in the taxonomy of large DNA phages which has been developed in such a way that existing families may unite phages with different gene layouts and phylogenies. The relationship between phylogeny and taxonomy is evolving and in future one might hope for important advancements that improve cross-family consistency in relation to phylogeny.

See also: Emerging and Reemerging Virus Diseases of Vertebrates; Evolution of Viruses; Origin of Viruses; Taxonomy, Classification and Nomenclature of Viruses; Virus Classification by Pairwise Sequence Comparison (PASC); Virus Databases; Virus Evolution: Bacterial Viruses; Virus Species.

Further Reading

Dolja VV and Koonin EV (eds.) (2006) Comparative genomics and evolution of complex viruses. *Virus Research* 117: 1–184.

Domingo E (2007) Virus evolution. In: Knipe DM, Howley PM, Griffin DE, et al. (eds.) *Fields Virology*, 5th edn., pp. 389–421. Philadelphia, PA: Wolters Kluwer, Lippincott Williams and Wilkins.

Domingo E, Webster RG, and Holland JJ (eds.) (1999) *Origin and Evolution of Viruses*. San Diego: Academic Press.

Fauquet CM, Mayo MA, Maniloff J, Desselberger U, and Ball LA (2005) *Virus Taxonomy: Eighth Report of the International Committee on Taxonomy of Viruses*. San Diego, CA: Elsevier Academic Press.

Felsenstein J (2004) *Inferring Phylogenies*. Sunderland, MA: Sinauer Associates, Inc.

Gibbs AJ, Calisher CH, and Garcia-Arenal F (eds.) (1995) *Molecular Basis of Virus Evolution*. Cambridge: Cambridge University Press.

Moya A, Holmes EC, and Gonzalez-Candelas F (2004) The population genetics and evolutionary epidemiology of RNA viruses. *Nature Reviews Microbiology* 2: 279–288.

Page RD and Holmes EC (1998) *Molecular Evolution. A Phylogenetic Approach.* Boston: Blackwell Publishing.

Salemi M and Vandamme AM (eds.) (2003) *The Phylogenetic Handbook. A Practical Approach to DNA and Protein Phylogeny.* Cambridge: Cambridge University Press.

Villarreal LP (2005) *Viruses and Evolution of Life.* Washington, DC: ASM Press.

Picornaviruses: Molecular Biology

B L Semler and K J Ertel, University of California, Irvine, CA, USA

© 2008 Elsevier Ltd. All rights reserved.

Glossary

***Cre* (*cis*-acting replication element)** First found in human rhinovirus genomic RNA and subsequently identified in other picornavirus genomes, the *cre* acts as template for the viral RNA polymerase 3D^{pol} to uridylylate VPg to VPg-pU-pU. Evidence suggests that this *cre* can function in *trans* as well.

IRES (internal ribosome entry site) An RNA sequence typically characterized by extensive nucleic acid secondary structure. The 40S ribosomal subunit of the cellular translation machinery interacts with RNA stem loops/sequence, and subsequently allows translation of downstream RNA sequence of the IRES. Translation, therefore, proceeds without recognition of a 5′ cap. Utilized by some virus families (including *Picornaviridae*) and cellular messenger RNAs.

Polyprotein In the context of the discussion of *Picornaviridae*, this refers to the long protein resulting from translation of the single open reading frame of picornavirus RNA. The polyprotein is processed by viral proteinases to yield mature viral proteins.

Positive-strand RNA A single molecule of picornavirus RNA that encodes functional viral protein when translated in the 5′–3′ direction. This is the 'sense' orientation of picornavirus RNA as it enters the cell that is also encapsidated in progeny virions.

RNP complex Ribonucleoprotein complex. Describing a stable interaction of RNA and protein(s), either *in vivo* or *in vitro*.

Uridylylation Refers to the addition of two uridylate residues to the VPg molecule by the picornavirus RNA-dependent RNA polymerase 3D^{pol} (and other viral proteins) using a viral RNA template.

VPg Virus protein, genome-linked. Also known as 3B, the function of VPg is to act as a protein primer for the picornavirus RNA-dependent RNA polymerase. Following uridylylation, VPg-pU-pU is covalently attached to the 5′ end of picornavirus RNAs (positive- and negative-strands).

Overview of Virus Family *Picornaviridae*

Introduction

Viruses belonging to the family *Picornaviridae* are small (Latin *Pico*) RNA (*rna*) viruses whose host range is typically restricted to mammals. Genera associated with *Picornaviridae* include erbovirus, teschovirus, kobuvirus, aphthovirus, cardiovirus, enterovirus, hepatovirus, parechovirus, and rhinovirus. The first three of these genera are relatively recent additions to the picornavirus family, and the last four contain pathogens that are the most extensively studied picornaviruses capable of infecting humans. Particularly, poliovirus of the enterovirus family is widely considered to be the 'prototypical' picornavirus, and perhaps the most feared by humans due to the potential for poliovirus infection to result in paralytic poliomyelitis. This debilitating affliction can result in paralysis of one or more limbs in an infected individual, and in rare cases even death. Human rhinovirus infection in humans results in the common cold, and is one of the most prevalent diseases throughout the world. While in the developed world the common cold is often seen merely as an inconvenience at worst, it is the most important cause of asthma exacerbations, and there is no effective vaccine for the virus nor any effective medical treatment for infection.

Due to the intense interest in *Picornaviridae* based upon the diseases associated with picornavirus infections, the virus family has received extensive scientific attention to understand the mechanisms of gene expression and replication of its members. In turn, insights into the manner in which members of the *Picornaviridae* propagate have allowed a better understanding of how these viruses cause disease and also how other unrelated virus families replicate within their own host cell systems. The aim of this article is to discuss the general mechanisms of gene expression and propagation of the *Picornaviridae* as well as to highlight some of the differences between the individual picornavirus family members.

Classification

From a disease perspective, individual virus species of the family *Picornaviridae* are grouped by genus based on their pathogenic properties and/or route of infection (**Table 1**). For example, the enterovirus genus includes poliovirus and coxsackievirus based upon the natural oral route of entry into the host and replication in gut tissue. The genus cardiovirus includes encephalomyocarditis virus (EMCV), which infects heart and nervous tissues in rodents. Based upon the ability of neutralizing antibody to recognize capsid antigens on the surface of the virion the species of *Picornaviridae* are further subclassified on the basis of serotype. This can range from a single serotype in the case of hepatitis A virus (HAV), or, in the case of rhinovirus, to ~100 identified serotypes.

In terms of molecular genetics, picornaviruses are broadly classified according to the internal ribosome entry site (IRES) in the 5′ noncoding region (NCR). Based upon similarity in RNA secondary structure and sequence studies, three distinct types of picornavirus IRES elements have been characterized. The genomes of enteroviruses (poliovirus and coxsackievirus) and rhinovirus contain type I IRESs, while type II IRESs are found in the genomes of cardiovirus, aphthovirus, erbovirus, and parechovirus. The HAV IRES is considered an outlier of the type II IRES and is classified as a type III IRES. The recently identified teschovirus and kobuvirus IRES elements are yet to be examined in enough detail for placement into a specific type. The function and characteristics of picornavirus IRES types are discussed in the section on 'Translation' of this article.

The other primary difference that distinguishes the picornavirus genera relates to structure and function of the proteins encoded by the polyprotein. For example, the genomes of enteroviruses, rhinoviruses, parechoviruses, and hepatoviruses do not code for a 'leader' (L) protein. Instead, many of these viruses encode a proteinase in the P2 region of the genome designated 2A. The 2A proteinase is structurally similar among the enteroviruses and rhinoviruses. Cardiovirus, aphthovirus, erbovirus, teschovirus, and kobuvirus genomes code for an L protein at the amino terminus of the uncleaved polyprotein, but only the aphthovirus L protein has been reported to be a proteolytically active proteinase. Parechovirus and HAV RNA genomes do not encode an L protein and the 2A regions of the genomes of these viruses are not proteolytically active.

Genome Structure and Features

All picornaviruses contain a single molecule of RNA of messenger RNA (mRNA) sense (hereafter referred to as positive-sense) within the capsid (**Figure 1**). The nucleotide (nt) length of the RNA genome found in virion particles is quite short compared to most other viruses, ranging from 6500 to 9500 nt in length. The 5′ end of the viral genome is covalently linked to the viral polypeptide VPg (virus protein, genome linked). The 5′ RNA terminus also contains RNA secondary structures important for viral RNA synthesis (see the section titled 'Viral RNA replication'). In comparison to cellular mRNAs, the picornavirus 5′ NCR is quite long often in excess of 700 nt and up to 1200 nt. Encompassing nearly 10% of the total RNA genome length, the 5′ NCR contains the picornavirus IRES, which allows initiation of translation of the viral genome. The 3′ NCRs of picornaviruses are, in contrast to many cellular mRNAs, quite short, ranging from 50 to 100 nt in length. Lastly, a homopolymeric poly(A) tract encoded by the viral genome follows the 3′ NCR and may participate in the formation of stable RNA secondary structure(s) with the heteropolymeric 3′ NCR. The poly(A) tract likely stabilizes and protects the viral RNA

Table 1 Members of *Picornaviridae*

Genus	Representative species	Associated disease
Aphthovirus	Foot-and-mouth disease virus (FMDV)	Foot-and-mouth disease (livestock)
Cardiovirus	Encephalomyocarditis virus (EMCV), Theiler's murine encephalomyocarditis virus (TMEV)	Encephalomyelitis, myocarditis (mice and livestock), demyelination in mice
Enterovirus	Poliovirus, coxsackievirus	Poliomyelitis; hand, foot, and mouth disease; myocarditis
Erbovirus	Equine rhinitis virus	Upper respiratory infection (horses)
Hepatovirus	Hepatitis A virus	Acute liver disease
Kobuvirus	Aichi virus	Gastroenteritis
Parechovirus	Human parechovirus	Gastroenteritis, paralysis, encephalitis, neonatal carditis
Rhinovirus	Human rhinovirus	Common cold
Teschovirus	Porcine teschovirus	Encephalitis, paralysis (swine)

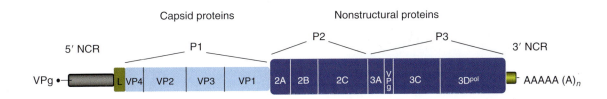

Figure 1 Organization of a typical picornavirus RNA genome. VPg is covalently attached to the 5′ terminus of genomic RNA. The 5′ NCR contains RNA sequences forming RNA structural elements used in virus RNA translation and replication. The coding region is divided into three sections consisting of the P1, P2, and P3 regions. The RNA region of P1 is translated and processed to virion capsid proteins. Some of the functions of the nonstructural proteins contained in the P2 and P3 regions are listed above. Following the P3 region, the 3′ NCR and poly(A) tract function in replication and RNA stability.

genome from intracellular degradation enzymes, much as the poly(A) tracts of cellular mRNAs protect their cognate RNAs. Despite the genetic and phenotypic differences that distinguish the members of the family *Picornaviridae*, all members utilize similar strategies of genome replication and production of infectious virus particles in their respective host-cell environments. Due to a positive-sense RNA genome, picornavirus RNA can immediately be translated into polyprotein upon entry into the cell cytoplasm, and does not require the inclusion of viral proteins such as polymerase in the capsid (as in the case of negative-strand RNA viruses). Moreover, nearly all picornaviruses inhibit cap-dependent translation of host cell mRNAs while utilizing a cap-independent mechanism to translate their own genome, reducing the competition for translation factors between the virus and host cell mRNAs in favor of the virus.

Picornavirus Entry and Gene Expression

Capsid

Picornavirus genomes are packaged into icosahedral virion structures consisting of 60 individual protomers formed by the structural proteins in the P1 coding region: VP1, VP2, VP3, and VP4. A single copy of VP1, VP2, and VP3 is organized into a triangular subunit on the surface of the capsid. Five of these subunits are organized around a fivefold symmetrical axis to form a pentameric protomer (**Figure 2**). Twelve of these pentameric subunits (a total of 60 triangular protomers) form the icosahedron, with twofold and threefold axes of symmetry located at the connection points. The small structural protein VP4 is not present on the outside of the mature virion; it is located in the interior of the capsid structure and interacts with the encapsidated RNA genome. During assembly, inclusion of the RNA genome coincides with the processing of a precursor polypeptide (VP0) into VP4 and VP2. The mechanism for the maturation of VP0 to VP4 and VP2 is currently unknown. What is known is that the viral proteinases 2A, L, and 3C/3CD do not appear to be involved, suggesting a cellular factor may be required for virion maturation or that viral RNA/capsid proteins effect this cleavage.

The virion capsid proteins VP1, VP2, and VP3 of enteroviruses and rhinoviruses protrude from the surface of the capsid structure at the axes of symmetry, forming the 'mesa' at the fivefold axes and the 'propeller' at the threefold axes (**Figure 2**). Observations of these extrusions using X-ray crystallography give the impression of clefts between the axes of symmetry, termed 'canyons'. The canyon region of *Picornaviridae* was initially proposed to be the site of virion binding to cell receptors prior to entry into the cytoplasm, based upon its predicted inaccessibility to neutralizing antibody. This hypothesis has been supported by cryoelectron micrograph studies of cell receptor bound to the canyon region of rhinovirus and coxsackievirus virion particles. While the canyon region of cardioviruses and aphthoviruses is not nearly as pronounced as in enteroviruses and rhinoviruses, neutralization studies investigating the virion capsid of foot-and-mouth disease virus (FMDV) indicate that the G–H loop of the VP1 capsid protein is important for cell receptor recognition.

A summary of some of the cellular molecules identified as cellular receptors for picornaviruses is listed in

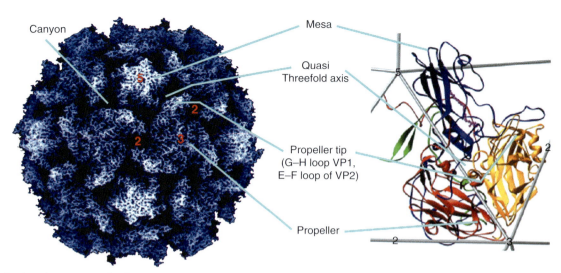

Figure 2 Atomic structure of poliovirus capsid. Left panel: A radially depth-cued rendering of the atomic model of the structure of the 160S poliovirus virion particle. Right panel: An expanded representation of a single protomer showing ribbon diagrams of VP1 (blue), VP2 (yellow), VP3 (red), and VP4 (green) overlaid on an icosahedral framework. The fivefold, threefold, and twofold axes are indicated by numbers. Cyan lines point to prominent surface features in the two panels. Reproduced from Bubeck D, Filman DJ, Cheng N, Steven AC, Hogle JM, and Belnap DM (2005) The structure of the poliovirus 135S cell entry intermediate at 10-angstrom resolution reveals the location of an externalized polypeptide that binds to membranes. *Journal of Virology* 79: 7745–7755, with permission from the American Society for Microbiology.

Table 2 Examples of cell receptors for picornavirus entry

Genus	Species	Cellular receptor
Aphthovirus	Foot-and-mouth disease virus	Integrin (strains A12), heparan sulfate (strain O1)
Cardiovirus	Encephalomyocarditis virus	VCAM-1
Enterovirus	Coxsackievirus B1-6	CAR, DAF (CD55)
	Poliovirus 1–3	CD155 (Pvr)
Hepatovirus	Hepatitis A virus	HAVcr-1
Rhinovirus	Major rhinovirus group	ICAM-1
	Minor rhinovirus group	LDL

VCAM, Vascular cell adhesion molecule; CAR, coxsackie-adenovirus receptor; DAF, decay-accelerating factor (CD55); Pvr, poliovirus receptor (CD155); HAVcr-1, hepatitis A virus cell receptor; ICAM-1, intercellular adhesion molecule; LDL, low-density lipoprotein.

Table 2. Some of the receptors, such as the rhinovirus major group receptor ICAM-1, have known cellular functions, while the normal cellular function of other virus receptors, such as the HAV receptor HAVcr-1, have not been determined. For poliovirus, entry of the virion RNA into the cell is not fully understood but may occur via a variation of receptor-mediated endocytosis; however, acidification of the endosome may not be necessary as release of the viral RNA is not pH dependent. Whether the mature virion is internalized wholly in a membrane-bound vesicle or whether the RNA exits at the plasma membrane surface is also unknown. Once bound to CD155, the virion capsid induces a conformational change and the internal VP4 capsid protein is extruded to the external surface to contact the plasma membrane along with the amino terminus of VP1 around the fivefold axis of symmetry, forming a channel by which the RNA has been proposed to enter the cytoplasm and begin the replicative cycle through translation of the RNA genome.

Translation of Genomic RNA

Once the virus has been uncoated and its RNA released into the host cell cytoplasm, the positive-strand RNA genome is translated. This viral mRNA contains *cis*-acting RNA sequence and structural elements distinct from those of host cell mRNAs. Eukaryotic mRNAs contain a 7-methyl guanosine cap that is attached to the nascent transcript in the nucleus during RNA processing. In addition to protecting the transcript from degradation, recognition of the cap structure by the eIF-4F complex serves as a scaffold for the assembly of other initiation factors and the 40S ribosomal subunit. Picornavirus genomes lack anything resembling a cap structure. Instead, the 5′ end of genomic RNA (as well as negative-strand RNA intermediates) is covalently bound to the viral protein VPg. This protein does not appear to have a role in translation of the polyprotein. Studies carried out with picornavirus mRNAs associated with translating polysomes have not detected VPg bound to viral RNAs, and further evidence indicates an unknown cellular factor as capable of cleaving the RNA–protein bond.

Translation of picornavirus positive-strand RNAs does not follow the conventional scanning model of most cellular mRNAs. Analysis of the first completely sequenced picornavirus, poliovirus, indicated that the 5′ NCR contains numerous AUG start codons in contexts favorable for initiation of translation that were not utilized by the virus. Further, biochemical analysis of the 5′ NCR determined that this region of viral RNA contains several RNA secondary structures with sufficient thermodynamic stability to most likely inhibit the 40S ribosomal subunit from scanning toward a start codon. Subsequent studies demonstrated that an IRES is utilized to translate the poliovirus RNA genome. All picornaviruses contain an IRES in their 5′ NCRs. IRES elements have also been identified in the genomes of other RNA viruses as well as in cellular mRNAs. An example of the mechanism of picornavirus RNA translation is shown in **Figure 3** for the poliovirus IRES.

In contrast to host cell cap-dependent translation, cap-independent translation initiation from an IRES does not use a cap structure at the 5′ end of the RNA to interact with the eIF-4F complex. In fact, many picornaviruses will actually proteolytically cleave or otherwise sequester eukaryotic initiation factors with high efficiency. The notable exception is HAV; cleavage of the eIF-4G component of the eIF-4F complex inhibits IRES activity, suggesting that an intact eIF-4F complex may interact with a site within the IRES to initiate HAV translation. While cleaved eIF-4G is rendered nonfunctional for cellular mRNA translation, eIF-4A and eIF-4B may participate with the cleaved portion of eIF-4G to bring the eIF-1A/eIF-2-met tRNA/eIF-3/40S ribosome complex to the viral IRES for enteroviruses, rhinoviruses, and aphthoviruses.

By convention, the RNA structure elements of the picornavirus type I IRES (enteroviruses and rhinoviruses) and HAV (denoted as having a type III IRES) are described by roman numerals, while the distinct 5′ NCR secondary structure elements of the type II IRES (including those of cardioviruses and aphthoviruses) are denoted by letters.

Chemical and enzymatic RNA structure probing aided by computer modeling studies have determined that the enterovirus and rhinovirus 5′ NCRs (approximately 700–800 nt in length) contain six stem–loop structures, with the region containing the IRES encompassing stem loops II through VI (**Figure 4**). Extensive genetic analysis of type I stem–loop structures involving point mutations, substitutions, or complete deletions of specific stem–loop structures indicates that specific RNA sequence elements

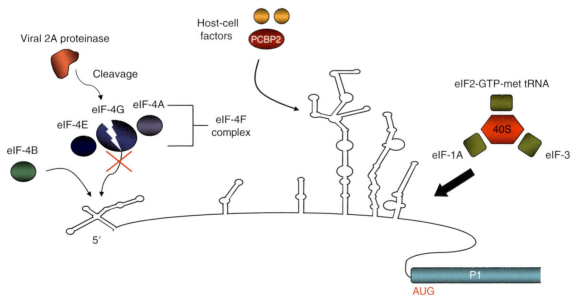

Figure 3 Cap-independent RNA translation of poliovirus RNA. Binding of cellular protein PCBP2 and other host cell factors to RNA secondary structures in the IRES element stabilizes the RNA, allowing association of the 40S ribosomal subunit with eIF-1A, eIF2-GTP-met, and eIF-3 bound. Following binding of the 60S subunit to form the complete 80S ribosome, initiation of RNA translation begins at the AUG start codon preceding the P1 coding region. Translation of the genomic RNA leads to the generation of the 2A proteinase, which will shut off host cell cap-dependent RNA translation through inactivation of the eIF-4F complex via cleavage of eIF-4G. Also note the absence of a cap structure at the 5′ end of the RNA.

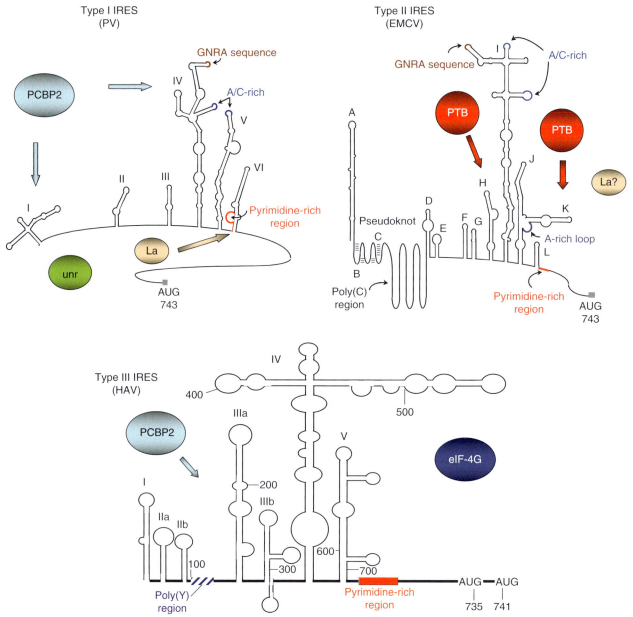

Figure 4 Comparison of IRES elements in the 5′ NCR of picornavirus RNA genomes. The predicted RNA secondary structure motifs associated with type I, II, and III IRES elements are represented by poliovirus, encephalomyocarditis virus, and hepatitis A virus, respectively. The secondary structure of the RNA sequence for each of the 5′ NCRs has been predicted by computer modeling and confirmed by RNA structure probing. Also depicted are some of the cellular factors known to bind to each particular IRES type, and their putative binding sites. Reproduced from Stewart SR and Semler BL (1997) RNA determinants of picornavirus cap-independent translation initiation. *Seminars in Virology* 8: 242–255, with permission from Elsevier.

are also important for IRES function. In particular, the poliovirus stem–loop IV contains a short region of single-stranded cytidines and a GNRA tetra-loop sequence important for cellular protein binding and efficient viral genome translation. Similar sequence and structural elements have also been predicted for the IRES of the closely related enterovirus coxsackievirus B.

The type II IRES elements of FMDV, EMCV, and the recently classified IRES of parechoviruses are bounded between stem-loops D through L (**Figure 4**) and are contained within a 5′ NCR of greater length than their enterovirus counterparts, from over 800 nt for EMCV to 1200 nt for FMDV. In addition to a greater number of RNA stem–loop structures within the 5′ NCR, some type II IRES elements also include a poly(C) region preceding stem–loop D that is missing in picornaviruses with a type I IRES (Theiler's murine encephalomyocarditis virus, a cardiovirus related to EMCV, does not contain a poly(C) tract). The function of this homopolymeric stretch of cytosine residues (ranging from 60 to over 400 nt) is unknown. Despite the structural differences compared to type I IRES elements, the type II IRES also contains a GNRA tetra-loop sequence in the stem–loop I, as well as A/C-rich regions that are found in type I IRES elements. The IRES of HAV is contained within stem loops IIIa to V, and although its stem–loop structures more closely resemble those of the type II IRES, the requirements for efficient IRES function are different than the type I or type II class of elements.

Picornavirus IRES elements operate to direct the assembly of initiation complexes using the host cell protein synthesis machinery. This assembly is mediated by the interactions between IRES RNA elements and cellular factors, including cellular proteins that do not appear to have a role in cap-dependent cellular translation. These RNA–protein interactions between viral RNA and host cell proteins have been identified as either being crucial for optimal translation efficiency specifically for one IRES type, or important for translation of more than one IRES type, perhaps acting to stabilize the RNA secondary structure that allows internal ribosome entry. For example, rabbit reticulocyte lysate (RRL) is a commonly used and efficient *in vitro* system for the translation of mRNA. However enterovirus, rhinovirus, and HAV IRES elements initiate translation aberrantly in RRL. In particular, the efficiency and fidelity of poliovirus translation in RRL can be increased significantly by the addition of crude extracts from HeLa cells or by high concentrations of the human La protein. In contrast, the type II IRES elements of EMCV, FMDV, and Theiler's virus RNA direct translation efficiently in RRL without the addition of La or other factors. Another cellular protein, polypyrimidine tract binding protein (PTB), interacts with poliovirus, human rhinovirus, EMCV, and FMDV IRES elements (**Figure 4**). The high-affinity RNA binding of PTB to the type II IRES elements of EMCV and FMDV suggests a role for the protein in directing viral RNA translation. Mutations introduced into the EMCV IRES designed to abrogate binding of PTB to the IRES result in a loss of IRES activity. **Table 3** lists some of the known interactions of cellular proteins with picornavirus IRES elements.

The cellular protein poly(rC) binding protein (PCBP) binds to the 5′ NCR (specifically stem–loop I and IV) of enterovirus and rhinovirus genomic RNAs and is necessary for the activity of the IRES element. Depletion of the PCBP isoform PCBP2 from HeLa cytoplasmic extracts drastically reduces the translation efficiency of poliovirus RNA, which can be restored with the addition of recombinant PCBP2. Intriguingly, while FMDV and EMCV IRES elements bind PCBP2, depletion of PCBP2 does not affect the ability of these IRESs to direct translation *in vitro*. A substantial amount of research has focused on the

Table 3 Examples of host proteins interacting with picornavirus 5′ NCRs

Cell proteins	Virus RNA binding
La autoantigen	PV, HRV, CVB, EMCV, FMDV
PCBP2	PV, HRV, CVB, FMDV, HAV
unr	PV, HRV
PTB	PV, HRV, EMCV, FMDV
ITAF-45	FMDV
eIF-4A	EMCV
eIF-4B	EMCV, PV, FMDV
eIF-4E	EMCV
eIF-4G	EMCV, FMDV
eIF-2	EMCV, PV

role of PCBP in poliovirus infection, as binding of PCBP (either as PCBP1 or PCBP2) to stem-loop I of poliovirus is necessary for RNA replication (described below).

Polyprotein Processing

Ribosome recognition and initiation within the picornavirus IRES allows the translation of the long polyprotein (approximately 250 kDa) encoded by the single open reading frame (ORF) encoded by picornavirus genomes. The viral polyprotein is proteolytically cleaved by the viral proteinases L (encoded by aphthoviruses), or 2A (encoded by enteroviruses and rhinoviruses), and 3C and 3CD (encoded by all picornaviruses) (**Figure 5**). Viral protein processing yields the precursor molecules P1, P2, and P3. The P1 region of the polyprotein is further processed to three or four distinct proteins, which are the structural components of the viral capsid, and in the case of cardioviruses, aphthoviruses, erboviruses, teschoviruses, and kobuviruses, the nonstructural protein L. The proteolytically processed P2 precursor contains proteins necessary for restructuring of cytoplasmic membranes and vesicles utilized in virus replication, for the shutoff of host cell translation, and for viral RNA replication. The P3 precursor is processed to produce major proteins necessary for RNA replication, including the RNA-dependent RNA polymerase 3D^{pol}, and the viral proteinases 3C and 3CD, both of which are responsible for the majority of viral polyprotein cleavage events. The translation stop codon for picornavirus genomes terminates synthesis of the polyprotein immediately after synthesis of the P3 polyprotein, at the carboxy terminus of the 3D^{pol} sequence.

Substrates of Viral Proteinases

Picornavirus Capsid Proteins

The P1 sequence encodes the capsid proteins VP4, VP2, VP3, and VP1. Initial processing of the viral polyprotein generates VP0 (a precursor to VP4 and VP2), VP3, and VP1. Parechoviruses and kobuviruses do not process the

Picornaviruses: Molecular Biology

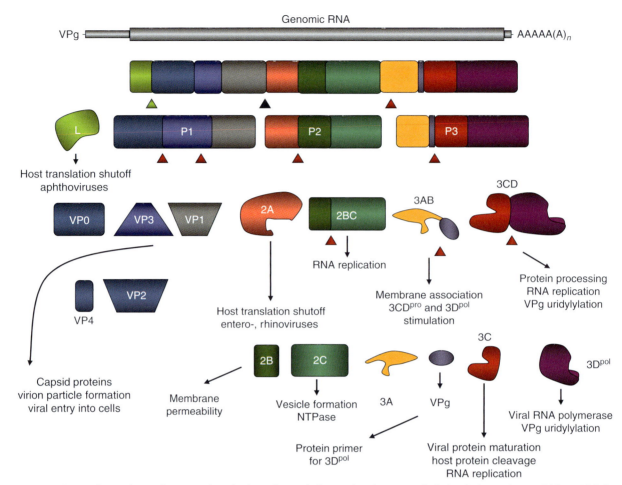

Figure 5 Picornavirus polyprotein processing. As the polyprotein is translated, autocatalysis by viral proteinases 2A/L or 3C/3CD results in the P1, P2, and P3 precursors. (The L protein of cardioviruses is not a proteinase and thus is not released by autocatalysis, and is instead released by 3C/3CD mediated cleavage from the P1 precursor.) These precursors are further processed into the virus structural proteins (P1) or nonstructural proteins (P2 and P3). The functions of the precursor and mature virus proteins are described above. The color-coded triangles correspond to specific cleavage events by the viral proteinases (green, L; black, 2A; red, 3C/3CD).

P1 capsid precursor in a manner analogous to how poliovirus processes VP0; therefore, complete processing of the capsid proteins results in only three proteins used in the formation of the virion capsid. For poliovirus, cleavage of the P1 region into the mature capsid components requires the proteolytic activity of 3CD. For all picornaviruses that process VP0 into VP4 and VP2, an unknown host or viral factor may effect this novel cleavage event.

Viral Proteins Involved in RNA Replication

Viral protein 2B is known to alter and increase the permeability of intracellular membranes including the endoplasmic reticulum. The 2B protein of coxsackievirus B3 (CVB3) contains a region of amino acids that resemble a cationic α-helix. Infection of tissue culture cells with a CVB3-containing mutations in the 2B α-helix predicted to disrupt the cationic character of the domain resulted in a decrease in virion progeny being released into the extracellular space, suggesting that 2B, possibly in conjunction with the viral proteins 2C/2BC, may facilitate virion release during later stages of picornavirus infection. While the precise function of picornavirus 2B is unknown, insertion mutations into the hydrophobic region of the poliovirus 2B produced a virus with defects in viral RNA synthesis that was incapable of rescue by complementation with wild-type virus, suggesting a *cis*-acting role for the 2B protein in picornavirus RNA replication.

During poliovirus infection, 2C and its precursor molecule 2BC are the viral factors responsible for induction of membranous vesicles that are rearranged by the 2B protein (see above) creating a vesicle scaffold for the viral replication complex to begin RNA synthesis. Both proteins are known to interact with the vesicles associated with the virus RNA replication complexes, and 2BC and 2C (but not 2B) induce the formation of these membranous vesicles when expressed alone in tissue culture cells.

The 2C coding region contains several putative structural domains that point to its role as a multifunctional protein in the picornavirus replication pathway. Poliovirus 2C protein is predicted to contain two amphipathic α-helices at its amino- and carboxy-termini and a cysteine-rich motif resembling a DNA zinc-binding finger domain. The middle of the coding region contains a domain with sequence elements found in RNA helicases; while helicase activity is yet to be reported, the 2C protein does have NTP-binding and hydrolysis functions. Although its precise function in RNA synthesis is currently unknown, the 2C protein of picornaviruses clearly has a role in viral RNA replication based upon studies using guanidine hydrochloride (Gu–HCl). In the presence of low concentrations of Gu–HCl (2 mM), picornavirus negative-strand RNA replication is specifically inhibited. Sequence analysis of poliovirus mutants resistant to the addition of Gu–HCl revealed lesions in the 2C coding region, indicating that poliovirus sensitivity to Gu–HCl mapped to the 2C protein. Poliovirus 2C and 2BC viral proteins also bind to the RNA stem–loop structure at the 3′ end of negative-strand RNA, suggesting that these proteins may also have a role in positive-strand RNA synthesis. Recent evidence has also suggested that poliovirus 2C may regulate the catalytic activity of the 3C proteinase to process viral precursor molecules.

The carboxy-terminal hydrophobic domain of the viral 3A protein anchors this small (approximately 130 amino acids) polypeptide and its precursor 3AB into cellular membranes. Mutations within the hydrophobic region predictably disrupt the interaction of the poliovirus 3AB precursor with cellular membranes, but more importantly they also disrupt efficient cleavage of viral P2-P3 or P2-3AB protein precursor molecules by the 3C. Thus, 3AB may provide a means of sequestering viral factors required for RNA synthesis and protein processing in the same cytoplasmic location as the vesicles formed from intracellular membranes, providing at the same time the VPg protein primer for RNA replication.

The viral proteinases 3C and 3CD cleave the polyprotein encoded by the genomic RNA to produce mature viral proteins. These proteinases also cleave cellular proteins, perhaps acting to disrupt host cell translation machinery or an antiviral response mounted by the cell. In addition, 3CD also participates in processes critical to viral RNA replication. Interaction of poliovirus 3CD, the precursor of 3C and the polymerase 3D^{pol}, is required to form an RNP complex with PCBP2 and the 5′ end of genomic RNA. This complex has been shown to be required for poliovirus RNA replication, possibly by facilitating negative-strand RNA synthesis via communication of the 5′ and 3′ termini of the genomic RNA. In addition, poliovirus 3CD enhances the rate of uridylylation of VPg from the *cre* element *in vitro*, suggesting that 3CD may assist the 3D^{pol} through either direct interaction or binding the RNA sequence of the *cre* to alter its conformation in a manner suitable to serve as template.

The 3D^{pol} of picornaviruses is the RNA-dependent RNA polymerase responsible for the synthesis of negative- and positive-strand RNAs. *In vitro* studies examining the requirements for poliovirus 3D^{pol} RNA synthesis determined that although the viral polymerase cannot initiate RNA synthesis without a primer sequence, it can elongate a nascent RNA strand following initiation without need of additional viral or cellular proteins. Interestingly, it has been proposed that 3AB may have a stimulatory role in the elongation step.

Host Cell Protein Cleavage and Translation Shutoff

In addition to their role in the maturation of viral proteins, the 2A (enteroviruses, rhinoviruses) and L (aphthoviruses) proteinases also have a role in influencing the translation activity of the host cell by cleaving factors involved in cap-dependent translation initiation. Both proteinases shut down host cell protein synthesis by cleaving the eIF-4G component of the eIF-4F initiation factor complex, resulting in an inability of the amino-terminal domain of eIF-4G to recognize and bind eIF-4E that is associated with the 7-methyl cap structure on the 5′ end of cellular mRNAs. Enterovirus and rhinovirus protein synthesis is upregulated as a consequence of this increase of free ribosomes that are available for internal ribosome entry. EMCV uses an alternate method to inhibit host cell translation by activating a cellular translational repressor, 4E-BP1, which binds cellular eIF-4E and therefore inhibits host cap-dependent translation.

Viral RNA Replication

Genomic RNA Elements Involved in Picornavirus RNA Replication

Like all RNA viruses, picornavirus replication uses a viral-encoded RNA-dependent RNA polymerase to specifically synthesize RNA from a viral RNA template. Picornavirus genomic RNA is released from the capsid into the cytoplasm without replication proteins (save for the viral VPg linked to the 5′ end); therefore, replication of the initial viral genome must follow translation of its coding region to generate the 3D^{pol} and other viral replication proteins. The polymerase uses the protein VPg as a primer to initiate RNA synthesis by catalyzing the addition of two uridylate residues to generate VPg-pU-pU. The initiation site of negative-strand RNA synthesis is at the 3′ poly(A) tract of the genomic positive-sense RNA. Viral negative-strand RNA exists in a duplex with its positive-sense RNA template. This double-stranded RNA intermediary is termed the replicative form, or RF. Negative-strand RNA in turn acts as template for the synthesis of positive-strand RNAs. The ratio of positive- to negative-strand RNA in infected cells has been observed to be approximately 50:1, suggesting that a single negative-strand RNA intermediate acts as a template for the production of multiple positive-strand RNAs. Newly synthesized viral positive-sense RNAs will be translated to produce additional viral proteins, used as additional template RNAs for negative-strand RNA synthesis, or packaged into virions for infection of other host cells (**Figure 6**). This section discusses the RNA elements and viral proteins utilized by picornaviruses and their functions in the replicative cycle, the contributions of cellular proteins and structures, and the unanswered questions regarding picornavirus RNA replication.

In addition to the IRES, picornavirus 5′ NCRs contain additional stable RNA secondary structural elements that are critical for replication of viral positive- and negative-strand RNAs. For entero- and rhinovirus genomic RNAs, this structure is termed stem–loop I or cloverleaf, so named for the cruciform secondary structure predicted to form within the first ∼100 nt of the RNA. The genomes of other picornaviruses including aphthoviruses and cardioviruses also contain an RNA structure at their 5′ termini, termed stem–loop A, which is analogous in function but not in structural conformation to the cloverleaf. For simplicity, the discussion below will primarily refer to the poliovirus stem–loop I RNA structure at the 5′ terminus. An insertion mutation in stem–loop I resulted in a deficiency in viral RNA synthesis. More importantly, a revertant virus containing a mutation in the genomic region encoding the viral proteinase 3C was isolated and found to have rescued the RNA synthesis defect. Binding of precursor 3CD to stem–loop I was confirmed genetically via mutations of stem–loop I nucleotides designed to destabilize the cruciform. Using electrophoretic mobility shift analysis, the RNA affinity of 3CD for stem–loop I was found to be increased when another protein was bound to the RNA cruciform. This protein was later identified as the cellular protein PCBP2, and it associated with stem–loop I and 3CD to produce a ribonucleoprotein

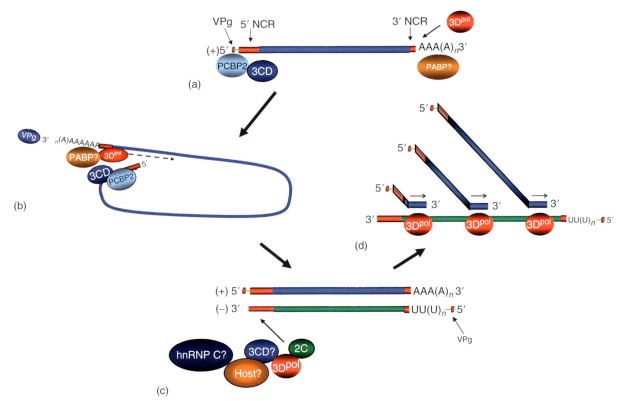

Figure 6 Picornavirus RNA replication. Shown here is a model of the general mechanism of RNA replication of picornaviruses, with the cellular and viral proteins depicted being specific for what is known for poliovirus. (a) The input genomic RNA (positive-sense) associates with viral and cellular factors (including PABP and PCBP2). Viral 3CD and cellular PCBP2 is proposed to interact with the RNA structure at the terminus of the 5′ NCR and contact PABP bound to the poliovirus poly(A) tract. The RNA-dependent RNA polymerase 3D^{pol} begins synthesis of a complementary negative-strand RNA. (b) The resulting double-stranded RNA intermediate (termed the replicative form or RF) forms part of the positive-strand RNA replication complex. (c) RNA replication complexes hypothesized to include host proteins (e.g., hnRNP C), viral proteins 3CD, 2C, and 3D^{pol}. Multiple initiations of 3D^{pol} RNA synthesis from the negative-strand RNA template result in the formation of the partially double stranded replicative intermediate, RI. (d) Positive-strand progeny RNAs will then re-enter the replication cycle for translation of additional viral proteins, or to serve as template for the replication of negative-strand RNA, or to be packaged into progeny virion particles.

(RNP) structure called the ternary complex. Formation of this RNA–protein complex has been confirmed *in vitro* and in poliovirus-infected cells. Depletion of PCBP2 from HeLa cell S10 extracts reduces levels of poliovirus RNA synthesis, supporting the hypothesis that PCBP2 is required for RNA replication. However, the question remained as to what the 3CD/PCBP2/stem–loop I ternary complex contributed to picornavirus RNA replication.

An intriguing possible answer to the role of the 3CD/PCBP2/stem–loop I complex involves another cellular protein, poly(A) binding protein (PABP). In the cell, PABP stabilizes and protects cellular mRNA transcripts by binding the poly(A) tract at the 3′ end of the mRNA. It also participates in translation through interactions with initiation factors at the 5′ end of the mRNA. *In vitro* binding studies using purified PABP and an RNA probe corresponding to the 3′ NCR and poly(A) tract of poliovirus correlated an ability of PABP to interact with this region with a minimal poly(A) tract. Furthermore, purified PABP could interact with the 3CD/PCBP2/stem–loop I ternary complex *in vitro*. PABP and poliovirus 3CD could be co-immunoprecipitated from extracts of poliovirus-infected HeLa cells, suggesting that PABP and the viral ternary complex may also interact during a poliovirus infection. These interactions would suggest that the 5′ and 3′ ends of positive-strand RNA are in close proximity through a 'protein–protein' bridge formed by viral and cellular factors. End-to-end communication of the positive-strand RNA could facilitate initiation of negative-strand RNA synthesis.

An internal RNA sequence within the picornavirus genome is also important in the picornavirus replicative cycle. This sequence has been termed the *cis*-acting replication element, or *cre*. First discovered in human rhinovirus 14 genomic RNA and later identified for other picornaviruses, the *cre* forms a short (50–100 nt) RNA hairpin structure in the positive-strand RNA of picornaviruses, with a conserved AAACAC sequence in the loop portion of the hairpin. Nucleotide mutations that disrupt the conserved sequence inhibit viral RNA synthesis and are therefore lethal to the virus. The position of the *cre* varies widely among picornaviruses; in enteroviruses, it is located within the 2C coding region, while in cardioviruses the *cre* maps to the VP2 coding region. The *cre* of aphthoviruses is unique among picornaviruses in that it is located outside the coding region in the 5′ NCR. The position of the *cre* varies even among serotypes of the same genus. For example, the human rhinovirus 2 *cre* is found in the 2A proteinase coding region, while human rhinovirus 14 *cre* is found in the VP1 RNA sequence. Studies of lethal mutations introduced in the poliovirus *cre* demonstrated that viral RNA replication could be rescued when a second, wild-type sequence was introduced elsewhere in the genome. The *cre* is considered to be a site at which viral replication proteins bind and uridylylate the VPg protein primer. The poly(A) tract at the end of the 3′ NCR was originally thought to be the template RNA used in VPg uridylylation; however, *in vitro* studies using synthetic VPg indicated that the poliovirus *cre* hairpin as a source of RNA template was much more efficient in stimulating the uridylylation of VPg than was the viral RNA poly(A) tract. Mutational analysis of poliovirus RNA has indicated that the *cre* may only be used in positive-strand RNA synthesis, and that the poly(A) tract is the primary source of template RNA for VPg uridylylation during negative-strand RNA synthesis.

Following the $3D^{pol}$ coding region, the 3′ NCR and the 3′ poly(A) tract are thought to be involved in RNA synthesis by acting as a *cis*-acting signal for negative-strand RNA replication. The poly(A) tract encoded by the viral genome may contribute to stable RNA secondary structures consisting of one to three stem–loop structures in the heteropolymeric 3′ NCR. The poly(A) tract likely stabilizes and protects the viral RNA genome from intracellular degradation enzymes, much as the poly(A) tracts of cellular mRNAs do. The poly(A) tract, in conjunction with bound cellular PABP, $3D^{pol}$, and other viral/cellular proteins may also function in negative-strand RNA synthesis. Human rhinovirus and poliovirus RNA genomes lacking the 3′ NCR (but leaving the genome encoded poly(A) tract intact) are still infectious in tissue culture, indicating the 3′ NCR is dispensable for viral replication. Mutant poliovirus lacking a 3′ NCR displays an RNA replication defect following infection of HeLa cells. This defect was shown to occur at the level of positive-strand RNA synthesis. Interestingly, the RNA replication defect exhibited by the poliovirus 3′ NCR deletion mutant was exacerbated in neuronal cells. Such a cell-specific defect suggested that the presence (or absence) of a cellular factor was important for proper poliovirus 3′ NCR function. The cellular protein nucleolin has been reported as being capable of interaction with the poliovirus 3′ NCR; however, its function in viral RNA replication is unknown. Viral proteins including the 3AB and 3CD protein have been reported to bind to the poliovirus 3′ NCR, and the $3D^{pol}$ of EMCV has been shown to bind to 3′ NCR sequences of the EMCV genome.

Synthesis of Positive- and Negative-Strand Viral RNA

Initiation of viral RNA synthesis requires the uridylylation of VPg by the viral polymerase $3D^{pol}$ and may also involve other viral proteins, including 3CD and the VPg precursor 3AB. Utilizing the 3′ poly(A) tract as a source of template, VPg is uridylylated to form VPg-pU-pU, and the $3D^{pol}$ synthesizes negative-strand RNA complementary to the genomic positive-sense RNA. The double-stranded RNA intermediate (RF) then serves as template for multiple initiation events by the RNA polymerase to synthesize positive-strand RNAs. In the infected cell, negative-strand RNA has only been observed either in a duplex with the positive-strand RNA or with several elongating positive-strand RNAs existing in a partially double-stranded RNA complex termed the replicative intermediate (RI). Similar to negative-strand RNA synthesis, positive-strand RNA synthesis requires the uridylylation of VPg and as noted earlier, *in vitro* studies have indicated that during synthesis of positive-strand RNAs uridylylation of VPg was more effective using a *cre* RNA hairpin as the source of template rather than the poly(A) tract. During positive-strand RNA synthesis, multiple positive-strand RNAs are initiated using a single negative-strand template; therefore, the more efficient *cre* acting as the site of uridylylation may be necessary to add VPg-pU-pU to newly synthesized positive-strand RNAs. RNA secondary structures within the 5′ and 3′ ends of positive- and negative-strand RNAs, respectively, may sufficiently destabilize base-pairing of the RNAs to allow the $3D^{pol}$ access to the negative-strand RNA template. The $3D^{pol}$ of poliovirus has also been reported to have duplex unwinding activity. Interestingly, the poliovirus 2C protein contains NTPase activity and has protein structural similarity to RNA helicases. It has also been shown to bind to the 3′ end of negative-strand RNA, suggesting a role in viral RNA replication. Perhaps $3D^{pol}$ and 2C form a complex capable of simultaneous unwinding and chain elongation of the RNA.

In addition to interacting with picornavirus genomic RNAs to promote virus translation and RNA replication, cellular proteins have also been shown to interact with negative-strand RNA intermediates. Cellular protein hnRNP C can interact with the RNA secondary structures formed by the 3′ end of poliovirus negative-strand RNA as well as with viral proteins 3CD and 2C. Complexes of hnRNP C and poliovirus replication proteins may promote initiation of positive-strand RNA synthesis via such RNA–protein complexes.

Unanswered Questions and Conclusions

Despite decades of research invested into the mechanisms by which members of the family *Picornaviridae* propagate their genome in infected cells, many questions remain unanswered. The discovery of the *cre* RNA structures helped address the mechanism of how VPg is primed for positive-strand RNA synthesis. However, how the negative strand acts as template for positive-strand RNA and what viral and/or cellular factors are necessary for RNA initiation are largely unanswered questions. The question pertaining to the role of cellular factors in this process is further complicated by the tissue tropism primarily determined by virus-specific cell receptor usage and by RNA structure/sequence differences among picornaviruses. Cap-independent RNA translation from the IRES element is known to occur for all picornaviruses, yet distinct differences exist even between IRES classes in the ability of the IRES to direct translation. For example, poliovirus and human rhinovirus utilize a type I IRES and share some similarities in the ability to bind cellular factors to promote translation; yet their tissue tropisms differ. What are the sequence and structural differences in the IRES element responsible for virus species-specific differences in translation, and what are the cellular proteins responsible for these differences?

One long-standing question for picornavirus researchers is how template utilization is controlled for translation versus viral RNA replication. Specifically, translating ribosomes on genomic RNA in the 5′ to 3′ direction would interfere with RNA polymerase synthesizing negative-strand RNA (proceeding along the genomic RNA in a 3′ to 5′ manner). Is there a molecular switch to turn off translation to allow RNA replication? Are viral and/or cellular factors involved in translation modified or relocated to permit RNA replication? Or is it an effect of increased levels of local concentrations of proteins that shifts the template RNA from translation to replication competency. Such mechanistic questions underscore the intricate molecular processes through which these pathogens have evolved to maximize the reproduction of their genomes. Study of these mechanisms of viral gene expression has provided insights into the design of vaccines and therapeutic drugs not only for picornaviruses, but also for other virus families.

See also: Rhinoviruses; Poliomyelitis.

Further Reading

Agol VI (2006) Molecular mechanisms of poliovirus variation and evolution. *Current Topics in Microbiology and Immunology* 299: 211–259.

Agol VI, Paul AV, and Wimmer E (1999) Paradoxes of the replication of picornaviral genomes. *Virus Research* 62: 129–147.

Bedard KM and Semler BL (2004) Regulation of picornavirus gene expression. *Microbes and Infection* 6: 702–713.

Belsham GJ (2005) Translation and replication of FMDV RNA. *Current Topics in Microbiology and Immunology* 288: 43–70.

Bubeck D, Filman DJ, Cheng N, Steven AC, Hogle JM, and Belnap DM (2005) The structure of the poliovirus 135S cell entry intermediate at 10-angstrom resolution reveals the location of an externalized polypeptide that binds to membranes. *Journal of Virology* 79: 7745–7755.

Hogle JM (2002) Poliovirus cell entry: Common structural themes in viral cell entry pathways. *Annual Review of Microbiology* 56: 677–702.

Jang SK (2006) Internal initiation: IRES elements of picornaviruses and hepatitis C virus. *Virus Research* 119: 2–15.

Semler BL and Wimmer E (eds.) (2002) *Molecular Biology of Picornaviruses*. Washington, DC: ASM Press.

Stewart SR and Semler BL (1997) RNA determinants of picornavirus cap-independent translation initiation. *Seminars in Virology* 8: 242–255.

Plant Antiviral Defense: Gene Silencing Pathway

G Szittya, Agricultural Biotechnology Center, Godollo, Hungary
T Dalmay, University of East Anglia, Norwich, UK
J Burgyan, Agricultural Biotechnology Center, Godollo, Hungary

© 2008 Elsevier Ltd. All rights reserved.

Introduction

One of the most striking biological discoveries of the 1990s is RNA silencing, which is an evolutionarily conserved gene-silencing mechanism among eukaryotes. Initially, it was identified in plants as a natural defense system thought to have evolved against molecular parasites such as viruses and transposons. However, it has become clear that RNA silencing also plays a very important role in plant and animal development by controlling gene expression.

RNA silencing operates through diverse pathways; however it uses a set of core reactions, which are triggered by the presence of long, perfect or imperfect double-stranded (ds) RNA molecules that are cleaved by Dicer or Dicer-like (DCL), an RNase III-type enzyme, into short 21–26 nucleotide (nt) long molecules, known as short interfering RNAs (siRNA) and microRNAs (miRNA). These small RNAs are then incorporated into a protein complex called RNA-induced-silencing complex (RISC) containing members of the Argonaute (AGO) family. RISCs are guided by the short RNAs to target RNAs containing complementary sequences to the short RNAs and this interaction results in cleavage or translational arrest of target RNAs. In fungi, nematodes, and plants, the silencing signal is thought to be amplified by RNA-dependent RNA polymerases (RDRs); however, RDRs have not been identified in other systems.

Recent discoveries have shown that the RNA-silencing machinery also affects gene function at the level of genomic DNA. The first observation of RNA-guided epigenetic modification was the RNA-directed DNA methylation (RdDM) of chromosomal DNA in plants resulting in covalent modifications of cytosins. RdDM was shown to require a dsRNA that was processed into short 21–24 nt RNAs and it also required some of the components of the silencing machinery such as RDR, DCL, and AGO proteins. RNA-silencing-mediated heterochromatin formation is also an epigenetic process and results in covalent modifications of histones (usually methylation of k9H3) and has been reported in fission yeast, animals, and plants.

It is now clear that RNA silencing is diverse and is involved in a variety of biological processes that are essential in maintaining genome stability, development, and adaptive responses to biotic and abiotic stresses.

Diverse Gene-Silencing Pathways in Plants

Most of the current knowledge about RNA silencing in plants has arisen from analysis of the model plant *Arabidopsis thaliana*. To date, there are several different, yet partially overlapping gene-silencing pathways known in plants. These pathways operate through the production of small RNAs and these small RNA molecules guide the effector complexes to homologous sequences where they exert their effect. Small RNAs are diverse and are involved in a variety of phenomena. They can be classified on the basis of their biogenesis and function as: microRNAs (miRNA), *trans*-acting siRNAs (ta-siRNA), natural *cis*-antisense transcript-derived siRNAs (nat-siRNA), heterochromatin siRNAs (hc-siRNA), and *cis*-acting siRNAs. In *Arabidopsis* the existence of multiple paralogs of RNA-silencing-associated proteins such as ten AGOs, six RDRs and four DCLs may explain the diversification of small RNAs and gene-silencing pathways.

MicroRNA Pathway

miRNAs are mainly 21 nt long and regulate endogenous gene expression during development and environmental adaptation. Important features of miRNAs are that one miRNA duplex is produced from one locus and that they are encoded by miRNA genes that are distinct from the genes that they regulate. Primary miRNAs (pri-miRNAs) are transcribed by RNA polymerase II (Pol II) and possess 5′ caps and poly(A) tails. These long single-stranded primary transcripts with an extensive fold-back structure are cleaved during a stepwise process in the cell nucleus by DCL1. DCL1 interacts with HYL1 (a double-stranded RNA binding protein) to produce the mature miRNA duplex (miRNA:miRNA*). Both strands of the miRNA duplex are methylated (specific to plants) at the 2′ hydroxyl group of the 3′ terminal ribose by HEN1 (an S-adenosyl methionoine (SAM)-binding methyltransferase). Methylation protects miRNAs from polyuridylation and probably from degradation at the 3′ end. The plant exportin-5 homolog HASTY exports the miRNA duplexes to the cytoplasm, where the miRNA strand is incorporated into the RISC complex and the miRNA* strand is degraded.

Plant miRNAs have near-perfect complementarity to their target sites and they bind to these sites to guide the cleavage of target mRNAs. AGO1 is part of the plant RISC, since plant miRNAs are associated with AGO1 and immuno-affinity-purified AGO1 cleaves miRNA targets *in vitro*.

In *Arabidopsis* miRNAs also have a role in RNA-directed DNA methylation, a mechanism generally associated with siRNA-mediated gene silencing. It was shown that PHABULOSA (PHB) and PHAVOLUTA (PHV) transcription factor coding sequences are heavily methylated several kilobases downstream of the miRNA-binding site. It was proposed that DNA methylation of PHB/PHV occurs *in cis* and depends on the ability of the miR165/166 to bind to the transcribed RNAs.

Currently, more than 100 miRNAs have been identified in *Arabidopsis*, but many more are likely to exist. The comparison of miRNAs from different plant species has shown that many of them are conserved; however, there are species-specific miRNAs. These 'young' miRNA genes were proposed to evolve recently and show a high degree of homology to the target genes even beyond the mature miRNA sequence. The list of the currently known plant miRNAs can be found in miRBase.

Trans-Acting siRNA Pathway

ta-siRNAs are 21 nt long endogenous siRNAs and they require components of both the miRNA and *cis*-acting siRNA pathways for their biogenesis. They are produced from noncoding (nc) TAS genes transcribed by Pol II. TAS ncRNAs contain an miRNA-binding site and are cleaved by an miRNA-programmed RISC. The production of cleaved TAS ncRNA therefore requires DCL1, HYL1, HEN1, and AGO1. The cleaved TAS ncRNA is converted into dsRNA by SGS3 (coiled-coil putative Zn^{2+} binding protein) and RDR6. This long dsRNA is then cleaved in a phased dicing reaction by DCL4 to generate 21 nt ta-siRNAs. The phase of the dicing reaction is determined by the initial miRNA cleavage site. Ta-siRNAs are methylated by HEN1 and guide mRNA cleavage through the action of AGO1 or AGO7 RISC. Some of the ta-siRNAs regulate juvenile-to-adult transition but other functions remain to be identified.

Natural *cis*-Antisense Transcript-Derived siRNA Pathway

Ten percent of the *Arabidopsis* genome contains overlapping gene pairs, also known as natural *cis*-antisense gene pairs. Nat-siRNA is a recently discovered class of endogenous siRNAs that are derived from natural *cis*-antisense transcripts regulating salt tolerance in *Arabidopsis*. It was reported that both 24 and 21 nt siRNAs (nat-siRNAs) were generated from a locus consisting of two overlapping antisense genes *P5CDH* and *SRO5*; *P5CDH* mRNA is expressed constitutively, whereas *SRO5* is induced by salt stress. When both transcripts are present, a dsRNA can form by the annealing of *P5CDH* and *SRO5* transcripts, which in turn initiates the biogenesis of a 24 nt nat-siRNA by the action of DCL2, RDR6, SGS3, and NRPD1A (RNA polymerase IVa subunit). This 24 nt nat-siRNA guides the cleavage of the constitutively expressed *P5CDH* transcript and sets the phase for a series of secondary, phased 21 nt nat-siRNAs produced by DCL1. The function of these secondary nat-siRNAs is unclear since the *P5CDH* transcript is also downregulated in their absence (in *dcl1* background) as well. However, in wt plants 21 nt nat-siRNAs mediate the cleavage of *P5CDH* transcript. The presence of many overlapping genes in eukaryotic genomes suggests that other nat-siRNAs may be generated from other natural *cis*-antisense gene transcripts.

Heterochromatin siRNA Pathway

The large majority of endogenous small RNAs is derived from transposons and repeated sequences. To explain their biogenesis, it is proposed that tandem repeats or multiple copies of transposable elements generate dsRNAs. Analysis of mutant plants suggests that these molecules are produced by one of the plant RNA-dependent RNA polymerases (RDR2). The long dsRNAs are cleaved by DCL3 to produce 24 nt long hc-siRNAs that are methylated by HEN1. The accumulation of hc-siRNAs also requires NRPD1a and NRPD2, putative DNA-dependent RNA polymerases collectively called PolIV. The PolIV complex is proposed to be involved in an amplification loop together with the *de novo* methyl-transferases (DRMs), the maintenance methyl-transferase (MET1), and the SWI2/SNF2 chromatin remodeling factor DDM1 that are also required for hc-siRNA production. The resulting hc-siRNAs are incorporated into an AGO4-containing complex which then guides DNA methylation and heterochromatin formation. An overview of these and other proteins with roles in *Arabidopsis* small RNA pathways is shown in **Table 1**.

cis-Acting siRNAs

Exogenous nucleic acid invaders such as viruses and transgenes trigger the production of *cis*-acting siRNAs, which are both derived from and target these molecular invaders. This pathway was the first RNA-silencing pathway identified in plants more than 15 years ago. Since this pathway is activated upon virus infection, it was proposed to be an RNA-based immune system against molecular parasites.

In plants, the *cis*-acting siRNA pathway can be used to induce sequence-specific silencing of endogenous mRNA

Table 1 Plant proteins involved in small RNA pathways

Gene	Alternative name	Protein name	Gene code	Biochemical activity and function	Pathway
AGO1	DND	ARGONAUTE 1	At1g48410	Ribonuclease activity; RNA slicer	miRNA ta-siRNA *cis*-acting siRNA
AGO4		ARGONAUTE 4	At2g27040	Unknown; RNA-directed DNA methylation	hc-siRNA
AGO7	ZIPPY	ARGONAUTE 7	At1g69440	Ribonuclease activity; RNA slicer	ta-siRNA
DCL1	CAF/SIN1/SUS1/EMB76	DICER-LIKE 1	At1g01040	RNase III; miRNA synthesis	miRNA nat-siRNA
DCL2		DICER-LIKE 2	At3g03300	RNase III; 22 nt or 24 nt siRNA synthesis	nat-siRNA
DCL3		DICER-LIKE 3	At3g43920	RNase III; 24 nt siRNA	hc-siRNA *cis*-acting siRNA
DCL4		DICER-LIKE 4	At5g20320	RNase III; 21 nt siRNA	ta-siRNA *cis*-acting siRNA
DDM1	SOM	DECREASE IN DNA METHYLATION 1	At5g66750	SWI2/SNF2-like chromatin remodeling enzyme; maintenance of CG DNA and histone methylation	hc-siRNA
DRM1		DOMAINS REARRANGED METHYLASE 1	At5g15380	Cytosine DNA methyltransferase; *de novo* DNA methylation and maintenance of asymmetric methylation of DNA sequences	hc-siRNA
DRM2		DOMAINS REARRANGED METHYLASE 2	At5g14620	Cytosine DNA methyltransferase; *de novo* DNA methylation and maintenance of asymmetric methylation of DNA sequences	hc-siRNA
HEN1		HUA ENHANCER 1	At4g20910	S-adenosyl methionine(SAM)-binding methyltransferase; miRNA and siRNA methylation	miRNA ta-siRNA nat-siRNA hc-siRNA *cis*-acting siRNA
HST		HASTY	At3g05040	Exportin; miRNA transport	miRNA

Continued

Table 1 Continued

Gene	Alternative name	Protein name	Gene code	Biochemical activity and function	Pathway
HYL1	DRB1	HYPONASTIC LEAVES 1	At1g09700	dsRNA biding; assisting in efficient and precise cleavage of pri-miRNA through interaction with DCL1	miRNA
MET1	DDM2/RTS2	METHYLTRANSFERASE 1	At5g49160	Cytosine DNA methyltransferase; maintenance of CG DNA methylation	hc-siRNA
NRPD1a	SDE4	NUCLEAR DNA-DEPENDENT RNA POLYMERASE IVa	At1g63020	DNA-dependent RNA polymerase; 24 nt siRNA production and siRNA-directed DNA methylation	nat-siRNA hc-siRNA
NRPD2a	DRD2	NUCLEAR DNA-DEPENDENT RNA POLYMERASE IVb	At3g23780	DNA-dependent RNA polymerase; 24 nt siRNA production and siRNA-directed DNA methylation	hc-siRNA
RDR2		RNA-DEPENDENT RNA POLYMERASE 2	At4g11130	RNA-dependent RNA polymerase; endogeneous 24 nt siRNA production	hc-siRNA
RDR6	SDE1/SGS2	RNA-DEPENDENT RNA POLYMERASE 6	At3g49500	RNA-dependent RNA polymerase; 21 nt and 24 nt siRNA biogenesis, natural virus resistance	ta-siRNA nat-siRNA cis-acting siRNA
SDE3		SILENCING DEFECTIVE 3	At1g05460	RNA helicase; Unclear	nat-siRNA cis-acting siRNA
SGS3	SDE2	SUPPRESSOR OF GENE SILENCING 3	At5g23570	Unknown; RNA stabilizer	ta-siRNA nat-siRNA cis-acting siRNA
WEX		WERNER SYNDROME-LIKE EXONUCLEASE	At4g13870	RNaseD 3′-5′ exonuclease; Unclear	cis-acting siRNA

through post-transcriptional gene silencing (PTGS). For this, sequences homologous to the endogenous gene are expressed ectopically in sense, antisense, or inverted-repeat orientation causing S-PTGS, AS-PTGS, and IR-PTGS, respectively. During IR-PTGS, IR transgenes direct the production of long dsRNAs which are processed to 21 and 24 nt siRNAs and methylated by HEN1. Twenty-one nt siRNAs are produced by DCL4 and guide AGO1 RISC to cleave the homologous endogenous mRNAs. Most likely, DCL3 produces 24 nt siRNAs that mediate DNA or histone modification at homologous loci. In *Arabidopsis* forward-genetic screens have identified many components of the S-PTGS pathway. During S-PTGS single-stranded transgene RNAs can be converted into dsRNAs by the combined action of RDR6, SGS3, SDE3 (putative DEAD box RNA helicase), and possibly WEX (RNase D 3′–5′ exonuclease). The resulting dsRNA is likely to be processed by DCL4 to 21 nt siRNAs, which are then methylated by HEN1. These siRNAs are incorporated into AGO1 RISC and guide the cleavage of homologous mRNAs. It is likely that this pathway also operates as an RNA-based immune system against viruses, where the dsRNA could be a replication intermediate or an internal secondary structure feature of the viral RNA. However, it is very likely that different viruses activate different siRNA pathways.

Gene Silencing and Antiviral Defense

The phenomenon that plants are able to recover from virus infection was first described almost 80 years ago. During recovery the initially infected leaves and first systemic leaves showed severe viral symptoms. However, the upper leaves were symptom free and became immune to the virus. As a consequence the plant became resistant to secondary infection against the inducing virus or close relatives. The discovery of recovery led to the concept of cross-protection. Cross-protection is a type of induced resistance and its basis is that prior infection with one virus provides protection against closely related viruses. Although recovery and cross-protection were described a very long time ago, it was not clear until recently how they operated. However, cross-protection demonstrated that plants harbor an adaptive antiviral defense system. During the last decade it became clear that this adaptive antiviral defense system was similar to RNA silencing and suggested that RNA silencing served as a natural defense system against viruses.

Successful virus infection in plants requires replication in the infected cells, cell-to-cell movement, and long-distance spread of the virus. However, plants try to protect themselves against virus infection and RNA silencing was evolved to detect and degrade foreign RNA molecules. The vast majority of plant viruses have RNA genomes with dsRNA replication intermediates and also with short imperfect hairpins in their genome that are ideal targets of the RNA-silencing mechanism. Moreover, plant viruses with DNA genome also serve as a target for the antiviral RNA-silencing pathway. In these cases, during replication the DNA is transcribed bi-directionally and overlapping transcripts form dsRNA and induce RNA silencing. It has become evident that plant viruses are potent inducers and targets of RNA silencing. In addition, as a part of co-evolution, plant viruses have developed silencing suppressor proteins to defend themselves against RNA silencing.

Many different types of plant viruses exist. They have different nucleic-acid content, genome organization, replication strategies, and have very different host ranges. However, it seems to be a common feature that the virus infection of plants results in the accumulation of viral siRNAs, yet the size profile of viral siRNAs varies depending on the virus. It has also been demonstrated that plants, mutant in different components of the RNA-silencing pathway, showed different responses to virus infection depending on the virus suggesting that different siRNA-mediated pathways are activated in plants in response to infection by different viruses.

Dicer-Like Proteins in Antiviral Defense

Arabidopsis encodes four DCL enzymes. DCL1 produces both miRNAs and siRNAs, while the three other DCLs produce only siRNAs. During the biogenesis of endogenous small RNAs DCL1 produces predominantly 21 nt miRNAs and 21 nt nat-siRNAs. DCL2 synthesizes a 24 nt nat-siRNA, DCL3 generates 24 nt hc-siRNA, and DCL4 produces 21 nt ta-siRNA and *cis*-acting siRNA. Analysis of *Arabidopsis dcl* mutants has revealed that there are some redundancies among the four DCL enzymes.

When a virus infects a plant, the virus-derived dsRNA or, as was shown with tombusvirus infection, the partially self-complementary structures of its genomic RNA, become substrates of DCLs and viral infection results in the accumulation of viral siRNAs. However, it has been shown that the size profile of viral siRNAs varies depending on the virus. Many viruses trigger the generation of 21 nt long viral siRNAs while others induce the production of both 21 and 24 nt siRNAs, with some viruses generating mostly 24 or 22 nt long viral siRNAs.

In *Arabidopsis* cucumber mosaic virus (CMV) siRNAs accumulate as 21 nt long species, tobacco rattle virus (TRV) produces virus-specific 21 and 24 nt siRNAs and turnip crinkle virus (TCV) produces only 22 nt long viral siRNAs. However, these viral siRNA size profiles were altered in different *Arabidopsis dcl* mutants. For example, in *dcl4* the accumulation of 21 nt CMV siRNAs was undetectable, but 22–24 nt viral siRNAs were present. Further analysis of *dcl* mutants has revealed that CMV 21 nt siRNAs are produced by DCL4, whereas the TRV 21 and 24 nt long siRNAs are generated by DCL4 and DCL3, respectively. However, the DCL3-dependent 24 nt long siRNAs are neither necessary nor sufficient to mediate defense against TRV. Thus, the DCL3-dependent RNA-silencing pathway alone cannot limit TRV infection. It was also shown that TCV 22 nt siRNAs are produced by DCL2 and they are sufficient to direct destruction of the virus. Further analysis has revealed that both DCL4 and DCL2 mediate TCV RNA silencing, with DCL2 providing redundant viral siRNA processing function, likely because TCV suppresses the activity of DCL4. It was concluded that viral siRNAs are produced by DCL4; however, DCL2 can substitute for DCL4 when the activity of the latter is reduced or inhibited by viruses. Furthermore, it was proposed that DCL4 might be the first line of antiviral defense, with DCL2-mediated activity coming to the fore when DCL4 is deactivated by a viral silencing suppressor protein. Whether all viral siRNAs are produced primarily by DCL4 from viruses with RNA genomes remains to be determined.

DNA viruses are also targets of RNA silencing, and virus-derived siRNAs have been detected in plants infected with DNA geminiviruses. *Arabidopsis* infected with Cabbage leaf curl begomovirus (CaLCuV), a geminivirus, generates 21, 22, and 24 nt viral siRNA of both polarities, with the 24 nt species being a predominant size-class. The accumulation of three different size-classes of viral siRNAs suggests that more than one DCL is involved in the biogenesis of these RNAs. Genetic evidence has shown that DCL3 is required to generate the 24 nt viral siRNAs and DCL2 is necessary for the production of a substantial fraction of the 22 nt viral siRNAs. Other yet unidentified DCLs or combination of DCLs generate the 21 nt and the remaining of 22 nt viral siRNAs. It has been shown that at least two RNA-silencing pathways are involved in DNA virus–plant interactions.

HEN1

HEN1 encodes an S-adenosyl methionine(SAM)-binding methyl-transferase that methylates the 2′-OH of the 3′-terminal nucleotide of miRNAs and siRNAs. It is suggested that all known endogenous small RNAs in *Arabidopsis* are methylated by HEN1. Infection of *Arabidopsis* with CaLCuV shows that the 3′-terminal nucleotide of viral siRNA is methylated by HEN1. However, mutations in HEN1 gene do not increase plant susceptibility to DNA viruses. Conversely, during the infection of a cytoplasmic RNA virus (oilseed rape mosaic tobamovirus), a major fraction of virus-derived siRNAs are not methylated in *Arabidopsis*. Moreover, it was also demonstrated that viral RNA-silencing suppressors interfere with miRNA methylation in *Arabidopsis*. However, the understanding of the exact biochemical process of viral siRNA methylation and its biological relevance together with the effect of viral silencing suppressors remains a challenge in plant virus interaction.

RISC

The existence of viral siRNA-programmed antiviral RISC that cleaves viral target RNA is widely accepted. However, in theory, DCL-mediated dicing of the viral genome would be sufficient to repress virus infection. Recently, it was shown that efficient accumulation of DCL-3-dependent (in *dcl2–dcl4* double mutant) TRV-derived 24 nt siRNAs was not sufficient to mediate viral defense against TRV. This may argue for an antiviral RISC to promote defense against TRV. In plants, AGO1 was shown to have RISC activity and it was physically associated with miRNAs, ta-siRNAs, and *cis*-acting siRNAs. However, the AGO1 slicer did not contain virus-derived siRNAs when *Arabidopsis* plants were infected with three different viruses (CMV, TCV, tobacco mosaic virus (TMV)). Unfortunately, the experiment did not examine more viruses and had not tested the effect of the viral suppressors on RISC assembly. Furthermore, no other viral siRNA-containing RISC complex has been identified, yet. However, the Arabidopsis genome encodes 10 AGOs and it is possible that some of them may have diversified to become the antiviral slicer of RISC.

The Role of RNA-Dependent RNA Polymerase 6 in Virus Infection

Plants contain many *RDR* genes, for example, the model plant *Arabidopsis* has six RDR paralogs. *Arabidopsis* RDR6 is necessary for ta-siRNA, nat-siRNA, and S-PTGS pathways and in some cases also in antiviral defense. A likely biochemical role of RDR6 in RNA silencing is to produce dsRNA that is cleaved by DCLs. *Arabidobsis rdr6* mutants showed hypersusceptibility to CMV, a cucumovirus, but not to TRV, a tobravirus, TMV, a tobamovirus, turnip vein clearing virus (TVCV), also a tobamovirus, TCV, a carmovirus, and turnip mosaic virus (TuMV), a potyvirus. In *Nicotiana benthamiana*, a species often used in plant virus research, RDR6 plays a role in defense against potato virus X (PVX), a potexvirus, potato virus Y (PVY), a potyvirus but is not involved in resistance against TMV, TRV, TCV, and CMV. The role of NbRDR6 in viral defense against PVX has been studied in detail. During PVX infection, NbRDR6 has no effect on replication, cell-to-cell movement, and virus-derived siRNA accumulation. However, NbRDR6 is implicated in systemic RNA silencing. In plants, there is a mobile silencing signal that spreads through the plasmodesmata and phloem. The exact nature of the silencing signal is unknown, but it is likely to be a 24 nt siRNA species. Both RDR6 and SDE3 are involved in the long-distance spread of the silencing signal generated during PVX infection. During infection NbRDR6 prevents meristem invasion by PVX and is required for the activity but not for the production of a systemic silencing signal. A model for NbRDR6 function in antiviral defense against PVX is that RDR6 uses the incoming silencing signal to produce dsRNA precursors of secondary siRNAs which mediate RNA silencing as an immediate response to slow down the systemic spread of the virus. However, it remains obscure why *rdr6* mutant plants are not hypersusceptible to all tested viruses.

SDE3

Arabidopsis SDE3 encodes a DEAD box putative RNA helicase. SDE3, in addition to RDR6, is involved in long-range but not cell-to-cell signaling of RNA silencing. *Arabidopsis* SDE3, like RDR6, is implicated in defense against CMV but has no effect on TMV, TCV, and TRV infection. It is very likely that they act together on the same RNA-silencing pathway during viral defense.

Many proteins have been identified with a role in RNA silencing. However, only a few have known function in viral defense. A simplified model summarizes the antiviral role of RNA silencing during virus infection (**Figure 1**). In the inoculated cell the entering viruses start to replicate using the virus-encoded replicase protein (Viral RdRP). During the replication cycle the virus forms dsRNA replication intermediates. The plant antiviral defense system recognizes the dsRNA and DCL4 uses it as a substrate to produce virus-derived 21nt siRNAs. The viral siRNAs may be methylated at the 3'-nucleotide by HEN1; however, the extent and general appearance of methylation on viral siRNAs remains undetermined. Virus-derived dsRNAs are diced mainly through DCL4. However, in some cases the invading virus-encoded silencing suppressor protein can repress DCL4 activity. In these cases other DCLs with redundant function become the major viral siRNA producer. Another source of virus-derived siRNAs is the partially self-complementary structures of the viral genomic RNA. These secondary structures can also be the substrate of a DCL. According

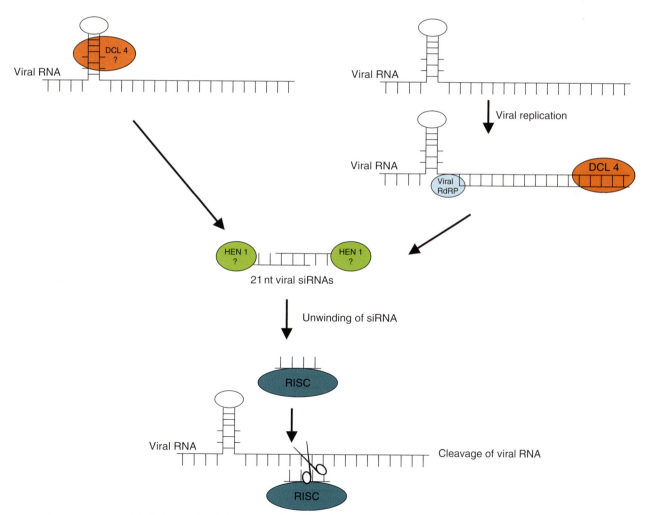

Figure 1 Simplified model of plant antiviral defence against RNA viruses in *Arabidopsis*.

to the size of these viral siRNAs, they may be DCL4 products. However, further investigation is needed to determine their exact origin. The viral siRNAs from both origins are unwound and one strand of the siRNA duplex is incorporated into RISC. The siRNA-loaded RISC recognizes and destroys complementary target viral RNAs. In some cases RDR6 and SDE3 are implicated in antiviral defense by slowing down the accumulation of certain viruses in systemically infected leaves.

Many viral genomes encode silencing suppressors that can inhibit different points of the plant defense system. They can repress HEN1 function or bind to the siRNA duplexes and sequester them from RISC assembly.

Environmental Influence on Gene-Silencing Pathways

It has long been known that plant–virus interactions are strongly modified by environmental factors, especially by temperature. Higher temperature is frequently associated with milder symptoms, low virus content of the infected plants, and rapid recovery from virus disease. In contrast, at cold air temperatures plants become more susceptible to virus infection, develop severe symptoms, and have higher virus content. The effect of temperature on virus-induced RNA silencing has been tested by infecting N. benthamiana plants with cymbidium ringspot virus (CymRSV) and Cym19stop virus (an RNA-silencing suppressor mutant of the same virus) and growing the infected plants at different temperatures. It has been found that CymRSV symptoms were attenuated at 27 °C and that the attenuated symptoms were associated with reduced virus level and that the amount of virus-derived siRNAs gradually increased with rising temperature. The CymRSV-encoded p19 protein acts as a viral suppressor of RNA silencing; thus, Cym19stop-infected plants recover from viral infection at a standard temperature (21 °C). However, at low temperature (15 °C) Cym19stop-infected plants display strong viral symptoms, and the level of virus-derived siRNAs is dramatically reduced. At low temperature, RNA silencing fails to protect the plants, even when the virus lacks a silencing suppressor. Therefore, RNA-silencing-mediated plant defense is temperature dependent and temperature regulates this defense pathway through the control of siRNA generation.

In N. benthamiana, NbRDR6 participates in the antiviral RNA silencing pathway that is stimulated by rising temperatures, suggesting that the function of NbRDR6 may be closely related to the temperature sensitivity of the RNA-silencing pathway. Importantly, it has also been demonstrated that temperature does not influence the accumulation of miRNAs, thus ensuring normal development at lower temperatures.

Viral Counterdefense: Silencing Suppressors

RNA silencing in plants prevents virus accumulation. To counteract this defense mechanism, most plant viruses express silencing suppressor proteins. More than a dozen silencing suppressors have been identified from different types of RNA and DNA viruses. These proteins probably evolved independently in different virus groups, because they are structurally diverse and no obvious sequence homology has been detected between distinct silencing suppressors. An interesting feature common to many viral silencing suppressors is that they were initially identified as pathogenicity or host range determinants. Suppressor activity has been identified in structural as well as nonstructural proteins involved in almost every viral function.

Viral silencing suppressor proteins operate through a variety of mechanisms. The molecular basis for suppressor activity has been proposed for several viruses, including p19 of tombusviruses, HC-Pro of potyviruses, p21 of closteroviruses, and B2 protein of flock house virus, a nodavirus infecting both plants and animals. Currently, the suppressor activity of p19 has been studied in most detail. The high-resolution crystal structures of two different tombusvirus p19 proteins, combined with molecular and biochemical data, have indicated precisely how silencing is blocked. It has been found that a tail-to-tail p19 homodimer specifically binds to siRNAs and recognizes their characteristic size, thus sequestering the products of Dicer and suppressing the silencing effect. Other studies have demonstrated that the molecular basis of silencing suppression of p21 and HC-Pro is also siRNA sequestration. B2 has been shown to bind to dsRNAs and inhibit siRNA formation. Analysis of many silencing suppressor proteins that bind dsRNA, either size selectively or size independently, has suggested that dsRNA binding is a general plant viral silencing suppression strategy which has evolved independently many times.

Viral infection leads to various symptoms, and many viral silencing suppressors have been previously described as pathogenicity determinants. Overexpression of viral silencing suppressors can result in developmental abnormalities in plants because they affect miRNA accumulation and function. Whether viral silencing suppressors inhibit the miRNA pathway because the miRNA and siRNA pathways share common components or because the miRNA pathway directly or indirectly influence virus infection is unknown.

Application of Virus-Induced Gene Silencing

Replicating plant viruses are both strong inducers and targets of the plant RNA-silencing mechanism.

This strong RNA-silencing-inducing ability of plant viruses is used in virus-induced gene silencing (VIGS). VIGS is an RNA-silencing-based technique used to reduce the level of expression of a gene of interest and study the function of the knocked-down gene. Full-length viral clones can be modified to carry a fragment of an endogenous gene of interest and these are known as VIGS vectors. DsRNA of the inserted fragment is generated during viral replication and mediates the silencing of the target gene. Both RNA and DNA viral genomes have been successfully developed into VIGS vectors and have been used in reverse genetics studies in many different plants. The most widely used VIGS vectors are based on TRV, TMV, and cabbage leaf curl geminivirus (CbLCV).

See also: Plant Resistance to Viruses: Engineered Resistance; Virus Induced Gene Silencing (VIGS); Viral Suppressors of Gene Silencing.

Further Reading

Akbergenov R, Si-Ammour A, Blevins T, *et al.* (2006) Molecular characterization of geminivirus-derived small RNAs in different plant species. *Nucleic Acids Research* 34: 462–471.

Baulcombe D (2004) RNA silencing in plants. *Nature* 431: 356–363.

Bouche N, Lauressergues D, Gasciolli V, and Vaucheret H (2006) An antagonistic function for Arabidopsis DCL2 in development and a new function for DCL4 in generating viral siRNAs. *EMBO Journal* 25: 3347–3356.

Brodersen P and Voinnet O (2006) The diversity of RNA silencing pathways in plants. *Trends in Genetics* 22: 268–280.

Chan SW, Henderson IR, and Jacobsen SE (2005) Gardening the genome: DNA methylation in *Arabidopsis thaliana*. *Nature Reviews Genetics* 6: 351–360.

Deleris A, Gallego-Bartolome J, Bao J, Kasschau KD, Carrington JC, and Voinnet O (2006) Hierarchical action and inhibition of plant Dicer-like proteins in antiviral defense. *Science* 313: 68–71.

Herr AJ (2005) Pathways through the small RNA world of plants. *FEBS Letters* 579: 5879–5888.

Qu F and Morris TJ (2005) Suppressors of RNA silencing encoded by plant viruses and their role in viral infections. *FEBS Letters* 579: 5958–5964.

Silhavy D and Burgyan J (2004) Effects and side-effects of viral RNA silencing suppressors on short RNAs. *Trends in Plant Science* 9: 76–83.

Szittya G, Silhavy D, Molnar A, *et al.* (2003) Low temperature inhibits RNA silencing-mediated defence by the control of siRNA generation. *EMBO Journal* 22: 633–640.

Vaucheret H (2006) Post-transcriptional small RNA pathways in plants: Mechanisms and regulations. *Genes and Development* 20: 759–771.

Vazquez F (2006) Arabidopsis endogenous small RNAs: Highways and byways. *Trends in Plant Science* 11: 460–468.

Voinnet O (2005) Induction and suppression of RNA silencing: Insights from viral infections. *Nature Reviews Genetics* 6: 206–220.

Voinnet O (2005) Non-cell autonomous RNA silencing. *FEBS Letters* 579: 5858–5871.

Wang MB and Metzlaff M (2005) RNA silencing and antiviral defense in plants. *Current Opinion in Plant Biology* 8: 216–222.

Relevant Website

http://microrna.sanger.ac.uk – miRBase::Sequences.

Plant Reoviruses

R J Geijskes and R M Harding, Queensland University of Technology, Brisbane, QLD, Australia

© 2008 Elsevier Ltd. All rights reserved.

Glossary

Open reading frame A sequence of nucleotides in DNA that can potentially translate as a polypeptide chain.
Icosahedral Having twenty equal sides or faces.
Capsid The protein shell that surrounds a virus particle.
Monocotyledon A flowering plant that has only one cotyledon or seed leaf in the seed.
Dicotyledon A flowering plant that has two cotyledons or seed leaves in the seed.
Transovarial Transmission from one generation to another through eggs.

Introduction

The family *Reoviridae* comprises a diverse group of viruses which can infect vertebrates, invertebrates, and plants. Despite their large host range, all members of the family *Reoviridae* share common properties including an icosahedral shaped virion and segmented double-stranded RNA (dsRNA) genome. The family *Reoviridae* consists of nine genera of which three genera, *Fijivirus*, *Oryzavirus*, and *Phytoreovirus* are plant-infecting reoviruses. These reoviruses generally replicate in both plant hosts and insect vectors. Infection of the insect vector is non-cytopathic and persists often throughout the life of the insect. Infection of the host plant is tissue specific and can cause severe disease. Fiji leaf gall disease, caused by Fiji disease virus (FDV), has caused yield losses of up to 90% in susceptible varieties of sugarcane in Australia. Rice ragged stunt virus (RRSV) is reported to reduce yield of rice by up to 100% in severe infections (generally 10–20%). Rice dwarf disease, caused by rice dwarf virus (RDV), can also cause significant losses as infected plants often fail to bear seeds. The genera of plant-infecting reoviruses are differentiated according to the number of genomic dsRNA segments and their electrophoretic profile, hosts, serological relationships, and capsid morphology (**Table 1**).

Taxonomy and Classification

Currently there are three genera of the family *Reoviridae* which are classed as plant-infecting reoviruses, *Fijivirus*, *Oryzavirus*, and *Phytoreovirus*. These reoviruses replicate both in plant hosts (except for one fijivirus: *Nilaparvata lugens* reovirus) and in their insect vectors (**Table 1**). Infection of the host plant is species specific, although the host range can often be extended under experimental conditions, and can produce various symptoms, including severe disease. The complete genome sequence has been obtained for a number of viruses and at least partial sequence information is now available for all plant reoviruses. This has allowed detailed comparisons within these genera and across all of the *Reoviridae*, thus providing a basis for the classification of these viruses into species and genera.

Within the genus *Fijivirus*, individual species have considerable similarities to the type species, Fiji disease virus. Classification into separate species is based on unique characteristics such as capacity to exchange genome segments, relatively high amino acid sequence similarity, serological cross-reaction, cross-hybridization of RNA or cDNA probes, host species, and insect vector species. In addition to these commonly used identifiers, analysis of the available genome sequences has assisted in identification of *Fijivirus* species. Gross genome characteristics for fijivirus members include a genome size of approximately 29 kbp and a characteristically low G+C content of 34–36%. Unique, and highly conserved, 5′ and 3′ terminal sequences are present in different plant reovirus species; in all RNA segments, the 3′ terminal trinucleotide is conserved across all species within the genus *Fijivirus* (**Table 2**). Inverted repeats are found adjacent to the terminal sequences and these differ from those of other plant reoviruses.

Members of the genera *Oryzavirus* and *Phytoreovirus* have significant similarity to type members rice ragged stunt virus and wound tumor virus, respectively.

Table 1 Characteristics of plant reoviruses

Genus	Virus	dsRNA genome segments	Host	Vector/s
Fijivirus	Fiji disease virus (FDV)	10	Monocot (Gramineae)	Planthoppers: *Perkinsiella saccharicida*, *P. vastatrix*, *P. vitiensis*
	Rice black-streaked dwarf virus (RBSDV)	10	Monocot (Gramineae)	Planthoppers: *Laodelphax striatellus*, *Ribautodelphax albafascia*, *Unkanodes sapporona*
	Maize rough dwarf virus (MRDV)	10	Monocot (Gramineae)	Planthopper: *Ribautodelphax notabilis*
	Pangola stunt virus (PaSV)	10	Monocot (Gramineae)	Planthoppers: *Sogatella furcifera*, *S. kolophon*
	Mal del Rio Cuarto virus (MRCV)	10	Monocot (Gramineae)	Planthopper: *Delphacodes kuscheli*
	Oat sterile dwarf virus (OSDV)	10	Monocot (Gramineae)	Planthoppers: *Javesella pellucidia*, *J. discolour*, *J.dubia*, *J.obscurella*, *Dicranotropis hamata*
	Garlic dwarf virus (GDV)	10	Monocot (Liliaceae)	Planthopper: Unknown
	Nilaparvata lugens reovirus (NLRV)	10	No plant host reported	Planthoppers: *Nilaparvata lugens*, *Laodelphax striatellus*
Oryzavirus	Echinochloa ragged stunt virus (ERSV)	10	Monocot (Gramineae)	Planthoppers: *Sogatella longifurcifera*, *S. vibix*
	Rice ragged stunt virus (RRSV)	10	Monocot (Gramineae)	Planthopper: *Nilaparvata lugens*
Phytoreovirus	Wound tumor virus (WTV)	12	Dicots	Leafhoppers: *Agallia constricta*, *A.quadripunctata*, *Agalliopsis novella*
	Rice dwarf virus (RDV)	12	Monocot (Gramineae)	Leafhoppers: *Nephotettix cincticeps*, *N. nigropictus*, *Recillia dorsalis*
	Rice gall dwarf virus (RGDV)	12	Monocot (Gramineae)	Leafhoppers: *Nephotettix cincticeps*, *N. nigropictus*, *N. virescens*, *N. malayanus*, *Recillia dorsalis*

Demarcation of species within the oryzaviruses and phytoreoviruses are primarily based on the ability to exchange genome segments although other characteristics, as mentioned for fijiviruses above, are also used. When available, genomic sequences are examined to reveal distinguishing features to support the classification. Oryzaviruses have a total genome size of approximately 26 kbp and specific 5′ and 3′ terminal sequences in all RNA segments: $^{5'}$GAUAAA—GUGC$^{3'}$. Phytoreoviruses have a total genome size of approximately 25 kbp, a G+C content between 38% and 48% and specific 5′ and 3′ terminal sequences, ($^{5'}$GG(U/C)A—(U/C)GAU$^{3'}$) in all RNA segments.

Virion Structure and Genome Organization

Fijivirus

The virions have a complex double icosahedral capsid construction and consist of a capsid, a core, and a nucleoprotein complex. Virions are fragile structures and readily break down *in vitro* to give cores. The outer capsid is 65–70 nm in diameter with 12 'A' type spikes located at the vertices of the icosahedron. The inner core is about 55 nm in diameter, with 12 'B' type spikes located at the vertices. The viral nucleic acid is located at the center of the virus particle, within the inner core capsid. Each virion contains a single full-length copy of the genome. Fijivirus genomes contain ten dsRNA genomic segments varying from approximately 1.8–4.5 kbp (**Figure 1**). The total genome is approximately 29 kbp with a low G + C content of 34–36%. Highly conserved unique 5′ and 3′ terminal sequences are found on all RNA segments (**Table 2**). Segment-specific inverted repeats are found adjacent to these terminal sequences. Segments 1–6, 8, and 10 are monocistronic, containing one open reading frame (ORF), while segments 7 and 9 each contain two ORFs. NLRV is the only fijivirus identified to date which differs from this structure, with one ORF on segment 7. The functions of proteins encoded by most ORFs are still unconfirmed; gene functions of segments 1–4 and 8–10 have been predicted based on protein expression studies or sequence similarities to related reoviruses (**Table 3**).

Oryzavirus

The virions have a double-shelled icosahedral capsid and consist of an outer capsid, an inner capsid, and a core. Virions are fragile and readily break down *in vitro* to give subviral core particles unless pre-treated. The outer capsid is 75–80 nm in diameter with 12 'A' type spikes located at the

Table 2 Conserved 5′ and 3′ sequences identified in fijiviruses

Virus	5′ conserved sequence	3′ conserved sequence[a]
Fiji disease virus	$^{5'}$AAGUUUUU—$^{3'}$	$^{5'}$—CAGCNNNN*GUC*$^{3'}$
Rice black-streaked dwarf virus	$^{5'}$AAGUUUUU—$^{3'}$	$^{5'}$—AGCUNN(C/U)*GUC*$^{3'}$
Maize rough dwarf virus	$^{5'}$AAGUUUUUU—$^{3'}$	$^{5'}$—U*GUC*$^{3'}$
Mal del Rio Cuarto virus	$^{5'}$AAGUUUUU—$^{3'}$	$^{5'}$—CAGCUNNN*GUC*$^{3'}$
Oat sterile dwarf virus	$^{5'}$AACGAAAAAAA—$^{3'}$	$^{5'}$—UUUUUUUUUAG*UC*$^{3'}$
Nilaparvata lugens reovirus	$^{5'}$AGU—$^{3'}$	$^{5'}$—GUU*GUC*$^{3'}$

[a]Italicized trinucleotide is conserved in all fijivirus sequences reported to date.

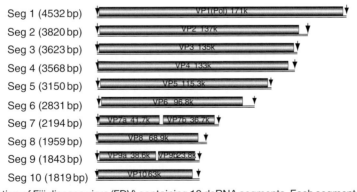

Figure 1 Genome organization of Fiji disease virus (FDV) containing 10 dsRNA segments. Each segment contains one ORF except for Seg 7 and Seg 9 which contain two ORFs. The arrows indicate the location of the 5′ and 3′ conserved sequences, respectively. Reproduced from Fauquet CM, Mayo MA, Maniloff J, Desselberger U, and Ball LA (2005) *Virus Taxonomy – Classification and Nomenclature of Viruses: Eighth Report of the International Committee on the Taxonomy of Viruses*. San Diego, CA: Elsevier Academic Press, with permission from Elsevier.

Table 3 Genome organization of FDV and predicted gene function

Segment	Size (bp)	Protein nomenclature	MW (kDa)	Predicted function (location)
S1	4532	VP1	170.6	RdRp (core)
S2	3820	VP2	137.0	Possible core protein (core)
S3	3623	VP3	135.5	Possible core protein (core)
S4	3568	VP4	133.2	Possible B spike (capsid)
S5	3150	VP5	115.3	Unknown
S6	2831	VP6	96.8	Unknown
S7	2194	VP7a	41.7	Unknown
		VP7b	36.7	Unknown
S8	1959	VP8	68.9	Possible NTP-binding (core)
S9	1843	VP9a	38.6	Structural protein (unknown)
		VP9b	23.8	Nonstructural
S10	1819	VP10	63.0	Outer capsid protein (capsid)

Reproduced from Fauquet CM, Mayo MA, Maniloff J, Desselberger U, and Ball LA (2005) *Virus Taxonomy – Classification and Nomenclature of Viruses: Eighth Report of the International Committee on the Taxonomy of Viruses*. San Diego, CA: Elsevier Academic Press, with permission from Elsevier.

five fold axis of the icosahedron. The core capsid is about 57–65 nm in diameter, with 12 'B' type spikes. The viral nucleic acid is located at the center of the viral particle, within the core capsid. The virus genome consists of ten dsRNA segments ranging in size from 1162–3849 bp (RRSV) with a total length of 26 kbp (**Figure 2**). Genome segments 1–3, 5, 7–10 of RRSV each contain a single ORF, while segment 4 contains two ORFs. Segment 8 encodes a polyprotein which is cleaved into two proteins. **Table 4** summarizes the organization of the RRSV dsRNA segments and the predicted function of the encoded proteins.

Phytoreovirus

The virions have a double-shelled icosahedral capsid construction and consist of an outer capsid, a core capsid, and a smooth core. Virions are approximately 70 nm in diameter with 12 spikes located at the fivefold vertices of the icosahedron and generally remain intact when purified. WTV, the type member, possesses three protein shells: an outer amorphous layer made up of two proteins, an inner capsid made up of two proteins, and a smooth core made up of three proteins that is about 50 nm in diameter. Each virion contains a single full-length copy of the genome. Phytoreoviruses have 12 segments of dsRNA which range in size from approximately 1–4.5 kbp with a total genome length of approximately 25 kbp (**Figure 3**) and a G+C content of 38–44%. Each segment of RDV contains a single ORF except for segment 12, which contains two ORFs. **Table 5** summarizes the organization of the RDV dsRNA segments and the putative function of the encoded proteins.

Replication and Gene Expression

The replication and gene expression of plant-infecting reoviruses is thought to be similar to that of other reoviruses. The best described of these is bluetongue virus (BTV), type member of the genus *Orbivirus*. If the BTV model is accurate for the plant-infecting reoviruses, replication occurs after virions (or viral cores) are delivered into the host cell. Replication is initiated when the viral capsid layer is removed and the core enters the cytoplasm of the cell. The viral genome (10–12 segments) remains packaged in the central cavity of the viral core to ensure host cell defense responses to dsRNA are not activated. The core is biochemically active with RNA-dependent RNA polymerase (RdRp), capping enzyme, and helicase enzyme. The viral core contains a number of channels, the largest of which is at the fivefold axis of the icosahedral structure. Smaller channels allow the entry of nucleotides into the core which are required for transcription. The large channel is located adjacent to the

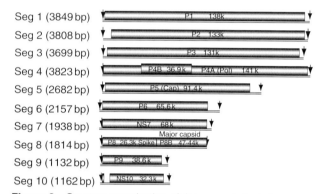

Figure 2 Genome organization of rice ragged stunt virus (RRSV) containing 10 dsRNA segments. Each segment contains one ORF except for Seg 4 which contains two ORFs. The arrows indicate the location of the 5′ and 3′ conserved sequences, respectively. Reproduced from Fauquet CM, Mayo MA, Maniloff J, Desselberger U, and Ball LA (2005) *Virus Taxonomy – Classification and Nomenclature of Viruses: Eighth Report of the International Committee on the Taxonomy of Viruses*. San Diego, CA: Elsevier Academic Press, with permission from Elsevier.

Table 4 Genome organization of RRSV and predicted gene function

Segment	Size (bp)	Protein nomenclature	MW (kDa)	Predicted function (Location)
S1	3849	P1	137.7	Virion associated (B spike)
S2	3810	P2	133.1	Inner core capsid (core capsid)
S3	3699	P3	130.8	Major core capsid (core capsid)
S4	3823	P4a	141.4	RdRp
		P4b	36.9	(unknown)
S5	2682	P5	91.4	Capping enzyme
S6	2517	P6	65.6	Unknown
S7	1938	NS7	68	Nonstructural protein (unknown)
S8	1814	P8	67.3	Precursor polyprotein/protease
		P8a	25.6	Spike
		P8b	41.7	Major capsid protein
S9	1132	P9	38.6	Vector transmission (spike)
S10	1162	NS10	32.3	Nonstructural protein

Reproduced from Fauquet CM, Mayo MA, Maniloff J, Desselberger U, and Ball LA (2005) *Virus Taxonomy – Classification and Nomenclature of Viruses*: Eighth Report of the International Committee on the Taxonomy of Viruses. San Diego, CA: Elsevier Academic Press, with permission from Elsevier.

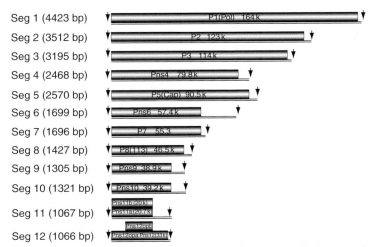

Figure 3 Genome organization of rice dwarf virus (RDV) containing 12 dsRNA segments. Each segment contains one ORF except for Seg 11 and Seg 12 which contain two ORFs. The arrows indicate the location of the 5′ and 3′ conserved sequences, respectively. Reproduced from Fauquet CM, Mayo MA, Maniloff J, Desselberger U, and Ball LA (2005) *Virus Taxonomy – Classification and Nomenclature of viruses*: Eighth Report of the International Committee on the Taxonomy of Viruses. San Diego, CA: Elsevier Academic Press, with permission from Elsevier.

replicase/transcriptase complex which has helicase activity for the unwinding and rewinding of the dsRNA genome during transcription of negative RNA strand. The newly formed positive strand mRNA molecules are modified to form a Cap1 structure by the guanylyltransferase, nucleotide phosphohydrolase, and transmethylase activity of the capping enzyme prior to the extrusion of mRNA, from the major pore, into the cytoplasm. These mRNA molecules released into the cytoplasm can be translated to produce viral proteins. Nonstructural viral proteins aggregate to form inclusion bodies or viroplasms. The viroplasm is the site of most of the mRNA production and assembly of core proteins. The mRNA molecules, one of each segment, are assembled with these viral proteins to form new virus core particles. Once a copy of the mRNA is inside a new viral core, the negative strand is synthesized completing the replication of the dsRNA genome. The complete viral core containing dsRNA then moves to the periphery of the viroplasm where capsid proteins are assembled to form the complete new viral particle.

Control of gene expression of a multisegmented genome is complex and not fully understood. Each genome segment contained within the viral core is associated with a single replicase/transcription complex, located adjacent to the major pore in the vertices of the icosahedron, and is transcribed separately to make full-length positive sense RNA copies. The location of the replicase/transcription complex also restricts the number of genome segments to a maximum of 12. These 10–12 mRNAs are produced in different molar amounts based largely on segment size

Table 5 Genome organization of RDV and putative gene function

Segment	Size (bp)	Protein nomenclature	MW (kDa)	Predicted function (location)
S1	4423	P1	170	RdRp (core)
S2	3512	P2	130	Capsid structural protein (outer capsid)
S3	3195	P3	110	Major core protein (core capsid)
S4	2468	Pns4	83	Nonstructural protein
S5	2570	P5	89	Guanylyltransferase (core)
S6	1699	Pns6	56	Nonstructural protein
S7	1696	P7	58	Nucleic acid binding protein (core)
S8	1427	P8	43	Major outer capsid protein
S9	1305	Pns9	49	Nonstructural protein
S10	1321	Pns10	53	Nonstructural protein
S11	1067	Pns11a	23	Nonstructural protein
		Pns11b	24	
S12	1066	Pns12	34	Nonstructural protein
		Pns12OPa	8	
		Pns12OPb	7	

Reproduced from Fauquet CM, Mayo MA, Maniloff J, Desselberger U, and Ball LA (2005) *Virus Taxonomy – Classification and Nomenclature of Viruses*: *Eighth Report of the International Committee on the Taxonomy of Viruses*. San Diego, CA: Elsevier Academic Press, with permission from Elsevier.

resulting in more copies of smaller mRNAs. This interaction between mRNA molecules of varying length and the replicase/transcription complex provides some control over the expression levels of individual virus genes. Translation of mRNA segments is largely independent of mRNA length although some segments are translated more efficiently resulting in a secondary method of control over expression. A third level of control results from the use of multiple or overlapping ORFs on one mRNA strand which are translated at different efficiencies. Lastly, some ORFs encode a polyprotein which must be processed to form functional proteins.

Distribution

Plant-infecting reoviruses are seasonally distributed as a result of plant host/crop cycles and presence of insect vector. Plant reoviruses have been isolated from every continent but some genera are more widespread than others. Fijiviruses are the most widely distributed, which is not surprising given that they are the most numerous. Fijiviruses occur in Africa, Europe, South America, Asia, Australia, and South Pacific Islands. Oryzaviruses have only been isolated from the Indian subcontinent and Asia while phytoreoviruses have been isolated from North America, Asia, and Africa.

Host Range and Virus Transmission

Fijiviruses

The genus *Fijivirus* contains eight species whose members infect a range of monocotyledonous plants of the families Gramineae and Liliacae. Common plant hosts include the Gramineae: *Avena sativa, Oryza sativa, Saccharum officinarum, Zea mays*, and the Liliacae: *Allium sativum*. However, this natural host range can be extended significantly by experimental virus infection. Virus is transmitted by delphacid planthoppers (**Table 1**). Virus can be acquired in juvenile stages, replicates in the vector, and, following a two week latent period, is transmitted to plants in a persistent manner. No transovarial transmission of virus has been reported. In addition to transmission by insect vectors, mechanical transmission of the virus to susceptible hosts has been achieved for some members with difficulty.

Oryzaviruses

The genus *Oryzavirus* contains two species whose members infect monocotyledonous plants of the family Gramineae. Common plant hosts include *Oryza sativa* and *Echinochloa crus-galli*. However, this natural host range can be extended by experimental virus infection to include other economically important species such as *Hordeum vulgare, Triticum aestivum*, and *Zea mays*. Virus is transmitted by delphacid planthoppers (**Table 1**). An acquisition period of 3 h is required followed by a 9-day latent period prior to transmission at all life stages in an intermittent manner. No transovarial transmission or mechanical transmission of virus has been reported.

Phytoreoviruses

The genus *Phytoreovirus* contains two species whose members infect monocotyledonous plants of the family Gramineae and one species which infects dicotyledonous plants. Common plant hosts include the Gramineae: *Oryza sativa* and the dicot – *Melilotus officinalis*. However, the natural host range of the dicot-infecting WTV can be extended significantly by experimental virus infection. Virus is transmitted by cicadellid leafhoppers (**Table 1**). Virus can be acquired after a short feeding period, replicates in the vector, and, following a 10–20 day latent period, is transmitted to plants throughout the life of the vector. Transovarial transmission of virus has been reported. Attempts to transmit the virus to susceptible hosts by mechanical methods have been unsuccessful.

Pathogenicity

The pathogenicity of plant reoviruses is particularly interesting as most viruses replicate in both insects and plant hosts. Most of these viruses do not appear to cause any disease in the insect host and pathogenicity of these viruses is restricted to the plant host. The pathogenicity of fijiviruses varies considerably. Fiji leaf gall disease (caused by FDV) has been reported to cause losses of up to 90% in susceptible sugarcane varieties, while NLRV has no known plant host and, therefore, no pathogenicity. Oryzaviruses can also cause important yield losses. Rice ragged stunt disease (caused by RRSV) has been reported to cause losses of 10–20% but sometimes as high as 100% in severe infections of susceptible varieties. The pathogenicity of phytoreoviruses is much milder although rice dwarf disease (caused by RDV) can be severe as infected plants often fail to bear seeds. There is currently little information on the molecular basis for pathogenicity and it is not known if different isolates of the same virus cause diseases of varying severity.

Diagnosis and Control

Diagnosis of plant reovirus infections can be done on the basis of symptoms or by use of molecular tests. Symptoms vary in different virus/host complexes but symptoms such as plant stunting, increased numbers of side shoots, and tumors or gall formation in phloem tissue are commonly observed. Given the variability in time to symptom expression and symptom severity, alternative tests are often used. Molecular and serological tests have been developed to assist in the diagnosis of viral infection in nonsymptomatic plant material and vector insects. Serological tests are usually in enzyme-linked immunosorbent assay (ELISA) format and rely on polyclonal antisera raised against virions or expressed viral proteins. Recently, molecular tests such as reverse transcriptase-polymerase chain reaction (RT-PCR), which are faster and more specific than serology, have become the most common method of diagnosis. Species-specific primers are now commonly available as increasing numbers of plant reovirus genomes are being sequenced.

Control strategies for plant reoviruses can be focused on either the host plant or insect vector. Plant-based control through breeding to develop resistant plant species is most commonly utilized in combination with removal of susceptible varieties and infected plants which provide a source of inoculum. This approach has provided robust control of RDV in rice and FDV in sugarcane. Genetic engineering of plant hosts has also been explored as an alternative control strategy. Pathogen-derived resistance approaches using either coat protein or other viral genes to control RDV, RRSV, and FDV have not proved as successful as those used to control other RNA plant viruses. However, this may improve in the future as more information on virus infection and replication becomes available providing new resistance targets. Control of insect vector numbers with insecticide has provided some additional disease control. Unfortunately, chemical control appears to be of limited use in cases of high vector pressure. The current combination of diagnosis and control measures is already relatively effective and has resulted in reduced disease incidence and impact.

Future

Although our understanding of plant reoviruses is increasing, many of the molecular and biological properties of these viruses are still unknown. The complete sequence information is now available for a number of these viruses which will allow production of cDNA probes to further elucidate the infection and replication processes in both plant and insect cells. The potential to produce infectious clones also holds promise for detailed studies of both plant and insect host ranges and methods of resistance employed by nonhost species. This information combined with knowledge gained from comparison to animal reoviruses may assist in further development of control strategies for diseases caused by these viruses in plants.

See also: Insect Reoviruses; Reoviruses: General Features; Reoviruses: Molecular Biology.

Further Reading

Bamford DH (2000) Virus structures: Those magnificent molecular machines. *Current Biology* 10: R558–R561.
Fauquet CM, Mayo MA, Maniloff J, Desselberger U, and Ball LA (2005) *Virus Taxonomy – Classification and Nomenclature of Viruses: Eighth Report of the International Committee on the Taxonomy of Viruses.* San Diego, CA: Elsevier Academic Press.
Hull R (2002) *Matthews Plant Virology.* London: Academic Press.
Mertens P (2004) The dsRNA viruses. *Virus Research* 101: 3–13.
Mertens PPC and Diprose J (2004) The Bluetongue virus core: A nano-scale transcription machine. *Virus Research* 101: 29–43.

Plant Resistance to Viruses: Engineered Resistance

M Fuchs, Cornell University, Geneva, NY, USA

© 2008 Elsevier Ltd. All rights reserved.

Introduction

Plant viruses can cause severe damage to crops by substantially reducing vigor, yield, and product quality. They can also increase susceptibility to other pathogens and pests, and increase sensitivity to abiotic factors. Detrimental effects of viruses can be very costly to agriculture. Losses of over $1.5 billion are reported in rice in Southeast Asia, $5.5 million in potato in the United Kingdom, $63 million in apple in the USA, and $2 billion in cassava worldwide. Also, grapevine fanleaf virus (GFLV) alone is responsible for losses of over $1.0 billion in grapevines in France.

A number of strategies are implemented to mitigate the impact of plant viruses. Quarantine measures and certification of seeds and propagation stocks limit the introduction of virus diseases in virus-free fields and areas. Cultural practices, rouging, control of vectors, and cross-protection based on mild virus strains or benign satellite RNA reduce the transmission rate of viruses in areas where epidemics can be problematic. However, the ideal and most effective approach to control viruses relies on the use of resistant crop cultivars. Host resistance genes have been extensively exploited by traditional breeding techniques. In recent years, cloning, sequencing, and functional characterization of some plant resistance genes have provided new insights into their structure and effect on virus multiplication at the cell and plant level. Notwithstanding, host resistance genes have been identified for a few viruses and only a limited number of commercial elite crop cultivars exhibit host resistance to viruses.

The advent of biotechnology through plant transformation and the application of the concept of pathogen-derived resistance (PDR) has opened new avenues for the development of resistant crop cultivars, in particular when resistant material with desired horticultural characteristics has not been developed by conventional breeding or when no host resistance sources are known. The past two decades have witnessed an explosion in the development of virus-resistant transgenic plants. Some crop plants expressing viral genes or other antiviral factors have been tested successfully in the field and a few have been commercialized. Extensive field evaluation experiments and commercial releases have demonstrated the benefits offered by virus-resistant transgenic plants to agriculture. Risk assessment studies have been conducted to address environmental safety issues associated with the release of transgenic plants expressing viral genes. Field experiments have provided a level of certainty of extraordinarily limited, if any, hazard to the environment. Also, studies on RNA silencing have provided new insights into the molecular and cellular mechanisms underlying engineered resistance to viruses in plants. Overall, the safe deployment of virus-resistant transgenic plants has become an important strategy to implement effective and sustainable control measures against major virus diseases.

Development of Virus-Resistant Transgenic Plants

Historical Perspectives

Over the past two decades, biotechnology expanded the scope of innovative approaches for virus control in plants by providing new tools to engineer resistance. The first approach to confer resistance to viruses in plants resulted from the development of efficient protocols for plant transformation and the application of the concept of PDR. This concept was initially conceived to engineer resistance by transferring and expressing a dysfunctional pathogen-specific molecule into the host genome in order to inhibit the pathogen. Therefore, a segment of the virus' own genetic material was hypothesized to somehow protect a plant against virus infection by acting against the virus itself.

The first demonstration of PDR in the case of plant viruses was with tobacco mosaic virus (TMV) in 1986. Transgenic tobacco plants expressing the TMV coat protein (CP) gene either did not display symptoms or showed a substantial delay in symptom development upon inoculation with TMV particles. Early experiments suggested a direct correlation between expression level of TMV CP and the degree of resistance. This type of PDR was known as CP-mediated resistance. It was successfully applied against numerous plant viruses. In the early 1990s, transgenic expression of the viral replicase domain was also shown to induce resistance. This approach was referred to as replicase-mediated resistance. Meanwhile, other virus-derived gene constructs, including sequences encoding movement protein, protease, satellite RNA, defective interfering RNAs, or noncoding regions, were also described to confer virus resistance in plants. It soon became evident that more or less any viral sequence could provide some level of resistance to virus infection when expressed in transgenic plants. Interestingly, in many cases, resistance was shown to require transcription of the virus-derived transgene rather than expression of a protein product. Subsequently, at the turn of the twenty-first century, tremendous progress has been made at unraveling the molecular and cellular mechanisms underlying virus resistance.

The discovery of RNA silencing as a key mechanism in antiviral defense in plants and regulation of gene expression in eukaryotes has been a major scientific breakthrough.

Resistance to viruses has also been achieved by transforming susceptible plants with antiviral factors other than sequences derived from viral genomes. Production of antibodies or antibody fragments, $2'-5'$ oligoadenylate synthase, ribosome inactivating proteins, double-stranded RNA (dsRNA)-specific RNases, dsRNA-dependent protein kinases, cystein protease inhibitors, nucleoprotein-binding interfering aptamer peptides, and pathogenesis-related proteins has been reported to protect plants against virus infection.

Over the past two decades, many plant species and crop plants have been successfully engineered for resistance against numerous viruses with diverse taxonomic affiliation and various mode of transmission. The majority of virus-resistant transgenic plants result from the application of PDR. Virus-derived transgene constructs include full-length, untranslated, and truncated coding and noncoding fragments, in sense or antisense orientation. The CP gene is the most commonly used viral sequence segment to engineer resistance.

Control of Virus Diseases

Virus resistance is evaluated in transgenic plants upon mechanical inoculation, agro-infiltration, grafting, or vector-mediated infection and expressed as immunity, restricted infection, delay in the onset of disease symptoms, or recovery. As an example, squash expressing the CP genes of zucchini yellow mosaic virus (ZYMV), watermelon mosaic virus (WMV), and cucumber mosaic virus (CMV) are highly resistant to mechanical inoculation with a mixture of these three viruses while control plants inoculated with only one of these three viruses are readily infected and exhibit severe symptoms (**Figure 1(a)**).

Some transgenic crop plants, such as cereal, vegetable, legume, flower, forage, and fruit crops, expressing virus-derived gene constructs or other antiviral factors have been tested under field conditions (**Table 1**). So far, a total of 58 virus species that belong to 26 distinct genera have been the target of field resistance evaluation in 37 different plant species.

In the USA, virus resistance represented 10% (1242 of 11 974) of the total applications approved for field releases between 1987 and early August 2006. Other target traits were herbicide tolerance (30%), insect resistance (26%), improved product quality (19%), agronomic properties (9%), and fungal resistance (6%). In the USA, field tests with virus-resistant transgenic crops started in 1988 and accounted for 13–37% of the approved releases from 1988 to 1999.

A high level of resistance to virus infection has been well documented for numerous transgenic crops expressing viral genes even under conditions of very stringent disease pressure and vector-mediated virus infection. For example, in the case of virus-resistant squash, transgenic cultivars expressing the CP gene of ZYMV and WMV are highly resistant to mixed infection by these two aphid-transmitted potyviruses (**Figure 1(b)**). Transgenic plants have a lush canopy and vigorous growth, whereas control plants show a chlorotic and distorted canopy, and stunted growth (**Figure 1(f)**). Transgenic plants produce fruits of marketable quality throughout the growing season (**Figure 1(h)**), whereas most fruits of control plants are malformed and discolored (**Figure 1(g)**), hence not marketable (**Figure 1(e)**). Remarkably, transgenic squash plants expressing the CP gene of ZYMV, WMV, and CMV exhibit identical levels of resistance to mechanical inoculation (**Figure 1(c)**) and aphid-mediated infection (**Figure 1(d)**) by these three viruses under field conditions in which no insecticides were sprayed to control aphid vector populations. Some of the 37 transgenic crops that have been field-tested or commercialized so far are also engineered for resistance to viruses that are not mentioned in **Table 1**. Similarly, in addition to the crops listed in **Table 1**, other virus-resistant transgenic plant species, including cabbage, cassava, grapevine, ryegrass, and petunia, among others, will likely be extensively field-tested in the future based on their promising performance under greenhouse conditions.

A few transgenic crops expressing viral CP genes have been commercially released in the USA and the People's Republic of China (**Table 1**). In the USA, virus resistance represented 10% (9 of 87) of the petitions granted for deregulation by the regulatory authorities until August 2006. Squash expressing the CP gene of ZYMV and WMV received exemption status in 1994 and was released thereafter. This was the first disease-resistant transgenic crop to be commercialized in the USA. Squash expressing the CP gene of ZYMV, WMV, and CMV has been deregulated and commercialized in 1995. Subsequently, numerous squash types and cultivars have been developed by crosses and back crosses with the two initially deregulated lines. The adoption of virus-resistant squash cultivars is steadily increasing since 1995. In 2004, the adoption rate was estimated to be 10% (2500 ha) across the country with an average rate of 20% in the states of Florida, Georgia, and New Jersey. Papaya expressing the CP gene of papaya ringspot virus (PRSV) was deregulated in 1998 and commercialized in Hawaii. PRSV-resistant papaya was the first transgenic fruit crop to be commercially released in the USA. Its adoption rate has been very high from the start of its release in 1998 because PRSV had devastated the papaya industry in Hawaii. Transgenic papaya cultivars were planted on more than half of the total acreage (480 ha) in 2004. Several potato cultivars expressing the replicase gene of potato leafroll virus (PLRV) or the CP gene of potato virus Y (PVY) were deregulated in 1998 and 2000.

158 Plant Resistance to Viruses: Engineered Resistance

Figure 1 Reaction of transgenic squash ZW-20 expressing the coat protein (CP) gene of zucchini yellow mosaic virus (ZYMV) and watermelon mosaic virus (WMV), and transgenic squash CZW-3 expressing the CP gene of ZYMV, WMV, and cucumber mosaic virus (CMV) to virus infection. (a) Resistance of transgenic CWZ-3 to mechanical inoculation with ZYMV, WMV, and CMV (upper right), and susceptibility of nontransgenic plants to single infection by mechanical inoculation of CMV (lower right), WMV (upper left), and ZYMV (lower left); (b) fields of transgenic ZW-20 (foreground) and nontransgenic plants (background) surrounded by a border row of mechanically inoculated nontransgenic plants; (c) transgenic ZW-20 (right) and nontransgenic (left) plants mechanically inoculated with ZYMV and WMV; (d) transgenic CZW-3 (right) and nontransgenic (left) plants subjected to aphid-mediated inoculation of ZYMV and WMV; (e) close-up of a fruit from transgenic ZW-20 (left) and fruits from nontransgenic plants (five fruits on the right), all subjected to aphid-mediated inoculation of ZYMV and WMV; (f) nontransgenic squash mechanically inoculated with ZYMV and WMV (left) and transgenic ZW-20 (right) exposed to aphid inoculation of ZYMV and WMV; and fruit production of (g) nontransgenic and (h) transgenic squash following aphid-mediated inoculation of ZYMV and WMV.

Table 1 Examples of virus-resistant transgenic crops that have been tested in the field or commercially released

Crop		Resistance to	
Category/common name	Scientific name	Virus	Genus
Cereals			
Barley	*Hordeum vulgare*	Barley yellow dwarf virus	*Luteovirus*
Canola	*Brassica napus*	Turnip mosaic virus	*Potyvirus*
Corn	*Zea mays*	Maize dwarf mosaic virus	*Potyvirus*
		Maize chlorotic dwarf virus	*Waikavirus*
		Maize chlorotic mottle virus	*Machlomovirus*
		Sugarcane mosaic virus	*Potyvirus*
Oat	*Avena sativa*	Barley yellow dwarf virus	*Luteovirus*
Rice	*Oryza sativa*	Rice stripe virus	*Tenuivirus*
		Rice hoja blanca virus	*Tenuivirus*
Wheat	*Triticum aestivum*	Barley yellow dwarf virus	*Luteovirus*
		Wheat streak mosaic virus	*Tritimovirus*
Ornamentals			
Chrysanthemum	*Chrysanthemum indicum*	Tomato spotted wilt virus	*Tospovirus*
Dendrobium	*Encyclia cochleata*	Cymbidium mosaic virus	*Potexvirus*
Gladiolus	*Gladiolus* sp.	Bean yellow mosaic virus	*Potyvirus*
Fruits			
Grapefruit	*Citrus paradisi*	Citrus tristeza virus	*Closterovirus*
Grapevine	*Vitis* sp.	Grapevine fanleaf virus	*Nepovirus*
Lime	*Citrus aurantifolia*	Citrus tristeza virus	*Closterovirus*
Melon	*Cucumis melo*	Cucumber mosaic virus	*Cucumovirus*
		Papaya ringspot virus	*Potyvirus*
		Squash mosaic virus	*Comovirus*
		Watermelon mosaic virus	*Potyvirus*
		Zucchini yellow mosaic virus	*Potyvirus*
Papaya[a]	*Carica papaya*	Papaya ringspot virus[a]	*Potyvirus*
Pineapple	*Ananas comosus*	Pineapple wilt-associated virus	*Ampelovirus*
Plum	*Prunus domestica*	Plum pox virus	*Potyvirus*
Raspberry	*Rubus idaeus*	Raspberry bushy dwarf virus	*Idaeovirus*
		Tomato ringspot virus	*Nepovirus*
Strawberry	*Fragaria* sp.	Strawberry mild yellow edge virus	*Potexvirus*
Tamarillo	*Cyphomandra betacea*	Tamarillo mosaic virus	*Potyvirus*
Walnut	*Juglans regia*	Cherry leafroll virus	*Nepovirus*
Watermelon	*Citrullus lanatus*	Cucumber mosaic virus	*Cucumovirus*
		Watermelon mosaic virus	*Potyvirus*
		Zucchini yellow mosaic virus	*Potyvirus*
		Papaya ringspot virus	*Potyvirus*
Forage			
Alfalfa	*Medicago sativa*	Alfalfa mosaic virus	*Alfamovirus*
Grass			
Sugarcane	*Saccharum* sp.	Sugarcane mosaic virus	*Potyvirus*
		Sugarcane yellow leaf virus	*Polerovirus*
		Sorghum mosaic virus	*Potyvirus*
Legumes			
Bean	*Phaseolus vulgaris*	Bean golden mosaic virus	*Begomovirus*
Clover	*Trifolium repens*	Alfalfa mosaic virus	*Alfamovirus*
Groundnut	*Arachis hypogaea*	Peanut clump virus	*Pecluvirus*
		Groundnut rosette virus	*Umbravirus*
Pea	*Pisum sativum*	Alfalfa mosaic virus	*Alfamovirus*
		Bean leafroll virus	*Luteovirus*
		Bean yellow mosaic virus	*Potyvirus*
		Pea enation mosaic virus	*Umbravirus*
		Pea seed-borne mosaic virus	*Potyvirus*
		Pea streak virus	*Carlavirus*
Peanut	*Arachis hypogaea*	Tomato spotted wilt virus	*Tospovirus*
		Groundnut rosette assistor virus	Unassigned
		Peanut stripe virus	*Potyvirus*
Soybean	*Glycine max*	Soybean mosaic virus	*Potyvius*
		Bean pod mottle virus	*Comovirus*
		Southern bean mosaic virus	*Sobemovirus*

Continued

Table 1 Continued

Crop		Resistance to	
Category/common name	Scientific name	Virus	Genus
Vegetables			
Cucumber	*Cucumis sativus*	Cucumber mosaic virus	*Cucumovirus*
		Papaya ringspot virus	*Potyvirus*
		Squash mosaic virus	*Comovirus*
		Watermelon mosaic virus	*Potyvirus*
		Zucchini yellow mosaic virus	*Potyvirus*
Lettuce	*Lactuca sativa*	Lettuce mosaic virus	*Potyvirus*
		Lettuce necrotic yellows virus	*Tenuivirus*
Pepper[b]	*Capsicum*	Cucumber mosaic virus[b]	*Cucumovirus*
		Tobacco etch virus	*Potyvirus*
		Potato virus Y	*Potyvirus*
Potato	*Solanum tuberosum*	Potato virus A	*Potyvirus*
		Potato virus X	*Potexvirus*
		Potato virus Y	*Potyvirus*
		Potato leafroll virus	*Polerovirus*
		Tobacco rattle virus	*Tobravirus*
		Tobacco vein mottling virus	*Potyvirus*
Squash[a]	*Cucurbita pepo*	Cucumber mosaic virus[a]	*Cucumovirus*
		Papaya ringspot virus	*Potyvirus*
		Squash mosaic virus	*Comovirus*
		Watermelon mosaic virus[a]	*Potyvirus*
		Zucchini yellow mosaic virus[a]	*Potyvirus*
Sugar beet	*Beta vulgaris*	Beet necrotic yellow vein virus	*Benyvirus*
		Beet western yellows virus	*Polerovirus*
Sweet potato	*Ipomea batatas*	Sweet potato feathery mottle virus	*Potyvirus*
Tomato[b]	*Solanum lycopersicum*	Beet curly top virus	*Curtovirus*
		Cucumber mosaic virus[b]	*Cucumovirus*
		Tobacco mosaic virus	*Tobamovirus*
		Tomato mosaic virus	*Tobamovirus*
		Tomato spotted wilt virus	*Tospovirus*
		Tomato yellow leaf curl virus	*Begomovirus*

[a]Commercially released in the USA.
[b]Commercially released in the People's Republic of China.

However, soon after their release, these potato lines were withdrawn from the market due to food processor rejection. In the People's Republic of China, tomato and pepper resistant to CMV through expression of the virus CP gene have been released.

Benefits of Virus-Resistant Transgenic Plants

Virus-resistant transgenic plants offer numerous benefits to agriculture, particularly in cases where genetic sources of resistance have not been identified or are not easy to transfer into elite cultivars by traditional breeding approaches due to genetic incompatibility or links to undesired traits. This is well illustrated for PRSV-resistant papaya and GFLV-resistant grapevine. In such cases, engineered resistance may be the only viable option to develop virus-resistant cultivars. Engineered resistance may also be the only approach to develop cultivars with multiple sources of resistance by pyramiding several virus-derived gene constructs. This is the case for squash resistant to CMV, ZYMV, and WMV. Commercial releases have demonstrated the stability and durability of the engineered resistance over more than a decade in the case of squash and papaya. Benefits of virus-resistant transgenic plants are also of economic importance because yields are increased and quality of crop products improved. For example, virus-resistant transgenic squash allowed growers to restore their initial yields in the absence of viruses with a net benefit of $19 million in 2004 in the USA. Also, after the release of PRSV-resistant papaya cultivars, papaya production has reached similar level than before PRSV became epidemic in Hawaii, with a $4.3 million net benefit over a 6-year period in Hawaii. Further, benefits are of epidemiological importance with virus-resistant transgenic plants limiting virus infection rates by restricting challenge viruses, reducing their titers, or inhibiting their replication and/or cell-to-cell or systemic movement. Therefore, lower virus levels reduce the frequency of acquisition by vectors and subsequent transmission within and between fields. Consequently, virus epidemics are substantially limited. Recently, it has been shown that transgenic squash resistant to ZYMV and WMV do not serve as virus source for secondary spread. In addition, benefits of virus-resistant transgenic plants are of environmental importance because chemicals directed to control virus vector populations are reduced or not necessary. Restricting the reliance on chemicals directed to arthropod, fungal, plasmodiophorid, and nematode vectors of plant viruses is important for sustainable agriculture. Finally, benefits of virus-resistant transgenic plants are of social importance. In Hawaii, growing papaya was not viable anymore prior to the release of PRSV-resistant transgenic papaya despite huge efforts to eradicate infected trees in order to limit the propagation of the virus. The impact of PRSV was so severe that some growers abandoned their farms and had to find new jobs. PRSV-resistant transgenic papaya saved the papaya industry and strengthened the social welfare of local communities in Hawaii. Altogether, benefits of virus-resistant transgenic crops are important not only to facilitate their adoption by growers but also to counterbalance real and potential environmental and health risks.

Mechanisms Underlying Resistance

Two types of mechanisms underlying engineered virus resistance have been described, one requiring expression of a transgene protein product (protein-mediated resistance) and the other depending on the expression of transgene transcripts (RNA silencing).

Protein-Mediated Resistance

Resistance to TMV in tobacco plants was initially related to the expression level of the viral CP. Resistance could be overcome by high doses of inoculum and was not very effective against virus RNA inoculation. Interference with an early step in the virus infection cycle, that is, disassembly of TMV particles, has been hypothesized to explain CP-mediated resistance, maybe by inhibition of viral uncoating or reduction of protein–protein interactions between transgene-expressed CP and challenge virus. It would be interesting to analyze whether RNA-mediated resistance is also active in transgenic plants for which the CP has been suspected as an elicitor of the engineered resistance.

RNA Silencing and Post-Transcriptional Gene Silencing

In the early 1990s, the use of untranslatable virus CP transgenes was shown to confer resistance to virus infection, indicating that resistance was RNA rather than protein-mediated. In addition, resistance was correlated to actively transcribed transgenes but low steady-state levels of transgene RNA, indicating a sequence-specific post-transcriptional RNA-degradation system. Remarkably, RNA degradation was shown to target transgene transcripts and viral RNA. This was the first indication of the occurrence of post-transcriptional gene silencing (PTGS) as a manifestation of RNA silencing, a mechanism that inhibits gene expression in a sequence specific manner.

RNA silencing operates through diverse pathways but relies on a set of core reactions that are triggered by dsRNAs, which are processed into RNA duplexes of 21–24 bp in length, called small interfering RNAs (siRNAs), by the enzyme Dicer and its homologs, which have RNAseIII, helicase, dsRNA-binding, Duf283, and PIWI-Argonaute-Zwille (PAZ) domains. One siRNA strand is incorporated into a large multi-subunit ribonucleoprotein complex called the RNA-induced silencing complex (RISC) upon ATP-dependent unwinding and guides the complex to degrade cellular single-stranded RNA molecules that are identical in nucleotide sequence to the siRNA such as challenge viral RNAs.

RNA silencing is a host defense mechanism against viruses that is triggered by dsRNA and targets the destruction of related RNAs. During virus infection, dsRNA intermediates of viral RNA replication, or dsRNA produced from viral RNAs by host polymerases, or partially self-complementary structures of viral genomic RNAs become substrates for the production of siRNAs by Dicer-like (DCL) enzymes. Plants produce at least four different variants of DCLs of which DCL2 and DCL4 provide a hierarchical antiviral defense system.

Viruses have the ability to not only trigger but also suppress RNA silencing as counterdefense mechanism. Indeed, viruses have evolved silencing-suppressor proteins that minimize PTGS. These proteins act in different ways and on different steps of the silencing pathways. Some viral suppressors bind to siRNA duplexes and prevent their incorporation into RISC. Others act directly on the enzymes or cofactors of this defense pathway. It seems that plants have duplicated DCLs as a response to the functional complexity of viral suppressors.

It is anticipated that RNA silencing mechanisms will be devised extensively in the future to create virus-resistant transgenic crop plants. Indeed, resistance can be very strong if transgene transcripts form a hairpin structure containing a substantial base-paired stem, if challenge viruses have extended nucleotide sequence identity with the transgene, and if infection with unrelated viruses carrying silencing-suppressor genes does not interfere with RNA silencing.

Environmental Safety Issues

The insertion and expression of virus-derived genes in plants has raised concerns on their environmental safety. Since the early 1990s, considerable attention has been paid to potential environmental risks associated with the release of virus-resistant transgenic crops. Potential risks relate to the occurrence and outcomes of heterologous encapsidation, recombination, and gene flow. However, it is important to keep in perspective that these phenomena are known to occur in conventional plants, in particular for heterologous encapsidation and recombination, in the case of co-infection. Therefore, it is critical to determine if they occur in transgenic plants beyond base line events in conventional plants.

Heterologous Encapsidation

Heterologous encapsidation refers to the encapsidation of the genome of a challenge virus by the CP subunits from another virus, upon co-infection or expressed in a transgenic plant containing a viral CP gene. Heterologous encapsidation is a particular case of functional complementation, that is, a transgene might help a challenge virus in another function. Since the CP carries determinants for pathogenicity and vector specificity, among other features, the properties of field viruses may change. For example, it is conceivable that an otherwise vector-nontransmissible virus could become transmissible through heterologous encapsidation in a transgenic plant and infect otherwise nonhost plants. Such changes in vector specificity will be a single generation, not a permanent, event because they will not be perpetuated in the virus genome progeny. Heterologous encapsidation is known to occur in conventional plants subjected to mixed virus infection. Therefore, it is not too surprising that it was documented with herbaceous transgenic plants in the laboratory. In contrast, heterologous encapsidation has not been found to detectable level in transgenic vegetable plants – expressing virus CP gene that were tested extensively in the field over several years at different locations. The only exception was with transgenic squash expressing the CP gene of WMV for which a low rate of transmission of an aphid non-transmissible strain of ZYMV was documented. However, transmission of this ZYMV strain did not reach epidemic proportion. Also, heterologous encapsidation is unlikely to occur in RNA-silenced plants since expression of the transgene does not result in the accumulation of a detectable protein product. Altogether, compelling evidence has shown that heterologous encapsidation in transgenic plants expressing virus CP genes is of limited significance and should be considered negligible in regard to adverse environmental effects.

Recombination

Recombination refers to template switching between viral transgene transcripts and the genome of a challenge virus. Resulting recombinant viruses may have chimerical genomic molecules consisting of a segment from the challenge viral genome and another segment from viral transgene transcripts. It is argued that recombinant viruses may have identical biological properties as their parental lineages or new biological properties such as changes in vector specificity, expanded host range, and increased pathogenicity. Since recombination alters the genome of challenge viruses, new properties of chimera viruses will be stably transmitted to and perpetuated within the virus progeny. Comprehensive studies have documented the occurrence of recombination in transgenic plants expressing viral genes. The stringency of selective pressure applied to the challenge virus has been shown to be a critical factor in the recovery of recombinant viruses. Conditions of high selective pressure enhance the creation of recombinant viruses. In contrast, limited, if any, recombinant viruses are found to detectable level under conditions of low or no selective pressure. It is important to keep in perspective that the latter conditions prevail under field conditions in which plants are infected by functional, not defective, viruses. So far, no recombination event has been found to detectable level in CP gene-expressing transgenic perennial plants that were tested in experimental fields over a decade. Therefore, the significance of recombination in transgenic plants expressing viral genes is very limited in regard to adverse environmental effects. Furthermore, recombination is unlikely to occur in RNA-silenced plants because transgene expression often results in no detectable RNA, especially so when noncoding sequences are used as antiviral constructs.

Gene Flow

Gene flow refers to the pollen-driven movement of *trans-genes* from a virus-resistant transgenic plant into a nontransgenic compatible recipient plant, for example, a wild relative. Hybrids resulting from gene flow can acquire and express virus-derived transgenes, and become resistant to the corresponding viruses. Subsequently, plants acquiring viral resistance traits can have a competitive advantage, exhibit increased fitness, and eventually become more invasive, maybe as noxious weeds. Movement of viral transgene constructs through pollen flow has been documented from virus-resistant transgenic squash into a wild squash relative under experimental field conditions. Hybrids between transgenic and wild squash exhibited increased fitness under conditions of intense disease pressure. In contrast, under conditions of low disease pressure, no difference was observed between hybrids and wild squash in terms of growth and reproductive potential. Since viruses do not limit the size and dynamics of wild squash populations in natural habitats, it is anticipated that gene flow with virus-resistant transgenic squash will be of limited significance. It remains to be seen if increased fitness will provide hybrids with a competitive edge that could eventually lead to enhanced weediness. Altogether, compelling evidence suggest that gene flow with virus-resistant transgenic squash should not be perceived more risky than the equivalent situation with virus-resistant conventional squash.

Gene flow can also occur from a virus-resistant transgenic plant into a compatible conventional plant. Although of negligible biological impact, this phenomenon can be essential for economical reasons such as organic production and export to countries that have not deregulated transgenic crops. Worth noting is the fact that the coexistence of transgenic and conventional papaya in spatiotemporal proximity is a reality in Hawaii.

Real versus Perceived Risks

It is important to discriminate perceived and real risks associated with virus-resistant transgenic plants. Field environmental safety assessment studies have provided strong evidence of limited or no environmental risks, indicating that issues associated with virus-resistant transgenic plants are substantially less significant than initially predicted. To fully grasp the significance of environmental risks, the situation in the absence of transgenic plants needs to be taken into account and considered as base line information. So far, there is no compelling evidence to indicate that transgenic plants expressing viral genes increase the frequency of heterologous encapsidation or recombination beyond background rates. Similarly, there is little evidence, if any, to infer that transgenic plants expressing viral genes alter the properties of existing virus populations or create new viruses that could not arise naturally in conventional plants subjected to multiple virus infection. Also, the consequence of gene flow is perceived identical with transgenic and conventional virus-resistant plants. Therefore, there seems to be increasing scientific evidence that, while initially perceived as major concern, heterologous encapsidation, recombination, and gene flow with transgenic plants expressing viral genes are not deemed hazardous to the environment. Nevertheless, a case-by-case approach is recommended to make sound decisions on the safe release of virus-resistant transgenic plants. Communicating scientific facts on the safety of transgenic plants expressing viral genes is critical to distinguish real and perceived risks, help counterbalance risks and benefits, and assist regulatory authorities in the decision-making process for the release of virus-resistant transgenic crops.

Conclusions

Engineered resistance has expanded the scope of innovative approaches for virus control by providing new tools to develop resistant crop cultivars and increasing opportunities to implement effective and sustainable management strategies. Many advances have been made on the development of virus-resistant transgenic plants over the past two decades. Since its validation with TMV in tobacco plants in 1986, the concept of PDR has been applied successfully against a wide range of viruses in many plant species. Studies on RNA silencing have provided new insights into virus–host interactions. They also shed light on the molecular and cellular mechanisms underlying engineered resistance in plants expressing virus-derived gene constructs. The exploitation of the sequence-specific antiviral pathways of RNA silencing has facilitated the design of transgene constructs for more predictable resistance. Broader spectrum resistance against multiple virus strains should also be achievable more easily by designing virus transgenes in highly conserved sequence regions and applying RNA silencing. It is anticipated that RNA silencing mechanisms will be devised more extensively in the future to create virus-resistant crop plants.

The past two decades have witnessed an explosion in the development of virus-resistant transgenic plants. Further, virus resistance has been extensively evaluated under field conditions. So far, 37 different plant species have been tested successfully in the field for resistance to 58 viruses. Despite remarkable progress, only a limited number of virus-resistant transgenic crops have been released so far for commercial use. Based on their efficacy at controlling virus diseases and high adoption rate by growers, more virus-resistant transgenic crops are likely to reach the market in the future.

There is no doubt that innovative and sustainable control strategies are needed for virus diseases to mitigate their impact on agriculture. Lessons from field experiments with various transgenic crops engineered for virus resistance and the commercial release of virus-resistant squash, papaya, tomato, and pepper have conclusively demonstrated that benefits outweigh by far risks to the environment and human health. A timely release and adoption of new virus-resistant transgenic crops is desirable, in particular in regions where viruses are devastating.

See also: Papaya Ringspot Virus; Plant Antiviral Defense: Gene Silencing Pathway; Plant Resistance to Viruses: Natural Resistance Associated with Dominant Genes; Plant Resistance to Viruses: Natural Resistance Associated with Recessive Genes; Plant Virus Diseases: Economic Aspects; Recombination; Recombination; *Tobamovirus*; Vector Transmission of Plant Viruses.

Further Reading

Baulcombe D (2004) RNA silencing in plants. *Nature* 431: 356–363.

Bendahmane M and Beachy RN (1999) Control of tobamovirus infections via pathogen-derived resistance. *Advances in Virus Research* 53: 369–386.

Fuchs M, Chirco EM, and Gonsalves D (2004) Movement of coat protein genes from a virus-resistant transgenic squash into a free-living relative. *Environmental Biosafety Research* 3: 5–16.

Fuchs M, Chirco EM, McFerson J, and Gonsalves D (2004) Comparative fitness of a free-living squash species and free-living x virus-resistant

transgenic squash hybrids. *Environmental Biosafety Research* 3: 17–28.
Fuchs M and Gonsalves D (2007) Safety of virus-resistant transgenic plants two decades after their introduction: Lessons from realistic field risk assessment studies. *Annual Review of Phytopathology* 45: 173–220.
Fuchs M, Klas FE, McFerson JR, and Gonsalves D (1998) Transgenic melon and squash expressing coat protein genes of aphid-borne viruses do not assist the spread of an aphid nontransmissible strain of cucumber mosaic virus in the field. *Transgenic Research* 7: 449–462.
Gonsalves C, Lee DR, and Gonsalves D (2004) *Transgenic Virus-Resistant Papaya: The Hawaiian 'Rainbow' was Rapidly Adopted by Farmers and Is of Major Importance in Hawaii Today.*
Gonsalves D (1998) Control of papaya ringspot virus in papaya: A case study. *Annual Review of Phytopathology* 36: 415–437.
Gonsalves D, Gonsalves C, Ferreira S, et al. (2004) *Transgenic Virus-Resistant Papaya: From Hope to Reality for Controlling Papaya ringspot virus in Hawaii*. http://www.apsnet.org/online/feature/ringspot.
Lindbo JA and Dougherty WG (2005) Plant pathology and RNAi: A brief history. *Annual Review of Phytopathology* 43: 191–204.
MacDiarmid R (2005) RNA silencing in productive virus infections. *Annual Review of Phytopathology* 43: 523–544.
Powell AP, et al. (1986) Delay of disease development in transgenic plants that express the tobacco mosaic virus coat protein gene. *Science* 232: 738–743.
Qu F and Morris TJ (2005) Suppressors of RNA silencing encoded by plant viruses and their role in viral infections. *FEBS Letters* 579: 5958–5964.
Sanford JC and Johnston SA (1985) The concept of parasite-derived resistance-deriving resistance genes from the parasite's own genome. *Journal of Theoretical Biology* 113: 395–405.
Tepfer M (2002) Risk assessment of virus-resistant transgenic plants. *Annual Review of Phytopathology* 40: 467–491.
Vigne E, Bergdoll M, Guyader S, and Fuchs M (2004) Population structure and genetic diversity within *Grapevine fanleaf virus* isolates from a naturally infected vineyard: Evidence for mixed infection and recombination. *Journal of General Virology* 85: 2435–2445.
Voinnet O (2005) Induction and suppression of RNA silencing: Insights from viral infections. *Nature Reviews Genetics* 6: 206–221.
Waterhouse PM, Wang MB, and Lough T (2001) Gene silencing as an adaptative defense against viruses. *Nature* 411: 834–842.

Plant Resistance to Viruses: Geminiviruses

J K Brown, The University of Arizona, Tucson, AZ, USA

© 2008 Elsevier Ltd. All rights reserved.

Glossary

Bipartite virus A virus having a genome comprising two segments of nucleic acid.
Consensus sequence An identical sequence, usually shared by a group of close virus relatives.
Hemiptera The insect order that contains plant feeding insects such as aphids, leafhoppers, planthoppers, and whiteflies.
Monopartite virus A virus having a genome comprising one segment of nucleic acid.
Reassortment The process by which nonhomologous viral components interact during an infection cycle. Naturally reassorted viruses are thought to be more fit than their homologous counterparts and so arise by natural selection.
Recombination The process by which nucleic acid sequences are exchanged between two or more nucleic acid molecules during replication.
Satellite A segment of nucleic acid that associates with a 'helper' virus and utilizes it for replication and transmission. Satellites typically share little to no sequence homology with the helper virus.
Vector A biological organism that transmits plant viruses from one host to another, usually an arthropod or nematode.

Introduction

Geminiviruses

For perspective, it is important to recognize that the family *Geminiviridae* was established only in 1978. This is an unusual family among plant viruses in that their genomes are circular, single-stranded DNA. When the group was originally established, fewer than 20 possible members were recognized. Even so, among the whitefly-transmitted group, now recognized as the genus *Begomovirus* (originally subgroup III), additional associated diseases had been described, but definitive characterization for most of them was delayed until tools for detecting and identifying them were developed.

Geminiviruses are composed of circular, single-stranded DNA genome that is packaged within a pair of quasi-isometric particles, hence, the name 'geminate', or twinned particles (**Figure 1**). Geminiviruses are divided into four genera based on the genome organization and biological properties. One of the most striking differences between the four genera is their restriction to either monocots or dicots (eudicots), and a highly specific relationship with the respective arthropod vector, treehoppers, leafhoppers, or a whitefly, all of which are classified in the order Hemiptera, suborder Homoptera, and which feed on phloem sap using a slender stylet.

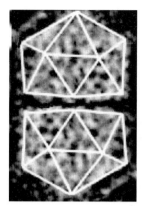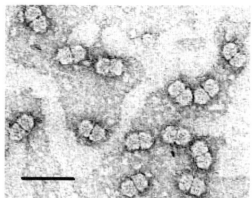

Figure 1 Typical geminate particles (~20 × 32 nm²) by transmission electron microscopy. Courtesy of C. Fauquet.

The four genera

Geminiviruses with monopartite genomes are transmitted by leafhopper vectors to monocotyledonous plants and are placed in the genus *Mastrevirus*. *Maize streak virus* is the type species. Geminiviruses with monopartite genomes that are transmitted to dicotyledonous plants by leafhopper vectors are classified in the genus *Curtovirus* with *Beet curly top virus* as the type species. The third genus, *Topocuvirus*, was recently established and has a single member, *Tomato pseudo-curly top virus*. Tomato pseudo-curly top virus (TPCTV) is monopartite, infects dicotyledonous plants, and has a treehopper vector. The fourth and largest genus, *Begomovirus*, houses the majority of the species in the family. Begomoviruses are transmitted exclusively by the whitefly *Bemisia tabaci* (Gennadius) complex to dicotyledonous plants, and *Bean golden yellow mosaic virus* is recognized as the type species. They can have either a monopartite or bipartite genome. For bipartite viruses, the two components are referred to as DNA-A and DNA-B. The B component is thought to have arisen from a duplication of DNA-A followed by the loss of all but the two genes involved in cell-to-cell movement and nuclear localization of the virus during replication.

Viral Genome and Protein Functions

Begomoviruses have a circular, single-stranded DNA (ssDNA) genome that replicates through double-stranded DNA (dsDNA) intermediates by a rolling-circle mechanism. The genomes of begomoviruses are approximately 2.6–2.8 kbp in size. Bipartite genomes share a common region (CR) of approximately 200 nt that is highly conserved among cognate components of a viral species, while the analogous region in monopartite viruses is referred to as the large intergenic region (LIR). The CR/LIR contain modular *cis*-acting elements of the origin of replication (*ori*) and promoter elements.

Four to six open reading frames (ORFs) capable of encoding proteins >10 kDa in size are present on the A component or monopartite genome. The viral capsid protein (CP) is encoded by the ORF (AV1), which is the most highly conserved begomovirus gene. The CP of begomoviruses is required for encapsidation of ssDNA and for whitefly-mediated transmission. The CP is also necessary for systemic spread of all monopartite and for some bipartite viruses, depending upon the degree of virus–host adaptation. The AC1 ORF encodes REP, a replication initiation protein, and specificity is mediated through sequence-specific interactions with *cis*-acting elements of the *ori*. The AC2 ORF encodes a transcription factor protein required for rightward gene expression and this protein also functions to suppress gene silencing imposed by the plant host. The AC3 ORF encodes a replication enhancer protein. Movement and systemic infection in bipartite viruses is accomplished by B component ORFs BV1 and BC1, which are responsible for nuclear transport of ssDNA and cell-to-cell movement functions, respectively. For monopartite viruses, several ORFs are involved in movement functions, including the analogous V1 (CP), V2, and the C4 ORF, which exert host-specific effects on symptom severity, virus accumulation, and systemic movement.

Satellite DNAs of Begomoviruses

Certain monopartite begomoviruses are associated with a nonviral circular, ssDNA referred to as a DNA-β type satellite. SatDNAs associate with a 'helper' begomovirus on which they depend for replication, encapsidation, and whitefly-mediated transmission. This type of satDNA comprises all nonviral sequences except for a 9 nt stretch that mimics the sequence of the viral origin of replication; hence, when replication initiates, the DNA-β is recognized as template and is replicated together with the begomoviral genome. In turn, the DNA-β satellite encodes a protein that has experimentally been shown to localize to the nucleus of the cell where it is thought to suppress host-induced gene silencing, an innate defense response that is intended to protect the plant from virus infection.

In the absence of the satellite, the 'helper' virus cannot sustain infection of the host and so if the helper virus becomes separated from its satellite, the virus would quickly become extinct. This relationship seems to imply that the virus itself has lost the capacity to be sufficiently pathogenic and so is dependent upon the satellite for its survival. It is interesting to speculate about the evolutionary significance of this type of interaction, in light of the hypothesis that after a time most pathogens and their hosts 'adjust' to one another, reaching a sort of evolutionary equilibrium. How these 'helper' begomoviruses became disabled and subsequently rescued by such a nonviral sequence is not known. And why they associate only with monopartite genomes, even though they have had ample opportunity for contact with bipartite begomoviruses, which also occur in the Old World, is not understood. It is thought by some that a monopartite, compared to bipartite, genome organization is the more primitive condition. However, it is not clear why satDNAs did not also co-evolve with the most ancient bipartite lineages.

Diversification and Host Shifts

There are over 389 complete geminiviral genome sequences, and at least 200 full-length DNA-β satellite sequences available in public databases, indicating the widespread nature of these viruses, discovered only recently. Geminiviruses appear readily adaptable to new hosts, and have proven remarkably capable of 'host-shifting'. Although the specific coding or noncoding regions of the genome that undergo change during host adaptation have not been identified, geminivirus genomes seem capable of undergoing quite rapid evolutionary change, for example, sometimes in a single growing season. The evidence that such rapid change occurs is taken from the observation that very few virus sequences in endemic species match identically with their closest relatives in a crop species, and many diverge by 10% or more. In only a very few instances have geminiviruses in an endemic species been found in a counterpart cultivated host, and then the virus is known to have emerged only recently as a pathogen of the crop plant. These observations suggest that the development of sustainable, resistant varieties to abate geminivirus diseases may prove difficult if these viruses are likewise capable of diversifying rapidly and overcome resistance genes. Moreover, the members of this virus family are well known for their ability to undergo recombination allowing them to diversify and evolve into more fit species. This phenomenon is best exemplified by the begomoviruses, some of which have been shown to contain sequence fragments that closely match sequences in several extant begomovirus genomes.

Diversification in bipartite viruses also appears to occur through reassortment of the DNA components based on sequence analysis. Under laboratory conditions, mixtures of certain heterologous DNA-A and DNA-B components have been shown to work cooperatively to infect a plant, and some of these 'pseudo-recombinants or reassortants' can infect host species that at least one of the parent viruses cannot. Because geminiviruses apparently do not 'cross-protect' against one another or undermine one another in the event of a mixed infection, it is clear that more than a species can infect the same plant. Experimental evidence has demonstrated that two begomoviruses can occupy the same nucleus in the plant host, a circumstance that would facilitate diversification processes and readily yield new and emerging viral species.

Symptomatology on Respective Major Hosts

Symptoms of infection caused by geminiviruses vary depending on whether the host is a monocot or a dicot, characterized as having distinct arrangements of vascular tissues and parallel or branching leaf vein patterns, respectively (**Figures 2(a)–2(d)**). Even so, geminivirus-infected plants exhibit unique symptom phenotypes that are often readily recognizable to the trained eye.

Monocots infected with mastreviruses often exhibit interveinal chlorosis that takes on the appearance of foliar streaking. Leaves may develop splotchy yellow or white patterns between the veins, and the veins may be distorted and cleared and plant vigor is reduced leading to stunted growth. Commonly, the symptoms develop in leaves closest to the growing tip when infection occurred.

A pattern of symptom development in the newest growth also is typical of begomoviruses, curtoviruses, TPCTV, and several unclassified, monopartite Old World geminiviruses. Symptoms in the leaves of dicots range from mosaic to yellow and bright yellow mosaics or sectored patterns, and leaf curling. Infected plants often develop shortened internodes and become dramatically stunted. Certain viruses have the effect of causing profuse flowering, while at the same time reducing fruit set and fruit size of those that are set. Certain geminiviruses also cause outgrowths on the top or underside of leaves, referred to as enations.

Economic Importance

Viruses of the family *Geminiviridae* are considered emergent pathogens because they have become recently increasingly important in a large number of agriculturally important crops ranging from cereals such as maize and wheat, to species grown for fiber, like cotton and kenaf, and to a number of annual and perennial ornamental plants. Other viruses in this family are primarily yield-limiting

Figure 2 (a) Bean plants infected with the New World, bipartite bean calico mosaic virus showing typical mosaic symptoms; (b) tomato plant infected with the Old World, monopartite tomato yellow leaf curl virus; (c) symptoms of the leafhopper-transmitted beet curly top virus infecting a pumpkin plant; (d) maize leaves showing characteristic symptoms of the Old World monopartite, monocot infected with maize streak virus. (c) Courtesy of R. Larsen. (d) Reproduced by permission of E. Rybicki.

pathogens of vegetable and root crop plants that are important as staples and for vitamins and nutritional variety. Yield losses in infected crops range from 20% to 100%, depending on the virus–host combination and the growth stage of the plant at time of infection. Such extensive reductions in crop productivity often result in conditions of malnutrition and in the loss of precious incomes, particularly in developing countries that have become reliant on exporting their crops.

Geminiviruses are distributed throughout the Tropics, subtropics, and mild climate regions of the temperate zones, occurring on all continents where food, fiber, and ornamental crops are cultivated. Geminiviruses also infect a wide range of endemic plant species, with which they have likely co-evolved, based on the evidence that they appear to cause little damage. This would not be surprising given their predicted longstanding association with uncultivated eudicot species. Thus the endemic species serve as a melting pot for native viruses that find their way into cultivated plants, particularly recently, as practices in many locations have shifted to monoculture production. In the subtropics, it is not uncommon to grow crops year-round. These practices enabled by extensive irrigation projects worldwide have resulted in sustained, high levels of virus inoculum and vector populations in the environment, and more frequent outbreaks of disease, particularly since the 1970s.

Molecular Diagnosis of the Virus and Its Homology with Other Viruses

For most plant viruses, serology has traditionally been the method of choice for identification. However, virus-specific antisera for geminiviruses are not generally available. This is in part because geminiviruses can be difficult to purify. Further, begomoviruses in particular are not good antigens, and a small amount of contaminating plant protein in the purified preparation outcompetes the viral epitopes during immunization. Even when antibodies are available, they are useful primarily for detection, not identification, owing to the high genus-wide conservation of the viral coat proteins. However, the greatest factor underlying the popular use of molecular diagnostics for begomoviruses is that they contain a circular, ssDNA genome that yields a dsDNA intermediate from replication, which is highly amenable to linearization with an endonuclease to yield a dsDNA fragment that can be cloned. Second, shortly after the geminiviruses were discovered, polymerase chain reaction (PCR) was invented and the two clearly went hand in hand.

Commonly, detection is accomplished using PCR primers designed to be broad spectrum (usually degenerate) or virus specific. Because substantial information on begomovirus genome organization and gene function is now available, as are numerous genomic and CP gene sequences, it is possible to scan partial or entire viral sequences and develop hypotheses regarding the prospective utility of a particular ORF or nontranslated region for inferring phylogenetic relationships. Sequences associated with functional domains of viral polypeptides, nucleotide sequence motifs or regulatory elements that are conserved within the genus or groups of species or strains, or sequences that are highly variable have been explored for this purpose.

Identification can be accomplished by determining the DNA sequence of the viral coat protein (tentative identification) and ultimately the complete DNA-A or monopartite DNA component. More recently the method, rolling-circle amplification, has been introduced as a non-sequence-specific approach for amplifying multimers of a viral genome, which can subsequently be cleaved to a full-length monomer using restriction endonucleases, allowing the sequence to be determined.

Sequence alignments are conveniently presented as cladograms to illustrate predicted relationships and distances can be calculated to estimate divergence. Phylogenetic analysis of the complete genome (monopartite) and A component (bipartite) viral sequences or the CP gene nucleotide sequences indicate strict clustering of begomoviruses by extant geographic origin, and not by monopartite or bipartite genome organization, crop of origin, or host range. Certain Old World viruses may be somewhat more divergent than the majority of New World viruses, perhaps because they have been more isolated by geographic barriers and/or patterns of human travel, suggesting fewer opportunities for interactions between viruses in different regions, while at the same time, great potential for localized diversity. All begomoviruses in the New World have bipartite genomes and many are distributed as overlapping geographic clusters, suggesting a lesser role of geographic barriers, fewer genotypes that have diverged in isolation, and, consequently, a more confounded evolution (**Figure 3**).

The genera in this family are discriminated based on numerical taxonomy in which a percentage nucleotide identity is used to classify them based on relatedness. Within each genus the working cutoff for species versus strains have been determined empirically by comparing the frequencies of percentage divergence across many sequences. For example, for the begomoviruses, a species is presently recognized when it is less than 89% identical to another, a value that has been shown to correlate nicely with biological differences. Species in the genera *Mastrevirus* and *Curtovirus* are classified as separate species when the genome sequence is found to share less than 75% and 89% nt identity, respectively.

All geminiviruses contain a conserved consensus sequence of *c.* 30 nt in length in this region. This sequence contains the T/AC cleavage site at which rolling-circle replication is initiated, and this conserved feature led to early speculation that recombination might occur frequently at this site. In addition, virus-specific, directly repeated sequences that bind the viral replicase-associated protein (REP) during replication initiation are also present in this region and are considered to provide useful clues with respect to the likelihood that the DNA-A and DNA-B components are cognate, or of the same virus. The directly repeated element in bipartite and most monopartite genomes contains two 4–5 nt directly repeated sequences separated by two to four 'spacer' nucleotides which act in a virus-specific manner to bind REP. As a result, these elements constitute informative indicators of viral genealogy useful in defining species and strains when traced through viral lineages. It is not entirely clear how much plasticity in sequence or spacing can be tolerated at this site without interfering with replication. Finally, these repeated sequences in the CR have been explored to make predictions about which bipartite genomes might feasibly interact *in trans* and reassort either with a close relative, or a more distantly related virus that harbors the same or similar iteron sequences and REP binding ability.

Recently Adopted Management Approaches

Management of diseases caused by geminiviruses requires a multifaceted strategy. This involves reduction of vector population levels in the field, typically by insecticide applications and/or encouragement of natural enemies in refuge areas such as predators or parasites of the vector. Sanitation in and around the field in which virus-infected plants, including endemic weeds, are removed after the crop is harvested and before it is planted are common practices. Implementation of a host-free period has proven relatively effective in reducing inoculum levels to a minimum so that when a crop is planted, it achieves near-mature growth before becoming infected with virus. Plants that become infected when they are partially mature usually experience less damage compared to virus infection early after planting. Implementation of a host-free period is accomplished by having a voluntary or an enforced period of time when no virus or vector hosts may be grown for a designated period of time usually ranging from 4 to 8 weeks, across an entire region. This is often an intangible goal given the requirement for complete cooperation. In all of the above examples, a single viruliferous vector can feasibly transmit virus to susceptible young plants, and so the most desirable control strategy is the use of resistant or tolerant crop varieties. For the most part, such varieties are not available. Recently, several resistant cassava, maize, and tomato varieties have been developed as the result of widespread efforts to provide genetically tolerant germplasm. There

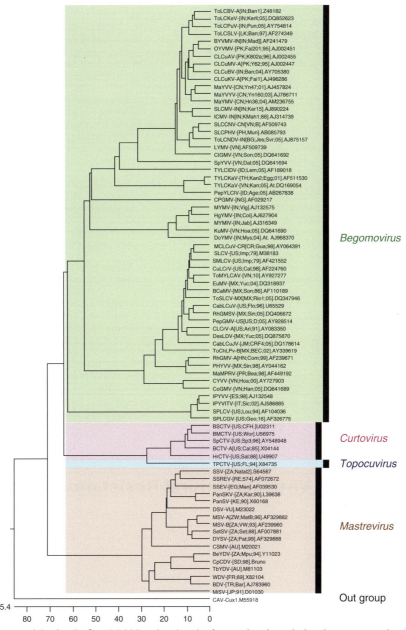

Figure 3 Phylogenetic tree of the family *Geminiviridae* showing the four major virus clades that serve to classify geminiviruses in one of four genera. The viruses are listed in the 8th ICTV Report. Courtesy of C. Fauquet.

also is an interest in developing and employing genetically engineered resistance using virus-derived resistance approaches such as RNA interference; however, the reluctance on the parts of many to accept this strategy has set back progress, which likely would have been brought to fruition for the control of at least certain geminivirus-incited crop diseases.

The recognition that begomoviruses are capable of rapid divergence employing multiple mechanisms underscores the need for accurate molecularly based methods that permit detection and tracking of biologically significant variants. Molecular approaches will continue to be combined to achieve knowledge about viral biology, ecology, and molecular and cellular aspects of the infection cycle. Databases containing genome sequences for geminiviruses now permit comparisons for establishing identity, taxonomic relationships, and interpreting disease patterns. Accurate molecular epidemiological information will assist plant breeding and genetic engineering efforts in developing disease-resistant varieties by enabling the selection of timely and relevant viral species (and variants) for germplasm screening and as sources of the most conserved viral transgene sequences, respectively, for sustainable disease control.

See also: African Cassava Mosaic Disease; Bean Golden Mosaic Virus; Beet Curly Top Virus; Cotton Leaf Curl Disease; Emerging Geminiviruses; Maize Streak Virus; Mungbean Yellow Mosaic Viruses; Tomato Leaf Curl Viruses from India; Tomato Yellow Leaf Curl Virus.

Further Reading

Arguello-Astorga GR, Guevara-Gonzalez RG, Herrera-Estrella LR, and Rivera-Bustamante RF (1994) Geminivirus replication origins have a group-specific organization of iterative elements: A model for replication. *Virology* 203: 90–100.

Bisaro DM (1994) Recombination in geminiviruses: Mechanisms for maintaining genome size and generating genomic diversity. In: Paszkowski J (ed.) *Homologous Recombination and Gene Silencing in Plants*, pp. 39–60. Dordrecht, The Netherlands: Kluwer.

Brown JK (2001) The molecular epidemiology of begomoviruses. In: Khan JA and Dykstra J (eds.) *Trends in Plant Virology*, pp. 279–316. New York: Haworth Press.

Brown JK (2007) *The Bemisia tabaci* complex: Genetic and phenotypic variability drives begomovirus spread and virus diversification. Plant Disease APSNet Feature Article December–January 2006–07. http://www.apsnet.org/online/feature/btabaci/ (accessed February 2008).

Brown JK and Czosnek H (2002) Whitefly transmitted viruses. In: Plumb RT (ed.) *Advances in Botanical Research*, pp. 65–100. New York: Academic Press.

Dry IB, Krake LR, Rigden JE, and Reizan AM (1997) A novel subviral agent associated with a geminivirus: The first report of a DNA satellite. *Proceedings of the National Academy of Sciences, USA* 94: 7088–7093.

Goodman RM (1977) Single-stranded DNA genome in a whitefly-transmitted virus. *Virology* 83: 171–179.

Hanley-Bowdoin L, Settlage SB, Orozco BM, Nagar S, and Robertson D (1999) Geminiviruses: Models for plant DNA replication, transcription, and cell cycle regulation. *Critical Reviews in Plant Sciences* 18: 71–106.

Harrison BD and Robinson DJ (1999) Natural genomic and antigenic variation in whitefly-transmitted geminiviruses (begomoviruses). *Annual Review of Phytopathology* 37: 369–398.

Hou YM and Gilbertson RL (1996) Increased pathogenicity in a pseudorecombinant bipartite geminivirus correlates with intermolecular recombination. *Journal of Virology* 70: 5430–5436.

Idris AM and Brown JK (2005) Evidence for interspecific recombination for three monopartite begomoviral genomes associated with tomato leaf curl disease from central sudan. *Archives of Virology* 150: 1003–1012.

Lazarowitz SG (1992) Geminiviruses: Genome structure and gene function. *Critical Reviews in Plant Sciences* 11: 327–349.

Padidam M, Sawyer S, and Fauquet CM (1999) Possible emergence of new geminiviruses by frequent recombination. *Virology* 265: 218–225.

Sanderfoot AA and Lazarowitz SG (1996) Getting it together in plant virus movement: Cooperative interactions between bipartite geminivirus movement proteins. *Trends in Cell Biology* 6: 353–358.

Saunders K, Norman A, Gucciardo S, and Stanley J (2004) The DNA beta satellite component associated with *Ageratum* yellow vein disease encodes an essential pathogenicity protein (BC1). *Virology* 324: 37–47.

Plant Resistance to Viruses: Natural Resistance Associated with Dominant Genes

P Moffett and D F Klessig, Boyce Thompson Institute for Plant Research, Ithaca, NY, USA

© 2008 Elsevier Ltd. All rights reserved.

Glossary

Avirulence gene Any pathogen gene whose protein product induces a resistance response in a plant possessing a matching resistance gene, thereby rendering the pathogen unable to establish an infection.

Gene-for-gene resistance A genetic paradigm whereby the outcome of a plant–pathogen interaction is determined by the genotype of both organisms. Plant disease resistance proteins confer resistance to pathogens possessing a corresponding avirulence gene.

Hypersensitive response A type of programmed cell death induced by a resistant plant in response to pathogen infection. Cell death is normally contained to infected cells.

NB-LRR protein Plant disease resistance proteins containing a central nucleotide-binding domain (NB) and C-terminal leucine-rich repeat (LRR) domain. Also known as NBS-LRR or NB-ARC-LRR proteins.

Resistance gene Plant genes which confer resistance to specific pathogens. Resistance genes confer resistance to pathogens possessing a corresponding avirulence gene.

Salicylic acid A phenolic compound synthesized by plants that plays a critical signaling role for activation and regulation of defense responses to pathogens. Salicylic acid also regulates other plant processes such as thermogenesis.

Virulence gene Any pathogen gene whose function contributes to increased pathogen proliferation or disease symptoms on an infected plant.

Introduction

In many plant–pathogen interactions, host plants possess polymorphic resistance to different strains or races of pathogen which is controlled by a single gene. Disease resistance controlled by recessive genes is often a passive form of resistance wherein the pathogen is unable to utilize host-cell machinery. Dominant resistance (R) genes in plants appear to actively recognize the presence of specific pathogens and initiate programmed responses to counteract infection. However, this resistance is dependent on the pathogen possessing a corresponding gene that matches the R gene. These pathogen avirulence (*Avr*) genes are also dominant over their cognate virulent (*avr*) alleles. Since the outcome of an attempted infection is dependent on the genotype of both host and pathogen (**Figure 1(a)**), this phenomenon is known as gene-for-gene resistance, also known as race-specific or cultivar resistance. Gene-for-gene resistance has been observed for many plant pathogens including fungi, oomycetes, bacteria, insects, nematodes, and viruses.

R Gene Products

Gene-for-gene resistance was first described over 50 years ago by H. H. Flor who observed that different cultivars of flax possessed dominant *R* genes that were specific to races of flax rust, caused by the fungus *Melampsora lini*, possessing corresponding dominant *Avr* genes. Although viruses and fungi have different genetics, the same concept applies to several virus resistance genes.

Genetically, *R* genes have been characterized for many years and have been incorporated into breeding programs of multiple crop plants. Since the cloning of the first *R* gene in 1993, a steady stream of *R* genes have been cloned from a number of plant species. The proteins encoded by these genes were found to belong to a small number of protein classes including transmembrane proteins possessing extracellar leucine-rich repeat (LRR) domains as well as a small number of intracellular serine/threonine kinases. By far, however, the most numerous type of *R* genes encode intracellular proteins containing a central nucleotide-binding (NB) domain as well as a C-terminal LRR domain. These NB-LRR proteins also possess a predicted signaling domain at the N-terminus. NB-LRR proteins can be placed into two classes; those which possess an N-terminal domain with homology to the Toll and interleukin receptors (TIR domain) of animals and those that do not (**Figure 1(b)**). The structure of the N-terminus of the latter class of NB-LRR proteins is not clear; however, many are predicted to form a coiled-coil structure and as such are commonly referred to as CC-NB-LRR proteins. CC-NB-LRR proteins are present in both monocots and dicots whereas TIR-NB-LRRs appear to be present only in dicots. Both types are present in gymnosperms.

As of 2007, over 50 *R* genes encoding NB-LRR proteins of known specificity have been cloned from plants belonging to various species. Sequencing projects have revealed that plant genomes contain large numbers of genes encoding NB-LRR proteins. The *Arabidopsis* genome contains 149 such genes whereas the rice genome encodes between 500 and 600. Many *R* gene loci are polymorphic with different alleles recognizing different pathogen races or even different pathogens. Furthermore, NB-LRR proteins are highly divergent both within and between species such that the *R* gene complement of a given species does not show strong similarity to the *R* gene complement of another unrelated species. Thus, *R* genes as a class of proteins likely possess a great deal of potential for recognition of pathogen diversity. Given the high degree of polymorphism, *R* genes recognizing a specific pathogen can often be identified by testing, for example, a large number of *Arabidopsis* accessions. For crop species, resistance sources are often identified in, and introgressed from, wild relatives. **Table 1** shows an incomplete list of a number of characterized dominant viral resistance genes. Although a number of these genes have not been cloned, they show characteristics typical of NB-LRR-encoding *R* genes such as dominant inheritance and the presentation of a hypersensitive response (HR) (see below).

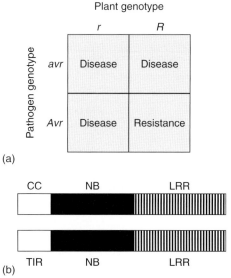

Figure 1 (a) Gene-for-gene resistance. Individual plant genotypes confer susceptibility (*r*) or resistance (*R*) to a given pathogen. This resistance is dependent on whether the pathogen possesses a matching avirulence (*Avr*) gene or not (*avr*). (b) Schematic diagram of the NB-LRR proteins encoded by virus resistance genes. The nucleotide-binding domain (NB) is represented as a black box and the leucine-rich repeat (LRR) domain is represented by a hatched box. These proteins are known as CC-NB-LRR or TIR-NB-LRR proteins depending on whether they possess an N-terminal coiled-coil (CC) or Toll and interleukin receptor (TIR) domain, represented by open boxes.

Table 1 Examples of plant–virus gene-for-gene relationships

R gene	Encoded protein	Plant	Origin	Virus	Avr determinant	Resistance response[a]
Molecularly characterized R genes						
N	TIR-NB-LRR	Tobacco	*N. glutinosa*	Tobamoviruses	Helicase/p50 subunit	HR
Rx	CC-NB-LRR	Potato	*Solanum tuberosum* ssp. *andigena*	PVX	CP	ER
Rx2	CC-NB-LRR	Potato	*Solanum acaule*	PVX	CP	ER
Sw-5	CC-NB-LRR	Tomato	*Solanum peruvianum*[b]	Tospoviruses (TSWV, GRSV, TCSV)	ND	ER, HR
HRT	CC-NB-LRR	Arabidopsis	Ecotype Dijon	TCV	CP	HR
RCY-1	CC-NB-LRR	Arabidopsis	Ecotype C24	CMV	CP	HR
Tm2	CC-NB-LRR	Tomato	*Solanum peruvianum*[b]	Tobamoviruses	30 kDa MP	ER
Tm2²	CC-NB-LRR	Tomato	*Solanum peruvianum*[b]	Tobamoviruses	30 kDa MP	ER
Rsv1	CC-NB-LRR[c]	Soybean	*Glycine max*	SMV	P3	ER, HR, SN
Ctv	CC-NB-LRR[c]	Trifoliate orange	*Poncirus trifoliata*	CTV	ND	ER
I	TIR-NB-LRR[b]	Bean	*Phaseolus vulgaris*	BCMV	ND	ER, HR, SN
Genetically characterized R genes						
Nb	ND	Potato	*Solanum tuberosum*	PVX	25 kDa MP	HR
Nx	ND	Potato	*Solanum tuberosum*	PVX	CP	HR
Ry	ND	Potato	*Solanum stoloniferum*	PVY	NIa protease	ER
L1	ND	Pepper	*Capsicum annum*	Tobamoviruses	CP	HR
L2	ND	Pepper	*Capsicum frutescens*	Tobamoviruses	CP	HR
L3	ND	Pepper	*Capsicum chinense*	Tobamoviruses	CP	HR
L4	ND	Pepper	*Capsicum chacoense*	Tobamoviruses	CP	HR
N'	ND	Tobacco	*Nicotiana sylvestris*	Tobamoviruses	CP	HR
Bdm1	ND	Bean	*Phaseolus vulgaris*	BDMV	BV1 protein	HR
TuRB01	ND	Brassica	*Brassica napus*	TuMV	C1 protein	ER
TuRB03	ND	Brassica	*Brassica napus*	TuMV	P3 protein	ER
TurB04	ND	Brassica	*Brassica napus*	TuMV	P3 protein	ER
TuRB05	ND	Brassica	*Brassica napus*	TuMV	C1 protein	HR
Cry	ND	Cowpea	*Vigna unguiculata*	CMV	2a protein	HR
Rym14^Hb	ND	Barley	*Hordeum bulbosum*	BaMMV/BaYMV	ND	ER
Rym16^Hb	ND	Barley	*Hordeum bulbosum*	BaMMV/BaYMV	ND	ER
Bdv2	ND	Wheat	*Thinopyrum intermedium*	BYDV	ND	ER
Plant–virus interactions resembling gene-for-gene resistance						
ND	ND	Tobacco	*Nicotiana tabacum*	TBSV	P19 protein	HR
ND	ND	*Nicotiana glutinosa*	*Nicotiana glutinosa*	TBSV	P22 MP	HR
ND	ND	*Nicotiana* spp.	*Nicotiana clevelandii*	CaMV	P6 protein	ER, HR, SN
ND	ND	Eggplant	*Solanum melongena*	Tobamoviruses	CP	HR

[a]Certain resistance genes condition more than one type of response depending on viral isolate, genetic background, or environmental conditions.
[b]Formerly known as *Lycopersicon peruvianum*.
[c]The *Rsv1*, *Ctv*, and *I* genes map to clusters of NB-LRR encoding genes. The individual genes conferring resistance have yet to be definitively assigned.
ND, not determined; CP, coat protein; MP, movement protein; HR, hypersensitive response; ER, extreme resistance; SN, systemic necrosis.
BaMMV/BaYMV, barley mild mosaic virus/barley yellow mosaic virus complex; BCMV, bean common mosaic virus; BDMV, bean dwarf mosaic virus; BYDV, barley yellow dwarf virus; CaMV, cauliflower mosaic virus; CMV, cucumber mosaic virus; CTV, citrus tristeza virus; GRSV, groundnut ringspot virus; PVX, potato virus X; PVY, potato virus Y; SMV, soybean mosaic virus; TBSV, tomato bushy stunt virus; TCSV, tomato chlorotic spot virus; TCV, turnip crinkle virus; TSWV, tomato spotted wilt virus; TuMV, turnip mosaic virus.

Molecular studies have shown that recognition specificity is largely determined by the LRR domain of NB-LRR proteins. For example, in certain cases, the recognition specificities of two similar NB-LRR proteins can be exchanged by swapping their respective LRR domains. Correspondingly, the LRR domain is the most polymorphic region of NB–LRR proteins. Bioinformatics studies have suggested that the LRR domains of *R* genes are under diversifying selection, suggesting that this class of genes is under selective pressure. There do not appear to be any obvious features that distinguish between those *R* genes that recognize, for example, bacteria versus those that recognize viruses. For example, the *HRT, RCY1,* and *RPP8* genes of *Arabidopsis* are different alleles of the same gene locus but recognize turnip crinkle virus (TCV), cucumber mosaic virus (CMV), and the oomycete *Hyaloperonospora parasitica*, respectively. Similarly, in potato, gene-duplication events have led to highly similar gene paralogues, *Rx* and *Gpa2*, which confer resistance to potato virus X (PVX) and to the nematode, *Globodera pallida*, respectively. Both CC- and TIR-NB-LRR proteins have been shown to specify resistance to viruses as well as other pathogen classes. Additionally, PVX can be engineered to express a bacterial *Avr* gene and a plant expressing the corresponding NB-LRR protein will be resistant to the recombinant virus. Thus, most NB-LRR proteins probably have the potential to evolve specificity to a variety of different pathogen types, allowing for maximum adaptability. As such, the mechanisms of recognition and initiation of disease-resistance responses are likely to be similar between R proteins that recognize viruses and R proteins that recognize other types of pathogens.

Avirulence Genes

Plant R proteins recognize the protein products of pathogen *Avr* genes, the latter being much more diverse in structure and function than the former. Many different types of pathogen proteins are either synthesized in the host cell by intracellular pathogens (viruses) or delivered into the cytoplasm by the various secretion systems of bacteria, fungi, or oomycetes, and many of these are recognized by NB-LRR proteins. In gene-for-gene resistance, an avirulence protein is defined simply as a protein that induces a resistance response in a plant with the appropriate *R* gene. Thus, the pathogen is unable to infect such a plant due to an active defense mechanism on the part of the plant rather than a lack of virulence on the part of the pathogen. In the absence of a corresponding *R* gene, Avr proteins generally act as virulence factors in that they often play a role in protecting the pathogen from host defense responses, actively inhibit host defenses, or are involved in replication of the pathogen. Pathogen *avr* alleles can overcome *R*-gene-mediated resistance either by mutation or by elimination of the appropriate *Avr* gene. However, this often leads to a loss of fitness. Given their small genomes, this is particularly acute for viruses as most viral genes are crucial for viral fitness and often have multiple functions. Thus, mutation may overcome recognition, but can have the side effect of making the viral protein ineffective in one or more of its functions.

Viral *Avr* genes may appear to share certain similarities; for example, a number of viral coat proteins (CPs) act as Avr determinants (**Table 1**). However, plant viruses have a limited repertoire of genes. Tobacco mosaic virus (TMV) has only three open reading frames (ORFs) (plus a read-through product) and each of these can act as an Avr determinant in the appropriate plant (see **Table 1**).

R Gene Responses to Viruses

The *N* gene has historically been one of the most studied antiviral resistance genes. Originating from *Nicotiana glutinosa*, the *N* gene was introduced into commercial tobacco cultivars via interspecific crosses as a means of conferring resistance to TMV. The *N* gene was among the first *R* genes to be cloned and encodes a typical TIR-NB-LRR protein. Like a number of TIR-NB-LRR-encoding genes, the *N* gene produces two different transcripts via alternative splicing, one encoding the full protein and a shorter version predicted to encode a protein lacking the LRR domain. The role of these truncated proteins is unclear; plants transgenic for versions of the *N* gene that cannot produce both splice forms are compromised in their resistance to TMV.

In addition to TMV, the *N* gene confers resistance to all other known tobamoviruses with the exception of a virus originally designated TMV-ob, and renamed obuda pepper virus (ObPV). A common strategy for identifying the Avr determinant of plant viruses is to engineer chimeras between virulent and avirulent viral strains in order to narrow down the region responsible for eliciting a resistance response. By exchanging genes between TMV and ObPV, the Avr determinant for the *N* gene was found to reside within the viral replicase protein. Furthermore, transient expression of the 50 kDa helicase domain of the TMV replicase (p50) in *N* plants was shown to be sufficient to induce a HR. The HR is a type of programmed cell death that is often associated with gene-for-gene resistance. In the interaction between TMV and tobacco plants possessing the *N* gene this is manifested as small necrotic lesions at the site of infection (**Figure 2(a)**). In this case the virus is able to replicate to a limited degree, spreading to cells surrounding the initially infected cell before the onset of cell death. As such, a macroscopically visible area undergoes cell death before the virus is fully contained. The HR is associated with *R*-gene-mediated resistance to other types of pathogens as well and its initiation requires only the presence of a matching pair of R and Avr proteins. **Figure 2(b)** shows the HR response resulting from the transient expression of an Avr protein in a plant leaf expressing the appropriate *R* gene.

Virus inoculation on a given plant can often result in an HR even though a strict gene-for-gene relationship has

Figure 2 (a) The hypersensitive response (HR). A tobacco leaf expressing the *N* gene has been mechanically inoculated with TMV. Necrotic lesions represent individual infection foci that have subsequently undergone cell death as part of the HR. A higher magnification of necrotic lesions is shown in the inset. (b) Responses of the *Rx* gene. A leaf of *Nicotiana benthamiana*, which is expressing the potato *Rx* gene from a transgene, has been mechanically inoculated with PVX on the left half of the leaf. The resulting resistance is characterized by a lack of viral accumulation or HR response and is known as ER. A higher magnification of the infected leaf is shown in the inset. Note that the only visible signs of infection are the slight leaf damage from the inoculation procedure. On the right side of the leaf, an *Rx*-dependent HR response has been induced by introducing PVX CP via *Agrobacterium*-mediated transient expression. The circle of grayness is caused by dying cells within the area expressing CP.

not been defined due to a lack of variability in either the host or the pathogen. The HR response often indicates that the resistance is due to the presence of an *R* gene. For example, eggplant undergoes a typical HR response when inoculated with tobamoviruses. A compatible interaction has not been seen in this interaction either because the corresponding *R* gene is present in all cultivars of eggplant, or simply because not enough genetic material has been tested. Regardless, this raises the possibility that some cases of nonhost resistance, a phenomenon where all accessions of a plant are resistant to all isolates of a pathogen, may in fact be controlled by the same mechanisms as gene-for-gene resistance.

The second viral resistance gene to be cloned was the *Rx* gene of potato which confers resistance to PVX and encodes a typical CC-NB-LRR-type protein. Using sequence swaps with the single resistance-breaking strain of PVX (strain HB), it was determined that the PVX CP is the Avr determinant for *Rx*. The *Rx* gene was introgressed into potato cultivars from the wild species *Solanum tuberosum* subsp. *andigena*. Additionally, the *Rx2* gene was introgressed from the wild species *Solanum acaule*. Despite their different origins and the fact that *Rx* and *Rx2* are located on separate chromosomes, the predicted gene products are highly similar and condition a similar response to PVX in potato. Unlike the *N* gene, *Rx* and *Rx2* confer what is known as extreme resistance (ER). ER is broadly defined as a lack of visible response to virus infection. In an ER response there is no necrosis and little or no virus accumulation can be detected (**Figure 2(b)**). *Rx*-mediated ER is functional at the single-cell level. When infectious viral RNA is introduced into protoplasts there is no apparent viral accumulation. However, some *R* genes that manifest ER, such as *Tm-2* and *Tm-2^2* (see below), do allow virus replication in protoplasts.

Although *Rx*- and *Rx2*-mediated resistance does not normally manifest as an HR, these proteins are capable of inducing an HR. When the Rx or Rx2 proteins are expressed together with the PVX CP an HR is rapidly induced (**Figure 2(b)**). Thus, Rx and Rx2 probably do not induce responses that are qualitatively different from other R proteins but may simply induce a very rapid response, eliminating the source of Avr protein before it can accumulate to levels sufficient to induce an HR.

An example of an *R* gene inducing both ER and HR responses is seen with the *Sw-5* gene of tomato. *Sw-5* encodes a CC-NB-LRR protein conferring resistance to several tospoviruses including tomato spotted wilt virus (TSWV), tomato chlorotic spot virus (TCSV), and groundnut ringspot virus (GRSV). When tomato plants possessing the *Sw-5* gene are mechanically inoculated with tospoviruses, both HR and ER responses can be seen depending on the strain of virus inoculated, suggesting that the HR and ER responses may represent a continuum of responses. These different responses conditioned by the same gene may occur because Sw-5 recognizes the Avr determinant (unknown) of different strains with varying efficiency. The occurrence of an *Sw-5*-induced HR can be affected by environmental conditions as well as genetic background. Cell type can also have an effect on the *Sw-5* response. Although *Sw-5* plants are largely immune to TSWV, infection can still occur from thrips carrying TSWV when they feed on the tomato fruit.

The effect of genetic background has been more extensively studied in the case of the *Arabidopsis HRT* (HR to TCV) gene. The *HRT* gene was discovered by inoculating a number of *Arabidopsis* ecotypes with TCV. The ecotype Dijon (Di) was found to condition an HR-type response to TCV and a line (Di-17) was selected with the most robust resistance. The *HRT* gene encodes a CC-NB-LRR protein and confers recognition of the TCV CP. Inoculation of TCV onto leaves of Di-17 plants results in a typical HR and containment of the virus. However, despite an HR in the initially infected plants, in ~10% of infected plants, the virus is able to move to, and replicate in, systemic tissues. This effect is much more dramatic when Di-17 is crossed with other ecotypes such as Col-0. In this genetic background, HRT still conditions an HR in the infected leaves but nearly all plants allow systemic spread of the virus. Genetic studies suggest that in order to be fully functional, HRT requires the presence of the recessive Di-17 allele of a gene termed *rrt*. Further studies have suggested that *HRT* requires elevated levels of salicylic acid (SA) in order to be fully functional and *rrt* may condition higher levels of this signaling molecule in Di-17.

The *Arabidopsis* ecotype C24 contains the *RCY1* gene which confers resistance to the Y strain of CMV. The *RCY1* and *HRT* genes are essentially different alleles of the same gene and encode proteins that are 91% identical. *RCY1* is also affected by genetic background and there appears to be a recessive gene present in ecotype Landsberg that abrogates *RCY1*-mediated resistance.

There are a number of other cases where an *R* gene is not able to contain a virus within the inoculated leaf. The *Rsv1* gene of soybean confers ER to all known American isolates of soybean mosaic virus (SMV) except for two. The G7d strain is resistance breaking and produces normal systemic infection in *Rsv1* soybeans. However, the G7 strain of SMV elicits an intermediate response. This strain is able to evade the initial resistance response and spread to systemic tissues. However, the build-up of virus in systemic tissues eventually leads to the initiation of cell death. This systemic necrosis (SN) is not sufficient to contain the virus, causing a trailing HR that eventually leads to the collapse of the plant. It is thought that the SN response is due to a weak recognition of the viral Avr determinant such that the HR is induced only once the virus has accumulated to high levels. Thus, the same responses (i.e., cell death) are induced as in an HR response, but too late to curtail virus spread.

Further evidence for quantitative differences between ER, HR, and SN responses is seen with the *I* locus of bean. The *I* gene has been mapped to a cluster of TIR-NB-LRR genes in bean and confers a defense response to nine different potyviruses, including bean common mosaic virus (BCMV). These responses range from ER to HR to SN depending on the viral isolate. At the same time, variation can be seen even when using the same viral isolate. At 23 °C plants heterozygous for the *I* gene (*I/I*) confer ER to BCMV whereas plants with only one copy of the *I* gene (*I/i*) confer an HR-type resistance response. At 34 °C, however, *I/i* plants undergo a SN response whereas *I/I* plants do so at a greatly reduced rate. Thus, the different responses to viruses by resistant plants likely represent a continuum of quantitatively different responses. The degree of response can be modulated by the presumed relative efficiency of recognition of the Avr determinant, gene dosage, temperature, and genetic background.

Mechanisms of Recognition

Given the relatively straightforward genetic relationship between *R* and *Avr* genes, it was originally proposed that this might represent a receptor–ligand interaction. Since LRR domains often act as protein–protein interaction domains, it was proposed that the variable LRR domains of NB–LRR proteins would bind to different Avr proteins and this binding would activate defense responses. However, initial difficulties in demonstrating such interactions led to the formulation of indirect models of recognition. The guard hypothesis states that, since many *Avr* genes are also virulence genes, R proteins would 'guard' the cellular targets of these virulence targets. In this scenario, it is the function of pathogen-encoded virulence factors to modify cellular proteins and this modification is in turn detected by R proteins. Thus, it is not the Avr protein *per se* that is detected, but rather its virulence activity. Several examples of such indirect recognition have been studied. For example, when expressed in *P. syringae*, the Avr proteins AvrB and AvrRpm1 are both recognized by the *Arabidopsis* CC-NB-LRR protein RPM1. RPM1 interacts physically with a cellular protein, RIN4, which in turn can interact with both AvrB and AvrRpm1. This latter interaction is thought to activate the RPM1 protein. The Avr protein AvrRpt2 from *P. syringae* is a cysteine protease and is recognized by the *Arabidopsis* CC-NB-LRR protein RPS2. RPS2 also binds RIN4 and becomes activated when RIN4 is cleaved by AvrRpt2. This activation can be mimicked by genetic ablation of the *RIN4* gene. In this case, RPS2 is constitutively activated in the absence of RIN4, thus the recognition event appears to be the proteolytic removal of a negative regulatory protein. Other probable examples of indirect recognition include recognition of AvrPphB from *P. syringae* by the *Arabidopsis* RPS5 protein through the PBS1 kinase as well as recognition by the tomato NB-LRR protein Prf of AvrPto from *P. syringae* due to the interaction of AvrPto with the Pto kinase.

At the same time, however, there is some indication that certain Avr/R protein pairs may interact directly. Yeast two-hybrid studies have shown Avr proteins interacting with the rice Pi-Ta, *Arabidopsis* RRS1, and the flax L NB-LRR proteins which confer resistance to rice blast fungus, *Ralstonia solanacearum,* and flax rust, respectively. Interactions have also been reported between the viral resistance protein N and its cognate elicitor p50. However, a great deal about these interactions remains unclear, such as whether the Avr proteins interact with the LRR domain, another part of the protein, or a combination thereof. Further studies may also demonstrate examples of indirect modes of recognition by NB-LRR proteins that recognize viruses.

Two traits desired by plant breeders related to recognition are broad-spectrum resistance and durability. Often, *R* genes are deployed that provide good resistance but are quickly overcome by new strains of a pathogen. The durability of an *R* gene may be affected by different aspects of the pathogen such as dispersal rates and mechanisms, occurrence of sexual reproduction, and variable repertoires of virulence factors. In the latter case, for example, phytopathogenic bacteria have relatively large repertoires of virulence factors which may be partially redundant and dispensable. However, since viruses cannot dispense with any of their limited number of genes, overcoming resistance must occur through sequence changes.

In the field, some viral *R* genes are very durable, such as *N*, *Rx*, and *Ry*. There are no strains of potato virus Y (PVY) that overcome *Ry* and although ObPV can overcome *N*, it is not widespread. Likewise *Rx*-breaking strains of PVX have reduced virulence and are found only in South America. Strains of PVX that overcome the *Nx* and *Nb* genes are more widespread, and TSWV strains that overcome *Sw-5* have been reported in several tomato-growing areas. An interesting case study in durability can be seen with the *Tm-2* and *Tm-2^2* genes. These genes were introgressed from different accessions of *Lycopersicum peruvianum* into tomato cultivars as a means of controlling TMV and tomato mosaic virus (ToMV). These genes confer resistance to nearly all strains of tobamoviruses through recognition of the viral movement protein (MP). Molecular cloning has revealed that *Tm-2* and *Tm-2^2* are different alleles of the same gene. These genes encode CC-NB-LRR proteins that differ by only four amino acids, two of which are in the LRR domain. Despite this similarity, *Tm-2^2* has proven to be more durable than *Tm-2*. Strains of ToMV that overcome *Tm-2* or *Tm-2^2* have mutations in different regions of the MP; some mutations can overcome *Tm-2*, but not *Tm-2^2*, and vice versa. The *Tm-2^2* gene may be more durable because the resistance-breaking mutations have a very negative impact on virus fitness as evidenced by reduced viral load in plants infected with these strains. Furthermore, the *Tm-2^2*-breaking strains cause very severe symptoms on *Tm-2^2* plants so that infected plants can be easily recognized and eliminated from the field or greenhouse.

Differential recognition by the *N'* gene of tobacco may represent a case where overcoming resistance does not impose a fitness cost. The *N'* gene confers resistance to several tobamoviruses including ToMV, but not TMV. Although *N'* has not been cloned, the Avr determinant has been shown to be the viral CP. Curiously, mutations can be introduced into the TMV CP that result in a gain of recognition by *N'*, but which appear to compromise CP function. It has been hypothesized that the recognition of TMV CP is normally masked from recognition by *N'*, and mutations that interfere with the tertiary or quaternary structure of the CP cause the recognition site to be exposed.

In some cases, viral Avr proteins are able to evade recognition by some resistance genes but not others. For example, the allelic pepper genes *L1–L4* also recognize the CP of tobamoviruses. Certain virus strains are able to overcome *L1*-mediated resistance, some overcome *L1* and *L2*, and some overcome *L1*, *L2*, and *L3*. However, the *L4* gene is very broad spectrum, controlling all known tobamoviruses.

Signaling Mechanisms

Studies with Rx, HRT, and N have shown that the different domains of NB-LRR proteins undergo physical intramolecular interactions with the LRR and N-terminal domains interacting with the NB domain. These interactions appear to condition an autorepressed state and alterations thereof are associated with activation of the protein. In the presence of PVX CP, the interdomain interactions of Rx are disrupted and the N protein has been shown to undergo self-association in the presence of p50. It is presumed that R proteins induce a signal transduction cascade leading to disease resistance by interacting with signal adaptor molecules. However, such molecules have yet to be identified. A number of proteins required for NB-LRR function have been identified. TIR-NB-LRR proteins, including N, have a general requirement for the EDS1 protein and often also require the EDS1-interacting protein PAD4. Both of these proteins have lipase signature domains although their exact function is not known. Given that the *N* gene is not effective above 32 °C, it is interesting to note that EDS1 expression levels have been shown in *Arabidopsis* to be greatly reduced at higher temperatures. Likewise, the barley MLA NB-LRR proteins accumulate to greatly reduced levels at high temperatures. These phenomena may at least partially explain the temperature sensitivity of some *R* genes.

A number of studies have shown that during a resistance response, kinase cascades are activated and there is a massive transcriptional reprogramming. Studies using virus-induced gene silencing (VIGS) have implicated several protein kinases, including MEK1, MEK2, NTF6, SIPK, WIPK, and NPK1, in *N* gene-mediated resistance. A requirement was also shown for the TGA and WRKY families of transcription factors as well as the individual transcription factors MYB1 and the TGA-interacting protein NPR1. These proteins and protein families have also been implicated in gene-for-gene resistance to other types of pathogens as well. Gene-for-gene resistance may also involve targeted proteolysis since silencing the expression of certain proteins involved in regulated protein degradation, including a subunit of the COP9 signalosome, SKP1 and COI1, also compromises *N*-mediated resistance. To date, the only genes whose silencing has been shown to break *Rx*-mediated resistance are *Hsp90* and *Sgt1*. Along with the Sgt1 binding protein Rar1, these gene products are required for the function of a number of NB-LRR proteins. The HSP90, SGT1, and RAR1 proteins are likely involved in chaperoning proteins in order to allow proper protein folding. Accordingly, these proteins are required for the accumulation of a number of NB-LRR proteins.

Ethylene plays a role in *R*-gene-mediated antiviral responses. *RCY1* functions at reduced efficiency in *Arabidopsis* containing the ethylene-desensitizing mutations *ein3* and *etr1*. The phytohormone SA also plays a role in *R*-gene-mediated resistance. The involvement of SA is often studied using SA biosynthetic mutants in *Arabidopsis* (such as *eds5*) or by using plants transgenically expressing the bacterial *nahG* gene which encodes a salicylate hydroxylase that degrades SA. The *eds5* mutation and *nahG* transgene both decrease the efficiency of *RCY1*-mediated resistance and completely abrogate *HRT*-mediated resistance.

When tobacco plants expressing *N* and *nahG* are infected with TMV, the resulting HR lesions are larger and the necrosis can spread beyond the infected leaf. Similarly, in *nahG* transgenic tomato, the normally ER response of *Tm-2^2* becomes a spreading HR phenotype in the inoculated leaf, although the virus does not spread systemically. The *Rx*-mediated response, however, is not affected by the *nahG* transgene. At the same time, SA has antiviral effects independent of *R*-gene-mediated resistance, such as causing the upregulation of an RNA-dependent RNA polymerase (RdRp) involved in RNAi-based antiviral responses. Thus, SA may enhance *R* gene responses rather than directly targeting virus and its requirement for resistance may depend on the strength of the *R*-gene-mediated response. In agreement with this, SA application results in reduced lesion size in the *N*-mediated response to TMV and an enhancement of *HRT*-mediated resistance. Cell death during the HR induces increased levels of SA. Thus, cell death may indirectly prime and/or augment *R*-gene-mediated antiviral responses. It is clear that cell death itself is not the only mechanism by which virus is eliminated as ER does not involve cell death, and in the *N*–TMV interaction, TMV can be detected outside the area of necrosis. Furthermore, in the interaction between cauliflower mosaic virus (CaMV) and *Nicotiana* spp., resistance is controlled by a single dominant gene. However, whether or not this resistance is accompanied by an HR is controlled by a separate gene. A clear understanding of how viruses are cleared from resistant plant cells will require further study.

See also: Plant Resistance to Viruses: Natural Resistance Associated with Recessive Genes; Plant Antiviral Defense: Gene Silencing Pathway; *Potexvirus*; *Tospovirus*; Cucumber Mosaic Virus; *Tobamovirus*; Virus Induced Gene Silencing (VIGS).

Further Reading

Belkhadir Y, Subramaniam R, and Dangl JL (2004) Plant disease resistance protein signaling: NBS-LRR proteins and their partners. *Current Opinion in Plant Biology* 7: 391–399.

Dangl JL and Jones JD (2001) Plant pathogens and integrated defence responses to infection. *Nature* 411: 826–833.

Harrison BD (2002) Virus variation in relation to resistance-breaking in plants. *Euphytica* 124: 181–192.

Hull R (2002) *Matthew's Plant Virology* 4th edn. San Diego, CA: Academic Press.

Kachroo P, Chandra-Shekara AC, and Klessig DF (2006) Plant signal transduction and defense against viral pathogens. *Advances in Virus Research* 66: 161–191.

Kang BC, Yeam I, and Jahn MM (2005) Genetics of plant virus resistance. *Annual Review of Phytopathology* 43: 581–621.

Martin GB, Bogdanove AJ, and Sessa G (2003) Understanding the functions of plant disease resistance proteins. *Annual Review of Plant Biology* 54: 23–61.

Ritzenthaler C (2005) Resistance to plant viruses: Old issue, news answers? *Current Opinion in Biotechnology* 16: 118–122.

Soosaar JL, Burch-Smith TM, and Dinesh-Kumar SP (2005) Mechanisms of plant resistance to viruses. *Nature Reviews Microbiology* 3: 789–798.

Plant Resistance to Viruses: Natural Resistance Associated with Recessive Genes

C Caranta and C Dogimont, INRA, Montfavet, France

© 2008 Elsevier Ltd. All rights reserved.

Glossary

Allele One of two or more forms that can exist at a single gene locus, distinguished by their differing effects on the phenotype.

Dominant allele An allele that expresses its phenotypic effect even when heterozygous with a recessive allele.

Paralog Paralogs are genes related by duplication within a genome. Orthologs (i.e., genes in different species that evolved from a common ancestral gene) retain the same function in the course of evolution whereas paralogs generally evolve new functions.

Pleiotropic mutation A mutation that has effects on several different characters.

Positional cloning Cloning a gene based simply on knowing its position in the genome without any idea of the function of that gene.
Quantitative trait loci (QTLs) QTLs correspond to genomic regions associated with the phenotypic variation of a quantitative trait.
Recessive allele An allele whose phenotypic effect is not expressed in a heterozygote.
TILLING A reverse genetics approach that relies on the ability of a special enzyme to detect mismatches in normal and mutant DNA strands when they are annealed. It can therefore detect single point mutations of the type induced by chemicals conventionally used in mutation breeding programs.
Virulence Genetic ability of a pathogen to overcome genetically determined host resistance and cause a compatible (disease) interaction.

Introduction

A century after the recognition that plant resistance to diseases can be genetically inherited, breeding for naturally occurring resistance genes remains the most effective and sustainable approach to the prevention of diseases caused by plant viruses. Genetic resistance (also called host resistance or cultivar resistance) occurs when some genotypes possess heritable resistance to a particular pathogen whereas other genotypes in the same gene pool, generally at the species level, are susceptible.

According to genetic criteria, resistance can be controlled by one, two, or several genes corresponding to monogenic, digenic, and polygenic resistance, respectively. Usually, mono- and digenic resistances segregate as qualitative traits (for instance, presence vs. absence of viral accumulation in inoculated leaves or presence vs. absence of symptoms), whereas polygenic resistances behave as quantitative traits controlled by several resistance genes named quantitative trait loci (QTLs). Currently, QTL mapping is a standard procedure in quantitative genetics and begins with the collection of genotypic (based on molecular markers) and phenotypic data from a segregating population, followed by statistical analysis to reveal all possible marker loci where allelic variation correlates with the phenotype.

Resistance can show dominant or recessive inheritance and this feature appears closely related to underlying molecular mechanisms. According to the plant phenotype, resistance genes control complete or partial resistance. Partial resistance includes tolerance defined as presence of the virus in all parts of the plant without symptoms and/or yield loss. Anyway, host resistance can occur by complete or partial interruption at any of a number of stages of the virus life cycle including entry into the cell, expression and replication of the viral genome, and cell-to-cell and long-distance movements.

Resistance may be overcome by the appearance of resistance-breaking strains compromising the durability of resistance genes deployed in culture. Genetic analyses comparing resistance-breaking and non-resistance-breaking viral strains have permitted the identification of viral components (i.e., avirulence genes) responsible for this process and produced significant results regarding resistance mechanisms and the molecular basis of compatible interactions between plants and viruses.

This article reviews phenotypic and molecular characteristics related to recessive virus resistance factors, that is, major genes and QTLs identified in the natural diversity of plants.

Survey of Recessive Resistance Genes

Recessive resistance genes are more prevalent for resistance to viruses than for resistance to fungi, bacteria, nematodes, or insects for which resistance is primarily a dominant trait. As described later in this article, this feature appears related to intrinsic properties of viruses. A survey of published recessive resistance against viruses identified in the natural diversity of plants is presented in **Table 1**. Recessive resistance to viruses is especially well documented within the dicots, chiefly Solanaceae, Cucurbitaceae, and Fabaceae, which comprise a number of economically important vegetables. Within monocot species, recessive resistances have been reported in barley, rice, and maize. For such genes, it is also important to keep in mind that the presence of dominant resistance genes that can mask recessive ones (both against the same virus), and/or the highly heterozygous genome structure of species such as *Vitis vinifera*, *Prunus* sp., and cultivated potatoes renders difficult the detection of recessive resistance.

Recessive inheritance of resistance has been described for various viruses which belong to very different viral genera, with RNA as well as DNA genomes (**Table 1**). However, the frequency of recessive resistance can be to some extent related to the virus family or genus. A large majority of recessive resistance genes which have been described to date concern resistance to viruses belonging to the genus *Potyvirus* in a wide range of plant species. *Potyvirus* is the largest genus of plant viruses and includes some of the most common and destructive viruses for a number of cultivated crops worldwide. It is remarkable that approximately 50% of resistance to potyviruses show recessive inheritance, while, for instance, resistance to tobamoviruses, which are also widespread in various crops, is usually monogenic dominant, with a few exceptions such as the *tm* gene for resistance to tobacco mosaic virus in common bean. More generally, ~70% of recessive resistances listed in **Table 1** are efficient against RNA

Table 1 Recessive resistances controlled by major genes or QTLs to plant viruses

Genus (family)	Virus	Host plant	Resistance gene(s) or loci	Resistance mechanism
RNA viruses with a 5′ VPg and 3′poly(A) tail				
Bymovirus (Potyviridae)	Barley mild mosaic virus	*Hordeum vulgare* (barley)	*rym4, rym5*	Complete resistance
			+ about 15 *rym* genes mapped on five chromosomes	From complete to partial resistance
	Barley yellow mosaic virus	*Hordeum vulgare* (barley)	*rym4, rym5*	Complete resistance
			+ about 15 *rym* genes mapped on five chromosomes	From complete to partial resistance
	Barley yellow mosaic virus-2	*Hordeum vulgare* (barley)	*rym-5*	Complete resistance
Potyvirus (Potyviridae)	Bean common mosaic virus	*Pisum sativum* (pea)	*rym1, rym11*	
		Phaseolus vulgaris (bean)	*bcm*	No systemic infection
			bc-1, bc-1², bc-2, bc-2², bc-3, bc-u (independent locus)	
	Bean yellow mosaic virus	*Phaseolus vulgaris* (bean)	*cyv*	No systemic infection
		Pisum sativum (pea)	*mo*	No symptoms
		Vicia mungo (mungbean)	2 genes	
	Bidens mottle virus	*Lactuca sativa* (lettuce)	*bi*	
	Blackeye cowpea mosaic virus	*Vigna unguiculata* (cowpea)	*bcm*	
	Celery mosaic virus	*Apium graveolens* (celery)	*cmv*	Resistance at a single cell level
	Clover yellow vein virus	*Phaseolus vulgaris* (bean)	*desc*	
		Pisum sativum (pea)	*cyv1*	
			cyv2 (independent locus)	
	Lettuce mosaic virus	*Lactuca sativa* (lettuce)	*mo1¹, mo1²*	Reduced accumulation
	Moroccan watermelon mosaic virus	*Cucumis sativus* (cucumber)	*mwm*	No symptoms
	Papaya ringspot virus-W	*Cucumis sativus* (cucumber)	*prsv-1*	Reduced accumulation
		Cucurbita moschata (squash)	*prv*	No symptoms
	Passionfruit woodiness virus	*Pisum sativum* (pea)	*pwv*	Complete resistance
	Pea seedborne mosaic virus	*Lens culinaris* (lentil)	*sbv*	Resistance at a single cell level
		Pisum sativum (pea)	*sbm1, sbm3, sbm4*	
			sbm2 (independent locus)	
	Peru tomato virus	*Lycopersicon esculentum* (tomato)	Monogenic	
	Pepper mottle virus	*Capsicum annuum* (pepper)	*pvr3*	Systemic movement
		Capsicum chinense (pepper)	*pvr1*	Resistance at a single cell level
	Pepper veinal mottle virus	*Capsicum annuum* (pepper)	*pvr2² + pvr6* (digenic)	Complete resistance
	Plum pox virus	*Arabidopsis thaliana*	*rpv1*	Restriction to long-distance movement
	Potato virus A	*Solanum tuberosum* subsp. *andigena*	*ra$_{adg}$*	Restriction of vascular movement

Continued

Table 1 Continued

Genus (family)	Virus	Host plant	Resistance gene(s) or loci	Resistance mechanism
	Potato virus Y	Capsicum annuum (pepper)	$pvr2^1$, $pvr2^2$	Resistance at a single cell level
			pvr3 (independent locus)	Complete resistance
			Several QTLs	Partial resistance
		Capsicum chinense (pepper)	pvr1 (allelic to pvr2)	Resistance at single cell level
		Lycopersicum hirsutum (tomato)	pot-1	Complete resistance
		Nicotiana tabacum (tobacco)	va	Cell-to-cell movement
	Tobacco etch virus	Capsicum annuum (pepper)	$pvr2^2$	Resistance at a single cell level
		Capsicum chinense (pepper)	pvr1	Resistance at a single cell level
		Lycopersicum hirsutum (tomato)	pot-1	Complete resistance
		Nicotiana tabacum (tobacco)	2 genes linked	Systemic movement
	Tobacco vein mottling virus	Nicotiana tabacum (tobacco)	va	Cell-to-cell movement
	Turnip mosaic virus	Brassica campestris (cabbage)	2 genes	No systemic infection
		Brassica rapa (turnip)	From 1 to 3 genes depending on strains	No systemic infection
	Watermelon mosaic virus	Cucumis melo (melon)	monogenic	Reduced accumulation and systemic movement
		Cucumis sativus (cucumber)	wmv-2	No symptoms at cotyledon stage
			wmv-3 + wmv-4 (digenic)	No symptoms at a true leaf stage
	White lupin mosaic virus	Pisum sativum (pea)	mo	No systemic infection
	Zucchini yellow mosaic virus	Pisum sativum (pea)	wlv	Reduced accumulation
		Cucumis sativus (cucumber)	zym	
	Zucchini yellow fleck virus	Citrullus lanatus (watermelon)	monogenic	No systemic infection
		Cucumis sativus (cucumber)	zyf	No systemic infection
Comovirus (Comoviridae)	Cowpea severe mosaic virus	Vigna unguiculata (cowpea)	Oligogenic (3 genes)	Complete resistance

RNA viruses with a 5′ VPg but no 3′ poly(A) tail			
Enamovirus (Luteoviridae)	Bean leafroll virus	Pisum sativum (pea)	lr	No symptoms
Polerovirus (Luteoviridae)	Beet western yellows virus	Lactuca sativa (lettuce)	bwy	Reduced accumulation
	Cucurbit aphid-borne yellows virus	Cucumis melo (melon)	cab-1 + cab-2 (digenic)	Complete resistance
Sobemovirus	Rice yellow mottle virus	Oryza sativa (rice)	rymv1	Complete resistance
			Several QTLs	Partial resistance
		Oryza glaberrima (rice)	rymv1	Complete resistance
RNA viruses with capped 5′ end and non-polyadenylated 3′ end				
Alfamovirus (Bromoviridae)	Alfalfa mosaic virus	Medicago sativa (alfalfa)	am-1	Restriction of systemic movement
Bromovirus (Bromoviridae)	Cowpea chlorotic mottle virus	Glycine max (soybean)	2 genes	No symptoms
Cucumovirus (Bromoviridae)	Cucumber mosaic virus	Cucumis melo (melon)	Polygenic with a major QTL	No systemic symptoms
		Cucurbita pepo (pumpkin)	2 unlinked genes	No systemic infection
		Capsicum frutescens (pepper)	Oligogenic (2–3 genes)	
Tobamovirus	Cucumber green mottle mosaic virus	Cucumis melo (melon)	Polygenic	No symptoms
Tospovirus (Bunyaviridae)	Tomato spotted wilt virus	Lycopersicon esculentum (tomato)	sw2, sw3, sw4	Partial resistance
RNA viruses with unmodified 5′ and 3′ ends				
Carmovirus (Tombusviridae)	Turnip crinkle virus	Arabidopsis thaliana	rrt + HRT (digenic)	Hypersensitive response
	Melon necrotic spot virus	Cucumis melo (melon)	nsv	Resistance at a single cell level
Luteovirus (Luteoviridae)	Barley yellow dwarf virus	Hordeum vulgare (barley)	ryd1	Partial resistance
Idaeovirus	Raspberry bushy dwarf virus	Rubus idaeus	Monogenic	Complete resistance
Other RNA viruses (not well-characterized genome)				
Umbravirus	Groundnut rosette virus	Arachis hypogaea (peanut)	2 genes	Escape to infection/no symptoms
	Rice tungro spherical virus	Oryza sativa (rice)	tsv-1	No systemic infection
			tsv-2 (independent loci)	
DNA viruses				
Begomovirus (Geminiviridae)	Bean golden yellow mosaic virus	Phaseolus vulgaris (bean)	bgm-1	Reduced symptoms
	Potato yellow mosaic virus	Lycopersicon pimpinellifolium	bgm-2 (independent loci) Bigenic	No systemic infection
	Tomato yellow leaf curl virus	Lycopersicon hirsutum	Oligogenic	Complete resistance
	Tomato chlorotic mottle virus	Lycopersicon esculentum (tomato)	Monogenic	No systemic infection
Curtovirus (Geminiviridae)	Beet curly top virus	Arabidopsis thaliana	Monogenic, two distinct loci in two ecotypes	Tolerance or no systemic infection
Tungrovirus (Caulimoviridae)	Rice tungro bacilliform virus	Oryza sativa (rice)	Polygenic	Tolerance, reduced virus accumulation

viruses with a viral protein genome-linked (VPg) bound to the 5′ end of their genome. Other recessive genes are involved in resistance to viruses with different genome organizations and different translation strategies. Resistance to several DNA viruses in the genus *Begomovirus*, viruses transmitted by whiteflies or planthoppers and responsible to very damageable diseases, also present monogenic recessive inheritance.

Resistance controlled by recessive genes involves various phenotypes from complete resistance expressed at the single cell level (as revealed by protoplast infection experiments) or during cell-to-cell or long-distance movements, to different levels of partial resistance, that is, reduced accumulation of the virus, reduced cell-to-cell or long-distance movement, and delayed symptoms or tolerance. Therefore, while most dominant resistances to viruses involve a hypersensitive response or an extreme resistance phenotype, recessive resistances show an important diversity of resistance phenotypes.

There are a number of examples of recessive resistance loci controlling resistance to different pathotypes of the same virus or distinct related viral species. One of the most remarkable examples has been described in pea (*Pisum sativum*), where recessive resistance genes against potyviruses have been mapped to two loci. The first locus includes *bcm* for resistance to bean common mosaic virus (BCMV), *cyv-1* for resistance to clover yellow vein virus (ClYVV), *mo* for resistance to pea mosaic virus, *sbm-2* for resistance to pea seedborne mosaic virus (PSbMV) pathotype L1, *pmv* for resistance to pea mosaic virus (PMV), and *wmv* for resistance to watermelon mosaic virus, while the second locus includes *sbm-1*, *sbm-3*, and *sbm-4*, which govern resistance to three PSbMV pathotypes, *cyv-2* for resistance to CYVV, and *wlv* for resistance to white lupin mosaic virus. Similarly in pepper, alleles at the *pvr2* locus control resistance to potato virus Y (PVY), tobacco etch virus (TEV), and pepper mottle virus (PepMoV). It is noticeable that all these viruses belong to the genus *Potyvirus*. Also in barley, a single recessive locus controls resistance to several bymoviruses, including barley mild mosaic virus, barley yellow mosaic virus, and barley yellow mosaic virus-2. The broad spectrum of these loci can be explained either by the action of one gene with a pleiotropic effect or by a cluster of tightly linked genes. In case of dominant resistance genes, clustering has been widely reported and was shown to correspond to duplicated or evolutionary related sequences of resistance genes with various resistance specificities against distinct pathogen groups (viruses, fungi, bacteria, etc.). Concerning recessive resistance genes against viruses and according to data obtained from the molecular cloning of these genes, broad-spectrum loci appear to result from the pleiotropic effect of a single gene (see next section).

Although most resistances with recessive inheritance are monogenic, there are several interesting examples where resistance alleles at two distinct loci are required to confer resistance. In *Phaseolus vulgaris*, resistance to several strains of BCMV requires the complementary effect of the *bc-u* locus and one at least of the strain-specific genes *bc-1*, *bc-2*, and *bc-3*. In *Capsicum*, resistance to the African potyvirus, pepper veinal mottle virus (PVMV), requires resistance alleles at the *prv2* and *pvr6* resistance loci. In melon, two complementary recessive genes are required to confer complete resistance to cucurbit aphid-borne yellows virus (genus *Polerovirus*). In the tomato-related wild species *Lycopersicon pimpinellifolium*, two recessive genes are also required to confer resistance to potato yellow mosaic virus (genus *Begomovirus*).

Relatively few QTLs for plant viral resistance have been genetically mapped and only in three cases they show a clear recessive inheritance. Several QTLs have been demonstrated to govern recessive resistance to PVY in pepper and to rice yellow mosaic virus (RYMV) in rice. In melon, quantitative resistance to cucumber mosaic virus (CMV) is conferred by a major-effect QTL explaining from 15% to 75% of the phenotypic variation depending on the strain, and also by several minor-effect QTLs. Phenotypically, virus resistance QTLs are usually revealed by a semi-quantitative scoring of symptom severity (that can be next used to calculate the area under the symptom progress curve –AUSPC) and/or a global evaluation of the virus content using enzyme-linked immunosorbent assays. However, these methods are difficult to develop on large-size progenies and do not permit accurate and quantitative resistance evaluations.

Molecular Structure and Mechanisms Underlying Recessive Resistance Genes

Fraser's Hypotheses

In 1990, R. S. S. Fraser proposed two hypotheses to explain the molecular mechanisms underlying recessive resistances to plant viruses. The first one, also called the 'negative model', suggests that resistance might be the consequence of the loss (null allele) or mutation of a specific host factor required by the virus to complete a particular step of its infectious cycle. Such resistance is likely to be passive, that is, it should never induce a hypersensitive nor a defensive response, and the dominant allele encodes a susceptibility factor. This hypothesis is in agreement with the fact that typical plant viruses encode relatively few proteins (between four and ten) and that they need therefore to reroute host cellular components to perform their infectious cycle. The compatibility between a plant and a virus therefore depends on complex interplays between functions encoded by the viral and host genomes. Conversely, the second hypothesis proposes that resistance might be the result of an active mechanism, involving a resistance allele encoding an inhibitor that interferes with the virus life cycle. In this case, the susceptibility allele encodes a dominant negative regulator of resistance. This hypothesis has been documented in the case of *mlo*, associated in barley with the recessive resistance against powdery mildew, caused by *Erysiphe graminis*.

The Key Role of Translation Initiation Factors in Recessive Resistance

The molecular cloning in 2002 of the first naturally occurring recessive resistance gene against a virus has validated the first hypothesis by demonstrating that mutations in the eukaryotic translation initiation factor 4E (eIF4E) are responsible of *pvr2*-mediated recessive resistance of pepper to PVY and TEV. Since then, although many host factors are required for plant virus interactions, all studies aimed at the characterization of recessive resistance genes from a range of plant species have so far only identified eIF4E and eIF4G proteins from the translation initiation complex.

In eukaryotic cells, translation initiation factor 4E binds to the m^7G cap as the first step in recruiting mRNA into the translation initiation complex. eIF4E is also associated with eIF4G to form the eIF4F complex. eIF4G is a scaffold for the other components of the complex and interacts with several other initiation factors like poly(A)-binding protein (PABP), eIF4A, a helicase for unwinding secondary structures, and eIF3, a multisubunit complex that binds to the 40S ribosomal subunit (**Figure 1**). In plants, two isoforms of eIF4F are present and appear to have complementary roles: eIF4F, containing eIF4F and eIF4G, and eIF(iso)4F, containing eIF(iso)4E and eIF(iso)4G. Although the two complexes are considered equivalent for *in vitro* translation of some mRNAs, they differ in their *in vivo* expression pattern and show some specificity for different capped cellular mRNAs. Plant genes encoding proteins from the eIF4F complex belong to small gene families. For instance, in *Arabidopsis thaliana*, three genes code for proteins of the eIF4E subfamily (i.e., *eIF4E1*, *eIF4E2*, and *eIF4E3*), one gene codes for eIF(iso)4E, one gene codes for eIF4G, and two genes code for proteins of the eIF(iso)4G subfamily (*eIF(iso)4G1* and *eIF(iso)4G2*). Although the structure of these gene families appears less complex in mono- than in dicotyledons, several loci have been identified in all species studied.

Among the eight natural recessive resistance genes against plant viruses which have been cloned in a range of plant species including both mono- and dicotyledons (pepper, lettuce, pea, tomato, barley, rice, melon), six code for eIF4E, one for its paralog eIF(iso)4E, and one for eIF(iso)4G (**Table 2**). All these resistances function against RNA viruses and the majority (five out of eight) against potyvirus infection. eIF4E has also been implicated in barley resistance to a bymovirus (a genus related to potyviruses) and in melon resistance to a carmovirus, and eIF(iso)4G has been implicated in rice resistance to a sobemovirus.

Characteristics of eIF4E-Mediated Resistance

Except for the *pvr6* gene, all resistances result from a small number of amino acid changes in the protein encoded by the recessive resistance allele. Concerning eIF4E-mediated resistance against potyviruses or related

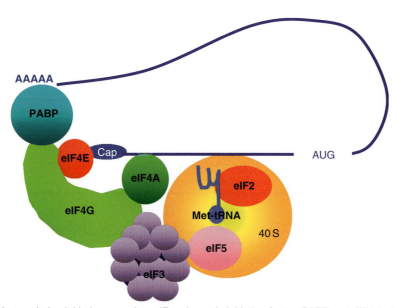

Figure 1 The eukaryotic translation initiation complex. eIF, eukaryotic initiation factor; PABP, poly(A)-binding protein; 40S, 40S ribosomal subunit. Reproduced from Robaglia C and Caranta C (2006) Translation initiation factors: A weak link in plant RNA virus infection. *Trends in Plant Science* 11(1): 40–45, with permission from Elsevier.

Table 2 Characteristics of cloned recessive resistance genes against plant RNA viruses

Genus	Virus	Gene/alleles	Species	Translation factor	Avr gene
Potyvirus	Potato virus Y	$pvr2^1$	Capsicum annuum	eIF4E1	VPg
	Potato virus Y and tobacco etch virus	$pvr2^2$	Capsicum annuum	eIF4E1	VPg
	Potato virus Y and tobacco etch virus and pepper mottle virus	pvr1	Capsicum chinense	eIF4E1	
Potyvirus	Pepper veinal mottle virus	pvr6 (+pvr2)	Capsicum annuum	eIF(iso)4E	
Potyvirus	Lettuce mosaic virus	$mo1^1$	Lactuca sativa	eIF4E1	CI, VPg, NIa region
	Lettuce mosaic virus	$mo1^2$	Lactuca sativa	eIF4E1	CI, VPg, NIa region
Potyvirus	Pea seedborne mosaic virus	sbm1	Pisum sativum	eIF4E1	VPg
Potyvirus	Potato virus Y and tobacco etch virus	pot-1	Lycopersicon hirsutum	eIF4E1	VPg
Bymovirus	Barley mild mosaic virus and barley yellow mosaic virus	rym4	Hordeum vulgare	eIF4E1	VPg
	Barley mild mosaic virus and barley yellow mosaic virus and barley yellow mosaic virus-2	rym5	Hordeum vulgare	eIF4E1	VPg
Sobemovirus	Rice yellow mottle virus	rymv1	Oryza sativa	eIF(iso)4G	VPg
Carmovirus	Melon necrotic spot virus	nsv	Cucumis melo	eIF4E1	3′ UTR

bymoviruses in pepper, tomato, lettuce, pea, and barley, most of the resistance-related changes correspond to nonconservative amino acid substitutions and are clustered in two neighboring regions at the surface of the predicted eIF4E three-dimensional structure. In melon, a single amino acid change mapped to the C-terminal part of eIF4E on an accessible region of the protein leads to resistance to MNSV. Amino acid substitutions responsible for resistance of rice to RYMV were also mapped to a surface location of eIF(iso)4G. Collectively these data argue in favor of resistance mechanism(s) mediated by subtle amino acid change(s) on the surface interaction of translation initiation factors.

Resistance genes against potyviruses and bymoviruses control multiple specificities targeted against distinct viruses of the same genus or pathotypes of the same virus (see section above). In pepper, lettuce, and pea, these resistance specificities were explained by the occurrence of small allelic series where distinct alleles were represented as a combination of distinct amino acid substitutions.

An intriguing aspect of eIF4E-mediated resistances is that they cover a diverse range of resistance phenotypes. Although most of them control complete qualitative resistance, they can also be quantitative and/or are components of polygenic resistance. For instance, in pepper, an eIF4E resistance allele at the pvr2 locus controls a partial resistance to some pathotypes of PVY and was initially mapped as a QTL. Similarly, in lettuce, two eIF4E resistance alleles, $mo1^1$ and $mo1^2$, appear associated with absence of accumulation of lettuce mosaic virus (LMV) at the cellular level, or to reduced accumulation of LMV and lack of symptoms, probably depending on other loci in the host genome. It was proposed that these distinct phenotypes might represent distinct outcomes of the same mechanism playing at quantitatively different intensities.

Among eIF4E-mediated resistances, the pvr6 resistance gene from pepper is of particular interest. It is involved in resistance to PVMV only when combined with some pvr2 resistance alleles (demonstrated to code for eIF4E), whereas pvr6 and pvr2 do not separately confer any resistance to PVMV. Moreover, pvr6 alone has no detectable effect on the potyvirus infection process. A recent study showed that the pvr6 resistance allele encodes a nonfunctional truncated eIF(iso)4E protein and that complete resistance to PVMV results from the combined effect of mutations in translation initiation factors eIF4E and eIF(iso)4E. Therefore, while most potyviruses require one specific eIF4E isoform to achieve infection, PVMV can use both 4E and (iso)4E isoforms.

A further common feature linking eIF4E-mediated resistance against potyviruses and also eIF(iso)4G-mediated resistance against RYMV is that regions of the viral protein named VPg (viral protein genome-linked; a protein that is covalently attached to the 5′ end of the viral RNA) have been mapped as avirulence determinants (i.e., involved in the overcoming of resistance) (**Table 2**). In this regard, resistance in melon to MNSV (lacking a VPg) differs because the viral avirulence determinant corresponding to eIF4E-nsv resistance gene was located at the 3′ untranslated region (3′ UTR) of the viral genome.

Natural versus Induced eIF4E-Mediated Resistance

Interestingly, *Arabidopsis thaliana* mutant lines (ethyl methane sulfonate (EMS)-induced mutations or transposon knockout) of the same genes from the translation initiation complex also exhibit complete or partial resistance phenotypes against RNA viruses, including several potyviruses, and also CMV and turnip crinkle virus (TCV). Similar to naturally occurring resistance alleles, single mutant alleles in *Arabidopsis* confer resistance to diverse related viruses in the genus *Potyvirus*. However, it is striking that a particular virus strain able to infect both *Arabidopsis* and crops appears not to be dependent on the same eIF4E isoforms in different hosts. For instance, TEV and LMV depend on eIF4E1 to infect pepper, tomato, or lettuce, whereas the same viruses depend on eIF(iso)4E to infect *Arabidopsis*.

Another feature linking analysis of natural and induced resistance is the ability of potyviruses to selectively recruit eIF4E isoforms. Indeed, disruption of *eIF(iso)4E* either by point mutations or by T-DNA insertion leads to *Arabidopsis*' lack of susceptibility to four different potyviruses (TEV, LMV, turnip mosaic virus (TuMV), and plum pox virus (PPV)), whereas disruption of *eIF4E1* leads to resistance to a fifth potyvirus, ClYVV. Recently, a crucial role of the eIF4G factors in potyvirus resistance/susceptibility has also been demonstrated by phenotyping *Arabidopsis* T-DNA insertion mutants that are null allele for each of the three *eIF4G* genes: the eIF(iso)4G1 mutant is resistant to PPV and LMV; the eIF4G mutant is resistant to ClYVV; and the double-mutant eIF(iso)4G1 X eIF(iso)4G2 is resistant to TuMV. Altogether these results indicate that eIF4G isoforms are selectively recruited by different potyviruses in a fashion that parallels the selective recruitment of eIF4E isoforms, suggesting the involvement of the whole eIF4F complex in potyvirus infection.

A very striking feature is that despite the key role of components of the eIF4F complex in *Arabidopsis*–RNA virus interactions, eIF4E- or eIF4G-mediated resistance was never identified in the natural diversity of this species. Several studies identified recessive resistance genes against viruses in different ecotypes of *Arabidopsis* (**Table 1**) but, up to now, none encodes eIF4F factors. Similarly, the systematic sequencing of *eIF4E* genes in a core collection of 54 ecotypes of *Arabidopsis* failed to identify signature amino acid substitutions previously demonstrated to be responsible for virus resistance in crops. Collectively, these data point toward a distinct mode of evolution of resistance against RNA viruses in this model species in comparison with crops.

Molecular Mechanism(s) Underlying eIF4E-Mediated Resistance

In link with their high frequency in the natural diversity of plant species, molecular mechanisms underlying eIF4E-mediated resistance have been studied most extensively for potyviruses. A key observation was the ability of eIF4E isoforms to bind the potyviral VPg in yeast two-hybrid and in *in vitro* binding assays. As described previously, the VPg was identified as the virulence determinant for several plant–potyvirus pairs. Moreover, amino acids within the central and C-terminal parts of the PVY VPg have undergone positive selection suggesting that the role of the VPg encoding region in virulence is determined by the protein and not by the viral RNA. A key role for the eIF4E–VPg interaction in the virus cycle is also supported by the demonstration that a mutation in the VPg that abolishes the interaction with eIF4E *in vitro* also prevents viral infection *in planta*. These data, together with signature amino acid substitutions identified at the surface of proteins encoded by resistance eIF4E alleles, argue in favor of a resistance mechanism mediated by disruption or impairing of a direct protein–protein interaction of eIF4E and VPg. This is supported by the recent demonstration that eIF4E proteins encoded by *pvr1* and *pvr2^2* resistance alleles from *Capsicum* failed to bind the VPg from TEV strains (by yeast two-hybrid and GST pull-down), whereas eIF4E proteins encoded by the susceptibility alleles interacted strongly. However, the fact that the outcome of the eIF4E–VPg interaction is the determinant of viral infectivity remains uncertain for some other plant–potyvirus pairs. This might be related to the methods used to detect eIF4E/VPg interactions and/or to the occurrence of distinct mechanism(s) not dependent on this sole interaction.

In spite of these data, the precise role of eIF4E and its interaction with the viral VPg has yet to be defined. In eukaryotes, most cellular RNAs contain both a 5′ cap and a 3′ polyA tail that act synergistically to stimulate translation (**Figure 1**). Most positive-sense single-stranded RNA viruses lack either one or both of these structures. For members of the family *Potyviridae*, which have a VPg protein covalently attached to the 5′ end instead of a cap structure, the VPg might function as a cap mimic and recruit the translation initiation complex to the viral RNA. Inconsistent with this hypothesis are the demonstrations that the 5′ UTRs of TEV and TuMV support internal initiation via the recruitment of eIF4G to an internal ribosome entry site (IRES) and that significant translation of potyviral RNA appears possible in the absence of the covalently bound VPg. However, it cannot be completely ruled out that the eIF4E–VPg interaction facilitates and/or enhances cap-independent translation of the viral genome. Several other possible roles for eIF4E during the potyviral life cycle have also been proposed. The eIF4E–VPg interaction might interfere with cap-dependent translation of host mRNA, freeing ribosomes for viral RNA translation. It might also play a role in genome replication in a manner similar to the replication of mammalian picorna-like viruses. In support of this model, potyviral VPg urydylation by the viral polymerase (NIb) has been demonstrated and the existence of complexes with VPg, NIb, eIF4E, and PABP has also been shown. For *sbm1*-mediated resistance to PsBMV, data also suggested that eIF4E facilitates cell-to-cell movement of the virus. Altogether, these results suggest that distinct mechanisms might be responsible of eIF4E-mediated resistance against potyviruses.

The distinctive features of eIF4E-mediated resistance against MNSV in melon (i.e., resistance controlled by a single amino acid change in the C-terminal part of eIF4E and avirulence determinant located at the 3′ UTR of the viral genome) associated with the fact that RNA of carmoviruses neither possesses a cap or a VPg at the 5′ end, nor a poly(A) tail at the 3′ end, strongly argue in favor of a resistance mechanism distinct from those against potyviruses. The authors of this work speculated that the MNSV avirulence determinant in the 3′ UTR interacts with eIF4E acting as a 5′ cap substitute, bringing the eIF4F complex to the 5′ UTR through a 5′–3′ RNA–RNA interaction. Indeed, they identified in the MNSV 3′ UTR, outside the avirulence determinant, nucleotide stretches with complementarities to nucleotides of a conserved loop in the MNSV 5′ UTR. This feature suggests that MNSV RNA initiation of translation might occur, as for other members of the family *Tombusviridae*, through a 5′–3′ RNA–RNA interaction that circularizes the message and involves eIF4E.

Conclusions

More than one hundred recessive resistance factors against plant viruses were identified in the natural diversity of crops and their wild relatives, among which a large number are currently used to control agricultural losses to viral diseases. The data shown in **Table 1** suggest strong biases toward resistance factors against RNA viruses (∼90%) and host responses that show monogenic inheritance (∼95%). Recent studies carried out to identify the molecular nature of recessive resistances genes against plant viruses have revealed a remarkable degree of conservation. All genes isolated and characterized to date encode eIF4E or eIF4G factors from the translation initiation complex. Their identification as essential susceptibility factors for RNA virus multiplication has enabled plant virology to make a significant step forward.

In future, the same type of recessive resistance genes is likely to be discovered in other plant families via candidate gene approaches, at least for viral genera for which translation initiation factors were demonstrated to play a key role in plant resistance (**Table 2**). More generally, the candidate gene approach should permit to examine the extent to which eIF4E and eIF4G factors, and also other factors from the translation complex, may serve as candidates for recessive resistance genes. It can also be anticipated that the molecular characterization of recessive resistances using approaches without presupposition on the molecular nature of the gene, such as positional cloning, will lead to the identification of new resistance targets that function as host susceptibility factors. Another promising approach is the identification of host proteins that interact with viral genes or gene products during infection through protein–protein interaction technologies and proteomics. Once the resistance genes are identified, it will be possible to discover novel alleles by surveying natural genetic variants using high-throughput specific marker technology. When such resistance alleles are not available in the natural diversity, it is possible to select plants carrying artificially induced mutations in a target gene through a reverse genetics technology termed TILLING (targeting induced local lesions in genomes). Overall, the characterization of new recessive resistance genes and probably of new susceptibility factors will provide plant biology with novel views on fundamental plant–virus interaction processes. Such knowledge will also significantly contribute to crop improvement for virus resistance through a diversification of genetic targets.

See also: Plant Resistance to Viruses: Engineered Resistance; Plant Resistance to Viruses: Natural Resistance Associated with Dominant Genes.

Further Reading

Colbert T, Till BJ, Tompa R, *et al.* (2001) High-throughput screening for induced point mutations. *Plant Physiology* 126: 480–484.

Diaz-Pendon JA, Truniger V, Nieto C, *et al.* (2004) Advances in understanding recessive resistance to plant viruses. *Molecular Plant Pathology* 5(3): 223–233.

Dreher TD and Miller AW (2006) Translational control in positive strand RNA plant viruses. *Virology* 344: 185–197.

Fraser RSS (1990) The genetics of resistance to plant viruses. *Annual Review of Phytopathology* 28: 179–200.

Kang B-C, Yeam I, and Jahn MM (2005) Genetics of plant virus resistance. *Annual Review of Phytopathology* 43: 18.1–18.41.

Maule AJ, Caranta C, and Boulton MI (2007) Sources of natural resistance to plant viruses: Status and prospects. *Molecular Plant Pathology* 8: 223–231.

Nieto C, Morales M, Orjeda G, *et al.* (2006) An eIF4E allele confers resistance to an uncapped and non-polyadenylated RNA virus in melon. *Plant Journal* 48: 1–11.

Robaglia C and Caranta C (2006) Translation initiation factors: A weak link in plant RNA virus infection. *Trends in Plant Science* 11(1): 40–45.

Thivierge K, Nicaise V, Dufresne PJ, *et al.* (2005) Plant virus RNAs. Coordinated recruitment of conserved host functions by (+)-ssRNA viruses during early infection events. *Plant Physiology* 138: 1822–1827.

Plant Rhabdoviruses

A O Jackson, University of California, Berkeley, CA, USA
R G Dietzgen, The University of Queensland, St. Lucia, QLD, Australia
R-X. Fang, Chinese Academy of Sciences, Beijing, People's Republic of China
M M Goodin, University of Kentucky, Lexington, KY, USA
S A Hogenhout, The John Innes Centre, Norwich, UK
M Deng, University of California, Berkeley, CA, USA
J N Bragg, University of California, Berkeley, CA, USA

© 2008 Elsevier Ltd. All rights reserved.

Introduction

The plant rhabdoviruses have distinctive enveloped bacilliform or bullet-shaped particles and can be distinguished based on whether they replicate and undergo morphogenesis in the cytoplasm or in the nucleus. Consequently, they have been separated into two genera, *Cytorhabdovirus* or *Nucleorhabdovirus*. More than 90 putative plant rhabdoviruses have been described although, in many cases, molecular characterizations necessary for unambiguous classification are incomplete or lacking. Recent analyses indicate that the eight sequenced plant rhabdoviruses have the same general genome organization as other members of the *Rhabdoviridae*, but that each encodes at least six open reading frames (ORFs), one of which probably facilitates cell-to-cell movement of the virus. Thus, plant rhabdoviruses have a number of similarities to members of other rhabdovirus genera, but they differ in several respects from rhabdoviruses infecting vertebrates.

Rhabdoviruses infect plants from a large number of different families, including numerous weed hosts and several major crops. Symptoms of infection vary substantially and range from stunting, vein clearing, mosaic and mottling of leaf tissue, to tissue necrosis. The most serious pathogens include maize mosaic virus (MMV), lettuce necrotic yellows virus (LNYV), rice yellow stunt virus (RYSV), also known as rice transitory yellowing virus (RTYV), eggplant mottled dwarf virus (EMDV), strawberry crinkle virus (SCV), potato yellow dwarf virus (PYDV), and barley yellow striate mosaic virus (BYSMV), which is synonymous with maize sterile stunt virus (MSSV), and wheat chlorotic streak virus (WCSV). A number of other rhabdoviruses also have disease potential that can be affected by agronomic practices, incorporation of genes for disease resistance, and control of insect vectors.

The spread of most plant rhabdoviruses is dependent on specific transmission by phytophagous insects that support replication of the virus, so their prevalence and distribution is influenced to a large extent by the ecology and host preferences of their vectors. Although some rhabdoviruses can be transmitted mechanically by abrasion of leaves, this mode of transmission does not contribute significantly to their natural spread due to the labile nature of the virion. Moreover, seed or pollen transmission of plant rhabdoviruses has not been described; thus, aside from vegetative propagation, direct plant-to-plant transmission is unlikely to be a major factor in the ecology or epidemiology of these pathogens.

This article focuses on recent findings concerning the taxonomy, structure, replication, and vector relationships of plant rhabdoviruses. More extensive aspects of plant rhabdovirus biology, specifically ecology, disease development and control, can be found in earlier reviews.

Taxonomy and Classification

The International Committee on Taxonomy of Viruses (ICTV) has used subcellular distribution patterns to assign plant rhabdoviruses to the genera *Cytorhabdovirus* and *Nucleorhabdovirus* (**Table 1**). Currently the ICTV has assigned eight virus species (BYSMV, *Broccoli necrotic yellows virus* (BNYV), *Festuca leaf streak virus* (FLSV), LNYV, *Northern cereal mosaic virus* (NCMV), *Sonchus virus* (SV), SCV, and *Wheat American striate mosaic virus* (WASMV)) to the genus *Cytorhabdovirus* and seven viruses (*Datura yellow vein virus* (DYVV), *Eggplant mottled dwarf virus* (EMDV), MMV, PYDV, RYSV, *Sonchus yellow net virus* (SYNV), and *Sowthistle yellow vein virus* (SYVV)) to the genus *Nucleorhabdovirus*. However, sufficient new information has been documented to justify provisional inclusion of the recently described maize fine streak virus (MFSV) and taro vein chlorosis virus (TaVCV) in the genus *Nucleorhabdovirus* (**Table 1**). Cereal chlorotic mottle virus (CCMoV) has also been provisionally included in the genus *Nucleorhabdovirus* based on its intracellular distribution and serology. The complete genomic sequences have been determined for three cytorhabdoviruses, LNYV, NCMV, and SCV and for five nucleorhabdoviruses, MFSV, MMV, RYSV, SYNV, and TaVCV. Phylogenetic analyses of these rhabdoviruses have confirmed their taxonomic classification. Most other plant rhabdoviruses have not been investigated in much detail beyond cursory infectivity studies, crude physicochemical analyses of virus particles, and electron microscopic observations of morphogenesis. Consequently, more than 75 putative rhabdoviruses await assignment to a genus (**Table 1**).

Table 1 List of plant rhabdoviruses and their host and vector specificity

Virus	Host	Vector
Cytorhabdovirus		
Barley yellow striate mosaic virus (BYSMV)	M	P
[Maize sterile stunt virus]		
[Wheat chlorotic streak virus]		
Broccoli necrotic yellows virus (BNYV)	D*	A
Festuca leaf streak virus (FLSV)	M	
Lettuce necrotic yellows virus (LNYV)	D*	A
Northern cereal mosaic virus (NCMV)	M	P
Sonchus virus (SonV)	D*	
Strawberry crinkle virus (SCV)	D*	A
Wheat American striate virus (WASMV)	M	L
[Oat striate mosaic virus]		
Nucleorhabdovirus		
Cereal chlorotic mottle virus (CCMoV)	M	L
Datura yellow vein virus (DYVV)	D	
Eggplant mottled dwarf virus (EMDV)	D*	L
[Pittosporum vein yellowing virus]		
[Tomato vein yellowing virus]		
[Pelargonium vein clearing virus]		
Maize fine streak virus (MFSV)	M	L
Maize mosaic virus (MMV)	M	P
Potato yellow dwarf virus (PYDV)	D*	L
Rice yellow stunt virus (RYSV)	M	L
[Rice transitory yellowing virus]		
Sonchus yellow net virus (SYNV)	D*	A
Sowthistle yellow net virus (SYVV)	D	A
Taro vein chlorosis virus (TaVCV)	M	
Unassigned Plant Rhabdoviruses		
Asclepias virus	D	
Atropa belladonna virus	D	
Beet leaf curl virus	D	LB
Black current virus	D	
Broad bean yellow vein virus	D	
Butterbur virus	D*	
Callistephus chinensis chlorosis virus	D	
Caper vein yellowing virus	D	
Carnation bacilliform virus	D	
Carrot latent virus	D	A
Cassava symptomless virus	D	
Celery virus	D	
Chondrilla juncea stunting virus	D	
Chrysanthemum vein chlorosis virus	D	
Clover enation (mosaic) virus	D	
Colocasia bobone disease virus	D	P
Coriander feathery red vein virus	D*	A
Cow parsnip mosaic virus	D*	
Croton vein yellowing virus	D*	
Cucumber toad skin virus	D	
Cynara virus	D*	
Cynodon chlorotic streak virus	M	P
Daphne mezereum virus	D	
Digitaria striate virus	M	P
Euonymus fasciation virus	D	
Euonymus virus	D	
Finger millet mosaic virus	M	P
Gerbera symptomless virus	D	
Gloriosa fleck virus	D	
Gomphrena virus	D*	
Gynura virus	D	
Holcus lanatus yellowing virus	M	
Iris germanica leaf stripe virus	M	
Ivy vein clearing virus	D*	
Kenaf vein-clearing virus	D	
Laburnum yellow vein virus	D	
Launea arborescens stunt virus	D	
Lemon scented thyme leaf chlorosis virus	D	
Lolium ryegrass virus	M	
Lotus stem necrosis	D	
Lotus streak virus	D	A
Lucerne enation virus	D	A
Lupin yellow vein virus	D	
Maize Iranian mosaic virus	M	P
Maize streak dwarf virus	M	P
Malva sylvestris virus	D	
Meliotus (sweet clover) latent virus	D	
Melon variegation virus	D	
Mentha piperita virus	D	
Nasturtium vein banding virus	D	
Papaya apical necrosis virus	D	
Parsely virus	D*	
Passionfruit virus	D	
Patchouli mottle virus	D	
Peanut veinal chlorosis virus	D	
Pigeon pea proliferation virus	D	L
Pinapple chlorotic leaf streak virus	M	
Pisum virus	D*	
Plantain mottle virus	M	
Poplar vein yellowing virus	D	
Ranunculus repens symptomless virus	D	
Raphanus virus	D*	
Raspberry vein chlorosis virus	D	A
Red clover mosaic virus	D	
Sainpaulia leaf necrosis virus	D	
Sambucus vein clearing virus	D	
Sarracenia purpurea virus	D	
Sorghum stunt mosaic virus	M	L
Soursop yellow blotch virus	D	
Soybean virus	D	
Triticum aestivum chlorotic spot virus	M	
Vigna sinensis mosaic virus	D	
Viola chlorosis virus	D	
Wheat rosette stunt virus	M	P
Winter wheat Russian mosaic virus	M	P

Names in brackets are synonymous to those immediately above. Host: D, dicot; M, monocot. (*) indicates ability to be mechanically transmitted. Vectors: A, aphid, L, leafhopper; LB, lacebug; P, planthopper. Blank spaces indicate that no insect vector has been identified.

Particle Morphology and Composition

Plant rhabdoviruses are normally bacilliform after careful fixation (**Figure 1(a)**) and estimates of their sizes range from 45 to 100 nm in width and 130 to 350 nm in length. The outer layer consists of 5–10 nm surface projections that appear to be composed of G protein trimers that penetrate a host-derived membrane (**Figure 1(b)**). The nucleocapsid core is composed of the genomic RNA, the nucleocapsid protein (N), the phosphoprotein (P), and the L polymerase protein (**Figures 1(a) and 1(b)**). Rhabdovirus virions also contain a matrix protein (M) that interacts with the G protein to stabilize the particle. A sixth protein (sc4) is associated with the membrane fractions of SYNV particles but the presence of an sc4 derivative has not been found in virions of other plant rhabdoviruses.

The overall chemical composition ($\sim$70 % protein, 2 % RNA, 20–25% lipid, and a small amount of carbohydrate associated with the G protein) of the plant and animal rhabdoviruses is similar. The minus-sense RNA genomes of plant rhabdoviruses, which range in size from $\sim$11 to 14 kb based on sedimentation, gel electrophoretic analyses, and genome sequencing, are slightly larger than those of most described animal rhabdoviruses. The lipids of plant and animal rhabdoviruses consist of fatty acids and sterols that are derived from sites of morphogenesis. The four sterols predominating in SYNV closely approximate sterols in the nuclear envelope, whereas those of NCMV, a cytorhabdovirus, are more typical of cytoplasmic membranes.

Genomic Structure and Organization

The consensus plant rhabdovirus genome deduced from the eight sequenced viruses is 3-ℓ-N-P-X-M-G-Y-L- **t**-5' (**Figure 1(c)**). The N, P, M, G, and L genes appear in the same order as in other rhabdoviruses and their encoded proteins are thought to be functionally similar to the five proteins of vesicular stomatitis virus (VSV). A variable number of genes at the X site have been found in each of the sequenced viruses and some of these appear to be involved in movement. The Y sites between the G and L genes encode short ORFs of unknown functions that are present in the nucleorhabdovirus, RYSV, and the cytorhabdoviruses, NCMV and SCV.

Rhabdovirus ORFs are separated by intergenic or 'gene-junction sequences' that provide vital regulatory functions during transcription and replication (**Figure 2(a)**). The gene-junction sequences can be grouped into three elements consisting of (1) a poly (U) tract at the 3' end of each gene, (2) a variable intergenic element that is not transcribed in the mRNAs, and (3) a short element complementary to the first 5 nt at the 5 start site of each mRNA. In general, the gene-junction sequences of each virus are highly conserved, and those of the plant rhabdoviruses share substantial relatedness and differ mostly at element II. Slightly more limited divergence is noted when comparing the genomes of other families within the order *Mononegavirales*, suggesting that these regulatory sequences have been stringently conserved.

The coding regions are flanked by 3 leader (ℓ) and 5 trailer (**t**) noncoding sequences that represent recognition signals required for nucleocapsid assembly and regulation of genomic and antigenomic RNA replication. These sequences have short complementary termini and small amounts of common sequence relatedness (**Figure 2(b)**). However the plant rhabdovirus ℓ RNAs differ in sequence from the ℓ and **t** sequences of vertebrate rhabdoviruses and are considerably longer than those of VSV. The transcribed ℓ RNA of SYNV is polyadenylated and differs in this respect from the ℓ RNA of VSV and other known rhabdoviruses.

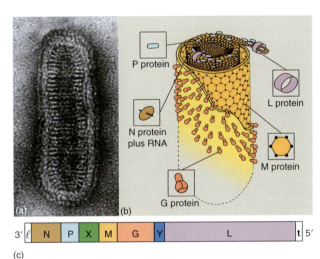

Figure 1 Electron micrograph, diagram, and genome organization of plant rhabdoviruses. (a) Transmission electron micrograph of a negative-stained virus showing the striated inner core, envelope, and glycoprotein spikes. (b) Architecture of the virus particle. The nucleocapsid core is composed of the minus-sense genomic RNA, the nucleocapsid protein (N), the phosphoprotein (P), and the polymerase protein (L). The matrix protein (M) is involved in coiling the nucleocapsid, attachment of the nucleocapsid to the envelope, and associations with the transmembrane glycoprotein (G). (c) Schematic representation of the negative-sense arrangement of genes encoded in the genomes of plant rhabdoviruses. The order of the genes is 3'-ℓ-N-P-X-M-G-Y-L-**t**-5', where ℓ represents the leader RNA, **t** represents the trailer sequence, X denotes putative movement and undefined plant rhabdovirus genes, and Y shows the location of open reading frames of unknown function in the genomes of several plant and animal rhabdoviruses. Reprinted, with permission, from the Annual Review of Phytopathology, Volume 43, © 2005 by Annual Reviews.

Properties of the Encoded Proteins

The most comprehensive biochemical analyses of the encoded proteins have been carried out with SYNV, LNYV, and RYSV. Overall, the plant rhabdovirus proteins have very little sequence relatedness to analogous proteins of animal rhabdoviruses, with the exception of

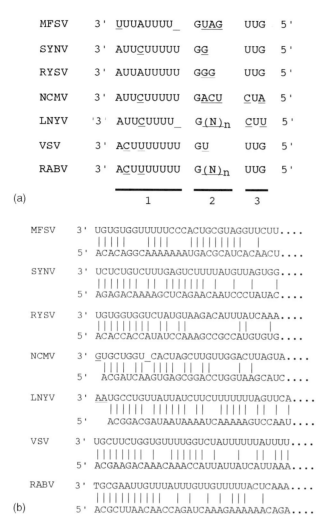

Figure 2 Comparisons of intergenic and terminal noncoding regions of rhabdovirus genomes. (a) Intergenic sequences separating the genes. (b) Complementary sequences at the 3′ and 5′ termini of the genomic RNAs. MFSV (maize fine streak virus), SYNV (sonchus yellow net virus), RYSV (rice yellow stunt virus), NCMV (northern cereal mosaic virus), LNYV (lettuce necrotic yellows virus), VSV (vesicular stomatitis virus), RABV (rabies virus). Modified from figures 2 and 3 of Tsai C-W, Redinbaugh MG, Willie KJ, Reed S, Goodin M, and Hogenhout SA (2005) Complete genome sequence and in planta subcellular localization of maize fine streak virus proteins. *Journal of Virology* 79: 5304–5314, with permission from American Society for Microbiology.

the L protein, which has conserved polymerase motifs common to those of most rhabdoviruses. A description of these proteins and their probable functions is outlined below.

The nucleocapsid protein (N)

The N protein functions to encapsidate the viral genomic RNA and is a component of the viroplasms and of the polymerase complex (**Figure 1**). The N genes of the nucleorhabdoviruses, SYNV, MMV, MFSV, TaVCV, and RYSV, and the cytorhabdoviruses, LNYV, SCV, and NCMV have been sequenced. The SYNV, MFSV, MMV, and RYSV nucleorhabdovirus N proteins exhibit short stretches of sequence similarity, suggesting that these four viruses are closely related. These regions of the nucleorhabdovirus N proteins are not significantly related to those of the cytorhabdoviruses and have no extensive relatedness to vertebrate rhabdovirus N proteins.

Experiments conducted in plant and yeast cells have shown that SYNV N protein contains a bipartite nuclear localization signal (NLS) near the carboxy-terminus that is required for nuclear import, and biochemical studies have shown that the protein interacts *in vitro* with importin α homologs. Related nuclear localization sequences are also present in the MFSV and TaVCV N proteins, but this signal is lacking in the N protein of MMV and RYSV. During transient expression in plant cells, the SYNV N protein forms subnuclear foci that resemble the viroplasms found in infected plants, and coexpression of the N and P proteins results in colocalization of both proteins to subnuclear foci (**Figure 3**). These foci require homologous interactions of the N protein that are mediated by a helix-loop-helix motif near the amino-terminus. Interestingly, the SYNV subnuclear foci are distinct from those of nucleolar marker proteins, whereas foci formed during interactions of the MFSV N and P proteins appear to colocalize to the nucleolus.

The phosphoprotein (P)

Direct experiments showing phosphorylation of the P protein are available only for SYNV. The amino-terminal half of the SYNV P protein is negatively charged, as is the case for the other rhabdoviruses. In SYNV, the P protein is phosphorylated *in vivo* at threonine residues and hence differs from the VSV P protein, which is phosphorylated at serine residues. No discernable sequence relatedness is evident between the P protein of SYNV and those of other rhabdoviruses. However, the plant rhabdovirus P proteins have similar hydrophilic cores and the cytorhabdovirus P proteins overall are hydrophilic with similar isoelectric points. In addition, there is nearly 50% sequence identity between the P proteins of MMV and TaVCV. Although these results suggest that the provisional assignments of the P proteins of other plant rhabdoviruses are probably correct, additional data need to be accumulated to obtain a consensus of the functional activities and the biochemical interactions of the putative plant rhabdovirus P proteins.

The P protein is a component of the viral nucleocapsid core and the polymerase complex. The SYNV P protein forms complexes *in vivo* with the N and L proteins that are analogous to N:P and P:L complexes found in VSV-infected cells, and hence the P protein probably functions in SYNV polymerase recycling. Biochemical experiments have shown that the solubility of the SYNV N protein is

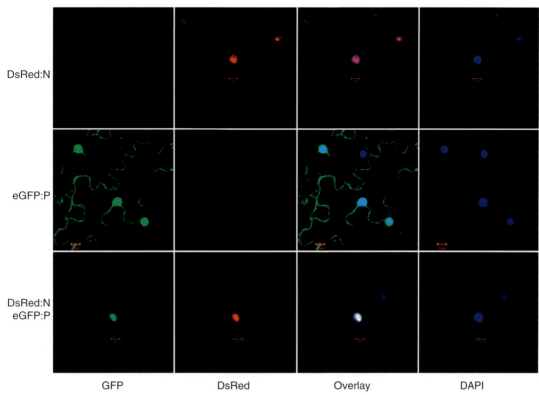

Figure 3 Subcellular localization of the N and P proteins of SYNV. The DsRed:N and eGFP:P fusion proteins were transiently expressed in *Nicotiana benthamiana* leaf tissue via infiltration with *Agrobacterium tumefacians* containing pGD vectors. The confocal micrographs show the subcellular localization of proteins at 3 days after infiltration. The top row depicts the individual expression of the DsRed:N fusion protein. The middle row shows fluoescence in cells expressing eGFP:P alone. The bottom row shows cells coexpressing the DsRed:N and eGFP:P proteins. Nuclei are identified by staining with DAPI (4′-6-diamino-2-phenylindole dihydrochloride). From Deng et al., unpublished.

increased during P protein interactions, so the P protein appears to have chaperone activity. Heterologous N:P protein complexes form by interactions of an internal region of the P protein with the amino-terminal helix-loop-helix region of N that overlaps the N:N protein binding site. The P protein also engages in homologous interactions that are mediated near the amino-terminus of the P protein. Hence, the SYNV P protein has functions similar to those of the P proteins of other well-characterized vertebrate rhabdoviruses.

Reporter gene fusions show that the SYNV P protein, when expressed alone, accumulates in both the nucleus and the cytoplasm (**Figure 3**). The central third of the SYNV P protein is required for nuclear import, but other regions of the protein affect the import efficiency. Recent experiments indicate that the SYNV P protein binds directly to human importin β derivatives *in vitro* and, since the N protein has an NLS site and interacts with importin α, the N and P proteins have different mechanisms for nuclear import. Interestingly, sequence analyses show that the SYNV P protein does not have a bipartite NLS, whereas the P proteins of RYSV and MMV have a bipartite NLS, and both P proteins have a pronounced nuclear localization pattern. Hence, due to these differences between the SYNV and the RYSV and MMV proteins, it is likely that the viruses have diverged in their strategies of nuclear localization. In SYNV and MFSV, coexpression of the P protein with the N protein of the same virus results in colocalization of the complexes to subnuclear foci characteristic of viroplasms. In SYNV, formation of these foci requires interactions of P with the amino-terminus of the N protein. However, the N and P interactions appear to be virus-specific because heterologous combinations of the SYNV and MFSV N and P proteins fail to form subnuclear foci, and the P protein continues to be expressed in both the nucleus and the cytoplasm.

The SYNV P protein also has a leucine-rich nuclear export signal located within the first third of the protein. Reporter proteins fused to the P protein are retained in the nucleus following treatment with Leptomycin B, an inhibitor of nuclear export. Interaction with the host nuclear export receptor XpoI provides further evidence for P protein nuclear export functions. These results and P protein mutagenesis experiments provide strong evidence suggesting that the SYNV P protein is involved in nuclear shuttling activities.

In addition to its role as a structural protein, the SYNV P protein shares many of the hallmarks of

RNA-silencing suppressor proteins. These characteristics include suppression of reporter gene silencing in transgenic plants and the ability to bind small interfering (siRNAs) and single-stranded RNAs *in vitro*. Together, these activities clearly point to key roles of the P protein in nucleocapsid structure, replication, countering innate host defenses, and possibly intercellular movement.

Position X proteins

Like other plant viruses, plant rhabdoviruses must encode proteins to assist in cell-to-cell movement of virus derivatives through the plasmodesmata and their systemic transport through the vascular system. Considerable evidence for a movement function has been accumulated for proteins encoded at position X between the P and M genes (**Figure 1(c)**). The predicted secondary structures of several plant rhabdovirus proteins, including SYNV sc4, LNYV 4b, RYSV P3, MMV P3, and MFSV P4, have a distant relatedness to the TMV 30 K superfamily structural motifs. Additional evidence for a role of the position X-encoded proteins in cell-to-cell movement is their association with host and viral membranes during transient expression. Unpublished evidence indicates that the sc4 protein is phosphorylated, as is also the case with the TMV 30 K movement protein. The movement hypothesis for genes occupying the X position has been reinforced by experiments carried out with RYSV. With this nucleorhabdovirus, the P3 protein is able to *trans*-complement cell-to-cell movement of a movement-defective potato virus X in *Nicotiana benthamiana* leaves. The P3 protein also interacts with the N protein, the major component of nucleocapsids; hence, it could possibly facilitate movement of nucleocapsids through these interactions.

Despite the persuasive evidence for a movement function of genes encoded at position X, considerable diversity appears in the X ORF(s) of several plant rhabdoviruses. For example, although the TaVCV gene X codes for a protein of a size similar to the other X protein genes, the protein has no obvious sequence similarity to proteins in the 30 K superfamily. MFSV encodes two proteins between the putative P and M proteins, and both have different localization patterns from sc4 of SYNV. In addition, four small ORFs reside between the P and M genes of NCMV. Thus, additional studies need to be undertaken to clarify the functional activities of the 'unusual' X proteins.

The matrix protein (M)

The M proteins of plant rhabdoviruses are basic and are thought to function in nucleocapsid binding and coiling, and interactions with the G protein. Sequence alignments of the M proteins of several plant rhabdoviruses have not revealed extensively conserved motifs. Unpublished studies suggest that the SYNV M protein is phosphorylated *in vivo* at both threonine and serine residues. When expressed ectopically, the M proteins of SYNV and MFSV localize in the nucleus. A central hydrophobic region of the M protein is thought to mediate membrane–lipid interactions with the G protein during morphogenesis. In addition to their roles in viral morphogenesis, preliminary experiments indicate that rhabdovirus M proteins have important roles in host–virus interactions because they appear to be able to inhibit host gene expression.

Position Y ORFs

NCMV, RYSV, and SCV contain a short ORF at position Y that separates the genes encoding the G and L proteins (**Figure 1(c)**). Small, nonvirion ORFs preceding the L gene are also found in the genomes of some animal viruses, but the products of these ORFs are either nonstructural or have not been detected in infected cells. The three predicted plant rhabdovirus Y proteins are small (<100 amino acids) and do not share obvious sequence identity with each other or with the nonvirion genes of the animal rhabdoviruses. However, short stretches of the Y ORFs have limited relatedness to other negative-strand RNA virus proteins, suggesting that these regions of the genome may have originated by gene duplication or recombination.

The RYSV P6 protein at position Y is predicted to contain an aspartic protease motif (DTG) and has five potential phosphorylation sites (S/T-X-X-D/E). *In vitro* phosphorylation assays using a GST:P6 fusion protein have shown that P6 is phosphorylated at both serine and threonine residues. Although P6 could not be detected in total protein extracts from infected leaf tissue, immunoblots of purified virus and protein extracts from viruliferous leafhoppers suggest that P6 is associated with virions, so it may have a structural role in infection.

The glycoprotein (G)

The G protein forms the glycoprotein spikes of rhabdovirus virions (**Figure 1**). The plant rhabdovirus G proteins do not have extensive similarity, but they are more closely related to each other than to the G proteins of several vertebrate rhabdoviruses. The plant rhabdovirus G proteins share putative N-terminal signal sequences, a transmembrane anchor domain, and several possible glycosylation sites. In addition, the SYNV G protein contains a putative NLS near the carboxy-terminus that could be involved in transit to the inner nuclear membrane prior to morphogenesis. Several glycosylation inhibitors interfere with N-glycosylation of the SYNV G protein, and tunicamycin treatment blocks SYNV morphogenesis, leading to accumulation of striking arrays of condensed nucleocapsid cores that fail to bud through the inner nuclear membrane. Thus, the G protein has a prominent role in morphogenesis, and the available evidence suggests that glycosylation is required for interactions of the protein with coiled nucleocapsids.

The polymerase protein (L)

The L proteins of plant rhabdoviruses are present in low abundance within nucleocapsids and in infected cells. The L proteins are the most closely related of the rhabdovirus-encoded proteins and are positively charged with conserved polymerase domains and RNA-binding motifs. The L protein of SYNV is required for polymerase activity, because antibodies directed against the GDNQ (polymerase) motif inhibit transcription. Alignment of the L protein sequence with polymerases of several other nonsegmented negative-strand RNA viruses reveals conservation within 12 motifs. Phylogenetic trees derived from L protein alignments indicate that the nucleorhabdoviruses and cytorhabdoviruses cluster together in two clades separated from the vertebrate rhabdoviruses. This suggests that the plant rhabdoviruses have diverged less from each other than from the vertebrate rhabdoviruses.

Polymerase Activity

A viral RNA-dependent RNA polymerase is activated after treatment of LNYV and BNYV cytorhabdovirus virions with mild nonionic detergents, and this activity cosediments with loosely coiled nucleocapsid filaments that are released from virions. The transcribed products are complementary to the genome, as expected of mRNAs. Thus, the described polymerases of these plant cytorhabdoviruses appear to be similar to the extensively studied polymerases of the vesiculoviruses.

In contrast, no appreciable polymerase activity is evident in dissociated preparations of SYNV or other nucleorhabdovirus virions that have been analyzed. However, an active polymerase can be recovered from the nuclei of plants infected with SYNV. Polymerase activity is associated with a nucleoprotein derivative, consisting of the N, P, and L proteins, that cosediments with SYNV nucleocapsid cores. The polymerase complex can be precipitated with P protein antibodies, but the activity of the complex is not inhibited by these antibodies. However, antibody inhibition experiments demonstrate that the L protein is required for polymerase activity. Kinetic analysis of transcription products also reveals that the complex is capable of sequentially transcribing a polyadenylated plus-sense leader RNA and polyadenylated mRNAs corresponding to each of the six SYNV-encoded proteins. Potential replication intermediates consisting of short incomplete minus-strand products homologous to the genomic RNA are also transcribed. These results thus provide a model whereby nucleorhabdovirus particles require polymerase activation by host components early in infection. In contrast, the polymerases of the cytorhabdoviruses appear to be present in an active form in virions and the released cores are capable of initiating primary transcription immediately upon uncoating *in vitro*.

Cytopathology and Replication

The plant rhabdoviruses vary profoundly in their sites of replication and morphogenesis, and those that replicate in the nucleus differ substantially from vertebrate rhabdoviruses that replicate and assemble in the cytoplasm. In plants, nucleorhabdoviruses replicate in the nucleus, bud in association with the inner nuclear membrane, and accumulate in enlarged perinuclear spaces formed between the inner and outer nuclear envelopes. Similar patterns normally occur in the majority of insect tissues, but MMV also buds in the outer membranes of the salivary glands and nerve cells of its leafhopper vector. Clearly, studies on insect cells need to be emphasized in future work to clarify aspects of insect transmission.

The limited evidence available indicates that the cytorhabdoviruses replicate in the cytoplasm, bud in association with the endoplasmic reticulum (ER), and accumulate in ER-derived vesicles (**Figure 4(a)**). Two slightly different variations in replication of LNYV and BYSMV have been proposed, based on extensive electron microscopic observations of infected cells. Indirect evidence suggests that a nuclear phase may be involved in LNYV replication because the outer nuclear membrane blisters and develops small vesicles that contain some virus particles. However, later in the life cycle, masses of thread-like viroplasms appear in the cytoplasm and these are located close to dense networks of the ER that appear during infection. These proliferated membranes form vesicles that may serve as sites for morphogenesis of the accumulating nucleocapsids. A similar scenario lacking a nuclear phase has been outlined for BYSMV. In this case, membrane-bound viroplasms appear in the cytoplasm and virus particles are found exclusively in association with cytoplasmic membranes that proliferate in close proximity to the viroplasms. Unfortunately, both of the cytorhabdovirus models have been derived solely from ultrastructural observations, and none of these studies has utilized specific antibodies to identify individual virus proteins, viral-specific probes for *in situ* hybridization, or modern techniques of cell biology to probe replication.

No direct information is yet available about the early entry and uncoating events, but a model for nucleorhabdovirus infection predicts that, after entry into the cell during vector feeding or mechanical transmission, rhabdovirus particles associate with the ER to release the nucleocapsid cores into the cytoplasm (**Figure 4(b)**). Released cores may then utilize the host nucleocytoplasmic transport

194 Plant Rhabdoviruses

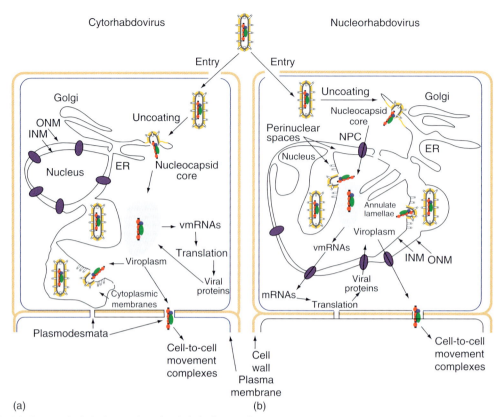

Figure 4 Models for cytorhabdovirus and nucleorhabdovirus replication in plant cells. Most rhabdoviruses are believed to enter plant cells during insect vector feeding and the nucleocapsid is thought to fuse with the endoplasmic reticulum (ER) and be liberated into the cytoplasm. Panels (a) and (b) provide contrasts between the cytorhabdovirus and the nucleorhabdovirus replication strategies, respectively. (a) Cytorhabdovirus replication model. The available information about cytorhabdovirus replication relies almost entirely on ultrastructural observations, and molecular or modern cytological information has not been obtained to extend these observations. After nucleocapsid release, primary and secondary rounds of transcription are followed by nucleocapsid accumulation in viroplasms to form dense masses that are associated with proliferated membrane vesicles. Morphogenesis occurs by budding of nucleocapsids into the ER. During the later stages of replication, large aggregates of bacilliform virions accumulate in pronounced vesicles that are thought to have originated from the ER. (b) Nucleorhabdovirus replication model. The nucleocapsid is thought to be transported into the nucleus through the nuclear pore complex and host components are thought to activate the nucleocapsids to initiate primary rounds of transcription to produce polyadenylated leader RNA and mRNAs for each of the viral proteins. The mRNAs are transported to the cytoplasm and translated, and the N, P and L proteins are transported through the nuclear pore complex into the nucleus. As the N, P, and L proteins increase in abundance, a switch occurs from primary transcription of mRNAs to a mode consisting of intermittent rounds of replication to produce antigenomic and genomic nucleocapsids, followed by secondary rounds of transcription to increase the pool of mRNAs. This phase of replication is regulated by a feedback mechanism that relies on the abundance of the core proteins for encapsidation of nascent leader RNAs. As replication progresses, the nuclei become greatly enlarged, and subnuclear viroplasms appear that consist of large masses of granular material that contain viral RNA and the N, P, and L proteins. Early during replication, some of the newly synthesized nucleocapsids are transported to the cytoplasm where they associate with movement proteins and are transported to other cells through the plasmodesmata. Late in replication, the M protein reaches sufficient concentration to coil the genomic nucleocapsids and mediate interactions with G protein patches at the inner nuclear membrane. During this process, virions undergo morphogenesis by budding through the inner nuclear membrane and accumulate in the perinuclear spaces. Reprinted, with permission, from the Annual Review of Phytopathology, Volume 43, © 2005 by Annual Reviews.

machinery to recognize karyophylic signals present on the N and P proteins to facilitate nucleocapsid entry into the nucleus. During the early stages of infection, the virion-associated polymerase is probably activated by host components to produce an active transcriptase that copies the genomic RNA into capped and polyadenylated mRNAs that are transported to the cytoplasm and translated. The translated N and P proteins are imported into the nucleus by separate mechanisms using host importin α and β proteins, respectively.

After entry into the nucleus, the N, P, and L proteins probably participate in multiple rounds of mRNA transcription, and antigenomic and genomic RNA replication. As replication proceeds, the viroplasms form discrete foci that appear near the periphery of dramatically enlarged nuclei. During the early stages of infection, small amounts of the nucleocapsids are postulated to be exported to the cytoplasm by interactions of the nuclear export signals on the P protein with host nuclear export receptor proteins. These exported nucleocapsids then interact with movement protein homologs to mediate transport through the plasmodesmata to adjacent cells. As infection progresses, the M protein accumulates in the nucleus and reaches concentrations sufficient to downregulate transcription and participate in coiling of minus-sense RNA nucleocapsid cores. The coiled cores then associate with G protein at sites on the inner nuclear envelope that are located in close proximity to the viroplasms. During budding, numerous enveloped virions accumulate in perinuclear spaces between the inner and outer nuclear envelope where they may be ingested during vector feeding.

A recent discovery that may shed new light on the processes of nucleorhabdovirus replication and maturation has been noted during infection of transgenic *N. benthamiana* plants that express green fluorescent protein (GFP) targeted to the ER. During infections with either SYNV or PYDV, a substantial proportion of the GFP appears to be redistributed to form spherules within the nuclei. In the case of SYNV, the spherules colocalized with foci formed by the N protein. A model to explain this phenomenon is that the spherules contacting the viroplasms in SYNV-infected plants are derived from the ER and become redistributed to the inner nuclear membrane to serve as sites for replication and virion maturation.

Vector Relationships, Distribution, and Evolution

Plant pathogenic rhabdoviruses are highly dependent on arthropod vectors for their distribution between plants. Although some plant rhabdoviruses have no known vector, most well-characterized members are transmitted by insects in which they also multiply, so it is possible that the plant rhabdoviruses radiated from a primitive arthropod. Plant rhabdoviruses are most commonly transmitted by aphids (Aphididae), leafhoppers (Cicadellidae), or planthoppers (Delphaccidae) (**Table 1**). An incompletely characterized putative rhabdovirus, beet leaf curl virus (BLCV), reportedly has a heteropteran beet leaf bug (*Piesma quadratum*) vector, but more extensive molecular and cytological analyses of this virus and other poorly characterized plant rhabdoviruses need to be carried out so their properties and vector relations can be clarified.

Vector–host relationships have profoundly affected plant rhabdovirus distribution and host range. For example, leafhoppers, planthoppers, and aphids are prevalent on both monocots and dicots, but the rhabdoviruses causing diseases of the Gramineae are all transmitted by leafhoppers or planthoppers. Except for PYDV and EMDV, which have leafhopper and planthopper vectors, respectively, dicot-infecting rhabdoviruses whose transmission has been investigated are transmitted by aphids. In all cases of insect transmission that have been carefully examined, rhabdoviruses are persistently transmitted in a propagative fashion, and in many cases, can be transmitted to vector progeny. Long latent periods are required before transmission occurs; insects often remain viruliferous throughout their lives, and transovarial passage has been observed through eggs and nymphs. In addition, strain-specific infection of tissue culture lines and explants combined with serological detection in vector cells provides unequivocal proof that rhabdoviruses replicate with high specificity in leafhopper and aphid vectors.

Several classical studies with PYDV in leafhoppers and SYVV in aphids, as well as recent studies with MMV in its planthopper vector *Peregrinus maidis*, have provided models for tissue-specific events in insect infection (**Figure 5**). After virus acquisition from plants, MMV initially accumulates in epithelial cells of the anterior part of the insect midgut and subsequently in nerve cells. Then, the virus appears in tracheal cells, hemocytes, muscles, and the salivary glands, and finally, in the fat cells, mycetocytes, and epidermal tissues. MMV infection is most extensive in the anterior portion of the gut, nerve cells, tracheal cells, and salivary glands of the planthopper. Based on this order of events, it is postulated that epithelial cells of the midgut are the first virus-entry sites,

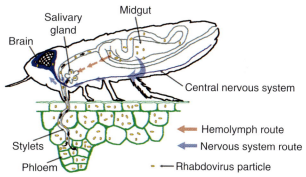

Figure 5 Events occurring during the infection cycle of rhabdoviruses in leafhopper vectors. Viruses are acquired during feeding on plant cells and move from the stylet to the midgut lumen of the digestive tract where they are hypothesized to invade epithelial cells by receptor-mediated endocytosis. From the epithelial cell layers, the virus moves into the nervous system, trachea, and hemolymph and spreads throughout the insect into the salivary glands and reproductive tissues. The salivary glands accumulate high levels of virus particles that are released by exocytosis, transported through the salivary canal, and transmitted to new host plants during subsequent feeding. Reproduced from figure 2a, Hogenhout SA, Redinbaugh MG, and Ammar E (2002) Plant and animal rhabdovirus host range: A bug's eye view. *Trends in Microbiology* 11: 264–271, with permission from Elsevier.

and that the virus quickly moves to the nervous system, trachea, and the hemolymph. From these tissues, MMV can be transmitted systemically to other tissues, including the salivary glands, which support high levels of virus accumulation.

MMV buds through the inner nuclear membrane in plant cells and has a similar pattern of morphogenesis in most cells of *P. maidis* tissues. However, MMV has also been observed to bud frequently from outer membranes in cells of nervous tissue and salivary glands of the planthopper, and similar observations have been made with other virus–vector combinations. Thus, the cellular budding site of MMV, and probably other rhabdoviruses, in insect hosts may be dependent on the cell type. Budding from outer cell membranes in salivary glands may be important because the process could allow release of virions into the saliva and permit introduction into plant cells during feeding.

Genetic experiments with PYDV have shown that highly efficient and inefficient leafhopper vectors can be selected. Continuous passage of PYDV by serial injection of insects can result in isolates that are unable to infect plants. Additional studies have shown that strains that have lost their capacity to be insect-transmitted can be recovered after protracted passage in plants. This phenomenon could provide a mechanism for evolution of vectorless rhabdoviruses, particularly in cases where infections were established in vegetatively propagated hosts.

Rhabdoviruses normally have the capacity to infect a greater range of plant hosts than the narrow range of species colonized by their insect vectors, because experimental host ranges usually can be extended considerably through mechanical transmission. In addition, plant rhabdoviruses have a wider insect host range than their natural insect vector hosts would indicate. For example, the majority of maize-feeding leafhoppers and planthoppers can acquire MFSV, but only one leafhopper, *Graminella nigrifrons*, can transmit this virus to maize. In addition, studies have shown that surrogate nonvectors of SCV injected with infected plant extracts can support replication and transmit virus to dicot hosts that do not support feeding by the native vector. Finally, cowpea protoplast infectivity experiments with the grass rhabdovirus FLSV and with SYNV show that both viruses are able to infect the legume protoplasts, but neither virus is able to infect cowpea plants. Together, these observations indicate that some plant rhabdoviruses have the ability to infect cells of several distantly related hosts, but that the natural host specificity is determined by (1) the insect vector feeding range; (2) the ability of the virus to move through the insect vector into the salivary glands and into the plant; and (3) systemic movement in the infected plant.

During evolution, plant rhabdoviruses faced two major challenges of a fundamentally different nature brought about by the necessity to alternately infect plants and insects. In each host, the virus must utilize different entry methods and accommodate distinct cellular and defense mechanisms. Rhabdovirus acquisition by the vector probably necessitates attachment to specific receptors at the surface of cells in the digestive system, followed by active invasion of the reproductive organs, fat bodies, and salivary glands. Very different barriers must be circumvented to establish systemic infections of plants. In order to establish a primary infection focus, the cell wall must first be breached by mouthparts of the insect, the virus must be introduced into the plant cell, where it uncoats and initiates the replication cycle. To establish systemic infections in plants, the virus must move from cell to cell through very small plasmodesmatal connections, enter the vascular system and spread throughout the plant. Therefore, plant rhabdoviruses are anticipated to have evolved a number of sophisticated mechanisms to circumvent the barriers to infection of their insect and plant hosts.

See also: Fish Rhabdoviruses.

Further Reading

Black LM (1979) Vector cell monolayers and plant viruses. *Advances in Virus Research* 25: 192–271.

Hogenhout SA, Redinbaugh MG, and Ammar E (2002) Plant and animal rhabdovirus host range: A bug's eye view. *Trends in Microbiology* 11: 264–271.

Jackson AO and Wagner JDO (1998) Procedures for plant rhabdovirus purification, polyribosome isolation, and replicase extraction. In: Foster G and Taylor S (eds.) *Plant Virology Protocols: From Virus Isolation to Transgenic Resistance*, vol. 81, ch. 7, pp. 77–97. Totowa, NJ: Humana Press.

Jackson AO, Dietzgen RG, Goodin MM, Bragg JN, and Deng M (2005) Plant Rhabdoviruses. *Annual Review of Phytopathology* 43: 623–660.

Krichevsky A, Kozlovsky SV, Gafney Y, and Citovsky V (2006) Nuclear import and export of plant virus proteins and genomes. *Molecular Plant Pathology* 7: 131–146.

Melcher U (2000) The 30K super family of viral movement proteins. *Journal of General Virology* 81: 257–266.

Sylvester ES and Richardson J (1989) Aphid-borne rhabdoviruses relationships with their vectors. In: Harris KF (ed.) *Advances in Disease Vector Research*, vol. 9, pp. 313–341. New York: Springer.

Tsai C-W, Redinbaugh MG, Willie KJ, Reed S, Goodin M, and Hogenhout SA (2005) Complete genome sequence and *in planta* subcellular localization of maize fine streak virus proteins. *Journal of Virology* 79: 5304–5314.

Tordo N, Benmansour A, Calisher C, *et al.* (2005) Family *Rhabdoviridae*. In: Fauquet CM, Mayo MA, Maniloff J, Desselberger U and Ball LA (eds.) *Virus Taxonomy: Eighth Report of the International Committee on Taxonomy of Viruses*, pp. 623–644. San Diego, CA: Elsevier Academic Press.

Plant Virus Diseases: Economic Aspects

G Loebenstein, Agricultural Research Organization, Bet Dagan, Israel

© 2008 Elsevier Ltd. All rights reserved.

Introduction

Estimates on crop losses due to pests and diseases are lacking, with respect to both up-to-date data and accuracy. Data from different countries obtained by a variety of methods should be taken only as a rough guide. Nevertheless, the available information sheds light on the tremendous loss to the food and fiber supply for the ever-growing world population.

A team of German crop scientists, with support of the European Crop Protection Association, published a comprehensive study of pest-induced crop losses on major food and cash crops. In the four principal crops (rice, wheat, barley, and maize) losses due to diseases ranged between 10% and 16% of potential production which translates into approx. US$64 × 10 billion in the years 1988–90. The losses increased to US$84 × 10 billion if four additional crops were included (potatoes, soybeans, cotton, and coffee) that together occupy half the world's cropland, with harvests worth US$300 billion in 1988–90. The study did not cover some important food crops of developing countries, such as cassava, millet, and sorghum, and horticultural crops, which are often heavily affected by virus diseases. The study found that pests and diseases accounted for pre-harvest losses of 42% of the potential value of output during 1988–90, with 15% attributable to insects and 13% each to weeds and pathogens. An additional 10% of the potential yield was lost post-harvest.

Yield losses due to plant virus infections are estimated to range between 10% and 15%, while losses due to fungal pathogens are estimated at 40–60%, the rest being caused by bacterial and phytoplasma pathogens and nematodes. Again, data should be viewed with caution, as many survey data are estimates based on educated guesses. On their face value these data do not agree with the previous data. Apparently, the selection of crops was different and fewer viral diseases were considered.

Yield losses caused by 11 different virus diseases in various crops, including bananas, cassava, cacao, maize, groundnuts varied from 0% up to 100%. In a survey conducted in eight African countries, cassava mosaic disease yield losses were estimated to reach 30–40%.

Economic losses are not only due to reduced vigor and growth resulting in yield losses, but may also affect product quality, as for example deformations in zucchini yellow mosaic virus (ZYMV)-infected squash or plum pox virus-infected peaches. In addition, major economic aspects to be considered are the costs to produce virus-tested propagation material, eradication programs, vector control, and breeding for resistance.

It seems that spread of viruses and subsequent economic losses are more prevalent in tropical and subtropical regions with continuous vegetation throughout the year. The continuous vegetation, annual and perennial, enables persistence of viruses and their vectors. In cooler areas winter temperatures are not favorable for annual vegetation and prevalence of vectors. Similarly, in regions with a dry hot summer, continuous herbaceous vegetation is interrupted which also reduces vectors and subsequent spread of plant viruses.

Economic losses due to viruses are often higher in vegetatively propagated crops such as citrus, cassava, and potato, than in seed-propagated ones. In many plants, viruses are not transmitted through seeds, a mechanism developed through evolution whereby the plant ensures that the next generation starts healthy. In vegetatively propagated crops, if cuttings, seed tubers, bud wood, bulbs, runners, etc. are taken from infected plants the virus will inevitably infect the next generation. This is in addition to spread of the virus by vectors, which may occur in both cases.

During the last decade whitefly transmitted viruses are on the increase in many crops, including vegetables, often becoming a major constraint.

This article is restricted to virus diseases in horticultural and plantation crops where virus diseases cause severe economic losses, many of them in developing countries.

Citrus Tristeza Virus

Citrus tristeza virus (CTV) is probably the most destructive citrus virus in the world. About 30 million trees on sour orange rootstocks were lost in Brazil and Argentina in the years 1940–60, 6.6 million trees in Venezuela in the 1980s, and an estimated 10 million trees in Florida and other Caribbean Basin countries. In these regions CTV was and is still being spread by the brown citrus aphid *Toxoptera citricida*. In Spain CTV killed about 10 million trees and several million more were lost in California, Israel, and other areas. In these areas the melon aphid, *Aphis gossypii*, is spreading CTV. When both vectors are present, as in Florida, spread of CTV is greatly enhanced. Yield losses have been documented in several places, for example, in Jamaica. At two locations totaling 1159 acres, losses over 5 years were more than US$4 million. These costs did not include the costs for removing dead trees and replanting. However, even though millions of trees in Brazil and Spain were lost it did not affect production of citrus in both countries in the long run. Thus, in Brazil production increased from 1.7 million t in 1960 to more than 20 million t in the 1990s of the previous century. Apparently, replanting new orchards on tolerant rootstocks, together with budwood control, and improved horticultural technologies made it possible to overcome the severe damages by CTV in the old plantations.

Various strains or isolates of CTV are known. Mild strains cause little if any symptoms on most commercial citrus and stem pitting on Mexican lime while severe strains cause decline and death on citrus grafted on sour orange rootstocks; seedling yellow strains cause stunting and yellowing of leaves while orange and grapefruit stem pitting strains will stem-pit sweet orange varieties and grapefruit, respectively. Stem pitting isolates are the most severe strains of CTV and cannot be controlled by the use of tolerant rootstocks.

CTV is a member of the monopartite genus *Closterovirus*, with a positive single-stranded RNA genome encapsidated in a flexuous particle about 2000 nm in length. The virus can be detected by ELISA or by biological indexing on Mexican lime.

The best way to control CTV and other graft-transmissible diseases of citrus is by a mandatory certification program. This program includes a 'clean stock program' for production and maintenance of pathogen-free propagation stock, and 'certification programs' for maintenance and distribution of virus-free material for commercial use. These publicly operated programs are generally self-supportive. Thus, in Florida a fee of US$2 is charged for registration of validated or parent trees.

Cross-protection with mild CTV strains is widely practiced in Brazil, Australia, and South Africa. The control of CTV strains causing stem pitting on sweet orange and grapefruit in Brazil is, by far, the largest and most successful use of cross-protection. By 1980, over 8 million trees of Pera sweet orange were cross-protected. Cross-protection to control CTV in Pera sweet orange in Brazil is still practiced.

An eradication program for tristeza is in operation in California. During 2000–01 the Central California Tristeza Eradication Agency surveyed approximately 31 400 acres and conducted up to 375 000 individual ELISA assays. This program delayed undoubtedly the spread of the disease but is expensive since it also includes the costs of removing infected trees. In Israel a tristeza suppression program was in operation between 1970 and 1977, whereby 300 000 tests were made using indicator plants and electron microscopy and 1700 CTV-infected trees were detected and removed. Between 1979 and 1980, 1.25 million trees were indexed by ELISA and only 0.13% of the tested trees were found to be infected. Apparently the eradication program delayed the outbreak of a tristeza pandemic in Israel by at least 15–20 years.

Cassava Mosaic Disease

Cassava is a low-cost carbohydrate food predominantly grown by resource-poor farmers in developing countries. Virus diseases seriously impede cassava production, especially in Africa, where it is grown in diverse climates, from coastal regions to semiarid zones of the Sahel. Virus diseases are regarded as the major constraint on cassava production.

Several similar but distinct whitefly transmitted geminiviruses cause cassava mosaic disease (CMD) in Africa. They can occur singly or in combination. The most important of them are African cassava mosaic virus (ACMV) and East African cassava mosaic virus (EACMV), ascribed to the genus *Begomovirus*. From 1988 to the present, a major pandemic of an unusually severe form of CMD is spreading throughout East and Central Africa, causing massive losses and affecting food supplies. A distinct strain of EACMV – the Ugandan strain – is the most significant threat to cassava production in Africa.

Yield losses with individual cassava cultivars in different African countries range from 20% to 95%, and continent-wide losses have been estimated at 12–23 million t of fresh tuberous roots per year, worth about US$1200–2300 million. Yield losses in Uganda at the peak of the recent EACMV-pandemic were immense and farmers had to abandon cultivation of cassava in the worst affected areas. It was assumed that each year an area of *c.* 60 000 ha of cassava yielding 600 000 t worth US$60 000 was lost due to CMD.

Symptoms of CMD are mainly green or yellow mosaic. Severe chlorosis is often associated with premature leaf abscission.

CMD is transmitted by the whitefly *Bemisia tabaci*. Virus dissemination between fields and over long distances is mainly through the use of infected stem cuttings as planting material. The disease is caused by several geminiviruses.

CMD can be detected by serological gel-diffusion tests, different ELISA and immunosorbent electron microscopy. Monoclonal antibodies enable cassava mosaic virus isolates to be differentiated though none exists so far for detecting specifically the Ugandan strain of EACMV. Polymerase chain reaction (PCR) with specific nucleotide sequence primers however detects all different cassava mosaic viruses, including EACMV.

For control of CMD, the main emphasis has been on the development of resistant varieties. Phytosanitation, involving the selection of cuttings from healthy plants, rouging of diseased plants, and production of virus-free tissue-cultured planting material are also practiced. Planting date can be adjusted to avoid times when populations of whiteflies are high. Use of insecticides to restrict spread of CMD by controlling the whitefly vector was not effective.

Virus Diseases of Potato

Virus diseases are a major constraint in potato production and often reduce yields by more than 50%. Due to their economic importance much research was devoted to them in the first half of the twentieth century, and new techniques such as serology, meristem cultures, and electron microscopy were developed to control them.

At least 37 viruses naturally infect cultivated potatoes, with potato virus Y (PVY), potato leafroll virus (PLRV), potato virus S (PVS), potato virus M (PVM), potato virus X (PVX), potato aucuba mosaic virus (PAMV), and potato mop-top virus (PMTV) being the most widespread and damaging ones. In addition, potato spindle tuber viroid (PSTV) and phytoplasma diseases are also causing major crop losses in potato.

PLRV is of great economic importance. It is widespread worldwide and in plants grown from infected tubers (secondary infection) yields may be reduced by 33–50%. Tubers from infected plants are small to medium sized. Even greater losses are observed when PLRV occurs together with PVX or PVY, reaching 40–70%. Yield losses due to PVY may reach 10–80%, especially if the virus occurs with PVX. A severe group of PVY strains designated PVY^{NTN} present in Europe, Japan, and the US cause a damaging disease in which tubers develop superficial rings that later are sunken and necrotic. They often become more conspicuous during storage, and can affect 90% of tubers of susceptible cultivars. At present there is no serological or other assay, which can distinguish between PVY^{NTN} and other PVY strains.

Yields of a crop in which all plants are secondarily infected with PVX and PVS will be 5–15% lower than normal. PVS and PVM are common in Eastern European countries. PVM can result in tuber yield losses of 40–75%. PVM occurs in complex with other viruses (especially PVS). PVS isolates may reduce tuber yield by 3–20%. Plants from infected stocks that have been propagated for several generations may suffer more severe losses. Thus, for example, potato yields in Kazakhstan are extremely low, averaging $9 t ha^{-1}$ during the years 1993–95. In a pilot scheme conducted by the Agricultural Institute of Astana, Kazakhstan, within a USAID project, microtubers obtained from meristems of virus-tested potatoes were used and resulted in a low infection rate in the elite seed tubers. These elite seed gave an increase in yield of about 90% when compared with yields of commercial fields in the vicinity.

Yields in former Soviet Union countries are low, averaging $11.26 t ha^{-1}$ during 1993–95 as there is no reliable system for providing certified tuber seeds, and farmers use seeds from their own fields. Low yields are to a great extent due to virus diseases carried over in the planting material. Thus, Loshitsky and Belorusky 3 N cultivars developed by the Belarusian Research Institute for Potato Growing (BRIP) yielded $30.0–50.0 t ha^{-1}$ in experimental fields, while Belarus' average potato yields range between 12.2 and $16.3 t ha^{-1}$. Supposing that only 20% of this difference is due to virus infection in the planting material, and extrapolating this over the total area of potato in the former Soviet Union countries of about 6 million ha, producing on an average 70 million t of potato an annual loss of about 18 million t seems to be a conservative estimate.

Costs of seed potato certification schemes should also be considered. The aim of these schemes is to produce true-to-type disease-free (mainly from viruses) potato seed tubers. These schemes nowadays often include *in vitro*-derived plantlets for the initial increase of pathogen-tested clonal selections, including production of minitubers. Prices for certified tuber seeds range between US$350 and 500 per ton while prices for ware potatoes range between US$170 and 180 per ton. If about 3 t seed potato is needed to plant 1 ha, this amounts to an additional expense of about US$450 when certified seeds are used compared to the farmer using his own crop for next year's planting. In fact the difference is presumably higher since the farmer might use the small, not marketable tubers for planting.

Virus Diseases of Sweet Potato

Sweet potato, *Ipomoea batatas*, is the seventh most important food crop in the world in terms of production. They are grown on about 9 million ha, yielding ∼140 million t, with an average yield of about $15 t ha^{-1}$. They are mainly grown in developing countries, which account for over 95% of world output. Sweet potato is a 'poor man's crop', with most of the production done on a small or subsistence level. Sweet potato produces more biomass and nutrients per hectare than any other food crop in the world. It is well suited to survive in fertile tropical soils and to produce tubers without fertilizers and irrigation and is one of the crops with an important role in famine relief. Thus, for example, across East Africa's semiarid, densely populated plains, thousands of villages depend on sweet potato for food supply. The Japanese used it when typhoons demolished their rice fields. Yields differ greatly in different areas or even fields in the same region. Thus, the average yield in African countries is about $4.7 t ha^{-1}$, with yields of 8.9, 4.3, 2.6, and $6.5 t ha^{-1}$ in Kenya, Uganda, Sierra Leone, and Nigeria, respectively. The yields in Asia are significantly higher, averaging $18.5 t ha^{-1}$. China, Japan, Korea, Thailand, and Israel have the highest yields with about 20, 24.7, 20.9, 12, and $33.3 t ha^{-1}$, respectively. In South America the average yield is $12.2 t ha^{-1}$, with Argentina, Peru, and Uruguay in the lead with 18, 11, and $10 t ha^{-1}$, respectively. For comparison, the average yield in the US is $16.3 t ha^{-1}$.

These differences in yields are mainly due to variation in quality of the propagation material. Sweet potatoes are vegetatively propagated from vines, root slips (sprouts) or tubers, and farmers in African and other countries often take vines for propagation from their own fields year after year. Thus, if virus diseases are present in the field they will inevitably be transmitted with the propagation material, resulting in a decreased yield. Often these fields are infected with several viruses, thereby compounding the effect on yields. In China, on average, losses of over 20% due to sweet potato virus diseases are observed mainly due to sweet potato feathery mottle virus (SPFMV) and sweet potato latent virus (SwPLV). The infection rate in the Shandong province reaches 5–41%. In countries where care is taken to provide virus-tested planting material, for instance in the USA and Israel, markedly higher yields of 16.3 and 30 t ha^{-1}, respectively, are obtained.

The sweet potato virus diseases (SPVDs) caused by the interaction of SPFMV and sweet potato chlorotic stunt virus (SPCSV) or sweet potato sunken vein virus (SPSVV) (possibly a variant of SPCSV) is the most important virus (complex) disease in East Africa. It can cause losses of 50–80%, especially in Uganda and Kenya, though in another study from Uganda losses were much smaller probably due to relatively high levels of virus resistance in their varieties. In a 3-year field study in Cameroon, SPVD reduced root yields by 56–90% in susceptible varieties. Yield reductions of 78% due to SPVD have been reported from field trials in Nigeria. In Israel in a 2-year field experiment, yield reductions of ~50% were observed in plots planted with SPVD-infected cuttings while infection by SPFMV or SPSVV alone were minor. Cucumber mosaic virus (CMV) can cause a complete failure of the crop when infected sweet potatoes carry SPSVV. However, CMV is unable to infect healthy sweet potato.

At present the best way to control virus diseases in sweet potato is to supply the grower with virus-indexed propagation material. Such programs are operating in Israel and in the Shandong province of China. In Israel, as a result of planting virus-tested material, yields increased at least by 100%, while in China increases ranged between 22% and 92%. The payoff to the farmer has been high and in Israel use of certified material, prepared by special nurseries, is common practice, costing ~3.0 US cents per cutting. Farmers buys about 30% of material needed to plant their fields and fill in the rest from cutting of vines grown from these plants. Yields are high and stable when virus-tested propagation material is used. In African countries such programs are operating only on a limited scale, because sweet potatoes are grown mainly as a food security crop, and not as a commercial one.

Breeding programs might be a future answer and such programs are in operation in Uganda, providing SPVD resistance. It will have to be seen if these improved cultivars will retain their resistance. Thus, several clones obtained from the International Potato Center (CIP) that were claimed to be resistant to SPFMV were found to be susceptible when Israeli and Ugandan virus isolates were tested.

Cacao Swollen Shoot Disease

The disease occurs in West Africa and was first reported in Ghana in 1936. Cacao swollen shoot disease is one of the most economically important viral diseases and has contributed to the drastic decline of cocoa production in Ghana. Since 1936, nearly 200 million cacao trees have been cut out from about 130 000 ha. From 1945 to 1950 yield of beans fell by 86% almost proportional to the killing of trees. In 1953 the 50 million infected trees represented a capital depreciation of £25 million since the discovery of the disease.

Cacao swollen shoot disease virus (CSSV) belongs to the genus *Badnavirus* and has bacilliform virions of 113 nm length and 28 nm width. A range of mealybug species, with *Planococcoides njalensis* being the most important one, transmit CSSV.

Attempts were made to control the disease by use of insecticides and biological control of the mealybugs, breeding for resistant cocoa cultivars, and protection with mild strains, but the main practice is by roguing of infected and their neighboring trees.

The eradication policy of infected trees enforced by the British colonial authorities in the Gold Coast (now Ghana) in the late 1940s met with serious opposition and was a major factor contributing to the rise of the political party led by Kwame Nkruma, who accused the authorities of attempting to destroy farmers' income.

Whitefly-Transmitted Geminiviruses in Tomato

These viruses such as tomato yellow leaf curl virus (TYLCV) have become a limiting factor in tomato production. In the USA whitefly-transmitted geminiviruses (WTGs) appeared in the early 1990s and losses are estimated to reach 20%, while in Cuba, Mexico, Guatemala, Costa Rica, and Brazil yield losses ranged between 30% and 100%. Losses in the Dominican Republic during 1989–95 were estimated at US$50 million.

Chemical control of the whitefly vector is a possibility, but the large whitefly populations make this option inefficient. Physical barriers such as greenhouses protected with fine net (50-mesh) screens are effective while 'floating barriers' of perforated polyethylene sheets stretched over tomato fields prevent the landing of whiteflies. These measures, however, increase the costs of production markedly. The best approach to control WTG is through the use of resistant or tolerant cultivars. Breeding programs are in progress, but so far no resistant or tolerant cultivars have been obtained.

Plum Pox Virus Sharka Disease of Stone Fruits

Plum pox symptoms were first observed in plums in Bulgaria between 1915 and 1918. Between 1932 and 1960 the disease moved north and east from Bulgaria into Yugoslavia, Hungary, Romania, Albania, Czechoslovakia, Germany, and Russia. The disease was observed mainly in plums and apricots and since the 1980s also in peaches.

Virus infection can cause considerable losses. About 100 million stone fruit trees in Europe are currently infected, and susceptible cultivars can result in 80–100% yield losses. In Eastern and Central Europe, sensitive plum varieties can exhibit premature fruit drop and bark splitting. Some sweet cherry fruits develop chlorotic and necrotic rings and premature fruit drop.

Plum pox does not kill trees but it makes the fruit unmarketable and reduces yield by 20–30%. The fruits drop before maturity and are unfit for use as they are bitter and unsweet. In Bulgaria, losses in 1968 were estimated to be 30 000 t. In the last three decades, the average fruit yield in the Czech Republic dropped by 80%, and the total number of plum trees was reduced from 18 million to 4 million. Emilia-Romagna region of Italy, between 1998 and 2002, 69 000 stone fruit trees were removed due to Sharka infection; and US$450 000 is being spent annually for their removal and replanting. In Spain since 1988, 1.5 million trees have been removed at a cost of US$17 million. In France, plum pox virus (PPV) is mainly present in the southeast part of the country. All trees showing symptoms are eliminated and if infection levels are above 10%, the whole orchard is destroyed. These methods have resulted in the destruction of about 27 000 trees in 1992 (of which 550 were found to be infected) and in 1993 100 ha (mainly peach) were eliminated. From 1973 to 1990, it is estimated that 91 000 trees were destroyed in France.

PPV is a potyvirus transmitted by aphids in a nonpersistent manner. However much of the spread in Europe can be attributed to movement of infected nursery material from the Balkans.

See also: Emerging and Reemerging Virus Diseases of Plants; Plant Resistance to Viruses: Engineered Resistance; Plant Resistance to Viruses: Natural Resistance Associated with Dominant Genes; Plant Virus Diseases: Fruit Trees and Grapevine; Plant Virus Vectors (Gene Expression Systems); Vector Transmission of Plant Viruses.

Further Reading

Loebenstein G and Thottappilly G (2003) *Virus and Virus-Like Diseases of Major Crops in Developing Countries.* The Netherlands: Kluwer Academic Publishers.

Orke EC, Dehne HW, Schonbeck F, and Weber A (1994) *Crop Production and Crop Protection: Estimated Losses in Major Food and Cash Crops.* Amsterdam: Elsevier.

Van der Zaag DE (1987) Yield reduction in relation to virus infection. In: de Bolas JA and van der Want JPH (eds.) *Viruses of Potatoes and Seed-potato Production*, pp. 146–150. Wageningen, The Netherlands: Pudoc.

Waterworth HE and Hadidi A (1998) Economic losses due to plant diseases. In: Hadidi A, Khetarpal RH, and Koganezawa H (eds.) *Plant Virus Disease Control*, pp. 1–13. St. Paul: American Pythopathological Society Press.

Relevant Website

http://faostat.fao.org – FAOSTAT, Food and Agriculture Organization of the United Nations (FAO).

Plant Virus Diseases: Fruit Trees and Grapevine

G P Martelli, Università degli Studi and Istituto di Virologia vegetale CNR, Bari, Italy
J K Uyemoto, University of California, Davis, CA, USA

© 2008 Elsevier Ltd. All rights reserved.

Introduction

Grapevine (*Vitis* spp.), rosaceous fruit tree (stone and pome fruits), and nut crop (walnut, hazelnut) varieties are propagated by grafting scions onto rootstocks which, in turn, are either clonally propagated or derive from seedlings. Both members (scion and rootstock) of the grafted plant are therefore liable to carry viruses if they come from infected sources, or can be infected in the field. Remarkably, the causal agents of a number of diseases of these different plant species share the same epidemiological behavior and/or taxonomic position as, for example, soil-borne nepoviruses and pollen-borne ilarviruses and cherry leafroll virus (CLRV).

Virus Diseases of Grapevines

Disease Symptoms and Yield Losses

There are four major virus diseases of grapevines, that is, infectious degeneration-decline, leafroll, rugose wood, and fleck, which differ symptomatologically and in the type of the causal agents.

Infectious degeneration affects European grapes (*Vitis vinifera*) and American rootstocks and is characterized by two distinct syndromes, 'infectious malformations' and 'yellow mosaic' caused by distorting and chromogenic virus strains, respectively. Leaves and shoots of vines infected by distorting virus strains are more or less severely malformed (**Figure 1(a)**), bunches are smaller and fewer than normal, and berries ripen irregularly, are small-sized, and set poorly. Yellow mosaic-affected vines show bright chrome yellow discolorations that may affect leaves (**Figure 1(g)**) shoots, tendrils, and inflorescences. Leaves and shoots show little malformation, but bunches are small and few. Symptoms of 'decline' resemble those of 'infectious malformations' but affected European grape varieties decline and may die. Crop losses due to these diseases can exceed 60–70%.

Leafroll elicits an early discoloration of the interveinal tissues of leaves of infected European vines which turn purple red in red-berried cultivars (**Figure 1(i)**) and yellowish in white-berried cultivars; both interveinal tissue colorations develop against a background of green primary veins. Discolorations are usually accompanied by downward rolling of the leaf margins and thickening of the blade. In white-berried cultivars of *V. vinifera*, the symptoms are similar, but the leaves become chlorotic to yellowish, instead of reddish. Bunches mature late and irregularly. Yield is decreased by 15–20% in average and rooting ability, graft take, and plant vigor are adversely affected as well as the quality of grapes and musts (sugar and protein content, aromatic profile, soluble solids, titratable acidity). Infection of American *Vitis* species and rootstock hybrids is symptomless, except for a variable decrease in vigor.

Rugose wood is a complex disease in which four different syndromes are recognized, that is, Rupestris stem pitting (RSP), Kober stem grooving (KSG), Corky bark (CB), and LN33 stem grooving (LNSG). Infected vines are less vigorous than normal, may show delayed bud opening in spring and a swelling above the bud union, which reflects a marked difference between the relative diameter of scion and rootstock. The bark above the graft union may be exceedingly thick and corky. Some vines decline and die within a few years from planting. The woody cylinder is marked by pits and/or grooves that can

Figure 1 Symptoms induced by: (a) Distorting strain of grapevine fanleaf virus (grapevine). (b) Double infections by apple mosaic virus (yellow banding) and prunus necrotic ringspot virus (shredding) (cherry). (c) American plum line pattern virus (plum). (d) Cherry rasp leaf virus (cherry). (e) Apple stem pitting virus (pear). (f) Apple stem grooving virus (Virginia crab). (g) Chromogenic strain of grapevine fanleaf virus (grapevine). (h) Plum pox virus (apricot). (i) Grapevine leafroll disease (grapevine). (j) Grapevine rugose wood disease (grapevine). (e, f) Courtesy of L. Giunchedi.

show on the scion, the rootstock (**Figure 1(j)**), or both, their severity varying with the scion/stock combination. Under cool and wet climates wood symptoms are milder or absent. No specific symptoms are seen on the foliage, although certain cultivars show rolling, yellowing, or reddening of the leaves similar to those induced by leafroll. Bunches may be fewer and smaller than normal and the crop is reduced on average by 20–30%.

Fleck is a complex consisting of several diseases ('fleck', 'asteroid mosaic', 'rupestris necrosis', and 'rupestris vein feathering') and viruses (grapevine redglobe virus, GRGV) that cause latent or semilatent infections in *V. vinifera* and most American *Vitis* species and rootstock hybrids. Only *Vitis rupestris* reacts to the different diseases with differential symptoms. Although the elusive nature of the complex hinders the assessment of its economic impact, adverse influence on vigor, rooting ability of rootstocks, and on graft take have been reported.

Geographical Distribution

All diseases are ubiquitous. There is not a single viticultural country where surveys were carried out in which one or more of the viruses involved in their etiology have not been found.

Causal Agents and Classification

Several different viruses are involved in the etiology of, or are associated with each single disease.

Infectious degeneration and decline are two diseases caused by nematode-borne viruses with isometric particles c. 30 nm in diameter and bipartite RNA genome 2.2–2.8 kb (RNA-1), and 1.5–2.4 kb (RNA-2) in size. Of the 16 viruses recovered from infected grapevines, 15 belong to different species of the genus *Nepovirus* (family *Comoviridae*): (1) subgroup A: *Arabis mosaic virus* (virus: ArMV), *Grapevine deformation virus* (GDefV), *Grapevine fanleaf virus* (GFLV), *Raspberry ringspot virus* (RpRSV), *Tobacco ringspot virus* (TRSV); (2) subgroup B: *Artichoke Italian latent virus* (AILV), *Grapevine Anatolian ringspot virus* (GARSV), *Grapevine chrome mosaic virus* (GCMV), *Tomato black ring virus* (TBRV); (3) subgroup C: *Blueberry leaf mottle virus* (BLMoV), CLRV, *Grapevine Bulgarian latent virus* (GBLV), *Grapevine Tunisian ringspot virus* (GTRV), *Peach rosette mosaic virus* (PRMV), *tomato ringspot virus* (ToRSV). The species *Strawberry latent ringspot virus* (SLRSV) is a member of the genus *Sadwavirus*. Infectious degeneration is caused by European nepoviruses (ArMV, GFLV, GCMV, RpRV, TBRV) and SLRSV, whereas the agents of decline disease are American nepoviruses (BLMoV, PRMV, TRSV, ToRSV).

The agents of leafroll are filamentous viruses with a monopartite RNA genome and a length of 1400–2200 nm, classified in the family *Closteroviridae*. Grapevine leafroll-associated virus 2 (GLRaV-2), the only representative of the genus *Closterovirus*, has a genome 15 528 nt in size. GLRaV-1 (17 647 nt), GLRaV-3 (17 917 nt), and GLRaV-5 are members of the genus *Ampelovirus*. GLRaV-4, GLRaV-6, GLRaV-8, and GLRaV-9 are tentative species in the same genus, whereas GLRaV-7 is unassigned to the family.

Viruses of the rugose wood complex have filamentous particles 730–800 nm long, encapsidating a monopartite RNA genome, and are classified in two genera of the family *Flexiviridae*. The genus *Vitivirus* comprises grapevine virus A (GVA) (7349 nt), the putative agent of 'Kober stem grooving', grapevine virus B (GVB) (7599 nt) and grapevine virus D (GVD); these two viruses may be involved in the etiology of 'Corky bark' rupestris stem pitting-associated virus (RSPaV) (8726 nt), a member of the genus *Foveavirus*, is the putative agent of the homonymous disease.

The family *Tymoviridae* comprises all agents associated with the fleck complex, that is, viruses with isometric particles c. 30 nm in diameter showing a prominent surface structure and a monopartite RNA genome. Grapevine fleck virus (GFkV, 7564 nt), the causal agent of fleck, and GRGV are both members of the genus *Maculavirus*, whereas grapevine asteroid mosaic-associated virus (GAMaV) and grapevine vein feathering virus (GVFV) are tentative members of the genus *Marafivirus*.

Transmission

All diseases are graft transmissible and persist in the propagating material which is responsible for their long-distance dissemination. Spread at a site varies with the disease.

Infectious degeneration and decline are two soil-borne diseases. Dorylaimoid nematodes have been identified as vectors of some of their agents. In particular, GFLV is transmitted by *Xiphinema index*; ArMV and SLRSV by *X. diversicaudatum*; ToRSV by *X. americanum sensu stricto*, *X. rivesi*, and *X. californicum*; PRMV by *X. americanum sensu lato* and *Longidorus diadecturus*; TRSV by *X. americanum sensu stricto*; TBRV by *L. attenuatus*; and RpRSV by *Paralongidorus maximus*. Seed transmission has been reported for PRMV, ToRSV, BLMoV, GCMV, and GFLV; however, grapevine is not seed-propagated in commercial viticultural practices.

Some of the viruses of the leafroll and rugose wood complexes are transmitted in a nonspecific semipersistent manner by pseudococcid mealybugs (i.e., *Planococcus ficus* (GVA, GVB, GLRaV-3), *Pl. citri* (GVA, GLRaV-3), *Pseudococcus longispinus* (GVA, GVB, GLRaV-3, GLRaV-5, GLRaV-9), *Ps. affinis* (GVA, GVB, GLRaV-3), *Ps. comstocki* (GVA, GLRaV-3), *Ps. calceolariae* (GLRaV-3), *Ps. viburni* (GLRaV-3), *Ps. maritimus* (GLRaV-3), *Heliococcus bohemicus* (GVA, GLRaV-1, GLRaV-3), *Phenacoccus aceris* (GLRaV-1)) and by soft scale insects (i.e., *Neopulvinaria innumerabilis*

(GVA, GLRaV-1, GLRaV-3), *Pulvinaria vitis* (GLRaV-1, GLRaV-3), and *Parthenolecanium corni* (GLRaV-1)). None of these viruses are seed-borne, contrary to RSPaV, which is pollen-borne and transmitted through seeds. None of the viruses of the fleck complex have known vectors.

Control

Use of clonally selected and sanitized propagative material is the best preventive method currently available for controlling all diseases. Heat therapy, meristem tip culture, and somatic embryogenesis are effective, though to different extents, for the elimination of parenchyma (nepoviruses) and of phloem-restricted (closteroviruses, vitiviruses, foveaviruses, maculaviruses) viruses. Restraining field re-infection of virus-free stocks is, however, difficult because no ultimate control of soil- or airborne vectors is possible. Attempts for introduction of transgenic resistance to some nepoviruses (GFLV, ArMV, GCMV), vitiviruses (GVA and GVB), and closteroviruses (GLRaV-2 and GLRaV-3) are in progress.

Virus Diseases of Stone Fruits

Disease Symptoms and Yield Losses

Stone fruits are affected by several diseases caused by recognized viruses and by unidentified graft-transmissible pathogens which, for brevity sake, are not addressed in this article.

Symptoms of virus infections depend on the pathogen, the host plant, and the environmental conditions. Variants of prunus necrotic ringspot virus (PNRSV) cause different diseases such as calico or bud failure in almond, necrotic ringspot or rugose mosaic in cherry, tatter leaf in cherry (**Figure 1(b)**) and peach, and mule's ear in peach. Variants of prune dwarf virus (PDV) cause yellows in sour cherry, blind wood and narrow leaf in sweet cherry, and gummosis in apricot. Co-infections by PNRSV and PDV cause peach stunt. Peach yields are drastically reduced (30–60%). Other viruses inducing yellow line pattern or vein netting symptoms in plum are Danish plum line pattern, a strain of PNRSV, European plum line pattern, a strain of apple mosaic virus (ApMV), and American plum line pattern virus (APLPV) (**Figure 1(c)**). In Mediterranean countries, mosaic symptoms in almond trees involve PNRSV, PDV, and/or ApMV.

Sharka, induced by plum pox virus (PPV), is characterized by chlorotic/necrotic ring pattern or mottling of the leaves, color breaks of flower petals, distortion of fruit shape, and rings or blotches of their surface (**Figure 1(h)**). Infected trees drop fruit prematurely. Because of this and fruit alterations, the crop can be completely lost. Peach mosaic virus (PcMV) and cherry mottle leaf virus (CMLV) cause delayed bud-break, stunted shoot growth, leaves with chlorotic spots and vein feathering and deformed rough skinned fruits (PcMV) or chlorotic spots over an uneven leaf surface with shredded leaf margins (CMLV). Cherry is resistant to PcMV, while CMLV causes latent infections in peach. Little cherry disease caused by little cherry virus 2 (LChV-2) affects shape and size of the fruits which lack deep color in dark varieties, and are flavorless. Canopy is light green and tree growth reduced. Black Beaut plums (*Prunus salicina*) infected by plum bark necrosis-stem pitting associated virus (PBNSPaV) are short-lived, exhibit gummosis and necrotic bark on branches and trunks, and severe stem pits.

Stem pitting is induced by different ToRSV strains in *Prunus*. In early spring, canopies of affected trees appear light green, progressing to general chlorosis, leaves are drooping and drop prematurely. The basal portion of trunks develops thick, spongy textured bark and pits and grooves on the woody cylinder, and trees decline. ToRSV-infected prune trees with a brown line at the graft union exhibit poor growth, sparse canopy, and in chronic infections develop limb dieback. PRMV causes rosettes of shoots with shortened internodes. Affected trees are dark green, stunted, and fruitless.

Other nematode-borne stone fruit diseases are: apricot bare twig and unfruitfulness caused by SLRSV, decline and death by cherry rasp leaf virus (CRLV), which also elicits the formation of enations on the underside of leaves (**Figure 1(d)**). In Europe, sweet cherry trees with rasp leaf disease are infected by ArMV, CLRV, raspberry ringspot virus (RpRSV), or SLRSV either alone or in combination with PDV. CLRV can also cause rapid decline of infected cherry trees on Colt rootstock. In Germany, TBRV-infected peach trees develop chlorotic spots and distortion of the leaves, dieback of scaffold branches, and tree death. Infections in peach by SLRSV are associated with willow leaf rosette (Italy), court-noué (France), and shoot dwarfing (Germany). In plum, RpRSV infections are associated with mild symptoms, but combined with PPV symptoms are severe and infected trees decline (Poland). Myrobalan latent ringspot virus (MyLRSV) induces stunting, sparse canopy, and decline in Myrobalan B plum rootstocks (France). Recently, a new putative cheravirus named stocky prune virus (StPV) was detected in stunted, low-yielding trees in France.

Apple chlorotic leaf spot virus (ACLSV) causes dark green, sunken mottle in leaves of the peach cv. Okinawa, is associated with a fruit disorder known as pseudopox in apricot and plum, and, in cherry, it induces fruit necrosis or fruit distortion when in combination with PNRSV. Bark split in plum and graft-incompatibility in apricot and peach have also been associated with infections by different ACLSV strains.

Apricot latent virus (ApLV), latent in apricot, elicits symptoms of asteroid spot disease in peach. Cherry green ring mottle virus (CGRMV) severely impacts sour cherry and Japanese flowering cherry trees, whereas cherry virus A (CVA) is always found in co-infections with other viruses; thus, specific symptoms have yet to be defined.

Geographical Distribution

Due also to symptomless infections in different *Prunus* species, ApMV, PDV, PNRSV, ACLSV, CGRMV, and, presumably, CVA and CLRV have now a worldwide distribution. PPV occurs through most of Europe, the Mediterranean region, and, more recently, became established in Chile, USA (Pennsylvania), Canada (Ontario), and India. PcMV is endemic in southwestern United States and in Northern Mexico. LChV-2 is present in North America, Europe, and Japan. PBNSPaV was identified in USA (California), Italy, Turkey, Morocco, Jordan, and Serbia. ToRSV and CRLV are endemic in North America and PRMV is restricted to Michigan (USA) and neighboring Ontario (Canada).

Causal Agents and Classification

Causal agents of stone fruit virus diseases belong to a wide array of genera and families, the properties of some of which, that is, the genera *Nepovirus* (*Comoviridae*), *Ampelovirus* (*Closteroviridae*), and *Foveavirus* (*Flexiviridae*), have been outlined above. This applies to the nepoviruses ToRSV, PRMV, ArMV, CLRV, RpRSV and TBRV, the ampeloviruses LChV-2 and PBNSPaV, and the foveaviruses ApLV and CGRMV.

ApMV, PNRSV, PDV, and APLPV (genus *Ilarvirus*, family *Bromoviridae*) have quasi-isometric particles 25–35 nm in diameter and a tripartite RNA genome 1.0–1.3 kb (RNA-1), 1.0–1.2 kb (RNA-2), and 0.7–0.9 kb (RNA-3) in size. CRLV, the type species of the recently established genus *Cheravirus*, has isometric particles c. 30 nm in diameter and a bipartite RNA genome 7030 nt (RNA-1) and 3315 nt (RNA-2) in size. ACLSV, PcMV, and CMLV (genus *Trichovirus*) have filamentous particles 720–760 nm in length and a 2.2–2.4 kb monopartite RNA genome. The capillovirus CVA has filamentous particles of undetermined length and a monopartite RNA genome 7383 nt in size while PPV, the only potyvirus known to infect stone fruits, has also filamentous particles c. 750 nm long, encapsidating a monopartite RNA genome 9741 nt in size.

Transmission

Ilarviruses and CLRV are pollen- and seed-borne, giving rise to infected seedlings. Horizontal spread of PNRSV and PDV requires also the sequential activities of honeybees (*Apis mellifera*) and thrips (*Thrips tabaci* or *Frankliniella occidentalis*). Spread via root grafts between neighboring trees have been reported in sweet cherry orchards in Washington (USA) for CLRV and in nurseries for ApMV. Several aphid species transmit PPV in a nonpersistent manner to account for localized spread, and movement in infected stocks for long-distance spread. The apple mealybug *P. aceris* vectors LChV-2. Similarly, the ampelovirus PBNSPaV is putatively vectored by mealybug species. PcMV is vectored by the bud mite, *Eriophyes insidiosus* and CMLV by *E. inaequalis*. Natural spread by vectors has not been confirmed for ACLSV, CGRMV, and CVA; however, root grafts are known to occur or suspected. ToRSV and CRLV are vectored by the dagger nematode, *Xiphinema americanum sensu stricto* and PRMV by *X. americanum sensu lato* and *Longidorus diadecturus*.

Control

Because infected orchard trees cannot be cured, effective control strategies are production and use of clean stocks and programs for pathogen assays to identify pretested propagation materials. To this aim, rules and regulations have been promulgated in various European and North American countries to enact some form of nursery tree fruit improvement schemes. Although planting clean stocks is an essential first step in prevention and control of virus diseases, in some instances additional steps may be required. With nepoviruses, diseased sites will require soil fumigation, use of resistant rootstocks, and/or broad leaf weed control.

Virus Diseases of Pome Fruits

Disease Symptoms and Yield Losses

Some of the viruses infecting stone fruit trees (ACLSV, ApMV, CRLV, and ToRSV) are also pathogenic to pome fruits. ApMV and tulare apple mosaic virus (TAMV) induce yellow mottling of the leaves which is outstanding on the spring vegetation but tends to fade away when the temperature rises. A yield reduction of 20–40% has been associated with infection by severe ApMV strains. ACLSV is the agent of different syndromes, such as *Malus platicarpa* dwarf, leaf deformation of several ornamental apple species, russet ring of apple fruits, and ring pattern mosaic of the leaves of pear and quince. Sensitive apple and quince cultivars can be heavily damaged; for example, yield of apples can be reduced in excess of 20%.

Apple stem pitting virus (ASPV) infections are latent in the great majority of commercial apple cultivars, whereas, in sensitive rootstocks, this virus elicits stem grooving, epinasty, that is, a marked curling of the leaves caused by necrosis of the main vein, and chlorotic/necrotic mottling of the leaves, followed by decline. Some of its strains are also responsible for diseases that affect the leaves (vein yellowing, necrotic spots) (**Figure 1(e)**) and fruits (stony pit) of pear, and the leaves (sooty ring spot), branches (necrotic grooves and bark necrosis), and fruits (deformation and stony pits) of quince.

Like ASPV, apple stem grooving virus (ASGV) infects latently ungrafted cultivars and rootstocks of apple, pear, and quince. However, a severe disease that leads to decline and death of the plants, known as 'apple junction necrotic pitting' or 'top working', develops when infected scions are used for grafting, especially on Virginia Crab, which reacts. Grafted plants react with stem grooving (**Figure 1(f)**), and necrosis of the tissues at the graft union (brown line) which leads to a rapid decline and death.

A somewhat similar disorder called 'union necrosis and decline' is induced by ToRSV in apples grafted on the rootstock MM.106. 'Flat apple', a disease induced by CRLV, is characterized by deformation of the fruits, reduced growth of lateral branches and upward rolling of the leaves.

Geographical Distribution

Except for TAMV, CRLV, and ToRSV, which are largely confined to the USA, all other pome fruit-infecting viruses (ApMV, ACLSV, ASGV, and ASPV) have a worldwide distribution.

Causal Agents and Classification

Affiliation and properties of ACLSV, ApMV, TAMV, CRLV, and ToRSV are as reported above. ASPV and ASGV are the type species of the genus *Foveavirus* and *Capillovirus*, respectively. Both viruses have filamentous particles c. 640 nm (ASGV) or c. 800 nm (ASPV) long, and a monopartite RNA genome 6495 nt (ASGV) and 9306 nt (ASPV) in size.

Transmission

ASPV, ASGV, and ACLSV have no known vector. Long-distance dissemination occurs via infected propagative material and nursery productions, whereas spread in apple orchards may take place through root grafts. The epidemiology of ilarviruses (ApMV, TAMV), the nepovirus ToRSV, and the cheravirus CRLV is as reported above.

Control

To control pome fruit viruses the same strategies used for stone fruit viruses can be implemented.

Virus Diseases of Walnut

Disease Symptoms and Yield Losses

Walnut blackline is a trunk union malady in English walnut scions propagated on rootstocks of several *Juglans* species. In California, blackline disease developed in English walnut trees propagated on seedlings of northern California black walnut, the hybrids Paradox and Royal, and Chinese wingnut (*Pierocarya stenoptera*).

In California, blackline diseased trees exhibit poor tree vigor and shoot growth, limb dieback, and yellow, drooped leaves. Rootstocks often produce sucker shoots. The main diagnostic symptoms are comprised of necrotic tissues embedded in bark tissue and into the woody cylinder at the scion-rootstock junction displayed as union blackline in trees of English walnut on black walnut rootstock or tissue necrosis extending downward from the scion-rootstock junction into Paradox rootstock. In Europe, grafted CLRV-infected trees develop a similar decline pattern. In addition, leaves develop chlorotic spots, rings, and yellow line patterns. Such leaf symptoms have not been observed in California.

Geographic Distribution

Walnut blackline disease has been reported in the USA (California and Oregon) and in Europe (Bulgaria, England, France, Hungary, Italy, Romania, and Spain).

Causal Agent and Classification

The incitant of walnut blackline disease is CLRV, genus *Nepovirus*.

Transmission

CLRV spreads by top-working trees with infected scions during cultivar conversion or in nature by pollen infecting healthy walnut trees during the flowering period. The virus is seed-borne and gives rise to infected seedlings.

Control

Use clean sources of scion wood and rootstocks. In orchards requiring supplemental pollination, pollen should be collected from CLRV-free trees.

Virus Disease of Hazelnut

Disease Symptoms and Yield Losses

Hazelnut mosaic diseased trees may develop leaves with chlorotic rings, flecking, and a variety of line patterns. Virus-infected trees may be symptomless also. Nut yields may be halved compared to production on healthy trees.

Geographical Distribution

The disease is reported in commercial orchards in Europe (Bulgaria, Italy, Turkey, Spain, Georgia) and likely occurs in other European countries. In the USA, hazelnut mosaic was detected in breeding lines and clonal germplasm importations, but not in commercial orchards.

Causal Agents and Classification

A complex of ilarviruses (ApMV, PNRSV, or TAMV) has been associated with diseased trees. In the USA, only ApMV has been identified in hazelnut trees.

Transmission

Hazelnut seeds harvested from ApMV-infected trees give rise to infected seedlings. ApMV is not known to be pollen-transmitted or vectored by insects. In contrast, PNRSV is known to be pollen- and seed-borne in several *Prunus* species.

Control

Propagate plants for planting from clean stocks.

See also: Nepovirus; Citrus Tristeza Virus; Flexiviruses; Tymoviruses; *Ilarvirus*; Plum Pox Virus.

Further Reading

Andret-Link P, Laporte C, Valat L, et al. (2004) Grapevine fanleaf virus: Still a major threat to the grapevine industry. *Journal of Plant Pathology* 86: 183–195.

Bovey R, Gärtel W, Hewitt WB, Martelli GP, and Vuittenez A (1980) *Virus and Virus-Like Diseases of Grapevines.* Lausanne: Editions Payot.

Desvignes JC, Boyé R, Cornaggia D, and Grasseau N (1999) *Maladies à Virus des Arbres Fruitiers.* Paris: Editions Ctifl.

Jones AL and Aldwinckle HS (eds.) (1990) *Compendium of Apple and Pear Diseases.* St. Paul, MN: APS Press.

Krake LR, Scott NS, Rezaian MA, and Taylor RH (1999) *Graft-Transmissible Diseases of Grapevines.* Collingwood, VIC: CSIRO.

Martelli GP (ed.) (1993) *Detection and Diagnosis of Graft-Transmissible Diseases of Grapevines.* Rome: FAO Publication Division.

Martelli GP and Boudon-Padieu E (2006) Directory of infectious diseases of grapevines. *Options Méditérranéens, Series B* 55: 11–201.

Mink GI (1992) Ilarvirus vectors. *Advances in Disease Vector Research* 9: 261–281.

Nemeth M (1986) *Virus, Mycoplasma, and Rickettsia Diseases of Fruit Trees.* Dordrecht, The Netherlands: Martinus Nijhoff.

Ogawa JM, Zehr EI, Bird GW, Ritchie DF, Uriu K, and Uyemoto JK (eds.) (1995) *Compendium of Stone Fruit and Diseases.* St. Paul, MN: APS Press.

Teviotdale BL, Michailides TJ, and Pscheidt JW (2002) *Compendium of Nut Crop Diseases in Temperate Zones.* St. Paul, MN: APS Press.

Walter B (ed.) (1997) *Sanitary selection of the grapevine. Protocols for Detection of Viruses and Virus-Like Diseases Les Colloques, No. 86.* Paris: INRA Editions.

Walter B, Boudon-Padieu E, and Ridé M (2000) *Maladies à virus, bacteries et phytoplasmes de la vigne.* Bordeaux: Editions Féret.

Wilcox WF, Gubler WD, and Uyemoto JK (eds.) (2007) *Compendium of Grape Diseases.* St. Paul, MN: APS Press.

Plant Virus Diseases: Ornamental Plants

J Engelmann and J Hamacher, INRES, University of Bonn, Bonn, Germany

© 2008 Elsevier Ltd. All rights reserved.

Glossary

Chlorotic symptoms Degradation or depletion of chlorophyll leads to light green coloration of plant parts. Characteristic patterns such as ring spots, line patterns, and mosaics occur during virus infections.

Deformations Virus-induced growth abnormalities leading to reduced growth of cells, twisted or curled leaves or a stunted appearance of the whole plant.

Flower breaking Discoloration of the flowers often leading to an uneven distribution or total loss (bleaching) of pigments.

Generative propagation Propagation via pollen and seeds, nonuniform daughter generations are obtained.

Micropropagation Art and science of plant multiplication *in vitro*. The process includes many steps – stock plant care, explant selection, and sterilization, and media manipulation – to obtain proliferation, rooting, acclimation, and growing on of liners.

Necrotic symptoms These symptoms occur when cells die, giving rise to dull grayish, brown, or black coloration of the respective plant tissues. Necrotic lesions are often surrounded by darkly stained cells.

Phylogenetic origin The evolutionary relatedness among groups of organisms (e.g., species and populations). Phylogenetics treats a species as a group of lineage-connected individuals over time.

Serological methods Diagnostic methods based on the specific reaction of antibodies raised in rabbits, mice, or chicken by immunization with plant viruses. Methods are, e.g., ELISA, direct immunoblots, Western blots, or lateral flow tests.

Subgenomic RNA Short copy of a genomic RNA often existing as a single additional translatable RNA for capsid protein within a viral capsid. One of the translation strategies of RNA-viruses to express their whole genetic information.

Vector transmission Most plant viruses are vectored by arthropods, often hemipteran insects. For persistently transmitted viruses, the virus is ingested, passes through the gut wall into the haemolymph, and then moves to the salivary glands where it can potentially be transmitted to other plants with the saliva. Long acquisition times of more than 1 h, a latency period of several days between uptake and release of the virus, and a long inoculation access of 1 h to several hours is characteristic. Persistently transmitted viruses, have two subclasses termed circulative if there is no multiplication in the insect vector and propagative if there is. For nonpersistently transmitted viruses, the virus is restricted to the tips of the insect stylet, where salivary duct and food canal coalesce. Nonpersistent viruses are efficiently transmitted after relatively brief (<5 min) acquisition and inoculation access periods. Semipersistent transmission is intermediate between nonpersistent and persistent. In this kind of transmission the virus moves to the foregut of the insect. Acquisition and inoculation access periods are longer than with nonpersistent transmission.

Vegetative propagation All kinds of plant propagation where parts of the plant are used to generate new plantlets, aiming uniform daughter generations.

Introduction

Virus diseases are of great economic importance in ornamentals because most of them are propagated vegetatively. This kind of plant propagation, either by tissue culture (micropropagation) or by scions, bulbs, rhizomes, or other tissues, is the most economic method to propagate plants maintaining their uniformity. Unfortunately, by this method viruses are propagated and spread from the mother stock to the next generation as well. After, for example, vegetatively propagated trailing petunias flooded the market in the 1990s, problems with tobamoviruses, mainly tobacco mosaic virus (TMV), arose quickly, compared to seed-propagated petunias.

The risk of virus infections in ornamental cultures is also raised by additional factors:

- introduction and establishment of novel virus vectors (e.g., *Thrips palmi*, *Frankliniella occidentalis*, *Bemisia tabaci*) and/or novel virus species into production areas;
- cross-breeding of novel, potentially virus-infected wild types of plants;
- introduction of novel exotic genera or species to broaden the range of ornamentals;
- cultural practices in closed irrigation and fertilization (nutrient solution) systems;
- worldwide trade with ornamentals; and
- production of plant material in countries representing different standards of production and/or other virus pressure.

As the number of ornamental species and varieties is very high and their phylogenetic as well as geographic origins vary considerably, the number of infecting agents, especially viruses, varies accordingly, giving rise to an immense range of virus–plant pathosystems. Viruses of ornamental plants have very different host ranges, many of them being ubiquitous, such as tomato spotted wilt virus (TSWV), impatiens necrotic spot virus (INSV), or cucumber mosaic virus (CMV). Some of them, however, are specialized and infect only certain species or genera of plants, such as, for example, pelargonium flower break virus (PFBV) or angelonia flower mottle virus (AFMoV). The following section gives an overview of the most common as well as some recently detected viral diseases of economically important ornamentals.

Virus Diseases in Diverse Ornamentals

Virus Diseases of Pelargonium

All pelargonias are often named geraniums. Pelargonias originated from the South African region. The main ornamental forms are obtained by cross-breeding of species, thus representing hybrids. Three forms of pelargonium are of main importance in floriculture: *Pelargonium grandiflorum* hybrids (pot plants), *P. peltatum* hybrids (ivy-leaved hanging pelargonias), and *P. zonale* hybrids (upright growing), the latter two forms being of highest economical importance as bedding or balcony plants. As pelargonias are usually propagated by cuttings or tissue culture, viral diseases endanger production and quality of that crop. In pelargonia, many virus symptoms evolving in late winter and early spring will disappear during summer and may reappear in the next season.

Specific pelargonium viruses

The most prevalent virus in Europe is by far PFBV, a virus assigned to the genus *Carmovirus*, family *Tombusviridae*. This icosahedral virus spreads via draining water, via contaminated pollen carried by thrips vectors, possibly by allowing virus from the surface of infected pollen to enter through feeding wounds, and most efficiently by vegetative propagation of infected plant material. Symptoms vary according to time of the year and cultural practice: pink flowering varieties show flower break (**Figure 1**), leaf symptoms vary from symptomless to light mottling and to ring spotting (**Figure 2**) as well as line patterns. The virus occurs more often in *P. zonale* than in *P. peltatum* hybrids (**Figure 3**).

Figure 1 Flower breaking on *P. zonale* induced by PFBV. By Joachim Hamacher.

Figure 2 Ring spots on *P. zonale* induced by PFBV. By Joachim Hamacher.

Figure 3 Leaf mottling on *P. peltatum* (PFBV). By Joachim Hamacher.

There has been considerable confusion about some pelargonium viruses causing very similar ring spot and line pattern symptoms. Investigations of these viruses on the molecular level solved the initial disarrangement: four different virus species have been characterized which are proposed to form a new genus of plant viruses named *Pelarspovirus*.

Pelargonium line pattern virus (PLPV) is a spherical virus, 30–32 nm in diameter, belonging to the family *Tombusviridae*, which has many traits of a carmovirus, but is different in one type of protein not expressed by carmoviruses and transcribes only one subgenomic RNA instead of two. It was reported initially in *P. zonale* from Great Britain. Most infections seem to be symptomless, but chlorotic rings and line patterns may develop (**Figure 4**). Investigations from Spain report that PLPV represents the most common symptomless virus disease of pelargonias. It can be transmitted mechanically and by vegetative propagation.

Pelargonium ring spot virus (PelRSV) produces symptoms of pronounced ring and line patterns resembling those induced by PLPV. PelRSV could, however, be clearly distinguished from PLPV by serology and nucleotide sequence as well as by the size of double-stranded RNAs (dsRNAs) produced in *Chenopodium quinoa*.

A further virus belonging to the family *Tombusviridae*, pelargonium chlorotic ring pattern virus (PCRPV) was isolated in Italy as geranium virus isolate 57. It was characterized and proposed to belong to the new virus genus mentioned above. This virus also leads to distinct chlorotic ring pattern symptoms (**Figure 5**).

Symptoms of pelargonium leaf curl virus, assigned to the genus *Tombusvirus*, were first reported in *P. zonale* from Germany as early as 1927. The symptoms are characterized by the appearance of small white-yellowish spots on leaves, which enlarge, becoming round flecks or stellate spots with

Figure 4 Chlorotic ring spots and line patterns induced by PLPV. By Joachim Hamacher.

Figure 5 Chlorotic ring patterns on *P. peltazonale* induced by PCRPV. By Joachim Hamacher.

Figure 7 Leaf crinkling and curling on *P. zonale* caused by PLCV. By Joachim Hamacher.

Figure 6 Whitish spots on PLCV-infected *P. zonale*. By Joachim Hamacher.

Figure 8 Yellow net vein symptoms on *P. zonale*, pathogen unknown. By Joachim Hamacher.

necrotic centers (**Figure 6**). In later infection stages leaf crinkling, leaf splitting, and plant degeneration or dwarfing occurs (**Figure 7**). Leaf symptoms will disappear in summer. The virus can be transmitted by water or nutrient solutions in soil-free substrates, mechanically, though with some difficulty, with plant sap, and most efficiently by cuttings from infected symptom-free sources.

Pelargonium zonate spot virus (PZSV), a virus now assigned to the new genus *Anulavirus*, family *Bromoviridae* has been reported from Italy (Apulia). PZSV causes concentric chrome-yellow bands in the leaves of *P. zonale*. Besides mechanical transmission, it can be transmitted by seeds and by pollen in *Nicotiana glutinosa* and *Diplotaxis erucoides*.

Pelargonium vein clearing virus (PelVCV) syn. eggplant mottled dwarf virus (EMDV), assigned to the genus *Cytorhabdovirus*, is reported from southern Italy. It causes mild vein clearing in *P. peltatum* and *P. zonale*. Transmission occurs via mechanical transmission and by vegetative propagation.

An often observed anomaly of pelargonium is named yellow net vein disease. It is graft transmissible and leads to obvious veinal chlorosis (**Figure 8**), persisting throughout the year. A causal virus or other agent has not been isolated or identified up to now.

Pelargonium nonspecific viruses

Some nepoviruses also infect pelargonias, mainly in North American varieties: tomato ringspot virus (ToRSV), tobacco ring spot virus (TRSV), and tomato black ring virus (TBRV) lead to ring spotting and chlorotic flecking and stippling (**Figure 9**). These viruses cause quarantine diseases because of their potential to infect fruit trees and their transmissibility by nematodes, pollen, and seeds.

A considerable threat for floriculturists are two tospoviruses: TSWV and INSV – the first infecting a great number of vegetables and ornamentals, the latter mainly ornamentals. TSWV infects *P. peltatum*, producing ring spots and line patterns (**Figure 10**). INSV infects pelargonias very

Figure 9 Chlorotic stippling of *P. zonale* induced by TBRV. By Joachim Hamacher.

Figure 11 Chlorotic flecking on *P. peltatum* induced by CMV (holes were punched in the left-hand leaf for analytical purposes). By Dietrich E. Lesemann.

Figure 10 Chlorotic line patterns and rings induced by TSWV on *P. peltatum*. By Joachim Hamacher.

rarely. Both viruses are transmitted by thrips, mainly the western flower thrips (*Frankliniella occidentalis*, Pergande).

CMV is a very ubiquitous virus with the widest known host range of all plant viruses, exceeding 1000 host plant species. It occurs with many different strains, may infect pelargonias and leads to pronounced flecking (**Figure 11**), asymmetry of leaves as well as breaking of the characteristic brown horseshoe zone in *P. zonale*. It can be transmitted by many aphid species in a nonpersistent manner.

Virus Diseases of Solanaceae

Petunia, calibrachoa

When vegetatively propagated trailing petunias (surfinias) came on the market, problems with viruses, unknown in seed-propagated petunias, arose. More than 150 different virus species may infect petunias, but only some of them infect petunias naturally. Petunias are vigorously growing plants requiring a balanced fertilization, water supply, and culture maintenance. If these requirements are not achieved, symptoms suspicious of viral origin may arise very soon. This has to be considered when monitoring the crops.

Specific petunia viruses

A new virus disease of petunias has been identified recently in Brazil. The pathogen has been named petunia vein banding virus (PetVBV) and is assigned to the genus *Tymovirus*. The virus occurred in mixed infections with another spherical virus (possibly PVCV) and produced local chlorotic and necrotic spots and systemic vein banding on petunias when inoculated as single pathogen.

Petunia vein clearing virus (PVCV), genus *Petuvirus*, eludes detection by immunological methods most time of the year, since its dsDNA genome is integrated in the host genome of some Petunia varieties. The genome expression is activated only in late winter/early spring or by stress. During symptom development, few icosahedral virus particles measuring about 45 nm in diameter can be observed by electron microscopy (**Figure 12**). Symptoms are veinal chlorosis with shrunken leaf veins (**Figure 13**) sometimes becoming necrotic.

Viruses nonspecific of petunias

The most frequently detected virus in trailing petunias is TMV. Symptoms vary according to variety, fertilization, and developmental stage of the plant. Veinal chlorosis (**Figure 14**) and necrosis (**Figure 15**), mottling (**Figure 16**), blistering with dark green areas as well as leaf curl and stunting of whole plants may appear. Flowers may also express color deviations as observed with bicolored petunia varieties (**Figure 17**). Tomato mosaic virus (ToMV)-infected plants show similar symptoms, although symptom expression is less severe. Mixed infections of both viruses occur as well. Two other tobamoviruses,

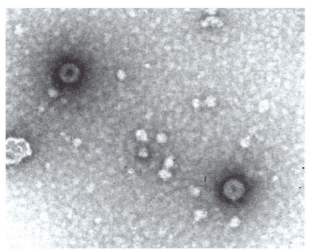

Figure 12 Spherical particles of PVCV from infected petunia showing darkly staining centers. Bar represents 50 nm. By Joachim Hamacher.

Figure 15 Necrotic leaf spots and veins of petunia caused by TMV. By Joachim Hamacher.

Figure 13 Veinal chlorosis of petunia leaves caused by PVCV. By Dietrich E. Lesemann.

Figure 16 Mottling of TMV-infected petunia leaves. By Joachim Hamacher.

Figure 14 Leaf vein chlorosis of young petunia plants induced by TMV. By Joachim Hamacher.

Figure 17 Flower breaking in bicolored petunia varieties with TMV infection. By Joachim Hamacher.

tobacco mild green mosaic virus (TMGMV) and turnip vein clearing virus (TVCV) have been detected in petunias, both leading to severe necrotic symptoms in petunias (**Figure 18**).

Transmission of tobamoviruses with contaminated plant sap is very easy and results in high infection rates, as petunias are vigorously growing plants with soft leaves and many hairs, which break when plants are handled.

Potato virus Y (PVY) is the type virus of the genus *Potyvirus* and often found in petunias. The main symptoms are mottling of leaves and color deviations (mottling) in purple flowering plants, also occurring in mixed infections with CMV (**Figure 19**). Another potyvirus infecting petunias, which induces flower breaking is turnip mosaic virus (TuMV). Potyviruses are aphid transmissible in a nonpersistent manner.

Tospoviruses may infect petunias locally but represent no real danger, as vegetative propagation will not transmit these viruses effectively. Nonflowering petunias, interestingly, may be used to monitor the occurrence of tospoviruliferous thrips, because they develop brown to black local lesions at the edges of thrips-feeding scars when infection has occurred.

Two viruses of the genus *Cucumovirus*, CMV and tomato aspermy virus (TAV) may infect petunias, leading to mosaic, flecking (**Figure 20**) or mottled leaves and TAV to flower deformation. Both viruses are aphid transmissible in a nonpersistent manner.

Alfalfa mosaic virus (AMV) produces marbling of leaves (**Figure 21**) and flower breaking in petunia.

Some nepoviruses (ArMV, TBRV, ToRSV, and TRSV) are reported to infect petunias, producing ring spots, line patterns, or tip necrosis. The symptoms will disappear in the course of infection.

Tobacco necrosis virus (TNV), tobacco rattle virus (TRV), and tobacco streak virus (TSV), were detected to infect petunias, but seem to appear only sporadically.

Some viruses naturally infecting petunias are of minor importance as they do not evoke obvious symptoms. *Calibrachoa mottle virus* (CbMV), assigned as a new virus species mainly infecting *Calibrachoa* sp., does not produce symptoms in petunias. Broad bean wilt virus 1 (BBWV1) is another latent virus of petunia.

Figure 18 Crippling of TMGMV-infected petunia scions (necrosis and deformations). By Joachim Hamacher.

Figure 19 Flower mottle of purple *Surfinia* caused by a mixed infection with PVY and CMV. By Dietrich E. Lesemann.

Figure 20 Bright yellow leaf spots of CMV-infected Surfinia. By Joachim Hamacher.

Figure 21 Marbling of petunia leaves induced by AMV. By Joachim Hamacher.

Figure 22 Chlorotic leaf blotching and mottling of *Calibrachoa* infected by CbMV. By Joachim Hamacher.

Figure 23 Chlorotic line patterns in ToMV-infected *Calibrachoa* leaves. By Joachim Hamacher.

Calibrachoa or 'Million Bells'

The name of calibrachoa goes back to the nineteenth century Mexican botanist and pharmacologist Antonio de Cal y Bracho. Million Bells are solanaceous bedding and balcony plants with growing popularity.

The most frequently detected virus of calibrachoas has been characterized recently, named calibrachoa mottle virus (CbMV) and has been tentatively assigned to the genus *Carmovirus*. Symptoms evolving after infection vary from light mottling to chlorotic blotching (**Figure 22**), but some varieties may remain symptomless.

Tobamoviruses play an important role in calibrachoas. ToMV is of greater importance than TMV, as it leads to pronounced chlorotic ring spots and line patterns (**Figure 23**), whereas TMV remains symptomless or induces only light mottling.

Some other viruses have been diagnosed in calibrachoas such ArMV, TRSV, BBWV1, INSV, PMMoV, RMV, CVB, PVY, and other potyviruses, as well as PNRV and TSV.

Virus Diseases of Balsaminaceae

Impatiens New Guinea hybrids and *I. walleriana* play a prominent role as bedding and balcony plants and may be hosts for quite a lot of virus diseases. The prevalent virus of *Impatiens* spp. is INSV. The symptoms observed after infection with INSV are necrotic spots (**Figure 24**), ring spots or mottling of leaves, and necrotic flecking of stems. On *I. New Guinea* hybrids the discolorations are often accompanied by foliar deformations (**Figure 25**). Besides vegetative propagation, virus transmission of tospoviruses within crops is mainly due to feeding of *Frankliniella occidentalis* (Pergande) the western flower thrips. Other thrips species may transmit as well but are of minor importance (**Table 1**). TSWV occurs by far less frequently on impatiens than INSV, but induces very similar necrotic symptoms.

Tobamoviruses are as well quite common in impatiens. TMGMV and TVCV have been diagnosed in *I. New Guinea* hybrids. The symptoms are very similar and comprise stunting, leaf deformation, or die back of young

Figure 24 Necrotic spots on leaves of INSV-infected *Impatiens New Guinea* hybrid (NGI). By Joachim Hamacher.

Figure 25 Leaf deformation and necrotic spots due to INSV infection of NGI. By Joachim Hamacher.

plants (**Figure 26**), leaf necrosis and deformation as well as foliar reddening on older plants. Pink or orange flowering varieties may exhibit flecking of flowers (**Figure 27**). Black line patterns, resembling symptoms induced by INSV or TSWV, have also been observed on some varieties. The viruses may be symptomless on mature plants. Both viruses may be transmitted by mechanical injury when handling the plants, by contaminated substrate and pots as well as by irrigation.

Leaf narrowing and rugged leaf borders as well as split petals and color breaking are typical symptoms of infections with CMV (**Figures 28** and **29**). Several aphid species may transmit the virus in a nonpersistent manner within seconds.

Further virus species infecting impatiens are TuMV, AMV (aphid transmissible in a nonpersistent manner as well), and TSV, a seed-borne ilarvirus, which probably may be transmitted with contaminated pollen via thrips feeding.

TuMV leads to dark areas at the base or tip of the leaves of some varieties of New Guinea hybrids as well as to asymmetry of the leaves. AMV-infected plants show slight leaf necrosis or cloudy yellowish discoloration of the leaves of green-leaved varieties (**Figure 30**). TSV is reported to occur quite frequently. It induces only slight mosaic on the lower side of the leaves or remains symptomless.

An hitherto not characterized virus on both species of cultivated impatiens occurs rather frequently in Europe and produces chlorotic to yellow concentric ring spots and line patterns (**Figures 31** and **32**). It is graft transmissible between *I. New Guinea* hybrids and *I. walleriana*. In plants exhibiting such symptoms, few spherical virus-like particles, with about 30 nm diameter could be observed by electron microscopy in newly infected tissue (**Figure 33**). A provisional name for the pathogen could be impatiens chlorotic line pattern agent.

Virus Diseases of Scrophulariaceae

Bacopa, Diascia, Nemesia, Angelonia

Bacopa (*Sutera cordata*), Nemesia (*N. fruticans* hybrids), Diascia (*D. vigilis*), and Angelonia (*A. angustifolia*) are genera of the family Scrophulariaceae, continuously gaining popularity.

Sutera cordata (diffusa), also known as *Bacopa*, is often infected by potyviruses. Lettuce mosaic virus (LMV) was detected by electron microscopy as well as by enzyme-linked immunosorbent assay (ELISA). Little and dented leaves, leaf curl, and marginal foliar necroses (**Figure 34**) were observed, but most frequently no symptoms were induced at all. Other filamentous viruses of bacopas are PVY and CVB. BBWV 1 was detected sporadically in bacopa, leading to mottling and flecking of leaves. Tobamoviruses, such as ToMV and TMV, tospoviruses (INSV and TSWV), as well as viruses belonging to the family *Bromoviridae* (AMV and CMV), have also been detected in bacopa.

Nemesia ring necrosis virus (NeRNV) has only recently been described as a new virus and assigned to the genus *Tymovirus*. Before identification as a new species, infections with the respective symptoms have been thought to be induced by scrophularia mottle virus (ScrMV), because of its strong immunological cross-reaction with that virus. NeRNV is widespread, leading to foliar concentric necrotic ring spots and line patterns as well as to sporadical flecking of flowers (**Figures 35** and **36**) in Nemesia, and in Diascia to black discoloration and line patterns, as well as dwarfing of the whole plant or parts of the plant (**Figures 37** and **38**).

Infections of nemesia with tospoviruses lead to necrotic areas and leaf narrowing (**Figure 39**), and in mixed

Table 1 Virus acronyms, species, and genera

Acronym	Virus/viroid species	Virus/viroid genus	Virus/viroid family
AFMoV	Angelonia flower mottle virus*	Carmovirus	Tombusviridae
AMV	Alfalfa mosaic virus	Alfamovirus	Bromoviridae
AnFBV	Angelonia flower break virus	Carmovirus	Tombusviridae
ApMV	Apple mosaic virus	Ilarvirus	Bromoviridae
ArMV	Arabis mosaic virus	Nepovirus	Comoviridae
BBWV 1	Broad bean wilt virus 1	Fabavirus	Comoviridae
BCMV	Bean common mosaic virus	Potyvirus	Potyviridae
BWYV	Beet western yellows virus	Polerovirus	Luteoviridae
BYMV	Bean yellow mosaic virus	Potyvirus	Potyviridae
CRSV	Carnation ringspot virus	Dianthovirus	Tombusviridae
CbMV	Calibrachoa mottle virus*	Carmovirus	Tombusviridae
ClYMV	Clover yellow mosaic virus	Potexvirus	Flexiviridae
CMV	Cucumber mosaic virus	Cucumovirus	Bromoviridae
CSNV	Chrysanthemum stem necrosis virus	tent. Tospovirus	Bunyaviridae
CSVd	Chrysanthemum stunt viroid	Pospiviroid	Pospiviroidae
CVB	Chrysanthemum virus B	Carlavirus	Flexiviridae
CymMV	Cymbidium mosaic virus	Potexvirus	Flexiviridae
CymRSV	Cymbidium ringspot virus	Tombusvirus	Tombusviridae
CypCSV	Cypripedium chlorotic streak virus	tent. Potyvirus	Potyviridae
CypVY	Cypripedium virus Y	Potyvirus	Potyviridae
EMDV	Eggplant mottled dwarf virus syn. Pelargonium vein clearing virus	Nucleohabdovirus	Rhabdoviridae
DMV	Dahlia mosaic virus	Caulimovirus	Caulimoviridae
DVNV	Dendrobium vein necrosis virus	tent. Closterovirus	Closteroviridae
INSV	Impatiens necrotic spot virus	Tospovirus	Bunyaviridae
LMV	Lettuce mosaic virus	Potyvirus	Potyviridae
LSV	Lily symptomless virus	Carlavirus	Flexiviridae
LVX	Lily virus X	Potexvirus	Flexiviridae
MNSV	Melon necrotic spot virus	Carmovirus	Tombusviridae
NeRNV	Nemesia ring necrosis virus*	Tymovirus	
OFV	Orchid fleck virus	Rhabdovirus	Rhabdoviridae
ORSV	Odontoglossum ringspot virus	Tobamovirus	
PCRPV	Pelargonium chlorotic ring pattern virus*	Pelarspovirus (proposed)	Tombusviridae
PelRSV	Pelargonium ring spot virus*	Pelarspovirus (proposed)	Tombusviridae
PelVCV	Pelargonium vein clearing virus syn. Eggplant mottled dwarf virus	Cytorhabdovirus	Rhabdoviridae
PetVBV	Petunia vein banding virus	Tymovirus	
PFBV	Pelargonium flower break	Carmovirus	Tombusviridae
PMMoV	Pepper mild mottle virus	Tobamovirus	
PLCV	Pelargonium leaf curl virus	Tombusvirus	Tombusviridae
PLPV	Pelargonium line pattern virus*	Pelarspovirus (proposed)	Tombusviridae
PNRSV	Prunus necrotic ringspot virus	Ilarvirus	Bromoviridae
PVCV	Petunia vein clearing virus	Petuvirus	Caulimoviridae
PVY	Potato virus Y	Potyvirus	Potyviridae
PZSV	Pelargonium zonate spot virus	Anulavirus (proposed)	Bromoviridae
ReTBV	Rembrandt tulip breaking virus	Potyvirus	Potyviridae
RMV	Ribgrass mosaic virus	Tobamovirus	
SLRSV	Strawberry latent ringspot virus	Sadwavirus	Comoviridae
SMV	Soybean mosaic virus	Potyvirus	Potyviridae
SrMV	Scrophularia mottle virus	Tymovirus	
TAV	Tomato aspermy virus	Cucumovirus	Bromoviridae
TBV	Tulip breaking virus	Potyvirus	Potyviridae
TBRV	Tomato black ring virus	Nepovirus	Comoviridae
TBSV	Tomato bushy stunt virus	Tombusvirus	Tombusviridae
TCBV	Tulip chlorotic blotch virus syn. Turnip mosaic virus	Potyvirus	Potyviridae
TEV	Tobacco etch virus	Potyvirus	Potyviridae
TMGMV	Tobacco mild green mosaic	Tobamovirus	
TMV	Tobacco mosaic virus	Tobamovirus	
TNV	Tobacco necrosis virus	Necrovirus	Tombusviridae
ToMV	Tomato mosaic virus	Tobamovirus	
ToRSV	Tomato ringspot virus	Nepovirus	Comoviridae
TRSV	Tobacco ringspot virus	Nepovirus	Comoviridae
TRV	Tobacco rattle virus	Tobravirus	
TSV	Tobacco streak virus	Ilarvirus	Bromoviridae

Continued

Table 1 Continued

Acronym	Virus/viroid species	Virus/viroid genus	Virus/viroid family
TSWV	Tomato spotted wilt virus	Tospovirus	Bunyaviridae
TulMV	Tulip mosaic virus	Potyvirus	Potyviridae
TuMV	Turnip mosaic virus	Potyvirus	Potyviridae
TVCV	Turnip vein clearing virus	Tobamovirus	
VanMV	Vanilla mosaic virus	Potyvirus	Potyviridae
VeLV	Verbena latent virus	Carlavirus	Flexiviridae

List of acronyms, virus names, and viral genera.
Names with an asterisk do not appear in the ICTV virus name list.
Adapted from Fauquet CM, Mayo MA, Maniloff J, Desselberger U, and Ball LA (eds.) (2005) *Virus Taxonomy: Eighth Report of the International. Committee on Taxonomy of Viruses.* San Diego, CA: Elsevier Academic Press.

Figure 26 Leaf necrosis and distortion of NGI caused by TMGMV. By Joachim Hamacher.

Figure 28 Leaf narrowing and curling of red-leaved NGI infected with CMV. By Joachim Hamacher.

Figure 27 Flower mottle of pink-flowering NGI caused by infection with TMGMV. By Joachim Hamacher.

Figure 29 Flower breaking and deformation of bicolored NGI induced by CMV. By Joachim Hamacher.

infections with NeRNV and TSWV to bleaching of flowers and necrotic flecks and ring spots (**Figures 40** and **41**).

A new virus disease of *Angelonia* sp. has recently been proved to be induced by a new putative member of the genus *Carmovirus*, for which the names AFMoV or angelonia flower break virus (AnFBV) respectively have been proposed. The virus (∼30 nm in diameter, **Figure 42**) leads to dark blotching of flowers (**Figure 43**). The virus occurs very frequently in angelonias.

Figure 30 Cloudy yellow mottling of green-leaved NGI infected with AMV. By Joachim Hamacher.

Figure 32 Marbling of *Impatiens walleriana* leaves infected with an unknown virus (impatiens chlorotic line pattern agent) after graft transmission from NGI. By Joachim Hamacher.

Figure 31 Chlorotic line pattern on leaves of NGI infected by an unknown virus (impatiens chlorotic line pattern agent). By Joachim Hamacher.

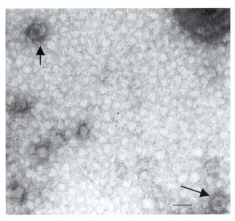

Figure 33 Electron micrograph of impatiens chlorotic line pattern virus-like particles (arrowheads) in plant sap of *I. walleriana* exhibiting symptoms of chlorotic line pattern. Bar represents 35 nm. By Joachim Hamacher.

Virus Diseases of Verbenaceae

Verbenas are traded as upright-, trailing-, and ground-covering forms (e.g., Tapien and Temari). Virus infections of verbenas occur frequently and are induced by a lot of virus species. AFMoV has only recently been detected to occur in varieties of verbenas. It may be latent or leads to distinct flecking of the plants, depending on the respective variety, and to degeneration when co-infected with INSV (**Figures 44** and **45**).

NeRNV has been reported from the UK to induce necrotic and chlorotic flecking, whereas clover yellow mosaic virus (ClYMV) leads to darkly staining and necrotic spots on the base of leaves as well as on stems in trailing verbenas.

Figure 34 LMV-infected bacopa with dented leaves and marginal necrosis. By Joachim Hamacher.

Plant Virus Diseases: Ornamental Plants 219

Figure 35 Blotches and ring spots on flowers of nemesia infected with NeRNV. By Joachim Hamacher.

Figure 36 Necrotic concentric ring spots on leaf of nemesia infected with NeRNV. By Joachim Hamacher.

Figure 37 Black rings and lines on NeRNV-infected diascia leaves. By Joachim Hamacher.

Figure 38 Dwarfed shoots of diascia infected with NeRNV. By Joachim Hamacher.

TSWV, INSV, CMV, BBWV1, AMV, and TMV have as well been diagnosed in verbenas. INSV and TSWV cause necrotic flecking, ring spotting, and line patterns (**Figure 46**). CMV and BBWV1 lead to mottling, necrosis, and little leaves (**Figures 47** and **48**). TMV causes yellow mottling, necrotic leaf borders, or die-back of leaves (**Figure 49**).

More viruses have been reported to infect verbenas: ApMV, TSV, ArMV, TRSV, ToRSV, CVB, VeLV, CRSV, MNSV, TEV, and a strain of BYMV, PNRV, RMV, and ToMV (for virus names see **Table 1**).

Figure 39 Necrotic symptoms and leaf narrowing on INSV-infected nemesia. By Joachim Hamacher.

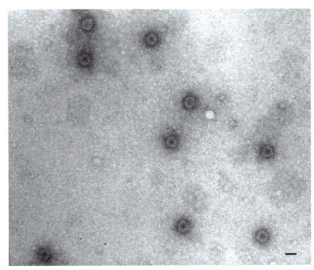

Figure 42 Electron micrograph of particles of AFMoV in plant sap from infected angelonia. Bar represents 30 nm. By Joachim Hamacher.

Figure 40 Flower bleaching of nemesia co-infected with NeRNV and TSWV. By Joachim Hamacher.

Figure 43 Darkly colored flecks on flowers (flower mottle) of angelonia infected with AFMoV. By Joachim Hamacher.

Figure 41 Necrotic ring spots and flecks on nemesia leaves co-infected with NeRNV and TSWV. By Joachim Hamacher.

Figure 44 Chlorotic spots on leaves of verbena co-infected with AFMoV and INSV. By Joachim Hamacher.

Figure 45 Necrotization of lower leaves of verbena co-infected with AFMoV and INSV. By Joachim Hamacher.

Figure 46 Necrotic lines and spots on TSWV-infected verbena. By Joachim Hamacher.

Figure 47 Necrotic spots and mottling on leaves of BBWV1-infected verbena. By Joachim Hamacher.

Figure 48 Light mottling and little leaves of BBWV1-infected verbena. By Joachim Hamacher.

Virus Diseases of Compositae

Osteospermum and dimorphotheca, dahlia, and chrysanthemum
Compositae or Asteraceae include plant species from all over the world

The South African species *Osteospermum* and *Dimorphotheca*, vernacular name, African daisy or cape daisy, are closely related genera and consist of about 70 (*Osteospermum*) and 2 (*Dimorphotheca*) species of evergreen shrubs, half shrubs, or annual plants. Osteospermums are relatively new to most gardeners. They have gained considerable popularity as summer bedding, balcony, and potted plants in the last decade. The frequently occurring virus disease in both species is caused by LMV which leads to foliar mottling, while some isolates also induce flower breaking. In some cases plants remain symptomless.

The tospoviruses TSWV and INSV may disturb plant growth, leading to a dwarfed appearance, deformed leaves, and chlorosis (**Figures 50** and **51**).

Infections with CMV also lead to deformation of plant parts: leaf narrowing and dwarfing in combination with chlorosis are typical symptoms (**Figure 52**).

Dahlias are native of Mexico and Latin America and belong to a genus with only 30 species. The very important bedding and cut flower forms go back to three species *D. coccinea* Cav., *D. pinnata* Cav., and *D. juarezii* hort. The latter most probably being already a garden form. The modern cultivars result from breeding in West European countries.

Dahlias are frequently infected by dahlia mosaic virus (DMV), a spherical dsDNA virus which mostly induces chlorotic oak leaf patterns (**Figure 53**) and other chlorotic symptoms and may induce stunting of susceptible varieties. The virus is spread via aphids in a nonpersistent or semipersistent way.

TSWV and INSV are among the commonly occurring viruses of dahlias. Symptoms may vary according to virus

Figure 49 TMV-infected verbena, yellow mottling of leaves, and necrotization of leaf margins. By Joachim Hamacher.

Figure 51 Leaf narrowing, chlorosis, and stunting of TSWV-infected osteospermum plant. By Joachim Hamacher.

Figure 52 CMV infection of osteospermum: leaf narrowing of plant parts. By Joachim Hamacher.

Figure 50 Dwarfing of TSWV-infected osteospermum (left), noninfected control (right). By Joachim Hamacher.

Figure 53 Chlorotic oak leaf pattern on DMV-infected dahlia. By Joachim Hamacher.

and variety but most often these viruses induce necrotic ring spots becoming concentric (**Figures 54** and **55**). Oak leaf patterns or discrete necrotic lines along the midrib may also appear in young leaves of certain cultivars.

Symptoms induced by CMV vary according to the developmental stage of the plant. They comprise light mosaic and typical leaf narrowing, also called 'fern leaf'. In some cultivars oak leaf patterns may prevail (compare **Figure 53**).

TSV appears rather frequently in dahlias. It does not induce typical or severe symptoms, but represents a potential danger for other crops.

The former genus *Chrysanthemum* encompasses about 200 species, originating predominantly from the Mediterranean region and Western Asia but some come from South Africa. *Chrysanthemum indicum* hybrids (*Dendranthema indicum* and *D. x grandiflorum*) originate from China to Japan. The genus *Chrysanthemum* is now differentiated into 14 genera, the economically most important of which are *Argyranthemum* (species *A. frutescens*), traded as potted flowering shrubs) and *Dendranthema* (species *Dendranthema x grandiflorum*), traded as cut plants or potted plants. *Chrysanthemum* is one of the leading ornamentals in the international market.

Dendranthema hybrids are most frequently infected by chrysanthemum stunt viroid (CSVd). The disease originated in the USA but reached Europe in the 1950s with infected cultivars exported to England. About 70% of the infected plants are systemically stunted with dwarfed leaves (**Figure 56**) and scrubby flowers, exhibiting floral bleaching of red colored cultivars. Color deviations of leaves are not always obvious, but may include pale, upright young leaves. Sometimes, leaf spots or flecks, often associated with leaf distortions ('crinkling'), are also observed. Symptoms are variable and highly dependent on environmental conditions, especially temperature and light. As CSVd is very stable *in vitro*, sap transmission during handling and cutting may occur. Detection of the viroid may be done by visual inspection, grafting onto indicator varieties, bidirectional electrophoresis, molecular hybridization, as well as reverse transcriptase polymerase chain reaction (RT-PCR) and nested RT-PCR, but not by immunological methods or electron microscopy.

A common virus of chrysanthemums is chrysanthemum virus B (CVB), a carlavirus. It leads to very mild leaf mottling or vein clearing in some cultivars; some infected varieties show slight loss of flower quality, and a few varieties sometimes develop brown necrotic streaks on the florets. Many cultivars become entirely infected, often without visible symptoms. CVB occurs in many other ornamental plants with rather long incubation periods and produces rather mild symptoms. Besides vegetative propagation, the virus can be transmitted in a nonpersistent manner by several aphids.

Figure 54 Chlorotic concentric rings and necrotic lines on TSWV-infected dahlia. By Joachim Hamacher.

Figure 55 Necrotic ring spots on dahlia infected with TSWV. By Joachim Hamacher.

Figure 56 Dwarfed and stunted chrysanthemum infected with CSVd (left) noninfected control (right). By Joachim Hamacher.

Figure 57 Necrotic flecks of TSWV-infected chrysanthemum leaves. By Rainer Wilke.

Figure 58 Tulips exhibiting flower breaking induced by TBV. By Franz J. Nienhaus.

Tospoviruses (TSWV and INSV) infect chrysanthemums quite often, inducing irregular chlorotic or necrotic spots (**Figure 57**), mild leaf deformations, primarily on young leaves, as well as browning or wilting of shoots. As tospoviruses are a major threat to growers, elimination of infected plants has to be executed early and rigidly.

TAV is fairly important in chrysanthemums, as it may cause heavy symptoms like breaking, dwarfing, as well as distortion of the flowers, and foliar chlorotic spots or ring spots. Most infected cultivars, however, do not show leaf symptoms.

Other virus diseases induced by AMV, BBWV1, BWYV, LMV, SMV, TMV, ToMV, ToRSV, and TSV have been reported for *Chrysanthemum* spp. but seem to be of minor importance.

Virus Diseases of Liliaceae

Tulipa, Lilium

Taxonomically, Liliaceae are assigned to the order Liliiflorae of the monocotyledonous plants. We concentrate on virus diseases of the genera *Tulipa* and *Lilium*, as they represent large genera with economically very important ornamentals.

Tulips originated from Central Asia and include about 100 different species. Tulips are propagated via bulbs, which develop at the base of the bulb of the previous year.

More than 20 different viruses are reported to infect tulips, the most important of which are TBV, TNV, and TRV.

Tulip breaking virus (TBV) is the most frequently encountered virus in tulips. It is assigned to the genus *Potyvirus* and can be transmitted by aphids (e.g., *Myzus persicae* and *Aphis fabae*) in a nonpersistent manner. TBV affects color breaking of flowers particularly in late-flowering pink, purple, and red cultivars, while white- and yellow-flowered cultivars are not affected. Breaking symptoms have been described as bars, stripes, streaks, featherings, or flames of different colors on petals (**Figure 58**). The color variation is caused by local fading, intensification, or accumulation of pigments in the upper epidermal layer after development of the normal flower color. Mottling or striping of the leaves also occurs. The infection causes loss of vigor and poor flower production. TBV played an important role in the Dutch 'tulipomania' in the seventeenth century, in that it increased the value of tulips with decorative flower breaking and led to wild speculations with astronomical prices for one variegated tulip bulb. At that time, the undesirable viral cause of the spectacular flower breaking was not yet known.

Tobacco necrosis virus D (TNV-D) causes the so-called Augusta disease, named after the variety in which it was first detected. The virus is transferred to the roots specifically adhering to the zoospores of the chytridiomycete *Olpidium brassicae*. This particularly happens when soil temperatures rise above 9 °C. Infected plants do not necessarily show symptoms, since the virus may be confined to the roots. The disease may suddenly appear when the plants are planted out during frost or when planted early at high temperatures after storage on a standing ground. Stunting and distortion of the shoot and leaves are the most severe symptoms but streaking or angular or elliptical spots are more typical symptoms. Fine necrotic lines in the tepals are of considerable value for differential diagnosis.

TRV (genus *Tobravirus*) is another common virus of tulips with an exceptionally wide host range of cultivated and wild plants. A vector transmission by soil-inhabiting nematodes of the genera *Trichodorus* and *Paratrichodorus* may spread the virus in the field. Symptoms of the disease include chlorotic flecks, oval lesions and streaks, appearing early in the season. Streaks of darker color may occur on tepals of red flowering varieties, whereas those streaks appear translucent in yellow or white flowers. Plants may become stunted and flowers sometimes deformed.

Tulip virus X (TVX) has been reported in tulips from Great Britain and from Japan. It causes chlorotic or necrotic streaks in leaves and streaks of intensified color mainly at the margins of tepals. The tepals may also become bleached or necrotic, resembling symptoms caused by TNV. TVX is a member of the genus *Potexvirus*. Mechanisms of spread within field or glasshouse plots are unclear but may rely on sap transmission.

Other viruses of tulips are nepoviruses: ArMV, TBRV, TRSV; a cucumovirus: CMV; several potyviruses: LMoV, ReTBV, TCBV, TulMV; a carlavirus: LSV; a tombusvirus: TBSV; and a tobamovirus: TMV (see **Table 1**).

Three viruses are common in lilies, these are lily mottle virus (LMoV), lily virus X (LVX), and lily symptomless virus (LVS).

LMoV is a potyvirus and was formerly thought to be a strain of TBV, but serological, host range, and molecular investigations have set it apart from TBV as an own species. It can be found wherever lilies are grown. Symptoms develop as chlorotic mottle to stripe-mosaic (**Figure 59**) and leaves may be twisted or show narrowing. When young plants are infected, severe yellowing of leaves or browning of veins in stems may occur. As LMoV is a potyvirus its transmission is affected by aphids in a nonpersistent manner. The virus can also infect tulips.

LVS does not induce specific symptoms in many cultivars, but reduced growth, small flowers, and lower bulb yields are recorded. Foliar vein clearing or intercostal light green stripes may develop as well. Other natural hosts are tulip and *Alstroemeria*. The virus can be transmitted nonpersistently by aphids.

LVX is a potexvirus, which normally does not lead to pronounced symptoms in most cultivars, but enhances symptoms of LSV. When symptoms appear, the plants show faint chlorotic spots and sometimes necrotic lesions. The virus can be sap-transmitted by mechanical inoculation onto herbaceous test plants.

Other viruses, reported to naturally infect lilies are ArMV, TRSV, ToRSV, CMV, TRV, TSWV, TMV, and SLRSV.

Virus Diseases of Orchidaceae

The family Orchidaceae contains about 800 genera with more than 25 000 species. Centers of origin are probably the Malaysian region with *c.* 12 000 species followed by the tropics of America with about 10 000 species. Genera with high economic importance are *Cattleya*, *Cymbidium*, *Dendrobium*, *Odontoglossum* hybrids, and *Phalaenopsis*.

Lawson summarizes 27 different viruses of orchids in a detailed review. The most important and often detected

Figure 60 Black line patterns on CymMV infected Cattleya leaves. By Dietrich E. Lesemann.

Figure 59 Intensive streaking of lily flower caused by co-infection of LMoV and LSV. By Martin Bucher.

viruses are cymbidium mosaic virus (CyMV) and odontoglossum ringspot virus (ORSV). Many rhabdoviruses or rhabdovirus-type viruses (nine different viruses) have been reported for orchids as well as potyviruses, such as BYMV, dendrobium mosaic virus syn, bean common mosaic virus (BCMV), cypripedium virus Y (CypVY), cypripedium chlorotic streak virus (CypCSV), TuMV and vanilla mosaic virus (VanMV), tospoviruses TSWV and INSV, a tentative closterovirus: dendrobium vein necrosis virus (DVNV), ToRSV, cymbidium ringspot virus (CyRSV), and CMV.

CyMV is a potexvirus and as such easily sap-transmissible. It infects a wide range of orchids and leads to necrotic flecks or streaks (**Figures 60** and **61**), black necrotic spots or line patterns (**Figure 62**) on *Cymbidium*, *Phalaenopsis*, and in mixed infections together with DVNV in *Dendrobium* (**Figures 63** and **64**). Flower necrosis induced in *Cattleya, Laelia, Cymbidium, Phalaenopsis, Epidendrum*, and *Vanda* has also been shown to be caused by CyMV (**Figure 65**).

ORSV (syn. TMV-O) is a tobamovirus and leads to a range of different symptoms in many orchid species (**Figures 66–68**). Flower breaking in violet *Cattleya* varieties with streaks of intensified pigmentation on tight buds have been observed as well as fine necrotic stripes (**Figure 69**). Symptoms on leaves comprise chlorotic and dark concentric ringspots in *Odontoglossum grande*. The virus is easily sap-transmissible and spread occurs via contaminated cutting tools.

Virus Control in Ornamentals

The ornamental industry is a very important branch of agriculture around the world with high amounts of investment and rising competition among growers. Outbreaks of diseases have shattered the ornamental industry more than once. Especially epidemics of viral or bacterial etiology may be disastrous for the reputation and standing of the enterprise struck by such diseases.

As direct chemical control strategies have proven to be ineffective in plants, only prophylactic measures to prevent or hinder infection and spread of viruses in the respective cultures are effective. Traditional resistance cross-breeding as a reliable measure to combat virus

Figure 61 Necrotic flecks on CymMV infected Phalaenopsis leaves. Dietrich E. Lesemann.

Figure 62 Cattleya leaf with black streaking caused by CymMV. By Dietrich E. Lesemann.

Plant Virus Diseases: Ornamental Plants 227

Figure 63 Cattleya flower with black streaking induced by CymMV. By Dietrich E. Lesemann.

Figure 64 Dendrobium leaves with black spots and chlorotic flecks caused by a co-infection of CymMV and DVNV. By Joachim Hamacher.

Figure 65 Dendrobium leaves with sunken necrotic spots and necrotic stripes caused by a mixed infection of CymMV and DVNV. By Joachim Hamacher.

diseases, however, is of minor importance, for on one hand cross-breeding of resistance genes is time consuming, and on the other hand the market demands a rapid change or enlargement of the range of varieties. Hygienic measures and, in many cases, meristem culture and thermotherapy are the essential prophylactic strategies to ensure virus-free cultures. The success of transgenic virus-resistant plants as a practicable alternative is strongly dependent on the legislative and the acceptance in the respective countries. To date the most promising strategy for the producer is the establishment of virus-free mother plants and elite stocks as well as continuous monitoring of production stocks. This demands staff skilled in horticultural production, as virus symptoms appear very varied and may be confounded with symptoms provoked by environmental stress or faulty cultural practices. Specialized diagnosis labs have to know the peculiarities of sampling and time of the year, in which virus testing is practicable, because virus concentrations vary considerably during the course of the year in many ornamental species. They tend to be highest in late winter/early spring in countries with moderate climate, and nearly disappear in summer and autumn, thus disabling secure diagnosis. To ensure virus-free production, the European and American plant protection organizations European and Mediterranean Plant Protection Organization (EPPO/OEPP) and US Department of Agriculture-Animal and Plant Health Inspection Service (USDA-APHIS), for example, have established continuously updated guidelines for the production of virus-free ornamentals. This is achieved by quarantine measures and well-regulated production requirements with highest phytosanitary standards. These include regular testing of candidate and elite stocks in specialized diagnostic labs that are either part of the horticultural enterprises or autonomous. Plant clinics and plant protection services contribute to the program as well. Routine testing is done by ELISA or comparable immunological tests. Molecular techniques with higher sensitivity than that of the conventional testing methods are gaining wider fields of application. Upon occurrence

Figure 66 Chlorotic and necrotic ring spots and line patterns on ORSV-infected cymbidium leaves. By Dietrich E. Lesemann.

Figure 68 Black streaking caused by ORSV on cymbidium leaves. By Dietrich E. Lesemann.

Figure 67 Chlorotic striping on cymbidium leaves infected with ORSV. By Dietrich E. Lesemann.

Figure 69 Chlorotic flecking on cattleya flowers infected with ORSV. By Dietrich E. Lesemann.

of symptoms (of unknown cause), electron microscopy may be of great help to clarify the viral etiology of observed symptoms. Regular visual inspections of production stocks (monitoring) ensure high quality and freedom from diseases of the products.

See also: Papaya Ringspot Virus; Plant Antiviral Defense: Gene Silencing Pathway.

Further Reading

Brunt A, Crabtree K, Dallwitz M, Gibbs A, and Watson L (1996) *Viruses of Plants*. In: *Descriptions and Lists from the VIDE Database*, 1484pp. Wallingford, UK: CAB International.

Daughtrey ML and Benson DM (2005) Principles of plant health management for ornamental plants. Annual Review of Phytopathology 43: 141–169.

Daughtrey ML, Robert L, Wick RL, and Peterson JL (eds.) (1995) *Compendium of Flowering Potted Plant Diseases*. St. Paul, MN: APS-Press.

Fauquet CM, Mayo MA, Maniloff J, Desselberger U, and Ball LA (eds.) (2005) *Virus Taxonomy: Eighth Report of the International Committee on Taxonomy of Viruses*. San Diego, CA: Elsevier Academic Press.

Hammond J (ed.) (2002) ISHS Acta Horticulturae 568: *X International Symposium on Virus Diseases of Ornamental Plants* (CD-ROM format only).

Hull R (ed.) (2001) *Matthews Plant Virology*. San Diego, CA: Academic Press.

Loebenstein G, Hammond J, Gera A, Derks T, and van Zaayen A (eds.) (1996) ISHS Acta Horticulturae 432: IX *International Symposium on Virus Diseases of Ornamental Plants*. (CD-ROM format only).

Loebenstein G, Lawson RH, and Brunt AA (eds.) (1995) *Virus and Virus-Like Diseases of Bulb and Flower Crops*, 543pp. New York: Wiley.

Mokrá V, Brunt AA, Derks T, and van Zaayen A (eds.) (1994) ISHS Acta Horticulturae 377: VIII *International Symposium on Virus Diseases of Ornamental Plants* (CD-ROM format only).

Pirone TP and Perry KL (2002) Aphids: Non-persistent transmission. Advances in Botanical Research 36: 1–19.

Ullman DE, Meideros RB, Campbell LR, Whitfield AE, Sherwood JL, and German TL (2002) Thrips as vectors of Tospoviruses. Advances in Botanical Research 36: 113–140.

Relevant Websites

http://www.dpvweb.net – Descriptions of plant viruses (Association of Applied Biologists).
http://www.agdia.com – List of ornamental plant samples of Agdia.
http://www.ebi.uidaho.edu – University of Idaho, VIDE Database.

Plant Virus Vectors (Gene Expression Systems)

Y Gleba, S Marillonnet, and V Klimyuk, Icon Genetics GmbH, Weinbergweg, Germany

© 2008 Elsevier Ltd. All rights reserved.

Glossary

Magnifection Process for heterologous protein expression in plants that relies on transient amplification of viral vectors delivered to multiple areas of plant body (systemic delivery) by *Agrobacterium*.

Noncompeting vectors Viral vectors capable of co-expressing recombinant proteins in the same cell throughout the plant body.

Introduction

Plant viral vectors have been and are still being developed to take advantage of the unique capability of viruses to extremely rapidly redirect most of the biosynthetic resources of a cell for the expression of a (usually) single nonhost protein – the viral coat protein (CP) (and hopefully, a protein of interest as well). Efficient viral vectors should allow to bypass the limitations of other expression methods, such as the low yield obtained with stable transgenic plants or standard transient expression methods, and the long time necessary for development of stable transgenic lines. Most of the progress achieved so far has been with RNA viruses, and the most advanced vectors have been built using just a handful of plant viruses such as tobacco mosaic virus (TMV), potato virus X (PVX), alfalfa mosaic virus (AMV), and cowpea mosaic virus (CPMV). Tremendous technical progress has been made over the past few years in the development of production processes and industrial plant hosts. In particular, novel developments include the design of vectors that are not just simple carbon copies of wild-type viruses carrying a heterologous coding sequence, but that, instead, have been 'deconstructed' by delegating some rate-limiting functions to agrobacteria or the plant host, thus allowing for a more efficient, versatile, controlled, and safe process.

Viruses have to carry out a number of processes to complete the viral replication cycle. These include host infection, nucleic acid amplification/replication, protein translation, assembly of mature virions, cell-to-cell spread, long-distance spread, reprogramming of the host biosynthetic processes, suppression of host-mediated gene silencing, etc. Viral vectors, in contrast, do not necessarily have to be able to perform all of these functions, but they have to perform a new function – high-level expression of a heterologous sequence.

Two different strategies can be used to develop a viral vector/host process.

Historically, the first approach that was developed, starting in 1990, consisted of designing vectors that were, in essence, wild-type viruses engineered to express an additional sequence – the gene of interest (approach also known as the 'full-virus' vector strategy). These vectors were expected to perform all of the functions that a virus normally undertakes, from the infection step to the formation of complete infectious viral particles. However, addition of a heterologous sequence reduced the efficiency of many of the normal steps of the replication cycle, such as high-level amplification and systemic movement. Moreover, these vectors did not necessarily provide the maximal level of expression for the protein of interest that one might have expected. Nevertheless, most of the practical results reported in the literature have been obtained with these 'first-generation' viral vectors.

The design of the next-generation viral vectors reflects an approach that admits inherent limitations of the viral replication process applied to heterologous gene expression. Rather than mimic the design of the wild-type virus, the goal is to 'deconstruct' the virus, and eliminate the functions that are either limiting (e.g., too species-specific, or rate-limiting) or undesired (such as the ability to create functional infectious viral particles), and rebuild the process, by either delegating the missing functions to the host (genetically modified to provide those functions in *trans*) or replacing them with analogous functions that are not derived from a virus (the 'deconstructed virus' vector strategy). In these 'second-generation' expression systems, elements of the viral machinery such as RNA/DNA amplification and cell-to-cell movement are integrated along with nonviral processes such as replicon formation via *Agrobacterium*-mediated delivery of T-DNAs encoding the viral vector or via activation from a plant chromosomal DNA encoding a pro-replicon or pro-virus.

First-Generation Virus Vectors

Under the 'full-virus' scenario, delivery of the replicons is provided by infecting the host with a mature viral particle or with a full copy of the viral DNA/RNA. The vector is essentially a functional virus that, in addition to the viral sequence, contains a coding sequence of interest under control of a strong viral promoter such as the CP subgenomic promoter (**Figure 1(a)**). Several efficient vectors have been made, primarily based on RNA viruses such as TMV and PVX. Vectors based on TMV were developed in several steps. Earlier versions simply used a duplicated CP subgenomic promoter for expression of the gene of interest, but this led to instability of the vector. Later versions replaced the duplicated subgenomic promoter by sequences derived from a different but phylogenetically related virus. Finally, different $3'$ untranslated sequences from a range of related viruses were screened to obtain a vector expressing the gene of interest efficiently.

An alternative expression strategy consists of expressing the protein of interest as a fusion to the CP (**Figure 1(a)**). In this case, only an essential part of the protein of interest, such as an immunogenic epitope, is fused to the CP since larger fusions usually eliminate viral particle formation and systemic movement ability.

Large-scale production with either type of vector can be achieved by spraying plants in the field or a greenhouse with a mixture of viral particles and carborundum. Depending on the efficiency of the vector and its ability to move systemically, 2–3 weeks are required for plants to achieve maximal transfection/expression.

During the last few years, improvements have been made to this first-generation vectors. Among the most interesting experiments coming from recent work is the improvement of viral vectors by directed evolution. It was found that cell-to-cell movement of the vectors described above was not as high as with wild-type viruses due to lower expression of the movement protein. The strategy employed was to use DNA shuffling to improve the movement protein gene. This work was very successful and resulted in mutant movement protein genes working better than the native version in the viral vector context, essentially in tobacco, where engineered vectors normally do not work very well. This work suggests that other parts of the viral vectors might be improved by DNA shuffling as well.

In general, first-generation vectors were very successful and were able to provide expression of heterologous protein at the levels close to 10% of total soluble protein or over 1 g recombinant protein kg^{-1} of fresh leaf biomass. Several pharmaceutical proteins made using such vectors have actually reached the stage of clinical trial.

However, despite these successes, first-generation viral vectors have several limitations. The yield that can be obtained with these vectors varies greatly depending on the protein to be expressed. In most cases, only relatively small proteins are expressed efficiently, any proteins larger than 30 kDa are usually poorly expressed. Moreover, yield of heterooligomeric proteins requiring two or more subunits expressed from the same vector are extremely low. With CP fusions, the yield can be as high as $2-4 \, g \, kg^{-1}$ of plant biomass; however, usually only short (up to 25 amino acids) epitopes are tolerated.

Host Improvement

To survive in nature, plants have naturally evolved resistance mechanisms against pathogen attacks. It is therefore expected that plants could be engineered to be more susceptible to viruses and therefore better hosts for expression of heterologous sequences using viral vectors. A dilemma is that plants susceptible to viral vectors and therefore good hosts for protein expression will also have poor agronomic characteristics.

One important element of plant defense mechanisms against viruses is the ability of plants to induce post-transcriptional silencing. A large amount of work in this area has shown that not only plants are able to defend themselves against viruses, but that viruses have counter-attacked by evolving suppressors of silencing. All these elements uncovered by basic research in virology offer elements that can be incorporated in expression strategies (with or without viral replication) to boost the level of protein expression. Another important field of research consists of approaches that aim at silencing/superactivating some plants genes that are, directly or indirectly, involved in viral replication or spread.

A typical host for viral research is *Nicotiana tabacum*, the host plant of the most-studied plant virus – tobacco mosaic virus. The most commonly used plant in virology is however the wild Australian species, *N. benthamiana*, which is unusually sensitive to a very wide array of viruses. Despite its poor agronomic characteristics, *N. benthamiana* can be grown very well in the greenhouse under controlled conditions and could therefore even be considered for industrial production.

One drawback of using plants as hosts for manufacturing therapeutic proteins is that plant post-translational modifications are not exactly identical to those made by animal/human cells (although, fortunately, not extremely different either). In particular, the enzymatic machinery

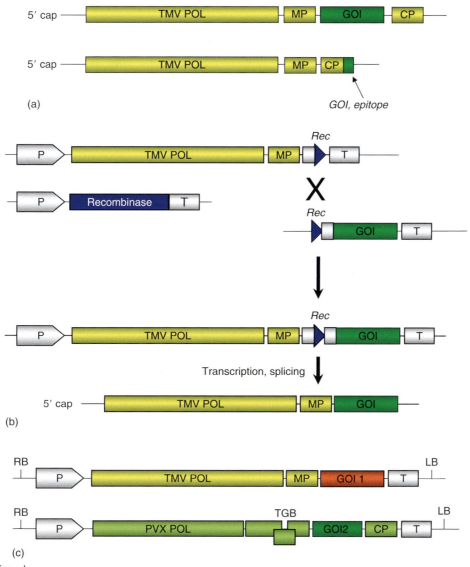

Figure 1 Continued

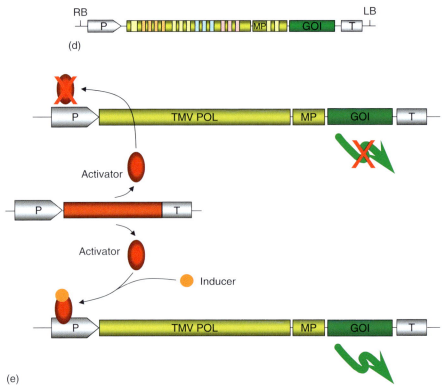

Figure 1 Plant viral vectors currently in use. (a) Typical first-generation RNA expression vectors based on TMV virus. (b) Second-generation vectors. Pro-vector system for rapid *in planta* assembly of viral replicons from T-DNAs delivered by *Agrobacteria*. (c) DNA vectors optimized for *Agrobacterium*-mediated delivery and expression of heterooligomers (IgG antibodies). (d) DNA vectors optimized for *Agrobacterium*-mediated delivery. (e) Systems relying on the regulated release of a replicon from a plant chromosome upon chemical induction. The viral components of the vector backbones are in yellow or in light green and include RNA-dependent RNA polymerase (TMV POL or PVX POL), movement protein (MP), coat protein (CP), triple gene block (TGB); the genes of interest encoding recombinant protein are in green or pink (GOI); the site-specific recombinase (REC) and its recognition sites (*Rec*) are in blue; multiple colored segments on the vector optimized for *Agrobacterium* delivery are plant introns; activator gene and its product are in red; promoter and terminator sequences are in gray.

responsible for N-glycosylation differs in plants and animals. More specifically, plant cells add some specific sugars (core-bound xylose and α-1,3-fucose), which might be immunogenic if protein containing these were administered to humans. In contrast, plant cells do not add other sugars such as terminal β-1,4-galactose residues or sialic acid, which are added to proteins made in animal cells. Fortunately, significant work has been made to engineer plant hosts, including tobacco and *N. benthamiana*, to provide more 'human-like' or even 'designer' post-translational modifications – in particular, low amounts of xylose and α-1,3-fucose as well as galactosyl terminal residues added to the core sugars. Such host plants will be useful for all vector types, whether of the first generation or vectors developed later.

Second-Generation Vectors

This strategy reflects attempts by engineers to part with the inherent limitations of the viral machinery, while keeping the useful viral elements. Some of the elements that can be used outside of the integrated virus system are the 'molecular machines' that provide for infectivity, amplification/replication, cell-to-cell movement, assembly of viral particles, suppression (shutoff) of the synthesis of plant proteins, silencing suppression, systemic spread, etc. Some of these elements are less efficient than others. For example, the ability of the virus to infect the host is low and requires some type of mechanical injury to the plant, or an insect as a vector. Systemic spread is a process that is usually very species-specific, easily impaired as a result of genetic manipulation. Replication/amplification ability, on the other hand, is a central element of viral vectors that is valuable as an 'amplifier'. Fortunately, replication is a relatively robust and species-independent mechanism.

Two basic applications, namely rapid expression in small or large quantities and large-scale industrial production of proteins in plants, each require different design of the expression strategies.

Vectors for Specific Application Areas

Plant Virus Vectors as Research Tools

Many different expression systems have been developed based on the backbones of entirely different viruses, and relying on different modifications of the core viral design. However, since the purpose of using such vectors is often rapid and high-throughput expression, and since only small (usually milligram) amounts of protein are usually required, many existing vectors do not contain the viral component(s) providing systemic movement.

An important and limiting step for the use of viral vectors is infection of the plants with the viral replicons. Use of *Agrobacterium* to deliver a copy of the viral vector encoded on the T-DNA ('agrodelivery') provides an excellent solution since *Agrobacterium* is an extremely efficient vector. All the steps that are necessary for the conversion of the 'agrodelivered' T-DNA into a functional DNA or RNA replicon have been shown to occur in plants. 'Agroinfection' has been used for many years. It is also often much more efficient than mechanical inoculation using viral particles, and is definitely more efficient than using DNA or RNA as infectious molecules. In case of RNA viruses, agroinoculation also represents a very inexpensive alternative to *in vitro* transcription to convert the DNA vector into an infectious RNA.

The use of agrodelivery for inoculation of RNA-viruses vectors also provides entirely new opportunities for vector engineering, since the vector is delivered to the plant cell as a DNA molecule, which can be manipulated prior to conversion into an RNA replicon. In particular, it has been shown that several T-DNAs transiently co-delivered from one or more agrobacteria can be efficiently recombined *in planta* by site-specific recombinases such as Cre or the φc31 integrase. The RNA obtained after transcription of the recombined DNA molecule can then be cleaned of the recombination sites by the nuclear RNA splicing machinery, provided that the recombination sites are engineered by flanking by intron sequences (**Figure 1(b)**).

Using such elements, a process has been developed that, in essence, allows simple and inexpensive '*in vivo*' engineering of RNA vectors by simply mixing various combinations of agrobacteria containing different components of a viral vector and co-infiltrating these mixtures into plants. Such an approach allows, for example, to rapidly assemble and express a variety of different protein fusions such as fusions of the protein of interest to targeting or signal peptides, binding domains, various coding sequences, purification domains, or cleavage sites. All of these expression experiments are done by simple mixing of pre-fabricated bacteria strains harboring the desired vector modules and a strain containing the gene-specific module. Depending on the viral vectors and the specific protein studied, milligram quantities of protein can be expressed in various plant compartments or as different fusions, within just a few days.

Vectors for Expression of Heterooligomeric Proteins

The first transgenic plants expressing full-size human antibodies were made around 1990. Despite these early positive results, low yields have prevented the widespread use of this technology for commercial applications. For example, although stably transformed plants express correctly folded and functional antibodies of immunoglobulin of the IgG and IgA classes, yields are generally very low (1–25 mg kg^{-1} of plant biomass). Moreover, the time necessary to generate the first gram(s) of such antibody material is longer than 2 years.

Transient systems, on the other hand, allow production of research quantities of material much faster. However, as for transgenic plants, transfection systems relying on *Agrobacterium*-mediated delivery of standard nonreplicating expresssion cassettes also provide only low yield of expressed proteins. The level of expression of such systems can be increased by using suppressors of silencing, but yields are still inferior to expression using viral vectors and are not readily scalable. Initial attempts to combine the speed of expression of transient systems with the high level of expression of viral vectors failed due to the inability of viral vectors to co-express two genes within one cell at high level; engineering of vectors for expression of two genes from one viral vector led to drastically reduced expression level of both genes; co-infiltration of two vectors expressing a separate gene also failed because competition between vectors resulted in either one or the other protein expressed in individual cells, but not of the two proteins within the same cell (which would be required for expression of a functional IgG antibody).

Recently a solution to this problem was found by cloning the two genes to be expressed in two separate viral vectors that were constructed on the backbone of two different noncompeting viruses such as, for example, TMV and PVX. Unlike vectors that are built on the same backbone, 'noncompeting viral vectors' were shown to be able to efficiently co-infect and replicate within the same cell. Therefore, high level of functional full-size monoclonal antibodies (Mabs) of the IgG class can now be obtained in plants by co-infiltrating two vectors containing the heavy- and the light-chain separately in each vector. This strategy was shown to work with different Mabs of the IgG1 and IgG2 classes; the molecules were found to be fully functional, and the first gram of material could be produced in less than 2 weeks after cloning in the vector (**Figure 1(c)**).

Expression Vectors for Industrial-Scale Protein Production: Transient Systems (Magnifection)

Industrial production requires viral amplification and expression in large numbers of plants (in a greenhouse or field) and, to obtain maximal yield, in as many tissues and leaves as possible within each plant. For achieving this goal, a first solution was provided by using vectors capable of systemic spread (the 'full-virus' strategy). Although successful, this strategy had many limitations: large inserts (larger than 1 kbp) could not be efficiently expressed, not only because of less-efficient amplification within each cell, but also because viral vectors containing large inserts are less able to move in systemic tissue without recombining; systemic vectors also never infect all harvestable parts of the plant (e.g., the lower leaves are usually not infected); the process is asynchronous as it invades different leaves in a sequential order. Moreover, for viral vectors designed to express protein fragments as CP fusions, only short epitopes (20 amino acids or less) could be successfully expressed until recently.

A simple technical solution was found to bypass these limitations; a viral vector lacking the ability to move systemically (with a simpler design, and more stable) is delivered to the entire plant using *Agrobacterium*-mediated delivery by vacuum-infiltration of the entire plant. This eclectic technology called 'magnifection' combines advantages of three biological systems: the speed and expression level/yield of a virus, the transfection efficiency and systemic delivery of an *Agrobacterium*, and the post-translational capabilities and low production cost of a plant. Such a process can be inexpensively performed on an industrial scale and does not require genetic modification of plants, and is therefore potentially safer (it also does not generate infectious virions as the full-vector strategy does) and more compatible with the current industrial infrastructure.

For magnifection to become an efficient tool, technical difficulties had however to be addressed. Indeed, initial vectors based on the backbones of plus-sense RNA viruses such as TMV could not be delivered to plant cells efficiently by *Agrobacterium* delivery. The first attempts were made with vectors built on the backbone of TMV and led to a rate of infectivity estimated at one successful transfection event per 10^8 agrobacteria. Later experiments using viral vectors based on different tobamoviral strains (infecting both crucifers and *Nicotiana* plants) were more successful, but clearly not optimal. Analysis of the early infection steps initiated by *Agrobacterium* delivery of T-DNAs encoding RNA viral vectors suggested that a bottleneck for the formation of active replicons may be the low ability of the primary transcript to leave the nucleus. The RNAs of cytoplasmic RNA viruses such as TMV normally never enter the plant cell nucleus at any point during the entire replication cycle and have therefore not evolved to be in contact with the cell RNA-processing machinery and to be exported to the cytosol. *In silico* analysis of the viral genome structure with programs designed to assess whether an unknown sequence is likely to be coding or noncoding (by comparison to expressed nuclear plant genes, using neural networks), showed that the viral RNA is unlikely to be recognized as a correct coding sequence and suggests that it will probably be processed abnormally or even degraded before it is able to reach the cytosol. By changing the codon usage of at least portions of the viral RNA, the viral genome can be modified to look more 'exon-like' (conversion of T-rich sequences to more GC-rich sequences, and removal of putative cryptic splice sites). Such changes have been useful to improve export of functional RNAs to the cytosol, but addition of plant introns improved the process even more significantly. The resultant synthetic T-DNA templates, when delivered as DNA precursors using *Agrobacterium*, provided efficient processing of the DNA information into active amplicons in almost all (>93%) cells of infiltrated tobacco (*Agrobacterium* infiltrated at optical density (OD) of 0.7), a 10^3-fold improvement compared with nonmodified vectors, and an up to 10^7-fold improvement over nonoptimized DNA templates reported in the first publications. Improved vectors lead to one successful infection event per 10–20 infiltrated agrobacteria (**Figure 1(d)**).

Using these vectors, a simple fully scalable protocol for heterologous protein expression in plants has been designed that does not require transgenic plants, but instead relies on transient amplification of viral vectors delivered to the entire plant using *Agrobacterium*. Entire plants are infiltrated (although the same procedure can be performed on detached mature leaves) with a highly diluted (up to 10^{-4} dilution) suspension of bacteria carrying a proviral amplicon on the T-DNA. The combination of vacuum infiltration/agroinfection can therefore be considered as replacing the conventional viral functions of primary infection and systemic movement. Viral vector-controlled amplification and cell-to-cell (short distance) spread is performed by the replicon, as is the case with standard vectors. Depending on the vector used, the host organism, and the initial density of bacteria, the magnifection process takes from 4 to 10 days and, for well-expressed proteins, results in level of up to 5 g recombinant protein kg^{-1} of fresh leaf biomass or over 50% of total soluble protein. Furthermore, since the viral vector lacks a CP gene (for tobamoviruses, the CP gene is needed only for plant infection or systemic movement, and is therefore not needed for magnifection), it can express longer genes (up to 2.3 kbp inserts or up to 80 kDa proteins). Infiltration of plants/detached leaves with bacteria can be performed in several ways, one simple process being vacuum infiltration by immersing whole aerial parts of plants in a bacterial suspension and applying a weak vacuum (approximately minus 0.8–1.0 bar) for 10–30 s. This process can be performed on a large scale and is expected to be commercially viable for production of up to 1 ton of recombinant protein per year. Industrial-scale production will require containment of infiltrated plants which contain agrobacteria (**Figures 2** and **3**).

Expression Vectors for Industrial-Scale Protein Production: Transgenic Systems

A radically different approach to establish an industrial expression system would be to produce a stably transformed plant that would contain an inactive or repressed copy of the viral replicon inserted on one of the host's chromosomes. Induction of viral replication could be activated at the time of the operator's choosing by a specific treatment. Obviously, establishing such a system would slow down the development phase by several months/years, in order to obtain stable transformants and select well-performing lines. However, this delay would be largely compensated by the ease of production that the system would provide and by the resulting low production costs.

Several strategies are possible for the design of an inactive replicon, and specific strategies will of course depend on the type of viral vector used. A viral vector might be encrypted by insertion of additional sequences within the vector, or inversion of part of the vector, and activation could be provided by regulated expression of a recombinase. Alternatively, a replicon could be kept inactive by silencing by the host plant, and activated by regulated expression of a silencing suppressor. Finally, a replicon could be kept in an inactive form by fusion to a promoter that is inactive in the absence of a specific activator or that is repressed in the presence or absence of a particular inducer.

The inducing treatment could be provided in many different ways. A simple strategy would be to induce release of the active replicon by genetic hybridization, which would combine in the same plant both the transgene locus encoding the replicon as well as the transgene(s) that control unencryption (e.g., a silencing suppressor or a recombinase that is developmentally regulated). Alternatively, hybridization could also be used to bring together genomes of multipartite viruses, thus providing for replication of RNAs that carry the transgene of interest but cannot replicate in the absence of the master RNA expressing the necessary RNA-dependent RNA polymerase.

A more versatile solution would be provided by spraying a small chemical inducer on an engineered plant. One technical challenge for establishing such a system is the design of an effective chemical switch that provides a tight control and can be effectively used for an industrial process. One specific issue in dealing with chemically inducible promoters for industrial purpose is the availability of small chemical molecules that are commercially available and safe. Among those available today, one should mention ethanol (used as a part of ethanol-inducible system), the commercial insecticide methoxyfenozide, which acts as an agonist for the ecdysone receptor, and tetracycline antibiotics.

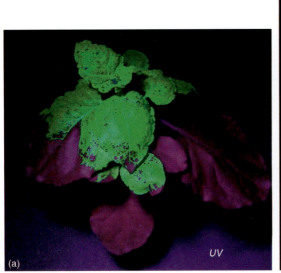 

Figure 2 Plant transfection with (a) first- and (b) second- generation vectors; GFP expression in *Nicotiana benthamiana*, photographed under UV light.

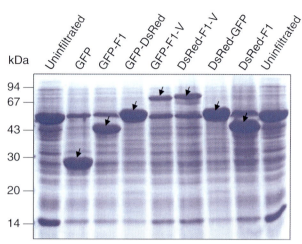

Figure 3 Expression of proteins and protein fusions using second-generation vectors; Coomassie-stained polyacrylamide protein gel showing protein profiles in crude extract from uninfected leaf tissue as well as tissues infected with genes encoding fluorescent protein GFP or protein fusions involving GFP or DsRed as fusion components. The arrow indicates the band with heterologous protein. Molecular weight ladder in kDa is shown on the left side.

Several groups are now working on the development of protocols relying on chemically inducible promoters for induction of both DNA and RNA viral vectors. As one example, an inducible geminivirus viral vector system based on bean yellow dwarf virus has been described. In this system, treatment with ethanol results in expression of the replicase, which induces release and replication of the gene-specific replicon and leads to expression of the gene of interest. Ethanol induction was reported to result in up to 80-fold increase of the mRNA levels and up to 10-fold increase of the translation product. In another example, an inducible system that relies on the estrogen-inducible promoter was shown to work efficiently in tobacco cell suspensions. Upon induction, a modified tomato mosaic virus expressing green fluorescent protein (GFP) instead of its CP amplified and expressed the gene of interest at high levels (10% of total soluble protein). In contrast, neither viral RNA nor GFP were detectable in uninduced cells (**Figure 1(e)**).

Despite these positive results, it is clear that the full potential of inducible viral vectors systems has still not been completely reached. Given the progress in this field, it is however expected that fully effective technology processes will be available within the next few years.

Vectors for Manufacturing 'Nanoscale' Materials

Due to their relatively simple macromolecular organization and very high accumulation titers, tobamoviruses provide an extremely cheap source of biopolymers than can be manufactured rapidly and under very simple conditions. TMV, the best-known tobamovirus, is also one of the most extensively studied viruses. It has a positive-sense RNA genome encoded in a single 6.4 kbp RNA molecule. The genome encodes four proteins, including the 17.5 kDa CP, the most abundant viral product and the only component of the TMV capsid. Over 2100 copies of CP fully protect the single-stranded viral RNA, resulting in rigid rod-shaped viral particles with a length of 300 nm and a diameter of 18 nm and a molecular mass of 40 000 kDa. The genomic RNA is packaged inside of a 2 nm-wide canal formed by the assembled CP capsid. Ninety-five percent of the mass of TMV particles consists of the CP. TMV accumulates to levels of up to $10\,g\,kg^{-1}$ of leaf biomass, and, therefore, the CP represents the most abundant individual protein that can be harvested from plants. The virions can be purified industrially using simple 'low-tech' protocols.

Due to the ability of the CP to polymerize *in vivo* and *in vitro* and the high stability and defined size of the assembled virions, the CP represents a potentially promising biopolymer feedstock for a number of applications in nanobiotechnology. The CP, the basic element of the 'viral polymer' (the viral particle) can be modified in two different ways.

One strategy consists of attaching novel chemical moieties to the viral particles *in vitro* by chemical modification of reactogenic groups exposed on the surface of the virus or in the inner cavity. One such reactogenic group, lysine, allows biotinylation of the capsid. In the case of the TMV CP, which does not contain a lysine, a randomized library was made and screened to introduce this desired amino acid at an externally located position, and this without affecting the formation of viral particles. The protein/peptide fusion component was then expressed independently as a streptavidin fusion, and used to decorate the TMV particles.

Another strategy for modification of the viral core monomer, the CP, consists of engineering the viral vector to express a protein fusion. Since the structure of many viral particles has been determined at atomic resolution, fusion proteins can be designed in such a way that the added sequences are predicted to lie at the surface of the assembled viral particle. For TMV, fusions can theoretically be made either at the N- or the C-terminal ends of the CP since both ends are exposed on the surface of the viral particle. For other viruses, short peptide epitopes can be inserted in surface-exposed loops. However, constraints such as the size or the pI of the inserted sequence are limiting the type of epitopes that can be successfully expressed using this strategy.

So far, extensive work by numerous groups to create new products based on protein fusions has met with limited success. Since the main goal of these studies was the design of new vaccines by surface display of immunotopes, only short inserts were extensively analyzed. For example, the limited size of peptides that could be fused to the CP of TMV without preventing virion assembly has restricted these systems to expression of 20-amino-acid (aa) or shorter peptide immunogens and of one peptide hormone only. However, it has recently been found that much longer polypeptides of up to 133 aa (for a fully functional fragment of protein A) can be displayed on the surface of TMV as C-terminal fusions to the CP, provided a flexible linker is being used. These nanoparticles coated with protein A allow purification of Mabs with a recovery yield of 50% and higher than 90% purity. The extremely dense packing of protein A on the nanoparticles confers an immunoadsorbent feature with a binding capacity of 2 g Mab per g. This characteristic, combined with the high level of expression of the nanoparticles (more than $3\,g\,kg^{-1}$ of leaf biomass), provides a very inexpensive self-assembling matrix that could meet industrial criteria for a single-use immunoadsorbent for antibody purification.

The field of nanomaterials built around plant viruses as scaffolds is still *in statu nascendi*, and we believe that it will offer multiple practical solutions not only to specific problems such as mentioned above for downstream processing of antibodies and other proteins, but also to entirely new technology processes that are difficult to imagine today.

Conclusions

Although there are currently no products or medicines in the market produced using viral vectors, compared to other expression platforms utilizing plants, this technology has made tremendous progress since its inception approximately 15 years ago (**Figure 4**), and it is simply a matter of time before products made with viral vectors become available. This progress has come as a result of the extensive studies that revealed the structure, functions, and molecular organization of plant viruses. In 1955, fascinated by the elegant simplicity of TMV's morphology, F. Krick supposedly exclaimed: 'Even a child could make a virus'. Today, some 50 years later, we tend to agree with him.

Acknowledgments

The authors wish to thank Dr. Anatoliy Girich, Icon Genetics, Halle, Germany, for the materials used in **Figure 3**.

See also: Virus Particle Structure: Nonenveloped Viruses; Virus Particle Structure: Principles.

	Nuclear transformation	Plastid transformation	*Agrobacterium*, transient	Virus, first generation	Virus, second generation
Yield	++	+++	+	++	+++
Protein yield	++	+++	+	++	+++
R&D speed	+	+	+++	+++	+++
Large genes, complex proteins	+++	+++	+++	+	++
Post-translational processing	+++	+	+++	+++	+++
Scalability	+++	+++	+	+++	+++

Figure 4 Strengths and weaknesses of different plant expression platforms. The platforms include nuclear transformation, plastid transformation, *Agrobacterium*-mediated transient expression, and the two viral expression platforms. The parameters that are most essential for expression of recombinant proteins, in particular biopharmaceuticals, include yield of recombinant protein, relative yield as a percentage of total soluble protein, speed of research and development supported by the platform, ability to express large genes and manufacture complex proteins, post-translational processing capabilities and scalability. The best performance is assigned three crosses, the worst is assigned one cross.

Further Reading

Canizares MC, Nicholson L, and Lomonossoff GP (2005) Use of viral vectors for vaccine production in plants. *Immunology and Cell Biology* 83: 263–270.

Donson J, Kearney CM, Hilf ME, and Dawson WO (1991) Systemic expression of a bacterial gene by a tobacco mosaic virus-based vector. *Proceedings of the National Academy of Sciences, USA* 88: 7204–7208.

Gleba Y, Marillonnet S, and Klimyuk V (2004) Engineering viral expression vectors for plants: The 'full virus' and the 'deconstructed virus' strategies. *Current Opinion in Plant Biology* 7: 182–188.

Ma JKC, Drake PMW, and Christou P (2003) The production of recombinant pharmaceutical proteins in plants. *Nature Genetics* 4: 794–805.

Mallory AC, Parks G, Endres VB, *et al.* (2002) The amplicon-plus system for high-level expression of transgenes in plants. *Nature Biotechnology* 20: 622–625.

Marillonnet S, Thoeringer C, Kandzia R, Klimyuk V, and Gleba Y (2005) Systemic *Agrobacterium tumefaciens*-mediated transfection of viral replicons for efficient transient expression in plants. *Nature Biotechnology* 23: 718–723.

Pogue GP, Lindbo JA, Garger SJ, and Fitzmaurice WP (2002) Making an ally from an enemy: Plant virology and the new agriculture. *Annual Review of Phytopathology* 40: 45–74.

Porta C and Lomonosoff GP (2002) Use of viral replicons for the expression of genes in plants. *Biotechnology & Genetic Engineering Reviews* 19: 245–291.

Smith M, Lindbo JA, Dillard-Telm S, *et al.* (2006) Modified tobacco mosaic virus particles as scaffolds for display of protein antigens for vaccine applications. *Virology* 348: 475–488.

Yusibov V, Shivprasad S, Turpen TH, Dawson W, and Koprowski H (1999) Plant viral vectors based on tobamoviruses. *Current Topics in Microbiology and Immunology* 240: 81–94.

Plum Pox Virus

M Glasa, Slovak Academy of Sciences, Bratislava, Slovakia
T Candresse, UMR GDPP, Centre INRA de Bordeaux, Villenave d'Ornon, France

© 2008 Elsevier Ltd. All rights reserved.

Introduction

Plum pox virus (PPV), the agent responsible for the Sharka disease, belongs to the genus *Potyvirus*. The natural host range of this virus is restricted to *Prunus* spp. (stone fruits and ornamental trees). The infection of susceptible genotypes results in characteristic foliar and fruit symptoms and premature fruit drop. The wide geographical distribution of PPV includes most of Europe and the Mediterranean region as well as some countries in Asia and North and South America, although with widely different incidence levels in different countries. PPV is transmitted nonpersistently by more than 20 aphid species, by grafting and vegetative multiplication of infected plants, but is not seed-borne. To date, six strains/groups of PPV have been identified based on biological, serological, and molecular properties (M, D, Rec, C, EA, and W). Although many diagnostic tools are available for the sensitive and/or specific detection of PPV, its uneven distribution in infected woody hosts and its low titer outside of the active growth period significantly complicate its detection. In regions free of PPV, strict quarantine measures are usually enforced. In the quasi-absence of resistant fruit tree varieties, a mix of prophylactic approaches including the use of virus-free propagation material, eradication of diseased trees, and vector control is generally used in an effort to control the virus in regions where it has not reached an endemic status.

Economical Importance

PPV is considered as the most detrimental viral pathogen of stone fruit crops (peach, apricot, plum, Japanese plum). During approximately a century of recognized existence, PPV had a devastating effect on the European stone-fruit industry, mainly in the central and south European countries. Fruit trees infected with PPV cannot be cured and are often eliminated as a consequence of disease eradication or containment efforts. Although the infected trees are usually not stunted and do not die, fruit yields can be severely affected. Besides foliar symptoms, the virus often severely damages fruits, so that they have decreased weight and sugar content, overall lower gustative quality and often become blemished and are frequently unsuitable for consumption or processing. Premature fruit drop can also be observed and may reach 80–100% in susceptible cultivars. Consequently, traditional susceptible cultivars have to be replaced by less susceptible or tolerant cultivars, which are often of lower agricultural or gustatory quality. Infected propagation material and nursery stock (rootstock, budwood, scions) is not marketable and has to be destroyed. Because of its high potential impact on stone fruit crops, plum pox virus is listed as a quarantine pathogen in many parts of the world. In Europe for example, PPV is listed in the EC Plant Health Directive (Annex II of the European Union council directive 2000/29/EEC).

History and Geographical Distribution

The sharka disease was observed for the first time around 1917 on plum cv. Kjustendil in the village of Zemen in Bulgaria, near the Yugoslavian border. The disease is named according to the characteristic symptoms on fruits (sharka means pox in Bulgarian). Initially described on plum, the virus was observed in 1933 on apricot, in the 1960s on peach, and in the 1980s on sour and sweet cherry.

After World War I, the virus progressively spread to a large part of the European continent and Mediterranean basin, probably mostly as a consequence of exchange of infected propagation material. In recent years, the virus has been reported from China, South America (Chile, Argentina), and North America (USA, Canada). However, the prevalence of the disease differs from region to region, from the endemic occurrence observed in the central and eastern European countries, where the virus is well established, to local and more limited incidence observed in other countries where the virus was only introduced recently or where strict phytosanitary control measures have been enforced and have successfully retained the virus under some level of control.

Host Range and Symptomatology

Natural host range is restricted to species of the genus *Prunus*, including cultivated stone fruits – plum (*P. domestica*), Japanese plum (*P. salicina*), apricot (*P. armeniaca*), peach and nectarine (*P. persica*), almond (*P. amygdalus*). Although not infected by the majority of PPV isolates, sweet cherry (*P. avium*) and sour cherry (*P. cerasus*) are also natural hosts for some specific PPV isolates.

Symptoms vary depending on virus isolate, host/cultivar susceptibility, physiological status and age of the host, and environmental conditions, such as temperature.

Depending on the host, the symptoms may affect leaves, flowers, fruits, and stones. In recently infected trees, symptoms tend to be restricted to only some parts of tree but tend to generalize with time. Under field conditions, the symptoms on *Prunus* plants infected with PPV are often masked late in the season or during the warm period of the growing season.

Mixed infection with other viruses such as prunus necrotic ringspot virus (PNRSV) or prune dwarf virus (PDV) may further increase the severity of symptoms.

Foliar symptoms on plum consist generally of pale green chlorotic rings, spots, or patterns. Susceptible cultivars develop shallow ring or arabesque depressions on fruits, sometimes with brown or reddish necrotic flesh and gumming. Tolerant plum cultivars show no symptoms on fruits.

Infected apricots develop chlorotic or pale-green rings and lines on leaves, light-colored depressed rings on fruits, which may be severely deformed. Stones are marked with typical discolored rings.

Symptoms on susceptible peach genotypes are pronounced vein clearing, small chlorotic blotches, and distortions of the leaves. Color-breaking symptoms on the petals are observed in some varieties. Pale rings or diffuse band are visible on the skin of the fruits. In general, the symptoms on peach tend to be less visible in comparison with those on plums and apricots. Infection of almond is often symptomless or with limited foliar symptoms.

Characteristic symptoms on cherries consist of pale-green patterns and rings on leaves. Fruits may be slightly deformed, with chlorotic and necrotic rings and notched marks.

Premature dropping of fruits is frequently observed in the most susceptible varieties of the various hosts, in particular plum and apricot. Depending on host (cultivar) susceptibility, the losses may reach up to 100%.

PPV also infects wild and ornamental trees, such as myrobalan (*P. cerasifera*), Japanese apricot (*P. mume*), nanking cherry (*P. tomentosa*), Canada plum (*P. nigra*), American plum (*P. americana*), dwarf flowering almond (*P. glandulosa*), and blackthorn (*P. spinosa*), which may act as local reservoirs of the virus.

The experimental host range of PPV is large, with over 60 reported host species (such as *Chenopodium foetidum*, *Nicotiana benthamiana* and *N. clevelandii*, *Pisum sativum*, *Ranunculus arvensis*, *Senecio vulgaris*, *Stellaria media*, etc.) in eight families. However, no epidemiologically significant contribution of weeds and annual herbaceous plants to the spread of plum pox virus has been reported.

In woody hosts, the virus often shows an uneven distribution within the tree and full systemic invasion of a tree may require several years. Long-distance viral movement is restricted in some resistant (apricot, peach) and hypersensitive (plum) genotypes. The irregular distribution and translocation of the virus in the tree and the low titer outside the active growth period may complicate the detection of the virus when methods of insufficient sensitivity are used.

Virion Structure and Genome Properties

The flexuous filamentous viral particles are approximately 750×15 nm. Viral particles are composed of the single-stranded genomic RNA encapsidated by a single type of capsid protein subunit. The genome is of positive polarity and is 9741–9795 nucleotides in length. It has a polyadenylated $3'$ end and a virus-encoded protein (VPg) covalently bound at its $5'$ end. The genomic organization is typical of potyviruses, with a single open reading frame encoding a large polyprotein precursor (3125–3143 amino acids, *c.* 355 kDa) that is proteolytically processed by three virus-encoded proteinases (P1, HC-Pro, and NIaPro) to yield as many as ten mature functional proteins.

Strains/Groups

Initially, different strains or groups of isolates have been described on the basis of symptomatology in various experimental hosts. The most developed analysis of this kind relied on symptoms on *C. foetidum*, an experimental local-lesion host. PPV isolates were classified as yellow, intermediate, and necrotic strains depending on the type of local lesions they induced. However, this phenotypic characterization proved to be at least to some extent under environmental control and no relationships could be later established between this biological property and serological/molecular properties.

In the late 1970s, the existence of two serogroups of PPV was demonstrated using agar double diffusion assays employing polyclonal antibodies and a purified, formaldehyde-treated, suspension of undegraded viral particles. These two serogroups were named M (based on the type isolate Marcus from peach in Greece) and D (based on the type isolate Dideron from apricot in southeastern France). Later it was shown that both strains differ also in the coat protein (CP) mobility in denaturing polyacrylamide gel electrophoresis (PAGE) and in the presence or absence of an *Rsa* I restriction site in the region encoding the CP C-terminus.

On the basis of recent molecular and serological analysis, six strains/groups of PPV isolates sharing common biological, serological, and molecular properties are now recognized. This classification has also been validated by comparisons of complete genomic sequences of isolates representative of each of these strains (**Figure 1**).

PPV-M (from the type isolate Marcus). The isolates of this group are present in many European countries, but absent from the Americas and China. They are often

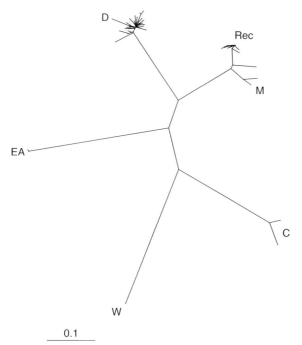

Figure 1 Unrooted phylogenetic tree computed using full-length coat protein gene sequences showing the relationships between isolates representative of all six known PPV strains.

Table 1 Origins and references of plum pox virus isolates for which complete genomic sequences are available

Isolate	Strain	Country of origin	Original host	Reference
PS	M	Yugoslavia	Peach	AJ243957
SK-68	M	Hungary	Plum	M92280
Dideron	D	France	Apricot	X16415
NAT	D	Germany	?	NC_001445
SC	D	?	?	X81083
PENN-1	D	USA	Peach	AF401295
PENN-2	D	USA	Plum	AF401296
PENN-3	D	USA	Peach	DQ465242
PENN-4	D	USA	Peach	DQ465243
Fantasia	D	Canada	Nectarine	AY912056
Vulcan	D	Canada	Peach	AY912057
48–922	D	Canada	Peach	AY912058
BOR-3	Rec	Slovakia	Apricot	AY028309
El Amar	EA	Egypt	Apricot	AM157175
El Amar	EA	Egypt	Apricot	DQ431465
SoC	C	Moldova	Sour cherry	Y09851
SwC	C	Italy	Sweet cherry	AY184478
W3174	W	Canada	Plum	AY912055

associated with rapidly spreading epidemics in peach but are less frequently found on plums. Usually, PPV-M isolates are transmitted efficiently by aphids.

PPV-D (from the type isolate Dideron). PPV-D isolates are widespread in all areas where PPV has been reported, including the recent outbreaks in Asia and South and North America. Isolates infect all the susceptible *Prunus* species excluding cherries, but PPV-D isolates are less frequently associated with spreading epidemics in peach.

PPV-Rec (from 'recombinant'). PPV-Rec isolates are derived from a single homologous recombination event between PPV-M and PPV-D isolates, with a cross-over located in the 3′ terminal part of the NIb coding region. This group of isolates was recognized only recently through the use of improved strain-typing methods. It was in fact found that a number of isolates originally described as PPV-M belong in fact to the PPV-Rec group. Isolates belonging to the PPV-Rec strain are widespread in several central and eastern European countries, frequently associated with plums and efficiently transmitted by aphids.

PPV-EA (from the type isolate El Amar) was originally isolated from apricots in the El Amar region of the Nile delta in Egypt in the late 1980s. It has also been observed in peach and in other regions of Egypt but has not been reported from outside this country thus far.

PPV-C (from 'Cherry') was first reported on sour cherry in Moldova and on sweet cherry in Italy. This group of cherry-adapted isolates seems sporadically present in some central and eastern European countries. PPV-C isolates are aphid transmissible and seem to be the only PPV isolates able to systemically infect cherry and *P. mahaleb*. PPV-C isolates are however also able to infect other *Prunus* species under experimental conditions.

PPV-W (from 'Winona') has been recently reported from a few infected plum trees in Canada. The PPV-W genome shows unique features, justifying its classification as a new strain but the epidemiology and other biological properties of this recently characterized strain remain to be determined.

Full-length genomic sequences have been determined for 18 PPV isolates representing each of the six recognized strains (**Table 1**). Moreover, partial sequences, focusing mainly on the 3′ terminal part of genome, are available for a large number of isolates, making PPV one of the genetically best studied potyviruses. The comparison of complete genomic sequences revealed up to 27.7% nucleotide divergence between representative isolates of the PPV-M, D, Rec, EA, C, and W strains (**Table 2**).

Recombination seems to have played a significant role in the evolutionary history of PPV. The PPV-Rec strain derives from a single recombination event involving isolates belonging to the PPV-D and PPV-M strains, with a recombination breakpoint in the C-terminus of the NIb gene. In addition, analysis of complete genomic sequences has recently demonstrated that the PPV-D and PPV-M strains themselves share an ancestrally recombined 5′ part of their genome (5′ noncoding region, P1, HC-Pro, and N-terminus of P3). Other recently

Table 2 Pairwise genetic distances calculated on the complete genomes of representative isolates of all six known PPV strains

	PPV-M	PPV-D	PPV-Rec	PPV-EA	PPV-C
PPV-D	0.127				
PPV-Rec	0.106	0.046			
PPV-EA	0.239	0.237	0.233		
PPV-C	0.257	0.266	0.260	0.277	
PPV-W	0.238	0.235	0.233	0.265	0.234

identified recombination events involve PPV-W, which shows a clear recombination signal with PPV-M in the region extending from the P1 coding region to the HC-Pro region and a divergent PPV-M isolate from Turkey (AY677114). It thus appears that at least four of the six recognized strains of PPV (among which the three most prevalent, PPV-D, -M, and -Rec) have recombination events in their evolutionary history. One remarkable consequence is that these four strains all share a similar genomic region corresponding to the C-terminus of their P1 and the N-terminus of their HC-Pro genes (genomic positions ~800–1450). On the other hand, PPV-C and PPV-EA appear to represent independent evolutionary lineages.

Assays that allow the discrimination between the various PPV strains include serological analysis with strain-specific monoclonal antibodies, restriction fragment length polymorphism (RFLP) analysis of polymerase chain reaction (PCR) fragments derived from the various genomic regions (CP, P3–6K1, and CI), PCR with strain specific primers and partial or complete sequence analysis. In addition, proper identification of recombinant isolates may require the use of several techniques targeting different parts of the viral genome or specific primers with binding sites located on both sides of the targeted recombination breakpoint. Strain-specific monoclonal antibodies have so far been obtained for isolates belonging to the PPV-M, PPV-D, PPV-EA, and PPV-C groups, but some isolates may show abnormal typing properties using either monoclonal antibodies or RFLP analysis techniques so that only PCR-based strain-specific assays or sequencing can provide unambiguous PPV strain identification.

Virus Spread

Natural spread of PPV occurs through aphids from neighboring infected reservoirs (cultivated, wild, or ornamental trees), while long-range movement of the virus is linked to international exchange of contaminated propagation material. The recent demonstration that aphids can acquire and transmit the virus from contaminated fruits indicates that fruit shipments could also represent a pathway for international virus movement. The virus is thus usually introduced in a new location by infected propagation material. Once established in the orchard or nursery, the virus is rapidly spread by aphid vectors. PPV is naturally transmitted by over 20 different aphid species in a nonpersistent manner (*Aphis craccivora, A. gossypii, A. hederae, A. spiraecola, Brachycaudus cardui, B. helichrysi, B. persicae, Myzus persicae, M. varians, Phorodon humuli, Rhopalosiphum padi*, etc.). Aphids generally transmit the virus over short distance, that is, next to infected sources, when migrating between plants and making test probes. The virus is transmitted in a nonpersistent manner so that aphids may acquire the virus very rapidly when probing infected plants with their stylets (piercing–sucking mouthpart). However, aphids can only transmit the virus for a short time after virus acquisition as the virus is retained in a viable state in the aphid's mouthparts for a period of only a few minutes to a few hours. The virus appears to be lost by the aphid the first time the aphid probes on the next plant and aphids do not remain infectious after moulting nor do they pass PPV onto their progeny.

The efficiency of transmission depends on the particular aphid species, virus isolate, and host species from which the virus is acquired and to which the virus is transmitted. As for other potyviruses, two viral proteins are known to be involved in the transmission mechanism, HC-Pro (helper component) and CP. The DAG amino acid motif is highly conserved in the N-terminus of the CP protein of aphid-transmissible PPV isolates and loss of transmissibility is correlated with its deletion.

Conflicting results have been reported in the past concerning seed transmission in some *Prunus* species, but recently more thorough studies have failed to demonstrate seed transmission of PPV in any of its woody hosts.

Ecology and Control

In the countries and regions where the virus is not yet present, strict quarantine measures need to be established to prevent the introduction of PPV through legal or illegal importation of infected fruit tree propagation material or fruits. In regions where the disease is present but still under control, a mix of prophylactic approaches is usually implemented, including the use of virus-free propagation material (rootstocks, budwoods), tight aphid vector control, regular inspection of orchards, and prompt eradication of infected plants. If the rate of infected trees in the prospected orchard exceeds a critical level (10–20% in most countries), whole orchards have to be removed.

In regions where eradication is no longer an option and where the spread of the disease is no longer under control (e.g., central and southeastern Europe), the cultivation of less susceptible or tolerant varieties is so far the only option that may allow to continue the production of susceptible fruit species. This practice, however, further contributes to the viral spread.

For PPV control in the long term, most European countries are placing strong emphasis on breeding programs aiming at the development of PPV-resistant varieties. Unfortunately, only a very limited number of natural sources of resistance to PPV have been identified within the *P. armeniaca* and *P. domestica* germplasm, and these are often characterized by low-quality fruits. In peach, despite extensive screening efforts, no resistant varieties are known but resistance has been identified in the wild relative *P. davidiana*. Breeding efforts are ongoing using these various resistance sources to introduce PPV resistance in commercially interesting varieties. Progress is however slowed by the long generation times, the length of the resistance tests, and the often polygenic nature of the resistances involved.

Development of genotypes with a hypersensitive response as an active defense mechanism against PPV is another promising way to produce resistant fruit trees, as was demonstrated for some plum varieties, but no hypersentivity sources are so far known in apricot and peach. In addition, it should be cautioned that in herbaceous crops, hypersensitive resistance to viruses has often proved to be nondurable as a consequence of genetic evolution of the virus.

Early attempts to use cross-protection with attenuated virus isolates, a technique successfully used to control some other potyviruses (zucchini yellow mosaic virus, papaya ringspot virus), have not met with success in the case of PPV.

Transgene-based resistance offers a complementary approach for the development of PPV-resistant stone fruit cultivars. Experimental herbaceous plants transformed with different regions of the PPV genome (i.e., P1, HC-Pro, CI, NIa, NIb, or CP) have been developed and have shown partial or complete resistance to PPV. A transgenic clone of *P. domestica* (C5), containing the PPV capsid protein gene, has been described as highly resistant to PPV in greenhouse and field tests, displaying characteristics typical of post-transcriptional gene silencing (PTGS) based resistance. Although this resistance can be partially overcome by graft inoculation, it provided effective protection against the natural spread of PPV in several field tests, even under conditions of high natural inoculation pressure.

See also: Papaya Ringspot Virus; Potato Virus Y; Potyviruses; Watermelon Mosaic Virus and Zucchini Yellow Mosaic Virus.

Further Reading

EPPO (2004) EPPO Standards PM 7/32 Diagnostic Protocol for Plum pox potyvirus. *EPPO Bulletin* 34: 247–256.

EPPO (2006) A review of plum pox virus. *EPPO Bulletin* 36: 201–349.

Glasa M and Candresse T (2005) Plum pox virus. AAB Description of Plant Viruses. No. 410. http://www.dpvweb.net/dpv/showdpv.php?dpvno=410 (accessed July 2007).

Kegler H, Fuchs E, Gruntzig M, and Schwarz S (1998) Some results of 50 years of research on the resistance to plum pox virus. *Acta Virologica* 42: 200–215.

López-Moya JJ, Fernández-Fernández MR, Cambra M, and García JA (2000) Biotechnological aspects of plum pox virus. *Journal of Biotechnology* 76: 121–136.

Martinez-Gomez P, Dicenta F, and Audergon JM (2000) Behaviour of apricot (*Prunus armeniaca* L.) cultivars in the presence of sharka (*Plum pox potyvirus*): A review. *Agronomie* 20: 407–422.

Poliomyelitis

P D Minor, NIBSC, Potters Bar, UK

© 2008 Elsevier Ltd. All rights reserved.

Introduction

A funerary stele from about 1300 BC currently in the Carlsberg museum at Copenhagen shows the priest Rom with the withered single limb and down-flexed foot typical of motor neuron destruction caused by poliovirus. It is considered to be the first documentary evidence for an infectious disease of humans. Two highly effective vaccines were developed in the 1950s, and a major program of the World Health Organization (WHO) is underway which may make poliomyelitis the second human disease to be eradicated globally after smallpox.

Pathogenesis and Disease

Poliomyelitis gets its name from the specificity of the virus for the motor neurons that form the gray matter (*polios* and *myelos*, Greek for gray and matter, respectively) of the anterior horn. Despite the obvious physical signs of muscle atrophy and motor neuron degeneration, few cases of poliomyelitis are identifiable in the literature before the end of the nineteenth century, and it is believed that it was extremely rare. However, at this time, the disease began to occur in large epidemics, initially in Scandinavia (particularly Sweden), and then in the USA, most commonly affecting young children (hence the alternative name of infantile paralysis). The legs are more commonly affected than the arms, and paralysis tends to occur in one limb rather than symmetrically.

Poliovirus occurs in three antigenically distinct serotypes designated 1, 2, and 3, such that infection with one serotype confers solid protection only against other viruses of that serotype. Infections primarily occupy the gut, and most are entirely silent, but in a small number of cases they lead to a systemic infection (the minor disease) 3–7 days post exposure, characterized by fever, rash, or sore throat. Depending on the strain or type of virus, usually less than 1% of all infections lead to the major disease or poliomyelitis which develops on average 7–30 days after infection. Spinal poliomyelitis resulting in lower limb paralysis or bulbar poliomyelitis in which the breathing centers are affected occurs when the lower or upper regions, respectively, of the spinal cord are affected. In the mid-twentieth century, about 10% of cases recovered without sequelae, 80% had permanent residual paralysis, and the remainder died. Encephalitis occurs but is rather rare. Meningitis, also known as abortive or nonparalytic poliomyelitis, occurs at a rate of about 5% of poliomyelitis cases.

Infection is mainly fecal–oral, although the virus can also be transmitted by the respiratory route from the throats of infected individuals. The primary site of infection in the gut is not known but may be the lymphoid tissues, specifically the Peyer's patches and tonsils or the mucosal surfaces in the gut or throat; infectious virus can be found in the local lymph nodes and in more distal mesenteric lymph nodes but it is not clear that the virus necessarily replicates there. The Peyer's patch-associated M-cells are believed by many to be the major infected cells in the gut.

Viremia may occur about 7 days after infection, corresponding roughly to the appearance of the minor disease. It is believed to seed sites including the peripheral and central nervous systems resulting in the major disease. Thus humoral antibodies should prevent poliomyelitis by blocking viremia; the protective effect of passively administered humoral antibodies was shown in the 1950s. This explains the change in the epidemiology of the disease at the start of the twentieth century as hygiene improved, and children were exposed to infection later in life when maternal antibody levels had declined to levels no longer able to confine the infection to the gut.

The Virus Genome

The virus contains a single strand of messenger sense RNA of about 7500 nt. A long highly structured 5′ noncoding region of about 740 nt which serves as an internal ribosomal entry site (IRES) is followed by a single open reading frame and terminates in a 3′ noncoding region of c. 70 nt followed by a polyadenylate tract. The 5′ end of the genome is covalently linked to a virus-encoded protein termed VPg. The single open reading frame encodes the structural proteins which make up the capsid (collectively termed the P1 region) followed by the regions encoding the nonstructural proteins P2 and P3. P1 is divided into VP1, VP3, and VP0 (VP2 plus VP4). P2 is cleaved into $2A^{pro}$, 2B, and 2C, and P3 into 3A, 3B (or VPg), $3C^{pro}$, and $3D^{pol}$ by virus-encoded proteases. $3D^{pol}$ is the viral polymerase, but all proteins in the nonstructural part of the genome play a role in RNA replication, which also depends on RNA structural elements and host cell membranes which are extensively rearranged in the course of infection.

The Virus Particle

The infectious virus particle consists of 60 copies each of VP1, VP2, VP3, and the smaller protein VP4; VP2 and VP4 are generated by autocatalytic cleavage of VP0, in the last stage of the maturation of the particle. The proteins are arranged with icosahedral symmetry such that VP1 is found at the pentameric apex of the icosahedron and VP2 and VP3 alternate around the pseudo-sixfold axes of symmetry of the triangular faces of the icosahedron. VP4 is located internally about the pentameric apex and is myristylated. The pentameric apex of the particle is surrounded by a dip or canyon into which the cellular receptor for poliovirus fits. The atomic structure of the virus was solved in 1985.

The Poliovirus Receptor

Although some strains of type 2 poliovirus are able to infect normal mice, only the higher primates and Old World monkeys are susceptible to infection by all serotypes as they possess the specific receptor site required, and this species restriction is one of the factors that makes eradication theoretically possible. The receptor site was identified in 1986 and is a three-domain membrane protein of the immunoglobulin superfamily termed CD155. It is necessary and sufficient for the infection of cells *in vitro*, and transgenic mice carrying the gene for the polio receptor can be infected by all poliovirus types, developing paralysis when infected with wild-type virus. Such mice have been developed as alternatives to monkeys in the safety testing of live polio vaccines.

The tissue distribution of the receptor site does not explain the targets of infection of the virus and it is possible that innate immunity plays a significant role in its highly specific tropism in normal individuals.

Classification

The members of the family *Picornaviridae* to which poliovirus belongs are small nonenveloped viruses of about

30 nm in diameter when hydrated, essentially featureless when examined by electron microscopy and containing a single strand of an RNA molecule of messenger sense. The family is currently split into nine genera on the basis of sequence similarities, as shown in **Table 1**.

The genus *Enterovirus*, of which poliovirus is the archetypal species, is further subdivided into eight species of which *A* to *D* and the *Poliovirus* infect humans. The classification is summarized in **Table 2**, with examples of the viruses within the species.

The sequences of the capsid region of the viruses within a species define a virus type while sequences of other regions of the genome do not. This suggests that viruses within a particular species can exchange genome segments more or less freely. On this basis, polioviruses should be assigned to the species *C enterovirus* but at the time of writing they remain a separate species because of the unique human disease they cause.

The comparison of the sequence of field isolates has proved a very valuable tool in the WHO program to eradicate polio. Viruses from the same geographical region tend to cluster in terms of their sequence largely independent of the year of isolation. Thus it is possible to identify a virus from a case as indigenous to the region where it was isolated or an importation. Similarly, as virus circulation is obstructed by effective vaccination programs, the variety of sequences found declines, indicating progress long before the disease is eradicated. Finally, it is possible to identify strains as wild type or vaccine derived. It has become clear that the rate at which the sequence of a vaccine or wild-type virus lineage drifts in an epidemic or during chronic infection of immunodeficient individuals unable to clear the virus is amazingly constant at about 2–3% silent substitutions per year, or 1% for all substitutions. Thus the comparison of viral sequences can date the common ancestor of two polioviruses accurately and therefore how long they have been circulating.

Vaccines

Inactivated polio vaccine (IPV) was developed by Salk. It consists of wild-type virus that has been treated with low concentrations of formalin sufficient to inactivate the virus without affecting its antigenic properties to any great extent. The vaccine was first licensed in 1955 in the USA and reduced the number of cases by over 99%. It has been shown to be capable of eradicating the disease in certain countries, notably Scandinavia and the Netherlands. However, there remained controversy over whether a nonreplicating vaccine could eradicate disease and the virus altogether in less-developed countries where exposure is more intense, and it was possible that live-attenuated vaccines imitating natural infection might give better protection against infection by wild-type viruses. The live-attenuated vaccines given by mouth (oral polio vaccine, OPV) that form the basis of the current eradication campaign were developed by Sabin and introduced in the early 1960s. The Sabin vaccines have been shown to be able to interrupt epidemics and to break transmission if used correctly, resulting in the eradication of the virus in entire regions and possibly eventually the world.

It was also known that the Sabin vaccines altered in the vaccine recipients, becoming more neurovirulent particularly in the case of the type 3 component, and that in rare instances, now estimated at about 1 case per 750 000 first-time vaccinees, the vaccine could cause poliomyelitis in recipients and their immediate contacts. Both IPV and OPV contain a single representative of each of the three serotypes.

Attenuation of the Sabin Vaccine Strains

The genome of poliovirus is of positive (messenger) RNA sense and therefore infectious. Complete genomes of the virulent precursors of the Sabin vaccine strains or isolates

Table 1 Genera within the family *Picornaviridae*

Genus
Enterovirus
Rhinovirus
Cardiovirus
Aphthovirus
Hepatovirus
Parechovirus
Erbovirus
Kobuvirus
Teschovirus

Table 2 Species within the genus *Enterovirus*

Species	Serotype
Poliovirus	1, 2, 3
Human enterovirus A	Coxsackievirus A2, 3, 5, 7, 8, 10, 12, 14, 16
	Enterovirus 71
Human enterovirus B	Coxsackievirus B 1–6, coxsackievirus A 9
	Human echovirus 1–7, 9, 11–21, 24–27, 29–33
	Human enterovirus 69
Human enterovirus C	Coxsackie virus A 1, 11, 13, 15, 17–22, 24
Human enterovirus C	Human enterovirus 68, 70
Bovine enterovirus	Bovine enterovirus 1, 2
Porcine enterovirus A	Porcine enterovirus 8
Porcine enterovirus B	Porcine enterovirus 9, 10

from vaccine-associated cases were cloned, sequenced, and mutations introduced or segments exchanged. Virus recovered from RNA transcribed from the modified plasmids was tested in monkeys or in transgenic mice carrying the poliovirus receptor. It was possible to identify two differences between Leon, the virulent precursor of the Sabin type 3 vaccine strain which would attenuate Leon or would de-attenuate the Sabin strain, one in the 5' noncoding region, the other in capsid protein VP3. Similarly, there were two major attenuating mutations in the type 2 strain, one in the 5' noncoding region, the other in capsid protein VP1. The story with type 1 was more complicated. Again there was one mutation in the 5' noncoding region but several throughout the capsid region. All three of the 5' noncoding mutations affect the highly structured region shown in **Figure 1**.

All 5' noncoding mutations have been shown to affect initiation of protein synthesis, all involve allowed but altered base pairing, and all revert or are suppressed in vaccine recipients when sequential isolates are made. Excretion of virus following vaccination with OPV continues on average for 4 or 5 weeks and the reversions observed occur early in the excretion period, usually within 1 or 2 days for the type 3 strain and within a week for the other two types. Other changes also take place, including the reversion or more usually direct or indirect suppression of the VP3 mutation in the type 3 strain.

In addition, the viruses recombine with each other at high frequency. The type 3 strain in particular is usually excreted as a recombinant from about 11 days post immunization with the part of the genome encoding the nonstructural genes from the type 1 or type 2 strains. Complex recombinants with portions from different serotypes are also common, although their selective advantage is not known. The adaptation of the virus to the gut is therefore rapid and subtle and by a variety of mechanisms including reversion, second site suppression, and suppression of the phenotype of poor growth by enhancing fitness by an entirely unrelated route, and may involve mutation or recombination. Once adapted, virus excretion persists typically for several weeks. The drift in the viral sequence over prolonged periods of time referred to above occurs in addition to these selected changes, and appears to be a consequence of a purely stochastic process.

The Global Polio Eradication Program

For many years, it was thought that OPV could not be effective in tropical countries, and despite large-scale use its impact was in fact minimal. The reasons put forward included interference by other enteric infections, and many other factors, including breast-feeding. The

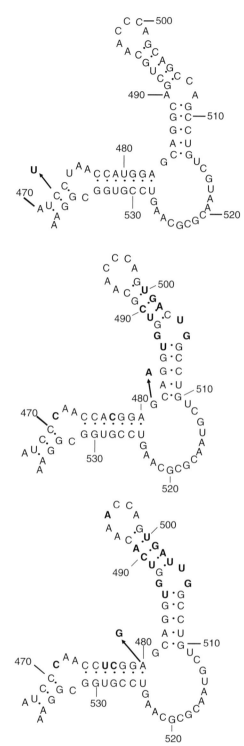

Figure 1 Structure of RNA domain in the 5' noncoding region of polio involved in attenuating the neurovirulence of the Sabin live-attenuated vaccine strains. Top: type 3; center: type 2; bottom: type 1. Differences from the type 3 sequence are shown in bold, illustrating the general conservation of the base-paired structures. Base changes involved in attenuation are shown by arrows for the three types. Note that for types 1 and 3, the changes result in weaker but allowed base pairing compared to the wild-type sequence.

most plausible explanations are that the vaccine used in routine vaccination programs was probably poorly looked after and therefore inactive, and the epidemiology of the disease was different in tropical and temperate climates. In countries of Northern Europe, poliomyelitis is highly seasonal, occurring essentially only in summer. Thus a routine vaccination campaign in which children are immunized at a set age is able to reduce the number of susceptibles during the winter when the virus is not freely circulating, so that circulation of the wild-type virus in the summer is impaired. Eventually, the wild-type virus dies out. In tropical climates, however, exposure is less seasonal, so that there is no respite during which immune populations can be built up by immunization; it remains a matter of chance whether a child is exposed first to vaccine or wild-type virus, and the impact on disease is correspondingly less. The strategy required is to immunize large proportions of the population at once. This approach was successfully followed in the 1960s in the USA in the southern states, where the climate is more tropical and routine immunization was less effective. However, it was not until the 1980s that the strategy was used in developing countries in South America when vaccine was given in mass campaigns termed National Immunization Days (NIDs). The success of the program led to a resolution in 1988 by WHO to eradicate polio from the world by the year 2000. While this has still not been achieved at the time of writing, there are only four countries from which poliovirus has never been eradicated: India, Pakistan, Afghanistan, and Nigeria. Northern India has proven extremely difficult and some children have contracted poliomyelitis after 10 or more immunizations with OPV. One approach that has been adopted is to use monovalent vaccine instead of the trivalent form containing all three serotypes. The rationale is that in the trivalent vaccine the different serotypes compete so that the most relevant serotype may not infect and immunize, whereas a monovalent type 1 vaccine used in an area where this is the problem serotype will be effective. There is evidence that this gives faster immune responses, and it was first used in Egypt where it eradicated the final type 1 strains. Monovalent vaccines are becoming the major tool in the current stages of eradication.

While there are few countries in which poliomyelitis has not been eradicated, many countries have suffered reintroductions. This is shown in **Figure 2**, which shows cases of the regions where polio has not been completely eliminated.

The first examples of this phenomenon were in central Africa. A decision was taken because of shortage of funds to concentrate immunization efforts on Nigeria which had many cases (**Figure 2(a)**). Thus immunization in the surrounding countries suffered, and at the same time resistance to immunization grew in northern Nigeria because of local concerns about supposed contaminants in the vaccine. During the period when immunization ceased, there was a resurgence in polio in Nigeria, which spread to adjacent countries where the immunization activities had been reduced (**Figure 2(b)**). As a result, polio spread from Nigeria across much of central Africa, being brought under control eventually by coordinated NIDs across most of the continent, an operation of unprecedented scale. Shortly after this, with polio still endemic in Nigeria, the annual pilgrimage to Mecca resulted in the introduction of polio from northern Nigeria into Yemen and Indonesia (**Figure 2(c)**). In 2006, polio got introduced into several African countries from northern India (**Figure 2(d)**).

The difficulties of eradication are hard to overestimate, and are compounded by the fact that most infections are entirely silent. It is clear that so long as one country remains a source of the virus the world remains at risk. This complicates the strategies to be followed once there is some confidence that the virus has in fact been eradicated in the wild, which are further complicated by the fact that OPV is a live vaccine derived from wild-type virus that is able to change in the infected vaccinee, from whom virus may be isolated for significant periods.

Vaccine-Derived Poliovirus

Vaccine-associated paralytic poliomyelitis (VAPP) cases were recognized from the first use of the Sabin vaccine strains, although their unambiguous identification required the molecular methods applied in the 1980s. They occur in primary vaccinees at a rate of about 1 per 750 000 but can also occur in contacts and those previously vaccinated. The occurrence of such cases, while rare, indicates that the vaccine strains can revert to virulence in recipients and that it is possible for the vaccine strains to spread from person to person.

In 2001, it was recognized for the first time that the Sabin vaccine strains could be the cause of outbreaks of poliomyelitis when about 22 cases were identified in Hispaniola, comprising Haiti and the Dominican Republic. Sequencing of the strains showed that they were very closely related to the type 1 Sabin vaccine strains and not to the previous endemic strains of which the last had been isolated 20 years earlier. Moreover, the sequence diversity indicated that the outbreak strains had been circulating unnoticed for c. 2 years, as they differed from the vaccine strain by about 2% overall. In addition, they were shown to be recombinant strains with a major portion of the nonstructural regions of the genome from viruses identified as species C enteroviruses unrelated to the vaccine strains. At least three different genomic structures were identified. Subsequently, it was reported that between 1988 and 1993 all supposed wild type 2 strains isolated from poliomyelitis cases in Egypt were in fact heavily drifted vaccine-related strains. Outbreaks in the Phillipines in 2001, two in Madagascar in 2002 and 2005, one in China in 2004, and one in Indonesia in 2005 have since been recorded. The most likely explanation is that vaccination programs become less vigorous once polio is eliminated from a country and other health issues take priority, so that while administration of OPV continues, coverage is less than 100% giving perfect conditions for the selection of transmissible strains. For reasons that are not understood, all such strains to date have been recombinants with unidentified species C enteroviruses. So far the outbreaks have been limited and easy to control; this was particularly shown in Indonesia where a type 1 outbreak from a

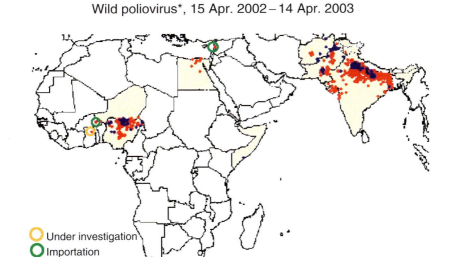

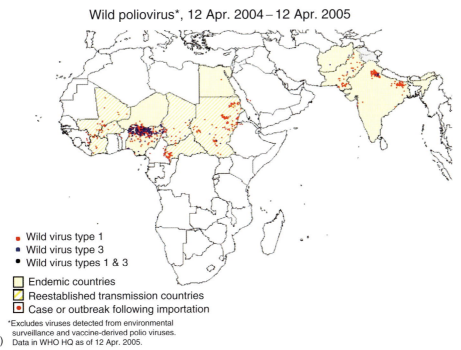

Figure 2 Continued

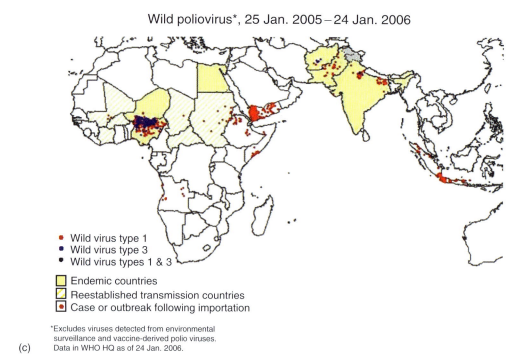

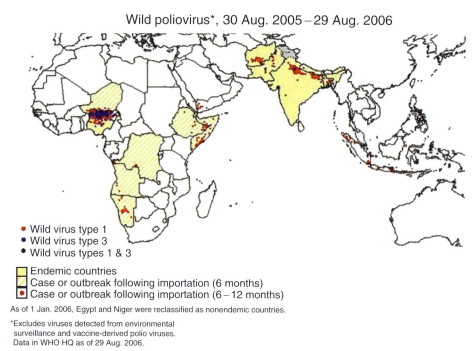

Figure 2 Occurrence of cases of poliomyelitis in remaining areas of infection from 2002 to 2006. Source: World Health Organization (WHO).

virus introduced from Nigeria occurred at the same time as an outbreak of circulating vaccine-derived poliovirus (CVDPV). The wild-type outbreak persisted longer. Virologically, there is no real reason why vaccine-derived viruses should not become as aggressive as wild-type viruses from which they derived, although this does not seem to have happened at the time of writing. The occurrence of such viruses is clearly a problem when cessation of vaccination is considered.

A second issue is the occasional case of long-term excretion of polioviruses by individuals deficient in humoral immunity. The viruses are termed immunodeficient vaccine-derived polioviruses (IVDPVs). Probably only a few percent of these hypogammaglobulinemic individuals will become long-term excreters of virus even if given OPV, and most stop shedding virus spontaneously after a period, which may be 2 or 3 years. Occasionally virus excretion may continue for much longer; in one instance isolates, were available for 12 years, and the sequence data and medical history of the patient suggest that virus has been shed for well over 20 years. The patient remains entirely healthy, although the virus is highly virulent in animal models. No treatment has been successfully applied to such individuals, although there are claims for success with an antiviral compound in one case. It is of interest that the mechanism by which virus is cleared is still not known. An infected individual can be expected to excrete virus for over 4 weeks and will continue to do so after an immune response is detectable. Immune-deficient patients can stop excreting virus spontaneously with no evidence of an immune response and while termination of excretion correlates in general with a rise in fecal IgA levels, this is neither necessary nor sufficient.

Cessation of Vaccination

OPV is the major tool of the eradication efforts that have been so successful in most of the world. While the remaining pockets of infection seem to be particularly intransigent, it is likely that polio can be eradicated possibly serotype by serotype by the increasing usage of monovalent OPV. In fact, the last known wild type 2 virus was isolated in October 1999, proving that eradication is possible in principle. This leads to the need to stop vaccinating while ensuring that the disease does not re-emerge. So far as CVDPVs are concerned the safest strategy might be to stop vaccinating abruptly after one last mass immunization, so that OPV is not continuously dripped into a population with overall declining immunity. The assumption would be that the virus dies out before susceptibles build up to a sufficient extent to maintain circulation. The strategy would require careful monitoring to ensure that the virus was not returning. IVDPVs are more difficult in that there is no known treatment and the only strategy currently may be to ensure that the populations around the cases have a high level of immunity, which clearly cannot involve the administration of OPV as this would risk starting the cycle all over again.

It is therefore likely that IPV will play a bigger role as time goes by, and many developed countries have already switched to its use, including the USA, Canada, and most of western Europe. IPV is both more expensive and difficult to administer than OPV and currently cannot be easily given in mass campaigns. It tends to be used in countries with higher incomes. However, it is possible that IPV could be combined with other vaccines such as diphtheria and tetanus for use in developing countries.

A remaining problem is that the production of IPV at present involves the growth of very large amounts of wild-type poliovirus. There have been at least two instances of virus escaping from such production facilities, although this may be less likely now that the issue has been recognized and containment at the major production sites greatly improved as a result. IPV production still poses a risk and if IPV were to be the only vaccine against polio production would probably have to increase in scale. Possibilities such as the use of the Sabin vaccine strains to make IPV are being examined as a possibly safer option, but raise complex issues of immunogenicity and yield as well as logistic concerns to do with the clinical trials of novel vaccine formulations which may be necessary.

The most likely current scenario for stopping vaccination seems to be that well-off countries and the better-off in developing countries will continue be immunized, but with IPV rather than OPV, while vaccination for the remainder will simply cease after a major last mass campaign.

See also: Smallpox and Monkeypox Viruses; Vaccine Production in Plants; Vaccine Safety; Vaccine Strategies.

Further Reading

Bodian D (1955) Emerging concept of poliomyelitis infection. *Science* 122: 105–108.
Fine PE and Carneiro IA (1999) Transmissibility and persistence of oral polio vaccine viruses: Implications for the global poliomyelitis eradication initiative. *American Journal of Epidemiology* 150: 1001–1021.
Hammon WD, Coriell LL, Wehrle PF, et al. (1953) Evaluation of Red Cross gammaglobulin as a prophylactic agent for poliomyelitis. Part 4: Final report of results based on clinical diagnosis. *JAMA* 151: 1272–1285.
Kew O, Morris-Glasgow V, Landarverde M, et al. (2002) Outbreak of poliomyelitis in Hispaniola associated with circulating type 1 vaccine-derived poliovirus. *Science* 296: 356–359.
Minor PD (1992) The molecular biology of poliovaccines. *Journal of General Virology* 73: 3065–3077.
Minor PD (1997) Poliovirus. In: Nathanson N, Ahmed R, Gonzalez-Scarano F, et al. (eds.) *Viral Pathogenesis*, pp. 555–574. Philadelphia: Lippincott-Raven Publishers.
Minor PD, John A, Ferguson M, and Icenogle JP (1986) Antigenic and molecular evolution of the vaccine strain of type 3 poliovirus during the period of excretion by a primary vaccinee. *Journal of General Virology* 67: 693–706.
Nkowane BU, Wassilak SG, and Orenstein WA (1987) Vaccine associated paralytic poliomyelitis in the United States: 1973 through 1984. *JAMA* 257: 1335–1340.
Paul JR (1971) *A History of Poliomyelitis.* New Haven, CT: Yale University Press.
Sabin AB (1956) Pathogenesis of poliomyelitis: Reappraisal in the light of new data. *Science* 123: 1151–1157.
Sabin AB, Ramos-Alvarez M, Alvarez-Amesquita J, et al. (1960) Live orally given poliovirus vaccine: Effects of rapid mass immunization on population under conditions of massive enteric infection with other viruses. *JAMA* 173: 1521–1526.

Relevant Website

http://www.polioeradication.org – The Global Polio Eradication Initiative.

Polydnaviruses: Abrogation of Invertebrate Immune Systems

M R Strand, University of Georgia, Athens, GA, USA

© 2008 Elsevier Ltd. All rights reserved.

Glossary

Fat body The main metabolic organ of insects and a primary source of numerous immune effector molecules including antimicrobial peptides.
Hemocoel The body cavity of the insect that contains blood and all internal organs.
Hemocytes Insect blood cells with diverse functions in immunity.
Oviposition The act of laying an egg by an insect.
Parasitoid An insect which is free-living as an adult but which develops as a parasite in or on the body of another invertebrate; parasitoids require only a single host to complete their immature development and almost always kill their hosts; most hosts of parasitoids are other insects and most parasitoids are in the order Hymenoptera (wasps, bees, and ants).

Introduction

The innate immune system of insects consists of both cellular and humoral components that provide defense against a diversity of invading parasites and pathogens. Parasitoid wasps form the largest group of multicellular parasites of insects with many species being of considerable economic importance as biological control agents of insect pests. There are an estimated 200 000 species of parasitoid wasps which may account for 10–20% of all insect species on earth. Almost all insects are attacked by some type of parasitoid but individual parasitoid species are usually specialists that parasitize only one or a few species of hosts. Insects have evolved potent immunological defense responses against parasitoid attack and parasitoids have evolved a diversity of counter-strategies to subvert these defenses. How these defense and counter-defense responses are regulated is not well understood across all insect and parasitoid species. However, it is known that a number of parasitoids rely on different types of microbial symbionts for survival in their hosts. The best characterized of these associations is the large number of parasitoid wasps that carry viruses in the family *Polydnaviridae*.

Polydnaviruses persist as stably integrated proviruses in the genome of their associated parasitoid species and replicate in the ovaries of female wasps in specialized cells that form a structure called the calyx (**Figure 1**). Virions accumulate in the lumen of the lateral oviducts to high density, creating a suspension of virus called calyx fluid. When a wasp lays an egg into a host, she injects a quantity of virus that infects different tissues including hemocytes. Polydnaviruses do not replicate in the wasp's host but expression of viral gene products confers protection from the host's immune system and alters other physiological processes that allow the parasitoid's offspring to successfully develop. Thus, a true mutualism exists between polydnaviruses and wasps as viral transmission depends on parasitoid survival, and parasitoid survival depends on infection of the host by the virus. Another entry in this volume discusses polydnavirus structure and genome organization. This article focuses on the function of polydnaviruses in abrogation of the insect immune response.

Polydnavirus Genera Differ in the Virulence Genes They Encode

Athough polydnaviruses share a similar life cycle, important differences also exist between the parasitoids that carry these viruses and the types of virulence genes different polydnaviruses encode. All members of the *Polydnaviridae* are enveloped and have circular, double-stranded DNA genomes that consist of multiple segments (15 to more than 30) of variable size (3 to >40 kbp) and abundance. However, the family is divided into two genera, *Bracovirus* and *Ichnovirus*, on the basis of their association with seven subfamilies of parasitoid wasps in the family Braconidae (Cardiochilinae, Cheloninae, Dirrhoponae, Mendesellinae, Khoikhoiinae, Miricinae, Microgastrinae; *c.* 17 500 species total) and two subfamilies of wasps in the Ichneumonidae (Campopleginae, Banchinae; *c.* 13 000 species total), respectively (**Figure 2**). Wasps in both families primarily parasitize larval stage Lepidoptera (moths and butterflies).

Comparative studies indicate that each polydnavirus carried by a given parasitoid species is genetically unique. Cumulative genome sizes for polydnaviruses are quite large and range from 180 kbp for microplitis demolitor bracovirus (MdBV), to 268 kbp for campoletis sonorensis ichnovirus (CsIV), and 568 kbp for cotesia congregata bracovirus (CcBV). In contrast, coding densities are very low and tend to be dominated by a small number of genes that have diverged into multiple variants. The majority of these gene family members are expressed in parasitized host insects but polydnaviruses are unable to replicate because their encapsidated genomes lack essential genes

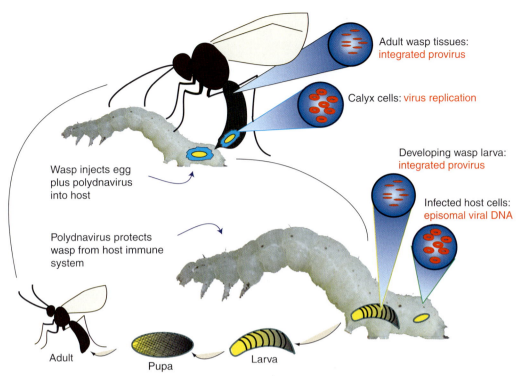

Figure 1 Parasitoid and polydnavirus life cycle. Polydnavirus DNA is integrated into somatic and germ cells of the associated wasp with replication restricted to calyx cells of the ovary. At oviposition, the wasp injects one or more eggs plus virions that infect hemocytes and other host tissues. Viral expression in the absence of replication produces gene products that protect the parasitoid's offspring from the host's immune system. The virus persists in its proviral form in the developing wasp larva. At the completion of wasp larval development the host dies. The wasp larva then pupates, emerges as an adult, and searches for new hosts.

such as polymerases and most structural proteins. Comparison of the fully sequenced MdBV, CsIV, and CcBV genomes to partial genomic sequences from other polydnaviruses identifies only one gene family (ankyrin genes) that is shared between bracoviruses (BVs) and ichnoviruses (IVs) (**Figure 2**). Two gene families are shared among known BVs (ankyrin genes, protein tyrosine phosphatase genes) while six gene families are shared among known IVs (ankyrin genes, innexin genes, cysteine-motif genes, repeat element genes, N-family genes, and polar-residue-rich genes). BVs and IVs from the same wasp genus encode similar gene families, whereas viruses associated with more distantly related wasps share lower levels of sequence similarly and exhibit greater differences in the types of genes and gene families they encode. For example, CcBV encodes cystatin, crv1-like, and C-lectin-related genes for which homologs are known from other cotesia BVs but not BVs from other Microgastrinae like MdBV and other microplitis BVs (**Figure 2**). Individual BVs and IVs also exhibit variation in the number of genome segments produced, the number of variants per gene family, and primary structure of individual genes.

That BVs share more similarities with one another than IVs and vice versa is fully consistent with the phylogeny of their associated wasps under conditions of Mendelian inheritance as proviruses (**Figure 2**). More importantly though, these phylogenetic data indicate that the association of BVs with braconids and IVs with ichneumonids most likely arose independently, and that similarities in the life cycle and genome organization of these viruses reflects convergence driven by their analogous roles in parasitism. Functionally, these patterns further suggest that BVs and IVs have evolved broadly different strategies for abrogating the immune system of host insects. The types of genes or number of variants of the same gene a given BV or IV encodes also likely reflects differences between parasitoids in how they interact with particular hosts or the range of host species they successfully parasitize.

The Insect Immune Response to Parasitoid Attack

A second key factor in understanding polydnavirus-mediated abrogation of the immune response is how insects respond to parasitoid attack in the absence of polydnavirus infection. Like most animals, the innate immune system of insects consists of a diversity of defense responses that are regulated by complex signaling pathways. The primary defense of most insects toward macroparasites, like parasitoids, is a cellular immune response called encapsulation which begins (phase 1) when the host

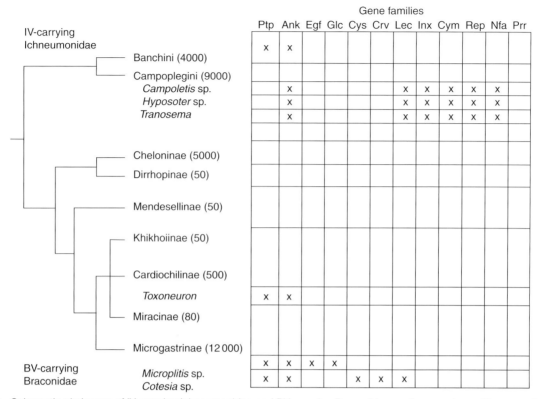

Figure 2 Schematic phylogeny of IV-carrying Ichneumonidae and BV-carrying Braconidae, and comparison of known polydnavirus gene families. The estimated number of species is indicated to the right of each wasp subfamily. Comparison of fully or partially sequenced viral genomes from selected wasp species identifies several shared gene families among IVs and BVs, respectively. Almost no genes, however, are shared between IVs and BVs. Gene family abbreviations: Ptp, protein tyrosine phosphatase; Ank, ankyrin; Egf, Egf motif gene; Glc, glycosylated mucin-like; Cys, cystatin; Crv, secreted Crv-like; Lec, C-lectin like; Inx, viral innexin; Cym, Cys-motif; Rep, repeat domain; Nfa, N-family; Prr, polar residue rich. No sequence data are currently available for IVs and BVs in the other wasp subfamilies presented.

recognizes the parasitoid egg or larva as foreign (**Figure 3**). This results in binding of a small number of hemocytes to the surface of the parasitoid and also stimulates proliferation and differentiation of hemocytes (phase 2). Recruitment of additional hemocytes to the target results in formation of a capsule comprised of overlapping layers of cells that bind to the parasitoid and one another. In phase 3, capsules often melanize due to activation of one or more pro-phenoloxidases (pro-POs). The parasitoid is then killed by asphyxiation and/or narcotizing compounds associated with the melanization pathway. Capsule formation usually begins 2–6 h after parasitism and is completed by 48 h.

Insects produce multiple hemocyte types that are morphologically and functionally distinct. In Lepidoptera, four hemocyte types are recognized (granular cells, plasmatocytes, spherule cells, oenocytoids), with capsules in most species comprised primarily of plasmatocytes. The model insect *Drosophila* in contrast produces three types of mature hemocytes (plasmatocytes, lamellocytes, and crystal cells) with capsules comprised primarily by lamellocytes. The absence of fully sequenced and well-annotated genomes has constrained identification of factors regulating capsule formation in Lepidoptera parasitized by most polydnavirus-carrying wasps. However, whole genome array and experimental studies have identified multiple genes of importance during encapsulation of parasitoids in *Drosophila*. Homologs of these genes are also likely involved in regulating encapsulation in Lepidoptera given the highly conserved nature of this defense response among insects generally.

Over the course of capsule formation in *Drosophila*, significantly different expression profiles are detected for more than 100 genes with functional annotations implicating several of these factors in capsule formation (**Figure 3**). During phase 1 of an encapsulation response, several genes under control of the Toll and JAK/STAT signaling pathways are upregulated that regulate hemocyte proliferation. Selected antimicrobial peptides under control of the Toll and imd pathways as well as proteins in the Tep and Tot families involved in pattern recognition, enzyme regulation, and stress responses are also differentially expressed. During phase 2, factors implicated in hemocyte adhesion include products of the *lectin-24A* and α*PS4 integrin* genes. *Drosophila* encodes several pro-PO genes but

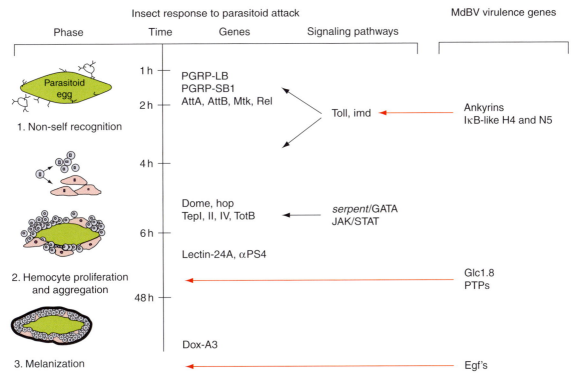

Figure 3 Overview of the host encapsulation response toward parasitoids in the absence of polydnavirus infection, and key virulence genes identified from MdBV involved in abrogation of the insect immune response. The two left-hand columns show the three major phases of capsule formation and the time elapsed since parasitoid attack. Selected insect genes and signaling pathways associated with the major phases of encapsulation are noted to the right of the time line. MdBV genes implicated in immune abrogation and their targets are noted at the far right.

melanization of capsules is associated specifically with upregulation of *proPO59/DoxA3* which is preferentially expressed by lamellocytes.

Polydnavirus Infection of Host Insects

Most polydnavirus-carrying parasitoids oviposit into the hemocoel of their host insect which is followed by the rapid entry of polydnaviruses into host cells and the release of viral DNA within nuclei. BVs and IVs infect several host tissues but host hemocytes are the most heavily infected cells in terms of the amount of detectable viral DNA. Other tissues typically infected by polydnaviruses is the fat body which also produces many immune factors with roles in encapsulation and antimicrobial defense. Unlike most other polydnavirus-carrying wasps, chelonine braconids oviposit into newly laid host eggs prior to differentiation of host tissues. Nonetheless, studies of chelonus inantius bracovirus (CiBV) indicate that most viral DNA is still detected in hemocytes with lesser amounts being present in fat body, nervous tissue, and the digestive system. Following infection of host cells, BVs and IVs persist in episomal form until the host is consumed by the developing parasitoid. Viral transcripts are expressed at near-steady-state levels by most polydnaviruses beginning 2–4 h after wasp oviposition and continuing until complete development of the wasp's progeny. However, viral expression is transient in a few systems or in the case of chelonine braconids is delayed until late in the life cycle.

Polydnavirus-Mediated Abrogation of the Host Insect Immune Response

The primary function of BVs and IVs is protection of the developing parasitoid from the host's immune system. This was first reported in 1981 for the ichneumonid *Campoletis sonorensis* and its host *Heliothis virescens*. These studies revealed that *Ca. sonorensis* was always encapsulated when CsIV was absent but was never encapsulated when CsIV was present. The protective effects of CsIV also required expression of one or more viral genes since inactivated virus could not prevent encapsulation of the wasp. Subsequent studies with other BV- and IV-carrying parasitoids produced similar outcomes indicating widespread protection of parasitoids by polydnavirus infection of host insects. Differences also exist among polydnaviruses in how they effect the insect immune response. Some polydnaviruses only impair the ability of the host's immune system to eliminate the wasp's offspring while others immunosuppress hosts more broadly. A few studies have also identified immune-related activities for polydnaviruses that do not involve gene expression.

Immunosuppressive Effects of BVs

BV-mediated abrogation of the insect immune system is currently best understood for MdBV which is carried by the microgastrine braconid *Microplitis demolitor*. In this system, MdBV broadly immunosuppresses host insects by blocking the ability of hemocytes to bind or phagocytize foreign targets, inducing apoptosis of granular cells, inhibiting inducible expression of antimicrobial peptide genes, and preventing hemolymph from melanizing. All of these effects manifest themselves within 12 h of MdBV infecting a host and also require expression of specific MdBV gene products. Genome-wide screens using RNA interference combined with other functional assays implicate four MdBV gene families in immune abrogation (**Figure 3**). Several of these MdBV genes also have conserved effects in immune cells of both their natural lepidopteran hosts and *Drosophila*. The first characterized anti-encapsulation effector from MdBV is Glc1.8, a cell surface mucin, that is expressed at high levels on the surface of infected host hemocytes. Glc1.8 prevents hemocytes from binding to parasitoid eggs or any other foreign surface. It also inhibits phagocytosis. These effects are due in part to interference with integrin-mediated adhesion but Glc1.8 likely also interacts with other molecules. Some phagocytic and adhesion pathways require tyrosine phosphorylation of several intracellular proteins that serve to link cell surface proteins to the actin cytoskeleton. One member of the MdBV protein tyrosine phosphatase (PTP) gene family dephosphorylates several proteins in insect hemocytes and significantly reduces phagocytosis. In *Drosophila* cells, this virally encoded enzyme also co-localizes to the tyrosine kinase homolog Dfak56, which is a component of focal adhesions associated with integrins that regulate adhesion and phagocytosis. This activity is similar to the bacterial pathogen *Yersinia pestis* which also produces a PTP that blocks phagocytosis by dephosphorylating essential proteins in mammalian macrophages. Encapsulation and phagocytosis by insect hemocytes involve multiple pattern recognition molecules and pathways. The inability of MdBV-infected hemocytes to bind or phagocytize any foreign target suggests that the activity of Glc1.8, a viral PTP, and possibly other factors together block several adhesion and phagocytosis pathways simultaneously.

The Toll and imd signaling pathways regulate expression of a diversity of immune genes following parasitoid attack or infection by other organisms. Both pathways involve specific nuclear factor kappa B (NFκB) transcription factors that are normally regulated by endogenous inhibitor κB (IκB) proteins. The 12-member ankyrin gene family of MdBV shares significant homology with insect IκBs. Two members of the family have also been experimentally shown to disrupt both Toll and imd signaling by binding to specific insect NFκBs and preventing them from translocating to the nucleus and interacting with κB sites in the promoters of target genes (**Figure 3**). This inhibits expression of known antimicrobial peptide genes but also likely suppresses other genes under NFκB regulation that regulate hemocyte proliferation, apoptosis, and other immune functions. An important question is why MdBV and other polydnavirus genes have diversified to form multiple gene variants. Because mammalian NFκB family members have cell-lineage-specific functions, one possibility for the MdBV ankyrin (IκB) family is that variants have functional activities that differ between important immune tissues like hemocytes and the fat body. Another possibility is that sequence variation among family members affects binding preferences for different insect NFκBs or confers binding properties for other host insect proteins.

The fourth MdBV effector belongs to the Egf gene family which encodes predicted proteins related to small serine protease inhibitors previously known only from nematodes. The pro-PO activation cascade of insects is regulated by a complex series of serine proteases that proteolytically activate prophenol oxidase (pro-PO) zymogen to phenol oxidase (PO). PO then hydroxylates monophenols to *o*-diphenols and oxidizes *o*-diphenols to quinones that polymerize to form melanin. The pro-PO cascade has not been fully characterized in any insect but is currently best understood in the lepidopteran *Manduca sexta*. Screening of MdBV Egf family members identified one member of the Egf gene family as a secreted protein that blocks melanization by inhibiting a specific serine protease in the *M. sexta* pro-PO activation cascade (**Figure 3**). Other members of the Egf family are likely also functional serine protease inhibitors but their target substrates are unknown. The picture that collectively emerges is that MdBV broadly immunosuppresses host insects by producing effector proteins that disrupt all three phases of the encapsulation process as well as other components of the insect immune system. These effectors preferentially also target signaling pathways rather than the gene products regulated by these pathways.

Most other studies on BV-mediated immunosuppression involve viruses from wasps in the genus *Cotesia*. Similar to MdBV, severe and permanent inhibition of hemocyte binding to foreign surfaces is reported for CmBV from *Cotesia melanoscela* and CkBV from *Cotesia kariyai*. Infection of *Pieris rapae* by CrBV from *Cotesia rubecula* also results in cytoskeletal alterations in hemocytes but these effects are only transient. Disruption of encapsulation by CcBV from *Cotesia congregata* is associated with both alterations to the cytoskeleton of host hemocytes and formation of aggregations rather than loss of adhesion. CcBV and other *Cotesia*

BVs also reduce melanization. Unlike other BV-carrying braconids, wasps in the genus *Chelonus* are all egg-larval parasitoids. Little is known about the role of chelonine BVs in immune suppression, but studies with CiBV indicate that viral infection of the host protects *Chelonus inanitus* larvae from encapsulation but does not suppress capsule formation of latex beads. The mechanism underlying this parasitoid-specific effect is unknown.

Conservation of the ankyrin and PTP gene families among all known BVs (**Figure 2**) suggests some family members function as inhibitors of NFκB signaling or phagocytosis as found for MdBV. At least some PTP family members from CcBV encode functional tyrosine phosphatases although the substrates and function of these enzymes in insects remain unclear. In contrast, no BVs from *Cotesia* species or other braconids outside the genus *Microplitis* encode homologs of the Glc or Egf gene families, suggesting that disruption of capsule formation or melanization must involve other effectors. Potential candidates include a gene (CrV1) originally identified from CrBV but that is conserved in other *Cotesia* BVs. This secreted protein causes alterations to the actin cytoskeleton of host hemocytes after endocytosis. *Cotesia* BVs also encode C-type lectins and a family of functional cysteine proteases (cystatins). Neither gene family has been experimentally linked to immunosuppression but some C-type lectins function as humoral pattern recognition receptors in insects while certain cystatins have immunomodulatory activities in vertebrates.

Immunosuppressive Effects of IVs

IVs are reported to have broadly similar immunosuppressive effects as BVs that include changes in the adhesive properties of host hemocytes and the number of hemocytes in circulation. In the CsIV system, hemocytes from noninfected hosts exhibit a decreased ability to attach to foreign surfaces when cultured in plasma from virus-infected hosts suggesting that secreted proteins of host or viral origin disrupt capsule formation. Hemocytes from CsIV-infected hosts are also unable to encapsulate eggs or other foreign targets, like glass, indicating that the host encapsulation response is broadly suppressed. In contrast, tranosema rostrale ichnovirus (TrIV) infection blocks encapsulation of *Tranosema rostrale* eggs but has no effect on encapsulation of other foreign targets.

IV infection also affects several elements of the humoral immune response including reduced levels of lysozyme expression in CsIV-infected hosts and suppression of melanization by several IVs. Serine protease inhibitor activity was isolated from the calyx fluid of *Venturia canescens* (=VcVLPs), though direct action on the serine proteases of the PO cascade was not shown. CsIV infection reduces melanization by inhibiting synthesis of enzymes in the melanization pathway at the post-transcriptional level. The resultant changes in enzyme levels alter substrate availability and composition, resulting in a reduced capacity of the host to produce melanin. In nonpermissive hosts of *Ca. sonorensis*, melanization reactions are only temporarily depressed with recovery of melanization occurring in hemolymph as ovarial and/or CsIV protein titers decline.

Few IV genes have been experimentally implicated in mediating any of the immune alterations described above. The CsIV Cys-motif proteins (VHv1.1, VHv1.4) are expressed in all infected tissues but the proteins are secreted and accumulate in the hemolymph where they bind to the surface of hemocytes. Experimental studies also suggest that VHv1.1 protein reduces but does not entirely block encapsulation of *Ca. sonorensis* eggs in hosts infected with a VHv1.1-expressing recombinant baculovirus. Other IV genes are indirectly implicated in immunosuppression because of homology to previously described BV genes or because they are highly expressed in key immune tissues like hemocytes or the fat body. IV-encoded ankyrin genes could potentially function as IκBs but differences in structural motifs between IV and BV ankyrins also suggest that they may have other biological activities. Products of the viral innexin genes from CsIV share high homology with insect innexin proteins that form gap junctions between many types of cells including hemocytes in capsules. CsIV innexin proteins are expressed at high levels in infected host hemocytes and also form functional gap junctions in *Xenopus* oocytes, suggesting they may alter normal gap junction function in a manner that disrupts encapsulation. The P30 ORF from HdIV is likewise predicted to encode a mucin-like protein with weak similarity to Glc1.8 from MdBV (see above) but its function in immunity is unclear.

Immunoevasive Activity of Polydnaviruses

While viral expression is usually required for defense against the host immune response, a few studies indicate that BVs and IVs passively protect parasitoids by coating the surface of the parasitoid egg. Surface features of the virions either prevent the parasitoid from being recognized as foreign by the host's immune system or inhibit binding of hemocytes. This activity also does not require expression of any viral genes. For example, eggs of *Co. kariyai* are coated by CkBV virions as well as other proteins secreted from ovarial cells. Two genes from *Co. kariyai* encode immunoevasive ovarial proteins (IEP1 and -2). These genes are not encoded within the CkBV genome but IEP1 and -2 are detected on the surface of CkBV virions. IEP proteins also confer protection to eggs from encapsulation and elimination of CkBV by hemocytes. Virions do not coat the eggs of *Toxoneuron nigriceps* or *Co. rubecula*, but the eggs of both are coated with ovarial proteins that also inhibit encapsulation. In the case of *Co. rubecula*, antiserum raised against CrBV cross-reacts with one of these proteins, Crp32, and when this protein is removed eggs are encapsulated by both nonparasitized and CrBV-infected hosts. Among ichneumonids, *V. canescens* produces virus-like particles (VLPs) morphologically similar to the IVs associated with other members of the Campopleginae, but these particles appear to lack any nucleic acid. Hosts parasitized by *V. canescens* also remain capable of mounting an encapsulation response against numerous foreign targets but are unable to encapsulate *V. canescens* because of VLPs and a hemomucin that coats the surface of their eggs. TrIV similarly coats the eggs of *Tranosema rosele* in a manner that also likely confers passive protection from encapsulation.

Future Perspectives

The importance of the insect immune system makes it a critical target for the survival of parasitoids and many other insect pathogens. Polydnaviruses usually abrogate the insect immune system by efficiently introducing immunosuppressive genes into hosts whose products confer protection to the associated parasitoid. The genetic diversity of immunosuppressive and other virulence factors produced by these viruses is also potentially enormous given that more than 30 000 species of parasitoid wasps carry polydnaviruses and that each polydnavirus–parasitoid association is genetically unique. As analysis of polydnaviruses increases, the number of identified gene products will continue to expand. Similarly, as our knowledge of the insect immune system increases, so will the ability to understand the activities of the effector molecules polydnaviruses produce and to use these products as tools for fundamental research or in insect pest management. Additional comparative genomic studies of polydnaviruses will also be essential in establishing the origin and diversification of these unique viruses as immunosuppressive pathogens of insects.

See also: Polydnaviruses: General Features; Viral Pathogenesis.

Further Reading

Brennan CA and Anderson KV (2004) *Drosophila*: The genetics of innate immune recognition and response. *Annual Review of Immunology* 22: 457.

Dupuy C, Huguet E, and Drezen J-M (2006) Unfolding the evolutionary story of polydnaviruses. *Virus Research* 117: 81.

Edson KM, Vinson SB, Stoltz DB, and Summers MD (1981) Virus in a parasitoid wasp: Suppression of the cellular immune response in the parasitoid's host. *Science* 211: 582–583.

Espagne E, Dupuy C, Huguet E, et al. (2004) Genome sequence of a polydnavirus: Insights into symbiotic virus evolution. *Science* 306: 286–289.

Gillespie JP, Kanost MR, and Trenczek T (1997) Biological mediators of insect immunity. *Annual Review of Entomology* 42: 611.

Godfray HJC (1994) *Parasitoids: Behavioural and Evolutionary Ecology*. Princeton, NJ: Princeton University Press.

Hoffmann JA (2003) The immune response of *Drosophila*. *Nature* 426: 33.

Lavine MD and Strand MR (2002) Insect hemocytes and their role in cellular immune responses. *Insect Biochemistry and Molecular Biology* 32: 1237.

Webb BA and Strand MR (2005) The biology and genomics of polydnaviruses. In: Gilbert LI, Iatrou K and Gill SS (eds.) *Comprehensive Molecular Insect Science*, vol. 6, 323pp. San Diego, CA: Elsevier.

Webb BA, Strand MR, Deborde SE, et al. (2006) Polydnavirus genomes reflect their dual roles as mutualists and pathogens. *Virology* 347: 160.

Wertheim B, Kraaijeveld AR, Schuster E, et al. (2005) Genome wide expression in response to parasitoid attack in *Drosophila*. *Genome Biology* 6: R94 (doi:10.1186/gb-2005-2005-6-11-r94).

Polydnaviruses: General Features

A Fath-Goodin and B A Webb, University of Kentucky, Lexington, KY, USA

© 2008 Published by Elsevier Ltd.

Glossary

Bracovirus Polydnavirus associated with braconid wasp.

Calyx cell Specific cell type in the female wasp reproductive tract.

Convergent evolution Organisms that are not closely related independently evolve similar traits.

Ecdysteroid Insect steroid hormone.

Ichnovirus Polydnavirus associated with ichneumonid wasps.

Polydnaviridae Family of DNA insect viruses associated with parasitic wasps.

Taxonomy and Classification

Polydnaviruses (PDVs) are an unusual group of insect viruses that exist in obligate mutualisms with certain parasitic wasps. There are two distinct genera recognized in the family *Polydnaviridae*: the bracoviruses (BVs) of the genus *Bracovirus* found only in the parasitic wasp family Braconidae and the ichnoviruses (IVs) of the genus *Ichnovirus* associated with the parasitoid family Ichneumonidae.

Virion Structure

Ichnovirus

IV virions are uniform in size (c. 85 × 330 nm), have the form of a prolate ellipsoid, and are individually surrounded by two unit-membrane envelopes (**Figure 1**). The inner envelope appears to be assembled *de novo* within the nucleus of infected ovarian calyx cells of the female parasitoid. The IV particles then bud through the plasma membrane of the calyx cells into the oviduct lumen, thereby acquiring the second envelope. IV virions have capsid dimensions large enough to support packaging of the entire genome but there are no data that show whether all or only some of the viral segments are actually present in a single virion. IV virions are complex containing 20–30 structural proteins detected in virions with molecular weights ranging from 10 to 200 kDa. Comparison of electron micrographs of IVs with that of other insect DNA viruses revealed that the IV virion morphology has some resemblance to that of ascoviruses. There are also similarities in their mode of transmission by parasitic wasps that may be indicative of an evolutionary relationship between these two virus groups.

Bracovirus

BV virions consist of enveloped cylindrical nucleocapsids of uniform diameter (30–40 nm) that are variable in length (8–150 nm) (**Figure 1**). BV nucleocapsids may possess long unipolar tail-like appendages of unknown function. The nucleocapsids are surrounded either individually or in groups by a single unit membrane and may be embedded in matrix proteins. There are some data to indicate that BV nucleocapsids encapsidate individual genome segments with the length of the capsid corresponding to the length of the segment. BV membranes are acquired in the nuclei of infected calyx cells of the wasp ovaries apparently by *de novo* membrane synthesis. In contrast to IV particles, mature BV particles are released into the oviduct lumen by lysis of calyx epithelial cells rather than by budding. BV virions are morphologically similar to the insect baculovirus and nudivirus virions but no homology has been detected between the structural genes of these insect viruses.

Genome Organization

PDV genomes are composed of multiple circular double-stranded DNA (dsDNA) segments which vary in size ranging from 2 to over 30 kbp (**Figures 2** and **3**). PDV genomes are unusual in that their genome sizes are comparatively large, ranging from 75 to over 500 kbp, but the gene density is one of the lowest reported for any virus. Analysis of the microplitis demolitor bracovirus (MdBV) and the campoletis sonorensis ichnovirus (CsIV) genomes by agarose gel electrophoresis, quantitative Southern blot hybridization, densitometry, and quantitative polymerase chain reaction (PCR) revealed that segment abundance is variable with some segments being present at a minimum of fivefold higher copy number than others. Furthermore, sequence analysis indicated that hypermolar segments contain large direct repeats that are suggestive of recombination events and consequent gene duplication. These large direct repeats are not present in low-abundance segments. The large repetitive DNA sequences of greater than 50 bp in length in PDV genomes comprise c. 15% of the genomic DNA in CsIV and MdBV. Transposable element sequences have been identified in both CsIV and MdBV, most of which are found in the noncoding regions of the genomes. Additional unusual characteristics of the PDV genome are the presence of several genes that harbor introns and some of

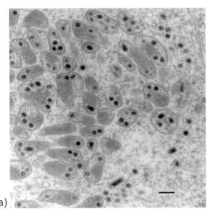 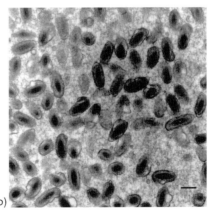

Figure 1 Virion morphology of BVs and IVs. Electron micrographs of the (a) protapanteles paleacritae bracovirus and the (b) hyposoter exiguae ichnovirus illustrate the morphological differences in members of the two PDV genera. Scale = 200 nm. Reprinted from Webb *et al*. (2005) *Polydnaviridae Virus Taxonomy* 8(1): 255–267, with permission from Elsevier.

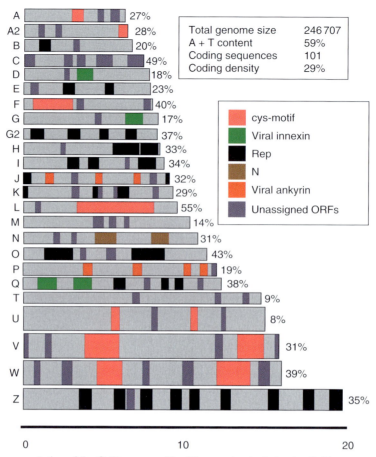

Figure 2 Diagrammatic representation of the CsIV genome. The 24 nonredundant circular CsIV genome segments are represented as linear molecules. To visualize segment size relationship within the genome, individual segments are listed from the smallest (A = 6.1 kbp) to the largest (Z = 19.6 kbp) segment with the segment name to the left. The percentage of coding DNA per segment is identified on the right. Wider bars represent high-copy, repetitive-type segments. ORFs and members of gene families have been annotated by color and located to individual genome segments. Gray regions represent noncoding DNA. Scale bar is in kilobases. Reprinted from Webb BA, Strand MR, Dickey SE, et al. (2006) Polydnavirus genomes reflect their dual roles as mutualists and pathogens. Virology 347: 160–174, with permission from Elsevier.

the PDV genes are organized into gene families. Furthermore, the encapsidated PDV genome lacks genes required for viral replication and virion assembly, suggesting that these factors reside in the genome of the parasitic wasp.

PDV gene expression has been investigated in most detail in the parasitized insect. Viral genes may be transiently expressed or members of viral gene families may be expressed in a tissue-dependent manner. Genes present on the most abundant segments are typically most highly expressed. Viral genes that are expressed only in the parasitic wasp, only in the parasitized lepidopteran host, or in both hosts have been described.

The genome organization of CsIV and MdBV are described in more detail in the next two sections.

CsIV Genome Organization

The CsIV genome is 246.7 kbp in size and is composed of 24 double-stranded, circular DNA segments ranging from 6.1 kbp (segment A) to 19.6 kbp (segment Z) (**Figure 2**). Some CsIV segments (U, V, W) undergo intramolecular recombination at intrasegmental repeats to generate smaller derivatives of the larger segments, a process referred to as segment nesting. It is believed that segment nesting provides a means for amplifying viral gene copy numbers. Interestingly, segment nesting in CsIV is associated with high-abundance segments and encodes only members of a single gene family, the cys-motif gene family. The coding density of the CsIV genome is 29% whereby individual genome segments have a coding density that ranges from 8% to 40% (**Figure 2**). Analysis of the CsIV genome identified 101 open reading frames (ORFs) whereby a total of 53 CsIV genes can be grouped into five gene families, namely the rep, cys-motif, vankyrin, innexin, and N gene families (**Figure 2**). The rep gene family is the largest gene family in the CsIV genome with 30 rep genes located on ten low-copy segments. The rep genes are characterized by a 540 bp sequence

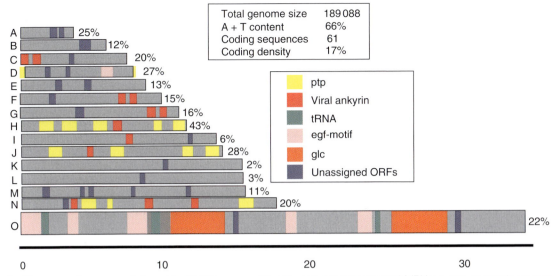

Figure 3 Diagrammatic representation of the MdBV genome. The 15 nonredundant circular MdBV genome segments are represented as linear molecules. To visualize segment size relationship within the genome individual segments are listed from the smallest (A = 3.6 kbp) to the largest (O = 34.3 kbp) segment with the segment name to the left. The percentage of coding DNA per segment is identified on the right. Wider bars represent high-copy, repetitive-type segment. ORFs and members of gene families have been annotated by color and located to individual genome segments. Gray regions represent noncoding DNA. Scale bar is in kilobases. Reprinted from Webb BA, Strand MR, Dickey SE, et al. (2006) Polydnavirus genomes reflect their dual roles as mutualists and pathogens. Virology 347: 160–174, with permission from Elsevier.

that is present in one to five copies but the function of the rep protein is unknown. The cys-motif gene family consists of ten genes on seven different segments harboring a conserved, cysteine-rich motif which is present in one or more copies in gene family members. Interestingly, all members of the cys-motif gene family contain introns whereas no spliced genes were found in the other CsIV gene families. All members of the cys-motif gene family possess signal peptides and share structural features encoding a novel cystine knot domain. Functional analysis of proteins encoding the cys-motif proteins VHv1.1 and VHv1.4 indicates that these proteins inhibit aspects of host development and cellular immunity. The vankyrin gene family is composed of seven members on segments P and J. Vankyrin genes encode for proteins containing truncated ankyrin repeat domains with sequence homology to the inhibitory domains of nuclear factor kappa B (NFκB) transcription factor inhibitors. Interestingly, the CsIV vankyrin proteins lack the N- and C-terminal regulatory regions required for dissociation and basal degradation of typical IκB (inhibitor of NFκB) suggesting a role for the vankyrin proteins in irreversible binding to modify or inhibit NFκB-mediated signal transduction cascades within the parasitized lepidopteran host. The four members of the vinnexin gene family were identified by sequence homology with invertebrate innexin gap junction proteins involved in cellular communication. CsIV vinnexins may function to alter gap junction proteins in infected host cells, thereby disrupting cell-to-cell communication required for defense responses. The smallest gene family in the CsIV genome is the N gene family with two structurally related members of unknown function located on segment N. The remaining 48 predicted CsIV genes are unique with no detectable homologs in the current protein databases.

Other IV genomes for which genome sequence is available indicate that there is considerable variation both in gene content and genome organization. However, most gene families appear to be present in ichneumonid genomes. The limited data available on banchine IVs indicate that this IV subgroup is more divergent.

MdBV Genome Organization

The MdBV genome is 189 kbp in size and is divided into 15 double-stranded, circular DNA segments that range from 3.6 kbp (segment A) to 34.3 kbp (segment O) (**Figure 3**). The average coding density of the MdBV genome is 17% with the coding density for individual segments ranging from 2% to 43% (**Figure 3**). Analysis of the MdBV genome identified 61 genes; 39 of those are grouped into five gene families, namely the ptp, vankyrin, tRNA, egf-motif, and glc gene families (**Figure 3**). The largest MdBV gene family is the ptp family with 13 members on four different segments encoding for proteins related to protein tyrosine phosphatases (PTPs). PTPs are known to regulate tyrosyl phosphorylation in numerous signal transduction pathways. The 12 vankyrin genes of the MdBV vankyrin gene family are located on seven different segments. They possess truncated ankyrin repeats and lack the C- and N-terminal control domains similar to the ones found in CsIV. The third gene family consists of seven nearly identical serine tRNAs on segment O. In addition, segment O harbors the two members of the glc gene family and five out of six members of the egf-motif gene family. The egf-motif genes share a cysteine-rich epidermal growth factor-like domain with similarity to the cysteine-rich domain of a unique serine proteinase inhibitor. The glc gene family is composed of two identical glc1.8 genes that encode for a mucin protein. The Glc1.8 protein localizes to the surface of hemocytes of MdBV-infected lepidopteran larvae where it blocks the encapsulation response and phagocytosis. The remaining 21 genes are unique and are predicted to encode for proteins of unkown function, not represented elsewhere in available databases. All members of the glc and egf-motif gene families contain introns while the other gene families do not contain any spliced genes. In contrast to the CsIV genome, segment nesting has not been observed in the MdBV genome.

Only one other BV genome sequence has been reported at this time. The cotesia congregatus bracovirus (CcBV) genome is more than twice as large as the MdBV genome and has a distinctive set of gene families with the ptp, tRNA, and vankyrin gene families present but apparently lacking the glc and mucin gene families. The CcBV genome is also reported to have a much higher number of spliced genes.

PDV Life Cycle and Replication

PDVs have an obligate association with parasitic wasps in the families Braconidae and Ichnomonidae in which the viral genome is stably integrated within the wasp genome. The PDV life cycle has been characterized as having two 'arms', namely the replication/transmission arm, in which PDV replication and transmission occurs in the viruses' wasp host and a pathogenic arm of the life cycle, in which viral gene expression occurs in the lepidopteran host to disrupt host physiological systems (**Figure 4**). Virus replication and assembly begin during the late stage of wasp pupal development and have been linked to changes in the wasp ecdysteroid titers that drive pupal development. In BV, proviral BV segments appear to be tandemly arrayed in the wasp genome and there is some evidence that larger chromosome regions are amplified before excision during viral replication, with individual viral segments then excised from amplified DNA for encapsidation in virions. Replication of the excised, circular form of viral segments was not detected. In IV, replication is thought to involve the excision of viral DNA segments from the wasp genome via site-specific recombination events followed by amplification of circular episomes in a rolling-circle-type mechanism. This model is attractive for the replication mechanism of IVs as their genome segments are not contiguous in the wasp genome. In all PDVs, replication occurs only from proviral DNA in specialized cells of the female oviduct, the calyx cells, and

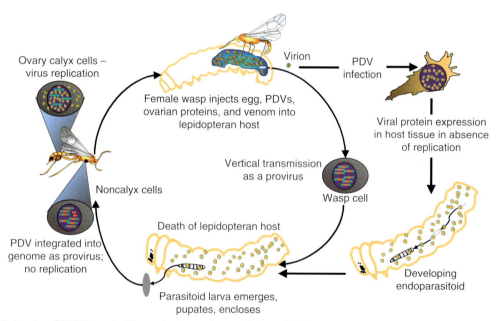

Figure 4 Polydnavirus (PDV) life cycle. The replication and transmission of PDVs and the life cycle of their associated wasp are shown. Viral DNA is transmitted as proviral DNA integrated into the wasp genome (curved arrows) and as circular DNA in virions (straight arrows). In calyx cells of the wasp ovaries and in infected lepidopteran cells, viral DNA is present as circular DNA, whereas in nonreplicative wasp cells the virus exists as a provirus and does not replicate. Reprinted from Webb BA, Strand MR, Dickey SE, et al. (2006) Polydnavirus genomes reflect their dual roles as mutualists and pathogens. *Virology* 347: 160–174, with permission from Elsevier.

only for transmission of virus to lepidopteran host insects. PDV virions accumulate to a high density in the oviduct lumen and are transferred to the lepidopteran host with wasp eggs, ovarian proteins, and venom during oviposition. In the lepidopteran host, the virus then infects hemocytes, fat body, and other tissues of the parasitized host. Viral gene expression can be detected as early as 1 h post parasitism and continues through parasitoid larval development. PDV gene expression induces physiological alterations in the lepidopteran host such as disruption of development and suppression of humoral and cellular immunity to permit the parasitoid's survival within host larvae and to mobilize host protein stores to redirect host resources for supporting parasitoid growth. The parasitoid completes its development inside the lepidopteran larva, pupates, and emerges as an adult wasp while the host larva dies. During parasitization, viral DNA is not only present in virions that infect lepidopteran cells but PDVs are also vertically transmitted in the proviral form within the chromosome of the wasp egg to ensure virus transmission to subsequent wasp generations.

Phylogenetic Information

PDVs are associated with two distinct lineages of ichneumonoid wasps. All BV-carrying wasps belong to the monophyletic braconid microgastroid lineage which arose 74 million years ago and contains more than 17 000 species whereas IVs are largely limited to the Campopleginae and Banchinae which consist of approximately 13 000 species. The absence of PDVs in the common ancestor to campoplegines and microgastroids is strongly indicative of an independent origin of IVs and BVs as are their distinctive modes of replication, virion structures, and genomes. Sequence data from different PDV genomes revealed that viruses within a genus have a number of genes in common whereas few genes are shared between IVs and BVs. Interestingly, the vankyrin gene family is the only gene family found to be in common between IVs and BVs but phylogenetic analysis of vankyrins from PDVs and mammalian and *Drosophila* IκBs revealed that IV and BV vankyrins cluster separately. Thus the data provide no clear evidence of shared ancestry between the two viral genera. The divergence in the viral genomes of closely related wasp species within the two PDV genera may be explained by the presence of transposon-derived sequences, large amounts of repetitive DNA, and segment nesting. The suite of unrelated genes in the IV and BV genomes implies that each genus has evolved different functional strategies to immunosuppress and alter the development of the parasitized host. However, similarities in their life cycle and similarities in genome organization between IVs and BVs suggest that common selection pressures among IVs and BVs have driven convergent evolution between the two PDV genera.

See also: Polydnaviruses: Abrogation of Invertebrate Immune Systems.

Further Reading

Beckage NE and Gelman DB (2004) Wasp parasitoid disruption of host development: Implications for new biologically based strategies for insect control. *Annual Review of Entomology* 49: 299–330.

Espagne E, Dupuy C, Huguet E, et al. (2004) Genome sequence of a polydnavirus: Insights into symbiotic virus evolution. *Science* 306: 286–289.

Kroemer JA and Webb BA (2004) Polydnavirus genes and genomes: Emerging gene families and new insights into polydnavirus replication. *Annual Review of Entomology* 49: 431–456.

Turnbull M and Webb B (2002) Perspectives on polydnavirus origins and evolution. *Advances in Virus Research* 58: 203–254.

Webb BA and Strand MR (2005) The biology and genomics of polydnaviruses. Comparative physiology and biochemistry. In: Gilbert LI, Iatrou K, and Gill SS (eds.) *Comprehensive Molecular Insect Science*, vol. 5, pp. 260–323. Amsterdam: Elsevier.

Webb BA, Strand MR, Dickey SE, et al. (2006) Polydnavirus genomes reflect their dual roles as mutualists and pathogens. *Virology* 347: 160–174.

Whitfield JB (2002) Estimating the age of the polydnavirus/braconid wasp symbiosis. *Proceedings of the National Academy of Sciences, USA* 99: 7508–7513.

Whitfield JB and Asgari S (2003) Virus or not? Phylogenetics of polydnaviruses and their wasp carriers. *Journal of Insect Physiology* 49: 397–405.

Polyomaviruses of Humans

M Safak and K Khalili, Temple University School of Medicine, Philadelphia, PA, USA

© 2008 Elsevier Ltd. All rights reserved.

JC Virus

JC virus (JCV) is classified into species *JC polyomavirus*, genus *Polyomavirus*, family *Polyomaviridae*. The JCV genome contains a small, double-stranded, closed circular DNA of 5130 bp. The capsid exhibits an icosahedral structure, approximately 45 nm in diameter. Structural and antigenic studies have indicated that JCV is closely related to two other polyomaviruses, BK virus (BKV) and simian vacuolating virus 40 (SV40). JCV is the etiologic agent of a fatal demyelinating disease of the central nervous system (CNS) known as progressive multifocal leukoencephalopathy (PML). Seroepidemiological data indicate that the overwhelming majority of the world's population (70–80%) is infected by JCV early in childhood without apparent clinical symptoms. The virus establishes a persistent infection in the kidneys (latent infection) and reactivates from latency under immunocompromised conditions. As well as the kidneys, hematopoietic progenitor cells, peripheral blood B lymphocytes, and tonsillar stromal cells have been shown to harbor JCV, suggesting that these sites could serve as additional sites for latent infection by JCV.

Recently, two new human polyomaviruses were also discovered. One was isolated from human respiratory tract samples by polymerase chain reaction (PCR) amplification and named KI polyomavirus (KIPyV). The other was isolated from a patient with symptoms of acute respiratory tract infection and named WU virus (WUV). Thus far, follow-up studies have not yielded evidence for associations between infections with KIPyV or WUV and respiratory disease.

JCV was first isolated from the brain tissue of a PML patient in 1971, and this opened new frontiers in polyomavirus research. A piece of brain tissue from a PML patient was used as a source of inoculum to infect primary cultures derived from human fetal brain, and the virus was then successfully cultivated from those long-term cultures. This was the first direct evidence that a neurotropic virus is associated with the development of PML. At the same time, a virus similar in structure was found in the urine of a patient who had undergone renal transplantation. Each virus was named with the initials of the donor, BK virus for the renal transplantation patient and JC virus for the patient with PML. Following the isolation of JCV, the oncogenic potential of the virus was demonstrated both in tissue culture and experimental animals. Particularly, animal model studies showed that JCV induces tumors in tissues of neural origin. In addition, the genome has recently been found to be present in a variety of human tumors.

JCV infects and destroys oligodendrocytes, which are the myelin-producing cells in the CNS, and indirectly causes the death of neurons in the white matter of the brain, because neurons are known to depend for survival on support provided by oligodendrocytes. Subsequently, the destruction of both oligodendrocytes and neurons in the CNS results in PML, a neurodegenerative disease. PML develops mostly in patients with underlying immunosuppressive conditions, including lymphoproliferative diseases, acquired immune deficiency syndrome (AIDS), and Hodgkin's lymphoma, although, in a small number of cases, PML also affects individuals lacking underlying disease. Before the AIDS epidemic, PML used to be a rare complication of middle-aged and elderly patients with lymphoproliferative diseases. However, it is now a commonly encountered disease of the CNS in patients of different age groups, and there is a noticeable increase in the incidence of PML in human immunodeficiency virus (HIV)-infected patients compared to noninfected individuals. The increased incidence of PML in AIDS patients suggests that HIV infection may directly or indirectly influence reactivation of JCV and therefore induction of PML. In support, recent reports indicate that the incidence of PML in HIV-seropositive patients reaches 5%, compared to 0.8% before the AIDS epidemic. Furthermore, reactivation of JCV in patients with multiple sclerosis (MS) or Crohn's disease treated with both interferon β1α and natalizumab, a selective adhesion-molecule blocker, suggests that such an immunosuppressive treatment may be a risk factor in induction of PML in MS and Crohn's disease patients.

Genome Organization: Regulatory and Coding Regions

The JCV genome is composed of bidirectional regulatory elements and coding regions (**Figure 1**). The regulatory region contains the origin of DNA replication (ori) and promoter/enhancer elements, and exhibits hypervariability in its sequence composition; that is, although the regulatory region of a prototype strain, Mad-1, is characterized by the presence of two 98 bp tandem repeats, there are considerable deviations from this in other strains. For example, the archetype strain contains only one copy of the 98 bp repeat with two insertions of 23 bp and 66 bp, and the Mad-4 strain contains two tandem repeats that are only 84 bp in length. The bidirectional nature of the regulatory elements may provide an advantage to the virus to transcribe its genes efficiently, by a mechanism that is currently unknown. The regulatory region of the archetype strain is thought to be more adaptive to a latent infection and, upon reactivation, undergoes a deletion/duplication process within its regulatory region, through which it apparently gains the ability to become a more virulent virus and infect oligodendrocytes. The coding sequences of JCV are divided into two regions, early and late. The early region encodes only regulatory proteins, including small tumor antigen (Smt-Ag), large tumor antigen (LT-Ag), and T′ proteins (T'_{135}, T'_{136}, and T'_{165}). The late region encodes a mixture of structural (VP1, VP2, and VP3) and regulatory (agnoprotein) proteins (**Figure 1**).

JCV is closely related to BKV and SV40 in the coding regions (70–80% identity). However, the regulatory regions are considerably diverged. This feature makes each virus unique with respect to replication in different cell types and tissues. In other words, the regulatory region largely determines the tissue-specific expression of gene products. LT-Ag also contributes to tissue-specific expression. Moreover, *in vivo* and *in vitro* transcription

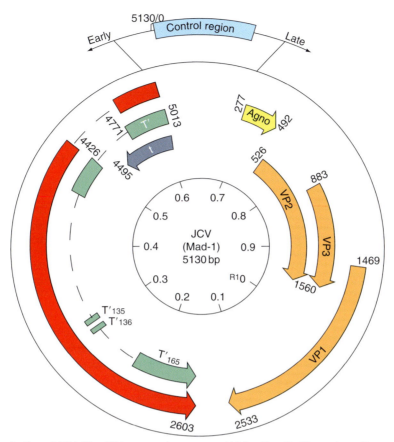

Figure 1 Genomic organization of JCV. The JCV genome is expressed bidirectionally. The early coding region encodes LT-Ag, Sm t-Ag, and T′ proteins (T'_{135}, T'_{136}, and T'_{165}). The late coding region encodes agnoprotein and the three structural proteins VP1, VP2, and VP3. The control (regulatory) region is located between the two coding regions and contains the origin of DNA replication (ori) and promoter/enhancer elements for the early and late promoters.

assays and cell fusion experiments have shown that tissue-specific cellular factors are critical for neurotropic expression of JCV. For instance, the early promoter was shown to be expressed well in glial cells but, when glial cells were fused with fibroblasts to form heterokaryons, early gene expression was significantly downregulated. This suggests that there are positively and negatively acting factors in glial and nonglial cells, respectively.

Regulatory Region

The regulatory region contains ori and multiple *cis*-acting regulatory elements that are involved in transcriptional regulation of the early and late regions. The regulatory region of the Mad-1 strain is shown schematically in **Figure 2**. The only region that shows substantial similarity to the corresponding regions in SV40 and BKV is ori, a 68 bp element located between a nuclear factor kappa B (NFκB) motif and the first 98 bp repeat. Each repeat contains a TATA box located on the early side of the repeat, and the first TATA box with respect to ori is involved in positioning of the transcription start sites for early genes. The second TATA box does not appear to have a similar function for the late genes. The other regulatory elements shown to be critical *cis*-acting elements for expression include Penta, AP-1, and NF-1 motifs that follow the TATA box. These elements serve as binding sites for several transcription factors and contribute to tissue tropism. The major *cis*-acting elements and transcription factors that bind to these regions are illustrated in **Figure 2**.

Cellular Transcription Factors Involved in JCV Gene Expression

Several studies have shown that many cell-specific and ubiquitous factors are involved in expression of the early and late promoters. These factors interact directly or indirectly with the *cis*-acting elements in the regulatory region and control gene expression. The regulatory proteins include LT-Ag and agnoprotein and have also been shown to play critical roles. LT-Ag, a multifunctional phosphoprotein, is not only involved in viral DNA replication but also late gene expression.

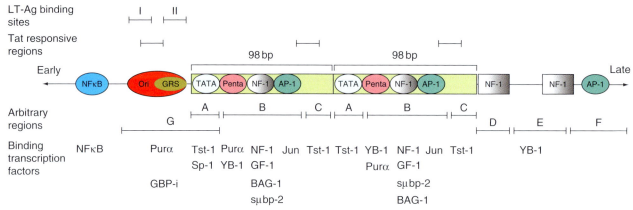

Figure 2 Regulatory region of JCV. The regulatory region contains the origin of DNA replication (ori) and promoter/enhancer elements. The two 98 bp tandem repeats are a characteristic of the Mad-1 strain. Promoter/enhancer elements that serve as binding sites for transcription factors are shown, and include the NFκB-binding element, the pentanucleotide element (Penta), the nuclear factor 1-binding element (NF-1), the GC-rich element (GRS), and the activating protein-1-binding element (AP-1). The transcription factors that have been shown to bind promoter/enhancer elements are shown at the lower part of the Figure. Large T antigen (LT-Ag)-binding sites and Tat-responsive elements are indicated. Arbitrary subregions A to G described in the text are indicated.

Along with cellular factors required for DNA replication, LT-Ag binds to ori, unwinds it, and initiates DNA replication in a bidirectional manner. With respect to the role of LT-Ag role in transcription, studies have shown that, whereas it autoregulates its own transcription from the early promoter, it robustly transactivates the late promoter, resulting in the expression of agnoprotein and capsid proteins.

A close inspection of the regulatory region sequences reveals the presence of multiple *cis*-acting elements to which transcription factors bind and regulate gene expression. These include NFκB, GRS, Penta, NF-1, AP-1, and 'AP-1-like' motifs. In addition, the regulatory region also contains LT-Ag-binding regions and elements responsive to the HIV transactivator, Tat (**Figure 2**). Furthermore, for the sake of simplicity in describing regulatory elements, the regulatory region is divided into arbitrary regions A to G. NFκB, a stress-inducible transcription factor, binds to the NFκB element and modulates gene expression from the early and late promoters. This factor represents a large family of transcription factors that are induced in response to a wide variety of extracellular stimuli, including phorbol ester, cytokines, and viral infection. p50 and p52 are two constitutively expressed subunits of NFκB that activate transcription from region D, whereas subunit p65 regulates transcription through the NFκB motif itself. The NFκB family members p50/p65 also influence transcription indirectly through a 23 bp element present in the regulatory regions of many JCV variants. Another inducible cellular factor, GBP-i, appears to target the GRS sequence present in region G and suppresses late promoter activity. Tst-1 is a member of the well-characterized tissue-specific and developmentally regulated POU family of transcription factors. This factor apparently interacts with distinct binding sites in regions A and C and regulates transcription from the early and late promoters. Interactions of cellular transcription factors with the viral regulatory proteins are also important. In this respect, interaction of Tst-1 with LT-Ag leads to synergistic activation of the early and late promoters. GC-rich sequences in Penta motif target the ubiquitously expressed transcription factor, Sp-1, which stimulates expression from the early promoter. The regulatory region also contains several NF-1-binding sites scattered across regions B, D, and E. NF-1 is important for both viral transcription and replication. Several factors interact with region B, including GF-1, YB-1, Purα, and AP-1. GF-1 is the human homolog of the murine Sμbp-2 protein and transactivates the early and late promoters. Two well-characterized cellular transcription factors, YB-1 and Purα, also interact with this region and regulate transcription and replication through interaction with LT-Ag. Another binding site for YB-1 is present in region E. BAG-1, a novel Bcl-1-interacting protein that is expressed ubiquitously in neuronal and non-neuronal cells, is a novel cellular transcription factor that regulates JCV promoters through the NF-1-binding site. Another transcription factor that interacts with region B is c-Jun, which is a member of the AP-1 family of transcription factors. Members of this family are known as the immediate–early inducible protooncogenes and are critical for the expression of many cellular and viral genes. c-Jun is phosphorylated during the JCV life cycle in an infection-cycle-dependent manner and interacts functionally with LT-Ag. In addition to its own regulatory proteins, the JCV genome is also crossregulated by regulatory proteins encoded by other viruses, including Tat and immediate-early transactivator 2 (IE2) from human cytomegalovirus (HCMV).

Life Cycle

Figure 3 highlights the chain of events during the polyomavirus life cycle. The infection cycle of JCV starts with the attachment of viral particles to cell surface receptors. The serotonin receptor $5HT_{2A}$ as well as α(2-6)-linked sialic acid are critical for the attachment of JCV to the cell surface, and the viral particle is then internalized by a process known as clathrin-dependent endocytosis. Following internalization, the particle is transported to the nucleus by an unknown mechanism. The next step is the uncoating of capsid proteins, which subsequently leads to the expression of early genes in the nucleus. Viral transcripts undergo splicing before they are transported to the cytoplasm and translated. Translation of early transcripts results in the production of early regulatory proteins, including LT-Ag, Sm t-Ag, and T' proteins (T'$_{135}$, T'$_{136}$, and T'$_{165}$). LT-Ag is then transported back to nucleus to initiate viral DNA replication. In the meantime, LT-Ag transactivates the late promoter for the production of regulatory agnoprotein and the three structural capsid proteins VP1, VP2, and VP3. Capsid proteins are also transported to the nucleus for virus assembly. In the encapsidation process, the capsid proteins are sequentially added to and arranged on the viral genome to package it into capsids. The final step is to release the virions from infected cells. The mechanism for this is not known, but it is thought that it takes place upon lysis of infected cells.

JCV Reactivation and Other Viral Infections

HIV infection and JCV reactivation

It has long been known that impaired cellular immunity is an exclusive predisposing condition for reactivation of JCV and onset of PML. Before the AIDS epidemic, lymphoproliferative disorders used to be the predominant underlying factor for reactivation of JCV from latency. In recent years, however, this trend has changed and a strong association of PML with AIDS has been established. The statistical data in this regard suggest that approximately 5–10% of AIDS patients eventually develop PML. This percentage is predicted to increase

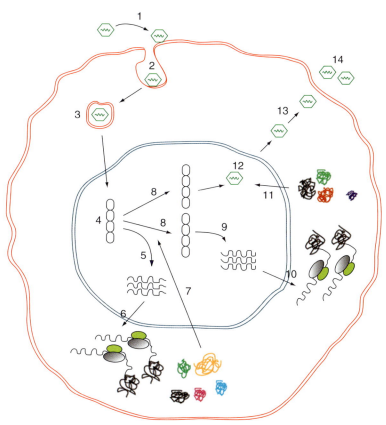

Figure 3 Life cycle of JCV. The steps are indicated by numbers: 1, adsorption of virus particles to cell surface receptors; 2, entry by clathrin-dependent endocytosis; 3, transport to the nucleus; 4, uncoating; 5, transcription of the early region; 6, translation to produce the early regulatory proteins, LT-Ag, Sm t-Ag, and T' proteins (T'$_{135}$, T'$_{136}$, and T'$_{165}$); 7, nuclear localization of LT-Ag; 8, replication of the viral genome; 9, transcription of the late region; 10, translation of late transcripts to produce agnoprotein and capsid proteins (VP1, VP2, and VP3); 11, nuclear localization of capsid proteins; 12, assembly of virus particles in the nucleus; 13, release of virions by an unknown mechanism; 14, released virions.

as more PML cases are evaluated in AIDS patients. In recent years, application of highly active antiretroviral therapies (HAART) to AIDS patients has had a significant effect on manifestations of AIDS-associated opportunistic infections, including HCMV infection and toxoplasmosis. The effect of HAART on the incidence of PML remains unclear, but there has been an indisputable increase in the frequency of PML since the inception of the AIDS epidemic. One simple explanation for this is that a higher degree and duration of cellular immunosuppression may pertain to HIV infection than to other immunosuppressive conditions. Another explanation is that alterations in the blood–brain barrier caused by HIV infection may render the brain more accessible to JCV infection. Direct activation of the JCV genome by HIV Tat protein or indirect activation by HIV-induced cytokines and chemokines could be other important factors for AIDS-related reactivation of JCV.

The degree and duration of immunosuppression in HIV infection may not be the only factors for the development of PML, because a substantial number of individuals with hematological malignancy, solid tumors undergoing chemotherapy, and/or underlying tuberculosis status post-organ transplantation all have a long period of immunosuppression but do not develop PML as often as HIV-infected individuals. What is special about HIV infection is that many arms of the host defense system are affected, as opposed to the more specific effects seen in other immunosuppressive conditions. It is also notable that PML typically manifests itself as a late complication of HIV infection. Perhaps a substantial loss in the number and function of CD4+ T cells during the late phases of AIDS (as well as other alterations, including deficiencies in chemotaxis, monocyte-dependent T-cell proliferation, Fc-receptor function, C3 receptor-mediated clearance, and oxidative burst response) contributes greatly to suppressive conditions of the immune system in AIDS patients, and this in turn facilitates JCV reactivation and the development of PML. Furthermore, infection of monocytes/macrophages by HIV results in the impairment of antibody-dependent, cell-mediated cytotoxicity, intracellular antimicrobicidal activity, and induction of interferon α secretion. Moreover, the circulating B cells decrease in number as the disease progresses and, in association with HIV infection, B cells secrete disproportional amounts of cytokines. Which of these deficiencies leads to predisposition to the development of PML remains unknown. However, it should be emphasized that on rare occasions PML may develop in the absence of any identifiable underlying immunosuppressive disorder.

The Tat protein is a potent transactivator of the HIV long terminal repeat (LTR). It mediates its transregulatory activity through a specific RNA sequence located in the leader of all HIV-1 RNAs, called the transactivation response (TAR) element. Several critical G residues in the TAR element are required for function. Tat also induces transcription from the JCV late promoter through several TAR elements in the regulatory region. The G residues required for the function of the HIV-TAR element are also conserved in the JCV-TAR element and play an important role in Tat-mediated activation of the late promoter. Thus, Tat may participate directly in activation of the JCV late promoter. It is also thought that Tat may indirectly influence JCV gene regulation through the induction of transcription factors such as NFκB and c-Jun/AP-1, both of which have been shown to interact directly with the JCV promoter. Taken together, the regulatory factors induced by HIV Tat or Tat-induced cytokines may activate JCV promoters directly or indirectly and thereby influence the viral life cycle.

Interaction of JCV with other human viruses

Initial infection by JCV is followed by a persistent infection in the kidney and perhaps in B cells, during which JCV appears to go through several stages: (1) a latent infection state of limited virus production; (2) an activated state, during which the virus causes cell lysis; and (3) a final tissue destruction state, which leads to the onset of PML. Although the precise mechanism(s) of viral reactivation processes is unknown, the immunosuppressed state of the patient is critical for the onset of PML. Moreover, infection of immunosuppressed individuals by other viruses may influence the reactivation process.

In addition to JCV, several human viruses, such as BKV, HCMV, and human herpesvirus 6 (HHV-6), are able to establish persistent infections in various tissues and organs. HCMV is highly prevalent in the human population and can infect a wide range of organs. It is commonly reactivated in bone marrow and renal transplantation patients. Particularly, it is often reactivated after renal transplantation in AIDS patients in the kidney, CNS, lymphoid organs, stromal cells, and CD34+ bone marrow progenitor cells. Recent reports indicate that HCMV infection positively affects the level of JCV DNA replication or possibly transcription. JCV is known to infect only glial cells, but this restricted cell specificity can be overcome by HCMV infection. Thus, fibroblasts, which are generally nonpermissive for the replication of JCV, but can become permissive if HCMV IE2 is provided to JCV-infected cells.

HHV-6 is also a ubiquitous virus, and is detected in close association with JCV in oligodendrocytes within PML lesions. This virus establishes a lifelong infection in various organs, including the brain, urogenital tract, lung, liver, and peripheral blood cells. There appears to be a high correlation between polyomavirus infection and HHV-6 infection. This suggests that polyomaviruses and HHV-6 may have a common host cell not only in the CNS, but also in peripheral organs, and co-infection may have an impact on JCV gene regulation and perhaps on the reactivation process.

BKV has been reported to infect human tissues, such as the urogenital tract, along with JCV. This has been verified by the detection of both viruses in the kidney and the urine. Such infections are commonly encountered after renal and bone marrow transplantation in HIV-infected people, pregnant women, and immunocompetent individuals. However, the influence of double infection on JCV replication remains unclear. Additionally, both viruses have been detected in the peripheral blood cells of healthy and immunocompromised individuals.

Lytic Infection versus Tumor Induction

PML

JCV lytically infects oligodendrocytes in the CNS, leading to a white matter disease in humans known as PML (**Figure 4**). Oligodendrocytes are members of the glial cell family in the CNS. The other cell types in the glial family include microglia, astrocytes, and neurons. The primary function of oligodendrocytes is to myelinate the axons that project from the neural cell bodies of the overlying cortex and to protect the neurons. Infection of oligodendrocytes results in the destruction of myelin insulating the neurons, creating sporadic plaques in different parts of the brain. Hence, demyelination occurs as a multifocal (rarely unifocal) process that can develop in any location in the white matter (**Figure 4(b)**). As a result of infection, demyelination can also occur in other regions of the CNS, including the brainstem and cerebellum. Destruction of oligodendrocytes initially leads to the development of microscopic lesions (**Figure 4(b)**), and, as the disease progresses, the demyelinated areas become enlarged and eventually may coalesce, making them visible on gross examination of cut sections (**Figure 4(a)**). In most cases, however, astrocytes have also been found to be abortively infected by JCV, exhibiting enlarged, lobulated nuclear structures. Lipid-laden macrophages frequently migrate to the areas of demyelination, perhaps phagocytizing the myelin breakdown products. In most cases, a large number of HIV-infected macrophages are also found within the necrotic lesions. However, it is not clear how these immune cells infiltrate into areas of demyelination. One possible explanation is that JCV infection may recruit HIV-infected macrophages into demyelinated areas or, alternatively, uninfected macrophages may become infected by HIV after they are recruited into the CNS.

Clinically, the most common signs and symptoms of PML at the time of presentation are visual deficit, which is the most common presenting sign accounting for 35–45% of cases, motor weakness, accounting for 25–33% of cases, and mental deficits (emotional lability, difficulty with memory and dementia), which is seen in approximately one-third of cases. PML usually progresses to death within 4–6 months, although, in occasional cases, clinical signs and symptoms appear to remain stable for a long period of time.

Oncogenic potential of JCV

JCV, BKV, and SV40 are all known to induce a variety of tumors in experimental animals. All also have the ability to induce neoplastic cell transformation in tissue culture. Following its isolation, JCV was demonstrated to induce tumors in experimental animals in tissues of neuronal origin, but the type of the tumor induced by JCV depends

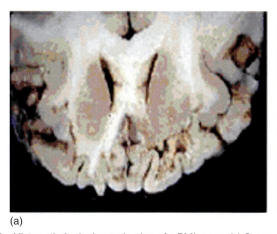
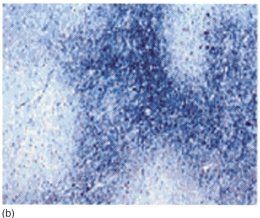

(a)   (b)

Figure 4 Histopathological examination of a PML case. (a) Gross appearance of PML in a coronal section of the brain. Multiple areas of cavitation are present in the subcortical white matter of the frontal lobes. (b) Staining for myelin demonstrates several areas of myelin loss in the white matter. Luxol Fast Blue, original magnification ×40. Reproduced by permission from Macmillan Publishers Ltd. *Journal of Clinical Pathology: Molecular Pathology* (Gallia G, Del Valle L, Line C, Curtis M, and Khalili K (2001) Concamitant progressive multifocal leukoencephalopathy and primary central nervous system lymphoma expression JC virus oncoprotein, large T antigen. *Journal of Clinical Pathology: Molecular Pathology* 54(5): 354–359), copyright (2001).

on the type and age of animal and the site of inoculation. For example, when the Mad-1 strain of JCV was inoculated intracerebrally and subcutaneously into newborn Syrian hamsters, more than 80% of animals developed medullablastomas, glioblastomas, or neuroblastomas. An entire biologically active JCV genome was isolated when cells from these tumors were co-cultivated with permissive glial cells. In contrast, when a similar group of animals was inoculated intraocularly with the same strain of JCV, they mostly developed abdominal neuroblastomas in several locations. It was also observed that tumors metastasized to the bone marrow, lymph node, and liver.

Interestingly, JCV is the only polyomavirus that induces tumors in nonhuman primates, such as monkeys. In order to mimic a case resembling PML in humans, owl and squirrel monkeys were inoculated with JCV subcutaneously, intraperitoneally, or intracerebrally. The animals developed tumors at different time intervals. For instance, one owl monkey developed a malignant cerebral tumor similar to astrocytoma in humans after 16 months. Another animal developed a malignant neuroblastoma 25 months after inoculation.

Transgenic mouse models have also been developed to mimic the acute demyelination observed in PML patients. A transgenic mouse was created by using the portion of the JCV genome that contains the promoter and coding regions for LT-Ag. Some of the offspring from this mouse exhibited a mild to severe tremor phenotype. Hypo- and dysmyelination were observed in the CNS but not in the peripheral nervous system (PNS), suggesting that expression of LT-Ag affects myelin formation in the CNS and not in the PNS. Further characterization of myelin formation in transgenic animals revealed that the level of myelin sheath wrapped around the axons was relatively low, although the expression level of proteolipid protein, myelin basic protein, and myelin-associated glycoprotein genes appeared to be normal at the RNA level. In contrast, the respective protein levels appeared to be reduced. The mechanism by which LT-Ag alters the levels of these proteins in transgenic mice remains unknown, but it has been suggested that LT-Ag may influence the rate of translation of mRNA for these genes or other cellular genes that negatively influence the maturation of oligodendrocytes. This may eventually alter myelin formation around the axons.

Our group also described the formation of different tumors in tissues derived from neuronal origin in transgenic mouse models. The LT-Ag coding region under the control of the regulatory region of the archetype strain was utilized to create these transgenic animal models. Histological and histochemical analysis of the tumor masses revealed no sign of hypomyelination in the CNS, which was a feature of this transgenic model. In contrast, cerebellar tumors resembling human medullablastomas were induced.

In addition to the induction of tumors in experimental animals, the JCV genome has also been detected in a variety of human tumors, which raises the possibility of involvement of JCV LT-Ag in tumor formation in humans. Richardson, who first described PML in 1961, reported an incidental case of an oligodendroglioma in a patient with concomitant occurrence of chronic lymphatic leukemia and PML. The association of PML with multiple astrocytomas was also reported in 1983. Similarly, another case in which a patient had a long history of immunodeficiency syndrome with PML was described, and numerous foci of anaplastic astrocytes were observed. The presence of viral particles in both oligodendrocytes and astrocytes, but not in neoplastic astrocytes, in the demyelinating lesions of PML foci was demonstrated by electron microscopy. The presence of a large number of dysplastic or dysmorphic ganglion-like cells, which showed the properties of neurons, was also described recently in the cerebral cortex of a patient with PML. Expression of JCV LT-Ag, but not capsid proteins, was detected in these cells.

In addition to cases of concomitant PML and cerebral neoplasm, JCV has been shown to be associated with human brain tumors in the absence of PML lesions. The detection of JCV DNA in the brain tumors of an immunocompetent patient with a pleomorphic xanthoastrocytoma has been reported. In another investigation, JCV DNA and expression of LT-Ag were detected in tumor tissue from an immunocompetent HIV-negative patient with oligoastrocytoma. These two cases demonstrated the association of JCV with brain tumors in immunocompetent non-PML patients, and further prompted attempts to establish the association of JCV with different types of brain tumors in humans. Analysis of multiple brain tumors for presence of the JCV genome revealed that 57.1% of oligodendrogliomas, 83.3% of ependymomas, 80% of pilocytic astrocytomas, 76.9% of astrocytomas, 62.5% of oligoastrocytomas, and 66% of anaplastic oligodendrogliomas contained JCV early gene sequences. Furthermore, JCV genomic DNA has been detected in tumor tissues of non-neural origin, including the gastrointestinal tract and solid non-neural tumors such as colorectal cancers.

The precise mechanism by which JCV induces tumors is not known, but the tumorogenic protein LT-Ag is known to play a major role in this process. LT-Ag from JCV, as well as from BKV and SV40, has been shown to target major cell cycle regulators, including the tumor suppressor protein p53 and the retinoblastoma (pRb) gene products. This targeting inhibits the functions of these two key regulators of the cell cycle and perhaps others. In fact, protein interaction studies have clearly showed complex formation between LT-Ag and cell cycle regulators including pRb, p53, and p107.

BK Virus

BKV belongs to the same genus as JCV, as species *BK polyomavirus*. Like that of other polyomaviruses, the BKV genome consists of a small, closed circular DNA approximately 5 kbp in size, with the size varying between strains. For example, the genome sizes of the DUN, MM, and AS strains are 5153, 4963, and 5098 bp, respectively.

The structural organization of the BKV genome resembles that of JCV (**Figure 1**) in that it contains bidirectional coding regions (early and late), which are regulated by the regulatory region. The early region encodes only the regulatory tumor antigens, Sm t-Ag and LT-Ag, and the late region is responsible for the capsid proteins (VP1, VP2, and VP3) and a small, basic regulatory protein, agnoprotein. As for JCV and SV40, LT-Ag is essential for viral DNA replication, but the function of Sm t-Ag in BKV regulation is unclear. LT-Ag also potently transactivates the late promoter. The capsid proteins form icosahedral shell structures (39–42 nm in diameter), into which the viral genome is packaged. The capsid proteins are critical for attachment of the virus to cell surface receptors. Although the function of the BKV agnoprotein is unclear, recent evidence from JCV suggests that it plays a role in viral DNA replication, transcription, and cell cycle regulation.

Infection and Associated Disorders

Like JCV, BKV has a worldwide distribution in the human population. Primary infection occurs during early childhood and is subclinical, though occasionally accompanied by mild respiratory illness or urinary tract disease. Little is known about the route of transmission, although induction of upper respiratory disease and detection of latent BKV DNA in tonsils suggest a possible oral or respiratory route. During primary infection, viremia occurs and the virus spreads to various organs, including kidneys, bladder prostate, uterine cervix, lips, and tongue, where it remains in a latent state. Reactivation from the latent state is mostly associated with the immunocompromised state of the individuals. Reactivated virus has been detected in the urine of renal and bone marrow transplant recipients undergoing immunosuppressive therapy and in the urine of pregnant women. Upon reactivation, the virus may cause interstitial nephritis and urethral obstruction in patients receiving renal transplants. BKV has surfaced as a significant pathogen in kidney transplant patients in recent years by association with nephropathy, better known as polyomavirus-associated nephropathy. BKV infection appears to be a serious problem in renal allograft recipients in the first 2 years after transplantation, if not treated properly. In addition, an association between hemorrhagic cystitis and BKV has been shown in bone marrow transplant recipients.

Oncogenicity of the Genome in Experimental Animals

Like JCV, BKV is also oncogenic in experimental animals, including young or newborn mice, rats, and hamsters. The route of inoculation is a significant factor in determining the types of tumors induced. For example, BKV induces tumors in high proportions when inoculated intracerebrally or intravenously, but is weakly oncogenic when inoculated subcutaneously. BKV induces tumors in hamsters in a variety of tissues and organs, including ependymoma, neuroblastoma, pineal gland tumors, fibrosarcoma, esteosarcoma, and tumors of pancreatic islets. Rats inoculated with BKV develop fibrosarcoma, liposarcoma, osteosarcoma, nephroblastoma, and glioma. In a similar setting, however, mice develop only choroid plexus papilloma.

LT-Ag is the primary protein of BKV responsible for tumor induction in experimental animals. The oncogenic ability of LT-Ag has also been tested in transgenic mice models. Like JCV LT-Ag, BKV LT-Ag is known to target and inhibit several key cell cycle regulatory proteins, including p53 and the pRb family members, pRb105, pRb107, and pRb130. BKV also causes renal tumors and hepatocellular carcinoma. In such studies, there appear to be differences among strains of BKV with respect to the ability of LT-Ag to cause tumors. For example, Gardner's strain appears to be more potent in transgenic mice than isolates such as the MM, BKV-IR, or RF strains.

BKV LT-Ag induces cell transformation in tissue culture, although to a lesser extent than SV40 LT-Ag. It has been proposed that a 'hit-and-run' mechanism is most likely to operate during this transformation process; that is, expression of LT-Ag is required for the initiation of a multistage process, but is not required after a certain stage of transformation is reached. For example, in one study it was observed that, although transfection of BKV DNA into human cells resulted in a transformed phenotype, it was absent from most new transformed clones.

Another mechanism by which human polyomavirus LT-Ag may cause transformation is via the induction of chromosomal structural alterations characterized by breaks, gaps, dicentric and ring chromosomes, deletions, duplications, and translocations. While the molecular mechanism of this clastogenic effect of BKV LT-Ag on host DNA is unknown, it is thought that it may reside in the ability of the protein to bind topoisomerase I or in its helicase activity, in which it may induce chromosomal damage while unwinding the strands of cellular DNA. Moreover, since LT-Ag targets and inactivates p53, this may lead to survival of damaged cells, increasing their chance of transformation and immortalization. As a result, the clastogenic and mutagenic activities of the LT-Ag of human polyomaviruses may disturb the crucial function of the genes that are important for the maintenance of genomic stability, such as oncogenes, tumor suppressor genes, and DNA repair genes.

Association of the Genome with Human Tumors

During late 1970s, BKV DNA was detected in a variety of human tumors and tumor cell lines, which prompted researchers to investigate the possible association of BKV with tumor induction. BKV was found to exhibit a specific oncogenic tropism for ependymal tissue, endocrine pancreas, and osteosarcomas in rodents. This led investigators to focus primarily on the presence of the BKV genome in such tumors. Southern hybridization studies showed that some pancreatic islet tumors, as well as some brain tumors, contain the BKV genome in a free, episomal state. BKV was even rescued from some of the tumors by transfection of human embryonic fibroblasts with tumor DNA.

The BKV genome was also reported in 46% of brain tumors of the most common histotypes, and was found to be integrated into chromosomal DNA in this particular study. Association of human tumors with immunocompromised conditions was also analyzed by Southern blotting, and the BKV genome was associated with Kaposi's sarcoma (KS) at low frequencies (20%).

Recently, normal and neoplastic human tissues, as well as tumor cell lines, were also examined for the presence of the BKV genome by PCR, utilizing primers for the early region. Nucleotide sequence analysis of PCR products revealed the presence of BKV-specific sequences in several brain tumor samples, one osteocarcinoma, two glioblastoma cell lines, one normal brain tissue, and one normal bone tissue specimen. In these studies, expression of the early region was demonstrated in some of the samples by reverse transcription (RT)-PCR. The presence of the BKV genome was also investigated in several different tumors, including urinary tract tumors and carcinomas of the uterine cervix, vulva, lips, and tongue. However, the data obtained from such studies were inconclusive because the proportion of positive samples in neoplastic tissues of the urinary and genital tracts and oral cavity was similar to that detected in corresponding normal tissues. BKV DNA has been shown to be present in a high proportion of KS cases, suggesting that BKV may be an important cofactor in KS.

Conclusion

Since the first cultivation of JCV from a PML patient and BKV from a renal transplant patient in 1971, we have learned much about the biology of both viruses. However, important aspects of the life cycle (viral entry, transport of viral particles to the nucleus, transcription, replication, and assembly and release of virions) remain to be elucidated. The more we investigate the biology of these viruses, the more complex their biology is revealed to be. Understanding the molecular mechanisms underlying the life cycle of these viruses will considerably enhance our view of their complex biology, which should then allow us to design effective therapeutics to intervene in the infection cycle at an early stage, before they cause more advanced disease.

See also: Human Cytomegalovirus: General Features; Human Herpesviruses 6 and 7; Polyomaviruses; Polyomaviruses of Mice; Simian Virus 40.

Further Reading

Berger JR and Concha M (1995) Progressive multifocal leukoencephalopathy: The evolution of a disease once considered rare. *Journal of Neurovirology* 1: 5–18.

Corallini A, Pagnani M, Viadana P, *et al.* (1987) Association of BK virus with human brain tumors and tumors of pancreatic islets. *International Journal of Cancer* 39: 60–67.

Del Valle L, Gordon J, Assimakopoulou M, *et al.* (2001) Detection of JC virus DNA sequences and expression of the viral regulatory protein T-antigen in tumors of the central nervous system. *Cancer Research* 61: 4287–4293.

Dorries K, Loeber G, and Meixensberger (1987) Association of polyomaviruses JC, SV40, and BK with human brain tumors. *Virology* 160: 268–270.

Gallia G, Del Valle L, Line C, Curtis M, and Khalili K (2001) Concamitant progressive multifocal leukoencephalopathy and primary central nervous system lymphoma expression JC virus oncoprotein, large T antigen. *Journal of Clinical Pathology: Molecular Pathology* 54(5): 354–359.

Gordon J, Del Valle L, Otte J, and Khalili K (2000) Pituitary neoplasia induced by expression of human neurotropic polyomavirus, JCV, early genome in transgenic mice. *Oncogene* 19: 4840–4846.

Hirsch HH (2005) BK virus: Opportunity makes a pathogen. *Clinical Infectious Diseases* 41: 354–360.

Kim J, Woolridge S, Biffi R, *et al.* (2003) Members of the AP-1 family, c-Jun and c-Fos, functionally interact with JC virus early regulatory protein large T-antigen. *Journal of Virology* 77: 5241–5252.

Lynch KJ and Frisque RJ (1991) Factors contributing to the restricted DNA replicating activity of JC virus. *Virology* 180: 306–317.

Major EO, Amemiya K, Tornatore CS, Houff SA, and Berger JR (1992) Pathogenesis and molecular biology of progressive multifocal leukoencephalopathy, the JC virus-induced demyelinating disease of the human brain. *Clinical Microbiological Reviews* 5: 49–73.

Monaco MC, Jensen PN, Hou J, Durham LC, and Major EO (1998) Detection of JC virus DNA in human tonsil tissue: Evidence for site of initial viral infection. *Journal of Virology* 72: 9918–9923.

Padgett BL, Zu Rhein GM, Walker DL, Echroade R, and Dessel B (1971) Cultivation of papova-like virus from human brain with progressive multifocal leukoencephalopathy. *Lancet* i: 1257–1260.

Raj GV and Khalili K (1995) Transcriptional regulation: Lessons from the human neurotropic polyomavirus, JCV. *Virology* 213: 283–291.

Richardson EP (1961) Progressive multifocal encephalopathy. *New England Journal of Medicine* 265: 815–823.

Sariyer IK, Akan I, Palermo V, Gordon J, Khalili K, and Safak M (2006) Phosphorylation mutants of JC virus agnoprotein are unable to sustain the viral infection cycle. *Journal of Virology* 80: 3893–3903.

Walker DL, Padgett BL, Zu Rhein GM, Albert AE, and Marsh RF (1973) Human papovavirus (JC): Induction of brain tumors in hamsters. *Science* 181: 674–676.

Polyomaviruses of Mice

B Schaffhausen, Tufts University School of Medicine, Boston, MA, USA

© 2008 Elsevier Ltd. All rights reserved.

Introduction

Murine polyomavirus, the virus that gives the name to the family *Polyomaviridae* and to the single genus (*Polyomavirus*) in the family, was discovered in the 1950s. Gross and Stewart observed that cell-free extracts from leukemic mice could induce neck tumors as well as leukemia in newborn mice. It soon became apparent that the leukemia-inducing activity could be separated from the activity inducing parotid tumors. Because the viral agent was found to induce a variety of solid tumors, the name 'polyoma' became attached to the virus. Epidemiology showed that the virus was widely disseminated in the mouse population.

Polyomaviruses have been valuable models for basic eukaryotic processes such as splicing and DNA replication. They have been especially important for studying growth regulation. The viruses require the apparatus of cellular DNA synthesis for their own replication. To meet this need they have evolved many different ways to intervene in cellular growth regulation. This intervention can cause tumors. Interestingly, although polyoma can be an extraordinarily potent tumor virus in certain laboratory strains of mice or in immunocompromised animals, the frequency of tumors in natural mouse populations is low despite the broad dissemination of the virus. Much of this resistance reflects the ability of T-cell immunity to prevent viral oncogenesis.

Virus Particles

Two major types of particles are found in lysates of infected cells: virions containing viral DNA and protein, and empty capsids that lack DNA and are noninfectious. Early physical characterization determined a molecular weight for virions of approximately 23 million Da with a sedimentation constant of 240S. The viral capsid is approximately 450 Å in diameter and possesses icosahedral symmetry. The viral coat consists of 72 pentameric capsomeres that are made up of VP1, the major viral structural protein. VP2 and VP3, which consists of the C-terminal 204 amino acid residues of VP2, represent the minor capsid proteins (**Table 1**). They appear to be found at the bottom of the central 'hole' in each capsomere. Although not required to assemble capsids, the infectivity of particles lacking VP2 or VP3 is dramatically reduced. The viral DNA is a closed, circular, supercoiled molecule of approximately 5300 bp. In virions, the DNA is associated with the cellular core histones H2A, H2B, H3, and H4, but not H1. Compared to host chromatin, the histones of the viral particle are much more highly acetylated, a modification known to be associated with gene activity.

Organization of the Viral Genome

The genome (**Figure 1**) is divisible into two almost equal parts. One half is used for the expression of the capsid proteins and the other is used for the expression of the three major early gene products. The early products are called T antigens (large T, middle T, and small T) because

Table 1 Polyomavirus (A2 strain) proteins and sizes (amino acid residues)

Early proteins (no of residues)		Late proteins	
Large T antigen	785	VP1	383
Middle T antigen	421	VP2	319
Small T antigen	195	VP3	204

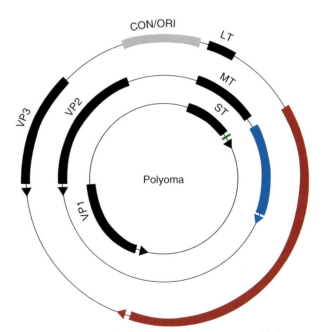

Figure 1 The organization of the polyoma genome. Early transcription of the T antigens (large T, LT; middle T, MT; small T, ST), late transcription of the capsid proteins (VP1, VP2, and VP3), as well as the origin of viral DNA replication (ORI), are regulated by the control (CON) region. The protein-coding sequences are indicated in colors to emphasize reading frame differences for the T antigens.

they were discovered using serum from tumor-bearing animals. Between the initiation codons for the late protein VP2 and the T antigens, there is a region of approximately 470 bp that contains control elements and the initiation sites for transcription and replication (**Figure 2**).

Genetic analysis has defined sequences necessary for viral DNA replication. There is a core origin of approximately 70 bp that includes an A+T-rich region on the late side and a highly purine-rich region (on one strand) on the early side. These regions flank a central 34 bp inverted repeat containing four pentanucleotide large T binding sites arranged as two head-to-head pairs. Replication initiated at this core origin requires a functional enhancer in *cis*.

Viral transcription and replication are regulated by the polyoma enhancer. **Figure 2** shows that the enhancer is located on the late side of the origin region. The enhancer has often been subdivided into two elements alpha (or A) and beta (or B) with overlapping function. It represents the binding sites for a series of cellular proteins that support viral transcription and replication. The most prominent binding sites are for PEA1 (AP1), PEA2 (runx), and PEA3 (ets family). Additional factors include polyomavirus enhancer B binding protein 1 (PEB1), EF-C, PED1, and c/EBP. Within the alpha element is a bipartite PEA1/PEA3 site of particular importance. Not surprisingly, given the number of cellular transcription factors that bind to it, there is redundancy built into the enhancer. The activity of the enhancer varies among cell types. It is clear that the enhancer structure can change in response to cellular environment or viral mutation. Ordinarily, polyoma grows poorly in embryonal carcinoma cell lines such as F9 cells. Viruses selected to grow on these cells were found to have alterations in the enhancer region. Hr-t mutants that lack middle T and small T were found to have enhancer alterations that contribute to their ability to grow. Mutants with deletions in the B enhancer domain grow poorly in neonatal, but almost normally in adult mice.

Not only can alterations in enhancer sequences affect behavior of the virus, but alterations on the early side of the control region can affect the tumor profile as well.

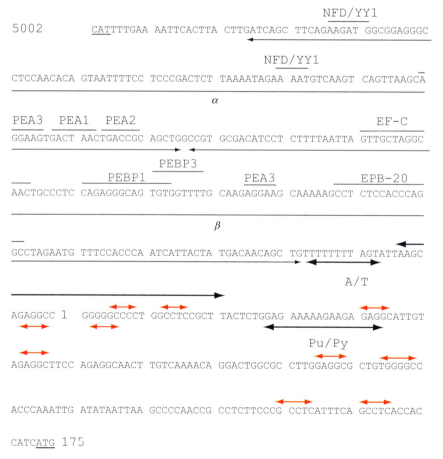

Figure 2 The origin/control region. The sequence between the initation codons for VP2 and large T is shown. The enhancer region is shown by the light arrows indicating the alpha (A) and beta (B) regions. Some transcription factor binding sites are indicated. The A+T-rich, purine-rich (one strand), and central palindrome regions of the core origin are shown by the thick arrows. Large T-binding sequences are indicated by red arrows.

Virus Infection

Infection by polyoma can have three outcomes: productive infection, which results in the production of more virus; 'abortive transformation' in which the infected cell temporarily assumes the transformed phenotype; and stable transformation, which permanently alters the cell to a transformed phenotype. In mouse cells, which are permissive for the virus, productive infection is the predominant response. In nonpermissive rat or hamster cells, abortive transformation is the most common response. Stable transformation is associated with integration of the viral DNA into host chromosomes, although episomal DNA has been demonstrated in tumors. The frequency of stable transformation is generally low and varies with the cell line.

A brief summary may be useful before dealing with the infectious cycle in more detail. The kinetics of infection depends on the multiplicity. In a high-multiplicity infection, early mRNA can be detected by polymerase chain reaction (PCR) as early as 6 h post infection. Large T can be observed by 8 h. Viral DNA replication begins 12–18 h post infection and continues for approximately 20 h. Late transcription is observed following viral DNA replication. Progeny virions begin to appear in the nucleus between 20 and 25 h post infection. Virus production plateaus around 40–48 h with the appearance of cytopathic effects.

Since the virus uses host DNA and RNA polymerases for its replication, it must enter the cell and reach the nucleus. Different polyomaviruses appear to have different entry processes upon infection. It has been known for 40 years that neuraminidase blocks polyoma infection. Polyoma binds sialic acid residues of glycoproteins and GD1a glycolipids with a critical sialic acid-α2,3-Gal structure. Interestingly, differences in the ability of virus strains to recognize different carbohydrate structures, thereby allowing interactions with 'pseudoreceptors', result in changes in virus spread in the mouse. After binding, the virus is internalized by caveolin-dependent and -independent pathways into the endoplasmic reticulum (ER). Successful infection occurs with conformational changes taking place in the ER determined by ER-localized oxidoreductase ERp29 and protein disulfide isomerase. Recent data indicating a role for Derlin2 in infection suggest that the quality control system of the ER is involved in transferring infecting virus to the cytoplasm. Nuclear localization sequences found in the capsid proteins and cellular histones are available to allow transport into the nucleus.

RNA Transcription and Processing

Viral transcription is divided into two types, early and late, that proceed in opposite directions from the viral control region. The early transcription unit that encodes the T antigens is active soon after infection begins. Differential splicing produces the three major early proteins, large T, middle T, and small T; a fourth spliced product, tiny T, has been observed, but its significance is uncertain. Late transcripts encoding the viral structural proteins are made early in infection, but they are not processed efficiently and have a short half-life. Late RNA does not accumulate until viral DNA replication proceeds, and late messages are generated for each of the three capsid proteins VP1, VP2, and VP3. Starting at that time, inefficient polyadenylation and termination generates large, multi-genome length, heterogeneous late mRNAs. These RNAs are processed by leader-to-leader splicing so that there are tandem repeats of a 57 bp late leader, which is thought to allow accumulation of stable late message. The accumulation of late message means that there are now sequences present that are antisense to the early transcript. Editing of double-stranded RNA (dsRNA) by the ADAR enzyme converts up to 50% of early RNA adenosines to inosines with a resulting decrease in levels of early protein synthesis. The basis for the switch that allows late leader splicing is not yet fully appreciated, but may well involve the organization of polyadenylation sites and a balance of their editing and cleavage.

Viral Transformation and Tumorigenesis

Investigating polyomavirus transformation has provided repeated insights into normal and abnormal cell behavior. Tyrosine phosphorylation and phosphatidylinositol 3-kinase (PI3-K) are two discoveries that came directly from polyoma middle T studies. The p53 tumor suppressor was discovered using SV40, the monkey polyomavirus. Studies on interactions of large T with the retinoblastoma (Rb) tumor suppressor family (pRb, p107, p130) have led to insights into the E2F family and its regulation. In addition, and in contrast to many oncogenic RNA viruses, polyoma early gene products are also required for productive infection. Large T is required for viral DNA replication. The limited host range of hr-t mutants shows that middle T and small T act to provide a cellular environment that is supportive for viral replication.

Viral transformation is carried out by the early region. As a result of the splicing described above, all four early proteins share a common domain of 79 amino acid residues. This common N-terminus represents a DnaJ domain. Proteins that contain DnaJ domains ordinarily function as cofactors for DnaK proteins that act as molecular chaperones. Small T shares additional 112 amino acid residues with middle T. Each T antigen has unique C-terminal sequences. In the case of small T, the unique C-terminus is only four amino acid residues. The result of this processing is that three proteins, large T, middle T, and small T, have different intracellular localization and different functions. Each of these proteins has the independent ability to affect cell growth and survival. Viral transformation results from T antigen association with, and regulation of, cellular signal-transducing proteins. Only large T is known to have intrinsic enzymatic activity, but even large T binds cellular proteins to regulate host function. Each of the three major T antigens will be discussed in turn. Because important insights have come from comparison studies with SV40, some mention will be made of SV40 small T and large T as well. SV40 has no direct counterpart to middle T.

Middle T

Middle T is key to polyoma transformation and tumor induction. Middle T is necessary and usually sufficient for transformation in cultured cells. The importance of middle T in tumor induction is clear. Viruses with middle T mutations show decreased tumorigenesis and changes in tumor profile. A number of transgenic models show tumor formation in response to middle T expression in the absence of either small T or large T. A mouse mammary tumor virus (MMTV)-middle T model of breast cancer has been especially well studied. Middle T has been a particularly useful model, because mutations that affect particular associations have usually had clearly identifiable phenotypes. In some instances, both in culture and in animals, middle T requires complementation from other viral oncogenes in order to transform. For example, in primary cells, polyoma middle T needs complementation by large T or nonpolyoma oncogenes for transformation. In REF52 cells, middle T also requires complementation with small T. While the role of middle T in transformation is clear, it is important to remember that middle T also plays an important role in polyoma infection. It can upregulate viral gene expression and viral DNA replication. Middle T has also been shown to regulate phosphorylation of the major capsid protein VP1.

Middle T functions as a kind of adaptor on which cellular signaling proteins are assembled (**Figure 3**). It might be viewed as a kind of constitutively active growth factor receptor. Middle T is associated with membranes; this membrane association is critical for transformation. Middle T binds the major cellular serine/threonine phosphatase PP2A. Cellular PP2A exists as a complex, either as two subunits, A and C, or as three subunits, A, B, and C. The B family is especially diverse, with many members, and appears to be involved in regulating targeting and activity. Both polyoma middle T and small T exist in complexes with the A and C subunits of the enzyme. Interestingly, middle T binds both Aα and Aβ forms of PP2A.

Association with PP2A allows middle Ts to bind and activate some of the src family tyrosine kinases (PTKs). C-src, c-yes, and c-fyn, but not c-hck, are all bound by middle T. Curiously, the PP2A does not have to be catalytically active for the PTK complex to form, suggesting that it may function as some kind of scaffold.

Tyrosine kinase activity is critical for transformation, so middle T mutants lacking associated tyrosine kinase activity are not able to transform. The c-src associated

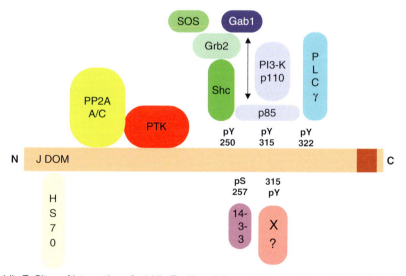

Figure 3 Polyoma middle T. Sites of interaction of middle T with cellular proteins are shown. Each of the signal transducers shown above the middle T sequence is known to be important for transformation. PP2A, protein phosphatase 2A; PTK, src family tyrosine kinase; PI3-K, phosphatidylinositol 3-kinase; X, missing partner at 315 indicated by genetics; PLCγ phospholipase Cγ1. There is also association between Gab1 and PI3-K. The red block represents the hydrophobic membrane attachment site. Small T shares with middle T the N-terminal 191 amino acid residues that include the J domain and PP2A binding site, but not the PTK binding site T and has only four unique residues at the C-terminus.

with middle T is activated, presumably, because it is lacking phosphorylation at Y527. In the complexes with PTK, middle T is phosphorylated on tyrosine residues. The major sites of phosphorylation are at residues 250, 315, and 322. These phosphorylations provide docking sites for signaling molecules. The initial picture of interactions was relatively straightforward. Each of the major sites represents a connection to a signal generator: 315 to PI3-K, 250 to Shc, and 322 to PLCγl. The picture is now more complicated from three points of view. First, there must be phosphorylation at minor tyrosines in the C-terminal half of middle T that can contribute to function. Second, there are clearly multiple connections to PI3-K, for example, mediated by more than one phosphotyrosine. Gab 1 protein, for example, binds middle T through its association with grb2 at 250 and can provide a connection to PI3-K. Third, multiple cellular targets seem to be reached through the single phosphotyrosine at 315. One is PI3-K, but the other remains to be identified.

Genetic analysis has been very useful in illuminating some of the signaling pathways of middle T. Mutation of 322, the PLCγl site, has modest effects in some transformation assays, but in low serum there is a substantial effect. Association with PLCγl is likely to be the basis for increased levels of IP3 observed in middle-T-transformed cells and for effects on PKC. Mutation of tyrosine 250 affects transforming ability and tumorigenesis, as do mutations just N-terminal to 250 (the NPTY motif). This represents a binding site for the adaptor Shc, and that binding leads to tyrosine phosphorylation of Shc. In turn, Shc binding and tyrosine phosphorylation are responsible for the association of Grb2. This in turn leads to the recruitment of SOS, a ras exchange factor that activates ras. Interestingly, while the 250 phosphorylation site seems to be especially important for transformation either in mouse cells or the transgenic middle T mammary tumor model, virus with a mutation at 250 is relatively efficient at forming tumors. Association with PI3-K, which is abolished by mutation at 315, is profoundly important for transformation in cultured rat fibroblasts and has a dramatic effect on the tumor profile in mice. PI3-Ks are broadly important enzymes that have been strongly linked to cancer. Downstream signaling from PI3-K includes both activation of Akt, a kinase important in preventing apoptosis, and activation of rac1, a G-protein involved in cytoskeletal organization and oxidative signaling.

Small T

In the past, small T has been studied less intensively than large T and middle T. However, increased attention has recently been focused on small T, since SV40 small T can play a role in transformation of human cells. It cooperates with hTert, SV40 large T, and ras in oncogene complementation assays. Transgenic SV40 small T can also contribute to mammary gland tumorigenesis.

Wild-type small T is not sufficient for transformation. Independent expression of small T in fibroblasts enables them to grow to a high cell density. However, small T can complement middle T for tumor induction and transformation. For example, small T can complement middle T for transformation of REF52 cells through its effects on arf-mediated activation of p53. Polyoma small T can resist the effects of p53-induced apoptosis. SV40 small T opposes apoptosis induced by large T. It also opposes Fas-mediated apoptosis of hepatocytes. Reports have suggested that Akt activity is induced by SV40 small T, which might account for these observations. Interestingly, in some cellular contexts both polyoma and SV40 small T also appear to be able to induce apoptosis.

Small T can be found in the nuclear and cytoplasmic compartments. Both SV40 and polyoma small T bind zinc. Like middle T, small T has been shown to bind PP2A. For SV40, the ability to displace specific B subunit family members such as B56γ seems important for its function.

Small T has been connected to a variety of cellular functions. For example, polyoma small T induces the membrane lectin agglutinability that is usually associated with cell transformation. SV40 small T can both disrupt tight junctions and disturb actin structure in epithelial cell monolayers. It has also been reported to perturb centrosome function. Most attention has focused on DNA replication and RNA synthesis. In serum-starved cells, polyoma small T can contribute to S phase induction in conjunction with large T. For example, it regulated the cyclin-cdk inhibitor p27. Small T can also affect viral DNA synthesis. Polyoma small T has also been implicated in virion assembly. For SV40, it has been shown that small T can transactivate or even repress various exogenous promoters. In many instances this has been related to the ability to bind PP2A. For SV40 small T, the interaction with PP2A stimulates the MAP kinase pathway, inducing cell proliferation. To cite just two recent examples of transcriptional activators, SV40 small T activates Sp1 and the FHL2 co-activator in a PP2A-dependent manner.

Large T

Polyomavirus large Ts have a dual role, acting not only directly in viral DNA replication and transcription, but also functioning to alter host-cell signaling. The role of large T in viral DNA replication has been studied extensively. SV40 provided the major model for establishing the mechanisms of cellular DNA replication, and its replication is still the best understood. In a productive infection, polyoma large T initiates viral DNA replication. It is thought to do this by forming a double hexamer at the two head-to-head pairs of binding sites in the origin region; it then can unwind the origin and recruit factors such as DNA polymerase and RNA polymerase. By analogy to SV40, it is also likely to participate in the elongation phase of DNA synthesis. Its role in replication is important in other contexts as well. In transformation, it is responsible for integration and excision of the viral genome and can also promote recombination. Polyoma large T possesses the biochemical activities that might be anticipated for a protein involved in DNA replication. It binds the polyoma origin region at GAGGC sequences. Like SV40 large T, it possesses helicase and ATPase activities. Large T associates with pol α-primase, and by analogy to SV40 is expected to associate with RNA polymerase and topoisomerase I.

Large Ts are directly involved in cell transformation. Large T is the major transforming protein of SV40. Polyoma large T does not transform by itself, and viruses that make only large T do not cause tumors. Much of the difference in phenotype comes from the obvious interaction of SV40 large T with p53. Although there is a recent indication that polyoma large T may interact with p53 phosphorylated on serine 18, polyoma tumors show no evidence of a p53 block seen in SV40 tumors, and cell lines from tumors retain a normal p53 response to DNA damage. Nonetheless, polyoma large T can cooperate with other oncogenes such as middle T or ras in the transformation of primary cells. Similar complementation can also be seen in tumorigenesis.

Polyoma large T has important effects on cell phenotype, presumably to prepare the cellular environment for viral replication. Large T immortalizes primary cells in a manner dependent on the binding site for the retinoblastoma susceptibility (pRb), p107, and p130 gene products. Large T prevents differentiation, either of myoblasts or preadipocytes. The ability of large T to block withdrawal of myoblasts from the cell cycle and to prevent differentiation is dependent on Rb binding. Large T can induce dramatic apoptosis. These effects also involve interactions with the Rb family.

Given these phenotypes, it is not surprising that large T affects cellular DNA and RNA synthesis. Large T induces cellular DNA synthesis, using both Rb-dependent and -independent mechanisms. In the case of SV40 large T, at least four separate functions that contribute to the induction of cellular DNA synthesis have been identified. These functions have been mapped to the binding domains for p53, pRb/p107, p300, and TEF-1, and at least in the case of pRb requires the J domain as well.

Large T is a transcriptional activator of cellular genes. The first target identified was dihydrofolate reductase (DHFR); since then, many others have been identified as targets, such as the thymidine kinase (TK), human heat shock protein 70 (hsp 70), DNA polymerase alpha (pol α), proliferating-cell nuclear antigen (PCNA), thymidylate synthase (TS), and cyclin A genes. Transactivation of the TK, pol α, PCNA, DHFR, and TS genes requires an intact pRb/p107 binding site on polyoma large T and is mediated via the cellular transcription factor E2F. The ability of large T to activate these E2F-responsive genes depends upon an intact N-terminal J domain that binds hsc 70. As shown most clearly for SV40 large T, the role of this chaperone function is to disrupt E2F-Rb family complexes. While Rb binding is one function, transactivation of cellular and viral promoters by polyoma large T can also occur in the absence of pRb/p107 binding. Similarly the cyclin A promoter can be activated, even when the E2F site is mutant. There are certainly multiple mechanisms that require a more complete understanding. In fact, studies on SV40 large T have suggested that it is a somewhat promiscuous activator. Both the nature of the TATA/Inr element and the upstream sequences can contribute to the activation. This has led to a description of SV40 large T as being a TBP-associated factor (TAF)-like protein. A potentially important mechanism is association with histone acetyltransferases. Large T can associate with p300/CBP-associated factor (PCAF), p300, and CREB-binding protein (CBP).

Figure 4 shows a current view of the anatomy of large T. Large T is a nuclear zinc-containing phosphoprotein of 785 amino acid residues. The zinc-binding motif is a C2H2 element that differs from that in middle T and small T. In large T, it promotes self-association. Large T can be divided into two major domains that exhibit independent function. The N-terminal domain primarily functions to stimulate the host cell, while the C-terminal domain functions primarily in DNA replication. There are additional subdomains, for instance, the DNA-binding domain can function autonomously. Large T functions both in self-association through the zinc-binding region and in association with cellular proteins, including members of the Rb family and the DnaK family of proteins. Large T function can clearly be regulated by its phosphoryla-

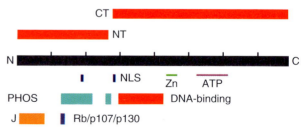

Figure 4 Large T antigen. The positions of the N-terminal (NT) and C-terminal (CT) domains are shown. The positions of the DNA-binding domain, zinc-binding element, ATP-binding domain, J domain, Rb-binding site, and nuclear localization sequences (NLS) are shown as bars. The regions containing phosphorylation sites are also indicated.

tion. Phosphorylation at threonine 278 by cyclin/cdk kinases is required for viral DNA replication.

Host Effects on Transformation and the Transformed Phenotype

Although the discussion of viral transformation and tumorigenesis has focused until now on the viral gene products, there are dramatic differences in the response of different inbred mouse strains to polyoma. Some strains such as C57BL/6J are quite resistant to tumor induction. Others such as C3H/BiDa are quite sensitive as neonates but develop resistance as adults. Much of this variation arises from the immune system. High tumor susceptibility can result, for example, from inheritance of a particular super antigen that affects the T-cell repertoire. Other mouse strains such as Ma/MyJ are resistant for reasons apparently related to spread of virus in the animal rather than immune mechanisms.

Transgenic middle T models point to the effects of host background on tumor behavior. The maternal genotype has a striking effect on tumor latency and on the likelihood of metastasis. Gene expression profiles resulting from middle T expression show obvious differences among strains.

See also: Simian Virus 40; Polyomaviruses of Humans; Polyomaviruses.

Further Reading

Michael JI and Eugene OM (2007) *Polyomaviruses.* In: Knipe D, Howley P, Griffin DE, et al. (eds.) *Fields Virology,* 5th edn., pp. 2263–2298. Philadelphia, PA: Lippincott Williams and Wilkins.

Stephen MD (2002) Polyoma virus middle T antigen and its role in identifying cancer-related molecules. *Nature Reviews Cancer* 2: 951–956.

Thomas LB (2001) Polyoma virus: Old findings and new challenges. *Virology* 289: 167–173.

Polyomaviruses

M Gravell and E O Major, National Institutes of Health, Bethesda, MD, USA

© 2008 Elsevier Ltd. All rights reserved.

Classification

Prior to the Seventh Report of the International Committee on Taxonomy of Viruses, polyomaviruses and papillomaviruses were classified in the family *Papovaviridae*. Commensurate with the release of this report, the genus *Polyomavirus* was removed from this family and elevated to independent status as the family *Polyomaviridae*. Capsid and genome size differences between polyomaviruses and papovaviruses were instrumental in prompting this change. Polyomavirus capsids are 40–45 nm in diameter and their genomes contain about 5000 bp, whereas papillomavirus capsids are 50–55 nm in diameter and their genomes contain 6800–8400 bp. Major differences in the replication cycles of polyomaviruses and papillomaviruses have also been described. The family *Polyomaviridae* currently comprises 13 members: four from monkeys, three from humans, two from mice, and one each from bird, hamster, rabbit, and cow. One virus, athymic rat polyomavirus, has been tentatively assigned to the family.

Morphological, Physicochemical, and Physical Properties

Members of the family *Polyomaviridae* have similar virion structures and the same general genome organization. Virions of polyomaviruses are nonenveloped, icosahedral particles composed of 360 copies of the major structural protein, VP1, and 30–60 copies of the minor structural proteins, VP2 and VP3. These molecules form 72 pentameric capsomers arranged in a skewed ($T = 7d$) lattice. Each capsomer is composed of five copies of VP1 and one copy of VP2 or VP3, each added to the internal cavity formed by association of the 5 VP1 molecules. The C-terminal ends of VP1 molecules extend to anchor neighboring capsomeres together. The VP2 and VP3 molecules are present within, but are not covalently linked to, the virion. In the course of virion assembly, occasional mistakes sometimes occur and aberrant capsid structures (such as empty particles, microcapsids, and tubular structures) are made.

Being nonenveloped, polyomaviruses are ether and acid resistant and relatively heat stable (50 °C, 1 h). However, they are unstable in 1 M $MgCl_2$. Virions have a sedimentation coefficient ($S20_w$) of 240S and an M_r of 2.5×10^7 Da. The buoyant densities of polyomaviruses are 1.2 and 1.34–1.35 g ml^{-1} in sucrose and CsCl gradients, respectively.

When the VP1 DNA sequence of the human polyomaviruses JC virus (JCV) or BK virus (BKV) is inserted into a baculovirus plasmid vector and expressed in insect cells as a recombinant gene, pentamers form resembling virion capsomeres. When these pentameric capsomere-like structures are purified and placed in a solution of physiological pH and ionic strength containing Ca^{2+} ions, directed self-assembly of pentamers into genome-free virions occurs. The capsids formed, which are called virion-like particles (VLPs), have the size, icosahedral symmetry and antigenicity of native virions. Practical use has been made of VLP production. JCV or BKV VLPs have been used in enzyme immunoassays (EIA) to measure titers of JCV- or BKV-specific antibodies elicited by infection with these viruses. Because of the greater sensitivity, specificity, and safety of these genome-free VLPs, their use in EIAs has now largely replaced hemagglutination assays as the preferred method to measure levels of JCV- or BKV-specific antibodies.

Genome

The genome in each polyomavirion contains a single, supercoiled molecule of closed circular, double-stranded DNA of about 5000 bp. It makes up about 10–13% by weight of the virion, the remainder being protein. The genome is composed of three functional regions: the genetically conserved early and late coding regions separated by the hypervariable regulatory region that contains the origin of DNA replication (ORI). Host cell histone proteins H2a, H2b, H3, and H4 associate with the supercoiled genome to form a mini-chromosome-like structure. Polyomavirus DNA synthesis occurs exclusively within the S-phase of the cell cycle. However, although cellular DNA replicates only once during each cellular S-phase, multiple cycles of viral genome replication can occur.

Proteins

Proteins produced during the course of polyomavirus replication are divided into those produced during the early and late stages of infection. The early proteins are nonstructural, and are called the large T and small t proteins (or antigens) based on their sizes; T (in either case) is derived from the word tumor. These T proteins are considered the master regulators because they stimulate cells to produce the enzymes and other factors required for cellular DNA replication, thereby setting up the conditions required for late events during viral DNA replication and virion assembly. The T proteins interfere with aspects of cell cycle regulation and cause cell transformation and sometimes tumorigenesis. They are translated from 2 to 5 mRNAs generated by alternative splicing from a common pre-mRNA.

The T protein genes are transcribed from one of the genomic DNA strands in the counterclockwise direction, and the late genes are transcribed in a clockwise direction from the other genomic strand from the opposite side of ORI. Polyomavirus replication occurs exclusively within the cellular S-phase. Binding of T protein to the hypophosphorylated retinoblastoma susceptibility protein (pRb) permits premature release of the E2F transcription factor, thus stimulating cell entry into S-phase. After recruitment of the host cell DNA polymerase complex to ORI, bidirectional replication is initiated. Activation of the late viral promoter by T protein and association with specific cellular transcription factors results in expression of late virus genes. These include the genes for the virion structural proteins, VP1, VP2, and VP3 and the nonstructural regulatory agnoprotein.

The three structural capsid proteins (VP1, VP2, and VP3) originate from a common precursor mRNA by alternative splicing. The VP2 precursor mRNA contains the coding sequence for the complete VP3 protein, and each protein is translated by virtue of different start codons in the same mRNA. More than 70% of the polyomavirus capsid protein is composed of VP1, the major structural protein. For the polyomaviruses simian virus 40 (SV40), mouse polyomavirus (mPyV), and JCV, the minor capsid proteins (VP2 and VP3) have been reported to be required for proper import of virion proteins to specific nuclear localization sites for assembly. VP2 and VP3 share an identical C-terminal sequence that contains the DNA-binding domain and the VP1-interacting domain. Disulfide bonds and calcium ions are required for maintenance of structural stability of virion capsids. Virion assembly occurs exclusively in the nucleus and has been linked to the presence of a higher calcium ion concentration in the nucleus than in the cytoplasm.

The nonstructural agnoprotein functions in the life cycle of polyomaviruses in several ways. In JCV replication, it has been implicated in regulating transcription and maturation. It has also been shown to interact functionally with YB-1, a cellular transcription factor, and to regulate JCV gene transcription negatively. More recently, it was reported to participate in JCV cell transformation by interfering with the ability of a DNA-dependent protein kinase repair complex, composed of a 470 kDa catalytic subunit and a K70/K80 heterodimer regulatory subunit, to repair breaks in double-stranded cellular DNA, thus inhibiting its role in cellular DNA repair.

Replication Cycle

Nuclear Entry

Major advances have been made in recent years in understanding mechanisms that polyomaviruses use to attach and penetrate cells in order to gain entry into the cytoplasm. All events, both early and late, required for polyomavirus replication occur in the nucleus. Although many steps in this process are now understood, the question of how, where, and in what state the virus minichromosome traverses the nuclear membrane is not fully understood. Intact polyomavirions have been seen inside the nucleus by electron microscopy prior to expression of T proteins, suggesting that the virions that initiate infection enter the nucleus in an undegraded state and are uncoated therein. However, partially uncoated virions have also been seen in the cytoplasm. Whether these partially uncoated intermediates are the entities that initiate virus DNA replication and virion assembly in the nucleus is uncertain. The site of entry and mechanism of virion entry into the nucleus is also not clearly defined. Virions in cytoplasmic vesicles and their fusion with the nuclear membrane have been observed, but whether this is the mechanism of virus genome entry into the nucleus is not clear. Two sites that have been implicated in virion entry into the nucleus are the endoplasmic reticulum and the nuclear pore complex.

Evidence that BKV enters the nucleus via passage through the endoplasmic reticulum follows. To enter the cytoplasm, BKV employs a caveolae-mediated endocytic pathway, which is generally slower than clathrin-mediated endocytosis. It is generally thought that cellular microtubules are involved in the shuttling of newly formed endocytic vesicles to various intracellular locations. Treatment of Vero cells with nocodazole does not prevent the endocytic uptake of BKV into the cytoplasm, but it does cause disassembly of the cellular microtubule network and inhibits movement of endocytic vesicles. BKV replication is inhibited if added to nocodazole-treated cells during the first 8 h of infection, which is the time period generally required for BKV to traverse the cytoplasm to the nucleus where replication begins. Microtubule disassembly prevents directional movement of endocytic vesicles, stopping them from passing through the endoplasmic reticulum to reach the Golgi. Treatment of LNCaP cells with brefeldin A, which blocks transport from the endoplasmic reticulum to the Golgi apparatus, prevents BKV infection of these cells. These results were interpreted to suggest that the route BKV uses for nuclear entry is via the endoplasmic reticulum.

Receptors and Modes of Cytoplasmic Entry

Polyomavirus entry into the cytoplasm has been studied in the greatest detail for mPyV, SV40, JCV, and BKV. The mechanisms of attachment and penetration are described below.

mPyV

This virus was the first polyomavirus for which sialic acids were implicated as an attachment receptor. The receptor studies were initiated in response to finding plaque size variants that produced small or large plaques. Upon sequencing the genomes of these size variants, it was found that the size differences in plaques resulted from a single amino acid change at VP1 position 92, in which glutamic acid was changed to glycine. Studies yielded evidence that the differences between small and large plaques originated from the capability of virus to bind to sialic acids. The small plaque isolates bound well to the straight chain $\alpha(2-3)$-linked sialic acids and the branched moieties containing both $\alpha(2-3)$- and $\alpha(2-6)$-linked sialic acids. In contrast, the large plaque variant bound well to only the straight chain $\alpha(2-3)$-linked structure and poorly to the branched structure. The single amino acid change from glutamic acid to glycine was shown to occur in a site responsible for mPyV hemagglutination of erythrocytes and receptor-binding to permissive cells. Because of this single amino acid change, the small plaque variant gained a receptor attachment advantage over the large plaque variant that translated to production of higher titers of infectious virus and virus having a higher cell transformation efficiency. These results illustrated that even a single amino acid change in a receptor site could produce major changes in virus yield and virulence.

Subsequent to implicating sialic acids as an attachment receptor for mPyV, the protein component of the receptor was found to be $\alpha4\beta1$ integrin. Endocytosis of mPyV has been shown to occur by either of two mechanisms: (1) by caveolae-mediated endocytosis, or (2) by an alternative clathrin, caveolin-1, dynamin-1 pathway. The target cell influences the pathway chosen. Recent evidence also has implicated gangliosides as a receptor for mPyV infection. Treatment with ganglioside GD1 of mouse cells lacking a functional receptor for permissive mPyV infection imparted susceptibility to infection to these cells.

SV40

Of the four polyomaviruses described herein, all except SV40 utilize a sialic acid component as receptor. By comparing the crystal structure of the SV40 capsid with that of mPyV, it was shown that the inability of SV40 to bind to sialic acid resided in truncation of an eight amino acid residue segment of an external loop on the capsid essential for binding to sialic acid.

The receptor by which SV40 initiates infection was found to be major histocompatability complex (MHC) class I molecules. Although these molecules mediate binding to cells, they neither envelop virions nor enter cells during the infectious process. Evidence suggests that after SV40 binds to MHC class I molecules, it sends an intracellular signal that facilitates its entry into cells by caveolae-mediated endocytosis. As with mPyV, gangliosides have been implicated as a receptor for SV40 attachment and cellular entry. GM1 gangliosides are important in initiating SV40 infection.

JCV

Evidence that sialic acid is a receptor for JCV attachment came from initial studies showing that treatment of red blood cells (RBCs) and permissive glial cells with crude neurominodase abrogated hemagglutination of RBCs and infection of permissive glial cells. However, treatment of these cell types with an $\alpha(2-3)$-specific neuraminidase did not cause inhibition, providing evidence that sialic acid might be a receptor, but not the $\alpha(2-3)$-linked sialic acid moiety. Results defining the specific sialic acid receptor came from several additional experiments. Binding of high concentrations of JCV to permissive glial cells was shown to block binding of the $\alpha(2-6)$-linked straight chain specific lectin, *Sambucus nigra* lectin (SNA). Use of specific O-linked and N-linked glycosylation inhibitors further showed that the JCV-specific receptor was an N-linked glycoprotein with a terminal $\alpha(2-6)$-linked sialic acid.

Chlorpromazine and the related compound clozapine both block clathrin-dependent endocytosis, and also JCV infection. These drugs belong to a class known as serotonin-dopamine inhibitors. Glial cells susceptible to JCV infection coexpress receptors for both serotonin and dopamine. To identify whether serotonin or dopamine is the receptor component required for JCV infection, glial cells permissive to JCV infection were treated with the dopamine antagonists bromocriptine and miniprine, and a dopamine agonist, pergolide. These agents did not block JCV infection and had only minimal influence on serotonin receptors. In contrast, treatment of glial cells with metoclopramide, chlorpromazine, or clozapine, all of which are antagonists of the $5HT_{2A}$ serotoninergic receptor, significantly inhibited JCV infection. Furthermore, monoclonal antibodies specific to $5HT_{2A}$ and $5HT_{2C}$ receptors inhibited JCV infection of glial cells, but monoclonal antibodies to the dopamine receptor did not. The specificity of these antibodies to JCV infection was further illustrated by their failure to inhibit SV40 replication in treated glial cells. This is as expected, since SV40 and JCV utilize different receptors for cell attachment and enter cells by different endocytic pathways.

HeLa cells are mostly refactory to JCV infection, but because they contain the N-terminal $\alpha(2-6)$-linked sialic acid receptor component on their surface, they bind JCV as well as permissive cells. When transfected with JCV genomic DNA, HeLa cells acquire the capacity to support JCV early gene expression. However, stable or transient transfection of HeLa cells with a $5HT_{2A}$ receptor-containing construct reversed their susceptibility to JCV infection. This illustrates the necessity of having both receptor components, N-terminal $\alpha(2-6)$-linked sialic acid and a $5HT_{2A}$ protein serotonergic component present on cells for susceptibility to JCV infection. Other than establishing the requirement for both cellular receptors, the mechanism by which $5HT_{2A}$ transfection into HeLa cell alters their susceptibility to JCV infection has not been explained.

After entry into the cell interior, polyomaviruses must enter the nucleus for replication and assembly. For nuclear entry, components of the cellular cytoskeleton have been shown to be important in intracellular movement of virus through the cell cytoplasm, as described above. After virion entry into the cell interior, an intact microtubule network has been reported to be important in the early phase of JCV infection. Furthermore, SVG cells (an astroglial cell line highly susceptible to JCV infection) was shown to require an intact actin cytoskeleton to maintain its susceptibility to JCV infection. It has been suggested that the actin cytoskeleton system does not participate directly in viral movement through the cytoplasm, but may be important in assembling the clathrin machinery necessary for JCV endocytosis. Participation of intermediate filaments is also required for permissive JCV infection of cells.

The etiological link between JCV and progressive multifocal leukoencephalopathy (PML) and the high concentrations of gangliosides in the human brain make them attractive candidates as JCV receptors. The sialic acids of glycoproteins and glycol lipids have been shown to bind to JCV VLPs at their VP1-binding sites. Pretreatment of JCV virions with the ganglioside GT1b also inhibited their ability to infect susceptible glial cells. This evidence was used to infer that GT1b ganglioside bound to JCV VP1 receptor sites, thereby blocking the ability of the virus to initiate infection of the highly JCV susceptible cell lines IMR-32 and SVG and suggesting that both glycoproteins and glycolipids may function as JCV receptors. JCV has been reported to have specificity for N-terminal $\alpha(2-6)$ sialic acids of glycoproteins in order to initiate infection, and has also been reported to bind JCV VLPs at various $\alpha(2-3)$ and $\alpha(2-6)$ structures. These results suggest that both glycoproteins and glycolipids may function as JCV receptors. The question of whether glycolipids are involved in JCV infection of the brain has not been resolved.

BKV

The role of sialic acids as a receptor for BKV entry was reported recently. The cellular receptor that BKV uses to initiate infection is a glycoprotein with an N-linked $\alpha(2-3)$ sialic acid. This was demonstrated by treatment of Vero cells with sialidase S, an enzyme that specifically removes $\alpha(2-3)$-linked sialic acid from glycoproteins and complex carbohydrates. Reconstitution of the asialo Vero cells by $\alpha(2-3)$-specific sialyltransferase restored susceptibility of these cells to infection by BKV, but restoration by an $\alpha(2-6)$-specific sialyltransferase did not. Evidence that the sialic acid was N-linked was obtained by showing that treatment with tunicamycin, an inhibitor of N-linked glycosylation, reduced BKV infection, whereas treatment with the O-linked glycosylation inhibitor benzylGalNac did not.

The protein component of the BKV receptor has not been determined. However, gangliosides have been implicated in binding to BKV and altering cell susceptibility to BKV infection. BKV binding to erythrocyte membranes has been investigated by use of sucrose flotation assays. It was shown that binding was to a neuraminidase-sensitive, proteinase K-resistant molecule. It was suggested that the terminal $\alpha(2-8)$-linked disialic acid motif, present on both GT1b and GD1b gangliosides, was responsible for inhibiting BKV binding to erythrocyte membranes. Furthermore, it was shown that LNCaP cells, which are normally resistant to BKV infection, became susceptible to BKV infection after treatment with gangliosides GT1b and GD1b. Also, it has been reported that recoating cells stripped of sialic acid and galactose with a mixture of all gangliosides derived from Vero cells restored their capacity to be infected by BKV. These results suggest that gangliosides have a role as receptors in BKV attachment and infection. The mechanisms that gangliosides employ in reversing cell susceptibility to BKV infection have not been determined.

In addition to glycolipids, the phospholipid bilayer has also been implicated in BKV hemagglutination and infection. Pre-incubating BKV with phospholipids decreases its ability to infect Vero cells, presumably because BKV binds to the exogenously added lipid, thus blocking its receptor sites from interacting with the host cell. Treating cells with either phospholipase A2 or D cleaves fatty acids, and reduces their permisssiveness to BKV infection and their ability to hemagglutinate RBCs. These parameters can be restored to the native state by adding back various preparations of phospholipids, such as L-α phosphatidylcholine and phospholipids derived from Vero cells. Hemagglutination titers of RBCs treated with phospholipase C are elevated compared to those of untreated controls, although the reason for these results is not completely understood. They point out that gangliosides are important to BKV hemagglutination and infection. Adding various gangliosides to cells has been shown to alter their susceptibility to BKV infection. Currently, four subtypes of BKV have been identified, although no clear-cut link has been established between any of these subtypes and increased BKV virulence. The receptor for BKV attachment has been linked to a region of VP1 between amino acid residues 61 and 83. This sequence aligns with a hydrophilic region that aligns with the the BC loop of the SV40 capsid. It has not been determined whether all of the BKV subtypes share the same amino acid receptor sequence for cellular attachment. As previously explained for mPyV, a single amino acid change in its receptor sequence translated to altered plaque size, hemagglutination activity, virus output, and virulence.

Host Cell Susceptibility

Although the human polyomaviruses JCV and BKV share about 75% DNA sequence homology, they differ widely in many biological properties, including host cell range, cellular receptors, cell entry mechanisms, tissue tropisms, and disease manifestations. A defining feature of these polyomaviruses is the hypervariability of their non-coding regulatory regions. The makeup of these regions affects levels of transcription and influences host cell range and tissue tropism. For an in-depth discussion of JCV variants, based on the nucleotide sequences of their noncoding region, refer to the work by Jensen and Major listed in the 'Further reading' section. The notion that polyomavirus replication is controlled at the intracellular, molecular level came from the similarity in results obtained for host cell susceptibility and virus output when native virions or transfected genomic DNA were used to infect various cell types. Results supporting this conclusion are presented below.

A family of DNA-binding proteins called NF-1, composed of four subtypes called A, B, C, and D (or X) has been linked with site-coding specific transcription of viral genes and JCV replication. Elevated levels of NF-1 class X (NF-1X) mRNA were expressed by brain glial cells that are highly susceptible to JCV infection, but not by nonsusceptible HeLa cells. $CD34^+$ precursor cells of the KG-1 line, when treated with phorbol ester 12-myristate 13-acetate (PMA), differentiated to cells with macrophage-like characteristics and lost susceptibility to JCV infection. To determine whether loss of JCV susceptibility in these cells was linked to reduced levels of NF-1X expression, reverse transcription-polymerase chain reaction (RT-PCR) was used to evaluate levels of mRNA of each of the four NF-1 subtypes found in PMA-treated KG-1 cells. Different levels of specific subtypes of mRNA were observed in the PMA-treated macrophage-like cultures compared with controls. Northern blot hybridization confirmed that the levels of NF-1X expressed by PMA-treated KG-1 cells were lower than those produced by untreated KG-1 cells. Use of gel mobility shift assays later confirmed this finding by showing the induction of specific NF-1-DNA complexes in KG-1 cells undergoing PMA treatment. These results suggested that the binding pattern of NF-1 class members may change in hematopoietic precursor cells, such as KG-1, as they undergo differentiation to macrophage-like cells. Transfection of PMA-treated KG-1 cells with an NF-1X expression vector restored their susceptibility to JCV infection. Transfection of PMA-treated KG-1 cells with NF-1 subtypes A, B, and C vectors failed to restore JCV susceptibility. These data collectively show the importance of NF-1X expression for JCV replication.

The importance of NF-1X expression for JCV replication has also been shown in multipotential human nervous system progenitor cells. By use of selective growth conditions, these progenitor cells were cultured from fetal brain tissue as undifferentiated, attached cell layers. Selective culture techniques yielded highly purified population of neurons or astrocytes. Infection with JCV virions or with a plasmid encoding the JCV genome demonstrated the susceptibility of astrocytes and, to a lesser degree, their undifferentiated progenitors. However, neurons remained nonpermissive for JCV replication. Expression of the NF-1X transcription factor was much higher in astrocytes than in neurons. Transfection of an NF-1X expression vector into progenitor-derived neuronal cells, prior to JCV infection, yielded JCV protein production. These results indicate that susceptibility to JC infection is regulated at the molecular level. Furthermore, they suggest that differential recognition of viral promoter sequences can predict lineage pathways of multipotential progenitor cells in the human nervous system.

Perspectives

Phylogenetic evidence indicates that polyomaviruses have evolved from a common ancestor. Despite their very similar morphological, physical, and chemical characteristics, events in the life cycle of closely related species can vary considerably. This article contains information of recent vintage on the variety of mechanisms involved in polyomavirus attachment, penetration and movement within the cell to reach the cell nucleus, where replication occurs. Our understanding of the role of sialic acid receptors in polyomavirus replication has been greatly expanded, especially those associated with glycoproteins. Recent work has also implicated various gangliosides as receptor molecules, and much new information should be available shortly on how they contribute to the infectious process.

Information is also presented on how levels of specific cellular transcription factors, such as NF-1X, can influence the infectability of specific cell types by JCV. Much new information has also been published on how the noncoding regulatory region of JCV changes from the archetype arrangement found in most urine isolates to those found in other tissues, such as lymphocytes, tonsils, and in brains of individuals with PML. Modifications to this promoter/enhancer structure can alter cellular host range and may be responsible for switching infection from the latent to lytic states. The importance of immunosuppression in JCV- or BKV-caused disease, and the predominant role of cellular immunity in their prevention, have been firmly established.

The tumorigenicity of JCV and BKV for animals has been widely known for many years. Their nucleic acids have been detected in tumor and surrounding tissue by sensitive PCR techniques. Also, polyomavirus T proteins that impair the cell-cycle regulatory functions of p53 and pRb have been detected by sensitive immunochemical techniques in human colon, gastric, brain, and pancreatic tumors. JCV and BKV persistently infect a high percentage of the human population worldwide. The range of cell types infectable by JCV or BKV has also increased in recent years, establishing the possibility that tumor cells could become infected after tumorigenesis begins.

See also: Polyomaviruses of Humans; Polyomaviruses of Mice; Simian Virus 40.

Further Reading

Cole CN and Conzen SD (2001) *Polyomaviridae*: The viruses and their replication. In: Fields BN, Knipe DM, Howley PM, et al. (eds.) *Fields Virology*, 4th edn., vol. 2, pp. 2141–2174. Philadelphia, PA: Lippincott Williams and Wilkins.

Eash S, Manley K, Gasparovic M, Querbes W, and Atwood WJ (2006) The human polyomaviruses. *Cellular and Molecular Life Sciences* 63: 865–876.

Gee GV, Dugan AS, Tsomaia N, Mierke DF, and Atwood WJ (2006) The role of sialic acid in human polyomavirus infections. *Glycoconjugate Journal* 23: 19–26.

Imperiale MJ and Major EO (2007) Polyomaviruses. In: Knipe DM, Howley PM, et al. (eds.) *Fields Virology*, 5th edn., vol. 2, pp. 2263–2298. Philadelphia, PA: Lippincott Williams and Wilkins.

Jensen PN and Major EO (2001) A classification scheme for human polyomavirus JCV variants based on the nucleotide sequence of the noncoding regulatory region. *Journal of Neurovirology* 7: 280–287.

Pomovirus

L Torrance, Scottish Crop Research Institute, Invergowrie, UK

© 2008 Elsevier Ltd. All rights reserved.

Glossary

Spore balls Resting spores (cystosori) of *Spongospora subterranea* found in lesions or pustules (powdery scabs) on potato tubers; a single cystosorus comprises 500–1000 resting spores aggregated to form in a ball that is partially hollow, traversed by irregular channels.

Spraing Virus disease symptoms of internal brown lines and arcs in potato tuber flesh.

Introduction

Pomoviruses have tubular rod-shaped particles and tripartite genomes; they are transmitted by soil-borne zoosporic organisms belonging to two genera (*Polymyxa* and *Spongospora*) in the family Plasmodiophoraceae. Pomoviruses have limited host ranges, infecting species in a few families of dicotyledenous plants. Agriculturally important hosts include potato and sugar beet. There are four member species: *Potato mop-top virus*, *Beet soil-borne virus*, *Beet virus Q*, and *Broad bean necrosis virus*. Beet virus Q (BVQ) and beet soil-borne virus (BSBV) can occur in mixed infections with the benyvirus beet necrotic yellow vein virus (BNYVV).

Taxonomy and Classification

The classification of tubular rod-shaped viruses transmitted by soil-borne plasmodiophorid vectors was revised in 1998 to establish four genera: *Furovirus*, *Pomovirus*, *Benyvirus*, and *Pecluvirus*. The revision was prompted by new virus sequence information that revealed major differences in genome properties (number of RNA species, sequence, and genome organization). The genera are not assigned currently to any family. Pomovirus is a siglum from *Potato mop-top virus*, the type species.

Physical Properties of Particles

Pomovirus particles are hollow, helical rods, 18–20 nm in diameter, comprising multiple copies of a single major coat protein (CP; *c.* 19–20 kDa). The CP gene is terminated by a UAG (or UAA in broad bean necrosis virus (BBNV)) stop codon that is thought to be suppressible, readthrough (RT) of which would produce a fusion protein of variable mass (54–104 kDa). One or a few copies of the RT fusion protein are present at the extremity of potato mop-top virus (PMTV) particles thought to contain the 5′ end of the virus RNA. Pomovirus particles are fragile and particle size distribution measurements are variable; PMTV particles have predominant lengths of 125, 137, and 283 nm. PMTV particles sediment as three components with sedimentation coefficients ($S_{20,w}$) of 126, 171, and 236 S.

Genome Properties

Complete genome sequences are available for three member species and an almost complete sequence (without the 5′ and 3′ untranslated regions (UTRs)) is available for BBNV. Pomovirus genomes comprise three species of positive-sense single-stranded RNA of *c.* 5.8–6, 2.8–3.4, and 2.3–3.1 kbp (**Figure 1**). RNA-1 encodes the replicase proteins. It contains a large open reading frame (ORF) that is interrupted by a UGA stop codon (ORF1) (or UAA in BVQ and BSBV); the sequence continues in-phase to encode an RT protein (204–207 kDa). The ORF1 protein (145–149 kDa) contains methyltransferase and helicase motifs while the RT domain contains the GDD RNA-dependent RNA polymerase (RdRp) motif. Phylogenetic analysis reveals that the RdRps of the pomoviruses and soil-borne wheat mosaic furovirus share between 50% and 60% sequence identity.

The 5′ UTR of pomovirus RNAs contains the starting sequence $GU(A)_{1-4}(U)_n$ (except BVQ RNA-1 which begins with AUA). The RNAs are probably capped at the 5′ end since the RNA-1 ORF contains methyltransferase motifs associated with capping activity. The terminal 80 nucleotides of the 3′ UTR can be folded into a tRNA-like structure that contains an anticodon for valine. Both BSBV and PMTV RNAs were shown to be valylated experimentally.

The virus movement proteins are encoded on RNA-2 of PMTV (RNA-3 in BSBV, BVQ, and BBNV). Three

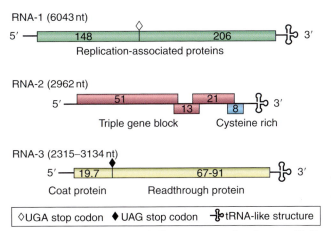

Figure 1 Diagram of the PMTV genome organization; boxes indicate open reading frames with the molecular masses of predicted protein products (kDa) indicated within.

5′ overlapping ORFs encode a conserved module of movement proteins known as the triple gene block (TGB). TGB movement proteins are found in the genomes of other rod-shaped viruses (hordei-, beny-, pecluviruses) and in monopartite filamentous viruses in the genera *Potexvirus* and *Carlavirus*. The TGB proteins (TGB1, TGB2, and TGB3) are named according to their position on the RNA and have molecular masses of 48–53, 13, and 20–22 kDa, respectively. The TGB1 contains a deoxyribonucleotide triphosphate (dNTP) binding site and helicase motifs in the C-terminal half typical of all TGB1 sequences and an extended N-terminal domain found in hordei-like TGB1s that do not share obvious sequence identities. TGB1 binds RNA and is thought to interact with genomic RNAs to facilitate movement. The sequence of the second TGB protein is the most conserved with BSBV, BVQ, and PMTV sharing 63–75% sequence identity and 49% identity with that of BBNV; there is little sequence identity among the TGB3 sequences. Analysis shows that TGB2 and TGB3 proteins contain two hydrophobic regions (predicted transmembrane domains) separated by a hydrophilic domain and these proteins are associated with intracellular membranes in infected plants.

In PMTV and BBNV, a fourth small ORF is predicted that encodes an 8 kDa cysteine-rich or 6 kDa glycine-rich protein, respectively, of unknown function whereas no such ORF is present in BSBV or BVQ. The 8 kDa cysteine-rich protein of PMTV is not needed for virus movement or infection of *Nicotiana benthamiana* and is not thought to function as silencing suppressor. Although a subgenomic RNA (sgRNA) that could encode the 8 kDa protein was detected in infected *N. benthamiana*, the protein is not readily detectable in extracts of infected leaves.

The CP and RT proteins are encoded on RNA-3 of PMTV (RNA-2 in BSBV, BVQ, and BBNV). The PMTV RNA-3 is of variable size, from 2315 to 3134 nt. Analysis of naturally occurring and glasshouse-propagated isolates revealed that deletions of c. 500–1000 nucleotides occur in both field and laboratory isolates and that they occur predominantly in the RT domain, particularly in the region toward the C-terminus. Deletions in this region are correlated with loss of transmission by the natural vector *Spongospora subterranea*. The first PMTV sequence to be published contained a shorter form of the CP–RT-encoding RNA and was designated as RNA-3, whereas in BSBV, BVQ, and BBNV the corresponding RNA is RNA-2. The RT domain of BVQ is shorter than the others and the RT coding sequence is followed by two additional ORFs for proteins of predicted mass of 9 and 18 kDa. Amino acid sequence comparisons reveal conserved motifs between these two proteins and domains in the C-terminal portions of BSBV and PMTV RT proteins which may indicate that they have arisen by degeneration of a larger ORF.

The occurrence of encapsidated deleted forms of RNA-2 and RNA-3 in natural and laboratory isolates of PMTV has been described and sequence analysis suggests that these PMTV RNAs contain sites that are susceptible to recombination possibly through a template switching mechanism. Variable base composition is found in natural isolates of BSBV in the sequence at the 3′ end of RNA-3 between the stop codon of the third TGB and the terminal tRNA-like structure.

Virus–Host Interactions and Movement

Cytoplasmic inclusions of enlarged endoplasmic reticulum (ER) and the accumulation of distorted membranes and small virion bundles can be seen by electron microscope examination of thin sections of BSBV- and BVQ-infected leaves. In PMTV-infected potato leaves, abnormal chloroplasts with cytoplasmic invaginations were seen in thin sections as well as tubular structures in the cytoplasm associated with the ER and tonoplast and in the vacuole.

PMTV does not require CP for movement and it is thought that TGB1 interacts with viral RNA forming a movement competent ribonucleoprotein (RNP) complex. Studies of transiently expressed PMTV TGB proteins fused to marker proteins such as green fluorescent protein or monomeric red fluorescent protein in epidermal cells of *N. benthamiana* have helped to elucidate events in intracellular trafficking and indicate that PMTV interacts with the cellular membrane recycling system. PMTV TGB2 and TGB3 were shown to co-localize on the ER (**Figure 2**) and in small motile granules that utilize the actin–ER network to reach the cell periphery and plasmodesmata (PD) and TGB3 contains a putative tyrosine sorting signal (Y-Q-D-L-N), mutation of which inhibits PD localization. TGB2 and TGB3 act together to transport GFP-TGB1 to the PD for movement into neighboring cells and TGB2 and TGB3 have the capacity to gate the PD pore. TGB2 co-localizes in endocytic vesicles with the Rab 5 ortholog Ara 7 (AtRabF2b) that marks the early endosomal compartment. Also, protein-interaction analysis revealed that recombinant TGB2 interacted with a tobacco protein belonging to the highly conserved RME-8 (receptor mediated endocytosis-8) family of J-domain chaperones, essential for endocytic trafficking.

Host Range, Geographical Distribution, and Transmission by Vector

Pomoviruses have a limited host range and are transmitted in soil by zoosporic plasmodiophorid vectors that have been classified as protists (**Figure 3**). Viruses that are transmitted by plasmodiophorid vectors include pomo-, peclu-, furo-, beny-, and bymoviruses and they are thought to be carried

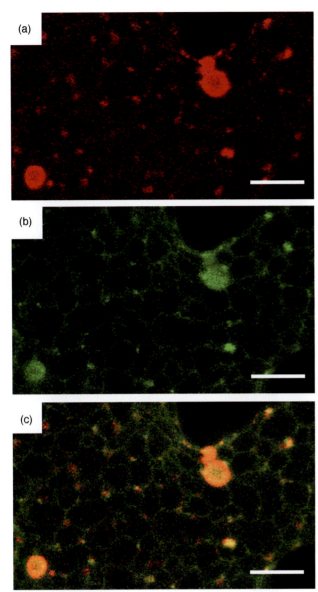

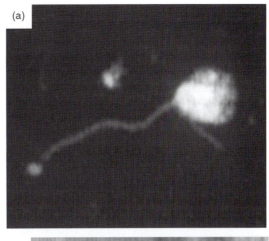

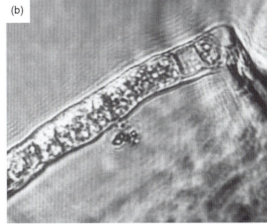

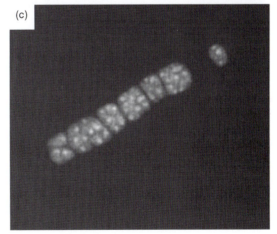

Figure 2 Confocal laser scanning microscope images of PMTV TGB2 and TGB3 fluorescent fusion proteins in epidermal cells of *Nicotiana benthamiana*. Monomeric red fluorescent protein (mRFP) tagged PMTV-TGB2 was transiently expressed with green fluorescent protein (GFP) tagged PMTV-TGB3, the proteins co-localized on membranes of the ER and in motile granules seen moving on the ER network. (a) Expression of mRFP-TGB2 (red channel); (b) expression of GFP-TGB3 (green channel); (c) merged image. Scale = 10 nm.

Figure 3 (a) Biflagellate zoopore of *Spongospora subterranea*. (b) Bright field and (c) fluorescence microscope images of zoosporangia in tomato root hair.

within the zoospores. The vector life cycle includes production of environmentally resistant thick-walled resting spores and viruliferous resting spores can survive in soil for many years.

BBNV has been reported only from Japan; it causes necrosis and stunting in broad beans and peas. It is mechanically transmitted by inoculation of sap to a few species including *Vicia faba*, *Pisum sativum*, and *Chenopodium quinoa*.

BVQ and BSBV are often found associated with BNYVV; BVQ is reported only from Europe whereas BSBV is found in sugar-beet-growing areas worldwide. No symptoms have been attributed to BSBV or BVQ alone in sugar beet. The viruses are thought to be transmitted in soil by *Polymyxa betae*. BVQ is mechanically transmissible only to *C. quinoa* and BSBV to members of the Chenopodiaceae.

PMTV is found in potato-growing regions of Europe, North and South America, and Asia; virus incidence is favored by cool, wet growing conditions. It is transmitted in soil by *S. subterranea*, also a potato pathogen which causes the tuber blemish disease powdery scab. PMTV can be transmitted mechanically to members of the Solanaceae and Chenopodiaceae.

Serological Relationships, Diagnosis, and Control

The viruses are serologically distinct. Distant relationships have been reported between BSBV and BVQ; PMTV and soil-borne wheat mosaic furovirus; and PMTV, BBNV, and tobamoviruses. PMTV, BSBV, and BVQ CPs contain a conserved sequence (SALNVAHQL) that reacts in Western blots with a monoclonal antibody (SCR70) produced against PMTV, but that is not exposed on intact particles. PMTV particles contain an immunodominant epitope at the N-terminus of the CP that is exposed at the surface along the sides of the particles and can be detected by monoclonal antibody SCR69 (**Figure 4**).

The viruses can be detected by serological tests (immunosorbent electron microscopy, enzyme-linked immunosorbent assay) and by assays based on the reverse transcriptase-polymerase chain reaction (RT-PCR) in leaves, roots, or tubers from naturally infected plants. PMTV is known to be erratically distributed in potato leaves and tubers and can move systemically in the absence of CP in potato leaves which raises a risk of false negative diagnosis. In addition, test plants grown in soil that has been previously air-dried can be used as indicators either by observing visual symptoms such as the PMTV-induced necrotic 'thistle-leaf'-shaped line patterns on leaves of *Nicotiana debneyi* or by conducting serological or RT-PCR analysis on the test plant roots or leaves.

PMTV causes an economically important disease affecting the quality of tubers grown for the fresh and processing markets. Tubers of sensitive potato cultivars that are infected from soil during the growing season develop spraing symptoms that include unsightly brown lines, arcs, or marks in the flesh sometimes accompanied by slightly raised external lines and rings (**Figure 5**). Potato plants grown from infected tubers display yellow markings or chevrons on the leaves and may

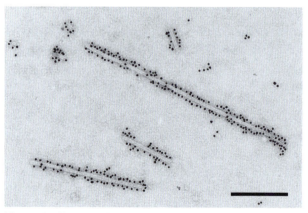

Figure 4 Electron micrograph of PMTV particles labeled with monoclonal antibody SCR69/gold conjugate.

Figure 5 Symptoms of PMTV in potato tubers cv. Nicola.

have shortened internodes (mop-top) producing cracked or malformed tubers; both tuber quality and yield can be affected. However, the virus does not infect all plants grown from infected tubers. Haulm and tuber symptoms vary markedly with cultivar, and some cultivars are symptomlessly infected. Environmental conditions also affect disease incidence and severity, and PMTV incidence was shown to increase with annual rainfall. Soil temperatures of 12–17 °C and high soil moisture at tuber initiation favor powdery scab incidence.

PMTV can be established at new sites by planting infected tubers and once established, PMTV is a persistent problem, as the resting spore balls of the vector *S. subterranea* are long-lived and resistant to drought and agrochemicals. Viruliferous spore balls are spread readily to new sites by farm vehicles, contaminated seed tubers, wind-blown surface soil; motile zoospores can be spread through contaminated irrigation or drainage water. In certain areas, potatoes have become infected by PMTV 18 years after potatoes were last grown (the longest period recorded). There is no effective practicable means to control *S. subterranea* but decreased severity of powdery scab can be achieved by application of fluazinam to soil. In addition, disinfection of tubers with chemicals such as formaldehyde decreases virus incidence, although the efficacy of this treatment depends on the level of soil infestation where the tubers are planted and a combination of tuber and soil treatments may be more effective.

The best prospect for virus disease control is development of resistant cultivars but there are no known sources of PMTV resistance in commercial potato cultivars and most of the commercially grown cultivars are also susceptible to *S. subterranea*. However, plants transformed with virus transgenes (CP and a mutated form of TGB2) have exhibited resistance to PMTV with decreased virus accumulation and incidence in tubers of plants grown in infested soil.

See also: *Benyvirus*; *Furovirus*; *Hordeivirus*; *Pecluvirus*; *Tobamovirus*.

Further Reading

Arif M, Torrance L, and Reavy B (1995) Acquisition and transmission of potato mop-top furovirus by a culture of *Spongospora subterranea* f. sp. *subterranea* derived from a single cystosorus. *Annals of Applied Biology* 126: 493–503.

Haupt S, Cowan GH, Ziegler A, Roberts AG, Oparka KJ, and Torrance L (2005) Two plant-viral movement proteins traffic in the endocytic recycling pathway. *Plant Cell* 17: 164–181.

Koenig R and Loss S (1997) Beet soil-borne virus RNA1: Genetic analysis enabled by a starting sequence generated with primers to highly conserved helicase-encoding domains. *Journal of General Virology* 78: 3161–3165.

Koenig R, Pleij CWA, Beier C, and Commandeur U (1998) Genome properties of beet virus Q, a new furo-like virus from sugarbeet determined from unpurified virus. *Journal of General Virology* 79: 2027–2036.

Lu X, Yamamoto S, Tanaka M, Hibi T, and Namba S (1998) The genome organization of the broad bean necrosis virus (BBNV). *Archives of Virology* 143: 1335–1348.

Pereira LG, Torrance L, Roberts IM, and Harrison BD (1994) Antigenic structure of the coat protein of potato mop-top furovirus. *Virology* 203: 277–285.

Reavy B, Arif M, Cowan GH, and Torrance L (1998) Association of sequences in the coat protein/read-through domain of potato mop-top virus with transmission by *Spongospora subterranea*. *Journal of General Virology* 79: 2343–2347.

Rochon D'A, Kakani K, Robbins M, and Reade R (2004) Molecular aspects of plant virus transmission by olpidium and plasmodiophorid vectors. *Annual Review of Phytopathology* 42: 211–241.

Sandgren M, Savenkov EI, and Valkonen JPT (2001) The readthrough region of potato mop-top virus (PMTV) coat protein encoding RNA, the second largest RNA of PMTV genome, undergoes structural changes in naturally infected and experimentally inoculated plants. *Archives of Virology* 146: 467–477.

Savenkov EI, Sangren M, and Valkonen JPT (1999) Complete sequence of RNA1 and the presence of tRNA-like structures in all RNAs of potato mop-top virus, genus *Pomovirus*. *Journal of General Virology* 80: 2779–2784.

Zamyatnin AA, Solovyev AG, Savenkov EI, et al. (2004) Transient coexpression of individual genes encoded by the triple gene block of potato mop-top virus reveals requirements for TGBp1 trafficking. *Molecular Plant Microbe Interactions* 17: 921–930.

Potato Virus Y

C Kerlan, INRA, UMR1099 BiO3P, Le Rheu, France
B Moury, INRA – Station de Pathologie Végétale, Montfavet, France

© 2008 Elsevier Ltd. All rights reserved.

Brief Description and Significance

Potato virus Y (PVY) was first recognized in 1931 as an aphid-transmitted member within a group of viruses associated with potato degeneration, a disorder known since the eighteenth century. PVY is the type species of the genus *Potyvirus*, one of the six genera in the family *Potyviridae*. PVY is naturally spread by vegetatively propagated material and by aphids in numerous species in a nonpersistent manner. Transmission by contact has also been reported. PVY has a wide host range and is highly variable with some host specificity. Genome sequences and reliable tools for detection and strain differentiation are available. Bioassays and serology have been largely developed.

PVY is one of the most damaging plant pathogens causing significant losses in four main crops around the world: potato, pepper, tomato, and tobacco. In surveys of viruses with worldwide economic importance, PVY was listed in the top-five viruses affecting field-grown vegetables. PVY was also found responsible for damages in petunias in Europe and in eggplant crops in India.

Efficient control strategies depending on the crop have been developed. However none of them seems capable to take into account PVY evolution and to suppress risks of new epidemics.

Viral Particle and Genome

PVY virions are nonenveloped, filamentous, flexuous rods, 730–740 nm long, 11–12 nm in diameter, with an axial canal 2–3 nm in diameter and helical symmetry. Assembly and disruption of PVY particles were studied in detail. Virions contain about 6% nucleic acid and a viral genome linked protein (VPg), but no lipid or other components.

The genome consists of one single-stranded, linear, positive-sense RNA molecule (3.1–3.2×10^6 Da) with a polyadenosine sequence at the 3' terminus. VPg is attached to the 5' end via a phosphate ester linkage to Tyr^{60}. A three-dimensional model structure of VPg was proposed. Complete genome sequences are available in databases for isolates from potato, tobacco, pepper, tomato, and *Solanum nigrum*. Hundreds of partial sequences of PVY isolates (more than 200 coat protein (CP) sequences) are also available.

PVY RNA is approximately 9700 nt in length excluding the poly(A) tail. As in all members of the genus *Potyvirus*, its single open reading frame is expressed as a large polyprotein (3061–3063 amino acids) autoproteolytically cleaved to yield ten functional proteins: P1 (284 aa), HC-Pro (456 aa), P3 (365 aa), 6K1 (52 aa), CI (634 aa), 6K2 (52 aa), VPg (188 aa), NIaPro (244 aa), NIb (521 aa), and CP (267 aa). There are two distal noncoding regions, 5'NTr (184 nt) and 3'NTr (from 326 to 333 nt) (**Figure 1**).

RNA synthesis is believed to occur in the cytoplasm. The replication complex comprises the proteins NIb, CI and VPg and possibly involves the proteins 6K1 and 6K2. The NIb protein is believed to be the RNA-dependent RNA polymerase.

The CP (29.95 kDa) consists of 267 amino acid residues. It comprises a core (218 aa), highly conserved in members of the genus *Potyvirus*, and two surface-exposed N-terminal (300 aa) and C-terminal (19 aa) regions which are not required for virus assembly and maintenance of infectivity.

The helper component protein (HC-Pro) is supposed to have a biologically active dimeric form (with a subunit molecular mass of 58 kDa). It is indeed capable in the yeast two-hybrid system of self-interaction that can be drastically reduced by mutations in its cysteine-rich region. HC-Pro is involved in PVY accumulation in tobacco and is a suppressor of post-transcriptional gene silencing (PTGS).

Transmission of PVY

PVY is naturally spread by vegetatively propagated plant organs (tubers, cuttings) and by aphids. PVY has also been reported to be spread by plant-to-plant contact,

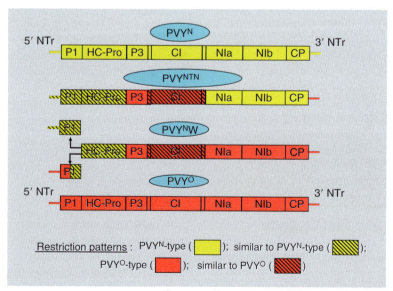

Figure 1 Schematic map of the PVY genome: strains PVY^O and PVY^N, variants PVY^{NTN} and PVY^{NW}. Reproduced from Glais L, Tribodet M, and Kerlan C (2002) Genetic variability in Potato Polyvirus Y (PVY): Evidence that PVY^{NW} and PVY^{NTN} variants are single to multiple recombinants between PVY^O and PVY^N isolates. *Archives of Virology* 147: 363–378, with permission from Springer-verlag.

for instance in tobacco and tomato crops in Southern America. It can also be transmitted by contact between sprouts of potato tubers during storage. Transmission by seed has been reported from *S. nigrum* and *Nicandra physaloides*. Transmission by pollen has never been proved for any host plant.

Unlike other potyviruses, PVY displays an unusually large range of aphid vectors. Aphids in 70 species, all in the family Aphidinae, were demonstrated to be able to transmit PVY, most of them with very low efficiencies compared to that of *Myzus persicae*. Apterae and alatae are vectors (**Figure 2**).

PVY is aphid-transmitted in a nonpersistent manner. Acquisition and inoculation periods are brief (a few seconds or minutes). Aphid stylets penetrate into the epidermal cell layer of the plants and puncture plant cell membranes. There is no discernible latent period. PVY does not pass through insect moults and retention of the virus in the aphid usually lasts not more than 1 or 2 h. However longer retention periods (up to 17 h in *Aphis nasturtii*) were reported. Prior starvation of the aphids increases the efficiency of transmission though it does not affect the number of electrically recorded membrane punctures during acquisition periods.

PVY transmissibility is determined by both HC-Pro and CP proteins. All aphid-transmissible PVY isolates contain the 'DAG triplet' (Asp–Ala–Gly) in the CP N-terminal domain. PVY isolates having the sequence DAGE are also aphid-transmissible, unlike those of tobacco vein mottling virus. PVY was the first virus for which it was proved that a virus-induced component of the plant sap is needed for aphid transmission. Effectively, HC-Pro of an efficiently transmitted PVY isolate was shown to mediate the transmission of nonaphid-transmitted PVY isolates and of other potyviruses. Its activity can be blocked by antisera specific to PVY HC-Pro but not by antisera to HC-Pros of other viruses. Monoclonal antibodies (MAbs) to PVY HC-Pro were also produced. Loss of HC-Pro activity correlates with nonretention of virions on aphid stylets. Aphid-transmissible and -nontransmissible isolates differ by one or two amino acid substitutions: Gly^{35} to Asp, Lys^{50} to Glu or Lys^{50} to Asn, Ile^{225} to Val, Ser^{355} to Gly. Lys^{50} is part of a conserved 'KITC motif' (Lys–Ile–Thr–Cys). Changes in or around this motif can result in losses of aphid transmissibility. Conversely the reverse mutation Glu^{50} to Lys restores the helper function of a defective HC-Pro.

Host Plants

The natural host range comprises plants in nine families. It includes potato (*S. tuberosum* ssp. *tuberosum*), several species of native potatoes in the Andes namely *S. andigena*, numerous *Solanum* wild species, pepper (*Capsicum* spp.), tobacco (*Nicotiana* spp.), tomato (*Lycopersicon* spp.), eggplant (*S. melongena*), ornementals (*Petunia* spp., *Dahlia* spp.), perennial plants (*Physalis virginiana*, *P. heterophyla*), and a number of annual or perennial self-propagating plants. Many region-specific hosts were recorded such as *Cotula australis* in New Zealand or *Sorbaria tomentosa* in Himalaya.

Weeds in the family Solanaceae are often potential inoculum sources for PVY infections in tomato and pepper crops: *S. nigrum* and *S. dulcamara* in many countries; *S. chacoense* and *P. viscosa* in Southern America; *S. gracile*, *S. aculeatissimum*, *P. angulata*, *P. ciliosa*, and *P. floridana* in Florida. *P. virginiana* and *P. heterophyla* (perennial ground cherries) are overwintering hosts in Northern America. *S. dulcamara* in Western Europe or *Datura* spp. in Mediterranean countries are potential inoculum sources in potato crops as well as volunteer potatoes. Infected pepper and tomato plants, and seed and volunteer potatoes were also identified as potential sources of inoculum in tobacco crops in the USA, Canada, and Italy.

Host specificity clearly stated in the past is currently under reinvestigation. Most PVY isolates appear able to infect common tobacco and tomato cultivars. Conversely most isolates naturally infecting potato are unable to infect pepper cultivars systemically while pepper isolates are usually not detected in potato fields.

PVY is easily sap-transmitted and can be transmitted by stem- and tuber-grafting. Its experimental host range comprises plants in 495 species in 72 genera of 31 families. It includes 287 species in the family Solanaceae (among which 141 *Solanum* species and 70 *Nicotiana* species), 28 species of Amaranthaceae, 25 species of Fabaceae, 20 species of Chenopodiaceae, and 11 species of Asteraceae. A large part of these plants comprises only local lesion hosts. *Datura stramonium*, formerly reported as a host plant, was demonstrated to be totally immune to all strains tested in 1980–90. Some PVY isolates can infect the model plant *Arabidopsis thaliana*.

Figure 2 Aphid vector of PVY: winged form of *Aphis nasturtii*. Photograph: B. Chaubet, INRA, France.

N. tabacum, *N. benthamiana*, *N. occidentalis* can be used as diagnostic species susceptible to all PVY strains (**Figure 3**). *N. tabacum* plantlets are often used as source and test plants for aphid transmission experiments. *N. tabacum* is a suitable host for virus purification. Yields of purification including a final step of cesium chloride gradient centrifugation usually vary from 10 to 25 mg kg^{-1} of tobacco leaves (**Figure 4**). Leaves of *N. tabacum* are the best virus-infected material to store. Antigenic properties can be retained for 1 year in freeze-dried crude extracts. Infectivity was reported to be preserved in freeze-dried material stored over calcium chloride at 4 °C for 15 years. However from our own experience numerous long-term stored isolates of PVY available in international collections are no longer infectious.

S. demissum Y, *S. demissum* PI 230579, and the hybrid *S. demissum* A6 are local lesion hosts. The 'A6 test' on detached leaves was commonly used in the past as a diagnostic tool for differentiating PVY from potato virus A (PVA), another potyvirus.

Cultivars of *N. tabacum*, potato and pepper, *Chenopodium amaranticolor* (**Figure 5**), *P. floridana*, some accessions of *S. brachycarpum* and *S. sparsipilum* are useful for distinguishing among PVY strains and pathotypes. Potato cultivars such as Bintje or Saco can be used to separate PVY from plants co-infected with PVA or potato virus X (PVX).

Serology

PVY is considered to be strongly immunogenic. Antisera with precipitin titers up to 1/4096 and MAbs have been produced in rabbits or mice immunized with purified virus preparations or synthetic peptides, or by using DNA-based immunization, or phage display antibody technology. Numerous sources of antibodies and serological detection kits are available.

ELISA (standard DAS-ELISA and related protocols), dot-blot immunobinding assay, immunosorbent electron microscopy, latex and virobacterial agglutination tests, immunodiffusion in agar gels were intensively studied for use in strain differentiation and virus detection from plant material and from aphids.

Polyclonal antisera do not discriminate among PVY strains. MAbs allow to separate two main serogoups broadly corresponding to potato strains PVY^O and PVY^C from one part and PVY^N from the other part (see below).

Figure 3 Typical mottle induced by a PVY isolate from potato on a leaf of *Nicotiana tabacum* cv *Xanthi* 15 days after mechanical inoculation. Photograph: C. Lacroix, INRA, France.

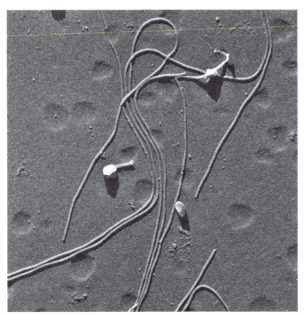

Figure 4 Purified suspension of PVY with a bacteriophage T$_4$ as internal calibration standard. Magnification 22 000×. Photograph: D. Thomas, CNRS-Rennes 1 University, France.

Figure 5 Host plants of PVY: local lesions induced by a PVY isolate from pepper on *Chenopodium amaranticolor* after mechanical inoculation. Photograph. L. Glais, INRA, France.

Specific MAbs to strain PVY^C or variants PVY^{NTN} and PVY^{N}W have also been produced, but their reliability still needs confirmation.

Relationships to Other Potyviruses

PVY is distantly serologically related to PVA and potato virus V (PVV), and to 17 other viruses in the genus *Potyvirus* including pepper mottle virus (PepMoV).

PVY, PepMoV, PVV, pepper yellow mosaic virus, pepper severe mosaic virus, wild potato mosaic virus and Peru tomato virus constitute a phylogenetic group distinguishable from other potyviruses including PVA.

Cytopathology

Virions are usually closely associated with inclusions in the cytoplasm of infected cells. They have also been observed within plasmodesmata or aligned on Golgi apparatus, endoplasmic reticulum, and around mitochondria. Most PVY strains induce type-IV inclusions (i.e., pinwheels, scrolls, and short curved laminated aggregates). They consist of a single, nonglycosylated protein (67 kDa), serologically unrelated to the CP protein, or to host proteins. PVY also induces cytoplasmic rod-like amorphous inclusions, but except two isolates from Brazil does not induce large crystalline cytoplasmic and nucleolar inclusions.

Cytopathology in PVY infections has recently been intensively studied in potato and tobacco leaves, notably changes in fresh matter content, photosynthesis, and other metabolic activities. P1 protein was detected in association with cytoplasm inclusions in tobacco cells. Amorphous inclusions appear to be the primary site of HC-Pro accumulation. NIa protein was proved to accumulate in both cytoplasm and nucleus. PVY CP, HC-Pro, and RNA were found within chloroplasts of tobacco leaves suggesting they may alter the chloroplastic function as also proved in transgenic tobacco plants expressing PVY CP.

PVY in Potato

PVY has become in the last decade the most important virus in most growing areas for seed, ware, and processed potatoes. Serious outbreaks were, reported in the 2000s. Tuber quality can be severely affected due to necrosis or defects for processing potatoes. Reduction in size and number of harvested tubers can result in losses up to 80%.

Symptoms consist of mild to severe mottle, often associated with crinkling of the leaves. Yellowing and necrosis (vein necrosis and necrotic spots) frequently occur in the lower leaves. Symptoms also include collapse and dropping of intermediate leaves, which remain clinging to the stem (leaf drop). Secondarily infected plants (when mother tubers are infected) are stunted with crinkled and smaller leaflets (**Figure 6**). Necrosis on and around leaf veins, on petioles, stems, and tubers may occur in numerous cultivars. Some necroses on stems and petioles are called stipple-streak disease. The potato tuber necrotic ringspot disease (PTNRD) is characterized by particular patterns on leaves (sometimes oak-leaf necrosis) (**Figures 7(a)** and **7(b)**) and stems, and superficial annular and arched necroses on tubers, first slightly protruding from tuber skin, then becoming dark brown with sometimes crackings of tuber skin (**Figures 7(c)** and **7(d)**).

Potato isolates have historically been divided into three main strain: PVY^O, PVY^N, and PVY^C. PVY^O and PVY^C are separated on the basis of hypersensitive reactions in potato cultivars bearing the resistance genes Ny_{tbr} and N_c (**Figure 8**). PVY^N differs from PVY^O and PVY^C in causing a veinal necrosis reaction in *N. tabacum* cv *Samsun* or cv *Xanthi* (**Figures 9(a)** and **9(b)**), but does not elicit any hypersensitive response in most potato cultivars. Two amino acids at positions 400 and 419 in the HC-Pro protein are involved in the induction of this necrotic reaction in tobacco. *P. floridana* is a third host for strain differentiation. PVY^C induces collapse and premature death of this plant. PVY^C was first characterized as non-aphid-transmissible, but many isolates of this strain were proved to be readily aphid-transmissible. Two former PVY^C isolates were later recognized as isolates of PVV.

Many unclassified variant isolates were reported in the past, notably isolates classified in a group called PVY^{An}. More recently PVY^Z and PVY^E were characterized as overcoming the resistance genes Ny_{tbr} and N_c. Two main variants have emerged for two decades: PVY^{NTN}, characterized by its ability to induce tuber necrosis, and PVY^{N}W which differs by its pathogenicity and its serotype O–C instead of serotype N. PVY$^{N:O}$ isolates

Figure 6 PVY on potato. Natural secondary infection by PVY in a potato field: crinkling, yellowing, and growth reduction of the leaflets. Photograph: V. Le Hingrat, FNPPPT, France.

Figure 7 PVY on potato. Symptoms associated with the potato tuber necrotic ringspot disease (PTNRD). (a) Yellowing and necroses on a basal leaf. (b) Necrotic oak-leaf pattern on an intermediate leaf. (c) Typical PTNRD symptoms on tubers of the cv *Monalisa*. (d) Atypical PTNRD induced by a PVY$^{N:O}$ isolate on a tuber of the cv *Yukon Gold*. Photographs: K. Charlet-Ramage, GNIS-INRA, France (a–c) and R. P. Singh Potato Research Centre, NB, Canada (d).

Figure 8 Hypersensitive reactions on potato cultivars 10 days after mechanical inoculation on cv *Desiree* containing the gene Ny_{tbr} inoculated by a PVY^O isolate (left) and on cv *Eersteling* containing the gene *Nc* inoculated by a PVY^C isolate (right). Photographs: C. Kerlan and J. P. Cohan, INRA, France.

described in the 2000s also share properties with both PVY^N and PVY^O and some of them induce tuber necrosis. PVY^O and PVY^N strains are distributed worldwide though PVY^N is yet a quarantine pathogen in Canada and the USA. The PVY^C strain is less frequent. The PVY^{NTN} variant has been identified in most potato-growing countries including the USA and Peru. The PVY^N-W variant has become prevalent in Poland and has been reported in several countries. PVY$^{N:O}$ was reported in Canada and the USA.

PVY potato isolates have been divided into three genetically distinct strains designated PVY^O, PVY^N, and PVY^{C1}. PVY^{C1} is a part of the genetically defined strain PVY^C or PVY^{NP} (nonpotato) which also includes isolates from pepper, tomato, and tobacco. PVY^{N}W, PVY$^{N:O}$, and the majority of PVY^{NTN} isolates are recombinants between

Figure 9 PVY indexing on *Nicotiana tabacum* cv *Xanthi* (15 days after mechanical inoculation). (a) Typical vein necrosis and leaf distortion induced by a PVY^N isolate. (b) Typical vein clearing without any leaf distortion induced by a PVYO isolate. Photographs: C. Lacroix, INRA, France.

PVY^N and PVY^O strains, with one to three recombination breakpoints (**Figure 1**). Several tuber necrosis-inducing isolates possess a PVY^N-type genome without any recombination breakpoint. Lastly, North American and European PVY^{NTN} isolates are separated on the basis of polymorphism in the 5′ NTr-P1 region.

Differences in aphid transmission between strains were reported. PVY^N isolates seem better transmitted than other PVY isolates with longer retention periods in aphids. Whatever the strain, *M. persicae* is clearly the most important vector in most potato-growing areas. However despite their low transmission efficiency, other species colonizing potatoes such as *A. nasturtii* (**Figure 2**), or visiting potatoes can also contribute to PVY spread. Some of these species, notably cereal aphids or the pea aphid, were thought to be involved in PVY epidemics in potato crops (in Europe and the USA) due to their high abundance.

Control methods are first based on control of seed potatoes, breeding for resistance and quarantine regulations, especially regarding PVY^N. Sophisticated schemes of seed production include monitoring of aphid vectors, treatments by mineral oils against aphid transmission, eradication of weeds, and post-harvest detection tests using large-scale ELISA. Numerous molecular assays were also developed for PVY detection in leaves, potato tubers, and aphids, including cDNA hybridization, various RT-PCR protocols, and microarray technology. Breeding for potato resistance to PVY takes into account resistance to infection (which includes resistance to virus and to vectors, and is largely used), hypersensitivity resistance (HR) and extreme resistance (ER). HR and ER are based on single dominant genes *Ny* and *Ry*, respectively. *Ry* genes (*Ry$_{sto}$*, *Ry$_{adg}$*), from *Solanum* wild species, map on chromosomes XI and XII, and confer broad-spectrum resistance, whereas HR genes *Nc*, *Ny$_{tbr}$* (which map on chromosome IV) (**Figure 10**) and *Nz* found in old potato cultivars, protect against PVY^C, PVY^O, and PVY^Z, respectively. The NIaPro protein of PVY is involved in the elicitation of *Ry*. The gene *Y-1* involved in the HR reaction has been cloned. More than 20 cultivars containing *Ry* have been produced and this resistance has proved to be quite durable. Marker-assisted selection for resistance to PVY has been developed for genes *Ry$_{adg}$*, *Ry$_{chc}$*, and *Ny$_{tbr}$*. Genetically engineered resistance has been intensively studied and used to generate resistant transgenic plants from commercial potato cultivars. CP-mediated resistance was described as ER with often a broad spectrum.

PVY in Pepper

PVY is the causal agent of major diseases and production losses in pepper crops. In some situations it can affect 100% of the plants and can be the most important disease. PVY affects pepper production worldwide, while other pepper-infecting potyviruses are mostly restricted to particular continents and are not present in Europe. Symptoms are mostly visible on the vegetative parts of the plants but rarely on fruits. Depending on the pepper genotype and on the virus isolate, they consist of mosaic and vein banding on leaves or on necrotic symptoms on leaves, petioles, and stems (**Figure 11**).

Pepper isolates of PVY belong to the same group as the C phylogenetic group of PVY in potato, although they could constitute a separate C subgroup. One recently described exception consists of a O:C recombinant strain inducing veinal necrosis in pepper in Italy. No PVY isolates from the N or O phylogenetic groups have been described as epidemic in pepper. Pepper PVY isolates have been classified according to their pathotype relative to the numerous resistance genes and alleles used to control PVY in pepper crops (see further).

The most widespread and efficient way to control PVY in pepper is through the growing of cultivars carrying resistance genes. The first resistance genes used to control PVY were the *pvr2^1* and *pvr2^2* alleles, which control

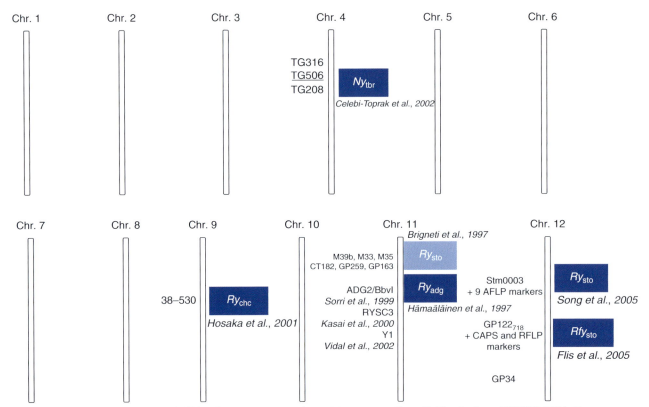

Figure 10 Genetic localization of PVY resistance genes on the potato genome. From S. Marchadour, FNPPPT-INRA, France (unpublished).

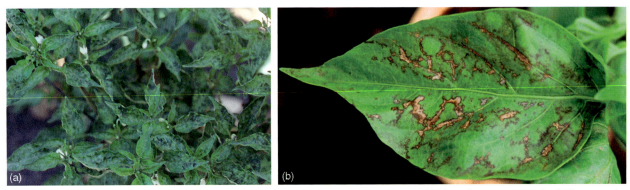

Figure 11 PVY infection in a pepper field: (a) severe crinkling and distortion of the leaves. (b) Necrotic patterns on a leaf. Photographs: P. Gognalons, INRA, France (a) and M. Pepelnjak Slovenia (b).

high-level resistance since no virus can be detected in inoculated organs of the plants. These genes have been exploited for decades in a large number of pepper cultivars and proved to be durable. However, PVY isolates virulent to the $pvr2^1$ allele were occasionally observed, especially in the Mediterranean basin and in tropical regions, while isolates virulent to the $pvr2^2$ allele are exceptional.

The three alleles $pvr2^+$ (susceptibility reference allele), $pvr2^1$ and $pvr2^2$ were used to classify the pepper isolates in three pathotypes: pathotype (0) isolates which can infect only plants with the $pvr2^+$ allele, pathotype (0,1) isolates which only infect plants carrying the $pvr2^+$ or $pvr2^1$ alleles and pathotype (0,1,2) isolates which are virulent toward all three $pvr2$ alleles.

It was shown that these alleles correspond to various copies of an isoform of the eukaryotic translation initiation factor 4E (eIF4E), and that amino acid substitutions in the central part of the viral protein genome-linked (VPg) of PVY determined virulence to both $pvr2^1$ and $pvr2^2$.

More recently, the dominant $Pvr4$ gene, originating from a *C. annuum* Mexican population, was introgressed into bell-pepper cultivars. This gene present in about 40% of European cultivars confers a high-level resistance to all tested isolates of PVY. This resistance shares common properties with hypersensitivity-based resistances. No isolate of PVY virulent to *Pvr4* has been described so far, indicating a high durability of the resistance.

Other pepper resistances to PVY have been described (notably polygenic resistances from *C. annuum* and *Pvr7* resistance from *C. chinense*).

Other control measures such as the use of oil sprays or physical barriers (polyethylene sheets and coarse nets) were also used for controlling spread of PVY in pepper crops in Israel and Florida.

PVY in Tomato

PVY can induce severe diseases in tomato crops but has long been considered a pathogen of secondary importance, inducing mild mosaic on the foliage only. Since the 1980s, new strains of PVY have arisen in Mediterranean countries, causing serious yield and quality loss of tomato fruits. These strains induce necrotic lesions on leaves and necrotic streaks on stems in all tomato varieties and frequently affected 100% of the plants, in greenhouses as well as in open fields. Importantly, combination of PVY infections with other viruses such as cucumber mosaic virus can induce very severe diseases in tomato crops.

Data about the genetic diversity of PVY isolates from tomato are scarce. Isolates belonging to the C/pepper group of strains and to the recombinant NTN group have been observed in tomato crops. In laboratory tests, most of the isolates belonging to all phylogenetic groups of PVY induce systemic infections in tomato.

There are few reports about PVY resistance in tomato and related species. Accession PI247087 of *L. hirsutum* was described highly resistant to PVY. The resistance is efficient toward all tested tomato or pepper isolates of PVY, inhibits the multiplication of the virus in the inoculated organs and is controlled by a single recessive gene (named *pot-1*) which maps to a region on chromosome 3 syntenic to the *pvr2* locus in pepper. *Pot-1* was shown to belong to the same family of genes as *pvr2*. Strains of PVY virulent to the *pot-1* resistance can be selected during laboratory tests, a property that is controlled by a single amino acid substitution in the VPg of the virus. The *pot-1* gene was introduced into genotypes of *L. esculentum*, but varieties carrying this gene are not widespread.

PVY in Tobacco

PVY was reported to be the most damaging virus in tobacco crops. It causes height reductions and modifies the chemical composition of cured leaves, especially the nicotine content. Yield losses of 14–59% were reported with incidence of up to 100%. Symptoms on leaves are usually mild mottling but particular chlorotic patterns and necroses, notably the veinal necrosis disease, may also occur (**Figures 12(a)** and **12(b)**).

Three strain groups, M^SM^R, M^SN^R, N^SN^R, have been identified according to their reaction in tobacco cultivars resistant or susceptible to the root-knot nematode (*Meloidogyne incognita*). The NIb protein was found to elicit the hypersensitive response in these resistant tobacco cultivars. VAM-B refers to a resistance-breaking group of isolates overcoming the gene *va* originally found in the genotype Virginia A mutant and characterized by deletions at the Va locus. The VPg protein was suggested to be involved in overcoming resistance conferred by *va*.

Breeding for resistance is the main control method against PVY in tobacco production. Many European and American commercial cultivars display noteworthy levels of recessive resistance. The gene *va* has been extensively

Figure 12 PVY infection on tobacco cv *Burley* in field trials in Southwestern France. (a) Chlorotic oak-leaf and halotic patterns. (b) Severe systemic necrosis. Photograph: D. Blancard, INRA, France.

utilized, though its introgression is often associated with decrease of yield and cured leaf quality. In countries where early infection frequently occurs, additional measures consist of protecting seedbeds by fleece, eradication of weeds, and isolation from potato, tomato and pepper fields. Pathogen-derived resistance has been extensively studied in tobacco, although more for investigating the mechanisms of protection than in order to use this strategy in commercial tobacco cultivars.

Prospects

Strains

Current distinctions within the species *Potato virus Y* is not totally clear. The exact relationship between the necrotic strains from potato and from tobacco (PVY^N, N^SN^R, M^SN^R) still must be defined precisely, as the distinction between PVY^{NTN} and other isolates of the PVY^N strain. Is there any PVY^N isolate unable to induce potato tuber necrosis whatever the conditions? Are $PVY^{N:O}$ isolates different to those previously referred as PVY^NW isolates? Molecular classification in most cases correlates only partially with biological classifications since they are based on neutral markers.

Evolution

The large variability and genomic diversity of PVY makes its evolutionary story fascinating. Links between mutation or recombination events and evolution of PVY are thoroughly studied. Current knowledge indicates that PVY was present in pre-Columbian America and might have followed different evolutionary pathways as a result of co-evolution with various solanaceous hosts.

See also: Vector Transmission of Plant VirusesVector Transmission of Plant Viruses; Plant Resistance to Viruses: Engineered Resistance; Plant Virus Diseases: Economic Aspects; Polyomaviruses of Mice; Potato Viruses; Tobacco Viruses.

Further Reading

Blancard D (1998) *Maladies du Tabac.* Paris: INRA.
Blancard D (1998) *Maladies de la Tomate.* Paris: INRA.
de Bokx JA and van der Want JPH (eds.) (1987) *Viruses of Potato and Seed-Potato Production,* Wageningen, The Netherlands: Pudoc.
Edwardson JR and Christie RG (1997) Potyviruses. In: *Florida Agricultural Experiment Station Monograph Series 18-II – Viruses Infecting Pepper and Other Solanaceous Crops,* pp. 424–524. Gainesville, FL: University of Florida.
Flis B, Hennig J, Strelczyk-Zyta D, Gebhardt C, and Marczewski W (2005) The $Ry\text{-}f_{sto}$ gene from *Solanum stoloniferum* for extreme resistance to potato virus Y maps to chromosome XII and is diagnosed by PCR marker $GP122_{718}$ in PVY resistant potato cultivars. *Molecular Breeding* 15: 95–100.
Glais L, Kerlan C, and Robaglia C (2002) Variability and evolution of *Potato virus Y* (PVY), the type-member of the *Potyvirus* genus. In: Khan JA and Dijkstra J (eds.) *Plant Viruses as Molecular Pathogens,* pp. 225–253. Binghamton, NY: The Haworth Press.
Kasai K, Morikawa Y, Sorri VA, Valkonen JPT, Gebhardt C, and Watanabe KN (2000) Development of SCAR markers to the PVY resistance gene Ry_{adg} based on a common feature of plant disease resistance genes. *Genome* 43: 1–8.
Loebenstein G, Berger P, Brunt AA, and Lawson RG (eds.) (2001) *Virus and Virus-Like Diseases of Potatoes and Production of Seed-Potatoes.* Dordrecht, The Netherlands: Kluwer Academic Publishers.
Milne RG (1988) The economic impact of filamentous viruses. In: Milne RG (ed.) *The Plant Viruses, Series 4: The Filamentous Plant Viruses,* pp. 331–407. New York: Plenum.
Moury B, Morel C, Johansen E, and Jacquemond M (2002) Evidence for diversifying selection in potato virus Y and in the coat protein of other potyviruses. *Journal of General Virology* 83: 2563–2573.
Moury B, Morel C, Johansen E, *et al.* (2004) Mutations in *Potato virus Y* genome-linked protein determine virulence toward recessive resistances in *Capsicum annuum* and *Lycopersicon hirsutum. Molecular Plant–Microbe Interactions* 17: 322–329.
Ruffel S, Dussault M-H, Palloix A, *et al.* (2002) A natural recessive resistance gene against potato virus Y in pepper corresponds to the eukaryotic initiation factor 4 E (eIF4E). *Plant Journal* 32: 1067–1075.
Ruffel S, Gallois J-L, Lesage M-L, and Caranta C (2005) The recessive potyvirus resistance gene *pot-1* is the tomato orthologue of the pepper *pvr2*-eIF4E gene. *Molecular Genetics and Genomics* 274(4): 346–353.
Shukla DD, Ward CW, and Brunt AA (1994) *The Potyviridae.* Wallingford, UK: CAB International.
Tribodet M, Glais L, Kerlan C, and Jacquot E (2005) Characterization of *Potato virus Y* (PVY) molecular determinants involved in the vein necrosis symptom induced by PVY^N isolates in infected *Nicotiana tabacum* cv. Xanthi. *Journal of General Virology* 86: 2101–2105.

Potato Viruses

C Kerlan, Institut National de la Recherche Agronomique (INRA), Le Rheu, France

© 2008 Elsevier Ltd. All rights reserved.

Glossary

Primary infection (primarily infected plants) Potato plants become infected during the growing season.
Secondary infection Infection results from an infected mother tuber.

Introduction

At least 38 potato viruses have been described, several of them being classified by ICTV as tentative species or possible strains of one species. They can be divided into three groups according to their importance and their distribution in potato-growing areas in the world. They include well-known viruses, and recently discovered, often poorly characterized viruses. Eight viruses are major pathogens causing severe damages in potato crops (**Table 1**). Many potato viruses occur only in Latin America. None of these viruses is economically important; some of them have been found only once in one cultivar (**Table 2**). The third group comprises viruses that occur in other parts of the world and are either of only local importance or without any significance (**Table 3**).

Potato Viruses: One Century

Viral diseases are thought to have become a threat to potato soon after its introduction in Europe. The phenomenon called potato degeneration was described in the eighteenth century. The leafroll disease was mentioned in the 1750s and the causal agent described in 1916 was one of the first viruses identified. The main species involved in the mosaic complex were identified in 1920s and 1930s, that is, potato virus M, potato virus X, potato virus Y, and potato virus A. Other important potato viruses were only discovered in the 1950s (tobacco rattle virus, potato virus S) or mid-1960s (potato mop-top virus). In the last four decades, an ever-increasing number of viruses were found infecting potato in Latin America, especially in Andean valleys where the potato originates from. Lastly, numerous viruses or strains more common in other host plants were found to infect potato occasionally. Some of them, for instance tomato spotted wilt virus, cause emerging diseases and could become important in the future. We may also point out that a contact-transmitted virus such as pepino mosaic virus, never detected in potato fields so far, was shown to be capable of infecting many important potato cultivars and hence could become a threat in the future.

Since the turn of the twentieth century, potato producers in many countries have recognized that certain potato-growing areas were better than others for the production of seed potatoes. Seed-potato production programs were developed, based during 1940–50 on clonal selection, then on basic seed production involving virus eradication by thermotherapy and meristem culture, and later on specific

Table 1 Major potato viruses

Species Acronym First mention	Genus	Economical importance[a]	Geographical distribution[a]	Spread[a]	Natural host range Other main hosts	Variability[a]
Potato virus Y PVY 1931	Potyvirus	Very high	Worldwide	Aphids NP[c], contact?	Wide Tobacco, tomato, pepper	Strains Y^O, Y^C, Y^N Variants Y^{NTN}, Y^NW, $Y^{N:O}$ Serotypes PVY^{O-C}, PVY^N
Potato leafroll virus PLRV 1916	Polerovirus	Very high	Worldwide	Aphids P[c]	Narrow Solanum spp., tomato, ulluco	Tomato yellow top and Solanum yellows strain
Potato virus X PVX 1931	Potexvirus	High	Worldwide	Contact	Narrow Tomato, pepper	Strain groups 1, 2, 3, 4 Strains HB, CP Serotypes PVX^O, PVX^A
Potato virus A PVA 1932	Potyvirus	Moderate	Worldwide[b]	Aphids NP[c]	Only potato	Four strain groups Three serotypes
Potato virus S PVS 1952	Carlavirus	Moderate	Worldwide	Aphids NP[c], contact	Narrow Pepino	Strains PVS^O, PVS^A
Potato virus M PVM 1923	Carlavirus	Moderate	Worldwide[b]	Aphids NP[c], contact	Narrow Mainly in Solanaceae	Strain PVM-ID
Tobacco rattle virus TRV 1946	Tobravirus	High	Worldwide	Nematodes	Weeds, flower bulbs, beet, tobacco, lettuce, spinach, etc.	Many strains
Potato mop-top virus PMTV 1966	Pomovirus	Rather high	Mainly in cooler climates	Fungus	Only potato	PMTV-S, PMTV-T

[a]In potato.
[b]Not found in Andean countries.
[c]Viruses are transmitted by aphids in a nonpersistent (NP), persistent (P) manner.
Species in bold italic letters: the virus (at least one strain) can induce tuber necrosis.

Table 2 Potato viruses only found in Latin America

Species Acronym First report	Genus	Economical importance[a]	Geographical distribution[a]	Spread[a]	Natural host range Other main hosts	Variability[a]
Andean potato latent virus APLV 1966	Tymovirus	Very low	Bolivia, Colombia, Ecuador, Peru	Contact, flea beetle, TPS[c]	Only potato and ulluco	Serotypes Hu, CCC, Col-Caj
Andean potato mottle virus APMoV 1977	Comovirus	Unknown	Chili, Ecuador, Peru, Brazil	Contact, flea beetle?	Only potato	Strains B, C, H
Arracacha virus B-oca strain AVB-O 1981	Nepovirus?	Unknown	Bolivia, Peru	TPS	Only potato and oca	Oca and potato strains
Potato deforming mosaic virus PDMV 1985	Begomovirus?	High in one cultivar	Southern Brazil	Whitefly	potato, S. chacoense, S. sisymbrifolium	Not reported
Potato rough dwarf virus PRDV 1995	Carlavirus	Very low	Argentina, Uruguay	Unknown	Only potato	Strain of PVP
Potato virus P PVP 1993	Carlavirus	Locally in two cultivars	Brazil	Aphids	Only potato	PRDV strain
Potato virus T PVT 1977	Trichovirus	Unknown	Bolivia, Peru	Contact, TPS	Mashua, oca, potato, ulluco	Not reported
Potato virus U PVU 1983	Nepovirus	None	Locally in Peru (Comas Valley)	Nematode	Unknown	None

Virus	Genus	Incidence (in potato)[a]	Location	Vector[b]	Host range	Strains
Potato yellow mosaic virus PYMV 1986	Begomovirus	Unknown	Venezuela	Whitefly	Narrow Tomato	Not reported
Potato yellow vein virus PYVV 1954	Crinivirus	Locally high	Colombia, Ecuador, Peru	Whitefly	Potato and several weeds	Not reported
Potato yellowing virus PYV 1991?	Alfamovirus?	Unknown	Peru, Chili, Bolivia	Aphid SP[b], TPS	Only potato	Not reported
Potato 14R virus P14R 1996	Tobamovirus?	None	Peru	Unknown	Only potato	None
Solanum apical leaf curl virus SALCV 1983	Begomovirus?	Very low	Peru	Unknown	Narrow Physalis peruviana, Solanum nigrum	Not reported
Sowbane mosaic virus SoMV 1997	Sobemovirus	Isolated once from a Mexican line	Unknown (Mexico?)	Unknown	Narrow Chenopodium spp.	None
Tobacco ringspot virus TRSV 1977	Nepovirus	Low	Peru	Nematode? TPS	Potato, arracacha, oca	TRSV-Ca & PBRSV
Wild potato mosaic virus WPMV 1979	Potyvirus	None	Peru	Aphid NP[b]	Narrow Pepino, tomato, wild potato	Not reported

[a] In potato.
[b] Viruses are transmitted by aphids in a nonpersistent (NP), persistent (P), and propagative (Pr) manner.
[c] TPS means true potato seeds.

Table 3 Viruses of limited significance to potato production

Species Acronym First report	Genus	Economical importance[a]	Geographical distribution[a]	Spread[a]	Natural host range Other main hosts	Variability[a]
Alfalfa mosaic virus AMV 1940	Alfamovirus	Only locally	Uncommon	Aphids NP[b], pollen, TPS[c]	Wide Clover, tobacco, tomato, pepper	Calico and tuber necrosing strains
Beet curly top virus BCTV 1954	Curtovirus	Only locally	Arid and semiarid regions	Leafhoppers	Bean, beet, spinach, tomato, pepper, many cucurbits	Not reported
Cucumber mosaic virus CMV 1958	Cucumovirus	Very low	UK, Egypt, India, Saudi Arabia	Aphids NP	Many cucurbits, pepper, tomato, etc.	Serotypes I, II
Eggplant mottled dwarf virus EMDV 1987	Nucleo-rhabdovirus	Very low	Iran	Aphids? contact?	Narrow Eggplant, tomato	Not reported
Potato aucuba mosaic virus PAMV 1961	Potexvirus	Low	Uncommon	Aphids NPH[b], contact	Narrow Clover, tomato, pepper, tobacco	Mild and severe strains
Potato latent virus PotLV 1996	Carlavirus	Unknown	North America	Aphid, whitefly?	Only potato	Not reported
Potato virus V PVV 1986	Potyvirus	Low	Bolivia, Peru, North Europe	Aphids NP	Narrow Tomato	Isolates Gl, AB, UF

Potato yellow dwarf virus PYDV 1 937	Nucleo-rhabdovirus	None	Canada, North USA, Saudi Arabia, South Russia?, Ukraine?	Leafhoppers P[b], Pr	Narrow Ox-eye daisy, periwinkle, clover	Two strains
Tobacco mosaic virus TMV 1977	Tobamovirus	None or low	China, India, Saudi Arabia, Andean countries	Contact	Wide Many crops and woody plants	Not reported
Tobacco necrosis virus TNV 1935	Necrovirus	Low	Europe, North America	Fungus	Bean, tulip, tobacco, cucumber, pea, lettuce	Strain D
Tobacco streak virus TSV 1981	Ilarvirus	Low	Brazil, Peru	Thrips?, TPS? pollen?	Wide	Not reported
Tomato black ring virus TBRV 1950	Nepovirus	Rather low	Europe	Nematodes, seeds, TPS	Very wide Grape, fruit trees, vegetables, weeds…	Potato bouquet and pseudo-aucuba strains, two serogroups,
Tomato mosaic virus TMV 1977	Tobamovirus	None	Hungary	Contact	Rather wide Pepper, tomato	TMV potato strain
Tomato spotted wilt virus TSWV 1938	Tospovirus	Locally high	South America, Australia, India, Europe, South Africa	Thrips	Very wide Tomato, pepper, pea, groundnut, soyabean, tobacco, etc.	Not clearly described

[a]In potato.
[b]Viruses are transmitted by aphids in a nonpersistent (NP), persistent (P), and propagative (Pr) manner; NPH means that aphid transmission requires a helper virus.
[c]TPS means true potato seeds.
Species in bold italic letters: the virus (at least one strain) can induce tuber necrosis.

multiplication systems such as propagation *in vitro* or by stem cuttings or from true potato seeds (TPSs). Planting of certified seed potatoes with low levels of virus infections has now become the main element in the fight against viral diseases. Breeding for virus resistance has been intensively used at least since 1925 for PLRV, and sources of resistance in many wild *Solanum* spp. have been exploited. Three types of resistance were described in the 1950s and reliable knowledge on their genetic background and inheritance was available in the 1970s. Now many resistance genes have been mapped in the potato genome, which has permitted breeding based on molecular markers. Lastly a remarkable catalog of drastic measures has been established in order to avoid the introduction of quarantine viruses. Technical guidelines for safe movement and utilization of potato germplasm have recently been published under authority of relevant international specialized institutions.

Main Traits of the Viral Diseases in Potatoes

Symptomatology and Damages

Secondary infection symptoms are often more severe than those of primary infections, resulting in emergence delays and growth reductions of plants. Combinations of viruses also result in severe symptoms. Common symptoms are mosaic, mottle, or other (bright or pale green) discolorations, crinkling, rolling, or distortion of leaflets, leaf dropping, stunting, and deformation of the whole plant. Various necrotic patterns may occur on leaves, stems, and tubers (both external and internal).

Size, number, aspect, and content of tubers can be affected. Yield losses greatly vary depending on virus, strain, and time of infection. They can reach up to 90%. Some infections can result in defects in processed products such as blackening of tuber flesh. Lastly viruses inducing tuber necrosis (strains of at least 11 species) can make tubers totally unmarketable as ware or processed potatoes.

Spread of Potato Viruses

Potato viruses are efficiently transmitted by aphids, either by species colonizing potatoes or by itinerant species migrating from other crops and weeds. PLRV and PVY are model viruses for studying the mechanisms of persistent and nonpersistent transmission by aphids. PLRV was the first virus of the family *Luteoviridae* for which passage of the virus through the intestine and not the hindgut was demonstrated. Other insects are vectors of less important viruses, notably leafhoppers, thrips, and whiteflies (**Tables 2** and **3**). PVX and some PVS strains are transmitted by contact between foliage or sprouts, between tubers when cutting, and also through farm implements. Several viruses are soil-borne, either transmitted by nematodes (TRV) or by fungi (PMTV, tobacco necrosis virus). Many Andean viruses can be transmitted by TPSs or by pollen. Inoculum sources are in most cases potato itself (volunteer or secondarily infected plants).

Control Methods

Control of seed potatoes

Production of certified seed potatoes nowadays is based on a catalog of measures more or less drastic and sophisticated depending on the country. It is undoubtedly a specialized and costly scheme involving the following main steps: (1) multiplication of virus-free prebasic and basic seeds under conditions where infection risks are minor, for example, first *in vitro* and in vector-proof greenhouses or plastic tunnels, then in isolated fields in areas known for low virus disease pressure; (2) field inspections involving roguing of diseased potatoes and eradication of all sources of inoculum; (3) treatments against vectors, notably insects; (4) specific growing techniques taking into account risks associated to the vector pressure such as early dates of plantation, haulm killing and use of barrier crops; (5) postharvest controls involving large-scale laboratory tests, commonly using ELISA and sometimes molecular techniques. The final aim is the quality grading of the commercial categories of seed potatoes, standards being dependent on each country but usually infection levels should not exceed 10% for all viruses together.

Resistance

Three types of resistance to viruses in potato are commonly distinguished: field resistance or resistance to infection, hypersensitivity (HR), and extreme resistance (ER). Resistance to infection is a polygenically inherited resistance appraised by the number of progeny tubers that are infected. It comprises resistance to virus itself and resistance to vectors. It is stated to be effective against all strains of the virus involved. Scores of resistance levels depend strongly on environmental conditions, time of infection, inoculum, and vector pressure. HR and ER are governed by single dominant genes coded N and R, respectively (**Figure 1**). ER is an enhanced form of HR, resulting in no obvious symptoms and no systemic infection. Conversely, HR is associated with local necrotic lesions on inoculated leaves but also in most virus–cultivar combinations with systemic necrotic reactions (spots, lines, arches) extending to lower and upper leaves, stems, and tubers. The virus though unevenly distributed can be detected in all these organs. However, except in special cases it is not transmitted to the daughter plants. N genes

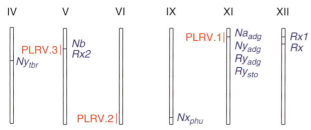

Figure 1 Genomic mapping of resistance genes (in blue) and QTL (in red) to four major viruses (PVY, PLRV, PVX, PVA) in potato. From S. Marchadour, FNPPPT-INRA, France (unpublished).

are strain-specific in contrast to *R* genes, each of which confers resistance to all strains of one or two viruses.

Breeding programs for resistance to infection to the main viruses have been well established in numerous countries producing potatoes. ER is also a frequently employed resistance, notably in countries where the use of certified seeds provides no satisfactory control. *R* genes providing protection against some major viruses (PVX, PVY, PVA, PVM) have been found in several wild *Solanum* species such as *S. stoloniferum*, *S. andigena*, *S. chacoense*, and *S. acaule*. They have been used in breeding programs since the mid-1940s and have appeared to provide sustainable resistance. *N* genes were present in the pool of old potato cultivars and hence have been easily incorporated into new cultivars. Potato is easily transformed by genetic engineering and pathogen-derived resistance against some major viruses (PVY, PLRV, PVX, PMTV) has been intensively studied, though without any significant applied use.

Quarantine measures

Quarantine measures taken against the risk of introducing foreign viruses (species or strains) in one state or continent have demonstrated their full efficiency. Indeed, so far no virus or strain listed as a quarantine pathogen is known to have spread outside its original region. Technical recommendations needed to test any exported or imported material (tubers, cuttings, true potato seeds, pollen, *in vitro* plantlets) have been established. These regulated viruses can be diagnosed by bioassays on selected indicator plants and, for the majority of them, by ELISA. Antibodies to potato virus U and potato yellow vein virus (PYVV) are not available. All potato viruses, except PYVV, potato-deforming mosaic virus and solanum apical leaf curl virus, are sap-transmissible. Molecular techniques or electron microscopy can also be used.

The Major Potato Viruses

Eight viruses are both widespread and damaging and, thus, of great concern in seed potato production (**Table 1**).

Potato Virus Y

PVY has become in the last decade the most important potato virus. It affects most cultivars and causes major yield losses of up to 80% and even more in the case of the potato tuber necrotic ringspot disease (PTNRD) (**Figure 2(a)**).

PVY isolates infecting potato have been divided into three strains (PVY^O, PVY^C, PVY^N), and many pathotypes and variants. Currently, PVY^O and PVY^N are the main widespread strains. PVY^C is less frequent (less than 5% of PVY isolates in France). Variants PVY^{NTN}, PVY^{N}W, and PVY$^{N:O}$ are also widespread. Most isolates of these variants are recombinants between PVY^O and PVY^N. Strains PVY^O, PVY^C, and PVY^N are differentiated on the basis of their reaction on *Nicotiana tabacum* and on potato cultivars carrying the hypersensitive genes Ny_{tbr} and Nc. PVY^{N}W isolates differ from other PVY^N isolates in their virulence and in belonging to serotype O-C.

Typical symptoms associated with PVY infections are mosaic, crinkle, necrotic patterns, and stunting of the plant in the case of PVY^O secondary infections (**Figure 2(b)**). Infection by PVY^N is often symptomless or associated with a mild mosaic (**Figure 2(c)**), only sometimes with a severe mottle. Yellowing and green halos can be observed on basal leaves (**Figure 2(d)**). Necroses (on or around leaf veins, on petioles, stems, and tubers) are related to hypersensitivity reactions in numerous cultivars (**Figure 2(e)**). Necrosis of basal leaves which remain clinging to the stem (leaf drop) are typical of PVY^O primary infections. PVY^C induces stipple-streaks on stems and brown spots around eyes of tubers. PVY^{NTN}, and some PVY^{N}W and PVY$^{N:O}$ isolates, can induce PTNRD.

Aphid transmission is undoubtedly the most important means of PVY spread in the fields. However, transmission by plant-to-plant contact has been consistently reported by several authors. Under experimental conditions, 70 aphid species were able to transmit PVY, *Myzus persicae* being the most efficient vector. This species and other aphids colonizing potato fields have long been seen as the main natural vectors. However, due to their abundance in potato fields during many PVY outbreaks, itinerant species multiplying in other crops (cereals, peas, etc.) have also been said to play a significant role in spread of PVY. PVY has a wide host range but besides potato no other host has so far been shown to act as a significant inoculum source, at least in the main potato-growing areas.

Control of PVY in seed fields requires use of mineral oil treatments to prevent aphid transmission. Forecasting and simulation models of PVY spread have been developed in many countries. At least 20 cultivars possess durable immunity to PVY though none of them is economically important. However, Ry_{sto} and Ry_{adg} genes are more and more introgressed into breeding programs.

Figure 2 (a) Secondary infection by PVY: mottle and crinkle of the leaflets; stunting of the plant. (b) Primary infection by PVY: mosaic on an apical leaf of cv. Bintje. (c) Primary infection by PVY: typical yellowing with green halos on a basal leaf of cv. Nicola. (d) Oak-leaf patterns and necrotic spots induced by PVY on leaflets of cv. Nicola. (e) PTNRD typical symptoms on tubers of cv. Nicola harvested in a field in Southern France. (a) Reproduced by permission of Y. Le Hingrat, FNPPPT, France. (b) Photograph: Christelle Lacroix, INRA, France. (c–e) Photographs: K. Charlet-Ramage, GNIS-INRA, France.

Potato Leafroll Virus

Although PLRV has been decreasing in incidence, it is yet a serious threat in most potato-growing areas. It can induce heavy yield losses, often more than 50%, sometimes up to 90%. Harvested tubers are small, whereas mother tubers often stay hard and unrotten. Spindling sprouts might occur in old cultivars. In some currently widespread cultivars, tuber quality is reduced due to the occurrence of necrotic spots distributed within the flesh (net necrosis) (**Figure 3(a)**). The virus is restricted to

Figure 3 (a) Net-necrosis induced by PLRV in a tuber of cv. Russet Burbank. (b) Secondary infection by PLRV: erect habit, plant stunting and pale yellow foliage. (a) Photograph: J. Martin, LNPV, France. (b) Reproduced by permission of Y. Le Hingrat, FNPPPT, France.

phloem. It induces formation of callose in sieve tubes blocking sugar transport and making the leaflets becoming hard and brittle. Cracking sounds resulting from squeezing such leaflets are a good indicator PLRV infections. It also induces phloem necrosis in petioles and stems. Photosynthesis and chlorophyll content are also reduced. Typical symptoms of secondary infection are an erect habit, plant stunting, and pale yellow foliage (**Figure 3(b)**). Leaflets roll upward, whereby lower leaves are dry, brittle, and leathery and may become necrotic. Primary infections are often latent, especially late infections. In most susceptible cultivars, apical leaves are erected and pale yellow (top yellowing) with sometimes purple or red margins.

PLRV is not transmitted by sap inoculation. At least ten aphid species were shown to be capable of transmitting it. PLRV does not multiply in the insect. Natural spread of PLRV in potato crops is mainly due to *M. persicae*, the most efficient vector though with large differences depending on aphid biotype and PLRV isolate. Other aphid species colonizing potatoes are poor vectors. However, *Aulacorthum solani* was shown to play a role in early infections in some PLRV outbreaks in Western Europe in the 1970s. *Capsella bursa-pastoris*, *Datura stramonium*, and *Physalis floridana* can act as virus reservoirs.

PLRV does not show any remarkable variability except that there have been reports on striking differences between isolates in aphid transmission rates. Differences between isolates can be established on the basis of symptom severity (stunting) in *P. floridana*. This host is also the common plant used to test the transmission efficiency of aphids.

In seed-potato producing countries, the use of efficient insecticide treatments has led to a considerable decrease of PLRV incidence. Many cultivars display high levels of field resistance. However, there is no commercial cultivar carrying *R* or *N* genes though ER and HR to PLRV were shown to be present in several pools of *Solanum* wild species. Two genetically engineered cultivars resistant to PLRV were registered in the USA but do not seem to be widespread.

Potato Virus X

PVX was in the past the most common virus in potatoes. It often induces a mild mosaic and mottle, sometimes severe mosaic, rugosity, crinkling of the leaflets, and also tuber necrosis in some cultivars. Yield reduction is usually low, often 10–15% but can be over 50% in the case of infections by severe strains or of mixed infections with PVY and PVA.

PVX has been separated into four strain groups (X^1, X^2, X^3, X^4) that differ in virulence to HR genes *Nx* and *Nb*. PVX^3 is the commonest strain at least in Europe. PVX^4 overcomes both *Nx* and *Nb* but is fortunately rare. PVX_{CP}, the common strain in Peru, belongs to group X^2. The strain HB discovered in Bolivia in 1978 overcomes these HR genes and also the *Rx* gene. Amino acid changes in its coat protein are responsible for its virulence. Both PVX_{HB} and PVX_{CP} belong to the serotype PVX^A which is distinct from the serotype PVX^O. Specific monoclonal antibodies and nucleic acid probes have been produced for diagnosis of isolates of each serotype. Reliable indicator cultivars for strain differentiation can be: Desiree or Pentland Crown (*nx*, *nb*) which are susceptible to all strains, King Edward (*Nx*, *nb*) resistant to X^1 and X^3, Bintje (*nx*, *Nb*) resistant to

X^1 and X^2, Maris Piper or Pentland Dell (Nx, Nb) resistant to X^1, X^2, and X^3. *Gomphrena globosa*, *C. amaranticolor*, and *C. quinoa* are other PVX indicator hosts.

PVX is efficiently controlled by use of certified seeds. Moreover, numerous important European and American cultivars (Ranger Russet, Saturna, Cara) harboring Rx_{adg} or Rx_{acl} are PVX immune. Numerous cultivars also contain Nx and Nb. PVX_{HB} is not a real threat due to its lack of fitness.

Potato Virus A

PVA is widespread except in some countries where it is rare (Poland) or absent (Andes). It is frequent in susceptible cultivars such as Desiree and Russet Burbank. In the last decade it has become more prevalent in some countries. Yield losses may be negligible or amount to 40%. Infection is often symptomless even in the most susceptible cultivars. Typical symptoms are a very mild mosaic, often transient or only discernible in cloudy weather conditions. Leaflets may be shiny, rough, and slightly rippled. Crinkling occurs in mixed infections with PVX. Combination with PVY also results in a severe disease.

PVA has a limited host range. Other hosts such as tobacco and cherry tomato do not play any significant role in PVA epidemiology. *M. persicae* is the main vector. *M. euphorbiae*, some *Aphis* species, and *Brachycaudus helichrysi* were reported as potential vectors.

PVA isolates have been divided into three serogroups using monoclonal antibodies and into four strain groups based on HR or systemic response in several potato cultivars (notably King Edward). Some isolates were proved to have lost aphid transmissibility. Mutations in proteins 6K2 and VPg were shown to be responsible for particular avirulence in *Nicandra physaloïdes*.

Virus concentration in infected plants is low and purification of virions is difficult. *N. tabacum* can be used as a propagation host. Virions are weakly immunogenic and specific antisera without any cross-reactivity with PVY are difficult to prepare. Commercial ELISA kits are available, though not always specific. RT-PCR protocols were also reported for the detection in potato leaves and dormant tubers. In the past *Solanum demissum* A6 and *S. demissum* A were frequently used for differentiating PVA from PVY.

A large part of old and current cultivars is either immune or hypersensitively resistant to PVA (notably the widespread cultivars Bintje, Eersteling, Spunta, Kennebec, Charlotte, Monalisa, Katahdin, King Edward).

Potato Virus S and Potato Virus M

PVS and PVM are distantly related to each other and to other potato viruses of the genus *Carlavirus* (PVP, PRDV; recently shown to be strains of the same species). They were widespread in the past infecting all plants of cultivars such as King Edward which were later virus-freed by meristem culture. They are currently quasi-eradicated in Western Europe but are yet widespread in Eastern Europe, notably in Poland, where PVS incidence may reach 100% in some cultivars. Yield losses do not exceed 15%, except in some Polish cultivars infected by PVM. Co-infections by both viruses are frequent. ELISA is routinely used for detection in leaves and tubers. Both viruses are controlled in seed-potato schemes.

PVS

Infections are usually symptomless though leaf mottling may be observed in particular genotypes (**Figure 4**). PVS^O and PVS^A are two strain groups which induce local and systemic reactions on *Chenopodium* spp., respectively. PVS^A can cause severe reactions in potato. Full-length genomes of at least three isolates were sequenced. A broad variability was shown throughout the whole genome. The 'Central European' variant (PVS-CP) which systemically infects *Chenopodium* spp. was genetically closely related to PVS^O but only distantly related to PVS^A.

Both PVS^O and PVS^A can be naturally transmitted by contact and by several aphid species, especially *M. persicae*, *Aphis frangulae*, *A. nasturtii*, *A. fabae*, and *Rhopalosiphum padi*. Natural host range is restricted to several species in the families Solanaceae and Chenopodiaceae. Tomato is immune to PVS^O, but symptomlessly infected by PVS^A. *Nicotiana debneyi* can be used for isolation from plants doubly infected with PVM. Specific anti-PVS^A monoclonal antibodies have been produced. Detection by radioactive and nonradioactive probes as well as by PCR was reported. Many cultivars

Figure 4 Infection by PVS: mottle observed on a plantlet from imported material tested in the Potato Quarantine Station in France. Photograph: J. Martin, LNPV, France.

harbor the Ns_{adg} gene and are efficiently protected. Conversely, resistance in cultivar Saco is polygenic and recessive.

PVM

PVM usually induces very slight mottles and mild abaxial rolling of leaflets (**Figure 5**). Main vectors are aphid species colonizing potato crops. Some isolates that are not transmitted by aphids have been described. PVM-ID is a serologically distinct strain. Potato is the main host naturally infected. Tomato (symptomless infection) and potato cultivar Saco are good hosts for propagation and separation from PVS. The main aphid vectors are *M. persicae*, *A. nasturtii*, and *A. frangulae*. Some strains may also be transmitted by contact. Monogenic dominant resistance was found in *S. gourlayi* (*Gm* gene), *S. megistacrolobum*, and *S. stoloniferum* (HR genes).

Tobacco Rattle Virus

TRV is a widespread virus infecting numerous weeds and crops. It causes a serious disease in potato, called spraing or stem mottle, mainly in areas with light and peaty soils. Spraing or corky ringspot denotes one type of necrotic arcing in the tuber flesh (**Figure 6(a)**) and on the tuber surface (**Figure 6(b)**). Stem mottle refers to the fact that only one single shoot may be stunted with leaves showing a mosaic and sometimes transient yellow chevrons. Spraing-affected tubers often give rise to virus-free progeny plants, sometimes to infected plants with symptomless daughter tubers. Lastly, some cultivars are symptomless carriers of TRV.

TRV particles are tubular straight rods varying in length depending on the isolate. Normal particle-producing isolates (called M-type) have two genomic RNAs and are readily transmitted by nematodes and by sap inoculation. NM-type isolates have only RNA-1, do not produce particles, are probably not transmitted by nematodes and are not easily sap-transmitted. Some of the best characterized strains were originally obtained from potato: PRN (potato ring necrosis) now used as the type strain although it seems to have lost its transmissibility by nematodes; Oregon strains, which include the variant Oregon Yellow; PSG, and PLB. Monoclonal antibodies to strain PLB have been produced. Complete genomic sequences of many

Figure 5 Infection by PVM: soft rolling of leaflets and pale green color on an infected plant (right) versus an healthy plant (left). Reproduced by permission of Y. Le Hingrat, FNPPPT, France.

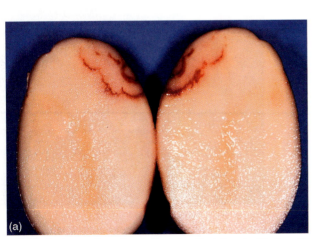
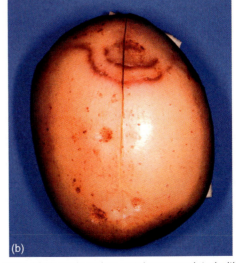

Figure 6 (a) Typical necrotic arches in the tuber flesh (spraing) induced by TRV. (b) Necrotic arches on tubers associated with spraing induced by TRV. (a, b) Photographs: J. Martin, LNPV, France.

potato strains are available in databases, namely for PLB and PSG.

At least 12 nematode species of the genera *Paratrichodorus* and *Trichodorus* are natural vectors, with a high specificity between virus strain and vector species. Adults and juveniles can transmit and retain virions for many months, particles being attached to the esophageal wall. Acquisition and inoculation periods can be short (1 h). TRV does not seem to multiply in its vector and to be transmitted through TPS.

Serological detection of TRV from potato is unreliable. Detection is based on the use of bait and indicator plants (*P. floridana*, *N. tabacum*, *C. quinoa*) but primarily on molecular techniques (RT-PCR and real-time PCR). However, none of these techniques is fully reliable.

Control methods are based on the use of certified seeds and of cultivars more or less resistant to spraing. Tolerant cultivars multiplying the virus without displaying any spraing cannot be recommended since they can contribute to spread of the virus. The use of nematicides for soil treatments is no longer permitted in most countries.

Potato Mop-Top Virus

PMTV can cause a serious disease in potato crops in Northern Europe, especially in cultivars such as Saturna. It is also present in the Andes, Japan, China, and Canada. PMTV induces necroses in potato tubers, both external and internal sinuous, often parallel, lines, also arcs (**Figure 7**). Blotching, skin cracking, and malformation can develop, mainly in cool conditions before harvest or during storage. Typical symptoms on leaflets after secondary infection are yellow or necrotic chevrons and blotches. Mop-top denotes bunched foliage on shortened stems. Virus does not invade all daughter tubers.

PMTV is vectored by the fungus *Spongospora subterranea* f. sp. *subterranea*, the causal agent of potato powdery also. Virus can persist in long-live resting spores of its vector. Strains differing in transmission and virulence have been reported. PMTV-T from Scotland is the type strain.

Potato is usually the only host plant. Many weed species such as *C. quinoa* may be natural hosts in Andean countries. *C. amaranticolor* and *N. debneyi* are local lesion hosts. *N. benthamiana* can be used as a bait plant and a suitable propagation host. ELISA, RT-PCR, and real-time PCR have been successfully used for detection, though results are inconsistent due to erratic distribution of the virus in infected plants. Virions are rod-shaped fragile particles similar to those of TRV. Polyclonal and monoclonal antibodies have been raised though the virus is difficult to purify. Antisera have also been produced using recombinant coat protein as antigen.

Control methods include breeding for tolerance to spraing, but no natural sources of resistance are available so far. Coat protein-derived resistance has been successfully tested in greenhouse conditions.

Viruses with Limited Significance

Many viruses of limited significance in potato may cause spectacular symptoms on potato foliage or tubers

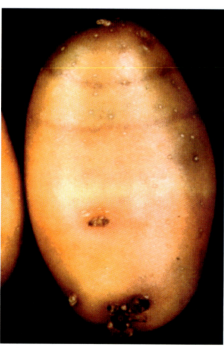

Figure 7 Necrotic sinuous lines induced by PMTV on a tuber of cv. Kerpondy. Photograph: C. Kerlan, INRA, France.

Figure 8 Calico symptoms induced by AMV on potato: bright yellow blotching of leaflets. Photograph: L. F. Salazar, CIP, Peru. Reproduced by permission of C. Jeffries, SASA, UK.

Figure 9 Infection by PAMV: bright yellow spots, flecks and blotches on cv. Ulster Premier. Reproduced by permission of C. Jeffries, SASA, UK.

Figure 10 Infection by TNV: necrotic lesions with crackings of the tuber skin. Reproduced by permission of C. Jeffries, SASA, UK.

Figure 11 External necroses on tubers induced by TSWV. Photograph: J. Martin, LNPV, France.

(**Figures 8–11**). Only TSWV was associated with significant losses in several countries with incidences sometimes reaching up to 90%. Its spread results from invasion of some species of thrips vectors.

See also: Bunyaviruses: General Features; History of Virology: Plant Viruses; *Carlavirus*; *Necrovirus*; *Nepovirus*; *Pomovirus*; *Potexvirus*; *Sobemovirus*; *Tobamovirus*; *Tobravirus*; *Tospovirus*; Vector Transmission of Plant Viruses; Luteoviruses; Plant Resistance to Viruses: Natural Resistance Associated with Recessive Genes; Plant Virus Diseases: Economic Aspects; Potato Virus Y; Potyviruses; Tymoviruses.

Further Reading

Brunt AA, Crabtree K, Dallwitz MJ, Gibbs AJ, and Watson L (eds.) (1996) *Viruses of Plants. Descriptions and Lists from the VIDE Database.* Wallingford, UK: CAB International.

de Bokx JA and Van der Want JPH (eds.) (1987) *Viruses of Potato and Seed-Potato Production,* 2nd edn. Wageningen, The Netherlands: Pudoc.

Delleman J, Mulder A, Peeters JMG, Schiper E, and Turkensteen LJ (eds.) (2005) *Potato Diseases.* Den Haag, The Netherlands: NIVAP Holland.

Gebhardt C and Valkonen J (2001) Organisation of genes controlling disease resistance in the potato genome. *Annual Review of Phytopathology* 39: 79–102.

Jeffries CJ (1998) *FAO/IPGRI Technical Guidelines for the Safe Movement of Germplasm, No. 19. Potato.* Rome: Food and Agriculture Organization of the United Nations/International Plant Genetic Resources Institute.

Loebenstein G, Berger P, Brunt AA, and Lawson RG (eds.) (2001) *Virus and Virus-Like Diseases of Potatoes and Production of Seed-Potatoes.* Dordrecht, The Netherlands: Kluwer.

Lorenzen JH, Meacham T, Berger PH, et al. (2006) Whole genome characterization of *Potato virus Y* isolates collected in the western USA and their comparison to isolates from Europe and Canada. *Archives of Virology* 151: 1055–1074.

Radcliffe EB and Ragsdale EB (2002) Aphid-transmitted potato viruses: The importance of understanding vector biology. *American Journal of Potato Research* 79: 353–386.

Ratke W and Rieckman W (1991) *Maladies et ravageurs de la pomme de terre.* Gelsenkirchen-Buer, Germany: Verlag Th. Mann.

Salazar LF (1966) *Potato Viruses and Their Control.* Lima, Peru: International Potato Center.

Singh RP (1999) Development of the molecular methods for potato virus and viroid detection and prevention. *Genome* 42: 592–604.

Solomon-Blackburn RM and Barker H (2002) A review of host-major gene resistance to potato viruses X, Y, A and V in potato: Genes, genetic and mapped locations. *Heredity* 86: 8–16.

Solomon-Blackburn RM and Barker H (2002) Breeding virus resistant potatoes (*Solanum tuberosum*): A review of traditional and molecular approaches. *Heredity* 86: 17–35.

Valkonen J (1994) Natural genes and mechanisms for resistance to viruses in cultivated and wild potato species (*Solanum* spp.). *Plant Breeding* 112: 1–16.

Valkonen J (1997) Novel resistances to four potyviruses in tuber-bearing potato species, and temperature-sensitive expression of hypersensitive resistance to *Potato virus Y*. *Annals of Applied Biology* 130: 91–104.

Potexvirus

K H Ryu and J S Hong, Seoul Women's University, Seoul, South Korea

© 2008 Elsevier Ltd. All rights reserved.

Glossary

Gene silencing A general term describing epigenetic processes of gene regulation, that is, a gene which would be expressed (turned on) under normal circumstances is switched off by machinery in the cell.

RNA interference (RNAi) A mechanism for RNA-guided regulation of gene expression in which double-stranded RNA inhibits the expression of genes with complementary nucleotide sequences.

Triple gene block (TGB) A specialized evolutionarily conserved gene module of three partially overlapping open reading frames involved in the cell-to-cell and long-distance movement of plant viruses.

Introduction

The genus *Potexvirus* (derived from *Potato virus X*) is one of nine genera in the family *Flexiviridae*. Most potexviruses are found wherever their hosts are grown. The genus *Potexvirus* contains a large number of members or tentative members. Host plants naturally infected with potexviruses may show mosaic, necrosis, ringspot, or dwarf symptoms or may be symptomless. The natural host range of individual viruses is usually restricted to a few plant species, although a few of the viruses can infect a wide range of plant species. Many members of the genus have been identified although only relatively recently. Most potexviruses are not transmitted by vertebrate, invertebrate, or fungal vectors; potato aucuba mosaic virus (PAMV), white clover mosaic virus (WClMV), and a few others are transmitted by aphids. The type species of the genus *Potexvirus* is *Potato virus X*. During the last few decades, potato virus X (PVX) has greatly contributed to our understanding of host resistance and gene silencing mechanisms, in particular because of the use of viral vectors in studying gene expression and RNA silencing in plants.

Taxonomy and Classification

The genus *Potexvirus* is one of nine genera in the family *Flexiviridae*. The type species of the genus is *Potato virus X*, and the name potexvirus was derived from the type species. The genus contains 55 members, of which 32 are definitive species and 23 tentative (**Table 1**). Some members listed as species and many members listed as tentative species have not been sequenced, and their taxonomic status is yet to be verified.

Different degrees of relationship exist among potexviruses, and phylogenetic analysis has revealed the presence of two major branches based on the composition of polymerase, coat protein (CP) and triple gene block (TGB) proteins.

Species demarcation criteria in the genus *Potexvirus* include the following: (1) members of distinct species have less than ~72% nucleotide or ~80% amino acid identity between their CP or replicase genes; (2) members of distinct species are readily differentiated by serology, and some virus strains can be differentiated with monoclonal antibodies; and (3) members of distinct species fail to cross-protect in common host plant species, and they usually have distinguishable experimental host ranges.

Virus Structure and Composition

Virions are flexuous filaments that are not enveloped, and measure about 470–580 nm in length and 13 nm in diameter, have helical symmetry and a pitch of 3.3–3.7 nm. Nucleocapsids are longitudinally striated. Virion M_r is about 3.5×10^6 with 5–7% nucleic acid content. Virions contain a monopartite, linear, single-stranded, positive-sense RNA that has a size range of 5.8–7.5 kbp. The 3′ terminus has a poly(A) tract and the 5′ terminus occasionally has a methylated nucleotide cap. CP subunits are of one type with a size in the range of 22–27 kDa. The genome is encapsidated in 1000–1500 CP subunits. The CPs of some strains of PVX and other viruses are glycosylated.

In some potexviruses such as bamboo mosaic virus (BaMV), a satellite RNA has been reported. The satellite RNA of 836 nt depends on BaMV for its replication and encapsidation. The BaMV satellite RNA (satBaMV) contains a single open reading frame (ORF) encoding a 20 kDa nonstructural protein.

Physicochemical Properties

Purified virion preparations sediment as one component with sedimentation coefficient of 115–130 S. The isoelectric point of the virion is about pH 4.4. The ultraviolet (UV) absorbance spectra of potexviruses have maxima at 258–260 nm and minima at 244–249 nm, with A_{max}/A_{min} ratios of 1.1–1.3, and the A_{260}/A_{280} ratio of purified

Table 1 Virus species in the genus *Potexvirus*

Species	Virus abbreviation	Accession number[a]
Definitive species		
Alternanthera mosaic virus	AltMV	NC_007731
Asparagus virus 3	AV-3	
Bamboo mosaic virus	BaMV	NC_001642
Cactus virus X	CVX	NC_002815
Caladium virus X	CalVX	AY727533
Cassava common mosaic virus	CsCMV	NC_001658
Cassava virus X	CsVX	
Clover yellow mosaic virus	ClYMV	NC_001753
Commelina virus X	ComVX	
Cymbidium mosaic virus	CYMMV	NC_001812
Daphne virus X	DVX	
Foxtail mosaic virus	FoMV	NC_001483
Hosta virus X	HVX	AY181252
Hydrangea ringspot virus	HdRSV	NC_006943
Lily virus X	LVX	NC_007192
Mint virus X	MVX	NC_006948
Narcissus mosaic virus	NMV	NC_001441
Nerine virus X	NVX	NC_007679
Opuntia virus X	OVX	NC_006060
Papaya mosaic virus	PapMV	NC_001748
Plantago asiatica mosaic virus	PlAMV	NC_003849
Plantago severe mottle virus	PlSMoV	
Plantain virus X	PlVX	
Potato aucuba mosaic virus	PAMV	NC_003632
Potato virus X	PVX	NC_001455
Scallion virus X	ScaVX	NC_003400
Schlumbergera virus X	SVX	AY366207
Strawberry mild yellow edge virus	SMYEV	NC_003794
Tamus red mosaic virus	TRMV	
Tulip virus X	TVX	NC_004322
White clover mosaic virus	WClMV	NC_003820
Zygocactus virus X	ZVX	NC_006059
Tentative species		
Allium virus X	AVX	AY826413
Alstroemeria virus X	AlsVX	NC_007408
Artichoke curly dwarf virus	ACDV	
Barley virus B1	BarV-B1	
Boletus virus X	BolVX	
Centrosema mosaic virus	CenMV	
Chenopodium mosaic virus X	ChMVX	NC_008251
Discorea latent virus	DLV	
Lychnis symptomless virus	LycSLV	
Malva veinal necrosis virus	MVNV	
Nandina mosaic virus	NaMV	
Negro coffee mosaic virus	NeCMV	
Paris polyphylla virus X	PPVX	DQ530433
Parsley virus 5	PaV-5	
Parsnip virus 3	ParV-3	
Parsnip virus 5	ParV-5	
Patchouli virus X	PatVX	
Pepino mosaic virus	PepMV	NC_004067
Rhododendron necrotic ringspot virus	RoNRSV	
Rhubarb virus 1	RV-1	
Smithiantha latent virus	SmiLV	
Viola mottle virus	VMoV	
Zygocactus symptomless virus	ZSLV	

[a]Accession number in GenBank.

preparation as 1.09–1.37. Physical properties of viruses in this genus are: thermal inactivation point 60–80 °C, longevity *in vitro* (several weeks to months), and dilution endpoint 10^{-5}–10^{-6}. Infectivity of sap does not change on treatment with diethyl ether. Infectivity is retained by the virions despite being deproteinized by proteases, phenols, or detergents.

The CPs of potexviruses are partly degraded during the purification and storage of virus preparations. The virions may dissociate by denaturing agents such as sodium dodecyl sulfate (SDS), urea, guanidine hydrochloride, acetic acid, and alkali.

Genome Structure and Gene Expression

The genome is a single-stranded RNA (ssRNA), 5.8–7.5 kbp in size and contains five ORFs, encoding the replicase, three putative protein components of movement proteins (MPs) called TGB, and the CP, from the 5'- to 3'- end in that order (**Figure 1**). Its genome organization resembles that of the genera *Allexivirus*, *Foveavirus*, *Mandarivirus*, and *Carlavirus*, but is distinguished from them by the presence of five ORFs and a small replication protein. The genome RNA is capped at the 5'-end and polyadenylated at the 3'-end. The 146–191 kDa viral replicase ORF is translated from the full-length genomic RNA. For expression of its 3' proximal viral genes, the virus utilizes at least two subgenomic RNAs. The genome structure of PVX (6435 nt) is represented in **Figure 1**. PVX RNA has 83 nt 5' untranslated region (UTR) at the 5'-terminus and 76 nt 3' UTR followed by a poly(A) tail at the 3'-terminus. The 5' UTR regulates genomic and subgenomic RNA synthesis and encapsidation, and the 3' UTR controls both plus- and minus-strand RNA synthesis. Potexvirus-infected plants contain double-stranded RNAs (dsRNAs) whose molecular mass corresponds to that of genomic RNA and two subgenomic RNAs. Although potexviruses typically have five ORFs, some potexviruses (cassava common mosaic virus (CsCMV), narcissus mosaic virus (NMV), strawberry mild yellow edge virus (SMYEV), and WClMV) have a sixth smaller ORF located within the ORF5, with no known protein product.

ORF1 encodes a 166 kDa polypeptide that is presumed to be the viral replicase. It contains motifs of methyltransferase, helicase, and RNA-dependent RNA polymerase. ORFs 2, 3, and 4 form the TGB encoding polypeptides of 25, 12, and 8 kDa, respectively, which facilitate the virus cell-to-cell movement. ORF5 encodes the 25 kDa CP. The CP is required for virion assembly and cell-to-cell movement. The 25 kDa TGBp1 protein contains an NTPase-helicase domain but is not involved in RNA replication.

Complete genomic sequences have recently been obtained for many potexviruses (see **Table 1**).

Infectious cDNA clones have been reported for some potexviruses, such as PVX, BaMV, cymbidium mosaic virus (CymMV), zygocactus virus X (ZVX), and WClMV. In particular, PVX has been widely used as a vector for the study of gene silencing and small RNAs as well as for expression of foreign genes in plants.

Viral Transmission

Potexviruses are spread by mechanical contact and horticultural and agricultural equipment. None of the viruses have known invertebrate or fungal vectors. Two potexviruses (PAMV and WClMV) are transmitted by aphids. A few members are not transmitted by mechanical inoculation. Seed transmission may occur with some potexviruses, but is not common.

Cytopathology

The cytoplasm of potexvirus-infected cells contains fibrous, beaded, banded, or irregular aggregates of virus particles. Virus particles, frequently in large aggregates, occur in the cytoplasm and occasionally in the nuclei. The distribution of potexviruses in the infected plant is not tissue specific. Inclusions such as cytoplasmic laminated inclusion components are present in infected cells.

Host Range

The natural host range of individual members is mostly limited, although some species can infect a wide range of experimental hosts. Some of the viruses cause serious damage to their hosts while others cause little damage to infected hosts.

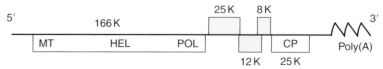

Figure 1 Genome organization of PVX, showing the typical genome structure present in the genus *Potexvirus*, family *Flexiviridae*. The 5' proximal large ORF encodes an RNA-dependent RNA polymerase (viral replicase), three overlapping ORFs encode the putative MPs (TGBs), and an ORF encodes CP. Motifs in the replicase are methyltransferase (MT), helicase (HEL), and RNA-dependent RNA polymerase (POL).

Symptomatology

Infection by most members of the genus *Potexvirus* is latent or causes only mild mosaic symptoms in the natural host. Symptoms vary cyclically or seasonally and may disappear soon after infection. Some viruses cause necrosis, ringspot, or dwarf symptoms in a wide range of plant species. If symptoms are evident, more severe symptoms such as mosaics appear in the early stages of infection. Some potexviruses such as PVX cause diseases that are of economic importance on their own; however, most of them are associated with more serious diseases when plants are co-infected with other viruses.

Serology

Virions are good immunogens. Most potexviruses are serologically related to several others, with the relationships varying from close to distant.

Geographical Distribution

Potexviruses are found wherever their hosts are grown, and so the geographical distribution of many species is restricted to only certain parts of the world.

Viral Epidemiology and Control

Most potexvirus-associated diseases are usually very mild or symptomless. The need for their control is often perceived as not important. However, crops such as potatoes, certain cactis, and some ornamental crops that may be infected by more damaging potexviruses require suitable control measures. Transgenic potato, tobacco, and orchid plants, which are resistant to infection by PVX and CymMV, have been developed.

See also: Allexivirus; *Capillovirus, Foveavirus, Trichovirus, Vitivirus*; *Carlavirus*; Flexiviruses; Plant Virus Vectors (Gene Expression Systems); Vector Transmission of Plant Viruses.

Further Reading

Adams MJ, Antoniw JF, Bar-Joseph M, *et al.* (2004) The new plant virus family *Flexiviridae* and assessment of molecular criteria for species demarcation. *Archives of Virology* 149: 1045–1060.

Fauquet CM, Mayo MA, Maniloff J, Desselberger U, and Ball LA (2005) In: *Virus Taxonomy, Classification and Nomenclature of Viruses, Eighth Report of the International Committee on the Taxonomy of Viruses*, 1101pp. New York: Academic Press.

Martelli GP, Adams MJ, Kreuze JF, and Dolja VV (2007) Family *Flexiviridae*: A case study in virion and genome plasticity. *Annual Review of Phytopathology* 45: 73–100 (online published).

Verchot-Lubicz J, Ye CM, and Bamunusinghe D (2007) Molecular biology of potexviruses: Recent advances. *Journal of General Virology* 88: 1643–1655.

Potyviruses

J J López-Moya, Instituto de Biología Molecular de Barcelona (IBMB), CSIC, Barcelona, Spain
J A García, Centro Nacional de Biotecnología (CNB), CSIC, Madrid, Spain

© 2008 Elsevier Ltd. All rights reserved.

Introduction

The existence of potyviruses as a specific group of plant pathogens with numerous members was recognized soon after the onset of virology as a scientific discipline. Historical evidence of symptoms caused in plants by potyviruses includes the color-breaking of infected tulips that was considered fashionable in the seventeenth century and abundantly reproduced in Dutch paintings from that period. Some viruses of the family were among the first plant viruses to be identified due to the importance of the diseases they cause, and relatively soon after potyviruses were grouped according to common characteristics. Electron microscopy provided a clear taxonomic criterion to allocate viruses into the group by showing the consistent presence of pinwheel-shaped cytoplasmic inclusions in plant cells infected by potyviruses. Further contributions to classification were provided by biological and serological studies, followed by deciphering of virus sequences. The finding of biological peculiarities in closely related viruses, with unusual vector organisms or genome composition, led to the current, although still provisional, classification of six genera in the family.

The large number of virus species within the family *Potyviridae* reveals its evolutionary success. Currently almost 200 definitive and tentative species are included in the genus of aphid-transmitted viruses, *Potyvirus*. Members of the family *Potyviridae* are distributed throughout the world, although each particular virus has a specific host range that limits its geographic distribution to the area potentially occupied by susceptible hosts. Generally speaking, host ranges are constrained to a limited number of natural and experimental hosts, although several potyviruses can infect a considerable number of plant species distributed in many botanical families and some members infect the most economically important crops, including grain, legumes, forages, vegetables, fruits, and ornamentals. The fact that several potyviruses are among the most damaging plant pathogens has fueled both the interest of researchers and the sustained attention of plant breeders and agronomists seeking practical solutions to the losses they cause in many crop species.

This article focuses on current knowledge about viruses in the family *Potyviridae*, with special emphasis on the more recent advances in taxonomy, evolution, diagnostics, functional and structural aspects of viral proteins, as well as host–virus interactions, including characterization of resistance genes and defense responses based on RNA silencing mechanisms. Present and future biotechnological applications of viruses within the family are also considered.

Taxonomy and Classification

The family *Potyviridae* comprises six genera of plant viruses, including the genus *Potyvirus* that is currently the largest genus of plant viruses. Other genera in the family are *Bymovirus*, *Ipomovirus*, *Macluravirus*, *Rymovirus*, and *Tritimovirus*.

Taxonomic standards for classification into the family include: (1) properties of virus particles: long flexuous filamentous (**Figure 1(a)**), (2) cytopathological manifestations: presence of pinwheel or scroll-shaped cylindrical cytoplasmic inclusions in infected plant cells (**Figure 1(b)**), and (3) genome structure and expression strategy: positive-sense ssRNA genomes, with 5' terminal proteins and 3' polyA tails, translated as large polyprotein precursors (see the section titled 'Properties of the genome').

Within the family, genera are differentiated according to biological and molecular criteria. Potyviruses are aphid-transmissible and possess a monopartite genome, characteristics that they share with macluraviruses, although particles of the latter are shorter in length. Other monopartite genera include whitefly-transmitted ipomoviruses, and mite-transmitted rymoviruses and tritimoviruses, further distinguished by their vector organisms belonging to the genera *Abacarus* and *Aceria*, respectively. Rymoviruses can also be distinguished by nucleotide similarity, being more closely related to potyviruses than tritimoviruses. Finally, bymoviruses have bipartite genomes that are encapsidated separately, and are transmitted by plasmodiophoraceous fungi of the genus *Polymyxa*. **Table 1** summarizes the family classification.

Phylogenetic analysis based on the alignment of genome regions has established standard thresholds of around

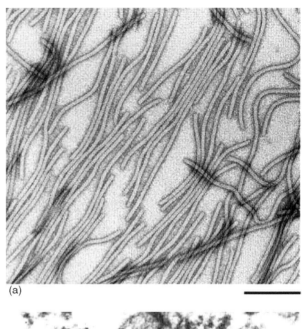

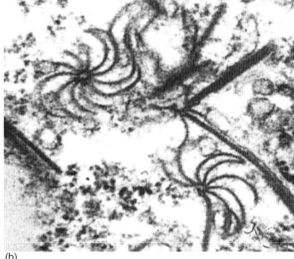

Figure 1 (a) Negative-stain preparation of purified tobacco etch potyvirus particles. (b) Thin section showing the typical pinwheel-shaped cytoplasmic inclusions present in cells of a *Nicotiana benthamiana* plant infected with plum pox potyvirus. Scale = 200 nm. Courtesy of D. López-Abella, CIB, CSIC.

50% and 75% of nucleotide identities (variable depending on the genomic region considered) as demarcation criteria for assigning viruses to genera and species, respectively. This approach has essentially confirmed the classification shown in **Table 1**. An example of phylogenetic analysis is shown in **Figure 2**. Subgrouping viruses, particularly between members of the large genus *Potyvirus*, is also contemplated as a possibility for future refinements in classification, and could follow host range restriction as biological demarcation criteria. Information about taxonomy and sequence of members of the family *Potyviridae* is available at several web-based databases (**Table 2**).

Table 1 Classification of members of the family *Potyviridae* into genera, with indication of type member, number of virus species, transmission vectors, and number of genomic RNAs

Genus	Type member	Number of species[a]		Genome	Vector[b]	Particle[c] length (nm)
		Definitive	Tentative			
Bymovirus	Barley yellow mosaic virus (BYMV)	6		Bipartite	Fungi	250–300
						500–600
Ipomovirus	Sweet potato mild mottle virus (SPMMV)	2	2	Monopartite	Whiteflies	900
Macluravirus	Maclura mosaic virus (MacMV)	3		Monopartite	Aphids	650–675
Potyvirus	Potato virus Y (PVY)	112	87	Monopartite	Aphids	700–900
Rymovirus	Ryegrass mosaic virus (RyMV)	4	1	Monopartite	Mites	700
Tritimovirus	Wheat streak mosaic virus (WSMV)	2		Monopartite	Mites	700

[a]Number of species according to the International Committee on Taxonomy of Viruses.
[b]Vector organisms responsible for transmission.
[c]Approximate length of virus particles.

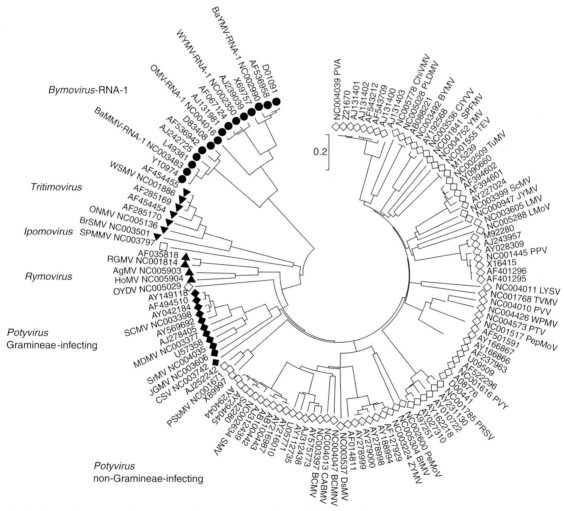

Figure 2 Phylogenetic analysis derived from the comparison of over 120 full-length viral genome sequences of 48 members of the family *Potyviridae*. The figure shows a neighbor-joining tree derived from a ClustalX alignment. Individual viral sequences, indicated by GenBank accession numbers and International Committee on Taxonomy of Viruses (ICTV) abbreviations, are grouped by genera and identified by symbols corresponding to potyviruses (diamonds, open or filled for viruses infecting non-gramineous or gramineous host plants, respectively), rymoviruses (triangles), one ipomovirus (open square), tritimoviruses (inverted triangles), and bymoviruses (circles). For bymoviruses, only the RNA-1 was used for comparison. Reproduced from Wang H, Huang LF, and Cooper JI (2006) Analysis on mutation patterns, detection of population bottlenecks, and suggestion of deleterious-compensatory evolution among members of the genus *Potyvirus*. *Archives of Virology* 151: 1625–1633, with permission from Springer-Verlag.

Table 2 Selection of resources available on the Internet (World Wide Web) for the family *Potyviridae*

Web page address (URL)	Contents and characteristics
http://www.ictvdb.rothamsted.ac.uk/Ictv/fs_potyv.htm	ICTV (International Committe on Taxonomy of Viruses): taxonomy structure, list of species
http://image.fs.uidaho.edu/vide/genus039.htm	VIDE (Virus Identification Data Exchange) project: nomenclature, host range, virion properties
http://www.danforthcenter.org/iltab/potyviridae/index.htm	Taxonomy, references, and sequence databases of members of the family
http://www.dpvweb.net/potycleavage/index.html	Analysis of the polyprotein cleavage sites

Virion Properties

Virion particles of viruses belonging to the family *Potyviridae* are flexuous rods constituted of protein (95%) and RNA (5%) (**Figure 1(a)**). Their size is about 11–15 nm wide, with lengths ranging from less than 700 nm in the case of macluraviruses to up to 900 nm in ipomoviruses (**Table 1**). Available data on particle structure suggest an helical assembly of identical coat protein (CP) subunits (about 2000) surrounding the nucleic acid, and a distribution of 7–8 subunits per turn has been suggested for the tritimovirus wheat streak mosaic virus (WSMV).

Molecular weight of CP subunits ranges between 30–40 kDa, with differences mainly due to a variable length of the N-terminal region. A more conserved internal CP core (about 220 amino acids) is probably involved in particle architecture. Both N- and C-termini are exposed at the surface of the particle. Superficially located residues might interact with other proteins during essential processes of the virus life cycle. As an example, the conserved DAG motif near the N-terminus of the CP is required for aphid transmission of tobacco vein mottling virus (TVMV). Recent studies have established the existence of host-dependent post-translational modifications, with probable regulatory functions during the viral cycle, in the CP of several potyviruses, including phosphorylation and glycosylation.

A recently identified peculiarity of potyviral particles is the presence of unusual structures at one of the ends of the long particles, probably associated to the 5′ end of the genomic RNA, in which the VPg and HC-Pro proteins were detected in some particles of purified potato virus Y (PVY) and potato virus A (PVA).

Serology and Diagnostics

Serological tools are the preferred diagnostic system of potyviruses. While the nonstructural viral proteins have limited use in serological diagnosis, the virus particles are usually strongly immunogenic. However, serological relationships among potyviruses are complex, with unexpected and inconsistent cross-reactivities that hindered their application in taxonomy. The cause of this is the presence of common epitopes in the conserved internal CP core, while specific epitopes map in the variable N-terminus, a surface-exposed region prone to degradation. New species-specific antibodies targeted against the immunodominant N-terminal region provided excellent serological detection tools applicable for many viruses.

The easy acquisition of sequence information after reverse transcription-polymerase chain reaction (RT-PCR) amplification is another way to deal with diagnosis. Degenerated primers able to amplify virtually any potyvirus have been described, for instance, targeted against conserved regions flanking the variable CP N-terminus. In combination with specific hybridization techniques or sequencing, the diagnosis is unequivocal. These molecular tools have been extensively used to identify new viruses infecting new hosts. Also, the combination of serological capture of particles with high sensitivity RT-PCR methods has resulted in robust detection systems. The application of modern molecular tools to diagnostics is being continually updated with new, more sensitive and specific approaches.

Properties of the Genome

The genome of monopartite potyviruses is an ssRNA molecule of + (messenger) sense. In all members studied so far, genomic RNA presents a 5′ terminal protein (VPg) and a 3′ polyA tail of variable length. Typical genome sizes range from 9.4 to 10.3 kbp. The genome comprises a single ORF coding for a long polyprotein (340–370 kDa) that generates mature products after ongoing an autoproteolytic processing cascade. The bipartite genome of bymoviruses is divided into two RNAs: a long RNA-1 homologous to the 3′ three-quarters portion of the monopartite potyvirus genome, and a short RNA-2 with partial similarity to portions of the 5′ region of the genome of monopartite potyviruses. Three virus-encoded proteinases are involved in the proteolytic processing of the potyviral polyprotein. The serine proteinase P1 and the cysteine proteinase HC-Pro are responsible for autocatalytic cleavages at their C-ends, and the serine proteinase NIaPro cleaves all other sites. Information on processing cleavage sites for all known viruses of the family is available on a web-based database (**Table 2**). The dynamics of the process might have regulatory implications for the sequential appearance and accumulation of intermediate and final products (**Figure 3**).

A typical potyvirus genome starts with a 5' noncoding region (NCR), less than 200 bp long. This leader acts as an enhancer of translation, and although the general mechanism is not fully understood, there is evidence that regulatory elements lie in this region. Studies in plum pox virus (PPV) showed that most of this region is dispensable for infectivity, although it contributes to viral competitiveness and pathogenesis, and that translation takes place by a cap-independent leaky scanning mechanism. In the case of the 5' leader of tobacco etch virus (TEV), the existence of an internal initiation site has been suggested, with the presence of an RNA pseudoknot domain that conferred cap-independent translation.

Another NCR of about 200 bp is located at the 3'-end of the genome before the polyA tail. Putative RNA structures in this region confer pathogenic properties in TVMV, and work with TEV and clover yellow vein virus (ClYVV) demonstrated the existence of *cis* acting elements necessary for infectivity.

Properties and Functions of Gene Products

The polyprotein coded in the genome of a potyvirus comprises the following gene products (from N-terminus to C-terminus): P1, HC-Pro, P3, 6K1, CI, 6K2, NIaVPg, NIaPro, NIb, and CP. For bymoviruses, RNA-2 encodes two proteins P2-1 and P2-2, while RNA-1 encodes the remaining products starting with P3. Besides the information provided in **Figure 3**, some characteristics of these products are given in this section.

The P1 protein is a serine protease, extremely variable in size (about 30–60 kDa). In TEV, it was shown to act as an accessory factor for genome amplification. This region contains important sequence differences among potyviruses. In the ipomovirus cucumber vein yellowing virus (CVYV) two duplicated P1-like serine proteinases are found, with the second one acting as a suppressor of gene silencing to compensate for the absence of HC-Pro in this virus.

The helper component HC-Pro is a multifunctional protein originally described as a factor required for aphid transmission ('HC' in the name of the protein stands for 'helper component' and refers to this function). Recently it was shown that the HC-Pro of tritimoviruses is also implicated in mite transmission. A modular distribution of functions has been proposed in this product: the N-terminus is essential for vector transmission, the central region is involved in suppression of silencing, and the C-terminal part is a papain-like cysteine proteinase ('Pro' in the name of the protein stands for 'proteinase' and refers to this function). HC-Pro can interact with RNA, and recent data indicate that it binds to double-stranded small interfering RNA (siRNA) molecules, a feature mechanistically coincidental with other silencing suppressors of viral origin. Structural studies with lettuce mosaic virus (LMV) and TEV have demonstrated HC-Pro in pairwise oligomeric forms. Interaction of HC-Pro with several host factors has been described.

In the case of bymoviruses, very limited information regarding roles for P2-1 and P2-2 is available. It has been postulated that P2-2 might play a role in vector transmission, since nontransmissible variants exhibited deletions in this region. However, sequence alignments indicate that P2-1, a putative cysteine proteinase, is more closely related to HC-Pro than P2-2.

Except for P1 and HC-Pro, all gene products are excised from the polyprotein by the action of the proteinase domain of NIa (nuclear inclusion a). The third gene product is P3, a protein of unknown functions, which is followed by a peptide called 6K1. Despite the presence of a typical cleavage sequence, cleavage between P3 and 6K1 of TEV does not occur *in vitro*, and excision does not seem to be required for viability of PPV. Immunology studies found both P3 and the precursor P3-6K1 in infected cells, and the latter is perhaps the functional product, although recently a tagged 6K1 product was also found in infected plants. Products of the P3-6K1 region play roles in host range definition and pathogenicity of several potyviruses.

The CI protein is the largest potyviral gene product. It forms very distinctive pinwheel-shaped cylindrical inclusions in the cytoplasm of infected cells, with high taxonomic value because they are unique to members of the family *Potyviridae* (**Figure 1(b)**). CI exhibits RNA helicase activity, and is supposed to act during RNA replication. Host factors interacting with CI of PPV and TEV have been recently identified. A second small peptide, 6K2, follows CI. This product has been implicated in virus replication, probably through a proposed anchoring capability which may serve to retain the replication complex in virus-induced membrane structures in the cytoplasm. Separation of 6K2 from NIa was also proposed to regulate nuclear targeting, and recent results in PVA reveals roles for 6K2 in movement and symptom induction.

The NIa protein has about 49 kDa, although it is subjected to an internal suboptimal cleavage to originate an N-terminal VPg (21 kDa) and a C-terminal proteinase fragment (28 kDa). VPg is the protein covalently attached to the 5' end of viral RNA. VPg can be uridylylated *in vitro* by the RNA replicase NIb, suggesting that it can be used as a primer during RNA replication as in the case of picornaviruses. VPg might be associated with host factors being a common pathogenicity determinant that also participates in long-distance movement. In PVA, VPg is translocated as a 'phloem protein' that specifically acts in companion cells to facilitate virus unloading. NIa is a

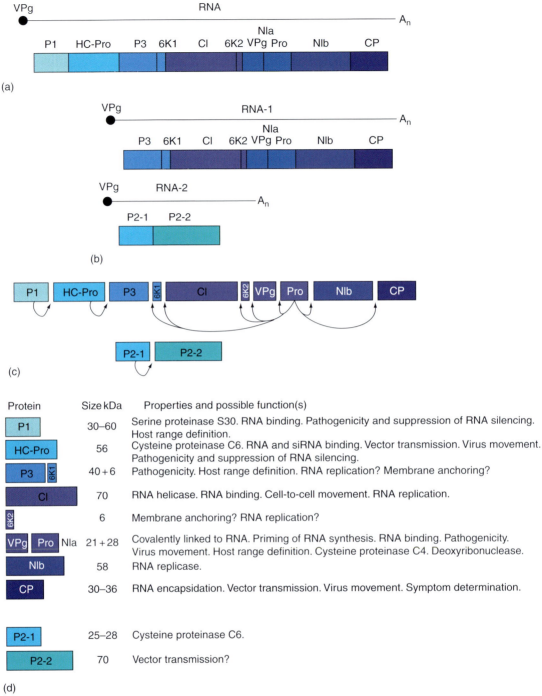

Figure 3 Representative genomic maps for viruses in the family *Potyviridae*. A monopartite virus characteristic of a potyvirus is shown in (a), while a bipartite bymovirus is depicted in (b). Genomic structures of ipomo-, maclura-, rymo-, and tritimoviruses are essentially similar to that of potyviruses. ssRNA genomes are shown as solid horizontal lines with VPgs represented by solid circles at each 5′ end, and polyA tails located at each 3′ end. Viral ORFs are depicted below as boxes divided in the different viral products, with names of gene products indicated. The proteolytic processing (rendering functional proteins or regulating their activity) of a typical potyvirus ORF is shown in (c), with indication of the three protease-specific cleavages sites indicated by arrows. Processing of the ORF encoded by the RNA-2 of bymoviruses is also presented. Properties of the different potyviral proteins are shown in (d). Abbreviation names, approximate size (kDa), and some properties and possible function(s) of each protein are indicated.

serine protease with a cysteine at the active site, responsible for the majority of cleavages in the polyprotein, acting in *cis* and *trans* at specific processing sites defined by characteristic heptapeptides. The protease domain of NIa shows nonspecific RNA binding activity, and was shown to exhibit nonspecific double-stranded DNA degradation activity in pepper vein banding virus (PVBV). NIa can form, together with NIb, inclusion bodies in the nucleus (hence the name of these proteins), where it is translocated responding to a bipartite signal sequence. The crystal structure of TEV NIa has recently been solved, providing clues on its mode of action.

NIb (nuclear inclusion b) is the second component of nuclear inclusions, being directed to the nucleus by specific signals in its sequence. NIb is the RNA-dependent RNA polymerase (RdRp) responsible for viral replication. Recruitment for the replication complex is postulated to occur via interaction with NIa. All NIb functions essential for RNA amplification may be provided in *trans* since NIb supplied in transgenic plants complement TEV deletion mutants.

The last product is CP, a protein with many roles in addition to the encapsidation of virus particles, being implicated in aphid transmission through interaction with HC-Pro, and in cell-to-cell and long-distance movement. The importance of maintaining the net charge of the CP N-terminus for infectivity has been shown in TVMV and zucchini yellow mosaic virus (ZYMV).

Replication and Propagation

Potyviruses replicate in the cytoplasm of infected cells, as schematically shown in **Figure 4**. After entering the cell, the viral genomic RNA must first be translated in the cytoplasm. In general, the early events during infection are still poorly understood. Disassembly of particles might

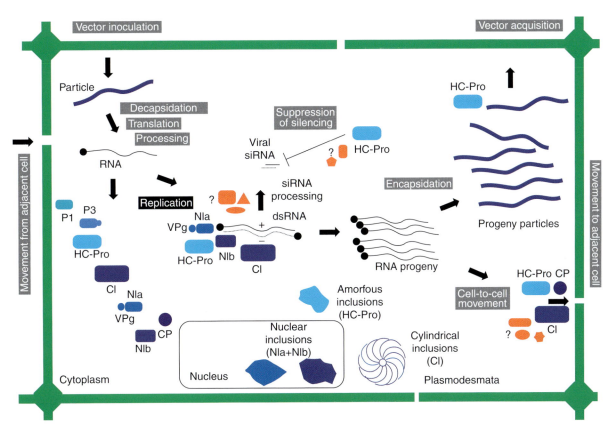

Figure 4 Schematic representation of different events during potyvirus infection of a plant cell. Colors and patterns of gene products match those displayed in **Figure 3**. The possible accumulation of viral proteins as inclusions in different compartments is also indicated. Putative host factors acting during the cycle are depicted as red color objects. The cycle begins (left upper corner) when the viral RNA enters the cell from an adjacent infected cell or a particle is initially inoculated by its vector and the genomic RNA undergoes decapsidation, translation, and processing to originate mature products. The replication complex is assembled with participation of NIb, CI, VPg, 6K2, and NIa, and probably other host-derived factors. The replication complex uses the genomic RNA (+ sense) to generate a complementary chain (− sense), which serves as a template for the synthesis of numerous genomic RNAs. Different mechanisms of plant defense might be activated during infection, including activation of the RNA silencing machinery that will produce virus-specific siRNA. A suppressor of RNA silencing, the HC-Pro in potyviruses, allows the virus to overcome this plant defense (indicated by the T-shaped symbol). After replication, the RNA progeny can move to adjacent cells through plasmodesmata, in a form not totally identified with the involvement of HC-Pro, CI, VPg, and CP, or it can be encapsidated and acquired by a vector organism to be transmitted again, in a process requiring HC-Pro.

occur co-translationally, and after translation the replication complex must be formed with participation of several viral products and probably host factors. This complex uses the genomic RNA (+ sense) as template to generate a complementary chain (− sense) through a dsRNA intermediate, and proceed with the asymmetric synthesis of numerous genomic RNAs. Virus specific siRNA accumulate during potyvirus infections, revealing the induction of a RNA silencing-mediated antiviral defense. Therefore, the virus needs to counterattack using a silencing suppressor, which in potyviruses was demonstrated to be the HC-Pro protein, although this function might be shared or displaced to other products in other genera. The need for several virus-encoded functions, including replication and silencing suppression, at very early stages of infection, points to the importance of viral RNA translation. Studies performed with pea seed-borne mosaic virus (PSbMV) served to identify a translation shut-off affecting many host proteins during infection. The potyvirus expression strategy through a polyprotein implies production of equimolar amounts of all gene products, giving rise to large excesses of some proteins which might remain soluble, be degraded or secreted, or end up in inclusion bodies. It is interesting that, whereas CI pinwheels are always formed, HC-Pro amorphous or NIa/NIb crystalline inclusions are present in some, but not all, potyviral infections.

Potyvirus RNA replication takes place in membranous structures probably derived mainly from the endoplasmic reticulum. NIb forms the core of the replication machinery, with involvement of CI, VPg, 6K2, and NIa. Other factors encoded by the virus or by the host might also be involved. For instance, NIa of turnip mosaic virus (TuMV) can interact with initiation factors eIF4E and eIF(iso)4E, and also with the PolyA binding protein. The fact that many potyvirus-specific resistance genes encode initiation factors points to the importance of these interactions.

Another essential process for virus infection is transport of the genomic RNA to adjacent cells. All tissues in the plant finally become invaded by the virus, with the probable exception of meristems, although a few seed-transmissible potyviruses might be capable of invading them. CP, but probably not virion formation, is essential for virus movement. In the process of local and systemic spread, again the silencing defense must be confronted by the virus in a race to avoid blockage by the transitivity diffusion of specific silencing signals.

The final step of the replicative cycle is spread of the virus to new plants, which rely on acquisition by the corresponding vector organisms and transmission. Two viral proteins, CP and HC-Pro, are involved in nonpersistent aphid transmission of potyviruses. Available data support a hypothesis in which HC-Pro serves as a reversible bridge to retain virus particles in aphid mouthparts. A conserved PTK domain in the central portion of HC-Pro could participate in binding to the DAG motif of the N-terminal region of CP, while a KITC domain at the N-terminal region of HC-Pro would be involved in binding to unknown structures of the aphid stylet. HC-Pro is also involved in semipersistent transmission of tritimoviruses by eryophid mites. Molecular information about the transmission of macluraviruses by aphids, bymoviruses by plasmodiophorids, rymoviruses by eryophid mites, and ipomoviruses by whiteflies is rather scarce.

Multifunctionality is observed in most potyviral proteins, illustrated, for instance, by HC-Pro. This capacity to participate in multiple processes suggests that potyvirus infection is not a consecutive succession of independent events, but a tightly regulated network of complex interactions between viral and host factors, still to be elucidated.

Pathogenicity

There is little information about how potyviruses cause diseases in their host plants. Several studies have served to identify sequences and/or products in the genome of potyviruses directly implicated in the production of symptoms. As mentioned above, symptom determinants are present in the 5′- and 3′-NCR. In addition, HC-Pro, P3, CI, 6K2, and VPg of different viruses have been described as determinants of pathogenicity, although single products were not always responsible for the different pathogenic responses.

As already mentioned, the P1-HC-Pro region is responsible for the synergistic effect in mixed infections of potyviruses with unrelated viruses, perhaps reflecting the capacity to interfere with RNA silencing-mediated defense. The discovery that HC-Pro is also able to affect miRNA regulated functions in plants provided a molecular explanation to some virus-induced symptoms. However, although it is tempting to regard interference with the metabolism of small RNAs in susceptible plants as a major element of pathogenicity, many other processes may also be affected. For example, specific features of the soybean mosaic virus (SMV) elicitor of the *Rsv1* resistance gene appears to be responsible for induction of either systemic mosaic or lethal systemic hypersensitive response in soybean. The fact that the HC-Pro protein of LMV targets and affects the proteasome might exemplify another way by which potyviruses can cause disease symptoms. Despite the fact that in most cases the mechanisms of pathogenicity remain uncertain, it is reasonable to conclude that the final macroscopic effects caused by potyviruses might be a combination of additive interference with several host functions.

Evolution

Members of the family *Potyviridae* are considered to belong to the picorna-like supergroup, characterized by the same genome expression strategy, and by a well-conserved set of replication-related proteins present at equivalent positions in the genome, which could be the result of cassette evolution.

Genera in the family show a close relationship, as indicated by the homologies between gene products and their conserved order in the polyprotein. Members of each genus within the family seem to be adapted to specific vector organisms. Moreover, the adaptation to particular host species might have contributed to speciation.

One intriguing issue is the amazingly large number of aphid-transmitted potyviruses. A combination of a very efficient transmission system and easy adaptation to new hosts must have contributed to this large expansion. In particular, vector transmission acting as bottlenecks might lead to speciation events. Together with recombination, switching events, radiation, and host and geographical adaptation are proposed as major traits of evolution. Existence of subpopulations able to differentiate and evolve independently within a single infected perennial plant was observed in PPV, and similar phenomena can happen in epidemics of viruses affecting annual hosts. Experimental evidence in PVA showed that recombination of nearly identical, phenotypically similar virus genomes can give rise to new viral strains with novel virulence and symptom phenotypes. Indications of recombination events, partial duplications, point mutations that apparently confer different host responses, as well as other factors, might help to explain the extraordinary variability observed among potyviruses.

Epidemiology and Control

Strategies currently applicable for potyvirus control are diverse, from cultural practices to the use of genetic resistance. Insecticide treatments against vectors are frequently considered unsatisfactory and have limited use because of the transmission type.

Severity of outbreaks is commonly related to abundance of initial foci of infection, dynamics of vector populations, and other factors, such as the presence of weeds acting as reservoirs of viruses. Human intervention is responsible in many cases for the introduction of emerging diseases into new territories, while vector organisms are mainly involved in propagation within particular regions. In a few potyviruses, seed transmission is also an important means of dissemination. A typical example of well-documented spread of a potyvirus over a territory and over time is the progressive emergence of Sharka disease, caused by PPV, in European countries during the twentieth century, and its recent diffusion to other continents to become nowadays a global pandemic.

The exploitation of pathogen-derived resistance and RNA silencing is yielding promising results for potyvirus control. Transgenic plants incorporating viral sequences were found to be resistant to several potyviruses. An example of success is provided by the engineering of papaya varieties resistant to papaya ringspot virus (PRSV) by expression of a viral CP transgene. Nowadays, RNA silencing is being further exploited for resistance by designing specific transgenes with hairpin structures, or even by direct application of dsRNAs. Cross-protection has also been related to RNA silencing mechanisms. Other approaches which are being explored include expression of ribozymes, plantibodies or inhibitors of proteinases. Stability and biosafety issues are under evaluation in all strategies.

The identification of potyvirus-specific resistance genes is leading to important advances. Characterization of new systems suggested that canonical resistance genes might be operating, for instance TIR-NBS-LRR class in PVY, or NBS-LRR class in SMV. As mentioned, mutations in eukaryotic initiation factors are frequently found as the responsible factor for recessive resistance against potyviruses. The generation of variability in candidate genes, using TILLING or equivalent platforms, could, therefore, be a promising strategy for generation of resistant plants.

Biotechnological Applications

Biotechnological uses of potyviruses are being pursued, both for the expression of foreign sequences in plants, and as a source of genetic elements and products of potential biotechnological utility. In this second aspect, extensive use of the TEV NIa protease to remove affinity tags from fusion proteins is a good example.

Regarding the use of potyviruses as expression vectors, several systems have been tested. Small peptides can be expressed fused to the CP, allowing the chimeras to be used as antigen presentation systems for immunization or diagnosis. Complete foreign genes have been expressed in vectors based on a quite large list of potyviruses which covers an important range of potential hosts. The seminal work with TEV pointed to the P1-HC-Pro junction as an adequate insertion site, but later new sites were exploited, such as the NIb-CP junction, or the P1 region. Recent reports also demonstrated the capacity to use potyviruses as double vectors expressing simultaneously two foreign proteins. The stability of the inserted genes seems to respond to characteristics of the foreign sequence still not completely defined.

The understanding of RNA silencing phenomena and the discovery of viral suppressors can serve as a technological basis to boost expression based on the simultaneous expression of a replicating virus plus a silencing suppressor.

Transgenic plants expressing HC-Pro are being used to improve the expression levels of viral vectors.

Concluding Remarks

Analysis of potyvirus molecular biology has been extraordinarily successful in many aspects, and is likely to continue and provide interesting data. The knowledge currently available about potyvirus replication, movement, and transmission should eventually permit new control strategies to be designed to interfere with these key stages in the virus life cycle. Structure resolution of virus components might also serve to explore new resistance strategies. Diagnostic tools are continually being improved, and future developments will certainly supply specific and sensitive means of virus identification. Current understanding of the selective involvement of host factors in resistance and pathogenesis is fuelling work on virus–plant interactions, a challenging research area. Besides, peculiarities of particular viruses are being unravelled, and will help to understand both generic features and specificities of viruses in the family. Finally, the biotechnological application of potyviruses is also a field in expansion, which will provide in the future more exciting developments.

See also: Bean Common Mosaic Virus and Bean Common Mosaic Necrosis Virus; Papaya Ringspot Virus; Plum Pox Virus; Potato Virus Y; Watermelon Mosaic Virus and Zucchini Yellow Mosaic Virus; Plant Resistance to Viruses: Engineered Resistance; Plant Resistance to Viruses: Natural Resistance Associated with Recessive Genes; Plant Antiviral Defense: Gene Silencing Pathway.

Further Reading

Adams MJ, Antoniw JF, and Fauquet CM (2005) Molecular criteria for genus and species discrimination within the family Potyviridae. *Archives of Virology* 150: 459–479.

Aranda M and Maule A (1998) Virus-induced host gene shutoff in animals and plants. *Virology* 243: 261–267.

Berger PH, Barnett OW, Brunt AA, et al. (2000) Family Potyviridae. In: Fauquet CM, Bishop DHL Carsten EB, et al. (eds.) *Virus Taxonomy: Seventh Report of the International Committee on Taxonomy of Viruses*, pp. 703–724. San Diego, CA: Academic Press.

Carrington JC, Kasschau KD, and Johansen LK (2001) Activation of suppression of RNA silencing by plant viruses. *Virology* 281: 1–5.

French R and Stenger DC (2003) Evolution of wheat streak mosaic virus: Dynamics of population growth within plants may explain limited variation. *Annual Review of Phytopathology* 41: 199–214.

Kasschau KD, Xie Z, Allen E, et al. (2003) P1/HC-Pro, a viral suppressor of RNA silencing, interferes with Arabidopsis development and miRNA function. *Developmental Cell* 4: 205–217.

López-Moya JJ, Fernández-Fernández MR, Cambra M, and García JA (2000) Biotechnological aspects of plum pox virus. *Journal of Biotechnology* 76: 121–136.

Ng JCK and Falk BW (2006) Virus–vector interactions mediating nonpersistent and semipersistent transmission of plant viruses. *Annual Review of Phytopathology* 44: 183–212.

Revers F, LeGall O, Candresse T, and Maule AJ (1999) New advances in understanding the molecular biology of plant/potyvirus interactions. *Molecular Plant Microbe Interactions* 12: 367–376.

Riechmann JL, Laín S, and García JA (1992) Highlights and prospects of potyvirus molecular biology. *Journal of General Virology* 73: 1–16.

Robaglia C and Caranta C (2006) Translation initiation factors: A weak link in plant RNA virus infection. *Trends in Plant Science* 11: 40–45.

Shukla DD, Ward CW, and Brunt AA (1994) *The Potyviridae*. Oxon, UK: CABI.

Urcuqui-Inchima S, Haenni A-L, and Bernardi F (2001) Potyvirus proteins: A wealth of functions. *Virus Research* 74: 157–175.

Voinnet O (2005) Induction and suppression of RNA silencing: Insights from viral infections. *Nature Reviews Genetics* 6: 206–220.

Wang H, Huang LF, and Cooper JI (2006) Analysis on mutation patterns, detection of population bottlenecks, and suggestion of deleterious-compensatory evolution among members of the genus Potyvirus. *Archives of Virology* 151: 1625–1633.

Poxviruses

G L Smith, P Beard, and M A Skinner, Imperial College London, London, UK

© 2008 Published by Elsevier Ltd.

Introduction

Poxviruses have been isolated from birds, insects, reptiles, marsupials, and mammals. The best known is variola virus (VARV), the cause of smallpox, an extinct disease that claimed millions of victims and influenced human history. All poxviruses have complex, enveloped virions that are large enough to be visible by the light microscope and contain double-stranded DNA (dsDNA) genomes with terminal hairpins linking the two DNA strands into a single polynucleotide chain. Poxvirus genes are transcribed by the virus-encoded RNA polymerase and associated transcriptional enzymes, which are packaged into the virion. Virus morphogenesis and entry have unique features, such as the possession of a thiol-oxidoreductase system to enable disulfide bond formation and morphogenesis in the cytoplasm, and a complex of several proteins for the fusion of infecting virions with the cell membrane. The large genome enables poxviruses to encode many virulence factors that are nonessential for virus replication in cell culture but which influence the outcome of infection *in vivo*. Diseases caused by poxviruses range from mild infections to devastating plagues, such as smallpox in man, mousepox in the laboratory mouse, and myxomatosis in the European rabbit.

Classification

The family *Poxviridae* comprises two subfamilies, the *Entomopoxvirinae* (**Table 1**) and *Chordopoxvirinae* (**Table 2**), whose members infect insects and chordates, respectively. The subfamily *Entomopoxvirinae* is divided into three genera: *Alphaentomopoxvirus*, *Betaentomopoxvirus*, and *Gammaentomopoxvirus*, which are typified by the species *Melolontha melolontha entomopoxvirus* (the virus name is abbreviated to MMEV), *Amsacta moorei entomopoxvirus 'L'* (AMEV), and *Chironomus luridus entomopoxvirus* (CLEV), respectively (**Table 1**). Alphaentomopoxviruses infect beetles (Coleoptera), betaentomopoxviruses infect butterflies and moths (Lepidoptera), and gammaentomopoxviruses infect mosquitoes and flies (Diptera). Compared to the *Chordopoxvirinae*, relatively few data are available for entomoxpoviruses and only two genomes have been sequenced. Nonetheless, these sequences revealed considerable divergence such that, although both sequenced viruses had originally been assigned to the same genus, *Melanoplus sanguinipes entomopoxvirus* (MSEV) was subsequently removed from this genus and the virus is now an unassigned member of the subfamily. Many features of the replication of entomopoxviruses are likely to be similar to those of chordopoxviruses, as exemplified by the orthopoxvirus *Vaccinia virus* (VACV). However, the morphology of entomopoxviruses differs from that of chordopoxviruses and there are also differences between the three entomopoxvirus genera (see below).

The subfamily *Chordopoxvirinae* is divided into eight genera: *Avipoxvirus*, *Capripoxvirus*, *Leporipoxvirus*, *Orthopoxvirus*, *Parapoxvirus*, *Suipoxvirus*, *Tanapoxvirus*, and *Yatapoxvirus* (**Table 2**). There are also several unassigned poxviruses that might form additional genera once their phylogenetic relationships have been established. For instance, the genome sequences of viruses isolated from Nile crocodiles (*Crocodylus niloticus*) and mule deer (*Odocoileus*

Table 1 Family *Poxviridae*, subfamily *Entomopoxvirinae*

Genus	Species[a] (abbreviation of virus name)	Genome accession no.	Genome size (bp)
Alphaentomopoxvirus	Anomala cuprea entomopoxvirus (ACEV)		
	Aphodius tasmaniae entomopoxvirus (ATEV)		
	Demodema boranensis entomopoxvirus (DBEV)		
	Dermolepida albohirtum entomopoxvirus (DAEV)		
	Figulus subleavis entomopoxvirus (FSEV)		
	Geotrupes sylvaticus entomopoxvirus (GSEV)		
	Melolontha melolontha entomopoxvirus (MMEV)		
Betaentomopoxvirus	Acrobasis zelleri entomopoxvirus 'L' (AZEV)	AF250284	232392
	Amsacta moorei entomopoxvirus 'L' (AMEV)		
	Arphia conspersa entomopoxvirus 'O' (ACOEV)		
	Choristoneura biennis entomopoxvirus 'L' (CBEV)		
	Choristoneura conflicta entomopoxvirus 'L' (CCEV)		
	Choristoneura diversuma entomopoxvirus 'L' (CDEV)		
	Choristoneura fumiferana entomopoxvirus 'L' (CFEV)		
	Chorizagrotis auxiliars entomopoxvirus 'L' (CXEV)		
	Heliothis armigera entomopoxvirus 'L' (HAVE)		
	Locusta migratoria entomopoxvirus 'O' (LMEV)		
	Oedaleus senigalensis entomopoxvirus 'O' (OSEV)		
	Operophtera brumata entomopoxvirus 'L' (OBEV)		
	Schistocera gregaria entomopoxvirus 'O' (SGEV)		
Gammaentomopoxvirus	Aedes aegypti entomopoxvirus (AAEV)		
	Camptochironomus tentans entomopoxvirus (CTEV)		
	Chironomus attenuatus entomopoxvirus (CAEV)		
	Chironomus luridus entomopoxvirus (CLEV)		
	Chironomus plumosus entomopoxvirus (CPEV)		
	Goeldichironomus haloprasimus entomopoxvirus (GHEV)		
Unassigned species in the subfamily	Diachasmimorpha entomopoxvirus (DIEV)		
Unassigned virus in the subfamily	Melanoplus sanguinipes entomopoxvirus (MSEV)	AF063866	236120

[a]Type species shown in bold.
Data from the International Committee on Taxonomy of Viruses (ICTV) (http://www.ncbi.nlm.nih.gov/ICTVdb/Ictv/index.htm) and the Poxvirus Bioinformatics Resource Center (http://www.poxvirus.org).

Table 2 Family *Poxviridae*, subfamily *Chordopoxvirinae*

Genus	Species[a] (abbreviation of virus name)	Genome accession no.	Genome size (bp)
Avipoxvirus	Canarypox virus (CNPV)	AY318871	359853
	Fowlpox virus (FWPV)	AF198100, AJ581527	288539, 266145
	Juncopox virus (JNPV)		
	Mynahpox virus (MYPV)		
	Pigeonpox virus (PGPV)		
	Psittacinepox virus (PSPV)		
	Quailpox virus (QUPV)		
	Sparrowpox virus (SRPV)		
	Starlingpox virus (SLPV)		
	Turkeypox virus (TKPV)		
	Crowpox virus (CRPV)		
	Peacockpox virus (PKPV)		
	Penguinpox virus (PEPV)		
Capripoxvirus	Goatpox virus (GTPV)	AY077836, AY077835	149723, 149599
	Lumpy skin disease virus (LSDV)	AF325528, AF409137	150773, 150793
	Sheeppox virus (SPPV)	AY077833, AY077834, NC_004002	150057, 149662, 149955
Leporipoxvirus	Hare fibroma virus (FIBV)		
	Myxoma virus (MYXV)	AF170726	161773
	Rabbit fibroma virus (RFV)	AF170722	159857
	Squirrel fibroma virus (SQFV)		
Molluscipoxvirus	**Molluscum contagiosum virus** (MOCV)	U60315	190289
Orthopoxvirus	Camelpox virus (CMLV)	AY009089, AF438165	202205, 205719
	Cowpox virus (CPXV)	AF482758, X94355, DQ437593	224499, 223666, 228250
	Ectromelia virus (ECTV)	AF012825,	209771, 207620
	Monkeypox virus (MPXV)	AF380138, DQ011156, AY753185, DQ011157	196858, 200256, 199469, 198,780
	Raccoonpox virus (RCNV)		
	Taterapox virus (TATV)	DQ437594	198050
	Vaccinia virus (VACV)	U94848, M35027, AF095689, AY243312	177923, 191737, 189274, 194711
	Variola virus (VARV)	L22579, X69198, Y16780, DQ441447	186103, 185578, 186986, 188251
	Volepox virus (VPXV)		
	Horsepox virus (HSPV)	DQ792504	212633
	Skunkpox virus (SKPV)		
	Uasin Gishu disease virus (UGDV)		
Parapoxvirus	Bovine papular stomatitis virus (BPSV)	AY386265	134431
	Orf virus (ORFV)	AY386264, DQ184476	139962, 137,820
	Parapoxvirus of red deer in New Zealand (PVNZ)		
	Pseudocowpox virus (PCPV)		
	Squirrel parapoxvirus		
	Auzduk disease virus		
	Camel contagious ecthyma virus		
	Chamois contagious ecthyma virus		
	Sealpox virus		
Suipoxvirus	**Swinepox virus** (SWPV)	AF410153	146454
Yatapoxvirus	Tanapox virus (TANV)		
	Yaba-like disease virus (YLDV)	AJ293568	144575
	Yaba monkey tumor virus (YMTV)	AY386371	134721
Unassigned viruses in the family	Nile crocodile poxvirus (CRV)	NC_008030	190054
	Mule deer poxvirus (DPV)	AY689436, AY689437	166259, 170560
	California harbor seal poxvirus (SPV)		
	Cotia virus (CPV)		
	Dolphin poxvirus (DOPV)		

Continued

Table 2 Continued

Genus	Species[a] (abbreviation of virus name)	Genome accession no.	Genome size (bp)
	Embu virus (ERV)		
	Grey kangaroo poxvirus (KXV)		
	Marmosetpox virus (MPV)		
	Molluscum-like poxvirus (MOV)		
	Quokka poxvirus (QPV)		
	Red kangaroo poxvirus (KPV)		
	Salanga poxvirus (SGV)		
	Spectacled caiman poxvirus (SPV)		
	Yoka poxvirus (YKV)		

[a]Type species shown in bold. Virus names in normal type (not italic) are tentative or unassigned.
Data from the International Committee on Taxonomy of Viruses (ICTV) (http://www.ncbi.nlm.nih.gov/ICTVdb/Ictv/index.htm) and the Poxvirus Bioinformatics Resource Center (http://www.poxvirus.org).

hemionus) in North America indicate that these will form additional genera. Within each genus there are distinct species and strains thereof. At least one genome sequence is available from each genus and in some cases large numbers of viruses have been sequenced. The orthopoxviruses have been studied most extensively and the poxvirus website lists genome sequences of 48 strains of VARV, 14 strains of VACV, and 9 strains of monkeypox virus (MPXV).

Virion Structure

Poxviruses have large virions with an oval or brick-shaped morphology. The best-studied poxvirus is VACV, which has dimensions of approximately 250 nm × 350 nm. Each infected cell produces two different forms of infectious virion called intracellular mature virus (IMV) and extracellular enveloped virus (EEV). These two forms differ in that EEV is surrounded by an additional membrane containing several virus proteins that are absent from IMV. The EEV form can be produced either by IMV budding through the plasma membrane or by exocytosis of a triple enveloped virus (see 'Morphogenesis', under 'Replication cycle').

The surface of the IMV particle contains surface tubules that have either an irregular appearance (in orthopoxviruses and avipoxviruses) or a basket-weave symmetry (in orf virus (ORFV), a parapoxvirus). The structure of the VACV IMV has been studied most intensively and there are differing views. Some investigators reported that the virus is surrounded by a double lipid membrane, whereas others propose that the outer layer comprises a single membrane surrounded by a protein coat. The evidence now strongly favors the one membrane model. The virus core has a dumbbell shape and there are lateral bodies of unknown function located in the core concavities. The virus core is surrounded by a palisade of protein spikes of 18 nm and the core wall contains pores through which virus mRNAs might extrude during early transcription. The virus genome is packaged in the core together with transcriptional enzymes and capsid proteins.

The structures of entomopoxvirus virions differ from those of chordopoxviruses. Virions of alphaentomopoxviruses have dimensions of approximately 350 nm × 450 nm and have a concave core with a single lateral body. The betaentomopoxvirus virions are a little smaller at about 250 nm × 350 nm with a cylindrical core and a lateral body that has a sleeve-like appearance. The gammaentomopoxvirus virions are 320 nm × 230 nm and are more like chordopoxviruses in having a biconcave core and two lateral bodies.

The general lack of symmetry of poxvirus virions contrasts with the icosahedral capsid of another family of large dsDNA viruses, the herpesviruses. Consequently, the structure of poxviruses is presently refractory to determination by current methods such as cryoelectron microscopy reconstruction and X-ray crystallography.

Genome Structure

Poxvirus genomes are dsDNA molecules that are linked at each terminus by a hairpin loop. The genome length varies from 134.4 kbp for bovine papular stomatitis virus, a parapoxvirus, to 359.9 kbp for canarypox virus, an avipoxvirus. When deproteinized, poxvirus genomes lack infectivity because the virus DNA-dependent RNA polymerase is required for transcription of poxvirus genes. The genomes of chordopoxviruses all show the same basic arrangement. Adjacent to each terminus there is an inverted terminal repeat (ITR) so that the sequence at one end of the genome is repeated at the other end in the opposite orientation. The length of the ITR varies considerably and may be less than 1 kbp and lacking protein coding sequences (e.g., VARV) or >50 kbp and containing numerous genes that are consequently diploid (e.g., some cowpox virus (CPXV) strains). Adjacent to each hairpin there is a short conserved sequence called the concatemer resolution sequence that is required for DNA replication.

The central region (approximately 100 kbp) of chordopoxvirus genomes encodes proteins that are conserved between chordopoxviruses and (mostly) essential for virus replication. Approximately equal numbers of genes from this region are transcribed leftward and rightward. In contrast, genes within the left and right terminal regions of the genome are transcribed predominantly toward the genome termini. These genes are more variable between viruses in both their type and number, and for the most part are nonessential for virus replication. Instead, these genes encode proteins that affect virus virulence, tropism, and interactions with the host immune system. Within chordopoxviruses the gene order in the central genome region is fairly well conserved, with two exceptions. First, in the avian viruses there are inversions and transpositions of blocks of genes. Second, in both avipoxviruses and crocodilepox virus (CRV) there are groups of inserted genes that are absent from other poxvirus genomes.

The entomopoxvirus genomes contain some of the genes conserved throughout the chordopoxviruses, but differ in that these genes are scattered throughout the genome and are not arranged centrally and in the specific order found in chordopoxviruses.

The nucleotide composition of poxvirus genomes varies considerably. The two sequenced entomopoxviruses are very rich in A + T (82.2% and 81.7%). Several chordopoxviruses are also (A + T)-rich, but less so than the insect viruses. For instance, sheeppox virus (SPPV) and VACV are 75% and 67% A + T, respectively. At the other end of the spectrum, the genome of molluscum contagiosum virus (MOCV) is 36% A + T. Despite these wide variations in base composition, the encoded proteins display considerable similarity.

Gene Content

Comparisons of sequenced chordopoxviruses have identified c. 90 genes that are present in every virus. These genes encode proteins that are essential for virus replication, such as capsid proteins and enzymes for transcription and DNA replication. All of these genes are located in the central region of the genome, and are likely to represent the core genome of an ancestral virus from which modern poxviruses have evolved. As poxviruses evolved and adapted to particular hosts, they seem to have acquired additional genes that are located predominantly in the terminal regions of the genome and give each virus its particular host range and virulence. For instance, many of the immunomodulatory proteins expressed by orthopoxviruses are absent from viruses of avian and reptile hosts and also from MOCV, which has a different set of immunomodulators.

If the entomopoxviruses are included in the above comparison, the number of conserved genes falls to fewer than 50. As more poxvirus genomes are sequenced the number of genes conserved in all poxviruses (the minimal gene complement) is likely to be reduced.

Phylogeny

Phylogenetic comparisons of poxviruses indicate that the entomopoxviruses are quite divergent from chordopoxviruses, but the exact relationships of most entomopoxviruses remain unknown owing to lack of sequence data. However, for the chordopoxviruses the situation is clearer. Genome sequences are available for at least one member of each genus, and a phlyogenetic tree (**Figure 1**) shows that the genera *Yatapoxvirus*, *Suipoxvirus*, *Leporipoxvirus*, and *Capripoxvirus* and deerpox virus (DPV)

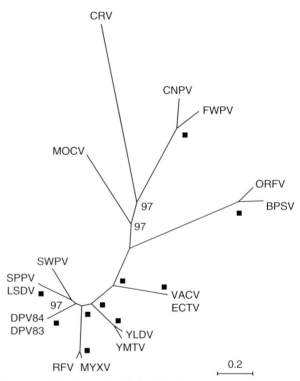

Figure 1 Phylogenetic relationships between chordopoxviruses based on a concatenated amino acid sequence alignment of 83 conserved proteins. Abbreviations: CRV, crocodilepox virus; CNPV, canarypox virus; FWPV, fowlpox virus; ORFV, orf virus; BSPV, bovine papular stomatitis virus; VACV, vaccinia virus; ECTV, ectromelia virus; YLDV, Yaba-like disease virus; YMTV, Yaba monkey tumor virus; MYXV, myxoma virus; RFV, rabbit fibroma virus; DPV, deerpox virus; LSDV, lumpy skin disease virus; SPPV, sheeppox virus; SWPV, swinepox virus; MOCV, molluscum contagiosum virus. Bootstrap values of greater than 70 are indicated at the appropriate nodes and black squares indicate values of 100. Reproduced from Afonso CL, Tulman ER, Delhon G, et al. (2006) Genome of crocodilepox virus. *Journal of Virology* 80: 4978–4991, with permission from the American Society for Microbiology.

cluster together. The orthopoxviruses are the nearest genus to this cluster, but have slightly larger genomes and other distinguishing features. The remaining genera, *Molluscipoxvirus*, *Parapoxvirus*, and *Avipoxvirus*, as well as the recently sequenced CRV are more divergent from the genera mentioned above. The avian poxviruses and CRV are most divergent and their genomes are distinguished by a lower degree of amino acid similarity, greater numbers of genes that are unique to that genus, and divergent gene order. The next most divergent genus is *Molluscipoxvirus*, which like VARV infects only humans but has very different immunomodulatory proteins compared to orthopoxviruses. Interestingly, there are some genes shared by only MOCV and CRV. Lastly, the parapoxviruses form a quite distinct genus.

Although only eight chordopoxvirus genera have been recognized so far by the International Committee for the Taxonomy of Viruses, the genomes of DPV and CRV are distinct from all other genera and so are likely to represent new genera.

Replication Cycle

Details of the poxvirus replication cycle have been obtained primarily by study of VACV and for greater detail the reader is referred to the article on that virus. Here only an overview is given.

Entry

VACV forms two distinct virion types (IMV and EEV), which are surrounded by different numbers of membranes. These different virions are also produced by viruses from several chordopoxvirus genera, and may be universal for poxviruses. IMV enters by fusion of its membrane with either the plasma membrane or the membrane of an intracellular vesicle following endocytosis. EEV entry is complicated by the second membrane that is shed on contact with the cell by a ligand-dependent nonfusogenic process. This places the IMV from within the EEV envelope against the plasma membrane and thereafter entry takes place as for free IMV. A remarkable feature of VACV fusion is the presence of a complex of nine proteins that are all required for fusion but not binding to the cell. In contrast, the fusion machinery of other enveloped viruses is often only a single protein. The fusion proteins of VACV are conserved in many other poxviruses, suggesting a common mechanism of entry. After a virus core has entered the cell, it is transported on microtubules deeper into the cell (see **Figure 2(d)** for an image of a virus core).

Gene Expression

Within minutes of infection the early virus genes are transcribed by the virus-associated DNA-dependent RNA polymerase within the core. The mRNAs are capped and polyadenylated, and are not spliced. They are extruded from the core and translated on host ribosomes. Early genes encode proteins that initiate DNA replication or combat the host response to infection (see 'Immune evasion' under 'Pathogenesis'). Once DNA replication has started, the intermediate class of genes is transcribed and, after expression of intermediate proteins, late transcription starts. Generally, late proteins encode structural proteins, which are expressed at higher levels than the early proteins involved in replication. Unusual features of late mRNAs are that they have a polyA tract just after the 5' cap structure as well as at the 3' end, and are heterogeneous in length due to a failure to terminate transcription at specific sites.

DNA Replication

DNA replication occurs in virus factories in the cytoplasm, and is catalyzed by the virus DNA polymerase and associated factors. The terminal hairpins and the adjacent concatemeric resolution sequence are essential for this process. Replication starts with the introduction of a nick near a terminal hairpin, followed by unfolding of the hairpin to enable self-priming from the free 3'-hydroxyl group. After extension to the genome terminus, separation of the parental and daughter strands, and refolding, DNA replication proceeds by strand displacement to form concatemeric molecules in either head-to-head or tail-to-tail configuration. Resolution of the concatemers occurs at the concatemer resolution sequence using the activity of a resolvase. Unit-length monomers are then packaged into virions. For additional details, see the article on VACV.

Morphogenesis

Virus assembly takes place in modified areas of the cytoplasm called factories or virosomes (**Figure 2**). The first structure seen by electron microscopy is crescent shaped, and is composed of host lipid and virus protein. This grows into immature virus (IV) in which the virus genome is packaged together with transcriptional enzymes. Once the virus membrane is sealed, the core condenses, some capsid proteins are cleaved proteolytically, and an IMV is formed. IMV may be released from cells following cell lysis either as free virions or within proteinaceous bodies. The latter are produced by some orthopoxviruses (including ectromelia virus (ECTV), raccoonpox virus, and some strains of CPXV), some avipoxviruses, and entomopoxviruses. These proteinaceous bodies are thought to give the virions extra stability outside the host until taken up by a new susceptible host. Interestingly, the proteinaceous body of entomopoxviruses is disassembled by high pH (as found in the insect gut) whereas the inclusion bodies of CPXV are disaggregated by low pH.

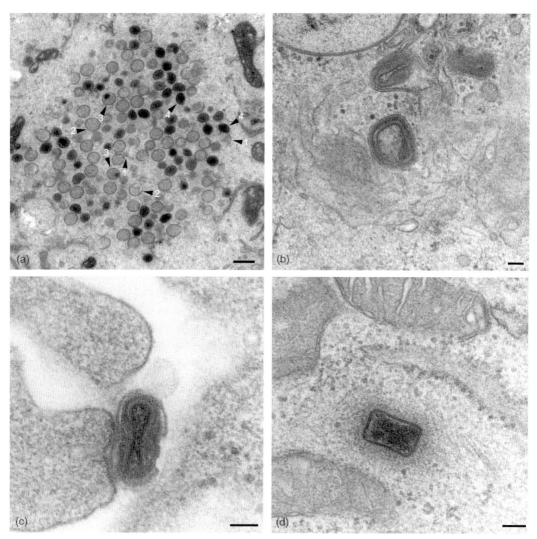

Figure 2 Electron micrographs of VACV. (a) Virus factory showing the different stages of VACV morphogenesis: (1) virus crescent; (2) immature virus, (3) immature virus showing stages of condensation and eccentric nucleoid formation, (4) IMV. (b) Wrapping of IMV to form IEV. The virion at top is partly wrapped while the virion in the center is completely wrapped. (c) CEV on the surface of an actin tail protruding from the cell surface. (d) Virus core within the cytosol shortly after infection. Scale = 500 nm (a), 100 nm (b–d). Reproduced by permission of Michael Hollinshead, Imperial College London.

Not all IMVs remain in the cell until cell lysis, as some are exported from the cell by two distinct pathways that each results in formation of EEV. The first pathway, exemplified by VACV, starts with transport of IMV away from factories on microtubules and the wrapping by a double layer of host membrane derived from the *trans*-Golgi network or early endosomes that has been modified by the insertion of virus proteins (**Figure 2(b)**). The wrapping process produces an intracellular enveloped virus (IEV), which contains three membranes, and this particle is transported to the cell surface on microtubules. At the cell surface, the outer membrane fuses with the plasma membrane to release a virion by exocytosis. The virion is called cell-associated enveloped virus (CEV) if it remains on the cell surface (**Figure 2 (c)**) and EEV if it is released. The majority of EEVs made by VACV are produced in this way.

The second pathway, illustrated by fowlpox virus (FWPV), is the budding of an IMV through the plasma membrane to release EEV.

In addition to being released as EEV, CEV can also induce formation of long cellular projections, which are driven by polymerization of actin (**Figure 2(c)**). These are important for efficient cell-to-cell dissemination of virus, because virus mutants that are unable to induce formation of actin tails form small plaques.

Pathogenesis

Diseases caused by poxviruses vary greatly and may be associated with high mortality or cause minor, local, or asymptomatic infections. Examples of systemic diseases caused by chordopoxviruses include smallpox, monkeypox, mousepox, camelpox, sheeppox, and myxomatosis.

Smallpox was the most serious human disease caused by a poxvirus and produced a systemic infection in which VARV spread sequentially through the body, culminating in the formation of skin lesions through which the virus was released to infect new hosts. A feature of smallpox that distinguished it from chickenpox (caused by varicella-zoster virus, a herpesvirus) was the centrifugal distribution of lesions (abundant on the face and limbs but less so on the trunk). The mortality rate associated with infection of immunologically naive subjects with variola major virus was 30–40%, whereas infection with variola minor virus caused only 1–2% mortality. That VARV was such a dangerous human pathogen may never be fully understood, but VARV contains many immunomodulatory proteins that subvert the host response to infection, especially the innate immune response (see below). Comparisons of the sequences of variola major and variola minor viruses revealed many differences, such that it was not possible to deduce which change(s) was (were) responsible for the large difference in virulence.

Human monkeypox is a systemic infection that is similar to smallpox but, unlike smallpox, the infection spreads very poorly from human to human. The virulence of MPXV depends on the virus strain. Viruses isolated from central Africa (e.g., Zaire) caused up to 10% mortality, whereas those from West Africa are less virulent. For instance, the zoonotic introduction of a West African strain into the USA in 2003 infected at least 37 people but no mortalities were recorded. The genomes of MPXV and VARV are significantly different, such that it is unlikely that MPXV would mutate into a VARV-like virus with efficient human-to-human transmission.

SPPV and goatpox virus (GTPV) are two very closely related members of the genus *Capripoxvirus*, and are of considerable veterinary importance because they are responsible for severe, systemic disease in sheep and goats. The viruses are endemic in parts of Africa, Asia, and the Middle East with a propensity of spreading to neighboring areas, as revealed by recent outbreaks in Vietnam, Mongolia, and Greece. The viruses are transmitted by direct contact, environmental contamination, or biting insects, and cause the most severe disease in young animals where the mortality can reach 100%. The affected animals exhibit typical poxviral disease with discrete, circumscribed cutaneous lesions most numerous in sparsely haired areas. Lesions are also found in the lung, gasterointestinal tract, and mucous membranes.

ECTV causes mousepox in inbred mouse colonies and can have very high mortality rates. For instance, in susceptible mouse strains the lethal dose 50 (LD50) of ECTV is less than one plaque-forming unit. Some features of mousepox have similarity with smallpox and consequently it is being used by some investigators as a model for smallpox. The natural host of ECTV is uncertain but is likely to be wild rodents.

Camelpox virus (CMLV) causes a severe systemic disease in camels that also has some features resembling smallpox. Sequencing of the CMLV genome showed that VARV and CMLV are closely related, although there remain many differences, especially within the terminal regions of the virus genome, so mutation of CMLV into a VARV-like virus is unlikely.

Myxomatosis is a disease caused by myxoma virus (MYXV) in the European rabbit (*Oryctolagus cuniculus*). However, the natural host for MYXV is the jungle rabbit (*Sylviagus brasiliensis*) or brush rabbit (*S. bachmani*), in which the outcome of MYXV infection is mild. MYXV infection in the European rabbit has been used as a model for poxvirus pathogenesis. MYXV virus is also famous as the virus used in Australia in the 1950s to control the European rabbit population, which increased from 13 in 1859 to more than 500 million early in the twentieth century. The spread of MYXV in Australian rabbits is an excellent example of host–pathogen evolution: very soon after release of the highly virulent virus into the rabbit population, the virus strains isolated from infected rabbits were of more modest virulence such that they did not kill all available hosts. Likewise, surviving rabbit populations were more resistant to the effects of virus infection.

Immune Evasion

Poxviruses encode many proteins that block or subvert the host response to infection. The genes encoding these proteins are mostly located within the terminal regions of the genomes and are nonessential for virus replication in cell culture. The proteins are directed predominantly against the innate response to infection, but target a wide range of host molecules both inside and outside the cell. Poxvirus intracellular proteins may block signaling pathways resulting from engagement of Toll-like receptors or receptors for cytokines or interferons. Additionally, intracellular proteins may inhibit the antiviral activity of interferon (IFN)-induced proteins or the pathways leading to induction of apoptosis. Extracellular proteins may inhibit the action of complement, cytokines, chemokines, or IFNs by binding these proteins in solution and preventing them from reaching their natural receptors on cells. Some poxviruses also subvert the cytokine or chemokine systems by encoding cytokines (e.g., interleukin-10, transforming growth factor beta) or chemokine mimics, or by the expression of chemokine receptors on the surface of infected cells. In general, the removal of individual immunomodulatory proteins has caused modest attenuation of virulence in one model or another, but the outcome is uncertain and in some cases an increase in virulence was reported. Very often, the attenuation deriving from loss of an immunomodulatory protein is modest compared to that caused by deletion of a gene involved in virus intracellular transport or release, but there are exceptional cases where loss of an immunomodulator has caused a several log increase in the virus LD50. The wide variety of immunomodulators expressed by poxviruses makes these viruses excellent models for studying virus–host interactions and provides a logical means by which the immunogenicity of recombinant poxviruses being considered as vaccine vectors can be improved.

Acknowledgments

G. L. Smith is a Wellcome Principal Research Fellow and P. M. Beard holds a Wellcome Trust Intermediate Clinical Fellowship.

See also: Capripoxviruses; Cowpox Virus; Entomopoxviruses; Fowlpox virus and other avipoxviruses; Leporipoviruses and Suipoxviruses; Molluscum Contagiosum Virus; Mousepox and Rabbitpox Viruses; Parapoxviruses; Smallpox and Monkeypox Viruses; Vaccinia Virus; Yatapoxviruses.

Further Reading

Afonso CL, Tulman ER, Delhon G, et al. (2006) Genome of crocodilepox virus. *Journal of Virology* 80: 4978–4991.

Buller RM, Arif BM, Black DN, et al. (2005) Poxviridae. In: Fauquet CM, Mayo MA, Maniloff J, Desselberger U, and Ball LA (eds.) *Virus Taxonomy: Eighth Report of the International Committee on Taxonomy of Viruses*, pp. 117–133. San Diego, CA: Elsevier Academic Press.

Fenner F, Henderson DA, Arita I, Jezek Z, and Ladnyi ID (1988) *Smallpox and Its Eradication.* Geneva: World Health Organization.

Fenner F, Wittek R, and Dumbell KR (1989) *The Orthopoxviruses.* London: Academic Press.

Mercer AA, Schmidt A, and Weber O (2007) *Poxviruses.* Berlin: Birkhäuser Verlag.

Moss B (2007) Poxviridae: The viruses and their replication. In: Knipe DM, Howley PM, Griffin DE, et al. (eds.) *Fields Virology*, 5th edn., pp. 2905–2946. Philadelphia, PA: Lippincott Williams and Wilkins.

Relevant Website

http://www.poxvirus.org – Poxvirus Bioinformatics Resource Center.

Prions of Vertebrates

J D F Wadsworth and J Collinge, University College London, London, UK

© 2008 Elsevier Ltd. All rights reserved.

Glossary

Codon 129 polymorphism There are two common forms of *PRNP* encoding either methionine or valine at codon 129; a major determinant of genetic susceptibility to and phenotypic expression of prion disease.

Conformational selection model A hypothetical model which explains transmission barriers on the basis of overlap of permissible conformations of PrP^{Sc} (prion strains) between mammalian species.

Molecular strain typing A means of rapidly differentiating prion strains by biochemical differences in PrP^{Sc}.

Prion The infectious agent causing prion diseases.

Prion incubation period The interval between exposure to prions and the development of neurological signs of prion disease; typically months even in rodent models and years to decades in humans.

Prion protein (PrP) A glycoprotein encoded by the host genome and expressed in many tissues but especially on the surface of neurons.

Prion strain Distinct isolates of prions originally identified and defined by biological characteristics which breed true in inbred mouse lines.

PRNP The human prion protein gene; mouse gene is designated *Prnp*.

Protein-only hypothesis The prions that lack a nucleic acid genome, are composed principally or solely of abnormal isomers of PrP (PrP^{Sc}), and replicate by recruitment of host PrP^{C}.

PrP Prion protein.

PrP^{C} The normal cellular isoform of PrP rich in α-helical structure.

PrP^{Sc} The 'scrapie' or disease-associated isoform of PrP which differs from PrP^{C} in its conformation and is generally found as insoluble aggregated material rich in β-sheet structure.

Subclinical infection A state where host prion propagation is occurring but which does not produce clinical disease during normal lifespan; essentially a carrier state of prion infection.

Transmission barrier This describes the observation that transmission of prions from one species to another is generally inefficient when compared to subsequent passage in the same host species

Introduction

The prion diseases are a closely related group of neurodegenerative conditions which affect both humans and animals. They have previously been described as the subacute spongiform encephalopathies, slow virus diseases, and transmissible dementias, and include scrapie in sheep, bovine spongiform encephalopathy (BSE) in cattle, and the human prion diseases, Creutzfeldt–Jakob disease (CJD), variant CJD (vCJD), Gerstmann–Sträussler–Scheinker disease (GSS), fatal familial insomnia (FFI), and kuru. While rare in humans, prion diseases are an area of intense research interest. This is first because of their unique biology, in that the transmissible agent appears to be devoid of nucleic acid and to consist of a post-translationally modified host protein. Secondly, because of the ability of these and related animal diseases to cross from one species to another, sometimes by dietary exposure, there has been widespread concern that the exposure to the epidemic of BSE poses a distinct and conceivably a severe threat to public health in the United Kingdom and other countries. The extremely prolonged and variable incubation periods of these diseases, particularly when crossing a transmission barrier, means that it will be some years before the parameters of any human epidemic can be predicted with confidence. In the meantime, we are faced with the possibility that significant numbers in the population may be incubating this disease and that they might pass it on to others via blood transfusion, blood products, tissue and organ transplantation, and other iatrogenic routes.

Aberrant Prion Protein Metabolism Is the Central Feature of Prion Disease

The nature of the transmissible agent in prion disease has been a subject of heated debate for many years. The understandable initial assumption that the causative agent of 'transmissible dementias' must be some form of virus was challenged by the failure to directly demonstrate such a virus (or indeed any immunological response to it) and by the remarkable resistance of the transmissible agent to treatments that inactivate nucleic acids. These findings led to suggestions that the transmissible agent may be devoid of nucleic acid and might be a protein.

Subsequently in 1982, Prusiner and co-workers isolated a protease-resistant sialoglycoprotein, designated the prion protein (PrP), that was the major constituent of infective fractions and was found to accumulate in affected brain. The term prion (from *pr*oteinaceous *in*fectious particle) was proposed by Prusiner to distinguish the infectious pathogen from viruses or viroids and was defined as "small proteinaceous infectious particles that resist inactivation by procedures which modify nucleic acids."

Initially, PrP was assumed to be encoded by a gene within the putative slow virus thought to be responsible for these diseases; however, amino acid sequencing of part of PrP and the subsequent recovery of cognate cDNA clones using an isocoding mixture of oligonucleotides led to the realization that PrP was encoded by a single-copy chromosomal gene rather than by a putative viral nucleic acid.

Following these seminal discoveries, a wealth of data has now firmly established that the central and unifying hallmark of the prion diseases is the aberrant metabolism of PrP, which exists in at least two conformational states with different physicochemical properties. The normal cellular form of the protein, referred to as PrP^C, is a highly conserved cell surface glycosylphosphatidylinositol (GPI)-anchored sialoglycoprotein that is sensitive to protease treatment and soluble in detergents. In contrast, the disease-associated scrapie isoform, designated as PrP^{Sc}, is found only in prion-infected tissue as aggregated material, partially resistant to protease treatment, and insoluble in detergents. Due to its physicochemical properties, the precise atomic structure of the infectious particle or prion is still undetermined but considerable evidence argues that prions are composed largely, if not entirely, of an abnormal isoform of PrP. The essential role of host PrP for prion propagation and pathogenesis is demonstrated by the fact that knockout mice lacking the PrP gene ($Prnp^{o/o}$ mice) are entirely resistant to prion infection, and that reintroduction of PrP transgenes restores susceptibility to prion infection in a species-specific manner.

Human Prion Diseases Are Biologically Unique

Human prion diseases are biologically unique and can be divided etiologically into inherited, sporadic, and acquired forms. Approximately 85% of cases of human prion disease occur sporadically as Creutzfeldt–Jakob disease (sporadic CJD) at a rate of roughly 1 case per million population per year across the world, with an equal incidence in men and women. The etiology of sporadic CJD is unknown, although hypotheses include somatic *PRNP* mutation, or the spontaneous conversion of PrP^C into PrP^{Sc} as a rare stochastic event. Polymorphism at residue 129 of human PrP (encoding either methionine (M) or valine (V)) powerfully affects genetic susceptibility to human prion diseases. About 38% of Europeans are homozygous for the more frequent methionine allele, 51% are heterozygous, and 11% homozygous for valine. Homozygosity at *PRNP* codon 129 predisposes to the development of sporadic and acquired CJD. Most sporadic CJD occurs in individuals homozygous for this polymorphism. This susceptibility factor is also relevant in the acquired forms of CJD, most strikingly in vCJD where all clinical cases studied so far have been homozygous for codon 129 methionine of the PrP gene *PRNP*. Additionally, a *PRNP* susceptibility haplotype has been identified indicating additional genetic susceptibility to sporadic CJD at or near the *PRNP* locus.

Approximately 15% of human prion diseases are associated with autosomal dominant pathogenic mutations in *PRNP*. How pathogenic mutations in *PRNP* cause prion disease is yet to be resolved, however, in most cases the mutation is thought to lead to an increased tendency of PrP^C to form a pathogenic PrP isoform. However experimentally manipulated mutations of the prion gene can lead to spontaneous neurodegeneration without the formation of detectable protease resistant PrP. These findings raise the question of whether all inherited forms of human prion disease invoke disease through the same mechanism, and in this regard it is currently unknown whether all are transmissible by inoculation.

Although the human prion diseases are transmissible diseases, acquired forms have, until recently, been confined to rare and unusual situations. The two most frequent causes of iatrogenic CJD occurring through medical procedures have arisen as a result of implantation of dura mater grafts and treatment with human growth hormone derived from the pituitary glands of human cadavers. Less frequent incidences of human prion disease have resulted from iatrogenic transmission of CJD during corneal transplantation, contaminated electroencephalographic (EEG) electrode implantation, and surgical operations using contaminated instruments or apparatus. The most well-known incidences of acquired prion disease in humans resulting from a dietary origin have been kuru that was caused by cannibalism among the Fore linguistic group of the Eastern Highlands in Papua New Guinea, and more recently the occurrence of variant CJD in the United Kingdom and some other countries that is causally related to human exposure to BSE in cattle. Incubation periods of acquired prion diseases in humans can be extremely prolonged, and it remains to be seen if a substantial epidemic of vCJD will occur within the UK and elsewhere.

PRNP codon 129 genotype has shown a pronounced effect on kuru incubation periods and susceptibility, and most elderly survivors of the kuru epidemic are heterozygotes. The clear survival advantage for codon 129 heterozygotes provides a powerful basis for selection pressure in the Fore. However, an analysis of worldwide haplotype diversity and allele frequency of coding and noncoding polymorphisms of *PRNP* suggests that balancing selection at this locus (in which there is more variation than expected because of heterozygote advantage) is much older and more geographically widespread. Evidence for balancing selection has been shown in only a few human genes. With biochemical and physical evidence of cannibalism on five continents, one explanation is that cannibalism resulted in several prion disease epidemics in human prehistory, thus imposing balancing selection on *PRNP*.

The Protein-Only Hypothesis of Prion Propagation

Despite extensive investigation, no evidence for a specific prion-associated nucleic acid has been found. Instead, a wide body of data now supports the idea that infectious prions consist principally or entirely of an abnormal isoform of PrP. PrP^{Sc} is derived from PrP^C by a post-translational mechanism and neither amino acid sequencing nor systematic study of known covalent post-translational modifications have shown any consistent differences between PrP^C and PrP^{Sc}. The protein-only hypothesis, in its current form, argues that prion propagation occurs through PrP^{Sc} acting to replicate itself with high fidelity by recruiting endogenous PrP^C and that this conversion involves only conformational change. However, the underlying molecular events during infection that lead to the conversion of PrP^C to PrP^{Sc} and how PrP^{Sc} accumulation leads to neurodegeneration remain poorly defined.

The most coherent and general model thus far proposed is that the protein, PrP, fluctuates between a dominant native state, PrP^C, and a series of minor conformations, one or a set of which can self-associate in an ordered manner to produce a stable supramolecular structure, PrP^{Sc}, composed of misfolded PrP monomers. Once a stable 'seed' structure is formed, PrP can then be recruited leading to an explosive, autocatalytic formation of PrP^{Sc}. Such a mechanism could underlie prion propagation and account for the transmitted, sporadic, and inherited etiologies of prion disease. Initiation of a pathogenic self-propagating conversion reaction, with accumulation of aggregated PrP, may be induced by exposure to a 'seed' of aggregated PrP following prion inoculation, or as a rare stochastic conformational change, or as an inevitable consequence of expression of a pathogenic PrP^C mutant which is predisposed to form misfolded PrP. It is now clear that a full understanding of prion propagation will require knowledge of both the structure of PrP^C and PrP^{Sc} and the mechanism of conversion between them.

Structure and Putative Function of PrP^C

PrP is highly conserved among mammals, has been identified in marsupials, birds, amphibians, and fish, and may be present in all vertebrates. It is expressed during early embryogenesis and is found in most tissues in the adult with the highest levels of expression in the central nervous system, in particular in association with synaptic membranes. PrP is also widely expressed in cells of the immune system. As a GPI-anchored cell surface glycoprotein, it has been speculated that PrP may have a role in cell adhesion or signaling processes, but its precise cellular function has remained obscure.

Mice lacking PrP as a result of gene knockout ($Prnp^{o/o}$ mice) show no gross phenotype; however, these mice are completely resistant to prion disease following inoculation and do not replicate prions. $Prnp^{o/o}$ mice do however show subtle abnormalities in synaptic physiology and in circadian rhythms and sleep. While the relative normality of $Prnp^{o/o}$ mice was thought to result from effective adaptive changes during development, data from $Prnp$ conditional knockout mice suggest this is not the case. These mice undergo ablation of neuronal PrP expression at 9 weeks of age and remain healthy without evidence of neurodegeneration or an overt clinical phenotype. Thus, acute loss of neuronal PrP in adulthood is tolerated and the pathophysiology of prion diseases appears to be unrelated to loss of normal PrP function in neurons.

Nuclear magnetic resonance (NMR) measurements and crystallographic determination of PrP from numerous mammalian species, including human PrP, show that they have essentially the same conformation. Following cleavage of an N-terminal signal peptide, and removal of a C-terminal peptide on addition of a GPI anchor, the mature PrP^C species consists of an N-terminal region of about 100 amino acids which is unstructured in the isolated molecule in solution and a C-terminal segment, also around 100 amino acids in length. The C-terminal domain is folded into a largely α-helical conformation (three α-helices and a short antiparallel β-sheet) and stabilized by a single disulfide bond linking helices 2 and 3. There are two asparagine-linked glycosylation sites. The N-terminal region contains a segment of five repeats of an eight-amino-acid sequence (the octapeptide-repeat region), expansion of which by insertional mutation leads to inherited prion disease. While unstructured in the isolated molecule, it seems likely that the N-terminal region of PrP may acquire coordinated structure $in\ vivo$ through coordination of either Cu^{2+} or Zn^{2+} ions.

Structural Properties of PrP^{Sc}

PrP^{Sc} is extracted from affected brains as highly aggregated, detergent insoluble material that is not amenable to high-resolution structural techniques. However, Fourier transform infrared spectroscopic methods show that PrP^{Sc}, in sharp contrast to PrP^C, has a high β-sheet content. PrP^{Sc} is covalently indistinguishable from PrP^C but can be distinguished from PrP^C by its partial resistance to proteolysis and its marked insolubility in detergents. Under conditions in which PrP^C exists as a detergent-soluble monomer and is completely degraded by the nonspecific protease, proteinase K, PrP^{Sc} exists in an aggregated form with the C-terminal two-thirds of the protein showing marked resistance to proteolytic degradation leading to the generation of N-terminally truncated fragments of di-, mono-, and nonglycosylated PrP (**Figure 1**). While there is no evidence for a specific prion-associated nucleic acid or other protein components, purified prion rods do however contain an inert polysaccharide scaffold.

Defining the precise molecular events that occur during the conversion of PrP^C to the infectious isoform of PrP is of paramount importance as this process is a prime target for therapeutic intervention. Direct $in\ vitro$ mixing experiments have been performed in an attempt to produce PrP^{Sc}. In such experiments, PrP^{Sc} is used in excess as a seed to convert PrP^C to a protease-resistant form, designated PrP^{Res}. While there are now many examples in the literature of conditions that generate PrP^{Res}, historically such reactions have not been able to demonstrate $de\ novo$ production of prion infectivity. Recently, however, a protein misfolding cyclic amplification system has demonstrated substantial amplification of PrP^{Res} and prion infectivity and may now provide the means to systematically investigate whether the generation of an infectious PrP isoform requires additional, as yet unknown, cofactors for the acquisition of infectivity.

The difficulty in performing structural studies on native PrP^{Sc} has led to attempts to produce soluble β-sheet-rich forms of recombinant PrP which may be amenable to NMR or crystallographic structure determination. Conditions have now been identified in which the PrP polypeptide can be converted between alternative folded conformations representative of PrP^C and PrP^{Sc}. At neutral or basic pH PrP adopts an α-helical fold representative of PrP^C and this conformation is locked by the presence of the native

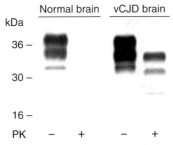

Figure 1 PrP analysis by immunoblotting. Immunoblot analysis of normal human brain and vCJD brain homogenate before and after treatment with proteinase K (PK). PrP^C in both normal and vCJD brain is completely degraded by proteinase K, whereas PrP^{Sc} present in vCJD brain shows resistance to proteolytic degradation leading to the generation of N-terminally truncated fragments of di-, mono-, and nonglycosylated PrP.

disulfide bond. Upon reduction of the disulfide bond, PrP^C rearranges to a predominantly β-sheet structure. This alternative conformation, designated β-PrP, is only populated at acidic pH with the PrP^C conformation predominating at neutral pH. Importantly, β-PrP shares overlapping properties with PrP^{Sc}, including partial resistance to proteolysis and a propensity to aggregate into fibrils. Success in producing disease in experimental animals that can be serially propagated following inoculation with PrP^{Sc}-like forms derived from recombinant PrP would not only prove the protein-only hypothesis, but would also provide an essential model by which the mechanism of prion propagation can be understood in molecular detail. In this regard, it has been recently reported that intracerebral injection of a β-sheet-rich fibrillar preparation of N-terminally truncated recombinant mouse PrP (comprising residues 89–230) into mice overexpressing PrP with the same deletion caused neurological disease after about 520 days.

Prion Disease Pathogenesis

Although the pathological consequences of prion infection occur in the central nervous system and experimental transmission of these diseases is most efficiently accomplished by intracerebral inoculation, most natural infections do not occur by these means. Indeed, administration to sites other than the central nervous system is known to be associated with much longer incubation periods, which in humans may extend to 50 years or more. Experimental evidence suggests that this latent period is associated with clinically silent prion replication in lymphoreticular tissue, whereas neuroinvasion takes place later. The M-cells in the intestinal epithelium appear to mediate prion entry from the gastrointestinal lumen into the body, and follicular dendritic cells (FDCs) are thought to be essential for prion replication and for accumulation of disease-associated PrP^{Sc} within secondary lymphoid organs. B-cell-deficient mice are resistant to intraperitoneal inoculation with prions probably because of their involvement with FDC maturation and maintenance. However, neuroinvasion is possible without FDCs, indicating that other peripheral cell types can replicate prions. The interface between FDCs and sympathetic nerves represents a critical site for the transfer of lymphoid prions into the nervous system; however, the mechanism by which this is achieved remains unknown. Distinct forms of prion disease show differences in lymphoreticular involvement that may be related to the etiology of the disease or to divergent properties of distinct prion strains. For example, the tissue distribution of PrP^{Sc} in vCJD differs strikingly from that in classical CJD, with uniform and prominent involvement of lymphoreticular tissues, with the highest amounts (up to 10% of brain concentrations) in tonsil. In contrast, in sporadic CJD, PrP^{Sc} has only been irregularly detected by immunoblotting in noncentral nervous system tissues at very much lower levels. Tonsil biopsy is used for diagnosis of vCJD and to date has shown 100% sensitivity and specificity for diagnosis of vCJD at an early clinical stage; tonsil is the tissue of choice for prospective studies investigating the prevalence of vCJD prion infection within the UK and other populations. The demonstration of extensive lymphoreticular involvement in the peripheral pathogenesis of vCJD raises concerns that iatrogenic transmission of vCJD prions through medical procedures may be a major public health issue. Prions resist many conventional sterilization procedures and surgical stainless steel-bound prions transmit disease with remarkable efficiency when implanted into mice. Disturbingly, cases of transfusion-associated vCJD prion infection have also now been reported. In contrast, there is no epidemiological evidence of transmission of classical CJD via blood transfusion or blood products.

Microscopic examination of the central nervous system of humans or animals with prion disease reveals typical characteristic histopathologic changes, consisting of neuronal vacuolation and degeneration, which gives the cerebral gray matter a microvacuolated or 'spongiform' appearance, and a reactive proliferation of astroglial cells (**Figure 2**). Demonstration of abnormal PrP immunoreactivity, or more specifically biochemical detection of PrP^{Sc} in brain material by immunoblotting techniques, is diagnostic of prion disease (**Figures 1** and **2**) and some forms of prion disease are characterized by deposition of amyloid plaques composed of insoluble aggregates of PrP. The histopathological features of vCJD are remarkably consistent and distinguish it from other human prion diseases with large numbers of PrP-positive amyloid plaques that differ in morphology from the plaques seen in kuru and GSS in that the surrounding tissue takes on a microvacuolated appearance, giving the plaques a florid appearance (**Figure 2**).

Prion Strains

A major problem for the 'protein-only' hypothesis of prion propagation has been to explain the existence of multiple isolates, or strains, of prions. Prion strains are distinguished by their biological properties: they produce distinct incubation periods and patterns of neuropathological targeting (so-called lesion profiles) in defined inbred mouse lines. As they can be serially propagated in inbred mice with the same *Prnp* genotype, they cannot be encoded by differences in PrP primary structure. Usually, distinct strains of conventional pathogen are explained by differences in their nucleic acid genome. However, in the absence of such a scrapie genome, alternative possibilities must be considered.

Support for the contention that prion strain specificity may be encoded by PrP itself was provided by study of two

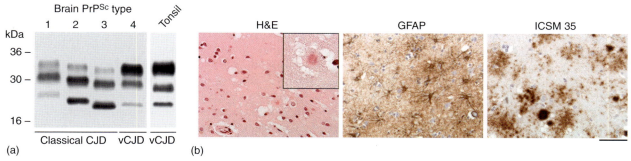

Figure 2 Characterization of disease-related prion protein in human prion disease. (a) Immunoblots of proteinase K-digested tissue homogenate with monoclonal antibody 3F4 showing PrP^{Sc} types 1–4 in human brain and type PrP^{Sc} type 4t in vCJD tonsil. Types 1–3 PrP^{Sc} are seen in the brain of classical forms of CJD (either sporadic or iatrogenic CJD), while type 4 PrP^{Sc} and type 4t PrP^{Sc} are uniquely seen vCJD brain or tonsil, respectively. (b) Brain from a patient with vCJD showing spongiform neurodegeneration following hematoxylin- and eosin staining (H&E), reactive proliferation of astroglial cells following staining with a monoclonal antibody recognizing glial-fibrillary acidic protein (GFAP), and abnormal PrP immunoreactivity following immunohistochemistry using anti-PrP monoclonal antibody ICSM 35 (ICSM 35). Scale (main panels) = 100 μm. Inset, high-power magnification of florid PrP plaques. Courtesy of Professor Sebastian Brandner.

distinct strains of transmissible mink encephalopathy prions which can be serially propagated in hamsters, designated 'hyper' and 'drowsy'. These strains can be distinguished by differing physicochemical properties of the accumulated PrP^{Sc} in the brains of affected hamsters. Following limited proteolysis, strain-specific migration patterns of PrP^{Sc} on Western blots are seen which relate to different N-terminal ends of PrP^{Sc} following protease treatment implying differing conformations of PrP^{Sc}. Distinct PrP^{Sc} conformations are now recognized to be associated with other prion strains and, similarly, different human PrP^{Sc} isoforms have been found to propagate in the brain of patients with phenotypically distinct forms of CJD.

The different fragment sizes seen on Western blots, following treatment with proteinase K, suggests that there are several different human PrP^{Sc} conformations, referred to as molecular strain types. These types can be further classified by the ratio of the three PrP bands seen after protease digestion, corresponding to N-terminally truncated cleavage products generated from di-, mono-, or nonglycosylated PrP^{Sc}. Four types of human PrP^{Sc} have now been commonly identified using molecular strain typing (**Figure 2**), although much greater heterogeneity seems likely. Efforts to produce an unified international classification and nomenclature of human PrP^{Sc} types has been complicated by the fact that the N-terminal conformation of some PrP^{Sc} subtypes seen in sporadic CJD can be altered *in vitro* via changes in metal-ion occupancy or solvent pH. Although agreement is yet to be reached on methodological differences, nomenclature and the biological importance of relatively subtle biochemical differences in PrP^{Sc}, there is strong agreement between laboratories that phenotypic diversity in human prion disease relates to the propagation of disease-related PrP isoforms with distinct physicochemical properties. Polymorphism at *PRNP* residue 129 appears to dictate the propagation of distinct PrP^{Sc} types in humans and it has now become clear that prion strain selection and the propagation of distinct PrP^{Sc} types may also be crucially influenced by other genetic loci of the host genome.

The hypothesis that alternative conformations or assembly states of PrP^{Sc} provide the molecular substrate for clinicopathological heterogeneity seen in human prion diseases (and that this relates to the existence of distinct human prion strains) has been strongly supported by transmission experiments to conventional and transgenic mice. Transgenic mice expressing only human PrP with either valine or methionine at residue 129 have shown that this polymorphism constrains both the propagation of distinct human PrP^{Sc} conformers and the occurrence of associated patterns of neuropathology. Biophysical measurements suggest that this powerful effect of residue 129 on prion strain selection is likely to be mediated via its effect on the conformation of PrP^{Sc} or its precursors or on the kinetics of their formation, as it has no measurable effect on the folding, dynamics, or stability of PrP^C. These data are consistent with a conformational selection model of prion transmission barriers and strongly support the 'protein only' hypothesis of infectivity by suggesting that prion strain variation is encoded by a combination of PrP conformation and glycosylation. These findings also provide a molecular basis for *PRNP* codon 129 as a major locus influencing both prion disease susceptibility and phenotype in humans.

The identification of strain-specific PrP^{Sc} structural properties now allows an etiology-based classification of human prion disease by typing of the infectious agent itself. Stratification of all human prion disease cases by PrP^{Sc} type will enable rapid recognition of any change in relative frequencies of particular PrP^{Sc} subtypes in relation to either BSE exposure patterns or iatrogenic sources of vCJD prions. This technique may also be applicable in determining whether BSE has been transmitted to other species thereby posing a threat to human health.

Neuronal Cell Death in Prion Disease

Although various mechanisms have been proposed to explain neuronal death in prion disease, the precise structure of the infectious agent and the cause of neuronal cell death in prion disease remains unclear. While PrP^C is absolutely required for prion propagation and neurotoxicity, knockout of PrP^C in adult brain and in embryonic models has no overt phenotypic effect, effectively excluding loss of of PrP^C function in neurons as a significant mechanism in prion neurodegeneration. Notably, however, there is also considerable evidence that argues that PrP^{Sc} and indeed prions (whether or not they are identical) may not themselves be highly neurotoxic. Consequently, it is now hypothesized that the neurotoxic prion molecule may not be PrP^{Sc} itself, but a toxic intermediate that is produced in the process of conversion of PrP^C to PrP^{Sc}, with PrP^{Sc}, present as highly aggregated material, being a relatively inert end product. The steady-state level of such a toxic monomeric or oligomeric PrP intermediate, designated PrP^L (for lethal), could determine the rate of neurodegeneration. Subclinical prion infection states may generate the toxic intermediate at extremely low levels below the threshold required for neurotoxicity. Recently, direct support for this hypothesis has been demonstrated by depleting endogenous neuronal PrP^C in mice with established neuroinvasive prion infection. This depletion of PrP^C reverses early spongiform change and prevents neuronal loss and progression to clinical prion disease despite the accumulation of extraneuronal PrP^{Sc} to levels seen in terminally ill wild-type mice. These data establish that propagation of non-neuronal PrP^{Sc} is not pathogenic, but arresting the continued conversion of PrP^C to PrP^{Sc} within neurons during scrapie infection prevents prion neurotoxicity. Importantly, this model also validates PrP^C as a key therapeutic target in prion disease.

Future Perspective

The novel pathogenic mechanisms involved in prion propagation are likely to be of wider significance and may be relevant to other neurological and non-neurological illnesses. Indeed, advances in understanding prion neurodegeneration are already casting considerable light on related mechanisms in other, commoner, neurodegenerative diseases such as Alzheimer's, Parkinson's, and Huntington's disease. While the protein-only hypothesis of prion propagation is supported by compelling experimental data and now appears also able to encompass the phenomenon of prion strain diversity, the goal of systematically producing prions *in vitro* remains. Success in producing disease in experimental animals that can be serially propagated following inoculation with PrP^{Sc}-like forms derived from recombinant PrP would provide an essential model by which the mechanism of prion propagation can be understood in molecular detail.

See also: Prions of Yeast and Fungi.

Further Reading

Caughey B and Baron GS (2006) Prions and their partners in crime. *Nature* 443: 803–810.

Collinge J (1999) Variant Creutzfeldt–Jakob disease. *Lancet* 354: 317–323.

Collinge J (2001) Prion diseases of humans and animals: Their causes and molecular basis. *Annual Review of Neuroscience* 24: 519–550.

Collinge J (2005) Molecular neurology of prion disease. *Journal of Neurology, Neurosurgery, and Psychiatry* 76: 906–919.

Collinge J, Sidle KCL, Meads J, Ironside J, and Hill AF (1996) Molecular analysis of prion strain variation and the aetiology of 'new variant' CJD. *Nature* 383: 685–690.

Griffith JS (1967) Self replication and scrapie. *Nature* 215: 1043–1044.

Hill AF and Collinge J (2003) Subclinical prion infection. *Trends in Microbiology* 11: 578–584.

Mabbott NA and MacPherson GG (2006) Prions and their lethal journey to the brain. *Nature Reviews Microbiology* 4: 201–211.

Mallucci G and Collinge J (2005) Rational targeting for prion therapeutics. *Nature Reviews Neuroscience* 6: 23–34.

Prusiner SB (1982) Novel proteinaceous infectious particles cause scrapie. *Science* 216: 136–144.

Prusiner SB (1998) Prions. *Proceedings of the National Academy of Sciences, USA* 95: 13363–13383.

Soto C (2004) Diagnosing prion diseases: Needs, challenges and hopes. *Nature Reviews Microbiology* 2: 809–819.

Weissmann C (2004) The state of the prion. *Nature Reviews Microbiology* 2: 861–871.

Wuthrich K and Riek R (2001) Three-dimensional structures of prion proteins. *Advances in Protein Chemistry* 57: 55–82.

Prions of Yeast and Fungi

R B Wickner, H Edskes, T Nakayashiki, F Shewmaker, L McCann, A Engel, and D Kryndushkin, National Institutes of Health, Bethesda, MD, USA

Published by Elsevier Ltd.

Glossary

Nonchromosomal gene A gene that segregates 4+:0 in meiosis and can be transferred by cytoplasmic mixing, in contrast to chromosomal genes that segregate 2+:2– in meiosis, and are not transferred by cytoplasmic mixing.

> **Nonsense suppressor tRNA** A mutant transfer RNA that recognizes a translational stop codon and inserts an amino acid thus allowing the peptide chain to continue.

Introduction and History

The word prion, meaning 'infectious protein' without need for a nucleic acid, was coined to explain the properties of the agent producing the mammalian transmissible spongiform encephalopathies (TSEs), although 25 years later there remains some debate if the TSEs are indeed caused by prions. The yeast and fungal prions were identified by their unique genetic properties which were unexpected for any nucleic acid replicon, but specifically predicted for an infectious protein.

[PSI] was described by Brian Cox in 1965 as a non-chromosomal genetic element that increased the efficiency of a weak nonsense suppressor transfer RNA (tRNA). [URE3] was described by Francois Lacroute as a nonchromosomal gene that relieved nitrogen catabolite repression, allowing expression of genes needed for utilizing poor nitrogen sources even when a good nitrogen source was available. The [Het-s] prion was described in 1952 by Rizet as a nonchromosomal gene needed for heterokaryon incompatibility in *Podospora anserina*. Each of these elements was later found to be a prion. The [PIN] prion was discovered in 1997 by Derkatch and Liebman in their studies of *de novo* generation of the [PSI] prion.

Genetic Signature of a Prion

Viruses of yeast and fungi generally do not exit one cell and enter another, but spread by cell–cell fusion, as in mating or heterokaryon formation. Infectious proteins (prions) should likewise be nonchromosomal genetic elements. To distinguish prions from nucleic acids, three genetic criteria were proposed (**Figure 1**): (1) If a prion can be cured, it can reappear in the cured strain at some low frequency. (2) Overproduction of the protein capable of being a prion should increase the frequency of the prion arising *de novo*. (3) If the prion produces a phenotype by the simple inactivation of the protein, then this phenotype should resemble the phenotype of mutation of the gene encoding the protein, which gene must be needed for prion propagation.

All three criteria were satisfied by [PSI] and [URE3], strongly indicating, perhaps proving, that they were prions. The [Het-s] prion of *P. anserina* and the [PIN] prion of *Saccharomyces cerevisiae* were likewise proved by application of the same genetic criteria, but because their prion form

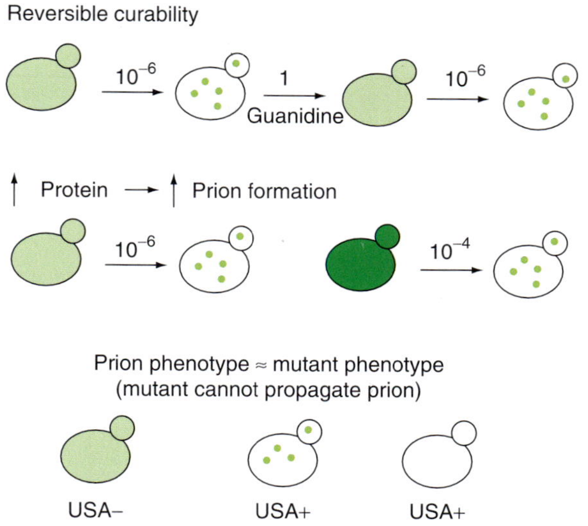

Figure 1 Genetic signature of a prion. Among nonchromosomal genetic elements, those with the three properties shown here must not be nucleic acid replicons and are almost certainly prions. However, only prions for which the prion form of the protein is inactive (such as [URE3] or [PSI]) will have property (3). The [Het-s] and [PIN$^+$] prions are active forms of the HET-s protein and Rnq1 protein, respectively and have properties (1) and (2), but not (3).

produces the phenotype, rather than the absence of the normal form, they do not satisfy criterion (3).

Self-Propagating Amyloid as the Basis for Most Yeast Prions

The finding that Sup35p, Ure2p, and HET-s were protease resistant and aggregated in prion-containing cells, that these proteins (and particularly their prion domains (**Figure 2**)) would form amyloid *in vitro* indicated that a self-propagating amyloid (**Figure 3**) was the basis of these prions. This was confirmed by the finding that the corresponding prions were transmitted by introduction of amyloid formed *in vitro* from the recombinant proteins (see below). However, in some cases, self-modifying enzymes have the potential to become prions (see below).

Chaperones and Prions

Chaperones of the Hsp40, Hsp70, and Hsp104 groups, as well as Hsp90 co-chaperones, have been found to be clearly involved in prion propagation. Millimolar concentrations of guanidine are known to cure each of the amyloid-based prions, and the mechanism of action of guanidine curing has been shown to be specific inhibition of Hsp104. It is believed that at least one function of these chaperones is to break large amyloid filaments into smaller ones which can then be distributed at cell division

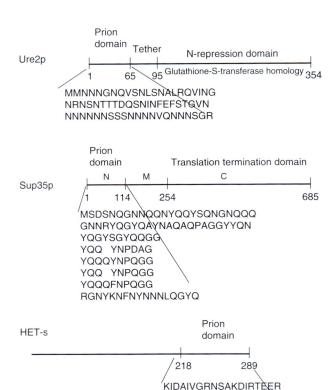

Figure 2 Prion domains. The prion domains of Ure2p, Sup35p, and HET-s are largely unstructured in the native form and in β-sheet in the prion form.

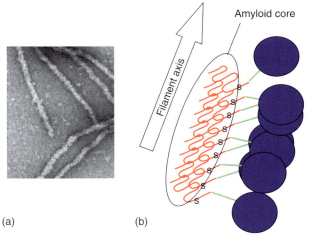

Figure 3 Amyloid. (a) Electron micrograph of amyloid of Ure2p. (b) Model of the structure of Ure2p amyloid. Red, prion domain; green, tether; blue, glutathione-S-transferase-like nitrogen regulation domain. Amyloid of recombinant HET-s, Sup35p, or Ure2p is infectious for yeast transmitting [Het-s], [PSI+], or [URE3], respectively. Amyloid of Ure2p is proposed to have the parallel in-register β-sheet structure shown based on the solid-state NMR studies of a fragment of the Ure2p prion domain, and the fact that the prion domain can be shuffled and still be a prion (see text).

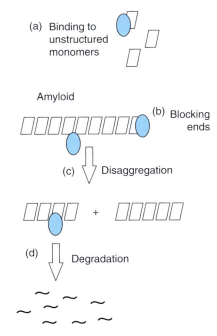

Figure 4 Chaperones and prions. Possible roles of chaperones in prion propagation are diagrammed.

to both daughter cells and insure the inheritance of the prion. Overexpression of some chaperones cure yeast prions, perhaps by solubilizing the filaments or perhaps by binding to the ends of filaments and preventing their elongation with new monomers. There is considerable specificity in which chaperone is needed for which prion and which chaperone can cure which prion by overexpression. The detailed mechanisms of chaperone action on prions (and amyloids in general) remain to be elucidated, but it is clear that they play an important role in these phenomena (**Figure 4**).

The Species Barrier and Prion Variants

Scrapie, a prion disease of sheep, only infects goats after a long incubation period, and subsequent goat-to-goat transmission has a much shorter incubation period. This is called the species barrier, and this barrier can in some cases be absolute, as appears to be the case between sheep and humans. The same phenomenon has now been documented in yeast, where [PSI] prions formed by the Sup35p of one species will not be transmitted to the Sup35p of another species, even though the other species' Sup35p can itself form a prion.

A single protein sequence can form several prions that are distinguishable, in yeast, by the intensity of their phenotype and the stability of their propagation. These are called 'prion variants' and are believed to reflect different amyloid structures. Paradoxically, a similar phenomenon, long documented in the mammalian TSEs, was used as an argument against the protein-only model. Elucidation of the structure of different prion variants, and the mechanism of their faithful propagation, in some cases across species barriers, remains an important problem.

The bovine spongiform encephalopathy epidemic in the UK has brought the species barrier and prion variant phenomena together. It is clear that the 'height' of the species barrier is a function of the prion variant. Collinge views the species barrier as a reflection of the degree of overlap of possible variants of the prion proteins of the two species. If they have few common amyloid conformers (prion variants) then the barrier will be high. If each sequence can adopt nearly all of the amyloid conformers of the other, there will be little species barrier.

Formation of Prions by Sup35p and Ure2p Homologs

The C-terminal domain of Sup35 is conserved in eukaryotes with a human homolog capable of complementing the S. cerevisiae protein. All Sup35 proteins have N-terminal extensions, however, with limited or no sequence homology between species. N-terminal sequences from some species related to S. cerevisiae are capable of forming a [PSI+]-like prion. Ure2p is limited to the ascomycete yeasts. As with Sup35p, Ure2 proteins have a conserved C-terminal domain and a variable N-terminal domain that in general is rich in Asn and/or Glu residues. Ure2p homologs of some Saccharomyces yeasts can propagate [URE3] in S. cerevisiae.

Prion Generation, and [PIN]: A Prion That Gives Rise to Prions

One of the lines of evidence that showed [PSI] was a prion of Sup35p was that overproduction of Sup35p increased the frequency with which [PSI] arises *de novo*. However, it was found that in some strains, overproduction of Sup35p did not yield detectable emergence of [PSI]-carrying clones. Another nonchromosomal genetic element, named [PIN] for [PSI]-inducibility, was found necessary. [PIN] is a self-propagating amyloid form of the Rnq1 (rich in Asn (N) and Gln (Q)) protein, and it promotes *de novo* generation of [URE3] as well as [PSI].

Transfection with Amyloid of Recombinant Proteins

Amyloid filaments formed *in vitro* from recombinant HET-s protein, Sup35p, or Ure2p can efficiently transform cells to the corresponding [Het-s], [PSI], or [URE3] prion. In some cases it was shown that the soluble form or nonspecific aggregates of the protein were ineffective. This argues that the respective amyloids are not by-products or a dead-end stage of these prions, but are themselves the infectious material. All infectious Ure2p amyloids are larger than about 40-mer size. Amyloid formed *in vitro* is capable, for at least [PSI] and [URE3], of transmitting any of several prion variants. This implies that the amyloids can have any of several structures, a fact demonstrated by solid-state nuclear magnetic resonance (NMR) for amyloid of the Alzheimer's disease peptide, Aβ.

Shuffling Prion Domains and Amyloid Structure

The prion domains (**Figure 2**) of Ure2p and Sup35p are quite rich in Asn and Gln residues, and nearly the entire sequence of Rnq1p, the basis of the [PIN] prion, is rich in these amino acids. However, many Q/N-rich proteins are not capable of being prions. Thus, it was assumed that specific sequences in the known prion domains were important for prion formation. The Sup35 prion domain has octapeptide repeats much like those in PrP, and deletion or duplication of these showed substantial effects on prion generation. In addition, single amino acid changes in the prion domain of Sup35p blocked prion propagation.

To critically test whether the Ure2p prion domain had sequences essential for prion development, the entire Q/N-rich region (residues 1–89) was randomly shuffled (without changing the amino acid content) and each of five shuffled sequences were inserted into the chromosome in place of the normal prion domain. Surprisingly, each of these five shuffled sequences could support prion generation and propagation, although one was rather unstable. Each protein with the shuffled sequence could also form amyloid (**Figure 3**) *in vitro*. This showed that it was the amino acid content of the Ure2p prion domain that determines prion formation, and that sequence plays only a minor role.

Similarly, five shuffled versions of the Sup35p prion domain were each inserted in place of the normal sequence. Again, all five shuffled versions allowed formation and propagation of a [PSI]-like prion. It is likely that the effects of deletion or duplication of the octapeptide repeats observed on prion formation or propagation were due to changes in the length or composition of the prion domain. It appears that repeats *per se* are not important for prion generation or propagation.

Shuffleable Prion Domains Suggests Parallel In-Register β-Sheet Structure

Amyloids are β-sheet structures, but there are at least three kinds of β-sheets. Antiparallel β-sheets have adjacent peptide chains oriented in opposite directions: N→C next to C→N. This results in pairing of largely nonidentical residues. A β-helix also involves pairing of largely nonidentical residues, although they are within the same peptide chain. A parallel β-sheet pairs identical residues if it is in-register, but in principle, one could have an out-of-register parallel β-sheet, in which case nonidentical residues would be bonded to each other.

Prion propagation (and amyloid propagation in general) is a very sequence-specific process. For example, a single amino acid change (at residue 138) can block propagation of scrapie in tissue culture. Humans are polymorphic at PrP residue 129 with roughly equal numbers of alleles encoding M and V. Either M/M or V/V individuals can get Creutzfeldt–Jakob disease (a human prion disease), but M/V heterozygotes cannot. Similarly, a single amino acid change in the prion domain of Sup35p can block propagation of [PSI] from the normal sequence, but the mutant Sup35p can itself become a prion nonetheless. Thus, if a prion amyloid has an antiparallel, parallel out-of-register, or β-helix structure, there must be some form of complementarity between bonded residues. Shuffling such a sequence would be expected to destroy the complementarity. In contrast, shuffling the sequence of a parallel in-register β-sheet would still leave identical residues paired. This suggests that prion domains that can be shuffled without destroying their prion-forming ability are forming parallel in-register β-sheets. Indeed Ure2p$^{10-39}$, a fragment of the prion domain, forms amyloid with a parallel in-register β-sheet structure, as does the Sup35 prion domain.

Biological Roles of Prions: A Help or a Hindrance?

In an attempt to discern whether yeast prions are an advantage or disadvantage to their host organism, cell growth of isogenic [PSI^+] and [psi^-] strains have been carried out under a variety of conditions. To what extent the various growth conditions tested represent the normal yeast habitat seems unknowable, although [psi^-] was an advantage under far more conditions than was [PSI^+].

An alternative approach was to compare the frequency with which [PSI^+] or [URE3] was found in wild strains to those of several 'selfish' RNA and DNA viruses and replicons known in *S. cerevisiae*. In any organism, an infectious element (such as a virus) may be widely distributed in spite of it causing disease in its host because the infection process overcomes and outraces the loss of infected individuals from negative selection. Certainly an infectious element, which is an advantage to its host, will quickly become widespread as selection and infection operate in the same direction. In fact, while the mildly deleterious RNA and DNA viruses and plasmids of yeast are easily found in wild strains neither [URE3] nor [PSI^+] was found in any of the 70 wild strains examined. This indicates that [URE3] and [PSI^+] produce disease in their hosts, and a rather more severe disease than the mild nucleic acid replicons.

The [Het-s] prion of *Podospora* appears to carry out the normal fungal function of heterokaryon incompatibility, thought to be a protection against the sometimes debilitating fungal viruses. Indeed, as one would expect for a prion with a function for the cell, 80% of wild *Podospora* isolates carry [Het-s], confirming its beneficial effects.

Unlike [PSI^+] and [URE3], [PIN^+] is found in wild strains at a frequency similar to that of the parasitic RNA and DNA viruses and plasmids. This suggests that [PIN^+] is at least not as severe a pathogen as are [URE3] and [PSI^+].

Enzyme as Prion

While most of the known prions involve amyloids, the word prion (infectious protein) is more general, requiring only that transmission be by protein alone. If an enzyme is made as an inactive precursor that needs the active form of the same enzyme for its activation, then such a protein can be a prion. The vacuolar protease B (Prb1p) of *S. cerevisiae* can be such a prion in a mutant lacking protease A, which normally activates its precursor. Cells initially carrying only the inactive precursor remain so unless the active enzyme is introduced. Once a cell has some active enzyme, the autoactivation process can continue indefinitely. It is likely that other examples of this type of phenomenon will be found among the many protein kinases, methylases, acetylases, and other modifying enzymes that are known. Indeed, a protein kinase of *P. anserina* appears to be able to become a prion in this manner, producing a nonchromosomal genetic element called 'crippled growth'.

The advent of yeast prions has propelled the prion field forward, and is giving us insight into the broader field of amyloids and how they interact with cellular components. There is already evidence for a number of new prions, mostly in yeasts and fungi, in part because they are so well suited to genetic studies, and in part because their frequent natural mating or heterokaryon formation (in fungi) results in complete mixing and exchange of cellular proteins.

See also: Prions of Vertebrates.

Further Reading

Aigle M and Lacroute F (1975) Genetical aspects of [URE3], a non-Mendelian, cytoplasmically inherited mutation in yeast. *Molecular and General Genetics* 136: 327–335.

Chan JCC, Oyler NA, Yau W-M, and Tycko R (2005) Parallel β-sheets and polar zippers in amyloid fibrils formed by residues 10–39 of the yeast prion protein Ure2p. *Biochemistry* 44: 10669–10680.

Chernoff YO, Lindquist SL, Ono B-I, Inge-Vechtomov SG, and Liebman SW (1995) Role of the chaperone protein Hsp104 in propagation of the yeast prion-like factor [psi$^+$]. *Science* 268: 880–884.

Coustou V, Deleu C, Saupe S, and Begueret J (1997) The protein product of the *het-s* heterokaryon incompatibility gene of the fungus *Podospora anserina* behaves as a prion analog. *Proceedings of the National Academy of Sciences, USA* 94: 9773–9778.
Cox BS (1965) PSI, a cytoplasmic suppressor of super-suppressor in yeast. *Heredity* 20: 505–521.
Derkatch IL, Bradley ME, Hong JY, and Liebman SW (2001) Prions affect the appearance of other prions: The story of [*PIN*]. *Cell* 106: 171–182.
Jung G, Jones G, and Masison DC (2002) Amino acid residue 184 of yeast Hsp104 chaperone is critical for prion-curing by guanidine, prion propagation, and thermotolerance. *Proceedings of the National Academy of Sciences, USA* 99: 9936–9941.
King CY and Diaz-Avalos R (2004) Protein-only transmission of three yeast prion strains. *Nature* 428: 319–323.
Maddelein ML, Dos Reis S, Duvezin-Caubet S, Coulary-Salin B, and Saupe SJ (2002) Amyloid aggregates of the HET-s prion protein are infectious. *Proceedings of the National Academy of Sciences, USA* 99: 7402–7407.
Masison DC and Wickner RB (1995) Prion-inducing domain of yeast Ure2p and protease resistance of Ure2p in prion-containing cells. *Science* 270: 93–95.
Paushkin SV, Kushnirov VV, Smirnov VN, and Ter-Avanesyan MD (1997) In vitro propagation of the prion-like state of yeast Sup35 protein. *Science* 277: 381–383.
Ross ED, Edskes HK, Terry MJ, and Wickner RB (2005) Primary sequence independence for prion formation. *Proceedings of the National Academy of Sciences, USA* 102: 12825–12830.
Tanaka M, Chien P, Naber N, Cooke R, and Weissman JS (2004) Conformational variations in an infectious protein determine prion strain differences. *Nature* 428: 323–328.
Wickner RB (1994) [URE3] as an altered *URE2* protein: Evidence for a prion analog in *S. cerevisiae*. *Science* 264: 566–569.
Wickner RB, Edskes HK, Ross ED, *et al.* (2004) Prion genetics: New rules for a new kind of gene. *Annual Review of Genetics* 38: 681–707.

Pseudorabies Virus

T C Mettenleiter, Friedrich-Loeffler-Institut, Greifswald-Insel Riems, Germany

© 2008 Elsevier Ltd. All rights reserved.

History

Although its taxonomic species name *Suid herpesvirus 1* testifies that the natural hosts of pseudorabies virus (PrV) are pigs, its symptoms were first described in cattle, and the virus was isolated for the first time from cattle, dogs, and cats. This is due to the fact that PrV infection in swine, particularly in older animals, may produce only innocuous respiratory symptoms or may go unnoticed altogether. However, in other susceptible species productive infection is invariably fatal and characterized by severe central nervous symptoms, a feature that prompted its designation as pseudorabies owing to the rabies-like clinical picture. A typical symptom in these species is extensive pruritus, resulting in the name 'mad itch' to describe the disease in cattle in the USA during the first half of the nineteenth century.

In 1902, the Hungarian physician Aládar Aujeszky reported isolation of the infectious agent from diseased animals (an ox, a dog, and a cat) and differentiated it from rabies. It could be passaged in rabbits, reproducing the typical symptoms. Guinea pigs and mice were also found to be susceptible, whereas chicken and doves were resistant. Thus, the illness has become widely known as Aujeszky's disease (AD). It was not until 1931 that Richard Shope established the identity of the 'mad itch' agent with an infectious agent widely present in domestic pig holdings in the USA. In Germany, Erich Traub was the first to cultivate PrV *in vitro* in organ explants in 1933. One year later, Albert Sabin published his findings of a serological relationship between PrV and herpes simplex virus (HSV), resulting in the inclusion of PrV in the herpesvirus group.

Taxonomy

PrV takes the formal species name *Suid herpesvirus 1* and belongs to the subfamily *Alphaherpesvirinae* of the family *Herpesviridae*. Originally on the basis of serological studies, and later confirmed by molecular biological analyses and comparison of the deduced amino acid sequences of homologous proteins, PrV was shown to be most closely related to bovine herpesvirus 1 (BHV-1) and equine herpesvirus 1 (EHV-1), and also to varicella-zoster virus (VZV). This prompted its assignment to the genus *Varicellovirus* within the *Alphaherpesvirinae*. Alphaherpesviruses are characterized by rapid lytic replication, a pronounced neurotropism with establishment of latency in sensory ganglia, and a broad host range. All these features apply to PrV.

The Virus Particle

PrV particles have a diameter of approximately 180 nm and exhibit the typical herpesvirus morphology. At the center is the genomic DNA enclosed in an icosahedral capsid. The capsid is surrounded by a structure designated the tegument, which is equivalent to the matrix of RNA viruses but significantly more complex.

The envelope, which contains virally encoded glycosylated and nonglycosylated proteins anchored in the lipid bilayer, encloses the capsid and tegument (**Figure 1**).

The Genome

The PrV genome consists of a linear, double-stranded DNA molecule. A complete sequence has recently been assembled from several partial sequences from six different PrV strains. It comprises 143 461 bp with more than 70 protein-coding regions, all of which exhibit homology to genes in related alphaherpesviruses (**Figure 2**). The genome contains a long (U_L) and short (U_S) unique region with the latter bracketed by inverted repeats (IR_S and TR_S), resulting in two isomeric forms of the genome in which U_S is inverted relative to U_L. This arrangement has been designated the class D herpesvirus genome structure. So far, three functional origins of DNA replication have been mapped, two in the inverted repeats and one in the middle of U_L. A fourth candidate located at the left genome end may not be functional. Compared to the genomes of other alphaherpesviruses, which are generally collinear in their gene arrangement, the PrV genome contains an inversion of approximately 40 kbp that encompasses genes homologous to the UL27 to UL44 genes of HSV (**Figure 2**). A similar inversion is also present in the genome of a distantly related avian alphaherpesvirus, infectious laryngotracheitis virus, but its biological significance is unknown.

A list of identified PrV genes and the functions of the encoded proteins is shown in **Table 1**. Wherever possible, PrV genes have been named after their homologs in HSV. However, the UL3.5 gene of PrV, which has homologs in other alphaherpesviruses such as VZV, BHV-1, and EHV-1, is absent from HSV, as are ORF1 and ORF1.2. PrV does not specify homologs of the UL56, UL45, US5, US10, US11, and US12 genes of HSV. Approximately half of the total number of PrV genes are considered 'nonessential', a status indicating that they are individually dispensable for viral replication, at least in cell culture. It is estimated that about half of the total number of viral proteins are located in the virion.

The Capsid

The icosahedral PrV capsid is composed of 162 capsomers. By analogy to the well-analyzed HSV-1 capsid, the capsomers consist of a total of 955 copies of the major capsid protein, the product of the UL19 gene (see below). Homologs of the HSV UL18 and UL38 proteins, which form triplexes connecting and stabilizing the capsomers, as well as the UL35 protein, which is located at the tips of the hexons, have been identified from the genome sequence. Also present are homologs of the UL6 portal protein which forms a dodecameric channel at one vertex for package and release of the viral genome. Several other proteins have been shown to be intimately associated with the capsid, such as the UL17 and UL25 gene products. Whereas the PrV UL17 protein may be present within the capsid shell, the UL25 gene product resides on the outside.

The Tegument

The herpesvirus tegument is a complex structure which, in the case of PrV, contains in excess of 15 viral proteins. It has become clear that tegument formation is an important step in the morphogenesis of the virion, and requires a network of partially redundant protein–protein interactions. It can be divided into capsid-proximal and envelope-proximal parts. The capsid-proximal tegument is composed of the UL36 gene product, the largest protein in PrV (3084 amino acid residues), which physically interacts with the UL37 gene product. From cryoelectron image reconstructions of herpesvirus particles, the UL36 gene product is thought to contact the capsid. The UL46, UL47, UL48, and UL49 gene products are easily stripped from the capsid together with the envelope, consistent

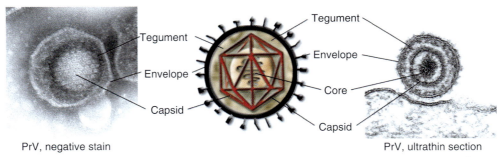

Figure 1 The PrV virion. A schematic diagram of the PrV virion is presented between an electron micrograph of a negatively stained PrV virion (left) and a thin-sectioned virus particle (right). The locations of the virion subcomponents (core, capsid, tegument, and envelope) are indicated. Spikes at the envelope represent viral glycoproteins.

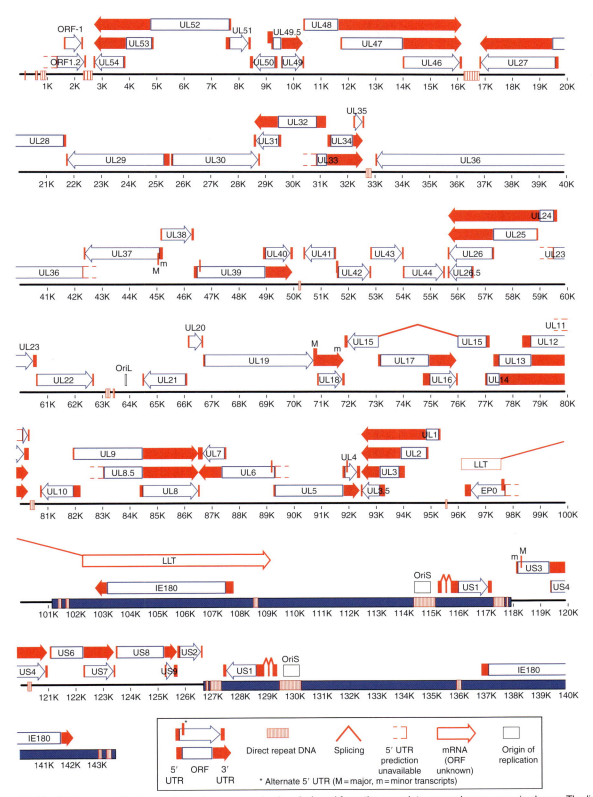

Figure 2 The PrV genome. The transcript and gene organization deduced from the complete genomic sequence is shown. The linear form of the PrV genome comprises the unique long sequence (U_L), the internal inverted repeat (IR_S), the unique short sequence (U_S), and the terminal inverted repeat (TR_S). The predicted locations of protein-coding regions (see **Table 1**), 5′ and 3′ nontranslated regions, DNA repeats, splice sites, and the origins of DNA replication are shown. Reproduced from Klupp BG, Hengartner C, Mettenleiter TC, and Enquist LW (2004) Complete, annotated sequence of the pseudorabies virus genome. *Journal of Virology* 78: 424–440, with permission from the American Society for Microbiology.

Table 1 PrV ORFs

Protein	ORF location[a]	Length (aa)	MW (kDa)	Alias	Function/property[b]	Virion subunit[c]
ORF1.2	1252–2259	335	35.3		Unknown	V (?)
ORF1	1636–2259	207	21.8		Unknown	V (?)
UL54	3815–2730r	361	40.4	ICP27	Gene regulation; early protein	NS
UL53	4833–3895r	312	33.8	gK	Viral egress; glycoprotein K; type III membrane protein	V (E)
UL52	7676–4788r	962	103.3		DNA replication; primase subunit of UL5/UL8/UL52 complex	NS
UL51	7663–8373	236	25.0		Tegument protein	V (T)
UL50	9333–8527r	268	28.6	dUTPase	dUTPase	NS
UL49.5	9257–9553	98	10.1	gN	Glycoprotein N; type I membrane protein; complexed with gM	V (E)
UL49	9591–10340	249	25.9	VP22	Interacts with C-terminal domains of gE and gM; tegument protein	V (T)
UL48	10404–11645	413	45.1	VP16/αTIF	Gene regulation (transactivator); egress (secondary envelopment); tegument protein	V (T)
UL47	11746–13998	750	80.4	VP13/14	Viral egress (secondary envelopment); tegument protein	V (T)
UL46	14017–16098	693	75.5	VP11/12	Unknown; tegument protein	V (T)
UL27	19595–16854r	913	100.2	gB	Viral entry (fusion); cell–cell spread; glycoprotein B; type I membrane protein	V (E)
UL28	21640–19466r	724	78.9	ICP18.5	DNA cleavage/encapsidation (terminase); associated with UL15, UL33, and UL6	pC
UL29	25315–21788r	1175	125.3	ICP8	DNA replication/recombination; binds single-stranded DNA	NS
UL30	25606–28752	1048	115.3		DNA replication; DNA polymerase subunit of UL30/UL42 complex	NS
UL31	29488–28673r	271	30.4		Viral egress (nuclear egress); primary virion tegument protein; interacts with UL34	pV (T)
UL32	30893–29481r	470	51.6		DNA packaging; efficient localization of capsids to replication compartments	pC
UL33	30892–31239	115	12.7		DNA cleavage/encapsidation; associated with UL28 and UL15	NS
UL34	31398–32186	262	28.1		Viral egress (nuclear egress); primary virion envelope protein; tail-anchored type II nuclear membrane protein; interacts with UL31	pV (E)
UL35	32241–32552	103	11.5	VP26	Capsid protein	V (C)
UL36	42314–33060r	3084	324.4	VP1/2	Large tegument protein; interacts with UL37 and UL19	V (T)
UL37	45111–42352r	919	98.2		Tegument protein; interacts with UL36	V (T)
UL38	45168–46274	368	40.0	VP19C	Capsid protein; forms triplexes together with UL18	V (C)
UL39	46470–48977	835	91.1	RR1	Nucleotide synthesis; large subunit of ribonucleotide reductase	V (?)
UL40	48987–49898	303	34.4	RR2	Nucleotide synthesis; small subunit of ribonucleotide reductase	NS
UL41	51498–50401r	365	40.1	VHS	Gene regulation (inhibitor of gene expression); virion host cell shut off factor	V (T)
UL42	51628–52782	384	40.3		DNA replication; polymerase accessory subunit of UL30/UL42 complex	NS
UL43	52842–53963	373	38.1		Unknown; type III membrane protein	V (E)
UL44	54029–55468	479	51.2	gC	Viral entry (virion attachment); glycoprotein C; type I membrane protein; binds to heparan sulfate	V (E)
UL26.5	56535–55699r	278	28.2	VP22a	Scaffold protein; substrate for UL26; required for capsid formation and maturation	pC
UL26	57273–55699r	524	54.6	VP24	Scaffold protein; proteinase; required for capsid formation and maturation	pC
UL25	58911–57307r	534	57.4		Capsid associated protein; required for DNA packaging and nuclear egress	V (C)
UL24	59519–59004r	171	19.1		Unknown; possible endonuclease	V (?)
UL23	59512–60474	320	35.0		Nucleotide synthesis; thymidine kinase	NS
UL22	60610–62670	686	71.9	gH	Viral entry (fusion); cell–cell spread; glycoprotein H; type I membrane protein; complexed with gL	V (E)
UL21	66065–64488r	525	55.2		Capsid associated protein; complexed with UL16	V (?)
UL20	66172–66657	161	16.7		Viral egress; type III membrane protein	V (?)
UL19	66744–70736	1330	146.0	VP5	Major capsid protein; forms hexons und pentons	V (C)
UL18	70896–71783	295	31.6	VP23	Capsid protein; forms triplexes together with UL38	V (C)
UL15 (Ex2)	73115–71979r	735	79.1		DNA cleavage/encapsidation; terminase subunit; interacts with UL33, UL28, and UL6	pC

ORF	Coordinates	Length (aa)	Other name	Function/property[b]	Localization[c]
UL15 (Ex1)	77065–75995r			DNA cleavage/encapsidation	V (C)
UL17	73166–74959	597		Tegument protein; complex with UL11 and UL21	V (T)
UL16	74986–75972	328		Unknown	?
UL14	77064–77543	159		Protein-serine/threonine kinase	V (T)
UL13	77513–78709	398		DNA recombination; alkaline exonuclease	?
UL12	78675–80126	483		Viral egress (secondary envelopment); membrane-associated tegument protein complex with UL16	V (T)
UL11	80084–80275	63		Viral egress (secondary envelopment); glycoprotein M; type III membrane protein; C-terminus interacts with UL49; inhibits membrane fusion in transient assays; complexed with gN	V (E)
UL10	81935–80754r	393	gM		
UL9	81934–84465	843	OBP	Sequence specific ori-binding protein	NS
UL8.5	83053–84465	470	OPBC	C-terminal domain of UL9	?
UL8	84462–86513	683		DNA replication; part of UL5/UL8/UL52 helicase/primase complex	NS
UL7	87479–86679r	266		Tegument protein; virion formation and egress	V (T)
UL6	89301–87370r	643		Capsid protein; portal protein; docking site for terminase	V (C)
UL5	89300–91804	834		DNA replication; part of UL5/UL8/UL52 helicase/primase complex; helicase motif	NS
UL4	91863–92300	145		Nuclear protein	NS
UL3.5	93150–92476r	224		Viral egress (secondary envelopment); membrane-associated protein	V (T)
UL3	93860–93147r	237		Nuclear protein	NS
UL2	94866–93916r	316	UNG	Uracil-DNA glycosylase	NS
UL1	95314–94844r	156	gL	Viral entry; cell-cell spread; glycoprotein L; membrane anchored via complex with gH	V (E)
EP0	97713–96481r	410	ICP0	Gene regulation (transactivator of viral and cellular genes); early protein	?
IE180 (IRS)	107511–103171r	1446	ICP4	Gene regulation; immediate-early protein	?
IE180 (TRS)	137091–141431				
US1 (IRS)	115995–117089	364	RSp40/ICP22	Gene regulation; immediate-early protein	?
US1 (TRS)	128607–127513r				
US3 (minor)	118170–119336	388	PK	Minor form of protein kinase (53 kDa mobility)	?
US3 (major)	118332–119336	334	PK	Viral egress (nuclear egress); major form of protein kinase (41 kDa mobility)	V (T)
US4	119396–120892	498	gG	Glycoprotein G (secreted)	secreted
US6	121075–122277	400	gD	Viral entry (cellular receptor binding protein); glycoprotein D; type I membrane protein; pC: present in intranuclear capsid precursor forms but not found in mature virion	V (E)
US7	122298–123398	366	gI	Cell-cell spread; glycoprotein I; type I membrane protein; complexed with gE	V (E)
US8	123502–125235	577	gE	Cell-cell spread; glycoprotein E; type I membrane protein; complexed with gI; C-terminus interacts with UL49	V (E)
US9	125269–125589	106	11K	Protein sorting in axons; type II tail-anchored membrane protein	V (E)
US2	125811–126581	256	28K	Tegument protein; prenylated	V (T)

[a]Numbering starts at +1 on the UL end of the genome. r indicates ORF encoded on reverse strand.
[b]Function/property as demonstrated for the PrV and/or HSV-1 homolog.
[c]V (C): virion capsid component; V (T): virion tegument component; V (E): virion envelope component; V (?): virion component of unknown subviral localization; pV: primary enveloped virion precursor component (not found in mature virion); NS: nonstructural protein; pC: present in intranuclear capsid precursor forms but not found in mature virion; ?: unknown.
Reproduced from Klupp BG, Hengartner C, Mettenleiter TC, and Enquist LW (2004) Complete, annotated sequence of the pseudorabies virus genome. *Journal of Virology* 78: 424–440, with permission from the American Society for Microbiology.

with a location in the envelope-proximal tegument. Correlating with these findings, the PrV UL49 gene product has been shown to interact with the intracytoplasmic C-termini of the gE and gM envelope proteins. Both parts of the tegument may be connected by the UL48 gene product. Tegument proteins enter the cell after fusion of the virion envelope and the cellular plasma membrane during entry, and prime the cell for virus production. The alphaherpesvirus UL48 gene products are strong transactivators of viral immediate-early (IE) gene expression, whereas the UL41 proteins possess endoribonucleolytic activity to degrade preexisting cellular mRNAs. Cellular proteins have also been detected in the PrV tegument, including actin, annexins, and heat shock proteins. Their biological role is unknown.

The Envelope

Receptor-binding proteins, as well as major immunogens, are located in the viral envelope. More than 10 envelope constituents have been identified in PrV. Most are modified by the addition of carbohydrate and thus are glycoproteins. Several type I, type II, and type III PrV glycoproteins have been described (**Table 1**). Since the early nomenclature of PrV glycoproteins was somewhat confusing, it has been agreed to name them after their HSV-1 counterparts. Several of these glycoproteins form complexes, such as homooligomeric glycoprotein B (gB) and heterodimeric gE/gI, gH/gL, and gM/gN. The discovery of gN, as well as the gM/gN complex which is conserved throughout the mammalian and avian herpesviruses, was first made in PrV. Nonglycosylated membrane proteins include the US9, UL20, and UL43 gene products (see **Table 1**). The nonstructural gG is proteolytically cleaved and released from infected cells.

The Replication Cycle

PrV is arguably the most intensively analyzed animal herpesvirus. It has become a major focus of molecular biological research on the basic mechanisms of herpesvirus biology.

In the replication cycle (**Figure 3**), PrV infection of host cells starts with interaction of envelope gC with cell surface heparan sulfate-containing proteoglycans. This interaction is beneficial to, but not essential for, the second step, which involves binding of the essential envelope protein gD to its cellular receptor, nectin. HSV-1 and BHV-1 also use heparan sulfate and nectin for attachment. However, gD-negative and even gC- and gD-negative infectious PrV mutants have been isolated, which indicates that infection can occur by other routes. These mutants harbor additional mutations in gB and gH.

Penetration (fusion of viral envelope and cellular plasma membrane) requires the essential proteins gB and gH/gL. These glycoproteins are conserved throughout the mammalian and avian herpesviruses, indicating a common mechanism for membrane fusion. After penetration, the capsid is transported via microtubules to the nuclear pore, where it docks and releases the viral genome into the cell nucleus through one vertex. Empty capsids may remain bound to the nuclear pore for a considerable time. The entire entry process can be bypassed *in vitro* by transfection of naked viral DNA.

In the nucleus, transcription of viral genes is initiated by expression of the major IE gene, resulting in the translation of a 180 kDa protein (IE180). Although IE180 has long been considered to be the only IE protein of PrV, studies using specific inhibitors have identified that the US1 (RSp40) mRNA is also expressed with IE kinetics. Like other herpesvirus IE proteins, IE180 is a potent transcriptional activator that transinduces the expression of viral early genes. Early genes encode enzymes involved in nucleotide metabolism (e.g., UL23 = thymidine kinase; UL2 = uracil-DNA glycosylase; UL39/UL40 = ribonucleotide reductase; UL50 = deoxyuridine triphosphatase) and DNA replication (UL30/UL42 = DNA polymerase and an associated factor; UL5/UL8/UL52 = helicase-primase complex; UL29 = single-stranded DNA-binding protein; UL9 = origin-binding protein), as well as two protein kinases (US3, UL13).

DNA replication, whether occurring exclusively via a rolling-circle mechanism or involving intra- and intermolecular recombination and branching, results in the formation of head-to-tail fused concatemers of the genome. Finally, late genes encoding primarily virion structural proteins are expressed and their gene products, after translation in the cytosol, are transported into the nucleus for capsid assembly and DNA packaging. Capsid assembly is morphologically similar in all herpesviruses: capsids containing the major capsid protein UL19, triplex proteins UL18 and UL35, hexon-tip protein UL35, and portal protein UL6 assemble autocatalytically around a protein scaffold consisting of the UL26 and UL26.5 gene products. Packaging occurs via the unique portal at one vertex comprising 12 molecules of the UL6 protein. Genome-length molecules are cleaved from concatemeric replication products during packaging, which requires the UL15, UL28, UL32, and UL33 gene products. The UL17 protein may accompany DNA into the capsid. The UL25 gene product is not required for cleavage/packaging but apparently stabilizes the capsid and is essential for triggering primary envelopment. It is located at the outside of the capsid.

Egress of herpesvirus capsids from the nucleus has been, and still is, a matter of debate. However, numerous findings in recent years, to which studies on PrV contributed significantly, have demonstrated that intranuclear capsids gain

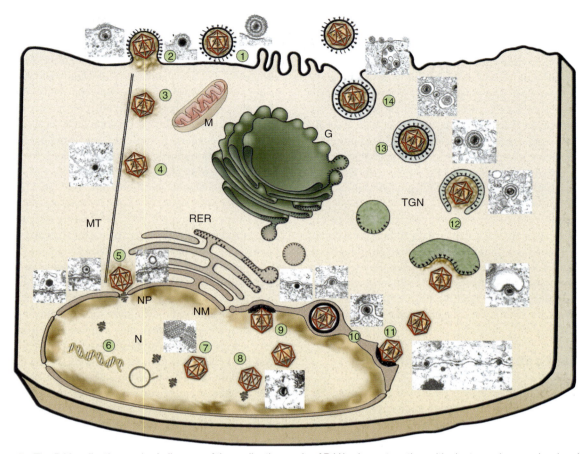

Figure 3 The PrV replication cycle. A diagram of the replication cycle of PrV is shown together with electron micrographs showing the respective stages. After attachment (1) and penetration (2), capsids are transported to the nucleus N (3) via interaction with microtubuli MT (4), docking at the nuclear pore NP (5) where the viral genome is released into the nucleus. Here, transcription of viral genes and viral genome replication occur (6). Concatemeric replicated viral genomes are cleaved to unit-length molecules during encapsidation (8) into preformed capsids (7), which then leave the nucleus by budding at the inner nuclear membrane NM (9) followed by fusion of the envelope of these primary virions located in the perinuclear cleft (10) with the outer nuclear membrane (11). Final maturation then occurs in the cytoplasm by secondary envelopment of intracytosolic capsids via budding into vesicles of the *trans*-Golgi network TGN (12) containing viral glycoproteins (black spikes), resulting in an enveloped virion within a cellular vesicle. After transport to the cell surface (13), vesicle and plasma membranes fuse, releasing a mature, enveloped PrV particle from the cell (14). RER, rough endoplasmic reticulum; M, mitochondrion; G, Golgi apparatus.

access to the cytoplasm by primary envelopment (i.e., budding at the inner leaflet of the nuclear membrane) followed by de-envelopment (fusion) at the outer leaflet. For primary envelopment, the conserved UL31 and UL34 proteins, which form a complex, have been shown to be important, though not always strictly essential, in all three subfamilies of herpesviruses. The UL31/UL34 complex is located in the nuclear membrane and recruits cellular protein kinase C, which phosphorylates and thereby dissociates nuclear lamins, allowing access of nascent capsids to the inner nuclear membrane. Primary enveloped virions in the perinuclear space also contain the UL31/UL34 complex, which constitutes part of the primary envelope (UL34 is a type II membrane protein) and tegument (UL31). The US3 protein kinase is present in primary and mature virions, whereas the UL31 and UL34 proteins are absent from mature virus particles. The mechanisms of de-envelopment are unclear. However, in the absence of the nonconserved and nonessential US3 kinase, primary enveloped virions accumulate in the perinuclear space, demonstrating the participation of this protein in nuclear egress.

Virion morphogenesis is completed in the cytoplasm by tegumentation, final envelopment, and transport of mature virus particles for release at the plasma membrane. Tegumentation apparently starts at two sites: the capsid and the future envelopment site. At the capsid, the conserved tegument proteins UL36 and UL37 interact. At the future envelopment site (i.e., at vesicles derived from the Golgi apparatus), the C-termini of the type I membrane protein gE and the type III membrane protein gM bind the UL49 tegument protein. Presumably, the UL48 protein links the two parts of the tegument, which drives budding of tegumented virions into the *trans*-Golgi vesicles containing viral glycoproteins to yield mature virions within a vesicle. PrV gM has been shown to relocate other viral and cellular proteins to the *trans*-Golgi and therefore may be involved in assembling the envelope proteins. The conserved UL11 gene product is thought to be involved in directing tegument proteins to the envelopment site. Finally, virion-containing vesicles move to the cell surface, a process in which the UL20 protein is involved, where plasma and vesicle membranes fuse, resulting in release of infectious particles. Apparently, virion gK inhibits an immediate re-fusion of released virions with the cell they just left.

Infectivity can be transmitted via direct cell-to-cell transmission as well as via free virions. Although several of the virion proteins required for penetration (gB, gH/gL) are also required for direct cell-to-cell spread, the mechanism remains enigmatic. In contrast to the situation with other alphaherpesviruses such as HSV, direct cell-to-cell spread of PrV does not require the receptor-binding gD molecule.

Clinical Features of Infection and Pathology

PrV is able to infect most mammals productively, with the exception of humans and other higher primates. However, primate and human cells are infectable in cell culture, and the reason for the natural resistance is not clear. Equids and goats are also rather resistant but may be infected experimentally. In addition, pseudorabies has been reported in many species of wild mammals, including wild boar, feral pigs, coyotes, raccoons, rats, mice, rabbits, deer, badgers, and coatimundi. It is so far not known whether these animals play a role in farm-to-farm transmission of PrV. In susceptible species other than porcines, infection is fatal and animals die from severe neuronal disorders.

After infection of the natural host, the clinical picture varies depending on the age of the animal, the virulence of the virus, and the route of infection. In nature, infection occurs predominantly oronasally, although genital transmission may also take place, especially in feral pigs. After replication in epithelial cells the virus gains access to neurons innervating the facial and oropharyngeal area, in particular the olfactory, trigeminal, and glossopharyngeal nerves. The virus spreads centripetally by fast axonal retrograde transport and reaches the cell bodies of infected neurons where either lytic or latent infection ensues (see below). PrV is disseminated viremically to many organs, where it replicates in epithelia, vascular endothelium, lymphocytes, and macrophages. In nonporcines, PrV is rather strictly neurotropic.

Neonates become prostrate and die quickly, often without nervous signs. In slightly older piglets, severe central nervous system (CNS) disorders are characterized by incoordination, twitching, paddling, tremors, ataxia, convulsions, and/or paralysis, whereas itching is only rarely present (**Figure 4**). Mortality in piglets up to 2–3 weeks of age may be as high as 100%, resulting in severe losses. Piglets at 3–6 weeks of age may still exhibit neurological signs and high morbidity, but mortality is usually reduced. Infection in older pigs induces primarily respiratory symptoms, such as coughing, sneezing, and heavy breathing, resulting from viral replication in, and destruction of, pulmonary epithelium. Despite the absence of overt nervous signs, virus gains access to neurons and remains latently established in the olfactory bulb, trigeminal ganglia, and brain stem or, after venereal transmission, in the sacral ganglia. PrV infection of pregnant sows may result in abortion or delivery of stillborn or mummified fetuses due to endometritis and necrotizing placentitis with infection of trophoblasts. In susceptible species other than swine, PrV infection is invariably fatal, sometimes after a rapid, peracute course without preceding overt clinical signs. Pruritus is a lead symptom of PrV infection in these species which, particularly in rabbits and rodents, may result in violent itching and automutilation. The death of mice, rats, cats, or dogs on farms is often a telltale sign of the presence of PrV prior to the appearance of symptoms in pigs.

Transmission occurs via virus-containing body fluids such as nasal and genital secretions, which gain access to epithelial surfaces within the respiratory or genital tract.

Figure 4 Neurological symptoms of PrV infection in piglets. The animals show ataxia (a), convulsions and paralysis (b) which ultimately lead to death.

Airborne transmission is efficient at short range, but long-range transmission covering several kilometers may also occur. Carnivores become infected by ingesting contaminated meat. After primary replication in epithelial cells, the virus enters the endings of sympathetic, parasympathetic or sensory and motor neurons innervating the area of primary replication. Infection probably occurs by the same mechanism as outlined above for cultured cells. Virus is transported in retrograde fashion to the neuronal cell body, where DNA replication and formation of progeny virions ensues. It is not clear whether complete virions or viral subassemblies are then transported to the synapse, or how transsynaptic transfer occurs. Depending on the virulence of the virus and the age and immune status of the host, infection may not proceed beyond the first neuronal level (i.e., ganglia directly innervating the affected peripheral site). However, virus may also spread to the brain resulting in ganglioneuritis and encephalitis. Lymphocytes can also become infected by PrV and this may help viral spread within the body, playing an important role in infection of the fetus. However, the percentage of infected cells in the blood is rather low, even during acute infection, and difficult to detect. A major target organ for latency in swine is the tonsils, and tonsil biopsies allow reliable detection of virus by molecular biological techniques or virus isolation.

There are no pathognomonic, gross lesions of AD. In piglets, there may be necrotizing tonsillitis, rhinotracheitis, or proximal esophagitis. Other lesions commonly seen include pulmonary edema, necrotizing enteritis, and multifocal necrosis of the spleen, lung, liver, lymph nodes, and adrenal glands. Histologically, PrV causes a nonsuppurative meningoencephalitis and paravertebral ganglioneuritis. The gray matter is especially affected, and infected neurons or astrocytes may present acidophilic intranuclear inclusions. The presence of viral antigen can be visualized by immunostaining and viral genomes can be detected by *in situ* hybridization. PrV infected cells usually show more or less extensive degeneration and necrosis due to lytic viral replication. Whether apoptosis induced by PrV infection also plays a role *in vivo* is unclear. A predominantly T-cell-mediated reaction of the immune system induces ganglioneuritis, polio- or panencephalitis with foci of gliosis contributing to the loss of neuronal function. The described extraneural lesions in pigs and acute myocarditis in carnivores might provide additional explanations for the fatal outcome of infections in which virus cannot be recovered from the brain.

Immunology

Live as well as inactivated vaccines induce efficient protective immunity against AD. Antibodies against a number of viral structural and nonstructural proteins have been detected in infected animals, and virus-neutralizing monoclonal antibodies have been isolated. Antibody responses are primarily directed against the major surface glycoproteins including gB, gC, gD, and gE as well as secreted gG. The most potent complement-independent virus neutralizing antibodies are directed against gC, gD and, to a lesser extent, gB, and subunit vaccines consisting of gB, gC, gD, as well as anti-idiotypic anti-gD antibodies and heterologous vectors expressing gC or gD, elicit protective immunity. In contrast, anti-gE antibodies require complement for neutralization, and anti-gG antibodies have no neutralizing ability at all. Antibodies against whole virus or specific for gB are used in diagnostic assays to detect PrV infection serologically. Major targets for cell-mediated immunity in pigs are primarily gC and, to a lesser extent, gB.

Although the numerous elaborate immune evasion mechanisms of beta- and gammaherpesviruses, including expression of virokines and viroceptors, have not been found in alphaherpesviruses, these viruses still interact with the immune system to evade its activity. Like other alphaherpesvirus gC proteins, PrV gC binds species-specifically to porcine complement component C3, and the gE/gI complex binds the Fc portion of porcine IgG. Secreted gG may bind chemokines, thereby impairing intercellular signaling. Moreover, infection of cells by PrV results in downregulation of major histocompatibility complex class I (MHC-I) antigen presentation, and envelope glycoproteins present at the plasma membrane of infected cells are internalized by as yet unknown factors, resulting in a paucity of antigens presented to the immune system at the cell surface. Recently, the PrV gN protein has been shown to inactivate the transporter which translocates processed peptides for loading onto MHC-I molecules in the endoplasmic reticulum. The combined action of these mechanisms may give the virus an edge over the immune system, facilitating establishment of latency and further virus spread.

Latency

Like other alphaherpesviruses, PrV has the capacity to become latent in neurons. During latency, the genome persists largely quiescently in a presumably circular form. Expression is restricted to one region, the latency-associated transcript (LAT) gene, which encompasses part of the inverted repeat and adjoining U_L region. The LAT gene, which encodes the large latency transcript (LLT), is located antiparallel to the genes encoding IE180 and EP0 (see **Figure 2**), and is transcribed into three different RNAs of 8.4, 8.0, and 2.0 kb. The 8.4 and 2.0 kb species are derived by splicing of a larger precursor. During latency, only the 8.4 kb RNA is produced from a separate promoter that is apparently active only under latent conditions, whereas the 8.0 and 2.0 kb species are also transcribed during lytic infection.

PrV encodes proteins which are able to suppress apoptotic cell death as a prerequisite for the establishment of latency. The US3 protein kinase has been demonstrated to mediate this function in porcine fetal trigeminal neurons. PrV establishes latent infections predominantly in neuronal tissues such as the trigeminal or sacral ganglia. However, tonsils have also been identified as sites of latency.

Epidemiology and Control

In the twentieth century, PrV has become a pathogen distributed worldwide with the exception of Australia, Canada, and the Scandinavian countries. In major swine-producing areas, PrV infection caused significant economic losses amounting to hundreds of millions of dollars, making it one of the most devastating pig diseases. Control and eradication of PrV infection in pigs relied on two strategies. In areas with a low prevalence of infection, serological screening and consequent elimination of seropositive animals resulted in the eradication of AD from countries such as the UK, Denmark, and East Germany. PrV infection can be diagnosed by detecting either the infectious agent (antigen detection by immunofluorescence or virus isolation) or viral DNA using polymerase chain reaction (PCR). The latter method is also suitable for detecting latent viral genomes. A PrV-specific immune response in live animals can be confirmed using various serological assays (e.g., virus neutralization, latex agglutination, or enzyme-linked immunosorbent assay (ELISA) systems based on complete virus particles or distinct viral antigens such as gB).

To reduce disease prevalence, vaccination with live-attenuated or inactivated vaccines has also been used. However, vaccination does not result in sterile immunity, and vaccinated animals may still be infected with and carry the virus, and these carriers are no longer identifiable by serological analysis. This problem has been solved by the advent of the so-called 'marker' vaccines. This novel concept provided a breakthrough in animal disease control, and serves as a blueprint for control of other infectious diseases. It was based on the finding that several immunogenic envelope glycoproteins of PrV, such as gC, gE, and gG (see above), are not required for productive replication and can be deleted from the viral genome without abolishing virus replication. These gene-deleted strains can be produced easily in conventional cell systems and can be administered as inactivated or modified-live vaccines. In fact, gene-deleted PrV strains were the first genetically engineered live-virus vaccines to be licensed. Thus, PrV has pioneered modern vaccinology. Whereas animals vaccinated with these vaccines do not mount an immune response to the missing gene product, wild-type virus infection results in seroconversion for the differentiating antigen. Serological assays (ELISA) have subsequently been developed that allow easy and sensitive detection of antibodies against these marker proteins, resulting in the identification of animals infected with wild-type virus, regardless of vaccination status (**Figure 5**). Thus, virus circulation can be reduced by vaccination, and infected animals that still harbor field virus can be identified subsequently and eliminated, resulting in cost-efficient eradication.

This breakthrough approach of 'differentiating infected from vaccinated animals' (DIVA) was pioneered in the field with PrV and is now widely accepted and practised

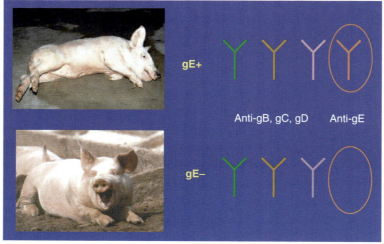

Figure 5 The principle of DIVA or marker vaccination. Whereas antibodies are produced against all immunogenic viral proteins after wild-type infection, antibodies against the missing gene product (circled) will not be formed after vaccination with a gene-deleted virus. The presence or absence of these antibodies is used to differentiate between infected and vaccinated animals. Reprinted with permission from Mettenleiter TC (2005) Veterinary viruses. *Nova Acta Leopoldina NF92* 344: 221–230.

with other infectious diseases, such as BHV-1 infection, classical swine fever, and foot-and-mouth disease. Its application resulted in eradication of PrV from heavily infected West Germany within 10 years, and also recently succeeded in eliminating PrV infection from pig herds in New Zealand and the USA. Although European wild boar and American feral pigs also harbor PrV, there is no epidemiological link between PrV in wild boar and domestic pigs in Europe since clearly different viral strains have been isolated from each. However, in the Southern USA infected feral pigs may represent a source of infection to domestic pig holdings.

PrV as a Tool in Neurobiology

Like other alphaherpesviruses, PrV exhibits a distinct neurotropism, invading the CNS via peripheral nerves. While wild-type strains of PrV may spread within the CNS both laterally and transsynaptically, attenuated PrV mutants have been identified which, under appropriate assay conditions, travel more or less exclusively along nerves and are transported transsynaptically. This property has prompted increasing use of PrV as a transneuronal tracer to label neuronal connections in experimental animal models, and has been useful in elucidating detailed neuroanatomical networks in mice and rats. The virus used most frequently in these studies is the Bartha strain of PrV, a modified-live vaccine strain which had been attenuated by the Hungarian veterinarian Adorján Bartha by multiple passages in embryonated chicken eggs and chicken embryo fibroblasts. Molecular biological analyses demonstrated that this strain carries several lesions compared to wild-type PrV: it lacks the gE, gI, and US9 genes, contains a mutation in the signal sequence for gC, specifies attenuating mutations in the UL21 gene, and expresses a UL10 gene product (gM) that is not glycosylated due to mutation of the N-glycosylation site. The glycoprotein deletion and the UL21 mutation have been shown to be most important for the observed attenuation. Recently, genetically engineered Bartha-derivatives expressing the marker proteins β-galactosidase or green fluorescent protein have been constructed and used in double-tracing studies. Moreover, mutants that express their markers only under specific conditions (e.g., in transgenic cells or animals expressing cre-lox recombinase under control of a tissue-specific promoter) have added further elegant possibilities for tissue-specific labeling.

Future Perspectives

PrV is a fascinating virus with several interesting properties. The availability of conventional and genetically engineered marker vaccines allows effective and cost-efficient disease control campaigns, which have been shown to result in the eradication of virus and disease from animal populations. Although PrV infection is still widespread, in particular in certain areas in Eastern Europe and Asia, concerted efforts could result in the elimination of the disease on a worldwide scale. Beyond its importance as the causative agent of a relevant animal disease, PrV is an ideal tool to study basic mechanisms of herpesvirus (molecular) biology and has the enormous advantage of an experimentally accessible natural virus–host system by infection of pigs. Moreover, its broad host range allows the use of other well-defined animal models for neuroanatomical, immunological, and molecular biological studies. Since PrV replicates exceedingly well in tissue culture, it is also well suited for detailed analysis of the requirements for (alpha) herpesvirus replication. Thus, PrV will remain under intensive scrutiny for sometime to come.

Acknowledgments

The author thanks Harald Granzow and Mandy Jörn for **Figures 1** and **3** and Jens Teifke, Thomas Müller, Hanns-Joachim Rziha, and Barbara Kluppa for helpful comments on the manuscript.

See also: Bovine Herpesviruses; Herpes Simplex Viruses: General Features; Herpes Simplex Viruses: Molecular Biology; Herpesviruses: General Features; Herpesviruses of Birds; Herpesviruses of Horses; Varicella-Zoster Virus: General Features; Varicella-Zoster Virus: Molecular Biology.

Further Reading

Enquist LW, Husak PJ, Banfield BW, and Smith GA (1999) Infection and spread of alphaherpesviruses in the nervous system. *Advances in Virus Research* 51: 237–347.

Granzow H, Weiland F, Jöns B, Klupp B, Karger A, and Mettenleiter TC (1997) Ultrastructural analysis of pseudorabies virus in cell culture: A reassessment. *Journal of Virology* 71: 2071–2082.

Klupp BG, Hengartner C, Mettenleiter TC, and Enquist LW (2004) Complete, annotated sequence of the pseudorabies virus genome. *Journal of Virology* 78: 424–440.

Mettenleiter TC (2000) Aujeszky's disease (pseudorabies) virus: The virus and molecular pathogenesis. *Veterinary Research* 31: 99–115.

Mettenleiter TC (2005) Veterinary viruses. *Nova Acta Leopoldina NF92* 344: 221–230.

Mettenleiter TC, Klupp BG, and Ganzow H (2006) Herpesvirus assembly: A tale of two membranes. *Current Opinion in Microbiology* 9: 423–429.

Pomeranz L, Reynolds AE, and Hengartner CJ (2006) Molecular biology of pseudorabies virus: Impact on neurovirology and veterinary medicine. *Microbiology and Molecular Biology Reviews* 69: 462–500.

Pseudoviruses

D F Voytas, Iowa State University, Ames, IA, USA

© 2008 Elsevier Ltd. All rights reserved.

Glossary

***env*-like gene** A gene found in some sireviruses with similarity to retroviral *env* genes: it occurs downstream of *pol* and often encodes a putative transmembrane protein.

Half-tRNA A tRNA fragment that encompasses the anticodon stem loop and primes reverse transcription of the hemiviruses.

Retrotransposon A mobile genetic element that replicates by reverse transcription but is not infectious.

Virus-like particle The nucleoprotein complex that serves as a replication intermediate for species in the family *Pseudoviridae*.

Introduction

Large-scale DNA sequencing projects have revealed that a significant fraction of most eukaryotic genomes is made up of mobile genetic elements. Among these are the long-terminal repeat or LTR retrotransposons, which share a similar genomic organization and life cycle with members of the family *Retroviridae*. The primary feature that distinguishes the retroviruses from the retrotransposons is that the former are infectious and encode an envelope gene that enables cell-to-cell transmission. The close relationship between the retroviruses and retrotransposons prompted the International Committee on Taxonomy of Viruses to provide a formal taxonomic description for the retrotransposons.

The retrotransposons are divided into the families *Pseudoviridae* (this article) and *Metaviridae* (see elsewhere in this volume). Pseudoviruses are particularly abundant in plant genomes, but are also found in diverse animal, fungal, and protist hosts. Members of the family *Pseudoviridae* are distinguished from those of the families *Metaviridae* and *Retroviridae* by a distinct organization of their *pol* genes: for pseudoviruses, reverse transcriptase (RT) precedes integrase (IN), and this order is reversed for members of related viral families. Phylogenetic analysis of RT amino acid sequences also clearly separates species of the family *Pseudoviridae* from those of the families *Metaviridae* and *Retroviridae*.

The family *Pseudoviridae* is often referred to as Ty1-copia retrotransposons after representative species from the yeast *Saccharomyces cerevisiae* (Ty1) and the insect *Drosophila melanogaster* (copia). For historic reasons, and because they are found in model experimental organisms, Ty1 and copia have been studied extensively at the molecular and biochemical level. Increasingly, research is also being conducted with the Ty5 hemivirus of *S. cerevisiae* and with representative pseudoviruses from plants, such as Tnt1 and Tto1 from tobacco (*Nicotiana tabacum*) and Tos17 from rice (*Oryza sativa*). Because Ty1 is the most intensely studied member of the family *Pseudoviridae*, much of the information in this article on virion properties and replication cycle is derived from research conducted with Ty1.

Taxonomy

The family *Pseudoviridae* is comprised of three genera: *Pseudovirus, Hemivirus,* and *Sirevirus* (**Table 1**). The genus *Pseudovirus* has the most species and the broadest host range. Although differences in RT amino acid sequences were originally used to distinguish the hemiviruses and pseudoviruses, as more species have been described, this criterion has proved to be less reliable (see below). Hemiviruses are also distinguished by the primer used for reverse transcription. Pseudoviruses and sirevirus use the 3'-OH of the acceptor stem of a host tRNA to prime DNA synthesis, whereas hemiviruses use a half-tRNA that corresponds to a portion of the tRNA anticodon stem loop. For copia, the tRNA fragment is generated by cleavage to release a 3'-OH that primes reverse transcription; however, the mechanism by which cleavage occurs remains obscure.

To date, the sireviruses are exclusively associated with plant hosts. Phylogenetic analyses indicate that the RT encoded by the sireviruses is distinct from those of the hemiviruses and pseudoviruses. An additional distinctive feature of many sireviruses is the presence of an open reading frame after *pol*. This gene is reminiscent of retroviral *env* genes both in terms of its location within the viral genome and the fact that it is often predicted to encode a transmembrane protein. Whether or not this *env*-like gene encodes a protein that mediates infectivity remains to be determined.

Virion Properties

Members of the family *Pseudoviridae* are generally thought to replicate through a noninfectious virus-like intermediate referred to as a virus-like particle or VLP. The most thoroughly characterized VLPs are those produced by

Table 1 Viruses in the family *Pseudoviridae*

Genus	Virus	Host species	Accession number
Pseudovirus	Art1	*Arabidopsis thaliana*	Y08010
	AtRE1	*A. thaliana*	AB021265
	Evelknievel	*A. thaliana*	AF039373
	Ta1	*A. thaliana*	X13291
	Melmoth	*Brassica oleracea*	Y12321
	Panzee	*Cajanus cajan*	AJ000893
	Tgmr	*Glycine max*	U96748
	BARE-1	*Hordeum vulgare*	Z17327
	Tnt1	*Nicotiana tabacum*	X13777
	Tto1	*N. tabacum*	D83003
	RIRE1	*Oryza australiensis*	D85597
	Retrofit	*O. longistaminata*	U72726
	Tp1	*Physarum polycephalum*	X53558
	Ty1	*Saccharomyces cerevisiae*	M18706
	Ty2	*S. cerevisiae*	M19542
	Ty4	*S. cerevisiae*	X67284
	Tst1	*Solanum tuberosum*	X52387
	WIS-2	*Triticum aestivum*	X63184; X57168 (LTR)
	Hopscotch	*Zea mays*	U12626
	Sto-4	*Z. mays*	AF082133
Hemivirus	Mosqcopia	*Aedes aegypti*	AF134899
	Tca2	*Candida albicans*	AF050215
	Tca5	*C. albicans*	AF065434
	1731	*Drosophila melanogaster*	X07656
	Copia	*D. melanogaster*	M11240
	Ty5	*S. paradoxus*	U19263
	Lueckenbuesser	*Volvox carteri*	U90320
	Osser	*V. carteri*	X69552
Sirevirus	Endovir	*A. thaliana*	AY016208
	SIRE1	*G. max*	AY205608
	ToRTL1	*Lycopersicon esculentum*	U68072
	Opie-2	*Z. mays*	U68408
	Prem-2	*Z. mays*	U41000

the pseudovirus Ty1 and the copia hemivirus. Although sometimes irregularly shaped, these VLPs are typically round to ovoid and have a mean radius of 30–40 nm. Ty1 VLPs are cytoplasmic, whereas those produced by copia are nuclear. When truncated forms of the Ty1 capsid protein (Gag) are expressed, the VLPs appear more icosohedral, as observed by cryoelectron microscopy (**Figure 1**).

Members of the family *Pseudoviridae* encode a *gag* gene, and Gag proteins are processed by a viral protease to yield mature proteins corresponding to retroviral capsid (CA) and nucleocapsid (NC). Like the retroviruses, many pseudoviruses encode a zinc finger motif in Gag that is part of NC. For Ty1, which lacks a zinc finger signature, the protein encoded by the corresponding region of *gag* appears to function like a retroviral NC.

Gag proteins are the primary structural components of the VLP. Inside the particle are packaged the Pol gene products – RT and IN – along with an LTR to LTR transcript of the viral genome (typically ∼5–9 kbp). This RNA is positive sense, capped and polyadenylated, and two RNAs are packaged within the particle (at least for the Ty1). Also found within the VLPs are various DNA intermediates – products of reverse transcription – which include full-length cDNA that represents the unintegrated form of the virus.

Genome Organization and Expression

Most members of the family *Pseudoviridae* encode a single open reading frame with similarity to both retroviral *gag* and *pol* genes (e.g., copia; **Figure 2**). Despite this single translation product, there appears to be a need, as in the retroviruses, to regulate relative amounts of Gag and Pol: for efficient replication, more structural Gag proteins are required to form the VLP relative to the catalytic Pol proteins. Work with the Ty5 hemivirus and the metavirus Tf1 suggests that differential protein stability may determine the proper Gag/Pol ratio in viruses that encode a single polyprotein. That is, at the onset of viral replication, Pol is degraded more quickly relative to Gag to achieve the proper stoichiometry. Some yeast pseudoviruses (e.g., Ty1, Ty2, and Ty4) encode pol in the +1 frame relative to *gag*, and Gag/Pol stoichiometry is regulated by ribosomal frameshifting (**Figure 2**). Several sireviruses also encode *gag* and *pol* on separate reading frames; however, the location of conserved stop codons suggests that ribosomal frameshifting is not employed, and ribosomal hopping, as described for bacteriophage T4 *gene 60*, has been proposed although not experimentally verified. For the hemivirus copia, *gag* is encoded on a spliced mRNA, and differential splicing determines proper Gag/Pol levels.

Gag amino acid sequences are poorly conserved among members of the family *Pseudoviridae*, although a few weakly conserved motifs have been described in addition to the zinc finger motif that characterizes NC. Gag proteins of sireviruses are considerably larger than those encoded by species in other genera, and the extended region of Gag interacts with components of the dynein microtubular motor, suggesting that it plays a role in transporting the virus throughout the cell.

The coding region for *pol* lies downstream of *gag* and directs the production of three enzymes: protease (PR), IN, and RT. PR lies at the Pol N-terminus and is required for release of the other enzymes from the Pol precursor and for processing Gag. Like the retroviral enzyme, PR is an aspartic protease and undoubtedly functions as a dimer. Downstream of PR lies the coding region for IN. At its N-terminus, IN bears a zinc finger motif followed by a catalytic core domain that is similar to the retroviral enzyme. Adjacent to the IN catalytic domain is a large motif (called GKGY) that is unique to INs encoded by members of the family *Pseudoviridae*. Additionally, the extreme C-terminus of IN is rich in simple sequence

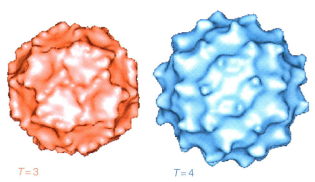

Figure 1 Virus-like particles produced by Ty1 of *S. cerevisiae*. Three-dimensional reconstruction of 1–346 $T=3$ and 1–408 $T=4$ particles calculated to 25 and 35 Å, respectively. Reprinted from Al-Khayat HA, Bhella D, Kenney JM, et al. (1999) Yeast Ty1 retrotransposons assemble into virus-like particles whose T-numbers depend on the C-terminal length of the capsid protein. *Journal of Molecular Biology* 292: 65–73, with permission from Elsevier.

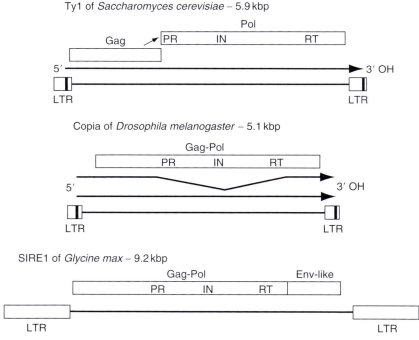

Figure 2 The genomic organization of representative viruses in the family *Pseudoviridae*. Ty1, copia, and SIRE1 are members of the genus *Pseudovirus, Hemivirus,* and *Sirevirus*, respectively. Transcript information is available for Ty1 and copia, and transcripts are depicted by arrows above the viral genomes. Note that the copia transcript is differentially spliced. Boxes on the ends of the viral genomes denote the LTRs. Black boxes within the Ty1 and copia LTRs denote the R region – sequences repeated at the 5′ and 3′ ends of the viral transcript. Encoded proteins are depicted as open boxes. Conserved amino acid sequences are labeled in Pol: protease (PR), integrase (IN), and reverse transcriptase/RNase H (RT). Note that Gag and Pol of Ty1 are on different reading frames, and a Gag-Pol fusion protein is expressed by ribosomal frameshifting. A single stop codon separates *gag-pol* from the *env*-like gene of SIRE1, and a Gag-Pol-Env-like polyprotein is synthesized by stop codon suppression.

motifs and shows considerable size heterogeneity. For the hemivirus Ty5, the IN C-terminus mediates target specificity through interactions with a specific heterochromatin protein (i.e., Sir4). This interaction tethers IN to heterochromatin and determines target site choice. The C-terminus of Pol encodes RT, and like the retroviral enzyme, RT has both RT and RNase H activity.

As described above, many sireviruses encode an additional gene downstream of *pol*. This coding region is often referred to as an *env*-like gene, both because of its location in the viral genome and because it typically encodes a protein with putative transmembrane domains, a hallmark of retroviral Env proteins. The *env*-like genes are often on separate reading frames, and for the SIRE1 virus, the *env*-like gene is separated from *pol* by a stop codon. Expression of the SIRE1 env-like gene is mediated by stop codon suppression, generating a Gag–Pol–Env-like fusion protein. The presence of an *env*-like gene, however, is not a universal feature of sireviruses, and the function of this protein in the life cycle of these elements remains to be determined.

The coding region for members of the family *Pseudoviridae* is flanked by two identical long terminal repeats that can range in size from a few hundred to a few thousand base pairs. The LTRs direct the expression of a genomic mRNA and often bear *cis*-acting sequences that regulate mRNA expression. The full-length genomic transcript has terminal redundancy like the retroviral transcript, and this redundancy is important in the replication life cycle of these elements (i.e., to facilitate strand exchanges during reverse transcription, see below).

Adjacent to the 5' LTR is a short *cis*-acting sequence called the primer binding site (PBS) that primes reverse transcription. For most members of the family *Pseudoviridae*, the PBS is complementary to a host initiator tRNA methionine. Importantly, and as mentioned above, complementarity of the PBS to the acceptor stem loop of a host tRNA is the distinguishing feature of the hemiviruses. Adjacent to the 3' LTR is a purine rich sequence (the polypurine tract, PPT) that serves as a primer for second strand DNA synthesis. There is considerable variation in the length of noncoding sequences between the priming sites and the coding regions, particularly for the sireviruses.

Replication Cycle

The mechanism by which viruses in the family *Pseudoviridae* replicate their genomes is principally derived from studies of the Ty1 pseudovirus of *S. cerevisiae*. It is generally assumed that other viruses in the family use a similar replication strategy, which as described below, is very similar to the replication mechanism employed by viruses in the family *Retroviridae*.

The replication cycle begins with the production of a full-length transcript that initiates in the 5' LTR and terminates in the 3' LTR. The LTRs are divided into three regions based on sequences present in the transcript: the U3 region includes sequences unique to the 3' end of the RNA; U5 includes sequences unique to the 5' end of the RNA, and the R regions are sequences repeated at both ends of the message. The first gene encoded on the full-length mRNA is *gag*, and as mentioned above, *gag* encodes the major CA as well as a small C-terminal peptide that is similar to retroviral NC proteins. CA and NC are derived from the Gag translation product through proteolytic processing by PR. Pol is expressed as a translational fusion with Gag, and in a few species, this occurs through ribosomal frameshifting. Pol is processed by PR to release mature IN and RT.

In addition to serving as a template for translation of Gag and Pol, the viral transcript also serves as a template for the synthesis of DNA through reverse transcription. The full-length mRNA is encapsidated in a precursor VLP consisting of unprocessed Gag and Gag–Pol proteins. Action by PR converts the particle into a mature VLP and likely activates reverse transcription.

The synthesis of viral cDNA from the template mRNA proceeds by a mechanism very similar to that of the retroviruses. The first step is the extension of cDNA from a tRNA primer that anneals to the PBS. The cDNA is extended to the 5' end of the mRNA generating minus ($-$)-strand strong-stop DNA (ssDNA). The ($-$) ssDNA is transferred to the 3' end of the full-length RNA where it pairs with the R region and then is extended to generate a nearly full-length ($-$)-strand cDNA. Plus ($+$)-strand synthesis begins at the RNase H resistant PPT adjacent to the 3' LTR. Extension yields a ($+$) ssDNA that is transferred to the left end of the ($-$)-strand DNA. Further extension yields a full-length cDNA. The double-stranded cDNA is imported into the nucleus where it is inserted into a chromosomal target site. Integration generates a 5 bp duplication of host target DNA. Some viruses, particularly in yeast, show strong target site preferences. In these cases, the integration machinery recognizes specific protein complexes at the target sites (e.g., transcription complexes, heterochromatin) to mediate target specificity.

Evolutionary Relationships

The taxonomic framework for the family *Pseudoviridae* is based on phylogenetic analyses of RT amino acid sequences encoded by family members (**Figure 3**). Early work indicated that members of the family *Pseudoviridae* form a monophyletic lineage among the diverse mobile genetic elements that encode RTs. The distinctiveness of the pseudovirus RT holds up regardless of the host from

Pseudoviruses

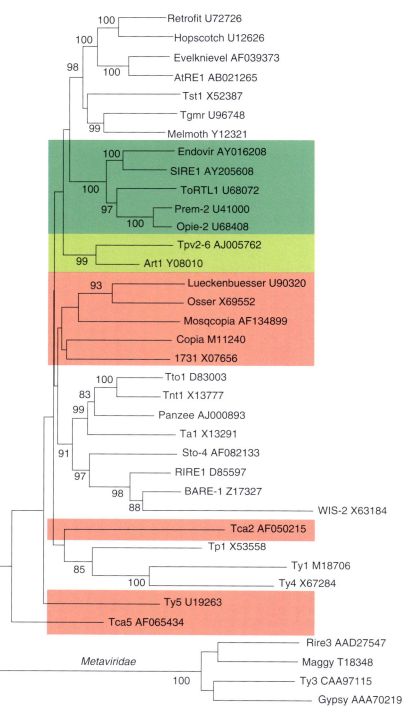

Figure 3 Phylogenetic relationships among members of the family *Pseudoviridae*. The tree is based on RT amino acid sequences and was performed using the neighbor-joining distance algorithm. RTs encoded by species in the family *Metaviridae* were used as an outgroup. The hemiviruses are shaded in pink and the sireviruses in green. Yellow denotes Tpv2-6 and Art1, which have not been formally assigned to the family *Pseudoviridae*.

which the species originate. The organization of *pol*, with IN preceding RT, is also a unique feature that identifies family members.

Designations for the genera *Pseudovirus* and *Hemivirus* were originally based on the primer used for (−) ssDNA synthesis (i.e., a full- or half-tRNA, respectively). The genera designations were underscored by phylogenetic analyses indicating that RTs encoded by members of these genera were distinctive. Recent genome sequencing efforts, however, have greatly expanded the ranks of the family *Pseudoviridae*. Analysis of the larger dataset indicates that the phylogenetic support for the original genera designations is weak (**Figure 3**). It may be that derived characters such as the use of particular primers is not an appropriate means for classification, and it is anticipated that the taxonomic structure of the family will need to be reconsidered.

Large-scale genomic sequencing efforts also revealed new lineages within the family *Pseudoviridae*, prompting the designation of the genus *Sirevirus*. The RTs encoded by the sireviruses are phylogenetically distinct from other members of the family, and sireviruses are found only in plant hosts. Although the original sireviruses described encode an additional *env*-like gene, some elements lack this coding sequence, and so distinctiveness of the RT amino acid sequence is the most reliable means of classification.

Genome sequencing efforts have also revealed species with *pol* gene organizations characteristic of the family *Pseudoviridae* and yet RTs that are distinct from the genera described to date (**Figure 3**). Among these are Tpv2 and its relatives, which currently are not officially assigned to the family *Pseudoviridae*. Future taxonomic revisions will have to address their candidacy for assignment to the *Pseudoviridae* and whether or not they define a new genus.

See also: Metaviruses; Retroviruses: General Features.

Further Reading

Al-Khayat HA, Bhella D, Kenney JM, *et al.* (1999) Yeast Ty1 retrotransposons assemble into virus-like particles whose T-numbers depend on the C-terminal length of the capsid protein. *Journal of Molecular Biology* 292: 65–73.

Boeke JD, Eickbush T, Sandmeyer SB, and Voytas DF (2000) Family Metaviridae. In: Van Regenmortel MHV, Fauquet CM, Bishop DHL, *et al.* (eds.) *Virus Taxonomy: Seventh Report of the International Committee on Taxonomy of Viruses*, pp. 358–367. San Diego: Academic Press.

Boeke JD, Eickbush T, Sandmeyer SB, and Voytas DF (2000) Family Pseudoviridae. In: Van Regenmortel MHV, Fauquet CM, Bishop DHL, *et al.* (eds.) *Virus Taxonomy: Seventh Report of the International Committee on Taxonomy of Viruses*, pp. 349–357. San Diego: Academic Press.

Doolittle RF, Feng DF, Johnson MS, and McClure MA (1989) Origins and evolutionary relationships of retroviruses. *Quarterly Review of Biology* 64: 1–30.

Eichinger DJ and Boeke JD (1988) The DNA intermediate in yeast Ty1 element transposition copurifies with virus-like particles: Cell-free Ty1 transposition. *Cell* 54: 955–966.

Flavell AJ, Dunbar E, Anderson R, Pearce SR, Hartley R, and Kumar A (1992) Ty1-*copia* group retrotransposons are ubiquitous and heterogeneous in higher plants. *Nucleic Acids Research* 20: 3639–3644.

Garfinkel DJ, Boeke JD, and Fink GR (1985) Ty element transposition: Reverse transcriptase and virus-like particles. *Cell* 42: 507–517.

Kikuchi Y, Ando Y, and Shiba T (1986) Unusual priming mechanism of RNA-directed DNA synthesis in *copia* retrovirus-like particles of *Drosophila*. *Nature* 323: 824–826.

Laten HM, Majumdar A, and Gaucher EA (1998) SIRE-1, a *copia*/Ty1-like retroelement from soybean, encodes a retroviral envelope-like protein. *Proceedings of the National Academy of Sciences, USA* 71: 6897–6902.

Mellor J, Malim MH, Gull K, *et al.* (1985) Reverse transcriptase activity and Ty RNA are associated with virus-like particles in yeast. *Nature* 318: 583–586.

Peterson-Burch BD and Voytas DF (2002) Genes of the *Pseudoviridae* (Ty1/*copia* retrotransposons). *Molecular Biology and Evolution* 19: 1832–1845.

Voytas DF and Boeke JD (2002) Ty1 and Ty5 of *Saccharomyces cerevisiae*. In: Craig NL, Craigie R, Gellert M, and Lambowitz AM (eds.) *Mobile DNA II*, 1st edn., 631–662. Washington, DC: American Society for Microbiology Press.

Voytas DF, Cummings MP, Koniczny A, Ausubel FM, and Rodermel SR (1992) *Copia*-like retrotransposons are ubiquitous among plants. *Proceedings of the National Academy of Sciences, USA* 89: 7124–7128.

Xiong Y and Eickbush TH (1990) Origin and evolution of retroelements based upon their reverse transcriptase sequences. *EMBO Journal* 9: 3353–3362.

Zou S, Ke N, Kim JM, and Voytas DF (1996) The *Saccharomyces* retrotransposon Ty5 integrates preferentially into regions of silent chromatin at the telomeres and mating loci. *Genes and Development* 10: 634–645.

Quasispecies

R Sanjuán, Instituto de Biología Molecular y Cellular de Plantas, CSIC-UPV, Valencia, Spain

© 2008 Elsevier Ltd. All rights reserved.

Glossary

Adaptive landscape A graphical representation of fitness as a function of the genotype, first used by the population geneticist S. Wright.

Antagonistic epistasis Genetic interaction that makes mutations have larger effects alone than in combination.

Antagonistic pleiotropy The situation in which a particular mutation that is beneficial in one environment becomes harmful in another environment.

Back-mutation Mutation that regenerates the ancestral genotype.

Biological fitness The number of descendants.

Degenerate quasispecies Quasispecies with more than a single fittest type.

Effective population size The number of individuals of a population that actually contribute to the next generation.

Epistasis Interaction between genetic loci, implying that mutations have nonindependent effects.

Error catastrophe A situation in which the mutation rate is too high to be effectively counteracted by selection and the population becomes a pool of randomly drifting sequences.

Error threshold The critical mutation rate that marks the transition from mutation–selection balance to error catastrophe.

Genetic drift Changes in the genetic composition of finite populations due to random sampling of genotypes between generations.

Master sequence The fittest and most abundant sequence in a nondegenerate quasispecies.

Muller's ratchet Accumulation of deleterious mutations in small populations due to genetic drift that becomes irreversible in the absence of recombination or back-mutations.

Multiplicative landscape An adaptive landscape in which there is no epistasis, that is, mutational effects are independent.

Mutation–selection balance The dynamic equilibrium between the generation of variability through deleterious mutation and its elimination through natural selection.

Mutational robustness Ability to tolerate mutations.

Phase transition A sudden change in the properties of a physical system in response to a change in one or more parameters.

Quasispecies A population at the mutation–selection balance composed by a master sequence and an ensemble of deleterious mutants.

Selection coefficient Denoted s, the relative fitness difference between a genotype and the wild type.

Sequence space A discrete space with as many dimensions as the genome length and four values on each dimension corresponding to the four possible nucleotides.

Wild type An arbitrary reference genotype, typically the most abundant one.

The Quasispecies Theory

Historical Context

The word quasispecies, borrowed from chemistry, refers to a population of quasi-identical molecules, by opposition to a molecular species, which is made of fully identical molecules. The term is probably unfortunate in the biological context because first, it could be erroneously thought of as if it made reference to the biological concept of species and second, biological populations are generally variable, making its usage redundant. For these reasons, some virologists have discarded it, while most keep making use of it to refer to a highly variable viral population. However, this is mainly a matter of semantics. Quasispecies constitute indeed a formal evolutionary theory first introduced in the early 1970s by Manfred Eigen and intended to capture the population dynamics of hypothetical replicons at the origin of life. These replicons were small and displayed a high rate of self-copying error due to the lack of proofreading mechanisms. Pioneer experiments with RNA bacteriophages showed high levels of genetic variability, suggesting error-prone replication and, later on, RNA viruses were confirmed to show high per-base mutation rates and short genomes. The quasispecies theory offered thus a relevant conceptual framework for RNA virus evolution and became a virological paradigm.

Scientific Context

Quasispecies theory is a population genetics theory. Population genetics, which took its first steps in the beginning of the twentieth century, addresses how changes in gene frequencies depend on measurable parameters such as selection coefficients, population sizes, mutation rates, or recombination rates, and provides a well-grounded theory for the study of evolution. Nevertheless, evolution is a very complex process and its mathematical treatment requires the use of simplifying assumptions. Choosing the appropriate assumptions is critical to build accurate models of evolution. This choice, which depends on the specific questions to be addressed, usually makes the difference between evolutionary models.

One central question in population genetics is how much variability can be stably maintained from the interplay between mutation and natural selection. In all populations, the continuous production of deleterious mutations through error-prone replication is counterbalanced by the elimination of these mutations through selection. When a population is sufficiently adapted to its environment, the observed variability is mainly determined by the interplay between the above two opposing factors which, in the long term, reach the so-called mutation–selection balance. Quasispecies theory mainly addresses this issue, though it had already been addressed decades before by the founders of population genetics. Since quasispecies come from the field of physical chemistry, their terminology is somewhat different, but most concepts are fully equivalent to those of population genetics (**Table 1**).

The Fundamental Model

The basic quasispecies model describes the population dynamics of macromolecular sequences maintained by error-prone replication. Mutations occurring during the copying process can modify the replication rate of the newly arisen sequences, allowing Darwinian selection to operate. The system can be described by a value matrix $\mathbf{W}$

Table 1 Conceptually equivalent terminology in quasispecies and population genetics theories

Quasispecies theory	Population genetics theory
Quasispecies	Population at the mutation–selection balance
Stationary state	Mutation–selection balance
Master sequence	Wild type
Superiority of the master	Average selection coefficient
Phase transition	Selective sweep
Excess production	Progeny number
Quality factor	Replication fidelity

in which each diagonal element W_{kk} equals the number of nonerroneous copies resulting from exact replication of sequence k and each off-diagonal element W_{kj} equals the number of copies of sequence k resulting from erroneous replication of sequence j. The stationary sequence distribution is called the quasispecies and consists of a master sequence along with a distribution of mutants. Although the slower replicators cannot sustain their abundance level by themselves, they are constantly recreated from other sequences that mutate into them. At the stationary state, the average biological fitness of the population equals the largest eigenvalue of $\mathbf{W}$ and the abundance of each sequence is given by the associated eigenvector.

In practice, given the huge dimension of $\mathbf{W}$ (equal to the genome length), exact solutions of the quasispecies equations cannot be found. If all sequences in a given error class (i.e., all one-error mutants, all two-error mutants, etc.) are assumed to have identical fitness, it becomes possible to estimate the frequency of each sequence as a function of the per-base mutation rate, its number of mutations, its relative fitness, and a recursively defined function that accounts for the fitness of all possible mutant intermediates. However, the fitness of all intermediates is not known in practice. The problem can be solved by assuming that mutations have independent fitness effects, or using population genetics terms, that there is no epistasis. This implies that the fitness of each mutant is simply the product of fitness values associated to each single mutant and thus, that the mean frequency of mutants carrying d mutations relative to that of the master (q_d/q_o), can be simply obtained by summing over all possible combinations of d mutations in a genome of length L. Using the binomial distribution, this can be written as $q_d/q_o = (1-\mu)^{-LBi(d|\mu,L)s^{-d}}$, where μ is the per-base mutation rate.

Assuming that L is large enough to be treated as infinite can be useful to illustrate the equivalence between quasispecies and population genetics models. The binomial distribution can then be substituted by a Poisson distribution of parameter $U = \mu L$, where U is the genomic mutation rate, and $(1-\mu)^L$ can be replaced by e^{-U}. Hence, $q_d/q_o = (U/s)^d/d!$ and, setting $q_o = e^{-U/s}$, it follows that

q_d is Poisson-distributed with parameter $\lambda = U/s$. Hence, the average number of mutations per genome is U/s and the mean relative fitness of the population is e^{-U}, which are classical population genetics results.

Common Assumptions

Although the quasispecies theory is very general in its basic formulation, some simplifications have to be done for the sake of tractability, going to the detriment of generality. Common assumptions of the original quasispecies models are that: (1) there is a single fittest sequence; (2) mutation rates are constant over time and genotypes; (3) mutants show invariant fitness regardless of the number of mutations they carry; (4) molecules are degraded at a constant rate; (5) the environment is constant; (6) the population size is infinite; and (7) there is no genetic recombination. All of these assumptions were made in most of the analyses originally published by Eigen and co-workers. The situation in which there is a single fittest sequence and all other genotypes have constant fitness can be represented by a two-class single-peak landscape, although real landscapes can be much more complex, with several local adaptive peaks separated by low-fitness regions (**Figure 1**).

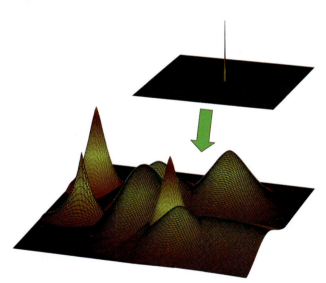

Figure 1 Two hypothetical adaptive landscapes. Sequence spaces are highly dimensional and need to be reduced to only two dimensions for visualization. The third dimension represents fitness, such that peaks in the landscape are local or global fitness maxima, whereas valleys are low-fitness regions. Above, the two-class single-peak landscape originally used by Eigen, in which there is a single fittest sequence and all other types show a lower, constant fitness value. A more complex landscape, with several local peaks, is represented below. Ruggedness appears as consequence of frequent epistasis of variable sign and intensity. The lower example provides a more realistic picture than the above two-class single-peak case, though it might still be much simpler than real landscapes.

Later on, some of the above assumptions were relaxed. For example, Peter Schuster and co-workers studied the case in which there is more than one fittest sequence and called it a degenerate quasispecies. The two master sequences can be located at distant regions of the sequence space separated by low-fitness regions, hence being effectively unconnected. Different generalizations of the two-class single-peak landscape have been studied, including multidimensional landscapes. A simple though useful case is the multiplicative landscape, in which epistasis is assumed to be null and hence fitness decays exponentially with mutation number. Others have also investigated quasispecies in variable environments or with finite population sizes.

Relevant Predictions

Although population genetics and quasispecies are fundamentally equivalent, the assumptions that are typically done in quasispecies models often differ from those made in classical population genetics. Importantly, different assumptions can lead to different predictions. Below, we focus on two relevant examples, concerning the frequency of mutations and the topology of adaptive landscapes.

The evolution of mutational robustness

The Haldane–Muller principle states that the average fitness of the population at the mutation–selection balance does not depend on the fitness effects of mutations or, more specifically, on selection coefficients. The reason is that the more deleterious a mutation is, the more it reduces the average population fitness but, on the other hand, the more its frequency is lowered by selection. These two factors cancel out and the average relative fitness of the population simply writes e^{-U} (see above). This result is based on neglecting back-mutations, which is justified for large genomes replicating at low per-base error rates. However, quasispecies were intended to capture the population dynamics of small replicons with high per-base mutation rates, implying that secondary and back-mutations could not be systematically neglected. The reason is that the probability of occurrence of back-mutations relative to forward mutations increases in short genomes and, at high genomic mutation rates, the evolutionary fate of any given mutant genotype can depend not only on its fitness effect, but also on secondary mutations, or on the influx of mutations regenerating the genotype from neighbors in sequence space.

When secondary and back-mutations are taken into account, the average fitness of the population does not anymore strictly depend on the deleterious mutation rate, but also on the average selection coefficient. In other words, the fitness of a genotype is not only determined by its own replication rate, but also by those of its mutational neighbors. Everything else being equal, the lower

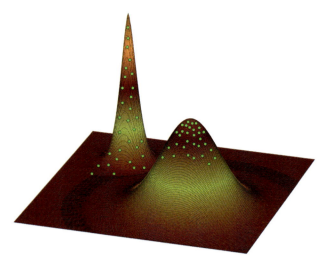

Figure 2 Representation of survival of the flattest in two-peak landscape. One population is located at a higher peak, which should confer it a selective advantage. However, the other population is located at a flatter peak, implying that neighbors in the sequence space have similar fitness and, thus, that the population is more robust to mutation. A prediction of the quasispecies theory is that, at high mutation rates, increased robustness can provide a selective advantage to the latter population.

the selection coefficient the higher the average fitness. A consequence is that, at high mutation rates, a slower replicator can potentially outgrow a faster competitor by virtue of its higher mutational robustness, a phenomenon called the 'survival of the flattest' (**Figure 2**). This effect rapidly vanishes as the mutation rate decreases and the genome length increases, but it can be magnified in small populations, in which genetic drift favors the accumulation of mutations. It should be noted that the statement that the average fitness is not a property of the master but also of the neighboring sequences does not mean that the genotypes constituting the quasispecies establish any kind of cooperation, complementation, or mutualism. Experiments showing cooperative interactions between genotypes should therefore not be considered as supportive of the quasispecies predictions.

The error catastrophe

The existence of an error catastrophe is probably the most widely known prediction of the quasispecies theory. An error threshold is the critical error rate beyond which Darwinian selection cannot further maintain the genetic integrity of the quasispecies. Below the threshold, the quasispecies is stably localized around one or more sequences, but if the mutation rate exceeds the threshold value, the variety of mutants that are in amounts comparable to the fittest types becomes very large and the population is said to enter into error catastrophe. According to the original definition given by Eigen, the threshold satisfies $W_{00} = \bar{w}_{k \neq 0}$, where W_{00} is the nonmutated progeny of the master and $\bar{w}_{k \neq 0}$ is the average fitness (mutated and nonmutated progeny) of all other sequences present in the population. If $W_{00} < \bar{w}_{k \neq 0}$, the master is overgrown by its own mutational cloud and, in the absence of back-mutation, goes extinct through error catastrophe (although assumptions about back-mutation do not have a bearing on the existence or magnitude of error catastrophes).

The existence of an error threshold, however, is not an obligate prediction of the quasispecies theory. For example, in a multiplicative landscape, the frequency of the master is given by $q_0 = e^{-U/s}$, a quantity that decreases as the mutation rate increases but never reaches zero, hence there is no error threshold. The original quasispecies model produces an error threshold because all mutants are assumed to have a constant fitness regardless of their mutational distance to the master. In this case, $\bar{w}_{k \neq 0}$ is a constant, $\bar{w}_{k \neq 0} = 1 - s$, and, therefore, an error threshold deterministically takes place when $L = \log \sigma_0 / \mu$, where σ_0 is the superiority of the master over all mutants. It must be noted that, even in this model, though there is an error threshold, no sharp transition in average fitness occurs. Average fitness smoothly decreases with increasing mutation rate until it reaches a plateau value equal to $\bar{w}_{k \neq 0}$. In general, whether there is an error threshold critically depends on the assumed adaptive landscape. Some generalizations of the two-class single-peak landscape predict error catastrophes, but, in general, the outcome mainly depends on the sign of epistasis. Error catastrophes have a greater probability to exist if epistasis is antagonistic, such that mutations tend to have progressively less fitness effects as they accumulate (notice that the two-class single-peak landscape represents an extreme form of antagonistic epistasis). However, recent work has shown that with null epistasis, there can be an error catastrophe if lethal mutations are taken into account. This is a complex theoretical issue which still remains to be fully resolved.

Experimental Evolution of RNA Viruses

RNA viruses are useful tools for experimentally addressing fundamental evolutionary issues while they still allow us to pay attention to the molecular aspects. Based on scientific but also historical reasons, the quasispecies theory is the *de facto* standard for interpreting evolutionary experiments with RNA viruses. RNA viruses are characterized by their error-prone replication and small genomes, which makes the quasispecies an *a priori* relevant theoretical approach. However, even if some predictions are specific to the quasispecies theory, most are equivalent to those made by previously proposed and simpler population genetics models. Furthermore, quasispecies do not provide a theoretical underpinning to some experimental aspects of viral evolution which are better explained by nondeterministic population genetics models. After three decades of experimental work, some general evolutionary properties of RNA virus populations have been elucidated, most of them being compatible with both quasispecies and classical population genetics. A few data, though, support the need to use the more complex quasispecies formulation.

Key Evolutionary Parameters of RNA Viruses

High per-base mutation rates

In pioneer evolutionary experiments performed in the late 1970s, RNA bacteriophage Qβ populations serially passaged in laboratory conditions were found to be highly variable at the nucleotide level, each viable phage genome differing on average in one to two positions relative to the reference sequence. It was suggested that these populations were in a dynamic equilibrium, with viable mutants arising at a high rate and being strongly selected against, an equilibrium known as the mutation–selection balance. From then on, many evolutionary experiments have been done with different model RNA viruses, including the bacteriophage φ6, the vesicular stomatitis virus (VSV), poliovirus-1, or foot-and-mouth disease virus (FMDV). It has been estimated that the mutation rate of RNA viruses is within the range 10^{-6}–10^{-4} substitutions per replication event, which is orders of magnitude higher than that of DNA-based organisms. Biochemical experiments with the avian myeloblastosis virus, VSV, and human immunodeficiency virus type 1 (HIV-1) have established that their polymerases lack $3'$ exonuclease activity, providing a basis for error-prone replication. Recent experiments have demonstrated that poliovirus-1 is capable of counteracting the mutagenic effect of base analogs such as ribavirin by evolving specific genotypic changes that confer increased replication fidelity. This suggests that the high mutation rate of RNA viruses is not only a consequence of fundamental biochemical restrictions, but also the product of evolutionary optimization.

Compacted genomes

Another seemingly general property of RNA viruses is their small genome size (e.g., 3569 nt for MS2 phage to 11 162 nt for VSV), with many examples of overlapping reading frames and multifunctional genes. In contrast to multicellular eukaryotes, noncoding regions occupy a small fraction of the genome, hence making silent mutations relatively infrequent. A consequence of genome compactness is that RNA viruses are extremely sensitive to deleterious mutations. For VSV, it has been estimated that single random nucleotide substitutions have average selection coefficients of 0.5 (fitness is decreased by 50%), at least an order of magnitude higher than in DNA-based organisms, and it has been estimated that up to 40% of random mutations inactivate the virus. A related consequence of genome compactness is that epistasis is antagonistic on average, because in such compact genomes, one or few changes are enough to disrupt most of the encoded functions, and hence additional mutations can only produce a comparatively smaller effect.

Rapid growth

RNA viruses critically rely on fast replication for survival. Their rapid infection cycles allow them to reach high titers before the onset of host defense mechanisms. In cellular cultures, phages complete an infection cycle in less than an hour, whereas mammalian viruses do it in a few hours. Each infected cell typically releases hundreds to few thousands of infectious particles and viruses rapidly reach population sizes of several billion particles per milliliter. Intuitively, high particle counts indicate that the number of available genotypic variants is high, which should allow Darwinian selection to operate more efficiently. However, it must be noted that although population sizes are huge, the relevant parameter in evolutionary terms is the effective population size, since it indicates the actual number of particles that contribute to the next generation. Demographic fluctuations, originated by transmission bottlenecks, host defense mechanisms, or antiviral treatments make the effective population considerably smaller than particle counts. For example, in typical passage regimes with mammalian lytic viruses, if 10^4 particles were inoculated into 10^5 cells, effective population sizes would be in the range of 10^4–10^5 particles, whereas particle counts would typically be in the order of 10^9.

Main Evolutionary Properties of RNA Viruses

Rapid adaptation dynamics

Similar to what is found with other microorganisms, a frequent observation in long-term evolution experiments is that RNA viruses show an initially fast adaptation that tends to decelerate over time. Such dynamics indicates that, after being placed in a new environment, populations are evolving from a region of low fitness toward an adaptive peak or plateau. High population sizes make it more likely that beneficial mutations are created and, together with fast replication rates, this explains the observed rapid adaptation. Also, elevated selection coefficients increase the probability that any given beneficial mutation becomes fixed in the population. Finally, high mutation rates produce high levels of genetic variability on which selection can operate, though it must be noted that, since most mutations are deleterious, increasing the mutation rate does not necessarily imply increasing the rate of adaptation. Maximal adaptation rates should indeed be reached at intermediate mutation rates, the optimal mutation value being mainly determined by the selection coefficient against deleterious mutations and the rate of recombination.

Evolutionary parallelisms and convergences

In several serial passage experiments with bacteriophages, plant viruses, and mammalian viruses, sequencing of lineages independently adapted to the same environment has revealed a large amount of evolutionary parallelisms and convergences at the genetic level, both synonymous and nonsynonymous. This pattern indicates that, upon facing identical selective pressures, viruses often find the same adaptive pathways. It is possible that in simple and compacted genomes, such as those of RNA viruses, the number of alternative evolutionary solutions is limited, or that, due to their elevated adaptive potential, RNA viruses are able of systematically finding the highest peak in the adaptive landscape. Both scenarios would indicate that viral evolution at high population sizes has an important deterministic component, as proposed by quasispecies models. However, it is also possible that the common ancestor was located near a local peak surrounded by low-fitness regions and that, since Darwinian selection always pushes populations uphill, these populations would have remained trapped in their local peak.

Fitness tradeoffs

Despite their remarkable adaptability, RNA viruses face some evolutionary constraints. For example, shifts between vertebrate and arthropod hosts impose drastic environmental fluctuations to arboviruses. Several studies have confirmed that adaptation to a novel host often decreases competitive ability in the former host. These tradeoffs can arise by antagonistic pleiotropy or antagonistic epistasis among beneficial mutations. Pleiotropy, epistasis, and hence fitness tradeoffs are expected to be common in compacted genomes with many overlapping functions. A particularly important example of tradeoff is that exerted on mutation rate. Most mutations are deleterious, and hence there is a short-term selective pressure for reducing mutation rates toward whatever limit is imposed by biochemical restrictions, but, on the other hand, mechanisms of replication fidelity should come at a kinetic cost and thus be selected against. Since RNA viruses critically rely on fast replication, the cost of proofreading functions in terms of replication rate might be particularly strong.

Transmission bottlenecks

Considering the fact that RNA viruses show large selection coefficients and high population sizes, it could be concluded that selection is the main factor driving their evolution. However, in nature, viral populations experience strong bottlenecks upon transmission and therefore, effective population sizes are well below particle counts. Low effective population sizes increase the strength of genetic drift, hence favoring the accumulation of deleterious mutations and potentially jeopardizing the survival of the population. This stochastic process is known as Muller's ratchet, named after the population geneticist who first anticipated it in the 1930s. The expected rate of fitness loss equals the product of the mutation rate and the average selection coefficient for deleterious mutations, thus making RNA viruses especially sensitive to Muller's ratchet. This prediction has been experimentally validated by performing serial plaque-to-plaque passages in a variety of RNA viruses, including bacteriophage $\phi 6$, VSV, FMDV, and HIV-1.

Is the Quasispecies Theory Relevant to the Evolution of RNA Viruses?

There has been some controversy as to whether quasispecies are relevant to RNA virus evolution. Detractors have sometimes argued that the theory overlaps with previous population genetics theory, whereas supporters have claimed that it goes beyond population genetics. Many virologists believe that the elevated genetic variability of RNA viruses proves their quasispecies nature, but the mere observation of genetic variability is a trivial consequence of mutation and hence can hardly be considered supportive of any biological theory. Most of the empirically inferred properties of RNA viruses are already well explained by classical population genetics, as for example the expected high adaptability of large populations, the most likely deleterious effects of mutation, the complex relation between mutation and adaptation rates, the existence and evolutionary implications of fitness tradeoffs and epistasis, or the genetic contamination of small populations throughout Muller's ratchet. To justify the necessity of the extra complexity introduced by the quasispecies theory, evidence for its specific predictions should be provided.

The evolution of robustness

RNA viruses are highly sensitive to mutation compared to more complex microorganisms. Genetic hypersensitivity and large population sizes make selection very efficient at purging deleterious mutations and hence promote the preservation of the wild type. Although this seems to be the predominant survival strategy among RNA viruses, quasispecies theory predicts that at high mutation rates and low population sizes, mutational robustness can be selectively advantageous. The first experimental evidence in favor of this prediction came from work with the bacteriophage $\phi 6$ showing that the evolution of a genotype can be influenced by the topology of the neighboring regions in the adaptive landscape and that robustness can be a selectable character. A more direct approach was undertaken with viroids, plant pathogens consisting of small, noncoding RNA molecules. Two different viroid species were competed in common host plants; one was characterized by fast population growth and genetic homogeneity, whereas the other showed slow population

Table 2 Differences between lethal mutagenesis and error catastrophe processes

	Lethal mutagenesis	Error catastrophe
Nature of the process	Population extinction through mutation accumulation	Delocalization of the quasispecies due to mutation
Name of the threshold	Extinction threshold	Error threshold
Key parameters	Mutation rate, fitness of the wild type, selection coefficient	Mutation rate and selection coefficient
Demography	Population size declines	No changes in population size are specified
Fate of the wild type	Not necessarily extinguished until the end of the process	Extinguished while other variants survive
Dependence on mutation rate	Extinction is more likely and occurs faster at higher mutation rates	Beyond the error threshold, further increases in mutation rate have no effect
Typically assumed effect of mutations on fitness	The fitness of mutant genotypes decays with mutation number	The fitness of mutant genotypes does not decay with mutation number
Mutational signature	No specific changes are required in the consensus sequence	The consensus sequence randomly drifts through time

growth and a high degree of variation. When the mutation rate was artificially increased using UVC radiation, the fitness of the slower replicator increased relative to that of the faster competitor, probably due to its higher mutational robustness, which might support the survival of the flattest hypothesis. Mutagens are especially suited for observing this quasispecies prediction, since they simultaneously increase mutation rates and decrease viral titers. Recent experiments with VSV populations indicate that base analogs such as 5-fluorouracil can provide a selective advantage to robust genotypes.

Error catastrophe and lethal mutagenesis

The consequences of artificially increasing error rates have been explored in cell culture experiments with a variety of RNA viruses, including VSV, HIV-1, poliovirus-1, FMDV, and lymphocytic choriomeningitis virus (LCMV). All of these studies have shown that mutagens are detrimental to viral fitness and, in some cases, lead to extinction. For example, in HIV-1, the addition of the base analog 5-hydroxydeoxycytidine can result in a loss of infectivity after 10–20 serial passages. Virologists, assuming that error catastrophe is a form of lethal mutagenesis, have often interpreted these experimental observations in terms of error catastrophe. However, the two concepts are not equivalent, because lethal mutagenesis implies extinction, whereas the error catastrophe is a shift in sequence space without a bearing on the survival of the population (**Table 2**). The results from lethal mutagenesis experiments are easily explained by classical population genetics, provided fitness is expressed in absolute rather than relative terms. A simple criterion for extinction is $e^{-U}R$, where R is the progeny number per infectious particle and e^{-U} equals the mean equilibrium fitness of the population (this expression simply states that if the absolute mean fitness is below one, population size will deterministically decline). Therefore, current experimental evidence does not support the error catastrophe prediction. More specific observations would be required (**Table 2**), although it must be noted that the error catastrophe is not a necessary prediction of the quasispecies theory because it critically depends on the assumed adaptive landscape.

See also: Antigenic Variation; Antiviral Agents; Evolution of Viruses; Nature of Viruses; Virus Evolution: Bacterial Viruses; Virus Species.

Further Reading

Bull JJ, Sanjuan R, and Wilke CO (2007) Theory of lethal mutagenesis for viruses. *Journal of Virology* 81: 2930–2939.

Domingo E, Biebricher CK, Eigen M, and Holland JJ (2001) *Quasispecies and RNA Virus Evolution: Principles and Consequences.* Austin, TX: Landes Bioscience.

Domingo E, Sabo D, Taniguchi T, and Weissmann C (1978) Nucleotide sequence heterogeneity of an RNA phage population. *Cell* 13: 735–744.

Eigen M, McCaskill J, and Schuster P (1988) Molecular quasi-species. *Journal of Physical Chemistry* 92: 6881–6891.

Elena SF and Lenski RE (2003) Evolution experiments with microorganisms: The dynamics and genetic bases of adaptation. *Nature Reviews in Genetics* 4: 457–469.

Hartl DL and Clark AG (2006) *Principles of Population Genetics*, 4th edn., Sunderland, MA: Sinauer.

Holmes EC and Moya A (2002) Is the quasispecies concept relevant to RNA viruses? *Journal of Virology* 76: 460–465.

Lenski RE (2000) Evolution, theory and experiments. In: Alexander M, Bloom BR, Hopwood DA, et al. (eds.) *Encyclopedia of Microbiology*, 2nd edn., pp. 283–298. New York: Academic Press.

Moya A, Holmes EC, and Gonzalez-Candelas F (2004) The population genetics and evolutionary epidemiology of RNA viruses. *Nature Reviews in Microbiology* 2: 279–288.

Sanjuan R, Cuevas JM, Furio V, Holmes EC, and Moya A (2007) Selection for robustness in mutagenized. *PLoS Genetics* 3(6): e93.

Wagner A (2005) *Robustness and Evolvability in Living Systems.* Princeton, NJ: Princeton University Press.

Wilke CO (2005) Quasispecies theory in the context of population genetics. *BMC Evolutionary Biology* 5: 44.

Wolf JB, Brodie ED III, and Wade MJ (2000) *Epistasis and the Evolutionary Process.* Oxford: Oxford University Press.

Rabies Virus

I V Kuzmin and C E Rupprecht, Centers for Disease Control and Prevention, Atlanta, GA, USA

© 2008 Elsevier Ltd. All rights reserved.

Glossary

Apoptosis One of the main types of programmed cell death which involves a series of biochemical events leading to morphological changes and death.
Chiropteran Pertaining to members of the order of flying mammals commonly called 'bats'.
Gliosis Proliferation of astrocytes in damaged areas of the central nervous system.
Hematophagous The habit of certain animals of feeding on blood.
Herpestidae A vertebrate family which includes mongooses.
Meninges The system of membranes which envelope the central nervous system. The meninges consist of three layers: the dura mater, the arachnoid mater, and the pia mater.
Neuronophagia Phagocytic destruction of nerve cells.
Paraesthesia A sensation of tingling, pricking, or numbness of a person's skin with no apparent physical effect.

Introduction

Rabies is fatal encephalitis caused by rabies virus and other lyssaviruses. This important zoonosis has been known for over 4000 years, but still causes more than 50 000 human deaths annually besides enormous economic losses, primarily developing countries of Asia and Africa where dog rabies is enzootic. Once symptoms appear, the disease is almost invariably fatal. Since the first successful use of rabies vaccine by Louis Pasteur in 1885, pre-exposure and post-exposure prophylaxis have been improved significantly, and canine rabies has been eliminated from Western Europe and large parts of North America. Nevertheless, the continuing burden of the disease in other parts of the world poses requirements for further development of affordable protective biologicals and treatment regimens for rabies.

Genome and Morphology

Rabies virus is classified as the type species of the genus *Lyssavirus*, family *Rhabdoviridae*, order *Mononegavirales*. The negative-sense single-stranded RNA (ssRNA) genome is *c.* 12 kbp in length (**Figure 1**) and encodes a short leader sequence (*c.* 50 nt), followed by five structural protein genes, separated by nontranscribed intergenic regions. The N gene consists of 1350–1353 nt and codes for the nucleoprotein. Due to its abundance and ease of the detection by fluorescent antibody and RNA-based methods, the nucleoprotein and the N gene are the most common targets of diagnostic tests, and antigenic and phylogenetic typing. The P gene consists of 891–915 nt and codes for the phosphoprotein. This protein is composed of two domains: the NH_2-terminal portion contains an L protein binding site, as well as a weak N protein binding site; the COOH-terminal portion contains a strong N protein binding site. The M gene consists of 606 nt and codes for the matrix protein. This protein is involved in later steps of virion formation, such as tightening of the ribonucleoprotein complex (RNP) into a compressed helical structure and the budding of nucleocapsids at cell membranes to obtain the viral envelope. The G gene consists of 1566–1599 nt and codes for the transmembrane glycoprotein. The glycoprotein interacts with receptors on the cell surface and is responsible for neutralizing antibody production. It is composed of four distinct domains: a signal peptide, ectodomain, transmembrane peptide, and cytoplasmic domain. The glycoprotein, and particularly the ectodomain, play a crucial role in rabies virus (RABV) pathogenicity. For example, one

Disclaimer: The findings and conclusions in this report are those of the authors and do not necessarily represent the views of the funding agency.

Figure 1 Structure of the rabies virus genome. N, nucleoprotein gene; P, phosphoprotein gene; M, matrix protein gene; G, glycoprotein gene (including SP, signal peptide; ECTO, ectodomain; TD, transmembrane domain; ENDO, endodomain); Ψ, large G-L intergenic region, sometimes referred to as a pseudogene; L, RNA-dependent RNA polymerase gene.

arginine residue at position 333 of the ectodomain is important for peripheral infectivity of RABV. Replacement of this amino acid in vaccine strains significantly attenuates pathogenicity. The L gene is 6381–6426 nt length and codes for the RNA-dependent RNA polymerase. This is a highly conserved polypeptide responsible for replication and transcription of the viral genome.

The RABV virion is bullet-shaped (**Figure 2**), 50–100 nm × 100–430 nm in dimensions, and composed of two structural units: an internal helical RNP, about 50 nm in diameter, and a lipid envelope which is derived from the host cytoplasmic membrane during budding. The RNP is comprised of the RNA genome and N nucleoprotein in tight association. The heavily phosphorylated phosphoprotein and polymerase are also bound to the RNP. The exact position of the matrix protein remains controversial, and may be either contained in the central channel of RNP or embedded into the inner layer of the virion membrane. Knobbed glycoprotein spikes, consisting of three glycosylated ectodomains and serving for binding of the virions to host cell receptors, protrude through the virion membrane.

Phylogeny and Evolution

RABV is the most broadly distributed member of the genus *Lyssavirus*, circulating in both carnivorous and chiropteran hosts. Phylogenetic lineages of RABV reflect the geographic distribution and association with a particular host species (**Figure 3**). In general, the RABV lineage is split into two large clusters. One includes viruses circulating worldwide among terrestrial carnivores (predominantly canids and mongoose). The second cluster is indigenous to the New World, and includes viruses circulating among raccoons, skunks, and a variety of bats. The first cluster includes the major 'cosmopolitan' canid RABV lineage, believed to have originated in Europe and widely disseminated as a consequence of global colonization during the sixteenth to nineteenth centuries. Presently, these viruses circulate in moderate latitudes of Eurasia, the Middle East, Africa, and the Americas. The majority of RABV vaccine strains, such as PV, ERA (SAD), PM, and HEP, belong to the 'cosmopolitan' RABV lineage. The lineage of Arctic and Arctic-like rabies viruses is monophyletic but, ancestrally, it is linked to the 'cosmopolitan' group. Viruses comprising the Arctic portion of this lineage are distributed in circumpolar

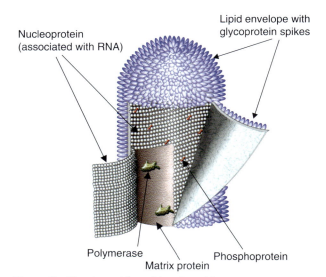

Figure 2 Structure of the rabies virus virion.

regions of Eurasia and North America, whereas distinct groups of Arctic-like rabies viruses circulate in some regions of the Middle East, and southern and eastern Asia. Representatives of some African virus lineages of Canidae and Herpestidae origin are more distantly related. The most divergent representatives of the Old World RABVs circulate in dogs in South Asia. Among the indigenous New World RABV lineages, the majority are tightly associated with particular bat species.

RABV is one of the most slowly evolving negative-stranded RNA viruses, with substitution rates of approximately 2×10^{-4} to 5×10^{-4} per site per year. No recombination has been reported in RABV genomes. Mutations occur due to the lack of proofreading and post-replication error correction by RNA polymerase. However, there is clear evidence of negative selection, as RABV and other lyssaviruses are subjected to strong constraints against amino acid substitutions, probably related to their unique pathobiology. There is some limited evidence that positive selection may have occurred at a few sites in the G gene, such as codons 156, 160, 183, and 370, but the role of these residues is not currently known. Although descriptions of a disease similar to rabies are known from 4000-year-old manuscripts, the origin of current RABV diversity has been estimated at 800–1500 years for separation of the major bat and terrestrial virus lineages, 1200–1800 years for the closest ancestor of

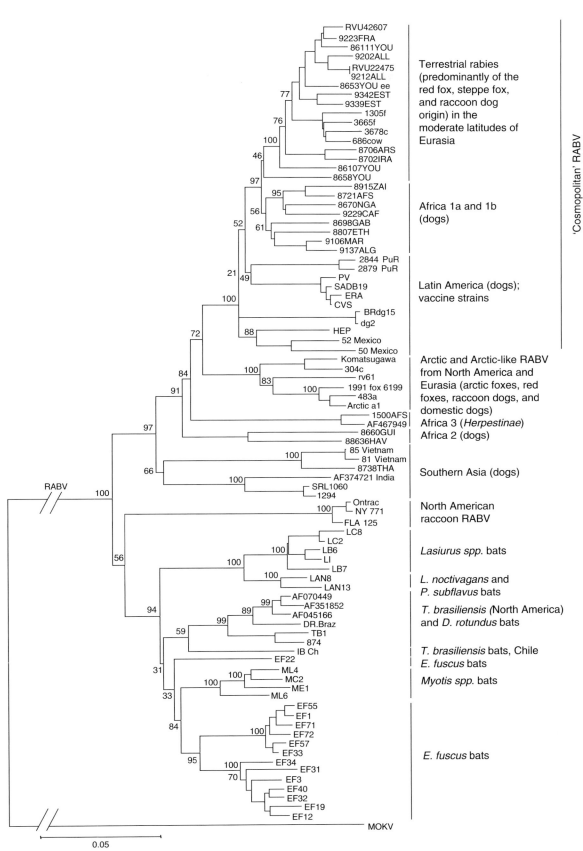

Figure 3 Phylogenetic tree of the major rabies virus lineages based on the sequences of the entire nucleoprotein gene. Bootstrap support values are presented for key nodes, and branch lengths are drawn to scale.

current American bat rabies viruses, and 280–500 years for the 'cosmopolitan' group.

Pathogenesis

After delivery into a wound, RABV can infect several types of cells and replicate at the inoculation site, as has been shown for skeletal muscle cells and fibroblasts. Attachment to cell membrane receptors is mediated by the glycoprotein spikes protruding from the virion membrane. Several types of putative receptors for RABV attachment have been suggested: nicotinic acetylcholine receptor, carbohydrate moieties, phospholipids, and gangliosides. Once bound, the virus enters the cell by endocytosis. Following a decrease in pH in the endosomal vesicle, the viral membrane fuses with the endosomal membrane, and the RNP is released into the cytoplasm. Because RABV RNA has negative polarity, it must be transcribed to produce the complementary positive-sense mRNA. This process is mediated by the viral RNA-dependent RNA polymerase. The RNP serves as a template for transcription and replication, and protects the RNA from nuclease activity. Translation of the viral mRNAs is ensured by the cellular protein synthesis machinery. All processes of transcription, translation, and replication take place in the cytoplasm. The glycoprotein is synthesized at the rough endoplasmic reticulum and delivered to the cytoplasmic membrane. The other viral proteins are expressed in the cytosol by free polyribosomes. At the final stage, transcription and replication are inhibited, the RNP becomes intensively condensed and assembled into mature nucleocapsids, which are subsequently delivered to the cell membrane for the budding of complete virions. During the budding process, virions acquire the lipid envelope and the glycoprotein which is embedded in the cell membrane.

After several replication cycles at the inoculation site, RABV penetrates peripheral nerves and spreads to the central nervous system (CNS) by retrograde axonal transport. Neuronal pathways shield the virus from host immune surveillance, resulting in the absence of an early antibody response. Once delivered to the CNS, the virus disseminates rapidly in the spinal cord, medulla, thalamus, pons, hippocampus, striatum, cerebellum, and cortex. The spread of infection within the CNS occurs from one neuron to another by both axonal and trans-synaptic transport. Neuropathological changes observed in the infected brain are relatively mild histologically and include gliosis, slight neuronophagia, and perivascular infiltration with inflammatory cells, with rare involvement of meninges. Occasionally, more severe brain damage occurs, such as spongiform lesions, extensive neuronal degeneration, and widespread inflammation. Functional alteration of the CNS is much more significant than morphological changes. Apoptosis as a response to RABV infection is an additional prominent factor of neuron damage. Generalized CNS dysfunction leads to a lethal outcome.

Reverse dissemination of virus from the CNS during the clinical period of rabies occurs along peripheral nerves. The RABV RNA may be detected in a variety of organs and tissues at the end of clinical period. However, infectious virus can be isolated from extraneural tissues only occasionally and in low titers. The exception is the salivary glands, where virus passes additional replication cycles and is released into the saliva to enable transmission.

Although a bite is the main method of transmission, non-bite exposures also may occur under unusual circumstances. For instance, two human rabies cases have been attributed to airborne exposure in laboratories, and another two cases have been attributed to airborne exposures in a bat-infested cave in Texas. Aerosol transmission has been demonstrated experimentally, only in very specific conditions with a highly concentrated viral aerosol. Several cases of human-to-human RABV transmission have occurred due to transplantation of cornea, liver, kidneys, and associated tissues obtained from donors who died of rabies. Exposure may occur by direct contact of saliva, nervous or other infected tissues from a rabid animal with mucous membranes, or from scratches. For example, several cases have occurred in trappers who skinned rabid animals without suitable protection.

Various mammalian species exhibit different susceptibilities to the variety of RABV variants, including mutual adaptation of virus and principal host. For example, canids are highly sensitive to homologous RABV variants, and develop the furious form of rabies with high titers of virus in salivary glands. These peculiarities ensure transmission of the infection to a critical number of susceptible individuals, before the death of the rabid animal. A very low level of seroprevalence has been detected in natural populations of foxes, indicating that most RABV contact events lead to a fatal infection. In contrast, a high level of seroprevalence has been detected among bats. These gregarious mammals demonstrate moderate to low susceptibility to RABV. In many cases, bats develop antibody rather than disease when they encounter RABV. In general, RABV evades immunological responses after entering into the nervous system. Development of antibodies could be attributed to peripheral virus activity rather than to CNS infection. There is contradictory evidence as to whether bats or other mammals can survive rabies. Initial observations made at the beginning of the twentieth century on vampire bats suggested that they can be 'asymptomatic carriers' of RABV. More recent reports from Spain have described some cases in which viral RNA was detected in extraneural tissues and oral swabs of naturally infected bats that were negative for rabies infection in brain tissue. Experimental studies of vampire bats have demonstrated intermittent shedding of RABV in saliva. However, a number of other experiments and field surveillance performed in Europe and the USA provided no evidence in support of a 'carrier' state: bats that developed rabies always died, and virus was detected only in those animals but not in healthy survivors. Some survivors do develop virus-neutralizing antibodies, suggesting a form of abortive infection. Most likely, this abortive infection occurs at the inoculation site. The susceptibility of other mammalian species, that do not serve as principal RABV hosts, is variable. Susceptibility of primates, including humans, is low to moderate, depending on the virus variant.

Indeed, rare cases of survival after the manifestation of clinical signs of rabies are occasionally registered in different animal species. However, these sporadic events cannot be taken in support for a theory of RABV persistence in vivo. At least six cases of human recovery after clinical rabies have been published. Five of the patients were vaccinated before clinical onset, and in one case no vaccination was performed. Rabies diagnosis in each case was based on a history of exposure, compatible symptoms, and increasing titers of anti-RABV antibodies in the serum and cerebrospinal fluid (CSF). However, no virus isolates were obtained from these patients.

Clinical Spectrum

The incubation period of rabies can vary from less than 10 days to more than 6 years. The reason for this variability is not yet clear. Some reports attribute long incubation periods to virus replication in the skeletal myocytes before entering the nerves, whereas short incubation periods may be associated with immediate penetration of the nerves and transfer to CNS. Most typical incubation periods after peripheral inoculation of natural hosts and humans vary from 3 to 14 weeks.

Prodromal symptoms are nonspecific: general malaise, fever, chills, sore throat, headache, nausea and vomiting, and sometimes diarrhea, anxiety, and irritability. Humans often suffer from a paresthesia or pain in the inoculation site, and the wound may become slightly inflamed. The prodromal period usually continues for 1–3 days before development of encephalitic symptoms.

Historically, two main clinical forms of rabies have been recognized, based on predominating symptoms: furious (encephalitic) or dumb (paralytic). However, mixed forms occur as well. When the disease is furious, animals or humans become agitated and aggressive. Insomnia, irritability, and anxiety are commonly observed. Other signs, such as pupillary dilation, altered phonation, aimless wandering, drooling of saliva, and muscle tremors and seizures may be noted. Humans often develop hallucinations and delirium. Some symptoms, which have been considered as 'classic' but are observed in 50% or less of patients, are hydrophobia (painful throat seizures at attempts to drink or even due to seeing or hearing running water), aerophobia, photophobia, and phonophobia (seizures in response to airflow, bright lights, and loud sounds, respectively). Subsequently, progressive pareses and paralysis appear. Sick animals or humans become comatose and die, usually due to respiratory failure.

Paralytic rabies is characterized by a greater prevalence of pareses and paralysis from the beginning of disease manifestations, whereas agitation and anxiety are moderate or absent. In part, the form of the disease may depend on the virus variant and animal species. For instance, vampire bat RABVs commonly cause paralytic rabies in cattle and bats. Canine RABV variants from both North America and Eurasia often cause furious rabies. Once symptoms appear, death occurs usually within 1–10 days. Ventilatory support may prolong survival to 3–4 weeks, but no effective treatment exists to date. On occasion, no rabies-specific symptoms can be observed, especially when the paralytic form of disease occurs. Rabies should be considered a possible cause in each case of encephalitis when the etiologic agent is unclear and the probability of lyssavirus exposure cannot be rejected.

Diagnosis

The direct fluorescent antibody test, performed on impressions of infected brain on glass slides, is the gold standard for rabies diagnosis. The brainstem and cerebellum are tissues of choice. Commercially available fluorescein-labeled anti-rabies antibodies, either polyclonal or monoclonal, react with the nucleocapsid of the whole spectrum of RABV variants described to date. The RNP in the infected neurons is condensed into 'inclusions' which are easily observed under the ultraviolet (UV) microscope. The indirect fluorescent antibody test, employing nonlabeled anti-nucleocapsid monoclonal antibodies (N-MAbs), is used commonly for typing of lyssaviruses. Among other methods for RABV antigen capture, some modifications of the enzyme-linked immunosorbent assay (ELISA), including an immuohistochemistry test, have been described and used in research laboratories worldwide.

RABV isolation can be attempted if replication-competent virus is needed for further investigations, or to confirm a negative result obtained by antigen-capturing methods. Isolation can be performed in laboratory mice, aged from suckling to 4 weeks old. Intracerebral inoculation is preferred because the susceptibility of mice to peripheral inoculation is usually 100–10 000 times less than susceptibility to intracerebral inoculation. Incubation periods in mice inoculated intracerebrally usually vary from 3 to 14 days, depending on the dose and particular properties of the isolate, but may be prolonged up to 6 weeks. Development of typical rabies signs, particularly paralysis and death, is characteristic. The inoculation result must be verified by detection of lyssavirus antigen in the brain of moribund mice. Among available cell cultures, those which derive from mammalian neurons are preferred (e.g., use of mouse neuroblastoma cell culture (MNA) is common in diagnostic laboratories). Susceptibility of MNA cells to RABV is similar to the susceptibility of suckling mice. The result can be detected after 48–72 h of incubation. The disadvantage of cell culture compared to mouse inoculation is related to the quality of the inoculum. Field specimens are often cytotoxic and so can be tested *in vitro* only after filtration or high dilution, reducing sensitivity of the test.

The reverse transcription-polymerase chain reaction (RT-PCR) test is a powerful adjunct diagnostic tool, particularly for antemortem diagnosis of rabies in humans. It is also useful for amplifying small quantities of RNA from infected tissues for use in genetic sequencing and phylogenetic analyses. PCR primer selection depends on the survey aim. For diagnostic purposes, when the virus variant is not known, the choice of primer and PCR regimen is a compromise between specificity and sensitivity. The N gene is targeted most commonly for diagnostic purposes because of its conservation, relative abundance in infected cells, and because it is well studied and represented in the public domain (GenBank). For phylogenetic comparisons, it is useful to amplify a variable region of a gene, but the primers should target flanking conserved regions to ensure specific annealing. For the N gene, the first 400 and the last 320 nucleotides are variable, whereas the space between them (*c.* 600 nt) is conserved. Nested and hemi-nested PCR methods have been developed for RABV. Although they are more sensitive than conventional RT-PCR, nonspecific amplification may still occur, and the result must be verified by molecular hybridization or sequencing of the PCR products. If used properly, and the result is verified, nested PCR is the most sensitive test among PCR-based methods available to date. A real-time PCR test for RABV has also been described. As for other real-time PCR applications, exceptional specificity has been demonstrated for RABV RNA detection. The real-time PCR test can detect highly degraded RNA because very short nucleotide chains (<100 nt) are amplified, and allows quantification of the load of viral RNA. A disadvantage of the specificity of the real-time method is the possibility that it could fail to detect genetically divergent viruses. Therefore, real-time PCR is especially suitable for experimental studies when the genetic sequence of the virus is known.

In some cases, particularly for human antemortem diagnosis, the detection of increasing titers of anti-RABV antibodies in serum, and particularly in the CSF, provides a suitable confirmatory diagnosis of rabies.

Epidemiology

RABV is global in distribution with the exception of Australia, Antarctica, and some isolated insular territories that are inaccessible to many viral hosts. Mammals and birds are susceptible to experimental infection, but only the former are relevant in the epidemiology of the disease. In the Old World, RABV circulates among terrestrial mammals (predominantly of the order Carnivora, families Canidae, and Herpestinae). Tropical Asia and Africa are endemic for dog rabies. These territories also encounter the greatest number of human rabies cases (40 000–60 000 annually). Nearly all occur after exposure to dogs. Dogs in Asia and Africa maintain circulation of different RABV lineages but the 'cosmopolitan' variant is distributed most broadly. Dog populations within moderate latitudes of Eurasia are more limited, and implementation of comprehensive vaccination programs has eliminated dog rabies, at least in developed countries. However, RABV still circulates in these countries among wild canids (e.g., red foxes, steppe foxes, raccoon dogs). Developed countries of western Europe are nearing eradication of terrestrial rabies in wildlife because of the implementation of oral rabies vaccination (ORV), but in other territories of the Palearctic, rabies is broadly distributed in wild canids. Some regions, such as conifer forests (taiga), are apparently free of the infection because of extremely low host population densities which are apparently incapable of maintaining RABV. The Arctic variant of RABV, harbored predominantly by Arctic foxes with involvement of other species, such as red foxes, wolves, and dogs, is distributed circumpolarly in the Arctic and subarctic zones.

In the New World, besides the Arctic and 'cosmopolitan' RABV variants introduced from Eurasia and circulating among canids, a number of indigenous RABV lineages are represented. Raccoon rabies emerged in the southeastern US during the 1950s and, to date, has occupied a broad territory along the eastern coast of North America. Skunks maintain circulation of distinct RABV lineages, such as 'south-central skunk', 'north-central skunk', and 'California skunk', and are involved often in the circulation of other lineages, originating from dogs, raccoons, and even bats. For example, a population of striped skunks in Arizona has maintained circulation of an RABV variant which is regularly isolated from big brown bats, indicative of a host shift. At least seven distinct RABV variants have been described to date among hematophagous and insectivorous bats in both North and South America. Migratory species, such as *Lasiurus* spp., *Tadarida brasiliensis*, etc., facilitate virus spread within broad territories along their migration pathways. The number of identified bat-associated RABV lineages will definitely increase with the improvement of surveillance systems, particularly in South America. Bat RABV variants (predominantly associated with the silver-haired bat, *Lasionycteris noctivagans*, and the eastern pipistrelle bat, *Pipistrellus subflavus*) have been responsible for nearly 90% of domestic human rabies cases reported in the USA during the last 10 years. Many of the bat exposure cases are 'cryptic' in that the circumstances of human exposure to bats frequently are peculiar and little attention is paid to rather small lesions caused by bat bites.

Mongooses also acquire and maintain circulation of distinct RABV variants in some regions. In some instances, they maintain circulation of viruses indistinguishable from those that circulate among dogs (the 'cosmopolitan' variant in the Caribbean and Indian dog variant in southern Asia). In Africa, mongooses harbor a specific RABV variant that circulates exclusively among herpestides of several species.

Prevention and Control

Rabies can be successfully prevented by immunization, not only prior to an exposure, but also after exposure. However, post-exposure prophylaxis (PEP) should commence soon as possible, before virus entry into magistral nerves and CNS, where it becomes largely inaccessible to the immune system due to the blood–brain barrier.

Two primary kinds treatment are used for PEP: vaccines, which lead to the development of active immunity; and anti-rabies immune globulins (RIGs), which provide passive immunity at early stages of immunization when immune response to the vaccine is limited.

Modern commercial human vaccines are pure, potent, safe, and efficacious against all RABV variants described to date. Most human vaccines are derived from the Pasteur virus strain, and are inactivated. Vaccines are manufactured using cell lines or primary cell cultures (such as human diploid cell, avian embryos, Vero, and BHK cells) from which virions are concentrated and partly purified to avoid allergic and adverse reactions. Older generations of nervous-tissue-derived vaccines (sheep or suckling mouse brain) are still used in some developing countries. Due to low potency, they must be administered in significantly higher doses and for longer courses of application. Adverse reactions are more common during vaccination.

Domestic animals are vaccinated against rabies to prevent virus circulation in the population, to eliminate important sources of human exposure, and to avoid economic losses. Most modern veterinary vaccines are inactivated, and potent in the ability to induce appropriate protection for 12–48 months after one dose. Vaccination of dogs and cats is recommended in all territories where contact with RABV is possible. Immunization regulations for other species of domestic and agricultural animals differ depending on the specific epizootic surroundings, geographic and economic circumstances, etc. Quarantine for animals imported from rabies-endemic territories is another measure implemented in some states or districts free of terrestrial rabies.

Outstanding progress in rabies prophylaxis in wildlife has been achieved by implementation of oral vaccination. Oral vaccines contain live-attenuated RABV or recombinant viruses (such as vaccinia virus) expressing the RABV glycoprotein. Wildlife oral vaccination campaigns have been conducted in North America and Europe. As a result, rabies epizootics in wild canids have been significantly reduced or eliminated in most West European countries, Canada, and in parts of the USA. The oral vaccination of raccoon populations, performed in the eastern USA, has prevented further westward spread. Due to very high population density of raccoons, oral vaccination is challenging in this species. To date, oral vaccination programs have not been applied to bats.

Further Reading

Baer GM (ed.) (1991) *The Natural History of Rabies*, 2nd edn. Boca Raton, FL: CRC Press.

Centers for Disease Control and Prevention (CDC) (1999) Human rabies prevention – United States, 1999. Recommendations of the Advisory Committee on Immunization Practices (ACIP). Morbidity and Mortality Weekly Report, vol. 48(no. RR-1), pp. 1–21. Atlanta, GA: CDC.

Jackson AC and Wunner WH (eds.) (2007) *Rabies*, 2nd edn. New York: Academic Press.

Krebs JW, Wilson ML, and Childs JE (1995) Rabies – Epidemiology, prevention, and future research. *Journal of Mammalogy* 76: 681–694.

Meltzer MI and Rupprecht CE (1998) A review of the economics of the prevention and control of rabies. Part 1: Global impact and rabies in humans. *Pharmacoeconomics* 14: 365–383.

Meltzer MI and Rupprecht CE (1998) A review of the economics of the prevention and control of rabies. Part 2: Rabies in dogs, livestock and wildlife. *Pharmacoeconomics* 14: 481–498.

Meslin F-X, Kaplan MM, and Koprowski H (eds.) (1996) *Laboratory Techniques in Rabies*, 4th edn. Geneva: World Health Organization.

Rupprecht CE, Dietzschold B, and Koprowski H (eds.) (1994) *Lyssaviruses*. Berlin: Springer.

Smith JS (1995) Rabies virus. In: Murray PR, Baron EJ, Pfaller MA, Tenover FC, and Yolken PH (eds.) *Manual of Clinical Microbiology*, pp. 997–1003. Washington, DC: ASM Press.

Tordo N, Charlton K, and Wandeler A (1998) Rhabdoviruses: Rabies. In: Collier LH (ed.) *Topley and Wilson's Microbiology and Microbial Infections*, pp. 666–692. London: Arnold.

Tsiang H (1993) Pathophysiology of rabies virus infection of the central nervous system. *Advances in Virus Research* 42: 375–412.

World Health Organization (1996) WHO recommendations on rabies post-exposure treatment and the correct technique of intradermal immunization against rabies. *WHO/EMC/ZOO/96.6*. Geneva: World Health Organization.

World Health Organization (2005) WHO expert consultation on rabies. First report. *WHO Technical Report Series*, vol. 931. WHO: Geneva.

Recombination

J J Bujarski, Northern Illinois University, DeKalb, IL, USA and Polish Academy of Sciences, Poznan, Poland

© 2008 Elsevier Ltd. All rights reserved.

Introduction

Genetic recombination of viruses could be defined as the exchange of fragments of genetic material (DNA or RNA) among parental viral genomes. The result of recombination is a novel genetic entity that carries genetic information in nonparental combinations. Biochemically, recombination is a process of combining or substituting portions of nucleic acid molecules. Recombination has been recognized as an important process leading to genetic diversity of viral genomes upon which natural selection can function. Depending on the category of viruses, recombination can occur at the RNA or DNA levels. Since these processes are different for DNA and RNA viruses, they are described separately.

Recombination in DNA Viruses

In many DNA viruses, genetic recombination is believed to occur by means of cellular DNA recombination machinery. Cellular DNA recombination events are of either homologous (general recombination) or nonhomologous types. The nonhomologous recombination events occur relatively rarely and are promoted by special proteins that interact with special DNA signal sequences. In general, homologous recombination events occur much more often and they are most commonly known as genetic crossing-over that happens in every DNA-based organism during meiosis.

The biochemical pathways responsible for DNA crossing-over are well established. General elements involved in general recombination include DNA sequence identity, complementary base-pairing between double-stranded DNA molecules, heteroduplex formation between the two recombining DNA strands, and specialized recombination enzymes. The best-studied recombination system of *Escherichia coli* involves proteins such as recA, and RecBCD, and it has led to a large amount of literature. Interestingly, related DNA recombination proteins have been characterized in eukaryotes, including yeast, insects, mammals, and plants.

Yet certain DNA virus species encode their own recombination proteins, and some of these viruses serve as model system with those to study the recombination processes. One of the best-known systems is of certain bacteriophages that recombine independently from the host mechanisms. These independent pathways are used for repairing damaged phage DNA and for exchanging DNA to increase diversity among the related phages. In Enterobacteria phage M13, high recombination frequency was observed within the origin of phage DNA replication in the *E. coli* host. There, the crossovers have occurred at the nucleotide adjacent to the nick at the replication origin, because of joining to a nucleotide at a remote site in the genome. These results implicated a breakage-and-religation mechanism of such apparently illegitimate cross-overs.

Importantly enough, many of these phage recombination mechanisms are analogous to the pathways operating in the host bacteria. For instance, Rec proteins of phages T4 and T7 are analogous to bacterial RecA, RecG, RuvC, or RecBCD proteins, while RecE pathway in the rac prophage of *E. coli* K-12 or the phage 1 red system influenced the studies of bacterial systems. A correlation of different stages of DNA recombination with transcription and DNA replication during Enterobacteria phage T4 growth cycle is shown in **Figure 1**.

Phage lambda (λ) has a recombination system that can substitute for the RecF pathway components in *E. coli*. The Enterobacteria phage λ moves its viral genome into and out of the bacterial chromosome using site-specific recombination. Based on crystal structures of the reaction intermediates, it is clear how the Enterobacteria phage λ integrase interacts with both core and regulatory DNA elements (**Figure 2**).

Recombination between viral DNA and host genes can lead to acquisition of cellular genes by DNA viruses. For instance, tRNA genes are present in Enterobacteria phage T4. Interestingly, these tRNA sequences contain introns suggesting that Enterobacteria phage T4 must have passed through a eukaryotic host during evolution. Similar viral–host recombination events were observed for retroviruses in eukaryotic cells.

Genetic recombination in DNA viruses is often studied using functional marker mutations. In single-component DNA viruses recombination occurs by exchanging DNA fragments, whereas in segmented DNA viruses, additional events rely on reassortment of the entire genome segments. This complicates the recombination behavior observed among mutants. One method of recombination analysis utilizes so-called conditional-lethal types, where the cells are infected with two variants and the recombinants are selected after application of nonpermissive conditions (two-factor crosses). This allows the mutants to be organized into complementation groups with the relative positions of mutations being placed on a linear map. Another method is called three-factor crosses. Here three

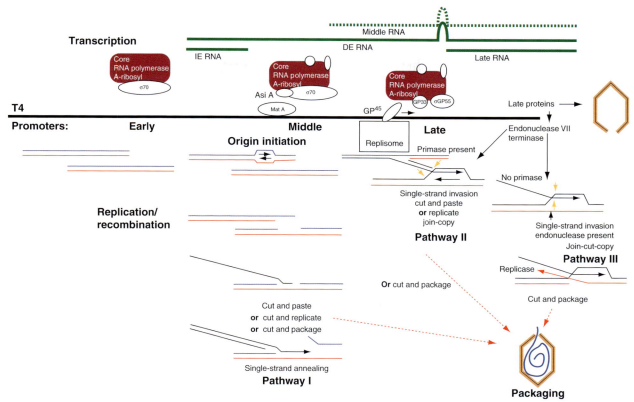

Figure 1 Diagram of the relationship between the Enterobacteria phage T4 transcriptional pattern and the different mechanisms of DNA replication and recombination. (a) Shows the transcripts initiated from early, middle, and late promoters by sequentially modified host RNA polymerase. Hairpins in several early and middle transcripts inhibit the translation of the late genes present on these mRNAs. (b) Depicts the pathways of DNA replication and recombination. Hatched lines represent strands of homologous regions of DNA, and arrows point to positions of endonuclease cuts. Reproduced from Mosig G (1998) Recombination and recombination-dependent DNA replication in enterobacteria phage T4. *Annual Review of Genetics* 32: 379–413, with permission from Annual Reviews.

mutations are employed, with crossing-over occurring between two mutations while the third mutation is not selected. This allows for determination of linkage relationships among mutants and of the order of marker mutations. Due to reassortment, both the two-factor and the three-factor crosses are of less use in segmented DNA viruses.

DNA viruses of eukaryotes also recombine their genomic material. For instance, herpes simplex virus (HSV) was found to support recombination while using pairs of temperature-sensitive mutants (two-factor crossings). In fact, a recombination-dependent mechanism of HSV-1 DNA replication has been described. The recombination frequency was proportional to the distance between mutations which suggested the lack of specific signal sequences responsible for the crossing-over. By using three-factor crossing, the HSV system involved two ts mutants and a syncytial plaque morphology as an unselectable marker. Similarly, in case of adenoviruses, the host range determined by the helper function of two mutations has been used as a third marker between ts mutants. Here, intertypic crosses between ts mutants have been identified based on segregation patterns and the restriction enzyme polymorphism.

Epstein–Barr virus (EBV) is a member of the family *Herpesviridae*, and it carries a long double-stranded genomic DNA, that shows a high-degree variation among strains. These variations include single base changes, restriction site polymorphism, insertions, or deletions. Based on tracking these mutations, it was found that some EBV strains arose due to DNA recombination.

Poxviruses represent the largest DNA viruses known (except those of algae and the mimivius). Homologous recombination was detected in the genome of vaccinia virus (VV), based on the high frequency of intertypic crossovers, the marker rescue, and the sequencing of recombinants. These processes could be both intra- and intermolecular, and they depend on the size of the DNA target. It has been suggested that either viral DNA replication itself or the activity of the viral DNA polymerase might participate in VV DNA recombination. Indeed, some VV proteins with DNA strand transfer activity have been identified.

The DNA genome of Simian virus 40 (SV40, *Papovaviridae*) was found to recombine in somatic cells. The artificially constructed recombinant circular oligomers were used to find high general recombination frequency

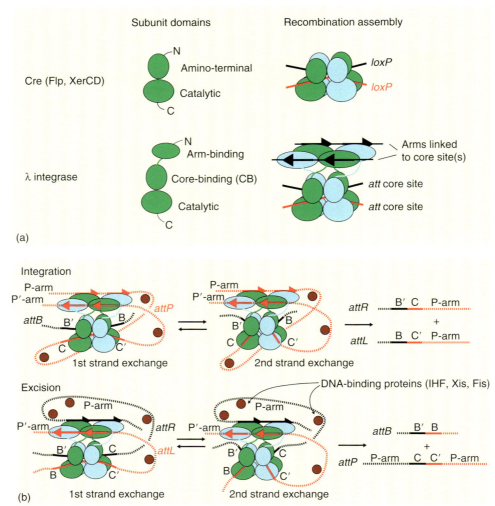

Figure 2 (a) Enterobacteria phage λ integrase compared to the simpler recombinases. Tyrosine recombinases such as Cre have two domains that bind the core recombination sites and carry out recombination on their own. Enterobacteria phage λ integrase has a third, amino-terminal 'arm binding' domain that binds to the arm region of the attachment site. The DNA complex cartoon for Enterobacteria phage λ integrase (lower right) represents the new crystal structures. (b) Integration and excision by Enterobacteria phage λ integrase. The first and second strand exchange cartoons represent the first and second halves of the recombination reaction, respectively. In the first half of integration, for example, Enterobacteria phage λ integrase brings attP and attB sites together and exchanges the first pair of strands to generate a Holliday junction intermediate. In the second half of the reaction, the Holliday intermediate has isomerized to form a distinct quaternary structure and exchange of the second pair of strands generates recombinant attL and attR products. Reproduced from Van Duyne GD (2005) Enterobacteria phage λ integrase: Armed for recombination. *Current Biology* 15: R658–R660, with permission from Elsevier.

of SV40 DNA. However, homologous recombination events were rare.

Among plant DNA viruses, genetic recombination was studied in case of geminiviruses and caulimoviruses. The geminiviruses carry a single-stranded DNA genome, composed of either one or two circular DNA molecules. Frequent intermolecular crossing-over events were observed by using mutant combinations. Homologous crossovers were detected to occur intramolecularly between tandem repeats of a geminivirus DNA using agro-infected tobacco plants. The mechanism may involve either homologous crossing-over events or copy-choice processes that rely on template switching by DNA replicase. Moreover, deletions, insertions, and more profound rearrangements have been detected in the geminivirus DNA. These are the illegitimate recombination processes that may involve aberrant breakage-and-religation events or errors in DNA replication, that could occur either inter- or intramolecularly.

Cauliflower mosaic virus (CaMV) belongs to a family of plant double-stranded (ds) DNA pararetroviruses that replicate via reverse transcription. A high recombination rate was observed during CaMV infection *in planta*. These crossovers could occur at the DNA level (thus in the nucleus) or at the RNA level (thus more likely during reverse transcription in cytoplasm). However, features such as recombinational hot spots and mismatch repair might indicate replicative (i.e., RNA) step, whereas mismatch repair can occur due to the formation of heteroduplex intermediates and thus suggest DNA recombination. These data further suggest that CaMV has the recombination mechanisms available at both steps of its life cycle. Recombination between CaMV variants and the CaMV transgenic mRNAs has been reported and this is believed to represent the RNA–RNA recombination events that happen during reverse transcription.

Recombination in RNA Viruses

RNA is the genetic material in RNA viruses, and a high mutation rate has been observed for the viral RNA genome. This likely occurs during RNA replication by means of action of an RNA-dependent RNA polymerase enzyme due to either replication errors or because of the replicase switching among viral RNA templates. The terms of classic population genetics do not describe RNA viruses. A better description of RNA viral populations is provided with a term 'quasispecies' that has been proposed to address a distribution of RNA variants in the infected tissue.

Many of the RNA viruses limit their life cycle to cytoplasm and thus the observed recombination events among RNAs of plus-stranded RNA viruses must occur outside the nucleus. In general, the RNA crossing-over processes are categorized as being either homologous or nonhomologous, but some earlier authors proposed that there are homologous, aberrant homologous, and nonhomologous RNA recombination types. Aberrant homologous recombination involves crossovers between related RNAs, but the crosses occur at not-corresponding sites leading to sequences insertions or deletions. More recently, mechanistic models were utilized to define the following RNA–RNA recombination classes: (1) The 'similarity-essential' recombination, where substantial sequence similarity between the parental RNAs is required as the major RNA determinant; (2) The 'similarity-nonessential' recombination does not require sequence similarity between the parental RNAs, although such regions may be present; and (3) There is the 'similarity-assisted' recombination where sequence similarity can influence the frequency or the recombination sites but additional RNA determinants are also critical.

Genetic RNA recombination has been described in many RNA virus groups. In particular, sequence data reveal RNA rearrangements reflecting RNA–RNA crossover events during RNA virus evolution. For instance, RNA rearrangements were demonstrated in the genomes of dengue virus-type I, flock house virus, hepatitis D virus, bovine viral diarrhea virus, and equine arthritis virus RNA. For plant RNA viruses, this has been demonstrated in potyviruses such as yam mosaic virus, sugarcane yellow leaf virus, and luteoviruses. Experimentally, RNA recombination has been shown to occur in picornaviruses, coronaviruses, or alphaviruses and in the following plant viruses: plum pox virus, cowpea chlorotic mottle virus, alfalfa mosaic virus, cucumber mosaic virus, tobacco mosaic virus, turnip crinkle virus (TCV), and tomato bushy stunt virus (TBSV). It has also been demonstrated in enterobacteria phage Qbeta, in negative RNA viruses, in double-stranded RNA viruses, and in retroviruses, as well as during formation of defective-interfering (DI) RNAs.

Recombination by reassortment was demonstrated for multisegmental animal RNA viruses, such as influenza virus, and in double-stranded reoviruses and orbiviruses. Specifically, the interpretation of two-factor crosses (using, e.g., ts mutants) in reoviruses turned out to be difficult due to recombination. The mutant sites cannot be ordered on a linear map and often no linkage between mutants could be detected.

Interestingly, there are examples of viral RNA recombination with host-derived sequences. These include the presence of uniquitin-coding region in bovine diarrhea virus, a sequence from 28S rRNA found in the hemagglutinin gene of influenza virus or a tRNA sequence in Sindbis virus RNA. Also, in plant viruses the host-derived sequences were found in potato leaf-roll virus isolates that carry sequences homologous to an exon of tobacco chloroplast. Chloroplast sequences were found in the actively recombining RNAs of brome mosaic virus (BMV). Several plant viruses were also confirmed to recombine with viral RNA fragments expressed in transgenic plants, including cowpea chlorotic mottle virus, red clover necrotic mottle virus, potato virus Y virus, and plum pox virus.

The existence of several RNA virus recombination systems has made possible the studies of the molecular mechanisms of RNA recombination. The majority of RNA recombination models predict copy-choice mechanisms, either due to primer extension (in flaviviruses, carmoviruses), at the subgenomic promoter regions (BMV, poliovirus), or by strand translocation (in nidoviruses). In retroviruses, there are three copy-choice mechanisms: (1) forced (strong stop) strand transfer, (2) pause-driven strand transfer, and (3) pause-independent (RNA structure-driven) strand transfer. However, in enterobacteria phage Qbeta a breakage-and-religation mechanism has been described. Details of some of these systems are discussed below.

The molecular mechanisms of the formation of both nonhomologous and homologous RNA recombinants have been studied using an efficient system of BMV. In order to increase recombination frequency, the BMV RNA3-based constructs were generated where the 3′ noncoding region was extended, while carrying partial deletions. This debilitated the replication of RNA such that the sequence got repaired by recombination with the sequences of other

BMV RNA segments. It appeared that short base-paired regions between the two parental BMV RNA molecules could target efficient nonhomologous crossovers. A proposed model predicted that the formation of local RNA–RNA heteroduplexes could function because they brought together the RNA substrates and because they slowed down the approaching replicase enzyme complexes.

These early studies also analyzed the molecular requirements of homologous recombination by inserting the BMV RNA2-derived sequences into the recombination vector. This revealed the accumulation of both precise and imprecise RNA2–RNA3 recombinants and that the recombination frequencies depended upon the composition of nucleotide sequences within the region of recombination. The crossovers tended to happen at stretches of GC-rich regions alternating with AU-rich sequences suggesting the RNA replicase switching between RNA templates. Elements capable of forming strand-specific, stem–loop structures were inserted at the modified 3′ noncoding regions of BMV RNA3 and RNA2 in either positive or negative orientations, and various combinations of parental RNAs were tested for patterns of the accumulating recombinant RNA3 components. This provided experimental evidence that homologous recombination between BMV RNAs more likely occurred during positive-rather than negative-strand synthesis.

True homologous recombination crossing-over has been observed among the RNA molecules of the same segments during BMV infections. By using nonselective marker mutations at several positions, it was demonstrated that RNA1 and RNA2 segments crossed-over at 5–10% frequency, whereas the intercistronic region in RNA3 supported an unusually high recombination frequency of 70%. The subsequent use of various deletion constructs has revealed that the high-frequency crossing-over mapped to the subgenomic promoter (sgp) region, and in particular to its internal polyA tract. Further studies have shown that sgp-mediated crossing-over has occurred at the minus-strand level (i.e., during plus-strand synthesis), most likely by discontinuous process, where the replicase complex detached from one strand and reinitiated on another strand. This process is most likely primed by a sg RNA3a intermediate, which prematurely terminates on the polyU (in minus strand) tract and re-anneals to this region on another minus RNA3 template (**Figure 3**). Also, it turned out that the frequency of crossing-over and the process of initiation of transcription of sg RNA4 were reversely linked, suggesting a competition between these two reactions.

The role of replicase proteins in BMV RNA recombination has been studied by using well-characterized 1a and 2a protein mutants. A ts mutation in protein 1a 5′ shifted the crossover sites indicating the participation of helicase domain of 1a. Likewise, mutations at several regions of 2a affected the frequency of nonhomologous recombination. The relationship between replication and recombination was studied by using BMV variants that

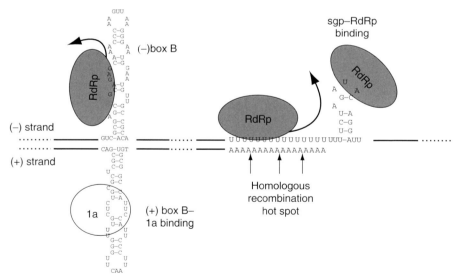

Figure 3 Model illustrating the synthesis of sg RNA3a in view of multiple functions of the intergenic region in (−) RNA3. The BMV RdRp enzyme complex (represented by gray ovals) migrates alongside the (−)-strand RNA template and pauses (represented by curved arrows) at the secondary structure or, most notably, at the oligoU tract, leading to the formation of subgenomic sgRNA3a. Yet another molecule of the RdRp enzyme binds to the sgp and initiates the *de novo* synthesis of sgRNA4. Also, the rehybridization of the sgRNA3a oligoA tail to the RNA3 (−) template can resume full-length copying which primes the observed RNA3–RNA3 recombination (5, 69). The (+) and (−) RNA strands are represented by thick lines and both the oligoU tract in the (−) strand template and the oligoA 3′ termini are exposed. The stem-and-loop structures adopted by the (+) and the (−) strands upstream to their oligoU (A) tracts are shown. The binding region to protein 1a via the box B of the stem–loop structure in (+) strands is shown. Reproduced from Wierzchoslawski R, Urbanowicz A, Dzianott A, Figlerowicz M, and Bujarski JJ (2006) Characterization of a novel 5′ subgenomic RNA3a derived from RNA3 of brome mosaic bromovirus. *Journal of Virology* 80: 12357–12366, with permission from American Society for Microbiology.

carried mutations in 1a and 2a genes. This revealed that the 1a helicase and the 2a N-terminal or core domains were functionally linked during both processes *in vivo* and *in vitro*. Also, it was shown that the characteristics of homologous and nonhomologous crossovers could be modified separately by mutations at different protein sites. All these studies confirmed the involvement of replicase proteins in recombination and supported the template-switching model.

More recently, the role of host factors in BMV recombination was addressed by using both yeast and Arabidopsis systems. In yeast, transient co-expression of two derivatives of BMV genomic RNA3 supported intermolecular homologous recombination at the RNA level but only when parental RNAs carried the *cis*-acting replication signals. The results implied that recombination occurred during RNA replication. In *Arabidopsis*, the use of gene-knock-out mutations in the RNA interference pathway revealed that BMV can recombine according to both the copy-choice template-switching and to the breakage-and-religation mechanisms.

The role of replicase proteins in RNA recombination has also been studied in other RNA viruses. For TCV, a small single-stranded RNA virus, a high-frequency recombination was observed between satellite RNA D and a chimeric subviral RNA C. The crossing-over most likely relied on viral replicase enzyme switching templates during plus-strand synthesis of RNA D which reinitiated RNA elongation on the acceptor minus-strand RNA C template. The participation of replicase proteins was demonstrated *in vitro*, where a chimeric RNA template containing the *in vivo* hot-spot region from RNA D joined to the hot-spot region from RNA C. Structural elements such as a priming stem in RNA C and the replicase binding hairpin, also from RNA C, turned to play key roles during recombination, probably reflecting late steps of RNA recombination such as strand transfer and primer elongation. The host factors related to the host-mediated viral RNA turnover have been found to participate in tombusvirus RNA recombination. The screening of essential yeast genes mutants identified host genes that affected the accumulation of TBSV recombinants, including genes for RNA transcription/metabolism, and for protein metabolism/transport. Suppression of TBSV RNA recombination was observed by the yeast Xrn1p 5′–3′ exoribonuclease, likely due to rapid removal of the 5′ truncated RNAs, the substrates of recombination. These 5′ truncated viral RNAs are generated by host endoribonucleases, such as the Ngl2p endoribonuclease.

Coronavirus RNAs were found to recombine between the genomic and DI RNA molecules. It was postulated that recombination has occurred due to the nonprocessive nature of the coronavirus RNA polymerase enzyme (**Figure 4**) and an efficient protocol for targeted recombination has been developed.

Similarly, in nodaviruses, the two-partite RNA viruses, recombination processes were found to occur between RNA segments at a site that potentially could secure base pairing between the nascent strand and the acceptor template. The recombination sites might have been chosen based on factors such as the similarity to the origin of replication or special secondary structures. A postulated model implies the polymerase to interact directly with the acceptor nodavirus RNA template.

A double-stranded RNA Pseudomonas phage Phi6 was hypothesized to recombine its RNA based on a copy-choice template switching mechanism, where the crossovers would have occurred inside the virus capsid structures at regions with almost no sequence similarity. Interestingly, the frequency of recombination was enhanced by conditions that prevented the minus-strand synthesis. Experiments were designed to reveal the effects of drift on existing genetic

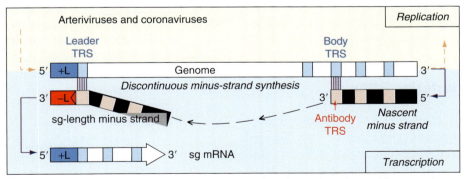

Figure 4 Models for discontinuous transcription from minus-strand sg-length templates in arteriviruses and coronaviruses. These viruses have a common 59 leader sequence on all viral mRNAs. Discontinuous extension of minus-strand RNA synthesis has been proposed as the mechanism to produce sg-length minus-strand templates for transcription. The replicase/transcriptase can attenuate at one of the body TRSs in the 39-proximal part of the genome, after which the nascent minus strand extends with the anti-leader ('L') sequence. Next, the completed sg-length minus strands serve as templates for transcription. Reproduced from Pasternak AO, Spaan WJM, and Snijder EJ (2006) Nidovirus transcription: How to make sense...? *Journal of General Virology* 87: 1403–1421, with permission from Society for General Microbiology.

variation by minimizing the influence of variation on beneficial mutation rate. The segmented genome of the pseudomonas phage Phi6 has allowed to present the first empirical evidence that the advantage of sex during adaptation increases with the intensity of drift.

The enterobacteria phage Qbeta, a small single-stranded RNA virus, could recombine both *in vivo* and *in vitro*. Here, the mechanism of recombination was not based on a template-switching by the replicase, but rather via a replicase-mediated splicing-type religation of RNA fragments. The system produced nonhomologous recombinants, whereas the frequency of homologous crossovers was low. These data suggested an RNA trans-esterification reaction catalyzed by a conformation acquired by enterobacteria phage Qbeta replicase during RNA synthesis. In summary, the results on various plus-strand RNA virus systems demonstrate the availability of a variety of template-switch mechanisms, the mutual-primer-extension on two overlapping RNA strands, the primer-extension on one full-length RNA strand, as well as both replicative and nonreplicative trans-esterification mechanisms where a piece of another RNA is added to the 3′ terminus of an RNA either by viral RdRp or by other enzymes (e.g., RNA ligase), respectively.

The recombination events in retroviruses contribute significantly to genetic variability of these viruses. The crossovers do occur by reverse transcriptase jumpings between the two genomic RNA molecules inside virion capsids. Apparently, the virally encoded reverse transcriptase enzymes have been evolutionarily selected to prone the jumpings between templates during reverse transcription. It turns out that the recombinant jumpings between RNA templates are responsible for both inter- and intramolecular template switchings and also for the formation of defective retroviral genomes. It has been found that the most stable interactions between two copies of retrovirus RNAs were within the 5′ nucleotides 1–754. There is experimental evidence demonstrating that the template 'kissing' interactions effectively promote recombination within the HIV-1 5′ untranslated region. The possibilities of recombination in retroviruses at the DNA level (of the integrated provirus sequences) were discussed earlier in this article.

Defective-Interfering RNAs

There is a variety of subviral RNA molecules that are linked to viral infections. Those derived from the viral genomic RNAs and interfering with the helper virus accumulation or symptom formation are called as DI RNAs. First reports (in 1954) about DI RNAs coexisting with viral infection was provided by Paul von Magnus with influenza virus. Thereafter, numerous both animal and plant viruses were found to generate DI RNAs. Naturally occurring DI RNAs have been identified in coronavirus infections. These molecules appear to arise by a polymerase strand-switching mechanism. The leader sequence of the DI RNAs was found to switch to the helper-virus derived leader sequence, indicating that helper virus-derived leader was efficiently utilized during DI RNA synthesis. Also, the leader switching likely occurred during positive-strand DI RNA synthesis, and the helper-virus positive-strand RNA synthesis tended to recognize double-stranded RNA structures to produce positive-strand DI RNAs. The parts of the coronavirus RNA required for replication and packaging of the defective RNAs were investigated, with both the 5′- and the 3′-terminal sequences being necessary and sufficient. The coronavirus DI RNAs have been utilized to study the mechanism of site-specific RNA recombination. This process relies on the acquisition of a 5′ leader that is normally used for production of numerous coronavirus sg RNAs. Also, these DI RNAs have been used as vehicles for the generation of designed recombinants from the parental coronavirus genome.

In case of plant viruses, tombusviruses and carmoviruses were found to accumulate DI RNAs, which maintain a consistent pattern of rearranged genomic sequences flanked by the 5′ and 3′ unchanged replication signals. In some cases, the base pairing between a partial nascent strand and the acceptor template can lead to the appearance of the rearranged regions in DI RNAs.

In addition to rearranged DI RNAs, some RNA viruses accumulate defective RNAs due to a single internal deletion in the genomic RNA of the helper virus. Such examples include beet necrotic wheat mosaic furovirus, peanut clump furovirus, clover yellow mosaic potexvirus, sonchus yellow net rhabdovirus, and tomato spotted wilt tospovirus. Features such as the ability to translate or the magnitude of the defective RNA seem to affect the selection of the best-fit sizes of DI RNAs during infection.

Another type of single-deletion DI RNAs are produced during broad bean mottle bromovirus (BBMV) infections from the RNA2 segment. A model has been proposed where local complementary regions bring together the remote parts of RNA2 which then facilitates the crossover events. Similar RNA2-derived DI RNAs have been reported to accumulate during the cucumovirus infections.

The closteroviruses, the largest known plant RNA viruses, form multiple species of defective RNAs, including the citrus tristeza virus defective RNAs that arise from the recombination of a subgenomic RNA with distant 5′ portion of the virus genomic RNA (**Figure 5**). Apparently, closteroviruses can utilize sg RNAs for the rearrangement of their genomes.

Negative-strand RNA viruses also form DI RNAs. For instance, in vesicular stomatitis virus (VSV), a rhabdovirus, the *cis*-acting RNA replication terminal elements participate in the formation of the 5′-copy-back DI RNAs, reflecting likely communication between distant portions of the VSV genome.

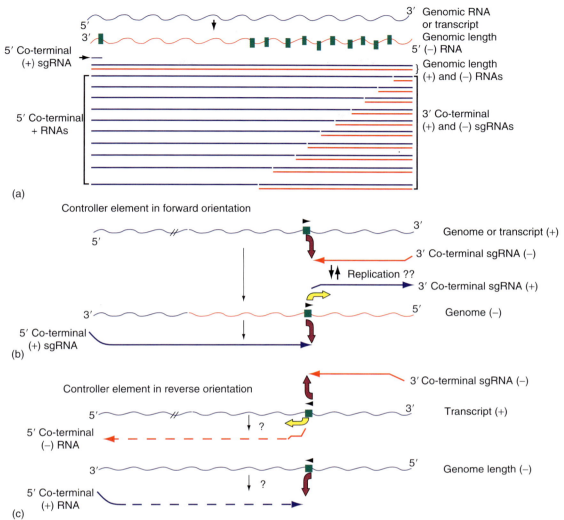

Figure 5 (a) The outline of different species of genomic RNA and 59 and 39 terminal sgRNAs potentially produced in CTV-infected cells. The positive-sense RNAs are shown in blue and the negative-sense RNAs are shown in red. The wavy line represents the genomic RNA or the plus-sense transcript (blue) and the genomic length minus-sense RNA (red) produced from the plus-sense RNA. The solid green boxes on the genomic negative-stranded RNA represent the sgRNA controller elements. The solid lines represent the full array of plus- and minus-stranded genomic and 39- and 59-terminal sgRNAs potentially produced during replication of CTV. (b, c) Models predicting the generation of 59- and 39-terminal positive- and negative-sense sgRNAs with the controller element present in normal and reverse orientation. One control region is shown for clarity. The wavy blue line represents the transcript (blue) containing the control region (green box) in normal and reverse orientation (the direction of the arrowheads above the controller element indicates the orientation of the controller element). The thick curved arrows represent the transcription termination (vertical direction, red) or promotion (horizontal direction, yellow). The solid blue lines with arrowheads represent the positive-sense 39- and 59-terminal sgRNAs and the solid red line with arrowhead indicates the 39-terminal negative-sense sgRNA. The dashed lines with arrowheads indicate the potential 59 terminal positive (blue)- and negative (red)-sense sgRNAs. Reproduced from Gowda S, Satyanarayana T, Ayllon MA, et al. (2001) Characterization of the cis-acting elements controlling subgenomic mRNAs of *Citrus tristeza virus:* Production of positive- and negative-stranded 39-terminal and positive-stranded 59 terminal RNAs. *Virology* 286: 134–151, with permission from Elsevier.

Summary and Conclusions

Genetic recombination is a common phenomenon among both DNA and RNA viruses. The recombination events have been observed based on natural rearrangements of the sequenced viral genomes. Also, experimental systems demonstrate the occurrence of recombination events that play important roles in securing the genetic diversity during viral infection. Different molecular mechanisms are involved in DNA versus RNA viruses. Many DNA viruses utilize host cellular machinery of general homologous recombination (such as meiotic crossing-over), whereas some encode their own proteins that are responsible for recombination. In addition, certain groups of DNA viruses support site-specific (nonhomologous) recombination events. In general, the virus DNA recombination mechanisms seem to involve post-DNA replication molecular events.

For RNA viruses the majority of known homologous and nonhomologous RNA recombination events appear to be integrally linked to RNA replication machinery. Various types of copy-choice (template-switching) mechanisms were proposed to describe the easy formation of RNA recombinants in numerous RNA virus systems. The roles of both special RNA signal sequences and viral proteins have been elucidated, reflecting the variety of the recombination strategies used by RNA virus groups. The involvement of host cell genes in RNA virus recombination has begun to get elucidated in several RNA viruses. Besides replicational copy-choice mechanisms, some RNA viruses use the breakage-and-religation mechanism where viral RNA gets regenerated by religation from RNA fragments, as shown experimentally for Enterobacteria phage Qbeta. New venues of RNA recombination research just emerge including our better understanding of the involvement of RNA cis-acting signals, the role of RNA replication, and the importance of cellular host genes such as RNA ribonucleases or RNA interference.

See also: African Cassava Mosaic Disease; Bean Golden Mosaic Virus; Brome Mosaic Virus; Cotton Leaf Curl Disease; Evolution of Viruses; Tomato Yellow Leaf Curl Virus.

Further Reading

Agol VI (2006) Molecular mechanisms of poliovirus variation and evolution. *Current Topics in Microbiology and Immunology* 299: 211–259.

Briddon RW and Stanley J (2006) Subviral agents associated with plant single-stranded DNA viruses. *Virology* 344: 198–210.

Chetverin AB (2004) Replicable and recombinogenic RNAs. *FEBS Letters* 567: 35–41.

Cromie GA, Connelly JC, and Leach DR (2001) Recombination at double-strand breaks and DNA ends: Conserved mechanisms from phage to humans. *Molecular Cell* 8: 1163–1174.

Figlerowicz M and Bujarski JJ (1998) RNA recombination in brome mosaic virus, a model plus strand RNA virus. *Acta Biochimica Polonica* 45(4): 847–868.

Galetto R and Negroni M (2005) Mechanistic features of recombination in HIV. *AIDS Reviews* 7(2): 92–102.

Gowda S, Satyanarayana T, Ayllon MA, et al. (2001) Characterization of the cis-acting elements controlling subgenomic mRNAs of *Citrus tristeza virus*: Production of positive- and negative-stranded 39-terminal and positive-stranded 59 terminal RNAs. *Virology* 286: 134–151.

Koonin EV, Senkevich TG, and Dolja VV (2006) The ancient Virus World and evolution of cells. *Biology Direct* 1(29): 1–27.

Masters PS and Rottier PJ (2005) Coronavirus reverse genetics by targeted RNA recombination. *Current Topics in Microbiology and Immunology* 287: 133–159.

Miller ES, Kutter E, Mosig G, Arisaka F, Kunisawa T, and Ruger W (2003) Bacteriophage T4 genome. *Microbiology and Molecular Biology Reviews* 67: 86–156.

Miller WA and Koev G (1998) Getting a handle on RNA virus recombination. *Trends in Microbiology* 6: 421–423.

Mosig G (1998) Recombination and recombination-dependent DNA replication in Enterobacteria phage T4. *Annual Review of Genetics* 32: 379–413.

Nagy PD and Simon AE (1997) New insights into the mechanisms of RNA recombination. *Virology* 235: 1–9.

Noueiry AO and Ahlquist P (2003) Brome mosaic virus RNA replication: Revealing the role of the host in RNA virus replication. *Annual Review of Phytopathology* 41: 77–98.

Pasternak AO, Spaan WJM, and Snijder EJ (2006) Nidovirus transcription: How to make sense...? *Journal of General Virology* 87: 1403–1421.

Poon A and Chao L (2004) Drift increases the advantage of sex in RNA bacteriophage Phi6. *Genetics* 166: 19–24.

Van Duyne GD (2005) Lambda integrase: Armed for recombination. *Current Biology* 15: R658–R660.

Weigel C and Seitz H (2006) Bacteriophage replication modules. *FEMS Microbiology Reviews* 30: 321–381.

White KA and Nagy PD (2004) Advances in the molecular biology of tombusviruses: Gene expression, genome replication, and recombination. *Progress in Nucleic Acid Research and Molecular Biology* 78: 187–226.

Wierzchoslawski R, Urbanowicz A, Dzianott A, Figlerowicz M, and Bujarski JJ (2006) Characterization of a novel 5′ subgenomic RNA3a derived from RNA3 of Brome Mosaic Bromovirus. *Journal of Virology* 80: 12357–12366.

Wilkinson DE and Weller SK (2003) The role of DNA recombination in herpes simplex virus DNA replication. *IUBMB Life* 55(8): 451–458.

Worobey M and Holmes EC (1999) Evolutionary aspects of recombination in RNA viruses. *Journal of General Virology* 80: 2535–2543.

Reoviruses: General Features

P Clarke and K L Tyler, University of Colorado Health Sciences, Denver, CO, USA

© 2008 Elsevier Ltd. All rights reserved.

Glossary

Caspase 3−/− mice Mice that do not express caspase 3.

CTL (cytotoxic T lymphocyte) A lymphocyte capable of inducing the death of infected somatic or tumor cells. CTLs express T-cell receptors (TcRs) that can recognize a specific antigenic peptide bound to class I MHC molecules, present on all nucleated cells, and a glycoprotein called CD8.

EHBA (extrahepatic biliary atresia) A progressive congenital disorder that destroys the external bile duct structure of the liver, impairing normal bile flow (cholestasis).

SCID mice Mice with severe combined immunodeficiency, that is, cannot make T or B lymphocytes.

History

In 1959, AB Sabin proposed the designation reovirus (respiratory enteric orphan) for a subgroup of respiratory and enteric viruses not known to be associated with human disease. These viruses had particular distinguishing characteristics including: (1) their size, which at ≤75 nm was larger than other known enteric viruses; (2) their capacity to produce cytoplasmic inclusions in monkey kidney cells in tissue culture; (3) their pathogenicity for newborn but not adult mice; and (4) their capacity to hemagglutinate human type O erythrocytes. The first reovirus, isolated from an aboriginal child in 1951 by NF Stanley and colleagues, was named hepatoencephalomyelitis virus. Later, in 1953, M Ramos-Alvarez and AB Sabin isolated the prototype virus for reovirus serotype 1 (reovirus serotype 1 strain Lang, T1L) from the stool of a baby named Lang. In 1955, Ramos-Alverez and Sabin also isolated the prototypes for reovirus serotype 2 from the stool of a child named Jones who had a summer diarrheal illness (reovirus serotype 2 strain Jones, T2J) and for reovirus serotype 3 from the stool of a child named Dearing (reovirus serotype 3 strain Dearing, T3D). A second prototype for reovirus serotype 3 was isolated from an anal swab from a baby named Abney (reovirus serotype 3 strain Abney, T3A) by L Rosen in 1957.

With the isolation and characterization of additional viruses, the acronym 'reo' was retained for the encompassing family of viruses, the *Reoviridae*, all members of which could be called reoviruses. The prefix ortho was added to the names of the initial isolates (orthoreoviruses) and their genus (*Orthoreovirus*) to distinguish them from other members of the family (see below). Despite these formal changes in nomenclature, the orthoreoviruses are still commonly referred to as reoviruses.

Taxonomy and Classification

Currently, the *Reoviridae* is the largest of all the six double-stranded (ds) RNA virus families and its members are also the most diverse in terms of host range (**Table 1**). The genomes of these viruses comprise 10, 11, or 12 segments of dsRNA, each encoding one to three proteins (usually one) on only one of the complementary strands. Their mature virions have characteristic sizes (60–85 nm excluding the extended fiber proteins that project from the surface of some members), no lipid envelope, and proteins arranged in two or three concentric layers that generally reflect icosahedral symmetry. A distinguishing feature of their replication cycles, also found in dsRNA viruses in other families, is the synthesis of viral mRNAs by virally encoded enzymes within the icosahedral particles.

Viruses in the 12 recognized genera of the family *Reoviridae* can be distinguished by differences in relative size, capsid number and structure, genome segment number, nature and number of structural proteins and patterns of reactivity with antisera (**Table 1**). Regions of significant sequence similarity across genus lines are few and of limited length; hence, exchanges of genetic material between genera are unlikely. Viruses in a subset of the genera have a turret-like projection from the innermost capsid layer (**Table 1**). This article focuses on the genus *Orthoreovirus*. The characteristics of other genera of the *Reoviridae* are discussed elsewhere in this encyclopedia.

The genus *Orthoreovirus* currently contains five species including *Mammalian orthoreovirus* (MRV) as species I, *Avian orthoreovirus* (ARV) as species II, *Nelson Bay orthoreovirus* (NBV) as species III, *Baboon orthoreovirus* (BRV) as species IV, and *Reptilian orthoreovirus* (RRV) as species V. All orthoreoviruses, except MRV, are fusogenic and induce syncytia.

Members of the genus *Orthoreovirus* have 10 segments of dsRNA contained in two concentric protein capsids of approximately 85 nm diameter (**Table 1**). Orthoreovirus genomes comprise three large (L1–L3), three medium (M1–M3), and four small (S1–S4) segments (**Table 2**). Most of the genome segments are mono-cistronic, except for S1 which is bi-cistronic in MRV, BRV, and RRV and tri-cistronic in ARV and NBV. Further details on the molecular biology of orthoreoviruses is described elsewhere in this encyclopedia.

Fusogenic Orthoreoviruses

The fusogenic orthoreoviruses include the species ARV, NBV, BRV, and RRV. Avian orthoreoviruses can be isolated from both domestic and wild birds and cause a variety of diseases including tenosynovitis (arthritis), a gastrointestinal maladsorption syndrome, and runting. Although some avian reoviruses can be adapted for growth in mammalian cells, or will grow in mammalian cells under certain conditions, there is little or no evidence for natural infection of mammals. Nelson Bay virus, isolated from an Australian flying fox, has traits intermediate between classical mammalian and avian orthoreoviruses. Although NBV was isolated from a mammal and replicates in mammalian cell cultures, it induces syncytia formation in those cultures and is thus fusogenic. Baboon reovirus, isolated from a colony in Texas, has similar properties to NBV, including replication and syncytia formation in mammalian cultures, but is distinguishable in other respects. Nucleotide sequence analysis indicates that NBV and BRV represent additional phylogenetic groups within the genus *Orthoreovirus*. Fusogenic isolates from snakes represent yet another distinct phylogenetic group within the genus.

Fusogenic reoviruses encode a unique group of small (95–140 amino acids) fusion-associated, small

Table 1 The family *Reoviridae*

Genera	No. of genome segments	No. of capsid layers	Hosts
Turreted[a]			
Orthoreovirus	10	2	Mammals, birds, reptiles
Aquareovirus	11	2	Fish, mollusks
Cypovirus	10	1[b]	Insects
Fijivirus	10	2	Plants, insects[c]
Oryzavirus	10	2	Plants, insects[c]
Myocoreovirus	11/12	2	Fungi
Idnoreovirus	10	2	Insects
Nonturreted			
Rotavirus	11	3	Mammals, birds
Orbivirus	10	3	Mammals, birds, arthropods[c]
Coltivirus	12	3	Mammals, arthropods[c]
Phytoreovirus	12	3	Plants, insects[c]
Seadornavirus	12	3	Mosquitoes[c], mammals, humans

[a]Two groups of genera are designated based on the presence or absence of turrets or spikes situated at the 12 icosahedral vertices of either the virus or core particle.
[b]Most cypovirus particles are characteristically occluded within a matrix of proteinaceous crystals called polyhedra.
[c]Serve as vectors for transmission to other hosts.
Modified from Nibert ML and Schiff LA (2001). Reoviruses and their replication. In: Knipe DM and Howley PM (eds.) *Fields Virology*, 4th edn., pp. 1679–1728. Philadelphia, PA: Lippincott Williams and Wilkins, with permission from Lippincott Williams and Wilkins.

Table 2 List of the dsRNA segments of mammalian orthoreovirus-3De (MRV-3De), their encoded proteins, and selected functions

Gene segment	Protein	Location	Function
L1	λ3 (Pol)	Core	RNA polymerase (Pol)
L2	λ2 (Cap)	Core Spike	Guanylyl transferase, methyl transferase (capping enzyme). 'Turret' protein
L3	λ1 (Hel)	Core	Inner capsid structural protein, binds dsRNA and zinc, helicase (Hel)
M1	μ2	Core	NTPase
M2	μ1	Outer capsid	Multimerizes with σ3. Cleaved to form μ1C and μ1N which assume $T=13$ symmetry in the outer capsid. μ1C is further cleaved to δ and φ during entry
M3	μNS μNSC	Nonstructural (NS)	Binds ssRNA and cytoskeleton. μNSC results from an alternate translational start site and has unknown function.
S1	σ1 σ1s	Outer capsid NS	Viral attachment protein Nonstructural, blocks cell-cycle progression
S2	σ2	Core	Inner capsid structural protein
S3	σNS	NS	ssRNA binding, genome packaging?
S4	σ3	Outer capsid	dsRNA binding, multimerizes with σ1, nuclear and cytoplasmic localization, translational control

Reproduced from Chappell JD, Duncan R, Mertens PPC, and Dermody TS (2005) *Othoreovirus*. In: Fauquet CM, Mayo MA, Maniloff J, Desselberger U, and Ball LA (eds.) *Virus Taxonomy: Eighth Report of the International Committee on Taxonomy of Viruses*, pp. 455–465. San Diego, CA: Elsevier Academic Press, with permission from Elsevier.

transmembrane (FAST) proteins. Three distinct members of the FAST protein family have been described; the homologous p10 proteins of ARV and NBV, and the unrelated p14 and p15 proteins of RRV and BRV, respectively. Unlike the well-characterized fusion proteins of enveloped viruses, the FAST proteins are nonstructural viral proteins and are not involved in viral entry into the cell. They appear to mediate cell to cell, rather than virus to cell, membrane fusion.

The fusogenic orthoreoviruses are less well characterized, by far, than the non-fusogenic mammalian reoviruses. The rest of this article focuses on mammalian reovirues.

Mammalian Reoviruses

Serotypes and Strains

The three major MRV serotypes (MRV-1, MRV-2, and MRV-3) represent numerous isolates including the early human reovirus prototype isolates T1L (now MRV-1La), T2J (now MRV-2Jo), and T3D (now MRV-3De). A fourth MRV serotype, Ndelle (MRV-4Nd) contains only one isolate. MRV serotype is determined by the cell attachment protein, σ1 (**Table 2**). The S1 genome segment, which encodes σ1, shows the greatest sequence diversity of all the genome segments (**Figure 1**), with only 26–49% identity between viruses belonging to different serotypes. In contrast, S1 sequence identity between viruses of the

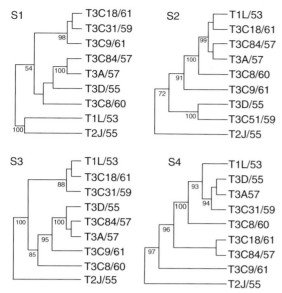

Figure 1 Phylogenetic trees indicating the potential evolutionary relationship and degree of diversity or relatedness between reovirus strains based on the nucleotide sequences of the reovirus S1, S2, S3, and S4 dsRNA segments. Each tree is rooted at the midpoint of its longest branch. Reproduced from Virgin HW, IVth, Tyler KL, and Dermody TS (1997) Reovirus. In: Nathanson N (ed.) Viral Pathogenesis, p. 669. Philadelphia, PA: Lippincott-Raven, with permission from Lippincott Williams and Wilkins.

same serotype is 86–99%, which is similar to the sequence identity of other genome segments between viruses belonging to different serotypes (**Figure 1**). This suggests that three versions of the S1 genome segment, corresponding to the three major MRV serotypes, arose from progenitors at different times and have subsequently diverged at a rate similar to that of other segments. Detailed phylogenetic analyses of the evolutionary relationship between a number of MRV strains from different serotypes based on the nucleotide sequence of their S1, S2, S3, and S4 gene segments are shown in **Figure 1**. Interestingly, the phylogenetic trees of individual MRV strains differ depending on the dsRNA segment chosen for comparison, indicating that reassortment of MRV gene segments occurs under natural circumstances.

MRV reassortants can easily be generated by co-infection of susceptible cells or mice with two distinct MRV serotypes. This capacity to exchange genetic material by genome reassortment during mixed infections to produce viable progeny virus strains can be used to identify individual species within the genus *Orthoreovirus*. In contrast to the evidence supporting reassortment of dsRNA segments between reovirus strains within a species, there is an absence of evidence of genetic recombination between either homologous or heterologous dsRNA segments.

Distribution and Host Range

Mammalian orthoreoviruses are ubiquitous in their geographic distribution. Studies of variation in the seasonal pattern of human infection are limited but an increased incidence of childhood illnesses associated with MRV-2 infection in the Northern Hemisphere summer (June–September) has been reported. In addition, many of the initial human reovirus isolates were from infants and children with summer diarrheal illnesses. Evidence of MRV infection has been found in animals of an enormous variety of species including humans, a wide variety of nonhuman primates, swine, horses, cattle, sheep, goats, dogs, cats, rabbits, rats, mice, guinea pigs, voles, bats, and a large number of marsupials.

Epidemiology

The majority of humans develop detectable serum antibodies against all three of the major MRV serotypes by late childhood. In a recent study of 272 serum specimens from young children, rapid loss of maternal antibody was detected between 0 and 6 months of age, seroprevalence was 0% in children 6–12 months of age, and then increased steadily throughout early childhood reaching 50% in children 5–6 years of age. The majority of cases of MRV infection in humans appear to be sporadic in nature, although outbreaks of infection caused by MRV-1 have been described. Age-related susceptibility to MRV infection has also been observed in both natural and experimental infection of animals. Calves, foals, piglets, and neonatal mice thus all appear more susceptible to MRV infection than their adult counterparts. Experimental studies in mice indicate that host immune status is another important factor in determining the nature and outcome of MRV infection. Immunocompetent adult mice develop an immune response but do not generally show clinical or pathological evidence of disease following reovirus infection. By contrast, after MRV infection, SCID mice develop prominent and often lethal hepatic disease. SCID mice and mice with targeted disruptions of the transmembrane exon of IgM (i.e., antibody and B-cell-deficient mice) also show altered patterns of viral clearance following peroral inoculation with MRV.

Transmission

Mammalian orthoreovirus transmission (horizontal spread) under natural circumstances involves respiratory aerosols and secretions, and fecal–oral transmission. In mice, there is an excellent correlation between the capacity of MRVs to grow in the intestine, the amount of virus subsequently shed in the stool, and the efficiency with which an infected animal transmits disease to its uninfected litter mates. The viral L2 gene, which encodes the core spike protein $\lambda 2$, is the primary determinant of the efficiency of viral transmission following peroral inoculation. Both the L2 (see above) and the S1 gene (which encodes the virus cell attachment protein, $\sigma 1$, and the small nonstructural protein, σ1s) influence growth and survival of reovirus in intestinal tissue (**Figure 2**).

Transmission is also influenced by the capacity of the virus to survive the environment after being shed from an infected host. Most MRVs are generally stable below room temperature although, at higher temperatures, strain-specific differences in thermostability become apparent. For example, MRV-1La has a half-life of 19 h at 37 °C, compared to 2.6 h for MRV-3De. MRVs are also stable in aerosols especially in the presence of high relative humidity. Viral outer capsid proteins appear to be the major determinants of virion stability.

Pathogenesis

The basic steps in the pathogenesis of mammalian orthoreovirus infection have been studied extensively in experimental animals, including mice and rats. After peroral or intratracheal inoculation, virions adhere to the surface of epithelial M (microfold) cells, which overlie collections of lymphoid tissue in the small intestine and bronchi that form part of the systems of gut-associated lymphoid tissue (GALT) and bronchus-associated lymphoid tissue (BALT). In the intestinal lumen, virions are partially digested by proteases to generate intermediate subviral particles (ISVPs). It appears that, at least in the intestine, ISVPs are the form of virus particles that bind to M cells. After binding, ISVPs and/or virions are transported across these cells to the underlying intestinal lymphoid tissue. Studies of intestinal infection suggest that replication may occur in macrophages within mucosal lymphoid tissue.

Spread of virus from the site of primary infection to distant tissues and organs, by means of the lymphatic system, blood stream, or by axoplasmic transport within neurons, results in systemic disease. MRV serotypes differ both in their capacity to generate and sustain viremia and the efficiency with which they utilize neuronal transport. Following footpad or intramuscular infection in neonatal mice, reovirus MRV-1La spreads to the central nervous system (CNS) primarily through the blood stream, whereas MRV-3De spreads predominantly through neural pathways. In this model the viral S1 gene determines both the pathway of spread in the infected host and the extent of extra-intestinal spread (**Figure 2**).

Depending on the viral strain, the route of inoculation and host factors such as age and immune status, MRVs can produce injury in a variety of target tissues. Among the most extensively studied targets of viral infection in murine model systems are the CNS, the lung, the heart, the hepatobiliary system, and the gastrointestinal tract (**Figure 2**). The specific pathology induced in these various organ systems is discussed extensively in the references included at the end of this article.

MRV strains often show striking differences in their pattern of organ and tissue tropism. For example, MRV-3De infects neurons and retinal ganglion cells, whereas MRV-1La infects ependymal cells and cells in the anterior lobe of the pituitary gland. Differences in tropism within the brain, the pituitary gland, and the retina are all determined by the viral S1 gene. Studies with monoclonal antibody resistant σ1 variants of MRV-3De indicate that a single amino acid substitution in this gene is sufficient to alter neurovirulence, CNS growth, and pattern of CNS tropism.

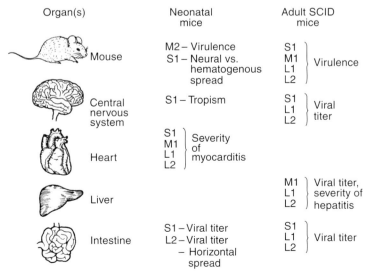

Figure 2 Reovirus dsRNA segments shown to have a role in determining organ-specific virulence in mice. Reproduced from Virgin HW, IVth, Tyler KL, and Dermody TS (1997) *Reovirus*. In: Nathanson N (ed.) *Viral Pathogenesis*, p. 669. Philadelphia, PA: Lippincott-Raven.

Attachment and Penetration

The cell attachment protein, σ1, consists of an elongated fibrous tail that inserts into the virion, and a virion-distal globular head. Four distinct and tandemly arranged morphologic regions within the σ1 tail have been designated (T(i)–T(iv)) based on proximity to the virion surface. A conserved surface at the base of the σ1 head domain of all three of the major MRV serotypes appears to determine virus binding to a serotype-independent receptor junction adhesion molecule-A (JAM-A). The σ1 protein of MRV-3De also contains a receptor binding domain in the T(iii) region of the tail that binds α-linked sialic acid. The relative importance of the JAM-A and sialic acid receptor binding domains of MRV-3De σ1 for efficient attachment and infection of host cells varies between different target cells. MRV-1La also binds a carbohydrate moiety but the nature of the glycosyl ligand remains uncertain. In contrast to MRV-3De σ1, the carbohydrate binding domain of MRV-1La σ1 has been mapped to tail region T(iv). Whereas there seems to be some flexibility on the binding of sialic acid and JAM-A for reovirus growth, both these receptors are required for the ability of reovirus to induce apoptosis in infected cells (see below).

In addition to receptor binding, β-1 integrin has recently been shown to facilitate reovirus internalization suggesting that viral entry occurs by interactions of reovirus virions with independent attachment and entry receptors on the cell surface.

Virus–Cell Interactions

Members of the MRV species induce apoptosis in cultured cells. MRV serotypes differ in this capacity with MRV-1La producing less apoptosis than MRV-3 strains. Serotype-specific differences in the capacity to induce apoptosis are determined by the S1 and M2 genome segments.

As noted above, the S1 genome segment encodes two proteins, the cell attachment protein (σ1) and a nonstructural protein (σ1s), which promotes G_2/M cell-cycle arrest. Several lines of evidence indicate that, at least in some cells in tissue culture, σ1 is the S1-encoded determinant of virus-induced apoptosis. First, apoptosis can be induced by UV-inactivated replication-incompetent virions, which lack σ1s. Second, apoptosis can be induced at nonpermissive temperatures by a variety of reovirus temperature-sensitive (ts)-mutants, which fail to synthesize σ1s in infected cells. Finally, the σ1s null-mutant MA, which fails to induce G2/M arrest in virus infected cells in tissue culture, retains the capacity to induce apoptosis, indicating that σ1s is not required for this process. Recent studies have, however, demonstrated that σ1s is a determinant of the magnitude and extent of reovirus-induced apoptosis *in vivo*, in both the heart and CNS.

The M2 gene encodes the major viral outer capsid protein μ1/μ1c. Apoptosis is inhibited following incubation of infected cells with MAbs directed against μ1 proteins and in cells infected with a temperature-sensitive (ts) membrane-penetration-defective M2 mutant. The μ1 protein is also sufficient to induce apoptosis in transfected cells. These observations support the role of the M2 genome segment in virus-induced apoptosis. In addition, recent studies suggest that binding of σ1 to JAM-A and sialic acid may be dispensable for virus-induced apoptosis and that the M2 gene segment is the only viral determinant of apoptosis when infection is initiated via Fc receptors.

As the M2 gene is a determinant of apoptosis, and as both anti-μ1 and anti-σ3 MAbs (which inhibit the virion-uncoating but not virus-cell attachment) can inhibit apoptosis, early events during virus entry, but subsequent to engagement with cellular receptors, appear to be required for apoptosis. This interpretation has subsequently been supported by experiments indicating that virus-uncoating but not replication is required for apoptosis.

MRV-induced apoptosis is associated with regulation of cellular MAPK signaling pathways, including c-Jun N terminal kinase (JNK) signaling, and transcription factors, including c-Jun and nuclear factor-kappa B (NF-κB). MRV-induced apoptosis also involves both the intrinsic and extrinsic apoptotic signaling pathways. Further details of reovirus-induced apoptotic signaling are provided in the references at the end of the article.

MRVs also induce apoptosis *in vivo* in the CNS and heart where virus infected and apoptotic cells co-localize to regions of viral injury. Virus-induced activation of caspase 3, injury, and viral load are diminished in the presence of chemical inhibitors of apoptosis or in caspase 3−/− mice (**Figure 3**). In addition, MRV-infected caspase 3−/− mice show increased survival compared to wild-type controls (**Figure 3**). These studies indicate that apoptosis is an important mechanism of virus-induced injury in the host and suggest that apoptosis inhibitors may provide useful antiviral therapies.

MRV infection is also associated with other cellular responses that influence virus growth and pathogenesis including: (1) increased expression of inducible NADPH-dependent nitric oxide (NO) synthase (iNOS) in the brains of MRV-infected mice, suggesting that NO may play an antiviral role during reovirus infection; and (2) increased phosphorylation of the eukaryotic initiation factor 2α which facilitates reovirus replication. Global expression analysis using microarrays indicates that, by 24 h following reovirus infection, the expression of 309 cellular genes (2.6% of the total number of genes present on the array) is altered in infected cells. Many of these genes are involved in cell-cycle regulation, apoptosis, and DNA repair. Further analysis of the 5′ upstream sequences of the most differentially expressed genes has revealed highly preserved sequence regions (modules) and higher-order patterns of modules (supermodules) containing binding sites for multiple transcription factors. This suggests a coordinated mechanism for virus-induced control of the expression of genes involved in similar biological processes.

Clinical Features and Infection

Human orthoreoviruses remain as much human orphan viruses as when they were first described. Human infection occurs during early childhood (see above) and is

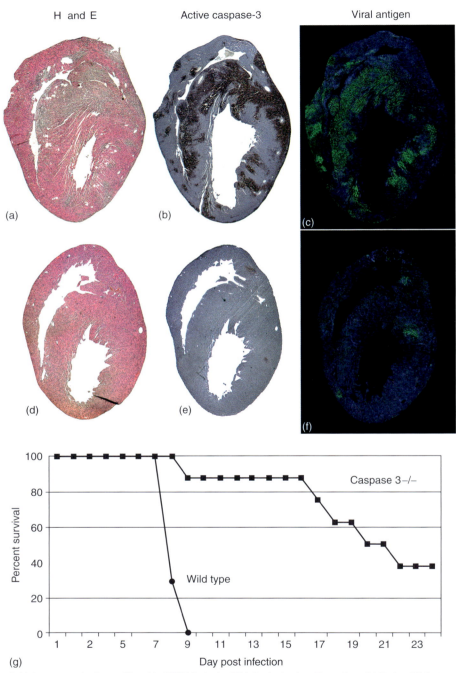

Figure 3 Myocardial injury, apoptosis, and load in MRV-3 strain 8B-infected mice. Two-day-old Swiss-Webster mice were infected with MRV-3 strain 8B followed by intraperitoneal administration of the pharmacologic caspase inhibitor Q-VD-OPH (50 mg kg^{-1} day^{-1}) or its diluent control on days 3–6 post infection. The animals were sacrificed on day 7. Consecutive sections were analyzed for histologic injury (H and E staining), active caspase 3 (brown diaminobenzadine staining), and virus antigen (fluorescent green staining). (a–c) Sections from Q-VD-OPH treated mice or (d–f) controls are shown. (g) Survival curves for MRV-3 strain 8B-infected caspase 3−/− animals and wild-type controls are also shown. Reproduced from DeBiasi RL, Robinson BA, Sherry B, et al. (2004) Caspase inhibition protects against reovirus-induced myocardial injury *in vitro* and *in vivo*. *Journal of Virology* 78: 11040–11050, with permission from American Society for Microbiology.

either asymptomatic or produces mild symptoms of upper respiratory or intestinal infections or, in some cases, exanthema with fever. The predominant symptoms observed in children during an outbreak of MRV-1La infection included rhinorrhea (81%), pharyngitis (56%), and diarrhea (19%), although the extent to which these were attributable exclusively to the MRV-1La infection itself is unclear. Over half the children shed virus in the stool for at least 1 week and 21% shed virus for at least 2 weeks. The longest reported duration of stool shedding was 5 weeks. Deliberate inoculation of adult human volunteers with MRVs produces similar patterns of infection to those that appear to occur under natural circumstances. Nasal inoculation of seronegative volunteers with MRV-1La, MRV-2Jo, or MRV-3De is associated with seroconversion and shedding of virus in the stool but is typically asymptomatic. Approximately one-third of MRV-1La-inoculated individuals do develop symptomatic infection (fever, headache, coughing, sneezing, rhinorrhea, and generalized malaise) lasting 4–7 days and beginning 24–48 h after viral challenge. Individuals challenged with MRV-3De sometimes develop mild rhinitis. In general, individuals with pre-existing antireovirus antibody prior to challenge with reovirus do not develop signs of clinical disease and do not shed significant amounts of reovirus in stools.

MRV-3De infection of mice produces a disease with clinical and pathological features that resemble human extrahepatoic biliary atresia (EHBA). However, attempts to link reovirus infection to human EHBA have produced conflicting results. Some studies show a higher frequency or higher titers of anti-reovirus antibodies in children with EHBA as compared to controls, whereas other studies do not. Similarly, some studies show that reovirus dsRNA can be amplified from patient tissues with increased frequency compared to controls, whereas other studies do not. Reovirus has not been directly isolated from pathological specimens obtained at biopsy, surgery, or autopsy from patients with EHBA; nor has reovirus been detected in the liver or biliary tissues of patients by immunocytochemistry.

One of the hallmarks of MRV infection in rodents is CNS disease. It is therefore not surprising that several case reports associating reovirus with CNS disease have appeared. Among the most convincing is a case of aseptic meningitis in a previously healthy 3-month-old. The child seroconverted and a serotype 1 MRV was isolated from cerebral spinal fluid (CSF) after inoculation onto green monkey kidney cells. In addition, a serotype 3 MRV strain was isolated from the CSF of a 6.5-week-old child with meningitis. This virus was capable of systemic spread in newborn mice after peroral inoculation and produced lethal encephalitis. Other rare reports of an association between MRV infection and human diseases including encephalitis, keratoconjunctivitis, and pneumonia exist. However, it is important to recognize that reoviruses are responsible for a vanishingly small percentage of the total number of cases of these various illnesses.

Immune Response

As noted above, both SCID mice and antibody and B cell deficient mice show increased susceptibility to MRV infection and diminished capacity to clear the virus. Similar results have been found with immunocompetent neonatal mice depleted of CD4+ and/or CD8+ T cells. This suggests that both B and T cell-mediated immune responses play a critical role in controlling MRV infection.

Following natural or experimental infection, the bulk of both the immunogobulin (Ig)A and IgG antibody response is directed against viral structural proteins and is not serotype-specific, as would be expected by the high degree of homology between proteins of viruses belonging to different serotypes. Serotype-specific antibody responses are directed against σ1 which is the least conserved of all the MRV proteins. The nature of the MRV-specific antibody response is influenced by the route of viral inoculation. Following peroral inoculation with MRV-1La, there is an increase in the number of reovirus-specific IgA-producing cells in intestinal Peyers patches and in the spleen. Enteric infection is also associated with the induction of IgG antibody, predominantly of the IgG2a and IgG2b subclasses. Variations in the dominant IgG antibody subclass are influenced both by the route of virus inoculation and the strain of mouse.

T cell responses are also induced during reovirus infection. Following peroral inoculation, MRV-1La-specific MHC-restricted cytotoxic T lymphocytes can be found in Peyer's patches and among the intraepithelial intestinal lymphocyte population. These cells are CD8+, bear the alpha/beta T cell receptor (TCR), are capable of MHC-restricted lysis of virus-infected target cells and increase dramatically after intestinal infection. Perforin, Fas-FasL, and TRAIL pathways are involved in intestinal lymphocyte cytoxicity against MRV-1La. Studies of Vβ TCR usage indicate that MRV infection is associated with oligoclonal expansion of specific TCR subpopulations. Serotype-specific MRV CTL responses are again directed against products of the S1 gene, whereas nonserotype-specific CTL responses are presumably directed against epitopes on other proteins or conserved epitopes within the S1-encoded proteins.

Both antibody and MRV-specific lymphocytes can protect mice against challenge with a variety of MRV strains and from infection by a variety of different routes. Passively transferred MRV-specific immune cells seem more effective than antibody in controlling viral replication at primary sites, whereas antibody may be more effective in controlling growth and spread of virus within certain tissues or organs, including the CNS.

Passive protection can be conferred by monoclonal antibodies specific for each of the viral outer capsid proteins (σ1, μ1C, σ3) and is associated with inhibited replication at primary sites, reduced viral spread to critical target tissues, and diminished growth and spread of virus within these tissues. Both CD4+ and CD8+ T cells are required for optimal protection following passive transfer of MRV-specific T cells. There is currently no evidence that intestinal intraepithelial γ/δ TCR+ T cells, as opposed to α/β TCR+ T cells, play a significant role in immunity to MRV infection.

In addition to humoral and cellular immune responses, cytokines and other mediators may play a role in modulating MRV infection. MRV-induction of interferon (IFN) depends on both the viral strain and the host cell and differs between mice of different strains. MRV-3De, for example, is a better inducer of IFN in mouse L-cell fibroblasts than MRV-1La. MRV-3De also induces higher levels of chemokine mRNA expression for TNFα and MIP-2 than MRV-1la in pulmonary cells *in vitro* and within the lung following *in vivo* infection.

MRVs are also susceptible to β-IFN, although strains differ strikingly in sensitivity, with MRV-3De being much more sensitive than MRV-1La. Differences in the levels of IFN-induced dsRNA-dependent protein kinase (PKR) may play a role in mediating effects of β-IFN on MRV replication. The σ3 protein inhibits the activation of PKR, by preventing its interaction with dsRNA. Differences in IFN sensitivity of reovirus strains may thus depend, in part, on their σ3 proteins. In cardiac myocyte cultures, reovirus induction of β-IFN is determined by viral core proteins and inversely correlates with the capacity of viruses to induce cytopathic effect *in vitro* and myocarditis *in vivo*. Depletion of β-IFN enhances the myocarditic potential of nonmyocarditic viral strains, suggesting a protective effect for β-IFN. These results contrast with studies of experimental reovirus serotype 2-induced murine diabetes, in which the severity of insulitis in mice correlates with increased expression of γ-IFN, and is ameliorated by administration of anti-γ-IFN antibody. This suggests that IFN induction may play a pathogenetic rather than a protective role in this setting.

See also: African Horse Sickness Viruses; Bluetongue Viruses; Coltiviruses; Insect Reoviruses; Orbiviruses; Plant Reoviruses; Rotaviruses; Seadornaviruses.

Further Reading

Barton ES, Chappell JD, Connolly JL, Forrest JC, and Dermody TS (2001) Reovirus receptors and apoptosis. *Virology* 290: 173–180.

Chappell JD, Duncan R, Mertens PPC, and Dermody TS (2005) Orthoreovirus. In: Fauquet CM, Mayo MA, Maniloff J, Desselberger U, and Ball LA (eds.) *Virus Taxonomy: Eighth Report of the International Committee on Taxonomy of Viruses*, pp. 455–465. San Diego, CA: Elsevier Academic Press.

Clarke P, Richardson-Burns SM, DeBiasi RL, and Tyler KL (2005) Mechanisms of apoptosis during reovirus infection. In: Griffin DE (ed.) *Role of Apoptosis in Infection. Current Topics in Microbiology and Immunology,* 289: pp. 1–24. Heidelberg: Springer.

Clarke P and Tyler KL (2003) Reovirus-induced apoptosis. *Apoptosis* 8: 141–150.

DeBiasi RL, Robinson BA, Sherry B, et al. (2004) Caspase inhibition protects against reovirus-induced myocardial injury *in vitro* and *in vivo*. *Journal of Virology* 78: 11040–11050.

Mertens PPC, Attoui H, Duncan R, and Dermody TS (2004) *Reoviridae*. In: Fauquet CM, Mayo MA, Maniloff J, Desselberger U, and Ball LA (eds.) *Virus Taxonomy: Eighth Report of the International Committee on Taxonomy of Viruses*, pp. 447–454. San Diego, CA: Elsevier Academic Press.

Nibert ML and Schiff LA (2001) Reoviruses and their replication. In: Knipe DM and Howley PM (eds.) *Fields Virology,* 4th edn., pp. 1679–1728. Philadelphia, PA: Lippincott Williams and Wilkins.

Tyler KL (2001) Mammalian reoviruses. In: Knipe DM and Howley PM (eds.) *Fields Virology,* 4th edn., pp. 1729–1746. Philadelphia, PA: Lippincott Williams and Wilkins.

Tyler KL and Oldstone MBA (eds.) (1998) *Reoviruses. Current Topics in Microbiology and Immunology,* 223 and 224. Heidelberg: Springer.

Virgin HW, IVth, Tyler KL, and Dermody TS (1997) *Reovirus*. In: Nathanson N (ed.) *Viral Pathogenesis,* p. 669. Philadelphia, PA: Lippincott-Raven.

Reoviruses: Molecular Biology

K M Coombs, University of Manitoba, Winnipeg, MB, Canada

© 2008 Elsevier Ltd. All rights reserved.

Glossary

EOP Efficiency of plating; a mathematical description of the relative capacity of temperature-sensitive virus mutants to replicate at a high, nonpermissive temperature compared to replication at a lower, permissive temperature.

Icosahedral A regular three-dimensional geometric shape consisting of 20 equilateral triangles and 12 vertices; the actual architecture of many 'spherical' viruses.

ISVP Intermediate (or infectious) subviral particle; an intermediate form of reovirus that lacks some proteins, compared to the intact virion, but contains

additional peptides compared to the innermost core; thought to represent the particle that crosses a cell membrane during viral entry.
PFU Plaque-forming unit; the basic unit used to indicate infectious titer for viruses capable of forming 'plaques' (individual localized regions of dead cells in a cell monolayer, each of which is formed by a single infectious virus).
Reassortant A hybrid virus that contains some genes from one parental virus and other genes from another parental virus; arise after co-infection of cells with two different members of the same virus species.
SCID mice Severe combined immunodeficient mice; so named because such animals lack both humoral-mediated and cell-mediated arms of immunity.
Top component A particle that contains all virus proteins but lacks detectable genomic material; so named because, being less dense than intact virions, it is found above virions in density gradients.

Introduction

Viruses comprising the species *Mammalian orthoreovirus* (MRV) are considered the prototype members of the genus *Orthoreovirus* in the virus family *Reoviridae*. Members of this family have a genome of 9–12 segments of double-stranded RNA (dsRNA) surrounded by 2–3 concentric, nonenveloped protein capsids (**Table 1**). Replication is exclusively cytoplasmic, viral uncoating is incomplete, and the inner capsid serves as the enzymatic complex that produces progeny mRNA. The family *Reoviridae* currently contains 12 genera, and inclusion of an additional genus (*Dinovernavirus*) has been proposed (**Table 2**). In addition to the genus *Orthoreovirus*, the family includes rotaviruses, agents responsible for a significant amount of viral gastroenteritis and numerous deaths annually worldwide, the economically important insect-vectored orbiviruses, and a variety of other viruses that infect animals, fungi, and plants. There have been increasing efforts to better understand other members of this genus, including avian orthoreoviruses and fish reoviruses. This review focuses upon MRV.

Reovirus Structure

The MRV particle comprises 10 dsRNA genome segments encased by two concentric protein capsids built from multiple, nonequivalent copies of eight different proteins (**Figure 1**). Five proteins make up the inner capsid (called 'core') that surrounds the dsRNA genome and transcribes it to produce progeny mRNA. Three additional proteins make up the outer capsid. In addition to the core and virus, an intermediate form, called the intermediate (or infectious) subviral particle (ISVP) is also naturally found. All three particle forms can be isolated *in vivo* or produced easily in the laboratory, and the protein compositions of each have been determined. The proteins, and their functions, are discussed more fully below. Three MRV serotypes (designated 1, 2, and 3) have been described. Prototype members of each serotype routinely used are type 1 Lang (T1L), type 2 Jones (T2J), and type 3 Dearing (T3D). In addition, a large number of field isolates that represent distinct clones have been identified. Some recent comparative sequence evidence supports establishment of a fourth serotype, represented by Ndelle virus.

Structure of the Genome

The complete nucleotide sequences of all ten genes of all three prototype MRV serotypes have been determined. The genome consists of three large segments (L1, L2, L3) ranging in size from 3854 to 3916 base pairs (bp), three medium segments (M1, M2, M3) ranging in size from 2203 to 2304 bp, and four small segments (S1, S2, S3, S4) ranging in size from 1196 to 1463 bp (**Figure 2** and **Table 3**), for a total aggregate genomic size of 23 606 bp for T1L, 23 578 bp for T2J, and 23 560 bp for T3D. In addition to the dsRNA genome, purified MRV particles contain large amounts (representing nearly 25% of total RNA) of short single-stranded oligoribonucleotides, which are thought to represent abortive gene transcripts. Every MRV dsRNA gene sequenced to date consists of an

Table 1 Characteristics of the *Reoviridae*

Structure
About 70–85 nm in diameter
Nonenveloped
Icosahedral
Multiple concentric protein capsids
    Innermost capsid serves as transcriptase complex
    Outermost capsid serves as gene-delivery system
Genome
Linear double-stranded RNA
9–12 gene segments
Total genome size 18–29 kbp
Segmented genome capable of assortment to produce hybrid reassortant viruses
Most gene segments monocistronic
Replication
Cytoplasmic
Proteolytic processing of intact virion to produce subviral particles
Uncoating is incomplete; innermost core capsid serves to transcribe mRNA

Table 2 Members of the *Reoviridae* family

Genus	Host Range	Prototype Virus
Turreted[a]		
Orthoreovirus	Vertebrates	Mammalian Reovirus
Mammalian	Mammals	*Reovirus 3*
Avian	Birds	*Avian Reovirus S1113*
Baboon	Baboon	*Baboon Reovirus*
Nelson Bay	Flying fox	*Nelson Bay virus*
Unclassified	Reptiles	
Aquareovirus	Vertebrates	*Golden shiner virus*
Cypovirus	Invertebrates	*Bombyx mori cypovirus 1*
Dinovernavirus[b]	Invertebrates	*Aedes pseudoscutellaris reovirus*
Entomoreovirus	Invertebrates	*Hyposoter exiguae reovirus*
Fijivirus	Invertebrates, plants	*Fiji disease virus*
Mycoreovirus	Fungi	*Mycoreovirus 1*
Oryzavirus	Invertebrates, plants	*Rice ragged stunt virus*
Nonturreted		
Coltivirus	Vertebrates, invertebrates	*Colorado tick fever virus*
Orbivirus	Vertebrates, invertebrates	*Bluetongue virus 10*
Phytoreovirus	Invertebrates, plants	*Wound tumour virus*
Rotavirus	Vertebrates	*Simian rotavirus SA11*
Seadornavirus	Vertebrates, invertebrates	*Banna virus*

[a]Turreted; innermost capsid (core) contains prominent spike projections at icosahedral vertices.
[b]Proposed membership within family.

open reading frame of variable length (1095 for S4 to 3867 for L2), bounded by a 5′ nontranslated region (NTR) of variable length (ranging from 12 bases in the S1 gene to 32 nucleotides in the S4 gene) and a 3′ NTR ranging from 35 bases in the L1 gene to 83 nucleotides in the M1 gene. Every gene contains a completely conserved consensus GCUA tetranucleotide at the extreme 5′ end of the gene and a completely conserved consensus UCAUC pentanucleotide (on the plus-sense strand) at the extreme 3′ end of the gene. Although still not known, the current belief is that signals that direct genomic assembly reside in one or both of the NTR. Pairwise comparisons of the T1L, T2J, and T3D NTR show generally high conservation. Most are of gene-specific uniform size (e.g., every L1 gene has a 16-nucleotide (nt) 5′ NTR and 35 nt 3′ NTR and every L3 gene has a 13 nt 5′ NTR and 63 nt 3′ NTR (**Figure 2**, left)) and some NTRs are completely conserved across all three serotypes. For example, the 13 nt L3 5′ NTR is identical in T1L, T2J, and T3D and the 18 nt M3 5′ NTR is 100% conserved across all three serotypes. A few genes show variability in the NTR. For example, the T3D L2 gene 5′ NTR has a 1 nt insertion compared to the T1L and T2J L2 5′ NTR and the T2J M1 3′ NTR has a 1 nt deletion compared to the T1L and T3D M1 3′ NTR.

Pairwise comparisons of gene identity show that, in most cases, T1L and T3D are more closely related to each other than either is to T2J (**Figure 2**, middle). For example, the percent identities of the T1L and T3D L1 genes are ∼95% whereas percent identity of the T2J L1 gene compared to either T1L or T3D is <80%. Similar patterns are seen in most other genes, although comparative values range from 78% for T1L:T3D L2 and 76% for T2J:T1L or T3D L2, to ∼98% for T1L:T3D M1 and ∼70% for T2J: T1L or T3D M1. Only the S1 gene shows a difference, with T1L and T2J being most similar to each other.

The segmented nature of the MRV genome allows genetic mixing if two different reoviruses infect the same cell. Progeny that contain mixtures of parental genome segments (called reassortants) can only arise if the two viruses belong to the same species. Genome segment reassortment may contribute to natural pathogenesis (as seen for rotavirus and influenza virus) and has served as a convenient genetic tool for understanding better reovirus gene/protein structures and functions.

Inner Capsid Structure

The reovirus inner capsid (core) contains the viral genome and is constructed from five proteins, called λ1, λ2, λ3, μ2, and σ2. The core has icosahedral symmetry and a diameter of ∼52 nm (excluding the spike 'turrets' which extend outward from each icosahedral fivefold axis (vertex) an additional 5.5 nm) (**Figure 1**). Cores serve as the metabolically active macromolecular 'machines' from which viral mRNA is transcribed, both during replication and *in vitro*. Detailed structural information has been provided from X-ray crystallographic studies of the core, which greatly aid in understanding each protein's structure and/or function.

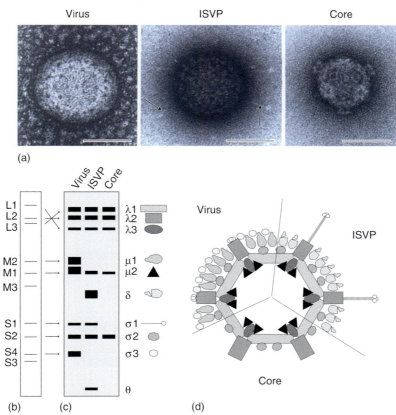

Figure 1 Structure, gene coding, and protein locations within T1L reovirus particle. (a) Electron micrographs of virus, ISVP, and core particles; arrows indicate protruding σ1 molecules on ISVP. (b) Cartoon of T1L dsRNA profile in SDS-PAGE with gene segments L1–S4 labeled on left. (c) Cartoon of protein profiles of virus, ISVP, and core in SDS-PAGE. Each protein is encoded by indicated gene segment in (b) (arrows). (d) Composite cartoon of reovirus virion, ISVP, and core, showing presumptive locations of various structural proteins, and their conversion or removal from each type of particle. Scale = 50 nm (a). Reproduced from Tran AT and Coombs KM (2001) Reoviruses. In: *Encyclopedia of Life Sciences*. New York: Wiley, with permission from John Wiley & Sons Ltd.

Full-length λ1 is a 1275-amino acid, 142 kDa protein encoded in the L3 genome segment. Protein λ1 is present in 120 copies within each reovirus core particle (**Table 3**). It is the major structural protein that forms the inner capsid. There is little serotypic variability in this inner capsid protein, implying fairly rigid structural requirements; pairwise protein identity analyses (**Figure 2**, right) show T1L and T3D λ1 share almost 100% identity (nine changes out of 1275 amino acids) and identity between T2J and either T1L or T3D is ~95%. The 120 λ1 proteins are organized as 60 asymmetric dimers in a $T = 1$ triangulation lattice to form a thin capsid shell (**Figure 1(d)**). The protein has a zinc-finger motif, a separate region that binds dsRNA, and ATPase activity, which probably is involved in transcriptional events.

Full-length λ2 is a 1289 amino acid (1288 aa for T2J), 144 kDa protein encoded in the L2 genome segment. Sixty copies of this protein are organized as distinctive pentameric turrets at each of the particle's 12 icosahedral vertices (**Figure 1**). This is one of the most variable of core proteins; T1L:T3D identity is ~92% and T2J has ~87% identity to T1L and T3D λ2 (**Figure 2**, right).

Protein λ2 binds glucose monophosphate (GMP) and S-adenosyl-L-methionine, possesses guanylyltransferase and methyltransferase activities, and serves to attach a type-1 ^{7}mG cap structure to nascent mRNA as the mRNA is extruded from transcribing cores. High-resolution structure determinations, coupled with mutagenesis studies, show λ2 to be a multidomain protein.

Full-length λ3 is a 1267 amino acid, 142 kDa protein encoded in the L1 genome segment. Cores contain 12 copies of this minor protein, which serves as the RNA-dependent RNA polymerase (RdRp). Pairwise protein identity values between any two clones is >90% (**Figure 2**, right). Current structural/functional data indicate that a single copy of this protein is located inside the core shell, slightly offset from center, at each of the icosahedral fivefold axes, located directly below the λ1 shell. Recent X-ray crystallographic structure determinations indicate this protein follows the basic 'right-hand' configuration of most nucleic acid polymerases, but also contains extra domains.

Full-length μ2 is a 736 amino acid, 83 kDa protein encoded in the M1 genome segment. This is the least understood of reovirus structural proteins; there is presently

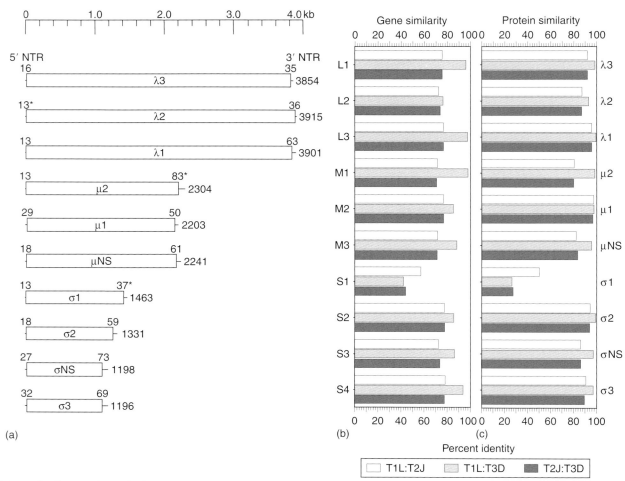

Figure 2 Gene and protein characteristics and similarities. Each of the ten double-stranded RNA genes of mammalian orthoreovirus are shown diagrammatically. (a) Genes, along with kilobase scale bar (top) are indicated. Rectangle corresponds to open reading frame and encoded protein is indicated within box. Total gene length indicated at right. Small numbers above each sequence at extreme left and right correspond to length of nontranslated region (NTR); * indicates variability in NTR length. (b) Pairwise comparisons in nucleotide identity between T1L:T2J, T1L:T3D, and T2J:T3D for each gene; genes indicated between panels A and B. (c) Pairwise comparisons in amino acid identity (indicated at far right) between T1L:T2J, T1L:T3D, and T2J:T3D for each encoded protein.

no high-resolution structure determination, and the precise function(s) remain unknown. The protein may represent an RdRp cofactor and it is present in 20–24 copies, possibly also directly under each of the core vertices. The T1L and T3D μ2 proteins share >98% identity, but the T2J μ2 protein shares only ~80% identity with T1L and T3D μ2. Genetic mapping experiments suggest that μ2 plays a role in determining the severity of cytopathic effect in cultured cells and the level of virus growth in various cells. The protein also is involved in myocarditis, in organ-specific virulence in SCID mice, in *in vitro* transcription of ssRNA, possesses nucleoside triphosphatase activity and binds RNA. It also plays important roles in virus inclusion formation and morphology, a process mediated, in part, by its level of ubiquitination.

Full-length σ2 is a 418 amino acid, 47 kDa protein encoded in the S2 genome segment. One hundred and fifty copies of the σ2 protein decorate the thin λ1 shell (**Figure 1**) and may act as 'clamps' to hold the shell together. Protein σ2 binds RNA and is required for assembly of core capsids. T1L and T3D σ2 share ~99% identity and T2J shows ~94% identity with T1L and T3D σ2.

Outer Capsid Structure

The reovirus outer capsid serves as a 'gene-delivery vehicle'; proteins within this layer are responsible for recognizing host cells and in permitting entry of the viral core into the cellular cytosol. The structures of all outer capsid proteins have been determined by X-ray crystallography. Proteins in the outer capsid are organized in a fenestrated $T=13(l)$ icosahedral lattice. This lattice is built from only 600 copies of each of two major proteins (μ1, also present as a carboxyl-terminal cleavage product μ1C, and

Table 3 Mammalian orthoreovirus gene and protein characteristics[a]

Gene	Serotype	Size (bp)[a]	Protein	Copy number[b]	Location	MW (Da)[a]	Size (aa)[a]	pI[a]	GenBank #	Functions
L1	T1L	3854	λ3	12	Core internal	142 372	1267	8.10	NC004271	RNA-dependent RNA polymerase
	T2J	3854				142 305	1267	8.17	NC004272	
	T3D	3854				142 287	1267	7.92	NC004282	
L2	T1L	3915	λ2	60	Core spike	143 957	1289	5.08	AF378003	Guanylyltransferase; methyltransferase
	T2J	3912				143 166	1288	4.96	AF378005	
	T3D	3916				144 082	1289	5.01	NC004275	
L3	T1L	3901	λ1	120	Core capsid	141 892	1275	5.92	AF129820	RNA binding; NTPase; RNA triphosphatase; helicase
	T2J	3901				142 043	1275	6.08	AF129821	
	T3D	3901				141 847	1275	5.98	AF129822	
M1	T1L	2304	μ2	24	Core internal	83 310	736	6.75	X59945	RdRp cofactor? NTPase; binds RNA
	T2J	2303				84 014	736	7.33	NC004254	
	T3D	2304				83 230	736	6.72	M27261	
M2	T1L	2203	μ1	600	Outer capsid	76 239	708	4.92	AF490617	Cell penetration; transcriptase activation
	T2J	2203				76 152	708	4.90	M19355	
	T3D	2203				76 339	708	4.91	U24260	
M3	T1L	2241	μNS		Nonstructural	80 174	721	5.92	AF174382	Binds RNA; inclusion formation
	T2J	2240				80 512	721	6.02	AF174383	
	T3D	2241				80 195	721	5.69	AF174384	
S1	T1L	1463	σ1	36	Outer capsid	51 481	470	4.97	M35963	Attachment protein; hemmaglutinin; type-specific antigen
	T2J	1440				50 482	462	5.18	M35964	
	T3D	1416				49 072	455	5.07	X01161	
S2	T1L	1331	σ2	150	Core nodule	47 115	418	8.30	L19774	Binds dsRNA
	T2J	1331				47 050	418	8.31	L19775	
	T3D	1331				47 166	418	8.30	L19776	
S3	T1L	1198	σNS		Nonstructural	41 207	366	6.34	M18389	Binds RNA; inclusion formation
	T2J	1198				41 338	366	6.58	M18390	
	T3D	1198				41 061	366	6.34	X01627	
S4	T1L	1196	σ3	600	Outer capsid	41 159	365	6.69	X61586	Binds ssRNA; role in assortment and replication
	T2J	1196				41 230	365	7.18	X60066	
	T3D	1196				41 164	365	6.84	K02739	

[a]From indicated GenBank nucleotide sequence.
[b]Within purified virion particle.

σ3), rather than 780 copies (13 × 60) as seen in nonturreted rotaviruses. This is because vertex positions that would normally contain the 'missing' 180 protein units are occupied by the core λ2 turrets.

Full-length μ1 is a 708 amino acid, N-terminal myristoylated protein encoded in the M2 genome segment. The protein folds into a four-domain structure; three lower predominantly α-helical domains (Domain I, II, and III) plus a fourth jelly roll β barrel head domain (Domain IV). When resolved by standard sodium dodecyl sulfate-polyacrylamide gel electrophoresis (SDS-PAGE), approximately 95% of virion μ1 appears as an ∼4 kDa amino-terminal peptide (called μ1N and not resolved in standard SDS-PAGE) and a 72 kDa carboxyl-terminal portion (called μ1C). Much of this μ1N/μ1C cleavage in purified virions, which takes place between Asn_{42} and Pro_{43}, was recently shown to be an artifact of sample preparation prior to SDS-PAGE. The Asn-Pro amino acid pair is found near the N-terminus in all MRV, avian orthoreovirus, and aquareovirus μ1 (and μ1 homolog) sequences determined to date, which suggests this μ1N/μ1C cleavage plays an important role in the replicative cycle. This protein alone accounts for approximately half of the total virion peptide mass, plays important roles in virion stability, and undergoes several specific cleavages during virus entry (discussed more fully below). Unlike the core proteins (discussed above), the T1L, T2J, and T3D μ1 proteins show approximately the same amount of identity to each other (∼97%) (**Figure 2**, right).

Protein σ1 is the cell attachment protein and serotype determinant. It is the most variable of reovirus proteins. The T1L and T2J σ1 proteins are more closely related (but share only ∼50% identity) whereas the T1L and T3D, and T2J:T3D σ1 proteins, share only ∼27% identity. The protein length varies with particular virus type; full-length T1L σ1 is a 470 amino acid, N-terminally acetylated protein encoded by the S1 genome segment, whereas the T2J σ1 is a 462 amino acid protein and the T3D σ1 is a 455 amino acid protein. The σ1 molecules have an overall stalk/knob structure similar to the adenovirus cell attachment fiber. The virion contains 36 copies of σ1, organized as trimers at each of the vertices.

Full-length σ3 is a 365 amino acid, 41 kDa N-terminally acetylated protein encoded in the S4 genome segment. The protein folds into a two-lobed structure, with the small lobe in contact with other major outer capsid protein μ1 and the large lobe exposed on the virion surface. Like the core proteins (discussed above), the T1L and T3D σ3 proteins share greatest identity (∼97%), whereas T2J shares ∼90% identity with the σ3 proteins of the other strains. The σ3 protein plays important roles in particle stability, in shutting down host macromolecular synthesis, and in downregulation of interferon-induced dsRNA-activated protein kinase (PKR), possibly by virtue of its intracellular distribution and capacity to interact with μ1C. The σ3 protein contains a zinc-finger motif and binds zinc, which appears important both for correct folding and stability, as well as a separate motif that binds dsRNA. Structural analyses indicate that three copies each of μ1 and σ3 form a heterohexameric $(μ1/σ3)_3$ aggregate. The three copies of μ1 are wound around each other to form a 'base', upon which three σ3 monomers sit. Two hundred such heterohexameric aggregates decorate the viral outer capsid, with the μ1 trimer base contacting the core and the σ3 monomers most distally located in the particle. This protein arrangement generates the ∼85 nm diameter fenestrated $T = 13(l)$ lattice. Proteolytic processing of σ3, which initiates in a central specific 'hypersensitive region' (HSR) and then proceeds bidirectionally toward the termini plays critical roles in virus entry and uncoating (discussed more fully below).

Reovirus Replication

Reoviruses infect a wide range of cells, both *in vitro* and *in vivo*. The virus usually infects specialized intestinal epithelial cells (M cells) that overlie Peyer's patches *in vivo*. Virus then migrates between and/or through the M cells into mucosal mononuclear cells in the Peyer's patch, and subsequently into a large number of extraintestinal sites, including heart, liver, and central nervous system. Elucidation of steps in MRV replication has been studied primarily in tissue-cultured mouse L929 fibroblast-like cells, although numerous studies have also been performed in a wide, and growing, range of other cells.

MRV replication is primarily cytoplasmic (**Figure 3**). There seems to be no significant nuclear involvement, although nonstructural virus protein σ1s (which is translated from an alternate S1 reading frame) targets the nucleus. The first step in virus replication is binding of virion to susceptible host cells (**Figure 3**, step 1), a process mediated by viral cell attachment protein σ1. The cellular protein(s) with which σ1 interacts is not completely known, but sialic acid appears to be an important component of the receptor(s). Several proteins have been identified as possible receptors, including junction-adhesion molecule. After initial binding, virus enters cells by either of the two mechanisms. Virus may be taken up by receptor-mediated endocytosis (**Figure 3**, step 2), and then converted into the ISVP by both acidification and proteolysis. A variety of agents that perturb either endosomal acidification (e.g., lysosomotropic agents such as ammonium chloride and chloroquine, or specific acidic protease inhibitors such as E-64), can inhibit infection by intact virions. Additional pharmacologic agents that inhibit a variety of other entry mechanisms (including methyl-β-cyclo-dextrin, which perturbs caveolae and lipid rafts, and chlorpromazine, which inhibits clathrin-mediated endocytosis) also reduce

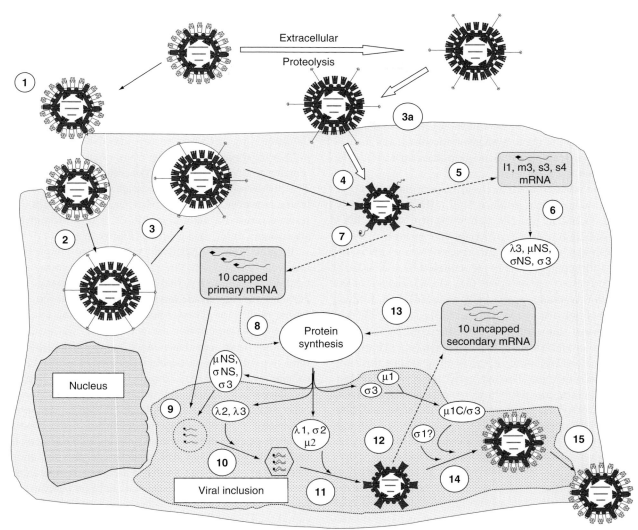

Figure 3 Model of MRV replication cycle. Details are provided in the text. Steps are (1) virus binding (upper left), (2) entry into endosomes where acid-mediated proteolysis occurs to remove outer capsid protein σ3, (3) membrane interaction to allow ISVP to escape endosome, (4) uncoating of ISVP to release transcriptionally active core particle, (5) initial 'pre-early' capped transcription, (6) initial 'pre-early' translation, (7) primary capped transcription, (8) primary translation, (9) assortment of mRNA segments into sets, probably mediated by σ3 and nonstructural proteins, (10) synthesis of negative RNA strands to generate progeny dsRNA (associated with accumulation of core proteins), (11) generation of transcriptase (replicase) complex, (12) secondary uncapped transcription, (13) secondary translation, (14) assembly of outer capsid, which halts transcription, and (15) release (lower right). Core proteins are shown in black and outer capsid proteins are shown in white; ISVP-specific modified proteins are shown in gray. Dashed arrows indicate transcription events leading to production of mRNAs (indicated in gray boxes), dotted arrows indicate translation events leading to production of proteins (indicated in white ovals), and black arrows indicate movement of proteins and viral complexes. An alternate entry mechanism for intermediate subviral particles (ISVPs), that are capable of directly penetrating membranes, is shown with white arrows at the top and in step 3a. Adapted from Tran AT and Coombs KM (2001) Reoviruses. In: *Encyclopedia of Life Sciences*. New York: Wiley, with permission from Wiley.

virus yields in cell culture. Accumulating evidence suggests that outer capsid protein σ3 must be proteolytically clipped, initially within a central HSR that is susceptible to a wide range of acidic, neutral, and basic proteases. Alternatively, the ISVP, which can be generated *in vitro* or extracellularly by a variety of intestinal proteases, appears capable of directly penetrating the cell membrane (**Figure 3**, step 3a). The acidification and protease inhibitors that block infection by intact virions do not block infection by ISVPs, making use of such inhibitors a convenient means for assessing outer capsid protein function. Cleavage, but not complete removal, of outer capsid protein σ3 appears to be a prerequisite for entry into the cytosol. Recent work has begun to generate, and use, a variety of novel particles that appear to represent additional intermediates along the virus → ISVP → core uncoating pathway. Final stages of membrane permeabilization and cell entry involve cleavage of μ1 to μ1N and μ1C, and release of the myristoylated μ1N peptide from particles. Upon entry into the cytosol, the remaining outer capsid proteins are removed, and the λ2 turrets undergo a dramatic conformational change that opens the λ2 channels further. Both alterations appear necessary for full activation of the viral transcriptase. Resultant core particles (**Figures 1** and **3**, step 4) constitute the final stage of uncoating, and incoming core particles persist throughout the remainder of the replication cycle.

These released cores serve as transcriptionally active RdRp machines that produce viral mRNA. Collectively, the λ2, λ3, and μ2 proteins manifest RdRp, helicase, RNA triphosphatase, methyltransferase, and guanylyltransferase activities to produce the viral mRNA, which is extruded through channels in the modified λ2 spikes. Early work suggested that only four (L1, M3, S3, and S4) of the ten dsRNA genes were initially transcribed to produce mRNA (**Figure 3**, step 5), and that the protein products of these initial transcripts act upon the core by unknown means to promote transcription of all ten genes (**Figure 3**, step 7). However, more recent work questions this 'cascade' interpretation. The ten 'early' transcripts, like parental genomes, contain methylated caps and lack poly(A) tails. The cap structures are provided to nascent mRNAs by the λ2 proteins as the mRNAs are extruded through the λ2 spikes. Viral transcripts are produced in quantities inversely related to transcript length; more 's' transcripts are produced than 'm' transcripts, and more 'm' transcripts are produced than 'l' transcripts. Viral 'early' mRNAs are then translated to produce the full complement of viral proteins (**Figure 3**, step 8). These proteins begin to coalesce and are directed into non-membrane-containing inclusions in the cytoplasm by the σNS, μNS, and/or μ2 proteins (**Figure 3**, step 9). The ten mRNA molecules are also used as templates for progeny minus-strand synthesis (**Figure 3**, step 10), a process most likely mediated by viral proteins. Eventual assembly of a complete progeny virion will require various structural proteins as well as a full complement of the ten dsRNA genes.

The ten different mRNA molecules are believed to be sorted (a process called 'assortment') to ensure that developing viral particles contain correct sets of genes. Viruses, like MRV, that contain multisegmented genomes, are thought to incorporate correct sets of genes by either a 'random' or 'specific' process. Because the random model predicts that only 1 out of every 2755 particles ($10!/10^{10}$) would be infectious, and because particle-to-PFU ratios approaching 2:1 have been described, it seems highly likely that assembly of the reovirus genome uses specific mechanisms. These mechanisms are currently being delineated, and recent evidence indicates regions of a few hundred nucleotides in the gene termini are involved in this process. The assortment process also probably uses nonstructural proteins σNS and μNS, both of which are translated in unusually large amounts and both of which bind single-stranded RNA (ssRNA). The σNS protein appears to perform its functions most efficiently as a multimer and binds nonspecifically most efficiently to ssRNA. Outer capsid protein σ3 also participates in assortment. Immunoprecipitation studies have indicated that proteins λ2 and λ3 are added to these complexes at the same time mRNA is copied into dsRNA. Once the mRNA is copied, the newly transcribed minus strand remains associated with it to generate progeny dsRNA. The plus-sense molecule is no longer available for translation. However, developing particles resemble cores that serve to transcribe additional mRNA, and so presumably also contain other core proteins (**Figure 3**, step 11).

Progeny core-like particles ('transcriptase complexes'), like incoming uncoated cores, are capable of synthesizing viral mRNA (**Figure 3**, step 12). These nascent replicase particles are responsible for the majority of transcription during replication. Transcripts produced during this second wave of transcription are not capped. The switch in translation from cap-dependence to cap-independence remains poorly understood, but is believed to involve intracellular distribution of σ3, its capacity to interact with μ1C, and interactions with several cellular interferon-regulated gene products, including PKR and RNase-L. Assembly of progeny virions requires condensation of the correct numbers of eight different viral proteins with one copy of each of the ten progeny dsRNA genes. *In vitro* 'recoating' studies and studies with assembly-defective, temperature-sensitive reovirus mutants have shed light on pathways by which viral proteins may associate to generate the double-capsid shell that will serve to protect the genome (**Figure 3**, steps 12–15). This is possible because many of the known reovirus temperature-sensitive mutants produce aberrant particles at the nonpermissive temperature (**Table 4**) and the precise genetic lesions (amino acid substitutions fit into available crystal structures) for many of them have been determined. For example, assembly of the core capsid shell requires only major core proteins λ1 and σ2, as indicated by expression studies and analyses of *tsC447*, a mutant defective in σ2. Reversion studies indicated that an $Asn_{383} \rightarrow Asp$ substitution, near the end of a long α-helix that spans residues $Thr_{350}–Asn_{386}$, is responsible for the inability of *tsC447* to assemble core particles at the nonpermissive temperature. The λ2 proteins then associate with the core shell particles, as implied by generation of 'spike-less' cores by *tsA279*, a double mutant that contains a defective λ2 protein. However, it currently is not known whether λ2 proteins form pentamers before associating with the core shell or whether they associate as monomers and then pentamerize. Addition of σ1 may take place at about this same time because of its intimate association with the λ2 spikes in mature particles. Outer capsid proteins μ1 and σ3 initially associate with each other, as determined both by studies with

Table 4 Characteristics of selected mammalian reovirus temperature-sensitive mutants

Mutant				Nonpermissive characteristics[a]			
Group	Clone	Gene	Protein	EOP[b]	dsRNA	Protein	Phenotype
Wild-type T3 Dearing				0.2	+	+	Normal
A	tsA201	M2	μ1	0.009	+	+	Normal
	tsA279	M2 and L2	μ1 and λ2	0.0002	+	−	Top component[c] Spike-less cores
B	tsB352	L2	λ2	0.00001	+	+	Cores
C	tsC447	S2	σ2	1×10^{-7}	−	−	Empty outer shells
D	tsD357	L1	λ3	0.00001	+	+	Top component
E	tsE320	S3	σNS	0.002	−	−	No inclusions
G	tsG453	S4	σ3	1×10^{-7}	+	+	Cores
H	tsH11.2	M1	μ2	1×10^{-7}	−	−	No inclusions
I	tsI138	L3	λ1	0.0001	−	−	No inclusions

[a]At nonpermissive temperature of 39.5 °C.
[b]EOP; Titer at 39.5 °C ÷ titer at 33.5 °C.
[c]Top component; genome-deficient double-shelled particles.

tsG453, a mutant defective in σ3 protein that forms core-like structures (**Table 4**) rather than ISVP-like structures, and by recoating experiments which have, so far, failed to attach only μ1 (in the absence of σ3) to cores. These proteins form heterohexamers ((μ1/3)$_3$) that then associate with the nascent core particles to complete the double capsid (**Figure 3**, step 14) and turn off transcription. Virus is released when infected cells lyse (**Figure 3**, step 15). The recent development of a 'reverse genetics system' for the reoviruses should prove extremely beneficial to complement some of the above-described transcapsidation and temperature-sensitive mutant analyses.

See also: African Horse Sickness Viruses; Aquareoviruses; Bluetongue Viruses; Coltiviruses; Enteric Viruses; Orbiviruses; Plant Reoviruses; Reoviruses: General Features; Rotaviruses.

Further Reading

Coombs KM (2006) Reovirus structure and morphogenesis. *Current Topics in Microbiology and Immunology* 309: 117–167.

Joklik WK (1998) Assembly of the reovirus genome. *Current Topics in Microbiology and Immunology* 233(1): 57–68.

Kobayashi T, Antar AAR, Boehme KW, et al. (2007) A plasmid-based reverse genetics system for animal double-stranded RNA viruses. *Cell Host and Microbe* 1: 147–157.

Mertens PPC, Attoui H, Duncan R, and Dermody TS (2005) *Reoviridae*. In: Fauquet CM, Mayo MA, Maniloff J, Desselberger U, and Ball LA (eds.) *Virus Taxonomy. Eighth Report of the International Committee on Taxonomy of Viruses*, pp. 447–454. San Diego, CA: Elsevier Academic Press.

Nibert ML and Schiff LA (2001) Reoviruses and their replication. In: Knipe DM and Howley PM (eds.) *Fields Virology*, 4th edn., pp. 1679–1728. Philadelphia, PA: Lippincott Williams and Wilkins.

Tran AT and Coombs KM (2001) Reoviruses. In *Encyclopedia of Life Sciences*. New York: Wiley, with permission from John Wiley & Sons Ltd.

Tyler KL (2001) Mammalian reoviruses. In: Knipe DM and Howley PM (eds.) *Fields Virology*, 4th edn., pp. 1729–1745. Philadelphia, PA: Lippincott Williams and Wilkins.

Replication of Bacterial Viruses

M Salas and M de Vega, Universidad Autónoma, Madrid, Spain

© 2008 Elsevier Ltd. All rights reserved.

Glossary

Lagging strand DNA strand synthesized as a series of 5′–3′ DNA fragments that are finally joined together to create an intact DNA strand.

Leading strand DNA strand whose synthesis proceeds continuously in the 5′–3′ direction as the parental duplex is unwound.

Sliding-back mechanism For linear chromosomes containing a terminal protein at their 5′-ends; initiation of DNA replication by which the DNA polymerase uses as template an internal 3′-end nucleotide, with the subsequent sliding of the initiation complex formed to recover the 3′-end terminal nucleotide information.

Okazaki fragment Short fragment of newly replicated DNA (with an RNA primer at the 5′-terminus) produced during discontinuous replication of the lagging strand.

O-some A first-stage initiation complex formed by the binding of lambda protein O to 100 bp of DNA close to the AT rich region of the DNA replication origin.

Processive polymerization Polymerization that takes place without dissociation of the DNA polymerase from the DNA.

Procesivity factor An accessory subunit of DNA polymerases that act to increase the processivity of polymerization.

Replisome The DNA-replicating structure at the replication fork consisting of two replicative DNA polymerases and a primosome.

Primosome A protein complex formed by a primase and a helicase that initiates synthesis of RNA primers on the lagging DNA strand during DNA replication.

Preprimosome For several replication systems, a protein complex assembled at the replication origin before the binding of the primase.

Transpososome The Mu transpososome refers to the Mu transposase protein complex and the three DNA segments bound by this protein complex.

σ replication mode For circular DNA, it refers to replication performed by a rolling circle mechanism.

θ (theta) replication mode For circular DNA, it occurs when two replication forks move in opposite directions from the replication origin, leading to a bidirectional replication mode.

Introduction

The requirement of a DNA/RNA molecule to prime DNA synthesis imposes replication strategies that avoid the loss of genetic information contained at the 5′-end of the lagging strand. Double-stranded (ds) DNA phages have developed different mechanisms to overcome such a problem by yielding head–tail concatemers, dependent on the presence of terminal redundancies, as in phages T4, T7, and SPP1, or on the circularization and further rolling circle replication, as occurs in phage λ. Several phages, such as φ29, have evolved to use a protein to prime DNA synthesis from each genome end, the priming protein becoming covalently linked to the 5′-ends of the genome. In other cases, as in phage Mu, replication of dsDNA depends on its capacity to be integrated into the host genome, replicating as a transposable element. The 5′-replication quandary does not exist in circular single-stranded (ss) DNA phages, such as φX174, which is replicated by a looped rolling circle to produce circular unit length genomes. In the case of ss- and dsRNA phages no such a problem exists since there is specific recognition by the RNA polymerase of the 3′-end of the template RNA. In addition, in phages λ, SPP1, or φX174, the replication of their genomes requires the presence of the host replication machinery, while others such as T4, T7, and φ29 use their own replication machinery.

In this article, we summarize the different replication strategies used by some of the well-studied phages.

Replication of dsDNA Phages

DNA Replication of Lamboid Phages

Bacteriophage λ has served as an important model system to elucidate part of the molecular mechanisms underlying eukaryotic and viral DNA replication. The virion particle contains a 48 502 bp linear dsDNA with a single-stranded protruding region, 12 bases long at each 5′-end, complementary to each other. Early after infection of its host *Escherichia coli*, the genome is circularized by means of the complementarity of the above-mentioned 5′-ends, and ligated by a host DNA ligase. Initiation of λ DNA replication requires the products of the early transcribed genes O and P, in addition to several functions from the host. Four homodimers of protein O bind to four 18 bp long binding sites at the λ replication origin. Through protein–protein interactions, the O dimers are brought together to form the O-some complex, promoting the further melting of an adjacent AT-rich region. The O-some prepares the λ replication origin for binding the host helicase DnaB. DnaB forms a heterodimer with λ protein P, which inhibits DnaB activity bringing it to the replication origin. Host proteins DnaJ, DnaK, and GrpE lead to disassembly of the preprimosome complex, releasing protein P and triggering replication fork movement. At early stages after infection, two DnaB copies unwind the DNA in opposite directions leading to a bidirectional θ (theta) replication mode. Such bidirectionality depends on the transcription activation from promoter P_R (transcriptional activation of *oriλ*), positively controlled by DnaA. At later stages, several DNA molecules start replicating following a σ mode (rolling circle replication) producing long concatemers of λ DNA containing more than 10 genome copies, which are cut at specific sites (*cos* sites) to yield unit length genomes with protruding (cohesive) 5′-ends that will be packaged into the virions. There is uncertainty about the process that governs the switch from θ to σ replication mode, the establishment of the σ replication mode being preceded by a round of unidirectional θ replication. Recent investigations point to the positive control of P_R promoter activity by bacterial DnaA in triggering such a switch. Thus, as the number of genome copies increases, free DnaA is titrated out due to its interaction with the DnaA-binding sites in λ DNA. It has been proposed that this may lead to inefficient *oriλ* transcriptional activation, leading to unidirectional θ replication followed by σ replication mode.

DNA Replication of Phage SPP1

SPP1 bacteriophage which infects *Bacillus subtilis* is one of the best-studied phages infecting Gram-positive bacteria. Its genome is dsDNA 44 007 bp long, partially circularly permuted with a terminal redundancy of 4%. SPP1 DNA replication depends partly on several replicative proteins from its host, such as the replicative DNA polymerase III and DnaG primase. As in the case of λ phage, SPP1 starts replication following a θ replication mode followed by a σ pathway. SPP1 protein gp38 binds to *ori*L inducing a local unwinding of an adjacent AT-rich sequence. The ssDNA generated is protected by gp36, the SPP1 ssDNA binding protein (SSB). As occurs with λ DNA replication, the helicase gp40 is loaded at the replication bubble by the helicase loader gp39 that inhibits the helicase activity of gp40. gp39 interacts with gp38 forming a heterodimer that is released from the replication origin, ceasing the inactivation of gp40 that now can exert its helicase activity. Later, the primase activity of the host DnaG is required to prime DNA synthesis by the host DNA polymerase III, the replication proceeding in only one direction. The formation of concatemeric molecules requires a switch from the θ mode to σ replication mode. Such a switch is mediated by gp38 which, after one or few rounds of θ-type replication, binds to *ori*L blocking progression of DNA replication. A subsequent nick either in the lagging or the leading strand will provide a 3′ ssDNA tail. In the latter case, a previous processing by gp34 5′–3′ exonuclease is needed. The 3′-OH end can pair with the open *ori*L of other DNA molecule and prime σ replication to generate concatemeric DNA molecules. Further encapsidation is started by cleavage at the *pac* sequence by the terminase small subunit gp1, followed by translocation of the concatemer to the interior of the procapsid.

DNA Replication of T4-Like Phages

The serologically related *E. coli* infecting bacteriophages T2, T4, and T6 are commonly called T-even phages. They have a dsDNA genome 170 kbp long whose termini contain repetitions of 3% of the genome. In addition, the genome of T-even phages contains glucosylated hydroxymethylcytosines that protect DNA from endonucleases and confer double-strand stability. The T-even phages encode for their own replication machinery, which makes them good candidates to study the general mechanism of DNA replication.

At early stages after infection, T4 DNA replication starts from only one replication origin. The T4 helicase/primase complex (gp41/gp61), loaded on to DNA by T4 gp59, moves processively in the 5′–3′ direction in lagging strand synthesis at the same time as the primase activity periodically synthesizes the RNA primers to initiate Okazaki fragment synthesis. Leading strand synthesis is initiated by an RNA molecule synthesized by a host RNA polymerase from early or middle promoters. DNA polymerase (gp43) catalyzes DNA synthesis of both strands assisted by gp45, a trimer that acts as a sliding clamp, holding the DNA polymerase tightly to the DNA. gp44/gp62 complex uses the hydrolysis of ATP to drive the binding of gp45 to DNA. Although T4 DNA replication onset depends on the replication origin, most of T4 DNA replication forks are initiated by using intermediates of recombination as DNA primers at random positions throughout the genome. Once the replication fork reaches the 3′-end, the single-stranded portion of the chain that is templating lagging strand synthesis invades an homology region in other DNA molecules because of the terminal redundancy of its ends, accomplishing a recombination-dependent DNA replication pathway called 'join-copy replication', that depends on genes expressed from early or middle promoters, and that is initiated from the invading 3′ DNA ends. This promotes the appearance of replicating DNA intermediates containing multiple covalently linked copies of the genome. When an endonuclease cuts at either of the invaded DNA strands, 'join-cut-copy recombination' is initiated from the 3′-ends to allow copying of single-stranded segments of an invading DNA. This pathway requires the action of either endo VII or terminase proteins, predominantely synthesized at late infection times, making the join-cut-copy the late pathway for DNA replication, since origin initiation of replication ceases during T4 development (**Figure 1**).

DNA Replication of T7-Like Phages

T7-like phages belong to the family *Podoviridae*. These phages code for a specific RNA polymerase resistant to rifampicine and specific for phage promoters. Phage T7 is the model member of this group of lytic phages. T7 contains a linear dsDNA 39 937 bp long with a terminal direct repetition 160 bp long, used to form concatemers during replication. This terminal repeat is different in both length and sequence in the different T7-like phages. T7 encodes for its own replication machinery and this fact has been used to study protein–protein interactions at the replication fork. Bacteriophage T7 genome is replicated as a linear monomer from replication origins in a bidirectional way at early stages after infection. The primary origin of replication is an AT-rich region (78% A+T) located downstream of φ1.1A and φ1.1B promoters. These promoters are used by T7 RNA polymerase to initiate synthesis of RNA to prime leading strand synthesis. However, alternative replication origins may exist since

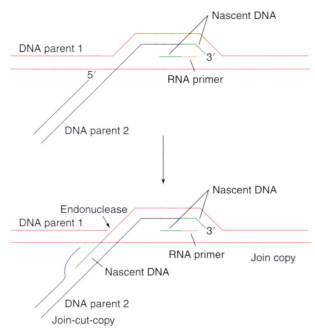

Figure 1 Initiation of DNA replication from intermediates of homologous recombination. Figure shows how ssDNA end of parent 2 invades the homologous dsDNA of parent 1. Reproduced from Mosig G, Gewing J, Luder A, Colowick N, and Vo D (2001) Two recombination-dependent DNA replication pathways of bacteriophage T4, and their roles in mutagenesis and horizontal gene transfer. *Proceedings of the National Academy of Sciences, USA* 98: 8306–8311, copyright (2001) National Academy of Sciences, USA, with permission from National Academy of Sciences.

in the absence of T7 primase, more than 20 RNA–DNA transition sites have been detected, scattered widely downstream from the $\phi 1.1$ promoter and mostly downstream from the $\phi 1.3$ promoter. In addition, secondary replication origins have been mapped close to the left genome end by using deletion mutants lacking T7 *ori*. To accomplish T7 genome replication, the phage synthesizes four replication proteins: (1) T7 RNA polymerase (gp1), which synthesizes the RNA primer to the leading strand; (2) T7 helicase/primase (gp4) which performs two activities. On the one hand, its C-terminal half contains a helicase activity that uses the hydrolysis of dTTP to unwind duplex DNA in a 5′–3′ direction with respect to the bound strand. This part of the protein physically interacts with T7 DNA polymerase. On the other hand, the N-terminal half contains a primase activity that catalyzes the synthesis of tetraribonucleotides to function as primers for Okazaki fragments synthesis; (3) T7 DNA polymerase (gp5) that contains both a 5′–3′ synthetic activity and a 3′–5′ exonuclease activity to proofread replication errors. T7 DNA polymerase is a distributive enzyme *per se*; however, *in vivo* it forms a complex with *E. coli* thioredoxin, a protein used by the polymerase as a processivity factor; (4) T7 ssDNA binding protein (gp2.5) that interacts physically with T7 DNA polymerase and primase/helicase. This protein is required for the coordinated synthesis of both leading and lagging strands of the replication fork. This SSB protein coats the ssDNA template of the lagging strand participating in mowing the polymerase from the end of one Okazaki fragment to the initiation site for the next.

Once the replication fork has reached the end of the linear molecule, the 3′-end of the strand used as template for lagging strand synthesis remains uncopied. By means of the terminal repeat sequences, several linear DNA molecules can anneal forming long concatemers containing from 10 to >100 genome equivalents. The concatemer will have a single copy of the terminal repeat between two genomes. Before the encapsidation of unit-length genomes, the terminal repeat sequence has to be duplicated. Many efforts have been done to elucidate the duplication mechanism. Two models have been proposed. One involves the transcription through the terminal repeat of the concatemer, promoting a displaced strand suitable to be cut by an endonuclease and generating a 3′-OH end that can be used to prime DNA synthesis and replicate the terminal repeat. A dsDNA break performed by the T7 terminase complex at the right part of the terminal repeat will provide another 3′-end that will complete duplication of the terminal repeat, allowing the genome to be packaged (**Figure 2(a)**). The presence of a palindromic sequence at the left part of the terminal repeat led to the proposal of an alternative model, invoking the formation of a cruciform structure followed by nicking by an unknown nuclease at a palindromic sequence, creating a hairpin primer for DNA polymerase, that once extended through the terminal repeat and into the genome will provide dsDNA that could be converted into mature left end. Primase initiated synthesis on the displaced strand will proceed through the terminal repeat, generating a mature right end of the next genome to be packaged (**Figure 2(b)**). Probably, both (or more) pathways could coexist to promote T7 DNA encapsidation and the yield of mature viral particles.

Bacteriophages with Terminal Protein at the 5′ DNA Ends

Bacteriophage $\phi 29$ which is a member of the family *Podoviridae* has a linear dsDNA 19 285 bp long, with a specific protein (product of the viral gene 3) covalently linked to the two 5′-ends called terminal protein (TP). The protein contains 266 amino acids and is linked to the DNA through a phosphoester bond between the OH group of serine 232 in the TP and 5′-dAMP.

Other *B. subtilis* phages related to $\phi 29$ that also contain linear dsDNA and TP of similar size are classified in three groups: (1) $\phi 15$, PZA, and PZE that belong to the $\phi 29$ group; (2) Nf, M2, and B103; and (3) GA-1. The DNA of

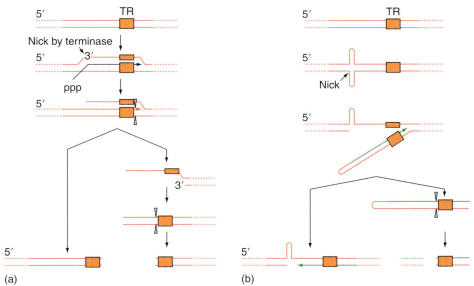

Figure 2 Mechanism for TR duplication. (a) Model involving the transcription through the terminal repeat of the concatemer. (b) The palindromic model. See text for details. (a) Adapted from Fujisawa H and Morita M (1997) Phage DNA packaging. *Genes to Cells* 2: 537–547, with permission from Blackwell Publishing. (b) Reproduced from Chung YB, Nardone C, and Hinkle DC (1990) Bacteriophage T7 DNA packaging III. A "hairpin" end formed on T7 concatemers may be an intermediate in the processing reaction. *Journal of Molecular Biology* 216: 939–948, with permission from Elsevier.

all these phages have a short inverted terminal repeat (ITR), six nucleotides long (AAAGTA) for φ29, φ15, PZA, and B103, eight nucleotides long (AAAGTAAG) for Nf and M2, and seven nucleotides long (AAATAGA) for GA-1, all of them showing a reiteration AAA at their 5' DNA ends. PRD1, a member of the family *Tectiviridae* of lipid-containing phages infecting *E. coli* and other Gram-negative bacteria, contains a linear dsDNA 14 925 bp long, whose 5'-termini are linked to a 28 kDa TP by a phosphoester bond between tyrosine 190 and 5'-dGMP. The DNA of PRD1 and related phages have a 110 bp long ITR with the reiteration 5' GGGG at the ends. The *Streptococcus pneumoniae* phage Cp-1 contains a 19 345 bp linear dsDNA with a TP of 28 kDa covalently linked to the 5' DNA ends by a phosphoester bond between the OH group of threonine and 5'-dAMP. Cp-1 has an ITR of 236 bp with the reiteration 5' AAA.

In all of the above-mentioned phages, the origins of replication are located at both ends of the linear genome and are specifically recognized by their cognate replication machinery, encoded by the phage, to initiate DNA replication.

Much work has been performed to elucidate the replication mechanism of these phages, the most studied case being bacteriophage φ29. DNA polymerase from phage φ29 forms, with a free TP, a heterodimer (**Figure 3(a)**) that specifically recognizes both replication origins. The DNA polymerase catalyzes the template-directed formation of a covalent linkage between dAMP and the OH group of serine 232 of the primer TP, giving rise to the TP–dAMP initiation complex, a reaction directed by the penultimate nucleotide of the 3' reiteration (3'TTT...5') and stimulated by the φ29-encoded protein p6. The TP–dAMP complex slides-back one position to recover the terminal nucleotide, the second 3'-terminal nucleotide acting again as template to direct the incorporation of the second dAMP residue. It has also been shown that phage GA-1 DNA initiates replication at the second nucleotide from the 3' DNA end. Since the TP–dAMP covalent complex is not a substrate of the 3'–5' exonuclease proofreading activity of the φ29 DNA polymerase, the sliding-back mechanism could provide a way to ensure the fidelity of the initiation reaction. Terminal reiteration also exists in PRD1 and Cp-1 DNAs. Indeed, it was shown that initiation of PRD1 and Cp-1 DNA replication occurs at the fourth and third 3'-terminal nucleotide of the template, respectively. In these cases, a stepwise sliding-back mechanism has been proposed. It is likely that internal initiation and a sliding-back mechanism is a feature of the genomes that initiate replication by protein priming.

The φ29 DNA polymerase/primer TP heterodimer does not dissociate after initiation or after sliding back. There is a transition stage in which the DNA polymerase synthesizes a five nucleotide-long DNA molecule while complexed with the primer TP, undergoes some structural changes during incorporation of nucleotides 6–9 (transition), and finally dissociates from the primer TP when nucleotide 10 is incorporated into the nascent DNA chain (elongation mode).

Once dissociated from the heterodimer, the same DNA polymerase molecule starts normal DNA elongation, catalyzing highly processive polymerization coupled to strand

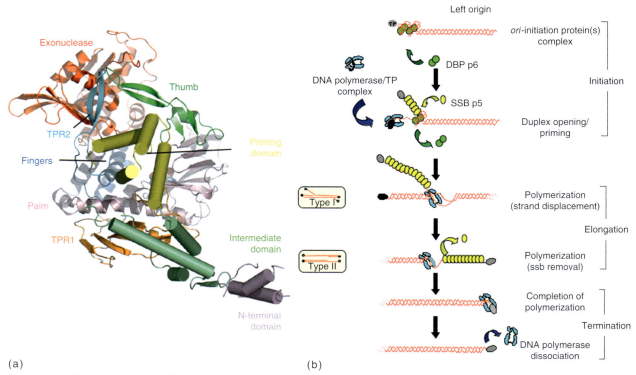

Figure 3 (a) Structure of the ϕ29 DNA polymerase-terminal protein heterodimer. Ribbon representation of DNA polymerase colored by subdomains and terminal protein shown with cylindrical helices. (b) Schematic representation of bacteriophage ϕ29 TP-DNA replication. (a) Adapted from Kamtekar S, Berman AJ, Wang J, et al. (2006) The ϕ29 DNA polymerase: protein-primer structure suggests a model for the initiation to elongation transition. *EMBO Journal* 25: 1335–1343, with permission from Nature Publishing Group. (b) With permission, from the Annual Review of Biochemistry, Volume 60 © 1991 by Annual Reviews.

displacement, and, therefore, complete replication of both strands proceeds continuously from each terminal-priming event. As the two replication forks move, DNA synthesis is initially coupled to strand displacement of long stretches of single-stranded ϕ29 DNA, producing type I replication intermediates (**Figure 3(b)**). When the two replication forks, moving in opposite directions, meet, a new type of replication intermediate (type II) is found. Electron microscopy analysis of ϕ29 replication intermediates *in vitro* showed that the viral protein p5 binds to the single-stranded portion of both type I and II molecules, thus acting as a SSB during ϕ29 DNA replication. Once replication of both strands is fulfilled, the two DNA polymerase molecules fall off the DNA to start initiation and replication of a new ϕ29 DNA molecule.

ϕ29 DNA replication takes place in close association with the bacterial membrane. Recent studies have identified p1 and p16.7 as membrane-localized, phage-encoded proteins likely to be involved in the membrane association of ϕ29 DNA replication. A multimeric p1 structure would provide an anchoring site for viral replisome through interaction with the primer TP, while p16.7 would recruit the ϕ29 DNA replication intermediates to the membrane by binding both the parental TP and the displaced ssDNA.

Bacteriophage Mu

E. coli phage Mu, a member of the family *Myoviridae*, is of great interest because it is both a bacteriophage and a transposon. The size of its linear dsDNA ranges between 37 and 42 kbp with a mean value of 36.7 kbp. The variability of its length is due to the presence of 50–150 bp and 0.5–3 kbp of host genome flanking the left and right ends of the viral genome, respectively. Once Mu has infected the host cell, the linear DNA is converted into a circular form, induced by the phage protein N that is noncovalently bound to Mu DNA. The early phage protein A (transposase) specifically inserts the viral genome in a random fashion into the host chromosome, through the genomic sequences *att*L and *att*R located at both ends of the viral DNA, following a 'nick-join-process' pathway by which the host sequences, still attached to Mu upon integration, are degraded shortly after, probably by gap repair mechanisms. Bacteriophage Mu replicates as a transposable element. Mu A protein remains bound to the Mu genome once inserted into the host genome, forming an oligomeric transpososome which promotes the transfer of the viral ends to other location of the host DNA, creating replication forks at each end. Host ClpX chaperonin reduces transpososome interaction with the DNA to promote the assembly of prereplisome. Later, primosome assembly protein PriA binds to the forked DNA structures and recruits PriB, DnaT, DnaB, and DnaC. PriA opens the dsDNA for DnaB binding that leads to the recruitment of DNA polymerase III holoenzyme to complete replisome assembly. In addition, DnaB attracts primase to catalyze lagging strand synthesis. The repetition of this integration process produces more than 100 copies of Mu DNA. Finally, these copies are cut at the *pac* sequence, located at the left end. The DNA is cut in a way that 50–150 bp of bacterial DNA at the left end are encapsidated in the phage particle. The total length of the encapsidated DNA depends on the size of the phage head, in a way that 0.5–3 kbp of host genome will flank the right end.

ϕX174 and Related Phages

ϕX174 and related a3, St1, and G4 bacteriophages are members of the viral family *Microviridiae*. All of them grow on various strains and species of *Enterobacteriaceae*, typically *E. coli*, *Salmonella*, and *Shigella* species. They contain a circular (+) ssDNA genome whose replication has been widely studied in phage ϕX174. The replication cycle can be divided into three stages.

Stage I. Once inside the host cell, the circular ssDNA is covered by the SSB protein, before starting the complementary (−)-strand synthesis. Different members of this group of phages exploit different host enzyme systems for complementary strand synthesis when ssDNA is complexed with SSB. Once covered by SSB, the assembly of preprimosome is carried out. The preprimosome is constituted by proteins PriA, PriB, PriC, DnaT, and DnaB. PriA recognizes the unique sequence *pas*, also called n', that forms a stem–loop structure. PriB and PriC act as stability and specificity factors. DnaT and DnaC load the host helicase DnaB. The preprimosome associates with the host primase DnaG to produce the primosome. Such a complex travels on ssDNA following a $5'$- to $3'$-direction, with the concomitant synthesis of short RNA molecules by DnaG to prime DNA synthesis by host DNA polymerase III holoenzyme. Host DNA polymerase I removes RNA primers and a DNA ligase ligates the different DNA fragments to produce a circular and supercoiled dsDNA (replicative form I; RFI). The stem loops of other phages such as G4, a3, and St-1 are directly recognized by DnaG primase without the need for auxiliary proteins.

Stage II. Phage protein A nicks between (+)-strand nucleotides 4305 and 4306 at the replication origin (30 bp long), releasing the superhelicity of the DNA molecule to give replicative form II (RFII) DNA molecules. Protein A creates a covalent ester linkage between a tyrosine residue and the $5'$-phosphate group of adenylic acid at position 4306 of viral (+)-strand. Host rep protein (helicase) forms a complex with protein A, unwinding the two strands of the duplex. In a coordinated way, DNA polymerase III holoenzyme uses the newly generated (+)-strand $3'$-OH to prime the synthesis of a new (+)-strand. The $5'$-end of the displaced strand travels with the replication fork in a 'looped rolling circle' way. Once the preprimosome assembly site on the displaced SSB-coated (+)-strand is available, synthesis of a new (−)-strand takes place as in stage I to give more RFI molecules that will be used as templates in further replication cycles. After one round of rolling circle synthesis, protein A cuts the newly generated replication origin, acting as a ligase to give circular (+) ssDNA molecules, protein A being transferred to the newly created $5'$-end setting the stage for a new round of replication.

Stage III. GpC protein binds to gpA/rep/ RFII complex, enabling them to serve as template in further RF replication rounds, forcing them to be used as template for the unique generation of (+) ssDNA molecules, that will be encapsidated later.

ss- and dsRNA Phages

The ssRNA coliphages form the family *Leviviridae*, of which the Qβ and MS2 bacteriophages are the best-studied members. Their infection depends on *E. coli* F-pili, normally used for bacterial conjugation. The genome of these phages is an ssRNA molecule with a length ranging from 3500 to 4200 nucleotides, depending on the genera. The viral (+) RNA molecule, once inside the cell, functions as messenger RNA for synthesis of phage proteins, and as template for viral replicase to multiply the viral genome. The high degree of secondary structure shown by the RNA plays a pivotal role in the fine-tuned coordination between translation (which proceeds in the $5'$–$3'$ direction of RNA) and replication (which advances in the opposite direction), which is required to prevent a head to head collision of ribosomes and phage replicase.

To replicate their genome, ssRNA phages make use of a replicase that is composed of four proteins, only one of which coded by the phage (the replicase or β-subunit). The other three are encoded by the host: ribosomal S1 (α-subunit), and translation elongation factors EF-Tu and EF-Ts (γ- and δ-subunits, respectively). The copy of (+) ssRNA into (−) ssRNA requires also the product of the host *hfq* gene, the host factor HF. The $3'$-end of (+) RNA is protected against host RNase E and other exonucleases by base pairing. This implies that the terminal nucleotides are also inaccessible to the ribosome and replicase, and suggests that S1 and HF are required. S1 is proposed to anchor the template on the polymerase in a standby complex, waiting for the occasional thermal breathing of the $3'$-region, such an opening being assisted by HF. The translation elongation factors bind the RNA to the polymerase. Under such conditions, replicase initiates (−) RNA synthesis at the penultimate $3'$-terminal C nucleotide, initially losing the ultimate A. Once replicase reaches the $5'$-end of the (+) RNA, the terminal

A nucleotide is recovered by an untemplated addition of an A at the 3′-end of the (−)-RNA. Although both (+)- and (−)-RNA chains are complementary, they do not anneal to form dsRNA. Such annealing is inhibited by the high degree of internal secondary structure formed in each ssRNA molecule. The (−)-ssRNA is used as a template to produce more (+)-ssRNA.

Members of the family *Cystoviridae* contain genomes composed of three dsRNA segments called S, L, and M. From this family, phages ø6, and more recently ø8, have been studied in great detail. Once the host cell is infected by these viruses, a transcriptionally active polymerase complex is released into the bacterial cytoplasm, where transcription of the dsRNA segments takes place. The L segment messenger RNA codes for proteins P1 (main component of the inner shell), P2 (RNA-dependent RNA polymerase), P4 (NTPase), P7 (involved in packaging and replication), and P14. The first four proteins form a polymerase complex which packages the three ssRNA (+)-strands. Such strands have an 18-base consensus sequence at their 5′-ends, and a *pac* sequence 200 bases long. Both the consensus and *pac* sequences are required and sufficient to package plus strands. The ssRNA(+) strands are packaged in a S, M, and L fashion. Once packaged within the viral particle, the polymerase complex synthesizes the minus RNA strand to convert the ssRNA into dsRNA.

See also: History of Virology: Bacteriophages; Icosahedral dsDNA Bacterial Viruses with an Internal Membrane; Icosahedral Enveloped dsRNA Bacterial Viruses; Icosahedral ssDNA Bacterial Viruses; Icosahedral ssRNA Bacterial Viruses; Icosahedral Tailed dsDNA Bacterial Viruses.

Further Reading

Au TK, Agrawal P, and Harshey RM (2006) Chromosomal integration mechanism of infecting Mu virion DNA. *Journal of Bacteriology* 188: 1829–1834.

Bamford DH (1999) Phage ø6. In: Granoff A and Webster RG (eds.) *Encyclopedia of Virology*, 2nd edn., pp. 1205–1208. San Diego, CA: Academic Press.

Barariska S, Gabig M, Wegrzyn A, et al. (2001) Regulation of the switch from early to late bacteriophage λ DNA replication. *Microbiology* 147: 535–547.

Chung YB, Nardone C, and Hinkle DC (1990) Bacteriophage T7 DNA packaging III. A "hairpin" end formed on T7 concatemers may be an intermediate in the processing reaction. *Journal of Molecular Biology* 216: 939–948.

Fujsawa H and Morita M (1997) Phage DNA packaging. *Genes to cells* 2: 537–547.

Jiang H, Yang J-Y, and Harshey RM (1999) Criss-crossed interactions between the enhancer and the *att* sites of phage Mu during DNA transposition. *EMBO Journal* 18: 3845–3855.

Kamtekar S, Berman AJ, Wang J, et al. (2006) The φ29 DNA polymerase: Protein-primer stucture suggests a model for the initiation to elongation transition. *EMBO Journal* 25: 1335–1343.

Martínez-Jiménez MI, Alonso JC, and Ayora S (2005) *Bacillus subtilis* bacteriophage SPP1-encoded gene 34.1 product is a recombination-dependent DNA replication protein. *Journal of Molecular Biology* 351: 1007–1019.

Miller ES, Kutter E, Mossig G, Arisaka F, Kunisawa T, and Rüger W (2003) Bacteriophage T4 genome. *Microbiology and Molecular Biology Reviews* 67: 86–156.

Molineux IJ (2006) The T7 goup. In: Calendar R (ed.) *The Bacteriophages*, 2nd edn., pp. 277–301. Oxford: Oxford University Press.

Mosig G, Gewing J, Luder A, Colowick N, and Vo D (2001) Two recombination-dependent DNA replication pathways of bacteriophage T4, and their roles in mutagenesis and horizontal gene transfer. *Proceedings of the National Academy of Sciences, USA* 98: 8306–8311.

Nakai H, Doseeva V, and Jones JM (2001) Handoff from recombinase to replisome: Insights from transposition. *Proceedings of the National Academy of Sciences, USA* 98: 8247–8254.

Ng JY and Marians KJ (1996) The ordered assembly of the φX174-type primosome. I. Isolation and identification of intermediate protein-DNA complexes. *Journal of Biological Chemistry* 271: 15642–15648.

Ng JY and Marians KJ (1996) The ordered assembly of the φX174-type primosome. II. Preservation of primosome composition from assembly through replication. *Journal of Biological Chemistry* 271: 15649–15655.

Salas M (1991) Protein priming of DNA replication. *Annual Review of Biochemistry* 60: 39–71.

Salas M (1999) Mechanisms of initiation of linear DNA replication in prokaryotes. *Genetic Engineering (New York)* 21: 159–171.

Taylor K and Wegrzyn G (1999) Regulation of bacteriophage λ replication. In: Busby SJW, Thomas CM, and Brown NL (eds.) *Molecular Microbiology*, pp. 81–97. Berlin: Springer.

Van Duin J and Tsareva N (2006) Single-stranded RNA phages. In: Calendar R (ed.) *The Bacteriophages*, 2nd edn., pp. 175–196. Oxford: Oxford University Press.

Replication of Viruses

A J Cann, University of Leicester, Leicester, UK

© 2008 Elsevier Ltd. All rights reserved.

Glossary

(+)-sense RNA (plus-sense RNA) A virus with a single-stranded RNA genome of the same polarity ('sense') as mRNA.

(−)-sense RNA (minus-sense RNA) A virus with a single-stranded RNA genome of the opposite polarity ('sense') as mRNA.

Assembly The stage of replication during which all the structural components come together at one site

in the cell and the basic structure of the virus particle is formed.
Attachment The binding of a virus particle to a specific receptor on the surface of a host cell.
Capsid A protein shell comprising the main structural unit of a virus particle.
Envelope A lipid membrane enveloping a virus particle.
Fusion protein The protein(s) on the surface of a virus particle responsible for fusion of the virus envelope with cellular membranes.
Gene expression An important stage of viral replication at which virus genetic information is expressed: one of the major control points in replication.
Genome replication The stage of viral replication at which the virus genome is copied to form new progeny genomes.
Matrix protein A structural protein of a virus particle which underlies the envelope and links it to the core.
Maturation The stage of viral replication at which a virus particle becomes infectious.
Molecular epidemiology The use of nucleotide sequence information to study the diversity and distribution of virus populations.
mRNA Messenger RNA, translated on ribosomes to produce proteins.
Nucleocapsid The core of a virus particle consisting of the genome plus a complex of proteins.
Penetration The stage of viral replication at which the virus genome enters the cell.
Polyprotein A long polypeptide encoding several mature proteins which are subsequently released by protease cleavage.
Receptor A specific molecule on the surface of a cell which is used by a virus for attachment.
Release The stage of viral replication at which virus particles escape the infected cell.
Tropism The ability of a virus to infect specific cell or tissue types.
Uncoating The stage of viral replication at which structural proteins are lost and the virus genome is exposed to the replication machinery.
Virions Structurally mature, extracellular virus particles.
Virus attachment protein The protein on the surface of a virus particle responsible for binding the receptor.

Unlike cellular organisms, which 'grow' from an increase in the integrated sum of their components and reproduce by division, virus particles are produced from the assembly of preformed components. Once manufactured, virus particles (virions) do not grow or undergo division. This alone makes the process of virus replication distinct from the growth of all other biological agents, and although the term 'grow' is sometimes used in the vernacular to refer to propagation of viruses, it is best to avoid this word when referring to the processes of virus replication.

Although this article will attempt to paint a general picture of the process of virus replication, the type of host cell infected by the virus has a profound effect on the replication process. There are many examples of viruses undergoing different replicative cycles in different cell types. However, the coding capacity of the genome determines the basic replication strategy used by different viruses. This strategy may involve heavy reliance on the host cell, in which case the virus genome can be very compact and need only encode the essential information for a few proteins, for instance, in parvoviruses. Alternatively, large and complex virus genomes, such as those of poxviruses, encode most of the information necessary for replication, and the virus is only reliant on the cell for the provision of energy and the apparatus for macromolecular synthesis, such as ribosomes. Viruses with RNA genomes have no apparent need to enter the nucleus, although during the course of replication, some do. DNA viruses, as might be expected, mostly replicate in the nucleus where host cell DNA is replicated and the biochemical apparatus necessary for this process is located. However, some viruses with DNA genomes (e.g., poxviruses) have evolved to contain sufficient biochemical capacity to be able to replicate in the cytoplasm, with minimal requirement for host cell functions.

Virus replication can be divided into eight stages, as shown in **Figure 1**. It should be emphasized that these are arbitrary divisions, used here for convenience in explaining the replication cycle of a theoretical, 'typical' virus. Regardless of their hosts, all viruses must undergo each of these stages in some form to successfully complete their replication cycle. Not all the steps described here are detectable as distinct stages for all viruses; often they blur together and appear to occur almost simultaneously. Some of the individual stages have been studied in great detail and a considerable amount of information is known about them. Other stages have been much harder to study, and less information is available.

Attachment

The attachment phase of replication comprises specific binding of a virus-attachment protein (or 'antireceptor') to a cellular receptor molecule. Virus receptors on cell surfaces may be proteins (usually glycoproteins) or carbohydrate residues present on glycoproteins or glycolipids. Some complex viruses (e.g., in the *Poxviridae* or *Herpesviridae*) use more than one receptor and therefore have alternative routes of uptake into cells. Most bacteriophage

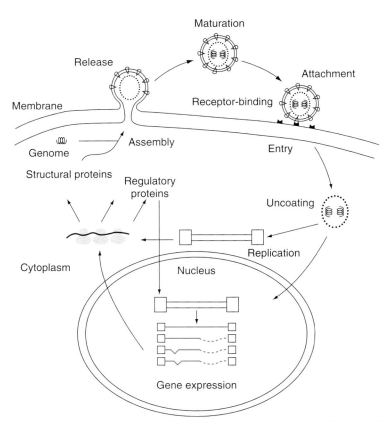

Figure 1 Schematic overview of a generalized scheme of virus replication. Reproduced from Cann AJ (2004) *Principles of Molecular Virology*, 4th edn. Amsterdam: Elsevier, with permission from Elsevier.

receptors are on the bacterial cell wall, although certain phages use cellular appendages (pili, flagella) as primary adsorption sites. Attachment is an automatic docking process and the kinetics of receptor binding are controlled by the chemical and thermodynamic characteristics of the molecules involved, that is, their relative concentrations and availability.

In most cases, the expression (or absence) of receptors on the surface of host cells determines the tropism of a particular virus, that is, the types of cell in which it is able to replicate. The attachment phase of infection therefore has a major influence on viral pathogenesis and in determining the course of a virus infection. Plant viruses must overcome different problems to animal viruses in initiating infection. The outer surfaces of plants are composed of protective layers of waxes and pectin, and each cell is surrounded by a thick wall of cellulose overlying the cytoplasmic membrane. No known plant virus uses a specific cellular receptor of the type that animal and bacterial viruses use to attach to cells and plant viruses must rely on mechanical breaks in the cell wall to directly introduce a virus particle into a cell.

Some virus receptors consist of more than one protein and multiple interactions are required for virus entry. An example of this is human immunodeficiency virus-1 (HIV-1), the primary receptor for which is the T-cell antigen, CD4. The binding site for the HIV-1 attachment protein (antireceptor), gp 120, has been mapped to the first variable region of CD4, although additional amino acids of the second variable domain also contribute toward binding. The sequences important for CD4 binding have also been mapped in gp120. Deletions in this region or site substitutions abolish CD4 binding. In addition to CD4, there is at least one accessory factor which is necessary to form a functional HIV-1 receptor. These factors have now been identified as a family of proteins known as β-chemokine receptors. Multiple members of this class of proteins have been shown to play a role in the entry of HIV-1 into cells, and their distribution in the body is the primary control for the tropism of HIV-1 for different cell types.

Occasionally, the specificity of receptor binding can be subverted by nonspecific interactions between virus particles and host cells. Virus particles may be taken up by cells by pinocytosis or phagocytosis. However, without some form of physical interaction which holds the virus particle in close association with the cell surface, the frequency of these events would be very low. In addition, the fate of viruses absorbed into endocytic vacuoles is usually to be degraded, except in cases where the virus particle enters cells by this route. On occasion, binding of antibody-coated virus particles to Fc receptor molecules on the surface of monocytes and other blood cells can result in virus uptake. The presence of antiviral antibodies can result in increased virus uptake by cells and increased pathogenicity, rather than virus neutralization, as would normally be expected. The significance of such mechanisms *in vivo* is not known.

Entry

Entry of the virus particle into the host cell normally occurs a short time after attachment of the virus to the receptor. Unlike attachment, cell entry is generally an energy-dependent process, that is, the cell must be metabolically active for this to occur. Three main mechanisms are observed:

1. Translocation of the entire virus particle across the cytoplasmic membrane of the cell. This process is relatively rare among viruses and is poorly understood. It is mediated by proteins in the virus capsid and specific membrane receptors.
2. Endocytosis of the virus into intracellular vacuoles. This is probably the most common mechanism of virus entry into cells. It does not require any specific virus proteins (other than those already utilized for receptor binding) but relies on the normal formation and internalization of coated pits (term to be explained) at the cell membrane. Receptor-mediated endocytosis is an efficient process for taking up and concentrating extracellular macromolecules.
3. Fusion of the virus envelope (where present) with the cell membrane, either directly at the cell surface or following endocytosis in a cytoplasmic vesicle. Fusion requires the presence of a specific fusion protein in the virus envelope, for example, influenza A virus hemagglutinin or the transmembrane glycoproteins of retroviruses. These proteins promote the joining of the cellular and virus membranes which results in the nucleocapsid being deposited directly in the cytoplasm. There are two types of virus-driven membrane fusions: pH-dependent and pH-independent.

The process of endocytosis is almost universal in animal cells and requires the formation of clathrin-coated pits which results in the engulfment of a membrane-bounded vesicle by the cytoplasm of the cell. At this point, any virus contained within these structures is still cut off from the cytoplasm by a lipid bilayer and therefore has not strictly entered the cell. As endosomes fuse with lysosomes, the environment inside these vessels becomes progressively more hostile as they are acidified and the pH falls, while the concentration of degradative enzymes rises. This means that the virus must leave the vesicle and enter the cytoplasm before it is degraded. There are a number of mechanisms by which this occurs, including membrane fusion and rescue by transcytosis. The release of virus particles from endosomes and their passage into the cytoplasm is intimately connected with (and often impossible to separate from) the process of uncoating.

Uncoating

Uncoating describes the events which occur after host cell entry, during which the virus capsid is partially or completely degraded or removed and the virus genome exposed, usually still in the form of a nucleic acid–protein complex. Uncoating occurs simultaneously with or immediately after entry and is thus difficult to study. In bacteriophages which inject their genome directly into the cell, entry and uncoating are the same process.

The removal of a virus envelope during membrane fusion is the initial stage of the uncoating process for enveloped viruses. Uncoating may occur inside endosomes, being triggered by the change in pH as the endosome is acidified, or directly in the cytoplasm. Entry into the endocytic pathway is a hazardous process for viruses because if they remain in the vesicle too long, they will be irreversibly damaged by low pH or lysosomal enzymes. Hence, some viruses have evolved proteins to control this process; for example, the influenza A virus M2 protein is a membrane channel which allows entry of hydrogen ions into the nucleocapsid, facilitating uncoating. The M2 protein is multifunctional, and also has a role in virus uncoating. In the picornaviruses, penetration of the cytoplasm by exit of virus from endosomes is tightly linked to uncoating. The acidic environment of the endosome causes a conformational change in the particle at around pH 5 that reveals hydrophobic domains not present on the surface of mature virus capsids. These hydrophobic patches interact with the endosomal membrane and form pores through which the RNA genome passes into the cytoplasm of the host cell.

The ultimate product of uncoating depends on the structure of the virus genome/nucleocapsid. In some cases, the resulting structure is relatively simple; for example, picornaviruses have only a small basic protein of approximately 23 amino acids covalently attached to the 5′ end of the RNA genome. In other cases, the virus core which remains is highly complex; for example, in the poxviruses uncoating occurs in two stages – removal of the outer membrane as the particle enters the cell and in the cytoplasm, followed by further uncoating as the core passes into the cytoplasm. In this case, the core still contains dozens of proteins and at least 10 distinct enzymes.

The structure and chemistry of the nucleocapsid determines the subsequent steps in replication. Reverse transcription can only occur inside an ordered retrovirus core particle and does not proceed to completion with the virus RNA free in solution. Eukaryotic viruses which replicate in the nucleus, such as members of the

Herpesviridae, *Adenoviridae*, and *Polyomaviridae*, undergo structural changes following penetration, but overall remain largely intact. This is important because these capsids contain nuclear localization sequences responsible for attachment to the cytoskeleton and this interaction allows the transport of the entire capsid to the nucleus. At the nuclear pores, complete uncoating occurs and the nucleocapsid passes into the nucleus.

Transcription and Genome Replication

The replication strategy of a virus depends, in large part, on the structure and composition of its genome. For viruses with RNA genomes in particular, genome replication and transcription are often inextricably linked, and frequently carried out by the same enzymes. Therefore, it makes most sense to consider both of these aspects of virus replication together.

Group I: Double-Stranded DNA

This class can be further subdivided into two as follows:

1. Replication is exclusively nuclear or associated with the nucleoid of prokaryotes. The replication of these viruses is relatively dependent on cellular factors. In some cases, no virus-encoded enzymes are packaged within these virus particles as this is not necessary, whereas in more complex viruses numerous enzymatic activities may be present within the particles.
2. Replication occurs in cytoplasm. These viruses have evolved (or acquired from their hosts) all the necessary factors for transcription and replication of their genomes and are therefore largely independent of the cellular apparatus for DNA replication and transcription. Because of this independence from cellular functions, these viruses have some of the largest and most complex particles known, containing many different enzymes.

Group II: Single-Stranded DNA

The replication of these virus genomes occurs in the nucleus, involving the formation of a double-stranded intermediate which serves as a template for the synthesis of new single-stranded genomes. In general, no virus-encoded enzymes are packaged within the virus particle since most of the functions necessary for replication are provided by the host cell.

Group III: Double-Stranded RNA

These viruses all have segmented genomes, as each segment is transcribed separately to produce individual monocistronic messenger RNAs. Replication occurs in the cytoplasm and is largely independent of cellular machinery, as the particles contain many virus-encoded enzymes essential for RNA replication and transcription since these processes (involving copying RNA to make further RNA molecules) do not normally occur in cellular organisms.

Group IV: Single-Stranded (+)-Sense RNA

These viruses can be subdivided into two groups.

1. *Viruses with polycistronic mRNA such as flaviviruses and picornaviruses.* As with all the viruses in this group, the genome RNA represents mRNA which is translated after infection, resulting in the synthesis of a polyprotein product, which is subsequently cleaved to form the mature proteins.
2. *Viruses with complex transcription such as coronaviruses and togaviruses.* In this subgroup, two rounds of translation are required to produce subgenomic RNAs which serve as mRNAs in addition to the full-length RNA transcript which forms progeny virus genomes. Although the replication of these viruses involves copying RNA from an RNA template, no virus-encoded enzymes are packaged within the genome since the ability to express genetic information directly from the genome without prior transcription allows the virus replicase to be synthesized after infection has occurred.

Group V: Single-Stranded (−)-Sense RNA

The genomes of these viruses can also be divided into two types.

- *Segmented.* The first step in the replication of these viruses (e.g., orthomyxoviruses) is transcription of the (−)-sense RNA genome by the virion RNA-dependent RNA polymerase packaged in virus particles to produce monocistronic mRNAs, which also serve as the template for subsequent genome replication.
- *Nonsegmented.* Monocistronic mRNAs for each of the virus genes are produced by the virus transcriptase in the virus particle from the full-length virus genome. Subsequently, a full-length (+)-sense copy of the genome is synthesized which serves as a template for (−)-sense progeny virus genomes (e.g., paramyxoviruses and rhabdoviruses).

Group VI: Single-Stranded RNA with DNA Intermediate

Retrovirus genomes are composed of (+)-sense RNA but are unique in that they are diploid and do not serve directly as mRNA but as a template for reverse transcription into DNA. A complete replication cycle involves conversion of the RNA form of the virus genetic material into a DNA form, the provirus, which is integrated into the host cell chromatin. The enzyme reverse transcriptase needs to be packaged into virus particles to achieve this conversion, as virus genes are only expressed from the DNA provirus and not from the RNA genome found in retrovirus particles of retroviruses.

Group VII: Double-Stranded DNA with RNA Intermediate

This group of viruses also relies on reverse transcription, but unlike the retroviruses, this occurs inside the virus particle during maturation. On infection of a new cell, the first event to occur is repair of the gapped genome, followed by transcription. As with group VI viruses, a reverse transcriptase enzyme activity is present inside virus particles, but in this case, the enzyme carries out the conversion of virus RNA into the DNA genome of the virus inside the virus particle. This contrasts with retroviruses where reverse transcription occurs after the RNA genome has been released from the virus particle into the host cell.

Assembly

During assembly, the basic structure of the virus particle is formed as all the components necessary for the formation of the mature virion come together at a particular site in the cell. The site of assembly depends on the pattern of virus replication and the mechanism by which the virus is eventually released from the cell and so varies for different viruses. Although some DNA virus particles form in the nucleus, the cytoplasm is the most common site of particle assembly. In the majority of cases, cellular membranes are used to anchor virus proteins, and this initiates the process of assembly.

For enveloped viruses, the lipid covering is acquired through a process known as budding, where the virus particle is extruded through a cell membrane. Lipid rafts are membrane microdomains enriched in glycosphingolipids (or glycolipids), cholesterol and a specific set of associated proteins. Lipid rafts have been implicated in a variety of cellular functions, such as apical sorting of proteins and signal transduction, but they are also used by viruses as platforms for cell entry (e.g., for HIV-1, SV40, and the rotaviruses), and as sites for particle assembly, budding and release from the cell membrane (e.g., in influenza A virus, HIV, measles virus, and rotaviruses).

As with the earliest stages of replication, it is often not possible to identify the assembly, maturation, and release of virus particles as distinct and separate phases. The site of assembly has a profound influence on all these processes. In general terms, rising intracellular levels of virus proteins and genomes reach a critical concentration and this triggers assembly. Many viruses achieve high levels of newly synthesized structural components by concentrating these into subcellular compartments known as inclusion bodies. These are a common feature of the late stages of infection of cells by many different viruses. Alternatively, local concentrations of virus structural components can be boosted by lateral interactions between membrane-associated proteins. This mechanism is particularly important in enveloped viruses which are released from the cell by budding (see above).

Maturation

Maturation is the stage of the replication cycle at which virus particles become infectious. This often involves structural changes in the newly formed particle resulting from specific cleavages of virus proteins to form the mature products or from conformational changes in proteins which occur during assembly (e.g., hydrophobic interactions). Protein cleavage frequently leads to substantial structural changes in the capsid. Alternatively, internal structural alterations, for example, the condensation of nucleoproteins with the virus genome, often result in changes visible by electron microscopy.

Proteases are frequently involved in maturation, and virus-encoded enzymes, cellular proteases or a mixture of the two may be used. Virus-encoded proteases are usually highly specific for particular amino acid sequences and structures, only cutting a particular peptide bond in a particular protein. Moreover, they are often further controlled by being packaged into virus particles during assembly and only activated when brought into close contact with their target sequence by the conformation of the capsid, for example, by being placed in a local hydrophobic environment, or by changes of pH or cation cofactor concentrations inside the particle as it forms. Retrovirus proteases are good examples of enzymes involved in maturation which are under tight control. The retrovirus core particle is composed of proteins from the *gag* gene and the protease is packaged into the core before its release from the cell on budding. During the budding process, the protease cleaves the gag protein precursors into the mature products – the capsid, nucleocapsid, and matrix proteins of the mature virus particle. Other protease cleavage events involved in maturation are less closely controlled. Influenza A virus hemagglutinin must be cleaved into two fragments (HA_1 and HA_2) to be able to promote membrane fusion during infection. Cellular trypsin-like enzymes are responsible for this process, which occurs in secretory vesicles as the virus buds into them prior to release at the cell surface; however, this process is controlled by the virus M2 protein, which regulates the pH of intracellular compartments in influenza virus-infected cells.

Release

For lytic viruses (most nonenveloped viruses), release is a simple process – the infected cell breaks open and releases the virus. The reasons for lysis of infected cells are not always clear, but virus-infected cells often disintegrate because viral replication disrupts normal cellular function, for example, the expression of essential genes. Many viruses also encode proteins that stimulate (or in some cases suppress) apoptosis, which can also result in release of virus particles.

Enveloped viruses acquire their lipid membrane as the virus buds out of the cell through the cell membrane, or into an intracellular vesicle prior to subsequent release. Virion envelope proteins are picked up during this process as the virus particle is extruded. This process is known as budding. As mentioned earlier, assembly, maturation, and release are usually simultaneous processes for viruses which are released by budding. The release of mature virus particles from their host cells by budding presents a problem in that these particles are designed to enter, rather than leave, cells. Certain virus envelope proteins are involved in the release phase of replication as well as in receptor binding. The best-known example of this is the neuraminidase protein of influenza virus. In addition to being able to reverse the attachment of virus particles to cells via hemagglutinin, neuraminidase is also believed to be important in preventing the aggregation of influenza A virus particles and may well have a role in virus release. In addition to using specific proteins, viruses which bud have also solved the problem of release by the careful timing of the assembly-maturation-release pathway. Although it may not be possible to separate these stages by means of biochemical analysis, this does not mean that spatial separation of these processes has not evolved as a means to solve this problem.

Further Reading

Cann AJ (2004) *Principles of Molecular Virology,* 4th edn. Amsterdam: Elsevier.

Freed EO (2004) HIV-1 and the host cell: An intimate association. *Trends in Microbiology* 12: 170–177.

Kasamatsu H and Nakanishi A (1998) How do animal DNA viruses get to the nucleus? *Annual Review of Microbiology* 52: 627–686.

Lopez S and Arias CF (2004) Multistep entry of rotavirus into cells: A Versaillesque dance. *Trends in Microbiology* 12: 271–278.

Moore JP, Kitchen SG, Pugach P, and Zack JA (2004) The CCR5 and CXCR4 coreceptors – central to understanding the transmission and pathogenesis of human immunodeficiency virus type 1 infection. *AIDS Research and Human Retroviruses* 20: 111–126.

Rossmann MG, He Y, and Kuhn RJ (2002) Picornavirus–receptor interactions. *Trends in Microbiology* 10: 324–331.

Schneider-Schaulies J (2000) Cellular receptors for viruses: Links to tropism and pathogenesis. *Journal of General Virology* 81: 1413–1429.

Reticuloendotheliosis Viruses

A S Liss and H R Bose Jr., University of Texas at Austin, Austin, TX, USA

© 2008 Elsevier Ltd. All rights reserved.

History

The reticuloendotheliosis viruses are a small group of avian retroviruses that are distinct from the avian leukosis and sarcoma viruses. The founding member of this family, reticuloendotheliosis virus strain-T (REV-T), was isolated in 1958 from lesions of a turkey that died with visceral reticuloendotheliosis. While other members of this family are also pathogenic, REV-T is unique in that it is acutely transforming due to its acquisition of the v-*rel* oncogene. Although periodic outbreaks of reticuloendotheliosis virus infections occur in commercial bird flocks they do not represent a major economic impact. The study of these viruses in the laboratory has provided insight into the mechanism of retroviral replication and has defined a role for Rel/NF-κB proteins in cancer.

Taxonomy and Classification

Reticuloendotheliosis viruses belong to the *Retroviridae* family of viruses. These single-stranded RNA viruses replicate through a DNA intermediate that is integrated into the host genome. The reticuloendotheliosis viruses belong to the subfamily *Orthoretrovirinae* and the genus *Gammaretrovirus*. Members of the reticuloendotheliosis virus family include REV-T, reticuloendotheliosis-associated virus (REV-A), Trager duck spleen necrosis virus (TDSNV), and chick syncytial virus (CSV). The reticuloendotheliosis viruses belong to a single interference group and can be divided into three distinct antigenic groups (REV-A, TDSNV, and CSV) based on their reactivity with monoclonal antibodies.

Reticuloendotheliosis viruses are more closely related to what have been referred to historically as mammalian type C and simian type D retroviruses than they are to other avian retroviruses. Analysis of their morphology by electron microscopy revealed that they exhibit morphology similar to that of mammalian type C retroviruses. Consistent with this classification, the reverse transcriptase encoded by reticuloendotheliosis viruses exhibits a preference for Mn^{2+} rather than Mg^{2+}. The envelope protein of reticuloendotheliosis viruses share 42% identity with the envelope proteins of simian type D retroviruses and some type C viruses, such as baboon endogenous virus (BaEV). However, the Gag and Pol sequences of reticuloendotheliosis viruses are homologous to those of type C retroviruses but are distinct from type D viruses. In addition, reticuloendotheliosis viruses utilize the same cellular receptor as mammalian type C (RD-114, BaEV, and HERV-W) and simian type D (SRV-1, -2, -3, -4, and -5) retroviruses. Their homology to mammalian retroviruses has led to the hypothesis that the reticuloendotheliosis viruses arose from a common ancestor of mammalian type C viruses that adapted to avian species.

Properties of the Virion

The morphology of reticuloendotheliosis viruses is similar for all members. The viral particles are enveloped, with surface projections that are approximately 6 nm long and an overall diameter of 110 nm. These projections are tapered, with a diameter of 10 nm at the top and 4 nm at the viral membrane. Each particle contains approximately 100 of these projections that are spaced 14 nm apart (center to center). The cores of mature virions are 67 nm in diameter. Reticuloendotheliosis virus particles exhibit a buoyant density of $1.16–1.18\,g\,cm^{-3}$ in sucrose gradients. The morphology and buoyant density of these viruses are comparable to other gamma retroviruses.

Properties of the Genome

The genomic RNA purified from reticuloendotheliosis virus particles forms a 60–70S complex, which can be dissociated into two identical 35S subunits. The entire genomes of two replication-competent members of this group, REV-A and TDSNV, have been sequenced. Their genomes share 94% identity and are approximately 8.5 kb in length (**Figure 1(a)**). The 5′ and 3′ ends of the genomic RNA contain unique sequences termed U5 and U3, respectively. An additional sequence, R, is also located at both ends and aids in viral replication. The virus contains the *gag*, *pol*, and *env* genes which encode the viral structural proteins, polymerase, and envelope proteins, respectively. In addition, an internal ribosome entry site (IRES) has been identified immediately upstream of *gag*. The replication-defective member, REV-T, was derived from REV-A and contains extensive deletions in the *gag* and *pol* genes (**Figure 2**). In addition, the *env* sequences have almost entirely been replaced with the v-*rel* oncogene in REV-T. These alterations result in a smaller genome of 5.7 kb.

Properties of the Proteins

The three genes encoded by replication-competent reticuloendotheliosis viruses encode polyproteins that comprise the virion structure, as well as those required for viral replication (**Figure 1(b)**). The *gag* gene product, pr60^{gag}, is a polyprotein that is subsequently cleaved to form the matrix protein (MA) p12, a protein with an unknown function pp18/pp20, the capsid protein (CA) p30, and the nucleocapsid protein (NC) p10. The Pol polyprotein encodes three products: protease (PR) p15, reverse transcriptase (RT) p84, and integrase (IN) p44. The *env* gene encodes two proteins present in the envelope of the virion: the surface glycoprotein (SU) gp90 and the transmembrane glycoprotein (TM) gp20. The replication-defective REV-T does not encode functional *gag*, *pol*, or *env* gene products. The v-*rel* oncogene expressed by this virus encodes a 59 kDa protein that is a member of the Rel/NF-κB family of transcription factors.

Replication

The replication of reticuloendotheliosis viruses employs a mechanism similar to that of other simple retroviruses. Reticuloendotheliosis viruses utilize a sodium-dependent neutral amino-acid transporter (RDR, SLC1A5, ATB^0) on the surface of susceptible cells as a receptor for infection. The binding of virions to the receptor is mediated by SU. While multiple domains throughout SU are required for this interaction, Asp192 is essential and may form a hydrogen bond with the receptor. TDSNV exhibits a pH-independent mechanism for entry that is likely to involve the direct fusion of the viral and cellular membranes. Once in the cytoplasm of the cell, the RT present in the uncoated viral core begins the process of copying the genomic RNA into a double-stranded cDNA. The first strand synthesis of viral DNA is primed by tRNA^{Pro} binding to sequences 3′ of the U5 region. RT synthesizes a

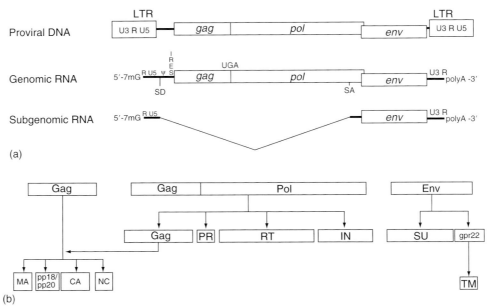

Figure 1 Structure of reticuloendotheliosis virus proviral DNA, RNAs, and proteins. (a) A schematic representation of the proviral DNA of a reticuloendotheliosis virus is shown at the top. The locations of the 5′ and 3′ LTRs are indicated. The relative locations of the U3, R, and U5 sequences of the LTR are also shown. Sequences encoding the gag, pol, and env genes are represented by white boxes. The env gene is offset from the pol gene to indicate that the coding sequences for these genes overlap. The structure of the viral genomic RNA is indicated in the middle. This RNA is capped at the 5′ end (7mG) and is polyadenylated (polyA) at the 3′ end. The locations of the packaging signal (ψ) and the internal ribosome entry site (IRES) at the 5′ end of the transcript are shown. Regions of the transcript that contain the U5, R, and U3 sequences are noted. The position of the amber stop codon (UGA) that is suppressed to form the gag-pol polyprotein is shown. The positions of the the splice donor (SD) and the splice acceptor (SA) utilized in the formation of the subgenomic RNA are indicated. A schematic of the viral subgenomic RNA is shown at the bottom. This transcript encodes env and lacks the sequences for ψ, IRES, gag and pol. (b) The protein products of the replication-competent reticuloendotheliosis viruses are shown. The Gag and Gag-Pol polyproteins are translated from the viral genomic RNA and Env is translated from the subgenomic RNA. The cleavage products of these polyproteins are indicated.

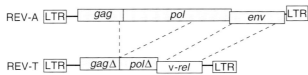

Figure 2 Structural differences between REV-A and REV-T. A schematic representation of the proviral DNA of the replication-competent REV-A is shown. The location of the 5′ and 3′ LTRs as well as the gag, pol, and env genes are indicated. The structural alterations which occurred during the formation of REV-T are indicated below. A nonhomologous recombination event occurred between the env sequences of REV-A and turkey c-rel, resulting in the formation of the v-rel oncogene (white box). The env sequences that encode for 11 N-terminal and 18 C-terminal amino acids of v-Rel are indicated by the gray boxes. The 3′ end of gag (gagΔ) and a large segment of pol (polΔ) are deleted in REV-T relative to REV-A. Due to the deletions of gag, pol, and env sequences, REV-T is replication-defective and requires that these proteins be supplied in trans by a helper virus, such as REV-A, or a packaging cell line.

cDNA encompassing U5 and R from the 5′ end of the viral genomic RNA. The viral RNA circularizes and the polymerase 'jumps' to the 3′ end of the genomic RNA, a process aided by the complementary R sequences. During synthesis of the cDNA the RNaseH activity of the RT degrades the viral RNA template. A portion of the viral RNA resistant to RNaseH activity serves as a primer for second strand DNA synthesis by RT. The synthesis of viral cDNA results in the duplication of the sequences located at the ends of viral genomic RNA, forming a pair of long terminal repeats (LTRs). Both circular and linear copies of viral DNA can be found in the cytoplasm of recently infected cells. The linear proviral DNAs (**Figure 1(a)**) are transported to the nucleus, where they are randomly integrated into the host genome by the virally encoded IN.

Two RNAs are transcribed from a promoter located in the 5′ LTR of the integrated proviral DNA, an 8.5 kb genomic-length RNA, and a 1.9 kb spliced, subgenomic RNA (**Figure 1(a)**). The genomic RNA encodes Gag and Gag-Pol proteins. Although the 5′ end of the genomic RNA has a 7mG cap, an IRES immediately upstream of gag is likely utilized for translation of gag. Synthesis of the Pol polyprotein requires the suppression of an amber stop codon at the 3′ end of gag. The resulting 200 kDa Gag-Pol protein is proteolytically processed by PR, releasing the Gag and Pol polypeptides.

The subgenomic viral RNA encodes the envelope proteins. These are synthesized as an intracellular precursor protein, gPr77^{env}, which is glycosylated with high mannose carbohydrates. This precursor is also myristylated, directing the envelope proteins to the cellular membrane. This precursor is further glycosylated with more complex carbohydrates to form gPr115^{env}. These complex carbohydrates are larger (14–17 kDa) than those found in the Env proteins of MuLV and RSV (3–8 kDa). gPr115^{env} is rapidly cleaved to form the envelope proteins gp90 and gPr22, which are incorporated into the virion.

Viral nucleocapsid assembly occurs in the cytoplasm of cells. This involves the formation of a ribonucleoprotein (RNP) complex composed of viral RNA dimers and 63 kDa Gag precursors. Dimerization of viral genomic RNA is mediated by the dimer linkage structure (DLS), which is located in a 180 nt region between the splice donor (SD) and *gag*. The 5′ end of this sequence forms a double hairpin structure that serves as the packaging signal for the encapsidation of viral RNA into the virion. The presence of this sequence between SD and *gag* ensures that only full-length genomic viral RNA is packaged into virions. The binding of the NC domain of Gag to the DLS facilitates the dimerization of the genomic RNA and is responsible for the specificity of viral RNA packaging. The myristylated MA of the Gag precursor directs the assembled viral nucleocapsids to the cell membrane, where they associate with the viral envelope proteins and eventually are released from the cell by budding.

The virus undergoes a final maturation step at the time of budding or shortly thereafter. The 63 kDa Gag precursor is cleaved by PR, resulting in the appearance of mature CA proteins in the virion. This temporally correlates with the condensation of virion cores from 73 nm during budding to 67 nm shortly after release from the cell. In addition, the transmembrane envelope protein gPr22 is further cleaved to form gp20, which has the fusagenic properties of TM. Release of progeny virus particles can be detected 24 h post infection.

Host Range and Transmission

Outbreaks of reticuloendotheliosis viruses have been reported in geographically diverse regions of the world, including the United States, Australia, and Israel. These viruses infect a wide range of avian species, such as chickens, turkeys, ducks, geese, pheasants, peafowl, prairie chickens, grouse, Japanese quail, and Hungarian partridges. Reticuloendotheliois viruses also infect and replicate in a variety of cultured cell types including fibroblasts and lymphoid cells from avian species and a dog osteosarcoma-derived cell line. The ability of these viruses to infect human cells is controversial; however, the more rigorous studies suggest productive infections do not occur. The failure of these viruses to replicate in certain mammalian cells is not due to an inability to infect these cells, but rather to a restriction in their replication cycle.

Reticuloendotheliosis viruses are most commonly transmitted horizontally within a flock, although vertical and germ line transmission of the virus has been reported. Mosquitoes have also been suggested as a vector for transmission. Reticuloendotheliosis viruses are able to integrate into the genomes of two DNA viruses, Marek's disease virus and fowlpox virus. Some of these integration events retain a full-length infectious provirus, suggesting that DNA viruses containing integrated reticuloendotheliosis provirus may provide a further mechanism for the transmission of reticuloendotheliosis viruses.

Immune Response

The immune response to infection by replication-competent reticuloendotheliosis viruses is dependent on the developmental stage at which the host is exposed to the virus. Infection of embryos induces a tolerant infection in which the bird exhibits a persistent viremia, but lacks antibody production against the virus. Tolerant infections are less common when birds are infected after hatching, and infection of day-old hatchlings generally results in the presence of virus-specific antibody 2–10 weeks post infection. The appearance of virus-specific antibodies in these birds temporally correlates with a decrease in the level of viremia. In addition to this humoral immune response, birds infected with reticuloendotheliosis virus exhibit a cell-mediated response against the infection. A cytotoxic T-lymphocyte (CTL) response mediated by CD4$^-$/CD8$^+$ $\alpha\beta$ T cells is observed in reticuloendotheliosis virus-infected birds 7 days after infection and persists for at least 21 days. CTLs mount the primary cell-mediated immune response against a viral infection, and lyse cells presenting viral antigens on their surface in the context of MHC class I molecules.

Pathogenesis

Non-Neoplastic Diseases

Reticuloendotheliosis viruses rapidly induce a severe, but transient immunosuppressed state in infected birds that likely plays an important role in the diseases caused by these viruses. This immunosuppression is characterized by a limited immune response to foreign antigens and a delay in the rejection of allogenic skin grafts in infected birds. Further, splenic lymphocytes from these birds fail to proliferate in response to T-cell mitogens and alloantigens *in vitro*. This immunosuppression does not appear to affect humoral immunity, since antisera specific to the infecting virus is produced shortly after infection. REV-T- and REV-A-mediated immunosuppression is due to the activation of a suppressor cell population in the spleen of infected birds. This suppressor cell population does not affect the cytotoxic responses of T cells, but rather suppresses their proliferation. The identity of the suppressor cell and the mechanism(s) by which it is activated are unknown.

The reticuloendotheliosis viruses induce a variety of diseases, including a visceral reticuloendotheliosis, spleen necrosis, hepatomegaly, splenomegaly, thymic and bursal atrophy, lymphoid nerve lesions, anemia, runting, and abnormal feather development. The mechanism(s) by which these diseases are induced is largely unknown. Syncytia formation has been observed in cultured cells infected with CSV and may contribute to some of the pathology. The runting syndrome observed in REV-A and REV-T infected chickens is not due to a reduced consumption of food, but rather to an alteration in metabolism. One such metabolic change that may at least partially account for runting is a decrease in phosphoenolpyruvate carboxykinase activity in the liver of infected birds. This enzyme is a key regulator of gluconeogenesis, and its reduced activity may be responsible for depletion of carbohydrate reserves in these birds. Runting and bursal atrophy in chickens have been linked to the structural genes of the virus. Studies performed by domain swapping with REV-A and the less pathogenic CSV demonstrated that both *gag* and *env* of REV-A are required for its full pathogenic effect.

It is possible that Gag proteins play an indirect role in pathogenicity. Infection of cultured cells with reticuloendotheliosis viruses induces a transient cytopathic effect that correlates with the appearance of large amounts of unintegrated and integrated viral DNA. The cells that overcome this cytopathic effect lack unintegrated viral DNA and have a reduced number of integrated proviruses. The transient nature of both the viral DNA accumulation and the cytopathic effect suggests that cell death occurs due to the high levels of this DNA or its encoded proteins. In some cell types, the MA portion of Gag provides a replicative advantage to REV-A relative to viruses containing the MA of less pathogenic reticuloendotheliosis viruses. It is possible that the higher level of virus production provided by the Gag of REV-A is cytotoxic to cells, resulting in some of these syndromes.

Neoplastic Diseases

Replication-competent reticuloendotheliosis viruses can induce both B-cell and T-cell lymphomas. CSV induces B-cell lymphomas after a relatively long latency period (6–10 months) that are indistinguishable from those induced by avian leukosis virus (ALV). Like ALV, these lymphomas are caused by insertional activation of c-*myc*. Proviral insertions are located upstream of the second exon of c-*myc*, with the vast majority lying between the first and second exons. The 5′ LTR of these proviral insertions are either deleted or rendered inactive by deletion of an enhancer sequence located 5′ of the *gag* sequence. T-cell lymphomas in CSV-infected birds have been observed after a relatively short latency period (6 weeks). These lymphomas are derived from thymic T cells and also develop from the insertional activation of c-*myc*. As in the B-cell lymphomas, these proviral integrations reside upstream of exon 2. However, 60% of these integrations are found 5′ of exon 1 and half of the proviruses insert in the opposite transcriptional orientation of c-*myc*. The induction of c-*myc* in many of these tumors appears to be through enhancer activation of a cryptic promoter in exon 1.

Transformation by REV-T

REV-T is the acutely transforming member of the reticuloendotheliosis virus family, inducing fatal lymphomas in experimentally infected birds after just 7–10 days. Large lesions in the spleen and smaller lesions in the liver are indicative of these lymphomas. REV-T transforms cells of various lymphoid lineages, including immature pre-B/pre-T cells, immature and mature B cells, T cells, myeloid cells, and predendritic cells. Although not a target cell for transformation *in vivo*, cultured chicken embryo fibroblasts (CEFs) are also readily transformed by REV-T. REV-T transformed cells are highly malignant and injection of only 100 non-virus-producing cells can induce a fatal lymphoma in young birds. In addition, transformed CEFs injected into the wing web of day-old chicks induces sarcomas at the site of injection. The ability of REV-T to transform cells is due to the presence of the v-*rel* oncogene. However, its transformation potential is not limited to avian cells. Transgenic mice that express v-*rel* under the control of the T-cell-specific *lck* promoter develop multiple lymphomas that are localized in enlarged spleens and livers.

The v-*rel* oncogene is formed through a recombination event between the envelope sequences of REV-A and turkey c-*rel* sequences. c-Rel is a member of the evolutionarily conserved Rel/NF-κB family of transcription factors. All members of this family contain a highly conserved 300 amino acid region located in the N-termini of these proteins termed the Rel homology domain (RHD). Rel/NF-κB proteins are normally sequestered in inactive complexes in the cytoplasm of cells by a family of inhibitory proteins referred to as IκBs. Upon proper extracellular stimulation, IκBs are phosphorylated, ubiquitinated, and ultimately degraded, allowing for the nuclear translocation of active Rel/NF-κB complexes. In the nuclei of cells, homo- and heterodimers of Rel/NF-κB proteins bind to a 10-bp DNA sequence (κB site) to regulate the expression of genes involved in the immune response, differentiation, proliferation, and stress responses. v-Rel transforms cells by evading the normal regulation of the Rel/NF-κB pathway, leading to the inappropriate activation or suppression of Rel/NF-κB regulated genes.

v-Rel complexes involved in transformation

The transformation potential of v-Rel is dependent on its ability to form dimers, bind DNA, and activate transcription. The κB binding complexes in the nuclei of v-Rel transformed cells consist of v-Rel homodimers and v-Rel heterodimers containing the endogenous Rel/NF-κB proteins c-Rel, NF-κB1 (p50), and NF-κB2 (p52). Analysis of mutants of v-Rel that are defective in homodimer and/or heterodimer formation with specific Rel/NF-κB family members revealed that the ability of v-Rel to dimerize and bind DNA with endogenous Rel/NF-κB proteins is not in itself sufficient for the transformation of chicken lymphocytes. Furthermore, studies from v-*rel* transgenic mice suggested that v-Rel heterodimeric complexes are not needed for transformation. Nuclear κB binding complexes in tumors from these mice consisted of only v-Rel homodimers and v-Rel/NF-κB1 heterodimers. v-Rel was still capable of inducing tumors when heterodimeric complexes were eliminated or reduced by mating v-*rel* transgenic mice with $nfkb1^{-/-}$ mice or *ikba* transgenic mice.

The characterization of two v-Rel mutants with an impaired ability to form homodimers demonstrated the importance of v-Rel homodimers in transformation. These mutants weakly transformed chicken splenic lymphocytes, and proviral DNA from cells transformed by these mutants contained multiple secondary mutations that enhanced or restored the ability of these v-Rel mutants to bind DNA as homodimers. Viruses expressing v-Rel mutants containing these secondary mutations transformed cells at levels comparable to or slightly less than viruses expressing wild-type v-Rel. These results suggested that a threshold level of DNA binding by v-Rel homodimers is required for transformation.

Mutations in v-Rel

The transduction of c-*rel* into the *env* sequences of REV-A resulted in the removal of sequences encoding the two N-terminal and 118 C-terminal amino acids of c-Rel. The N-terminus of v-Rel contains 11 N-terminal amino acids encoded by the 5' end of the *env* gene. The 18 C-terminal amino acids of v-Rel are encoded by out-of-frame *env*-derived sequences. v-Rel has also acquired 14 amino acid substitutions and three deletions relative to turkey c-Rel. The majority of these mutations occur in sequences outside of the RHD. In addition, amino acid substitutions have occurred in the N- and C-terminal *env*-derived

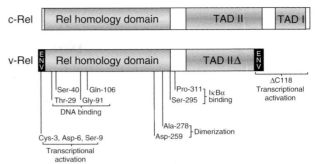

Figure 3 Structural differences between c-Rel and v-Rel. The Rel homology domain is shown as a shaded box. The two transcriptional activation domains of c-Rel (TAD I and TAD II) are similarly indicated. v-Rel is missing TAD I and contains a deletion in TAD II (TAD IIΔ). The 11 N-terminal and 18 C-terminal *env*-derived amino acids (ENV) of v-Rel are indicated by black boxes. Amino acids in v-Rel that differ from c-Rel or the Env of REV-A that alter the biochemical activities of v-Rel are noted. Their influence on DNA binding, transcriptional activation, dimerization, or IκBα binding is also indicated. The serine at amino acid 40 noted in this figure is only found in S2A3 v-Rel, a naturally occurring B-cell tropic variant of v-Rel. Amino acid differences between v-Rel and c-Rel that have no defined effect on the function of v-Rel are not shown.

sequences. The results of extensive studies evaluating the structural and biochemical differences between v-Rel and c-Rel indicate that these mutations modulate the DNA binding, dimerization, IκBα association, and transactivation properties of v-Rel relative to c-Rel (**Figure 3**). By far, the most significant contribution to the transformation potential of v-Rel is the deletion of the 118 C-terminal amino acids of c-Rel, which have transactivation and cytoplasmic retention properties. However, the high transformation potential of v-Rel is not the result of any one mutation or single group of mutations, but rather the functional cooperation of these mutations with one another.

v-Rel target genes

The major focus of v-Rel research in recent years has been the identification and characterization of v-Rel target genes. Although more than 50 genes have been identified with altered expression in cells expressing v-Rel (**Table 1**), few have been analyzed for their role in transformation. What is clear from the study of these genes is that v-Rel transforms cells, at least in part, by efficiently inducing the expression of genes that regulate cell proliferation. One of these, IRF-4, is a member of the interferon regulatory factor family of transcription factors and is elevated in cells expressing v-Rel. Experiments involving the overexpression of IRF-4 with v-Rel and inhibition of IRF-4 expression by antisense technology have identified a key role for IRF-4 in the transformation of both lymphocytes and fibroblasts by v-Rel. IRF-4

Table 1 Genes that exhibit altered expression in cells expressing v-Rel

Functional group	Genes induced or suppressed by v-Rel
Rel/NF-κB/IκB family members	c-rel, nf-kb1, nf-kb2, ikba
Transcription factors	c-fos, c-jun, fra-2[a], HMG-14b, IRF-1, IRF-3, IRF-4, IRF-8, IRF-10, c-myb[a], NAP-1[a], STAT-1
Translation factors	eIF2α[a]
Structural proteins	Actin[a], ARP, β-tubulin[a], CAP-23, δ-crystallin, type I collagen α[a], vimentin
Signal transduction	TC10, rhoC, rac1, sh3bgrl[a], sh3bgrl2, sh3bgrl3
Inhibitors of apoptosis	ch-IAP1, NR13, TERT, TR
Cytokines	IFN1, IL-6[b], IL-8, mip-1β, ctca, TNF-α[b]
Cell surface proteins	DM-GRASP[c], IFNaR1, IFNaR2, IL-2R, MHC Class I and II, p75, Sca-2
Metabolism	iNOS, mitochondrial cytochrome b, OAS, ODC-antizyme, p40^{phox}, p47^{phox}
Kinases and phosphotases	JAK1, PP2A
Chaperones	GRP-78

[a]Downregulated in cells expressing v-Rel.
[b]Upregulated in T-cell lymphomas from v-rel transgenic mice, not tested in avian cells.
[c]Upregulated in T-cell lymphomas from v-rel transgenic mice and avian cells transformed by v-Rel.

regulates the expression of components of the interferon signal transduction pathway, and is likely to contribute to v-Rel-mediated transformation by limiting the induction of this antiproliferative pathway. DM-GRASP, an immunoglobulin superfamily adhesion molecule, is also upregulated in v-Rel transformed fibroblasts and lymphoid cells and functions to promote cell proliferation. This was demonstrated by the ability of a monoclonal antibody specific for DM-GRASP to inhibit the proliferation of v-Rel transformed B-cells.

Although enhanced proliferation is a critical step in the malignant transformation of cells, it is not sufficient. Transformation also requires the suppression of apoptotic pathways, and v-Rel induces the expression of a number of genes with known anti-apoptotic activities, including ch-IAP1 and telomerase. ch-IAP1, a member of the inhibitor-of-apoptosis family, is one of the few known direct transcriptional targets of v-Rel. When temperature-sensitive v-Rel transformed cells were placed at the nonpermissive temperature, the expression of ch-IAP1 was reduced and the cells rapidly underwent apoptosis. Ectopic expression of ch-IAP1 in these cells protected them from apoptosis at the nonpermissive temperature, demonstrating that the induction of ch-IAP1 by v-Rel contributes to the immortalization of cells. In addition, v-Rel activates telomerase during transformation by inducing the expression of two of its components, telomerase RNA (TR) and telomerase reverse transcriptase (TERT). Telomerase has a well-defined role in the maintenance of telomere length, and has been more recently demonstrated to protect cells against apoptosis. v-Rel transformed cells rapidly undergo apoptosis when transfected with siRNAs specific for TERT or treated with a chemical inhibitor of telomerase, defining an important anti-apoptotic role for the elevated telomerase activity in v-Rel transformed cells.

Many of the genes regulated by v-Rel participate in diverse signal transduction pathways in normal cells. Although a number of these genes have been implicated in transformation, the mechanism(s) by which they contribute to transformation remains to be elucidated. Members of the AP-1 family are differentially regulated by v-Rel and the overexpression of *supjun-1*, a dominant negative mutant of AP-1 activity, dramatically decreases the transformation efficiency of v-Rel. The chemokine macrophage inflammatory protein-1β (MIP-1β/CCL4) is strongly induced by v-Rel in a variety of cell types, and its overexpression cooperates with v-Rel and c-Rel in promoting the growth of fibroblasts in soft agar. TC10 is a member of the Rho family of small GTPases and its expression and activity are induced by v-Rel. Both TC10 and a gain-of-function mutant synergistically cooperated with v-Rel to transform fibroblasts. Moreover, co-expression of a dominant-negative mutant of TC10 with v-Rel dramatically decreased the transformation of cells by v-Rel. In contrast to the abundance of genes that are upregulated by v-Rel, few genes have been identified that are downregulated by this oncogene. One gene that is strongly downregulated in v-Rel transformed cells encodes SH3BGRL, a member of the SH3BGR (SH3 domain binding glutamic acid-rich) family of proteins. Overexpression of SH3BGRL in primary splenic lymphocytes dramatically inhibits the ability of v-Rel to transform these cells. Furthermore, an intact SH3-binding domain in SH3BGRL is required for this function. Although the precise functions of SH3BGR family members are unknown, these results indicate that the suppression of an SH3-mediated signaling pathway is important for v-Rel-mediated transformation. SH3BGRL is the only downregulated v-Rel target gene to date with a demonstrated role in transformation.

The v-Rel target genes discussed above invariably exhibit a more dramatic change in expression in response to v-Rel than c-Rel. However, the ability of v-Rel to less efficiently activate the expression of certain genes is also critical for its transformation potential. Both v-Rel and c-Rel directly activate the transcription of *ikba*. However, since v-Rel lacks the most potent transactivation domain of c-Rel, it induces *ikba* expression more slowly and to lower levels than does c-Rel. This prevents the rapid relocalization of v-Rel from the nucleus to the cytoplasm than is observed in cells expressing c-Rel.

The identification of v-Rel target genes has highlighted the importance of this viral oncoprotein as a model system to study the role of Rel/NF-κB proteins in cancer. Many v-Rel target genes, including those with demonstrated roles in v-Rel-mediated transformation, exhibit altered regulation in human cancers linked to deregulated Rel/NF-κB activity. The recent availability of chicken DNA microarrays will greatly expand our knowledge of the pathways altered by v-Rel during transformation and likely provide additional insight into the mechanisms by which Rel/NF-κB proteins participate in neoplastic disease.

See also: Retroviral Oncogenes; Simian Retrovirus D.

Further Reading

Gilmore TD (1999) Multiple mutations contribute to the oncogenicity of the retroviral oncoprotein v-Rel. *Oncogene* 18: 6925–6937.

Witter RL and Fadly AM (2003) Reticuloendotheliosis. In: Saif YM, Larry R, McDougald LR, Barnes JH, Glisson R, and Swayne D (eds.) *Diseases of Poultry*, 11th edn., pp. 517–535. Ames, IA: Iowa State University Press.

Retrotransposons of Fungi

T J D Goodwin, M I Butler, and R T M Poulter, University of Otago, Dunedin, New Zealand

© 2008 Elsevier Ltd. All rights reserved.

Glossary

Retrotransposition The process by which a retrotransposon replicates.

Retrotransposon A eukaryotic mobile genetic element that can replicate via the reverse transcription of an RNA intermediate and the insertion of the resulting DNA into the host genome.

Reverse transcriptase A DNA polymerase that can use both RNA and DNA as a template.

Introduction

Retrotransposons are mobile genetic elements. They are generally regarded as selfish or parasitic entities and appear as inserts within the genomes of their hosts. They can spread through the host genome by copying their RNA transcripts into DNA and inserting these new DNA copies back into the host chromosomes. As this process, retrotransposition, is replicative it results in an increase in the copy number of the retrotransposons. Over time this can result in retrotransposons reaching very high copy numbers and making up a large proportion of the genome. As an example, there are more than 500 000 copies of the L1 retrotransposon in the human genome and these make up about 17% of human DNA.

Much of what is known about retrotransposons and their interactions with their hosts has been derived from studies with fungi. Fungi are a diverse group of eukaryotes of great ecological, economic, medical, and scientific importance. They range from tiny obligate intracellular pathogens, such as *Encephalitozoon cuniculi*, to large, multicellular organisms that can have a mass in excess of 100 kg, such as *Bridgeoporus nobilissimus*. They include such organisms as the bakers' and brewers' yeast *Saccharomyces cerevisiae*, genetic model organisms such as *Neurospora crassa*, plant pathogens such as the causal agent of rice blast disease *Magnaporthe grisea*, human pathogens such as *Candida albicans* and *Cryptococcus neoformans*, sources of antibiotics such as *Penicillium chrysogenum*, and mycorrhiza-forming fungi such as *Glomus intraradices*. The fungi are classified into six divisions: Ascomycota, Basidiomycota, Chytridiomycota, Glomeromycota, Microsporidia, and Zygomycota. The great majority of well-known fungi, such as *S. cerevisiae*, *C. albicans*, and *N. crassa* are ascomycetes. The next largest group is the basidiomycetes, which includes such organisms as the button mushroom *Agaricus bisporus*, the pathogenic yeast *C. neoformans*, as well as rusts and smuts. The chytrids are unusual fungi that produce flagellated motile spores. This group includes the frog pathogen *Batrachochytrium dendrobatidis*. The zygomycetes are characterized by zygospores formed during sexual reproduction. They include *Rhizopus oryzae*, which is commonly found on decaying vegetable matter and can cause fatal infections in humans. The glomeromycetes include the arbuscular mycorrhizal fungi such as *G. intraradices* and *Gigaspora margarita*. These species form symbiotic associations with the roots of many land plants, the fungus supplying the plant with nutrients obtained from the surrounding soil, in exchange for carbohydrates. The last fungal division, the microsporidia, consists of obligate intracellular parasites of animals. These are atypical fungi in that they lack mitochondria and peroxisomes. They include the human pathogen *Encephalitozoon cuniculi*.

Much of the contribution of fungi to our knowledge of retrotransposons has come from the yeast *S. cerevisiae*. The Ty elements found in this species were among the first retrotransposons to be identified and characterized. Indeed, the term retrotransposon was coined following the demonstration that Ty elements replicate (transpose) through an RNA intermediate, in a series of steps essentially the same as that for retroviruses. The widespread use of *S. cerevisiae* as a model for many features of eukaryotic life, due to its ease of culture and experimental manipulation, rapid growth, readily accessible genetics, and small genome size, meant that many tools were available to assist in the analysis of Ty elements. Numerous features that have made *S. cerevisiae* attractive for the study of retrotransposons are also shared with other fungi. In particular, the general small size of fungal genomes has meant that many complete genome sequences, from an evolutionary diverse array of fungi, have been determined in recent years and many more will become available in the near future. Analyses of these genome sequences are ongoing but have already contributed much to our understanding of retrotransposons. Most importantly, the analysis of a diverse range of fungi has uncovered many types of retrotransposons, and interesting interactions between the retrotransposons and their hosts, that are not observed in *S. cerevisiae*, making it clear that study of a wide range of species will be essential for an in-depth understanding of fungal retrotransposons. Here we will outline many of the features of retrotransposons, with special emphasis on the fungal elements. We also describe the impact of these elements on the fungal genome and the ways in which the interactions between the retrotransposons and their hosts have evolved to permit the long-term survival of the elements while minimizing the deleterious impact on their hosts.

Types of Retrotransposons

Retrotransposons come in a wide variety of forms with diverse sequences and diverse replication mechanisms (**Figures 1** and **2**). The one feature common to all retrotransposons is a sequence encoding reverse transcriptase (RT), the enzyme responsible for copying the RNA transcripts into DNA. The RTs of all known retrotransposons are homologous, indicating that they share a common origin. Phylogenetic analyses based upon alignments of RT sequences (**Figure 2**), together with comparisons of other structural and mechanistic features, have been used to classify retrotransposons into various groups. At the broadest level many retrotransposons can be classified into one of two groups. Members of the first group consist of a protein-coding internal region flanked by generally noncoding long terminal repeats (LTRs) and are known as LTR retrotransposons. The Ty elements of *S. cerevisiae* are all LTR retrotransposons. A typical LTR retrotransposon is 5–6 kbp long and contains two protein-coding open reading frames (ORFs), *gag* and *pol* (**Figure 1**). The *gag* ORF encodes a protein, Gag, which forms the major structural component of a virus-like particle (VLP) within which the reverse transcription reactions take place. The *pol* ORF encodes a polyprotein bearing various enzymatic domains: aspartic protease (PR), the RT, ribonuclease H (RH), and integrase (IN). PR is responsible for processing the initial *gag* and *pol* ORF products (see further ahead) into their functional domains. RT copies the RNA transcripts into DNA, assisted by RH. IN is involved in the insertion of the nascent DNA copy into the host genome. The LTRs flanking the protein-coding regions contain signals regulating transcription of the element, and, due to their redundant nature, they allow a full-length DNA copy of the element to be constructed from somewhat less-than-full-length RNA transcripts (see further ahead). LTR retrotransposons are themselves subdivided into further groups, named the Ty1/copia, Ty3/gypsy, and BEL groups, after founding elements from fungi and insects. The members of these groups can be distinguished from each other by sequence comparisons and the order in which the enzymatic domains appear within the *pol* gene. In Ty1/copia elements the *pol* domain order is PR-IN-RT-RH. In BEL and Ty3/gypsy elements (with a few exceptions) the order is PR-RT-RH-IN.

The second major group of retrotransposons are distinguished from LTR retrotransposons by the absence of LTRs, by differences in the set of encoded proteins, and a completely different replication mechanism (see further ahead). This group of retrotransposons has been given various names, including non-LTR retrotransposons, poly-A retrotransposons, long interspersed nuclear elements (LINEs), and retroposons. However, none of these names is ideal: the 'non-LTR' term refers to a feature all these elements lack, rather than one they all share; the 'poly-A' term refers to a run of A residues found at the $3'$ end of many of the first elements of the group to be characterized, but subsequently found to be lacking in many other elements; the LINE terminology is not ideal, as many of the members of this group are not interspersed, but are found only in specific regions of the genome. Finally, the term 'retroposon' has been used in the literature to refer to numerous different groupings of retrotransposons and retrotransposon-like elements, so its usefulness has been lost. It has recently been suggested, however, that the elements of this group be referred to as the target-primed (TP) retrotransposons, after their insertion mechanism (target-primed retrotransposition – see further ahead), which is distinct from that of LTR retrotransposons and is likely to be shared by all members of the group. This is the terminology used in this article.

A typical TP retrotransposon consists of one or two protein-coding regions flanked by $5'$ and $3'$ untranslated

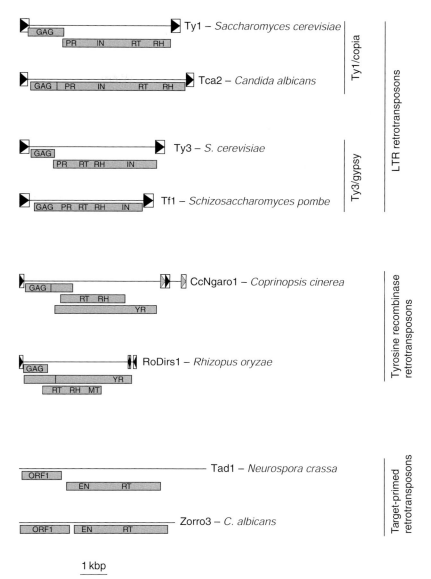

Figure 1 Structures of different types of retrotransposons. The lengths of the elements are indicated by the horizontal lines. Repeat sequences are shown as boxed triangles. Open reading frames are depicted as shaded boxes, with the approximate positions of the various protein-coding domains indicated: EN, endonuclease; IN, integrase; MT, putative methyltransferase; PR, protease; RH, ribonuclease H; RT, reverse transcriptase; YR, tyrosine recombinase. The groups to which the elements belong are shown on the right.

regions (**Figure 1**). All elements encode an RT and an endonuclease (EN). The TP retrotransposons can be divided into 15 or more clades, defined by sequence comparisons and structural distinctions. In members of what are believed to be the older clades, the EN-coding sequence lies downstream of RT and the EN is similar in sequence to certain restriction endonucleases. In the more recently evolved elements, the EN-coding sequence lies upstream of RT and is related to eukaryotic apurinic/apyrimidinic (AP) endonucleases. An RH-coding sequence also makes a sporadic appearance among these latter elements. In the elements with two ORFs, RT and EN are always encoded by the downstream ORF. The upstream ORF, ORF1, is generally poorly conserved. The function of the protein product of ORF1 is unknown, although in some elements ORF1p has been shown to have RNA-binding and nucleic acid-chaperone activities.

In addition to these two major categories of retrotransposons, several other types of retrotransposons have been described. These elements are generally rarer than the LTR and TP retrotransposons and so have not been characterized in as much detail. The first of these additional groups, the tyrosine recombinase (YR) retrotransposons, bear RT- and RH-coding sequences related to

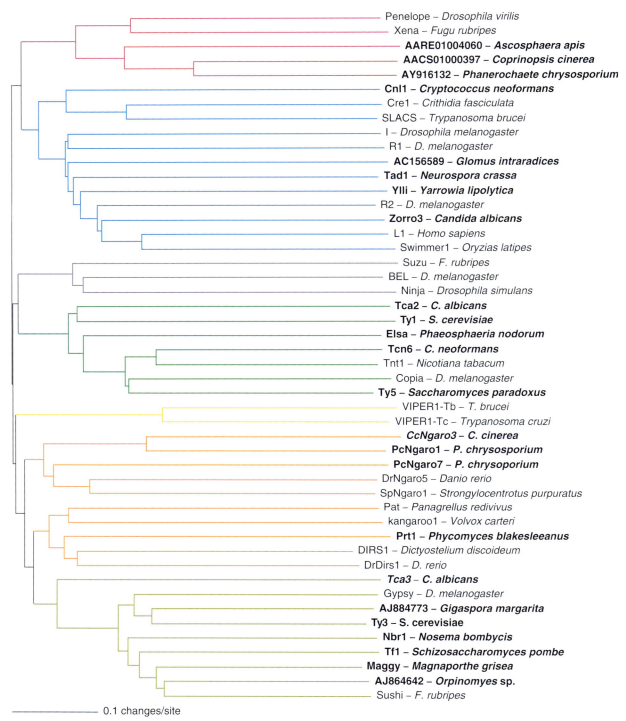

Figure 2 Relationships among retrotransposons. This tree is based on an alignment of RT sequences from a wide variety of retrotransposons. The various groups are indicated in different colors: red, Penelope-like retrotransposons; blue, target-primed retrotransposons; purple, BEL-like LTR retrotransposons; green, Ty1/copia LTR retrotransposons; yellow, VIPER elements; orange, tyrosine recombinase retrotransposons; khaki, Ty3/gypsy LTR retrotransposons. Elements from fungi are in boldface.

those of LTR retrotransposons, but they lack coding sequences for PR and IN. Instead they encode a tyrosine recombinase, related to the recombinase of bacteriophage lambda (**Figure 1**). Like LTR retrotransposons, they also bear flanking repeats, but these are distinct from typical LTRs in that they are either 'split direct repeats' or have a complex inverted-repeat structure. The DIRS1 element from the slime mold *Dictyostelium discoideum* was the first YR retrotransposon to be described. The second group, the Penelope-like elements (PLEs; named after an element from *Drosophila virilis*), are similar to TP retrotransposons in that they encode an RT and an EN. The EN of PLEs is, however, unrelated to that of TP retrotransposons, but is similar to the Uri endonucleases found in group I introns. PLEs are often found associated with variable terminal repeat sequences in both direct and inverted orientations. Their replication mechanism is not well understood, but may be an unusual form of target-primed retrotransposition. The final category of retrotransposons, the vestigial interposed retroelement (VIPER)-like elements, have been found in only a very small number of species (trypanosomes) to date. They bear RT and RH genes somewhat similar to those of LTR retrotransposons, but otherwise very little is known regarding their conserved features.

Distribution of Retrotransposons

Retrotransposons have only been found in eukaryotic genomes (with the exception of a few rare elements integrated into the genomes of eukaryotic viruses). While RT-encoding elements (group II introns and retrons) do appear in prokaryotes, these elements are not mobile genetic elements in the same sense as the eukaryotic retrotransposons. Among eukaryotes, retrotransposons have been found in almost all genomes that have been examined in detail. However, the diversity of retrotransposons found within particular species varies enormously. For instance, examples of nearly every type of known retrotransposon have been identified in the genome of the zebrafish *Danio rerio*, while only one active retrotransposon, the L1 TP retrotransposon, is known in humans and no retrotransposons of any sort are found in the complete genome sequence of the malaria parasite *Plasmodium falciparum*. Similarly, the distribution of different types of retrotransposons also displays great variation. For instance, members of the Ty3/gypsy class of LTR retrotransposons appear in plants, animals, fungi, and various protists, whereas BEL-like LTR retrotransposons have only been found in animals to date.

Within the fungi, retrotransposons are widespread and abundant (**Table 1**). Most have been found in ascomycetes and basidiomycetes. Only a few have been identified in zygomycetes, glomeromycetes, chytrids, and microsporidia: this is, however, likely to be simply the result of there being less data available for species from these divisions, rather than low numbers of retrotransposons, as those species that have been examined often contain abundant elements. Of the LTR retrotransposons, members of the Ty1/copia group have been identified in a large number of ascomycetes, many of the basidiomycetes that have been examined, and also in one species of chytrid (*Orpinomyces* sp.). None has yet been identified in zygomycetes, glomeromycetes, or microsporidia. Ty3/gypsy elements have been identified in all fungal divisions. They are abundant in many ascomycetes and basidiomycetes and have also been found in microsporidia (*Nosema bombycis* and *Spraguea lophii*), glomeromycetes (*Glomus intraradices* and several species of Gigaspora), chytrids (*Orpinomyces* sp.) and zygomycetes (*R. oryzae*). In many fungi, Ty3/gypsy elements are more abundant than the Ty1/copia elements, suggesting that the former elements have been somewhat more successful in colonizing and persisting in fungal genomes. No BEL-like LTR retrotransposons have been identified, despite the large amounts of sequence data available, suggesting that this group is absent from fungi.

Target-primed retrotransposons are also widespread in fungi with numerous elements again found in ascomycetes and basidiomycetes and examples also identified in zygomycetes (*R. oryzae*), glomeromycetes (*G. intraradices* and some Gigaspora species), and chytrids (*Orpinomyces* sp.). None has yet been identified in microsporidia. Despite their widespread occurrence, TP elements are notably absent from several species for which complete genome sequences are known. In particular, they are absent from *S. cerevisiae* and another model yeast, *Schizosaccharomyces pombe*. This absence from model species has meant that the contribution of fungi to our understanding of the replication mechanisms and other features of TP retrotransposons has been much less than that for LTR retrotransposons.

YR retrotransposons have a more sporadic distribution in fungi than LTR and TP retrotransposons. Examples have been identified in basidiomycetes (*Phanerochaete chrysosporium* and *Coprinopsis cinerea*) and zygomycetes (*Phycomyces blakesleeanus* and *R. oryzae*). None has yet been identified in other divisions. For ascomycetes at least, with the vast amount of sequence data available, this suggests that they are either absent from this division or extremely rare. Sequences related to Penelope retrotransposons are also rare in fungi, but nevertheless, such sequences are evident in at least three fungal divisions: basidiomycetes (*P. chrysosporium* and *C. cinerea*), zygomycetes (*R. oryzae*), and ascomycetes (*Ascosphaera apis*). No elements similar to VIPER have been detected in fungi to date.

Overall, the distribution of the various types of retrotransposons in fungi parallels their distribution throughout eukaryotes in general; that is, TP retrotransposons and members of the Ty1/copia and Ty3/gypsy groups of LTR retrotransposons are common occurrences and found in most species. The types of retrotransposon that are rare in general, such as the YR and Penelope-like retrotransposons, are also rare in fungi. Nevertheless, these latter elements are still widespread, in the sense that they are found in species from several different divisions. Their widespread distribution and a general high level of diversity in their sequences, suggests that these elements have a long

Table 1 Types of retrotransposons identified in various fungal phyla

Phylum	Type of retrotransposon	Examples (host)[a]	Accession no.
Ascomycota	LTR–Ty1/copia	Ty1 (*Saccharomyces cerevisiae*)	M18706
		Tca2 (*Candida albicans*)	AF050215
		Elsa (*Phaeosphaeria nodorum*)	AJ277966
	LTR–Ty3/gypsy	Ty3 (*S. cerevisiae*)	M34549
		Tf1 (*Schizosaccharomyces pombe*)	M38526
		Maggy (*Magnaporthe grisea*)	L35053
	Target–primed	Tad1 (*Neurospora crassa*)	L25662
		Zorro3 (*C. albicans*)	AF254443
	YR	—	
	Penelope	Unnamed (*Ascosphaera apis*)	AARE01004060
Basidiomycota	LTR–Ty1/copia	Tcn6 (*Cryptococcus neoformans*)	Retrobase[b]
	LTR–Ty3/gypsy	MarY1 (*Tricholoma matsutake*)	AB028236
	Target-primed	Cnl1 (*C. neoformans*)	Retrobase[b]
	YR	CcNgaro1 (*Coprinopsis cinerea*)	AACS01000194
	Penelope	Unnamed (*Phanerochaete chrysosporium*)	AY916132
Chytridiomycota	LTR–Ty1/copia	Unnamed (*Orpinomyces* sp.)	AJ864659
	LTR–Ty3/gypsy	Unnamed (*Orpinomyces* sp.)	AJ864642
	Target-primed	Unnamed (*Orpinomyces* sp.)	AJ864661
	YR	—	
	Penelope	—	
Glomeromycota	LTR–Ty1/copia	—	
	LTR–Ty3/gypsy	Unnamed (*Gigaspora margarita*)	AJ884773
	Target-primed	Unnamed (*Glomus intraradices*)	AC156589
	YR	—	
	Penelope	—	
Microsporidia	LTR–Ty1/copia	—	
	LTR–Ty3/gypsy	Nbr1 (*Nosema bombycis*)	DQ444465
	Target-primed	—	
	YR	—	
	Penelope	—	
Zygomycota	LTR–Ty1/copia	—	
	LTR–Ty3/gypsy	Unnamed (*Rhizopus oryzae*)	AACW02000049
	Target-primed	Unnamed (*R. oryzae*)	AACW02000090
	YR	Prt1 (*Phycomyces blakesleeanus*)	Z54337
	Penelope	Unnamed (*R. oryzae*)	AACW01000082

[a]A dash indicates that no element of a particular class has been identified in the phylum to date.
[b]Retrobase (http://biochem.otago.ac.nz).

history in fungi, probably dating back to the last common ancestor of all fungi. In this case, their current sporadic distribution is likely to be the result of the frequent loss of these elements from evolving lineages. It is not clear why these elements should be more frequently lost than TP and LTR retrotransposons. One possibility is that they might not replicate so efficiently and so never reach such high copy numbers and are thus more prone to stochastic loss through deletion or point mutation.

Retrotransposon Replication

LTR Retrotransposons

The replication of LTR retrotransposons is well understood due to its extensive similarities with retroviral replication and the thorough characterization of the Ty elements of *S. cerevisiae*. While specific features may vary from element to element, the general features of Ty1 element replication likely apply to all LTR retrotransposons. The first step in Ty1 replication is transcription of an integrated element. The Ty1 LTRs contain regulatory sequences which cause transcription to begin within the left LTR sequence, proceed all the way through the internal region, and terminate within the right LTR, resulting in a terminally redundant RNA. This RNA is then exported to the cytoplasm where it is translated. In Ty1 (and in all other LTR retrotransposons) the *pol* gene is located downstream of *gag* and is translated at a much lower level. This downregulation of *pol* is achieved by having it positioned in a different translational reading frame from the *gag* ORF, with the 5′ end of *pol* having a small overlap with the 3′ end of *gag*. All translation begins with the *gag* ORF and usually stops at the *gag* termination codon. The *pol* ORF is translated only when the ribosome undergoes a rare, programmed frameshift near the end of the *gag* ORF. When this occurs, a Gag–Pol fusion protein is produced. Next, the abundant Gag proteins and the rare Gag–Pol fusion proteins assemble into a hollow VLP. The N-terminus of the Gag protein lies on the exterior of the particle and the C-terminus lies on the inside. The fusion of the Pol protein to the C-terminus of Gag in the Gag–Pol fusion ensures that the Pol enzymes are packaged into the interior of the particle. Also packaged inside the particle are the retrotransposon RNA (usually two copies) and a tRNA (initiator Met tRNA) which will act as a primer during the reverse transcription step. The walls of the particle are porous to nucleotides and other small molecules.

The first step in the reverse transcription process is the annealing of the primer tRNA to a complementary sequence in the retrotransposon RNA (the minus-strand primer-binding site (PBS)) that lies just downstream of the left LTR. The 3' OH group of the tRNA is then used as a primer by the RT to initiate minus-strand DNA synthesis. Minus-strand DNA synthesis then proceeds to the end of the molecule, the 5' end of the mRNA, where it temporarily halts. The RNA in the RNA/DNA hybrid is then removed by RH. This allows the nascent minus-strand DNA to anneal to the complementary LTR sequence at the 3' end of the RNA. Minus-strand DNA synthesis then resumes and proceeds to the 5' end of the template. Next, the RNA of the RNA/DNA hybrid is again degraded by RH. A short purine-rich segment of RNA immediately upstream of the right LTR, known as the polypurine tract, is, however, resistant to removal by RH. This sequence remains bound to the minus-strand DNA and its 3' OH group is used by RT to prime plus-strand DNA synthesis. This continues as far as the 5' end of the minus-strand DNA template. The nascent plus-strand is then dissociated from the 5' end of the minus-strand DNA, possibly by displacement by further plus-strand DNA synthesis initiated from a second poly-purine tract lying in the central region of the element. The displaced plus-strand DNA can then anneal to the 3' end of the minus-strand DNA at their complementary sequences. A full-length double-stranded DNA copy of the retrotransposon can then be formed by extension of each strand to the end of its template and removal of any RNA nucleotides remaining from the primer sequences. Finally, in a series of steps, which are not so well understood, the new double-stranded DNA molecule associates with IN, exits the VLP, and enters the cell nucleus. Here, the DNA/protein complex associates with the chromosomal DNA and IN mediates the insertion of the retrotransposon into the host genome, completing the replication process.

While most of these steps are likely to be essentially the same in most LTR retrotransposons, many minor variations are known. For instance, in elements such as Tf1 and Tf2 from *S. pombe* (and closely related elements present in other fungi and also in vertebrates) the primer for minus-strand DNA synthesis is not a tRNA, but rather the 5' end of the elements own mRNA which is complementary to the PBS. The two sequences anneal to each other and the resulting structure is recognized by the element's RH and cleaved to produce a free 3' OH which is then used as a primer. As another example, in the element Tca2 from *C. albicans*, the *gag* and *pol* ORFs are in the same phase and separated by a stop codon, rather than being separated by a frameshift (**Figure 1**). Expression of the Gag–Pol fusion protein in this element presumably occurs via the occasional suppression of the *gag* stop codon. In other elements, for example, Tf1 and Tf2 (**Figure 1**), the *gag* and *pol* ORFs are fused into a single ORF and translated together. The correct stoichiometry of the Gag and Pol proteins is subsequently obtained by the preferential degradation of Pol.

Target-Primed Retrotransposons

The replication of TP retrotransposons is not as well understood as that of LTR retrotransposons, at least partly due to the lack of a suitable model system in an experimentally tractable microorganism. Nevertheless, much has been learned from the study of TP elements in insect and mammalian systems. Replication begins with transcription initiated from an internal promoter in the 5' untranslated region. Transcription proceeds to the 3' end of the element to produce a full-length RNA. The RNA is exported to the cytoplasm where it is translated. In the case of elements with two ORFs, the first ORF is translated much more efficiently than the second ORF. Following translation, the proteins remain associated with the RNA to form a RNA/protein complex. The complex moves to the nucleus and associates with the chromosomal DNA. The EN produces a single-stranded nick in the host chromosome and the resulting 3' OH group is used as a primer by RT to synthesize minus-strand DNA directly into the insertion site. The following steps are not well understood but presumably include the nicking of the second strand of the target site by EN, the use of the resulting 3' OH group to prime plus-strand synthesis, the completion of both strands of DNA synthesis, and the sealing of the ends of the retrotransposon to the host chromosomes. Although fungal elements have, as yet, contributed little to our understanding of TP retrotransposition, several TP elements have recently been identified in yeasts that may be developed into useful experimental systems for analyzing the process in more detail. These elements include Zorro3 from *C. albicans* and Ylli from *Yarrowia lipolytica*.

Other Retrotransposons

Very little is known about the replication of the other types of retrotransposons and most of this is based upon analyis of sequence data rather than direct experimental evidence. Briefly, replication of the YR retrotransposons is thought to proceed via an RNA which is copied into a circular, double-stranded DNA by the actions of RT and RH. The circular DNA is then integrated into the host chromosome by recombination mediated by the tyrosine recombinase. It is likely that PLEs integrate via a mechanism related to target-primed retrotransposition. Nothing is known about the replication of the VIPER-type of retrotransposon. The recent identification of YR and Penelope-like retrotransposons in fungi, and the likelihood that more will be identified in the near future as more fungal genome sequence data are obtained, may soon permit the development of systems to characterize these elements in more detail.

Retrotransposons and Fungal Genomes

Retrotransposons can have many and varied effects on the genomes of their hosts. They can make up a significant proportion of the host genome. In humans, the L1 TP retrotransposon alone makes up about 17% of the genome. In some plants, up to as much as 90% of the genome is made up by retrotransposons. Obviously, such an abundance of retrotransposons will have a profound effect on the functioning of the genetic material. In the fungi that have been analyzed in depth, the proportion of the genome made up by retrotransposons is generally lower than the figures quoted above. Figures for fungi include 3.1% for *S. cerevisiae*, 1.3% for *C. albicans*, 1.1% for *S. pombe*, 5.4% for *M. grisea* (all ascomycetes), and 7.6% for *Microbotyrum violaceum* (a basidiomycete). It should be noted that many of the fungi that have been analyzed in depth have been chosen in part because of their small genome size (10–50 Mbp). However, fungi with much larger genomes exist. For instance, among the glomeromycetes the genome size of *Gigaspora margarita* has been estimated at ~740 Mbp and that of *Scutellospora gregaria* at >1000 Mbp. It is possible that fungi with such large genomes might contain very large numbers of retrotransposons. On the other hand, some fungi have much lower numbers of retrotransposons. For instance, the microsporidian *E. cuniculi* has a very small genome (~2.9 Mbp) which contains no identifiable retrotransposons at all. Likewise, the yeast *Pichia farinosa* appears to contain no retrotransposons, despite these elements being abundant in closely related species.

In addition to making up a significant fraction of many fungal genomes, retrotransposons can act as powerful mutagens. For instance, they can insert into or adjacent to genes thus altering gene structure and/or regulation. They can create new copies of genes via the occasional accidental copying of gene transcripts into DNA and the insertion of these back into the genome. The resulting sequences will often be nonfunctional (retropseudogenes) but occasionally a new copy will be inserted in an intact form, adjacent to a promoter sequence where it will be expressed and perform some useful function. Retrotransposons can move their flanking sequences around the genome, potentially creating new arrangements of exons in genes and/or altering promoter sequences. The retrotransposon sequences themselves may occasionally be adopted to perform functions useful for the host. In addition, by providing regions of sequence similarity dispersed throughout the genome, retrotransposons can promote recombination events between otherwise nonhomologous regions, leading to chromosomal rearrangements. For example, the breakpoints of many chromosomal translocations in *S. cerevisiae* have been found to coincide with Ty elements. While the majority of mutational events associated with retrotransposons are likely to be neutral or deleterious to the hosts, occasional events will result in beneficial rearrangements. This suggests that the relationship between retrotransposons and their hosts should not be considered as strictly the same as that of parasite and host, but more as one with mutual benefits. In this regard, it is also of interest to note that, since the long-term survival of a retrotransposon is dependent on the long-term success of its host, many retrotransposons have developed strategies that act to minimize the damage they cause. For instance, some retrotransposons direct their integration to specific areas of the genome where they are least likely to cause deleterious mutations. As an example, the integration of Ty3 occurs very precisely 1–4 bp upstream of tRNA genes, areas where the insertion of the retrotransposon is unlikely to have a negative impact. Likewise, Ty5 integration is directed to areas of silent chromatin.

Not only can retrotransposons have profound effects on their hosts, but the hosts can also have profound effects on their retrotransposons. This is usually in the form of mechanisms to eliminate active retrotransposons, thus minimizing their potential mutagenic effects. Active retrotransposons can be eliminated in various ways. For instance, in addition to random point mutations, LTR retrotransposons can be inactivated by recombination between the two LTRs of a single element. This results in the internal region being excised and just a single LTR remaining at the original insertion site. In some species retrotransposons can be silenced by RNA interference or through methylation of repeat sequences. One of the most interesting mechanisms for eliminating active retrotransposons to have been identified in fungi is a process called

RIP (repeat-induced point mutation). This was first found in *N. crassa* and similar processes have since been identified in a variety of other filamentous ascomycetes. RIP acts during the sexual cycle and efficiently identifies and mutates repeat sequences that are greater than about 400 bp in length and share more than about 80% sequence identity. In *N. crassa* RIP produces a large number of C-to-T and G-to-A mutations in both copies of the repeated sequence. RIP progressively mutates the repeated sequences over successive sexual cycles until they no longer are sufficiently similar to be recognized (i.e., <80% identity). This process very effectively inactivates retrotransposons. Its efficiency is demonstrated by the finding that in the strain of *N. crassa* whose genome was sequenced, numerous retrotransposon relics were identified but not a single active element remains. RIP in *N. crassa* is one of the more extreme examples of the extents to which fungi may go to eliminate active transposable elements. Its actions not only eliminate active retrotransposons but also appear to prevent gene duplication of any sort, thus eliminating this pathway as a means for evolving new gene functions. This is illustrated by the finding that in *N. crassa* the most closely related paralogous genes were duplicated >200 million years ago, likely predating the origins of RIP. Although RIP very effectively inactivates retrotransposons, an active element, the Tad1 TP retrotransposon, has nevertheless been identified in some strains of *N. crassa*. It is thought that active copies of this element have evaded RIP by persisting in strains that have not recently gone through a sexual cycle and by occasionally spreading via anastomoses between different asexual lineages. Analyses of these active Tad elements may have revealed a previously unsuspected importance of the asexual phase in the biology of *N. crassa*.

One of the most interesting features that becomes apparent when the sets of retrotransposons in different fungal genomes are compared is the great variation. Different species vary in the proportion of the genome made up by retrotransposons (as outlined above), in the diversity of retrotransposons that they harbor (e.g., in *S. pombe* the only known retrotransposons are two very closely related members of the Ty3/gypsy group of LTR retrotransposons, whereas species such as *C. albicans* or *C. neoformans* harbor a much wider range of elements), in the distribution of elements along the chromosomes (e.g., in *S. pombe* the elements seem to be fairly evenly distributed along the chromosomes, whereas in *C. neoformans* elements were found to be greatly concentrated in putative centromeric regions), and in whether or not there are many active elements (e.g., *S. cerevisiae* contains numerous intact elements, whereas nearly all elements in *N. crassa* are heavily mutated). In most cases the reasons for the variation have not been determined. No doubt some of the variation is simply due to chance, but the remainder likely reflects fundamental differences between the host species and the elements that they contain. Determining the reasons behind the variation will lead to important insights into the evolution of fungal genomes.

Uses of Fungal Retrotransposons

Fungal retrotransposons have been put to numerous uses. For instance, Ty elements have been employed as insertional mutagens and gene-tagging systems in *S. cerevisiae*. As another example, retrotransposons have been employed in population genetics studies of various species, such as the ascomycete *M. grisea* and the basidiomycete *Chondrostereum purpureum*. Probably the most important uses of fungal retrotransposons, however, are as models for elements from other species. Fungi are eukaryotes and contain many of the same cellular features as higher organisms, including man. However, the rapid growth, simple lifecycle, and ease of experimental manipulation of many fungi mean that it is possible to use them for studying aspects of eukaryotic cells that would be difficult or impossible in other organisms. The fact that fungi contain retrotransposons closely related to elements of profound importance, such as retroviruses and TP retrotransposons, makes them ideal for studying many aspects of retrotransposons and the retrotransposition process.

See also: Metaviruses; Polyomaviruses of Mice; Retrotransposons of Plants; Retrotransposons of Vertebrates; Retroviruses: General Features.

Further Reading

Daboussi M-J and Capy P (2003) Transposable elements in filamentous fungi. *Annual Review of Microbiology* 57: 275–299.

Galagan JE and Selker EU (2004) RIP: The evolutionary cost of genome defense. *Trends in Genetics* 20: 417–423.

Goodwin TJD and Poulter RTM (2000) Multiple LTR-retrotransposon families in the asexual yeast *Candida albicans*. *Genome Research* 10: 174–191.

Goodwin TJD and Poulter RTM (2001) The diversity of retrotransposons in the yeast *Cryptococcus neoformans*. *Yeast* 18: 865–880.

Kim JM, Vanguri S, Boeke JD, Gabriel A, and Voytas DF (1998) Transposable elements and genome organization: A comprehensive survey of retrotransposons revealed by the complete *Saccharomyces cerevisiae* genome sequence. *Genome Research* 8: 464–478.

Roth J-F (2000) The yeast Ty virus-like particles. *Yeast* 16: 785–795.

Retrotransposons of Plants

M-A Grandbastien, INRA, Versailles, France

© 2008 Elsevier Ltd. All rights reserved.

Glossary

Env Envelope protein of retroviruses and errantiviruses, containing a transmembrane domain and a host receptor-binding domain, and mediating formation of infectious virions and targeting of host receptor cells.

Env-like Coding domains possessing some features similar to those of the env domains of retroviruses (such as a transmembrane domain), but for which no similar function has been shown as yet; env-like domains are found in addition to the gag-pol domain of many plant LTR retrotransposons.

Gag (group-specific antigen) Nucleic acid-binding protein forming the structural core of the retrovirus virion core and of the cytoplasmic VLP of retrotransposons.

Int Integrase function ensuring the insertion of the double-stranded DNA daughter copy in the genome; also sometimes referred to as endonuclease (endo).

LTR (long terminal repeat) Long sequence found repeated in the same orientation at both ends of retroviruses and LTR retrotransposons and containing promoter and regulatory sequences involved in transcription, that starts in the 5' LTR and terminates in the 3' LTR.

PBS (primer binding site) Short sequence found downstream to the 5' LTR, and involved in the priming of the (−)-DNA strand synthesis during reverse transcription.

PPT (polypurine tract) Short AG-rich sequence found upstream to the 3' LTR, and involved in the priming of the (+)-DNA strand synthesis during reverse transcription.

Prot Protease protein involved in the cleavage of the functional proteins units from the (gag) pol polyprotein produced by the retrovirus or the retrotransposon.

RT Reverse transcriptase protein ensuring the synthesis of the double-stranded daughter DNA copy from the RNA template produced by the integrated provirus or retrotransposon; an associate ribonucleaseH (RNaseH) function ensures the concomitant degradation of the RNA template.

VLP (virus-like particle) Cytoplasmic intermediate analogous to the virion core of retroviruses; the VLP is constituted by structural gag proteins associated to two copies of the RNA template, together with enzymatic proteins (RT, int); the reverse transcription of the RNA template is thought to occur within the VLP.

Introduction and General Classification

Transposable elements are ubiquitous mobile DNA sequences found in both prokaryotic and eukaryotic genomes. For a long time, they were considered as 'parasite' DNA, but have now been shown to be major components of plant genomes, where they can represent up to 80% of the bulk of large cereal genomes. Transposable elements, with a few notable exceptions, encode the functions involved in their mobility and are classified into two main classes with radically different transposition mechanisms. Class I elements or retrotransposons transpose via an RNA intermediate that is reverse-transcribed into a daughter copy DNA and re-inserted into the genome, while class II elements or DNA transposons move directly from DNA to DNA. Retrotransposons thus belong to the large class of reverse-transcribing elements (or retroelements) that multiply by transferring their genetic information from RNA to DNA, and are close relatives to viral entities such as animal retroviruses or pararetroviruses infecting both animal and plant hosts. However, in contrast to true infectious viruses, which propagate only functional genomes, retrotransposons are vertically transmitted intrinsic components of host genomes. Their integrated genomes are thus submitted to evolutionary drift, with various mutations and restructurations destroying their functionality, and defective copies can still be amplified, sometimes at surprisingly high levels, via functional related copies. Plant retrotransposons are thus found in a tremendous variety of structural variants, including highly defective and deleted versions, making their classification a difficult task. In addition, retrotransposons are composed of modules that frequently show incongruent phylogenies, and this intrinsic chimerical nature often obscures classifications based on phylogenies.

There are two main categories of retrotransposons: LTR retrotransposons that contain two long terminal repeats (LTRs), and non-LTR retrotransposons often referred to as retroposons. All functional copies encode as basic modules the structural and functional proteins required for the retrotransposition cycle (summarized in **Figure 1**). These include a structural RNA-binding protein (gag or gag-like),

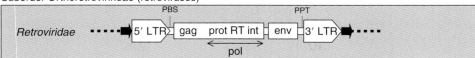
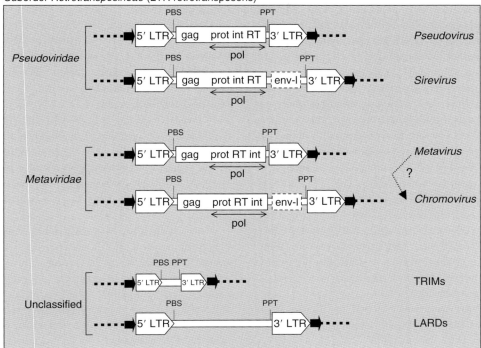
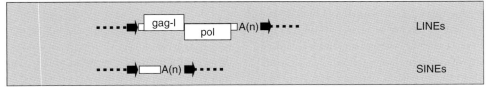

Figure 1 Structural diversity of retrotransposons found in plant genomes. The structure of a retrovirus has been provided for reference. The gag and pol domains are encoded in one or two frames, depending on the element, and have been represented here as a single ORF for simplification. Family names are indicated on the left and genus names on the right. LTR, long terminal repeat; gag, nucleic acid-binding protein forming the structural core of the retrovirus virion core and of the cytoplasmic VLP of retrotransposons; prot, protease protein involved in the cleavage of the functional proteins units from the (gag) pol polyprotein; int, integrase function ensuring the insertion of the double-stranded DNA copy in the genome; RT, reverse transcriptase ensuring the synthesis of the double-stranded DNA copy from the RNA template; env, envelope protein ensuring infectiosity of the retrovirus; env-l, env-like coding domain possessing structural features similar to env, but for which no similar function has been shown yet (a dotted box framing the env-like domain indicates that elements within the genus may or may not contain the domain); PBS, primer binding site; PPT, polypurine tract. Arrows flanking elements represent the target site duplication. The question mark linking the genera *Metavirus* and *Chromovirus* indicates that the definition of *Chromovirus* as an additional genus within the family *Metaviridae* has not been included in the official classification, with the result that many chromoviruses are still considered to be metaviruses.

and a pol domain encoding the reverse transcriptase (RT) ensuring the synthesis of the double-stranded daughter DNA copy from the RNA template, as well as in most cases a protein ensuring insertion of the double-stranded DNA daughter copy in the genome (referred to as integrase for most elements). A number of plant LTR retrotransposons contain an additional coding domain currently termed env-like, although, as discussed below, functional and structural analogies to retroviral env proteins remain to be demonstrated.

LTR retrotransposons have been further classified in Ty1/*copia*-type elements and Ty3/*gypsy*-type depending on the order of the coding domains. Non-LTR retrotransposons include LINEs (long interspersed nuclear elements), which are elements carrying coding sequences, and SINEs (short interspersed nuclear elements), small noncoding elements of a few hundred base pairs that exploit the transposition machinery of LINEs to ensure their amplification. LTR retrotransposons are by far the most abundant transposable elements in plant genomes, while LINEs appearsomewhat less represented, in contrast to mammalian genomes. SINEs are also abundant in plant genomes, although their small size prevents them from playing a large structural role.

The continuous discovery of new structural variants of these major categories has made it necessary to revise and refine previous classifications. Retrotransposons have been included in a recent classification proposed for all reverse-transcribing elements, and based on viral nomenclature with type species forms defined for some genera (**Table 1**). All retroelements have been defined as belonging to class *Retroelementopsida*, presently divided in two orders, order *Retrovirales* that encompasses retroviruses, pararetroviruses, and LTR retrotransposons, and order *Retrales* that contains non-LTR retrotransposons and bacterial retrons. While order *Retrales* has not been further organized at the present time, order *Retrovirales* has been divided in several suborders, families, and genera.

Suborder *Orthoretrovirinae* contains the family *Retroviridae*, or true retroviruses, of which no example is known in plants at the present time.

Suborder *Pararetroviridae* contains two families of pararetroviruses, one of which (*Caulimoviridae*) is represented in plants. Pararetroviruses represent sensibly different lineages of retroelements, and are thought to derive from preexisting viruses having acquired a reverse transcription function, presumably from retrotransposons of the family *Metaviridae*. Many plant genomes also contain vertically transmitted integrated endogenous pararetroviral sequences (EPRVs), which represent integrated derivatives of pararetroviruses, although the infectious counterpart has not always been characterized. Plant pararetroviruses are not discussed here.

Suborder *Retrotransposineae* contains LTR retrotransposons, and has been divided in family *Pseudoviridae* (Ty1/*copia* retrotransposons) and family *Metaviridae* (Ty3/*gypsy* retrotransposons), based on the organization of the pol domain.

The Life Cycle of LTR Retrotransposons

High Similarities with Intracellular Steps of the Retroviral Cycle

LTR retrotransposons represent by far the most abundant plant retrotransposons. They share striking similarities with animal retroviruses, both in their structure (**Figure 1**) and in their replication cycle (**Figure 2**). Like integrated retroviral forms (termed proviruses), LTR retrotransposons are terminated by two LTR sequences. As the reverse transcription process generates two identical LTRs, with mutations accumulating subsequently, the rate of divergences between LTRs of a given copy is often used as a molecular clock to date its insertion. The gag and pol coding domains are found between the LTRs. Other typical features include short sequences known as PBS (primer binding site) and PPT (polypurine tract). The PBS is found immediately downstream the 5′ LTR, and is involved in priming of the (−)-DNA strand synthesis during reverse transcription. It is generally complementary to the 3′ end of a tRNA recruited for the priming of DNA synthesis (tRNA^{met} for most *Pseudoviridae*, more variable for *Metaviridae*). The PPT is a short AG-rich sequence found immediately upstream to the 3′ LTR, and involved in the priming of the (+)-DNA strand synthesis. Inserted copies are typically bounded by short direct repeat of the target host site, generated upon insertion, and usually 5 bp long for plant LTR retrotransposons.

Transcription begins in the 5′ LTR, ends in the 3′ LTR, and produces both the RNA template used for reverse transcription and the mRNA(s) that will be translated into proteins, after cleavage of the gag-pol polyprotein by the protease (prot) function. The gag protein binds to the RNA template and forms a cytoplasmic virus-like particle (VLP) analogous to the retroviral virion core. Together with the RNA template, VLPs encapsulate the RT that will produce the double-stranded DNA copy, and the integrase (int) that will be involved in transfer of the linear DNA copy to the nucleus and its insertion in the genome. This amplification cycle includes multiple steps of control, with transcription being absolutely crucial for retrotransposition. As a consequence, the mode of amplification of a particular element will be conditioned by its transcriptional regulation.

The amplification cycle of LTR retrotransposons is thus closely related to the intracellular steps of the retroviral cycle, with strong similarities in the reverse transcription and integration processes, as well as similar involvement of a cytoplasmic encapsulated intermediate. It has to be noticed that the amplification of non-LTR retrotransposons is sensibly different, with reverse transcription occurring at the integration site. The reality of the retroviral-type cycle of LTR retrotransposons has been demonstrated in yeast, but not yet in plants. However, VLPs have been detected for the barley BARE1 element. In addition, it has been demonstrated very recently that plant elements of the genus *Sirevirus* (see below) encode unusually large Gag domains extended in their C-terminus, and that this Gag extension binds to LC8/LC6 dynein proteins. Although the biological significance of this interaction remains to be determined, it is very similar to the binding of some human retroviruses'

Table 1 Classification of retroelements

	Type species	A few examples in Viridiplantae[a]
Order *Retrovirales*		
Suborder *Orthoretrovirineae*		
Family *Retroviridae* (retrovirus)		Not yet
Suborder *Pararetrovirineae* (pararetrovirus)		
Family *Hepadnaviridae*		Not yet
Family *Caulimoviridae*		CaMV, BSV, integrated EPRVs (banana BSV; petunia PVCV, tobacco TVCV, *Nicotiana* NtoEPRV, and Ns EPRV)
Suborder *Retrotransposineae* (LTR retrotransposons)		
Family *Pseudoviridae* (Ty1/*copia* retrotransposons)		
Genus *Pseudovirus*	Ty1	Ta1 (*Arabidopsis* X1329), Tnt1 (tobacco X13777), Tst1 (potato X52387), Tto1 (tobacco D83003), WIS2 (wheat X63184), BARE1 (barley Z17327), Hopscotch (maize U12626), Tto1 (tobacco D83003), Tos17 (rice D85876), RIRE1 (rice D85597), Art1 (arabidopsis Y08010), Retrofit (rice U72726), Melmoth (kale Y12321), Tgmr (soybean U96748), Stonor (maize AF082134), Panzee (pigeon pea AJ000893), Tpv2 (bean AJ005762), LERE1 (tomato AF275345), Retrolyc1 (*S. peruvianum* AF228701), AtRE1 (*Arabidopsis* AB021265), TLC1 (*S. chilense* AF279585), Toto1 (tomato AF220602), Angela (wheat DQ666286)
Genus *Hemivirus*	Copia	Osser (*Volvox* X69552)
Genus *Sirevirus*	SIRE1	SIRE1 (soybean AY205608), ToRTL1 (tomato U68072), Endovir (*Arabidopsis* AY016208) Opie-2 (maize U68408), PREM2 (maize U41000), Osr7 (rice AP002538), Osr8 (rice AC021891)
Family *Metaviridae* (Ty3/*gypsy* retrotransposons)		
Genus *Errantivirus*	gypsy	Not yet
Genus *Metavirus*	Ty3	Athila (*Arabidopsis* X81801), Cyclops (pea AJ000640), RIRE8 (rice AB014740), RIRE2 (rice AB030283), BAGY2 (barley AJ279072), Cinful (maize AF049110), Grande1 (maize X97604), Piggy1 (pea AY299398), Ogre (pea AY299398), Diaspora (soybean AF095730) Genus *Chromovirus?*: del1 (lilium X13886), IFG7 (pine AJ004945), Reina (maize U69258), RIRE3 (rice AB014738), RIRE7 (rice AB033235), Galadriel (tomato AF119040), tekay (maize AF448416), Beetle1 (beet AJ539424)
Genus *Semotivirus*	BEL	Not yet
Genus ?	DIRS	Not yet
Unclassified		LARDs: Sukkula (barley AY054378), Squiq (rice AY355293), Spip (rice AY355292), Dasheng (rice) TRIMs: (potato AJ276865), Katydid (*Arabidopsis*), mini-Toto1 (tomato X58273)
Order *Retrales*		
Suborder *Retroposineae* (non-LTR retrotransposons)		
	L1	Cin4 (maize Y00086), del2 (lilium Z17425), Isabelle (maize AF326781), Karma (rice AB081316), Bali1A (rapeseed AF525305)
	Alu	TS (tobacco D17453), S1 (rapeseed L76840), Au (*Aegilops* AB046134)
Suborder *Retronineae* (bacterial retrons)		

[a]Elements have been classified according to year of publication or sequence submission to Genbank databases. Note that in several cases, only partial sequences are available. Extensive data on various plant elements can be found at: Repbase, a database of repetitive DNA elements of all organisms, the TIGR Plant Repeat Database and the Triticeae-specific TREP database. Retroelements were classified according to Hull R (2001) Classifying reverse transcribing elements: A proposal and challenge to the ICTV. *Archives of Virology* 146: 2255–2261; and Boeke JD, Eickbush T, Sandmeyer SB, and Voytas DF (2004) *Pseudoviridae/Metaviridae* In: Fauquet CM, Mayo MA, Maniloff J, Desselberger U, and Ball LA (eds.) *Virus Taxonomy: Eighth Report of the International Committee on Taxonomy of Viruses*, pp. 397–420. San Diego, CA: Elsevier Academic Press.

Gag protein to LC8 proteins to ensure intracellular movement of virion cores along microtubules. Such observations reinforce the strong similarities between the life cycle of retroviruses and LTR retrotransposons.

The major difference between retroviruses and LTR retrotransposons resides in the infectious potentialities of the former, mediated by the env function that allows them to produce extracellular virions and infect host

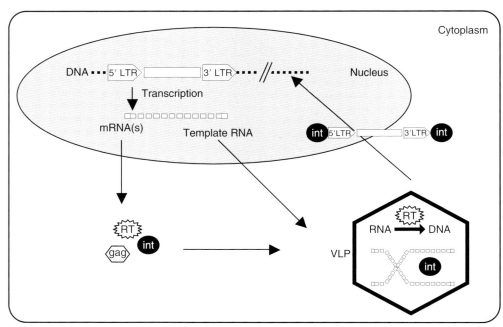

Figure 2 The amplification cycle of LTR retrotransposons. Transcription begins in the 5′ LTR and ends in the 3′ LTR, and produces both the RNA template used for reverse transcription and mRNAs translated into proteins. The gag protein binds to the RNA template and forms a cytoplasmic virus-like particle (VLP). Together with the RNA template, VLPs encapsulate the RT that will produce the double-stranded DNA copy, and the integrase (int) involved in transfer of the linear DNA copy to the nucleus and its insertion in the genome.

receptor cells. *Bona fide* retrotransposons, on the other hand, are not infectious and their amplification cycle is exclusively intracellular. However, we shall see later that this assumption should be treated with care and that the delimitation between retroviruses and retrotransposons is increasingly fuzzy, possibly even in plants.

It is now considered as certain that retroviruses are derivatives of some LTR retrotransposon *Metaviridae* lineages. LTR retrotransposons are themselves derived, with LINE-like ancestors, from the assembly of various modules together with an RT-like function derived from ancestral cellular functions used for the transition of genetic information from RNA to DNA at the dawn of life. Retroviruses have evolved from LTR retrotransposons through transduction of env domains ensuring their ability to exit the cell. Although the process can be reversed (e.g., with vertebrate endogenous retroviruses having lost their env functionality and reversed to a restricted intracellular life cycle), retroviruses can be seen as the ultimate evolutionary step in freeing of cellular genetic information outside of the cell, and they basically represent LTR retrotransposons that have succeeded in life. It is not clear whether this process has occurred only once during the evolution of *Retroviridae*; however, it is now increasingly clear that a number of their LTR retrotransposons' left-behind cousins have repeatedly attempted to join them in acquiring an extracellular life and that some of them have truly succeeded, albeit not yet in plants.

Evolution as Quasispecies-Like Populations

Other similarities of retrotransposons with retroviral characteristics include their high level of sequence variability. This has been best illustrated for elements related to the tobacco Tnt1 element, which are found in many species of the family Solanaceae, and evolve via a high variability of the LTR region carrying regulatory features (U3 region). Species of the genus *Nicotiana* contain Tnt1 elements that are classified into several groups (termed subfamilies) which are closely related, except in their U3 regions that differ strongly. Members of each subfamily are detected within each species, and their U3 divergence thus predates the *Nicotiana* radiation. However, the respective proportions of each subfamily differ in each species, indicating that each host species has preferentially amplified specific subfamilies. Similar evolutionary patterns have been detected in other Solanaceae species, such as tomato, pepper, or aubergine, which contain elements closely related to *Nicotiana* elements, except in the U3 region, and which can also be further subdivided in subfamilies based on the nature of the U3 region only. Interestingly, the different tobacco Tnt1 subfamilies are each expressed in response to different stimuli, within the same host genome. The very ancient Tnt1 element has thus evolved in Solanaceae a large range of highly related progeny populations that have gained new regulatory sequences and new expression patterns.

This strategy is highly similar to the evolutionary patterns of RNA viruses and retroviruses, generating populations of closely related but different genomes referred to as quasispecies. This variability has been attributed to the high-error-prone process of reverse transcription, due the lack of proofreading repair activity of RNA polymerases and reverse transcriptases. This allows viral populations to evolve very rapidly when environmental conditions change and endow retroviruses with high adaptive capacity. Thus, LTR retrotransposons similarly evolve continuums of closely related sequences, which can be defined as quasispecies-like. This evolutionary pattern allows them to vary their amplification conditions, as regulatory controls in the U3 will determine the conditions in which the element will be transcribed and amplified. The preferential amplification of specific subfamilies in each host species suggests that optimal patterns of amplification may have been selected and maintained in each host genome, possibly depending on its own evolutionary, reproductive, and environmental history.

The Different Genera of LTR Retrotransposons

The Family *Pseudoviridae* (Ty1/*copia* Retrotransposons)

Pseudoviridae are differentiated from all other LTR retrotransposons and from retroviruses by an unusual organization of the pol domain, with the int domain placed before the RT domain (**Figure 1**). This confirms that the different LTR retrotransposon families have evolved by independent acquisition of modular functions. Historically, Ty1/*copia* retrotransposons have been the first LTR retrotransposons characterized in higher plants, and the best-known plant LTR retrotransposons belong to this family, including the tiny handful of elements for which mobility has been demonstrated. *Pseudoviridae* have been classified in the genera *Pseudovirus* (type species Ty1 of yeast), *Hemivirus* (type species *copia* of Drosophila), and the recent genus *Sirevirus* (type species, SIRE1 of soybean).

Pseudoviruses and hemiviruses differ by the fact that the former use a full $tRNA^{met}i$ to prime synthesis of the (−)-DNA strand, while the latter use a half $tRNA^{met}i$. So far, no hemivirus has been referenced in plants, and all LTR retrotransposons usually – and incorrectly – referred to as *copia*-type retrotransposons in fact classify in the genus *Pseudovirus*. Sireviruses have so far been identified only in plants. They are phylogenetically well separated from the two other genera, and display some very interesting characteristics. In particular, several members of the genus have acquired additional ORFs carrying additional functions that present some similarities to the env domain of retroviruses, such as a transmembrane domain (TMD). This is, for instance, the case with SIRE1, ToRTL1, and Endovir (**Table 1**). A number of other elements classified as sireviruses, such as Opie-2, PREM-2, Osr7, or Osr8 (**Table 1**), do not carry additional ORFs with typical env-like features. However, all of them carry a significant amount of noncoding DNA after the end of the pol domain, from a few hundred base pairs to over 1 kbp, while members of the genera *Pseudovirus* and *Hemivirus*, such as Tnt1, Ty1, or *copia*, separate the pol domain from the 3′ LTR by a few dozen nucleotides or less. This suggests that the env-less sireviruses may also have contained similar additional ORFs that have decayed over evolutionary times.

The Family *Metaviridae* (Ty3/*gypsy* Retrotransposons)

The family *Metaviridae* is characterized by an organization of the pol domain identical to that of retroviruses, with the int protein encoded at the end of pol, after the RT domain (**Figure 1**). Metaviruses appear to be basal to both retroviruses and pseudoviruses, and the family *Metaviridae* therefore encompasses a large variety of members, whose classification is probably not at its final stage yet. Although metaviruses have been characterized in plants more recently than pseudoviruses, they have been found to be more abundant in several plant genomes.

The family *Metaviridae* comprises the genera *Errantivirus* (type species *gypsy* of drosophila), *Metavirus* (type species Ty3 of yeast), and *Semotivirus* (type species BEL of drosophila) (**Table 1**). To these three genera are often added the DIRS retrotransposons, originally described from *Dictyostelium discoidum*, that show differences with other metavirus elements. DIRS do not carry typical LTRs, that is, direct repeats in the same orientation, and do not encode an int. Instead, they are bounded by inverted long repeats and integration in the host genome is ensured by a tyrosine-recombinase. In spite of these differences, they are more related to metaviruses, both in terms of RT sequence and organization of the pol domain, with the tyrosine-recombinase placed downstream to the RT domain. They could represent chimerical elements having acquired an RT domain from some metaviruses, and are placed by several authors as a putative additional genus (still unnamed) in this family. No DIRS, semotivirus, or errantivirus has yet been characterized in plants, with the totality of so-called *gypsy*-like retrotransposons of plants classified in the genus *Metavirus* (**Table 1**).

Errantiviruses are grouped together on the basis of the presence of an additional env-like domain, which ensures a real envelope function for the two infectious *gypsy* and ZAM elements of drosophila. *Gypsy* and ZAM are the first examples of true retroviruses outside of the vertebrate subphylum. However, their phylogenies place them quite apart from retroviruses, and they are more closely related to other LTR retrotransposons of the family *Metaviridae*. They do presumably represent recent progresses of LTR retrotransposons toward an extracellular life cycle.

The separation of errantiviruses and metaviruses has been interpreted by many authors as based on the lack of an env or env-like domain in metaviruses. However, the situation is more complex. Many plant elements presently classified in the genus metavirus indeed do not encode for additional ORFs after the pol domain (such as del1, IFG7, Reina, RIRE3, RIRE7, Galadriel, tekay, and Beetle-1; **Table 1**). In contrast to env-less sireviruses mentioned above, these elements do not show significant stretches of additional sequences between the pol domain and the 3' LTR (only a few dozens at the most, often very few), indicating that they may never have contained any additional ORF. However, a large number of plant elements presently classified in the genus *Metavirus* do contain an additional ORF with typical env-like features (such as Athila, Cyclops, RIRE8, RIRE2, BAGY2, Cinful, Grande1, Piggy1, or Ogre; **Table 1**). The two groups of elements cluster separately in most phylogenetic studies, indicating that they could be considered as two different genera of the family *Metaviridae*. An additional genus, *Chromovirus*, has recently been proposed within the family *Metaviridae*, based on the presence of a chromodomain in the integrase. However, it has not been included in the current official classification. Interestingly, all plant metaviruses devoid of additional coding domains seem to be chromoviruses. It is therefore likely that the classification of *Metaviridae* will be reconsidered in future, to separate the two groups of metaviruses and recognize the existence of the chromodomain. Interestingly, the soybean Diaspora element is devoid of any additional DNA between pol and the 3' LTR; however it clusters with env-containing elements, suggesting that its original env-like domain has been deleted. Therefore, the presence/absence of an env-like domain would not be a relevant criterion in the redefinition of genera within the family *Metaviridae*.

Unclassified LTR Retrotransposons

A number of LTR retrotransposons have recently been discovered that cannot be classified into any of the above genera due to their lack of coding sequences. The TRIMs (terminal repeat retrotransposons in miniature) consist of small LTRs framing a short central domain of noncoding sequences, while the LARDs (large retrotransposon derivatives) are larger defective derivatives that contain a large internal domain without significant similarities to retrotransposon coding domains. Both TRIMs and LARDs share many of the structural characteristics of LTR retrotransposons, such as RT priming sites PBS and PPT, and are highly amplified, suggesting efficient transactivation by functional retrotransposons. For several LARDs, sequence identities have been found with LTRs of canonical LTR retrotransposons (e.g., Dasheng with RIRE2, Squip with RIRE8, Spip with RIRE3), suggesting that they have been derived from these elements, and may eventually be classified together. Similarly, some TRIM-like small elements are derived from canonical LTR retrotransposons, such as the tomato mini-Toto1 element, derived from the Toto1 pseudovirus. In addition, a number of intermediate situations are frequently found, such as TRIM derivatives containing larger additional unknown DNA, a possible intermediate situation between TRIMs and LARDs. The existence of TRIMs and LARDs and various intermediates indicates that plant genomes contain a large range of retrotransposons with structural features much more diverse and complex than previously thought. As long as the features required in *cis* for amplification are conserved (e.g., LTRs, PBS, and PPT, as well as a few internal sequences involved in packaging), and as long as the LTR promoter remains transcriptionally active to ensure the production of an RNA template, these defective or atypical elements can be amplified to high levels using the structural and enzymatic retroviral functions provided by their original functional parent, or possibly by other related LTR retrotransposons.

Env-Like Functions in Plants?

Although plant retrotransposons with additional ORFs of the env type have been detected, the potential role of such a function in infectivity remains very hypothetical in plants. Retroviral env glycoproteins carry a number of typical structures, such as a TMD mediating formation of infectious virions and a host receptor-binding domain, mediating targeting of host receptor cells. No significant homology to any retroviral env gene has been shown for any plant element; however, such identification is highly unlikely due to the intrinsic high variability shown by animal retroviral env proteins. As the definition of env-like functions in plant retrotransposons is mostly based on the presence of TMDS, the issue of the naming of these additional ORFs has been hotly debated. TMDs can also be found in various proteins, and the env denomination has a specific meaning related to anchoring of the protein in the membrane envelope of extracellular virion particles and to recognition of receptors on target cells to ensure viral infection after membrane fusion. Quite likely, such mechanisms are rarely used by plant viruses for host-to-host transmission, due to the cell wall barrier. Plant retrotransposons carrying additional ORFs related to functions used by most plant viruses to ensure host-to-host transmission, for example, via vector insects, have not yet been found. However, examples of enveloped viruses do exist in plants, and retrotransposon-encoded env-like proteins could be involved in anchoring of VLPs to cellular membranes, playing a role in cell-to-cell transmission via plasmodesmata and possibly in transmission by insects.

Furthermore, conservation of env-like domains has been observed between elements from different plant genera within both the *Metaviridae* and the *Pseudoviridae* (SIRE1 of soybean and Endovir of *Arabidopsis*). In addition, conserved splice acceptor sequences are detected 5' of env-like domains of several plant metaviruses, and production of an env-like spliced subgenomic RNA has been demonstrated for the barley BAGY-2 element, a mechanism similar to those used by retroviruses to ensure production of the env protein. In contrast, sireviruses such as SIRE1 use stop codon suppression to express env-like protein. The existence of such specific expression mechanisms strongly suggests that env-like domains are not mere incidental transduction of cellular functions, as shown for the maize Bs1 element, and that acquisition of additional coding domains with TMDs, whether in a quest or not for extracellular life, may play an important role in the plant LTR retrotransposon life cycle. In the search of a better definition, and in spite of lack of evidence for any retroviral env-type function, it has been proposed to maintain the env denomination due to its vernacular long-term use: for example, the nucleocapsid protein used to build up the VLP core particle is commonly termed gag in all retrotransposons, by analogy to the gag 'group-specific antigen' region of retroviruses, although such functional denomination does not make sense in plants.

A striking observation is that LTR retrotransposons generally have been very efficient in acquiring env-like domains, and have done so many times. For instance, additional ORFs with env-like features have been observed both for Ty1/*copia* and Ty3/*gypsy* retrotransposons, suggesting that they have been acquired independently during evolution. In addition, the *Metaviridae* contain clear examples of acquisition of env function from different sources, for example, from baculoviruses for gypsy of drosophila, and from phleboviruses and herpesviruses for the semotiviruses Cer of *Caenorhabditis elegans* and Tas of *Ascaris lumbrocoides*, respectively. This indicates that LTR retrotransposons have repeatedly been able to hijack viral functions leading to extracellular autonomy, and there is no conceivable reason for plants to be an exception.

The Nomenclature Issue

Rules for naming individual elements also continue to be hotly debated. Earlier retrotransposons have been named according to authors' personal fancies (e.g., Athila, Cyclops, Opie, Ogre) or based on the plant species they were isolated from, but following no defined rule (e.g., Tnt1 for Transposon of Nicotiana Tabacum, BARE-1 for Barley Retrotransposon, WIS-2 for Wheat Insertion Sequence). An official nomenclature has recently been proposed for all retroelements, with three letters used for the species, up to three (exceptionally four) for the element and 'V' for virus. In this nomenclature, the yeast Ty1 and drosophila *copia* elements have been renamed SceTy1V and DmeCopV, respectively, and Tnt1 and BARE1 have been renamed NtaTnt1V and HvuBV, respectively. However, this nomenclature has so far not been implemented by the plant scientific community, most likely because it does not solve the major problem arising from the presence of highly related elements in different plant species or genera, which is a general rule in plants.

In addition, most transposable element populations, and especially retrotransposons, are generally composed of populations of closely related sequences. Defining whether a particular copy is related to an already known element, or is different, is thus often a difficult task. A consensus position has been to consider that copies that show above 75% identity over most of their length (not taking into account deletions or insertions of unrelated material) are variants of the same element. However, many elements have been discovered and named independently. For instance, the wheat WIS2 and Angela elements are quite similar to the barley BARE1 element, with internal nucleotidic sequences 75–80% identical between BARE1 and WIS2 and over 80% identical between BARE1 and Angela. Similarly, *Solanum* subsection *Lycopersicon* counterparts of the tobacco Tnt1 element have been named Retrolyc1 (first identified in *S. peruvianum*) and TLC1 (identified in *S. chilense*), although Retrolyc1 and Tnt1 nucleotide sequences are over 85% identical and TLC1 and Retrolyc1 are 93% identical. This highlights the difficulties inherent to a nomenclature based on host species names.

A proposition was made to replace species-based names by first names, a trend that seems to be developing at present, at least in the plant field. More specifically, it has been proposed that female first names should be used for retroelements and male first names for class II DNA transposons, although this particular proposal is not really implemented. Such type of nomenclature should alleviate the problem of the presence of related elements in different hosts, provided authors make sure their favorite first name has not been used yet. Attempts to unify the nomenclature (as well as the general classification) of transposable elements of Triticeae are currently coordinated by Dr. Thomas Wicker, who curates the TREP database for Triticeae repeats. This should lead to the redefinition of consensus guidelines that could be fruitfully followed by the entire scientific community for transposable elements of all plant genera.

See also: Caulimoviruses: General Features; Caulimoviruses: Molecular Biology; Endogenous Retroviruses; Metaviruses; Movement of Viruses in Plants; Pseudoviruses; Reoviruses: General Features; Retrotransposons of Fungi; Retrotransposons of Vertebrates; Retroviruses: General Features; Vector Transmission of Plant Viruses.

Further Reading

Boeke JD, Eickbush T, Sandmeyer SB, and Voytas DF (2004) *Pseudoviridae/Metaviridae*. In: Fauquet CM, Mayo MA, Maniloff J, Desselberger U, and Ball LA (eds.) *Virus Taxonomy: Eighth Report of the International Committee on Taxonomy of Viruses*, pp. 397–420. San Diego, CA: Elsevier Academic Press.

Casacuberta JM, Vernhettes S, Audéon C, and Grandbastien M-A (1997) Quasispecies in retrotransposons: A role for sequence variability in Tnt1 evolution. *Genetica* 100: 109–117.

Grandbastien M-A, Audeon C, Bonnivard E, et al. (2005) Stress activation and genomic impact of Tnt1 retrotransposons in Solanaceae. Special Issue: Retrotransposable Elements and Genome Evolution. *Cytogenetic and Genome Research* 110: 229–241.

Havecker ER, Gao X, and Voyas DF (2004) The diversity of LTR retrotransposons. *Genome Biology* 5: 225.

Havecker ER, Gao X, and Voytas DF (2005) The sireviruses, a plant-specific lineage of the Ty1/*copia* retrotransposons, interact with a family of proteins related to dynein light chain 8. *Plant Physiology* 139: 857–868.

Hull R (2001) Classifying reverse transcribing elements: A proposal and challenge to the ICTV. *Archives of Virology* 146: 2255–2261.

Kalendar R, Vicient CM, Peleg O, Anamthawat-Jonsson K, Bolshoy A, and Schulman AH (2004) Large retrotransposon derivatives: Abundant, conserved but nonautonomous retroelements of barley and related genomes. *Genetics* 166: 1437–1450.

Kordis D (2005) A genomic perspective on the chromodomain-containing retrotransposons: Chromoviruses. *Gene* 247: 161–173.

Lucas H, Yot P, Lockhart BEL, and Capy P (2004) Taxa of viruses, virus-like and subviral agents infecting Poaceae, class '*Retroelementopsida*'. In: Lapierre H and Signoret PA (eds.) *Virus and Virus Diseases of Poaceae (Graminae)*, pp. 279–303. Paris: INRA Editions.

Malik HS, Henikoff S, and Eickbush TH (2000) Poised for contagion: Evolutionary origins of the infectious ability of invertebrate retroviruses. *Genome Research* 10: 1307–1318.

Peterson-Burch BD, Wright DA, Laten HL, and Voytas DF (2000) Retroviruses in plants? *Trends in Genetics* 16: 151–152.

Vicient CM, Kalendar R, and Schulman AH (2001) Envelope-class retrovirus-like elements are widespread, transcribed and spliced, and insertionally polymorphic in plants. *Genome Research* 11: 2041–2049.

Wicker T, Sabot F, Hua-Van A, et al. (2007) A unified classification system for eukaryotic transposable elements. *Nature Previews Genetics* 8: 973–982.

Witte CP, Le QH, Bureau T, and Kumar A (2001) Terminal-repeat retrotransposons in miniature (TRIM) are involved in restructuring plant genomes. *Proceedings of the National Academy of Sciences, USA* 98: 13778–13783.

Wright DA, Daniel F, and Voytas DF (2001) Athila4 of *Arabidopsis* and Calypso of soybean define a lineage of endogenous plant retroviruses. *Genome Research* 12: 122–131.

Relevant Websites

http://www.girinst.org – Genetic Information Research Institute (GIRI).
http://www.tigr.org – The TIGR Plant Repeat Databases, J. Craig Venter Institute (JCVI).
http://wheat.pw.usda.gov – The Triticeae Repeat Sequence Database (TREP), GrainGenes, Agricultural Research Service, USDA.

Retrotransposons of Vertebrates

A E Peaston, The Jackson Laboratory, Bar Harbor, ME, USA

© 2008 Elsevier Ltd. All rights reserved.

Glossary

Ancestral retrotransposon A retrotransposon present in the genome of a common ancestor of two or more host groups. Also referred to as ancestral repeat.

Apurinic/apyrimidinic Endonuclease enzyme that catalyzes the cleavage of a phosphodiester bond in a DNA molecule.

Autonomous retrotransposon A retrotransposon encoding proteins required for its reverse transcription and transposition.

Clade A group of organisms consisting of a single common ancestor and all its descendents.

DDE transposases A class of transposase enzymes containing a highly conserved amino acid motif, aspartate–aspartate–glutamate (DDE), required for metal ion coordination in catalyzing integration of retrotransposon cDNA into the host DNA, thus also known as integrase.

Exaptation In broad terms, a feature conferring evolutionary fitness on an organism but which was originally nonfunctional or designed for some other function in the organism. Thus, the genes of a retrotransposon newly inserted in the host genome maybe, over evolutionary time, co-opted for function in the host; in this new role they are called exaptations.

Homoplasy A structure arising in two or more species as the result of a convergent evolution, and not as a result of common descent which indicates that the feature in one species is homologous to that in the other.

Lineage-specific retrotransposon

A retrotransposon introduced into the genome of one but not another host grouping after their evolutionary divergence.

Nucleotide substitution rate The rate at which single nucleotide mutations occur within regions of a genome not subject to selection. One way to

> estimate this is by comparison of the consensus sequences of ancestral repeats with their remnant sequences in different host genomes.
> **Y-transposase, also known as tyrosine recombinase** A class of transposase enzymes which can use a conserved tyrosine to cut and rejoin its DNA substrates by a 3' phosphotyrosine linkage; used by some retrotransposons to integrate their cDNA into the host DNA, and thus sometimes called integrase.

History

Transposable elements are defined segments of DNA which replicate and move to other loci within the genome by a variety of mechanisms. The vast majority of these mobile elements in vertebrates are retrotransposons, which replicate by means of an RNA intermediate that is reverse-transcribed into DNA and inserted in a new location within the genome. Retrotransposons are found in all vertebrate genomes, but have been intensively studied in relatively few.

Retrotransposons were initially detected as discrete fragments of genomic DNA that rapidly reannealed after denaturation, and were recognized to be repetitive sequence elements interspersed within genomic DNA. The advent of DNA analysis by restriction enzyme fragmentation led to the identification of discrete families of repetitive elements whose members shared particular sets of internal restriction sites. The most highly repeated long and short sequences were named long interspersed repetitive elements (LINEs) and short interspersed repetitive elements (SINEs), respectively.

The LTR elements, a third general type of retrotransposon resembling the integrated form of proviruses, was similarly discovered. These endogenous retroviruses and retrovirus-like elements are generally restricted to an intracellular life cycle and vertical transmission through the germline and, unlike 'true' retroviruses, are not infectious. However, notable exceptions blur this distinction.

The advent of whole genome sequencing and sequence analysis is rapidly providing detailed pictures of vertebrate retrotransposon landscapes, such as those of a pufferfish (*Takifugu rubripes*), the chicken (*Gallus gallus*), the laboratory mouse (*Mus musculus*), human (*Homo sapiens*), dog (*Canis lupus familiarus*), and others. New lineage-specific retrotransposons of vertebrates such as primate-specific SVA (SINE-R, VNTR, Alu), are coming to light, as well as the discovery within selected vertebrate lineages of ancient retrotransposons, such as the Penelope-like elements (PLEs) that are present in the genomes of metazoans, fungi, and amoebozoans. Analysis of retroelement phylogenies can assist in disentangling evolutionary relationships of their hosts.

Nomenclature

Different structural and functional classification schemes for retrotransposable elements have been proposed. No scheme has been generally accepted by both the virology and the retroelement communities, and at the level of individual elements, nomenclature is frequently based on historical tradition and can be confusing. A simple classification scheme based primarily on genetic structure is used here, with acknowledgment that, with the exception of reverse transcriptase, coding domains are not uniformly retained across different groups. Detailed discussion of classification and individual elements is outside the scope of this article, but the subject can be pursued through references listed in Further Reading.

Retrotransposon Structural and Functional Features

Retrotransposons are usually subdivided into two structural superfamilies based on the presence or absence of terminal directly repeated sequences of several hundred nucleotides, the LTRs. The LTR-retrotransposon superfamily includes elements resembling integrated or endogenous proviruses, the non-LTR superfamily includes LINEs, SINEs, and SVA elements. In each superfamily, there are autonomous elements, which encode reverse transcriptase (RT) as well as a variable set of other proteins necessary for replication and transposition of the element. The superfamilies also include nonautonomous elements whose open reading frames (ORFs) usually encode no proteins, or putative proteins lacking homology to other transposable element proteins and lacking known function. Nonautonomous elements always have an autonomous partner to provide the necessary proteins *in trans*. A third group of retrotransposable elements, the PLEs, were originally described in invertebrates but have more recently also been reported in vertebrates. PLEs have distinct genomic and transcript structural features which preclude a neat fit in either the LTR or non-LTR superfamilies; consequently, this group will be treated independently in this article.

Fundamental structural features of non-LTR and LTR retrotransposons are illustrated schematically in **Figure 1**. A useful repository of consensus sequences of retrotransposons from different animal species, and their classification is curated by the Genetic Information Research Institute.

LTR Superfamily

Four major groups of LTR retrotransposons have been identified in vertebrates to date: Ty3/*gypsy*; BEL; the vertebrate retrovirus group; and Ty1/*copia*. The hepadnaviruses, a fifth group encoding RT similar to LTR retrotransposons but which are circular and lack LTRs, are discussed elsewhere in this encyclopedia. The LTR retrotransposon groups are distinguished from one another by amino acid sequence comparison of their homologous enzymatic domains, primarily RT, and by features specific to one or two groups. Nonautonomous retrotransposon families are distinguished from one another by DNA sequence comparisons of their internal domain and of their LTRs.

The DIRS1 (Dictyostelium repeat sequence 1) subclass LTRs are either inverted repeats, or split direct repeats. In addition, these elements lack the typical integrase and protease coding domains, containing instead a domain encoding a bacteriophage lambda recombinase-like protein also known as tyrosine-recombinase or Y-transposase. These features of DIRS1 make its inclusion in the LTR superfamily problematic, although its RT has homology to the more typical elements of the superfamily.

Structure

LTR elements are typically constructed as two direct LTRs flanking an internal sequence of variable coding content (**Figure 1**). The LTRs are necessary and sufficient for promoter activity and transcription of the retroelement. Functional features common to LTRs include: 5′ TG and 3′ CA dinucleotides necessary for integration into the host genome; Pol II promoter elements and a transcription start site; and a polyadenylation and cleavage signal.

In typical autonomous LTR retrotransposons, the internal sequence contains a variable but small number of ORFs. Most elements contain an ORF including Gag (group-specific antigen) and Pol (polymerase) domains, and some additionally contain an ORF for Env (envelope). Gag encodes a structural polyprotein integral to the formation of cytoplasmic virus-like particles in which reverse transcription takes place. Pol encodes several enzymatic activities: (1) protease required to cleave the translated products of the gag–pol transcript into their functional forms; (2) RT for transcribing the retrotransposon RNA into double-stranded cDNA; (3) ribonuclease H (RNase H) for processing the RNA template prior to plus strand cDNA synthesis; and (4) integrase incorporating an aspartate (D), aspartate, glutamate (E) (DDE)-type

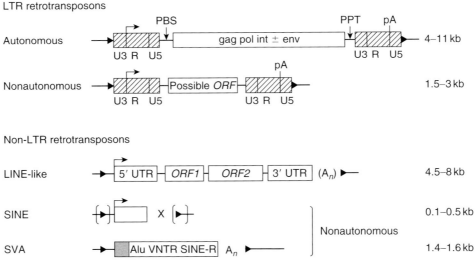

Figure 1 General structure of selected retrotransposons. A typical LTR retrotransposon is flanked by target site duplications (black arrowheads) and its LTR domains (hatched) contain a pol II promoter, transcription start site and a polyadenylation and cleavage signal. The internal domain of autonomous elements contains ORFs encoding homologs of retroviral gag and pol genes, and occasionally an additional env gene homolog. The short ORF of nonautonomous elements does not encode these proteins, if any. LINE-like elements terminate in family-specific 3′ UTRs with or without poly (A) tails, the poly (A) tail (A_n) indicated here is typical of L1. Flanking target site duplications typical of the L1 clade may or may not be present in other LINE-like clades. The autonomous LINE-like elements contain an internal pol II promoter and ORFs encoding structural and reverse transcriptase (RT) genes required for retrotransposition. SINEs contain an internal pol III promoter but lack an ORF, and may or may not be flanked by target site duplications (in parentheses), depending on their specific LINE-like partner. SVA consists of a 5′ variable number hexameric repeat (gray box), followed by an Alu-like element, a VNTR domain, the SINE-R element derived from the human endogenous retrovirus HERVK10, and a poly (A) tail. SVA elements are thought to be mobilized *in trans* by L1 proteins, and are flanked by target site duplications typical of L1. Right angled arrow, transcription start site; pA, polyadenylation and cleavage signal; PBS, primer binding site for initiating minus strand cDNA synthesis; PPT, polypurine tract used for initiation of plus strand cDNA synthesis. Element sizes are approximate ranges. The diagrams are not to scale.

transposase activity. The integrase sequence of the chromovirus genus of the Ty3/*gypsy* group additionally encodes a chromodomain, a domain found in chromatin complex structural proteins or chromatin remodeling proteins. Further information regarding retroviral genes and structure can be found elsewhere in the encyclopedia. The Pol domain order is usually protease, RT, RNase H, integrase, except for the Ty1/Copia group and the Gmr1-like elements of the Ty3/Gypsy group in which integrase is upstream of RT.

In addition to coding sequence, the internal sequence contains a 5′ primer binding site (PBS) for first-strand DNA synthesis and a 3′ polypurine tract which serves as the primer binding site for second-strand DNA synthesis. These two features are conserved in the internal sequence region of nonautonomous LTR retrotransposons.

Only a few LTR retrotransposons in any genome are intact, a majority of them being inactivated by different mutations. These may be due to intrinsically error-prone, reverse transcription by RT, to mutation associated with the nucleotide substitution rate of the specific genome, to mutation associated with methylation of CpG dinucleotides of the element, to insertional mutation by other transposable elements, or to recombination and deletion. Solitary LTRs usually far outnumber full-length elements in the genome and are understood to be the result of homologous recombination between the 5′ and 3′ LTRs of a full-length element, with excision and loss of the intervening sequence. Over time, older transpositionally active elements of a retrotransposon lineage are inactivated by mutation, becoming functionally extinct. Younger functionally intact members of the lineage continue the retrotransposition activity, or if none exist then that lineage becomes extinct. Transposition-incompetent elements in the genome can be viewed as molecular fossils of past transposition activity.

Transposition mechanism

Most LTR-retrotransposons are believed to replicate and transpose using a complex process very similar to that of infectious retroviruses, as covered elsewhere in this encyclopedia. In the DIRS1 group, the presence of a Y-transposase and lack of target site duplications suggests a different mechanism, involving RT-mediated synthesis of a closed circular cDNA, and insertion into the target using Y-mediated recombination.

Non-LTR Superfamily

Retrotransposons included in this superfamily are the autonomous LINEs and nonautonomous SINEs and SVA elements. One clade of LINEs is the LINE-1 clade (L1), and to minimize confusion, the general category of 'LINE' retrotransposons will be referred to hereafter as LINE-like.

LINE-like retrotranposon structure

Multiple clades of LINE-like retrotransposons have so far been recognized on the basis of RT and other protein sequence comparisons. At least nine clades are represented in vertebrate genomes. Common structural features of LINE-like retrotransposons include a 5′ untranslated region (UTR) containing a (G+C)-rich promoter region, followed by two nonoverlapping open reading frames (ORF1 and ORF2), and a 3′ UTR.

However, different clades are distinguished by many structural differences. A brief description of the 3′ UTR and ORF differences between the first three identified vertebrate LINE-like clades (LINE-1 (L1), CR1, and RTE) is given below to convey a sense of the structural features distinguishing LINE-like retrotransposons.

Variable-sized target-site duplications flank full-length elements of the L1 and RTE lineages, but not the CR1 lineage. The 3′ UTR of L1 characteristically contains an AATAAA polyadenylation signal and terminates in a 3′ poly (A) tail. In contrast, the 3′ UTR of CRI elements contains no A or AT-rich regions and terminates in a [(CATTCTRT) (GATTCTRT)$_{1-3}$] motif, while the very short 3′ UTRs of the RTE clade are dominated by A/T-rich trimer, tetramer, and/or pentamer repeats.

In the L1 clade, ORF1 encodes a single-stranded, nucleic acid-binding protein with nucleic acid-chaperone activity. A similar activity is proposed for the ORF1 product of CR1, while the 43-amino-acid ORF1 of RTE is thought to be too short to encode these components. In most LINE-like clades, ORF2 encodes a large protein with an N-terminal endonuclease and a C-terminal RT domain. As in most LINE-like clades, the endonuclease is of the apurinic/apyrimidinic type (AP) in L1, CR1, and RTE elements. A site-specific endonuclease typifies a minority of clades. A third domain in the ORF2-encoded protein, a COOH-terminal cysteine-rich domain, is conserved in mammalian L1 elements.

As with LTR retrotransposons, most LINE-like retrotransposons in a genome are transpositionally disabled. For example, of the half million or so L1 copies in the human genome, less than 1% are intact full-length elements. In all LINE-like lineages, 5′ truncation is an exceptionally common mutation, likely reflecting the rarity of successful full-length element reverse transcription and insertion. An interesting speculation is that, by providing opportunity for L1 to acquire novel 5′ ends, 5′ truncation may be an evolutionary strategy for L1 to evade suppression of transcription by the host.

LINE-like retrotransposon transposition

LINE-like retrotransposons are thought to all use a target-primed mechanism of transposition, reverse transcribing their RNA directly on the new chromosomal integration site, although the details likely vary between the different clades and are not well understood for all. The human L1 is perhaps the best-studied vertebrate LINE-like retrotransposon and is used here to outline the process. Genetic evidence and recent cell culture experiments suggest the following model (**Figure 2**). Following transcription, nuclear export, and translation of the L1 mRNA in the cytoplasm, L1 proteins preferentially associate with their encoding transcript ('*cis* preference') forming a cytoplasmic ribonucleoprotein particle. Upon access of the particle to the nucleus, L1 endonuclease nicks the host DNA at a loose consensus sequence (3'-AA/TTTT-5') in the minus strand. The L1 transcript poly (A) tail base pairs with the nicked target and the target thus primes first strand synthesis using L1 RT with L1 mRNA as template. A nick in the target plus strand, offset from the minus-strand nick, is made by an unknown enzyme, possibly L1 endonuclease. Second strand synthesis proceeds using the first strand as template, perhaps primed by microhomology-mediated priming between the first strand and the 3' end of the plus strand. Host enzymes are thought to be involved in completion of synthesis of both new strands and DNA repair linking them to the host DNA, creating the flanking, perfect target-site duplications typical of L1. L1 uses host ribonuclease H to degrade L1 RNA template, but other clades of LINE-like retrotransposons encode their own RNase H.

SINE structure

SINEs are small nonautonomous retrotransposons, usually derived from different cellular structural RNAs, such as 7SLRNA or 5S rRNA; however, most eukaryotic SINE families are derived from tRNAs. A SINE generally consists of a 5' structural RNA-like region containing an internal polymerase III promoter, a region unrelated to structural RNA, and a 3' end derived from a LINE-like retrotransposon in the same genome. An exception is Alu elements, whose 3' ends are not shared with L1, their autonomous partner, unless one counts the poly (A) tail. SINEs are usually flanked by target site duplications typical of the element from which their 3' end is derived.

SVA structure

SVA elements, exclusively found in hominoid primates, are chimeric elements named for their principal components, SINE-R, VNTR, and Alu. A 5' (CCCTCT) hexamer repeat is followed by an antisense Alu sequence and then, multiple copies of a variable number tandem repeat (VNTR). The 3' portion consists of the primate-specific SINE-R, derived from an LTR-retrotransposon thought to be human endogenous retrovirus K-10 (HERVK-10), and finally a polyadenylation and cleavage signal and poly (A) tail.

Transposition of SINE and SVA

Experimental evidence from several vertebrate species indicates that, as long suspected, the 3' end of SINE transcripts is recognized *in trans* by the cognate LINE-like machinery for transposition. For example, transcripts of Alu elements, and the mouse SINEs, B1, B2, and ID, are recognized and transposed by L1 machinery, whereas UnaSINE1, an eel SINE, is retrotransposed by UnaL2. Similarly, the characteristics of SVA insertions strongly suggest they use the L1 machinery for retrotransposition.

Penelope-Like Elements

Structure

In a few species of vertebrates, a limited number of elements with intact ORFs resembling the Penelope element of *Drosophila virilis* have been described. The elements are flanked by short target site duplications, and usually consist of LTRs flanking anINT. The LTR sequences do not resemble those of LTR retrotransposons, and are thought

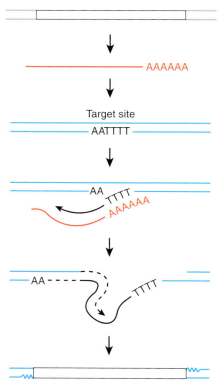

Figure 2 Retrotransposition mechanism of L1 elements. Polyadenylated L1 transcripts (red line) are transported to the cytoplasm, packaged in their own translation products, and then transported back to the nucleus. In the nucleus, L1 endonuclease nicks the minus strand of the new insertion site (blue line) at the target consensus sequence 5' TTTT/AA 3'. The T-rich 3'-OH minus strand primes reverse transcription and first strand synthesis (black line). A nick in the plus strand, staggered in position to the minus strand nick, exposes 3' plus strand target DNA which is thought to prime L1 second strand transcription (black broken line) using the first strand as template, after RNA degradation by RNase H and strand switching by RT. Finally, host enzymes are thought to participate in repairing the gaps in the host DNA, creating target-site duplications flanking the L1 (zig-zag lines indicate duplicated host DNA).

to represent tandem arrangement of two copies of the element with variable 5′ truncation of the upstream copy. The upstream LTR may be preceded by an inverted LTR fragment. The single ORF includes an N-terminal domain containing a conserved DKG amino acid motif, followed by the RT domain, a variable length linker sequence thought to contain a nuclear localization signal, and an endonuclease domain. The endonuclease is of the GIY-YIG type, otherwise unreported in eukaryotes.

Transposition

It is not clear how PLEs are transposed. The presence of introns in genomic copies of some PLEs found in invertebrates, and their absence from cDNA of the element, argues against an L1-like 'cis preference' action of PLE proteins. Other possibilities include unconventional transposition of full-length unspliced mRNA, or use of a DNA template for transposition.

Evolutionary Features

Origin

Phylogenetic analyses of autonomous retrotransposable elements have historically relied on amino acid sequence alignments of RT, the only protein coding domain common to all elements. The evidence indicates that LINE-like retrotransposons are divided into 17 or more clades, many of which have wide distribution in eukaryotes. These analyses, and abundant representation of LINE-like retrotransposons in basal eukaryote genomes, support the origin of ancestral LINE-like retrotransposons close to the emergence of the eukaryotic crown group in the Proterozoic eon. Combined analyses of RT, RNaseH, and endonuclease domains suggest that ancestral LINE-like retrotransposons evolved from group II introns in genomes of eubacteria, and fungal and plant organelles, and originally possessed a single ORF for RT and a site-specific endonuclease. Early branching lineages such as L1 acquired a second ORF (ORF1), and the site-specific endonuclease was replaced by a relatively non-site-specific AP endonuclease probably acquired from the host DNA repair machinery. Later branching lineages also acquired an RNaseH domain, most likely from their eukaryotic host, but elements within these lineages have not all retained this domain. Most evidence suggests that LINE-like retrotransposon transmission is strictly vertical through the germline. Evidence supporting horizontal transmission of RTE from snakes to ruminants has been reported, although the mechanism is unknown, and the topic remains controversial.

Phylogenetic studies of LTR elements, together with their absence from some basal eukaryotic genomes, suggest they arose more recently than LINE-like retrotransposons, although this is not a settled topic. It has been speculated that LTR retrotransposons arose as a chimera between a non-LTR retrotransposon carrying RT and RNaseH and a DNA transposon carrying an integrase domain. The acquisition of additional ORFs distinguishes many clades. Evidence suggests the vertebrate retrovirus group acquired a second RNaseH domain, and the primary sequence of the first degenerated, maintaining some structural information but not the catalytic site. The acquisition of Env, enabling infectious transfer is a striking feature of some lineages within the vertebrate retroviral group, although the origin, potentially ancient, of Env in this group is obscure. Strong phylogenetic evidence from invertebrate genomes indicates that other LTR element groups independently acquired different Env-like genes from infectious viruses, but vertebrate representatives of these are yet to be described. The evolutionary success of the Chromovirus clade of the Ty3/*gypsy* group, found in plants and fungi as well as vertebrates, has been attributed to its acquisition of the chromodomain. As with LINE-like retrotransposons, LTR elements have had varying success in colonizing vertebrate genomes, although all major LTR groups are represented by an active element in at least one species of vertebrates (**Table 1**).

New autonomous endogenous LTR elements are acquired through horizontal transfer of exogenous elements, which can then invade the genome and be transmitted through the germline, blurring the distinction between exogenous and endogenous elements. An example of current interest is the ongoing infectious epidemic and endogenization of the Koala retrovirus. However, some genera of the vertebrate retrovirus group, such as the lentiviruses, appear incapable of generating endogenous elements. This appears to be the result of Env mutations that disable their ability to infect germ cells. New nonautonomous elements arise through recombination, and all are transmitted in the germline. In general, LTR elements seem to be active within a genome over short evolutionary scales relative to LINE-like lineages.

As previously mentioned, SINEs appear to have arisen from structural RNA sequences. The primate Alu, mouse B1, and related families were derived from 7SL RNA. However, almost all other SINEs are derived from tRNA, and are placed as an evolutionarily older family than the 7SL-derived group. A new SINE-like family of diverse low-copy-number species- or lineage-specific retrotransposons derived from small nucleolar RNA was recently described in vertebrates.

Construction of a RT-based PLE polygeny is problematic since PLE RT differs from all of other retroelement RTs and more closely resembles telomerase. This, together with their distinct structure, has led to general agreement that PLEs form a separate group from the LTR and non-LTR retrotransposons. A relatively early origin for PLEs is supported by grouping of PLEs found in many eukaryotic genomes. Degenerate, and full-length PLEs

Table 1 Retrotransposon clades identified in some vertebrate genomes

Type of retrotransposon	Cartilaginous fish[a,b]	Teleost fish[c] (Tn, Tr, Dr)	Mammals[d] (Hs, Mm, Cf)	Birds (chicken)[e]
Non-LTR retrotransposons[f]	2.15	0.78, 1.32, 0.39	20.42, 19.2, 16.49	6.5
Restriction-enzyme like				
NeSL	n.a.	0.08, 0.01, <0.01	n.d., n.d., n.d.	n.d.
R2	n.a.	n.d., n.d., <0.01	n.d., n.d., n.d.	n.d.
R4	n.a.	Fossils,[e] 0.09, n.d.	n.d., n.d., n.d.	n.d.
Apurinic/apyrimidinic				
R1	*	n.d., n.d., n.d.	n.d., n.d., n.d.	n.d.
L1/TX1	*	0.03, 0.06, 0.02	16.9, 18.8, 14.5	n.d.
RTE/Rex3[g]	n.a.	0.18, 0.39, 0.2	n.d., n.d., n.d.	n.d.
L2/Maui	***	0.04, 0.53, 0.11	3.22, 0.38, 1.84	0.1
L3/CR1[h]	***	n.d., n.d., n.d.	0.31, 0.05, 0.15	6.4
Rex1/Babar	***	0.45, 0.25, 0.05	n.d., n.d., n.d.	n.d.
I/Bgr	n.a.	0.01, fossils, 0.01	n.d., 0.01, n.d.	n.d.
Jockey	*	n.d., n.d., n.d.	n.d., n.d., n.d.	n.d.
LOA	*	n.d., n.d., n.d.	n.d., n.d., n.d.	n.d.
LTR retrotransposons	0.1	0.12, 0.30, 0.40	8.29, 9.87, 3.25	
Vertebrate retroviruses	*	0.03, 0.09, <0.01	8.29, 9.87, 3.25	1.3
TY1/Copia	n.a.	0.02, 0.01, <0.01	n.d., n.d., n.d.	n.d.
TY3/Gypsy[j]	*	0.06, 0.17, 0.13	n.d., n.d., n.d.	n.d.
BEL	*	Fossils, 0.02, 0.01	n.d., n.d., n.d.	n.a.
DIRS1[h]	n.a.	0.02, 0.01, 0.25	n.d., n.d., n.d.	n.a.
Penelope-like elements	n.a.	0.06, 0.09, <0.01	n.d., n.d., n.d.	n.a.
SINEs[h]	1.36		13.14, 8.22, 9.12	Fossils

[a]The estimated percentage of the genome occupied by the respective elements is shown. Where numeric estimates are unavailable, *indicates detected at low frequency, ***indicates detected at high frequency. There are some differences in calculation of the estimates as indicated below.
[b]*Callorhincus milii*. The estimates are based on analysis of 18 Mb of random sequence from this fish; copy number is uncertain and additional retrotransposon clades may be identified when the complete sequence is available. From Venkatesh B, Tay A, Dandona N et al. (2005) A compact cartilaginous fish model genome. *Current Biology* 15: R82–R83.
[c]Teleost fish considered here include *T. nigroviridis*, *T. rubripes*, and *D. rerio* (Tn, Tr, Dr respectively); the percentage is the percentage of RT gene-containing sequence from whole genome shotgun sequences. From Volff JN, Bouneau L, Ozouf-Costas C, and Fischer C (2003). Diversity of retrotransposable elements in compact pufferfish genomes. *Trends in Genetics* 19: 674–678.
[d]Mammals here include human, mouse and domestic dog (Hs, Mm, Cf, respectively) From Waterston RH, Lindblad-Toh K, Birney E, et al. (2002) Initial sequencing and comparative analysis of the mouse genome. *Nature* 420: 520–562; Lander ES, Linton LM, Birren B, et al. (2001) Initial sequencing and analysis of the human genome. *Nature* 409: 860–921; Kirkness EF, Bafna V, Halpern AL, et al. (2003) The dog genome: Survey sequencing and comparative analysis. *Science* 301: 1898–1903.
[e]From Hillier LW, Miller W, Birney E, et al. (2004) Sequence and comparative analysis of chicken genome provide unique perspectives on vertebrate evolution. *Nature* 432: 695–716.
[f]Non-LTR retrotransposons are subdivided into a phylogenetically older group encoding a restriction enzyme-like endonuclease, and a younger group with an apurinic/apyrimidinic endonuclease.
[g]Members of this clade detected in reptiles.
[h]Members of this clade detected in reptiles and amphibians.
[i]Fossils of this clade detected in reptiles, potentially active elements detected in amphibians.
n.a. indicates data not known. n.d. indicates the element was not detected after searching. Fossils are elements deemed to be extinct, having lost the means of retrotransposition or lacking evidence of recent retrotransposition.

have been reported from fungi and a wide range of invertebrates. In vertebrates, they have so far been found in teleost fish, sharks, and amphibia, not always as degenerate molecular fossils. It is possible that some are still active in the fish *Tetraodon nigroviridis* and *Danio rerio*.

Retrotransposons as Phylogenetic Markers

At any point in time, relatively few copies of a retrotransposon are capable of replication and retrotransposition. As these copies accumulate mutations, the mutations are inherited by subsequent members of the retroelement lineage in the host genome. Once the host lineage splits, new insertions will occur independently in each descendant lineage. Since stably integrated elements are identical by descent, and the probability of parallel independent insertions into a genome is low, retrotransposons can be considered to be homoplasy-free characters, offering unique utility as markers to study evolution of host species. Retrotransposons have been used in several species of fish, mammals, birds, and reptiles, to clarify phylogenetic relationships. As suggested above, these analyses rest on the assumption that, in general, retrotransposons integrate randomly into genomes, with an exceptionally low probability that an element would independently insert in orthologous positions in two species. Another assumption is that, unless removed by segmental deletion, an insertion remains in its locus once fixed in the genome, eventually becoming unrecognizable as a result of accumulated mutations. Evidence that these assumptions do not always hold indicates that, as with analyses based on other genome elements, care is required in the conduct and interpretation of phylogenetic analyses based on retroelements.

Retrotransposon Effects on the Genome

Transposable elements have profoundly affected the structure and function of vertebrate genomes in many different ways.

Retrotransposon Content

Vertebrate genomes differ markedly in the total quantity and diversity of retrotransposons they contain, and the evolutionary trajectory of different elements. Genome size has been correlated to the quantity of transposable elements in the genome. Retrotransposons occupy approximately 35–50% or more of mouse, human, and domestic dog genomes. The chicken genome is approximately 39% of the size of mouse and human genomes, but only about 8% of the genome is recognizable retrotransposable elements. In the very compact genomes of smooth pufferfish *T. nigroviridis* and *Takifugu rubripes*, roughly 12% the size of mouse and human genomes, retrotransposable elements occupy less than 5% of the DNA. The enormous retrotransposon copy number accumulation in species with large genomes suggests that these species lack some constraint on retrotransposon activity that is present in animals with small genomes. It has been suggested that retrotransposons physically organize the genome through higher order chromatin structuring, provision of dispersed regulatory units, and other means. Thus, differences in genome retrotransposon content could significantly affect the operation of different genomes.

The pufferfish genomes, considering their small size, contain a remarkable diversity of retrotransposons in comparison with mammals. In mammalian genomes, 4 LINE-like clades have been identified (L1, L2, RTE, and CR1) with L1 the major currently active element, while in pufferfish genomes seven clades have been identified (NeSL, R4, L1, RTE, I, L2, Rex1) most of which have been recently active. Within the L1 clade alone, a single lineage has dominated L1 activity in mouse and human genomes since the mammalian radiation and comprises about 20% of the genome, whereas multiple L1 lineages predating the mammalian L1 emergence are active in several fish species, although present in very low copy numbers. All the major groups of LTR retrotransposons are represented in pufferfish genomes, as are PLEs. In contrast, only the vertebrate retrovirus group (endogenous retroviruses) and a few molecular fossils of the Ty3/Gypsy and DIRS1 groups are present in the mouse and human genomes, and there is no evidence for the presence of PLEs. The zebrafish, *D. rerio*, with a larger genome than the pufferfish, also hosts a great variety of retrotransposons. Whether these obvious differences in retrotransposon evolutionary biology between the three fish genomes and the mouse/human genomes represent the general case between teleosts and mammals is as yet unknown. Birds, as exemplified by the chicken, are different again. A single LINE-like retrotransposon, CR1, comprises about 90% of all identified chicken retrotransposons, L2/MIRs and endogenous retroviruses equally comprise the remainder. Curiously, the chicken genome lacks SINEs although it contains faint remnants of ancient SINEs pre-dating the bird-mammal split. A variety of retrotransposons have been reported from reptile genomes, some revealing lineage-specific retrotransposons such as the Sauria SINE derived from a LINE-like element of the RTE clade.

Mammalian genomes *per se* can differ markedly from one another in their retrotransposon content and activity, reflecting the evolutionary trajectory of different elements. For example, LINE, SINE, and LTR elements occupy similar percentages of mouse and human genomes. However, endogenous retroviruses are almost extinct in humans, while multiple families of endogenous retroviruses are active in rodents. The L1 lineage appears to be still active in most mammalian genomes, but recent evidence indicates its extinction in several tribes of sigmodontine rodents at or after their divergence from the earliest extant genus. As would be predicted, L1 extinction was linked to extinction of B1, a rodent SINE thought to be transposed by L1. Unexpectedly, vigorous expansion of an endogenous retrovirus, MysTR, to very high copy numbers unprecedented in any other endogenous retrovirus group, was also linked to L1 extinction. Whether there is any relationship between L1 activity and endogenous retrovirus activity is unknown.

Genomic Distribution of Retrotransposons

Diverse patterns of retrotransposon distribution in genomes are dependent on the type of element and the host genome. In the mouse and human, retrotransposons are generally dispersed widely through the genome. In contrast, retroelements strongly cluster in heterochromatic gene-poor regions in *T. nigroviridis*, similar to the distribution in *Drosophila*. Whether this extremely uneven distribution is specific to small genomes, or teleosts, is unknown, and how and why it might arise is the subject for some speculation.

At a smaller scale, much variation is evident within genomes. The density of retrotransposons varies among different chromosomes in individual vertebrate species, and in mammals is usually highest on the X and Y chromosomes. A higher density of L1 on X chromosomes than autosomes is also observed in the Ryuku spiny rat, *Tokudaia osimensis*, in which both males and females have an XO karyotype, arguing against the hypothesis that evolutionary selection of a high density of L1 on X is due to involvement of L1 in X-inactivation. Although LTR elements are distributed more or less uniformly in mouse and human genomes, L1 occurs at much higher density in gene-poor, AT-rich regions and the human SINE, Alu, occurs at high density in GC-rich regions. Closer analysis demonstrates young Alus preferentially occur in AT-rich regions, whereas in older Alus a stronger bias toward GC-rich regions emerges. One of several possible explanations for this skewed distribution of LINEs and Alus is selective targeting of L1 and Alu to AT-rich regions, and subsequent positive selection for Alus in gene-rich GC-rich regions. A significant antisense bias observed for many older intact LTR elements and solitary LTRs located in mammalian introns, is thought to arise from negative selection of sense-oriented elements bearing strong splice acceptor or donor sites, or strong transcriptional regulatory function.

Effect on Genes

Through their specific exaptation for use by the host, or through incorporation within genes, retrotransposons significantly contribute to gene evolution, and some examples follow. Sequences from LTR elements alone occupy about 1.5% of mouse and 0.8% of human genes, and genes containing these elements tend to be newly evolved genes. Indeed, LTRs drive developmentally regulated expression of cellular genes in early mouse embryos, perhaps an evolving example fitting with the hypothesis that randomly distributed retrotransposons provide a means to set up, over evolutionary time, co-ordinated transcriptional regulatory circuits. An interesting example of exaptation is the independent selection by sheep, primate, and rodent lineages of Env expression from lineage-specific endogenous retroviruses for function in placental syncytiotrophoblast morphogenesis. At least one ultra-conserved sequence in mammalian genomes has proved to be an ancient SINE whose multiple insertions, predating the divergence of amniotes and amphibians, have been exapted for use in transcriptional regulation or as a conserved exon in multiple unrelated genes.

New insertions can also disrupt normal gene function through interference with transcriptional regulation, through physical alteration of transcripts by aberrant splicing or premature termination, and through exon shuffling by inadvertent transduction of non-retrotransposon sequences. While L1 generates processed pseudogenes in the human genome, the paucity of processed pseudogenes in the chicken genome indicates that CR1 does not, suggesting that its retrotransposition machinery does not recognize mRNA. Thus, different vertebrate LINE-like retrotransposons may differ in their effects on different genomes.

Retrotransposons also act as substrates for recombination, fostering genomic instability, and are involved in both duplications and deletions within genomes, and other structural rearrangements. For example, in the human genome, SVA elements are reported to be associated with about 53 kbp of genomic duplications in the human genome, including duplication of entire genes and the creation of new genes. Recently, recombination hot spots in mouse and human genomes were linked with a sequence, CCTCCCT, found in an ancient nonautonomous LTR element (THE1) in humans. The LTR element itself is not recognizable in mouse, likely having been mutated beyond recognition by the higher nucleotide substitution rate in this species. Finally, illegitimate recombination between Alu elements in primate genomes has been linked with occasional genomic deletions. However, the effect of different retrotransposons on genome integrity may vary according to the retrotransposon type and host species. For example, high karyotypic variation in sigmodontine rodent species with extinct L1 compared with those maintaining active L1 supports the notion that L1 proteins may be important for DNA break–repair in these animals. Interestingly, a single L1 element, L1_MM, was enriched in regions of low recombination activity in the mouse, and L1 is underrepresented in human recombination hot-spots.

Host Responses to Retrotransposons

The inherent tendency of retrotransposons to amplify their copy number creates mutagenic insertions potentially harmful to the host, leading to the view of retrotransposons as genomic parasites. An adaptive response by the host would selectively encourage the evolution of repressive mechanisms directed against retroelements, and this in turn would exert selective pressure on retroelements to resist repression. It has been suggested that evolution of the APOBEC3 family of cytidine deaminases, cellular inhibitors of retrotransposition of LINE-like, SINE and LTR retrotransposons, may have been driven in part as a genome response to invasion of mammalian genomes by retrotransposons. In addition to random mutation of active elements within the genome, other strategies to inhibit retrotransposon proliferation include transcriptional silencing through epigenetic modifications to chromatin and DNA, and post-transcriptional silencing through RNA interference. From time to time, retroelements escape repression and undergo expansion in the affected genome. Thus, continued activity of a retroelement in the genome, such as of endogenous retroviruses in the mouse, may indicate that a host is lagging in the evolutionary race to control the genome invader. Some retrotransposons have been astoundingly successful in certain genomes, for example, there are over 1×10^6 Alu copies in the human genome, consisting of one family with about 20 subfamilies. This, together with the low frequency of pathogenic effects for many retroelements, and the adoption by hosts of many elements for their own biology, suggest the controversial idea that many retrotransposons have formed or are forming a symbiotic rather than parasitic relationship with their hosts.

Acknowledgments

This work was supported in part by VSPHS NIH (R01HD037102).

See also: Retroviruses: General Features.

Further Reading

Aparicio S, Chapman J, Stupka E, et al. (2002) Whole-genome shotgun assembly and analysis of the genome of *Fugu rubripes*. Science 297: 1301–1310.

Craig NL, Craigie R, Gellert M, and Lambowitz AM (eds.) (2002) *Mobile DNA II*. Herndon, VA: ASM Press.

Curcio MJ and Derbyshire KM (2003) The outs and ins of transposition: From mu to kangaroo. Nature Reviews Molecular Cell Biology 4: 865–877.

Eickbush TH and Furano AV (2002) Fruit flies and humans respond differently to retrotransposons. Current Opinion in Genetics and Development 12: 669–674.

Furano AV, Duvernell DD, and Boissinot S (2004) L1 (LINE-1) retrotransposon diversity differs dramatically between mammals and fish. Trends in Genetics 20: 9–14.

Goodwin TJ and Poulter RT (2001) The DIRS1 group of retrotransposons. Molecular Biology and Evolution 18: 2067–2082.

Hedges DJ and Deininger PL (2006) Inviting instability: Transposable elements, double-strand breaks, and the maintenance of genome integrity. Mutation Research: Fundamental Mechanisms of Mutagenesis doi:10.1016/j.mrfmmm.2006.11.021.

Hillier LW, Miller W, Birney E, et al. (2004) Sequence and Comparative analysis of chicken genome provide unique perspectives on vertebrate evolution. Nature 432: 695–716.

Kazazian HH, Jr. (2004) Mobile elements: Drivers of genome evolution. Science 303: 1626–1632.

Kordis D (2005) A genomic perspective on the chromodomain-containing retrotransposons: Chromoviruses. Gene 347: 161–173.

Kirkness EF, Bafna V, Halpern AL, et al. (2003) The dog genome: Survey sequencing and comparative analysis. Science 301: 1898–1903.

Lander ES, Linton LM, Birren B, et al. (2001) Initial sequencing and analysis of the human genome. Nature 409: 860–921.

Malik HS and Eickbush TH (2001) Phylogenetic analysis of ribonuclease H domains suggests a late, chimeric origin of LTR retrotransposable elements and retroviruses. Genome Research 11: 1187–1197.

Ostertag EM and Kazazian HH, Jr. (2001) Biology of mammalian L1 retrotransposons. Annual Review of Genetics 35: 501–538.

Piskurek O, Austin CC, and Okada N (2006) Sauria SINEs: Novel short interspersed retroposable elements that are widespread in reptile genomes. Journal of Molecular Evolution 62: 630–644.

Roy-Engel AM, Carroll ML, El-Sawy M, et al. (2002) Non-traditional Alu evolution and primate genomic diversity. Journal of Molecular Biology 316: 1033–1040.

van de Lagemaat LN, Medstrand P, and Mager DL (2006) Multiple effects govern endogenous retrovirus survival patterns in human gene introns. Genome Biology 7: R86.

Venkatesh B, Tay A, Dandona N, et al. (2005) A compact cartilaginous fish model genome. Current Biology 15: R82–R83.

Volff JN (ed.) (2005) *Cytogenetic and Genome Research, Vol. 110: Retrotransposable Elements and Genome Evolution*. Basel: S Karger AG.

Voff JN, Bouneau L, Ozouf-Costas C, and Fischer C (2003) Diversity of retrotransposable elements in compact Pufferfish genomes. Trends in Genetics 19: 674–678.

Waterston RH, Lindblad-Toh K, Birney E, et al. (2002) Initial sequencing and comparative analysis of the mouse genome. Nature 420: 520–562.

Weber MJ (2006) Mammalian small nucleolar RNAs are mobile genetic elements. Public Library of Science Genetics 2: e205doi:10.1371/journal.pgen.0020205.

Relevant Website

http://www.girinst.org – Genetic Information Research Institute, Mountain View.

Retroviral Oncogenes

P K Vogt and A G Bader, The Scripps Research Institute, La Jolla, CA, USA

© 2008 Elsevier Ltd. All rights reserved.

Glossary

Chimeric transcript An mRNA that encodes a fusion protein derived from two individual and originally separate genes.

Subtractive hybridization A method to identify differentially expressed mRNAs; the method is based on hybridizing a 'tester' mRNA population with a 'driver' mRNA population and selectively eliminating tester–driver hybrids.

Historical Perspective

Retroviruses that carry an oncogene induce neoplastic transformation of cells in culture and rapidly cause tumors in the animal. Early studies with Rous sarcoma virus in chicken embryo fibroblasts showed a controlling influence of the retroviral genome on the properties of the transformed cell and suggested that the virus carries oncogenic information. The isolation of temperature-sensitive mutants of Rous sarcoma virus firmly established a dominant role of viral genetic information in the process of virus-induced carcinogenesis. These mutants of Rous sarcoma virus encode an unstable oncoprotein and are able to transform chicken embryo fibroblasts at a low, permissive temperature but fail to induce oncogenic changes at an elevated, nonpermissive temperature. Yet the virus is able to propagate under both permissive and nonpermissive conditions, demonstrating that virus viability and replication are not affected by the mutation and that oncogenicity is governed by genetic information that is distinct from viral replicative genes. Cells transformed by temperature-sensitive Rous sarcoma virus under permissive conditions become normal in morphology and growth behavior if the cell cultures are switched to the nonpermissive temperature. Thus, viral genetic information is required for the initiation as well as for the maintenance of the transformed cellular phenotype. A physical correlate to these genetic experiments was found in studies with transformation-defective Rous sarcoma virus. These spontaneously emerging variants of the virus fail to induce transformation in cell culture but are able to generate viable progeny virus that remains transformation defective. The transformation-defective viruses contain genomes that are about 20% smaller than the genome of oncogenic Rous sarcoma virus. The missing genetic information is of obvious importance for oncogenicity. Subtractive hybridization of DNA transcripts from transformation-defective and transformation-competent viral genomes generates a specific cDNA probe for the oncogenic sequences in Rous sarcoma virus. This probe hybridizes to cellular DNA, revealing the presence of sequences that are homologous to the transforming information of Rous sarcoma virus in the genome of all vertebrate cells. Retroviral oncogenes are therefore derived from the cell genome. The cellular versions of retroviral oncogenes are also referred to as proto-oncogenes or, more commonly, as c-*onc* genes; the corresponding viral versions are called v-*onc* genes. The discovery of oncogenes owes much to the exceptional genetic structure of Rous sarcoma virus. This virus is unique among oncogene-carrying retroviruses in that it is replication competent, containing a full set of viral genes plus the oncogene. Loss of the oncogene in transformation-defective variants still leaves a viable virus, greatly facilitating molecular and genetic analyses.

Taxonomy of Oncogene-Carrying Retroviruses

Retroviruses are broadly divided into two categories, viruses with simple genomes and viruses with complex genomes. Simple retroviral genomes contain four coding regions with information for virion proteins. These regions are referred to as *gag*, which directs the synthesis of matrix, capsid, and nucleoprotein structures; *pro*, generating the virion protease; *pol*, containing the information for the reverse transcriptase and integrase enzymes; and *env*, which encodes the surface and transmembrane components of the envelope protein. Complex retroviral genomes contain additional information for regulatory nonvirion proteins that are translated from multiply spliced mRNAs.

In taxonomic terms, retroviruses form a family that consists of the subfamilies *Orthoretrovirinae* and *Spumaretrovirinae*. Oncogenes are carried by only two genera of the orthoretroviruses. These are the alpharetroviruses, representing the avian leukosis complex of viruses, and the gammaretroviruses, encompassing murine, feline, and primate leukemia and sarcoma viruses. This restriction of oncogenes to two viral genera probably reflects a set of conditions that have to be met for the acquisition of cellular sequences by a retroviral genome. These conditions include absence of cytotoxicity and efficient viral integration and replication. Complex retroviruses may not tolerate the insertion of cellular sequences into the viral genome because some functions of the regulatory nonvirion proteins that would be displaced by a cellular insert can probably not be complemented in *trans*. These are all hypothetical reasons for the occurrence of oncogenes in just a few retroviruses; mechanistic explanations of these restrictions are currently not available.

Acquisition of Cellular Sequences by Retroviral Genomes

The acquisition of a cellular oncogene by a retrovirus is a rare event that occurs during viral passage in an animal but is seldom seen in cell culture. Because *de novo* acquisition of cellular sequences cannot be reproduced with significant frequency in cultured cells, the molecular mechanism of acquisition has to be reconstructed using information derived from the structure of viral genomes and from the virus life cycle. The life cycle of retroviruses contains two steps that leads to genetic recombination. (1) Retroviruses are diploid and thus can form heterozygous viral particles. Recombination between distinct but related viral genomes encased in the same particle occurs very frequently and probably results from copy-choice events that occur during reverse transcription. (2) Integration of the provirus into the cellular genome produces a recombinant between virus and cell (**Figure 1**). These two recombinational activities of retroviruses can explain the incorporation of cellular sequences into the viral genome. The first step in this acquisition consists of the integration of a provirus containing a single 5′ (left-hand) long terminal repeat (LTR) into the oncogene proper or into the immediate upstream vicinity of a cellular oncogene. This provirus can then produce a chimeric RNA transcript that starts in the viral LTR, continues by read-through into the cellular gene, and terminates with the poly A stretch of the cellular gene. Alternatively, such a chimeric transcript can be generated by a splicing event that uses a viral splice donor and a cellular splice acceptor in joining upstream viral to downstream cellular sequences. Chimeric RNAs of this type would be incorporated into virions. In the subsequent cycle of infection, reverse transcription could effect a second recombination event, generating a junction between the 3′ region of the cellular sequences and a part of the viral genome carrying the 3′ terminal repeat sequences that are essential for the efficient production of proviral DNA. In this model, the first step of recombination in acquiring cellular sequences is the integration of the provirus, which is DNA-based recombination. The second step occurs during reverse transcription of a heterozygote particle and is RNA-based recombination. It is possible, however, to envisage a second recombination step that is also DNA based. It requires the integration of another provirus immediately downstream of the cellular oncogene to serve as donor of the necessary 3′ (right-hand) LTR for the new provirus that now carries an insertion of cellular sequences. Currently available experimental evidence can be adduced to support either model for the second recombination step in the acquisition of a cellular oncogene. Specific oncogene acquisitions may in fact occur by either mechanism.

Retroviruses Carrying an Oncogene Are Replication Defective

The incorporation of cellular sequences into the retroviral genome occurs at the expense of viral sequences that are displaced in the process. With one notable exception, Rous sarcoma virus, transducing retroviruses lack one or several essential viral genes. Typical deletions extend from within the viral *gag* gene into the *env* gene, eliminating the 3′ portion of *gag*, all of the *pol* gene, and part of *env*. Such defective viruses can infect cells, integrating into the cellular genome, producing the oncoprotein, and inducing neoplastic transformation, but they are unable to synthesize infectious progeny virus. For infectious virus production, they require co-infection of the same cell with a closely related helper retrovirus that in *trans* provides those replicative functions that are missing from the defective transforming virus. Helper retroviruses contain a complete set of viral genes but have not incorporated any cellular oncogenic sequences. They do not induce oncogenic transformation when replicating in cultured cells. In animal infections, they can cause tumors by insertional mutagenesis after extended latent periods. These tumors result from transcriptional upregulation of a cellular oncogene by promoter activities of a provirus integrated nearby.

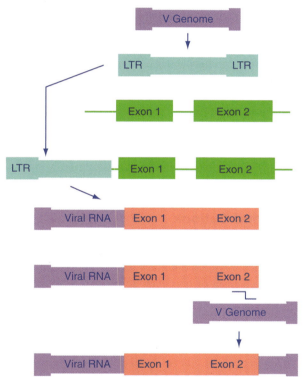

Figure 1 A hypothetical mechanism for the acquisition of a cellular oncogene by a retroviral genome. The genome of a retrovirus without oncogene is transcribed into DNA and, in a first recombinational event, is integrated upstream of a cellular oncogene containing two exons. The right-hand long terminal repeat (LTR) and adjacent sequences of the viral genome are lost during the integration process. Transcription from the integrated proviral LTR generates a chimeric mRNA by read-through or aberrant splicing. This mRNA contains cell-derived oncogene and viral information. The chimeric mRNA is packaged into viral particles together with wild-type viral genomes. In the next round of infection the second recombinational event occurs during reverse transcription and adds 3′ viral terminal sequences to the DNA transcript, facilitating the production and integration of a functional provirus that contains a cell-derived oncogene. Green: DNA, red and purple: RNA.

Oncoproteins Are Often Fusion Proteins

As a consequence of the initial integration event that leads to oncogene acquisition, viral oncogenes usually code for fusion proteins, consisting of an N-terminal portion derived from the virus and a C-terminal component representing the oncoprotein proper. The viral sequences generally include the N-terminus of the Gag polyprotein. The initial recombination event also often eliminates short lengths from the N-terminus of the cellular oncoprotein, except when viral sequences are spliced onto the cellular gene. Fusion of the cellular oncoproteins to viral *gag* sequences can have important consequences that contribute to the gain of function associated with the viral oncogene. The efficiency of protein translation and the stability of the fusion protein are often increased. Gag sequences also provide an affinity for the plasma membrane which can be critical in activating the oncogenic potential of the protein.

Functional Classes of Oncogenes

Retroviral oncogenes code for components of cellular growth-regulatory signals (**Table 1**). The major functional categories include growth factors, receptor and nonreceptor tyrosine kinases, serine-threonine and lipid kinases, adaptor proteins, hormone receptors, and a variety of transcriptional regulators. Cellular growth signals are propagated from the cell periphery to the nucleus. Therefore, the nuclear oncoproteins coding for transcriptional regulators are the ultimate effectors of oncogenicity, converting the signal into a pattern of gene expression that is the basis of the oncogenic phenotype of the cell. Because of this pivotal role of nuclear oncoproteins in specifying neoplastic properties, oncogenic transformation can be viewed as a case of aberrant transcription. However, there is also abundant evidence for a critical role of protein translation in oncogenesis. Oncogenesis induced by the PI3K pathway in particular depends on differential translation of specific growth-promoting proteins.

Retroviral Transduction of a Cellular Oncogene Results in Gain of Function

Compared to their cellular counterparts, retroviral oncoproteins show an increase in activity. Some of this gain of function is purely quantitative: the viral promoter assures highly efficient transcription of the oncogene. There are several oncogenes for which mere overexpression is sufficient to turn them into effective agents of neoplastic transformation. These code for wild-type oncoproteins that deregulate cellular growth by their virus-mediated abundance. However, many oncoproteins carry specific mutations that are responsible for the gain of function. These mutations remove or inactivate domains of the oncoprotein that effect negative regulation or may enhance specific enzymatic activities of the oncoprotein by other mechanisms including conformational change, improved substrate affinity, or change in cellular localization. They may stabilize the protein or alter the spectrum of downstream targets. Thus both quantitative and qualitative changes resulting from viral transduction and the associated mutation of the oncogene contribute to the gain of function.

Cooperation of Oncogenes

A few retroviruses carry two oncogenes. These oncogenes are derived from distinct cellular loci situated at distant positions within the host genome. Retroviruses with two oncogenes include Avian erythroblastosis virus R (AEV-R), carrying v-*erbA* and v-*erbB*, Avian myeloblastosis–erythroblastosis virus E26, encoding *myb* and *ets* as a single Gag-myb-ets protein product, and Avian myelocytoma virus MH2, which contains the v-*myc* and v-*mil* (raf) genes. Each of these single oncoproteins can induce neoplastic transformation on its own. However, viruses with two oncogenes are more potent in transformation, indicating that the two oncoproteins cooperate. Whereas primary avian cells can be readily transformed by viruses carrying a single oncogene, in mammalian cells oncogenic transformation generally requires the cooperation of two or more oncogenes. The reasons for this difference in cellular susceptibility to oncogene action are not known.

Coda and Outlook

There are still significant gaps in our knowledge of retroviral oncogenes. We do not know and cannot reproduce the exact mechanism of oncogene acquisition by a retroviral genome. We have only tentative explanations for the failure of some groups of retroviruses to capture cellular sequences. We have no idea why certain growth-promoting cellular genes have not shown up as retroviral oncogenes. Have not enough retroviruses been studied or is there an active mechanism that excludes certain genes? Does some cryptic homology between provirus and oncogene determine the spectrum of genes that can be incorporated? On a more basic level, there is evidence that oncogenes can induce tumors by activating the transcription of specific microRNAs. Are there also rapidly oncogenic retroviruses that carry a micro-RNA gene as an oncogene?

Despite these puzzling questions, at a fundamental level the nature and workings of retroviral oncogenes are well understood and have provided important insights into the mechanisms of virus-induced carcinogenesis. The discovery that all retroviral oncogenes are derived from cellular information has greatly expanded the significance of these genes. Originally seen as viral

Table 1 Classes of retroviral oncogenes

Functional groups and oncogenes	Identity and function of cellular homolog	Retrovirus
Growth factor		
sis	Platelet–derived growth factor (PDGF)	Simian sarcoma virus
Receptor tyrosine kinases		
erbB	Receptor of epithelial growth factor (EGF)	Avian erythroblastosis virus
fms	Receptor of colony-stimulating factor 1 (CSF-1)	McDonough feline sarcoma virus
sea	Receptor of macrophage-stimulating protein (MSP)	Avian erythroblastosis virus S13
kit	Hematopoietic receptor of stem cell factor (SCF)	Hardy–Zuckerman 4 feline sarcoma virus
ros	Orphan receptor tyrosine kinase	Avian sarcoma virus UR2
mpl	Hematopoietic receptor of thrombopoietin	Mouse myeloproliferative leukemia virus
eyk	Closest homolog of mammalian c-*mer*; c-Mer ligands include anticoagulation factor protein S and the growth arrest-specific gene product Gas6	Avian retrovirus RPL30
Hormone receptor		
erbA	Thyroid hormone receptor	Avian erythroblastosis virus
G proteins		
H-*ras*	GTPase; MAPK signal transduction	Harvey murine sarcoma virus
K-*ras*	GTPase; MAPK signal transduction	Kirsten murine sarcoma virus
Adaptor protein		
crk	Adaptor protein containing SH2 and SH3 domains; PI3K/Akt signal transduction	Avian sarcoma virus CT10
Nonreceptor tyrosine kinases		
src	MAPK and PI3K/Akt signal transduction	Rous sarcoma virus
yes	Src family kinase; signal transduction	Avian sarcoma virus Y73
fps	Cytokine receptor signaling; the *fps* and *fes* oncogenes are derived from the same cellular gene	Fujinami poultry sarcoma virus
fes	Cytokine receptor signaling; the *fps* and *fes* oncogenes are derived from the same cellular gene	Gardner–Arnstein feline sarcoma virus
fgr	Src family kinase; signal transduction	Gardner–Rasheed feline sarcoma virus
abl	Cytoskeletal signaling and cell cycle regulated transcription	Abelson murine leukemia virus
Serine/threonine kinases		
mos	Regulator of cell cycle progression; required for germ cell maturation; activates MAPKs	Moloney murine sarcoma virus
raf	MAPKKK, MAPK signal transduction	Murine sarcoma virus 3611
akt	PI3K/Akt signal transduction	Murine retrovirus AKT8
Lipid kinase		
p3k	PI 3-kinase; PI3K/Akt signal transduction	Avian sarcoma virus 16
Transcriptional regulators		
jun	bZIP protein of AP-1 complex; homo- and heterodimer with AP-1 family members; cell cycle progression	Avian sarcoma virus 17
fos	bZIP protein of AP-1 complex; heterodimer with AP-1 family members; cell cycle progression	FBJ murine osteogenic sarcoma virus
myc	bHLH-ZIP protein; heterodimer with Max; cell cycle progression	Avian myelocytoma virus MC29
myb	HTH protein; development of hematopoietic system	Avian myeloblastosis virus

Continued

Table 1 Continued

Functional groups and oncogenes	Identity and function of cellular homolog	Retrovirus
ets	HTH protein; myeloid and eosinophil differentiation	Avian myeloblastosis–erythroblastosis virus E26
rel	p65 NF-κB subunit; survival pathways	Avian reticuloendotheliosis virus
maf	bZIP protein; homo- and heterodimers with various bZIP proteins; differentiation of various tissues	Avian musculoaponeurotic fibrosarcoma virus
ski	Adaptor protein for various transcription factors; chromatin-dependent transcriptional regulation; muscle differentiation	Avian Sloan-Kettering retrovirus
qin	Avian homolog of mammalian brain factor 1 (BF-1/FoxG1); forkhead/winged helix (FOX) protein; monomer; neuronal differentiation	Avian sarcoma virus 31

Abbreviations: Gas6, growth arrest-specific gene 6; MAPK, mitogen-activated protein kinase; MAPKKK, mitogen-activated protein kinase kinase kinase; SH2, Src homology domain 2; SH3, Src homology domain 3; PI3K, phosphoinositide 3-kinase; AP-1, activator protein 1; bZIP, basic region leucine zipper; bHLH-ZIP, basic region helix–loop–helix leucine zipper; HTH, helix–turn–helix; NF-κB, nuclear factor kappa B; FBJ, Finkel–Biskis–Jinkins.

pathogenicity genes, they have become universal effectors of oncogenicity. Viruses have been demoted to just one of several instruments that can activate these genes; mutation, overexpression, and amplification are some of the others. The study of oncogenes, initially a somewhat esoteric part of virology, has grown to determine the course of cancer research during the past three decades and has contributed immensely to our understanding of cancer in general. As targets of specific inhibitors, oncoproteins are now revolutionizing cancer treatment. Gleevec, directed at the BCR-ABL oncoprotein in chronic myelogenous leukemia, and Iressa, inhibiting mutants of the epithelial growth factor receptor in non-small cell lung cancer, have dramatically proven the promise of therapy targeted to oncoproteins.

Acknowledgments

This work was supported by grants from the National Cancer Institute. This is manuscript number 18 394 of The Scripps Research Institute.

See also: Feline Leukemia and Sarcoma Viruses.

Further Reading

Bister K and Jansen HW (1986) Oncogenes in retroviruses and cells: Biochemistry and molecular genetics. *Advances in Cancer Research* 47: 99–188.

Duesberg PH and Vogt PK (1970) Differences between the ribonucleic acids of transforming and nontransforming avian tumor viruses. *Proceedings of the National Academy of Sciences, USA* 67: 1673–1680.

Hughes SH (1983) Synthesis, integration, and transcription of the retroviral provirus. *Current Topics in Microbiology and Immunology* 103: 23–49.

Land H, Parada LF, and Weinberg RA (1983) Tumorigenic conversion of primary embryo fibroblasts requires at least two cooperating oncogenes. *Nature* 304: 596–602.

Martin GS (1970) Rous sarcoma virus: A function required for the maintenance of the transformed state. *Nature* 227: 1021–1023.

Schwartz JR, Duesberg S, and Duesberg PH (1995) DNA recombination is sufficient for retroviral transduction. *Proceedings of the National Academy of Sciences, USA* 92: 2460–2464.

Stehelin D, Guntaka RV, Varmus HE, and Bishop JM (1976) Purification of DNA complementary to nucleotide sequences required for neoplastic transformation of fibroblasts by avian sarcoma viruses. *Journal of Molecular Biology* 101: 349–365.

Stehelin D, Varmus HE, Bishop JM, and Vogt PK (1976) DNA related to the transforming gene(s) of avian sarcoma viruses is present in normal avian DNA. *Nature* 260: 170–173.

Swanstrom R, Parker RC, Varmus HE, and Bishop JM (1983) Transduction of a cellular oncogene: The genesis of Rous sarcoma virus. *Proceedings of the National Academy of Sciences, USA* 80: 2519–2523.

Tam W, Hughes SH, Hayward WS, and Besmer P (2002) Avian bic, a gene isolated from a common retroviral site in avian leukosis virus-induced lymphomas that encodes a noncoding RNA, cooperates with c-myc in lymphomagenesis and erythroleukemogenesis. *Journal of Virology* 76: 4275–4286.

Temin HM (1960) The control of cellular morphology in embryonic cells infected with Rous sarcoma virus *in vitro*. *Virology* 10: 182–197.

Toyoshima K and Vogt PK (1969) Temperature sensitive mutants of an avian sarcoma virus. *Virology* 39: 930–931.

Varmus HE (1982) Form and function of retroviral proviruses. *Science* 216: 812–820.

Varmus HE (2006) The new era in cancer research. *Science* 312: 1162–1165.

Vogt PK (1971) Genetically stable reassortment of markers during mixed infection with avian tumor viruses. *Virology* 46: 947–952.

Vogt PK (1971) Spontaneous segregation of nontransforming viruses from cloned sarcoma viruses. *Virology* 46: 939–946.

Wang LH (1987) The mechanism of transduction of proto-oncogene c-src by avian retroviruses. *Mutation Research* 186: 135–147.

Relevant Website

http://www.ncbi.nlm.nih.gov – Virus Databases Online (ICTVdB Index of Viruses), National Center for Biotechnology Information.

Retroviruses of Insects

G F Rohrmann, Oregon State University, Corvallis, OR, USA

© 2008 Published by Elsevier Ltd.

Glossary

Baculovirus F protein The envelope fusion protein encoded by many baculoviruses that is related to the env protein of insect retroviruses.
Errantivirus A genus of insect retroviruses of the family *Metaviridae*.
Semotivirus A genus of the family *Metaviridae* containing retrotransposons and retrovirus-like elements encoding an env-like ORF.
TniTedV A virus of *Trichoplusia ni* called Ted.

Introduction to Retroelements

Retroelements are a diverse group of genetic sequences that have the capacity to be transcribed into RNA, reverse-transcribed into DNA, and inserted into a new site in a genome. A major category of these elements has long terminal repeat (LTR) sequences that are directly repeated at the 5′ and 3′ termini of the complete cDNA. These structures contain signals essential for RNA transcription and cDNA synthesis. LTR-retrotransposons are also distinguished from all other retroelements by sequence features within the reverse transcriptase region (see below) and by the nature of primers employed by this enzyme. Reverse transcriptases of LTR retroelements use the 3′ end of an annealed tRNA to prime reverse transcription. In contrast, non-LTR retrotransposons use the 3′ end of DNA to prime reverse transcription.

Major Divisions of LTR Retroelements

There are two major lineages of LTR-containing retrotransposons that have been differentiated by the domain order in their reverse transcriptases (**Figure 1**). They are the *Pseudoviridae* which have a domain order of proteinase, integrase, reverse transcriptase-RNase H (PR, IN, RT-RH), and the *Metaviridae* (PR, RT-RH, IN). The *Metaviridae* (from Greek *metathesis*: transposition) includes three genera the semotiviruses (from Latin *semoti*: distant, removed) which refers to their distant relationship to other genera of the *Metaviridae*, and the metaviruses and errantiviruses (from Latin *errans*: to wander). Three other lineages including the family *Retroviridae*, the *Caulimoviridae* (a family of DNA plant viruses with an RNA intermediate), and DIRS retrotransposons are also thought to have been derived from the family *Metaviridae* based on the relatedness of the reverse transcriptase sequences.

The Origin of Insect Retroviruses: 'Capture' of an Envelope Fusion Protein from the *Baculoviridae*

Individual members of all three genera of the family *Metaviridae* may contain an env-like open reading frame (ORF); however, both the errantiviruses and a lineage within the semotiviruses are specific to insects and appear to have originated by recombination events in which their progenitors incorporated an *env* gene from a baculovirus. The phylogenetic relationship between these different lineages of envelope proteins is shown in **Figure 2**. The two major groups of insect retroviruses are the *gypsy*-like errantivirus lineage and the *Kanga* and *Roo* semotivirus lineages. Although errantivirus and the semotivirus lineages resemble retroviruses, they have not been included within the family *Retroviridae*, because they are distinct lineages, and evidence that they are infectious is limited (see below).

Errantivirus Infectivity

The most intensively studied errantivirus is the *gypsy* element from *Drosophila melanogaster* (designated drosophila melanogaster Gypsy virus (DmGypV)) and most research has been focused on this virus. The reasons for questions regarding the infectivity of errantiviruses stem from the inability to manipulate them by employing commonly used virological techniques. Although they produce virus-like particles (vlps), the virus has not been titered, plaqued, or neutralized with antibodies, and no evidence of cytopathogenic effects on cells exposed to the virus have been documented. However, indirect evidence has been presented suggesting that DmGypV is infectious for *Drosophila*. These data were obtained by feeding a strain of *Drosophila* that lacks active DmGypV transposition with either purified vlps from insects with tranpositionally active DmGypV, or extracts derived from such insects and then documenting increased levels of transposition in the recipient insects. Similarly, it was observed that the DmGypV could be transmitted between cells in culture.

The Role of env in Errantivirus Infectivity

The evidence implicating env in errantivirus infectivity is varied. In one study, it was found that a preparation of two monoclonal antibodies against DmGypV env mixed with the vlp fraction reduced the number of insertion events in insect feeding experiments. In addition, evidence suggests that an integrated Moloney leukemia virus-luciferase construct pseudotyped with a DmGypV envelope is infectious for *Drosophila* cells. This suggested that DmGypV env is capable of mediating infection of *Drosophila* cells. Although DmGypV may be infectious, their infectivity appears to be very limited. It has been suggested that since they are adapted for integration into the cell genome, they no longer require propagation via infection. However, this does not explain why they have retained, conserved, and continue to express an *env* gene. Since envelope proteins often play major roles in the virulence of other viruses, the errantivirus *env* proteins may allow the viruses to be infectious and spread between different organisms, but this infectivity may be restricted by features of their env proteins. It may also facilitate spread between cells or tissues within an organism, although evidence has been presented that in some instances, at least, env is not required for viral movement between cells within organisms or between cultured cells.

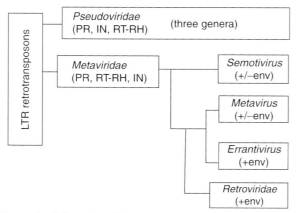

Figure 1 Schematic relationship of families of LTR retroelements based on the phylogeny of their reverse transcriptase domain. The genera of the *Pseudoviridae* are not shown. Two categories of the *Metaviridae*, the *Caulimoviridae* and DIRS retrotransposons are not included. Insect retroviruses comprise the genus *Errantivirus* and are also found in the genus *Semotivirus*. Abbreviations: PR, proteinase; IN, integrase; RT, reverse transcriptase; RH, RNase H. Not to scale.

Errantivirus Distribution

Five different categories of errantiviruses have been found in the sequenced genome of *D. melanogaster* and

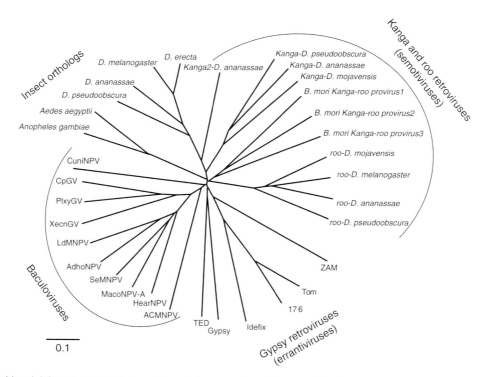

Figure 2 Neighbor-joining phylogenetic tree of baculovirus envelope fusion proteins, insect retrovirus env proteins, and orthologous proteins from insects. The tree was derived from an alignment of central domains of the proteins. Adapted from Malik HS and Henikoff S (2005) Positive selection of iris, a retroviral envelope-derived host gene in *Drosophila melanogaster*. *PLoS Genetics* 1: 429–443, with permission from HS Malik.

they range from a single full-length copy of DmeGypV, to 29 partial and 18 full-length copies of element Dme297V. DmeZamV was not found in this sequence indicating the variability of errantivirus distribution between *D. melanogaster* strains. Errantiviruses have also been identified in a number of other insect species. They are divided into two phylogenetic groups based on a primer binding site of either tRNA^{Ser} or tRNA^{Lys}, which are predicted to be used for the initiation of cDNA synthesis.

Semotivirus Distribution

Although several retrovirus-like lineages of the semotiviruses have been reported, they are distinct from the errantiviruses (**Figure 2**). An apparent insect specific semotivirus lineage with an env ORF related to the baculovirus F gene became evident when the genomes of *Drosophila* species other than *D. melanogaster* became available. The additional species include *D. ananassae, pseudoobscura, mojavensis,* and *erecta.*

Additional Relationships of Errantiviruses to Baculoviruses

The relatedness of the errantivirus env and the baculovirus F proteins may reflect more than a fortuitous recombination event between these two viruses. The errantivirus TniTedV is a mid-repetitive element (about 50 copies per genome) in *Trichoplusia ni* and is capable of transposition from the insect into the baculovirus genome. It encodes a full complement of retrovirus ORFs including an *env* gene. A key feature of the relationship that may have led to the capture of a baculovirus F gene by a pre-errantivirus retrotransposon involves the ability of baculoviruses to express genes at very high levels. This feature appears to be due at least in part to the fact that they encode an RNA polymerase capable of high levels of transcription. This polymerase recognizes a novel promoter sequence (DTAAG) that is found in the TniTedV LTR as a palindrome. When TnTedV is integrated into the viral genome, mRNA is expressed from the LTR promoters at high levels late in the baculovirus infection cycle. Therefore, integration into a baculovirus genome may reflect a strategy to exploit baculovirus late gene expression to express the integrated retrotransposon/retrovirus genome at high levels. This could result in the production of a mixture of retrovirus particles and occluded baculoviruses containing integrated retroviruses and could provide two methods of escape from an insect with a fatal baculovirus infection – they could survive either by integration into baculovirus genomes, or as infectious virus particles. The evolution of this relationship between a baculovirus and a primordial LTR-type retrotransposon provides a clear pathway, via DNA recombination, for the transposable element to incorporate the baculovirus F homolog into its genome, thereby converting it into an infectious retrovirus.

Features of Baculovirus F and Errantivirus env Proteins

The baculovirus F and errantivirus env proteins appear to be members of a group of envelope fusion proteins common to many vertebrate viruses (**Figure 3**). Although these proteins, with the exception of baculovirus F and insect retrovirus env protein, lack convincing sequence relatedness, it has long been suggested that a number of envelope fusion proteins from a variety of disparate viruses are related. This is based on their features including that cleavage is often required for their activation and the fact that the fusion protein complex is membrane associated via a transmembrane domain. In addition, the membrane-associated peptide contains a hydrophobic fusion peptide sequence downstream of the cleavage site followed by predicted coiled-coil domains that are involved in forming hairpin-like structures that are important in virus–cell fusion. Such structures have been characterized in fusion proteins from viruses as diverse as filoviruses, retroviruses, orthomyxoviruses, and paramyxoviruses. Evidence suggests that baculovirus F proteins are members of this group, and it has been demonstrated that they require cleavage for activation. Errantivirus env proteins also have similar features consistent with their being members of this group.

Cleavage of Baculovirus F and Errantivirus env Proteins

One of the most striking features of the comparison of the baculovirus F and errantivirus env proteins is the conservation of the region which includes the polybasic furin-like cleavage site and the sequence immediately downstream (**Figure 3**). This region is conserved in all baculovirus F proteins except those that use GP64 as their fusion protein. Cleavage at this site is required for the baculovirus F proteins to carry out cell fusion. Similarly, the homologous region in DmGypV env has been shown to be required for cleavage. DmGypV env has also been shown to localize to cell membranes which is a prerequisite for many envelope proteins. Therefore, the *errantivirus env* genes encode many of the motifs associated with envelope fusion proteins including the furin-like cleavage site, which is critical for activation of other fusion proteins.

Just downstream of the furin cleavage motif is a 21-amino-acid hydrophobic sequence that is the most

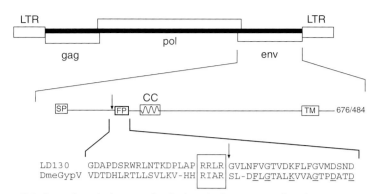

Figure 3 Comparison of predicted envelope fusion protein of a baculovirus to a predicted env protein of an errantivirus. Two representative proteins are shown: LD130 from the *Lymantria dispar* nucleopolyhedrovirus and DmeGypV env from the gypsy errantivirus of *D. melanogaster*. The diagram shows the predicted signal peptide (SP), cleavage site (arrow), fusion peptide (FP), coiled-coil region (CC), and transmembrane domain (TM). The numbers at the end are the lengths of LD130 and DmeGypV env, respectively. Beneath the diagram is the sequence surrounding the cleavage site. The boxed sequence is the furin cleavage signal. The underlined amino acids in the fusion peptide region are conserved between many baculovirus envelope fusion (F-type) proteins and errantivirus env proteins.

conserved region shared between baculovirus F and errantivirus env proteins and has the features of a fusion peptide.

A Cellular Homolog of the Insect Retrovirus *env* and *Baculovirus F* Genes

A cellular homolog (called cg4715 in *D. melanogaster*) of the insect retrovirus *env* gene has been identified in the *Sophophora* subgenus of *Drosophila* (**Figure 2**). It is preserved in the same location relative to adjacent genes in the different *Drosophila* species and its phylogeny corresponds to that of the host insects indicating that it is inherited in a strictly vertical manner rather than through horizontal transfer. The cg4715 lineage is most closely related to the *env* gene from a branch of insect retroviruses and phylogenic analyses suggest that it was incorporated by at least two independent events; once into a member of the *Drosophila* and once into a mosquito lineage (**Figure 2**). In addition, in some *Drosophila* there are two adjacent cg4715 genes indicating that a gene duplication event occurred. The second gene represents a distinct lineage that is characterized by the presence of a single predicted intron suggesting that it is spliced. In contrast to the cg4715 homologs found in the *Drosophila* lineage, those present in mosquitoes appear to contain a functional furin cleavage site. It has been suggested that insect recruitment of cg4715 occurred because its endogenous expression might protect insects from infection by retroviruses and baculoviruses that share a homologous env protein. This may result from its interacting with and blocking retrovirus receptors on the cell surface or by acting as dominant negative inhibitors. In the latter scenario, cg4715 would interact with and form multimers with the env protein of invading retroviruses and prevent infection because of their inability to be cleaved. Therefore, the retrovirus env proteins would be trapped in defective multimeric complexes.

See also: Ascoviruses; Baculoviruses: Molecular Biology of Granuloviruses; Baculoviruses: Molecular Biology of Nucleopolyhedroviruses; Baculoviruses: Molecular Biology of Sawfly Baculoviruses; Metaviruses; Oryctes Rhinocerous Virus; Ustilago Maydis Viruses.

Further Reading

Boeke JD, Eickbush T, Sandmeyer SB, and Voytas DF (2004) Metaviridae. In: Fauquet CM, Mayo MA, Maniloff J, Desselberger U, and Ball LA (eds.) *Virus Taxonomy: Eighth Report of the International Committee on Taxonomy of Viruses*, pp. 359–367. San Diego, CA: Elsevier Academic Press.

Boeke JD, Eickbusch T, Sandmeyer SB, and Voytas DF (2004) Pseudoviridae. In: Fauquet CM, Mayo MA, Maniloff J, Desselberger U and Ball LA (eds.) *Virus Taxonomy: Eighth Report of the International Committee on Taxonomy of Viruses*, pp. 663–672. San Diego, CA: Elsevier Academic Press.

Eickbush TH and Malik HS (2002) Origins and evolution of retrotransposons. In: Craig NL, Craigie R, Gellert M and Lambowitz AM (eds.) *Mobile DNA II*, pp. 1111–1144. Washington, DC: ASM Press.

Kaminker J, Bergman C, Kronmiller B, et al. (2002) The transposable elements of the *Drosophila melanogaster* euchromatin: A genomics perspective. *Genome Biology* 3(12), doi: 10.1186/gb-2002-3-12-research0084.

Malik HS and Henikoff S (2005) Positive selection of iris, a retroviral envelope-derived host gene in *Drosophila melanogaster*. *PLoS Genetics* 1: 429–443.

Malik HS, Henikoff S, and Eickbush TH (2000) Poised for contagion: Evolutionary origins of the infectious abilities of invertebrate retroviruses. *Genome Research* 10: 1307–1318.

Pearson MN and Rohrmann GF (2002) Transfer, incorporation, and substitution of envelope fusion proteins among members of the *Baculoviridae*, *Orthomyxoviridae*, and *Metaviridae* (insect retrovirus) families. *Journal of Virology* 76: 5301–5304.

Terzian C, Pélisson A, and Bucheton A (2001) Evolution and phylogeny of insect endogenous retroviruses. *BMC Evolutionary Biology* 1(3), doi:10.1186/1471-2148-1-3.

Retroviruses of Birds

K L Beemon, Johns Hopkins University, Baltimore, MD, USA

© 2008 Elsevier Ltd. All rights reserved.

History

Avian retroviruses have been studied for 100 years. Avian leukosis virus (ALV) was discovered in 1908 by Ellermann and Bang, and the related Rous sarcoma virus (RSV) was isolated by Peyton Rous in 1911. Numerous isolates of ALV and of transforming viruses, which cause sarcomas and a variety of hematopoietic neoplasms, were reported in the decades that followed. Progress in understanding their nature was very slow until the development of cell culture assays in the late 1950s and the use of genetic and cell biological approaches to study replication and transformation. The discovery of reverse transcriptase (RT) in 1970 and of the origin and mechanism of action of viral oncogenes in the decade following led to an explosion of research activity. In addition to the genetic information of ALV needed for replication, RSV has a transduced oncogene, *src*, that enables it to induce sarcomas rapidly *in vivo* and to transform cells in culture. Src was the first retroviral oncogene to be characterized and the first tyrosine kinase to be identified. Currently, avian retroviruses provide a robust system for studies of retroviral gene expression and of virion assembly *in vitro*, as well as for studies of oncogenesis.

Taxonomy and Classification

The avian leukosis viruses comprise a single genus *Alpharetrovirus*, of the family *Retroviridae* and the subfamily *Orthoretrovirinae*. Although they share structural and biological characteristics with the mammalian C-type gammaretroviruses (such as murine leukemia virus), these two groups are not closely related. All ALVs are closely related to one another sharing considerable sequence and antigenic identity. Isolates are differentiated by subgroup (i.e., receptor utilization) and the presence or absence of oncogenes.

Distribution, Host Range, and Propagation

ALVs are endemic in flocks of domestic chickens (*Gallus gallus*) worldwide, and natural infections seem limited to this species, within which they are of some economic importance. Related endogenous viruses are found in ring-necked and golden pheasant (but not other related birds, such as turkeys), but exogenous viruses have not been isolated from other species. ALV will replicate efficiently in species closely related to the chicken such as quail, turkeys, and pheasants, but less so in more distant species such as ducks. RSV of some subgroups can transform mammalian cells and induce tumors in mammals, but with greatly reduced efficiency, and virus replication in mammalian cells cannot be reproducibly observed. The restriction of the virus to avian species is due to a lack of suitable receptors for most subgroups as well as to blocks to viral gene expression. The rare transformants that arise in RSV-infected mammalian cells often display rearrangements in proviral DNA that relieve this block.

A variety of cell types from gallinaceous birds (including chickens, turkeys, and quail) can be used to propagate ALVs and their relatives. Primary and secondary fibroblast cultures or cell lines (DF-1) are most commonly used, as well as lines derived from quail tumors (QT6). To avoid problems associated with frequent recombination, it is advisable to use cells that do not contain related endogenous proviruses, such as cells from species other than chickens or from chickens bred to contain no such proviruses.

Properties of the Virion

Like all retroviruses, ALVs are transmitted as enveloped virions of about 100 nm diameter, derived by budding from the host cell membrane. Lipids derived from the host plasma membrane are present in the viral envelope and make up 35% of the virion weight. Within the retrovirus family, they are defined as having a C-type morphology. Small, dispersed spikes project from the surface of the virion; these consist of trimers of the two *env*-encoded proteins, SU (surface) and TM (transmembrane). The internal core of the virion is of uncertain symmetry in mature virions but appears in electron micrographs as a centrally located, roughly spherical structure about 30 nm in diameter. Immature virions seen during or shortly after budding have a more open, spherical core structure, substantially larger in diameter than the processed one. The core comprises about 1500 copies each of the four gag-encoded proteins (as well as protease) and about 100 copies each of RT and integrase.

Properties of the Genome

The ALV genome consists of a homodimer of positive-sense, single-stranded (ss) RNA about 7500 bp in length. Transforming viruses, in which an oncogene has been inserted, have genomes varying in length from about 3.2 kbp (for UR2 virus) to about 9.3 kbp (for nondefective Rous sarcoma virus). In most of these viruses, the oncogene has replaced some of the normal genome, leading to genetically defective virus, which requires co-infection with helper ALV for replication. The genome is modified and processed by cell machinery. It contains a 5′m7GpppGm capping group and a 3′ poly (A) sequence, as well as some internal m6A residues. In *in vitro* translation systems, it is capable of serving as mRNA for the *gag-pro* and *gag-pro-pol* gene products. As with all retroviruses, the order of genes is 5′-*gag-pro-pol-env*-3′.

Important noncoding regions found near the end of the genome are necessary to provide signals for virus replication. These include an 18–21 base sequence (R) repeated at each end as well as unique sequences U3 (*c.* 250 bp) near the 3′ end and U5 (*c.* 80 bp) near the 5′ end which are duplicated in the long terminal repeat (LTR) during reverse transcription. The LTR contains sequences controlling transcription initiation and polyadenylation. Adjacent to these are the sites for initiation of reverse transcription: the primer binding (PB) sequence next to U5. Between PB and the beginning of *gag* is an approximately 300 bp leader region, which contains signals important for the dimerization and packaging of the genome into virions. The direct repeat (DR) sequences flanking the *src* gene in RSV are necessary for cytoplasmic accumulation and packaging of full-length viral RNA. ALV has one copy of the DR sequence. Other regulatory sequences are within coding sequences. The negative regulator of splicing (NRS) within the *gag* gene suppresses splicing and also promotes polyadenylation. ALVs with mutations in the NRS can lead to rapid-onset lymphomas *in vivo*, characterized by readthrough and splicing from the viral genome into downstream oncogenes or onco-miRs.

Properties of the Proteins

The virion contains nine proteins, the products of four coding regions. The Gag proteins constitute the major structural components and are sufficient to form recognizable virions if expressed alone. The Gag-Pro precursor is processed during release of virus into four Gag proteins: MA (matrix, about 19 kDa) which interacts with the cell membrane; p10, a 10 kDa protein of unknown function and location; CA (capsid, about 27 kDa) which forms the core shell structure; and NC (nucleocapsid, about 12 kDa), an RNA-binding protein necessary for specific encapsidation of genome RNA. The Gag-Pro precursor also contains the 15 kDa PR (protease) peptide necessary for processing all internal virion proteins. The Pol reading frame is expressed as a fusion protein with Gag and Pro, processed to yield RT (usually present as a heterodimer of 98 and 66 kDa reflecting partial processing by PR), and integrase (IN, about 32 kDa); these are the two enzymatic activators necessary for synthesis and integration of the DNA provirus. The *env* gene encodes the Env precursor (Pr95) which is processed as a membrane protein and cleaved by host cell proteases to yield the SU glycoprotein, which has an apparent molecular weight of about 85 kDa, about half of which is due to the provision of *c.* 14 N-linked carbohydrate side chains, and the TM glycoprotein, which has an apparent molecular weight of about 37 kDa. The SU and TM products remain as a disulfide-bonded heterodimer with SU containing the activity necessary for receptor binding and TM-mediating fusion with the cellular membrane.

Physical Properties

Virions of ALV have an equilibrium density in sucrose solutions of about 1.16–1.18 g ml^{-1} and a sedimentation coefficient of about 600S. They are quite labile and are readily inactivated by extremes of pH, as well as by heat or mild detergent treatments. They are somewhat radiation resistant, perhaps reflecting the recombinational repair capability provided by the dimeric genome.

Replication

Replication of ALV is like that of other retroviruses, and this group of viruses provided some of the important early models for studying the process. Entry of the virion follows interaction with a specific receptor on the cell surface. Genetically, at least ten subgroups (A–J) have been identified on the basis of distinct receptor recognition. The presence of receptors for specific subgroups is polymorphic among birds. Three unlinked genetic loci (*Tv-a*, *Tv-b*, and *Tv-c*) for ALV receptors have been genetically identified in chickens. The dominance of susceptibility over resistance alleles at each of these loci implies that they encode the receptor directly. The *Tv-b* locus has several alleles, controlling susceptibility to subgroups B, D, and E. Receptors for ASLV subgroups A, B, D, and E have been cloned. The Tv-a receptor resembles a portion of the receptor for low-density lipoprotein and is unrelated to other known retroviral receptors. The receptors for B, D, and E are all in the tumor necrosis factor receptor family. Chickens have two alleles that can act as receptors for these viruses. Entry of the virion core into the cell is by fusion of viral and cellular membranes, perhaps following endocytosis.

Once within the cytoplasm of the infected cell, the process of reverse transcription within the poorly defined core structure copies the ssRNA genome into a molecule of double-stranded (ds) DNA. This process – which varies little from that of other retroviruses – includes a series of 'jumps' from one end of the template to the other. The product is a dsDNA molecule, which differs from the genomic RNA by the presence at either end of the LTR. The LTR contains sequences necessary for DNA integration and for synthesis and processing of viral RNA.

Integration of viral DNA into more or less random sites in the cell genome is accomplished by the IN protein which has entered the cell with the virion and remains with the DNA in an ill-defined structure. The process of integration leads to the insertion of the viral DNA into cell DNA in the same general organization as both genome and unintegrated DNA. Integrated ALV DNA is characterized by the loss of two bases from each end of the viral sequence and the duplication of six bases of cell DNA at the integration site.

Transcription of the provirus into genomic and mRNA is mediated by cellular RNA polymerase II directed to the correct initiation site by promoter and enhancer sequences in the LTR. The strength of the enhancer elements is a major factor distinguishing pathogenic from nonpathogenic ALV isolates. Unlike some other retroviruses, there is no apparent role of virus-encoded proteins in regulating the transcription process. Processing of the viral transcripts includes addition of poly(A) following a canonical signal (AAUAAA) in the RNA derived from the 3' LTR and splicing of the fraction of the transcripts destined to become mRNA for the *env* gene. The splicing removes most of the *gag*, *pro*, and *pol* sequences, leaving the beginning of *gag* fused to *env*.

Translation of the full-length RNA leads to two products: The Gag-Pro precursor of about 76 kDa and the Gag-Pro-Pol precursor of about 180 kDa. Synthesis of the latter molecule is made possible by a −1 translational frameshift about 5% of the time, bypassing the termination codon at the end of Pro. Assembly of the precursors is at the cell surface, and is coincident with budding, implying a simultaneous association of the precursors with the genome, with the cell membrane, and with one another. Release of the immature particle (characterized by a hollow, symmetrical core which almost fills the virion) is rapidly followed by cleavage of the Gag-Pro and Gag-Pro-Pol precursors to yield the finished proteins. This cleavage is accompanied by condensation of the core into its mature form. Since the PR protein embedded in the Gag-Pro precursor contains only onehalf of the active site, dimerization of this domain is necessary for cleavage to occur. This requirement probably helps to delay cleavage until the appropriate time.

Once infected, the host cell is usually not killed by virus replication. A strong superinfection resistance due to blockage or loss of viral receptors develops soon after infection and prevents accumulation of proviruses by reinfection. In some cases, weak or slow development of superinfection resistance is associated with a cytopathic interaction of the virus with its host cell.

Transformation

A unique characteristic of ALV and a few other retroviruses is their ability to incorporate certain host sequences into their genome and alter the function of these protooncogenes to generate oncogenes. The presence of an oncogene renders the virus capable of inducing malignant transformation of cells in culture and one of a variety of malignant and rapidly fatal diseases in birds. At least 20 distinct cell sequences have been incorporated by ALV into a very large number of distinct isolates. RSV, which contains *src*, is the prototype oncogene-containing virus. Other notable oncogene-containing ALV variants include avian myeloblastosis virus (AMV; containing *myb*); avian myelocytomatosis virus-29 (MC-29; *myc*); avian erythroblastosis virus (AEV; *erb*-A and *erb*-B); Fujinami sarcoma virus (FSV; *fps*); and University of Rochester sarcoma virus-2 (RU-2; *ros*). Study of the genetic alterations that distinguish these oncogenes from proto-oncogenes, and the enzymatic and physiological function of the proteins they encode has been a keystone of modern cancer research. Incorporation of oncogenes into the virus genome is usually at the expense of some viral genes and co-infection of a cell with a wild-type (helper) ALV is thus necessary to provide viral proteins for replication of oncogene-containing viruses. RSV, which is a replication-competent transforming virus, is the exception.

Endogenous Viruses

Another unique feature shared by ALV and a few other retrovirus groups is their ability to become established in the germline and inherited stably as endogenous proviruses. Naturally occurring endogenous proviruses form a distinct lineage of ALVs, showing a specific host range (subgroup E) for which many domestic chickens lack receptors (a phenomenon known as xenotropism), and a reduced replication capacity and pathogenicity relative to exogenous viruses. Endogenous viruses are usually expressed at a very low rate, due largely to methylation of CpG residues in the proviral DNA, and are often (but not always) defective in sequence.

Genetics

Strain differences among ALV isolates are primarily in host range and are encoded by differences within the central portion of SU; other parts of the genome, with the exception of the U3 end of the LTR, are quite highly conserved. Like other retroviruses, ALVs exhibit very high rates of homologous recombination – a consequence of the diploid genome and the 'jumping' mechanism of reverse transcription. The latter also permits relatively high rates of nonhomologous recombination, leading to frequent (but not usually lethal) rearrangements of the genome as well as the occasional acquisition of foreign sequences such as oncogenes.

Evolution

Amino acid sequence relationships reveal a common origin of all retroviruses, but the ALV group forms a divergent branch, with its closest relative being the mouse mammary tumor virus. Whereas the recent spread of viruses among chickens is probably due largely to human intervention, the virus group is of considerable antiquity, since distantly related endogenous viruses are widespread in the genomes of avian and even mammalian species. The closely related endogenous viruses seem to be recent introductions derived by germline infection with exogenous virus since they are found only in *Gallus gallus*, and not in other species of *Gallus*, although they do appear in more distantly related pheasants.

Transmission and Tissue Tropism

Transmission of virus is principally vertical by infection of the offspring through virus secreted into the egg. Indeed, high titers of ALV are often detectable in commercial hen's eggs. Horizontal spread of virus is naturally much more rare, requiring close contact, but virus can be readily spread from infected birds via contaminated needles during vaccination or through vaccines prepared from infected eggs or cell cultures.

All isolates of ALV replicate efficiently in fibroblast cultures and in the bursa. Tropism for other tissues varies among isolates and is determined by both *env* and LTR sequences.

Pathogenicity

ALVs induce a wide spectrum of disease in naturally or experimentally infected animals. The prototypic disease induced by ALV is a B-cell lymphoma arising in the bursa of Fabricius starting a few months after infection and spreading to the liver and other organs during its course. Other malignancies, including erythroleukemia, sarcoma, and others, are not uncommon depending on the strain of virus and bird and the time and route of inoculation. The malignant diseases induced by viruses, which do not contain oncogenes, are the consequence of insertional activation of cellular proto-oncogenes (such as *c-myc* in the case of lymphoma, *c-erb*-B in erythroleukemia, and others). In addition to malignancies, these viruses also induce hemangiomas, osteopetrosis, and wasting diseases.

In some cases, an immune response against infected cells may be important; in others, cytopathic effects of the sort noted above may play a significant role.

Acquisition of oncogenes by ALVs greatly alters the nature and course of the disease. Infection of newly hatched chicks with AMV, for example, can lead to their death from myeloblastic leukemia in as few as 5 days. Moribund animals display enormously elevated myeloblast counts and a level of viremia sufficient to render the plasma noticeably turbid. Similarly, birds inoculated with RSV develop rapidly growing, usually fatal, sarcomas at the site of injection in a few weeks. It should be noted that the oncogene-containing viruses are not efficiently transmitted from one animal to another due to their rapid pathogenicity. In most cases, they have probably arisen in the animal from which they were isolated and would have died out if not brought into the laboratory.

Not all members of this group are highly pathogenic. RAV-0 (an endogenous virus) can infect susceptible chickens and induce viremia, but disease is rare and occurs only after a long latent period. The reduced virulence is probably an important feature of viruses inherited in the germline.

Immune Response

In infected birds, the only significant immune response is the appearance of type-specific neutralizing antibodies, which apparently recognize the regions of Env involved in receptor recognition. Group specific responses against Env or other proteins are not usually observed in infected chickens, although inoculation of virus into mammals induces antibodies capable of recognizing all virion proteins in the absence of subgroup-specific reactivity. The limited immune response observed in infected chickens has been attributed to the presence of endogenous proviruses whose expression (even at a low level) can induce tolerance to antigens in common with infecting virus. Indeed, it has been suggested that induction of tolerance might be a desirable feature for the animal, since it could prevent or limit immunopathological sequelae of infection. Postinfection immune response seems to be of little consequence in preventing subsequent malignant disease, since the cells, which will eventually form the tumor, are probably infected quite soon after infection, and the long latency reflects the necessity for subsequent rare events (such as mutations in other genes) rather than a continuing period of virus replication.

Prevention and Control

ALV-induced disease is a cause of some economic loss to the poultry industry in the United States, and occasional more serious epizootics (such as a recent outbreak of hemangioma in Israel) due to ALV have occurred. Control of infection is generally by detection and culling of infected individuals. No useful vaccination strategy has been developed. In principle, it should be possible to virtually eliminate the disease by breeding the appropriate *Tv-a* and *Tv-b* alleles into commercial strains; in practice, this has not been done very often. A more recent strategy is to introduce defective proviruses encoding envelope protein into the germline of birds; these can block infection by inducing superinfection resistance.

Future

Although of economic importance to the poultry industry, the value of ALV and the related oncogene-containing viruses to science has been far greater. The study of these viruses as models will continue to illuminate fundamental aspects of retrovirus biology. Continued searches for new transforming viruses and selected retroviral integration sites are likely to yield novel and important oncogenes. The goal of eradication of ALV disease from commercial chickens is attainable with present technology; its realization is largely a matter of economic considerations.

See also: Endogenous Retroviruses; Retroviruses: General Features.

Further Reading

Coffin JM, Hughes SH, and Varmus HE (1997) *Retroviruses.* Coid Spring Harbor, NY: Cold Spring Harbor Laboratory.

Flint SJ, Enquist LW, Racaniello VR, and Skalka AM (2004) *Principles of Virology, Molecular Biology, Pathogenesis, and Control of Animal Viruses,* 2nd edn. Washington, DC: ASM Press.

Goff SP (2006) *Retroviridae*: The retroviruses and their replication. In: Knipe DM, Howley PM, Griffin DE, et al. (eds.) *Fields Virology,* 5th edn., pp. 1871–1940. Philadelphia, PA: Lippincott Williams and Wilkins.

Retroviruses: General Features

E Hunter, Emory University Vaccine Center, Atlanta, GA, USA

© 2008 Elsevier Ltd. All rights reserved.

Glossary

Lenti From Latin *lentus*, 'slow'; refers to the slow development of pathology associated with lentivirus infections.

Retro From Latin *retro*, 'backward'; refers to the activity of reverse transcriptase and the transfer of genetic information from RNA to DNA.

Spuma From Latin *spuma*, 'foam'; refers to the vacuolated morphology of spumavirus infected cells.

Introduction

The family *Retroviridae* contains a large and diverse group of viruses that infect vertebrates. They are enveloped viruses that undergo a unique replication cycle, which clearly distinguishes them from other viruses. Virions generally contain a single-stranded RNA genome that upon introduction into the target cell is converted to a double-stranded DNA (dsDNA) copy by a process termed reverse transcription. This DNA version of the viral genome is then integrated into the chromosomal DNA of the cell, allowing the virus to persist and produce progeny for as long as the cell lives and providing a mechanism for lifelong infection in the vertebrate host.

The retrovirus family includes two human pathogens, human immunodeficiency virus (HIV), the causative agent of acquired immune deficiency syndrome (AIDS), and human T-lymphotropic virus 1 (HTLV-1), which induces T-cell lymphomas and degenerative nervous system disease in man. The family also includes important pathogens of horses, cows, sheep, cats, and rodents, where members induce cancers, anemias, arthritis, immunodeficiencies, and degenerative disease.

Retroviruses are an ancient group of viruses with archival evidence in the form of endogenous genomes pointing to infections that date back tens of millions of years. They have provided new tools and insights into molecular biology, have yielded clues to the basis of cancer, and continue to impact mankind through their pathogenic potential.

History

Although in 1904 equine infectious anemia was the first retroviral disease to be described, it was not recognized until much later that this was the result of viral infection. Four years later, however, Ellerman and Bang showed that chicken leukosis, a form of leukemia, was caused by a virus, and 3 years after that (1911) Peyton Rous reported cell-free transmission of sarcomas in the chicken. This virus, Rous sarcoma virus (RSV), named after its discoverer, together with the avian leukosis viruses (ALVs), is representative of the genus *Alpharetrovirus*. In part because these observations were made in birds, the scientific community did not immediately appreciate the importance of these oncogenic virus discoveries and it was 25 years before John Bittner identified the first mammalian retrovirus. He demonstrated that mouse mammary tumors were caused by a milk-transmitted, filterable agent. Then in 1957, Ludwik Gross described the development and serial cell-free passage of a highly potent strain of mouse leukemia virus. This was followed over the next two decades by the discovery and isolation of many oncogenic retroviruses from mice, cats, cows, and nonhuman primates.

It was also in the 1950s (1954) that Sigurdsson described visna, a neurological disease in sheep that slowly and progressively induced paralytic symptoms in its host. This led the authors to put forward the concept of slow viral infections and to the nomenclature that now describes members of the genus *Lentivirus* (derived from Latin: *lentus*, slow).

Spumaviruses are unusual in that they were not isolated from a specific disease state but rather from cell cultures derived from a healthy monkey contaminated by a simian member of this genus in 1954. In cell culture, the viruses induce a characteristic foamy appearance in the cytoplasm of the cell and this led to the nomenclature of the genus (derived from Latin: *spuma*, foam).

The first human retrovirus to be described in 1980 was HTLV-1, a member of the genus *Deltaretrovirus*, which induces a cutaneous T-cell lymphoma in a small number of individuals infected years earlier with the virus. It was only 3 years later that human immunodeficiency virus type-1 (HIV-1 − initially called lymphadenopathy-associated virus (LAV)) was isolated from patients with early manifestations of AIDS by Montagnier and his co-workers, and in 1984 that the link between this lentivirus (initially termed HTLV-III) and AIDS was conclusively established by Gallo and his colleagues. Since then numerous nonhuman primate, feline, and bovine members of the genus *Lentivirus* have been described, and they provide powerful animal models for the study of HIV/AIDS.

The archival evidence of previous retroviral infections in the form of endogenous proviruses that are found in high copy number in the genome of several mammalian species, including human, argues for multiple instances of widespread retroviral infections in mammals. Endogenous proviruses can vary from intact genomes capable of producing complete virus (modern endogenous viruses) to highly mutated variants that have evolved with the host and date back tens of millions of years (ancient endogenous retroviruses).

Taxonomy and Classification

The family *Retroviridae* contains two subfamilies: the *Orthoretrovirinae* and the *Spumaretrovirinae*. Differentiation between subfamilies is based on morphological characterization, replication differences, and variation in the expression and function of viral proteins. Six genera are present in the *Orthoretrovirinae*: *Alpharetrovirus*, which includes the avian leukosis and sarcoma viruses; *Betaretrovirus*, which includes mouse mammary tumor virus (MMTV) and Mason–Pfizer monkey virus (M-PMV); *Gammaretrovirus*, which contains the mammalian 'C-type' retroviruses, including rodent and primate leukemia and sarcoma viruses and the avian reticuloendotheliosis virus (REV); *Deltaretrovirus*, which includes bovine leukemia virus (BLV) an HTLV-1; *Epsilonretrovirus*, comprising the fish retroviruses, including the walleye dermal sarcoma virus (WDSV); and *Lentivirus*, which include HIV-1 and HIV-2, as well as viruses from ungulates such as visna-maedi and equine infectious anemia virus (EIAV). The subfamily *Spumaretrovirinae* is comprised of a single genus, *Spumavirus*, which contains the prototype foamy virus (PFV; originally named human foamy virus but more recently shown to be of chimpanzee origin). Members of this genus have been isolated from cats, cows, horses, and a variety of nonhuman primates but not from humans. Spumaviruses have replication characteristics intermediate between the orthoretroviruses and members of the *Hepadnaviridae*. Their genomic organization is similar to other retroviruses and reverse transcription of the viral genome is a prerequisite for infection, assembly and budding occur on intracellular membranes, DNA transcription from the RNA genome occurs prior to virus release from the cell, and the viral (reverse) transcriptase is translated independently from a spliced mRNA and requires the viral genome for incorporation (**Figure 1**).

Genome Structure and Organization

The viral genome is genetically diploid, consisting of two linear, positive-sense, single-stranded RNAs that range in size from 7 to 11 kbp depending on the species in question. The RNA monomers are held together in a (70S) dimer through hydrogen bonds in a $5'$-located dimer linkage structure. Each monomer of RNA is polyadenylated at the $3'$ end and has a cap structure (type 1) at the $5'$ end. The purified virion RNA is not infectious and must be converted into a dsDNA provirus via a process of reverse transcription in order for the virus life cycle to proceed.

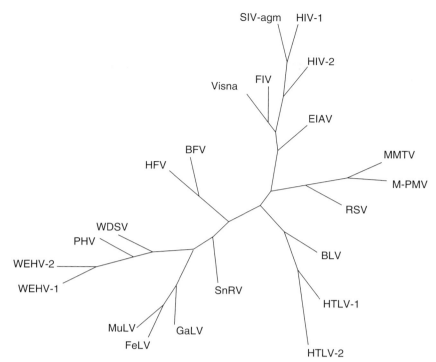

Figure 1 Phylogenetic relationships: phylogenetic analysis of conserved regions of the pol genes of retroviruses. An unrooted neighbor-joining phylogenetic tree was constructed based on an alignment of the amino acid residues of reverse transcriptase genes of several retroviruses. Courtesy of Quackenbush S and Casey J. Reproduced from *Virus Taxonomy: Seventh Report of the International Committee on Taxonomy of Viruses*, Elsevier/Academic Press, with permission.

Each RNA monomer has a specific tRNA molecule base-paired to a region (termed the primer binding site) near the 5′ end of the genomic RNA. This tRNA acts as the primer for DNA synthesis on the RNA template.

Infectious viruses have a minimum of four genes that encode the virion structural and replication proteins in the order: 5′-*gag–pro–pol–env*-3′. The *gag* gene encodes the viral structural proteins that are translated initially as a gag polyprotein precursor; *pro* encodes the viral aspartyl proteinase (PR), which mediates polyprotein cleavage during virus maturation; *pol* codes for the reverse transcriptase enzyme (RT) and the viral integrase (IN). RT is responsible for converting the viral RNA genome into a dsDNA provirus, which must then be integrated into the chromosomal DNA of the target cell by the IN protein (**Figure 2**).

Members of the genera *Deltaretrovirus*, *Epsilonretrovirus*, *Lentivirus*, and *Spumavirus* encode additional genes encoding nonstructural proteins important for the regulation of gene expression and for virus replication. The primate lentiviruses appear to have the most complex set of six accessory genes that include *tat*, which encodes a 'transactivator of transcription', and '*rev*', which codes for a protein (Rev) that transports unspliced and partially spliced viral RNA transcripts out of the nucleus. Genes encoding proteins with similar transcription enhancing and RNA transport functions are found in the deltaretroviruses (*tax* and *rex*) and spumaviruses (*tas*).

Some members of the alpharetroviruses and the gammaretroviruses carry cell-derived sequences that are important in pathogenesis. These cellular sequences are either inserted in a complete retrovirus genome (e.g., some strains of RSV), or in the form of substitutions for deleted viral sequences (e.g., murine sarcoma virus). Such deletions render the virus replication defective and dependent on replication-competent helper viruses for production of infectious progeny. In many cases, the cell-derived sequences form a fused gene with a viral structural gene that is then translated into one chimeric protein (e.g., gag-onc protein).

Virus Morphology

The virions are spherical, enveloped particles 80–100 nm in diameter, that for the most part derive their lipid bilayer from the plasma membrane of an infected cell. Inserted into the viral envelope are glycoprotein surface projections, which are about 8 nm in length and have the appearance of knobbed spikes. An internal core encapsidates the viral ribonucleoprotein and associated replicative enzymes. This structure has a spherical appearance in

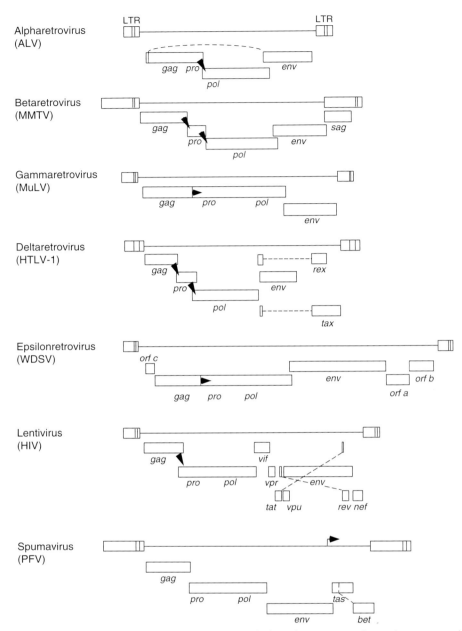

Figure 2 Genomic organization of retroviruses: the different prototypical provirus genomes for each genus are shown indicating the positions of the LTRs and encoded structural genes (*gag*, *pro*, *pol*, *env*) and certain other nonstructural genes (e.g., *tax* and *rex* in the deltaretroviruses) as well as their reading frames (ribosomal frameshift or ribosomal readthrough sites: arrowheads). LTR, long terminal repeat.

members of the genera *Alpharetrovirus*, *Gammaretrovirus*, and *Deltaretrovirus* of the subfamily *Orthoretrovirinae* and in members of the subfamily *Spumaretrovirinae*. It is spherical or rod shaped for members of the genus *Betaretrovirus*, and has a truncated cone shape in virions from the genus *Lentivirus*. Retroviruses have a characteristic buoyant density of $1.16-1.18\,\mathrm{g\,cm^{-3}}$ and a sedimentation coefficient (S_{20w}) of approximately 600S in sucrose (**Figure 3**).

Two distinct morphogenic pathways exist. Members of the genera *Alpharetrovirus* and *Gammaretrovirus*, which assemble their immature capsids at the plasma membrane, were historically classified as C-type viruses based on electron microscopy, and members of the genus *Lentivirus* are also assembled via this pathway. In contrast, members of the genus *Betaretrovirus* assemble immature capsids (previously termed A-type particles) in the cytoplasm, which are then enveloped at the plasma membrane with either a B-type (MMTV) or D-type (M-PMV) morphology. Members of the subfamily *Spumaretrovirinae* also assemble immature capsids in the cytoplasm, but these are enveloped primarily at intracellular membranes.

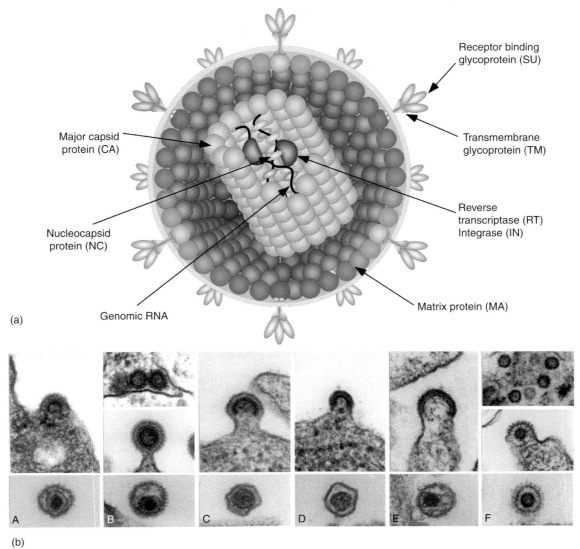

Figure 3 Virion structure and morphology. (a) Schematic cartoon (not to scale) shows the inferred locations of the various structures and proteins. (b) A, *Alpharetrovirus*: avian leukosis virus (ALV), type 'C' morphology; B, *Betaretrovirus*: mouse mammary tumor virus (MMTV), type 'B' morphology; C, *Gammaretrovirus*: murine leukemia virus (MLV); D, *Deltaretrovirus*: bovine leukemia virus (BLV); E, *Lentivirus*: human immunodeficiency virus 1 (HIV-1); F, *Spumavirus*: human foamy virus (HFV). The bar represents 100 nm. Reproduced from *Virus Taxonomy: Seventh Report of the International Committee on Taxonomy of Viruses*, Elsevier/Academic Press, with permission.

Viral Proteins

The surface of the virion is studded by envelope glycoprotein (Env) spikes comprised of three copies each of two envelope proteins: SU (surface) and TM (transmembrane). These individual components are synthesized as part of a single Env precursor, which is encoded by the viral env gene and is proteolytically cleaved during intracellular transport and prior to assembly into the virus. The number of Env trimers/virion can vary from an estimated low of 7–14 for HIV to a high of more than 150 for primate foamy viruses. SU acts as the receptor-binding component and TM as a membrane-spanning anchor that mediates virus and cell membrane fusion during virus entry.

For members of the *Orthoretrovirinae*, virions contain 3–6 internal, nonglycosylated structural proteins (encoded by the *gag* gene). These are, in order from the N-terminus, (1) MA (matrix protein), (2) a protein, frequently phosphorylated, that in some viruses plays a role in viral budding (3) CA (capsid protein), (4) NC (nucleocapsid protein), and (5) a small C-terminal protein, found in some viruses, that can play a role in assembly and/or budding. These proteins are translated as a single gag polyprotein precursor, which is cleaved by a virus-encoded aspartyl proteinase to the mature products. The MA is often modified with a myristic acid moiety that is covalently linked to the N-terminal glycine. This modification combined with basic residues in the MA domain form a bipartite signal that facilitates intracellular transport of the gag precursor and its association with the plasma membrane. Viruses of the genera *Alpha-*, *Beta-*, and *Gammaretrovirus* encode a protein between MA and CA, which contains one or two motifs that recruit host factors (ESCRT proteins) necessary for successful release of the budding virus from the cell. In primate lentiviruses, these motifs are encoded in a protein at the C-terminus of gag. The CA domain of the Gag precursor plays a key role in immature capsid assembly as well as in the formation of the mature viral core, which encapsidates the viral genome and replicative enzymes. Coating of the viral RNAs by NC appears to facilitate the compact packing of the viral genome into the core.

For members of the subfamily *Spumaretrovirinae*, the Gag protein is only cleaved once near the C-terminus and there are no mature cleavage products analogous to MA or NC of the *Orthoretrovirinae*. Gag also lacks features of other retroviral Gag proteins such as the addition of myristic acid at the N-terminus, the Cys-His boxes of NC, and the major homology region (MHR) of CA. Instead there are three glycine–arginine-rich (GR) boxes near the C-terminus, which are likely involved in assembly and/or RNA binding.

In all retroviruses three nonstructural, enzymatic proteins are incorporated into virions. These are the aspartyl protease (PR, encoded by the *pro* gene), the reverse transcriptase (RT, encoded by the *pol* gene), and integrase (IN, encoded by the *pol* gene). PR is required for cleavage of the Gag precursor, an obligate step in maturation and for virus infectivity. In some viruses a dUTPase (DU, role unknown) is also present. Proteins constitute about 60% of the virion dry weight.

Retrovirus Replication Cycle

Attachment and Penetration

Entry into the host cell is mediated by an interaction between the virion glycoproteins and specific protein receptors at the surface of the host cell. This interaction, sometimes aided by low pH in endosomal vesicles, induces a conformational change in the Env trimer that allows the fusion peptides of the TM proteins to be inserted into the target cell membrane and additional rearrangements of TM, which bring the viral and cell membranes together. Fusion of the viral envelope with the plasma membrane then occurs. Retroviral receptors are cell-surface proteins. For HIV, both the CD4 protein, which is an immunoglobulin-like molecule with a single transmembrane region, and a chemokine receptor (CCR5 or CXCR4), which span the membrane seven times, are required for membrane fusion. The receptors for ecotropic murine leukemia virus (MLV), amphotropic MLV, and gibbon ape leukemia virus (GALV) as well as M-PMV are involved in the transport of small molecules. These transporters have a complex structure with multiple transmembrane domains. For the ALVs two receptors have been identified: that for subgroup A viruses is a small molecule with a single transmembrane domain, distantly related to a cell receptor for low-density lipoprotein while that for subgroup B viruses is related to the TNF-receptor family of proteins.

Once the viral membrane has fused with that of the target cell, the viral core is exposed to the intracellular environment of the cytoplasm. It is not clear whether a specific uncoating event occurs or whether initiation of nucleic acid transcription is sufficient to partially disassemble the viral core. Subsequent early reverse transcription events are carried out in the cytoplasm in the context of a nucleoprotein complex derived from the mature capsid.

Reverse Transcription

Replication of the viral genome starts with reverse transcription by RT of virion genomic RNA into cDNA. This process was so termed because at the time of its discovery, it reversed the accepted flow of genetic information in the cell. The 3′ end of the genome-associated tRNA acts as a primer for synthesis of a negative-sense cDNA transcript (**Figure 4(b)**). Because the primer binding site is located near the 5′ end of the genome, the initial short product (minus-strand strong stop) must transfer to the 3′ end of the genome through duplicated sequences (R) that are present at each end of the viral RNA. The transferred transcript can now prime continued cDNA synthesis (**Figure 4(c)**).

The RT enzyme is structured such that the DNA synthetic active site is located in close proximity to an RNase H active site that specifically digests the RNA template strand present in the newly formed RNA–DNA hybrid. Reverse transcription thus involves the concomitant synthesis of DNA and digestion of the viral RNA. Short undigested (purine-rich) RNA products of this process act to prime virus-sense cDNA synthesis on the negative-sense DNA transcripts (**Figure 4(d)**). One of these polypurine primers (PP) located near the 3′ end of the genome initiates a second (plus-strand strong stop) strand transfer, this time using homologies in the tRNA primer-binding site (PB, **Figure 4(e)**). Both strand transfers result in duplication of 5′ and 3′ genomic RNA sequences respectively, so that in its final form the single linear dsDNA transcript derived from the diploid viral genome contains long terminal repeats (LTRs) composed of sequences from the 3′ (U3) and 5′ (U5) ends of the viral RNA that now flank the R sequence (**Figure 4(f)**). A high frequency of recombination is observed during the process of reverse transcription when, following co-infection of a cell, two genetically distinct RNAs are packaged into

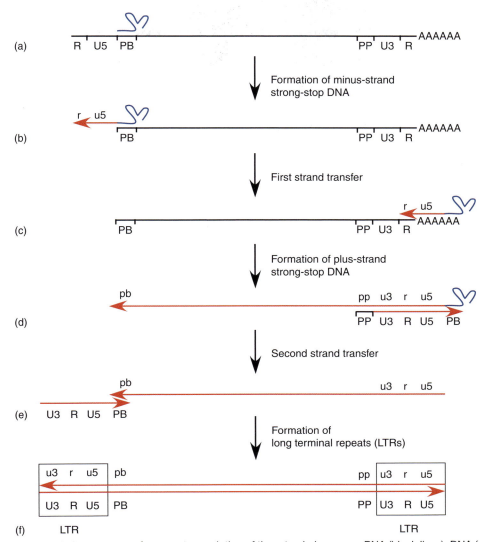

Figure 4 Reverse transcription: process of reverse transcription of the retroviral genome. RNA (black lines), DNA (red lines).

a retrovirus. This appears to reflect the frequent transfer of the elongating RT from one template RNA to the other, and implies that the RNAs are packaged in the core in a fashion that facilitates this.

Integration

The linear, double-stranded, greater-than-genome-length DNA product of reverse transcription complexed with the viral integrase is called the pre-integration complex (PIC). It must be transported into the nucleus in order for the viral DNA to be integrated into the chromosomal DNA of the host to form an integrated provirus. For most retroviruses, nuclear membrane breakdown during cell division is required for entry of the PIC, but the lentivirus PIC can be transported into the nucleus via nuclear pores, allowing proviral integration in nondividing cells. Integration is mediated by the viral IN protein. The ends of the virus DNA are joined to cell DNA, following the removal of two nucleotides from the ends of the linear viral DNA. Integration generates a short duplication of cell sequences at the integration site, the length of which is virus specific. Proviral DNA can integrate randomly at many sites in the cellular genome with no specific sequence being targeted, although there does appear to be a preference for integration in or near actively transcribed genes. Once integrated, a sequence is generally incapable of further transposition within the same cell. The map of the integrated provirus is collinear with that of unintegrated viral DNA. Integration appears to be a prerequisite for virus replication.

Genomic and mRNA Transcription

The integrated provirus is transcribed by cellular RNA polymerase II into virion RNA and mRNA species in response to transcriptional signals in the U3 region of the viral LTRs. In some genera, virus-encoded transactivating proteins also regulate transcription. Transcription starts at the beginning of the 5′ R region and proceeds to the end of the 3′ R′ sequence. Signals in the U5 region of the 3′ LTR promote cleavage and polyadenylation of the RNA at this site. There are several classes of mRNA depending on the virus and the genetic organization of the retrovirus. With the exception of the spumaviruses, an mRNA comprising the whole retroviral genome serves for the translation of the *gag*, *pro*, and *pol* genes. This results in the formation of polyprotein precursors, which are the source of the structural proteins, protease, RT, and IN, respectively. A smaller mRNA consisting of the 5′ end of the genome spliced to sequences from the 3′ end of the genome that include the *env* gene and the U3 and R regions, is translated into the precursor of the envelope proteins. In viruses that contain additional genes, various additional forms of spliced mRNA are also made; however, all these spliced mRNAs share a common sequence at their 5′ ends. Spumaviruses are unique in that they make use of an internal promoter (IP) located in the *env* gene upstream of the accessory protein reading frames. Most primary translational products in retrovirus infections are polyproteins, which require proteolytic cleavage before becoming functional. The *gag*, *pro*, and *pol* products are generally produced from a nested set of primary translation products. For *pro* and *pol*, translation involves bypassing translational termination signals by ribosomal frameshifting or by readthrough at the *gag-pro* and/or the *pro-pol* boundaries (**Figure 1**).

Assembly and Release of Virions

Immature capsids assemble either at the plasma membrane (a majority of the genera) or at intracytoplasmic particles (*Betaretrovirus* and *Spumavirus*). Little is known about the intracellular targeting of Gag precursor proteins within the cell, although the myristic acid modification at the N-terminus of Gag and positively charged amino acids in the MA domain appear to provide a bipartite signal to initiate budding at the plasma membrane. For the betaretroviruses, a short (cytoplasmic targeting/retention) signal (CTRS) in the MA domain of Gag interacts with components of the dynein motor so that translating polysomes are transported to the pericentriolar region (microtubule organizing center) of the cell, where immature capsid assembly occurs. Mutations in the CTRS result in plasma membrane assembly of immature capsids. A similar mechanism for intracellular targeting and assembly of Gag appears to be utilized by the spumaviruses.

Most retroviruses are released from the cell by a process of budding from a region of the plasma membrane where viral glycoproteins must also be targeted. There is evidence in several genera that interaction of Gag and Env components occurs at an intracellular location prior to co-localization at the budding site. This process of budding, however, does not appear to require the Env proteins since expression of Gag alone is sufficient for release of virus particles. The final pinching-off and release of virus requires the complex cellular machinery (ESCRT) normally involved in multivesicular body formation, which is recruited to the site of budding by sequence motifs in Gag. Polyprotein processing of the internal proteins occurs concomitant with or just subsequent to release of virus from the cell and is accompanied by maturation of the virion. This includes morphological changes that include condensation of the viral RNA into an electron-dense ribonucleoprotein core and the acquisition of infectivity.

Pathogenesis

Members of the family *Retroviridae* establish persistent lifelong infections in their hosts – a reflection of their replication cycle and their ability to insert a copy of the proviral genome into the chromosome of a target cell. Because retroviruses are in general noncytopathic to their host cells and function as effective parasites that siphon off only a small percentage of the macromolecular machinery, continued production of progeny viruses over the lifespan of the cell is thus the norm. Moreover, in the context of the vertebrate host, this means that curing infection is effectively impossible, since a single retrovirus-infected cell can be the source of systemic infection.

Viruses from several retroviral genera are capable of inducing tumor formation in their natural hosts. It was the 'acute transforming' retroviruses, exemplified by RSV, that provided the key to our initial understanding of how retroviruses induce cancer in their hosts. These viruses, members of the genera *Alpha-* and *Gammaretrovirus*, have transduced cellular genes (now known as proto-oncogenes) that function in the signal transduction pathways involved in growth factor upregulation of cell proliferation. With the exception of certain RSV isolates, these acutely ransforming retroviruses are generally replication defective, since the inserted oncogene replaces replicative genes, and require a helper virus to provide the missing replicative functions. Nevertheless, they are capable of rapidly inducing a variety of cancers in their hosts. The replication-competent WDSV, a member of the genus *Epsilonretrovirus*, also induces cell transformation through expression of a cell-derived gene, although in this case it is related to the cyclin family of regulators.

For the so-called 'chronic transforming' members of the genera *Alpha-*, *Beta-*, and *Gammaretrovirus*, persistent infection with the high numbers of associated viral integration events eventually results in the insertional activation of cellular oncogenes. This was first defined for ALV where, in birds with this form of leukemia, frequent integrations just upstream of the *c-myc* oncogene in malignantly transformed cells were found to result in its unregulated expression. Related mechanisms that deregulate cellular oncogene expression have been described for MMTV, MuLV, and other vertebrate 'leukemia' viruses. The human pathogen HTLV-1 induces T-cell-derived tumors in a fraction of infected individuals but this appears in part to be the result of *trans*-activation of cellular genes such as Il-2 and Il-2 receptors as well as inactivation of cell-cycle regulators such as p53 by Tax, which leads to unregulated proliferation of T-cells.

Members of the genus *Lentivirus*, including visna-maedi virus, EIAV, caprine arthritis encephalitis virus (CAEV), and HIV-1, generally establish persistent infections that progressively impose a defined pathology on their host. It was the progressive central nervous system degeneration in sheep induced by visna virus that led to the term 'slow infections' and to the nomenclature for this genus. The human pathogen HIV-1 is typical of members of the genus; although it is cytopathic in CD4+ T-lymphocytes, it establishes a persistent disease through constant cycles of infection of lymphocytes and macrophages in lymphoid tissues. Generally, it takes several years for depletion of the immune system to progress to a level where opportunistic infections and cancers, which are the hallmark of AIDS, develop. Interestingly, the simian immunodeficiency viruses (SIVs) do not appear to cause disease in their natural hosts, where presumably there has been co-evolution of virus and host to reach a nonpathogenic equilibrium.

See also: AIDS: Disease Manifestation; AIDS: Global Epidemiology; AIDS: Vaccine Development; Endogenous Retroviruses; Retroviral Oncogenes.

Further Reading

Desrosiers R (2007) Nonhuman lentiviruses. In: Knipe DM, Howley PM, Griffin DE, *et al.* (eds.) *Fields Virology,* 5th edn., pp. 2215–2241. Philadelphia, PA: Lippincott Williams and Wilkins.

Freed EO and Martin MA (2007) HIVs and their replication. In: Knipe DM, Howley PM, Griffin DE, *et al.* (eds.) *Fields Virology,* 5th edn., pp. 2107–2185. Philadelphia, PA: Lippincott Williams and Wilkins.

Goff SP (2007) *Retroviridae*: The retroviruses and their replication. In: Knipe DM, Howley PM, Griffin DE, *et al.* (eds.) *Fields Virology,* 5th edn., pp. 1999–2069. Philadelphia, PA: Lippincott Williams and Wilkins.

Kuritzkes DR and Walker B (2007) HIV-1: Pathogenesis, clinical manifestations, and treatment. In: Knipe DM, Howley PM, Griffin DE, *et al.* (eds.) *Fields Virology,* 5th edn., pp. 2187–2214. Philadelphia, PA: Lippincott Williams and Wilkins.

Lairmore MD and Franchini G (2007) Human T-cell leukemia viruses types 1 and 2. In: Knipe DM, Howley PM, Griffin DE, *et al.* (eds.) *Fields Virology,* 5th edn., pp. 2070–2105. Philadelphia, PA: Lippincott Williams and Wilkins.

Linial M (2007) Foamy viruses. In: Knipe DM, Howley PM, Griffin DE, *et al.* (eds.) *Fields Virology,* 5th edn., pp. 2245–2262. Philadelphia, PA: Lippincott Williams and Wilkins.

Linial ML, Fan H, Hahn B, *et al.* (2004) *Retroviridae*. In: Fauquet CM, Mayo MA, Maniloff J, Desselberger U, and Ball LA (eds.) *Virus Taxonomy: Eighth Report of the International Committee on Taxonomy of Viruses,* pp. 421–440. San Diego, CA: Elsevier Academic Press.

Rhinoviruses

N W Bartlett and S L Johnston, Imperial College London, London, UK

© 2008 Elsevier Ltd. All rights reserved.

Glossary

Afebrile Without a fever.
Bronchopulmonary dysplasia Chronic lung disease of infancy that follows mechanical ventilation and oxygen therapy for acute respiratory distress after birth in premature newborns.
Ciliated Cells with hair-like structures (cilia) on the surface.
Cytopathology Cell damage.
Endocytosis Uptake of material into the cell by membrane-bound vesicles.
Endosome Membrane-bound vesicle formed during endocytosis.
Heterotypic Of different types.
Nasal mucosa Mucous membrane lining the nasal cavity.
Nasopharynx The portion of the pharynx extending from the posterior nares to the level of the soft palate.
Nim Neutralizing immunogenic site.
Otitus media Infection and inflammation of the middle ear space and ear drum.
Peak expiratory flow The maximum flow at the outset of forced expiration.
Rhinorrhoea The free discharge of a thin nasal mucus.
Sinusitis Inflammation of the sinus.
TBP TATA-binding protein, part of RNA polymerase II transcription system.

History

Rhinoviruses are the most common infectious agents of humans and most frequent cause of the common cold (acute nasopharyngitis), a mild disease of the upper respiratory tract. More recently, their role in acute exacerbations of asthma and other airway disease has been highlighted, implicating the virus in illnesses significantly more severe than the common cold. Hieroglyphs representing the cough and common cold date back to ancient Egypt. In the fifth century BC Hippocrates gave a description of the disease and in the first century Pliny the Elder suggested 'kissing the hairy muzzle of a mule' as therapy for colds. The common cold was also known among the ancient American Indian, Aztec, and Maya civilizations. It was not until first Walter Kruse and then Alphonse Donchez in the early part of the twentieth century demonstrated that viruses (filtered material from the nasal secretions from a cold sufferer) caused the common cold. The first isolations of human rhinovirus (HRV) were reported by two laboratories in the late 1950s: Price in 1956 and Pelon and co-workers in 1957. With advances in tissue culture techniques that matched temperature and pH conditions to those found in the nose, identification of serologically distinct rhinoviruses increased rapidly such that by 1987 100 serotypes had been identified (**Table 1**). Efforts in the 1960s and 1970s to develop vaccines based on inactivated or attenuated viruses were unsuccessful due to the large number of rhinovirus serotypes and ineffective mucosal immunization. Lack of a practical small animal model for rhinovirus infection has meant that experimental human challenge models have been used to study virus pathogenesis. Although it has been long recognized that rhinovirus infections are the most frequent cause of the common cold, recent epidemiologic studies using sensitive polymerase chain reaction (PCR) techniques and data from human studies has highlighted the importance of rhinoviruses as precipitants of serious respiratory illnesses. This is especially evident in the context of persons suffering chronic airway diseases such as asthma and chronic obstructive pulmonary disease (COPD).

Taxonomy and Classification

The family *Picornaviridae* (pico = small + RNA) contains three genera of human pathogens that are structurally and genetically closely related: enterovirus, parechovirus, and rhinovirus. Rhinoviruses have been isolated from humans and cattle, with HRVs comprising by far the largest group. The HRV genus currently consists of 102 serotypes (HRV-1A to HRV-1B and HRV-2 to HRV-100, Hanks). Designation of serotypes is based on antibody neutralization, with absence of cross-reactivity of polyclonal antisera with defined serotypes constituting designation of a distinct serotype. (It should be noted that extensive serotyping has not been performed since the 1980s and several rhinovirus isolates await classification and may represent new serotypes.) The three bovine rhinoviruses (BRV-1–3) also await classification. All serotypes (except HRV-87) segregate into two genetic clusters or species based on nucleotide sequence across the VP4/VP2 interval: human rhinovirus A (HRV-A; 76 serotypes) and human rhinovirus B (HRV-B; 25 serotypes). Analysis of nucleotide sequences from HRV87 demonstrated that it is an enterovirus, most closely related to human enterovirus 68 (species D). The rhinovirus genus can also be classified into two groups according to receptor tropism, with approximately 90% of HRV (major receptor group) exploiting the intercellular adhesion molecule 1 (ICAM-1, CD54) for cell binding. The remainder (minor receptor group) use members of the low-density lipoprotein (LDL) receptor family. HRV-87 uses an as yet unknown sialoprotein for cell-surface attachment. The rhinovirus genus can also be divided into two groups according to the variability in susceptibility to capsid-binding drugs. The pattern of susceptibility has served as an alternative method of classification of HRV serotypes into two groups, A and B.

Virion Structure and Physical Properties

Rhinoviruses, typical of the family *Picornaviridae*, are small (approximately 30 nm, molecular mass of 8.5×10^6 kDa) icosahedral particles (**Figure 1**) composed of 60 copies of each of the four capsid proteins VP1–VP4 with molecular masses of 32, 29, 26, and 7 kDa, respectively. The capsid proteins are symmetrically arranged into protomers which contain one copy of VP1, VP3, and a precursor (VP0) in which VP2 and VP4 are covalently linked. Protomers, arranged around a fivefold axis, form pentamers and 12 pentamers form the icosahedral capsid shell. Capsid precursors encapsidate viral genomic RNA

Table 1 Species[a] and receptor binding[b] grouping of human rhinovirus serotypes (excluding HRV-87)

Human rhinovirus A	Human rhinovirus B
1A, **1B**, **2**, 7, 8, 9, 10, 11, 12, 13, 15, 16, 18, 19, 20, 21, 22, 23, 24, 25, 28, **29**, **30**, **31**, 32, 33, 34, 36, 38, 39, 40, 41,43, **44**, 45, 46, **47**, **49**, 50, 51, 53, 54, 55, 56, 57, 58, 59, 60, 61, **62**, 63, 64, 65, 66, 67, 68, 71, 73, 74, 75, 76, 77, 78, 80, 81, 82, 85, 88, 89, 90, 94, 95, 96, 98, 100	3, 4, 5, 6, 14, 17, 26, 27, 35, 37, 42, 48, 52, 69, 70, 72, 79, 83, 84, 86, 91, 92, 93, 97, 99

[a]Species classification is based on genetic clustering of capsid (VP4/VP2) gene nucleotide sequences.
[b]Viruses shown in bold type bind to the LDL receptor (minor group); the remainder bind to ICAM-1 (major group).

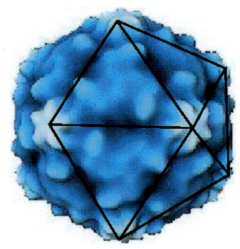

Figure 1 Human rhinovirus 14 solved by cryoelectron microscopy and image reconstruction. The fivefold axis of symmetry is superimposed on the image. Courtesy of The Big Picture Book of Viruses.

to form 150S provirions. Maturation cleavage of VP0 to VP2 and VP4 is the final step of assembly and yields infectious mature virions. VP1, VP2, and VP3 structurally constitute most of the capsid. They are similar in structure and have eight-stranded, antiparallel β-barrel motifs differing mainly in the loops and elaborations which join or project from the β-strands. VP1 is the most external and dominant structural protein and contains most of the motifs known to interact with cellular receptors and neutralize monoclonal antibodies. VP4 is smaller, has an extended structure, and lies at the RNA–capsid interface in close association with the RNA core and functions as an anchor to the virus capsid. A deep cleft, called the canyon, surrounds the fivefold axes of icosahedral symmetry and encloses the ICAM-1 binding site in the major receptor group viruses. Amino acids at the base of the canyon are more conserved than those on the protruding rim of the canyon and surface of the virion, which are more prone to substitutions and contain the binding sites for neutralizing antibody. Beneath this canyon, within VP1, lies a pore that leads to a hydrophobic pocket occupied by a pocket factor that is likely to be a fatty acid. The minor group LDL receptor binds to the star-shaped dome on the fivefold axis of the virion.

Rhinoviruses have a buoyant density in CsCl of 1.38–1.42. The property that distinguishes them from the closely related enteroviruses is their acid lability with inactivation occurring below pH 6. In contrast most rhinoviruses are thermostable, surviving for days at 20–37 °C. The ability to survive for extended periods of time in the environment is likely to be an important factor in their spread. Several serotypes (3–12, 15, 18, and 19) are relatively stable at 50 °C for 1 h. The lack of a lipid membrane enables rhinoviruses to resist 20% ether, 5% chloroform, and sodium deoxycholate solutions. Alcohol/phenol disinfectants are effective virucidal agents.

Properties of the Rhinovirus Genome

The nonenveloped virion encapsidates a genome composed of a single-stranded positive-(messenger) sense RNA molecule of 7100–7400 nucleotides. This RNA molecule functions directly as a message, encoding a single open reading frame containing approximately 2150 codons flanked at both termini by untranslated regions (UTRs) (**Figure 2**). At the 5′ terminus the 600-nucleotide-long UTR is covalently linked to a small, virus-encoded protein, VPg which initiates viral RNA synthesis. The UTRs have important secondary and tertiary structures; the first 100 nucleotides of the 5′-UTR forms a clover leaf structure that binds viral and host proteins forming a nucleoprotein complex required for viral RNA replication. The RNA molecule is directly translated from a type I internal ribosome entry site (IRES) also located in the 5′-UTR which directs recruitment of host-cell ribosomes and cap-independent translation of viral proteins. Downstream of the 5′-UTR is the capsid-coding region (P1), followed by the nonstructural protein encoding regions P2 and P3. The 3′ terminus of the genome contains a short (approximately 40 nucleotides) UTR and terminates with a poly-A tail. The single ∼250 kDa polyprotein encoded by the rhinovirus genome is processed both during and after translation into mature viral proteins by a sequence of cleavages executed by virus-encoded proteases.

Properties of Rhinovirus Proteins

The rhinovirus proteins are numbered 1A(VP4), 1B(VP2), 1C(VP3), 1D(VP1), 2A, 2B, 2C, 3A, 3B, 3C, 3D, according to their physical location in the unprocessed polyprotein. The polyprotein is proteolytically cleaved to yield four capsid (structural) proteins and ten nonstructural proteins: seven mature proteins and three intermediate proteins with functions distinct from their cleavage products (2BC, 3AB, and 3CD^{pro}). Three of the products of polyprotein processing function as proteases (^{pro}) while 3D^{pol} is the virally encoded RNA-dependent RNA polymerase. All of the nonstructural proteins are required for replication of the viral RNA. In addition, some nonstructural proteins alter host-cell function. The VPg protein (3B) acts as a primer for the initiation of RNA synthesis. Proteins 2A^{pro} and 3C^{pro} are cysteine proteases involved in processing the viral polyprotein. The first cleavage is catalyzed by 2A^{pro} which cleaves only at tyrosine–glycine bonds and performs the primary cleavage between the capsid and nonstructural precursors at the junction of the C terminus of VP1 and its own N terminus releasing

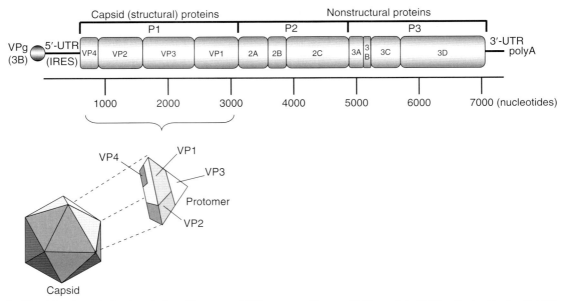

Figure 2 The rhinovirus single-stranded positive-sense RNA genome. The small VPg protein (3B) is attached to the 5′-UTR which encodes a type I internal ribosome entry site (IRES). The genome is polyadenylated (polyA) at the 3′ terminus. The single polyprotein encoded by the genome is cleaved by proteases to yield individual capsid and nonstructural proteins.

P1. The 2A^{pro} also mediates shutdown of host-cell translation by cleaving cellular eIF-4G, a key factor in cap-dependent translation. The 3C^{pro} catalyzes most of the subsequent cleavage events on the picornaviral polyprotein and along with the 3CD^{pro} cleaves the polyprotein at glutamine–glycine bonds. Rhinovirus 3C^{pro} may also play a role in virus-induced shutoff of host-cell transcription (by RNA polymerase II) by cleavage of the transcription factor TBP. It is thought that 3C enters the nucleus via a nuclear localization signal within the 3CD precursor. Experiments with the 2B protein of enteroviruses indicate that this viral protein is a 'viroporin' increasing plasma membrane permeability and inhibiting secretory pathways. The 3A protein may also modulate the secretory functions of cells affecting surface expression of host major histocompatibility complex (MHC) class I.

Life Cycle

Entry, Replication, and Assembly

The major group HRVs utilize ICAM-1 to attach to cells. ICAM-1 is a cell-surface glycoprotein and a member of the immunoglobulin (Ig) protein superfamily. Once bound to the receptor the viral particle is internalized by receptor-mediated endocytosis. Replication of rhinoviruses takes place in the cytoplasm of host cells. Uncoating and release of the viral RNA into the cytoplasm occurs after acidification of the late endosome. Acidification triggers conformational changes in the capsid centered on the fivefold axis, where a channel through the capsid and the endosomal membrane opens allowing RNA release.

The positive-(message) sense RNA genome is translated into a single large polyprotein from the IRES located in the 5′-UTR of the rhinovirus genome. The IRES forms a complex secondary structure which can direct ribosomes to the polyprotein start AUG, thus initiating cap-independent translation. This allows the synthesis of viral proteins while cap-dependent translation of cellular proteins is shut off. The polyprotein is processed into functionally active proteins through a sequence of cleavages performed by virus-encoded proteases (described in the previous section). The cloverleaf secondary structure of the 5′-UTR in the viral genome binds viral (3C or 3CD precursor) and host (poly (A)-binding protein) proteins to form a nucleoprotein complex required for RNA replication, which is catalyzed by newly synthesized 3D^{pol} (viral RNA-dependent RNA polymerase), the most highly conserved polypeptide among members of the family *Picornaviridae*. Initially primed by the terminally bound VPg protein, the viral RNA polymerase uses the genomic positive-sense RNA as a template for synthesizing negative-sense copies which in turn act as templates for positive-sense RNA synthesis. Some newly synthesized positive-sense RNA copies act as messages, whereas others are packaged into virus particles. The viral polymerase has a mutation rate of one every 2200 bases (approximately four mutations per transcript) and is therefore at the threshold for genetic maintenance. This characteristic in conjunction with the ability to undergo viral recombination is probably a significant reason why rhinoviruses are such efficient pathogens of humans.

There is still much not known about the assembly of virus particles. Capsid assembly is via the precursor 5S protomers and 14S pentamers already described and takes place in association with membranes. RNA is thought to be packaged into pre-formed 80S capsids although the molecular mechanism by which this process occurs is not well understood. Maturation to yield infectious virus relies on cleavage of VP0 into VP2 and VP4. Productive infection induces apoptosis (programmed cell death) and cell lysis facilitating release of virus from the cytoplasm.

Host Range, Propagation, and Detection

HRVs exhibit a high degree of species specificity due to the inability to bind nonhuman ICAM-1 on the cell surface. In addition, there appears to be a block to viral replication in nonpermissive cells. As a result efficient growth occurs only in human and some primate cells. An attempt to adapt a serotype for a mouse infection model had limited success and currently no small-animal HRV infection model has been reported. Nonhuman cells can be made permissive for infection if manipulated to express human ICAM-1 on the cell surface. Primary cells such as human embryonic kidney, bronchial epithelial, tonsil, and continuous human cell lines such as HeLa, H292, and HEP-2 can support growth of HRVs. The most commonly used cells for rhinovirus growth are the WI-38 strain and the MRC-5 strain of diploid fibroblasts, foetal tonsil cells, and HRV-sensitive HeLa (e.g., Ohio HeLa). Growth of virus in cell monolayers is usually detected by the appearance of cytopathic effect (CPE), which initially appears as foci of rounded up cells. HRVs can be plaqued on a number of cell lines with most techniques employing a semisolid overlay. Many HRV serotypes, particularly group B members, do not grow well in tissue culture. Nevertheless virus culture in susceptible cell lines has been the 'gold standard' for laboratory diagnosis of respiratory virus infections. However tissue culture techniques are generally laborious, time consuming, and insensitive when compared with more recent PCR-based assays. For this RNA is extracted from potentially infected samples and copied to cDNA by reverse transcription. PCR using HRV-specific primers targeting conserved viral sequences, such as the 5′-UTR, is then performed to determine the presence of viral genetic material in the original sample. PCR techniques have advanced further with the development of quantitative real-time PCR (qRT-PCR). This technique is more sensitive than conventional PCR and enables the number of viral RNA copies in a sample to be measured. In addition to direct detection of HRV by culture or PCR, antiviral antibodies can be measured by serological methods such as enzyme-linked immunosorbent assay (ELISA). Serological assays are diagnostically useful only when the HRV serotype is already known (i.e., for research purposes such as experimental human infections).

Serological Relationship and Antigenic Variability

With over 100 immunologically distinct serotypes currently recognized HRVs exhibit remarkable antigenic variability. The existence of a large number of HRV serotypes is in contrast to two other medically important picornaviruses, the polioviruses (three serotypes) and hepatitis A virus (a single serotype). While the development of virus-specific neutralizing antibody correlates with protection from disease, anti-HRV antibodies are highly serotype specific, rarely exhibiting cross-serotype neutralizing activity. Several serotypes can be grouped on the basis of low-level immunological cross-reactivity with hyperimmune rabbit serum (HRV-36, -58, and -89, for example), but it is not known whether this plays any role in cross-serotype protection. Mutagenesis studies have identified four neutralizing immunogenic sites along the edge of the capsid canyon. Nim-1a and Nim-1b on VP1 are located above the canyon, and Nim-2 on VP2 and Nim-3 on VP3 are positioned below. Structural analyses of neutralizing antibodies complexed with virus indicate that effective neutralizing activity depends on bivalent antibody binding that both blocks receptor interaction and stabilizes the virus capsid. Typically the most effective neutralizing antibodies bind with high affinity and span the canyon receptor site. In contrast to antibody responses, T-cell responses may be cross-reactive against numerous HRV serotypes, with the major T-cell epitopes buried within the viral capsid where amino acid sequences are more conserved.

Epidemiology

The common cold is almost certainly the commonest illness affecting mankind, occurring in all populations and ages throughout the year. The illness that results from infection is a major cause of morbidity and in turn lost productivity through absenteeism from work or school. Infection rates are highest among infants and children who may be infected up to 12 times a year. Infection rates decrease with age with adults infected on average 2 to 5 times per year. Infections exhibit a seasonal pattern and are more prevalent in autumn and in late spring. It is clear that school attendance is a major contributing factor to seasonality of infection. HRVs cause between 35% and 60% of common colds. While cold exposure does not seem to be a contributing factor to susceptibility to HRV infection, factors such as age and family structure are clearly important. Infections increase significantly from the second year of life, peak around the age of 6, and decline thereafter. The family unit is a major site for spread of HRV in modern societies. Most often an infected child introduces the infection into the family with other siblings and the mother most at risk of secondary infection, presumably because of increased exposure. Schools and day-care centers are also sites of high transmission due to overcrowding, low immunity, and the unhygienic habits of children. Rhinoviruses have also been shown to spread among university students and boarding school residents. HRVs also cause a significant number of afebrile respiratory illnesses in military populations. All populations are affected and observations indicate that multiple HRV serotypes circulate within a population at any given time, and prevalent serotypes change from year to year.

Pathogenesis of Rhinovirus-Induced Disease

Transmission

Two routes are likely to be important for person to person spread of rhinovirus infection via virus-contaminated respiratory secretions: direct hand–surface–hand contact and aerosol inhalation. Rhinoviruses are able to survive on environmental surfaces for several hours at ambient temperature. HRVs were recovered from up to 90% of hands of persons with a cold and from a range of environmental objects such as door knobs, dolls, cups, and glasses. Transfer of virus through touching such objects occurs in seconds. Subsequent rubbing of the nose or eyes with infected hands can result in direct inoculation. Despite compelling evidence supporting direct transmission, demonstrating this in natural circumstances has been less convincing. In contrast there is good evidence suggesting inhalation of aerosols is the major route of HRV infection.

The incubation period preceding virus shedding is 1–4 days, usually 2–3 days. Virus in nasal secretions peaks 2–4 days post infection and remains high for 7–10 days, although low levels may persist for as long as several weeks. Atopic asthmatic subjects may have impaired virus clearance and increased virus load, factors that correlate with increased severity of illness. High virus titers in nasal secretions, increased symptoms, time spent in contact, and social factors such as crowding and poor hygiene contribute to increased transmission of virus.

Pathogenesis and Clinical Features

The nose is the main portal for HRV entry into the body and the primary site of replication is the epithelial surface of nasal mucosa. The virus is then transported to the posterior nasopharynx by ciliated epithelial cells. Clinical signs of illness are generally limited to common cold-like symptoms. HRV replication may also occur in the lower airways and this may actually be a common occurrence following infection of the upper airways. Unlike influenza or adenovirus infections HRV infection produces little, if any cytopathology of the nasal mucous membrane. The absence of histopathology in infected nasal mucosa initially led to the suggestion that mediators of inflammation produced by infected cells in the airway are responsible for cold symptoms. HRV infection *in vitro* and *in vivo* induces production of numerous pro-inflammatory and immune mediators including interferons (IFN-α/β, IFN-λ), chemokines that attract neutrophils (CXCL8/IL-8, CXCL5/ENA-78, CXCL1/GROα), esoinophils (Eotaxin/CCL10, CCL5/RANTES) or lymphocytes (CCL5/ RANTES, CXCL10/IP-10), pro-inflammatory cytokines (IL-1β, IL-2, IL-4, IL-11, IL-12, IL-13, IL-16, IFN-γ, and TNFα), growth and differentiation factors (IL-6, G-CSF, GM-CSF, IL-11), cell adhesion molecules ICAM-1 and vascular cell adhesion molecule (VCAM), and respiratory mucins. The presence of an array of pro-inflammatory mediators is supportive of an immune rather than cytopathic mechanism of rhinovirus-induced illness (**Figure 3**).

Rhinovirus infections are usually relatively trivial, producing symptoms of the common cold including rhinorrhoea, sneezing, nasal discharge and obstruction, coughing, and sore throat. Fever and malaise are less commonly seen than in other respiratory virus infections. Sleep patterns are often disrupted affecting mood and mental functioning. Symptoms appear after a 24–48 h incubation period, peak 2–3 days later, and last for 5–7 days in total, but may persist for up to 2–4 weeks. Severity of symptoms resulting from HRV infections is highly variable ranging from a barely apparent illness to a non-influenza flu-like disease. The most common complications associated with infection of the upper respiratory tract in children are acute otitis media and sinusitis. Rhinoviruses can also infect the lower respiratory tract and thereby account for lower respiratory symptoms and can cause serious and debilitating disease, particularly in young children, the elderly, the immunocompromised, and patients with chronic disorders such as cystic fibrosis and bronchopulmonary dysplasia. In patients hospitalized with respiratory problems rhinovirus infection has been associated with pneumonia in infants, wheezing in asthmatic children, exacerbations of chronic bronchitis, asthma and COPD, and congestive heart failure in older adults. HRV infection is the most frequent causative agent of asthma exacerbations. Recent studies have demonstrated that 80–85% of asthma exacerbations in children are associated with respiratory virus infections, with rhinoviruses accounting for two-thirds of these infections or 50% of all exacerbations (**Figure 4**).

Immune Response and Immunity

Serum antibody specific to the infecting HRV develops between 7 and 14 days after inoculation, but initially antibody-free volunteers may take up to 21 days to be

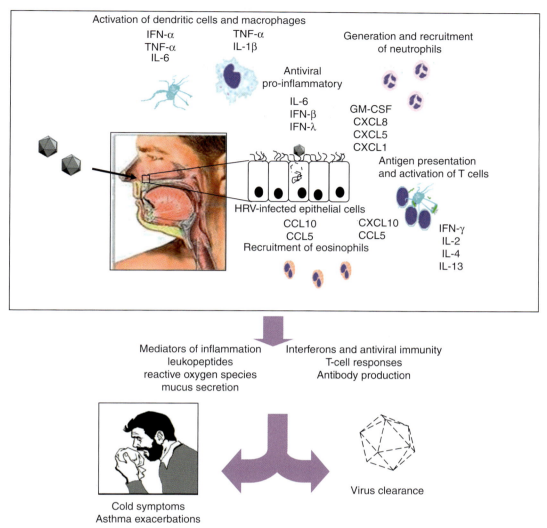

Figure 3 Mechanism of rhinovirus-induced illness. Infection of nasal epithelium stimulates production of a range of pro-inflammatory mediators causing immunopathology usually associated with cold symptoms, or exacerbations of asthma in susceptible individuals.

detected. Virus-specific IgG and IgA antibodies remain low for the first week post infection, peaking approximately a week later. The dominant serum antibody response to infection is IgG1, followed by IgG3, IgG4, and IgG2; IgA1 is the dominant IgA subclass. IgG antibodies stay at high levels for at least a year, while IgA levels decline slowly, but remain detectable for the same period. IgA is the dominant immunoglobulin in nasal secretions, becoming detectable 2 weeks post inoculation, peaking 1 week later. Serum and secretory antibody persist for several years after infection, although levels decline. Given that HRV-specific antibodies appear relatively late in the infection, humoral immunity is likely to be not essential for recovery from viral illness but may be involved in final viral clearance. Preexisting antibodies are likely to be important in protection from reinfection with the same serotype. Mechanisms of antibody-mediated virus inactivation might include virus aggregation and complement activation, blocking receptor binding and stabilization of capsid to prevent uncoating.

Cellular immunity is also activated following HRV infection. In contrast to the highly serotype-specific humoral response, virus-specific lymphocytes may be activated by several serotypes indicating shared viral epitopes, but their role in subsequent protection is not known.

Prevention and Control

Currently there are no effective strategies for prevention or treatment of rhinovirus common colds. The large number of serotypes and lack of heterotypic humoral immunity is a major issue precluding the use of conventional vaccine approaches. Vaccine enhancement of virus-specific cell-mediated immunity may be a more promising approach since T-cell epitopes are more conserved.

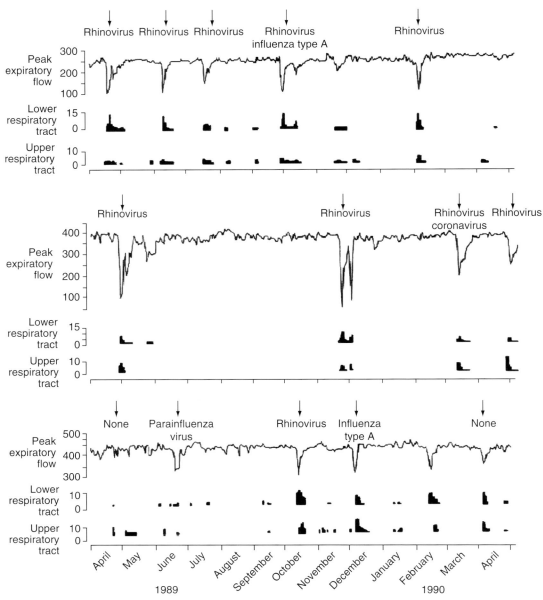

Figure 4 Association of marked reductions in lung function (peak expiratory flow) and exacerbations of lower respiratory tract symptoms with rhinovirus infection for three asthmatic children. Reproduced from SL Johnston, PK Pattemore, G Sanderson, et al. (1995) Community study of role of viral infections in exacerbations of asthma in 9–11-year-old children. *British Medical Journal* 310(6989): 1225–1229, with permission from BMJ Publishing Group.

The lack of a vaccine or specific therapies and the commonness of the illness have resulted in the emergence of numerous nonspecific therapies for the common cold. These include ascorbic acid, zinc gluconate, echinacea, and inhalation of hot humidified air. The efficacy of any nonspecific measure in prevention or treatment of the cold is yet to be generally accepted. Most work has been invested into the development of chemotherapeutic approaches. These include pharmacological antiviral agents that indirectly inhibit virus replication such as IFN-α. Studies demonstrated that IFN-α was effective in preventing the onset of cold symptoms when administered prophylactically. However it had little to no effect when given after infection and was often associated with various side effects. A large number of compounds designed to inhibit virus uncoating and/or cell binding and entry have been studied. So far an effective drug free of side effects is yet to emerge from this research. One such drug, Pleconaril, was submitted for approval to market as a treatment for the common cold in adults. Evidence showed that Pleconaril reduced the duration of cold symptoms if taken within the first 24 h of a cold. Unfortunately, the drug was not approved due to interactions in test subjects with the oral contraceptive and associated side effects.

Another approach has involved blocking virus binding to cells using a soluble form of the receptor. A number of reports have demonstrated that soluble ICAM-1 inactivates the virus and is effective at inhibiting viral entry. Despite encouraging results with soluble ICAM-1 no development is currently ongoing for these agents due to problems with formulation and delivery.

See also: Common Cold Viruses; Picornaviruses: Molecular Biology.

Further Reading

Contoli M, Message SD, Laza-Stanca V, *et al.* (2006) Role of deficient type III interferon-lambda production in asthma exacerbations. *Nature Medicine* 12(9): 1023–1026.

Couch RB (2001) Rhinoviruses. In: Knipe DM, Howely PM, Griffin DE, *et al.* (eds.) *Fields Virology*, 4th edn., pp. 777–797. Philadelphia, PA: Lippincott Williams and Wilkins.

Edwards ME, Kebadze T, Johnson MW, and Johnston SL (2006) New treatment regimes for virus-induced exacerbations of asthma. *Pulmonary Pharmacology and Therapeutics* 19: 320–334.

Heikkinen T and Jarvinen A (2003) The common cold. *Lancet* 361(9351): 51–59.

Johnston SL, Patterson PK, Sanderson G, *et al.* (1995) Community study of role of viral infections in exacerbations of asthma in 9–11-year-old children. *British Medical Journal* 310(6989): 1225–1229.

Papadopoulos NG and Johnston SL (2004) Rhinoviruses. In: Zuckerman AJ, Bantavala JR, Griffiths PD, and Schoub BD (eds.) *Principles and Practice of Clinical Virology,* 5th edn., pp. 361–377. West Sussex, England: Wiley.

Wark PA, Johnston SL, Bucchieri F, *et al.* (2005) Asthmatic bronchial epithelial cells have a deficient innate immune response to infection with rhinovirus. *Journal of Experimental Medicine* 201(6): 937–947.

Ribozymes

E Westhof and A Lescoute, Université Louis Pasteur, Strasbourg, France

© 2008 Elsevier Ltd. All rights reserved.

Glossary

Nucleolytic Qualifies ribozymes which undergo cleavage and ligation following a nucleophilic attack of the 2′ hydroxyl group on the adjacent 3′ phosphate group.
Transesterification Transfer of a phosphoryl group.
Transpeptidation Formation of a peptide bond.
Concatemer Multiple copies of a DNA sequence arranged end to end in tandem.

Introduction

The discovery in the 1980s of RNA molecules with catalytic activity has revolutionized molecular biology. Ribozymes, molecules able to catalyze various chemical reactions, are widespread in biology. Large- and small-size ribozymes can be distinguished. The large ribozymes comprise the group I introns, the group II introns, and ribonuclease P (RNaseP). A step of maturation of pre-mRNAs in some bacteria or in nuclear eukaryotic rRNAs implicates the self-splicing of group I introns which use a guanosine molecule as a cofactor to catalyze a two-step transesterification reaction. The crystal structures of group I introns show that the catalytic core is conserved and stabilized by long-range interactions that involve different peripheral elements. At the same time the structures of the specificity domain of two RNaseP RNAs, responsible for the maturation of pre-tRNA, have been solved and show that the global architecture of the molecule is conserved partly due to the long-range interactions. Most importantly, it was established that the rRNA 23S is responsible for catalyzing the transpeptidation reaction during protein synthesis. Among the simplest catalytic RNA molecules are the small nucleolytic RNA species. The small ribozymes include the Varkud satellite ribozyme (VS ribozyme), the hairpin ribozyme, the hepatitis delta virus (HDV), and hammerhead ribozymes. They carry out reversible cleavage and ligation reactions in the presence of physiological concentrations of magnesium ions. Ribozyme activity is necessary during the replication of some plant pathogen RNA genomes.

The Chemical Reactions

The reaction catalyzed by small nucleolytic ribozymes takes place at a specific phosphodiester bond by a transesterification reaction involving the 2′-hydroxyl group for cleavage and the 5′-hydroxyl group for ligation. The cleavage reaction proceeds by activation of the 2′-hydroxyl adjacent to the scissile phosphate. The 2′-hydroxyl group performs a nucleophilic attack on the adjacent phosphorus atom, which leads to a pentacoordinate transition state. After capturing a proton, the 5′-oxygen atom leaves leading to the formation of a cyclic 2′,3′-phosphate group (**Figure 1**). Ligation is the reverse reaction of cleavage and involves: (1) activation of the 5′-hydroxyl group by a base; (2) attack of the 2′,3′-cyclic phosphate; and (3) liberation and protonation of the 2′-hydroxyl group. Magnesium ions are generally necessary for the native folding of RNA molecules. Whether magnesium ions participate actively or indirectly in the catalysis reaction is still a matter of debate. During recent years, it has become clear, however, that the nucleic acid bases themselves can play an active role in acid–base catalysis. Nowadays, the crucial roles for efficient catalysis of tertiary contacts between peripheral elements, even in small ribozymes, have been firmly established.

The Ribozymes of Plant Viroids and Virusoids

Replication by the Rolling-Circle Mechanism

Two of the autocatalytic RNAs, the hammerhead and hairpin ribozymes, were discovered first in some viroids and virusoids responsible for several economically important infectious diseases of plants. These pathogens are small circular single-stranded RNAs, without a capsid, possess no open reading frame (ORF) and the RNAs are classified as noncoding RNAs (**Figure 2**). Their replication takes place (1) in the cytoplasm with a helper virus in the case of virusoids; (2) in an autonomous way (without helper virus) in chloroplasts, or in the nucleus in the case of viroids. The replication of these circular genomes is made according to a 'rolling-circle' mechanism, which involves the copying of the circular genome positive (+) strand by an RNA polymerase of the host or helper virus to give a multimeric negative-strand RNA (−) (concatemer) (**Figure 3**). In the case of the viroid family *Pospiviroidae*, the synthesized linear concatemers of the (−) strands are used as matrix for the synthesis of (+) concatemers; the circularization of those which form unit genomes seems to require an RNA ligase as indicated by the presence in some instances of a 2′ phosphomonoester and a 3′,5′ phosphodiester bond.

In the family *Avsunviroidae*, the (−) concatemer undergoes an autocleavage reaction involving a hammerhead ribozyme. The mechanism of circularization of the (−) strand to be used as a matrix is still controversial. Indeed, two hypotheses have been proposed: either the intervention of an RNA ligase or the ribozyme catalyzes the opposite reaction and ligates the RNA. In the end, the (−) strand is used as matrix for the synthesis of (+) concatemers by the mechanism of rolling circle. After cleavage by the hammerhead ribozyme, the (+) strands are circularized. Thus, only the RNA intermediaries are implicated in this mechanism of replication since the viroid genome does not code for proteins. Thus, the various biological properties of the viroids, such as the ability to identify the host, depend exclusively on the sequence and the structure of their RNAs.

The hammerhead ribozyme plays a fundamental role in the replication of these pathogenic RNA genomes. Yet, its role is not limited to this function as it has been identified in genomes of the triton (*Notophtalamus viridescens*), in parasitic schistosomes, and in cave crickets (*Dolichopoda* sp.). Two hammerhead ribozymes have been found also in the *Arabidopsis thaliana* genome.

The Hairpin Ribozyme

The hairpin ribozyme is involved in the processing of RNA molecules generated in the replication cycles of various species. The tobacco ringspot virus satellite RNA (sTRSV), for example, carries a hammerhead ribozyme on the (+) strand and a hairpin ribozyme on the (−) strand. The crystal structure of the hairpin ribozyme shows that it is comprised of two important structural elements: a four-way helical junction which is necessary for the optimal function of the ribozyme and two internal loops within the interacting helices (**Figure 4**). The minimal ribozyme (i.e., without the other two arms) catalyzes the cleavage reaction but requires 3 orders of magnitude higher concentration of magnesium cations (Mg^{2+}) than the natural form. Single-molecule studies show that the natural ribozyme exists in three states: an undocked form (where helices do not contact each other), an intermediate state (with helices docked but without contact between the internal loops), and a docked form (with intimate contacts between the loops). Although the junction is not essential to the hairpin ribozyme activity, it greatly enhances the rate of folding of the ribozyme in an active form at physiological concentration of Mg^{2+} ions. These auxiliary elements play the same role as in the hammerhead ribozyme and appear to act as 'folding enhancers'. However, there is no evidence for the direct implication of metal ions in the chemistry of the hairpin ribozyme.

The Hammerhead Ribozyme

Classification

The natural hammerhead ribozymes can be classified into two main types according to the helix numbering which carries the RNA 3′ and 5′ ends. In type 1, the 5′ end is carried by helix I which always presents an internal loop. Helix II presents in some instances an internal loop before the terminal loop. Helix III does not present a terminal loop but possibly an internal loop. In type 3, the 5′ end is carried by helix III. Helices I and II present sometimes an internal loop preceding the terminal loop. Ribozymes where the 5′ end is carried by helix II have, so far, never been observed in any genomes.

Structure and folding

The structure of the catalytic core of the hammerhead ribozyme was determined by crystallography using an RNA/DNA hybrid as well as in the 'all RNA' form with a methyl group on the 2′-hydroxyl of the site of cleavage

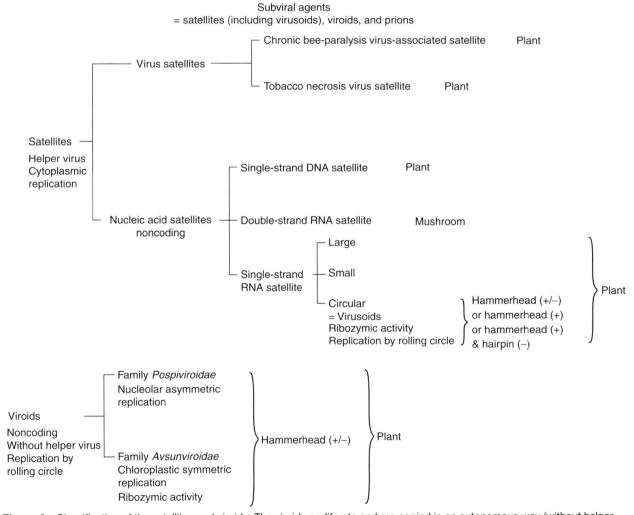

Figure 1 Reactions occurring during self-cleavage of the nucleolytic ribozymes like the hammerhead, the hairpin, the Varkud, or the Hepatitis delta virus ribozymes. The adjacent 2′-hydroxyl group is activated for nucleophilic attack. In a concomitant way, a proton is given to the leaving 5′-oxygen group.

Figure 2 Classification of the satellites and viroids. The viroids proliferate and are copied in an autonomous way (without helper virus). Their circular RNA genomes do not have an open reading frame. The virusoids are satellite RNAs with circular single-stranded genomes. They do not code for protein, contrary to the satellite genomes of viruses which code for capsid proteins.

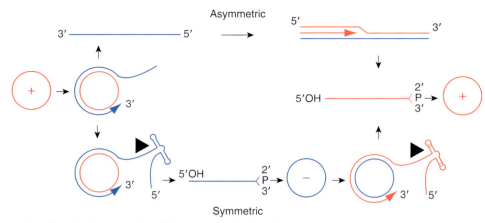

Figure 3 Mechanism of replication using the rolling-circle mechanism. There are two main ways for replication: an asymmetrical way (top) which requires only one rolling circle as seen in the family *Pospiviroidae* and a symmetrical way which requires two rolling circles specific to the family *Avsunviroidae*. The (+) strands are drawn in red and the (−) strands in blue. The hammerhead ribozymes in the cleavage of the concatemers utilize only the symmetrical way.

which prevents the cleavage reaction. The two practically identical structures show a catalytic core composed of 11 nucleotides essential for catalysis, organized in a three-way junction from which the helices I, II, and III leave. Helix II is stacked on helix III, while helix I is parallel to helix II. Helix II, composed of Watson–Crick base pairs, is stacked on three non-Watson–Crick base pairs.

Many structural studies of the ribozyme in solution have shown the importance of Mg^{2+} ions for folding. Indeed, in absence of divalent ions, the core of the ribozyme is not structured and the helices are equidistant from each other, revealing a disorganization of stacking, whereas in the presence of divalent ions the ribozyme folds in two stages to an active structure. At a low concentration, helices II and III are stacked but the angle between helices I and II is very large and the catalytic core is not structured; the ribozyme is inactive. At higher concentration, helix I reorients to a position near helix II and the ribozyme is then in an active conformation. However, the majority of the studies on hammerhead ribozyme folding were carried out on minimal ribozymes presenting the conserved and presumably essential elements, that is, 13 nucleotides of the catalytic core of which 11 cannot be mutated without causing loss of activity, and three helices I, II, and III with variable length and termination. These minimal ribozymes are active in an optimal way at high concentrations of Mg^{2+} ions (>10 mM) but almost inactive at physiological concentration of Mg^{2+} (0.1–0.3 mM).

However, recent studies showed that the tertiary contacts between the terminal loops of helices I and II, peripheral to the catalytic core and nonessential for catalysis, are essential for the activity of hammerhead ribozymes in physiological conditions (low concentration of Mg^{2+} ions). The role of the peripheral structures and their interactions in the catalytic activity of the hammerhead ribozymes is expected from previous observations

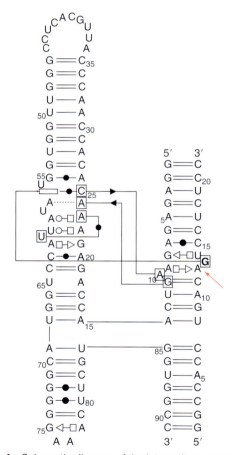

Figure 4 Schematic diagram of the interactions present in the three-dimensional crystal structure of the hairpin ribozyme (PDB ID 1HP6). The red arrow indicates the cleavage site.

on longer ribozymes such as group I introns. The progressive deletion of the peripheral area of group I introns results in increasing requirements of salts and ultimately in loss of activity. The various categories of group I intron all have the same catalytic core, although the outlying areas and the tertiary interactions differ and they all contribute to the stability of the catalytic core. This suggests that during evolution, several folding solutions were found which ensured the optimal catalytic activity of the ribozyme. In the same manner, the peripheral area of the natural hammerhead ribozymes evolved so as to optimize the folding and the catalytic activity for biological needs under natural conditions of Mg^{2+} ion levels. The tertiary interactions between these regions reduce the conformational space accessible to the molecule and facilitate the folding of the ribozyme, thereby increasing its catalytic activity. It is clear that the catalytic core of the ribozyme had to undergo structural changes, which explains the increase in its catalytic efficiency although the precise nature of the changes is unknown. The structure of the full-length *Schistosoma mansoni* hammerhead ribozyme reveals the structural basis for the effectiveness of natural ribozymes and explains most of the apparent inconsistencies between the structure of the minimalist ribozymes and the biochemical data.

Role of Peripheral Elements in the Mechanism of Ligation

Like any enzyme, the hammerhead and hairpin ribozymes are able to carry out the reaction opposite to cleavage, that is, ligation. However, in majority of conditions, cleavage is favored at least 100-fold compared to ligation. The various stages for ligation are (1) activation of the 5′-hydroxyl by a base, (2) attack of the 2′,3′-cyclic phosphate, and (3) release and protonation of 2′-hydroxyl, restoring the 3′,5′-phosphodiester bond.

The Hairpin Ribozyme

The studies of the dynamics of the hairpin ribozyme made it possible to determine the conformational kinetic properties of the reactions of ligation and cleavage. They show that after cleavage the four-way junction of the ribozyme moves into an undocked state. The reaction of cleavage is definitely slower than the reaction of ligation, but ligation is slower than helical unstacking. The product is thus normally released before ligation takes place. However, if the product of cleavage remains bound to the enzyme, the ribozyme alternates between docked and undocked state before ligation takes place. These experiments also show that the nature of the product is a factor influencing the speed of ribozyme unfolding. Indeed, the presence of a 2′,3′-cyclic phosphate increases the unfolding of the ribozyme compared to a product having either a 3′-phosphate or 3′-hydroxyl. Therefore, the product has important effects on structural dynamics of the ribozyme, and the loss of covalent continuity in the backbone is not the only factor increasing the undocking of the ribozyme. The undocking promoted by the presence of cyclic phosphate, involves detachment of the product. This is in agreement with the biology of the ribozyme in its natural environment where it cleaves the RNA concatemer (produced by rolling circle) and releases a monomeric 2′,3′-cyclic phosphate product. In the case of the hairpin ribozyme (−) strand of the satellite RNA 'sTRSV', the helix formed by the product and the enzyme, with a length of six base pairs, has a weak dissociation constant. Single-molecule spectroscopy shows that docking–undocking events can take place several times before the ligation does take place. The equilibrium constant of cleavage and ligation shows a significant bias toward ligation with an internal equilibrium constant of $K_{int} = k_L/K_C = 34$. This preference for ligation ensures that during virus replication a certain number of circular RNAs will be maintained which will be used as matrix for the synthesis of the concatemer. On the other hand, the rate of undocking is greater than the rate of ligation which allows cleavage of concatemers because of the rapid undocking that follows cleavage.

The Hammerhead Ribozyme

For the hammerhead ribozyme, it has been shown that the kinetic constant of ligation ($k_{-2} = 0.008$ min^{-1}) is 100 times lower than the constant of cleavage ($k_2 = 1$ min^{-1}); the equilibrium constant $K_{eq} = k_2/k_{-2}$ thus has a value of 125 (**Figure 5**). This is explained by the more favorable thermodynamic state of the enzyme/product complex than of the complex enzyme/substrate which favors the formation of the latter by driving the reaction toward cleavage. The tertiary interactions, by limiting the accessible conformational space (i.e., by stabilizing the folding of the complex enzyme/product), increase the ligation activity of the ribozyme. The same phenomena are observed for the hairpin ribozyme, since the presence of the native four-way junction with four helices moves the internal equilibrium toward binding and increases the rate of folding.

The Mechanism of Monomer Circularization after Replication

The peripheral elements together with the catalytic core are not only involved in the reaction of cleavage but also in the reaction of ligation. Each monomeric copy of the genome, product of the replication by rolling circle and cleavage by the hammerhead ribozyme, must be circularized to be used in turn as a replication matrix. Little is known about the *in vivo* circularization of the monomer. The viroid RNAs belonging to the family *Pospiviroidae* are probably ligated by an RNA ligase as indicated by the presence of a 2′-phosphomonoester and a 3′,5′-phosphodiester bond at the site of ligation. On the other hand,

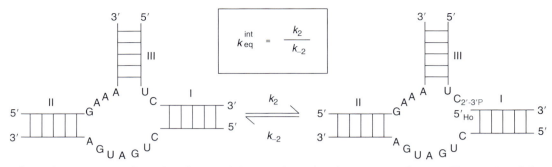

Figure 5 Secondary structure and reaction of the catalytic core of a minimal hammerhead ribozyme. The reaction of cleavage is defined by the constant of cleavage k_2 and the reaction of ligation is defined by the constant of ligation k_{-2}. The definition of the equilibrium constant is framed.

information on the ligation of viroid RNAs in the family *Avsunviroidae* is limited. In the case of the circularization of the genome of peach latent mosaic viroid (PLMVd), a 2′,5′-phosphodiester bond has been observed *in vitro* and *in vivo*. Such a bond would prevent cleavage of the ribozyme and would ensure the maintenance of an essential circular RNA matrix by the rolling-circle mechanism.

Yeast and Human Satellite Viruses

The *Neurospora crassa* versus Ribozyme

The VS RNA is transcribed from the mitochondrial DNA of *Neurospora crassa*. The VS RNA contains a ribozyme of around 150 nt in length acting in the processing of replication intermediates. There is only limited information available on its structure. The VS ribozyme comprises five helices organized in two three-way junctions (2–3–6 and 3–4–5 junctions) to form a Y shape. A pseudoknot is formed between the terminal loop of the substrate helix and the helix V of the ribozyme (**Figure 6**). The scissile phosphate is within an internal loop of the substrate stem–loop. The catalyzed reaction requires divalent Mg^{2+} cations.

The Human HDV

The HDV is a satellite virus of hepatitis B virus (HBV) on which it is dependent for its replication cycle. The genome of HDV is a 1700-nt-long circular single-stranded RNA. Its replication is RNA directed without DNA intermediates. Like certain pathogenic subviral RNAs that infect plants, HDV RNA replicates by a rolling-circle mechanism. Both genomic (+) and complementary (−) strands of RNA contain a single ribozyme of about 85 nt. The HDV ribozyme folds into a compact structure comprising a double pseudoknot and its disruption results in a marked loss of activity (**Figure 7**). The X-ray structures of the two states of the ribozyme (pre- and post-cleaved states) reveal a significant conformational change in the RNA

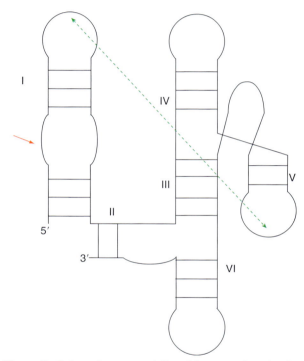

Figure 6 Schematic representation of the secondary structure of the VS ribozyme. The pseudoknot is indicated by the green arrow. The red arrow indicates the cleavage site.

after cleavage and point to the role of a divalent metal ion in the cleavage mechanism.

Conclusions

The hammerhead and hairpin ribozymes were initially studied in their minimalist form, which resulted in weak activities due to the presence of numerous conformations in dynamic exchange, most of them inactive, which required high concentrations of divalent ions for activity. However, the peripheral elements are essential for optimal activity in native conditions. These peripheral regions, by interacting with each other, facilitate and stabilize folding into a single active structure. They are, therefore, necessary

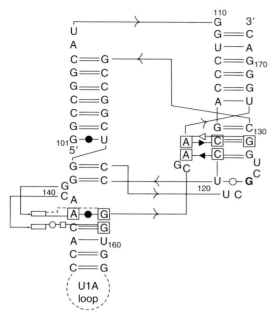

Figure 7 Schematic diagram of the interactions present in the three-dimensional crystal structure of HDV ribozyme after self-cleavage (PDB ID 1DRZ).

mimicking the cap of mRNA. Our present knowledge underlines the fundamental links existing between the folding pathway, the selection and stabilization of a single native state by tertiary interactions between peripheral elements and the catalytic activity of ribozymes.

See also: Icosahedral Enveloped dsRNA Bacterial Viruses; Icosahedral ssRNA Bacterial Viruses; Interfering RNAs; Satellite Nucleic Acids and Viruses.

Further Reading

Blount KF and Uhlenbeck OC (2005) The structure–function dilemma of the hammerhead ribozyme. *Annual Review of Biophysics and Biomolecular Structure* 34: 415–440.
Daros JA, Elena SF, and Flores R (2006) Viroids: An Ariadne's thread into the RNA labyrinth. *EMBO Reports* 7: 593–598.
Doudna JA and Cech TR (2002) The chemical repertoire of natural ribozymes. *Nature* 418: 222–228.
Gesteland RF, Cech TR, and Atkins JF (eds.) (2006) *The RNA World*. New York: Cold Spring Harbor Laboratory Press.
Khvorova A, Lescoute A, Westhof E, and Jayasena SD (2003) Sequence elements outside the hammerhead ribozyme catalytic core enable intracellular activity. *Nature Structural and Molecular Biology* 10: 708–712.
Lafontaine DA, Norman DG, and Lilley DM (2002) The global structure of the VS ribozyme. *EMBO Journal* 21: 2461–2471.
Leontis NB and Westhof E (2001) Geometric nomenclature and classification of RNA base pairs. *RNA* 7: 499–512.
Lilley DM (2004) The Varkud satellite ribozyme. *RNA* 10: 151–158.
Lilley DM (2005) Structure, folding and mechanisms of ribozymes. *Current Opinion in Structural Biology* 15: 313–323.
Martick M and Scott WG (2006) Tertiary contacts distant from the active site prime a ribozyme for catalysis. *Cell* 126: 309–320.
McKay DB (1996) Structure and function of the hammerhead ribozyme: An unfinished story. *RNA* 2: 395–403.
Westhof E (2007) A tale in molecular recognition: The hammerhead ribozyme. *Journal of Molecular Recognition* 20: 1–3.

in physiological conditions although they are not directly implied in the catalysis. These results put into question the conclusions drawn previously from data obtained with molecular systems that had been simplified and reduced to the extreme. Full understanding of the chemical and biological actions of ribozymes is certainly not yet achieved, however, and surprises are still lurking. Recently, a ribozyme structurally derived from group I introns was shown to catalyze the formation of a $2',5'$-linkage leading to the formation of a lariat structure

Rice Tungro Disease

R Hull, John Innes Centre, Colney, UK

© 2008 Elsevier Ltd. All rights reserved.

Glossary

Agroinoculation Infection of a plant by injecting *Agrobacterium* containing a modified Ti plasmid into which an infectious copy of the viral genome has been inserted. In the case of pararetroviruses, this infectious copy comprises the sequence that can express the more-than-full-length RNA transcript required for the reverse transcription stage of viral replication.
Pararetrovirus A virus that replicates by reverse transcription but differs from retroviruses in that it encapsidates the DNA phase of replication and that virus replication does not require integration of the viral genome into the host chromosome.
Semipersistent transmission Virus–vector interaction in which the virus is acquired rapidly on feeding on an infectious plant, and then is transmitted by relatively short feeds on a healthy host plant. The virus interacts with specific sites in the vectors' mouthparts or foregut.

Rice Tungro Disease

The characteristic symptoms of rice tungro disease had been reported from several Southeast Asian countries for many years before they were recognized in 1963 to be of viral etiology. During and since the 1960s, there have been several severe outbreaks of the disease mainly associated with the major agronomic improvements in rice cultivation resulting from the 'green revolution'. The disease is found in all countries of South and Southeast Asia and in the south and southeast of China (**Figure 1**). The importance and distribution of the disease is evidenced by the variety of local names for it: *accep na pula* (Philippines), *mentek* (Indonesia), *penyakit merah* (Malaysia), yellow orange leaf (Thailand), leaf yellowing (India). In the early 1990s, it was estimated that tungro caused annual losses in excess of $1.5 billion without taking account of the cost of insecticide to control the vector. In parts of India, the disease causes losses of up to 50% of the expected crop yield.

In many rice cultivars, tungro infection is evidenced by a marked stunting of plants, discoloration of the leaves, and reduced tiller number. The leaves of tungro-infected plants are yellow or orange (**Figure 2**) with the discoloration starting at the tip and extending down toward the base; infected leaves are slightly rolled inward. Tolerant rice cultivars usually show only partial discoloration which may not be apparent on younger leaves. The disease is transmitted in the semipersistent manner by several leafhopper species with the rice green leafhopper, *Nephotettix virescens* (**Figure 3**), being the most important vector. It is also transmitted by *Nephotettix cincticeps*, *Nephotettix nigropictus*, *Nephotettix malayanus*, and *Recilia dorsalis*; these other vectors may be of relative importance in certain situations. The distribution of the disease is circumscribed by the distribution of *N. virescens* (**Figure 1**).

It was not until 1978 that it was recognized that tungro is caused by a complex of two viruses, rice tungro bacilliform virus (RTBV) and rice tungro spherical virus (RTSV) (**Figure 4**). When these viruses are separated, RTBV causes moderate symptoms in rice but is not leafhopper transmitted. RTSV, on the other hand, is leafhopper transmitted but causes few, if any, symptoms; in some Japanese rice cultivars, it caused a disease termed 'waika'. In the disease complex, both viruses are leafhopper transmitted, RTSV effecting the transmission of RTBV as well as itself. RTSV enhances the symptoms of RTBV in infected plants giving rise to the severe symptoms.

Most host range studies on the tungro viruses are unreliable as the association of the two viruses was not then recognized. Host range studies are also constrained by the ability of the vector species to transmit the virus to the plant species tested. RTSV appears to have a restricted host range limited to members of the Poaceae and Cyperaceae (*Echinochloa crus-galli*, *Echinochloa glabrescens*, *E. colona*, *Leptochloa chinensis*, *Leersia hexandra*, *Oryza sativa*, *Panicum repens*, *Cyperus rotundus*); RTSV causes few, if any, symptoms in these hosts.

The severity of the disease has encouraged extensive molecular studies on both tungro viruses even though they have been difficult to work with as they are in very low concentration in the plant and are not mechanically transmissible. RTBV is transmissible by agroinoculation.

Rice Tungro Bacilliform Virus

RTBV is a pararetrovirus belonging to a monospecific genus, *Tungrovirus*, in the family *Caulimoviridae*. It has bacilliform particles of about 130 nm length (though there can be longer ones in some isolates) and 30 nm diameter (**Figure 5**). The particle structure is based on a $T = 3$ icosahedron cut across its threefold axes, the tubular portion being made up of rings of hexamer subunits with a repeat distance of about 10 nm.

The viral coat protein comprises one major species of M_r 37 000; this may be processed to give a product of M_r about 33 000 and possibly others. The molecular mass of the coat protein has been measured by mass spectrometry and the N- and C-terminal sequences have been determined. This protein contains two specific amino acid sequence motifs, $CXCX_2CX_4HX_4C$ characteristic of reverse transcribing virus *gag* proteins and $CX_2CX_{11}CX_2CX_4CX_2C$ unique to 'badnaviruses'.

The particles contain a circular double-stranded DNA genome of about 8 kbp which has one discontinuity at a specific position in each strand. The genomes of several isolates have been sequenced and the virus has been shown to have four open reading frames (ORFs), all on one strand (**Figure 6**). The start codon of ORF1 is ATT but the function of its product of M_r 24 000 is unknown. ORF2 encodes a product of M_r 12 000 which interacts with coat protein and is possibly involved in virus assembly. The polyprotein encoded by ORF3 is processed by an aspartate protease to give: N-terminal protein(s) thought possibly to be cell-to-cell movement protein(s), viral coat protein, aspartate protease, and the replicase comprising reverse transcriptase and RNaseH activities. The ORF4 product of M_r of 46 000 has no known function.

As with all plant pararetroviruses, the genome is transcribed asymmetrically to give a more-than-genome-length RNA, which is the template for reverse transcription and is also the template for the translation of the first three ORFs. The fourth ORF is translated from an mRNA spliced from the more-than-genome-length transcript (**Figure 6**).

There are two major strains of RTBV, one from Southeast Asia and the other from the Indian subcontinent. These strains have about 70% sequence similarity and the Indian strain can be distinguished from the Asian

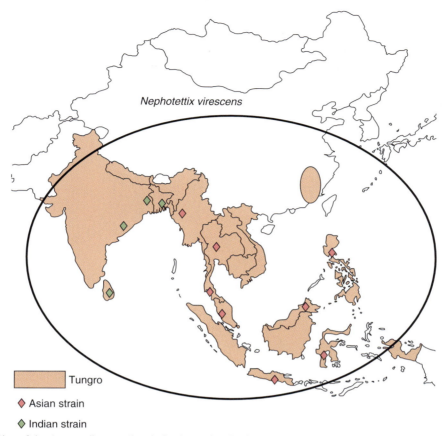

Figure 1 Distribution of rice tungro disease; the circle shows the distribution of the main leafhopper vector.

Figure 2 Field symptoms of rice tungro disease.

Figure 3 The rice green leafhopper, *Nephotettix virescens*.

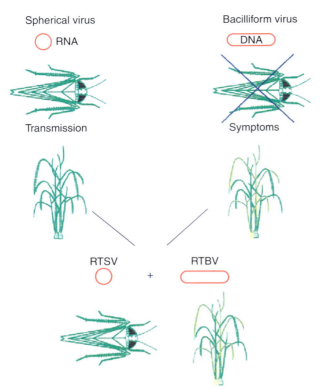

Figure 4 Involvement of RTBV and RTSV in tungro disease. Reprinted, with permission, from the *Annual Review of Phytopathology*, Volume 34 © 1996 by Annual Reviews.

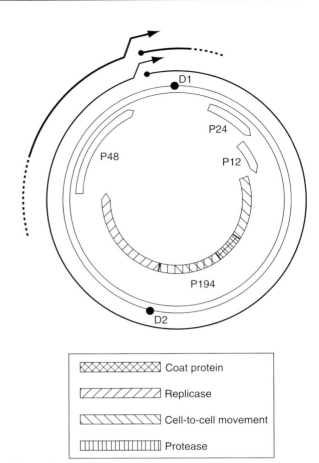

Figure 6 Genome organization of RTBV. The double complete circles represent the dsDNA genome, the positions of the discontinuities (D1 and D2) being shown. Inside the double circles, the arced boxes represent the ORFs, the size of the product being indicated in kDa. Outside the double circles, the positions of the 35S RNA transcript and the spliced mRNA for ORF4 are shown, the arrowheads marking the 3′ ends and the dotted lines the splice. The infills of the boxes represent the functions of the ORF products. Reproduced from Hull R (2001) *Matthews' Plant Virology*, 4th edn., figure 6.4. London: Academic Press, with permission from Elsevier.

Rice Tungro Spherical Virus

RTSV belongs to the genus *Sequivirus* in the family *Sequiviridae* and has isometric particles of about 30 nm diameter (**Figure 5**). RTSV is limited to vascular tissues where it is restricted to the phloem cells. Virus particles are scattered in the cytoplasm or are embedded in nonenveloped electron-dense granular inclusions. Particles also occur as crystalline aggregates often in vacuoles. Small vesicles containing fibers are found in the cytoplasm, usually along the wall of infected cells.

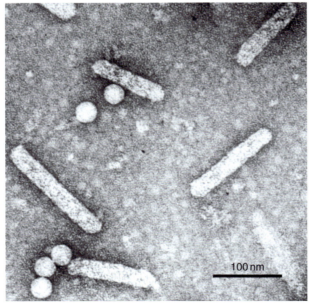

Figure 5 Particles of RTBV and RTSV negatively stained in uranyl acetate.

strain by a deletion of *c.* 65 nt in a noncoding region. Within field isolates, there is microvariation which might be expected from a replication mechanism which does not involve proof reading.

The capsid is made up of three coat protein species (CP1, CP2, and CP3) that are apparently in equimolar amounts, and is thought to have a similar $T = 1$ icosahedral architecture to that of picornaviruses. Identification of cleavage sites by N-terminal sequencing shows that

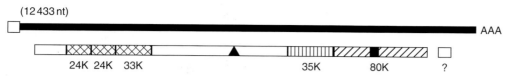

Figure 7 Genome organization of RTSV. The single line represents the (+)-strand RNA genome. The boxes show the coding regions with the final gene products after processing identified for those that are known. The infills represent the function of the gene products – see **Figure 6** for details of these. Reproduced from Hull R (2001) *Matthews' Plant Virology*, 4th edn., figure 6.4. London: Academic Press, with permission from Elsevier.

CP1 is 22.9 kDa and CP2 is 22.3 kDa; the C-terminus of CP3 has not been identified. In Western blots of crude extracts from infected plants, an antiserum to CP3 identifies a band at 33 kDa and several in the range 40–42 kDa; the nature of these larger bands is unknown.

Each RTSV particle contains one molecule of single-stranded (+)-strand RNA of about 12.2 kbp. The RNA encodes a large polyprotein of about 393 kDa and has two short ORFs at the 3′ end (**Figure 7**). The polyprotein is processed to give various products including the three coat protein species, a cysteine or 3C-type protease, and an RNA-dependent RNA polymerase at its 3′ end. Functions of other products have not yet been determined. No functions have been ascribed to the products of the 3′ short ORFs and there is no evidence that these are expressed.

RTSV strain Vt6 overcomes the resistance to the type strain in rice cultivar TKM6. Several serological variants of RTSV have been reported. Molecular techniques reveal much microvariation in the sequences encoding CP1 and CP2 but there is no evidence for major geographic strains that reflect those of RTBV.

Control

There are three major approaches to controlling tungro: by the manipulation of various cultural practices such as planting time, having a crop-free period, and seedbed protection; invasion of young susceptible rice plants by viruliferous leafhoppers can be avoided. The leafhoppers can be controlled by spraying insecticide but this has not proved to be particularly effective. There have been various programs to breed resistance to tungro but most have proved of limited success. The International Rice Research Institute has screened more than 40 000 accessions of rice for resistance to tungro viruses and have found two basic types of resistance, that to the vector and that to the virus. Resistance to the vector has proved to be not very durable and has broken down in less than 1 year in some places. Resistance to each of the viruses is primarily tolerance reducing virus replication and/or reducing symptoms and transmission efficiency. As noted above, resistance to RTSV has been overcome by a new strain of the virus. However, these forms of resistance are being incorporated into breeding programs as resistance to RTSV can reduce the spread of the disease and resistance to RTBV can mitigate losses due to reduced symptoms. There are also attempts to use transgenic approaches to conferring protection against these two viruses but so far nothing has proved to be fully effective.

See also: Cereal Viruses: Rice.

Further Reading

Anjaneyulu A, Satapathy MK, and Shukla VD (1994) *Rice Tungro*. New Delhi: Oxford and IBH Publishing Co. Pvt. Ltd.

Hull R (1996) Molecular biology of rice tungro viruses. *Annual Review of Phytopathology* 34: 275–297.

Hull R (2001) *Matthews' Plant Virology*, 4th edn., London: Academic Press.

Rice Yellow Mottle Virus

E Hébrard and D Fargette, IRD, Montpellier, France
G Konaté, INERA, Ouagadougou, Burkina Faso

© 2008 Elsevier Ltd. All rights reserved.

Introduction

Rice yellow mottle virus (RYMV) is a member of the genus *Sobemovirus*. RYMV is present only on the African continent. It was first reported in Kenya in 1966 and in Côte d'Ivoire in 1974 in the irrigated rice fields. Since the early 1990s, RYMV is present everywhere in sub-Saharan Africa and in Madagascar where rice is grown. It affected all types of rice cultivations, including lowland, upland, rain-fed, floating, and mangrove rice. RYMV regularly induces severe yield losses to rice production ranging from 25% to 100%. In some regions, when epidemics are recurrently very severe, farmers abandoned their fields and eradicated the tropical forests in order to plant new rice fields. Highly susceptible cultivars have been eliminated by the disease. RYMV is now ranked as the main biotic threat to rice production in Africa.

Virion Properties

RYMV has icosahedral particles of 25 nm in diameter. The virions contain a single coat protein (CP) of 29 kDa, a genomic RNA (gRNA), and one subgenomic RNA (sgRNA) molecule. The capsid contains 180 copies of the CP subunit arranged with $T=3$ symmetry (**Figure 1**). The structure of RYMV was determined by X-ray crystallography at 2.8 Å resolution and compared to the structure of southern cowpea mosaic virus (SCPMV) and sesbania mosaic virus (SeMV). Sobemovirus CP subunits are chemically identical but structurally not equivalent. Three types of CP subunits termed A, B, C are related by quasi-threefold axes of symmetry and are involved in different inter-subunit contacts. In the C-type subunits, a longer part of the N-terminus is ordered (residues 27–49), forming an additional β-strand named βA arm. The analysis of molecular determinants involved in the $T=3$ assembly of SeMV demonstrated the major role of βA arm. Sobemovirus particles are stabilized by divalent cations, pH-dependent protein–protein interactions, and salt bridges between protein and RNA. Upon alkali treatment in presence of chelators, the capsid shell swells and becomes sensitive to enzymes and denaturants. In these conditions, RYMV is more stable than SCPMV. This property is likely due to the 3D swapping of the RYMV βA arm around the distal quasi-sixfold axes as found with SeMV whereas the SCPMV βA arm makes a U-turn and is localized around the nearby quasi-sixfold axes. Another reason for the better stability of RYMV is likely to be the strong RNA–protein interactions resulting from the presence of ordered RNA within the capsid shell. RYMV particles exist in three forms with different stability (1) an unstable swollen form dependent on basic pH but lacking Ca^{2+}, (2) a transitional compact form dependent on acidic pH, but also lacking Ca^{2+}, and (3) a stable compact form that is pH independent and contains Ca^{2+}. The compact form is highly infectious and probably required for virus movement and transport by vector.

Recently, molecular diversity of the CP was analyzed in relation to the capsid three-dimensional structure in order to identify which amino acids are involved in the differential recognition by certain monoclonal antibodies. The residues in position 178 and 180, despite their internal localization in the capsid, can modify the antigenic reactivity, with Mabs G and E allowing serotypes Sr3 and Sr5 to be differentiated.

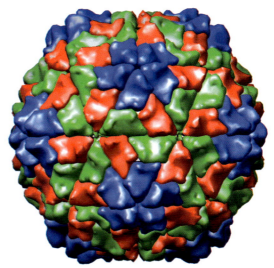

Figure 1 RYMV capsid structure. The capsid comprised 180 copies of one single type of polypeptide arranged in $T=3$ quasi-equivalent symmetry. The icosahedral asymmetric unit contains three subunits: A (in blue), B (in red), and C (in green). Each subunit is involved in different inter-subunit contacts. This image was automatically generated from ViPER virus capsid PDB file 1F2N using the MultiScale extension to the Chimera interactive molecular graphics package. Reproduced from Rice yellow mottle virus In: *Characterization, Diagnosis and Management of Plant Viruses*, Vol. 4, Ch. 2, pp. 31–50, 2008. Houston, TX: Studium Press, with permission from Studium Press.

Organization of the Genome

The genomic RNA is one single-stranded messenger-sense molecule, 4450 nucleotides in size. The 5′ terminus of the RNAs has a genome-linked protein (VPg), and lacks a poly(A) tail. RYMV often encapsidates, in addition to its genomic RNA, a viroid-like satellite RNA (satRNA) that is dependent on a helper virus for replication. RYMV satRNA has 220 nt and is the smallest naturally occurring viroid-like RNA known today. The RYMV satellite RNA is not involved in pathogenicity.

Like other sobemoviruses, the RYMV genome is organized in 4 overlapping open reading frames (ORFs). Two noncoding regions at the RNA extremities contain 80 and 289 nt, respectively. The ORF1 and the ORF4 are localized at the 5′ and 3′ extremities, respectively. An intergenic region of 54 nt separates the first two ORFs. The two coding regions ORF2a and ORF2b overlapped, ORF2b being situated in the −1 reading frame within ORF2a (**Figure 2**).

The ORF1 encodes a protein P1 which is dispensable for replication but is required for infectivity in plants. RYMV P1 exists in two forms of 18 and 19 kDa which probably result from degradation or post-translational modifications. P1 is required for cell-to-cell movement of the virus. Moreover, RYMV and cockfoot mottle virus

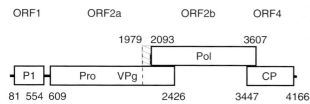

Figure 2 RYMV genomic organization. Positions of the ORFs are indicated in nucleotides. P1, proteinase (Pro), VPg, RNA-dependent RNA polymerase (Pol), and CP are labeled. The dotted line at nucleotide 1979 represents the frameshifting signal. The fusion point of the polyprotein P2a+b is unknown, and an AUG codon present at the beginning of the ORF2b (nucleotide 2093) is indicated by the vertical line. Reproduced from Rice yellow mottle virus In: *Characterization, Diagnosis and Management of Plant Viruses*, Vol. 4, Ch. 2, pp. 31–50, 2008. Houston, TX: Studium Press, with permission from Studium Press.

(CoMV) P1 are suppressors of post-translational gene silencing in the nonhost plant *Nicotiana benthamiana*. The two P1 are able to suppress the initiation and the maintenance of silencing but the suppression of systemic silencing is stronger with RYMV P1 than with CoMV P1.

The polyproteins encoded by the RYMV ORF2 are presumably translated from leaky translation. The start codon of ORF2 appeared to be in a more favorable context than ORF1. The ORF2b is translated as a polyprotein fused with ORF2a after a -1 programmed ribosomal frameshifting mechanism. This phenomenon happened with *in vitro* efficiency from 26 to 29% for the CoMV. The presence of a heptanucleotide slippery sequence UUUAAAC and a predicted stem–loop structure are necessary. The ORF2a contains sequence motifs for the proteinase and the VPg and the ORF2b encodes the RNA-dependent RNA polymerase (RdRp). RYMV proteinase contain the consensus sequence $H(X_{32-35})[D/E](X_{61-62})TXXGXSG$ characteristic of serine proteinases, they cleave between E–T and E–S amino acid residues.

RYMV VPg is not absolutely required for infectivity but *in vitro* transcripts required capping to be infectious. The putative and known sobemovirus VPgs are characterized by the presence of a conserved W[A/G]D sequence followed by a D- and E-rich region. The position of sobemovirus VPg, like polerovirus VPg, differs from the genome arrangement VPg proteinase–polymerase characteristic of many RNA virus species including members of the families *Picornaviridae* and *Comoviridae*. The localization of the polymerase was predicted from the presence of the GDD motif and surrounding conserved motifs characteristics of RdRp. Little is known about the replication signals needed for initiation of plus and minus-strand synthesis in sobemoviruses.

The ORF4 is translated from the subgenomic RNA and encodes the CP (which appeared as a doublet of 28–29 kDa). RYMV CP is dispensable for replication but required for movement to long distance and perhaps from cell to cell. Encapsidation is likely to be essential for long distance movement. This movement was suggested to occur during the cell differentiation to vessels. The N-terminal sequence contains a putative bipartite nuclear localization signal (NLS). This highly basic region is thought to interact with viral RNA and to stabilize the virion.

Relationships of the Species with Other Taxa

The species *Rice yellow mottle virus* belongs to the genus *Sobemovirus* which is not assigned to any family. The genus contains 13 definitive species including *Southern bean mosaic virus*, the type species. The sobemovirus CPs are related to those of necroviruses (*Tombusviridae*), whereas the proteinase, VPg, and polymerase are related to those of poleroviruses and enamoviruses (*Luteoviridae*).

Biological Properties

The natural host range of RYMV is restricted to a few members of the Poaceae family, principally *Eragrostidae* and *Oryzae* sp. The most commonly cited species are *Oryza sativa*, *O. longistaminata*, and *Echinocloa colona*. A few additional Poaceae plants have been infected experimentally.

RYMV particles are present in all plant parts including roots and seeds. Particles are present in large amount in mesophyll cells, and in vascular tissues mainly in xylem and associated parenchyma cells. Particles were found individually, as aggregates, or as crystalline forms in the cytoplasm and the vacuoles of infected cells. In addition, the cytoplasm of infected cells contains fibrillar material located either in vesicles or distributed in diffused patches. Virus particles are also found in mature xylem, as well as inside the primary wall.

Cytological changes were found in the chloroplasts of infected mesophyll cells where the starch grains decreased in size and number. The chloroplasts form invaginations containing mitochondria, peroxysomes, and virus particles. Electron-dense materials were observed in the nuclei and associated with fibrillar elements in cytoplasmic vesicles. The most dramatic changes induced by RYMV occurred in the cell walls of parenchyma and mature xylem cells causing disorganization of the middle lamellae wall. From infected cells, a viral ribonucleoprotein complex moves from cell to cell to reach the vascular bundle sheath. Viral replication, encapsidation, and storage in vacuoles occur mainly in vascular parenchyma cells. The calcium linkage from pit membranes to virus particles during xylem differentiation could contribute to disruption of these membranes and facilitate systemic virus transport. In the upper leaves, virions are suspected to cross the pit membrane to infect new vascular cells and virions spread out by cell-to-cell movement through plasmodesmata. At this stage of infection, most replication occurs in mesophyll and vascular cells, and most of the virions accumulate in large crystalline patches.

Vector transmission is mainly due to chrysomelidae vectors, but other biting insects including the grasshopper *Conocephalus* have been reported to occasionally transmit the virus. Transmission is not persistent. The virus is present within the seed, although RYMV is not seed-transmissible. Abiotic transmissions through plant residues, irrigation water, gutation water, direct contact between plants, and contamination by infected agricultural tools have been observed. Experimentally, mechanical transmission is easily achieved.

Diagnostic and Identification

RYMV diagnosis was first based on symptoms. However, many factors such as mineral deficiencies can induce symptoms of yellow mottle on rice. Therefore, several tools for specific detection of RYMV were developed. First, polyclonal antibodies (PAbs) directed against different isolates were used in double-diffusion tests and in direct antibody sandwich enzyme-linked immunosorbent assay (DAS-ELISA). No cross-reactions with other sobemoviruses was observed, demonstrating the high specificity of these antibodies. Western-blot analysis can be used to detect RYMV with PAbs. These antibodies were also used in immunolocalization techniques. Immunoprinting tests derived from the direct tissue blotting technique were developed to analyze the distribution of the virus in entire leaves. Moreover, cytological detection of RYMV CP in rice tissues can be performed by immunofluorescence microscopy after glutaraldehyde/paraformaldehyde fixation and labeling with secondary antibodies linked to fluorescein. Polyclonal antibodies directed against P1 produced in *E. coli* are also available, and were used in Western-blot analyses. Monoclonal antibodies were also developed to study RYMV diversity. Indirect triple antibody ELISA (TAS-ELISA) with strain-specific MAbs allows the major serotypes to be differenciated.

In parallel to the serological methods, molecular tools were developed. The first full-length sequence was obtained after RNA extraction from purified virus. Several specific primers were defined to perform RT-PCR amplification and to produce infectious transcripts. Northern and Southern blot methods are available using a DNA probe obtained by PCR amplification of the CP gene (ORF4) from the infectious clone CIa. *In situ* hybridization tests of the RNA in rice tissues were also performed. In this case, the probe was DNA, obtained from the full-length genome of the clone CIa labeled randomly with digoxigenin-UTP. RT-PCR amplification of the CP gene followed by direct DNA sequencing progressively has replaced other molecular methods of analyses.

Epidemiological Aspects

Typical symptoms of RYMV are a mottle and a yellowing of the leaves. However, streaks, necrosis, and whitening are sometimes observed with some cultivars and under specific growing conditions. Infection resulted in stunting of the plant, reduced tillering, poor panicle exertion, and sterility. Death of the plants of susceptible cultivars occurs after early infection. Severe yield losses ranging from 20% to 100% after RYMV infection have been reported in several countries. Most cultivars currently grown are susceptible to RYMV. Heavy infection is associated with reduced fertility.

RYMV is an emergent disease. It is thought that the virus was originally harbored in wild Poaceae and transferred only recently to cultivated rice. The rapid and intense spread of the virus was associated to the change of agricultural practices in response to increasing food demand. Very productive but susceptible *O. sativa* cultivars have been introduced from Asia which allows two crop cycles a year resulting in the maintenance of the inoculum all over the year. RYMV propagation probably occurs as follows: (1) a virus source made of volunteers, of wild Poaceae or plant residues act as primary inoculum; (2) a direct contamination of the new rice fields occur after mechanical transmission by man or biological infection by insects; (3) rice beds can also be infected by insect vector; (4) the new rice field originating from these infected seed-beds are contaminated by man during replanting; (5) further propagation of the virus is done by man, animals, wind, and irrigation water (**Figure 3**).

The epidemiology has been little studied and there are even uncertainties about the respective importance of biotic and abiotic transmission. Under these conditions, precise forecasting of RYMV infection is not possible, although growing intensification of rice cultivation in Africa will no doubt favor RYMV spread, unless durable resistant cultivars are introduced. Phytosanitary measures are sometimes advised although they are often economically not practicable and their impact to reduce virus spread is unknown. They include protection of seedbeds by nets, disinfection of tools used at replanting, destruction of volunteers and rice residues. Chemical control of the vectors is not economically feasible, is ecologically dangerous, and is most unlikely to be successful considering the large number of species involved in transmission.

Several breeding programs have been conducted in order to select and breed resistant cultivars. The genotype and the phenotype of two kinds of natural resistance

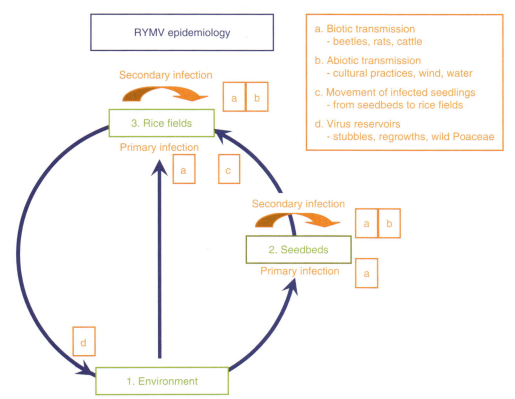

Figure 3 Descriptive model of RYMV epidemiology. 1. Environment. RYMV is present in the environment in rice stubble and regrowths and in wild Poaceae. 2. Rice seedbeds. Primary infection occurred via biotic transmission (beetles, rats, cows) whereas cultural practices contribute to secondary infection in seedbeds 3. Rice fields. Primary infection occurred via biotic transmission from the environment, and also when transplanting infected seedlings from seedbeds. Both biotic and abiotic transmissions contribute to secondary spread in rice fields. After harvesting, RYMV is perpetuated in contra-season in the environment through rice stubbles and regrowths and transmitted to wild Poaceae.

have been characterized. Partial resistance is encountered in cultivars of *O. japonica* species with a delayed virus multiplication and symptom expression. Partial resistance is polygenic and a major quantitative trait locus has been identified on chromosome 12. High resistance has been identified in a limited number of accessions of *O. glaberrima* and *O. sativa indica* species. No symptoms are apparent and virus content is most often undetectable. This resistance is monogenic and recessive. The high resistance is controlled by the recessive gene *Rymv1* that maps on chromosome 4 and encodes the translation initiation factor eIF(iso)4G. Different alleles were identified in the sativa-resistant varieties and the *O. glaberrima* accessions. Compared to susceptible varieties, they are characterized by a point mutation or a small deletion in the conserved domain of the gene. The high resistance was efficient against representative isolates of the main virus strains. However, high resistance was overcome experimentally. The VPg was identified as the virulence factor. A single point mutation was sufficient to break the resistance. Transgenic plants with portions of the ORF2 expressing partial resistance have also been produced. This transgenic resistance is thought to involve a gene silencing mechanism. However, the transgenic lines showed a less effective, partial, and temporary resistance compared to natural resistances.

Diversity and Evolution

The diversity of RYMV was assessed by studying isolates from several countries where the disease has been observed. Isolates were serologically typed with monoclonal antibodies. Their CP was sequenced. Several isolates representative of the geographic and molecular diversity were fully sequenced. RYMV is a highly variable virus and analyses of the geographic distribution of the genetic diversity elucidated the process of evolution and of dispersal of the virus. RYMV showed a high level of population structure marked at the continental scale with three subdivisions: East Africa, Central Africa, and West Africa. The highest diversity was observed in East Africa, with a pronounced peak in Eastern Tanzania, and a decrease from the east to the west of the continent. This pattern suggests a westward expansion with a succession of founder effects and subsequent diversification phases. Accordingly, genetic diversity would be adversely affected by recurrent bottlenecks occurring along the route of colonization, resulting in the lowest diversity in the extreme west. This, together with accumulation of *de novo* mutations postdating population separation, provides an explanation for the genetic differences among strains across Africa.

See also: Cereal Viruses: Rice; Emerging and Reemerging Virus Diseases of Plants; *Sobemovirus*.

Further Reading

Abubakar Z, Ali F, Pinel A, et al. (2003) Phylogeography of rice yellow mottle virus in Africa. *Journal of General Virology* 84: 733–743.

Albar L, Bangratz-Reyser M, Hébrard E, et al. (2006) Mutations in the eIF (iso)4G translation initiation factor confer high resistance of rice to rice yellow mottle virus. *Plant Journal* 47: 417–426.

Bonneau C, Brugidou C, Chen L, et al. (1998) Expression of the rice yellow mottle virus P1 protein in vitro and in vivo and its involvement in virus spread. *Virology* 244: 79–86.

Brugidou C, Holt C, Yassi A, et al. (1995) Synthesis of an infectious full-length cDNA clone of rice yellow mottle virus and mutagenesis of the coat protein. *Virology* 206: 108–115.

Fargette D, Konate G, Fauquet C, et al. (2006) Molecular ecology and emergence of tropical plant viruses. *Annual Review of Phytopathology* 44: 235–260.

Fargette D, Pinel A, Abubakar Z, et al. (2004) Inferring the evolutionary history of rice yellow mottle virus from genomic, phylogenetic and phylogeographic studies. *Journal of Virology* 78: 3252–3261.

Fargette D, Pinel A, Halimi H, et al. (2002) Comparison of molecular and immunological typing of isolates of rice yellow mottle virus. *Archives of Virology* 147: 583–596.

Hébrard E, Pinel-Galzi A, Bersoult A, et al. (2006) Emergence of a resistance-breaking isolate of rice yellow mottle virus during serial inoculations is due to a single substitution in the genome-linked viral protein VPg. *Journal of General Virology* 87: 1369–1373.

Hébrard E, Pinel-Galzi A, Catherinot V, et al. (2005) Internal point mutations of the capsid modify the serotype of rice yellow mottle sobemovirus. *Journal of Virology* 79: 4407–4414.

Konaté G, Traoré O, and Coulibaly M (1997) Characterization of rice yellow mottle virus isolates in Sudano-Sahelian areas. *Archives of Virology* 142: 1117–1124.

Pinel A, Traoré O, Abubakar Z, Konaté G, and Fargette D (2003) Molecular epidemiology of the RNA satellite of rice yellow mottle virus in Africa. *Archives of Virology* 148: 1721–1733.

Qu C, Liljas L, Opalka N, et al. (2000) 3 D domain swapping modulates the stability of members of an icosahedral virus group. *Structure* 8: 1095–1103.

Sorho F, Pinel A, Traoré O, et al. (2005) Durability of natural and transgenic resistances in rice to rice yellow mottle virus. *European Journal of Plant Pathology* 112: 349–359.

Traoré O, Pinel A, Hébrard E, et al. (2006) Occurrence of resistance-breaking isolates of rice yellow mottle virus in West and Central Africa. *Plant Disease* 90: 256–263.

Traoré O, Traoré M, Fargette D, and Konaté G (2006) Rice seedbed as a source of primary infection by rice yellow mottle virus. *European Journal of Plant Pathology* 115: 181–186.

Rift Valley Fever and Other Phleboviruses

L Nicoletti and M G Ciufolini, Istituto Superiore di Sanità, Rome, Italy

© 2008 Elsevier Ltd. All rights reserved.

Glossary

Arbovirus Any virus of vertebrates biologically transmitted by infected hematophagous arthropods.

Aseptic meningitis Inflammation of the covering of the brain (meninges) caused by a virus.

Epizootic A disease affecting a large number of animals at the same time within a particular region or geographic area.

Phlebotomus fever viruses or sandfly fever viruses Viruses that cause disease of brief duration characterized by sudden onset fever, headache, pain in the eyes, malaise, and leukopenia and are transmitted by the bite of infected sandflies.

Phlebovirus A genus in the family Bunyaviridae containing many viruses, of which the best known are Rift Valley fever virus and sandfly fever viruses.

Sandfly Any of various small biting two-winged flies of the families Psychodidae, Simuliidae, and Ceratopogonidae.

Transovarial transmission Vertical transmission of a virus from mother to offspring.

Zoonosis A disease of animals that can be transmitted to humans.

Introduction

The genus *Phlebovirus* is one of the five genera of the family *Bunyaviridae*, in which are included more than 300 virus species. The five genera (*Orthobunyavirus, Hantavirus, Nairovirus, Phlebovirus, Tospovirus*) are grouped in one family, primarily because they share structural characteristics, all bear a tripartite RNA genome of negative polarity, and all have a roughly similar protein-coding pattern within each genome segment. It has been demonstrated that Uukuniemi virus, the prototype virus of a group of tick-borne viruses, and its relatives have the same ambisense coding strategy as do phleboviruses. In addition, a high degree of similarity occurs in proteins of Uukuniemi virus and in those of some phleboviruses. As a consequence, all viruses previously included in the genus *Uukuvirus* are now classified in the genus *Phlebovirus*. Therefore, in the genus *Phlebovirus*, viruses previously known as phlebotomus fever viruses and uukuviruses are now placed.

The first description of what was most likely phlebotomus fever occurred at the time of the Napoleonic wars, when a similar disease was reported as 'Mediterranean fever'. The disease was first described with considerable accuracy by Pick in 1886, who termed it 'dog fever' (*hundsfieber*), probably due to the marked signs and symptoms of conjunctivitis, which resembled the eyes of a bloodhound. At the same time in Italy, the disease was already known as pappataci fever, suggesting a possible link with sand flies. In Yugoslavia, epidemics of the disease occurred each summer among newly arrived Austrian troops stationed along the Adriatic coast.

Rift Valley fever virus (RVFV) was first isolated in 1930 during investigation of an epidemic with high mortality rates among sheep on a farm in the Rift Valley in East Africa. In retrospect, epizootics due to this virus had been identified as early as 1912. Epizootics were reported to have occurred in many sub-Saharan countries after 1940 and the vector-borne origin of the disease was definitively proved in 1948.

RVFV is one of the most important viral zoonoses in Africa. Transmission of the virus to humans occurs via arthropod vectors, aerosols of blood or amniotic fluid of infected livestock, or by direct contact with infected animals. RVFV in humans is manifested by a broad spectrum of infections, from asymptomatic infection to a benign febrile illness, to a severe illness (approximately 1–3% of cases) that can include retinitis, encephalitis, and hemorrhagic fever. In addition to the human illness, disability, and suffering, RVFV outbreaks can result in devastating economic losses when livestock in an agricultural society are affected.

RVFV was first confirmed outside Africa in September 2000. An outbreak occurred in southwestern coastal Saudi Arabia and neighboring coastal areas of Yemen. RVFV isolated from the floodwater mosquito *Aedes vexans arabiensis* during the outbreak was closely related to strains from Madagascar (1991) and Kenya (1997), suggesting that the virus was imported through infected mosquitoes or livestock from East Africa.

From blood samples taken during an epidemic occurring in Italy among Allied troops during World War II, Sabin isolated two serologically distinct agents, sandfly fever Sicilian virus and sandfly fever Naples virus. Since then, many distinct viruses have been isolated from sandflies or humans in both the Old World and the New World and, on the basis of serological relationships, classified in the phlebotomus fever group (**Table 1**).

In 1971, in Italy, Toscana virus (TOSV) was isolated from the sandfly, *Phlebotomus perniciosus*. The virus was shown to be antigenically related to sandfly fever Naples virus, and antibody to TOSV was shown to occur at a relatively high prevalence of antibodies in healthy humans. TOSV has since been shown to cause human disease and has been associated with acute neurologic disease.

Morphology, Structure, and Strategy of Replication

The prototype phlebovirus is RVFV. Similar to other bunyaviruses, the virion is spherical or pleiomorphic, depending on the method used for fixation, with a diameter of 90–100 nm. The virion consists of a core containing the genome and its associated proteins, which are in turn surrounded by an envelope composed of a lipid bilayer containing equivalent numbers of two glycoproteins.

The genome consists of three single-stranded RNA segments: large (L, 6.5–8.5 kbp), medium (M, 3.2–4.3 kbp), and small (S, 1.7–1.9 kbp). Sequence studies demonstrate that the three segments are identical on the 3′-end (UGUGUUUC) and a complementary 5′-end. The S RNA produces many copies of the nucleocapsid protein (20–30 kDa) and the L RNA produces a few copies of the large (L) protein (150–250 kDa), which is a transcriptase.

Table 1 Viruses of the genus *Phlebovirus*

Virus species	Viruses
Bujaru virus	Bujaru, Munguba
Candiru virus	Candiru, Alenquer, Itaituba, Nique, Oriximina, Turuna
Chilibre virus	Chilibre, Cacao
Frijoles virus	Frijoles, Joa
Punta Toro virus	Punta Toro, Buenaventura
Rift Valley fever virus	Rift Valley fever, Belterra, Icoaraci
Salehebad virus	Salehebad, Arbia
Sandfly fever Naples virus	Sandfly fever Naples, Karimabad, Tehran, Toscana
Uukuniemi virus	Uukuniemi, EgAN 1825–61, Fin V 707, Grand Arbaud, Manawa, Murre, Oceanside, Ponteves, Precarious Point, RML 105355, St. Abbs Head, Tunis, Zaliv Terpeniya
Unassigned	Aguacate, Anhanga, Arboledas, Arumowot, Caimito, Chagres, Corfou, Gabek Forest, Gordil, Itaporanga, Odrenisrou, Pacui, Rio Grande, sandfly fever Sicilian, Saint-Floris, Urucuri

The glycoproteins, together with a nonstructural protein (NS_M), are translated from an mRNA complementary to the M segment as a precursor which is post-translational to G_N and G_C proteins. The S segment codes for two proteins. The N protein is read from a subgenomic mRNA complementary to the 3′-end segment of the viral RNA; the second, a nonstructural protein (NS_S, 29–37 kDa), is read from a subgenomic virus-sense mRNA species corresponding to the 5′-half of the viral RNA (**Figure 1**). This strategy of replication is called 'ambisense', and it is utilized also by members of the genus *Tospovirus* as well, and of viruses of the family *Arenaviridae*.

The maturation of phleboviruses occurs in intracellular smooth membranes, principally in the Golgi complex. Exceptionally, RVFV matures on the cell surface of infected rat hepatocytes. The glycoproteins accumulate in the Golgi complex, causing vacuolization, but the Golgi complex remains functional. Virions bud into Golgi vesicles, which are then transported to the cell surface where the particles are released by exocytosis.

The Agents

Rift Valley Fever

RVF was first recognized as a viral zoonosis in Kenya in 1930. Since then, several massive epizootics affecting domestic livestock have been reported in widely separated parts of Africa. In 1977–78, RVFV appeared for the first time in Egypt in an epizootic epidemic of unprecedented size. It was estimated that in some areas along the Nile, 25–50% of sheep and cattle were infected, and that there were as many as 200 000 human cases and at least 600 deaths. Several types of severe and sometimes fatal illnesses were observed in humans, including hepatitis with hemorrhagic manifestations, meningoencephalitis, and retinitis. During this epidemic, RVFV was isolated from *Culex pipiens* mosquitoes; however, arthropods of many species have also been shown to be potential vectors.

During epizootics of RVFV, disease occurs first in animals, then in humans. RVFV produces severe disease in domestic animals, sheep being more susceptible than cattle, goats least susceptible. A greater than 90% rate of mortality is observed among infected lambs, and 20–60% among adult sheep. Pregnant ewes (90–100%) abort within a few days after infection.

RVFV can be transmitted to humans by contact with tissues or blood of infected animals during slaughter, autopsy, or disposal of infected animals. The possibility of transmission of the virus via such an airborne route among abattoir or laboratory workers has been recognized. Vector-borne transmission, particularly at the beginning of epizootic phase, is important.

The most common form of illness associated with RVFV infection in humans is the acute febrile disease (see **Table 2**). The febrile period lasts 4–7 days; patients then recover completely within 2 weeks. During the epidemic of RVFV in 1977–78 in Egypt, 5–10% of patients experienced a typical febrile illness, followed by decreased visual acuity after 7–20 days. Hemorrhagic RVFV, which occurs in about 1% of cases, was not documented until a 1975 epizootic in South Africa. Patients infected during this epidemic experienced the sudden onset of a febrile illness, which was followed 2–4 days later by jaundice and hemorrhagic manifestations. Hematemesis, melena, gingival bleeding, and petechial and purpuric skin lesions were common. In fatal cases, death occurred within a week of the onset of jaundice. Shock and hepatic insufficiency represented the most probable cause of death. Mortality rates range from 0.2% to 14%. The exact incidence of each of the severe clinical syndromes complicating RVFV during the epidemics has not been established. These rates could be

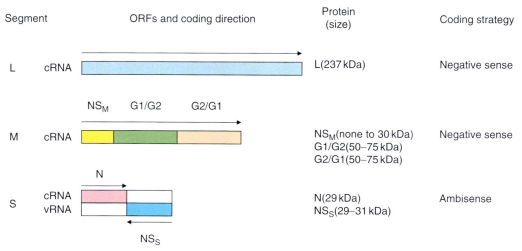

Figure 1 Genome organization and coding strategy of phleboviruses.

Table 2 Phleboviruses infecting humans

Virus	Evidence of infection	Isolation from humans	Geographic distribution	Arthropod association
Alenquer	Febrile illness	Blood	Brazil	Unknown
Arboledas	Antibody		Colombia	*Lutzomyia* spp.
Arumowot	Antibody		Africa	*Culex antennatus, Culex rubinotus, Mansonia uniformis*
Bujaru	Antibody		Brazil	Unknown
Cacao	Antibody		Panama	*Lutzomyia trapidoi*
Candiru	Febrile illness	Blood	Brazil	Unknown
Chagres	Febrile illness	Blood	Panama, Colombia	*Lutzomyia trapidoi, Lutzomyia ylephiletor*
Corfou	Antibody		Greece	*Phlebotomus major*
Gabek Forest	Antibody		Africa	Unknown
Gordil	Antibody		Africa	Unknown
Karimabad	Antibody		Iran	*Phlebotomus* spp.
Punta Toro	Febrile illness	Blood	Panama	*Lutzomyia trapidoi, L. ylephiletor*
Rift Valley fever	Febrile illness Hemorrhagic fever	Blood, CSF	Africa, Arabian peninsula	*Aedes caballus, Aedes circumluteolus, Aedes lineatopennis, Aedes vexans arabiensis, Culex theileri, Culicoides* spp., *Eretmapodites chrysogaster, Mansonia africana*
Saint-Floris	Antibody		Africa	Unknown
Salehabad	Antibody		Iran	*Phlebotomus* spp.
Sandfly fever Sicilian	Febrile illness	Blood	North Africa, South Europe, Central Asia	*Phlebotomus* spp.
Sandfly fever Naples	Febrile illness	Blood	North Africa, South Europe, Central Asia	*P. papatasi, P. perfiliewi*
Toscana	Febrile illness Meningitis	CSF	Italy, Spain, Portugal, Cyprus, France, Greece	*P. perniciosus, P. perfiliewi*

influenced by susceptibility of the human population, the virulence of various RVFV strains, and population dynamics of the vector.

Sandfly Fever Viruses

Classical phlebotomus fevers caused by sandfly fever Naples and sandfly fever Sicilian viruses are mild, self-resolving, flu-like illnesses characterized by sudden onset, fever, frontal headache, low back pain, generalized myalgia, retro-orbital pain, conjunctival injection, photophobia and malaise, and, in some cases, nausea, vomiting, dizziness, and neck stiffness (see **Table 2**). Fever is always present, ranging from 38 to 41 °C. The incubation period is 2–6 days, and the duration of disease varies from a few hours to four days (phlebotomus fever is also known as 'a 3-day fever'). Patients completely recover within 1–2 weeks. No deaths associated with phlebotomus fever have been reported but the large numbers of cases that can occur during epidemics, particularly during wars, make these illnesses potentially incapacitating to entire populations, and therefore are important.

Toscana Virus

TOSV was initially isolated from *P. perniciosus* in central Italy in 1971. The virus was subsequently isolated from *Phlebotomus perfiliewi* (but never from *Phlebotomus papatasi*) in other areas of Italy and from the brains of Kuhl's Pipistrelle bats (*Pipistrellus kuhlii*) captured in areas where the insect vectors were present. The virus also has been isolated from humans. Transovarial transmission has been demonstrated experimentally by viral isolation from male *Phlebotomus* spp. Venereal transmission from infected males to uninfected females has also been demonstrated. For many years, the known distribution of TOSV was limited to Italy and Portugal. More recently, the geographic distribution of the virus has been extended to France, Spain, Slovenia, Greece, Cyprus, and Turkey, according to results from viral isolation and serologic surveys.

TOSV is neurovirulent, a characteristic it shares with RVFV. Clinical cases with signs and symptoms ranging from aseptic meningitis to meningoencephalitis caused by TOSV are observed annually during the summers (see **Table 2**). All studies reported that the highest risk of acquiring TOSV is in August, which corresponds with peak sandfly activity.

Seroprevalence studies suggest that a substantial proportion of infections likely result in asymptomatic infections or cause only mild illnesses. Only severe cases, those involving CNS disease, require hospitalization.

After an incubation period ranging from a few days to 2 weeks, disease onset is intense, with headache, fever, nausea, vomiting, and myalgias. Physical examination may show neck rigidity Kernig signs, poor levels of consciousness, tremors, paresis, and nystagmus.

In most cases, cerebrospinal fluid (CSF) contains more than 5–10 cells with normal content of sugar and protein in it, or leukopenia. The mean duration of the disease is 7 days, and the outcome is usually favorable. However, a small number of severe cases with unusual symptoms have been reported. No sequelae have been described.

A study including subjects of a high-risk, professionally exposed population reported a seroprevalence of 70%, but without neurological symptoms. These data confirmed that TOSV infection can occur with either mild or no symptoms and suggest the frequent presence of TOSV infection in regions where the vector occurs.

Diagnostic Procedures

Diagnosis of phlebovirus infections has been classically attempted by isolating the agent or demonstrating seroconversion in paired acute- and convalescent-phase serum samples. Several molecular techniques have been developed and used to demonstrate the presence of the virus or viral RNA in patient samples.

Virus Recovery and Detection

The recovery of phleboviruses from acutely ill phlebotomus fever patients is rare, given that viremias associated with disease are transient (24–36 h) and most patients do not seek medical care within this period.

The most common method of phlebovirus isolation is intracranial inoculation of suckling mice. However, several studies have demonstrated that Vero cells (from kidney tissue of an African green monkey) are more sensitive than are newborn mice for isolation of phleboviruses from wild-caught sandflies. Similar results have been obtained with RVFV.

Identification of isolates can be made with serological methods, such as immunofluorescence, neutralization in mice, or by plaque reduction (PRNT) using hyperimmune sera. Identification can also be performed by complement fixation (CF) or hemagglutination-inhibition (HI); however, care should be taken in areas where more than one phlebovirus circulates, as these latter tests do not discriminate between certain antigenically related viruses. For example, sandfly fever Naples, Toscana, and Tehran viruses are indistinguishable by CF, yet easily identified by PRNT.

Notable diagnostic success in TOSV infections was achieved through molecular reverse transcription-polymerase chain reaction (RT-PCR) techniques. This approach has shown the occurrence of different TOSV variants.

Serologic Tests

Serologic diagnosis can be performed with many different methods. Most of the work in the past has been done with HI. PRNT provides more clear-cut results, but this test is time consuming, and is not more useful for diagnosis than are other tests when a single serum sample is available. However, the PRNT is the most common assay for assessing antibody titer.

Enzyme-linked immunosorbent assays for IgG and IgM antibodies have been established in the case of many phleboviruses; diagnosis is obtained in most cases by detection of specific IgM antibody. The use of immunoenzymatic techniques has been very useful in the developing tests in which the recombinant nucleoprotein expressed in *Escherichia coli* is used as the specific antigen.

Epidemiology

Epizootics of RVFV have been reported in many African countries. Prior to 1977, it was believed that the infection was confined to sub-Saharan Africa. However, the epizootic/epidemic in Egypt in 1977–78 occurred in areas outside the recognized range of the virus. Retrospective studies indicated that RVFV had not been endemic in Egypt prior to the epidemic. The epidemic occurred after the progressive implementation of the Aswan dam project, which regulated the irrigation of the Nile delta. The dam resulted in an increase in the number of breeding places of mosquitoes and also facilitated the transport of animals from Sudan to northern Egypt. From 1973 to 1976, an epizootic outbreak of RVFV occurred in Sudan, which extended from south to north and possibly reached Egypt in 1977.

RVFV was firstly detected outside Africa in September 2000. In southwestern coastal Saudi Arabia and neighboring coastal areas of Yemen, an epizootic occurred with >120 human deaths and major losses in livestock populations from disease and required slaughter. Most RVF activity was associated with flooded wadi agricultural systems; no cases were reported in the mountains or in dry sandy regions, where surface water does not accumulate long enough to sustain mosquito breeding.

Representative phleboviruses have been isolated in Southern Europe, Africa, Central Asia, and the Americas.

Many phleboviruses occur in tropical and subtropical regions, but some important ones are active in temperate areas (e.g., sandfly fever Naples virus, sandfly fever Sicilian virus, and TOSV in Southern Europe and in the Middle East). Most phleboviruses (66%) have been isolated in Central and Southern America and are associated with sandflies of the genus *Lutzomyia*. On the contrary, the majority of sandfly fever viruses isolated in the Old World always are associated with *Phlebotomus* spp. sandflies.

In general, each serotype has a unique distribution, with limited geographical overlap. However, there are indications of simultaneous circulation of different phleboviruses in the same sandfly population. Sandfly fever Naples and sandfly fever Sicilian viruses have the largest geographical distributions, paralleling that of their vector *P. papatasi*. Reported isolations as well as results of serologic studies indicate that sandfly fever Naples and sandfly fever Sicilian viruses are present in the Mediterranean coastal regions of Europe and North Africa, the Nile valley, most of southwest Asia, areas adjacent to the Black and Caspian Seas, and Central Asia as far as Bangladesh.

Despite the preponderance of phleboviruses in the New World, the incidence of phlebotomus fever, and in general of human infections, in the Americas is relatively low. On the contrary, the Old World phleboviruses, such as sandfly fever Naples virus, sandfly fever Sicilian virus, and TOSV, are relatively widely distributed and are well-known causes of epidemics of human infections. In some rural human communities in endemic areas of the Old World, the prevalence of phlebovirus infection is quite high among the indigenous population. For example, results of serosurveys aimed at defining the extent and prevalence of infection with different serotypes among indigenous human populations indicated prevalences as high as 62% for Karimabad virus in some provinces of Iran and 59% for sandfly fever Sicilian and 56% for sandfly fever Naples viruses in Egypt.

A similar serosurvey for the presence of antibodies to sandfly fever Sicilian virus, sandfly fever Naples virus, and TOSV was done in 1977 in Italy. These seroepidemiological studies indicated that both sandfly fever Sicilian and sandfly fever Naples viral infections decreased after the 1940s, probably as the result of insecticide spraying during malaria eradication campaigns. On the contrary, age-specific antibody rates suggested that TOSV, which is transmitted by *P. perniciosus* and *P. perfiliewi*, is endemic in Italy. A high infection rate (24.8%) was observed among residents of the region where the virus was first isolated. Similar results have been found in other Mediterranean countries (Spain, France, Greece, Cyprus).

Because illnesses due to phlebotomus fevers have been of considerable military and historical interest, much of the early research on phlebotomus fevers was performed by military physicians and epidemiologists. In 1984, cases of sandfly fever Sicilian virus infections were documented in Cyprus among Swedish United Nations soldiers. A follow-up study revealed that 11 of 298 soldiers seroconverted to sandfly fever viruses (mostly Sicilian) during a 6 month stay in Cyprus in the summer of 1985. Infection due to sandfly fever viruses (mostly sandfly fever Sicilian virus) was diagnosed in Swedish tourists contracting a febrile illness associated with their travel during 1986–88 to the Mediterranean region. In contrast, a serological survey among Cypriots revealed a higher antibody prevalence to the sandfly fever Naples virus than to sandfly fever Sicilian virus and TOSV.

Clinical cases with signs and symptoms ranging from aseptic meningitis to meningoencephalitis caused by TOSV are observed annually in central Italy during the summer. The incidence of cases of meningitis due to TOSV is directly related to yearly differences in the density of the vector sandflies, which is greatly influenced by variations of climatic conditions. Most cases have been reported in residents or travellers in central Italy or Spain, and sporadically from other Mediterranean countries, such as Portugal, Cyprus, France, and Greece.

In Spain, TOSV is one of the three leading causes of meningitis. A large study conducted in different regions of Spain showed the presence of IgG antibodies to TOSV (26.2%), sandfly fever Naples virus (2.2%), and sandfly fever Sicilian virus (11.9%).

Studies of people living on the Ionian Islands and on the western mainland of Greece showed a seroprevalence for TOSV of 60% and 35%, respectively. It is also reported that in Cyprus 20% of the healthy population had IgG antibody to TOSV.

Therapy

The mild nature of phlebotomus fevers has not encouraged studies on the possibility of treating the disease with specific antiviral compounds. Rather, treatment of phlebotomus fever patients is symptomatic. Due to the severity of the disease, several studies have been done both *in vitro* and *in vivo* in an effort to devise possible treatments of RVFV infections. Two different approaches have been evaluated: the use of an antiviral compound such as ribavirin or its derivatives, or the use of interferon or interferon inducers.

High doses of ribavirin have been used successfully to treat experimental infections with RVFV in several strains of mice. However, at low drug doses, treatment failures occur, resulting in death due to either hepatitis or subsequent encephalitis. Better results have been obtained when animals were treated with liposome-encapsulated ribavirin.

Experimental data and epidemiological experience suggest that the major utility of interferon in RVFV infections could most likely be in early postexposure prophylaxis (e.g., laboratory accidents, high-risk exposures). The success of interferon treatment initiated after the appearance of clinical symptoms has not been assessed. However, if maximal levels of circulating interferon in humans are obtained before early viremia, as is seen in rhesus monkeys, it is possible that interferon treatment begun at the first appearance of symptoms could affect the course of illness by limiting the incidence of future complications, particularly hemorrhagic fever. In addition, laboratory data and anecdotal information suggest that ribavirin and convalescent plasma may be beneficial in treating hemorrhagic fevers due to RVFV.

Prevention and Control

Vaccines

To date, there are no vaccines available against phlebotomus fever viruses. Given the large number of serotypes that can infect humans, and the lack of cross-immunity between them, to be practical a phlebotomus fever vaccine would have to be polyvalent. Moreover, in view of the relatively benign nature of the disease, vaccination against phlebotomus fever is not cost-effective or advisable as a general public health policy.

As RVFV is primarily of veterinary importance and human epidemics follow livestock epizootics, the immunization of susceptible animals is the most effective means of controlling RVFV infection.

Formalin-inactivated RVFV vaccines have been used for many years in South Africa, Egypt, and Israel to immunize sheep and cattle but they require multiple inoculations to ensure lasting protection. This type of vaccine is recommended for nonendemic areas as well as for livestock to be exported from enzootic areas.

An inactivated RVFV vaccine, developed for human use, is recommended for use to immunize at-risk laboratory and field researchers, or other people at high risk of infection. No clinical cases have been reported after vaccination; however, a few subclinical infections have been detected.

Two live RVFV vaccine candidates have been developed. One is a minute plaque variant and the other is an extensively mutagenized strain, MP-12, passaged in the diploid human lung cell line, MRC-5. Experimental studies in animals have demonstrated that a single dose of MP-12 RVFV vaccine is immunogenic and non-abortogenic in pregnant ewes. It can, therefore, be used safely to protect pregnant ewes and newborn lambs during an RVFV outbreak without increasing mortality due to vaccination.

RVFV MP12 vaccine was developed in cells certified for human vaccine production, and this vaccine may be employed for protecting both humans and livestock against epizootic RVFV infection. Because it is attenuated, this vaccine may be used to prevent an impending RVFV outbreak in an area where the virus is nonendemic, without the risk of spread to the environment.

Prevention of Transmission

During the Egyptian epidemic of RVFV in 1977–78, insecticides were used both inside and outside human and animal shelters to reduce the adult populations of mosquito vectors. However, it was not demonstrated that the use of these chemicals actually decreased the incidence of the disease. In sub-Saharan Africa, where RVFV outbreaks are thought to occur following flooding of vector mosquito breeding habitats, the possibility of using encapsulated formulations of insecticides to control the larval stages has been considered.

Measurements of green leaf vegetation dynamics recorded by advanced very high resolution radiometer instruments on polar-orbiting, meteorological satellites were used to derive ground moisture and rainfall patterns in Kenya and monitor resulting flooding of mosquito larval habitats likely to support RVFV vector mosquitoes (*Aedes* spp. and *Culex* spp.). These data could be used by local authorities for implementations of specific mosquito control measures to prevent transmission of RVFV to susceptible vertebrate hosts.

At present, the only method effective in controlling phlebotomus fevers is to reduce human contact with the vectors that carry their etiologic agents. Insecticides are extremely effective in the control of peridomestic sandfly species, but are of little value for sylvan species. Mechanical means (e.g., protective clothing, bed nets and screening of windows) and use of repellents, such as diethyltoluamide, can also be used to prevent human–vector contact.

See also: Bunyaviruses: General Features; Replication of Viruses.

Further Reading

Baldelli F, Ciufolini MG, Francisci D, et al. (2004) Unusual presentation of life-threatening TOSV meningoencephalitis. *Clinical Infectious Diseases* 38: 515–520.

Balkhy HH and Memish ZA (2003) Rift Valley fever: An uninvited zoonosis in the Arabian peninsula. *International Journal of Antimicrobial Agents* 21: 153–157.

Braito A, Corbisiero R, Corradini S, et al. (1997) Evidence of Toscana virus infections without central nervous system involvement: A serological study. *European Journal of Epidemiology* 13: 761–764.

Charrel RN, Gallian P, Navarro-Mari JM, et al. (2005) Emergence of Toscana virus in Europe. *Emerging Infectious Diseases* 11(11): 1657–1663.

Ciufolini MG, Maroli M, Guandalini E, Marchi A, and Verani P (1989) Experimental studies on the maintenance of Toscana and Arbia viruses (*Bunyaviridae: Phlebovirus*). *American Journal of Tropical Medicine and Hygiene* 40: 669–675.

Flick R and Bouloy M (2005) Rift Valley fever virus. *Current Molecular Medicine* 5: 827–834.

Geisbert TW and Jahrling PB (2004) Exotic emerging viral diseases: Progress and challenges. *Nature Medicine* 10: S110–S121.

Nicoletti L, Verani P, Caciolli S, *et al.* (1991) Central nervous system involvement during infection by the phlebovirus Toscana of residents in natural foci in central Italy (1977–1988). *American Journal of Tropical Medicine and Hygiene* 45: 429–434.

Nicoletti L, Ciufolini MG, and Verani P (1996) Sandfly fever viruses in Italy. *Archives of Virology Supplement* 11: 41–47.

Sánchez Seco MP, Echevarría JM, Hernández L, Estévez D, Navarro-Marí JM, and Tenorio A (2003) Detection and identification of Toscana and other phleboviruses by RT-PCR assays with degenerated primers. *Journal of Medical Virology* 71: 140–149.

Torres-Velez F and Brown C (2004) Emerging infections in animals – potential new zoonoses? *Clinics in Laboratory Medicine* 24: 825–838.

Weaver SC (2005) Host range, amplification and arboviral disease emergence. *Archives of Virology Supplement* 19: 33–44.

Relevant Website

http://www.cdc.gov – CDC Special Pathogens Branch, Centers for Disease Control and Prevention.

Rinderpest and Distemper Viruses

T Barrett, Institute for Animal Health, Pirbright, UK

© 2008 Elsevier Ltd. All rights reserved.

Glossary

Cytopathic effect Alterations in the microscopic appearance of cultured cells following virus infection.

Enzootic A disease constantly present in an animal community in a defined geographic region.

Epizootiology The study of disease epidemics in animal populations.

Hemagglutinin Any substance that causes red blood cells to agglutinate; many viruses possess a hemagglutinin protein in the outer envelope.

History

Rinderpest and its related viruses form a distinct group of paramyxoviruses, the genus *Morbillivirus*. In addition to measles virus (MV), the type virus of the genus, they include important animal pathogens: rinderpest virus (RPV), which infects many species of large ruminant; peste des petits ruminants virus (PPRV), a similar disease of small ruminants; canine distemper virus (CDV), found mainly in carnivores; phocid distemper virus (PDV), in seals; and the cetacean morbillivirus (CeMV), which is found in whales, dolphins, and porpoises. Of the animal viruses, rinderpest is the most important economically. It was also known as cattle plague in the past because of the high mortality associated with infections and the speed with which it spreads in naive cattle populations, which made it an easily recognizable disease. It is one of the oldest documented plagues of domestic livestock. Aristotle (384–322 BC) described a disease in cattle, *struma*, that had all the characteristics of rinderpest and other descriptions date to the fourth century when invasion by Huns into Europe resulted in outbreak of highly contagious disease which had all the characteristics of rinderpest. Steppe cattle of Central Asia are thought to be the original source of rinderpest and 'steepe murrain' was the old English name for the disease. Panzootics of rinderpest were also brought to western Europe with the Mongol invasions in the thirteenth century. In 1709, the disease again entered Europe through Venetian trade with the east and by 1714 had spread as far west as Britain. The economic and social consequences which followed in the wake of these plagues led to the establishment of the first veterinary schools – the first at Lyon, France, in 1762. Other European countries followed France's example and set up their own veterinary schools. Subsequent brief reintroductions into Europe in the early twentieth century, and into South America and Australia through the import of infected cattle led to the establishment of the Office Internationale des Epizooties (OIE), a body that functions as the World Organisation for Animal Health, to deal specifically with animal diseases in relation to international trade.

Rinderpest was introduced into Africa with disastrous consequences in the late 1880s with cattle imported from India to feed Italian troops in fighting a colonial war in Abyssinia (now Ethiopia). The subsequent panzootic spread to nearly all parts of the African continent within a period of 10 years, reaching South Africa by 1897. The devastation that followed its path as it swept across the African continent wiped out 90% of the domestic cattle and wild buffalo (*Syncerus caffer*). A number of other large wild ruminant species were also highly susceptible to the virus and died in large numbers during this so-called 'Great African Pandemic'.

A similar plague of small ruminants, peste des petits ruminants (PPR), was first scientifically documented in West Africa in the early 1940s and is also known as *kata* in that region. At first, it was thought that PPRV was a variant of RPV adapted to small ruminants but it was subsequently shown to be a genetically distinct virus with an independent epizootiology in areas where both viruses co-circulated.

Canine distemper, which infects many terrestrial carnivore species, is also a disease with a long history. In the eighteenth century, Edward Jenner studied its neurological effects following infection in dogs. In 1988, a distinct but closely related virus, phocine distemper virus (PDV), was isolated from seals found dying in large numbers along the beaches of Northern Europe and which showed clinical signs similar to distemper in dogs. Both harbor (*Phoca vitulina*) and gray (*Halichoerus grypus*) seals were affected and the epizootic resulted in the deaths of more than 20 000 seals. In 1990, large numbers of striped dolphins (*Stenella coeruleoalba*) in the Mediterranean Sea were found dying from a similar infection and a virus was isolated from sick animals. This virus has since been associated with mass die-offs among whales, porpoises, and dolphins (order *Cetacea*) and a fatal epizootic in bottlenose dolphins (*Tursiops truncatus*) from the northwestern Atlantic Ocean in the 1980s. A morbillivirus was also isolated from diseased porpoises (*Phocoena phocoena*) found along the coasts of Ireland and the Netherlands between 1988 and 1990 during the epizootic of PDV in seals. This virus was later found to be very closely related to the dolphin virus and genetically quite distinct from all other morbilliviruses. The dolphin and porpoise virus isolates are now commonly referred to as CeMV, and these virus infections have the potential to cause severe disease and threaten the ecology of many marine mammal species.

Taxonomy and Classification

As indicated above, these antigenically closely related viruses are classified in the genus *Morbillivirus* within order *Mononegavirales*, family *Paramyxoviridae*, subfamily *Paramyxovirinae*. The name is derived from the diminutive (*morbilli*) of the Latin word *morbus* meaning 'little plague', to distinguish measles from deadlier diseases such as smallpox. The paramyxoviruses are large enveloped pleimorphic particles which bud from the surface of infected cells and the different species are indistinguishable in the electron microscope. They vary in diameter from 300 to 500 nm. The lipid envelope is derived from the host cell membrane as the virions bud from the cell and it encloses a ribonucleoprotein (RNP) core which contains the genome encapsidated by the nucleocapsid protein giving it a characteristic 'herring bone' appearance. Unlike other paramyxoviruses, the morbilliviruses generally lack neuraminidase but a highly substrate-specific neuraminidase activity has been reported for PPRV and RPV. Only MV and PPRV viruses can hemagglutinate red blood cells reproducibly.

Geographic Distribution

Until the mid-1980s, rinderpest was found in much of tropical Africa and in western and southern Asia. Control and eradication program were established in the mid-1980s and this effort, based on mass vaccination campaigns in the endemic regions, has succeeded in eliminating the disease from Asia and the Middle East; the virus is now almost certainly confined to a defined region of eastern Africa known as the southern Somali ecosystem, a region in eastern Africa comprising southeast Ethiopia, southern Somalia, and northeast Kenya.

PPR is enzootic in West Africa and on the Indian subcontinent and is now spreading from Afghanistan into Central Asia (**Figure 1**). It first appeared as a recognized disease in India in 1988 and has subsequently been found as far east as Bangladesh and north as far as Nepal. Epizootics regularly occur in the Middle East and Turkey through import of infected animals. Turkey's proximity to Southern Europe poses a threat to small ruminants in that region. The widespread distribution of PPRV in southwest Asia suggests that the virus had been present on the continent for some considerable time before it was identified in India. Its presence was most likely masked by rinderpest since all morbillivirus-like diseases in small ruminants in India before 1988 were considered to be caused by rinderpest infections. With the subsequent elimination of rinderpest from Asia and the development of more accurate diagnostic techniques it has now become easier to chart the spread of PPRV in the region.

CDV is found in all but the hottest and most arid regions of the globe. The development of attenuated live vaccines in the late 1950s greatly reduced the incidence of the disease in domestic dogs in developed countries. However, many wildlife species are susceptible to CDV infection and can act as reservoirs for the virus. CDV has also been found in Lake Baikal seals (*Phoca sibirica*) and Caspian Sea seals (*Phoca caspica*) and it poses a threat to the survival of the latter species which is highly endangered. There is also serological evidence for infection in Antarctic seals and, as there are no native carnivores on that continent, unvaccinated sledge dogs are blamed for introducing the virus.

The geographic distribution of PDV is not fully defined but, based on serological evidence, it appears to be an endemic infection in some species of seal in the northern Atlantic and Arctic oceans. These are considered to be the most likely source of PDV for the European

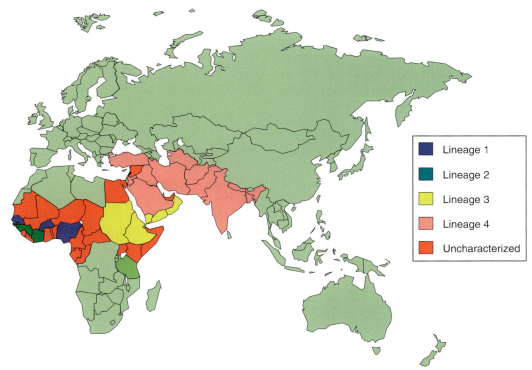

Figure 1 The distribution of PPRV. Where known, the different lineages are color-coded. There are only serological or capture enzyme-linked immunosorbent assay (ELISA) data available for countries shown in red.

seal epizootics. The CeMV appears to have a worldwide distribution and there is serological evidence for the presence of this virus in diverse species from all of the world's major oceans.

Host Range Virus

Each morbillivirus is generally able to cause serious disease only in one order of mammals, the exception being CDV. All cloven-hoofed animals (Artiodactyla) are thought to be susceptible to infection with RPV but the disease is not manifest in all. In the case of cattle, Indian and African breeds (*Bos indicus*, zebu) are more resistant than European (*Bos taurus*). The virus can also infect a range of wild ungulates but disease progression depends on the innate resistance of the species concerned. Some, such as kudu (*Tragelaphus imberbis*), eland (*Taurotragus* spp.), giraffe (*Giraffa* spp.), and wildebeest (*Connochaetes* spp.), are highly susceptible to the virus and died in large numbers during the first African pandemic. Others, mainly small antelope species, proved to be more resistant (**Table 1**).

The full host range of PPRV is unknown but, in addition to sheep and goats, several species of antelope have been fatally infected by contact with infected sheep. Outbreaks of PPRV have been reported in game reserves and zoos where the mortality was 100% in some species.

Table 1 Susceptibility of wildlife to rinderpest

Very high	Buffalo, eland, kudu, wart hog
High	Giraffe, bushbuck, bush pig, sitatunga, Uganda cob, bongo, wildebeest
Moderate	Reedbuck, topi, gemsbok, blesbok, bontbok, oribi, impala, springbok
Low	Waterbuck, dukier, orynx, Grant's gazelle, dikdik, hartebeest
Very low	Thomson's gazelle, hippopotamus, gerenuk

Based on data from Plowright W (1982) The effects of rinderpest and rinderpest control on wildlife in Africa. In: Edwards MA and McDonald U (eds.) *Symposia of the Zoological Society of London, No. 50: Animal Diseases in Relation to Animal Conservation*, p. 1. London: Academic Press.

Goats are generally considered to be more sensitive to PPRV infection than sheep. Indian buffaloes (*Bulbalus bulbalis*) have also been reported to have died from PPRV infections. Cattle have been found in West Africa which were seropositive for PPRV, with up to 80% prevalence in some herds, but there is no evidence that it can cause disease in cattle.

CDV can infect most carnivores but in some it may result in only a mild or subclinical infection, for example, in domestic cats. It causes severe disease in all members of the Canidae (dog, wolf, fox), Mustelidae (ferret, weasel, mink), Procyonidae (raccoon, panda), as well as in collared peccaries (*Tayassu tajacu*, order Artiodactyla). More recently, CDV has been shown to be responsible for high mortalities in both wild and captive big cats and in hyenas (*Crocuta crocuta*). Failure to recognise the disease earlier in these species may have been due to a lack of awareness of a possible viral etiology and/or the availability of diagnostic tools to detect the virus. Outbreaks of CDV in Siberian and Caspian seals have extended its host range to include these species. PDV is known to infect many species of seal in the North Atlantic and Arctic oceans but its full host range is unknown. There is also serological evidence that terrestrial carnivores in Canada, including polar bear (*Ursus maritimus*), lynx (*Fellis lynx*), and wolves (*Canis lupus*), have been infected with the virus and also with CeMV. CeMV infections have been described in a variety of cetaceans and there are serological data indicating infection in many more.

Virus Propagation

All known morbilliviruses can be propagated on Vero cells, which lack the ability to produce interferon, but it generally requires several blind passages to adapt the virus to these cells. This adaptation can alter the receptor-binding characteristics of the virus and often attenuates it for the natural host. Clinical isolates of rinderpest that retain their pathogenicity can best be grown on primary bovine kidney cells or transformed lymphoid cell lines such as B95a cells. PPRV is normally grown on lamb kidney cells and primary mitogen-stimulated dog or ferret macrophages are the most suitable cells for the isolation of CDV. Alternatively, Madin–Darby canine kidney (MDCK) cells have been shown to be useful for CDV isolation from infected tissues. Primary seal kidney cells were initially used to isolate PDV. Virus is most readily obtained from tissues such as mucosal lesions, lymph nodes, or by co-cultivation of washed buffy coat from infected animals with suitable tissue culture cells. Lung tissue is also a good source of virus for PPRV, CDV, and PDV isolation. Typical cytopathic effects such as cell elongation, cell rounding, the formation of stellate cells, and syncytia can be observed 3–12 days post infection of the cell cultures although several blind passages may sometimes be necessary before cytopathic changes are observed.

Host range is to some extent, but not completely, determined by the expression of a suitable receptor on the host cell. The CD150 molecule, also known as signal lymphocyte activating molecule or SLAM, is thought to be the main cellular receptor for wild-type RPV, CDV, and MV isolates. SLAMs specific to their respective host species may be general morbillivirus receptors. CDV replicates better and causes extensive cell fusion in Vero cells expressing dog SLAM (Vero DST) compared to normal Vero cells and these are a better cell line for propagation and titration of CDV isolates. The expression of SLAM on cells of lymphoid origin, activated lymphocytes, immature thymocytes, macrophages, and mature dendritic cells explains the strong lymphotropism of these viruses. Tissue culture-adapted viruses can use other receptors such as CD46. Alterations in receptor recognition could possibly be a means whereby these viruses could widen their host range in future.

Properties of the Genome

The morbillivirus genomes consist of a single strand of negative-sense RNA just under 16 kbp in length. They are organized into six contiguous, nonoverlapping, transcription units which encode the six structural proteins, namely the nucleocapsid (N), the phospho (P), the matrix (M), the fusion (F), the hemagglutinin (H), and the large (L) proteins, the latter being the viral RNA-dependent RNA polymerase. They have highly conserved sequences at their 3′ (leader) and 5′ (trailer) terminal extremities that act as promoters for transcription and replication (**Figure 2**). The complete promoter elements include sequences that extend into the untranslated region at the start of the N gene open reading frame (ORF) and the untranslated region at the end of the L gene ORF. These regions contain all the *cis*-acting signals necessary for primary transcription as well as for the production of a full-length positive-sense RNA genome copy required for the production of new genome RNA. There are semi-conserved start–stop sequence motifs at the start and end

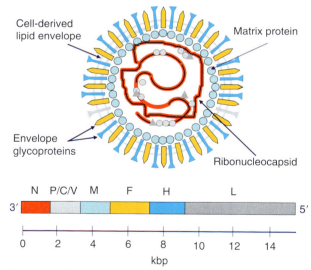

Figure 2 Cartoon showing the structure of morbillivirus virions. The outer envelope is shown with two projecting glycoproteins (H and F) and the inner helical nucleocapsid containing the genome RNA encapsidated with the N protein and the P and L proteins associated with it. The matrix protein is shown as a ring of circles underneath the virus envelope. The linear order of genes in the genome RNA from 3′ to 5′ is also shown.

of each mRNA transcription unit: (UCCU/C) at each transcription start; and a sequence rich in U residues, which signals the polyadenylation of the mRNAs, at the end. Between the end of one transcription unit and the start of the next is an intergenic triplet (usually GAA) which is not translated into mRNA. Downstream mRNA synthesis depends on termination of the upstream mRNA. The N, P, and L proteins, along with the genome RNA, constitute the transcription/replication unit of the virus, the RNP core. The F and H glycoproteins are embedded in a lipid envelope which is derived from the host cell during the budding process. The nonglycosylated M interacts with cytoplasmic domains of the membrane-associated F and H proteins and also with the nucleocapsid RNPs formed in the cytoplasm during replication and brings together the two components that make up the budded virion. The M protein is essential for efficient virus budding to occur.

Morbilliviruses also produce two nonstructural proteins (C and V) encoded in the P gene transcription unit. The first of these, the C protein, is translated by ribosomes that scan past the first AUG codon and start at the second which is located about 20 nt downstream. This protein is in a different reading frame to the P protein and bears no antigenic relationship to it. The second nonstructural protein, the V protein, in derived by alternative transcription of an mRNA from the P transcription unit by which a nontemplated G residue added to approximately 50% of the P mRNAs. The extra Gs are inserted at a specific, highly conserved, sequence (5′ UUAAAAAGGG[G]CACAG), known as the editing site, positioned about halfway along the P protein ORF. This so-called editing process is a property of the virus polymerase as it does not occur in artificial transcription systems. Translation of this mRNA produces the V protein which is a chimeric protein consisting of the N-terminus of the P protein with a new, shorter C-terminus rich in cysteine residues derived from template sequence in the third reading frame. The V mRNA is also capable of translating the C protein as its coding region is located before the editing site in P. The nonstructural proteins have functions in controlling transcription and replication and also are involved in virus evasion of the host's innate immune responses.

Morbilliviruses, like some other paramyxoviruses, have a strict genome length requirement in that they should be divisible by 6 (the 'rule of six'). This requirement can be explained by the fact that each N protein monomer associates with exactly 6 nt and that efficient transcription and/or replication can only occur if the RNA genome is encapsidated by the N protein in its entirety. Reverse genetics systems have been established for MV, RPV, and CDV. This enables virus to be 'rescued' from a DNA copy of its genome and this copy can be manipulated to make virus mutants which can then be used to determine the functions of various proteins and sequence motifs and to study the molecular basis of host range and pathogenicity.

Evolution

The close antigenic relationship and sequence similarity of morbilliviruses indicate that they all evolved from a common ancestor; however, the details of their evolutionary history are unclear. The phylogenetic relationship between the different morbilliviruses, based on the sequence of the P proteins, is shown in **Figure 3**. RPV, PPRV, CDV, and CeMV are equidistant from each other. It appears, therefore, that no existing morbillivirus represents the ancestral virus. MV is genetically more closely related to RPV and PDV to CDV, suggesting that they evolved more recently from their respective common ancestors. A factor that must be considered in their evolutionary history is that animals which have recovered from morbillivirus infections are immune for life and so fairly large populations are required to ensure a constant supply of naive hosts needed to maintain these viruses in circulation. The minimum population size that satisfies this requirement has been estimated to be about 300 000 for MV, and so this virus could not have existed before settled human populations became large enough. Herds

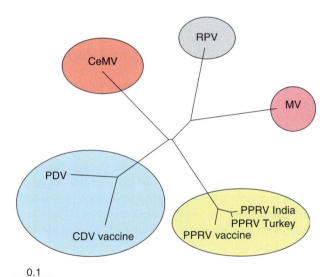

Figure 3 Radial tree showing the morbillivirus phylogenetic relationships based on their P protein sequences. The P proteins were aligned using ClustalW (BLOSUM) matrix program. A neighbor-joining radial tree was then generated from this alignment using ClustalX v1.83 and TreeView v1.6.6. Three different PPRV sequences were included to show the extent of variation within each morbillivirus type. The PPRV vaccine and PPRV Turkey are the most distant viruses in the group. Access numbers: RPV (X98291); PPRV (AY560591, X74443, AJ849636); CDV (AF305419); PDV (D10371); DMV (AJ608288); MV (AB012948).

of ruminants roaming the steppes of Central Asia, the historic source of rinderpest, would, however, have been able to maintain a morbillivirus in circulation. It is probable that when human communities became large enough, rinderpest, or a rinderpest-like infection, was passed to them from their domesticated cattle. RPV and PPRV are equidistant from the putative ancestral virus and so PPRV did not evolve directly from RPV and must have a long independent evolutionary history in small ruminants.

Carnivores preying on infected ruminants may similarly have become infected with the progenitor morbillivirus which subsequently evolved in these species to become CDV. CDV infects a wide variety of carnivores and so there is potentially a large reservoir of virus in susceptible wildlife species. PDV most probably evolved quite recently from CDV. Seals have many opportunities to become infected with CDV by contact with terrestrial carnivores such as wolves, foxes, dogs, and polar bears. Arctic seals populations, unlike European seal populations, are large enough to enable a morbillivirus to be maintained and subsequent evolution would select a virus more adapted to replicate efficiently in seals.

The two CeMVs that have been isolated from dolphins and porpoises are very closely related to each other, as related as different lineages of either RPV or PPRV, and can be considered to constitute one virus species. CeMV is antigenically closest to PPRV but is equidistant for the putative ancestral virus by sequence analysis and so this virus also has had a long independent evolutionary history.

Serological Relationships and Variability

All the morbilliviruses are antigenically related, the F, M, and N proteins being the most highly conserved across the group. The two virus-coded glycoproteins, the H and F, are embedded in the virus envelope and neutralizing antibody responses are generated to both of these proteins. The H protein, responsible for attachment of the virions to the host cell surface receptor molecule and is therefore a determinant of host range, is the least cross-reactive of the morbillivirus proteins. While it is possible to differentiate strains of each morbillivirus by monoclonal antibody and sequence analysis, these variations do not result in different serotypes and for vaccination purposes each morbillivirus has only one serotype.

Epizootiology

Rinderpest and PPR are normally introduced into new regions by importation of live animals. Transmission through infected meat and meat products is considered to be a very low risk due to the highly labile nature of the viruses. Historically, rinderpest outbreaks have been associated with wars and civil conflict when there is uncontrolled movement of people and troops, often bringing live animals from enzootic regions as food. Sheep and goats, and possibly other ruminants, may show only mild disease or subclinical infection with RPV but nevertheless they can then pass the infection to in-contact cattle. Trade in live animals is also another factor that has been responsible for long-distance spread of rinderpest.

Epizootics of rinderpest in naive populations are generally very severe and mortality rates can exceed 90% and affect all age groups, the classic characteristics of cattle plague. The last outbreak of this nature occurred in Pakistan in 1994. In endemic areas, disease is less severe and most often affects animals less than 2 years of age as older animals are generally immune due to previous exposure or vaccination. Newborns are protected for up to 9 months by maternal antibody. The rinderpest virus strain in circulation in the last enzootic focus in Africa belongs to African lineage 2 (**Figure 4(a)**) and is unusual in that it causes only mild disease in cattle. It becomes clinically apparent when it spreads to highly susceptible wildlife species such as buffalo, eland, and kudu. These have suffered very high mortalities when infected with this strain of virus during outbreaks in Kenyan national parks and because of this wildlife are considered excellent sentinels for incursions of rinderpest. They are closely monitored for evidence of reintroduction of the virus from its only known enzootic focus in the southern Somali ecosystem.

PPRV shows a similar epizootiology and can also cause very high mortalities in naive populations of sheep and goats. Introductions into disease-free areas are usually traceable to movement of infected animals from enzootic regions. Phylogenetic analysis on PPRV strains reveals that there are four distinct lineages in circulation (**Figure 4(b)**). PPRV infections in wildlife have not been studied in any detail but the severity of disease when it occurs in captive small ruminants infected by contact with domestic livestock makes it unlikely that there is a significant wildlife reservoir of the virus.

CDV causes periodic epizootics in domesticated dogs and wild carnivores and it remains a problem in poor urban areas where there are many stray dogs and vaccination is not carried out rigorously. In enzootic areas, the density of the susceptible population is an important factor in determining the frequency of epizootics and can also affect its maintenance and spread. Again, disease is most often seen in young animals after maternal antibody has waned. A variety of wild carnivore species are highly susceptible to CDV infections and presence of the virus can be a high-risk factor for some endangered species.

The source of PDV in the two epizootics that occurred in European seals is thought to have been either Canadian harp seals (*Phoca groenlandica*) or Arctic ringed seals (*Phoca hispida*). Retrospective analysis of seal sera from Canada

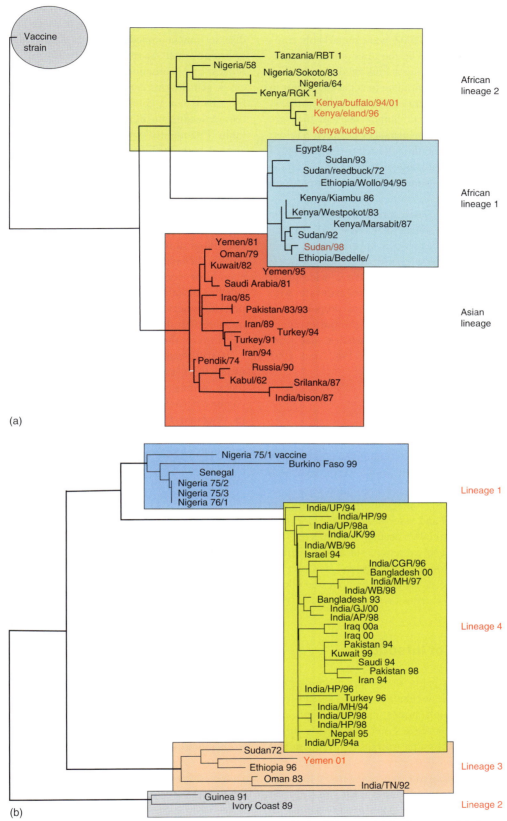

Figure 4 A computer-generated tree showing the relationship between (a) the African and Asian lineages of rinderpest virus and (b) the currently circulating PPRV lineages. The trees were generated using the PHYLIP program using partial sequence data from the F protein gene. Only RPV African lineage 2 remains extant and the outbreaks in Kenya since 1994 are highlighted in red. The vaccine strain of rinderpest dates to the early part of the twentieth century and was used as the outgroup in the rinderpest and PPRV analysis.

dating back to the early 1980s showed that they were positive for virus-specific antibodies. The two European PDV epizootics, in 1988 and 2002, were remarkably similar and this may be explained by the susceptibility, migratory patterns, and/or breeding habits of the seals. The gray seal is the most likely candidate to act as the vector in the transmission of PDV between Arctic and European seals and between colonies of harbor seals as they are known to move much greater distances between haulout sites. Another important factor is the relative resistance of gray seals to the disease, possibly enabling PDV to circulate in that population without necessarily causing high mortality.

The pilot whale (*Globicephala* spp.) is considered to be the most likely endemic source of CeMV and also the vector for its transmission to other species. Pilot whale populations have the characteristics required for both; they move in large groups (pods), have a widespread pelagic distribution, and are known to associate with many different cetacean species. A high proportion of pilot whales sampled in the mid-1990s showed evidence of infection and over 90% of pilot whales that were involved in mass strandings between 1982 and 1993 were seropositive for morbillivirus.

Transmission

All morbilliviruses are extremely labile in the environment and are inactivated by heat, ultraviolet (UV) light, and chemicals that alter pH or destroy their lipid envelopes. Therefore, although highly contagious, they require close contact between infected and susceptible animals for their transmission which normally occurs via infected air droplets. All secretions and excretions potentially harbor virus and, along with fomites, can also be a source of infection. In the case of seals, contact at haulout sites allows the proximity required for aerosol transmission.

Clinical Features of Infection

Morbillivirus infections begin in the upper respiratory tract and, after a variable incubation period spread from the local draining lymph nodes via the lymph and blood to other lymphatic tissues and then to the upper and lower respiratory tracts, gastrointestinal mucosa and in some cases the brain. With virulent strains, there is a marked leucopenia leading to a deficiency in the immune system. The incubation period ranges from 3 to 6 days in natural infections or following experimental parenteral inoculation. With less virulent strains of RPV, the incubation period can extend to 15 days and in many cases there may be no clinical signs but the animals seroconvert. The cell (epithelial and lymphoid) tropisms explain the pathological signs most associated with morbillivirus-induced disease: mucocutaneous lesions, severe infection in the gastrointestinal tract; and destruction of the lymphoid organs and consequent immunosuppression. In some cases, skin lesions can be associated with RPV infection but this is not a common finding. Hyperkeratosis of the cornea leading to blindness is a striking feature seen in kudu. Pneumonia is less marked in RPV than in PPRV and other morbillivirus infections.

In CDV infections, the incubation period can range from 1 to 6 weeks. Animals then show an initial febrile response and develop clinical signs commonly seen in other morbillivirus infections: mucopurulent nasal and ocular discharges, vomiting, diarrhoea, and pneumonia. Neurological signs such as convulsions, tremors, and seizures or behavioral changes are seen and they can develop acutely or weeks or months later or they may follow a subclinical infection. Recovered dogs frequently show persistent nervous tics or involuntary movements of one or more legs. In some cases, a hyperkeratosis (hard pad) develops on the foot pads.

Seals infected with either CDV or PDV also show the usual signs associated with a morbillivirus infection, fever, serous or mucopurulent oculonasal discharge, conjunctivitis, dyspnea, diarrhea, lethargy, and abortion in pregnant females. Bronchopneumonia is the most marked pathological feature in aquatic and marine mammal infections (CDV, PDV, and CeMV) which severely affects the animal's ability to dive and forage for food and quickly results in loss of body condition and a reduced blubber thickness. During the epizootic in the Mediterranean in 1990–91, many striped dolphins (*Stenella coeruleoalba*) were also found to be in poor body condition and loss of fat stores led to decreased buoyancy. Skin lesions, necrosis of the buccal mucosa, lymphodepletion, and hemoconcentration were commonly observed. Neurological signs similar to those found in CDV infections are a feature of both PDV and CeMV infections.

Pathology and Histopathology

The severity of gross pathological lesions observed in RPV and PPRV infections, like the clinical signs observed, is related to the virulence of the virus strain involved. Erosions and ulcerations are found in the upper respiratory, urinogenital, and digestive tracts. In the small intestine there is necrosis and destruction of Peyer's patches. Destruction of the epithelial lining of the gut is responsible for the severe bloody diarrhea seen in acute cases and the packed cell volume can be increased by 40–65%. In the cecum, colon, and rectum of animals infected with RPV and PPRV, so-called 'zebra' or 'tiger' stripes are often found and result from the distension of blood vessels packed with erythrocytes. Dehydration also causes changes in hematology and blood chemistry. A nonsuppurative encephalitis with central nervous system (CNS) degeneration occurs during infections with CDV, PDV, and CeMV, but no CNS involvement has ever been reported in ruminants infected with either RPV or PPRV. Immunosuppression associated with morbillivirus infections can lead secondary bacterial infections which may complicate both the clinical and pathological findings and latent or concurrent infections can be activated.

Histologically, the morbilliviruses show a strong tropism for epithelial and lymphoid cells and all lymphoid organs are affected with damage to the mesenteric lymph nodes, the gut-associated lymphoid tissue, the lymphoid follicles of tonsils, lymph nodes, spleen, and mucosa-associated lymphoid tissues, where severe destruction of the B- and T-cell areas is seen in infections with virulent strains. Mild strains induce less extensive lymphoid destruction and mucosal lesions and in these animals tissue samples show unremarkable histopathological changes in the gastrointestinal tract and the lymphoid tissues and this may account for the reduced ability for these strains to transmit by contact.

Immune Response

There is a strong cell-mediated component in the response to morbillivirus infection and immunosuppressed individuals are known to be at extreme risk from MV. Neutralizing antibodies are generated only in response to the H and F glycoproteins and vaccines containing purified H or F proteins are effective only if administered with a strong cytotoxic T-cell-stimulating adjuvant, for example, Quil A (ISCOM vaccines). Inactivated whole virus vaccines give only a poor and short-lived protection. Both the H or F proteins can confer immunity to disease as poxvirus recombinant vaccines expressing either of these glycoproteins can confer fairly long-term (up to 3 years) immunity to clinical disease. In addition, RPV recombinants can also confer cross-immunity to PPR disease in the absence of cross-neutralizing antibodies. All morbillivirus infections appear to give lifelong protection against disease but it is not clear if subclinical reinfection can occur.

Prevention and Control

In 1711, during a prolonged epizootic of rinderpest in Italy, the Pope's physician, Giovani Lancisi, promulgated rules to deal with the disease in cattle. He insisted on movement controls on all animals in the areas affected, the slaughter of diseased and in-contact cattle, and their burial in lime. He also introduced the idea of quarantine and his policies were backed by strong legal enforcement with severe punishments for transgressors, principles which are still applied today to control animal diseases. This approach, along with import restrictions on cattle from the East, succeeded in controlling the disease in Europe and by the beginning of the twentieth century western Europe was free of enzootic rinderpest. However, these conditions are not always easy to impose in developing countries.

The existence of only one serotype for each of the morbilliviruses, the absence of persistence of infectious virus, and lifelong immunity after recovery from the initial acute infection suggest that outbreaks of these viruses should be easy to control. In addition, the morbilliviruses need close contact for infection to occur, are labile and do not survive long in the environment, and disinfection of infected premises is fairly straightforward. These characteristics of the virus, and the availability of a safe and effective vaccine, were the main drivers for the decision to try to control rinderpest following its resurgence in the early 1980s. International rinderpest control campaigns were begun in the late 1980s, and in 1992 the United Nations Food and Agriculture Organisation (FAO) recommended the global eradication of rinderpest as an internationally coordinated program. In 1994, this became the Global Rinderpest Eradication Programme (GREP), a time-bound program to eliminate rinderpest from the world by the year 2010.

All this effort has succeeded in eradicating the virus from Asia and has eliminated it from most of Africa. The last outbreak of rinderpest occurred in India in 1995 and in Pakistan in 2000. In Africa, the last confirmed case caused by African lineage 1 virus was in southern Sudan in 1998 and so this has probably been eliminated in the field, while the last confirmed case of African lineage 2 was in buffalo in Meru National Park in 2001. Meru is in northeastern Kenya, within the southern Somalia ecosystem, and the major obstacle to completing the GREP by the projected date of 2010 is the continuation of conflict in Somalia.

It was feared that even if rinderpest were eliminated from the domestic cattle populations wild ruminants might act as a reservoir of infection, but history shows that when the disease is eliminated from cattle it disappears from surrounding wildlife, as evidenced by its disappearance from Tanzania, South Africa, and southern Kenya, areas of the continent with many wildlife species. Nevertheless, during outbreaks, rinderpest-infected wildlife can help spread the disease over large distances.

In contrast, only limited resources have been directed to solving the problem of PPRV although a very effective live-attenuated vaccine is also available. From an economic and social perspective, PPR is now considered to be of great importance as it threatens small-ruminant production, the mainstay of many subsistence farmers in much of the developing world.

CDV vaccines have been very effective in controlling infections in domestic dogs; however, not much can be done to prevent infections in wild carnivores, seals, dolphins, and whales. Even if good vaccines were available, vaccination of wild animal populations is logistically very difficult. There are also ethical issues to take into account, such as the potential for uncontrolled spread of vaccine, which may not be attenuated for all species, and the disturbance caused to the animals which may also be harmful.

Diagnosis

Rapid and accurate diagnosis is the key to success in controlling morbillivirus outbreaks in domestic livestock. Clinical signs, however, are not always clear enough to make a confirmatory diagnosis, even when severe clinical signs are evident, as other viruses can mimic those commonly seen in morbillivirus infections. For example, rinderpest and bovine viral diarrhea viruses are often confused and PPRV can be mistaken for pasteurellosis or other microbial pneumonias. Confirmatory laboratory diagnosis is therefore essential.

Simple and rapid diagnostic tests such as capture enzyme-linked immunosorbent assay (ELISA) and reverse-transcriptase polymerase chain reaction (RT-PCR) have been developed for RPV and PPRV in recent years, and, since these are easy to use for analyzing large numbers of samples, these are now favored for virus detection. With RT-PCR, the DNA product can be sequenced and used for phylogenetic analysis that can be used to identify the strain of virus involved and the potential source of virus entering a new region.

For serological detection of morbillivirus antibodies, the ELISA format is favored and the success of the rinderpest eradication campaign depended to a large extent on the ability to seromonitor large herds following vaccination. The test is based on the use of a highly specific monoclonal antibody directed against the H protein of RPV in a competitive ELISA format. A companion test for the detection of PPRV antibodies is also available.

Vaccination

The first attenuated rinderpest vaccines were produced in the 1930s when the virus was adapted to replicate in goats (caprinized) and rabbits (lapinized). In the 1940s, a chick embryo-(avianized) adapted vaccine was produced. These vaccines were widely used but they had drawbacks in that they were not fully attenuated and could cause disease in more susceptible breeds. In the early1960s, a tissue culture-adapted strain of RPV, the 'Plowright vaccine', was developed by multiple passage of the virus in primary bovine kidney cells. The vaccine is relatively easy to produce, safe for use in all cattle breeds, and does not spread by contact. It is highly effective in preventing disease and the immunity induced in vaccinated cattle proved to be lifelong. This vaccine has been used successfully since the early 1960s for the control of rinderpest. A similar vaccine to control PPR was produced in the late 1980s by multiple passages of the virus on Vero cells and this is now being used to control the disease in parts of Africa and Asia.

Egg-adapted (Onderstepoort vaccine) and canine tissue culture-attenuated (Rockborn) vaccines for CDV were produced in the 1950s and are still widely used to vaccinate dogs against distemper. Immunity lasts for several years following vaccination of dogs with either vaccine. The distemper vaccines are not attenuated for all carnivores; for example, some species of ferret are extremely susceptible to CDV and develop disease on vaccination.

All morbillivirus vaccines, like their wild-type progenitors, are extremely fragile and heat-labile and so it is expensive and logistically difficult to store and use them in hot climates. The establishment of an effective cold chain for the delivery of vaccine was an essential feature of the success of the rinderpest eradication campaign as was follow-up seromonitoring studies to determine the level of herd immunity and the effectiveness of vaccination teams.

Future Prospects

Rinderpest is currently on the verge of global eradication and if successful it will be the first veterinary virus disease to have been eradicated globally and the second after smallpox. This goal must not be forgotten and the mild strain of the disease which may still persist in Somalia is of great concern. The worst case scenario is that such a clinically mild strain could move unnoticed into other regions of Africa where, since mass vaccination has ceased, there is a vast naive cattle and wildlife population which could become reinfected. This would provide an excellent opportunity for the virus to evolve to become more pathogenic for cattle. There is a high probability that this could happen as the virus most closely related to this strain was isolated from a giraffe in Kenya in 1962 and is highly pathogenic in cattle.

There is also a major gap in our understanding of the ecology of morbillivirus disease in wild animals, especially in marine and aquatic mammals, and this is an area of research which should be encouraged.

Many questions also remain concerning the host range determinants, biology, molecular biology, and pathogenesis of the morbilliviruses which hopefully can be addressed using reverse genetics.

See also: Border Disease Virus; Bovine Viral Diarrhea Virus; Epidemiology of Human and Animal Viral Diseases;

Measles Virus; Paramyxoviruses of Animals; Viral Pathogenesis.

Further Reading

Appel M (1987) Canine distemper virus. In: Apple MJ (ed.) *Virus Infections of Carnivores*, p. 133. Amsterdam: Elsevier.

Barrett T (2001) Morbilliviruses: Dangers old and new. In: Smith GL, McCauley JW, and Rowlands DJ (eds.) *Society for General Microbiology, Symposium No. 60: New Challenges to Health – Threat of Virus Infection*, 155pp. Cambridge: Cambridge University Press.

Barrett T, Pastoret P-P, and Taylor WP (eds.) (2005) *Biology of Animal Infections: Rinderpest and Peste des Petits Ruminants Virus*, 341pp. London: Elsevier.

Barrett T and Rima BK (2002) Molecular biology of morbillivirus diseases of marine mammals. In: Pfeiffer CJ (ed.) *Molecular and Cell Biology of Marine Mammals*, pp. 161–172. Melbourne, FL: Krieger Publishing Company.

Kock RA, Wambua JM, Mwanzia J, et al. (1999) Rinderpest epidemic in wild ruminants in Kenya 1993–97. *Veterinary Record* 145: 275–283.

Plowright W (1982) The effects of rinderpest and rinderpest control on wildlife in Africa. In: Edwards MA and McDonald U (eds.) *Symposia of the Zoological society of London, No. 50: Animal Diseases in Relation to Animal Conservation*, 1pp. London: Academic Press.

Rotaviruses

J Angel and M A Franco, Pontificia Universidad Javeriana, Bogota, Republic of Colombia
H B Greenberg, Stanford University School of Medicine and Veterans Affairs Palo Alto Health Care System, Palo Alto, CA, USA

© 2008 Elsevier Ltd. All rights reserved.

Glossary

Antigenemia Presence of viral antigen in blood.
CFTR Cystic fibrosis transmembrane conductance regulator, a chloride channel localized in the apical membrane of epithelial cells implicated in secretory diarrhea.
Genotype Specific genetic makeup of one or more viral genes determined by sequence comparison.
Intussusception Pathological event in which the intestine acutely invaginates upon itself and becomes obstructed, followed by local necrosis of gut tissue.
Serotype Significant differences in the antigenic composition of the neutralizing antigens, VP4 and VP7 in the case of rotavirus.
Transcytosis Active transport by which polymeric IgA and IgM antibodies are transported from the basolateral to the lumen of the intestine by the polymeric Ig receptor.

Introduction

Using electron microscopy, rotaviruses were discovered as the etiological agents of epizootic diarrhea of infant mice (EDIM) in 1963 and of calf scours in 1969. Using this same technique, Ruth Bishop identified a rotavirus (RV) in intestinal biopsies of children with diarrhea in 1973. Since then, rotaviruses have been recognized as the most important cause of severe gastroenteritis in children worldwide and an important pathogen of the young of many animals.

Notwithstanding that the overall global mortality from childhood diarrhea decreased in the last 20 years, recent studies suggest that the proportion of hospitalizations attributable to RV-induced diarrhea may have increased during the same time period. For this reason, the burden of RV-related deaths was recently revised, and it is estimated that RV causes around 611 000 childhood deaths every year. More than 80% of these deaths occur in developing countries of sub-Saharan Africa, and South Asia. Worldwide, it is estimated that nearly every child under 5 years of age will have an episode of RV diarrhea, one in five will require medical attention, one in 65 will be hospitalized, and approximately one in 293 will die from RV disease. Although deaths due to RV are rare in developed countries, the incidence of viral infection is the same as in developing countries and health costs associated with RV disease are considerable. In the United States, for example, it has been estimated that 58 000–70 000 rotavirus-associated hospitalizations occur each year and cost-effectiveness studies clearly justify the use of RV vaccines in that country.

Rotaviruses are very well adapted to their host: they replicate very efficiently, sterilizing immunity is not developed and, despite an important host range restriction, many animal hosts exist (**Table 1**). These characteristics help to explain the high viral prevalence, and suggest that the prevention of severe disease is an appropriate goal for vaccination.

Morphology

Rotaviruses were given their name because, when examined by classical electron microscopy, they appear as wheel (*rota*)-shaped, 70 nm particles. However, by cryoelectron microscopy (a method that permits visualization of the viral spikes), the viral diameter is 100 nm (**Figure 1**). Using this method, the virus particle has been shown to be formed by three concentric layers of proteins: the core comprises viral structural protein 2 (VP2), the RNA-dependent RNA polymerase (VP1), guanyl tranferase (VP3), and the viral genome (**Figure 1**), the intermediate layer is formed by structural protein VP6, the most abundant and most antigenic viral protein, and the external layer comprises 780 copies of glycoprotein VP7 and 60 viral spikes formed by VP4. The surface of the virion has three types of pores that penetrate into the interior of the capsid. These channels seem to be important during viral replication, allowing exchange of compounds in aqueous solution to the inside of the capsid and the export of nascent RNA transcripts.

Several RV structural proteins and nonstructural proteins (NSPs) have been crystallized, permitting the initiation of detailed molecular studies of viral physiology. By this method, the viral spikes have been shown to consist of VP4 trimers, the structure of which rearranges upon trypsin cleavage (a process that enhances viral infection) and probably again on entry into the cell. These changes resemble the conformational transitions of membrane

Table 1 Virus-associated factors that contribute to high viral prevalence and reinfections

Characteristic	Comments
Natural infection does not generate sterilizing immunity	Goals of vaccination are to decrease severe disease but not to prevent infection
Multiple animal hosts exist	Eradication does not seem feasible
Short incubation period (1–2 days)	Does not allow time for the recall of high levels of immune effector mechanisms
The entry cell is the same as the cell used for viral replication	Does not allow time for the recall of high levels of immune effector mechanisms
Virus is excreted in high quantities (up to 10^{11} pfu g^{-1} of feces)	High levels of viral dissemination in the environment
Up to 30% of children excrete antigen up to 57 days after onset of diarrhea	High level of viral dissemination
High rate of viral mutation and gene reassortment	May permit the virus to evade the immune system. Currently unproven
Over 50% of infections are asymptomatic	RV well adapted to the human host

Reprinted from Franco MA, Angel J, and Greenberg HB (2006) Immunity and correlates of protection for rotavirus vaccines. *Vaccine*, 24: 2718–2731, with permission from Elsevier.

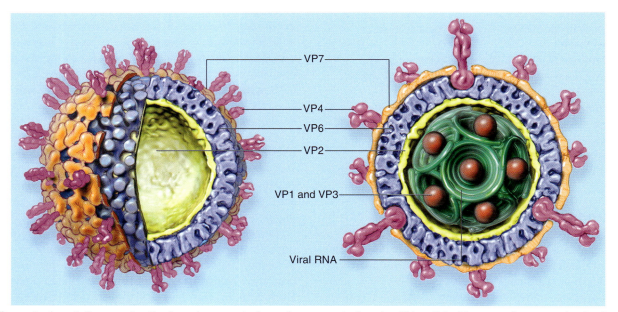

Figure 1 An artist's reconstruction based on cryoelectron microscopy studies of an RV particle. Shown are the seven structural proteins, and the viral RNA. Reproduced with permission from Andrew Swift, Swift Illustration.

fusion proteins of enveloped viruses. The crystal structure of the viral hemaglutinin VP8 (VP4 is cleaved by trypsin into VP8 and VP5) that contains several virus-neutralizing epitopes has also been determined. Details of the characteristics and function of the viral proteins are described elsewhere in this encyclopedia.

Classification and Epidemiology

Rotaviruses are classified in the genus *Rotavirus* of the family *Reoviridae* that comprises icosahedral, nonenveloped viruses with segmented, double-stranded RNA (dsRNA) genomes. Based primarily on epitopes in VP6, rotaviruses are classified serologically into seven groups (A–G). These serologically distinct groups are also very distinct genetically and genome segment reassortment does not occur between serogroups. According to current taxonomic classification, serogroups A–E each correspond to a different RV species. Serogroups F and G are currently regarded as tentative species. Most human pathogens fall into groups A, B, and C. The information presented in this chapter is mostly limited to group A rotaviruses which are, by far, the most common cause of severe diarrhea in humans. Among group A rotaviruses, antigenic differences in VP6 have also been identified and used to establish viral subgroups, primarily for epidemiological studies. Serotypes within each serogroup are defined by epitopes in VP7 (glycoprotein, G types) and VP4 (protease, P types) that induce neutralizing antibodies. As the genes encoding these two proteins segregate separately during genome segment reassortment, a binary serotyping system has been developed to identify isolates.

Genotypes, determined by nucleic acid sequence similarity of genes encoding VP7 and VP4, and serotypes, determined by antigenic similarity in VP4 or VP7, as tested by neutralization assays, are generally equivalent for VP7. Currently, 15 G types have been described. G1, G2, G3, G4, and G9 constitute more than 92% of all G serotypes of humans detected globally and appear to be equally virulent. For VP4, there is no direct relationship between genotypes and serotypes and, therefore, a dual system P classification is in use (P genotype numbers are denoted in brackets and P serotypes without brackets). At least 23 P genotypes P[1]–P[23] and 14 serotypes have been described. More than 91% of circulating human RV strains express the P[8] and P[4] genotypes. These genotypes correspond to two subtypes of P1 serotype (P1A and P1B, respectively) that share some cross-reactive epitopes.

Overall, strain variability is less than would be expected from a random association of G and P genotypes since human RV strains belonging to G1, G3, and G4 serotypes are preferentially associated with P[8], while G2 serotype strains are most frequently associated with P[4] genotypes. Importantly, cross-hybridization studies using labeled viral RNA have shown that, in general, viruses that express P[8] form a different genogroup than viruses that express P[4].

In addition to these more prevalent strains, rotaviruses with unusual serotypes currently circulate and can arise sporadically in developed and particularly in developing countries. Several serotypes may coexist within a community, but, in temperate climates especially, each season is usually dominated by a single serotype that may change from season to season. The global distribution of rotaviruses varies over continents: G1P[8] represents over 70% of RV infections in North America, Europe, and Australia, but only about 30% of the infections in South America and Asia, and 23% in Africa. These differences in geographic distribution can probably be explained by variations in sanitary and climatic conditions and/or closer contact of individuals with animal rotaviruses in areas with more RV diversity.

Animal and human group A rotaviruses can undergo genome segment reassortment *in vitro* and *in vivo*. Reassortment originates in cells simultaneously infected with two different rotaviruses and results in progeny viruses that have a combination of genes from each parental strain. In several cases, human rotaviruses have been found that have genes of animal RV origin, adding a further dimension to strain diversity. However, rotaviruses have an important host range restriction in that, in general, humans are infected only with human rotaviruses. This restriction has been exploited for the development of several viral vaccines.

In the temperate zones of the world (mostly developed countries), RV infection occurs primarily during epidemic peaks in the cooler months of the year. A yearly wave of rotaviral illness spreads over the United States; it begins in the southwest in November and terminates in the northeast in March. A similar phenomenon has recently been reported to occur in Europe. This pattern is not seen in countries within 10° of latitude from the Equator (mostly developing countries) where epidemics occur year-round. No clear explanation is available for these temporal patterns of viral prevalence.

Rotaviruses are highly contagious and spread easily but, unlike certain human bacterial infections that occur disproportionately in developing countries, the prevalence of RV infection is the same in developed and developing countries, implying that sanitation and hygiene are not effective measures for disease control.

The Viral Genome and Viral Replication

The RV genome comprises 11 dsRNA segments that code for six structural proteins and five or six nonstructural proteins. The size of segments varies from 0.6 to 3.3 kbp. Each segment contains a single long open reading frame

(ORF), with the exception of segments 9 and 11, each of which may contain two ORFs. The 5′ and 3′ ends of the RNA segments have noncoding regions that differ between rotaviruses from groups A, B, and C. These sequences are important in transcription, replication, and reassortment of the virus genome.

The viral RNA itself is not infectious. For this reason, the engineering of recombinant rotaviruses has been very difficult, and this has impeded functional analysis of the viral RNA noncoding sequences and of the viral proteins. This problem has been partially solved by the development of a cell-free system that supports the synthesis of dsRNA from exogenous mRNA, and by gene silencing using small interfering RNA (siRNA) to specifically inhibit translation of viral proteins. In addition, a reverse genetics system for introduction of site-specific mutations into the dsRNA genome of infectious RV has been recently developed (see *recommend readings*). Evaluation of the utility of this method is eagerly awaited by scientists in the field.

Rotaviruses are highly variable. Three mechanisms for generating genetic diversity have been identified: point mutation, genome segment reassortment, and recombination. Rotaviruses have high rate of mutation and it has been estimated that, on average, at least one mutation occurs during each genome replication. Reassortment of genome segments also occurs at high frequency during mixed infections with two or more rotaviruses, both *in vitro* and *in vivo*. Natural reassortment *in vivo* appears to influence the serotypic diversity in humans, especially in less developed countries (see below). Recombination and related rearrangements of viral RNA segments (e.g., partial gene duplication or deletions) probably play a minor role in generating viral diversity but may be important in longer-term viral evolution.

RV entry into host cells is a multistep process and several molecules have been identified as RV receptors or co-receptors. However, the process has been shown to vary in different viral strains. For example, for several animal RV strains, the first binding step involves the interaction of VP8 with sialic acid, while some human RV strains appear to bind initially to GM1 ganglioside. As a second step, RV binds to the integrin $\alpha 2\beta 1$ in an interaction mediated by the integrin-binding motif DGE in VP5. In addition to these two interactions, integrins $\alpha v\beta 3$ and $\alpha x\beta 2$, and the heat shock protein hsp70, have also been shown to be involved at a later step of rotavirus cell entry. It seems that the association of some of these molecules with rafts is important for viral entry.

Although it was initially proposed that rotaviruses enter by direct penetration, current models favor the hypothesis that virus entry is by endocytosis. Inside the cell, VP5 seems to induce a size-selective membrane permeabilization of the putative viral endosome that facilitates the transit of Ca^{2+} from the vesicle. When Ca^{2+} concentrations in the endosome are lowered, a disassembly of VP4 and VP7 is postulated to occur. This event permeabilizes the membrane of the endosome and the viral transciptase (viral particles without VP4 and VP7) gains access to the cytoplasm and begins to synthesize viral mRNAs. The mRNAs produced by the transcriptase are exact copies of each genome segment, with a 5′-terminal type 1 methylated cap structure and without a 3′-terminal poly(A) sequences.

The viral mRNAs are translated, giving rise to the structural and nonstructural proteins necessary to complete the viral replication cycle. NSP3 is reported to shut off the synthesis of cellular proteins and induces the preferential translation of viral proteins. These proteins accumulate in the cytoplasm in an electron dense region called the viroplasm, where the viral genome is replicated and the assembly of progeny double-layered particles takes place. The mechanism by which one viral particle assembles with each and only one of the 11 RNA segments is unknown. Synthesis of the dsRNAs occurs following the packaging of viral mRNAs into intermediate precursors of double-layered particles.

Assembled double-layered particles interact with NSP4, that has been synthesized by ribosomes associated with the endoplasmic reticulum (ER), and bud into the ER lumen. In this organelle, the double-layered particles acquire a transient lipid membrane. Then, in a very poorly understood mechanism (probably related to the high Ca^{2+} levels of the ER), the viral particles acquire VP7 and lose the transient enveloping membrane, giving rise to the triple-layered particle. Recent experiments suggest that VP4 is acquired by the viral particle in a compartment outside the ER.

The physiological mechanism of exit of the mature triple layer viral particle from the cell is unknown. In polarized Caco-2 cells (intestinal epithelial cells derived from a human colon adenocarcinoma), RV is released before cell death using a vesicle-associated vectorial transport system to the apical pole. However, rotaviruses are lytic viruses, and could also exit the cell after cell lysis. The mechanism of cell death induced by RV is not completely understood. Results in polarized Caco-2 cells and *in vivo* studies in mice suggest that it is by viral induced cell apoptosis.

Pathogenesis

Important viral antigenemia and some level of viremia are observed in the initial phase of RV-induced diarrhea in children and animals. In mice, extraintestinal viral replication occurs commonly during homologous and some heterologous infections. Also in this model, the level and location of extraintestinal replication varies between RV strains and replication can occur in several leukocytes subsets. However, the clinical relevance of these findings is still unclear and, in children and in animals, the bulk of RV replication most likely occurs in the mature villus tip cells of the small bowel.

RV-induced diarrhea probably occurs by multiple mechanisms that vary, depending on the animal species analyzed (**Table 2**). Pathological findings in the small intestine of children with RV diarrhea include: shortening and atrophy of the intestinal villi, enterocyte vacuolization with distended cisternae of the endoplasmic reticulum, and mononuclear infiltration in the intestinal lamina propia. However, in children a direct relationship between the extent of histopathology and disease has not been demonstrated. This finding suggests that RV diarrhea can occur without important enterocyte death.

In pigs, RV infection induces an intestinal lactase deficiency and increased lactose in feces induces an osmotic diarrhea. Elevated levels of lactose are also observed in the feces of some RV infected children. Lactase deficiency could be due to RV-induced enterocyte destruction or to alteration in the synthesis or metabolism of disaccharidases. In mice, two mechanisms of diarrhea, which do not involve enterocyte destruction, have been identified. NSP4, and a derived peptide of NSP4, induce diarrhea in mouse pups but not adult animals, making it the first viral enterotoxin described. However, its role in RV-induced diarrhea in other species has not been characterized. Also in mice, RV has been shown to activate the intestinal autonomous neural system and increase the secretion of water and electrolytes, as well as the intestinal motility. Racecadotril, an inhibitor of the enteric nervous system, is somewhat efficacious in treating RV-induced diarrhea in children, suggesting that this mechanism plays at least some role in human diarrhea.

It is possible that the three mechanisms (malabsorption, NSP4 enterotoxicity, and enteric neural system stimulation) may play a major or minor role in the pathophysiology of RV-induced diarrhea, depending on the species and the time point after the onset of disease.

Diagnosis, Clinical Characteristics, and Treatment

Initial efforts to isolate wild-type rotaviruses in tissue culture were not very successful. For this reason, early clinical studies characterized rotaviruses by isolating viral RNA from feces and then analyzing the RNA by electrophoresis in polyacrylamide gels. Using this method, rotaviruses were classified depending on the pattern of migration of their RNA segments (electropherotypes). At present, this method seems of limited value because a relationship between electropherotypes and virulence or serotypes is not generally apparent. Currently, diagnosis is commonly conducted using commercial ELISAs that detect viral antigen (mostly VP6) in the feces. Rotavirus is shed in very large amounts making the ELISA a highly effective and accurate diagnostic assay. For epidemiological studies, human RV strains present in feces can be grown in cell culture in MA104 (African green monkey kidney) and Caco-2 cells using trypsin for enhancement of viral growth during the culture. Trypsin cleaves VP4 and increases infectivity in culture. Genotype characterization of RV strains is mainly by reverse transcription-polymerase chain reaction (RT-PCR). RV detection by electron microscopy is only conducted in research laboratories, especially when other viral pathogens are being investigated.

Rotaviruses are highly contagious since approximately 10^{11} particles per gram of feces are excreted, and they are very resistant to ambient conditions. In addition, viral excretion in most children lasts for up to 10 days and, in some children, may extend for 2 months after onset of infection. Rotaviruses are mainly transmitted by an oral–fecal route although, in some cases, a respiratory route has been suggested.

In developing countries, the peak incidence of RV disease occurs in children between 6 and 11 months of

Table 2 Mechanisms of RV-induced diarrhea (may vary according to the animal species studied)

Mechanism	Comments
Action of NSP4 as a toxin induces a secretory diarrhea	Only demonstrated in rodents. Non-CFTR mediated. Occurs early in infection prior to cell death
RV stimulates the enteric nervous system (ENS) inducing a secretory diarrhea and increased intestinal motility	Drugs that inhibit the ENS are useful to treat RV diarrhea. Occurs early in infection prior to cell death
Altered metabolism of disaccharidases and other enterocyte membrane proteins induces malabsorptive/osmotic diarrhea	Occurs early in infection prior to cell death
Enterocyte death contributes to malabsorptive/osmotic diarrhea	Late mechanism. In polarized intestinal epithelial cell lines and *in vivo* in murine enterocytes, RV infected cells seem to die by apoptosis

Reprinted from Franco MA, Angel J, and Greenberg HB (2006) Immunity an correlates of protection for rotavirus vaccines. *Vaccine*, 24: 2718–2731, with permission from Elsevier.

age. In contrast, in developed countries, the highest incidence is observed in older children (2 years old). This difference is probably related to differences in sanitation in the different settings. Notwithstanding this variation, RV incidence is similar in both developing and developed countries but mortality is mainly observed in developing countries, presumably due to limited access to appropriate health care. The relative protection of infants younger than 2 months of age, which occurs worldwide, could be related to the presence of protective maternal antibodies. Up to 50% of adults caring for children with RV diarrhea can become infected and, of these, 50% develop disease which is generally mild.

The primary clinical syndrome caused by RV infection is acute gastroenteritis. After a short incubation period of 48 h or less, children frequently present with vomiting that lasts for 1–2 days. The vomiting is often accompanied with fever (37.9 °C or greater). Subsequently, or at the same time, a watery diarrhea appears and, if it is not treated, frequently induces dehydration. It is estimated that up to 50% of RV infections in children are asymptomatic but some of these may represent second or third exposures.

Children attending day care institutions are at high risk of developing RV-induced diarrhea. Although RV infections in neonatal care units have been classically described as asymptomatic, probably due to the presence of maternal antibodies, severe symptomatic outbreaks have also been described. RV strains that induce nosocomial infections in neonatal nurseries are generally different from those circulating in the community. RV infection can cause severe and prolonged disease in children with primary immunodeficiencies, some of whom develop a systemically disseminated infection. Acquired immunodeficiency also predisposes to severe RV disease in bone marrow- and liver-transplanted children. The role of RV-induced disease in immunosuppressed adults with HIV seems, at present, less important.

RV diarrhea is self-limited and treatment is aimed at reducing symptoms until the immune response resolves the infection. Children with mild diarrhea are treated by oral rehydration. Those presenting moderate to severe dehydration may require intravenous rehydration. In cases of severe disease, treatment with probiotics, preparations that contain antibodies against RV, Racecadotril (an inhibitor of the enteric nervous system), and Nitasoxanide (a drug of unknown mechanism of action) have been shown to accelerate resolution of the disease. However, it is not yet clear whether these interventions truly provide significant advantages to ill children.

Immune Response

Immunity to RV is incompletely understood and animal models have been important in the acquisition of our currently available knowledge. Pepsin and gastric acid seem to be important host defense factors against RV infection, since these factors inactivate rotaviruses in adult mice but not in suckling mice. In addition, the innate immune response and interferons, in particular, seem to mediate an antiviral effect, and viral mechanisms for evading this response involving NSP1 have recently been suggested. However, mice lacking T and B cells become chronically infected with RV, suggesting that the adaptive immune system is essential for viral elimination. Also, children lacking T and B cells, or only B cells have been shown to shed virus chronically.

Antibodies seem to be the principal mechanism that mediates protection against viral reinfection. In agreement with the fact that most viral replication occurs in the intestine, the localization of virus specific B cells in the intestine seems important for their capacity to mediate protection. It is postulated that local IgA antibodies can mediate expulsion of rotaviruses inside the enterocytes and exclusion (to avoid *de novo* infection of enterocytes) of rotaviruses in the gut lumen. Neutralizing antibodies to VP4 and/or VP7 can block enterocyte infection directly when present in the gut lumen (exclusion). In mice, anti-VP6 non-neutralizing polymeric IgA may bind virus VP6 during transcytosis from the basolateral membrane of enterocytes to the gut lumen and 'expulse it'. These antibodies may also inhibit RV transcription intracellularly. In addition, antibodies to NSP4 may block diarrhea, but not infection, by blocking the enterotoxic property of the molecule. However, the antiviral effects of antibodies against VP6 and NSP4 have only been shown to be protective in mice. In piglets, a model that is probably closer to humans than mice, the presence of antibodies against VP4 and/or VP7 seems necessary for protection.

Although local antibodies appear to be the principal mechanism that protects against viral infection, T cells also directly mediate antiviral immunity, at least in mice. CD4+ T cells are also essential for the development of more than 90% of the RV-specific intestinal IgA, and thus their presence seems critical for the establishment of protective long-term memory responses. Moreover, murine RV-specific CD8+ T cells are involved in the timely resolution of primary RV infection and can mediate short-term partial protection against reinfection.

In children, primary RV infections are generally the most severe, with severity decreasing as the number of reinfections increases. Complete protection against moderate-to-severe illness is achieved after two natural RV infections, whether symptomatic or asymptomatic. In agreement with the results of animal models mentioned previously, total serum RV IgA, induced by a primary infection, seems to be the best but imperfect correlate of protection against subsequent reinfections. Primary RV infection induces homotypic neutralizing antibodies to VP7 and VP4 and some level of heterotypic immunity, but the role in protection of these antibodies is still controversial. In this respect, it is noteworthy that an attenuated G1P[8] human RV vaccine has been shown to induce excellent protection against severe diarrhea caused by P[8] rotaviruses, independently of their G type. This vaccine can also induce protection against G2P[4] rotaviruses, but this protection is somewhat lower than against the P[8] viruses. Altogether, these findings suggest a role for neutralizing antibodies directed against VP4 in protection. However, since, as mentioned previously, P[4] viruses belong to a separate genogroup from most P[8] viruses, differences in immunity to other non-neutralizing viral proteins could also be the explanation for the relatively lower protection rate induced by the vaccine against P[4] viruses.

The lack of understanding of RV immunity limits our capacity to develop new RV vaccines. Recent studies have tried to characterize RV-specific lymphocytes in humans to better understand human immunity. Due to the relative inaccessibility for study of the human intestinal immune tissue, cells involved in the RV immune response have been studied in peripheral blood. It is hypothesized that at least a fraction of these lymphocytes are recirculating to and from the intestine, and thus may reflect intestinal immunity. In blood of children with acute RV infection, plasmablasts and plasma cells (secreting mainly RV specific IgM) are found, and these express homing receptors, that allow them to return to the intestine. In the convalescence phase, RV-specific B cells are mainly memory cells, some of which express intestinal homing receptors. In addition, RV-specific CD4+ and CD8+ T cells have been characterized. In children and adults with acute infection, both types of T cells that secrete gamma interferon circulate, but in unexpectedly low frequencies. It is speculated that the limited induction of RV-specific T cells can be related to the occurrence of delayed viral clearance and symptomatic reinfections in a subset of individuals.

Viral Vaccines

The first commercially available RV vaccine was a simian-human reassortant tetravalent vaccine (RotaShield). Although this vaccine was highly effective, it was withdrawn from the market 1 year after its introduction due to its temporal association with low levels of intussusception (1 in 10 000 children). Two new RV vaccines have been recently licensed for use in many countries worldwide. These vaccines appear to be safe and not cause intussusception. The vaccine produced by Merck (RotaTeq^{TM}) contains mono-reassortants of a bovine virus with G1, G2, G3, G4, and P1A[8] human RV genes. The vaccine produced by GlaxoSmithKline (GSK, Rotarix) is a human-attenuated G1P1A[8] virus. In trials that involved over 60 000 infants, both of these vaccines have been shown to be safe and to provide over 70% protection against any RV diarrhea and over 98% protection against severe RV infection. Importantly, both vaccines reduced the rates of all gastroenteritis-related hospitalizations of any cause by over 40%. Despite these encouraging results, it is still undetermined if these vaccines will work in very poor developing countries in Africa and Asia in which children can be malnourished and in which atypical RV strains may circulate frequently. Moreover, the intussusception associated with RotaShield was predominantly seen in children older than 2 months, and the children studied in the safety evaluation of the two new vaccines were mostly 2 months of age. Thus, the safety of the new RV vaccines in children of more than 2 months of age has not been fully established and the development of second-generation vaccines may be desirable. A promising strategy to drastically reduce the risk of intussusception is to administer the vaccine to neonates, in whom intussusception almost never occurs. Current studies in animal models are focused on the development of second generation nonreplicating RV vaccines, such as DNA vaccines, and recombinant RV proteins or virus-like particles.

Further Reading

Franco MA, Angel J, and Greenberg HB (2006) Immunity and correlates of protection for rotavirus vaccines. *Vaccine* 24(5): 2718–2731.

Franco MA and Greenberg HB (2002) Rotaviruses. In: Richman DD, Hayden FG, and Whitley RJ (eds.) *Clinical Virology*, 2nd edn., pp. 743–762. Washington, DC: ASM Press.

Komoto S, Sasaki J, and Taniguchi K (2006) Reverse genetics system for introduction of site-specific mutations into the double-stranded RNA genome of infectious rotavirus. *Proceedings of the National Academy of Sciences, USA* 103(12): 4646–4651.

Rothman KJ, Young-Xu Y, and Arellano F (2006) Age dependence of the relation between reassortant rotavirus vaccine (RotaShield) and intussusception. *Journal of Infectious Diseases* 193(6): 898.

Ruiz-Palacios GM, Perez-Schael I, Velazquez FR, et al. (2006) Safety and efficacy of an attenuated vaccine against severe rotavirus gastroenteritis. *New England Journal of Medicine* 354(1): 11–22.

Santos N and Hoshino Y (2005) Global distribution of rotavirus serotypes/genotypes and its implication for the development and implementation of an effective rotavirus vaccine. *Reviews in Medical Virology* 15: 29–56.

Vesikari T, Matson DO, Dennehy P, et al. (2006) Safety and efficacy of a pentavalent human-bovine (WC3) reassortant rotavirus vaccine. *New England Journal of Medicine* 354(1): 23–33.

Rubella Virus

T K Frey, Georgia State University, Atlanta, GA, USA

© 2008 Elsevier Ltd. All rights reserved.

Glossary

Birth defects Malformations or abnormalities developed during gestation that are apparent at, or soon after, delivery.

Congenital infection Infection of the fetus during gestation as a result of virus passage through the placenta.

Elimination program Vaccination program designed to eliminate 'indigenous' sources of a pathogen from a geographic region and thus restrict sources of the pathogen to imports from other regions.

Enveloped virion Virus particle with a lipid bilayer membrane, or envelope, surrounding the capsid or core. The envelope is usually derived from host cell membranes and contains virus-specified glycoproteins decorating its outer surface.

Icosahedral capsid In the virus particle, the proteinaceous shell surrounding the virus genome – in this case, with symmetry of an icosahedron (appears quasi-spherical or round in electron micrographs).

Live, attenuated vaccine A vaccine formulation using an infectious variant of the virus that has been attenuated or weakened by repeated passage in cell culture or alternate hosts. Such vaccines recapitulate the infection, including induction of a complete adaptive immune response, but do not cause disease.

Persistent infection Infection characterized by continued presence of the virus, often despite the induction of an adaptive immune response.

Plus-strand RNA virus Virus with a single-stranded RNA genome that is of messenger RNA sense and can be directly translated to produce virus-specified proteins.

Serodiagnosis Detection or diagnosis of infection through immune status, usually the presence of IgM antibodies or a rise in IgG antibodies in paired sera collected at the time of symptoms and convalescence (post-recovery).

Systemic infection Infection that spreads from local site entry via the lymphatic system and blood stream to one or more target internal organs or tissues.

Introduction

Rubella virus (RUBV) infection causes a benign disease known as rubella or German measles that can result in profound birth defects if contracted *in utero*. Norman Gregg, an Australian ophthalmologist, first reported the association of congenital cataracts as a consequence of gestational rubella in 1941, establishing RUBV as a major teratogen. The last major epidemic of rubella to impact the United States occurred in 1964–65, resulting in 20 000 congenital rubella syndrome (CRS) cases, and subsequently live, attenuated vaccines were developed and applied in the US, Canada, Japan, and several European countries by 1969. These programs have been successful in greatly reducing the incidence of both rubella and CRS in these countries, particularly the US, in which indigenous rubella was declared eliminated in 2004. Currently, ~50% of countries have national rubella vaccination programs, but the majority of the world's population is not covered and rubella thus remains a worldwide challenge.

Classification

RUBV is a member of the family *Togaviridae* and the only member of the genus *Rubivirus*. The other togavirus genus, *Alphavirus*, contains 26 members, all of which are arthropod-borne. In common with the alphaviruses, the rubella virion consists of a single-stranded, plus-sense genomic RNA contained within a quasi-spherical core or capsid, composed of a single virus-specified protein, C, which is surrounded by a lipid envelope in which are embedded two virus-specified glycoproteins, E1 and E2. Unlike alphaviruses, RUBV has no invertebrate vector and the only known natural reservoir is humans. RUBV also is antigenically distinct from the alphaviruses. Phylogenetic analysis of E1 gene nucleotide sequence of worldwide RUBV isolates has led the World Health Organization to propose a standardized taxonomy for RUBVs which consists of two clades, 1 and 2, and genotypes within each clade. Clade 1 contains seven genotypes and is distributed worldwide. Clade 2 contains three clades and appears to be restricted to Asia.

Host Range and Virus Propagation

RUBV has no known natural host other than humans. No reliable animal model exists for the study of RUBV pathogenesis. RUBV replicates in a number of laboratory cell culture lines; however, in most of these no cytopathic effects (CPEs) are routinely observed. Continuous cell lines commonly used to propagate RUBV include Vero (African green monkey kidney), RK-13 (rabbit kidney), and BHK-21 (baby hamster kidney). In all cell lines in which RUBV replicates, persistent infections are readily established and maintained, whether or not the cell line exhibits a functional interferon system, and persistent RUBV infection in cell culture cannot be cured by the inclusion of neutralizing antibodies in the culture medium because virus budding occurs at intracytoplasmic locations. Thus, RUBV is highly adapted for persisting infection.

Properties of the Virion

Rubella virions are 60–70 nm spherical particles composed of an electron-dense core separated from the lipid envelope by an electron-lucent zone. Rubella virions exhibit a marked degree of pleomorphism (**Figure 1**). The virion has a density of 1.18–1.19 g ml^{-1}, whereas isolated capsids have a density of 1.44 g ml^{-1}. The C protein (~34 kDa), which comprises the capsid, is present as disulfide-linked homodimers. Although presumed to be icosahedral, the symmetry of the RUBV capsid has not been solved.

The virion spikes are formed by two virion glycoproteins, E1 and E2. E1 has a molecular weight of 59 kDa whereas E2 is a heterogeneous species ranging from 44 to 50 kDa due to differential glycosylation. Both E1 and E2 appear to be primarily in the form of heterodimers which are easily disrupted by routine preparation techniques. The higher-order architecture of the virion spikes is entirely unclear. E1 is more exposed on the virion surface than is E2, contains both the viral hemagglutinin and receptor site, and is also immunodominant in terms of the humoral response.

Rubella virions are stable at physiological pH values and can be frozen at temperatures below −20 °C for years, without loss of infectivity. Live, attenuated vaccine virus is stored in lyophilized form. Rubella virions are susceptible to most commonly used inactivating agents, such as formaldehyde, UV light, and lipid solvents.

Genomic Organization

The RUBV genomic RNA is 9762 nucleotides in length and contains a 5′ terminal cap structure and a 3′ terminal poly(A) tract. A distinctive feature of the genomic RNA is that it contains 30% guanine residues and 39% cytosine residues, the highest G+C content of all RNA viruses. The genome contains two long, nonoverlapping open-reading frames (ORFs) plus untranslated regions (UTRs) at the 5′ and 3′ ends and between the ORF's (**Figure 2**). The 5′ proximal ORF, or nonstructural proteins ORF (NS-ORF), encodes a 2116-amino-acid product that is proteolytically cleaved into two products, 150 kDa (P150) and 90 kDa (P90), which are at the N- and C-termini of the ORF, respectively. The cleavage is mediated by a papain-like cysteine protease located at the C-terminus of P150. P150 and P90 are responsible for virus RNA replication. By computer-assisted comparisons with other viruses, P150 contains a domain predicted to have methyl/guanylyltransferase activity (responsible for forming the cap structure at the 5′ end of the genomic and subgenomic RNAs) in addition to the protease, whereas P90 contains both a helicase domain and an RNA-dependent RNA polymerase (or replicase) domain. The 3′ proximal ORF, or structural protein ORF (S-ORF), encodes a 1063-amino-acid product that is proteolytically processed into the virion proteins by a cell protease, signal endopeptidase. The order of the virion protein genes within the ORF is 5′ C-E2-E1 3′. The S-ORF is translated from a subgenomic RNA synthesized in infected cells and containing the sequences from the start site through the 3′ end of the genome. An infectious clone for RUBV has been developed. An infectious clone is a cDNA copy of the viral RNA contained in a plasmid in which it is placed adjacent to an RNA polymerase promoter. Since RUBV has a plus-sense genome, *in vitro* transcripts from the plasmid will initiate virus replication following transfection into susceptible cells. The infectious clone allows for studies of the effects of site-directed mutagenesis on the RUBV genome.

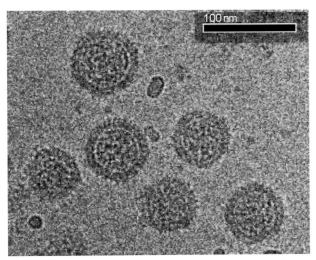

Figure 1 Cryoelectron micrograph of rubella virions. Unlike conventional negative staining, the virions have a uniform structure when visualized by this technique; however, pleomorphism of particle diameter and the gap between the core and envelope remains apparent. Courtesy of Tao Sun, Yumei Zhou, Michael Rossmann, and Teryl Frey (unpublished data).

Intracellular Replication Cycle

The receptor for RUBV on the surface of susceptible cells has not been identified. Following attachment to the receptor, the virus is taken into the cell by receptor-mediated endocytosis. In the reduced pH environment of the endocytic vesicle, fusion between the viral envelope and the vesicular membrane occurs, releasing the capsid and genomic RNA into the cytoplasm of the cell.

The genomic RNA is translated to produce the NS protein precursor which is cleaved into P150 and P90 (**Figure 3**). These proteins then use the genomic RNA

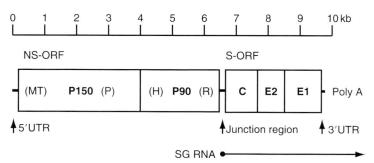

Figure 2 Coding strategy of the RUBV genome. Shown is a schematic representation of the RUBV genomic RNA with untranslated regions (UTRs) drawn as solid black lines and coding regions (ORFs) as open boxes (NS-ORF, nonstructural protein ORF; S-ORF, structural protein ORF). Within each ORF, the coding sequences for the proteins processed from the translation product of the ORF are delineated and, in addition, within the NS-ORF, the locations of motifs associated with the following activities are indicated: methyl/guanylyltransferase (MT), protease (P), helicase (H), and RNA-dependent RNA polymerase (R). The sequences encompassed by the subgenomic RNA (SG RNA) are also shown. The scale at the top of the diagram is in kilobases.

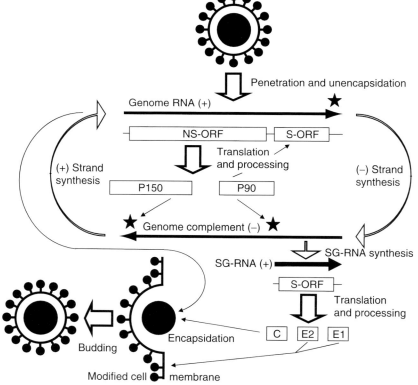

Figure 3 Replication strategy of RUBV. The plus-sense genome and subgenomic RNA are represented by solid black arrows indicating plus polarity; beneath each, the ORFs that they contain are shown as open boxes. The minus-sense genome RNA complement, represented as a dotted arrow, is used solely as a template for the two plus-sense RNA species. Putative cis-acting sequences on each RNA, which are recognized by the virus RNA-dependent RNA polymerase to initiate synthesis of complementary RNAs, are marked with stars. The general functions of the virus proteins are indicated by arrows (e.g., P150 and P90 functioning as the RNA-dependent RNA polymerase by interacting with cis-acting sequences on the viral RNA species and synthesizing complementary strands).

as a template for synthesis of a genome-length, minus-sense RNA. The genome-length, minus-sense RNA is then used as the template for synthesis of both the genomic RNA and the subgenomic RNA. Synthesis of the subgenomic RNA is initiated by internal recognition of sequences on the genome-length, minus-sense RNA template. Host-cell proteins are likely involved in the replication process and, interestingly, recent evidence indicates that the C protein is as well. RNA synthesis is asymmetric in infected cells in that more of the plus-sense species than the minus-sense RNA is produced. The uncleaved NS protein precursor is active in minus-strand RNA synthesis, while the cleaved P150/P90 complex appears to be active in only plus-strand RNA synthesis. Thus the activity of the NS protease through mediating this cleavage is important in regulating plus- and minus-sense RNA synthesis. RUBV RNA synthesis occurs in cytopathic vacuoles of lysosomal origin, in infected cells.

In the structural protein ORF, E2 and E1 are immediately preceded by hydrophobic signal sequences that function to direct translation of secreted and membrane-associated proteins into the lumen of the endoplasmic reticulum (ER). Therefore, following translation of the C sequences within the ORF, the E2 signal sequence mediates association of the translation complex with the ER. C–E2 cleavage is mediated by signal endopeptidase, or signalase, which functions in the lumen of the ER to remove signal sequences from secreted and membrane-associated proteins; unlike the proteins of the alphaviruses, RUBV C protein does not have autocatalytic protease activity. Following cleavage, the E2 signal sequence remains associated with C. Similarly, the E1 signal sequence maintains the association of the translational complex with the ER, signalase mediates the E2–E1 cleavage, and the E1 signal sequence remains attached to E2. Soon after synthesis, heterodimerization of E2 and E1 occurs in the lumen of the ER. The three-dimesional folding of E1 appears to be a complicated process requiring intramolecular disulfide-bond formation by all 20-Cys residues in the ectodomain of the protein. Both E1 and E2 acquire high-mannose glycans in the ER; E1 contains three potential glycosylation sites and E2 contains four and all appear to be utilized. The sites of O-glycosylation of the E2, which accounts for the size heterogeneity of E2, is not known. E1 contains an ER retention signal that is only overridden once conformational folding is complete, after which the E1–E2 heterodimer migrates to the Golgi. In the Golgi the N-glycans of both E2 and E1 are modified to complex form, although modification is not complete and the extent of modification on both proteins is heterogeneous. Modifications of the O-glycans on E2 also occur and E2 contains a Golgi retention signal, indicating that the Golgi is the preferred site of viral budding in infected cells. However, late in infection E1 and E2 migrate to the cell surface and budding can also occur at this site in some cell lines.

RUBV capsid morphogenesis occurs in association with cell membranes. The association of RUBV capsid protein with membranes is probably mediated by the E2 signal sequence, which is retained at the COOH terminus of C. In fact, C may associate with the E2–E1 heterodimer and migrate as a passenger on the cytoplasmic side of vesicles transporting E2–E1 from the ER to the Golgi and among the Golgi stacks. Unlike the alphaviruses, whose capsids accumulate in infected cells, RUBV capsids only become visible in association with deformed, thickened membranes that appear to be in the early process of budding. A putative encapsidation signal has been localized near the 5′ terminus of the genomic RNA. The C protein is phosphorylated and phosphorylation/dephosphorylation by cell enzymes is proposed as the regulator of the process by which the genome is unencapsidated following entry (phosphorylation) and encapsidated later in infection following replication (dephosphorylation). Interestingly, in cells in which the complete S-ORF is expressed in the absence of genomic RNA, virus-like particles form and are secreted and these have the same morphology and isopycnic density as do virions.

RUBV replicates in the cytoplasm of the infected cell. None of the virus proteins exhibit any involvement with the nucleus during infection and RUBV infection does not appear to inhibit cell macromolecular synthesis in any grossly detectable manner; however, perturbations of specific macromoleclar products and induction of specific genes may occur. Microspically, RUBV-infected cells appear similar to uninfected cells; however, rearrangements of cellular cytoskeletal elements and organelles such as mitochondria have been reported. RUBV reportedly inhibits growth in primary human cell cultures in part due to an inhibition of mitosis; however, the virus has no reproducible effect on the growth of stable cell lines. In those cell lines which exhibit CPE (Vero and RK-13 cells), cell death is due to apoptosis.

Genetics and Evolution

The RUBV genome is extremely stable. Currently, the genomes of 19 independent strains of RUBV from eight genotypes, five in clade 1 and three in clade 2, have been sequenced in their entirety and all are nearly identical in terms of the size of the genome as well as the coding and noncoding regions within the genome. The only exceptions are 1–2 nucleotide deletions occasionally encountered in the junction region. Maximal observed distance at the nucleotide level is 5% among clade 1 viruses, 7.5% among clade 2 viruses, and 9% between viruses in the two clades. Across the 19 genomic sequences, 78% of the nucleotides are conserved, explaining the low level of variability. A unique feature of RUBV evolution is that changes to G and C are selected for, indicating an adaptive advantage of the high G+C content of the genome.

Because there are no known close relatives of RUBV, the origin of the virus prior to its introduction into the human population is unknown. Except for short stretches at the 5′ end of the genome and at the subgenomic promoter site, RUBV and the alphaviruses share no nucleotide homology. Thus these two genera are only distantly related. RUBV and the alphaviruses belong to the 'alphavirus-like superfamily' of plus-sense RNA viruses, which includes a large number of plant viruses as well as human hepatitis E virus, the sole member of the genus *Hepevirus*. Within this superfamily, computer-assisted phylogenetic analysis of the NS proteins indicates that RUBV is more closely related to hepatitis E virus than to the alphaviruses and this dissimilarity is borne out by differences in the order of motifs in the NS-ORF. Thus, it is hypothesized that the evolution of the togaviruses may have been more complicated than simple divergence from a common ancestor and probably involved recombination between progenitors of the current alphaviruses, RUBV, hepatitis E virus, and, possibly, certain plant viruses.

Serologic Relationships and Variability

RUBV is monotypic and immunological characterization of diverse strains, including both clade 1 and 2 viruses, has only revealed subtle antigenic differences which map to C or E2. As might be anticipated from the lack of serologic cross-reaction with the alphaviruses, there is no homology at the amino acid level between RUBV and alphaviruses within the virion proteins.

Epidemiology

Historically, RUBV was endemic worldwide. Over the past 35 years, vaccination programs have curtailed this distribution, as discussed below. Before the advent of vaccination programs, rubella was considered a disease of middle childhood in temperate zones, with seasonal peak occurrence in the spring and epidemics at 5–9-year intervals. However, in tropical zones the highest infection rates were in children under 5 years of age. RUBV is not as transmissible as is measles virus and even during epidemics, susceptibles are spared. Thus, infection of adolescents and young adults, the population at risk for CRS, in endemic areas is not uncommon.

Transmission and Tissue Tropism

RUBV is transmitted between individuals by aerosol. The epithelium of the buccal mucosa provides the initial site for virus replication and the mucosa of the upper respiratory tract and nasopharyngeal lymphoid tissue serve as portals of virus entry. The virus is then spread by local lymphatics, which seed regional lymph nodes where further virus replication occurs. After an incubation period of 7–9 days, virus appears in the blood. Viremia ceases with the onset of detectable rubella-specific antibody, shortly after the rash appears 2–3 weeks post infection. Patients are most infectious immediately preceeding and during the rash phase; virus generally disappears from nasopharyngeal secretions within 4 days of appearance of the rash. Congenitally infected infants shed virus for 3–6 months following birth and are a source of transmission during that period. Reinfection with RUBV occurs, usually without clinical illness or virus shedding. There are a small number of cases in which RUBV reinfection of pregnant women with well-documented immunity has resulted in CRS.

During pregnancy, placental tissues are very susceptible to infection. Placental infection results in scattered foci of necrotic syncytiotrophoblast and cytotrophoblast cells and evidence of damage to vascular endothelium. Following placental infection, virus can spread to the fetus but this does not always occur and RUBV is more often recovered from placental tissue than from fetal products of conception. Once fetal infection occurs, virus spreads throughout the fetus and almost any organ may be infected. Severe fetal damage is only associated with infection during the first trimester of pregnancy; the rate of CRS is >50%, 25%, and 10% when infection occurs during the first, second, and third months, respectively. This is due to a combination of an apparent decline in the efficiency of placental transfer after the first trimester and a reduction in the ability of the virus to inflict fetal damage after this time of gestational development.

Clinical Features of Infection

Rubella acquired in childhood or early adulthood is usually mild and it is estimated that up to 50% of rubella infections are clinically inapparent. Symptomatic rubella encompasses combinations of maculopapular rash, lymphadenopathy, low-grade fever, conjunctivitis, sore throat, and arthralgia. The rash is the most prominent feature and appears following an incubation period of 16–20 days. The rash begins as distinct pink maculopapules on the face that then spread over the trunk and distally onto the extremities. The maculopapules coalesce and the rash fades over several days. An associated posterior cervical and suboccipital lymphadenopathy is also characteristic. Infrequent complications include thrombocytopenia and post-infectious encephalitis. Acute polyarthralgia and arthritis following natural RUBV infections of adults are common and occur more frequently and with greater severity in women than in men. Joint involvement is usually transient; however, chronic arthritis, persisting or recurring over several years, has been reported.

The clinical manifestations of CRS apparent at birth vary widely, most frequently including thrombocytopenic purpura ('blueberry muffin syndrome'), intrauterine growth retardation, congenital heart disease (patent ductus arteriosus or pulmonary artery or valvular stenosis), psychomotor retardation, eye defects (cataract, glaucoma, retinopathy), hearing loss, and hepatomegaly and/or splenomegaly. Nearly 80% of CRS children show some type of neural involvement, particularly neurosensory hearing loss.

Most clinical manifestations of congenital rubella are evident at or shortly following birth and some are transient. However, recognition of retinopathy, hearing loss, and mental retardation may be delayed for several years in some cases. Progressive consequences of congenital rubella have become increasingly appreciated as CRS children from the 1964 epidemic have been followed longitudinally. These predominantly involve endocrine dysfunction (diabetes mellitus, which ultimately affects 40% of CRS patients, and thyroid dysfunction). A rare, fatal neurodegenerative disease, progressive rubella panencephalitis (PRP), described in CRS patients, bears superficial resemblance to subacute sclerosing panencephalitis associated with measles virus. Subsequently, PRP cases have been reported in individuals infected postnatally.

Pathogenesis, Pathology, and Histopathology

There is limited information on the pathogenesis of uncomplicated rubella because of the benign nature of the illness. With respect to the complications that can accompany acute rubella, the postinfectious encephalitis is thought to be autoimmune in nature since RUBV cannot be isolated from cerebrospinal fluid or brain at autopsy. Interestingly, however, extensive inflammation and demyelination are not observed. In a few cases of rubella arthritis, the presence of RUBV in synovial fluid and/or cells has been demonstrated and therefore it is assumed that virus persistence is involved. However, considering the age and sex factors in the incidence of arthritis, it seems likely that immunopathological mechanisms also play a role. In one study, human leukocyte antigen (HLA) class II haplotypes predisposing adult women to arthritis and arthralgia following rubella vaccination were found to also correlate with predisposition to arthritis and arthralgia regardless of vaccination status.

Following fetal infection, virus can be isolated from practically every organ of abortuses or infants who die soon after birth. However, only 1 in 10^3 to 1 in 10^5 cells are infected and it is not known how such a low infection rate leads to the profound birth defects exhibited in CRS. Affected organs are routinely small for gestational age and contain reduced numbers of cells. Considering the inhibitory effect of RUBV on primary cells, it is thought that virus infection early in organogenesis inhibits cell division, leading to both retardation and alteration in organ development. Virus persistence continues after birth, as evidenced by virus shedding, which generally ceases within 6 months of age. Whether virus persistence continues beyond cessation of shedding and plays a role in the delayed and progressive manifestations of CRS is not known.

Histologically, affected organs of CRS patients show a limited number of well-recognized malformations, with noninflammatory histopathology predominating. Particularly apparent are vascular lesions and focal destruction in tissue bordering these lesions. These lesions are likely to be due to virus replication in the vascular endothelium and damage to neighboring tissue may play a role in the pathogenesis of CRS. The neuropathology of CRS is of interest not only because of the defects manifest shortly after birth, but also because some CRS patients develop schizophrenia-like symptoms later in life. CRS brains are generally free of gross morphological malformations, with a common tendency toward microcephaly. Vascular damage, leptomeningitis, decreased number of oligodendroglial cells, and alteration of white matter are observed. Recently, magnetic resonance imaging of a group of CRS adults with schizophrenia-like symptoms revealed specifically reduced cortical gray matter and enlargement of the ventricles, which were not previously observed aspects of CRS-induced neuropathology. Interestingly, the comparative finding that non-CRS schizophrenia patients exhibit a pattern of brain dysmorphosis similar to that found in CRS patients with schizophrenia-like symptoms supports the hypothesis that schizophrenia is developmental in nature (there is some evidence for a viral trigger to schizophrenia).

Immune Response and Serodiagnosis

Following acute infection, anti-RUBV IgM antibodies are detectable at the time of rash onset and for a month or two afterwards, so that serodiagnostic testing for the presence of IgM is the current primary means for diagnosis of acute RUBV infection; in the succeeding weeks anti-RUBV antibodies in all immunoglobulin classes appear. The dominant early and persistent IgG response is in the IgG_1 subclass and antibodies of this class persist indefinitely after natural infection of otherwise healthy individuals. The majority of the antibody response is directed to the E1 glycoprotein, with proportionally lesser amounts of the response directed at E2 or C. Although neutralizing and complement-fixing antibodies are induced as well, the classical assay for the presence of anti-RUBV antibodies has been hemagglutination inhibition (HAI) and the current standard level of 10 IU ml^{-1} is based on an HAI titer of roughly 8. Because of the importance of serodiagnostic testing for rubella, a worldwide commercial market for rubella tests exists and a number of commercial laboratories offer such kits, most of which are based on latex agglutination or enzyme immunoassay.

RUBV-specific cellular immune responses are measurable within 1–2 weeks of onset of illness. Major histocompatibility complex (MHC)-restricted CD4+ epitopes have been mapped to all three of the virus structural proteins; however, CD8+ epitopes have thus far only been mapped to the C protein. HLA class I and II alleles both positively and negatively associated with antibody and lymphoproliferative responses following rubella vaccination have recently been determined.

Following fetal infection, the fetus produces IgM antibody, detectable at 18–20 weeks of gestation, and maternal IgG antibody crosses the placenta. Both types of antibody exhibit virus-neutralizing activity *in vitro*; however, neither is sufficient to resolve virus infection during gestation. As discussed above, the intracellular maturation of virus probably shields it from antibody. After birth, the presence of IgM or a lack of decline of IgG titer are both considered diagnostic of fetal infection. CRS infants exhibit various degrees of impairment in the cellular immune response to RUBV and it is thus a deficiency in this arm of the immune response that allows the virus to persist. Considering the cellular immune deficiency in CRS infants, it is curious that detectable virus persistence ends relatively shortly after birth. The means by which virus persistence is cleared under these conditions is not understood.

Diagnosis

The common symptoms of acute rubella, lymphadenopathy, erythematous rash, and low-grade fever, are nonspecific and easily confused with illnesses caused by other common pathogens. A definitive diagnosis of rubella requires detection of the presence of IgM antibodies or a rise in IgG antibodies in paired acute-phase and convalescent-phase serum samples. Virus is also readily isolated from saliva or nasopharyngeal washings at the time of rash onset. Reverse transcription-polymerase chain reaction (RT-PCR) assays have been developed to detect virus in saliva, blood, or urine. Generally, virus isolation is only done to collect specimens for molecular epidemiological analysis. Diagnosis of acute rubella is of utmost importance during early pregnancy. There is no intervention for congenital rubella other than abortion; however, as discussed above, maternal infection does not always lead to CRS, resulting in an extremely difficult decision. RT-PCR can be used to detect virus RNA in amniotic fluid or chorionic villi, but is rarely employed. Diagnosis of CRS in a newborn is initially made on the basis of symptoms and confirmed in the laboratory by serological testing for the presence of IgM antibodies, by lack of decrease of IgG antibodies in serum specimens, or by detection of virus or virus RNA. As stated above, CRS infants shed virus and therefore are a source of contagion.

In countries with vaccination programs, the need for detection of acute rubella is rare. However, serodiagnosis for determination of immune status is routine for pre-pregnancy planning, employment in medical facilities and, in some states in the US, to obtain a marriage license. Serodiagnosis can also be used in lieu of proof of vaccination, which is required for school enrollment. Individuals found to be seronegative are subsequently vaccinated, which is done postpartum in the case of a woman who is pregnant at the time of testing.

Prevention and Control of Rubella

As discussed above live, attenuated vaccines were developed and placed in use by 1970. The vaccine used in most countries is the RA 27/3 vaccine. In Japan, five live, attenuated vaccine strains were developed and are currently in use. Additionally, at least one Chinese vaccine strain is currently in use in China.

Attenuated RUBV vaccines cause subclinical infection with transient viremia in susceptible patients. Natural transmission of vaccine virus has not been reported. The RA27/3 vaccine strain produces seroconversion in greater than 95% of recipients. Vaccine-induced titers are lower than those induced by natural infection but appear to last indefinitely. The RUBV vaccine is generally administered to children in trivalent form with the measles- and mumps-attenuated vaccines (MMR) but inclusion of the recently licensed varicella vaccine in a tetravalent vaccine is being strongly considered.

The RUBV vaccines have been among the most successful in terms of induction of immunity with an absence of side effects. However, two issues have arisen concerning rubella vaccination. The first is that the vaccine virus can cross the placenta and infect the fetus. However, US and Israel registries of deliveries to women inadvertently vaccinated during early pregnancy revealed no reported congenital abnormalities. Similar findings were reported in subsequent studies conducted in other countries in conjunction with mass vaccination campaigns. Nevertheless, vaccination during pregnancy is contraindicated and is deferred until postpartum. Second, is the occurrence of arthralgia and arthritis following vaccination. Joint complications are nonexistent in children with the currently used rubella vaccines; however, transient arthralgia and arthritis is reasonably common among adult female vaccinees. There have also been reports of chronic arthritis and related neurological involvement following vaccination of adult women. Although these complications are consistent with complications that can accompany natural rubella in adult females, recent studies have shown that the incidence of such vaccine-related complications is low and cannot be statistically differentiated from the incidence of similar symptoms in control, unvaccinated populations.

Since the inception of rubella vaccination in 1969, the US has employed a strategy of universal vaccination at 15 months of age augmented with vaccination of seronegative 'at-risk' individuals (women planning pregnancies, healthcare workers) which was successful in bringing the incidence of rubella and CRS to low levels by 1988. A resurgence of rubella and CRS that occurred between 1989 and 1991 among foci of unvaccinated individuals led, in part, to administration of a second dose of MMR vaccine when the recipients were between 5 and 10 years of age. After 2000, the incidence of rubella fell to fewer than 25 cases per year and in 2004 and it was concluded that indigenous rubella had been eliminated from the US. Vaccination programs have also been in place in Europe and Japan since the development of live, attenuated vaccines; however, developing countries were slower to institute rubella vaccination programs. However, efforts among developing countries have accelerated since 2000, particularly in conjunction with ongoing measles elimination efforts, and currently roughly 50% of countries worldwide have national rubella vaccine programs. The World Health Organization Regions of the Americas and Europe have set goals for elimination of rubella and CRS by 2010.

Despite the recent increase in intensity of rubella vaccination efforts worldwide, most of Asia and Africa are not currently included and thus well over half of the world's population is not covered. The cost of the vaccine, the general mildness of the disease, and the nature of national public health infrastructures are deterrents to vaccination programs in developing countries. Moreover, estimates of CRS incidence are difficult to obtain because the three hallmark symptoms of CRS, deafness, blindness, and mental retardation, are not uniformly present and are not readily diagnosable in newborns. Surveys in developing countries show that the rate of CRS in developing countries is the same as that in developed countries prior to vaccination and a recent economic analysis concluded that the cost/benefit ratio for rubella vaccination was similar to that for hepatitis B vaccination, a pathogen that is universally recognized to impart a societal load.

Nevertheless, the success of worldwide rubella vaccination appears dependent on piggy-backing on global measles vaccination programs. In addition to its efforts to initiate rubella vaccination programs in developing regions, the WHO established a global Measles–Rubella Surveillance Network in 2003. Interestingly, rubella vaccination and the concomitant reduction in rubella simplifies measles surveillance which requires diagnosis of rubella because of the similarity of symptoms of the two diseases. A major modification in measles/rubella vaccination may be forthcoming in the form of aerosolized vaccines. Aerosol administration mimics the natural route by which these two viruses are contracted and preliminary studies show that aerosolized vaccine uptake and efficacy is as good as or better than the current injection route. Use of aerosolized vaccines would reduce vaccination costs by eliminating needles and the concomitant need for trained personnel for administration. Finally, it is emphasized that rubella reduction in developing countries imparts benefit to developed countries, in which most rubella outbreaks are due to importation, by decreasing such outbreaks and thus easing control efforts. However, discontinuation of rubella vaccination in developed countries would require eradication.

Future

Because of its clinical similarity to any of a diverse group of clinical diseases, RUBV will remain a fascinating pathogen. As an example, the incidence of diabetes in CRS patients is the best statistically direct association between a specific human virus and a specific autoimmune disease. The mechanism of viral involvement, if any, in each of these diseases is not fully understood. Elucidating the mechanism of RUBV-induced birth defects could also provide understanding into teratogenesis by other infectious agents. Unfortunately, our present understanding of disease mechanisms in the RUBV-related syndromes is hindered by the lack of a suitable animal model system that fully mimics the infection seen in humans, making development of an animal model a research priority. RUBV is taxonomically unique and appears to have evolved as a recombinational hybrid of distantly related ancestor viruses. Thus, investigation of the molecular biology of RUBV likely will reveal novel replication strategies and yield insight into virus evolution. The greatest challenge concerning RUBV is the current and forthcoming elimination efforts in most regions of the world.

See also: Central Nervous System Viral Diseases; Evolution of Viruses; Hepatitis E Virus; Measles Virus; Togaviruses: Alphaviruses; Togaviruses: General Features.

Further Reading

Best JM, Castillo-Solorzano C, Spika JS, et al. (2005) Reducing the global burden of congenital rubella syndrome: Report of the World Health Organization steering committee on research related to measles and rubella vaccines and vaccination, June 2004. *International Journal of Infectious Diseases* 192: 1890–1897.

Frey TK (1994) Molecular biology of RUBV. *Advances in Virus Research* 44: 69–160.

Hobman TC and Chantler JK (2007) RUBV. In: Knipe DM and Howley PM (eds.) *Fields Virology,* 5th edn., pp. 1069–1100. Philadelphia: Lippincott Williams and Wilkins.

Law LJ, Ilkow CS, Tzeng WP, et al. (2006) Analyses of phosphorylation events in the RUBV capsid protein: Role in early replication events. *Journal of Virology* 80: 6917–6925.

Ovsyannikova IG, Jacobson RM, Vierkant RA, Jacobsen SJ, Pankratz VS, and Poland GA (2005) Human leukocyte antigen class II alleles and rubella-specific humoral and cell-mediated immunity following measles–mumps–rubella-II vaccination. *International Journal of Infectious Diseases* 191: 515–519.

Plotkin SA (2006) The history of rubella and rubella vaccination leading to elimination. *Clinical Infectious Diseases* 43(supplement 3): S164–S168.

Reef SE, Frey TK, Theall K, et al. (2002) The changing epidemiology of rubella in the 1990s: On the verge of elimination and new challenges for control and prevention. *JAMA* 28(7): 464–472.

Tzeng WP, Matthews JD, and Frey TK (2006) Analysis of RUBV capsid protein-mediated enhancement of replicon replication and mutant rescue. *Journal of Virology* 80: 3966–3974.

Zhou Y, Ushijima H, and Frey TK (2007) Genomic analysis of diverse RUBV isolates. *Journal of General Virology* 88: 932–941.

Sadwavirus

T Iwanami, National Institute of Fruit Tree Science, Tsukuba, Japan

© 2008 Elsevier Ltd. All rights reserved.

Glossary

Natsudaidai A popular sour citrus grown in some parts of Japan.

Satsuma A major mandarin-type citrus grown in Japan and some other Asian countries.

Introduction

The genus *Nepovirus* includes viruses which are transmitted by nematodes with polyhedral particles. Typical nepoviruses have a broad host range, a single coat protein, and a bipartite single-stranded RNA genome and are transmissible through seed. Sadwaviruses have been previously considered atypical and tentative members of the genus *Nepovirus*. Like typical nepoviruses, sadwaviruses have polydedral virus particles which contain two species of genomic RNA, but can be distinguished on the basis of the genome organization, sequence homologies, and the number of coat proteins. Some sadwaviruses are transmitted by aphids.

The derivation of the name comes from the virus species, *Satsuma dwarf virus*, the type member of the genus *Sadwavirus*. The major disease caused by satsuma dwarf virus (SDV) is stunting, accompanied by the presence of boat-shaped leaves, in Satsuma mandarin (*Citrus unshiu* Marc.).

Taxonomy and Classification

Satsuma dwarf virus, *Strawberry latent ringspot virus*, and *Strawberry mottle virus* are the definite members of the genus *Sadwavirus*. Lucerne Australian symptomless virus (LASV), Rubus Chinese seed-borne virus (RCSV), and Black raspberry necrosis virus (BRNV) are tentative members (**Table 1**).

Virus species are demarcated on the basis of type of biological vector, if known, host range, absence of serological cross-reaction, absence of cross-protection, and difference in amino acid sequence (less than 75% in the large coat protein and the proteinase-polymerase region).

The genus *Sadwavirus* has not been assigned to any virus family.

Strains and Synonyms

Citrus mosaic virus (CiMV), natsudaidai dwarf virus (NDV), and navel orange infectious mottling virus (NIMV) are distantly related strains of *Satsuma dwarf virus*. Typical symptoms of CiMV-infected Satsuma mandarin are spotting and blotching of the fruit rind. NDV induces vein clearing and mottling on new leaves of nastudaidai (*Citrus natsudaidai*). NIMV produces persistent chlorotic spots on leaves of sweet orange. The strains of *Satsuma dwarf virus* share around 80% sequence homology at the amino acid level.

Strawberry latent ringspot virus has a few synonyms, including rhubarb virus 5. Some isolates from olive, peach, raspberry, and grapevine were serologically distinguishable from the type strain, but sequence diversity among these isolates and strains has not been studied.

Geographic Distribution

SDV occurs mostly in Japan, but has also been found in China and Turkey. Strawberry latent ringspot virus (SLRSV) is distributed throughout Europe and the USA, as well as Australia and New Zealand. Strawberry mottle virus (SMoV) is probably spread worldwide, as well as BRNV. Occurrence of LASV and RCSV is limited to Australasia and China, respectively.

Virion Properties

Particles of sadwaviruses are icosahedral, about 26–30 nm in diameter (**Figure 1**). Some particles are penetrated

Table 1 Species in the genus *Sadwavirus*[a]

Species and strains	Abbreviation	RNA-1 accession number	RNA-2 accession number
Satsuma dwarf virus			
Citrus mosaic virus	CiMV		D64079 (partial)
Natsudaidai dwarf virus	NDV		AB032750 (partial)
Navel orange infectious mottling virus	NIMV		AB000282 (partial)
Satsuma dwarf virus	SDV	AB009958	AB009959
Strawberry latent ringspot virus			
Strawberry latent ringspot virus	SLRSV	NC 006964	NC 006965
Strawberry mottle virus			
Strawberry mottle virus	SMoV	AJ311875	AJ311876
Tentative species in the genus			
Black raspberry necrosis virus	BRNV	NC 008182	NC 008183
Lucerne Australian symptomless virus	LASV		
Rubus Chinese seed-borne virus[a]	RCSV		

[a]Species names are given in italic script; strain names and tentative species names are in Roman script.

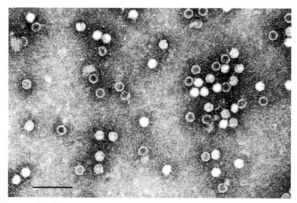

Figure 1 Electron micrograph of negatively stained virus particles of SDV. Scale = 100 nm.

by negative stain in the electron microscope. Typically, particles of sadwaviruses are of three types, named according to the relative rates of sedimentation of purified preparations. Top component (T) particles are empty shells. Middle component (M) and bottom component (B) particles contain genomic RNA. The buoyant densities in CsCl are about 1.43 and 1.46 g cm^{-3}, respectively. SLRSV has only B particles with buoyant density about 1.46 g cm^{-3}. Sadwavirus particles contain two sizes of RNA molecules. SDV has M and B particles that contain either of the different RNAs. In contrast, SLRSV has only B particles. Some particles contain one larger RNA, others contain two molecules of smaller RNA.

Genome Organization

The genome of sadwaviruses consists of two molecules of positive-sense single-stranded RNAs called RNA-1 and RNA-2. The sizes of RNA-1 and RNA-2 are about 7 and 4–5.5 kb, respectively. The 3′ termini of both RNAs are polyadenylated. In the case of SLRSV, both RNAs are linked at the 5′ termini to a small protein (VPg). VPgs of other sadwaviruses are unknown.

Both RNA-1 and RNA-2 have a single large open reading frame (**Figure 2**). The general genome organization is similar to that of como- and nepoviruses. The polyprotein encoded by RNA-1 contains domains for helicase, protease, and RNA-dependent RNA polymerase, whereas the polyprotein encoded by RNA-2 has movement protein and coat proteins. The amino acid sequences of the N-termini of polyproteins encoded by RNA-1 and RNA-2 of SDV are highly conserved. The feature is observed in the polyproteins of tomato ringspot virus in the genus *Nepovirus*. The function of these regions is unknown.

Sequence Comparisons

Comparison among the sequences of sadwaviruses of different species at nucleotide and amino acid levels shows few marked similarities. Some regions show about 20–40% sequence identities, and other parts are less similar.

Phylogeny

Complete genomic nucleotide sequences of SDV, BRNV, SLRSV, and SMoV have been determined. Partial sequences of CiMV, NDV, and NIMV are available (**Table 1**). There is little information on the nucleotide sequences of other sadwaviruses. To reveal relationships, the RNA-dependent RNA polymerase region, which is the most conserved among sadwa-, como-, nepo-, and other plant picorna-like viruses, has been used for phylogenetic comparisons. Alignment of the most conserved regions of the RNA-dependent RNA polymerase region shows that all plant picorna-like viruses are separated to

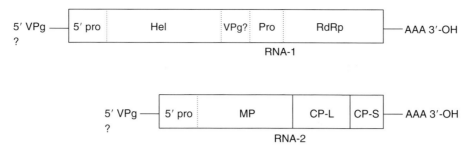

Figure 2 Genome organization of SDV. The boxes represent polyproteins. The vertical solid lines within the box show sites where cleavages occur in the polyprotein and the dashed lines indicate sites where cleavages are presumed to occur. Abbreviations: VPg, genome-linked viral protein; 5′ pro, 5′ protein; Hel, helicase; Pro, protease; RdRp, RNA-dependent RNA polymerase; MP, movement protein; CP-L and CP-S, large and small components of coat protein; AAA, poly(A).

three distinct groups: the first and second groups include comoviruses and nepoviruses, respectively, whereas the third group consists of SDV (and closely related CiMV, NDV, NIMV), SMoV, apple latent spherical virus (ALSV, which is a member of a newly established genus *Cheravirus*), and viruses of the family *Sequiviridae*, which have monopartite RNA geneome.

Satellites

SLRSV has strains that contain satellite RNAs. The SLRSV satellite encodes a polyprotein, which shows little homology with other known proteins. There is no evidence as to the function of the polyprotein. Other sadwaviruses do not contain a satellite RNA.

Host Range and Virus Propagation

Natural host of SDV, CiMV, NDV, and NIMV is limited to woody plants, whereas SMoV, LASV, and RCSV are confined to herbaceous plants. BRNV infects *Rubus* spp. SLRSV has a wide natural host range that includes both woody and herbaceous plants. All sadwaviruses are readily transmitted by mechanical inoculation to herbaceous plants of several families, including Chenopodiaceae, Fabaceae, and Solanaceae. Common experimental hosts used for the propagation of sadwaviruses are *Chenopodium quinoa*, *Cucumis sativus*, and *Physalis floridana*. SMoV propagates very poorly in any of the experimental herbaceous hosts, and purification of virus particles is difficult.

Epidemiology

SLRSV is transmitted by the soil nematode, *Xiphinema diversicaudatum*, and the transmission from plants to plants is slow, reflecting the restricted mobility of the nematode. SDV, CiMV, NDV, and NIMV are disseminated in a similar manner, and soil transmission is suspected, although vectors have not yet been identified. SMoV and BRNV are transmitted by aphids. All sadwaviruses but SMoV are seed-borne, and dissemination in seeds is important for long-distance movement of some sadwaviruses. Indeed, RCSV was first found in the UK from imported seeds from China.

Hosts of SDV, CiMV, NDV, and NIMV (citrus), BRNV (raspberry), SMoV (strawberry), SLRSV (strawberry and fruit trees) are vegetatively propagated, and movement of contaminated budwoods and tubers plays an important role in the long-distance dissemination. For example, occurrence of SDV in China and Turkey is obviously related to unchecked budwoods from Japan.

Cytopathology

Tubules which contain virus particles arranged in a single file are formed in the cytoplasm of infected cells. A file of virus particles is sometimes observed in a plasmodesma between two cells. Virus particles have also been detected in the vacuoles of infected cells in crystalline forms. These structures are also commonly observed in cell infected with comoviruses and nepoviruses.

Prevention and Control

Control measures consist of establishment of healthy stocks, chemical controls of vectors (nematode for SLRSV, aphids for BRNV and SMoV), and introduction of tolerant cultivars when available. Distributions of virus-free materials are important for vegetatively propagated crops (citrus for SDV, strawberry for SMoV and SLRSV, fruit trees for SLRSV). Several diagnostic tools including enzyme-linked immunosorbent assay (ELISA) and reverse transcriptase-polymerase chain reaction (RT-PCR) are available for most of the sadwaviruses. Chemical fumigation of citrus field to prevent soil transmission of SDV is partially effective at best. Many strawberry cultivars are symptomless hosts for either SMoV or SLRSV. Mixed infection with other strawberry viruses induces apparent foliar symptoms and decline of tubers.

See also: Nepovirus; Sequiviruses.

Further Reading

Halgren A, Tzanetakis IE, and Martin RR (2007) Identification, characterization, and detection of black raspberry necrosis virus. *Phytopathology* 97: 44–50.

Iwanami T, Kondo Y, and Karasev AV (1999) Nucleotide sequences and taxonomy of satsuma dwarf virus. *Journal of General Virology* 80: 793–797.

Le Gall O, Iwanami T, Karasev A, et al. (2005) Genus *Sadwavirus*. In: Fauquet CM, Mayo MA, Maniloff J, Desselberger U, and Ball LA (eds.) *Virus Taxonomy: Eighth Report of the International Committee on Taxonomy of Viruses*, pp. 799–802. San Diego, CA: Elsevier Academic Press.

Mayo MA and Robinson DJ (1996) Nepoviruses: Molecular biology and replication. In: Harrison BD and Murant AF (eds.) *The Plant Viruses*, vol. 5, pp. 139–185. New York: Plenum.

Murant AF, Jones AT, Martelli GP, and Stace-Smith R (1996) Nepoviruses: General properties, diseases, and virus identification. In: Harrison BD and Murant AF (eds.) *The Plant Viruses*, vol. 5, pp. 99–137. New York: Plenum.

Thompson JR, Leone G, Lindner JL, Jelkmann W, and Schoen CD (2002) Characterization and complete nucleotide sequence of strawberry mottle virus: A tentative member of a new family of bipartite plant picorna-like viruses. *Journal of General Virology* 83: 229–239.

Satellite Nucleic Acids and Viruses

P Palukaitis, Scottish Crop Research Institute, Invergowrie, UK
A Rezaian, University of Adelaide, Adelaide, SA, Australia
F García-Arenal, Universidad Politécnica de Madrid, Madrid, Spain

© 2008 Elsevier Ltd. All rights reserved.

Glossary

Satellite-like RNA A subviral genome dependent upon another virus for replication and encapsidation, but is required for vector transmission of the helper virus.

Satellite RNA or satellite DNA A subviral genome dependent on a helper virus for both replication and encapsidation.

Satellite virus A subviral genome dependent on a helper virus for its replication, but encoding its own capsid protein.

Virus-associated nucleic acid A subviral genome that depends on another virus for encapsidation and transmission, but not for its replication.

Introduction

Satellites of viruses constitute a heterogeneous collection of subviral agents. Satellites can be differentiated from other subviral nucleic acids, such as subgenomic (sg) RNAs, defective (D) or defective interfering (DI) RNAs, and viroids, by their molecular, biological, and genetic nature. Unlike sg RNAs, D RNAs, and DI RNAs, satellites and viroids have little or no sequence similarity to any known virus. Whereas viroids are replicated by host polymerases, RNA satellites are replicated by the polymerase of a virus, referred to as the helper virus (HV). Satellites have been found associated with DNA and RNA viruses, and in the latter case, with both ssRNA and dsRNA viral genomes, the satellites being of the same nucleic acid type as the HV. While, in general, the HV can exist independent of the satellite, there are exceptions where a satellite contributes to the transmission of the HV and thus is referred to as being satellite-like. In addition, some viral RNAs are found in association with other viral genomes on which they depend for their encapsidation and transmission, but not their replication. These viral RNAs are referred to here as virus-associated RNAs. Satellites are divided basically into two main groups; those that encode their own capsid protein (CP) are called satellite viruses, while those that require their HV for both replication and encapsidation are referred to as satellite RNAs or DNAs. There are also satellites of satellites, in which some satellite RNAs are replicated by the HV, but are encapsidated by the CP of a satellite virus. In addition, in the case of the carmovirus turnip crinkle virus (TCV), which replicates both satellite RNAs and DI RNAs, it also produces a chimeric RNA (referred to as satC) that is in part a DI RNA and in part a satellite RNA (satD). While the vast majority of satellites are found in association with plant viruses, a few have been found in association with animal or fungal viral genomes. Most plant viruses do not contain satellites associated with them, but as some satellites can significantly affect the disease induced by the HV, the presence of satellites has important consequences for viral-induced diseases.

History

The use of the term satellite as a subviral agent was first conceptualized in 1962 by Kassanis, to describe the relationship of a 17-nm diameter viral particle found in association with some isolates of the 26-nm diameter necrovirus tobacco necrosis virus (TNV). The smaller particle was dependent on TNV for its accumulation, but was serologically unrelated to TNV. It became known as the satellite virus of TNV (now known as tobacco necrosis satellite virus, TNSV). A few other satellite viruses have been described since then, but, in general, they are rare. By contrast, satellite RNAs are more common. The first satellite RNA was described by Schneider in 1969, in association with the nepovirus tobacco ringspot virus (TRSV), the satellite RNA being encapsidated by the CP of the HV. The symptoms induced by TRSV were attenuated dramatically by the presence of the satellite RNA. This is not always the case, with some satellite RNAs having no effect on either the HV accumulation or disease induced by the HV, and a few satellite RNAs exacerbating the HV-induced disease. Most of these satellite RNAs contain ssRNA genomes, while a few contain genomes of dsRNA. Some of the satellite RNAs with ssRNA genomes are translated to produce proteins, which may or may not be required for their replication, depending on the particular satellite RNA. Satellite DNAs were first described in 1997 by Dry and colleagues, in association with the geminivirus tomato leaf curl virus (TLCV). It was isolated from field-infected tomato plants in northern Australia. Satellite DNAs also are encapsidated by the CP of their HV. Since then, satellite DNAs referred to as satellite DNA β have been described in association with many geminiviruses, principally from Asia.

Geographical Distribution

Since satellite RNAs are only found associated with their specific HVs, they are limited to the distribution range of the HV and the vectors of the HV. On the other hand, not all isolates of the HV have satellites associated with them, further delimiting the distribution of satellites. Nevertheless, as many of the HVs are distributed worldwide, so are many of the satellites. The distribution of satellite DNA β has so far been linked to the geminiviruses of the Old World, an exception being DNA β in honeysuckle that appears to have been distributed through vegetative plant material.

Classification

There is no correlation between the taxonomy of their HV and the presence of satellites. Satellites are associated with viruses belonging to at least 17 genera, and with only a limited number of species within those genera. Satellites do not constitute a homogeneous group in terms of their nucleic acid type, size, sequence, structure, or translatability. Most satellites are composed of RNA, but some consist of DNA. Most satellites are linear molecules, but some RNA satellites and all of the DNA satellites are circular in structure. Some satellites may encode proteins, while many do not. Thus, satellites are classified primarily into categories based on the features of the above properties, vis-à-vis their nucleic acid form and their genetic capacity. All satellites are grouped into the following categories, first differentiated on the basis of whether they are satellite viruses (**Table 1**) or satellite nucleic acids (**Table 2**), but

Table 1 Satellite viruses

Helper virus/satellite virus	Particle size (nm)	CP[a]	Satellite RNA size (nt)	Accession no.
Subgroup 1				
Chronic bee-paralysis virus (CBPV)/ CBPV-associated satellite virus (CBPVA)	17	NR[b]	~1100 (three species)	
Subgroup 2				
Necrovirus				
Tobacco necrosis virus (TNV)/satellite TNV (STNV)	17	~21 600	1239	J02399
Sobemovirus				
Panicum mosaic virus (PMV)/satellite PMV (SPMV)	16	~17 500	824–826	M17182
Tobamovirus				
Tobacco mosaic virus (TMV)/satellite TMV (STMV)	17	~17 500	1059	M24782
Nodavirus				
Macrobranchium rosenbergii nodavirus (MrNV)/extra small virus (XSV)	15	~17 000	796	AY247793
Unassigned				
Maize white line mosaic virus (MWLMV)/ satellite MWLMV (SMWLMV)	17	23 961	1168	M55012

[a]CP, capsid protein MW.
[b]NR, not reported.

Table 2 Satellite nucleic acids

Helper virus/satellite	Satellite size	Encoded protein (aa/kDa)	Accession no.
ssDNA satellites (circular)			
Ageratum yellow vein virus (AYVV)/AYVV satellite DNA β	1347	118/13.7	AJ252072
Bhendi yellow vein mosaic virus (BYVMV)/BYVMV satellite DNA β	1353	140/	AJ308425
Cotton leaf curl Bangalore virus (CLCuBV)/CLCuBV satellite DNA β	1355	118/	AY705381
Cotton leaf curl Gezira virus (CLCGV)/CLCGV satellite DNA β	1348	117/13.6	AY077797
Cotton leaf curl Multan virus (CLCMV)/CLCMV satellite DNA β	1349	118/13.7	AJ298903
Eupatorium yellow vein virus (EuYVV)/EuYVV satellite DNAβ	1356	116/13.5	AJ438938
Honeysuckle yellow vein virus (HYVV)/HYYV satellite DNA β	1344	116/13.5	AJ316040
Malvastrum yellow vein virus (MYVV)/MYVV satellite DNA β	1350	118/	AJ786712
Tobacco curly shoot virus (TCSV)/TCSV satellite DNA β	1354	118/	AJ421484
Tomato yellow leaf curl China virus (TYLCCV) /TYLCCV satellite DNA β	1336	118/	AJ421621
Tomato leaf curl virus (TLCV)/TLCV satellite DNA	682	None	U74627
dsRNA satellites (linear)			
L-A ds RNA virus of *Saccharomyces cerevisiae*/M satellite RNAs	1801 bp	NR	U78817
	0.5–1.8 kb	NR	
Trichomonas vaginalis T1 virus (TVTV)/TVTV satellite RNA	497 bp	NR	U15991
	~700 bp	NR	NR
	~1700 bp	NR	NR
Ophiostoma novo-ulmi mitovirus 3a (OnuMV3a)/OnuMV3a S-dsRNA	738–767 bp	4 small ORFs	AY486119
Large ssRNAs satellites			
Benyvirus			
Beet necrotic yellow vein virus (BNYVV)/BNYVV RNA 5	1342-1447 nt	~26 kDa	U78292
Nepovirus			
Arabis mosaic virus (ArMV)/ArMV large satellite RNA	1104 nt	~39 kDa	D00664
Chicory yellow mottle virus (CYMV)/CYMV large satellite RNA	1145 nt	~39 kDa	D00686
Grapevine Bulgarian latent virus (GBLV)/GBLV satellite RNA	~1500 nt	NR	NR
Grapevine fanleaf virus (GFLV)/GFLV satellite RNA	1114 nt	~37 kDa	D00442
Myrobalan latent ringspot virus (MLRV)/MLRV satellite RNA	~1400 nt	~45 kDa	NR
Strawberry latent ringspot virus (SLRV)/SLRV satellite RNA	1118 nt	~36 kDa	X69826
Tomato black ring virus (TBRV)/TBRV satellite RNA	1372-1375 nt	~48 kDa	X05689
Potexvirus			
Bamboo mosaic virus (BaMV)/BaMV satellite RNA	836 nt	~20 kDa	L22762
Small, linear ssRNAs satellites			
Carmovirus			
Turnip crinkle virus (TCV)/TCV satellite RNA	194 nt, 230 nt, 356 nt		X12749
Cucumovirus			
Cucumber mosaic virus (CMV)/CMV satellite RNA	333–405 nt		M18872
Peanut stunt virus (PSV)/PSV satellite RNA	393 nt		Z98198
Necrovirus			
Beet black scorch virus (BBSV)/BBSV satellite RNA	615 nt		AY394497
Tobacco necrosis virus (TNV)/TNV small satellite RNA	620 nt		NR
Nepovirus			
Chicory yellow mottle virus (CYMV)/CYMV small satellite RNA	457 nt		NC006453
Sobemovirus			
Panicum mosaic virus (PMV)/PMV satellite RNA	350 nt		NR
Tombusvirus			
Artichoke mottled crinkle virus (AMCV)/AMCV satellite RNA	~700 nt		NR
Cymbidium ringspot virus (CymRSV)/CymRSV satellite RNA	621 nt		D00720
Carnation Italian ringspot virus (CIRV)/CIRV satellite RNA	~700 nt		NR
Pelargonium leaf curl virus (PLCV)/PLCV satellite RNA	~700 nt		NR
Petunia asteroid mosaic virus (PAMV)/PAMV satellite RNA	~700 nt		NR
Tomato bushy stunt virus (TBSV)/TBSV satellite RNA	612 nt, 822 nt		AF022788
Umbravirus			
Pea enation mosaic virus (PEMV)/PEMV satellite RNA	717 nt		U03564
Circular ssRNAs satellites			
Polerovirus			

Continued

Table 2 Continued

Helper virus/satellite	Satellite size	Encoded protein (aa/kDa)	Accession no.
Cereal yellow dwarf virus (CYDV-RPV)/CYDV-RPV satellite RNA	322 nt		M63666
Nepovirus			
Arabis mosaic virus (ArMV)/ArMV satellite RNA	300 nt		NC001546
Tobacco ringspot virus (TRSV)/TRSV satellite RNA	359 nt		M14879
Sobemovirus			
Lucerne transient streak virus (LTSV)/LTSV satellite RNA	324 nt		X01984
Rice yellow mottle virus (RYMV)/RYMV satellite RNA	220 nt		NC003380
Solanum nodiflorum mottle virus (SNMV)/SNMV satellite RNA	377 nt		J02386
Subterranean clover mottle virus (SCMV)/SCMV satellite RNA	332 nt, 388 nt		M33000
Velvet tobacco mottle virus (VTMoV)/VTMoV satellite RNA	365–366 nt		J02439

Table 3 Satellite-like and virus-associated nucleic acids

Helper virus/satellite-like/virus-associated nucleic acid	Size of nucleic acid	Encoded protein
Satellite-like ssRNAs[a]		
Benyvirus		
Beet necrotic yellow vein virus (BNYVV)/BNYVV RNA 3	1754 nt	~25 kDa
BNYVV/BNYVV RNA 4	1467 nt	~31 kDa
Umbravirus		
Groundnut rosette virus (GRV)/GRV satellite RNA	895–903 nt	
Virus-associated nucleic acids[b]		
Hepadnavirus		
Hepatitis B virus (HBV)/hepatitis delta virus (HDV)	1979 nt	~22 kDa (δAg-S) ~24 kDa (δAg-L)
Luteovirus		
Beet western yellows virus (BWYV) /	2843 nt	~85 kDa, plus others
BWYV-associated RNA (BWYVaRNA)		
Carrot red leaf virus (CRLV) /		
CRLV-associated RNA (CLRVaRNA)	2835 nt	~85 kDa, plus others
Begomovirus		
Cotton leaf curl virus (CLCuV)/		
CLCuV-associated DNA 1 (CLCuVaDNA 1)	1376 nt	~33 kDa
Nanovirus		
Banana bunchy top virus (BBTV)/BBTV-associated DNAs S1, S2, and S3	~1100	
Faba bean necrotic yellows virus (FBNYV)/FBNYV-associated DNAs C1, C7, C9, and C11	~1000	~33 kDa
Milk vetch dwarf virus (MDV)/MDV-associated DNAs C1, C2, C3, and C10	~1000	~33 kDa
Subterranean clover stunt virus (SCSV)/SCSV-associated DNAs C2 and C6	~1000	~33 kDa

[a]Dependent on a helper virus for replication and encapsidation, but is essential for vector transmission of the helper virus.
[b]Dependent on a helper virus for encapsidation, but not for replication.

having in common the fact that they are not required for the replication of the HV.

1. Satellite viruses:
   - subgroup 1 – chronic bee-paralysis virus-associated satellite virus
   - subgroup 2 – satellites that resemble tobacco necrosis virus
2. Satellite nucleic acids:
   - ssDNA satellites
   - dsRNA satellites
   - ssRNA satellites: (1) subgroup 1 – large, ssRNA satellites; (2) subgroup 2 – small, linear, ssRNA satellites; (3) subgroup 3 – circular, ssRNA satellites.

Some RNAs previously described as satellite RNAs contribute to the natural means of transmission of the HV and thus are considered satellite-like RNAs, rather than true satellites (**Table 3**). In other cases, there are nucleic acids that are dependent on the HV for encapsidation, but not replication. These subviral agents are considered here as virus-associated nucleic acids (**Table 3**) and will be described and differentiated below.

General Properties and Effects of Satellites

Satellite Viruses

Best characterized of the satellite viruses (**Table 1**) are the satellites that resemble TNSV, all of them depending, for their replication, on plant viruses with ssRNA genomes. All satellite viruses in this subgroup contain an ssRNA of 800–1200 nt. Particles are isometric, 17 nm in diameter, with a $T = 1$ symmetry, built of 60 protein subunits. The particle structure has been determined at high resolution by X-ray diffraction for TNSV, tobacco mosaic satellite virus (TMSV), and panicum mosaic satellite virus (PMSV), with ~80% of the encapsidated RNA in stem–loop structures. Particle structure differs from that of their HV. In spite of the different structure of CP and virus particle, the particles of both the satellite and HV may share important properties; for example, the particles of TNSV and TNV are able to bind specifically to the zoospores of the vector fungus *Olpidium brassicae*. The 1000–1200 nt RNA of TNSV does not have a methylated cap structure or a genome-linked protein (VPg) in its 5′ end. Unlike most plant virus RNAs it has a phosphorylated 5′ end. The 3′ termini may share a structure with that of the HV genomic RNA (gRNA); for instance, in both tobacco mosaic virus and TMSV, the 3′ termini form tRNA-like structures and are aminoacylable with histidine. *Cis*-acting sequences necessary for RNA amplification have been mapped at the 5′ and 3′ nontranslated regions (NTRs) of PMSV RNA as well as within the CP open reading frame (ORF), and are conserved in a D RNA, which is maintained by PMSV. A translational enhancer domain has been mapped in the 3′ NTR of TNSV. In addition to the ORF encoding the CP, some satellite viruses contain other ORFs; it remains unclear if the encoded products have any role *in vivo*.

Interference with the accumulation of the HV has been described for TNSV. Satellite viruses can also modify the symptoms induced by the HV, as is the case with PMSV, which in co-infection with panicum mosaic virus (PMV) enhances the mild symptoms of PMV to cause a severe mosaic and chlorosis. Symptom induction is due to the PMSV CP, and a chlorosis-inducing domain has been mapped. In addition to its structural and symptom-inducing functions, the CP of PMSV is involved in systemic movement, binds PMV particles, and counters the effects of post-transcriptional gene silencing suppressors.

The particles of TNSV may contain a noncoding satellite RNA of about 620 nt, which depends on TNV for replication and on TNSV for encapsidation. A satellite RNA with similar dependence relationships has been described for PMV and PMSV. These systems are good examples of the complexity of dependence relationships in satellitism.

The other subgroup of satellite viruses contains satellites found associated with chronic bee-paralysis virus (CBPV). The satellites consist of three RNA species of about 1.1 kbp, which can be encapsidated either in 17-nm isometric particles by the satellite-encoded CP, or by the CP of the HV, CBPV. The satellites interfere with CBPV replication.

Satellite Nucleic Acids

ssDNA Satellites

All the DNA satellites reported to date are associated with a single virus genus, *Begomovirus*, in the family *Geminiviridae* (**Table 2**). This group of over 100 viruses is transmitted by whiteflies and their geminate particles encapsidate either one or two ssDNA species, each of about 2700 nt, as well as satellite DNAs that may occur. The first satellite DNA was found in association with TLCV and is a single-stranded (ss), circular DNA molecule of 682 nt, which can be supported for replication by a number of begomovirus species. It lacks any significant ORF or identifiable promoter element and does not contribute to the infection of TLCV. More recently, a large group of ss, satellite DNAs termed DNA β has been isolated. They are related to TLCV satellite DNA but are about twice the size of the TLCV satellite DNA and encode a protein known as βC1. The C1 protein is expressed from a single complementary-sense transcript with conserved regulatory elements.

The search for the causal agents of Ageratum yellow vein disease led to the discovery of the first DNA β satellite because it was required for disease symptom expression. Subsequently, a similar role was established for cotton leaf curl virus (CLCuV) DNA β, which is part of a disease complex causing major crop losses in Pakistan. Some geminivirus satellite DNAs reported as DNA β are of similar size to TLCV satellite DNA and lack a βC1 gene. Geminivirus satellite DNAs therefore can be considered a single group in which some species are defective for encoding a protein.

Geminivirus satellite DNAs can exert a drastic effect on the symptoms produced by their HV. The pathogenesis is mediated by the βC1 protein. The mechanism of βC1-mediated pathogenesis is not clear; however, the protein causes a drastic disease-like phenotype when expressed transgenically. This severe effect on host plants suggests changes to growth pattern and is accompanied by vein thickening, enations, and development of leaf-like structures. The protein has been demonstrated to be a suppressor of gene silencing, raising the possibility that it may control host functions by microRNA regulation.

Geminivirus satellite DNAs do not show a strict affinity for HV. TLCV satellite DNA is supported for replication by viruses as diverse as tomato yellow leaf curl virus, African cassava mosaic virus, and even beet curly top virus, which belongs to a different genus. In another case, the DNA β associated with Ageratum yellow vein disease was found to be maintained in experimental plants infected with Sri Lankan cassava mosaic virus (SLCMV), which has a bipartite genome. This interaction altered the host range of SLCMV to include Ageratum. Interestingly, the satellite could substitute for DNA B component implying a functional similarity. Other evidence supporting the role of geminivirus satellite DNAs in movement has also been obtained. The DNA A component of tomato leaf curl New Delhi virus (TLCNDV) is not capable of systemic infection in the absence of DNA B and remains confined to the sites of inoculation where it is capable of replication; however, CLCuV DNA β can substitute DNA B of TLCNDV to restore systemic infection in tomato. Surprisingly, systemic infection of DNA A component alone, accompanied by symptoms, could also be mediated by transient expression of the βC1 protein.

dsRNA Satellites

A number of examples of dsRNA satellites have been described, but only two are recognized officially at this time (**Table 2**): one group is associated with the yeast *Saccharomyces cerevisiae* and is designated the M satellites of L-A dsRNA virus; the other group of three dsRNAs is associated with the Trichomonas vaginalis T1 virus (TVTV) of the eponymous protozoan. Both HVs are in the family *Totiviridae*. In the case of the three TVTV satellite RNAs of 497, ~700, and ~1700 bp, the ds satellite RNAs are maintained only by some isolates of TVTV, which both replicate and encapsidate these dsRNAs. The L-A dsRNA virus M satellites consist of several dsRNAs varying from 1.0 to 1.8 kbp in size. These dsRNA satellites are dependent on genes of the HV for both replication and encapsidation. The M1 dsRNA satellite encodes a toxin which kills other yeasts not harboring this dsRNA. In its prototoxin form, this protein provides immunity to the yeast secreting the toxin. Other L-A dsRNA viruses have different M satellites associated with them, all of them also encoding toxin and immunity systems.

All of the putative dsRNA satellites were found with fungi, in association with dsRNA viruses in the families *Hypoviridae*, *Narnaviridae*, *Partitiviridae*, and *Totiviridae*. One of these dsRNA satellites, from Ophiostoma novo-ulmi mitovirus 3a (OnuMV3a) in the genus *Mitovirus*, of the family *Narnaviridae* consists of a group of dsRNAs of 738–767 bp, designated OnuMV3a S-dsRNA (**Table 2**). These dsRNA satellites did not affect the hypovirulence associated with their HV in the fungus *Sclerotinia homeocarpa*. Their nucleotide sequences indicate that these OnuMV3a S-dsRNAs did not have the coding capacity for their own replicase, and thus presumably depend on the replicase of the HV.

ssRNA Satellites

Subgroup 1: Large ssRNA satellites

Subgroup I satellites have messenger RNA properties. These satellite RNAs are between 0.8 and 1.5 kb in size and encode nonstructural proteins that are expressed *in vivo* (**Table 2**). Most satellites in this subgroup are associated with nepoviruses. Large ssRNA satellites of nepoviruses share the 5′ and 3′ structural features of the gRNAs of the HV: a 3′-terminal poly(A) sequence and a 5′-terminal VPg that, in the analyzed instances, is indistinguishable from that on the HV genome and thus, is encoded by it. The encoded nonstructural protein of different satellite RNAs are basic proteins of 36–48 kDa. For the satellite RNAs of tomato black ring virus (TBRV), grapevine fanleaf virus (GFLV), and arabis mosaic virus (ArMV), it has been shown that the encoded proteins are needed for the replication of the satellite RNA. Most large satellite RNAs of nepoviruses do not seem to have an effect on the accumulation or pathogenicity of the HV. However, this may depend on the experimental system, as the large satellite RNA of ArMV was shown to modulate the symptoms of the HV depending on the species of host plant.

Large satellite RNAs (1342–1347 nt) have been found to be associated with many isolates of the benyvirus beet necrotic yellow vein virus (BNYVV). These satellite RNAs, designated BNYVV RNA 5, encode a protein of 26 kDa, which is responsible for intensification of the rhizomania disease induced by infection of BNYVV RNAs 1 and 2, plus its two satellite-like RNAs 3 and 4.

Another large satellite RNA has been found to be associated with the potexvirus bamboo mosaic virus (BaMV), the only satellite RNA encapsidated into rod-shaped particles. BaMV satellite RNA encodes a 20-kDa protein that is expressed *in vivo* but is not needed for satellite RNA replication. Interestingly, this protein shares significant sequence similarity with the structural protein of the PMSV, and binds cooperatively to RNA, with a preference for the satellite RNA. The presence of BaMV satellite RNA significantly reduces the accumulation of BaMV RNA. *Cis*-acting sequences required for BaMV satellite RNA replication have been mapped at the 5′ NTR, and comprise a stem–loop structure that is conserved among BaMV satellite RNA variants.

Subgroup 2: Small, linear ssRNA satellites

This subgroup contains ss satellite RNAs associated with seven genera of plant viruses (**Table 2**), although many of the members are associated with nepoviruses. There are no circular forms of these satellite RNAs present in infected cells, although dimer or higher multimer forms may be detected in virions and/or in infected cells. These satellite RNAs vary in size from ~200 to ~800 nt and do not appear to encode proteins, although isolates of some satellite RNAs contain nonconserved ORFs. These RNAs appear to have highly ordered secondary structures, which is probably responsible for the high stability and infectivity of these RNAs, as well as their biological properties. Most of the small ss satellite RNAs either have no effect on or reduce the accumulation of their HV. This reduction in HV accumulation often, but not always, results in a decrease in the disease symptoms induced by the HV. Moreover, in a few cases, the presence of the satellite RNA reduced or had no effect on the HV accumulation, but intensified the disease induced by the HV. One of the latter, a tomato necrosis-inducing satellite RNA of the cucumovirus cucumber mosaic virus (CMV), could even induce necrosis in tomato independent of the HV, if expressed in the complementary-sense orientation within the genome of virus vector but not when expressed as a transgene; however, in the case of other satellite RNAs of CMV, different strains of CMV and the related HV tomato aspermy virus (TAV) have been used to show that the HV makes a contribution to the disease syndrome associated with the satellite RNA. The basis for satellite RNA-mediated attenuation of disease symptoms is not clear, but may not be due simply to reduction of the HV titer, since CMV satellite RNAs supported by TAV show an attenuation of symptoms, but no effect on the accumulation of the HV. Similarly, the reduction of HV level may not be due simply to competition between the HV and the satellite RNA for the same replicase, since data from TCV and CMV satellite RNAs indicate that satellite RNAs may affect the suppression of RNA silencing function of their associated HV, leading to a reduction in the titer of the virus due to RNA silencing. A general model for satellite- and viroid-mediated intensification of disease symptoms has been proposed, caused by silencing inducing RNAs generated from such subviral RNAs acting on specific host mRNAs.

Subgroup 3: Small, circular ssRNA satellites

This subgroup of satellite RNAs contains members that vary in size from 220 to 388 nt, and are associated with three genera of plant viruses (**Table 2**). The satellite RNAs exist in circular as well as linear forms in infected plant cells, but only the circular satellite RNA form is encapsidated by the helper sobemovirus, while only the linear satellite RNA form is encapsidated by the helper polerovirus and nepoviruses. Multimeric, linear forms of these satellites are produced in infected cells and, in some cases, these are also packaged into virus particles. The multimeric forms are generated via a rolling-replication mechanism and the unit-length linear as well as circular forms are generated via ribozymes. There is no mRNA activity associated with these various satellite RNAs, and they all have a high degree of secondary structure. The satellite RNAs either have no effect on (sobemovirus satellites) or reduce (nepovirus and polerovirus satellites) the accumulation of their HV. In addition, the presence of satellite RNAs either reduces (nepovirus and polerovirus satellites) or exacerbates (sobemovirus satellites) the symptoms induced by the helper virus. The molecular basis for symptom attenuation or intensification is unknown.

Satellite-Related Nucleic Acids

A number of subviral nucleic acids have been described that have in some instances been referred to as satellite RNAs or DNAs. However, as these nucleic acids either contribute to the infection cycle of the HV, or are not dependent upon an HV for their replication, they are better described as satellite-like or virus-associated nucleic acids, respectively (**Table 3**).

Satellite-like ssRNAs

This group of subviral RNAs includes the satellite RNAs of groundnut rosette virus (GRV) and RNAs 3 and 4 of BNYVV. In the case of satellite RNAs of GRV, they are required for transmission of GRV by the luteovirus groundnut assistor virus. How these satellite RNAs of 895–903 nt affect transmission is unknown, since they do not code for any functional proteins. Some of the satellite RNAs of GRV also affect accumulation of the HV and the symptoms induced by the HV, causing either attenuation due to severely reduced replication of the HV, or exacerbation, producing unique symptoms. In the case of BNYVV, where only RNAs 1 and 2 are essential for replication, movement within leaves and particle assembly, RNA 3 (1774 nt) is required for spread of the virus into roots and RNA 4 (1467 nt) is essential for virus transmission by the soil-fungus vector of BNYVV. Thus, BNYVV RNAs 3 and 4 are considered satellite-like, in that they contribute to the natural infection cycle of the virus. By contrast, BNYVV RNA 5 is a true satellite, in that it is not required for any phase of the infection cycle of BNYVV, although its encoded 26 kDa protein in combination with the 25 kDa protein encoded on RNA 3 intensifies symptom expression of the disease rhizomania.

Virus-associated nucleic acids

This group of subviral agents does not fit the above definition of satellites, but as they are peripheral to but dependent on an HV for their encapsidation, they have sometimes been referred to as satellite-like RNAs or DNAs. However, as a functional distinction exists between these subviral nucleic acids and those of the satellite-like agents described above, which are dependent on an HV for their replication, these subviral agents are collectively referred to as (virus-) associated nucleic acids.

Hepatitis delta virus

Hepatitis delta virus (HDV) is a subviral agent of the human pathogen, hepatitis B virus (HBV). HDV is replicated in the nucleus by the host DNA-dependent RNA polymerase II, but is dependent upon the envelope proteins of HBV for its encapsidation and transmission, and is thus not a satellite. The HDV (1679 nt) RNA folds into an unbranched rod-like structure similar to that of viroids, and in common with the circular ssRNA satellites, contains ribozymes that are involved in the processing and maturation of this RNA during replication. The HDV RNA encodes two forms of a protein designated the delta antigen (δAg). The smaller, 22 kDa form (δAg-S), which is a nuclear phosphoprotein, is required for replication of HDV RNA, while the 24 kDa larger form (δAg-L), which contains an additional 19 aa at the C-terminus and is produced late in infection, is required along with δAg-S for assembly of the HDV RNA into HBV envelope particles. δAg-L also acts as an inhibitor of HDV replication. HDV intensifies the severity of liver disease caused by HBV.

Luteovirus-associated RNAs

Some isolates of two luteoviruses, beet western yellows virus and carrot red leaf virus (CRLV) were found to have associated with them additional RNAs of size 2.8 kb. These associated RNAs encoded their own viral replicases, but required the respective luteoviruses for their movement, encapsidation, and transmission. These RNAs, designated BWYVaRNA (2843 nt) and CRLVaRNA (2835 nt), have similar genome organizations, are about 50% similar in sequence, and both intensify the disease caused by their HV. CLRVaRNA together with CLRV and the umbravirus carrot mottle virus induced carrot motley dwarf disease.

Begomovirus- and nanovirus-associated DNAs

Circular ssDNA species of about 1300 nt, unrelated to geminivirus genomes, have been isolated from plants infected with some Old World begomoviruses. These molecules, named DNA 1, resemble a DNA component of nanoviruses. CLCuV-associated DNA1 is an example (**Table 3**). They encode their own replicase-associated protein (Rep) and replicate independent of the HV. In the absence of a nanovirus CP which allows aphid transmission, the virus-associated DNAs are encapsidated in helper geminiviruses and are transmitted by their whitefly vector. Nanoviruses themselves may also contain virus-associated DNAs. These are about 1.0 or 1.1 kbp in size and capable of autonomous replication but require the respective nanovirus for their movement, encapsidation, and transmission by aphids.

Replication and Structure

Satellite DNAs

The replication of geminivirus satellite DNAs mimics that of the HV. Unlike RNA viruses, geminiviruses do not encode a polymerase and depend on host DNA polymerases for replication. The only geminivirus-encoded protein essential for replication is Rep. Geminivirus satellite DNAs utilize the HV Rep and contain the conserved sequence for DNA nicking by the Rep. Replication takes place by a combination of rolling circle and recombination dependent mechanisms and the DNA is encapsidated in the HV virion. The replication is independent of βC1 expression.

Sequence comparisons at the nucleotide and amino acid levels reveal a number of conserved structural features, despite a significant level of sequence divergence. These include an A-rich region, a sequence of about 80 nt referred to as the satellite conserved region, and putative stem–loop structure containing a mononucleotide motif, also present in geminivirus genomes.

Satellite RNAs

A general property of satellite RNAs is that their replication depends on the replication machinery of the HV, which involves virus- and host-encoded factors. Hence, replication of satellite RNAs depends on interactions with both the HV and the host plant. Satellite RNAs may or may not share structural features at their 5′ and 3′ termini in common with the HV RNA. For instance, the large satellite RNAs of the nepoviruses have an HV-encoded VPg at the 5′ end and a poly(A) tail at the 3′ end, as do their HV RNAs, but the satellite RNAs of CMV have a 5′ cap structure, as do the CMV RNAs, but unlike the HV, the satellite RNAs do not have a tRNA-like structure that can be valylated at the 3′ end. Structural differences and, often, differences in the replication process between the satellite RNA and the HV RNA, indicate that the replication machinery of the HV may need to be adapted to replicate the satellite RNA, in ways which are not completely understood. The HV replication complex could be modified by satellite-encoded factors, as has been hypothesized for the large satellite RNAs of nepoviruses, or by unidentified host factors. It should be pointed out that in the best characterized systems, the efficiency of replication depends on the host plant. For CMV satellite RNAs, replication efficiency also depends on the strain of HV. Similarly, while the expression of the HV RNA-dependent RNA polymerase was enough for the replication of cereal yellow dwarf virus (CYDV) satellite RNA in the homologous host, this was not the case for cymbidium mosaic virus satellite RNA in a heterologous host system, in which, however, a DI RNA was amplified.

Replication has been studied best in the small noncoding satellite RNAs. For the small linear satellite RNAs of TCV and CMV, multimeric forms of both positive (arbitrarily defined as the encapsidated sense) and negative sense are found in infected tissues. The junction between monomers can be perfect or have deletions. Circular forms are not found, and replication does not proceed through a rolling-circle mechanism. Regulatory sequences for RNA replication have been mapped in detail in the satellite RNAs of TCV. In the hybrid satC molecule, a hairpin structure is a replication enhancer and has a role in depressing the accumulation of the HV. Also, structural motifs favoring recombination have been analyzed extensively in satD and satC of TCV.

The replication of small, circular satellite RNAs is by a rolling-circle mechanism; upon, infection, linear, multimeric forms of the negative strand are synthesized. The multimeric plus strand is then synthesized on this template or on negative strand circular monomers, depending on the satellite RNA. The positive strand multimer is cleaved autocatalytically and ligation occurs for those satellite RNAs that are encapsidated as a circular form. Hammerhead and hairpin ribozyme structures are found in these satellite RNAs monomers or dimers, which catalyze hydrolysis or hydrolysis and ligation, respectively. Other structures required for replication and encapsidation have been mapped in CYDV satellite RNA.

For one circular satellite RNA, a DNA counterpart has been shown to occur in the host genome, as a retroid element. HDV, which depends on HBV for encapsidation, shares structural features and replication mechanisms with the small, circular satellite RNAs.

Structural analyses have been done mostly with small satellite RNAs. Models for the *in vitro* secondary structure have been proposed for several satellite RNAs (e.g., CMV satellite RNA, TCV satellite RNAs, TRSV satellite RNA, and CYDV satellite RNA) based on nuclease sensitivity and chemical modification of bases data. For CMV satellite RNA, *in vivo* models have also been proposed. All analyzed satellite RNAs have a high degree of secondary structure, with more than 50% and up to 70% of bases involved in pairing. The high degree of secondary structure could explain the high stability of these molecules as well as the very high infectivity reported for some of them, such as CMV satellite RNA and TCV satellite RNA. The high degree of secondary structure may also be related to their biological activity, which for these noncoding molecules must depend on structural features. As has been detailed in other sections, sequences, structural domains, and tertiary interactions involved in pathogenicity and replication have been well characterized for some small satellite RNAs.

Sequence Variation and Evolution

Sequence variation and evolution has been analyzed most for CMV satellite RNA. Early experiments under controlled conditions showed a high variability and genetic plasticity. Populations started from cDNA clones rapidly evolved to a swarm of sequences whose master sequence changed upon passage on different hosts. Analysis of genetic variants from natural populations showed again a high diversity for CMV satellite RNA. In natural populations, satellite RNA diversification and evolution proceeded by mutation accumulation and by recombination. Constraints to genetic variation were analyzed, and were related to the maintenance of base pairing in secondary structure elements. Population diversity of CMV satellite RNA was higher than for the HV, and the population structures of CMV and the satellite RNA were not correlated. These analyses also showed that CMV satellite RNA behaved in nature as a molecular hyperparasite spreading epidemically on the CMV populations. The incidence of CMV satellite RNA in CMV populations has been shown to be low, except during episodes of epidemics of tomato necrosis. High diversity of natural populations has also been shown for PMSV RNA and for BaMV satellite RNA. For BaMV satellite RNA, the diversity was highest in the NTRs. Conversely, the population of rice yellow mottle virus satellite RNAs over sub-Saharan Africa showed little sequence diversity. The incidence of these three satellite RNAs in the analyzed populations of the respective HV was, in all cases, high.

Phylogenetic analyses divide geminivirus satellites broadly into two categories, those isolated from host plants of the family Malvaceae and the rest isolated from other plants, mostly in the Solanaceae. Compared to satellite RNAs, satellite DNAs do not exhibit a wide diversity. They are structurally similar, and lack strict specificity for HV species. Geminivirus satellite DNA sequences analyzed exhibit a minimum overall similarity of about 47% and 37%, at the nucleotide level and amino acid level, respectively. To date, over 120 DNA β sequences have been reported in data banks; however, many of these are members of the same species because their overall sequence similarity is well above 90%. Some examples of distinct geminivirus satellites that have been adequately characterized are listed in **Table 2**.

There has been much speculation on the origin of satellite RNAs. For CMV satellite RNA and TCV satellite RNA, it has been suggested that they could have been generated out of small sequences synthesized upon the HV RNA as a template by the HV polymerase. This hypothesis is no longer sustained by those that proposed it. Whatever the mechanism of generation of the satellite RNAs, it seems that the satellitism as a phenomenon, that is, the dependence on a certain virus for replication, has evolved independently several times. This is suggested by the lack of correlation between satellite and HV taxonomy, and by the existence of subviral nucleic acids with different degrees of dependence on an HV. For instance, an evolutionary line from nondependent viruses to satellites such as BaMV satellite RNA through satellite viruses could have proceeded by size and information content reduction. A phylogenetic relationship between viroids and small circular satellite RNAs has also been proposed.

Expression of Foreign Sequences from Satellite Vectors

The satellite RNAs of BaMV and BNYVV, as well as the satellite DNA of tomato yellow leaf curl China virus isolate Y10 (TYLCCNV-Y10), all of which express proteins not required for the replication or spread of the satellite or HV, have been used as expression vectors of foreign sequences. In the case of the satellite RNA of BaMV, the expression of such sequences and the accumulation of the satellite RNA were reduced considerably in systemically infected leaves. Nevertheless, this satellite expression system has been useful for studying cis- and trans-acting replication signals. The 26 kDa protein encoded by RNA 5 of BNYYV has been replaced with the sequence encoding the green fluorescent protein, which was expressed in both the inoculated leaves and systemically infected leaves. The small size of the satellite DNA of TYLCCNV-Y10 precludes use of this system for expression of most genes, but the system has been used to express plant gene segments inducing RNA silencing in several plants species. The TMSV system also has been used to express plant gene sequences, in place of its CP, resulting in RNA silencing of those plant genes, but not of the satellite virus or its HV. The modified TMSV expression vector was able to spread through the plant in the absence of its CP.

Satellite-Mediated Control of Viruses

The ability of some satellite RNAs to attenuate the symptoms induced by their HV has been utilized to develop strategies for disease control. In one strategy, the satellite RNA of CMV has been used in combination with mild strains of CMV as a 'vaccine', to pre-inoculate tomato plants and ameliorate symptoms induced by the replicating challenge virus (through the attenuating ability of the vaccine satellite RNA), in fields and greenhouses. Moreover, the vaccine could also prevent or reduce the frequency of infection of both the HV (through cross-protection by the HV of the vaccine strain) and of satellite RNAs that exacerbate the disease caused by the HV (through cross-protection by the vaccine satellite RNA). In a second strategy, the satellite RNAs of CMV, TRSV, and GRV have been expressed in transgenic plants, and provided either resistance to infection by the respective HV, or tolerance to the disease induced by the HV and/or related pathogenic satellites.

See also: Beta ssDNA Satellites.

Further Reading

Briddon RW and Stanley J (2006) Subviral agents associated with plant single-stranded DNA viruses. *Virology* 344: 198–210.

Dry IB, Krake LR, Rigden J, and Rezaian MA (1997) A novel subviral agent associated with a geminivirus: The first report of a DNA satellite. *Proceedings of the National Academy of Sciences, USA* 94: 7088–7093.

Gosselé V, Fache I, Meulewaeter F, Cornelissen M, and Metzlaff M (2002) SVISS – A novel transient gene silencing system for gene function discovery and validation in tobacco plants. *Plant Journal* 32: 859–566.

Palukaitis P and García-Arenal F (2003) Cucumoviruses. *Advances in Virus Research* 62: 241–323.

Simon AE, Roossinck MJ, and Havelda Z (2004) Plant virus satellite and defective interfering RNAs: New paradigms. *Annual Reviews in Phytopathology* 42: 415–437.

Tao X and Zhou X (2004) A modified satellite that suppresses gene expression in plants. *Plant Journal* 38: 850–860.

Vogt PK and Jackson AO (eds.) (1999) *Satellites and Defective Viral RNAs.* Heidelberg: Springer.

Seadornaviruses

H Attoui, Université de la Méditerranée, Marseille, France
P P C Mertens, Pirbright Laboratory, Woking, UK

© 2008 Elsevier Ltd. All rights reserved.

Glossary

Arthralgia Joint pain.
Cerebral palsy Disability resulting from damage of the brain before, during, or shortly after birth and manifested as muscular and speech disturbances.
Hemiplegia Total or partial paralysis of one side of the body resulting from an affliction of the motor centers in the brain.
Myalgia Muscle pain.
Viremia The presence of virus in the blood.

Introduction

The family *Reoviridae* includes 12 recognized genera and 3 proposed genera (**Table 1**). The genera *Coltivirus*, *Seadornavirus*, and *Orbivirus* include arboviruses that can cause disease in animals (including humans).

In 1991, the International Committee for the Taxonomy of Viruses (ICTV) formally recognized the genus *Coltivirus* (sigla from Colorado tick virus) containing the 12-segmented double-stranded RNA (dsRNA) animal viruses that were previously classified within the genus *Orbivirus*. *Colorado tick fever virus* is the prototype species of the genus *Coltivirus* and includes strains of several tick-borne viruses (including isolates of Colorado tick fever virus (CTFV) from humans and ticks; California hare coltivirus (CTFV-Ca) from a hare in northern California; and Salmon River virus (SRV) from humans in Idaho). The genus *Coltivirus* also contains a second distinct species: *Eyach virus* (EYAV) isolated from ticks in Europe.

The seadornaviruses are 12-segmented mosquito-borne dsRNA viruses that were initially considered as tentative species within the genus *Coltivirus*. However, based on their antigenic properties and nucleotide sequence comparisons to the coltiviruses and other members of the *Reoviridae*, their taxonomic status was reassessed. They are now formally recognized as members of the genus *Seadornavirus* (sigla: from *South East Asian Dodeca RNA virus*; genus recognized in 2000), which includes the type species *Banna virus* (BAV), isolated from humans; two other species, *Kadipiro virus* (KAD) and *Liao ning virus* (LNV); and several unclassified isolates from mosquitoes. The seadornaviruses have been implicated in a variety of pathological manifestations in humans, including flu-like illness and neurological disorders, as described below.

Historical Overview of Seadornaviruses

The first seadornaviruses were reported in 1980 and 1981, when numerous isolates of 12-segmented dsRNA viruses were made from pooled homogenates of mosquitoes in Indonesia and were initially designated as JKT-6423, JKT-6969, JKT-7043, and JKT-7075. These original viruses were recently renamed as Banna virus (BAV) and Kadipiro virus (KDV). Shortly after the initial identification of BAV in Indonesia, a similar virus was isolated from the serum and cerebrospinal fluid (CSF) (2 isolates) and sera (25 isolates) of human patients with encephalitis in Yunnan province in the south of China and BAV was reported as a causative agent of human viral encephalitis. Seadornaviruses have subsequently been isolated on several occasions from humans, cattle, pigs, and mosquitoes and are vectored by *Anopheles*, *Culex*, and *Aedes* species. BAV is now regarded as endemic in southeast Asia, particularly in Indonesia and China.

Table 1 Twelve recognized genera and the three proposed genera of family *Reoviridae*

	Type species			No. of segments	Hosts of the member viruses
Genus					
1. Idnoreovirus	*Diadromus pulchellus reovirus*			10	Insects
2. Cypovirus	Cypovirus type 1			10	Insects
3. Oryzavirus	Rice ragged stunt virus			10	Plants
4. Fijivirus	Fiji disease virus			10	Plants
5. Orthoreovirus	Mammalian orthoreovirus			10	Primates, ruminants, bats, birds, reptiles
6. Orbivirus	Bluetongue virus			10	Ruminants, equids, rodents, bats, marsupials, birds, humans
7. Rotavirus	Rotavirus A			11	Mammals, birds
8. Aquareovirus	Aquareovirus A			11	Aquatic animals
9. Mycoreovirus	Mycoreovirus-1			11 or 12	Fungi
10. Phytoreovirus	Wound tumor virus			12	Plants
11. Coltivirus	Colorado tick fever virus			12	Mammals including man
12. Seadornavirus	*Banna virus*	Isolates:	BAV-Ch BAV-In6423 BAV-In6969 BAV-In7043	12	Humans, porcine, cattle
	Kadipiro virus	Isolates:	KDV-Ja7075		
	Liao ning virus	Isolates:	LNV-NE9712 LNV-NE9731		
Proposed genera					
1. Dinovernavirus	*Aedes pseudoscutellaris reovirus*			9	Insects
2. Mimoreovirus	*Micromonas pusilla reovirus*			11	Algae
3. Cardoreovirus	*Eriocheir sinensis reovirus*			12	Crabs

Figure 1 The known world distribution of seadornaviruses (shaded areas): BAV and LNV in China and BAV and KDV in Indonesia.

Distribution and Epidemiology

BAV was initially isolated from mosquitoes and mammalian species in 1980–81, between latitudes of 39° N and 38° S, in southeast Asia (**Figure 1**). BAV was subsequently isolated from humans in the Yunnan province of southern China (Xishuangbanna prefecture in 1987). A virus identified as BAV was also isolated in 1987 from patients with fever and flu-like manifestation, in Xinjiang province of western China province. In 1992, two 12-segmented dsRNA viruses were isolated from serum of patients with an unknown fever in Mengding county of Yunnan province. Eight further isolates of antigenically Banna-like viruses were made from the sera of 24% of patients with an unidentified fever and viral encephalitis in Xinjiang province of northwest China in 1992. Twelve-segmented dsRNA viruses that are described as antigenically related to BAV have also been isolated from humans in various other provinces in China, including Beijing, Gansu, Hainan, Henan, and Shanshi and consequently BAV is now classified as a BSL3 arboviral agent.

Four viruses initially identified as JKT-6423, JKT-6969, JKT-7043, and JKT-7075, were isolated from *Culex* and *Anopheles* mosquitoes in central Java in Indonesia. The first three isolates were isolates of BAV, while JKT-7075 represents the only known strain of the *Kadipiro virus* species.

Recently, isolates of two distinct serotypes of *Liao ning virus* (LNV) were obtained from *Aedes dorsalis* mosquito in the Liaoning province of north eastern China.

Vectors, Host Range, and Transmission

BAV is the only seadornavirus that has been isolated from humans. Experimentally, seadornaviruses have also been shown to infect and replicate in adult mice and can be detected in the infected mouse blood from 3 to 5 days post infection. LNV replicates readily in several mammalian cell lines with massive lytic effect and kills adult mice. KDV and LNV have only been isolated only from mosquitoes. Indeed seadornaviruses are thought to be transmitted by mosquitoes and have been isolated from *Culex vishnui, Culex fuscocephalus, Anopheles vagus, Anopheles aconitus, Anopheles subpictus*, and *Aedes dorsalis*.

BAV, KDV, and LNV all occur in tropical and subtropical regions, where other mosquito-borne viral diseases, including Japanese encephalitis (JEV) and dengue, are endemic. This suggests that conclusive diagnosis and differentiation of these diseases may be difficult, based on clinical signs alone. Indeed several cases were diagnosed as JE in China (Henan province, Naijing city, and Fujian province in summer–autumn season of 1994), despite the absence of detectable JEV, or JEV-specific antibodies. However, a proportion of these patients (7/98) did show a 4- to16-fold rise in anti-BAV immunoglobulin G (IgG) by enzyme-linked immunosorbent assay (ELISA), indicating a BAV infection. A further 130 sera (from 1141 patients diagnosed with JE or viral encephalitis) from Chinese health institutes were also positive for BAV-specific immunoglobulin M (IgM).

In the summer of 1987, 16 BAV-like isolates were recovered from sera of 53 cattle (30% of the tested animals) and 4 isolates recovered from sera of 13 pigs (30% of the tested animals) collected in a slaughterhouse in the Dai nationality autonomous prefecture of the Xishuangbanna area of Yunnan province, where BAV has been previously isolated from humans.

BAV is regarded as 'a common cause of viral encephalitis and fever in humans during summer–autumn in

China. Its high isolation rate and wide geographical distribution make BAV an important public health problem in China. It is likely that seadornaviruses are also endemic in other countries within Southeast Asia.

Virion Properties, Genome, and Replication

Seadornavirus particles, like those of many of the 'unturreted' reoviruses, are approximately spherical in appearance with icosahedral symmetry. The intact virion is 60–70 nm in diameter and has three concentric capsid layers, containing a total of seven structural proteins. The double-layered core-particle contains five proteins (VP1, VP2, VP3, VP8, and VP10) and is approximately 50 nm in diameter, with a central cavity that contains the 12 segments of the viral genome.

VP1 (the polymerase) and VP3 (the capping enzyme) form the transcriptase complex that is present within the core as a dimer of VP1 and a monomer of VP3. This complex is located at each of the 12 fivefold axes of the viral particle. The subcore layer is composed of 120 copies of VP2 and represents the $T=2$ layer. The outer layer of the core is made of 780 copies of VP8 and represents the $T=13$ layer. Based on sequence comparison to other viruses of the *Reoviridae* and in particular the rotaviruses, the seadornavirus VP10 proteins are thought to be anchored at the core surface. The outer capsid layer, which is composed of only two proteins (VP4 and VP9) (**Figure 2**), completely surrounds the core particle.

The BAV genome encodes five nonstructural proteins which are the VP5, VP6, VP7, VP11, and VP12. VP7 contains sequences which are found in various protein kinases. VP12 contains a dsRNA-binding domain and might play a role in circumventing cell defense mechanisms, possibly acting as an anti-PKR component.

Negative staining and electron microscopy have shown that the surface of the virus particles has a well-defined capsomeric structure and icosahedral symmetry. The outer surface also has spikes that are similar to those of the rotaviruses (**Figure 3(a)**). The virus core has a smooth outline that is typical of the nonturreted members of the family *Reoviridae* (**Figure 3(b)**).

Like the orbiviruses, the seadornaviruses are unstable in CsCl and readily lose their outer coat proteins. However, intact virus particles have been purified from infected tissue culture supernatant by ultracentrifugation on colloidal silica gradients (Percoll). Purified virus particles are stable at 4 °C and approximately pH 7.0 for up to 3–4 months. However, increased acidity decreases their infectivity, which is abolished at pH 3.0. The virus is more stable as an unpurified cell culture lysate at 4 °C, which is a convenient way for medium-term storage. For longer periods of storage, viruses are stable at −80 °C, and infectivity can be further conserved by addition of 50% fetal calf serum.

The infectivity of seadornaviruses is decreased considerably by heating to 55 °C. Treatment with sodium dodecyl sulfate (SDS) disrupts the virus particle and consequently abolishes infectivity completely. Infectivity is very significantly reduced by treatment with butanol,

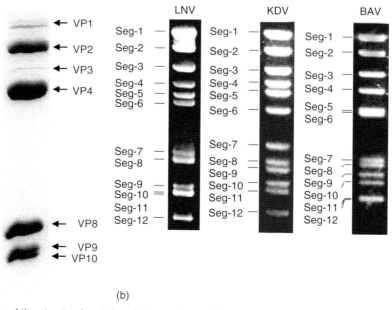

Figure 2 The PAGE profile of the structural proteins of Banna virus and the AGE profile of the genomes of seadornaviruses. (a) The PAGE profile of the structural proteins of Banna virus separated on a 10% polyacrylamide gel: VP4 and VP9 are the outer coat proteins. (b) The AGE profile of the genomes of Liao ning virus (LNV), Kadipiro virus (KDV), and Banna virus (BAV) run in a 1% agarose gel. The distinct electropherotypes of the three species allow to distinguish them.

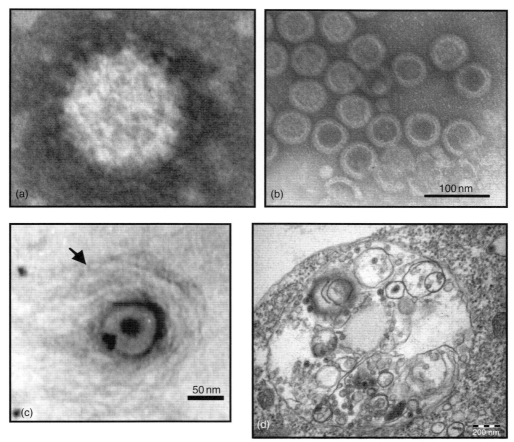

Figure 3 Electron micrographs of Banna virus particles. (a) A whole particle with spike proteins at the surface; (b) double-layered purified cores with smooth outlines; (c) particles surrounded by temporary envelope structure; and (d) thin section of infected C6/36 cells with virus particles seen within vacuole structures.

although organic solvents such as Freon 113 or Vetrel XF do not affect virion structure or infectivity and have been used successfully during purification of virus particles from cell lysates.

During replication, core-like particles are found within vacuole-like structures in the infected cell cytoplasm (**Figure 3(c)**), and are thought to be involved in virion morphogenesis. These structures appear similar to those found in cells infected with rotaviruses, where progeny particles bud into the endoplasmic reticulum, acquiring the outer coat components and generating mature virus particles.

Seadornaviruses replicate in a number of mosquito cell lines, including C6/36 and AA23 (both from *Aedes albopictus*), A20 (*Aedes aegypti*), and Aw-albus (*Aedes w albus*). Over 40% of the progeny virus particles are liberated in the culture medium (as determined by titration), prior to cell death and massive cytopathic effect (CPE) (fusiform cells), which (with BAV, LNV and KDV) is evident at 48–72 h post infection. Infected cells do not lyse initially and the virus is released by budding, acquiring a temporary membrane envelope in the process (**Figure 3(d)**). However, cell lysis does occur late in infection, as a result of cell death. Intracellular radiolabeling of viral polypeptides has shown that label is incorporated predominantly into viral polypeptides, even in absence of inhibitors of DNA replication such as actinomycin D, demonstrating 'shut off' of host-cell protein synthesis.

Like the other members of the genus, LNV can replicate in several mosquito cell lines, but it is the only seadornavirus that readily replicates in a variety of transformed or primary mammalian cells, including BHK-21, Vero, BGM, Hep-2, and MRC-5, leading to massive cell lysis after 48 h post infection.

The seadornavirus genome consists of 12 segments of dsRNA that are identified as Seg-1 to Seg-12 in the order of decreasing molecular weight and order of migration during agarose gel electrophoresis (AGE). The full-length genome sequence of BAV, KDV, and LNV has been determined. The genome comprises approximately 21 000 bp and the segment length ranges between 3747 and 862 bp. The genomic RNA of BAV and LNV show 6–6 electrophoretic profiles in 1% agarose gel electrophoresis (**Figure 2**). The genome of KDV migrates in 6-5-1 profile. Each genome segment encodes a single protein with an open reading frame spanning almost the whole length of the segment (**Figure 4**).

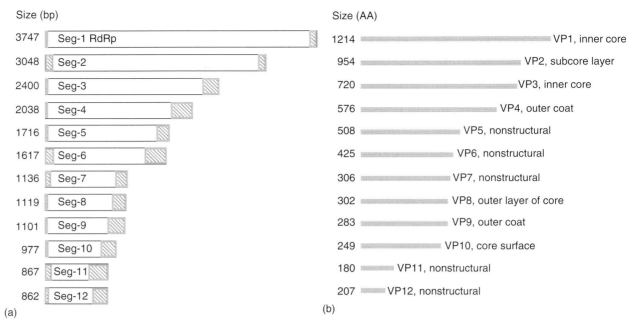

Figure 4 (a) Organization of the genome segments of BAV and (b) their putative encoded proteins. Shaded areas in (a): 5' and 3' NCRs. See www.iah.bbsrc.ac.uk/dsRNA_virus_proteins/Bannavirus-Proteins.htm.

Table 2 Correspondence between seadornavirus genes (relative to the type species Banna virus), putative functions, and protein copy numbers per intact virus particle

BAV	LNV (%AA identity with BAV)	KDV (%AA identity with BAV)	Function (location)	Protein copy number per particle
Seg-1 (VP1, Pol)	Seg-1 (VP1, Pol) (41)	Seg-1 (VP1, Pol) (42)	RNA-dependent RNA polymerase (core)	24
Seg-2 (VP2, T2)	Seg-2 (VP2, T2) (27)	Seg-2 (VP2, T2) (33)	$T=2$ protein, nucleotide binding (core)	120
Seg-3 (VP3, Cap)	Seg-3 (VP3, Cap) (37)	Seg-3 (VP3, Cap) (38)	Capping enzyme (core)	12
Seg-4 (VP4)	Seg-4 (VP4) (32)	Seg-4 (VP4) (32)	(Outer coat)	~330
Seg-5 (VP5)	Seg-6 (VP6) (26)	Seg-6 (VP6) (27)	(Nonstructural)	0
Seg-6 (VP6)	Seg-5 (VP5) (26)	Seg-5 (VP5) (26)	NTPase (nonstructural)	0
Seg-7 (VP7)	Seg-7 (VP7) (27)	Seg-7 (VP7) (29)	Protein kinase (nonstructural)	0
Seg-8 (VP8, T13)	Seg-8 (VP8, T13) (23)	Seg-9 (VP9, T13) (21)	(Core)	780
Seg-9 (VP9)	Seg-10 (VP10) (18)	Seg-11 (VP11) (23)	Cell attachment (outer coat)	~310
Seg-10 (VP10)	Seg-9 (VP9) (24)	Seg-10 (VP10) (24)	(Core)	~260
Seg-11 (VP11)	Seg-12 (VP12) (37)	Seg-12 (VP12) (34)	(Nonstructural)	0
Seg-12 (VP12)	Seg-11 (VP12) (35)	Seg-8 (VP8) (26)	dsRNA-binding (nonstructural)	0

BAV, Banna virus, KDV, Kadipiro virus, and LNV, Liao ning virus. The copy number was determined by radiolabeling of the proteins using ^{35}S methionine. The encoded putative proteins are indicated between brackets followed by the nomenclature that is used in the *Eighth Report of the ICTV* and presented on the website www.iah.bbsrc.ac.uk/dsRNA_virus_proteins/protein-comparison.htm.

Antigenic and Phylogenetic Relationships between Seadornaviruses

The earliest data that was generated concerning the relationships between different seadornaviruses were obtained by RNA cross-hybridization analyses, which identified BAV and KDV as members of distinct genogroups. Antigenic relationships between different seadornaviruses were also investigated using polyclonal mouse immune-sera. BAV from southern China and Indonesia, LNV from northeast of China, and KDV from Indonesia, which are classified as distinct species, show no cross-reaction in neutralization tests. Subsequent comparisons of nucleotide and amino acid sequences confirmed that BAV, KDV, and LNV represent three distinct virus species, with AA identities between homologous proteins of 24–42% (**Table 2**). A neighbor-joining tree based on an alignment of the most highly conserved seadornavirus protein, VP1 (the viral polymerase), shows BAV, KDV, and LNV as three distinct phylogenetic groups, confirming their status as distinct species (**Figure 5(a)**).

Antigenic variation between different BAV strains was also investigated using immune-sera, and identified two distinct serotypes, that did not cross-neutralize. These have been named as BAV serotypes A and B. A seroneutralization epitope was identified on the outer coat protein VP9. Sequence analysis of Seg-9 showed only 40% amino acid identity between BAV serotypes A and B, identifying two genotypes: genotype A (represented by isolates BAV-Ch (China) and BAV-In6423 (Indonesia)); and genotype B (represented by isolates BAV-In6969 and BAV-In7043 (Indonesia)) (which match with virus serotypes A and B).

Seg-7, which codes for nonstructural protein VP7, also shows a significant variation, between the genotypes/ serotypes, with only 70% amino-acid-sequence identity between types A and B. The amino-acid-sequence identity (VP1–VP12) between different BAV isolates within a single serotype ranged from 83% to 100%, while sequence identity between different serotypes ranged from 73% to 95% (excluding VP9). This level of genetic divergence is comparable to that observed between serotypes in other insect-transmitted arboviruses of family *Reoviridae*, in particular the orbiviruses. The two identified serotypes of LNV have nucleotide sequence identities ranging between 81% and 90%, while amino acid identity ranging between 80% and 96%. The lowest AA identity (80%) was found in the proteins of the outer capsid VP10 (the homolog of BAV VP9), which defines virus serotype in the seadornaviruses. A neighbor-joining tree, comparing the sequences of cell-attachment outer capsid proteins of

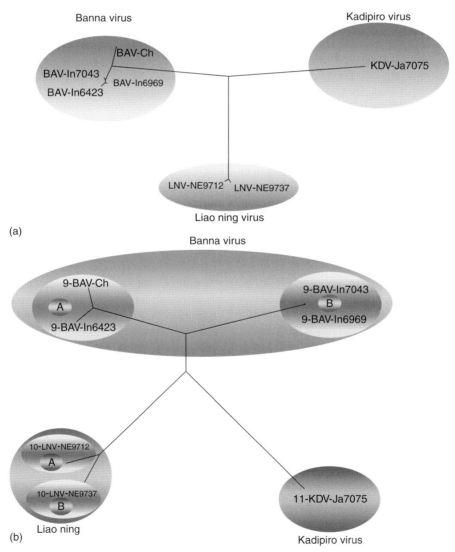

Figure 5 Phylogenetic tree for seadornaviruses based on (a) the polymerase gene and (b) on the cell-attachment (and seroneutralization) protein. The phylogenetic comparison based on the cell-attachment (and seroneutralization) protein distinguishes the the two genotypes (serotypes) of BAV and also those of LNV.

BAV (VP9), KDV (VP11), and LNV (VP10) clearly identifies the two serotypes of BAV (A and B) and distinguishes the A and B serotypes of LNV (**Figure 5(b)**).

All 12 segments of BAV, KDV, and LNV have conserved sequences, located at their 3′ and 5′ termini. The motifs 5′-GUAU^A/$_U$^{A}/$_U$A^A/$_U$^{A}/$_U$^{A}/$_U$-3′ and 5′-C^C/$_U$GAC-3′ (+ve strand) were found in the 5′ noncoding region (NCR) and the 3′ NCR of BAV, respectively, while in KDV, these motifs were 5′-GUAGAA^A/$_U$^{A}/$_U$^{A}/$_U$^{A}/$_U$-3′ and 5′-^{C}/$_U$GAC-3′ (+ve strand) and in LNV they were 5′-GUUAU^A/$_U$^{A}/$_U$^{A}/$_U$-3′ and 5′-^{C}/$_U$^{C}/$_U$GAC-3′ (+ve strand), respectively. For the BAV and KDV viruses, the 5′ and 3′ terminal trinucleotides of all segments are inverted complements. LNV has a difference in position 3 of the 5′ ends and therefore only the 5′ and 3′ terminal dinucleotides of all segments are inverted complements.

The genetic relation between BAV, LNV, and KDV is reflected in their morphological characteristics. Electron microscopic analysis showed that the three viruses are morphologically identical.

Within the family *Reoviridae*, the most conserved gene between different genera encodes the RNA-dependent RNA polymerase (RdRp). Values for amino acid identity of lower than 30% can be used to distinguish members of distinct genera. A tree constructed from the alignment of polymerase sequences of representative members of the family *Reoviridae* (**Table 3**) is shown in **Figure 6**. Calculations of amino-acid-identity values based on this alignment show that the coltivirus polymerases exhibit a maximum of 15% identity with those of the seadornaviruses, confirming the status of *Coltivirus* and *Seadornavirus* as a distinct genera. In contrast amino acid identities as high as 28% were detected between the seadornaviruses and the rotaviruses (members of a distinct genus of 11-segmented viruses also within the family *Reoviridae*).

Evolutionary Relationships between BAV and Rotaviruses

A sequence comparison of the structural proteins of BAV (VP1, VP2, VP3, VP4, VP8, VP9, and VP10) to those of other members of the *Reoviridae*, have shown similarities between VP9 and VP10 of BAV and the VP8* and VP5* subunits of the outer coat protein VP4 of rotavirus A, with amino acid identities of 21% and 26%, respectively. This suggests that an evolutionary jump has occurred between the two genera, involving the expression of two BAV proteins (VP9 and VP10) from two separate genome segments rather than the generation of VP8* and VP5* by the proteolytic cleavage of a single gene product (as seen in the cleavage of rotavirus VP4). VP3 of BAV, which is the guanylyltransferase of the virus also exhibits significant identity (21%, AA identity) with the VP3 (guanylyltransferase) of rotavirus.

Structural Features and Relation to Rotaviruses

Electron micrographs of purified, intact BAV particles show striking similarities to those of the rotaviruses. Numerous protein spikes were observed on the surface of BAV, reminiscent of those of rotaviruses. This morphological resemblance has not previously been reported between rotaviruses and other members of family *Reoviridae*. The evolutionary relationship between the seadornaviruses and rotaviruses was further confirmed when the atomic structure of BAV VP9 was determined by X-ray crystallography. VP9 is a trimeric molecule that is held together by an N-terminal helical bundle (**Figure 7**). The N-terminal tail of the BAV VP9 monomer is reminiscent of the coiled-coils structures of the HIV gp41 protein, and the carboxy terminal of the VP5* of rotavirus (**Figure 7**). In contrast the monomer of BAV VP9 has a head domain made mainly of β-sheets (**Figure 7**), which shows significant structural similarities to rotavirus VP8*. However, VP8* does have a sialic acid-binding domain, which is absent from the head domain of the VP9 of BAV. This might explain why rotavirus infectivity can be decreased when cells are pretreated with sialidases although BAV infectivity, to C6/36 mosquito cells, is not altered by a similar treatment.

Functional Studies

Expressed BAV VP9 was used in competition assays with intact virus particles, in an attempt to decrease their infectivity, as demonstrated with the outer capsid proteins of some other reoviruses (e.g., the avian orthoreoviruses). Surprisingly, however, pretreatment of cells with VP9 increased the infectivity of BAV by 10–100 times. VP9 is thought to act as a membrane fusion protein and the N-terminal coiled-coils may be responsible for such an activity. This may explain why soluble trimeric VP9 can increase viral infectivity. VP9 trimers not only bind receptor but also initiate endocytosis, perhaps by receptor oligomerization. The initiation of endocytosis facilitates penetration by virus particles present at or near the cell surface and thereby increases their infectivity.

BAV VP9 also carries the virus seroneutralization epitopes. Anti-VP9 antibodies are highly neutralizing and define the two serotypes A and B of BAV. Animals immunized with VP9 failed to replicate the virus when challenged with the homologous type. Animals immunized with VP9 of the heterologous serotype showed unaltered virus replication and viremia. Similar results were obtained in cell cultures. VP9 could therefore be used as a vaccine subunit for immunization. Sequence analysis of BAV VP3 suggests that it is the viral capping enzyme (CaP).

This has been confirmed experimentally by incubating virus particles with α-^{32}P GTP. VP3 was covalently

Table 3 Sequences of RdRp used in phylogenetic analysis (**Figure 6**) of various members of family *Reoviridae*

Species	Isolate	Abbreviation	Accession number
Genus *Seadornavirus* (12 segments)			
Banna virus	Ch	BAV-Ch	AF168005
Kadipiro virus	Java-7075	KDV-Ja7075	AF133429
Liao ning viurus	LNV-NE9712	LNV-NE9712	AY701339
Genus *Coltivirus* (12 segments)			
Colorado tick fever virus	Florio	CTFV-Fl	AF134529
Eyach virus	Fr578	EYAV-Fr578	AF282467
Genus *Orthoreovirus* (10 segments)			
Mammalian orthoreovirus	Lang strain	MRV-1	M24734
	Jones strain	MRV-2	M31057
	Dearing strain	MRV-3	M31058
Genus *Orbivirus* (10 segments)			
African horse sickness virus	Serotype 9	AHSV-9	U94887
Bluetongue virus	Serotype 2	BTV-2	L20508
	Serotype 10	BTV-10	X12819
	Serotype 11	BTV-11	L20445
	Serotype 13	BTV-13	L20446
	Serotype 17	BTV-17	L20447
Palyam virus	Chuzan	CHUV	Baa76549
St. Croix river virus	SCRV	SCRV	AF133431
Genus *Rotavirus* (11 segments)			
Rotavirus A	Bovine strain UK	BoRV-A/UK	X55444
	Simian strain SA11	SiRV-A/SA11	AF015955
Rotavirus B	Human/murine strain IDIR	Hu/MuRV-B/IDIR	M97203
Rotavirus C	Porcine Cowden strain	PoRV-C/Co	M74216
Genus *Aquareovirus* (11 segments)			
Golden shiner reovirus	GSRV	GSRV	AF403399
Grass Carp reovirus	GCRV-873	GCRV	AF260511
Chum salmon reovirus	CSRV	CSRV	AF418295
Striped bass reovirus	SBRV	SBRV	AF450318
Genus *Fijivirus* (10 segments)			
Nilaparvata lugens reovirus	Izumo strain	NLRV-Iz	D49693
Genus *Phytoreovirus* (10 segments)			
Rice dwarf virus	Isolate China	RDV-Ch	U73201
	Isolate H	RDV-H	D10222
	Isolate A	RDV-A	D90198
Genus *Oryzavirus* (10 segments)			
Rice ragged stunt virus	Thai strain	RRSV-Th	U66714
Genus *Cypovirus* (10 segments)			
Bombyx mori cytoplasmic polyhedrosis virus 1	Strain I	BmCPV-1	AF323782
Dendrlymus punctatus cytoplasmic polyhedrosis 1	DsCPV-1	DsCPV-1	AAN46860
Lymantria dispar cytoplasmic polyhedrosis 14	LdCPV-14	LdCPV-114	AAK73087
Genus *Mycoreovirus* (11 or 12 segments)			
Rosellinia anti-rot virus	W370	RaRV	AB102674
Cryphonectria parasitica reovirus	9B21	CPRV	AY277888
Genus Mimoreovirus (11 segments)			
Micromonas pusilla reovirus	MPRV	MPRV	DQ126102
Genus Dinovernavirus (9 segments)			
Aedes pseudoscutellaris reovirus	ApRV	ApRV	DQ087277
Genus Cardoreovirus (12 segments)			
Eriocheir sinensis reovirus	Isolate 905	EsRV	AY542965

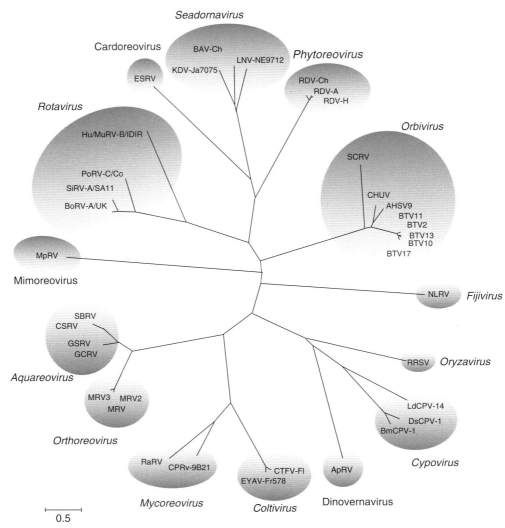

Figure 6 Phylogenetic relations among members of the *Reoviridae* based on the sequences of their putative RNA-dependent RNA polymerases (RdRps). Neighbor-joining phylogenetic tree built with available polymerase sequences (using the Poisson correction or gamma-distribution algorithms) for representative members of 12 recognized and 3 proposed genera of family *Reoviridae*. The abbreviations and accession numbers are those provided in **Table 3**.

labeled with α-^{32}P derived from GTP. The reaction was found to be dependent on the presence of divalent ions such as magnesium. The recombinant VP3 also covalently bound α-^{32}P GTP and exhibited an inorganic pyrophosphatase activity, possibly as a detoxifying mechanism which removes the inorganic pyrophosphate accumulated during RNA and Cap structure synthesis by the virus-core-associated enzymes. Similar properties were described for the capping enzyme VP4 of Bluetongue virus (genus *Orbivirus*).

Clinical Features and Diagnostic Assay for Seadornaviruses

As mentioned earlier, the only seadornavirus that has been isolated from humans and associated with human disease (to date) is BAV. Humans infected with BAV develop flu-like manifestation, myalgia, arthralgia, fever, and encephalitis. Reports of children born to infected mothers (as revealed by serological assays) show various manifestations including cerebral palsy, hemiplegia, delay of development, and viral encephalitis.

A diagnostic serological assay was developed for different serotype of BAV, based on outer coat protein VP9 (which is responsible for cell-attachment and sero-neutralization). Molecular diagnostic assays have also been developed for BAV and KDV. These are RT-PCR-based assays, which have been validated using an infected murine model in which viral RNA could be detected in blood as early as 3 days post infection. These RT-PCR assays can also be used to distinguish genotypes A and B of BAV, based on distinct amplicons of different lengths, that are obtained by specific primers at different locations in

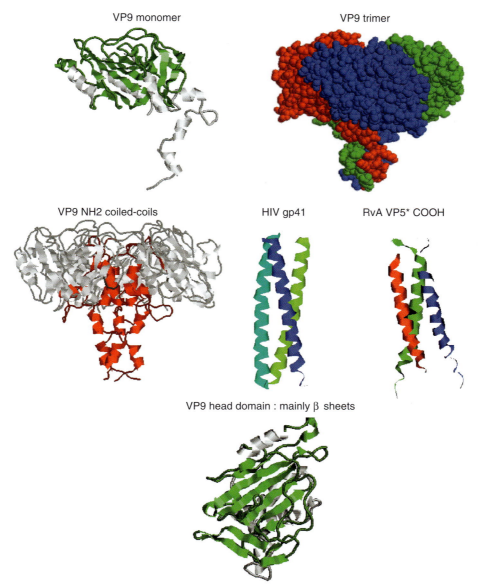

Figure 7 The structure of the VP9 outer capsid protein of Banna virus determined by X-ray crystallography at a resolution of 2.6 angstrom. The NH2 terminal of the BAV VP9 is organized in the form of coiled-coils that are similar to those found in fusion active proteins such as HIV gp41 and VP5* of rotavirus. RvA, Rotavirus A.

Seg-9. A quantitative PCR assay has also been developed for BAV based on the sequence of Seg-10. A standard RT-PCR assay was recently developed for LNV based on the sequence of Seg-12, which can be used to detect viral RNA in infected mouse blood.

Severity of Infection in Animal Model and Diversification of Genome Sequence after Replication in Permissive Animals

Banna, Kadipiro, and Liao ning viruses all replicate in mice after intraperitoneal injection. The viral genome can be detected in mouse blood 3 days until 5–7 days post injection. Clearance of the viremia is accompanied by the appearance of antibodies. When animals were injected on a second occasion with the same virus, BAV and KDV failed to replicate, demonstrating the immune status of the mice. However, LNV replicated in the immunized mice causing death with a severe hemorrhagic syndrome. An analysis of the PCR amplicons from the infected mice blood showed very few changes in the sequences of the BAV and KDV genomes. However, there was considerable diversification in the genome of LNV, as demonstrated by analyses of the sequence of Seg-12, leading to formation of a diverse quasispecies. This might explain why LNV can continue to replicate in the immunized mice, as diversification of the sequence might help the virus to 'escape' the immune system.

Treatment and Immunity

There is no specific treatment for BAV infection. Patients infected with BAV have shown a fourfold rise in the anti-BAV antibody titers in paired sera tested by ELISA, showing that an immune response is developed to the virus infection. Mice experimentally infected with BAV developed viremia. The clearance of the virus from the blood circulation occurred concomitantly with the appearance of anti-BAV antibodies. Based on the mouse model, immunization with a recombinant expressed VP9 is likely to provide a significant level of protection against an initial infection with the homologous virus type. Immunization with VP9 could therefore provide a significant level of protection in high-risk areas and it might form the basis for an effective subunit vaccine.

See also: Human T-Cell Leukemia Viruses: Human Disease; Reoviruses: General Features.

Further Reading

Attoui H, Billoir F, Biagini P, et al. (2000) Complete sequence determination and genetic analysis of Banna virus and Kadipiro virus: Proposal for assignment to a new genus (*Seadornavirus*) within the family *Reoviridae*. *Journal of General Virology* 81: 1507–1515.

Attoui H, Charrel R, Billoir F, et al. (1998) Comparative sequence analysis of American, European and Asian isolates of viruses in the genus *Coltivirus*. *Journal of General Virology* 79: 2481–2489.

Attoui H, de Lamballerie X, and Mertens PPC (2005) *Coltivirus, Reoviridae*. In: Fauquet CM, Mayo MA, Maniloff J, Desselberger U, and Ball LA (eds.) *Virus Taxonomy: Eighth Report of the International Committee on Taxonomy of Viruses*, pp. 497–503. San Diego, CA: Elsevier Academic Press.

Attoui H, de Lamballerie X, and Mertens PPC (2005) *Seadornavirus, Reoviridae*. In: Fauquet CM, Mayo MA, Maniloff J, Desselberger U, and Ball LA (eds.) *Virus Taxonomy: Eighth Report of the International Committee on Taxonomy of Viruses*, pp. 504–510. San Diego, CA: Elsevier Academic Press.

Billoir F, Attoui H, Simon S, et al. (1999) Molecular diagnosis of group B coltiviruses infections. *Journal of Virological Methods* 81: 39–45.

Brown SE, Gorman M, Tesh RB, and Knudson DL (1993) Coltiviruses isolated from mosquitoes collected in Indonesia. *Virology* 196: 363–367.

Chen BQ and Tao SJ (1996) Arbovirus survey in China in recent ten years. *Chinese Medical Journal (Engl)* 109: 13–15.

Mertens PPC, Attoui H, and Bamford DH (eds.) (2002) Identification of comparable proteins of the dsRNA viruses. In: *The RNAs and Proteins of dsRNA Viruses*. www.iah.bbsrc.ac.uk/dsRNA_virus_proteins/protein-comparison.htm (accessed May 2007).

Mertens PPC, Duncan R, Attoui H, and Dermody TS (2005) *Reoviridae*. In: Fauquet CM, Mayo MA, Maniloff J, Desselberger U, and Ball LA (eds.) *Virus Taxonomy: Eighth Report of the International Committee on Taxonomy of Viruses*, pp. 447–454. San Diego, CA: Elsevier Academic Press.

Mohd Jaafar F, Attoui H, Bahar MW, et al. (2005) The structure and function of the outer coat protein VP9 of Banna virus. *Structure* 13: 17–29.

Mohd Jaafar F, Attoui H, Gallian P, et al. (2004) Recombinant VP9-based enzyme-linked immunosorbent assay for detection of immunoglobulin G antibodies to Banna virus (genus *Seadornavirus*). *Journal of Virological Methods* 116: 55–61.

Mohd Jaafar F, Attoui H, Mertens PPC, et al. (2005) Structural organisation of a human encephalitic isolate of Banna virus (genus *Seadornavirus*, family *Reoviridae*). *Journal of General Virology* 86: 1141–1146.

Tao SJ and Chen BQ (2005) Studies of coltivirus in China. *Chinese Medical Journal (Engl)* 118: 581–586.

Relevant Website

http://www.iah.bbsrc.ac.uk – The RNAs and Proteins of dsRNA Viruses.

Sequiviruses

I-R Choi, International Rice Research Institute, Los Baños, The Philippines

© 2008 Elsevier Ltd. All rights reserved.

Glossary

Helper component A viral-encoded protein that mediates the transmission of viruses by vectors.

Semipersistent transmission A mode of vector-mediated transmission of plant viruses in which viruses are usually acquired within few minutes, and retained up to several days by vectors. Viruses do not multiply in vectors and often require a helper component during semipersistent transmission.

Introduction

Sequiviridae is a relatively newly recognized family of viruses, although the viruses classified into this family had been reported already in the early 1970s. Viruses belonging to the family *Sequiviridae* were often referred to as 'plant picorna-like viruses' due to their similarities to picornaviruses in virion morphology and genome structure. They have a monopartite single-stranded RNA (ssRNA) genome encapsidated in isometric particles. They infect plants and are usually transmitted by insect vectors in nature. Special attention has been given to two of the members, maize chlorotic dwarf virus (MCDV) and rice tungro spherical virus (RTSV), since they are involved in major diseases of important cereal crops. Chlorotic dwarf disease is considered to be the second most important viral disease of maize in the USA, and tungro disease is the most serious threat to rice production in South and Southeast Asia. One of the unique biological features found in these viruses is that most of them act as or require a helper virus for transmission by insect vectors. The name of the family comes from the Latin word *sequi* which means to follow or accompany, in reference to the dependent insect transmission of parsnip yellow fleck virus (PYFV), one of the type members in this family.

Taxonomy

The family *Sequiviridae* is divided into two genera, *Sequivirus* and *Waikavirus*. At present the genus *Sequivirus* has two species *Parsnip yellow fleck virus* (type species) and *Dandelion yellow mosaic virus*. Three species, *Rice tungro spherical virus* (type species), *Maize chlorotic dwarf virus*, and *Anthriscus yellows virus* are classified as members of the genus *Waikavirus* (**Table 1**). In addition, lettuce mottle virus is considered as a tentative member of the genus *Sequivirus*.

Viruses belonging to the genus *Sequivirus* infect mesophyll and epidermal cells and are able to be transmitted by mechanical inoculation and insect vectors, while those belonging to the genus *Waikavirus* are usually limited in phloem tissue and transmitted only by insect vectors. Sequiviruses are dependent on a helper virus for their transmission by insect vectors. Waikaviruses are independently transmitted by insect vectors and presumably encode a helper component in their genomes. The viruses in the family *Sequiviridae* are primarily classified on the basis of their biological and physical characteristics, but conspicuous differences are also found in their genome features. The genomes of MCDV and RTSV are approximately 12 kb in length and polyadenylated at the 3′ end, whereas that of PYFV is about 10 kb and devoid of the 3′ poly(A) region.

Properties of Virions

Viruses belonging to *Sequiviridae* have nonenveloped isometric particles of approximately 30 nm in diameter (**Figure 1**). The sedimentation coefficients of the virions are 153–159S for sequiviruses and 175–183S for waikaviruses. The buoyant density of PYFV virions in CsCl is $1.49\,g\,ml^{-1}$, and that of waikaviruses is $1.51–1.55\,g\,ml^{-1}$. Genome sequences and immunodetection of the viruses belonging to the family *Sequiviridae* indicate that the virions consist of three capsid proteins (CPs). The sizes of CPs range from 22 to 35 kDa, depending on virus species and isolates. The virion of RTSV (Philippine-type strain A) consists of three proteins of 22.5 (CP1), 22 (CP2), and 33

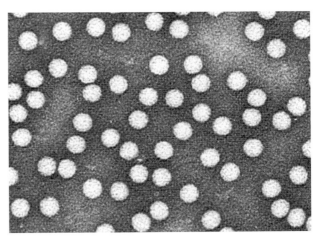

Figure 1 Virions of RTSV. The estimated size of particles is approximately 30 nm.

Table 1 Virus members in the family *Sequiviridae*

Genus/virus	Genome size (kb)	Geographical distribution	Major natural host	Transmission vector
Sequivirus				
Parsnip yellow fleck virus	9.9	Europe	Parsnip, hogweed, cow parsley	Aphids (*Cavariella aegopodii, C. pastinacae*)
Dandelion yellow mosaic virus	10.0[a]	Europe	Lettuce, dandelion	Aphids (*Acyrthosiphon solani, Myzus ornatus, M. ascalonicus, M. persicae*)
Waikavirus				
Rice tungro spherical virus	12.2	Asia	Rice, *Oryza* species	Green leafhoppers (*Nephotettix virescens* and four other species)
Maize chlorotic dwarf virus	11.8	USA	Maize, Johnson grass	Deltocephaline leafhopper (*Graminella nigrifrons*)
Anthriscus yellow virus	10.6[a]	Eurasia, UK	Cow parsley	Aphid (*Cavariella aegopodii*)

[a]Estimated size (nucleotide sequence not determined).

(CP3) kDa. The predicted molecular masses for the CP of MCDV (Tennessee (TN) isolate) are 22, 23, and 31 kDa, and those of PYFV (P-121 isolate) are 22.5, 26, and 31 kDa.

Genome Structure

The genomes of viruses belonging to the family *Sequiviridae* are positive-sense, monopartite ssRNA. The length of genomes varies significantly, ranging from approximately 10 kb for sequiviruses to 12 kb for waikaviruses. The genomes contain a large open reading frame (ORF) encoding a polyprotein presumably proteolytically processed by viral-encoded protease(s) during translation (**Figure 2**). The large ORF in the genome of PYFV (P-121 isolate) putatively encodes a polyprotein consisting of 3027 amino acid residues with the predicted molecular mass of about 336 kDa. The large ORF in the genomes of waikaviruses encodes a polyprotein of approximately 400 kDa, which has about 3440–3470 amino acid residues. In addition to the ORF for the polyprotein, short ORFs were also identified near the 3′ and the 5′ ends of MCDV and RTSV genomes. Two RNA species which seemingly correspond to subgenomic transcripts from the short ORFs locating near the 5′ end of the RTSV genome (strain A) were detected from infected plants. However, the lengths and locations of the short ORF in the genomes of waikaviruses vary considerably among viruses and isolates, and the presence of products translated from these ORFs was not confirmed in plants. The 5′-untranslated region (UTR) in the genome of PYFV (P-121 isolate) is about 0.28 kb in length, while those of waikaviruses are longer, about 0.43 kb in MCDV and 0.52 kb in RTSV. The 5′-UTR in the genomes of MCDV and RTSV has several AUG sequences upstream of the putative polyprotein start site, and appeared to form extensive secondary structures. As observed in the genome of picornaviruses, such secondary structures may serve as the internal ribosomal entry site to avoid the interference on translation by upstream AUG sequences. No stable secondary structures are recognized in the 5′-UTR of the PYFV genome; however, several short stretches of pyrimidines (UCUCUY) are present in the region. The 3′-UTR in the genomes of waikaviruses are polyadenylated and unusually long. The length of 3′-UTR in the genomes of MCDV and RTSV are approximately 1.0 and 1.2 kb, respectively, provided that the short ORFs near the 3′ end of genomes are not translated. The 3′-UTR in the genome of PYFV is approximately 0.5 kb in length. Unlike the genomes of waikaviruses, that of PYFV is not polyadenylated, but a part of 3′-UTR is likely to form a stem–loop structure which is similar to that found in the genomes of flaviviruses.

Properties of Viral Proteins

Based on the protein sequences predicted from the large ORF, and the actual sequences of the N-termini of CPs, it was predicted that the polyproteins of viruses belonging to the family *Sequiviridae* are cleaved into at least seven proteins through proteolytic maturation. It appears that the N-terminal half of the polyprotein contains regions for a leader protein and three CPs. Protein regions showing similarities to helicase, protease, and RNA-dependent RNA polymerase (RdRp) of picornaviruses and comoviruses are recognized in the central to carboxyl (C)-terminal regions of the polyproteins. The arrangement of functional domains in the polyproteins of the viruses belonging to *Sequiviridae* shows significant similarity to that of picornaviruses, indicating that the strategies of genome replication and expression for the viruses of *Sequiviridae* might be analogous to those of picornaviruses.

Comparative sequence analysis indicated that the polyproteins of the viruses belonging to the family *Sequiviridae*

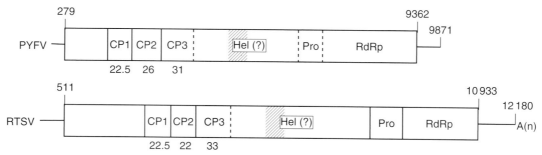

Figure 2 Genome structure of PYFV (P-121 isolate) and RTSV (strain A). The ORFs for polyprotein are indicated with rectangles. Horizontal lines at both ends of the ORFs represent the 5′- and 3′-UTR of the genome. Solid vertical lines dividing the ORF correspond to the positions of proteolytic cleavage sites in the polyproteins when translated, while the positions corresponding to predicted cleavage sites are indicated with dotted lines. Positions of translation start and stop codons for the polyproteins and the length of the entire genomes are indicated with the numbers above the genomes. Numbers below the regions encoding CP1–CP3 are the predicted molecular weight (kDa) for the respective CP. Shadowed areas represent the approximate region encoding NTP-binding domain. The regions in the ORF are shown with Hel (?) for putative NTPase/helicase, Pro for 3C-like cysteine proteinase, and RdRp for RNA-dependent RNA polymerase. A(n) indicates the polyadenylation at the 3′ end of the RTSV genome.

contain a protease region similar to the 3C cysteine protease of picornaviruses and other viral proteases resembling the 3C protease such as the 24 kDa protease of cowpea mosaic virus (CPMV) and the NIa protease of tobacco etch virus. For example, considerable similarity was found between the region delimited by the amino acid residues 2643 and 2853 in the polyprotein of RTSV (strain A) and a region in the 24 kDa protease of CPMV. Immunodetection using an antiserum raised against a protein containing the putative protease region of RTSV indicated that the size of mature protease in infected plants is approximately 35 kDa. The sequence context of cleavage site for the 3C cysteine protease and the results from *in vitro* translation from partial genomic RNA templates defined the region of the protease in the RTSV polyprotein to be from amino acid residues 2527–2852. The RTSV protease presumably acts *in cis* to cleave itself from the adjacent protein regions, but *trans*-cleavage by the RTSV protease was also observed to occur between the protease and the putative helicase regions *in vitro*. Based on the sequence alignment among the 3C-type proteases, amino acid residues such as His^{2680}, Glu^{2717}, Cys^{2811}, and His^{2830} in the RTSV polyprotein were predicted to constitute the catalytic triad or substrate-binding pocket of the cysteine protease. Substitutions of these amino acid residues abolished or drastically reduced the proteolytic activity, substantiating their critical roles in the cysteine protease. Conserved amino acid residues which may constitute the catalytic triad were also identified in the protease regions of MCDV and PYFV, although the involvement of the individual residues in the protease activity has not been experimentally demonstrated.

The C-terminal region in the polyproteins of viruses belonging to the family *Sequiviridae* shows extensive similarity in amino acid sequence to RdRp of picornaviruses and comoviruses. The RdRp region of RTSV (strain A) was defined in the region between amino acid residues 2853 and 3473 with the conserved YGDD motif at amino acid residues 3270–3273. The RdRp region of MCDV (isolate TN) has the conserved YGDD motif at amino acid residues 3238–3241. In addition, motifs such as DYSXFDG (amino acid residues 3129–3135 in the MCDV polyprotein) and $PSGX_3TX_3NS$ (amino acid residues 3189–3200) were identified to be conserved among the RdRp regions of MCDV, CPMV, and tomato black ring virus (TBRV). Multiple alignment of the RdRp region of PYFV (isolate P-121) with those of CPMV, TBRV, and poliovirus showed that they share several conserved motifs including sequences YGDD (amino acid residues 2629–2632 in the PYFV polyprotein), $PSGX_3TX_3NS$ (amino acid residues 2580–2591), and FLKR (amino acid residues 2684–2687).

The central region flanked by the CP and the protease regions in the polyproteins of viruses belonging to the family *Sequiviridae* contains sequence motifs GX_4GKS and DD, which are conserved among proteins with NTP-binding domain. The NTP-binding domain of the central polyprotein regions shows extensive similarity to the corresponding domains in the 58 kDa protein of CPMV and the 2C proteins of picornaviruses. The 58 kDa protein and the 2C protein presumably function as a nucleoside triphosphatase (NTPase)/helicase required for the initiation of negative-strand RNA synthesis. Antibodies specific to protein segments from the central region of the MCDV polyprotein reacted with three protein species in extracts from infected plants. The sizes of the proteins detected were smaller than predicted from the intact central region, indicating that the protein in the central region might be processed into smaller proteins.

It appeared that the polyproteins of viruses belonging to *Sequiviridae* have a region for putative leader protein(s) at the N-terminal region. Although the predicted sizes of the putative leader proteins are about 40 kDa in PYFV, and 70–78 kDa in waikaviruses, the sizes of proteins detected with the antisera specific to the putative leader proteins are significantly smaller than predicted. An antiserum specific to the putative RTSV leader protein detected a protein of about 32 kDa in extracts from infected plants. Meanwhile, proteins with apparent sizes of 50, 35, and 25 kDa in the extracts from plants infected with MCDV (severe (S) strain) reacted with the antibody specific to the putative MCDV leader protein. These results suggest that the putative leader proteins of waikaviruses may be processed post-translationally in plants. The function associated with the leader protein region is still unclear. However, such leader protein found in the polyprotein of aphthoviruses in the family *Picornaviridae* has a protease activity which cleaves itself autocatalytically from the polyprotein. It was proposed that the 35 kDa protein of MCDV is generated from the leader protein region through autoproteolysis at the putative cleavage site ALVRLFHGSAE (amino acid residues 150–160), while the 25 kDa protein results from the cleavage at Q^{445}/S^{446} and Q^{686}/S^{687} by a cysteine protease. The 25 kDa protein may function as the helper component in the insect transmission of MCDV since the protein was observed to accumulate in vector insects after feeding on plants infected with MCDV. This observation is consistent with the result from a serological blocking experiment showing that the helper component is not the virion or CP.

Three consecutive regions of CPs are present in the N-terminal half of the polyproteins of the viruses belonging to the family *Sequiviridae*. The proteolytic cleavage sites for the RTSV CPs determined by N-terminal amino acid sequencing were mapped to Q^{644}/A^{645}, Q^{852}/S^{853}, and Q^{1055}/D^{1056} for the junctions between leader protein/CP1, CP1/CP2, and CP2/CP3, respectively. The context of these cleavage sites suggests that the CPs are processed *in trans* by a cysteine protease. However, the cleavage at these sites with the RTSV protease was not detected *in vitro*. The difference in the reactivity among the CPs of

RTSV with the antibody raised against the virus particles indicated that the 33 kDa CP (CP3) is the major antigenic determinant on the surface of particles, although the structural details of virus particles have not been elucidated yet. The size of CP3 of RTSV (Philippine isolate) in the crude extract from infected plants detected by the antibody specific to CP3 appeared to be 40–42 kDa, markedly larger than that detected from purified virus preparation or that expected from the genome sequence. It is likely that the size of CP3 in infected cells is larger due to post-translational modification, but the modified moiety appeared to be cleaved off probably by the treatment with cellulolytic enzymes during virion purification.

Phylogenetic Relationships

Comparison of amino acid sequences revealed that the sequence similarities of the MCDV (isolate TN) polyprotein to those of RTSV (strain A) and PYFV (isolate P-121) were 51% and 35%, respectively. Overall similarity in genome organization and the presence of several conserved motifs in the polyproteins indicate that the viruses of the family *Sequiviridae* are closely related to picornaviruses and comoviruses. The NTP-binding domain appeared to be the most conserved region in the polyproteins among the viruses belonging to the family *Sequiviridae*, picornaviruses and comoviruses. The NTP-binding domain in the polyprotein of MCDV (TN isolate) showed significant sequence similarity to those of viruses such as RTSV (strain A, 79%), PYFV (isolate P-121, 55%), CPMV (46%), hepatitis A virus (HAV, 47%), and poliovirus (45%). The regions of RdRp also show significant similarities to one another among members of the family *Sequiviridae*, picornaviruses and comoviruses. For instance, the sequence similarities in the RdRp region of MCDV (TN isolate) to those of related viruses are 75% for RTSV, 50% for PYFV, 48% for CPMV, 44% for TBRV, and 40% for HAV. Phylogenetic analysis based on the sequences of NTP-binding domains indicates that PYFV and RTSV are not noticeably similar to each other compared to their relatedness to picornaviruses and comoviruses. However, phylogenetic analysis based on the regions of RdRp suggests that PYFV and RTSV are more closely related to each other than to picornaviruses and comoviruses. Unlike the region of RdRp and NTP-binding domain, the CPs of viruses belonging to the family *Sequiviridae* show no sequence similarities to those of comoviruses, and only limited similarities to those of some picornaviruses. The 26 kDa CP of PYFV and the 22.5 kDa CP of RTSV contain amino acid sequences resembling those in VP3 of encephalomyocarditis virus and human rhinovirus 14. The closer relatedness of nonstructural protein regions in members of the family *Sequiviridae* to those of comoviruses, and the similarity in the CP sequences among PYFV, RTSV, and some picornaviruses imply that the family *Sequiviridae* may fit taxonomically between *Comoviridae* and *Picornaviridae*.

Variation of Isolates and Strains

Isolates of PYFV are largely divided into two groups. One is the parsnip serotype which includes isolates from parsnip, celery, and hogweed, while those from carrot and cow parsley belong to the other group, the *Anthriscus* serotype. The two groups of isolates are distinguishable by reciprocal immunodiffusion tests with antisera raised against isolates belonging to either group. In addition to the difference in natural hosts, the artificial inoculation of test plants with the respective isolates showed evident difference in host ranges between the two groups of isolates, although minor differences in host range and symptoms on certain plants were also observed among the isolates within each serotype.

Examination of nucleotide sequences of the RTSV CP revealed broad genotypic variation among and within geographic isolates, although the relationship of genotypic variation with the pathogenicity is not understood yet. Isolates of RTSV from the Philippines and Malaysia show about 95% sequence similarity, while those from Bangladesh and India differ from the Philippine isolate by about 15%. The CP3 of the Indian isolate was distinguishable from that of the Philippine isolate in electrophoretic mobility and the response to cellulolytic enzyme. Genotypic survey on the CP sequences of RTSV field isolates collected from various sites in the Philippines and Indonesia indicated that a high degree of genetic diversity exists among the field isolates, and that infections with mixed genotypes in single sites are not uncommon. Phylogenetic analysis based on the CP sequences of the RTSV field isolate suggested that the clustering of genotypes found in the Philippines sites was significantly different from that found in the Indonesia sites, indicating geographic isolation of RTSV populations. Strain Vt6 of RTSV was found to possess enhanced virulence, showing infectivity to some rice cultivars which the type strain A may not be able to infect. Amino acid sequence of strain Vt6 was approximately 95% identical to that of strain A, with greater dissimilarity in the leader protein region and the putative small ORF found near the 3′ end.

Few isolates of MCDV with distinctive biological and genotypic characteristics have been reported. The S isolate of MCDV was observed to produce more pronounced symptoms than the type (T) isolate. The mild (M1) isolate usually exhibits mild symptoms by itself, but it develops severe symptoms by synergistically interacting with other MCDV isolates. The deduced amino acid sequences of isolates S and T show 99.5% identity, while that of isolate M1 has only 61% identity to that of isolate T. In fact, antisera raised against isolate T react strongly with isolate

S, but not with isolate M1. Isolate TN of MCDV is also significantly divergent from isolate T, showing only 60% of amino acid sequence identity. The low levels of amino acid sequence identity among the isolates of MCDV raised the possibility that they may represent distinct virus species.

Interactions between Viruses

Viruses belonging to the family *Sequiviridae* are often detected in plants infected with other viruses, for instance, cow parsley infected with PYFV and anthriscus yellows virus (AYV), and rice plants infected with rice tungro bacilliform tungrovirus (RTBV) and RTSV. Such mixed infections appear to be the outcome of the dependent insect transmission of one virus on the other. Incidences of mixed infection in lettuce with dandelion yellow mosaic virus and lettuce mosaic virus were also reported, but their relationships in aphid-mediated transmission are still unclear.

PYFV is transmitted by mechanical inoculation, but AYV is not. Both viruses are also transmitted by aphids in a semipersistent manner. However, the aphid-mediated transmission of PYFV is dependent on AYV. PYFV is transmitted by aphids from plants infected with both viruses or by aphids previously fed on plants infected with AYV. Aphids seem to retain PYFV and AYV for up to 4 days. Even though AYV cannot infect parsnip, PYFV can be transmitted to parsnip by aphids fed on other plants infected with AYV and PYFV. Therefore, AYV apparently acts only as the helper virus for aphids to acquire PYFV, and is not necessary for the infection process of PYFV.

Rice tungro disease is caused by the interaction between RTSV and RTBV. Both viruses are transmitted by green leafhoppers (GLHs) in a semipersistent manner. GLHs transmit RTSV and RTBV simultaneously or singly from source plants infected with both viruses. RTSV is independently transmitted by GLHs, but RTBV can be transmitted by GLHs which feed on plants infected with RTSV. Although GLHs retain RTSV for only 3–4 days, their ability to acquire and transmit RTBV may persist for 7 days. Neutralization of RTSV-viruliferous GLHs with anti-RTSV immunoglobulin markedly reduced the ability to transmit RTSV but still retained the ability to acquire and transmit RTBV. These observations indicate that RTSV is the helper virus for the GLH-mediated transmission of RTBV, but the helper function is associated with factors other than RTSV virions or CP. Rice plants infected with RTBV alone exhibit symptoms such as stunted growth, yellow to yellow-orange discoloration of leaves, and reduced tillering. The symptoms become more severe when plants are simultaneously infected with RTBV and RTSV, despite the fact that RTSV alone does not cause conspicuous symptoms except occasional slight stunted growth (**Figure 3**). Such synergistic effects of RTSV on

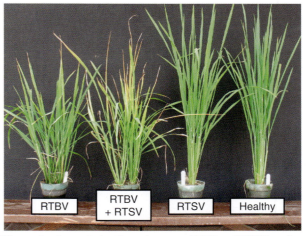

Figure 3 Synergistic effects between RTSV and RTBV on the symptom development in rice. Rice plants shown (left to right) were infected with RTBV alone (RTBV), both RTBV and RTSV (RTBV + RTSV), RTSV alone (RTSV), or not infected (healthy). Plants infected with both RTBV and RTSV were stunted and showed severe discoloration of leaves.

symptom development are also observed when RTSV co-infects plants with viruses such as rice grassy stunt virus and rice ragged stunt virus.

See also: Cereal Viruses: Maize/Corn; Cereal Viruses: Rice; Picornaviruses: Molecular Biology; Rice Tungro Disease; Vector Transmission of Plant Viruses.

Further Reading

Azzam O, Yambao MLM, Muhsin M, McNally KL, and Umadhay KML (2000) Genetic diversity of rice tungro spherical virus in tungro-endemic provinces of the Philippines and Indonesia. *Archives of Virology* 145: 1183–1197.

Chaouch-Hamada R, Redinbaugh MG, Gingery RE, Willie K, and Hogenhout SA (2004) Accumulation of maize chlorotic dwarf virus proteins in its plant host and leafhopper vector. *Virology* 325: 379–388.

Elnagar S and Murant AF (1976) Relations of the semi-persistent viruses, parsnip yellow fleck and anthriscus yellows, with their vector, *Cavariella aegopodii*. *Annals of Applied Biology* 84: 153–167.

Hemida SK and Murant AF (1989) Host ranges and serological properties of eight isolates of parsnip yellow fleck virus belonging to the two major serotypes. *Annals of Applied Biology* 114: 101–109.

Isogai M, Cabauatan PQ, Masuta C, Uyeda I, and Azzam O (2000) Complete nucleotide sequence of the rice tungro spherical virus genome of the highly virulent strain Vt6. *Virus Genes* 20: 79–85.

Reddick BB, Habera LF, and Law MD (1997) Nucleotide sequence and taxonomy of maize chlorotic dwarf virus within the family Sequiviridae. *Journal of General Virology* 78: 1165–1174.

Shen P, Kaniewska M, Smith C, and Beachy RN (1993) Nucleotide sequence and genomic organization of rice tungro spherical virus. *Virology* 193: 621–630.

Turnbull-Ross AD, Mayo MA, Reavy B, and Murant AF (1993) Sequence analysis of the parsnip yellow fleck virus polyprotein: Evidence of affinities with picornaviruses. *Journal of General Virology* 74: 555–561.

Severe Acute Respiratory Syndrome (SARS)

J S M Peiris and L L M Poon, The University of Hong Kong, Hong Kong, People's Republic of China

© 2008 Elsevier Ltd. All rights reserved.

Glossary

Bioavailability A measurement of the proportion of the orally administered dose of a therapeutically active drug that reaches the systemic circulation and is available at the site of pathology.

Coryza A runny nose.

Desquamation Shedding of epithelialium.

Lymphopenia Reduction of lymphocytes in the circulating blood below the normal range for age.

Myalgia Muscle pain.

Pathognomonic Characteristic and diagnostic of a particular disease.

PEGylated An adjective for describing molecules conjugated with polyethylene glycol (PEG).

Radiological abnormalities Atypical findings observed by medical imaging procedures (e.g., chest X-ray).

History

From November 2002 to January 2003, cases of an unusually severe atypical pneumonia were being observed in Guangdong Province, China. The disease was characterized by the lack of response to conventional antibiotic therapy and the occurrence of clusters of cases within a family or healthcare setting. In retrospect, these were the first known cases of the disease that was later to be called severe acute respiratory syndrome (SARS). Through January, the numbers of cases of this unusual 'atypical pneumonia' continued to increase with examples of 'super-spreading incidents' that were to punctuate the course of the subsequent SARS epidemic. Between 16 November and 9 February, 305 cases were identified, one-third of them in healthcare workers (**Table 1**).

On 21 February 2003, a 65-year-old doctor working in a hospital in the city of Guangzhou, the provincial capital of Guangdong, arrived in Hong Kong and checked into Hotel M. He had treated patients with 'atypical pneumonia' in Guangzhou and had been ill himself since 15 February. His 1-day stay on the ninth floor at this hotel led to the infection of at least 17 other guests or visitors, some of whom traveled on to Hanoi, Toronto, Vancouver, Singapore, USA, Philippines, Guangzhou, and Australia. Five of these secondary cases initiated clusters of infection in Hanoi, Singapore, Toronto and two clusters of infection within Hong Kong. This was the most significant single event in the global spread of SARS, and arguably the most dramatic known event in the global spread of any infectious disease. However, because the secondary cases had largely dispersed outside of Hong Kong, this cluster of cases remained 'invisible' until the epidemiological linkages were reconstructed in mid-March.

Between 26 February and 10 March, disease outbreaks were recognized in the Hanoi-French Hospital in Vietnam and in Prince of Wales Hospital in Hong Kong. Dr. Carlo Urbani, a World Health Organization (WHO) communicable diseases expert stationed in Vietnam, examined the first cases of the disease outbreak in Hanoi and provided WHO with the first case descriptions of this new disease. Later, Dr. Urbani was himself one of the victims who succumbed to this disease. On 12 March, the WHO issued a Global Health Alert regarding an atypical pneumonia that was a particular risk to healthcare workers. Subsequently, Singapore and Toronto also reported clusters of cases. On 15 March, the WHO issued a Travel Advisory. The new disease was named SARS and a preliminary case definition was provided. The WHO set up virtual networks of virologists, clinicians, and epidemiologists to rapidly collate, evaluate, and disseminate information about the new disease.

Within weeks, SARS had spread to affect 8096 patients in 29 countries across five continents with 744 fatalities, an overall case–fatality rate of 9.6%. Healthcare facilities served as a major amplifier of infection, constituting 21% of all reported cases.

By 21–24 March, the etiological agent of SARS was identified to be a novel coronavirus, subsequently termed SARS coronavirus (SARS CoV). Serological tests demonstrated that the human population had no prior evidence of infection with SARS CoV, indicating that this virus had newly emerged in humans and implying a likely zoonotic origin.

Early case detection and isolation of infected individuals reduced and interrupted SARS CoV transmission across the world. By 5 July 2003, the WHO announced that all chains of human transmission of SARS were broken and the outbreak was at an end. This was indeed a historic triumph for global public health. Although SARS was subsequently to re-emerge to cause limited human disease (and in one instance, limited human-to-human transmission) as a result of laboratory escapes and zoonotic transmission from the live game animal markets of Guangdong in December 2003–January 2004 (**Table 1**), the human outbreak of SARS had been controlled.

Table 1 A chronology of events in the emergence of SARS

Date	Key events
16 November 2002	45-year-old man in Foshan city, Guangdong Province, mainland China becomes ill with fever and respiratory symptoms and transmits the disease to four other relatives.
10 December 2002	35-year-old restaurant chef working in Shenzhen is admitted to Heyuan City People's Hospital. Transmits disease to eight healthcare workers.
January 2003	Pneumonia outbreaks in Guangzhou (capital city of Guangdong Province). These include number of healthcare workers infected through the care of patients with the disease.
11 February 2003	Guangdong health authorities report an outbreak of respiratory disease in Guangdong with 305 cases and five deaths, one-third of the cases being in healthcare workers caring for patients with the disease. Cases were reported from Foshan, Heyuan, Zhongshan, Jiangmen, Guangzhou, and Shenzhen municipalities of Guangdong Province.
21 February 2003	A 65-year-old doctor from Guangdong arrives and checks in at Hotel M in Hong Kong. His stay of 1 day at this hotel leads to the infection of at least 17 other guests or hotel visitors who initiate clusters of infection within Hong Kong, Vietnam, Singapore, and Toronto.
26 February 2003	A 48-year-old 'Hotel M contact' is admitted to Hanoi-French Hospital in Vietnam and is the source of an outbreak there. Seven healthcare workers were ill by 5 March.
1 March 2003	A 22-year-old 'Hotel M contact' is admitted to Tan Tock Seng Hospital, Singapore. She will pass on infection to 22 close contacts.
4 March 2003	A 26-year-old Hotel M contact admitted to Prince of Wales Hospital, Hong Kong. His illness is relatively mild and is not categorized as severe pneumonia. He transmits infection to 143 persons including 4 members of his family, 67 healthcare workers or medical students, and 30 other patients.
5 March 2003	A 78-year-old 'Hotel M contact' dies at home in Toronto, Canada. Four family members are infected. They are the source for the subsequent Toronto outbreak.
5–10 March 2003	Outbreaks are recognized in Hanoi and Hong Kong.
12 March 2003	WHO issues a Global Alert about atypical pneumonia in Guangdong, Hong Kong, and Vietnam that appears to place healthcare workers at high risk.
13–14 March 2003	Singapore and Toronto report clusters of atypical pneumonia. In retrospect, both groups have an epidemiological link to Hotel M. One of the doctors who had treated develops symptoms while traveling and is quarantined on arrival in Germany on 15 March.
15 March 2003	WHO has received reports of over 150 cases of this new disease, now named severe acute respiratory syndrome (SARS). An initial case definition is provided. Travel advisory issued.
17 March 2003	WHO multicenter laboratory network on SARS etiology and diagnosis is established.
21–24 March 2003	A novel coronavirus is identified in patients with SARS.
12 May 2003	The genome sequence of the SARS coronavirus is completed.
June 2003	A virus related to SARS CoV is detected in civets and other small mammals in live game-animal markets in Guangdong.
5 July 2003	Lack of further transmission in Taiwan, the last region to have SARS transmission, signals the end of the human SARS outbreak.
September 2003	Laboratory-acquired SARS coronavirus infection in Singapore.
December 2003–January 2004	Re-emergence of SARS infecting humans from animal markets in Guangdong. Laboratory-acquired SARS coronavirus infections in Taiwan.
February 2004	Laboratory-acquired SARS leads to community transmission in Beijing and Anhui in China.

Adapted from Peiris JSM, Guan Y, Poon LLM, Cheng VCC, Nicholls JM, and Yuen KY (2007) Severe acute respiratory syndrome (SARS). In: Scheld WM, Hooper DC, and Hughes JM (eds.) *Emerging Infections* 7, p. 23. Washington, DC: ASM Press, with permission from ASM Press.

Virology

SARS Virus

SARS coronavirus is a member of the genus *Coronavirus* within the family *Coronaviridae* and the order *Nidovirales*. Coronaviruses are classified on genetic and antigenic characteristics into three groups and SARS CoV is presently regarded as a group 2b coronavirus. It is an enveloped, positive-sense, single-stranded RNA virus with a genome size of approx 29.7 kbp. The virus particle is approximately 100–160 nm in diameter with a distinctive corona of petal-shaped spikes on the surface which is comprised of the spike glycoprotein (S). The S protein is in a trimeric form on the viral surface. It has an N-terminal variable subdomain (S1) which contains the motifs responsible for receptor binding. A more conserved subdomain (S2), which contains heptad repeats and a coiled-coil structure, is important in the membrane fusion process. The S1–S2 subdomains remain in a noncleaved form in the intact SARS CoV virion and cleavage is believed to occur within the endocytic vesicle during the viral entry process. The envelope also contains a transmembrane glycoprotein M and in much smaller amounts, an envelope (E) protein. The M protein is a triple-spanning membrane protein and has a key role in coronavirus assembly. The hemagglutinin-esterase (HE) glycoprotein, found in some group 2 coronaviruses, is absent in SARS CoV. The nucleocapsid protein (N) interacts with the viral genomic RNA to form the viral nucleocapsid. Viral replication complexes are believed to be localized within double-membraned vesicles or autophagosomes.

SARS CoV Genome

The genome of SARS CoV is that of a typical coronavirus. The viral genomic RNA has at least 14 open reading frames (ORFs) (**Figure 1**). The genome codes for 16 nonstructural proteins (nsp1–16), 5 structural proteins, and 7 accessory proteins. The genomic RNA encoding the replicase gene functions as mRNA to generate polyproteins 1a and 1ab. The translation of ORF1b is directed by a −1 ribosomal frameshift (RFS) signal that contains a nucleotide slippery sequence (5′-UUUAAAC-3′) and an RNA pseudoknot. By contrast, the structural and accessory proteins are products derived from subgenomic RNA (sgRNA 2–9) which are synthesized by discontinuous RNA transcription. Translated products from ORF2 (S), ORF4 (E), ORF5 (M), and ORF9a (N) are viral structural proteins as described above. Recently, it was reported that the protein encoded by the ORF3a, which is able to interact with S and contains ion channel activity, is also a structural protein. However, the full function of this protein is yet to be determined.

The polyproteins 1a and 1ab generated from the replicase gene are cleaved by a papain-like proteinase (part of nsp3) and a 3C-like proteinase (nsp5) to generate 16 nonstructural proteins (**Figure 1(b)**). Nsp12 is a primer-dependent RNA-dependent RNA polymerase (RdRp), whereas nsp8 is a noncanonical RdRp (nsp8) synthesizing primers utilized by nsp12. In addition, eight nsp7 and eight nsp8 subunits are able to form a hexadecamer with a hollow, cylinder-like structure. RNA-binding studies and the overall architecture of this nsp7–nsp8 complex suggest that it might encircle RNA and confer processivity of nsp12. The nsp9 is a single-stranded RNA-binding protein and is able to interact with nsp8. The nsp13 is a helicase and unwinds duplex RNA (and DNA) in a 5′-to-3′ direction. The nsp3, nsp14, nsp15, and nsp16 have been shown to have ADP-ribose 1′-phosphatase, 5′-to-3′ exonuclease, endoribonuclease, and 2′-O-ribose methyltransferase activities, respectively. These four proteins are distantly related to cellular enzymes involved in RNA metabolism. These observations may be relevant to viral RNA processing. The nsp10 contains two zinc finger motifs and is suggested to be a regulator of vRNA synthesis. The biological functions of nsp1, nsp2, nsp4, nsp6, and nsp11 are largely unknown. The nsp1 is reported to induce chemokine dysregulation and host mRNA

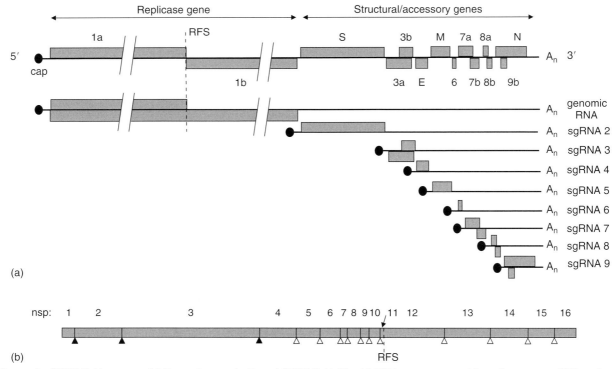

Figure 1 SARS CoV genome. (a) Genomic organization of SARS CoV. The 14 ORFs are expressed from the genome RNA and a nested set of subgenomic mRNA (sgRNA 2–9) that all have a common leader sequence derived from the 5′ end of the genome. The genomic RNA and all sgRNA contain a 5′ cap and a polyadenylated tail at the 3′ end. (b) Domain organization of the proteins for ORF1ab. Black and white arrow heads represent the sites cleaved by papain-like and 3C-like proteinases, respectively. The ribosomal frameshift (RFS) site is highlighted by a broken line.

degradation. The nsp2 is dispensable for virus replication. The nsp4 and nsp6 each contain a putative transmembrane domain.

Apart from the ORFs encoding the replicase and structural proteins, the viral genome contains additional ORFs that code for accessory proteins (3b, 6, 7a, 7b, 8a, 8b, and 9b). Genetically modified recombinant viruses without these accessory ORFs have been shown to be replication competent in cell cultures, indicating that the accessory ORFs may not be essential for virus replication *in vitro*. However, recombinant viruses with deletions in these regions are attenuated, suggesting that these proteins might have functions that are important for viral replication and pathogenesis *in vivo*. The accessory proteins from ORF3b and ORF7a induce apoptosis in transfected cells. There is also evidence suggesting that the 7a protein is incorporated into virions. The protein encoded in ORF6 has been shown to inhibit the nuclear import of STAT-1 and function as an interferon antagonist in infected cells. These properties might relate to virus virulence. Interestingly, comparative sequence analysis of SARS CoV isolated from palm civets (see below) and humans showed that all animal isolates contained a 29-nucleotide (nt) sequence which is absent from most human isolates obtained in the later phase of the SARS outbreak. As a result, the ORF8 in these human SARS CoVs encodes 8a and 8b proteins, whereas the corresponding ORF in the animal isolates encodes a single protein, known as the 8ab protein. These proteins from the animal and human ORF8 have differential binding affinities to various SARS CoV structural proteins. Furthermore, the expression of E can be downregulated by 8b but not 8a or 8ab in infected cells. These observations may suggest that the 29-nt deletion might modulate the replication or pathogenesis of the human SARS CoV. The crystal structure of the 9b protein suggests that it might be a lipid binding protein but its function is yet to be identified. Overall, these accessory proteins may play roles in viral replication and pathogenesis.

Ecology and Animal Reservoir

Until the end of January 2003, 39% of patients with SARS in Guangdong had handled, killed, or sold wild animals or prepared and served them as food. However, such risk factors were found in only 2–10% of cases from February to April 2003 when the virus had adapted to efficient human-to-human transmission. Thus, the early epidemiological evidence pointed to the live game animal trade as a potential source of the SARS CoV. SARS-like coronaviruses were identified in a number of small mammalian species sold in the live game animal markets in Guangdong, including the palm civet (*Paguma larvata*), raccoon dog (*Nyctereutes procyonides*), and the Chinese ferret badger (*Melogale moschata*). A high proportion of individuals working in these markets were observed to have developed antibodies to SARS CoV, although none of them had a history of the disease. Viruses isolated from the re-emergent SARS cases in Guangdong in December 2003–January 2004 were more similar to those found in civets in these markets, rather than to viruses causing the global outbreak in early 2003. These observations strongly implicated the live game animal trade as the interface for interspecies transmission of a precursor animal SARS-like coronavirus to humans.

SARS CoV can be shed for weeks in experimentally infected palm civets but many of the other species appear to clear the virus rapidly. While civets in live animal markets were often observed to be positive for SARS-like coronavirus RNA, civets tested in the farms that supply these markets and those caught in the wild rarely have evidence of infection. Thus, palm civets were believed not likely to be the natural reservoir of the precursor SARS CoV (see below). More recently, group 2b coronaviruses related to SARS CoV have been identified in *Rhinolophus* bats in Hong Kong and mainland China. Such bats are also sold live in these game animal markets. It is now believed that these or related bat coronaviruses may be the precursor from which SARS CoV originated (see below).

Phylogeny

SARS CoV and the SARS-like civet and bat coronaviruses form a distinct phylogenetic subgroup (2b) within the group 2 coronaviruses (**Figure 2**). Genetic and phylogenetic analysis indicates that the viruses associated with the early phase of the human SARS outbreak are more closely related to the viruses found in palm civets and other small mammals in the live game animal markets in Guangdong. The genomes of viruses in the early phase of the human outbreak in 2003 were observed to be under strong positive selective pressure, suggesting that the virus was rapidly adapting in a new host. Furthermore, virus in civets was also found to be under strong positive selective pressure, supporting the view that civets were not the natural host of the precursor SARS-like coronavirus. The search for the precursor of SARS CoV led to the discovery of a number of novel coronaviruses in bats which are related to group 1 and group 2 coronaviruses. Some of these bat coronaviruses are genetically related to SARS CoV (group 2b) and are likely to be the direct or indirect precursor of SARS CoV (see below).

Interestingly, considered overall, the recently discovered group 1 and group 2 (including SARS CoV-like) bat coronaviruses appear to be in evolutionary stasis while many other mammalian coronaviruses still appear to be under evolutionary selection pressure, raising the intriguing possibility that bats may in fact be the precursors, not only of SARS CoV, but also of most other mammalian coronaviruses.

Virus Receptors

The functional receptor for SARS CoV on human cells is the angiotensin-converting enzyme 2 (ACE-2) which

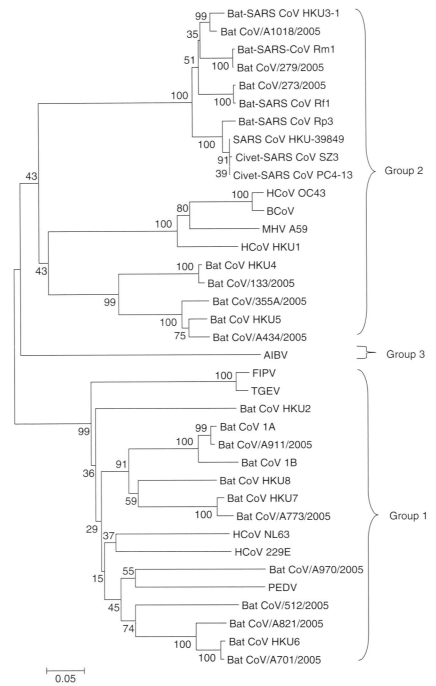

Figure 2 Phylogenetic analysis of RNA sequences coding for the RNA-dependent RNA polymerase (partial sequence). The phylogenetic tree was constructed by the neighbor-joining method and bootstrap values were determined with 1000 replicates. Human SARS CoV (GenBank accession AY278491.2), SARS CoVs isolated from palm civets in 2003 (AY304486.1) and 2004 (AY613948.1) and bat CoVs [Bat CoV 1A (DQ666337.1), Bat CoV 1B (DQ666338.1), Bat CoV HKU2 (DQ249235.1), Bat-SARS CoV HKU3–1 (DQ022305), Bat CoV HKU4 (DQ249214.1), Bat CoV HKU5 (DQ249217.1), Bat CoV HKU6 (DQ249224.1), Bat CoV HKU7 (DQ249226.1), Bat CoV HKU8 (DQ249228.1), Bat-SARS CoV Rp3 (DQ071615), Bat-SARS CoV Rm1 (DQ412043), Bat-SARS CoV Rf1 (DQ412042), Bat CoV/A434/2005 (DQ648819.1), Bat CoV/A701/2005 (DQ648833.1), Bat CoV/A773/2005 (DQ648835.1) Bat CoV/A821/2005 (DQ648837.1), Bat CoV/A970/2005 (DQ648854.1) Bat CoV/A911/2005 (DQ648850.1), Bat CoV/A1018/2005 (DQ648795.1), Bat CoV/133/2005 (NC_008315), Bat CoV/273/2005 (DQ648856), Bat CoV/279/2005 (DQ648857), Bat CoV/355A/2005 (DQ648809.1), Bat CoV/512/2005 (DQ648858)] were aligned with references sequences as indicated. Reference sequences are: transmissible gastroenteritis virus, TGEV (DQ811789); HCoV 229E (AF304460); HCoV NL63 (AY567487); HCoV-OC43 (AY391777); HCoV-HKU1 (DQ415903 HKU1); porcine epidemic diarrhea virus, PEDV (AF353511); avian infectious bronchitis virus, AIBV (AY646283); mouse hepatitis virus, MHV (AY700211); bovine coronavirus, BCoV (AF220295); feline infectious peritonitis virus, FIPV (AY994055).

binds the receptor-binding motif (amino acid residues 424 to 494) of the SARS CoV spike (S) protein. While the human SARS CoV S protein binds efficiently to both human and civet ACE-2, the civet-like SARS CoV S protein binds efficiently to ACE-2 from civets but poorly to human ACE-2. The spike protein of the bat SARS-like coronavirus lacks the ACE-2 receptor-binding motif and is therefore unlikely to bind to human ACE-2.

These findings explain the increased human transmissibility of SARS CoV in the later stages of the SARS outbreak, the observation that human SARS CoV efficiently infects civets under experimental conditions, and the failure of civet SARS CoV or bat SARS-like CoV to replicate productively in primate (Vero-E6, FRhK4) cells that support replication of human SARS CoV. This finding also explains the poor virulence and transmissibility of re-emergent SARS in December 2003–January 2004 when humans are believed to have been infected with a civet-like SARS CoV.

Other cell-surface molecules such as L-SIGN, DC-SIGNR, DC-SIGN (CD209), and L-SECtin may serve as binding receptors but do not appear to be functional viral receptors in the absence of ACE-2. They may, however, promote cell-mediated transfer of the virus to other susceptible target cells. On the other hand, binding to L-SIGN appears to lead to proteasome-dependent viral degradation and it may function as a scavenger receptor (see below).

Human Disease

Transmission

Respiratory droplets are the major source of infectious virus for transmission of SARS. However, aerosol exposure has probably contributed to disease transmission, at least in some defined instances where aerosol-generating procedures (e.g., nebulizers, high-flow oxygen therapy, intubation) have been used. The unusual stability of SARS CoV also suggests that contaminated surfaces and fomites may contribute to disease spread. As SARS CoV is present in feces and urine (and possibly other body secretions), these body fluids may also play a part in disease transmission. The largest single outbreak of SARS at the Amoy Gardens apartment block in Hong Kong, where over 300 individuals were infected from a single index case, is believed to have been caused by aerosols generated from infected body secretions (e.g., feces).

The estimated incubation period for SARS is 2–14 days. During the 2003 outbreak, the majority of cases did not transmit disease at all and only a few patients accounted for a disproportionately large number of secondary cases. Host factors may have played a role in these super-spreading events but, in many cases, there was a unique combination of host factors and environmental circumstances that facilitated transmission. In contrast to the high transmission rates in these super-spreading events and within hospitals, there was less evidence of secondary transmission within the family or within households (e.g., 15% in Hong Kong). Notwithstanding the 'super-spreading phenomenon' that has characterized SARS, the basic reproduction number (Ro) of SARS is estimated to range from 2 to 4.

Seroepidemiological studies of contacts of SARS patients (both adults and children) have revealed that asymptomatic infection was uncommon. The absence of large numbers of asymptomatic transmitters and the paucity of transmission during the first 5 days of illness explain the success of the public health measures of aggressive case detection and isolation in interrupting transmission of human-adapted SARS CoV and the control of the global disease outbreak. These features of SARS have been attributed to the observation that, unlike many other acute viral respiratory infections, SARS transmission has mostly occurred only after the fifth day of illness. This is, in turn, probably related to the low viral load in the upper respiratory tract during the early phase of the illness (see below).

Clinical Features

As the clinical features of SARS are not pathognomonic, a contact history and virological evidence of infection are important for confirmatory diagnosis. SARS typically starts with myalgia and loose stools around the time of onset of fever without coryza or sore throat (seen in 70% of patients). The upper respiratory manifestations are less commonly observed. Radiological abnormalities have been observed in >60% of cases at initial presentation and preceded lower respiratory tract symptoms in approximately 41% of patients.

Children have had much milder illness than adults and mortality rates progressively increase with age. Some patients, particularly those with progressive lower respiratory tract involvement have had a watery diarrhea. Other extrapulmonary manifestations included hepatic dysfunction and a marked lymphopenia involving both B, T (CD4 and CD8 subsets), and natural killer (NK) cells. High serum levels of chemokines (interleukin 8 (IL-8), CCL2, and CCl10) and pro-inflammatory cytokines (IL-1, IL-6, IL-12) have been observed.

The overall case–fatality rate was 9.6% and the terminal events were severe respiratory failure associated with acute respiratory distress syndrome (ARDS) and multiple organ failure. Age, presence of co-morbidities, and viral load in the nasopharynx and serum during the first 5 days of illness correlated with an adverse prognosis.

Autopsy findings of those who died in the first 10 days of illness were diffuse alveolar damage, desquamation of pneumocytes, and hyaline membrane formation. Viral

RNA was detected by quantitative polymerase chain reaction (PCR) at high copy number in the lung, intestine and lymph nodes, and at lower levels in spleen, liver, and kidney. In lung biopsy tissue or in autopsy tissue of patients dying in the first 10 days after disease onset, viral antigen and viral nucleic acid were demonstrated by immunohistochemistry and *in situ* hybridization methods respectively, in alveolar epithelial cells and to lesser extent in macrophages. A few unconfirmed studies have also reported the detection of virus particles or viral RNA in multiple organs but these findings require independent confirmation.

Laboratory Diagnosis

Highly sensitive and specific real-time PCR assays for detection of viral RNA remain the best choice for early SARS diagnosis. Viral RNA has been detected in respiratory specimens, feces, serum, and urine. Specimens from the lower respiratory tract such as endotracheal aspirates have higher viral load than those from the upper respiratory tract and are better diagnostic clinical specimens. As viral load is low during the first 5 days of disease, a negative PCR result from specimens collected at this time does not exclude the diagnosis. Testing multiple specimens improves the detection rate of SARS. Virus culture on Vero E6 or FRhK-4 cells and viral antigen detection tests are much less sensitive than reverse transcriptase PCR (RT-PCR) for detecting the virus. While viral RNA remains detectable in the respiratory secretions and feces for many weeks after the onset of illness, specimens rarely yield a virus isolate after the third week of illness.

Sero-conversion by immunofluoresence or neutralization occurs during the second week of illness and can provide reliable retrospective diagnosis. Enzyme-linked immunoassays using inactivated whole virus or recombinant antigens are convenient alternatives for serological screening, but any positive results must be confirmed by the more specific immunofluoresence or neutralization tests.

Pathogenesis

The primary mechanism of lung damage appears to be due to infection of type 1 and type 2 pneumocytes which are key target cells of the virus. Type 2 pneumocytes are important in the repair of lung injury and infection of these cells can potentially impair the regenerative responses of the lung and aggravate the respiratory impairment.

Whereas mice deficient in NK, T or B lymphocytes display similar kinetics of viral replication to normal mice, infection of mice with defects in the STAT1 signaling pathway results in more prolonged viral replication and more severe disease. These findings indicate the importance of innate immune responses in the control of infection, at least in the mouse. Infection of epithelial cells, macrophages, and myeloid dendritic cells fails to induce a type 1 interferon response although other interferon response genes are activated. Viral proteins expressed from ORF3b, ORF6, and the N gene have interferon antagonist effects *in vitro*. In contrast, macrophages and dendritic cells respond to infection *in vitro* with strong chemokine responses, including those (e.g., CCL10) that are elevated in the serum of SARS patients, and macrophage-chemoattractant chemokines (CCL2). This may explain the predominantly macrophage infiltrate in the lung.

There is evidence of viral replication within intestinal epithelial cells but there is minimal cellular infiltrate or disruption of intestinal architecture and the pathogenesis of diarrhoea in SARS remains unclear.

Treatment

As SARS emerged as a disease of unknown etiology, empirical therapeutic options were initially tested including broad spectrum antivirals and immunomodulators such as ribavirin, intravenous immune globulin, type 1 interferon, SARS convalescent plasma, and corticosteroids. However, in the absence of controlled clinical trials, no conclusions can be drawn on the efficacy of these interventions.

Anti-SARS CoV activity *in vitro* has been demonstrated for several therapeutics already in clinical use for other conditions, including lopinavir–nelfinavir, glycyrrhizin, baicalin, reserpine, and niclosamide. There are contradictory reports on the *in vitro* activity of ribavirin, interferon beta, and interferon alpha. In summary, taking into account bio-availability of these compounds and *in vitro* data, interferon alpha $n1/n3$, leukocytic interferon alpha, interferon beta and nelfinavir appear to be worthy of animal studies and randomized placebo-controlled clinical trials if SARS was to return.

A clinical trial of lopinavir 400 mg with ritonavir 100 mg orally every 12 h (added to an existing regimen of ribavirin and corticosteroid therapy) appeared to provide clinical benefit compared to historical controls. However, the lack of concurrent controls makes it difficult to draw conclusions. Similarly, a limited clinical trial of 13 patients using interferon alfacon-1 treatment showed a trend toward improved radiological and clinical outcomes, but without achieving statistical significance.

Studies in primate models have demonstrated prophylactic or therapeutic benefit from PEGylated recombinant interferon alpha-2b and from small interfering RNA therapy. More recently, screening of combinatorial chemical libraries *in vitro* has identified potential inhibitors of the viral protease, helicase, and spike protein-mediated entry.

Animal Models

Experimental SARS CoV infection leads to virus replication in a number of animal species including nonhuman primates (e.g., cynomolgous and rhesus macaques, African green monkeys, and marmoset monkeys), mice (BALB/c, C57/BL6), Golden Syrian hamsters, ferrets, and cats. Only some of these develop pathological lesions in the lungs (cynomolgus macaques, ferrets, hamsters, marmosets, aged BALB/C mice).

Interestingly, whereas SARS CoV replicates in the lung of both young and aged (12–14 months) BALB/c mice, only aged mice manifest clinical symptoms and histological evidence of lung pathology. This is reminiscent of disease in humans in which children have mild illness (see above). Furthermore, few animal models reproduce the gastrointestinal manifestations of the illness.

While the ideal animal model for understanding SARS pathogenesis is lacking, those that support viral replication (with or without clinical disease) are adequate for evaluating the efficacy of vaccines.

Vaccines and Immunity

A wide range of strategies have been explored for development of SARS vaccines. These have included: inactivated whole virus vaccines; subunit vaccines including baculovirus expressed S1 subdomain or the complete trimeric spike protein of the virus expressed in mammalian cells; DNA vaccines expressing S (full-length and fragments), N, M, or E proteins; and vectored vaccines based on modified vaccinia Ankara (MVA) virus, vesicular stomatitis virus, adenoviral vectors carrying S, M, or N proteins, and attenuated parainfluenza virus type 3 vectored vaccines carrying S, E, M, and N proteins. Neutralizing antibody responses and, where appropriate, cell-mediated immune responses have been measured as correlates of immunity. Some of these vaccines have been evaluated in experimental models by challenging with infectious SARS CoV.

Trials in hamsters of attenuated parainfluenza virus type 3-vectored vaccines individually expressing SARS CoV S, E, M, and N proteins have indicated that only the S protein construct elicits neutralizing antibody and protects against experimental challenge. Furthermore, passive transfer of serum containing S protein neutralizing antibody has been shown to be sufficient to induce protective immunity in mice. It is concluded that neutralizing antibody to the S protein is an important correlate of protection. The receptor-binding determinant of the S1 subdomain is an immuno-dominant epitope and a critical determinant for virus neutralization.

As antibody can enhance rather than protect against the coronavirus disease feline infectious peritonitis, antibody-dependent enhancement has been a concern for SARS-Co vaccine development. However, no evidence of vaccine-enhanced disease has been observed to date, with two possible exceptions. There is a report that a modified vaccinia Ankara virus S protein vaccine has led to hepatitis in vaccinated ferrets but this has not been independently confirmed. There is also a report that S protein antibody elicited by a subunit vaccine enhances entry of pseudo-particles carrying S spike into lymphoblastoic cell lines which lack ACE-2 and are not normally permissive to infection. However, in the challenge experiments in hamsters, the vaccine did not induce protection and there is no evidence of disease enhancement.

Passive immunization with human monoclonal antibodies to the S protein has been successful at protecting mice and ferrets from experimental challenge by reducing viral load in the lung but not in the nasopharynx.

Most of these active and passive immunization studies have evaluated protection from challenge using the homologous human-adapted SARS CoV. However, a newly emergent SARS outbreak will probably arise from the animal reservoir and it is therefore important to investigate cross-protection against animal SARS-like CoV. As none of the civet or bat SARS CoV has yet been successfully grown *in vitro*, the cross-reactive neutralizing antibody response has been studied using lentiviruses pseudotyped with CoV S protein from a civet virus (SZ3), a civet-like virus causing re-emergent SARS in humans in December 2003 (GD03), and from a human SARS CoV (Urbani-strain) isolated from the major human SARS outbreak in 2003. The viruses pseudotyped with human Urbani virus S protein were neutralized by antibodies to the civet SARS-like virus but pseudotypes with the civet-like S protein were not neutralized by antibodies to the human SARS CoV (Urbani). On the contrary, antibody to the Urbani virus appeared to enhance the infectivity of the GD03 and SZ3 pseudotyped viruses. These findings appear to reflect receptor usage of these viruses as it has been shown that GD03 and SZ3 bind poorly to human ACE-2 (see above). The development of vaccines that can prevent re-emergence of SARV CoV from its zoonotic reservoir remains a challenge.

Conclusion: Will SARS Return?

Like many recent emerging infectious diseases that threaten human health, SARS was a zoonosis. The SARS CoV that was responsible for the global outbreak in 2003 was well adapted to bind to human ACE-2 and was efficiently transmitted human-to-human. Laboratories remain a potential source of infection from such viruses and, as occurred in February 2004, laboratory escape can lead to a community outbreak.

The SARS-like coronavirus found in civets (and other mammals) in live game-animal markets is very closely related to SARS CoV, but it binds inefficiently to the human ACE-2 receptor (see above). Consequently, when human infection with the civet SARS-like CoV occurred in December 2003–January 2004, there was no human-to-human transmission and clinical disease was mild. While SARS-like coronaviruses have been found in bats, they are genetically distinct to SARS CoV and the bat SARS-like CoV S protein appears unable to bind to human or civet ACE-2. Thus, it is likely that re-emergence of a virus capable of causing human disease from this source probably requires extensive adaptation in an intermediate host (e.g., small mammals such as civets). While it is difficult to assess the likelihood of SARS re-emergence, this possibility cannot be excluded.

The rapid expansion of the live game-animal trade and the development of large markets in southern China which house a diversity of wild and domestic animal species were probably important in facilitating the emergence of SARS CoV. It is therefore possible that, like Ebola, SARS may re-emerge at intervals in the future. However, a number of epidemiological characteristics of SARS (see above) should allow it to be contained by public health interventions, once the disease is diagnosed. Indeed, the chain of community transmission arising from a laboratory escape of SARS CoV in February 2004 was contained by such public health measures and community transmission was aborted. However, if the dynamics of transmission of a re-emergent virus are different, and particularly if transmission occurs earlier in the illness and there are more asymptomatic infections, the options for control and the ultimate consequences may be very different. It remains important, therefore, to understand better the ecological and viral factors that predispose to interspecies transmission and the emergence of animal viruses with efficient competence for transmission in humans. Attention should be directed toward the adaptation strategies and the ecological factors that are important in determining interspecies transmission, rather than focus on the disease itself (i.e., SARS). Efforts to understand better the molecular basis for interspecies transmission that led to the genesis of SARS CoV will help us to prepare better for the next emerging infectious disease challenge; whether this comes from SARS CoV, avian influenza H5N1, or a yet unknown virus.

See also: Coronaviruses: General Features; Coronaviruses: Molecular Biology.

Further Reading

Chan JCK and Taam Wong VCW (2006) *Challenges of Severe Acute Respiratory Syndrome.* Singapore: Elsevier.

Cinatl J, Michaelis M, Hoever G, Preiser W, and Doerr HW (2005) Development of antiviral therapy for severe acute respiratory syndrome. *Antiviral Research* 66: 81.

de Haan CAM and Rottier PJM (2006) Hosting the severe acute respiratory syndrome coronavirus: Specific cell factors required for infection. *Cellular Microbiology* 8: 1211.

Gillim-Ross L and Subbarao K (2006) Emerging respiratory viruses: Challenges and vaccine strategies. *Clinical Microbiology Reviews* 19: 614.

Gorbalenya AE, Enjuanes L, Ziebuhr J, and Snijder EJ (2006) Nidovirales: Evolving the largest RNA virus genome. *Virus Research* 117: 17.

Li W, Wong SK, Li F, et al. (2006) Animal origins of the severe acute respiratory syndrome coronavirus: Insight from ACE-2 protein interactions. *Journal of Virology* 80: 4211.

May RM, McLean AR, Pattison J, and Weiss RA (2004) Emerging infections: What have we leant from SARS? *Philosophical Transactions of the Royal Society of London, Series B* 359: 1045.

Peiris JSM, Guan Y, Poon LLM, Cheng VCC, Nicholls JM, and Yuen KY (2007) Severe acute respiratory syndrome (SARS). In: Scheld WM, Hooper DC, and Hughes JM (eds.) *Emerging Infections* 7, p. 23. Washington, DC: ASM Press.

Peiris JSM, Yuen KY, Osterhaus ADME, and Stohr K (2003) The severe acute respiratory syndrome. *New England Journal of Medicine* 349: 2431.

Peiris M, Anderson L, Osterhaus A, Stohr K, and Yuen KY (2005) *Severe Acute Respiratory Syndrome.* Oxford: Blackwell.

Perlman S and Holmes KV (eds.) (2006) The *Nidoviruses*: Towards control of SARS and other nidovirus diseases. *Advances in Experimental Medicine and Biology*, vol. 581. 617pp. New York: Springer.

Stockman LJ, Bellamy R, and Garner P (2006) SARS: Systematic review of treatment effects. *PLoS Medicine* 3: e343.

World Health Organization(2003) A multicentre collaboration to investigate the cause of severe acute respiratory syndrome. *Lancet* 361: 1730.

World Health Organization, Western Pacific Region (2006) *SARS. How a Global Epidemic was Stopped.* Geneva: WHO Press.

Zhong NS and Zeng GQ (2003) Our strategies for fighting severe acute respiratory syndrome (SARS). *American Journal of Respiratory and Critical Care Medicine* 168: 7.

Shellfish Viruses

T Renault, IFREMER, La Tremblade, France

© 2008 Elsevier Ltd. All rights reserved.

Glossary

Aquaculture Cultivation of aquatic animals or plants.
Bivalve Marine or freshwater mollusks having a soft body with plate-like gills enclosed within two shells hinged together.
Gills Respiratory organ of aquatic animals that breathe oxygen dissolved in water.
Hatchery A place where eggs are hatched under artificial conditions.
Hemocyte Any blood cell especially in invertebrates.

Larva The immature free-living form of most invertebrates which on hatching from the egg is fundamentally unlike its parent and must metamorphose.
Mantle A protective layer of epidermis in mollusks that secretes a substance forming the shell.
Mollusk Invertebrates having a soft unsegmented body usually enclosed in a shell.
Nursery A place for the cultivation of juveniles under controlled conditions.
Shellfish Aquatic invertebrates belonging to the crustacean or mollusk families.
Velum Membrane of mollusk larvae that allows swimming activity.

Introduction

A natural abundance of shellfish was common in many areas of the world until the early twentieth century. However, industrial and urban development and population growth in coastal areas, coupled with extreme harvest pressure, appear to have contributed to a stready decline in natural shellfish populations. This decline in wild harvests, together with a greater demand for seafood from an increasing world population, have driven the development of technology for the intensive management and cultivation of shellfish. As a result, global shellfish production, the greatest proportion of which is bivalves, was estimated to be 10 732 000 metric tons in the year 2000. However, as husbandry practices have developed, the significant impact of infectious diseases on productivity and product quality has been increasingly recognized. Numerous examples worldwide have demonstrated that entire shellfish industries in coastal areas are susceptible to diseases and that the production of healthy shellfish is a key to the economic viability of mollusk farming.

The study of shellfish diseases is a relatively young science and the discovery of viruses in marine mollusks is a fairly recent event. Viral diseases have seriously affected the aquaculture industry during the last decades. Viral pathogens are often highly infectious and easily transmissible, and are commonly associated with mass mortalities. Viruses interpreted as members of the families *Iridoviridae*, *Herpesviridae*, *Papovaviridae*, *Reoviridae*, *Birnaviridae*, and *Picornaviridae* have been reported as associated with disease outbreaks and causing mortality in various mollusks. However, there is currently a lack of information concerning the occurrence of mollusk viruses worldwide, and the basic method for identification and examination of suspect samples is still predominantly histopathology. This technique enables the identification of cellular changes associated with infection but does not provide conclusive identification of mollusk viruses unless completed by other methods such as transmission electron microscopy. Moreover, as there is a lack of marine mollusk cell lines and, since invertebrates lack antibody-producing cells, the direct detection of viral agents remains the only possible approach to diagnosis.

As filter feeders, bivalves may also bioaccumulate viruses from humans and other vertebrates, acting as a transient reservoir. The consumption of raw or undercooked shellfish can result in human disease and contamination of shellfish cultivated in coastal marine waters by microorganisms that are pathogenic to humans is a public health concern worldwide. The association of shellfish-transmitted infectious diseases with sewage pollution has been well documented since the late nineteenth and early twentieth centuries. Human enteric viruses, including rotaviruses, enteroviruses, and hepatitis A virus, are the most common etiological agents transmitted by shellfish. These enteric viruses are associated with several human diseases ranging from ocular and respiratory infections to gastroenteritis, hepatitis, myocarditis, and aseptic meningitis. Many of these viruses are transmitted by the fecal–oral route and are highly prevalent in locations with poor sanitation. There is considerable literature on the implications for human health. However, this is not the subject of the present article.

Irido-Like Viruses

Hosts and Locations

Infections by irido-like viruses have been reported in oysters in France and in the USA. Two distinctive conditions have been associated with mass mortalities in adult Portuguese oyster, *Crassostrea angulata*, along French coasts: gill necrosis virus disease and hemocytic infection virus disease. A viral infection similar to the latter was reported in the Pacific oyster *Crassostrea gigas* during summer mortalities in the Bay of Arcachon (Atlantic coast, France) long after the disappearance of the Portuguese oyster. A third type of irido-like virus, the oyster velar virus (OVV), has been reported from hatchery-reared larval Pacific oysters on the west coast of North America (Washington State, USA).

Disease Manifestations and Epizootiology

Gill necrosis virus disease is regarded as the primary cause of disease outbreaks and mortalities that occurred in the late 1960s among Portuguese oysters on the Atlantic coast of France. The disease appears to have affected up to 70% of oyster populations with maximum losses reported in 1967. Losses subsequently declined and survivors recovered from the disease. The first gross signs were small perforations in the center of yellowish discolored zones of tissue on gills and labial palps. Further development and extension of the lesions resulted in larger and deeper ulcerations. In advanced stages, total destruction of affected gill filaments was observed. Yellow or green pustules also developed on the adductor muscle and mantle.

In 1970, high mortality rates were again reported in *C. angulata* oysters in France. Mortality was first observed in the basin of Marennes Oleron (Altantic coast) and in Brittany. The high mortality rates that occurred during this epizootic led to almost total extinction of French *C. angulata* by 1973. The disease affected adult oysters. No distinctive clinical signs were noted (e.g., no gill lesions). Histological observations included an acute cellular infiltration consisting of atypical virus-infected hemocytes. Pacific oysters seemed to be resistant to the vius and subsequently replaced *C. angulata* in France. However, a morphologically similar virus was reported from Pacific oysters during an outbreak of summer mortality in the Bay of Arcachon (Atlantic coast, France) in 1977. Although affected Pacific oysters exhibited virtually no gross signs, the presence of atypical cells interpreted as infected hemocytes and degeneration of connective tissues were reported in infected animals.

Oyster velar virus disease (OVVD) of the Pacific oyster occurred in the USA from mid-March through mid-June each year from 1976 to 1984, suggesting that the expression of the disease may be related to particular environmental conditions. When cultured at 25–30 °C, mortalities in oyster larvae greater than 170 μm in shell length typically begin at about 10 days of age. The infection results in the sloughing of ciliated velar epithelial cells and detachment of infected cells from the velum. Other cells lose cilia and infected larvae become unable to move normally.

Descriptive Histopathology

Histologically, gill necrosis virus disease is characterized by tissue necrosis with massive hemocytic infiltration around the lesions. The most distinctive lesion is the occurrence of giant polymorphic cells which may be up to 30 μm in size and contain large fuchsinophilic granules in the cytoplasm. In some giant cells, a voluminous basophilic inclusion (5–15 μm) occupies the greatest part of the cytoplasm in which finer basophilic granules (0.4–0.5 μm) are also present. Electron microscopy indicates that the inclusions are the viroplasm, and the fine granules are large viral particles. The most characteristic histological lesion of hemocytic infection virus disease is an acute cellular infiltration with the presence of atypical hemocytes in the connective tissues. Basophilic intracytoplasmic inclusion bodies are found in atypical blood cells in which irido-like virus particles can be observed in ultrathin sections. OVVD disease manifests histologically by the presence of intracytoplasmic inclusion bodies, 1.2–4 μm in diameter, located most commonly in the ciliated velar epithelium. The presence of DNA in the inclusion bodies is suggested by a positive Feulgen and Rosenbeck reaction.

Viruses

Mature icosahedral virions (380 nm diameter) are scattered throughout the cytoplasm of infected cells (**Figure 1(a)**). The outer shells of virions appear to consist of two trilaminar layers. The electron-opaque core (250 nm diameter) is limited by a three-layered fringe of definite width and surrounded by a layer of dense material

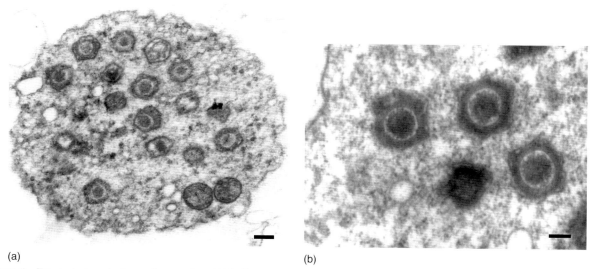

Figure 1 Transmission electron micrographs of irido-like particles infecting *Crassostrea angulata* oysters. (a) Intracytoplasmic irido-like virus particles in an infected *C. angulata* cell (gill necrosis virus disease). Scale = 200 nm. (b) Irido-like virus particles from *C. angulata*. Virions are icosahedral in shape with a central electron-dense core, surrounded by an electron-lucent zone followed by another dense layer. Two unit membranes separated by a clear zone enclose the particle. Scale = 100 nm.

(**Figure 1(b)**). Morphogenesis takes place in the cytoplasm. Oyster irido-like viruses have not been isolated from infected tissue and have not been characterized biochemically. However, the presence of viral DNA was demonstrated by histochemical techniques including acridine orange staining and the Feulgen and Rosenbeck reaction. The characteristic morphology and cytoplasmic localization of these large DNA viruses suggest that they may eventually be classified as members of the family *Iridoviridae*. However, no molecular characterization has yet been conducted and there remains a need for definitive demonstration of viral etiology for the reported diseases.

Herpesviruses

Hosts and Locations

Herpes-like virus infections have been identified in various marine mollusk species throughout the world, including the USA, Mexico, France, Spain, the UK, New Zealand, Australia, and Taiwan. The first description of a virus morphologically similar to members of the family *Herpesviridae* in a bivalve mollusk was reported in 1972 in the eastern oyster, *Crassostrea virginica*. Since then, a wide host range has been reported for herpes and herpes-like viruses infecting bivalve species, including the Pacific oyster *C. gigas*, the European oyster *Ostrea edulis*, the Antipodean flat oyster *Ostrea angasi*, the Chilean oyster *Tiostrea chilensis*, the Manila clam *Ruditapes philippinarum*, the carpet shell clam *Ruditapes decussatus*, the Portuguese oyster *C. angulata*, the Suminoe oyster *C. ariakensis*, and the French scallop *Pecten maximus*. It is noteworthy that recently a herpes-like virus has also been observed by transmission electron microscopy in the gastropod mollusk *Haliotis diversicolor supertexta* in Taiwan associated with high mortality rates.

Disease Manifestations and Epizootiology

Herpesvirus and herpes-like virus infections have been associated with high mortalities of hatchery-reared larvae and juveniles stages of several bivalve mollusk species. Observations by transmission electron microscopy indicate that larvae exhibit generalized infections, whereas focal infections usually occur in juveniles. Although viral infections have also been observed in adult bivalves, they are apparently less sensitive than younger stages. Infected larvae exhibit velar and mantle lesions. They swim weakly in circles and shortly before death settle at the bottom of the tanks. Infected juveniles exhibit sudden high mortalities in a short period of time (less than 1 week) often during the summer. Histologically, lesions are confined to connective tissues. Fibroblast-like cells exhibit abnormal cytoplasmic basophilia and enlarged nuclei with marginated chromatin. Other cell types including hemocytes and myocytes show extensive chromatin condensation. Peculiar patterns of chromatin, ring-shaped or crescent-shaped, are also observed suggesting that apoptosis may occur. Viral DNA and proteins have been detected in asymptomatic adult oysters. Like other herpesviruses, the *C. gigas* herpesvirus seems to be capable of long-term persistence in the infected host. The pathogenicity of the virus for the larval stages of *C. gigas* has been demonstrated by experimental transmission to axenic larvae. Attempts to reproduce symptoms experimentally in juveniles and adult oysters have so far been inconclusive.

Virus Ultrastructure

Based primarily on virion morphology and aspects of morphogenesis and genome organization, Ostreid herpesvirus 1 (OsHV-1) is currently classified as an unassigned member of the family *Herpesviridae*. Particles present in the nucleus are circular or polygonal in shape. Empty particles are presumed to be capsids; others containing an electron-dense toroidal or brick-shaped core are interpreted as nucleocapsids (**Figure 2(a)**). Capsids and nucleocapsids are scattered throughout the nucleus in infected cells (**Figure 2(b)**). An electron–lucent gap of approximately 5 nm with fine fibrils is observed between core and capsid. Digital reconstruction of the OsHV-1 capsid based on cryoelectron microscopic images indicates an icosahedral structure with a triangulation number of $T = 16$, which is an architecture unique to herpesviruses. Prominent external protrusions at the hexon sites, and a relatively flat and featureless appearance of the inner surface, reported for OsHV-1 capsids, are also characteristic features of herpesviruses. Extracellular particles are usually enveloped with a trilaminar unit-membrane and measure 100–180 nm in diameter (**Figure 2(c)**). Tegument between the outer membrane and the capsid shell of enveloped particles is either absent or minimal (**Figure 2(c)**).

Genome Structure and Organization

Virus particles have been purified from fresh infected *C. gigas* larvae and the entire OsHV-1 genome has been cloned and sequenced (GenBank accession number AY509253). The total genome size is 207 439 bp. The overall genome organization is TR_L-U_L-IR_L-X-IR_S-U_S-TR_S (**Figure 3**) in which TR_L and IR_L (7584 bp) are inverted repeats flanking a unique region (U_L, 167 843 bp), TR_S and IR_S (9774 bp) are inverted repeats flanking a unique region (U_S, 3370 bp), and X (1510 bp) is located between IR_L and IR_S. A somewhat similar genome structure has been reported for certain vertebrate herpesviruses (e.g., herpes simplex virus and human cytomegalovirus). A small proportion of OsHV-1 genomes either lacks the X-sequence or contains an additional X-sequence at the left terminus.

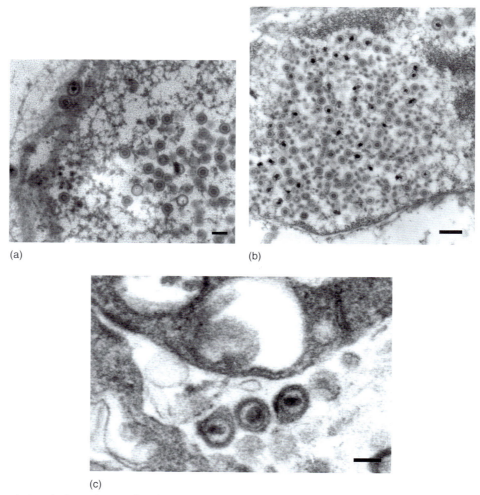

Figure 2 Transmission electron micrographs of ostreid herpesvirus 1 (OsHV-1) infecting Pacific oyster larvae. (a) Intranuclear spherical or polygonal virus particles; some particles appear empty and other contain an electron-dense core. Scale = 100 nm. (b) Nucleus of an infected interstial cell containing empty capsids and nucleocapsids. Scale = 200 nm. (c) High magnification of extracellular enveloped particles. Scale = 100 nm.

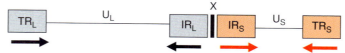

Figure 3 General genome organization of ostreid herpesvirus 1 (OsHV-1). TR_L and IR_L are inverted repeats flanking the unique region U_L. TR_S and IR_S are inverted repeats flanking the unique region U_S.

Since herpesvirus genomes are packaged into capsids from head-to-tail concatemers, this minor genome form may result from rare cleavage of concatemers at X–TR_S rather than at IR_L–IR_S. Moreover, approximately 20–25% of genomes contain a 4.8 kbp region of U_L in inverse orientation. The two orientations of U_L and U_S are present in approximately equimolar amounts in viral DNA, giving rise to four genomic isomers. This is also a feature of the vertebrate herpesvirus genomes with similar structures and results from recombination between inverted repeats during DNA replication. The genome termini are not unique but a predominant form is apparent for each. The IR_L–IR_S junction is also not unique, but the predominant form corresponds to a fusion of the two termini if each possesses two unpaired nucleotides at the 3′-end. Unpaired nucleotides are characteristic of herpesvirus genome termini.

Detailed analysis of the OsHV-1 genome sequence indicates that there are 124 unique open reading frames (ORFs). Owing to the presence of inverted repeats, 12 ORFs are duplicated resulting in a total of 136 genes in the viral genome. These numbers include several fragmented genes, each of which is counted as a single ORF. It is not yet known if splicing contributes to further elaboration of gene expression. A total of 38 genes shares species is accentuated by the lack of specific chemotherapies and vaccines. Improved knowledge and understanding of shellfish viruses is needed to develop new tools for disease control.

See also: Iridoviruses of Invertebrates.

Further Reading

Arzul I, Renault T, Lipart C, and Davison AJ (2001) Evidence for inter species transmission of oyster herpesvirus in marine bivalves. *Journal of General Virology* 82: 865–870.

Barbosa-Solomieu V, Dégremont L, Vazquez-Juarez R, et al. (2005) Ostreid herpesvirus 1 detection among three successive generations of Pacific oysters (*Crassostrea gigas*). *Virus Research* 107: 47–56.

Chang PH, Kuo ST, Lai SH, et al. (2005) Herpes-like virus infection causing mortality of cultured abalone *Haliotis diversicolor supertexta* in Taiwan. *Diseases of Aquatic Organisms* 65: 23–27.

Chou HY, Chang SJ, Lee HY, and Chiou YC (1998) Preliminary evidence for the effect of heavy metal cations on the susceptibility of hard clam (*Meretrix lusoria*) to clam birnavirus infection. *Fish Pathology* 33: 213–219.

Comps M and Duthoit JL (1979) Infections virales chez les huîtres *Crassostrea angulata* (Lmk) et *C. gigas* (Th.). *Haliotis* 8: 301–308.

Davison AJ, Trus BL, Cheng N, et al. (2005) A noval class of herpesvirus with bivalve hosts. *Journal of General Virology* 86: 41–43.

Farley CA, Banfield WG, Kasnic JRG, and Foster WS (1972) Oyster herpes-type virus. *Science* 178: 759–760.

Le Deuff R-M and Renault T (1999) Purification and partial genome characterization of a herpes-like virus infecting the Japanese oyster, *Crassostrea gigas*. *Journal of General Virology* 80: 1317–1322.

Lees D (2000) Viruses and bivalve shellfish. *International Journal of Food Microbiology* 59: 81–116.

Lipart C and Renault T (2002) Herpes-like virus detection in *Crassostrea gigas* spat using DIG-labelled probes. *Journal of Virological Methods* 101: 1–10.

McGeoch DJ, Rixon FJ, and Davison AJ (2006) Topics in herpesvirus genomics and evolution. *Virus Research* 117: 90–104.

Meyers TR and Hirai K (1980) Morphology of a reo-like virus isolated from juvenile American oysters (*Crassostrea virginica*). *Journal of General Virology* 46: 249–253.

Rasmussen LPD (1986) Virus-associated granulocytomas in the marine mussel, *Mytilus edulis*, from three sites in Denmark. *Journal of Invertebrate Pathology* 48: 117–123.

Renault T and Novoa B (2004) Viruses infecting bivalve molluscs. *Aquatic Living Resources* 17: 397–409.

Renault T, Le Deuff R-M, Lipart C, and Delsert C (2000) Development of a PCR procedure for the detection of a herpes-like virus infecting oysters in France. *Journal of Virological Methods* 88: 41–50.

Shrimp Viruses

J-R Bonami, CNRS, Montpellier, France

© 2008 Elsevier Ltd. All rights reserved.

Glossary

Cephalothorax The shrimp head, containing the main organs, hepatopancreas, stomach, foregut and midgut, gonads, heart, gills.

Epizootic An epidemic in animal populations. Rapid spreading of a disease.

Hepatopancreas Organ of the digestive tract with secretion–absorption functions and located in the cephalothorax (head) of shrimp. Also called the digestive gland.

Postlarvae Stage of shrimp development, after the larval stages.

Introduction

To date, more than 20 viral diseases have been reported in shrimp and prawns. Most of the described viruses are related, but often only on the basis of morphological characteristics to known virus families. Two of the most important pathogens of shrimp have been sufficiently characterized to be accepted by the International Committee on Taxonomy of Viruses (ICTV) as members of new virus families within the classification and nomenclature of viruses. Yellow head virus (YHV), together with closely related gill-associated virus (GAV), have been classified as members of the new genus *Okavirus* in the new family *Roniviridae*. White spot syndrome virus (WSSV) has been classified as the only known member of the new genus *Whispovirus* in the new family *Nimaviridae*. A third major pathogen, Taura syndrome virus (TSV), has been accepted for classification in the family *Dicistroviridae*. Although not yet officially accepted by the ICTV, six other shrimp viruses have been sufficiently well characterized to be considered as possible members of four existing virus families: *Totiviridae* (infectious myonecrosis virus, IMNV), *Nodaviridae* (macrobrachium rosenbergeii nodavirus, MrNV), *Parvoviridae* (infectious hypodermal and hematopoietic necrosis virus, IHHNV, and hepatopancreatic parvovirus, HPV) and *Baculoviridae* (BP-type and MBV-type viruses).

History

Three periods can be defined in the recent history of shrimp virus discovery: the first was during the 1970s, in which observations of viral agents were made by 'chance'; the second period commenced in the 1980s, in which evidence of viral pathogenic agents was obtained as a result of disease observations or mortalities occurring in shrimp farms; and finally, since the early 1990s with the development of molecular biology, more structured and intensive approaches to the descriptions of the pathogens and the associated diseases were undertaken.

The first virus of Crustacea was reported in 1966 by Vago in the Mediterranean crab *Portunus depurator*. The virus appeared to be related to members of the *Reoviridae*. In shrimp, occluded bacilliform particles were reported in 1973 by Couch during investigations on the effect of pollutants in pink shrimp (*Penaeus duorarum*). The agent was reported to be closely related to members of the *Baculoviridae*.

In 1981, Sano and colleagues reported the detection of a 'nonoccluded baculovirus' in kuruma shrimp (*Penaeus japonicus*) during mortalities in hatchery-reared shrimp in two prefectures in Japan. Two years later (1983), Lightner reported a transmissible disease named infectious hypodermal and hematopoietic necrosis (IHHN) as the Pacific blue shrimp (*Penaeus stylirostris*) as the source of mortalities in ponds. While at first thought to be related to members of the *Picornaviridae*, IHHNV was later assigned to the *Parvoviridae* after investigations on the structure of the genome. Since that time, reports of new viruses have increased in parallel with the rapid development of shrimp farming globally and, to date, more than 20 viruses have been reported in shrimp (**Table 1**).

Table 1 Known shrimp viruses

Genome	Virus family[a]	Acronym	Virus name	Host	Genomic data
ssRNA	*Dicistroviridae*	TSV	Taura syndrome virus	*P. vannamei*	Complete
	Roniviridae	YHV	Yellow head virus	*P. monodon*	Complete
		GAV	Gill-associated virus		Complete
	Bunyaviridae	MoV	Mourilyan virus	*P. monodon*	Partial
				P. japonicus	
	Totiviridae	IMNV	Infectious myonecrosis virus	*P. vannamei*	Complete
	Rhabdoviridae	RPS		*P. stylirostris*	n.d.
				P. vannamei	
	Togaviridae	LOVV	Lymphoid organ vacuolisation virus	*P. vannamei*	n.d.
	Nodaviridae	MrNV/XSV	Macrobrachium rosenbergii nodavirus/extra small virus	*M. rosenbergii*	Complete
		LSNV		*P. monodon*	n.d.
dsRNA	*Birnaviridae*	IPN-like virus	Infectious pancreatic necrosis-like virus	*P. japonicus*	n.d.
	Reoviridae	Reo-Pj		*P. japonicus*	n.d.
		Reo-Pm		*P. monodon*	n.d.
		Reo-Pv		*P. vannamei*	n.d.
		PBRV		*Palaemon* sp.	n.d.
ssDNA	*Parvoviridae*	IHHNV	Infectious hemocytic and hematopoietic virus	*P. stylirostris*	Complete
		HPV	Hepatopancreatic parvovirus	*P. semisulcatus*	Complete
		HPV-like		*P. merguiensis*	n.d.
				M. rosenbergii	n.d.
		SMV		*P. monodon*	n.d.
		LPV		*P. monodon*	n.d.
				F. merguiensis	n.d.
				P. esculentus	n.d.
dsDNA	*Iridoviridae*			*Protrachypene precipua*	n.d.
	Nonoccluded bacilliform virus	BMNV or (PjNOB)	Baculoviral midgut gland necrosis virus	*P. japonicus*	Partial
		PHRV			n.d.
		CcBV		*Crangon crangon*	n.d.
	Baculoviridae	PvNPV or (BP-type)	Single-nucleocapsid polyedrosis virus	*P. duorarum*	n.d.
		PmNPV or (MBV-type)	Single-nucleocapsid polyedrosis virus	*P. vannamei*	Partial
		MbNPV	Single-nucleocapsid polyedrosis virus	*P. monodon*	Partial
				Metapenaeus bennettae	n.d.
	Nimaviridae	WSSV	White spot syndrome virus	Penaeids and numerous other crustaceans	Complete

[a]Only TSV, GAV, YHV, IHHNV, and WSSV have been formally classified to date.

The major developments in shrimp virus research commenced in the early 1990s following the development and application of molecular tools. Cloning and sequencing of genome fragments allowed construction of specific and sensitive tools for detection and diagnosis. These DNA-based methods, such as the polymerase chain reaction (PCR) and dot-blot hybridization, have led to the use of commercial diagnostic kits for selection of healthy shrimp and the production of specific pathogen-free (SPF) breeding stock. SPF shrimp have proven to be particularly useful, not only in limiting or preventing production losses due to disease on farms, but also as experimental animals for studies on shrimp viruses.

Impact of Viral Diseases on Farmed Shrimp Production

The rapid progress in shrimp virology during the past two decades is clearly related to the high commercial value of these farmed crustaceans. This has significantly increased our understanding of viruses infecting only invertebrates which previously had been confined primarily to insect viruses. However, in contrast to most insect viruses in which investigations were oriented primarily toward biological control of insect pests and their potential use as gene expression vectors, shrimp virus research has focused on disease prevention.

It is difficult to know the impact of viral diseases on shrimp populations in the natural environment. Diseased shrimp are rarely found, essentially due to predation and cannibalism when animals appear weakened. However, under farming conditions, where the density of the shrimp population can easily exceed 300 000 postlarvae per hectare, the consequence of a disease outbreak and associated mass mortalities can be disastrous. In 2003, total world production from shrimp farming was more than 1.6 million metric tons, representing a value of almost US$ 9000 million and accounting for about 25% of marketable shrimp production (capture and farming). Viral disease has been a major problem for this large and expanding industry and the economic and socioeconomic impacts have been severe. For example, a report commissioned by the World Bank in 1996 estimated the annual global cost of disease (primarily viral) to shrimp farming was US$ 3000 million or 40% of production capacity at that time. In China and Thailand alone, annual production losses due to WSSV have reached US$ 500–1000 million. In Ecuador, the emergence of WSSV in 1999 resulted in a 75% drop in annual production and 130 000 industry workers were reported to have lost their jobs. The production records of these major shrimp-producing countries exhibit depressions coinciding with the sequential emergence of major viral pathogens. In contrast, in Brazil, which was free of the major diseases until 2003, shrimp production constantly increased. Nevertheless, Brazil has also experienced significant production losses since that time with the emergence of infectious myonecrosis caused by a new toti-like virus.

Factors Responsible for Viral Disease Emergence in Shrimp

Disease emergence in shrimp aquaculture can be attributed to three primary factors: (1) the common practice in the shrimp farming industry of introducing each season fresh wild broodstock collected from the natural environment as the source of seed (postlarval shrimp) for farms; (2) national and international trade in broodstock and commercial seed (postlarvae) to be transferred to ponds until they reach the marketable size; and (3) the culture of shrimp in earthern ponds in high densities.

The use of wild brooders in hatcheries commonly occurs without adequate knowledge of their health status or adequate quarantine. This often results in the introduction of pathogens into the farming system and it is a practice that persists even though SPF shrimp, bred in biosecure facilities, are now readily available in many countries.

Trade in broodstock and seed also commonly occurs without checking for at least the known pathogens. This practice was responsible for the rapid worldwide spread of the IHHNV, the introduction of TSV to the Eastern Hemisphere, and the rapid spread of WSSV from the Eastern to the Western Hemisphere. The practice of farming in brackish water earthern ponds creates an environment in which new pathogens may be encountered and disease spread rapidly once outbreaks occur. As a result, short explosive epizootics may, in few days, wipe out the whole shrimp population of a farm.

Environmental and host factors also play an important role in the emergence of shrimp disease. Poor-quality pond conditions (e.g., high salinity, pH, or nitrogen) can cause physiological stress in shrimp leading to increased susceptibility to disease. The shrimp culture medium (brackish water) allows very efficient transfer of free viruses, either as free infectious agents or in pathogen vectors. As ectotherms, shrimps are susceptible to water temperature shifts that can stimulate the replication of viruses that commonly persist in shrimp as unapparent infections. Shrimp also lack the sophisticated adaptive immune mechanisms (antibodies, cytokines, T-cells) deployed by vertebrates to combat viral infections and so have less capacity to survive disease.

Characteristics of Some Shrimp Viruses

Shrimp Baculoviruses

Two distinct groups of viruses, sharing numerous characters with members of the family *Baculoviridae*, have been described in shrimp. They have been improperly called BP-type viruses (BP, baculovirus penaei) and MBV-type viruses (MBV, monodon baculovirus) on the basis of the morphology of their occlusion bodies (OBs). The BP-type group exhibits tetrahedral OBs; the MBV group exhibits rounded OBs. It is classified as a tentative species (*Penaeus monodon* NPV, PemoNPV) in the genus *Nucleopolyhedrovirus* of the family *Baculoviridae*.

General characteristics

All baculoviruses detected in shrimp develop in the nuclei of digestive epithelial cells (hepatopancreas and mid-gut) with the release of OBs within feces. They are bacilliform in shape and comprise a trilaminar envelope surrounding a rod-shaped nucleocapsid. Virions may be occluded in paracrystalline occlusion bodies. Only small fragments of the double-stranded DNA (dsDNA) genome have been cloned and sequenced for a few shrimp baculoviruses. For the BP-type virus, Penaeus duorarum single nuclear polyhedrosis virus (PdSNPV), the genome has been shown by electron microscopy to be a circular structure with an estimated size of 75×10^6 kDa (113–115 kbp). There appear to be different morphotypes of BP-type baculoviruses which were distinguished based on differences in particle size; two different BP genotypes have been reported in the same shrimp tissue section by *in situ* hybridization.

Polyhedra

In contrast to insect baculoviruses, shrimp baculovirus OBs are unenveloped and comprise large subunits (SuOBs) of polyhedrin which are icosahedral in shape (**Figure 1**). The SuOBs are always associated in triplets. They are organized in two different crystalline forms: the BP-type viruses form parallel rows in all three dimensions (**Figure 2(a)**); the MBV-type viruses assemble four triplets to form a 'rosette' (hollow sphere) as the building block of the OB (**Figure 2(b)**). MBV-type OBs are more sensitive to thaw–freezing but each OB type disaggregates spontaneously in CsCl gradients. Compared to insect baculovirus polyhedrin subunits, both BP- and MBV-type SuOBs are larger in dimensions (15–22 nm compared to 6.5–9 nm diameter) and molecular mass (52–58 kDa compared to $\sim$30 kDa).

There is evidence of antigenic cross-reactivity between PdSNPV and MBV SuOBs, but this requires confirmation as, unexpectedly, PdSNPV SuOBs were also found to cross-react with both polyhedrin and granulin from the insect baculoviruses Autographa californica nuclear polyhedrosis virus and and Trichoplusia ni granulosis virus, respectively.

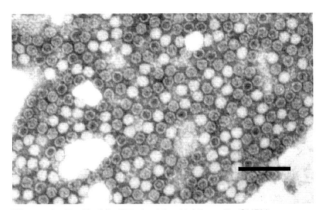

Figure 1 Purified SuOBs (polyhedrin) from PmSNPV (MBV-type virus). Note the full and empty virus-like particles. The particles were negatively stained with 2% phosphotungstic acid (PTA). Scale = 100 nm. Reproduced from Bonami: JB, Aubert H, Mari J, Poulos BT, and Lightner DV (1997) The polyhedra of the occluded baculoviruses of marine decapod crustacea: A unique structure, crystal organization, and proposed model. *Journal of Structural Biology* 120: 134–145, with permission from Elsevier.

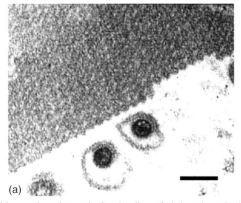 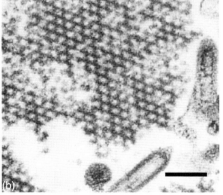

Figure 2 Ultrathin sections in occlusion bodies of shrimp baculoviruses; (a) PvSNPV (BP-type virus) and (b) PmSNPV (MBV-type virus). The two types of crystal (bullet-like arrangement and rosette formation) are evident. Scale = 100 nm (a, b). Reproduced from Bonami JB, Aubert H, Mari J, Poulos BT, and Lightner DV (1997) The polyhedra of the occluded baculoviruses of marine decapod crustacea: A unique structure, crystal organization, and proposed model. *Journal of Structural Biology* 120: 134–145, with permission from Elsevier.

Viral infection cycle

All life stages of the pink shrimp (*P. duorarum*), white shrimp (*Penaeus setiferus*), and the brown shrimp (*Penaeus aztecus*) are susceptible to BP-type baculovirus, while only larval and postlarval stages of the Pacific white shrimp (*Penaeus vannamei*) and Pacific blue shrimp (*P. stylirostris*) have been observed to be infected. In the black tiger shrimp (*Penaeus monodon*), postlarvae, juveniles, and adults have been observed to be infected with MBV.

For PdSNPV in *P. duorarum*, the major steps in the infection cycle have been described including the appearance of naked nucleocapsids in the cytoplasm of epithelial digestive cells, nucleoprotein release at nuclear pores, followed by all morphogenesis steps in infected nuclei and the development of tetrahedral OBs. Cellular pathological changes include nuclear hypertrophy and perinuclear membranous proliferation. Nuclear hypertrophy associated with increasing size of OBs leads to the release of the nuclear contents into the digestive lumen of hepatopancreatic tubules and gut, and the excretion of OBs in the feces. The infection cycle is completed by oral ingestion and OB dissolution in the digestive tract of the new host. A similar infection cycle is hypothesized for MBV-type baculoviruses. In addition to horizontal transmission, vertical transmission is suspected, based on the presence of OBs in very young postlarvae a few days after hatching.

Geographic distribution and host range

BP-type virus infection is widespread in farmed and wild penaeid shrimp in the Americas. The known distribution is from the Gulf of Mexico in the north through the Caribbean, the east coast of South America as far south as the State of Bahia in Central Brazil. On the Pacific coast, the distribution ranges from Mexico to Peru. BP-type virus has also been observed in wild shrimp from Hawaii. Several penaeid shrimp species have been reported to be infected, both in aquaculture facilities and in the wild environment. These include *P. duorarum*, *P. aztecus*, *P. vannamei*, *P. setiferus*, *P. stylirostris*, *P. penicillatus*, *P. schmitti*, *P. paulensis*, *P. subtilis*, *Trachypenaeus similis*, and *Protrachypene precipua*.

MBV-type virus infections are widely distributed in penaeid shrimp of the Eastern Hemisphere, particularly in the Indo-Pacific countries of China, India, Indonesia, Philippines, Malaysia, Thailand, Sri Lanka, and Australia where the virus is enzootic in wild stocks. The virus has also been reported in Kuwait, Oman, Israel, Italy, and West Africa (Kenya and Gambia). It has been observed in imported penaeid shrimp in Tahiti and Hawaii, Mexico, Ecuador, Brazil, Puerto Rico, and some states of the USA. The known host range includes *P. monodon*, *P. merguiensis*, *P. semisulcatus*, *P. indicus*, *P. penicillatus*, *P. esculentus*, and *P. vannamei*.

Shrimp Parvoviruses

Two viruses sharing the primary characteristics of members of the family *Parvoviridae* have been reported in penaeid shrimp. IHHNV causes a systemic infection of multiple organs of ectodermal and mesodermal origin as and is described in the Pacific blue shrimp (*P. stylirostris*). It is associated with runt and deformity syndrome (RDS) in Pacific white shrimp (*P. vannamei*) and occurs commonly as a low-level persistent infection in apparently healthy black tiger shrimp (*P. monodon*). HPV causes an infection of the digestive system (epithelial cells of hepatopancreas and mid-gut) in black tiger shrimp *P. monodon* and in the fleshy prawn (*P. chinensis*). HPV isolates from *P. monodon* (HPVmon) and *P. chinensis* (HPVchin) represent two distinct genetic lineages. A third parvo-like virus, spawner-isolated mortality virus (SMV) infects *P. monodon* but is less well characterized.

IHHNV and HPV are each nonenveloped icosahedral viruses, 22 nm in diameter and containing a linear single-stranded DNA (ssDNA) genome. IHHNV virions have been reported to contain four polypeptides as detected by sodium dodecyl sulfate polyacrylamide gel electrophoresis (SDS-PAGE) and silver staining (74, 47, 39, and 37.5 kDa). For HPVchin, only one virion structural protein has been observed (54 kDa) but analysis of HPVmon has revealed a doublet protein band (57 kDa major band and 54 kDa minor band). In addition to differences in tissue tropism, the viruses display differences in cytopathology. During HPV infection, cellular lesions correspond to typical densonucleosis in insects with the formation of enlarged, densely stained, and Feulgen-positive nuclei. In IHHNV infections, lesions are discrete, often difficult to detect, and display characteristic eosinophilic intranuclear Cowdry type A inclusion bodies.

Genomic organization

The IHHNV genome comprises 4075 nucleotides (according to a recent GenBank submission) and contains three long open reading frames (ORFs) on the complementary (positive) strand (**Figure 3**). The three ORFs are referred to as left, mid, and right. The left ORF is in a different reading frame (+1) to mid- and right ORFs. As for other parvoviruses, noncoding termini at each end of the genome contain palindromic sequences. The left ORF encodes a 666 amino acid protein with high homology with putative major nonstructural protein (NS-1) of mosquito brevidensoviruses. The mid-ORF overlaps the left ORF and encodes a 363 aa protein of unknown function (possibly an NS-2 protein). The right ORF overlaps the left ORF by 62 nt and encode a 329 aa protein. By analogy to other parvoviruses, it has been speculated that the right ORF encodes the capsid proteins but this is not consistent with the four structural proteins reported in purified virions. The negative strand also contains an ORF with a

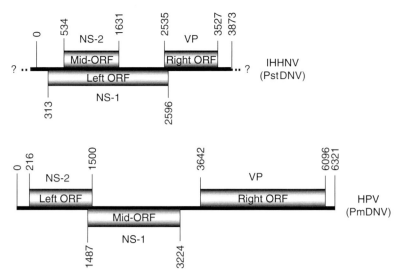

Figure 3 Genome organization of IHHNV and HPV. IHHNV (PstDNV) is according to Shike H, Dhar AK, Burns JC, et al. (2000) Infections hypodermal and haematopoietic necrosis virus of shrimp is related to mosquito brevidensoviruses. *Virology* 277: 167–177.

coding capacity of 134 amino acids. Further work is required to define the IHHNV expression strategy.

The HPV genome comprises 6321 nucleotides containing three long ORFs in the complementary (positive) strand and palindromic noncoding termini that form hairpin-like structures, as found in other parvoviruses (**Figure 3**). ORF1 (left ORF) encodes a 428 amino acid putative nonstructural protein-2 (NS-2) of unknown function. ORF2 (mid-ORF), in −1 frame relative to ORF1 encodes a 579-amino-acid polypeptide that contains conserved replication initiator motifs, NTP-binding, and helicase domains similar to NS-1 of other parvoviruses. The right ORF (ORF3), which is in the same frame as ORF1, encodes the 818-amino-acid capsid protein (VP protein).

Taxonomic position

In terms of morphology and genome structure and organization, both IHHNV and HPV share similarities with members of the family *Parvoviridae*. Genomic analysis has clearly indicated that IHHNV is closely related to mosquito densoviruses and it has recently been classified as a tentative species (*Penaeus stylirostris densovirus*), in the genus *Brevidensovirus*. Similarly, HPV was recently proposed as new member of subfamily *Densovirinae* to be designated as the species *Penaeus monodon densovirus*.

Geographic distribution and host range

HPV was first observed in wild and farmed penaeid shrimp in Australia, China, Korea, Philippines, Indonesia, Malaysia, Kenya, Kuwait, and Israel. It has also been reported subsequently from several locations in North and South America (Pacific coast of Mexico and El Salvador). Natural infections have been reported in *P. merguiensis*, *P. semisulcatus*, *P. chinensis*, *P. esculentus*, *P. monodon*, *P. japonicus*, *P. penicillatus*, *P. indicus*, *P. vannamei*, and *P. stylirostris*.

IHHNV is widely distributed in the Americas (Brazil, Ecuador, Central America, Mexico, Peru, southeast USA), the Central Pacific (Guam, Hawaii, Tahiti, and New Caledonia), and Asia and the Indo-Pacific area (Indonesia, Malaysia, Philippines, Thailand, and Australia). Natural infections have been reported in *P. stylirostris*, *P. vannamei*, *P. occidentalis*, *P. californiensis*, *P. monodon*, *P. semisulcatus*, and *P. japonicus*. It is likely that infection with IHHNV or a similar virus also occurs in other penaeid shrimp species.

Recent investigations have revealed evidence of the integration of IHHNV-related sequences within the genomes of *P. monodon* populations from Africa (Madagascar, Tanzania, Mozambique) and Australia. The integrated sequences vary by up to 14% from epidemic IHHNV and do not appear to be associated with the formation of complete or infections virions.

Infectious Myonecrosis Virus

In 2002, a new disease associated with low mortalities was reported in the Pacific white shrimp *Penaeus vannamei* from northeastern Brazil. Clinical signs included necrotic areas in muscles, resulting sometimes in a whitish and opaque appearance to the tail. The causative agent was named infectious myonecrosis virus (IMNV) on the basis of the clinical signs.

General properties

In infected animals, virions accumulate in the cytoplasm of target cells (muscular cells, hemocytes, and connective tissue cells) forming basophilic inclusions. IMNV has a density of $1.366\,\text{g}\,\text{ml}^{-1}$ in cesium chloride. It is a nonenveloped icosahedral virus, 40 nm in diameter (**Figure 4**) with a genome consisting of a single segment of double-stranded RNA (dsRNA), 7560 bp in length. The capsid contains a single major polypeptide of 106 kDa.

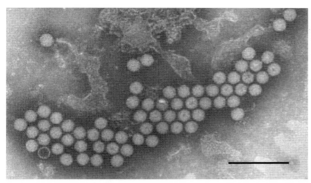

Figure 4 Purified IMNV virions. The particles were negatively stained with 2% phosphotungstic acid (PTA). Scale = 200 nm. From Poulos BT, Tang KFJ, Pantoja CR, Bonami JR, and Lightner DV (2006) Purification and characterization of infectious myonecrosis virus of penaeid shrimp. *Journal of General Virology* 87: 987–996.

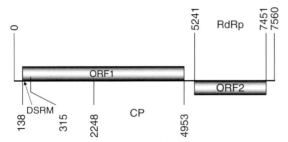

Figure 5 Genome organization of IMNV.

Genomic organization

The IMNV genome organization is illustrated in **Figure 5**. Two large nonoverlapping ORFs in different reading frames (ORF1 and ORF2) are flanked by 3′ untranslated region (UTR) and 5′ UTR of 109 and 135 nt, respectively, and are separated by a noncoding region of 287 nt. ORF1 contains a dsRNA-binding motif (DSRM) of 60 amino acids in the N-terminal region. The translated sequence of 901 amino acids corresponds approximately to the size determined in SDS-PAGE for the major capsid protein (106 kDa). ORF2 encodes an RNA-dependent RNA polymerase (RdRp) with significant sequence similarity to members of the family *Totiviridae*.

Taxonomic position

Phylogenetic analysis of IMNV RdRp clusters this virus within members of the genus *Giardiavirus* in the family *Totiviridae*. The genome structure and organization are also consistent with classification in the *Totiviridae*. However, as known giardiviruses infect only flagellated protozoan parasites (*Giardia lamblia* and *Trichomonas vaginalis*), further consideration of the properties of IMNV will be required before formal classification can be obtained.

Geographic distribution and host range

The disease was, until recently, restricted to farmed *P. vannamei* in northeastern Brazil. However, recently the virus and the disease have been reported at several sites in Indonesia where infected *P. vannamei* postlarvae have been introduced from Brazil. *Penaeus stylirostris* and *P. monodon* shrimps are susceptible to experimental infection with IMNV.

MrNV/XSV Complex

White tail disease occurs in farmed giant freshwater prawns (*Macrobrachium rosenbergii*) and is named after the obvious clinical signs in diseased postlarvae. Two viruses, Macrobrachium rosenbergii nodavirus (MrNV) and a very small virus-like particle named XSV (extra small virus), are each found in diseased prawns. The disease was first observed in a hatchery in Guadeloupe Island in French West Indies where abnormal and sudden mortalities had been recorded since 1994. Losses were variable in intensity (5–90% cumulative mortality), depending on the broodstock used to generate the postlarvae. The first gross sign of the disease is the presence of whitish postlarvae 2–3 days after emergence. The prevalence of opaque and milky postlarvae increases dramatically 1–2 days later and changes are particularly obvious in abdomen (tail). The mortality rate reaches a maximum 5 days after the first observation of gross signs.

In diseased prawns, the most affected tissues are the striated muscles in the abdomen and cephalothorax and the connective tissues of all organs. The cytoplasm of infected cells contains discrete, pale to darkly basophilic inclusions, ranging from <1 to 40 μm in diameter. Muscles exhibit multifocal areas of hyaline necrosis of the fibers, with moderate edema.

General properties of isolated particles

Two types of particles are observed in the cytoplasm of infected cells (**Figure 6**). The larger particles are nonenveloped, icosahedral in shape, and 26–27 nm in diameter, with a density in CsCl of 1.27–1.28 g ml^{-1}. The genome consists of two segments of linear single-stranded RNA (ssRNA) of 3202 (RNA1) and 1250 nt (RNA2). The capsid contains a single major protein of 43 kDa (CP43). The properties of this virus are similar to those of nodaviruses and it has been named Machrobrachium rosenbergii nodavirus (MrNV).

The smaller particles are nonenveloped, icosahedral, and 15 nm in diameter, and have a density in CsCl of 1.325 g ml^{-1}. The genome is composed of a single segment of linear ssRNA 796 nt in length. The capsid comprises coat proteins of 16 kDa (CP16) and 17 kDa (CP17). Because the particles are very small, this agent has been called extra small virus (XSV).

Genomic organization

The organization of the MrNV genome is shown in **Figure 7**. RNA1 and RNA2 each encodes a long ORF

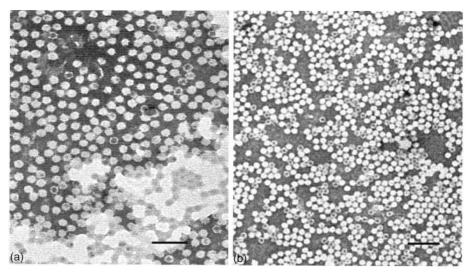

Figure 6 Purified MrNV and XSV particles. The particles were negatively stained with 2% phosphotungstic acid (PTA). Scale = 50 nm. (a, b) Reproduced from Bonami JR, Shi Z, Qian D, and Sri Widada J (2005) White tail disease (WTD) associated virions and characterization of MrNV as a new type on nodavirus. *Journal of Fish Diseases* 28: 23–31, with permission from Blackwell Publishing.

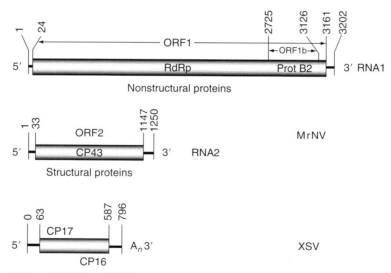

Figure 7 Genome organizations of MrNV and XSV.

flanked by terminal noncoding regions, and lacks a 3′ poly(A) tail. The coding capacity is 1046 and 371 amino acids, equivalent to 122 and 43.5 kDa, for ORF1 and ORF2, respectively. Similar to other nodaviruses, RNA1 contains two long ORFs encoding nonstructural proteins: ORF1 encodes the 1046-amino-acid (~122 kDa) A protein which contains motifs characteristic of the RdRp; and ORF1b, located at the 3′ end of the segment, encodes the 134-amino-acid (~13 kDa) B protein. ORF2 encodes the 371-amino-acid (~43.5 kDa) capsid protein (CP43). The coding assignment of RNA2 has been demonstrated by N-terminal amino acid sequencing of CP43. As for all other nodaviruses, a nonencapsidated RNA3 of 453 nucleotides has also been detected. The MrNV RNA3 is longer than that of any other nodavirus RNA3 (~380–400 nt) and is likely to be a subgenomic mRNA encoding ORF1b.

The 796 nt XSV genome contains a single ORF located between nucleotides 63 and 587, and a short poly(A) tail. A potential polyadenylation signal (AAUAAA) is located 6 nt upstream of the poly(A). The ORF encodes both capsid proteins (CP16 and CP17). Amino acid sequence analysis indicates that CP16 is synthesized by internal initiation at a second methionine residue 11 amino acids downstream of the first initiation codon. The two forms of the capsid protein are synthesized in an equimolar ratio.

Taxonomic position

No formal taxonomic classification of MrNV has yet been assigned. The particle morphology and genome organization are similar to those of members of the family *Nodaviridae*. Amino acid sequence alignments of the RdRp indicate highest homology with members of the genera *Alphanodavirus* (insect nodaviruses) and *Betanodavirus* (fish nodaviruses). Sequence similarities are higher with alphanodaviruses than with betanodaviruses. However, coat protein sequence alignments indicate that the distance between MrNV and alphanodaviruses is similar to those separating alphanodaviruses and betanodavirues. This suggests that MrNV may represent a possible third genus within the family.

The appropriate taxonomic classification of XSV is currently unclear. Sequence similarity searches have not indicated any significant homology with known virus genomes. As the monocistronic genome encodes only the capsid proteins, it is possible that the RdRp required for replication and transcription may be provided by MrNV. This agent meets the primary criteria of satellite viruses and appears more similar to the known satellite viruses infecting plants (e.g., tobacco necrosis satellite virus-like) than those infecting insects (e.g., chronic bee paralysis-associated satellite virus).

MrNV/XSV relationships

MrNV and XSV are each always found associated in white tail disease infections in *M. rosenbergii* in all hatcheries and farms where disease is reported. Analysis by quantitative (real-time) polymerase chain reaction (PCR) has indicated variable proportions of MrNV RNA and XSV RNA in different isolates. However, nothing is known to date about their respective roles in the disease or its severity.

Geographic distribution and host range

White tail disease was first recognized in a hatchery in Guadeloupe Island and was then reported in Martinique Island and in the Dominican Republic. It is suspected that infected postlarvae were introduced to Puerto Rico.

In China, the disease was recorded in Zhejiang, Jiangsu, Shanghai, Guangxi, and Guangdong Provinces and, more recently, it has been reported in Taiwan. In India, the disease has been reported in Andhra Pradesh and Tamil Nadu on the east coast and in Kerala on the west coast where *M. rosenbergii* farming was developed to replace marine (*P. monodon*) culture following the dramatic effects of the WSSV epidemic.

To date, white tail disease has only been reported in the giant freshwater prawn *M. rosenbergii*. However, marine shrimp (*P. monodon*, *P. indicus*, and *P. japonicus*) are susceptible to experimental infection, suggesting a possible role as reservoir. No disease has been reported in marine shrimp species.

Control of Viral Diseases in Shrimp

As invertebrates, marine shrimp and freshwater prawns lack the acquired immune systems that exist in vertebrates. Consequently, a disease control based on vaccination cannot be used. Some antiviral defense reactions (innate immunity) have been reported in Crustacea but the mechanisms by which these occur are largely unknown and prevention appears as the most useful method for control. For this reason, specific and sensitive DNA-based tools for early detection are now available for most of the viral diseases. These tools allow selection of healthy broodstock and a good knowledge of the health status of postlarval seed introduced into ponds. The availability of these tools has also led to the establishment of biosecure breeding programs for commercial production of specific pathogen-free (SPF) broodstock. The more widespread use of SPF stock will significantly reduce the risk of future emergence of devastating viral diseases in shrimp aquaculture.

See also: Baculoviruses: General Features; Giardiaviruses; Insect Viruses: Nonoccluded; Nodaviruses; Parvoviruses of Arthropods; Satellite Nucleic Acids and Viruses; Taura Syndrome Virus; Totiviruses; White Spot Syndrome Virus; Yellow Head Virus.

Further Reading

Bonami JR, Aubert H, Mari J, Poulos BT, and Lightner DV (1997) The polyhedra of the occluded baculoviruses of marine decapod crustacea: A unique structure, crystal organization, and proposed model. *Journal of Structural Biology* 120: 134–145.

Bonami JR, Mari J, Poulos BT, and Lightner DV (1995) Characterization of the HPV, a second unusual parvovirus pathogenic for penaeid shrimp. *Journal of General Virology* 76: 813–817.

Bonami JR, Shi Z, Qian D, and Sri Widada J (2005) White tail disease of the giant freshwater prawn, *Macrobrachium rosenbergii*: Separation of the associated virions and charaterization of MrNV as a new type of nodavirus. *Journal of Fish Diseases* 28: 23–31.

Bonami JR, Trumper B, Mari J, Brehelin M, and Lightner DV (1990) Purification and characterization of the infectious hypodermal and haematopoietic necrosis virus penaeid shrimps. *Journal of General Virology* 71: 2657–2664.

Couch JA (1991) *Baculoviridae*. Nuclear polyhedrosis viruses. Part 2: Nuclear polyhedrosis viruses of invertebrates other than insects. In: Adam JR and Bonami JR (eds.) *Atlas of Invertebrate Viruses*, pp. 205–226. Boca Raton, FL: CRC Press.

Lightner DV (1988) Diseases of cultured penaeid shrimp and prawns. In: Sindermann CJ and Lightner DV (eds.) *Diseases Diagnosis and Control in North American Marine Aquaculture*, pp. 8–127. Amsterdam: Elsevier.

Lightner DV (ed.) (1996) *A Handbook of Shrimp Pathology and Diagnostic Procedures for Diseases of Cultured Penaeid Shrimp*, 304pp. Baton Rouge, LA: World Aquaculture Society.

Mari J, Bonami JR, and Lightner DV (1993) Partial cloning of the genome of the IHHNV, an unusual parvovirus pathogenic from penaeid shrimp. Diagnosis of the disease using a specific probe. *Journal of General Virology* 74: 2637–2643.

Pillai D, Bonami JR, and Sri Widada J (2006) Rapid detection of macrobrachium rosenbergii nodavirus (MrNV) and extra small virus (XSV), the pathogenic agents of white tail disease of *Macrobrachium* rosenbergii (De Man), by loop-mediated isothermal amplification. *Journal of Fish Diseases* 29: 275–283.

Poulos BT, Tang KFJ, Pantoja CR, Bonami JR, and Lightner DV (2006) Purification and characterization of infectious myonecrosis virus of penaeid shrimp. *Journal of General Virology* 87: 987–996.

Sahul Hameed AS, Yoganandhan K, Sri Widada J, and Bonami JR (2004) Studies on the occurrence of macrobrachium rosenbergii nodavirus and extra small virus-like particles associated with white tail disease of *M. rosenbergii* in India by RT-PCR detection. *Aquaculture* 238: 127–133.

Shike H, Dhar AK, Burns JC, et al. (2000) Infectious hypodermal and haematopoietic necrosis virus of shrimp is related to mosquito brevidensoviruses. *Virology* 277: 167–177.

Sri Widada J, Richard V, and Bonami JR (2004) Characteristics of the monocistronic genome of extra small virus (XSV), a virus-like particle associated with macrobrachium rosenbergii nodavirus (*Mr*NV): Possible candidate for a new species of satellite virus. *Journal of General Virology* 85: 643–646.

Sukhumsirichart W, Attasart P, Boonsaeng V, and Panyim S (2006) Complete nucleotide sequence and genomic organization of hepatopancreatic parvovirus (HPV) of *Penaeus monodon*. *Virology* 346: 266–277.

Zhang H, Wang J, Yuan J, et al. (2006) Quantitative relationship of two viruses (MrNV and XSV) in white tail disease of *Macrobrachium rosenbergii* de Man. *Diseases of Aquatic Organisms* 71: 11–17.

Sigma Rhabdoviruses

D Contamine and S Gaumer, Université Versailles St-Quentin, CNRS, Versailles, France

© 2008 Elsevier Ltd. All rights reserved.

Glossary

Permissive allele A host allele which allows or participates in viral propagation.
Restrictive allele A host allele which opposes viral propagation.
Stabilized lineage for a virus A female or male host lineage, with infected primordial germinal cells, which transmits the virus to its progeny.

Introduction

Sigma virus (SIGV) is a rhabdovirus that naturally infects fruit flies (*Drosophila* spp.). SIGV infection causes increased sensitivity to carbon dioxide (CO_2) gas, which is anesthetic for flies. Whereas uninfected flies recover rapidly from the effects of CO_2 exposure upon return to a normal atmosphere, flies infected with SIGV remain irreversibly paralyzed. This effect has been observed among both wild flies and laboratory strains of as many as 13 *Drosophila* species and in three of these (*D. melanogaster*, *D. affinis*, and *D. athabasca*) it has been shown to be the consequence of infection.

The discovery of SIGV originates from 1937 when a study by P. L'Héritier and G. Tessier described a CO_2-sensitive strain of *D. melanogaster*. While this gas is commonly used as an anesthetic in fly genetics, these flies were irreversibly paralyzed when exposed to a CO_2-rich atmosphere. The paralysis was specific to CO_2, and dependent on the gas concentration and temperature during exposure. For example, irreversible paralysis appears at 10 °C with CO_2 concentrations higher than 50% while CO_2 concentrations must reach 75% to induce irreversible paralysis at 16 °C.

Transmission of this characterstic appeared to be heritable but could not be linked to any chromosome. The agent was considered to be cytoplasmic and was named Sigma. Its infectivity was demonstrated by inoculation of resistant flies with sensitive-fly extracts. Inactivation by X-ray irradiation indicated the sensitive volume diameter to be 42 nm and filtration through a 180-nm membrane eliminated 99% of infectivity. In 1965, Sigma was finally described as a particle of 70×140 nm similar to vesicular stomatitis virus (VSV) or rabies virions. Thereafter, studies on SIGV focused on its hereditary transmission among flies. More recently, publications on SIGV have been centered on the population genetics of the SIGV–*Drosophila* linkage and descriptions of the defense mechanisms developed by *D. melanogaster*.

Classification

SIGV is currently classified as an unassigned member of the family *Rhabdoviridae*. The virion morphology, genome organization, and sequence relationships of several structural proteins are clearly consistent with its assignment as a rhabdovirus. However, it is not sufficiently closely related to other rhabdoviruses to be classified in any existing genus. It is closer to vesiculoviruses than to other rhabdoviruses according to phylogenetic studies (**Figure 1**) and biological properties (i.e., the CO_2 symptom induced by both vesiculoviruses and SIGV).

Virion and Genome Structure

The SIGV virion is a spiked and enveloped bullet-shaped particle of approximately 75×140–200 nm containing a helical nucleocapsid (**Figure 2**). The genome is a

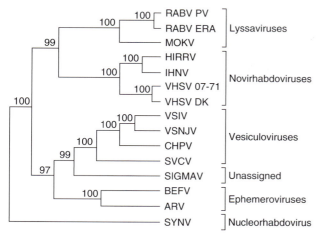

Figure 1 Phylogenetic tree of 15 rhabdovirus glycoproteins. The values adjacent to the branches indicate the boostrapping confidence limits. SIGV is not assigned to any genus of the *Rhabdoviridae* but the most conserved proteins (including the G protein) are most closely related to those of the vesiculoviruses. Reproduced from Björklund HV, Higman KH, and Kurath G (1996) The glycoprotein genes and gene junctions of the fish rhabdoviruses spring viremia of carp virus and hirame rhabdovirus: Analysis of relationships with other rhabdoviruses. *Virus Research* 42: 65–80, with permission from Elsevier.

negative-sense single-stranded RNA. It contains five genes arranged in the same order as in other rhabdoviruses ($3'$-N-P-M-G-L-$5'$) and encodes proteins with significant levels of sequence identity to the corresponding proteins of rhabdoviruses. For example, the P protein of SIGV is acidic and the distribution of charges is similar to other rhabdovirus P proteins. The charge and size of the M protein of SIGV are also similar to the M proteins of vesiculoviruses and the primary domains (basic domain, proline-rich domain, hydrophobic domain) are found in other rhabdovirus M proteins. The SIGV genome also contains an additional gene (X) between P and M genes (**Figure 3**). The X gene encodes a putative protein of 298 amino acids that is of unknown function, but which contains three conserved domains found in reverse transcriptases. Another unusual feature of the sigma genome is the 33 nt overlap of the M and G genes. The putative transcription initiation sequence CAACANG (+ sense) sequence is found at the beginning of the P, X, M, G, and L genes. The putative transcription termination/polyadenylation sequence $CAUG(A)_7$ (+ sense) terminates the N, P, X, M, and G genes. Phylogenetic analyses based on the N and G proteins indicate that SIGV clusters with vesiculoviruses (**Figure 1**).

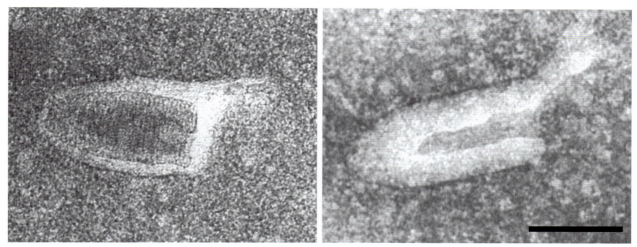

Figure 2 SIGV virions as observed by negative-contrast electron microscopy. The membrane fragment evident on the top right-hand side of the virions in each illustration is frequently but not always observed. Scale = 100 nm.

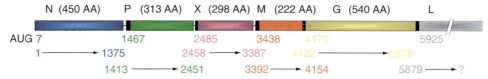

Figure 3 Genome organization of SIGV. The single-strand RNA genome encodes six proteins (N, P, X, M, G, and L) for which the size are indicated. The X protein is also named PP3 (protein product 3). The genome sequence has been obtained for all regions except the L gene which has only been partially sequenced. By comparison with other rhabdoviruses, the SIGV L gene is estimated to be approximately 6000 nt in length, resulting in a complete genome sequence of approximately 12 000 nt. The locations of mRNA transcripts and the position of putative initiation codons (AUG) are indicated. The arrows indicate the direction of transcription and the numbers indicate the transcription initiation and termination points. The M and G transcripts overlap from nucleotides 4122–4154.

CO_2 Sensitivity

The molecular basis of CO_2 sensitivity is unknown. However, the CO_2-induced paralysis of wings and legs correlates with viral invasion of thoracic ganglia that are involved in the nervous control of locomotion. Following abdominal inoculation of flies, several cycles of replication are necessary for SIGV to reach the thoracic ganglia and induce CO_2 sensitivity. Following injection of one infectious dose at 25 °C, an incubation period of 20 days is required for expression of the symptom and, as the virus dose is increased, the mean incubation period decreases. Moreover, a dose of virus that sensitizes flies to CO_2 in 10 days when injected into the abdomen, leads to CO_2-induced paralysis in only 3 days if injected into the thoracic ganglia.

The paralysis symptom is specific to CO_2 as other gases (e.g., N_2, H_2, CO, propane, and volatile acids) do not induce similar effects. This characteristic distinguishes SIGV from other viruses infecting Drosophila such as the entomobirnavirus drosophila X virus which induces sensitivity to both CO_2 and N_2 in what appears to be a general anoxia. There is evidence, however, that several vesiculoviruses, although not naturally infecting drosophila, induce CO_2-specific sensitivity in flies similar to that induced by SIGV. Each vesiculovirus tested to date, including vesicular stomatitis Indiana virus (VSIV), Cocal virus, pike fry rhabdovirus, and spring viremia of carp virus, is capable of replication in *D. melanogaster* and induces sensitivity to CO_2 but not to N_2 or propane. In contrast, the lyssavirus rabies virus (CVS strain) and two novirhabdoviruses of fish (infectious hematopoeitic nectosis virus and vesicular hemorrhagic septacaemia virus) do not replicate and do not induce CO_2 symptoms in *D. melanogaster*. The specificity of CO_2-induced paralysis in flies infected with SIGV or vesiculoviruses suggests that the underlying molecular mechanism responsible for the symptom is likely to be similar. Nevertheless, CO_2 symptoms vary for different vesiculoviruses. For example, Cocal virus induces CO_2-dependent paralysis when the virus titer in the nervous system reaches the maximum. Later in infection, although the virus titer in the entire fly remains constant, the titer decreases in the thoracic ganglia and CO_2 sensitivity disappears. It has also been observed that, following infection of flies with VSIV, a proportion of the exposed population displays a delayed CO_2 symptom while the remainder of the CO_2-sensitive flies are immediately paralyzed. The delayed sensitivity correlates with a more rapid invasion of the cephalic ganglia, than the thoracic ganglia and results in death 2–3 days after exposure, presumably due to paralysis of the mouthparts. Direct injection of SIGV into the cephalic ganglia also leads to this delayed sensitivity and lethality.

Host Range

SIGV has been reported to trigger CO_2 sensitivity in 15 of the 16 *Drosophila* species tested to date, the exception being *D. repleta*. However, some *D. melanogaster* genotypes are also resistant to SIGV replication, and so CO_2 sensitivity, and the observed resistance of *D. repleta* may not be valid for all genotypes.

One of the criteria used to demonstrate virus replication is the recovery of a virus yield that is in excess of the original infecting dose. This criterion is very stringent. Indeed, the virus yield from infected flies decreases soon after infection due to the entry of the nucleocapsids into the cytoplasm of infected cells, and there is a period of latency before replication generates new progeny virus. Therefore, a less-stringent criterion for SIGV replication is the observation of a higher virus yield than can be recovered during the latent period. Using these criteria, of 13 Diptera species (*Phormia terranovae*, *Ceratitis capitata*, *Musca domestica*, *Calliphora erythrocephala*, *Sarcophaga argyrostoma*, *Glossina mortisans*, *Aedes albopictus*, *Aedes aegypti*, *Aedes detritus*, *Anopheles stephensi*, *Toxorhynchites amboinensis*, *Culex pipiens*, *Culex quinquefasciatus*) that have been tested, only the fleshfly (*Phormia terranovae*) and the mosquito (*Aedes albopictus*) appeared incapable of supporting SIGV replication. As for *D. repleta*, there may be genotypic variation in the susceptibility of these insects and it is possible that susceptibility can be increased by SIGV strain adaptation. This is supported by the observation that SIGV can be adapted to replication in *A. albopictus* cell cultures by passage in the mosquito *Toxorynchites amboinensis*. The selected variants multiply and induce CO_2 sensitivity in *A. albopictus*. Adaptation of SIGV to replication in restrictive genotypes of *D. melanogaster* has also been observed. SIGV has not been observed to replicate in tests conducted in nine insect species representing five orders (Blattaria, Orthoptera, Lepidoptera, Hymenoptera, Hemiptera) other than Diptera or in cultured vertebrate cells.

Virus Transmission

SIGV is not transmitted between insects by casual contact and no vector of horizontal transmission has been identified. In nature, SIGV transmission appears to be exclusively vertical and, although 100% prevalence of infection can be achieved in laboratory strains, only a proportion of natural populations of flies are infected.

There is no evidence of integration of the SIGV genome into host chromosomes and the virus appears to remain exclusively in the cytoplasm. SIGV establishes a stable infection in oogonia, and between 10 and 40 viral genomes can be detected per oogonium according to the virus strain. Thereafter, there is an accumulation of genomes as oocyte volume increases. This balance reflects an autoregulation of viral infection in which cellular proteins may play a role and allows virus transmission during cell division. The nature of the process also implies low cytopathogenicity of SIGV in oocytes. Furthermore, if viral genomes segregate stochastically following cell division, relatively rare events of completely asymmetrical distribution would lead to the presence of uninfected individuals in the progeny of an infected female.

In a lineage stabilized for SIGV infection, some females will not transmit the virus to 100% of their progeny and infected females are less fertile. This should lead to a rapid elimination of the virus if the males were not able to transmit the infection. The efficiency of viral transmission by males can be as high as 90% in the progeny of a cross between uninfected females and males infected with a recently isolated strain of wild virus. In *D. melanogaster*, the male progeny of such a cross does not transmit the virus to the next generation. However, the female progeny can transmit SIGV to the next generation and some of their daughters are the source of stabilized lineages. These transmission rules are the roots of the invasive character of the virus and allow the virus to maintain itself in nature.

In *D. affinis* and *D. athabasca*, SIGV transmission follows similar rules with the exception that the male progeny of an uninfected female crossed with an infected male can transmit the virus to the next generation.

Virus Replication Cycle

The replication cycle of SIGV has been studied *in vivo* following injection of a viral extract into the abdomen of flies. There is a rapid reduction in recoverable infectivity following inoculation such that after 1 h, only 1% of the infecting dose can be recovered and new viral production commences between 24 and 48 h post infection. Studies of the viral replication cycle have been conducted using both wild-type viruses and temperature-sensitive (*ts*) mutants. Heat–shock experiments have identified three groups of *ts* mutants corresponding to three phases in the replication cycle. The first group of mutants identified a transiently thermolabile complex comprising the viral genome and viral proteins. The half-life of thermosensitivity is 4 h with a maximum span of 9 h post infection. The molecular basis of thermoresistance has not been determined. However, similar studies in vesiculoviruses suggest that this phase could include the steps that precede replication of the viral genome such as the release of the ribonucleoprotein complex from the endosome and primary transcription and protein synthesis.

A second group of *ts* mutants is affected in a replication phase that has a half-life of 9 h, commencing 7 h post infection and terminating at 14 h post infection. This phase corresponds to genome replication since hereditary transmission of these mutants is interrupted at the restrictive temperature. Late functions are also altered, as have been observed in molecular studies of vesiculoviruses which have shown that the N, P, and L proteins are essential for genome replication and secondary transcription.

A third group of *ts* mutants is affected in the late phase of the viral cycle. At the restrictive temperature, these mutants can be hereditarily transmitted but no infectious particles are produced and CO_2-induced paralysis of the host is not observed. The proteins modified in these mutants could be the G, M, or X proteins. Since the N, P, and L proteins are involved in viral replication and the X protein is absent in vesiculoviruses, CO_2 sensitivity must be due to G, M, or both G and M proteins.

At 20 °C, the viral replication cycle can vary in length between 25 and 90 h post infection in different individuals with an average of 60 h. The cycle length varies according to temperature and is much faster at 25 °C, as is the metabolism of flies. The host genotype also influences the cycle length. In more permissive genotypes, the replication cycle is faster and in the most permissive genotypes, the longest cycle is only 48 h in duration.

Host Immunity

A reduced fertility that has been observed in female flies infected experimentally with SIGV should be a strong selective pressure in nature. Surprisingly, poor fertility is not evident in natural populations. A loss of fertility is observed in crosses between infected females from a natural population and uninfected males from laboratory strains. These results suggest that the viral infection cycle is controlled by the host genome. Indeed, seven such genes are known: *ref(1)H*, *ref(2)M*, *ref(2)P*, *ref(3)G*, *ref(3)O*, *ref(3)D*, and *ref(3)V*. Each is polymorphic with alleles segregating into two categories. Permissive alleles allow the viral infection cycle to proceed while restrictive alleles restrict virus cycling. The only exception is the *ref(3)V* gene for which the restrictive allele blocks hereditary transmission of SIGV from stably infected males.

The most intensively studied of the *ref* genes is *ref(2)P*. Restrictive *ref(2)P* alleles modify the rules for hereditary transmission of SIGV. The frequency of uninfected flies in the progeny of infected parents increases with the number of restrictive alleles in the genome of the female. Stabilized mothers that are homozygous for a restrictive allele do not display CO_2 sensitivity even though they remain capable of virus transmission. Their progeny are transmission defective but may be CO_2 sensitive, depending on the genotype. Moreover, when stabilized males are crossed with uninfected females, the proportion of progeny that is infected is twofold lower when the mother is heterozygous for *ref(2)P* (permissive/restrictive) and zero when the mother is homozygous restrictive for *ref(2)P*. These restrictive alleles are often encountered in natural populations and do not appear to be counter-selected in uninfected drosophila. When the populations are infected with currently observed virus strains, which are sensitive to this defense system, *ref(2)P* restrictive alleles are favored until their frequency reaches 0.3. At this frequency, the sensitive strain of virus is eliminated. Thereafter, the lack of counter-selection of restrictive alleles maintains their frequency around 0.3, which protects the population from any new invasion by a sensitive virus.

Three functional domains have been identified in both the Ref(2)P protein and its mammalian homolog, p62/sequestosome-1. Two protein–protein interaction motifs are found: an amino-terminal PB1 (Phox and Bem 1) domain and a more central ZZ zinc finger. At the other end of both proteins there is a carboxy-terminal UBA (ubiquitin-binding area) domain. The Ref(2)P PB1 domain mediates the interaction with the drosophila atypical protein kinase C (DaPKC) and p62 binds mammalian aPKCs. The physiological function of Ref(2)P remains unknown even though it is essential for male fertility in some specific genotypes.

A comparison of susceptibility to the SIGV between flies that are homozygous for permissive alleles of *ref(2)P* and flies *ref(2)P*$^{-/-}$ has shown that permissive Ref(2)P protein is required for the virus to multiply at highest efficiency. For SIGV strains that are the most sensitive to *ref(2)P* alleles, a 16-fold higher dose is required for infection of a *ref(2)P*$^{-/-}$ genotype and a 10 000-fold higher dose is required for infection of a homozygous restrictive genotype than to infect homozygous permissive flies.

A comparison of the 15 sequenced alleles of *D. melanogaster ref(2)P*, among which three are restrictive, and the reference sequence of *D. simulans ref(2)P* indicate that the ancestral gene was permissive. Three mutations affecting the PB1 domain of the protein are necessary and sufficient to convert a permissive *ref(2)P* allele into a restrictive allele. Both permissive and restrictive alleles with sequence variations in the PB1 domain have been shown to form a monophyletic group that shows less internal variability than the group of ancestral permissive alleles. This suggests that the three variations in the PB1 domain that affect susceptibility to SIGV infection have emerged relatively recently. The high frequency of these variants (up to 50% of the observed alleles) indicates that their existence provides a selective advantage in drosophila populations.

The interaction between SIGV and the *ref(2)P* gene is highly specific. None of the other viruses that replicate in *D. melanogaster*, such as drosophila X virus and vesiculoviruses, is sensitive to restrictive *ref(2)P* alleles. A unique amino acid change in the PB1 domain is sufficient to suppress the restrictive character of an allele. In a viral population, genotypes with a capacity to replicate efficiently in a restrictive environment exist, even if the virus has never been exposed to restrictive alleles. For this reason, the elimination of sensitive viruses in a natural fly population by restrictive *ref(2)P* alleles is often associated with a new infection of the population by adapted viruses. Therefore, one could doubt that restrictive alleles provide a real defense system against SIGV but *ref(2)P* is not the only *ref* gene.

Each of the *ref* genes impacts independently on the viral replication cycle. For example, if a particular viral strain requires a tenfold higher dose to successfully infect flies that are restrictive at the *ref(2)P* locus, and a tenfold higher dose to infect flies restrictive at the *ref(2)M* locus, a 100-fold higher dose is required to overcome the flies that are restrictive for both *ref* genes. Adapted viruses can be isolated from a clone sensitive to *ref(2)P* restrictive alleles and studied for their sensitivity to *ref(2)M* restrictive alleles. As shown in **Figure 4**, the viral mutants

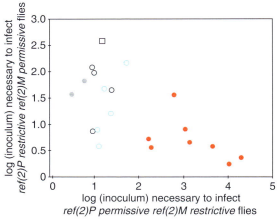

Figure 4 Virus clone distribution according to the inoculum size necessary to infect different fly genotypes. The reference *Drosophila melanogaster* strain was homozygous permissive for both *ref(2)P* and *ref(2)M*. The size of inoculum necessary to infect these flies was arbitrarily defined as 1.00. The y-axis values are the logarithm of the smallest inoculum capable of infecting flies homozygous for both a permissive *ref(2)M* allele and homozygous for a restrictive *ref(2)P* allele. The x-axis values are the logarithm of the smallest inoculum capable of infecting flies homozygous for both a permissive *ref(2)P* allele and a restrictive *ref(2)M* allele. The square represents the parental virus while circles represent viral mutants adapted to *ref(2)P* restrictive alleles. Open circles symbolize mutants that are not significantly different from the parental virus when assayed in *ref(2)M* restrictive flies. Blue circles depict the results observed for thermosensitive mutants, while gray and red circles show the observation made for mutants that are slightly adapted to both *ref* genes and mutants that are more sensitive to *ref(2)M* restrictive alleles than the parental clone, respectively. Among the mutants that are at least 20 times more adapted to *ref(2)P* than the parental clone, three are thermosensitive and thus would not survive in nature. The majority (seven out of eight) of the other 'adapted mutants' are more sensitive to *ref(2)M* restrictive alleles.

most adapted to *ref(2)P* restrictive alleles become more sensitive to restrictive alleles of *ref(2)M*. Increasing the frequency of *ref(2)M* restrictive alleles can also contribute to the elimination of viral mutants, only allowing viruses adapted to both *ref(2)P* and *ref(2)M* restrictive alleles to invade a fly population. It appears that the various *ref* genes cooperate to defend the fly population against the virus. For example, if the frequency of *ref(2)P* restrictive alleles was around 0.3, the frequency of *ref(2)M* restrictive alleles could rise while viruses sensitive to *ref(2)M* would disappear.

There are still SIGV strains in nature which are sensitive to *ref(2)P* restrictive alleles. In addition, a virus adapted to *ref(2)P* with an increased sensitivity to *ref(2)M* (i.e., equivalent to the red circles of **Figure 4**) is yet to be found in nature. Moreover, analysis of *ref(2)P* allele genealogy suggests that, if all the viruses present in a region were adapted to *ref(2)P* restrictive alleles, a new mutation in the PB1 domain of the protein would be selected to generate a new restrictive allele. Therefore, Ref(2)P remains as an active defensive mechanism against natural SIGV infections. The role of the other Ref proteins in host immunity remains unknown.

Concluding Remarks

In earlier studies, the detailed rules of vertical transmission of SIGVs in *D. melanogaster* populations illustrated how vector insects such as mosquitoes or sandflies can become the reservoirs of a virus. Currently, the two major interests in SIGV research are the description of the host–virus interactions in natural populations together with the development of an improved understanding of defensive mechanisms mediated by the *ref* genes. This immunity has certain similarities to innate immunity since its components are hereditary, but they differ because of the high specificity of protection and the memory effect that is generated in host populations. Strictly speaking, Ref(2)P may be part of the innate immune response through its interaction with DaPKC and a more extensive study could reveal that other *ref* genes also form part of a complex innate host immune system in flies.

See also: Chandipura Virus; Vesicular Stomatitis Virus; Fish Rhabdoviruses.

Further Reading

Björklund HV, Higman KH, and Kurath G (1996) The glycoprotein genes and gene junctions of the fish rhabdoviruses spring viremia of carp virus and hirame rhabdovirus: Analysis of relationships with other rhabdoviruses. *Virus Research* 42: 65–80.

Carpenter JA, Obbard DJ, and Maside X (2007) The recent spread of a vertically transmitted virus through populations of *Drosophila melanogaster*. *Molecular Ecology* 16: 3947–3954.

Carré-Mlouka A, Gaumer S, Gay P, et al. (2007) Control of sigma virus multiplication by the *ref(2)P* gene of *Drosophila melanogaster*. An *in vivo* study of the PB1 domain of Ref(2)P. *Genetics* 176: 409–419.

Fleuriet A and Periquet G (1993) Evolution of the *Drosophila melanogaster*–sigma virus system in natural populations from Languedoc (southern France). *Archives of Virology* 129: 131–143.

Landès-Devauchelle C, Bras F, Dezélée S, and Teninges D (1995) Gene 2 of the sigma rhabdovirus genome encodes the P protein and gene 3 encodes a protein related to the reverse transcriptase of retroelements. *Virology* 213: 300–312.

Teninges D, Bras F, and Dezélée S (1993) Genome organization of the sigma rhabdovirus: Six genes and a gene overlap. *Virology* 193: 1018–1023.

Teninges D and Bras-Herreng F (1987) Rhabdovirus sigma, the hereditary CO_2 sensitivity agent of *Drosophila*: Nucleotide sequence of a cDNA clone encoding the glycoprotein. *Journal of General Virology* 68: 2625–2638.

Wang Y, Cowley JA, and Walker PJ (1995) Adelaide River virus nucleoprotein gene: Analysis of phylogenetic relationships of ephemeroviruses and other rhabdoviruses. *Journal of General Virology* 76: 995–999.

Wayne ML, Contamine D, and Kreitman M (1996) Molecular population genetics of *ref(2)P*, a locus which confers viral resistance in *Drosophila*. *Molecular Biology and Evolution* 13: 191–199.

Simian Alphaherpesviruses

J Hilliard, Georgia State University, Atlanta, GA, USA

© 2008 Elsevier Ltd. All rights reserved.

History

Members of the subfamily *Alphaherpesvirinae* in the family *Herpesviridae* have been identified in Old World monkeys and apes and in New World monkeys. It appears that nearly all host species have co-evolved with one or more alphaherpesviruses, and so it comes as no surprise that new agents in this group continue to be discovered. Indeed, the increased use of nonhuman primates in biomedical research has facilitated the discovery and characterization of such viruses. As assays improve, more virus isolates from nonhuman primates have been identified, and subsequently have been differentiated from closely related viruses by molecular and immunological means. The benefit of enhanced techniques for virus differentiation is exemplified by the early reports in the 1970s that simian agent 8 (SA8; *Cercopithecine herpesvirus 2*) was not only found in vervets but also in baboons. It was not until nearly two decades later that the virus isolated from baboons in a captive colony was recognized to be herpesvirus papio 2 (HVP-2; *Cercopithecine herpesvirus 16*). Although improved technology enables differentiation of closely related herpesviruses, it remains a time-consuming task to find species-specific viruses, characterize the associated pathogeneses, and verify the natural hosts. Like human alphaherpesviruses, nonhuman primate alphaherpesviruses fall into the genera *Simplexvirus* and *Varicellovirus*.

Simplexviruses are readily isolated because they replicate rapidly in cell cultures from a variety of mammals. They produce a distinct cytopathology as quickly as 24–48 h following infection. Examples of this cytopathology and some of the differences among simplexviruses are shown in **Figure 1**. The property of rapid growth is one of the initial criteria for establishing that an isolate is an alphaherpesvirus. These viruses are most frequently isolated from mucosal sites or from necropsy samples, and occasionally from fomites. With sequence analysis readily available, they can be differentiated easily from close counterparts endemic in related species of Old and New World monkeys, apes, and humans. The zoonotic potential of these viruses is of particular interest, because at least one, B virus (*Cercopithecine herpesvirus 1*), can be rapidly lethal in humans when antiviral therapy is initiated too late.

Frequently, the identity of the natural host of a virus isolate has been established by the use of seroepidemiological assays. The natural host often shows 80–100% seroprevalence in the wild by the time the animals reach sexual maturity. Most data in this regard have been summarized from studies of vervets, baboons, and macaques in the wild, although there are excellent studies that have focused on apes and New World monkeys. The assays, however, upon which seroprevalence are based are not always uniform, and may utilize antigens that are not necessarily unique to the specific virus for which antibody is reactive, resulting in difficulty identifying the natural host of a particular virus. Nonetheless, the plethora of assays has afforded investigators the opportunity to establish candidates for the natural host, and, as assays improve, the identity of the natural host can be narrowed.

Owing to the close evolutionary relationships between monkeys, most notably within Old World groups and New World groups, the viruses that have co-evolved in each host group share significant similarities genotypically and phenotypically. Although genes are arranged mostly in a collinear layout in viruses from both groups of nonhuman primates, viruses from Old World and New World monkeys differ from each other more than viruses from within each monkey group. This point is best illustrated by studies evaluating immunological cross-reactivity among nonhuman primate alphaherpesviruses between groups and within groups. Within either of the Old World and New World groups, extensive genetic conservation renders classical serological assays using virus or infected cell lysates of relatively limited value in identifying the natural host, due to significant cross-reactivity of immunogenic epitopes. An assay for detection of HVP-2 will cross-react sufficiently strongly with mangabey or langur herpesviruses, such that differentiation of the antibody specificity will be nearly impossible using a virus- or infected

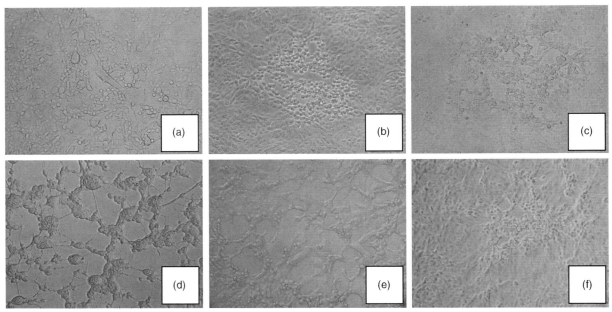

Figure 1 Vero cells in culture (ATCC-081) 24 h after infection with (a) B virus (E2490), (b) HSV-1 (KOS), (c) HVP-2 (SWF), (d) mangabey herpesvirus (EM-GS), (e) SA8 (Hull), or (f) langur herpesvirus (BZ).

cell-based immunoassay. With the advent of recombinant viral proteins, differentiation assays for human alphaherpesviruses have improved, but conservation of the major antigenic proteins can still cloud identification of endemic populations in nonhuman primates. The strategies that differentiate alphaherpesviruses in humans have not proved to be entirely successful when extrapolated to identifying alphaherpesviruses in nonhuman primates. The best example of this is the confusion caused by the finding that SA8 was endemic in baboons as well as vervets, which was still manifest decades after the first publication citing the identification of this agent in baboons. It is now appreciated that the viruses infecting vervets and baboons are distinct alphaherpesviruses, SA8 and HVP-2, respectively. Identification of the natural host of simian varicella virus (SVV) provides a similar example, since both macaques (Asian) and vervets (African) host similar, if not identical, variceloviruses.

Classification and Evolution

The current list of nonhuman primate alphaherpesviruses is shown in **Table 1**. Nonhuman primate alphaherpesviruses belong to genus *Simplexvirus* (with human herpes simplex virus (HSV) types 1 (HSV-1) and 2 (HSV-2)), with the exception of SVV (*Cercopithecine herpesvirus 9*), which belongs to genus *Varicellovirus* (with human varicella-zoster virus, VZV). The marked genetic similarities between the nonhuman primate viruses and their counterparts in humans strongly support a co-evolutionary history.

The genomes of the Old World monkey simplexviruses have been shown to exist in four isomeric forms, like those of HSV (type E genome structure). The DNA sequences of B virus, SA8, mangabey herpesvirus, langur herpesvirus, and HVP-2 are similar in size (151–157 kbp) to those of their human counterparts, with somewhat greater G+C contents. The genetic contents of these viruses are very similar to that of HSV-1 and HSV-2, except that they lack the gene encoding the neurovirulence protein ICP34.5. The variceloviruses have a type D genome structure, and the genetic content of SVV is very similar to that of VZV, differing in only a couple of genes at the left genome terminus.

In accord with the relationships among their hosts, simplexviruses of New World primates are more distantly related to those of Old World primates, and the genomes are smaller in size with relatively less complex isomeric organization. The evolutionary distances are apparent from the fact that antisera prepared against the major HSV proteins cross-react with homologous proteins from Old World, but not New World, simplexviruses. Thus far, no variceloviruses have been found to be endemic in New World monkeys.

Virion Structure

Nonhuman primate alphaherpesviruses are similar in structure to the other members of the subfamily *Alphaherpesvirinae*. By electron microscopy, virus particles consist of capsids sheathed by a tegument and surrounded by an envelope. The icosahedral capsids are assembled in the infected cell nucleus and finally enveloped in the cytoplasm. Unenveloped capsids can be found in both the cytoplasm and nucleus, and are most abundant in the latter. Temporal studies of the kinetics of virus replication suggest that the envelope is acquired in a two-step process, similar to that of HSV. The envelope is rich in virus glycoproteins, which, as judged from antibodies induced during infection, are highly immunogenic. Sequencing and immunological cross-reactivity studies indicate that glycoprotein B is the most highly conserved glycoprotein.

Table 1 Nonhuman primate alphaherpesviruses

Host	Common virus name	Official species name
Old World		
Vervet (African green monkey)	Simian agent 8 (SA8)	Cercopithecine herpesvirus 2
Baboon	Herpesvirus papio 2 (HVP-2)	Cercopithecine herpesvirus 16
Macaque	B virus	Cercopithecine herpesvirus 1
Macaque	Simian varicella virus (SVV) (also known as Patas delta, Liverpool vervet and Medical Lake macaque herpesvirus)	Cercopithecine herpesvirus 9
Langur	Langur herpesvirus	Currently unnamed
Mangabey	Mangabey herpesvirus	Currently unnamed
New World		
Spider monkey	Spider monkey herpesvirus (also known as herpesvirus ateles)	*Ateline herpesvirus 1*
Marmoset	Marmoset herpesvirus (also known as herpesvirus saimiri and, in other platyrrhine monkeys, herpesvirus tamarinus)	*Saimiriine herpesvirus 1*

Physical Properties and Epidemiology

Each of the nonhuman primate alphaherpesviruses has features in common with its human counterparts with respect to host range, replication kinetics, transmission, tissue tropism, pathogenecity, histopathology, and induced immune responses. The exception is B virus, which can infect humans, often resulting in a fatal zoonotic infection when left untreated. Each virus is predominantly found in one type of monkey, and recent experience indicates that antibody cross-reactivity with isolates from other species should be evaluated carefully in order to avoid deducing wrongly that a virus is endemic in multiple monkey types. As mentioned above, this is particularly apparent in the case of SA8 and HVP-2, but also features in case studies of alphaherpesviruses from Asian and African monkeys, as well as greater and lesser apes. Although apes are likely to have co-evolved with their own respective alphaherpesviruses, investigators have not yet found herpesviruses other than HSV-1 and HSV-2 in these animals.

Transmission of nonhuman primate alphaherpesviruses from animal to animal occurs as a result of biting, scratching, and splashing activities that contaminate susceptible mucosal epithelial cells. Seroprevalence of virus-induced antibodies increases with the onset of sexual maturity, but infected infants have also been identified, albeit infrequently. Infection is associated with few or no apparent symptoms, whether the virus is latent or replicating actively. Periodic reactivations from latency have been associated with virus shedding from mucus membranes, and this is the time period during which transmission is greatest. Except for B virus, none of the nonhuman primate alphaherpesviruses appears to be transmissible to humans under natural circumstances. Evidence for transmission of human viruses to apes has been substantiated in certain circumstances, as mentioned above, but nonhuman primates are not usually found to be infected with human alphaherpesviruses. When transmission of Old World simplex- or varicelloviruses to New World primates occurs, infection is readily apparent with high morbidity and frequent mortality.

Virus Replication

Nonhuman primate alphaherpesviruses appear to follow the same pattern of biosynthetic activities leading to virus replication and assembly as do their human counterparts, with some exceptions noted. The nonhuman primate simplexviruses replicate in a manner similar to HSV, with a replication cycle of approximately 18 h. Cytopathic effects in cultured cells are also similar, with the exceptions of B virus and mangabey herpesvirus, which induce cell fusion between infected cells and also with neighboring uninfected cells. **Figure 1** shows representative cytopathic effects of B virus versus other alphaherpesviruses in nonhuman primate cells. SVV replicates with kinetics similar to VZV over an interval of 48–72 h in cell culture and, like VZV, remains mostly cell associated.

Following adsorption of a nonhuman primate alphaherpesvirus to a susceptible cell, de-enveloped capsids are released into the cytoplasm and proceed to the nuclear membrane by mechanisms that are poorly understood, but which probably involve the cell's cytoskeleton components. Microarray studies have revealed that host gene remodeling in B virus-infected cells begins within the first hour post infection, and within 3 h post infection the events are clearly distinguishable from those transpiring in HSV-infected cells. Nonetheless, the outcome at the level of the infected cell in culture is the same for both human and nonhuman primate alphaherpesviruses – productive virus replication. The temporal cascade of protein synthesis is conserved, with immediate early, early, and late expression. The major difference between human and nonhuman primate alphaherpesviruses is that the US11 protein is produced as an immediate early protein in the nonhuman primate viruses and as an early protein in HSV-1 and HSV-2. This protein in nonhuman primate viruses prevents phosphorylation of protein kinase R, which in turn prevents phosphorylation of eIF-2α, thus blocking apoptosis in the infected cell and enabling virus replication.

Immune Response to Infection

Nonhuman primates infected by their respective alphaherpesviruses induce humoral antibodies generally within 7–14 days following the onset of acute replication. Antibody titers, however, are not apparent in all animals within this period. Titers can also wax and wane depending on the intervals between reactivated infections. Nonetheless, intermittent virus shedding makes reliance on virus isolation or polymerase chain reaction (PCR) impractical; thus, detection of antibodies is used as the current indicator of whether an animal is infected. Antibodies induced in Old and New World monkeys are cross-reactive with the alphaherpesviruses infecting animals within each group, but there is no apparent cross-reactivity between the two groups. There is little evidence that Old World monkeys can be infected by New World monkey viruses, but New World monkeys generally succumb to viruses from Old World monkeys or humans. Cross-species housing is generally avoided.

The antibodies most commonly produced are against the major glycoprotein (glycoprotein B). There is no clear-cut evolution of the humoral immune response, but the antibodies most observed include those reactive with glycoprotein B, glycoprotein C, glycoprotein D, glycoprotein G, and the major capsid protein. Having said this, antibody profiles from different animals are quite distinct, as in the case of human antibody profiles induced against HSV.

Future Perspectives

Identification and characterization of simplex- and varicelloviruses from nonhuman primates afford the opportunity to learn more about the origins, evolutionary processes, and pathogenic properties of these viruses. Thus, study of a particular member provides insights on the virus as an individual species and as a representative of a lineage of related agents. Perhaps even more important is the development of understanding on how to approach novel viruses, whether newly discovered or emerging. Knowledge of these viruses in relation to their natural hosts often cannot be extrapolated to their effects in foreign hosts, and the responses of a non-natural host can influence pathogenesis in unexpected ways. This is illustrated by B virus infection of humans, the most dramatic example of cross-species transmission among the herpesviruses. Current attention focused worldwide on emerging viruses and viruses that invade foreign hosts places an emphasis on primate alphaherpesviruses that will teach investigators for decades to come.

See also: Herpes Simplex Viruses: General Features; Herpes Simplex Viruses: Molecular Biology; Herpesviruses: General Features; Simian Gammaherpesviruses; Varicella-Zoster Virus: General Features; Varicella-Zoster Virus: Molecular Biology.

Further Reading

Cohen JI, Davenport DS, Stewart JA, et al. (2002) Recommendations for prevention of and therapy for exposure to B virus (*Cercopithecine herpesvirus 1*). *Clinical Infectious Diseases* 35: 1191–1203.

Davison AJ and Clements JB (1997) Herpesviruses: General properties. In: Mahy BWJ and Collier LH (eds.) *Topley and Wilson's Microbiology and Microbial Infections,* 9th edn., pp. 309–323. London: Arnold.

Eberle R and Hilliard J (1995) The simian herpesviruses. *Infectious Agents Disease – Reviews Issues and Commentary* 4: 55–70.

Gray WL, Gusick NJ, Ek-Kommonen C, Kempson SE, and Fletcher TM, III (1995) The inverted repeat regions of the simian varicella virus and varicella-zoster virus genomes have a similar genetic organization. *Virus Research* 39: 181–193.

Weigler J (1992) Biology of B virus in macaque and human hosts: A review. *Clinical Infectious Diseases* 14: 555–567.

Whitley RJ and Hilliard J (2007) Cercopithecine herpesvirus 1 (B Virus). In: Knipe DM and Howley PM (eds.) *Fields Virology,* 5th edn., vol. 2, p. 2889. Philadelphia: Lippincott Williams and Wilkins.

Simian Gammaherpesviruses

A Ensser, Virologisches Institut, Universitätsklinikum, Erlangen, Germany

© 2008 Elsevier Ltd. All rights reserved.

Introduction

The *Gammaherpesvirinae* is a large subfamily of the family *Herpesviridae*. Although gammaherpesviruses usually cause limited disease upon primary infection of their natural hosts, several are relevant tumor viruses of the hematopoietic system and form an important chapter of viral oncology. The first clearly identified human herpesvirus, Epstein–Barr virus (EBV; species *Human herpesvirus 4*), is the prototype of genus *Lymphocryptovirus* (whose members are referred to as lymphocryptoviruses or γ1-herpesviruses) and the cause of infectious mononucleosis. Homologs of EBV had been recognized for decades in various Old World primates, and have been found recently in several species of American monkeys. They may serve to develop models for pathogenesis or treatment of human lymphoproliferative diseases and cancers that are caused by EBV, such as B-cell lymphomas and other lymphoproliferative syndromes, nasopharyngeal carcinomas, and, possibly, gastric cancer.

The second genus of gammaherpesviruses, *Rhadinovirus* (whose members are referred to as rhadinoviruses or γ2-herpesviruses), is distinct biologically and molecularly. The prototypic member of this group, herpesvirus saimiri (HVS; species *Saimiriine herpesvirus 2*), and herpesvirus ateles (HVA; species *Ateline herpesvirus 2* and ateline herpesvirus 3) were detected as T-lymphotropic viruses in neotropical primates and raised primary interest from the fact that they cause fulminant T-cell lymphomas in numerous primate species as well as in rabbits. The related animal pathogens alcelaphine herpesvirus 1 and ovine herpesvirus 2 cause malignant catarrhal fever, a T lymphoproliferative disease of ruminants. Although no exact correlates of these T-cell tumors exist in human pathology, HVS strains of subgroup C are capable of transforming human and simian T lymphocytes to continuous growth in cell culture. This provided for the first time a reliable means of immortalizing human T lymphocytes in cell culture, a useful tool for T-cell immunology. These viruses have been used as expression vectors for gene transfer in T lymphocytes and have facilitated study of the mechanisms of episomal persistence in the T-cell system. Further interest in the rhadinoviruses arose when the first human member of this genus was recognized. This virus was found to be strongly associated with all forms of Kaposi's sarcoma (KS), as well as with multicentric Castleman's disease and primary effusion lymphoma (PEL). Since DNA from this virus is regularly found in all KS forms, specifically in the spindle cells of KS, it was also termed KS-associated herpesvirus (KSHV; species *Human herpesvirus 8*). Viral membrane-associated oncoproteins Stp and Tip, which act on T-lymphocyte signaling, were defined in HVS, though it is far less clear which of several candidate genes encode the relevant oncoproteins of KSHV.

For many years, research on lymphotropic simian herpesviruses focused on the tumorigenic T-lymphotropic rhadinoviruses, especially HVS. Then the discovery of KSHV prompted research on B-lymphotropic agents and led to the description of rhesus rhadinovirus (RRV; species *Cercopithecine herpesvirus 17*) and several closely related rhadinoviruses in various Old World primates, although these are only loosely associated with pathogenicity or tumor induction. New World primate EBV-like viruses were discovered recently, and an increasing number of DNA sequences from additional, new gammaherpesviruses are being amplified from diverse host species using degenerate polymerase chain reaction (PCR) techniques that target strongly conserved herpesvirus genes, such as that encoding DNA polymerase. A provisional compilation of the better defined primate gammaherpesviruses is represented in **Table 1**, and more extensive information is available in the website of the International Committee on Taxonomy of Viruses (ICTV).

Herpesvirus Saimiri and Herpesvirus Ateles

This section focuses on the basic biology, gene content, and viral mechanisms of oncogenic transformation of HVS and HVA, and their possible applications as T-cell vectors and in cell-based immunotherapy. These gammaherpesviruses must not be confused with two alphaherpesviruses isolated from the same host species, designated as species *Saimiriine herpesvirus 1* and *Ateline herpesvirus 1*, respectively.

History, Host Range, Transmission, and Pathology

HVS was originally isolated by Melendez and others from captive monkeys of various species, but it soon became clear that this virus is found regularly only in squirrel monkeys (*Saimiri sciureus*), whose natural habitat is South American rainforests. Squirrel monkeys are usually infected via saliva within the first two years of life. The virus does not cause disease or tumors, and establishes lifelong persistence. In other New World primate species such as tamarins (*Saguinus* spp.), common marmosets (*Callithrix jacchus*), or owl monkeys (*Aotus trivirgatus*), infection with HVS causes acute peripheral T-cell lymphoma within less than 2 months after experimental intramuscular or intravenous infection. Intramuscular injection of purified virion DNA can also cause disease in susceptible primates.

HVS strains are classified into three subgroups (A, B, and C) depending on pathogenic properties and on sequence divergence in the left-terminal nonrepetitive region of the genome (**Figure 1**). The major representative strains are the prototypic A11 for subgroup A; B S295C and B-SMHI for subgroup B; and C488 and C484 for subgroup C.

Viruses of HVS subgroups B and C are considered to be the least and most oncogenic, respectively. Tamarins are susceptible to viruses of all subgroups, whereas subgroup B viruses are not able to cause disease in adult common marmosets. Strain C488 causes acute peripheral T-cell lymphoma within only a few weeks in common marmosets or cottontop tamarins (*S. oedipus*). A similar fulminant disease is induced in Old World rhesus and cynomolgus monkeys (*Macaca mulatta* and *M. fascicularis*, respectively) by large intravenous doses of C488. Similar to the situation in New World primates, the disease in cynomolgus monkeys is designated as a pleomorphic, peripheral T-cell lymphoma or a pleomorphic, T-lymphoproliferative disorder. A high-titer infection in New Zealand white rabbits results in tumor induction, but pathogenicity has not been reported in rodents. HVS can be isolated from the peripheral blood cells of persistently infected squirrel monkeys or diseased tamarins, presumably from infected T cells, by co-cultivation with permissive owl monkey kidney (OMK) cells. HVS replicates productively in, and induces cell lysis of, OMK cells and some primary mesenchymal cultures established from marmosets, and less efficient replication is possible in Vero (African green monkey) cells.

HVA can be isolated at a high frequency from spider monkeys (*Ateles* spp.). Strain 810 from *A. geoffroyii* is a member of species *Ateline herpesvirus 2*, whereas strain 73 and related strains (87, 93, and 94) from *A. paniscus* are isolates of ateline herpesvirus 3. HVA replicates in OMK cells, but remains mostly cell-associated with syncytia formation. As a result, supernatants of such cultures have low, unstable virus titers.

Like HVS, HVA is not pathogenic in its natural host, but causes acute T-cell lymphomas in various New World primate species, including cottontop tamarins and owl monkeys. The pathological changes are similar to those observed after HVS infection. In addition, HVA transforms

Table 1 Primate gammaherpesviruses

Species	Common name(s) and abbreviation(s)	Host	Associated pathogenicity
Genus *Rhadinovirus*			
Human herpesvirus 8	Kaposi's sarcoma-associated herpesvirus (KSHV)	Human (*Homo sapiens*)	Kaposi's sarcoma, multicentric Castleman's disease, primary effusion lymphoma
NA[a]	Chimpanzee rhadinovirus	Chimpanzee (*Pan troglodytes*)	Unknown
NA	Gorilla rhadinovirus	Gorilla (*Gorilla gorilla*)	Unknown
Cercopithecine herpesvirus 17	Rhesus rhadinovirus (RRV), *Macaca mulatta* rhadinovirus	Rhesus macaque (*M. mullata*)	B-cell hyperplasia?
NA	Retroperitoneal fibromatosis-associated herpesvirus (RFHV, RFHVMn, RFHVMm)	Southern pig-tailed macaque (*M. nemestrina*), rhesus macaque (*M. mullatta*)	Retroperitoneal fibromatosis?
NA	*Macaca nemestrina* rhadinovirus 2 (MnRRV)	Southern pig-tailed macaque (*M. nemestrina*)	Unknown
Saimiriine herpesvirus 2[b]	Herpesvirus saimiri (HVS)	Squirrel monkey (*Saimiri sciureus*)	T-cell lymphoma in other neotropical monkey species
Ateline herpesvirus 2	Herpesvirus ateles (HVA)	Spider monkey (*Ateles paniscus*)	T-cell lymphoma in other neotropical monkey species
NA	Herpesvirus ateles strain 73 (HVA), ateline herpesvirus 3	Spider monkey (*A. paniscus*)	T-cell lymphoma in other neotropical monkey species
Genus *Lymphocryptovirus*			
Human herpesvirus 4[b]	Epstein–Barr virus (EBV)	Human (*Homo sapiens*)	B-cell lymphoma, nasopharyngeal lymphoma, Hodgkin's disease
Pongine herpesvirus 1	Herpesvirus pan, chimpanzee lymphocryptovirus	Chimpanzee (*Pan* sp.)	Unknown
Pongine herpesvirus 2	Orangutan herpesvirus	Orangutan (*Pongo* sp.)	Unknown
Pongine herpesvirus 3	Gorilla herpesvirus	Gorilla (*Gorilla* sp.)	Unknown
Cercopithecine herpesvirus 12	Baboon herpesvirus, herpesvirus papio	Baboon (*Papio* sp.)	Spontaneous B-cell lymphoma (and in immunosuppressed animals)
Cercopithecine herpesvirus 14	African green monkey EBV-like virus	African green monkey (*Chlorocebus aethiops*)	Unknown
Cercopithecine herpesvirus 15	Rhesus EBV-like herpesvirus, rhesus lymphocryptovirus	Rhesus macaque (*M. mullatta*)	Spontaneous B-cell lymphoma (and in immunosuppressed animals)
NA	Cynomolgus EBV-like virus, *Macaca fascicularis* gammaherpesvirus (herpesvirus MF1, A4, TsB-B6, Si-IIA-EBV)	Cynomolgus monkey (*M. fascicularis*)	Spontaneous B-cell lymphoma (and in immunosuppressed animals)
Callitrichine herpesvirus 3	Marmoset lymphocryptovirus	Common marmoset (*Callithrix jacchus*)	Spontaneous B-cell lymphoma
NA	Gold-handed tamarin lymphocryptovirus (SmiLHV1)	Gold-handed tamarin (*Saguinus midas*)	Unknown
NA	Squirrel monkey lymphocryptovirus (SscLHV1)	Squirrel monkey (*S. sciureus*)	Unknown
NA	White-faced saki lymphocryptovirus (PpiLHV1)	White-faced saki (*Pithecia pithecia*)	Unknown

[a]NA, species not assigned by ICTV.
[b]Type species of the genus.

T cells of certain New World monkey species (such as cottontop tamarin) in culture, yielding cytotoxic T-cell lines. Human T cells are not susceptible to transformation with various HVA strains, but could be transformed by a recombinant HVS C strain in which the HVS oncogenes were replaced by the HVA oncogene Tio.

Transformed T-cell lines have been derived from HVA-infected tamarins and cultivated continuously for several years. Whereas in most cases virus particles were found initially, virus production was frequently lost after prolonged culture. The episomal DNA is heavily methylated in such nonproductive cell lines, and rearrangements or

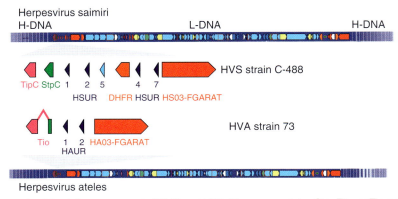

Figure 1 Gene arrangements at the left genome end of HVS and HVA. The oncoproteins Stp, Tip, or Tio are encoded at the variable left terminal region of the coding L-DNA. Stp, saimiri transformation-associated protein of the respective subgroup A, B, or C; Tio, two-in-one-protein of HVA; Tip, tyrosine kinase-interacting protein of HVS subgroup C; HSUR or HAUR, HVS or HVA-encoded URNA; HS03-/HA03-FGARAT, formylglycineamide ribotide amidotransferase ORF3.

large deletions are evident in the viral genomes. Marmoset and tamarin T cells can be transformed by HVS to stable T-cell lines *in vitro* and are designated as semipermissive, since virus particles are released, although to lower titers than from OMK cells.

Genome Properties, Replication, and Gene Content

The term 'rhadino' viruses was coined from the ancient Greek word ραδινοσ for fragile, because the viral genomic DNA breaks upon isopyknic centrifugation in CsCl gradients into two classes of highly differing densities. The L DNA (low density, low G + C content) contains the viral protein-coding genes, and the repetitive H DNA (high density, high G + C content) from the genome termini is noncoding. The intact viral (M) genome has intermediate density. Two strains of HVS, strain A11 and the highly oncogenic subgroup C strain C488, have been sequenced. The A11 H DNA consists of multiple tandem repeats of 1444 bp (70.8% G + C), and the unique L DNA comprises 112 930 bp (34.5% G + C). The size of the M DNA genome is variable owing to different numbers of H DNA repeats attached to both ends of the linear virion genome. In strain C488, the L DNA comprises 113 027 bp, and it is flanked by arrays of two distinct repeat unit types of 1318 and 1458 bp, the shorter representing the longer with 140 bp deleted. The packaged M genome of C488 is approximately 155 kbp in size, with a range of 130–160 kbp owing to variable numbers of terminal H DNA repeats. The HVS L DNA contains at least 76–77 protein-coding open reading frames (ORFs) and encodes 5–7 U RNAs (termed HSURs) (**Figure 1**).

HVA strain 73 has a similar genome structure to HVS, with a slightly shorter L DNA of 108 409 bp (36.6% G + C), and H DNA consisting of multiple tandem repeats of 1582 bp (77.1% G + C). The HVA L DNA contains 73 ORFs and only two genes for U RNA-like transcripts (termed HAURs). HVA does not encode ORF12, vIL17, vCD59, or vFLIP homologs, but the genes encoding superantigen (SAG), cyclin, and G-protein-coupled receptor (GPCR) are conserved. Thus, HVA may be an ancient variant of HVS that has either collected a smaller set of cell-homologous genes or has secondarily lost several genes.

In all gammaherpesviruses, the genes that are conserved among the herpesvirus subfamilies are arranged in blocks. Flanking or interspersed among the blocks are other genes, most of which do not occur in the other subfamilies. Among these are transforming oncogenes and viral homologs of cellular genes, which are described below. Most genes are well conserved between different HVS strains, but there is pronounced sequence variation near the left end of the HVS L DNA and in the region of the R transactivator gene (*orf50*) and the adjacent glycoprotein gene (*orf51*), a region that is also highly variable among other rhadinoviruses.

The replication mechanism of rhadinoviruses has not been investigated in much detail, and is generally considered to follow that of other herpesviruses. The lytic origin of DNA replication (OriLyt) in HVS strain A11 has been mapped to an untranslated region upstream of the thymidylate synthase gene. A putative latent origin of DNA replication (OriP) in the left-terminal region of the L DNA in strain C484 was reported to mediate episomal maintenance, but is not conserved between different HVS strains and is not required for viral replication or episomal persistence. Thus, although HVS persists in transformed human T cells as stable nonintegrated episomes at high copy number, OriP and the viral factors involved remain unidentified. Histone modification of the HVS C488 episome in human T cells has been analyzed, and bears similarities to that of KSHV episomes in B cells.

Infection of tissue culture cells by HVS is asynchronous, and hence the assignment of HVS genes to the immediate early (IE) phase of infection is based mostly on experiments using cycloheximide to inhibit protein synthesis. The IE57 post-transcriptional regulator encoded by *orf57* appears to be the sole regulatory viral IE gene. It codes for a nuclear phosphoprotein of 52 kDa with structural and functional homology to herpes simplex virus ICP27 and EBV BMLF1. IE57 stimulates the expression of unspliced, and represses the expression of spliced, transcripts, has been shown to redistribute nuclear components of the splicing machinery, and is involved in nuclear RNA export. A strong viral transactivator function was mapped to the delayed early gene *orf50*, the homolog of the EBV R transactivator gene. Owing to differential splicing and promoter usage, this gene codes for a full-length protein (ORF50A) and a smaller, C-terminal variant (ORF50B). The transactivation domain resides in the C-terminal region of these proteins and binds to the TATA-binding protein in the basal transcription complex. Although IE57 is highly conserved between subgroups A and C, the *orf50* region is divergent. Neither HVS nor HVA encodes a homolog of bZip/Zta of EBV or KSHV.

The HVS ORF73 protein of strains A11 and C488 localizes to the host cell nucleus, and, like the latent nuclear antigen (LANA) of KSHV, can associate with host cell chromosomal DNA. The A11 ORF73 protein can associate with cellular p32 and binds to GSK-3β. Although not detectable by northern blotting of RNA from C488-transformed human T cells, *orf73* transcripts are detectable by reverse transcription-polymerase chain reaction (RT-PCR). The C488 ORF73 protein can downregulate the *orf50A* and *orf50B* promoters, and this prevents ORF50-mediated activation of viral replication gene promoters. This suggests that the HVS ORF73 protein, and its homologs in the other rhadinoviruses, can block initiation of the lytic replication cascade, thereby controlling the transition between latency and lytic replication.

Sequestered Cellular Genes

Rhadinoviruses such as HVS and KSHV contain several intronless genes that are homologous to cellular genes; in this context, a role for reverse transcription during putative capture of these genes might be speculated upon. A few of these cellular gene homologs are unique to specific viruses, and some are common to several rhadinoviruses (and to lymphocryptoviruses, including EBV). This suggests that successful uptake of cellular genes is a rather infrequent event during herpesvirus evolution. Most of these cellular homologs can be categorized into two major groups: (1) genes related to cellular growth control or nucleotide metabolism, and (2) genes that modulate innate or adaptive immune functions, including apoptosis. For example, HVS *orf72* codes for a functional viral cyclin D, and homologs related to nucleotide metabolism include a dihydrofolate reductase (DHFR; *orf2*) and a functional thymidylate synthase (TS; *orf70*). Both *orf3* and *orf75* encode large tegument proteins that share similarity with formylglycineamide ribotide amidotransferase (FGARAT). It is thought that these enzymes may possibly augment the free nucleotide pools and could thus facilitate DNA synthesis and virus replication.

Oncogenic Signaling and Transformation

The Stp Oncoproteins

The HVS oncogenes required for induction of T-cell leukemia and T-cell transformation *in vitro* reside in the variable region at the left end of the L DNA (**Figure 1**). Subgroup A and B strains have a single gene termed *stpA* or *stpB* (saimiri transformation-associated protein of subgroup A or B strains), and subgroup C strains carry *stpC* (*stp* of subgroup C strains) and *tip* (tyrosine kinase-interacting protein). The proteins StpA and StpB share limited sequence homology with StpC, but are structurally unrelated to Tip. Although *stpA* and *stpC/tip* are not required for viral replication, deletion of either *stpA*, *stpC*, or *tip* abolishes transformation by HVS *in vitro* and pathogenicity *in vivo*. *stpA*- or *stpC*-transfected rodent fibroblasts form foci *in vitro* and induced tumors in nude mice. *stpA*-transgenic mice develop polyclonal peripheral T-cell lymphomas, and an *stpC* transgene induces epithelial tumors.

stpC and *tip* are transcribed into a single bicistronic mRNA from a common promoter directed toward the left end of the L DNA, with *tip* situated downstream from *stpC*. Transcription of *stpC/tip* is regulated similarly to IE genes in human T cells, and no obvious viral factors seem to be involved. The *stpC/tip* promoter carries euchromatic histone modifications in C488-transformed human T cells.

The 102-residue StpC phosphoprotein has an N terminus of 17 mostly charged residues, and the C terminus contains a hydrophobic region that probably serves as an anchor to perinuclear membranes. In between are 18 collagen tripeptide repeats of the form $(GPX)_m$, which may mediate multimerization of the protein. StpA and the less efficiently transforming StpB bind to, and are phosphorylated by, the nonreceptor tyrosine kinase Src. StpC interacts with the small G-protein Ras and stimulates mitogen-activated protein (MAP) kinase activity. Both StpA and StpC interact with tumor necrosis factor receptor-associated factors (TRAFs), leading to nuclear factor kappa B (NFκB) activation.

The Tip Oncoprotein

The subgroup C-specific 40 kDa Tip phosphoprotein has been shown to co-precipitate with the T-cell-specific nonreceptor Src family tyrosine kinase p56/Lck in C488-transformed T cells. Tip-transgenic mice develop T-cell proliferations. Tip has an N-terminal glutamate-rich region, duplicated in some strains, followed by one or two serine-rich regions, a bipartite kinase-interacting domain, and a C-terminal hydrophobic domain that anchors the molecule at the inside of the plasma membrane. The kinase-interacting domain consists of nine residues with homology to the C-terminal regulatory regions of various Src kinases (CSKH), and a proline-rich SH3-domain-binding sequence (SH3B). Several tyrosine residues, three of which are conserved between all strains investigated, are substrates for Lck. Tyrosine residue 127 (Y127) is the major tyrosine phosphorylation site of Tip (strain C488), but this modification does not enhance Lck binding in T cells. Recombinant viruses expressing mutations in Tip show that the strong Lck binding mediated by cooperation of the SH3B and CSKH motifs is essential for transformation of human T cells by C488, whereas Tip Y127 is required for transformation in the absence of exogenous interleukin-2, suggesting its involvement in cytokine signaling pathways.

Tip binding to Lck modulates the kinase activity and could result in an altered substrate specificity, contributing to the abrogation of ZAP70 phosphorylation. This dysregulation may further link Tip-bound Lck to alternative downstream effectors. In addition, the implication of Tip Y114 with constitutively active signal transducers and activators of transcription (STATs), especially STAT3, and the role of STATs in growth regulation and oncogenesis in multiple cell types, suggest a central role for Tip-induced STAT activity in viral T-cell transformation. However, recombinant HVS C488 expressing Tip with a tyrosine-to-phenylalanine mutation at residue 114 was able to transform primary human T lymphocytes in the absence of STAT1 or STAT3 activation. Tip is further associated with lipid rafts, and this is essential for the T-cell receptor (TCR) and CD4 downregulation but not for inhibition of TCR signal transduction and activation of STAT3 transcription factor. The activation of Lck and the inhibition of T-cell signaling by Tip may represent two different aspects of the same function, since the activation of Lck by Tip might trigger negative feedback mechanisms, such as apoptosis, in stably transfected Jurkat cells expressing high levels of Tip.

The Tio Oncoprotein

A spliced gene with two exons is located at the junction between H DNA and the left-terminal L DNA in HVA strain 73. The encoded protein shares local similarity with StpC and Tip of HVS subgroup C strains, and was therefore termed 'two in one' (Tio). Tio is expressed in HVA-transformed simian T cells, and is bound to, and phosphorylated by, the Src family tyrosine kinases Lck or Src. Phosphorylation of Tio at Y136 is required for successful transformation of human T cells. These cells are also transformed by recombinant HVS C488 in which *stpC* and *tip* has been replaced by a *tio* cDNA transcribed from a heterologous promoter. Furthermore, Tio induces NFκB signaling via direct interaction with TRAF6.

Growth Transformation of Human T Cells by Rhadinoviruses

Human T-cell growth transformation by HVS subgroup C strains has provided a reproducible technique for generating T-cell lines, and has opened up a new research direction linking T-cell biology, signal transduction pathways, and viral transforming functions. Infection of cord or peripheral blood mononuclear cells, thymocytes, or established human T-cell clones by C488 results in T-cell lines that grow continuously without restimulation by antigen or mitogen and do not require the presence of feeder or antigen-presenting cells. Many HVS subgroup C strains are able to transform human T cells, though to a varying extent; C488 is often preferred, as it achieves dependable growth transformation. Recombinant HVS C488 in which *stpC* and *tip* have been replaced by HVA *tio* can offer increased efficiency of human T-cell transformation along with a decreased requirement for IL-2. HVS C488 carrying mutations in Tip is being investigated for an expanding range of T-cell phenotypes.

The resulting polyclonal T-cell lines display the irregular morphology of T blasts. They carry nonintegrated HVS genomes in high copy numbers, have a normal karyotype, and are not tumorigenic in nude or severe combined immune-deficient (SCID) mice. The phenotype of HVS-transformed T cells is remarkably stable for many months in culture. It corresponds to that of mature, activated CD4+ CD8− or CD4− CD8+ T cells, usually with αβ-type (less frequently γδ-type) T-cell receptors. Transformed lines derived from established T-cell clones show the phenotype and human leukocyte antigen-restricted of the parental T cells. Cellular responses after CD3, CD4, or IL-2 receptor stimulation or antigen contact can be measured by signal transduction parameters, by proliferation, or, most reliably, by interferon-γ production. Transformation of cytotoxic T lymphocytes (CTLs) is rather inefficient, but may be increased by optimized protocols for prestimulation and culture of CTLs.

Transformation by HVS C488 has, in many cases, been the only way to cultivate and amplify T cells from patients with primary human immune deficiencies, including genetic T-cell defects involving the CD3γ chain, IL-2Rγ chain, CD95/Fas, IL-12R, major histocompatibility complex class II, Wiskott–Aldrich syndrome, or CD18/LFA-1. HVS-transformed human CD4+ T cells provide a productive system for T-lymphotropic viruses such as human herpesvirus 6 and human immunodeficiency virus (HIV) types 1 and 2, including primary clinical and macrophage-tropic HIV isolates.

Although most HVS-transformed New World monkey T lymphocytes produce infectious viral particles, HVS-transformed human T-cell lines maintain an intact viral genome but do not shed infectious virus. Production of infectious particles is also not induced by specific or nonspecific stimulation of the cells, using phorbol esters, nucleoside analogues, or other drugs that can reactivate viruses such as EBV or KSHV. Many macaque T-cell lines have been shown to shed very low amounts of virus particles, in contrast to their human counterparts, and the infusion of HVS-transformed autologous T cells into donor macaques did not cause disease. The reinfused T cells persisted for extended periods and the animals were protected against challenge with HVS C488. This is a relevant observation, since macaques are a common model for the situation in humans, and HVS-transformed simian T lymphocytes are similar to their human counterparts in many characteristics, including retained antigen specificity and presentation.

Alterations

StpC and Tip are the only viral proteins that have been demonstrated regularly in HVS-transformed human T cells, and yet their expression alone or together in a lentiviral background is not sufficient to transform primate T cells. The HSURs are expressed abundantly in a similar way to the small, noncoding RNAs (EBERs) of EBV, but deletion of all the HSUR genes does not influence virus replication or T-cell transformation. Viral transcription other than that of the bicistronic *stpC/tip* genes is rarely detected in human T cells; it is restricted to *ie14/vsag*, and few others at extremely low abundance (*orf57*/IE57, *orf50*/RTA, *orf70*/TS, *orf71*/vFLIP, *orf72*/vCyclin, and *orf73*/LANA). Some of these increased after stimulation with phorbol ester. Many other viral genes, such as the weakly transcribed *orf71* and *orf72*, have been shown by deletion analysis not to be required for T-cell transformation.

Compared to parental, untransformed T cells, a few cellular and biochemical alterations have been detected consistently in HVS-transformed T-cell lines: CD2 and its ligand CD58 are both expressed at high densities on the cell surface and there is hyper-responsiveness to CD2 ligation. Since withdrawal by limiting dilution halts the growth of HVS-infected human T cells, IL-2 induction by CD2–CD58 contact likely contributes to the transformed phenotype of HVS-transformed human T cells. Furthermore, subcloning of HVS-transformed cells is not possible. The Src family protein tyrosine kinase $p53/56^{Lyn}$ is usually expressed in B cells. Lyn is also found in HVS-transformed T-cell lines, similar to HTLV-1-immortalized T cells, but is not activated by HVS Tip. HVS-transformed T cells secrete high amounts of the Th1 cytokine interferon gamma (IFN-γ), and Th2-skewed T cells or Th2 clones shift toward a Th1 or Th0 profile. Many transformed clones also secrete large amounts of chemokines, such as MIP-1α and MIP-1β, and CCL1/I-309, which may protect HVS-transformed T cells from apoptosis via CCR8. IL-26, a new IL-10 cytokine family member, was discovered due to its over-expression in HVS-transformed T cells. IL-26 may influence T-cell interaction with epithelial cells *in vivo* but seemingly does not contribute to HVS-mediated T-cell transformation.

Gene Transfer

HVS vectors are attractive for gene transfer into T cells, since the functional phenotype of transformed T lymphocytes is maintained and the T cells can be simultaneously expanded by transformation. They may even be considered for therapeutic redirection of human T-cell antigen specificity, as tools for experimental cancer therapy applications. However, replication-deficient vector variants are necessary, and a number of biosafety aspects remain to be clarified. Remarkably, ganciclovir administration does not prevent pathogenesis by HVS expressing a TK suicide gene, and tumor induction is even more rapid than with a wild-type HVS control.

Genetic alteration of HVS-transformed cells can be achieved by transduction with retroviral or lentiviral vectors, or by using recombinant HVS. Efficient infection and occasionally limited productive replication of HVS have been observed in various human cell types, including human bone marrow stroma cells, primary fibroblasts, and hematopoietic precursors. Although foreign genes were first inserted into the genome of HVS more than two decades ago, the reconstitution of virus from overlapping cosmids and engineering of bacterial artificial chromosomes has greatly facilitated mutational analysis of the HVS genome and expression cloning in HVS. This includes attenuated, nononcogenic vectors deleted in the transformation-associated left-terminal region of L DNA that harbors the HVS oncogenes. Episomally persisting herpesvirus vectors, based either on replication defective viruses or on amplicons, are currently regarded as a promising alternative that can avoid side effects of integration, which is now a major concern in the field of gene transfer.

Rhesus Rhadinovirus and Related Old World Primate Rhadinoviruses

Natural Occurrence and Pathology

The discovery of KSHV as the first human rhadinovirus in 1994 greatly stimulated the search for rhadinoviruses in other Old World primates. Serological studies using KSHV-derived antigens indicated that a related herpesvirus may exist in rhesus monkeys. This led to the isolation of RRV by co-cultivation of lymphocytes from seropositive rhesus monkeys with rhesus fibroblasts by Desrosiers and colleagues in 1997. RRV seems to be very widespread in captive monkeys and, in contrast to KSHV, can be propagated efficiently in cell culture. Although there exists no clear disease association for RRV in infected healthy macaques, there is one report concerning rhesus macaques that were immunosuppressed by previous infection with simian immunodeficiency virus (SIV). In this case, RRV infection resulted in a multifocal lymphoproliferative disease resembling multicentric Castleman's disease. However, this had not been noticed previously in numerous studies of SIV-infected macaques of unknown, but presumably mostly positive, RRV infection status.

Using degenerate PCR of the DNA polymerase gene, DNA fragments of rhadinovirus origin have been identified in various Old World primates, including African green monkey, chimpanzee, gorilla, and mandrill. Phylogenetic analysis of short sequences has revealed that Old World primate rhadinoviruses probably segregate into two groups: one that is more closely related to KSHV and another that is more closely related to RRV.

Genome Structure and Replication

Analysis of the genome sequences of two independent strains has revealed that RRV is indeed more closely related to KSHV than to the prototypic rhadinovirus HVS. The RRV genome organization is essentially collinear with that of KSHV. It has (at least) 79 genes, 67 of which are homologous to genes found in both KSHV and HVS. Of the remaining 12 genes, 8 are similar to KSHV genes (see below). Interestingly, *orf2*/DHFR is in the same position as in HVS, and different from that in KSHV. The RRV OriLyt is located in the same region as in HVS and KSHV, between *orf69* and *orf71*/vFLIP. The functions of viral transactivators and regulatory proteins resemble those of KSHV. The gene content of RRV is similar to that of KSHV, but contains only one *vMIP* gene and lacks *K3* and *K5*. The genes encoding CCPH and vIL-6 are conserved in RRV, and eight genes are present with homology to the family of viral interferon regulatory factors (*vIRF-1* through *vIRF-8*). Several large DNA viruses have been shown to encode micro-RNAs (miRNAs), including EBV and the rhadinoviruses RRV and KSHV, and miRNA genes are evolutionary conserved in at least the lymphocryptoviruses. However, the role of miRNA and RNA interference in the viral context is controversial. Given the specificity of RNA interference, it remains to be determined whether this can have a role in herpesviral transformation of foreign hosts. An interesting speculation is that herpesviral miRNAs may act as specificity factors that initiate heterochromatin assembly of the latent viral genome.

Rhadinoviruses from Retroperitoneal Fibrosis

In an approach using a degenerate PCR technique, Rose and co-workers identified fragments of a herpesvirus DNA polymerase gene in tissue specimens from retroperitoneal fibromatosis (RF) from macaque species, *M. nemestrina* (six cases) and *M. mulatta* (one case). RF is a rare disease occurring in immunesuppressed macaques that consists of aggressively proliferating fibrous tissue with a high degree of vascularization; thus, it somewhat resembles KS. Earlier transmission studies indicated that an infectious agent may be involved in RF pathogenesis. Sequence comparisons indicate that the DNA polymerase and adjacent genes of these two potentially novel rhadinoviruses, tentatively termed RFHVMm and RFHVMn, are related more closely to KSHV than RRV. Attempts to isolate the viruses on cultured cells have so far been unsuccessful. The possible coexistence of two different rhadinoviruses in the same host animal is also indicated, since in one study all RF-diseased macaques harbored RRV DNA (and all were coinfected with simian retrovirus 2 and/or SIV). In these animals, RFHV DNA was present at significantly higher copy numbers in the RF tumors.

Lymphocryptoviruses of Old and New World Primates

Gammaherpesviruses closely related to EBV have been recognized in several species of Old World primates from the mid-1970s. The genome of rhesus lymphocryptovirus (abbreviated to rhesus LCV; species *Cercopithecine herpesvirus 15*) has been sequenced. Until recently, the paradigm was that the lymphocryptoviruses are restricted to Old World primates, including humans. However, a virus related to EBV was isolated from common marmosets, both from healthy animals and animals with spontaneous B-cell lymphomas. Related lymphocryptoviruses have also been detected in several other New World primate species (**Table 1**). The new lymphocryptovirus (marmoset LCV; species *Callitrichine herpesvirus 3*) has an EBV-like genome structure, and determination of the genome sequence has shown that several Old World primate lymphocryptovirus-specific genes are absent. Specifically, homologs of EBV BCRF1/vIL10, BARF1/CSF-1R, BARF0, the EBERs, and several other genes of unknown function have not been detected, and marked divergences exist in LMP-1, LMP-2, EBNA-LP, EBNA-2, and the EBNA-3 family. Also, the organization of the putative marmoset LCV OriP-region is clearly distinct from that in Old World primate lymphocryptoviruses.

In vitro transformation of B cells by human and simian Old World primate lymphocryptoviruses seems to be mostly restricted to the natural host or closely related species. However, experimental T-cell tumors can be induced in rabbits following infection by cynomolgus LCV or baboon LCV.

Conclusion and Perspective

Viruses of the gammaherpesvirus genera *Lymphocryptovirus* and *Rhadinovirus* can be found in New World and Old World primates, including humans. Although several members of both genera are closely associated with viral oncogenesis, the simian viruses do not generally provide a straightforward model for multifaceted human diseases. The direct transforming action of viral oncogenes, as well as a chronic inflammatory reaction that may be affected by KSHV-encoded or KSHV-induced cytokines or angiogenic factors, may contribute to the genesis of KS. The KSHV-related simian rhadinoviruses do not provide a corresponding animal model as yet. Historically, interest in the rhadinoviruses has focused on the long-established prototype, HVS. Although a comparable virus-associated, acute, peripheral, pleomorphic T-cell lymphoma is not known yet in humans, this disease, which is induced reproducibly by HVS within weeks, can serve as an experimental model for general tumor development. The ability of certain HVS strains to transform human T lymphocytes to stable proliferation in culture provides a valuable tool for laboratory studies of T-cell immunology, including inherited and acquired immunodeficiency. In addition to their use as an immunological and biochemical T-cell model, HVS-transformed T cells can provide a source for the purification of specifically overexpressed cytokines or chemokines from culture supernatants. A detailed analysis of differential gene expression will lead to identification of signaling pathways that lead to lymphocyte transformation by herpesviral oncoproteins. Those involving STAT, nuclear factor of activated T cells (NFAT), and NFκB may be particularly important, perhaps providing hints to the roles of the respective pathways in human (nonviral) T-cell malignancy. Furthermore, rhadinovirus-transformed T-cell lines can be valuable tools in screening for specific drugs that target these pathways.

The side effects of retroviral integration have shown the requirement for efficient nonintegrating vectors. Recombinant rhadinoviruses can deliver foreign genes into primary human mesenchymal cells and T lymphocytes; this may prepare the ground for future therapeutic applications of persisting rhadinoviral vectors in adoptive immunotherapy. Safety considerations will prevent the use of unconditionally transforming rhadinovirus vectors, and require the development of novel rhadinovirus-based T-lymphotropic episomes, including conditional/attenuated or amplicon vector systems. Analysis of viral genome episomal modification in host T cells and the detection of genomic insulating regions can provide markers for the selection of regions suitable for the insertion of transgenes into the viral backbone.

Acknowledgments

Original work included in this review article was supported by the Deutsche Forschungsgemeinschaft (SFB643 TP A2, EN423/2-1), the Wilhelm Sander Stiftung, Bavarian International Graduate School of Science (BIGS), and the Interdisciplinary Center for Clinical Research (IZKF) at the University of Erlangen-Nuremberg.

See also: Epstein–Barr Virus: General Features; Epstein–Barr Virus: Molecular Biology; Herpesviruses: Discovery; Herpesviruses: General Features; Immune Response to viruses: Antibody-Mediated Immunity; Kaposi's Sarcoma-Associated Herpesvirus: General Features; Kaposi's Sarcoma-Associated Herpesvirus: Molecular Biology; Murine Gammaherpesvirus 68; Simian Alphaherpesviruses.

Further Reading

Alberter B and Ensser A (2007) Histone modification pattern of the T cellular *Herpesvirus saimiri* genome in latency. *Journal of Virology* 81: 2524–2530.

Bruce AG, Bakke AM, Bielefeldt-Ohmann H, et al. (2006) High levels of retroperitoneal fibromatosis (RF)-associated herpesvirus in RF lesions in macaques are associated with ORF73 LANA expression in spindleoid tumour cells. *Journal of General Virology* 87: 3529–3538.

Cai X, Schäfer A, Lu S, et al. (2006) Epstein–Barr virus microRNAs are evolutionarily conserved and differentially expressed. *PLoS Pathogens* 2: e23.

Cho NH, Feng P, Lee SH, et al. (2004) Inhibition of T-cell receptor signal transduction by tyrosine kinase-interacting protein of herpesvirus saimiri. *Journal of Experimental Medicine* 200: 681–687.

Cullen BR (2006) Viruses and microRNAs. *Nature Genetics* 38: S25–S30.

Damania B (2004) Oncogenic gamma-herpesviruses: Comparison of viral proteins involved in tumorigenesis. *Nature Reviews Microbiology* 2: 656–668.

Ensser A (2006) Transformation by herpesviruses: Focus on T-cells. *Future Virology* 1: 109–121.

Ensser A and Fleckenstein B (2004) Herpesvirus saimiri transformation of human T lymphocytes. In: Coligan JE, Bierer BE, Margulies DH, et al. (eds.) *Current Protocols in Immunology*, pp. 7.21.1–7.21.10. New York: Wiley.

Ensser A and Fleckenstein B (2005) T-cell transformation and oncogenesis by γ2-herpesviruses. *Advances in Cancer Research* 93: 91–128.

Ensser A, Thurau M, Wittmann S, et al. (2003) The genome of herpesvirus saimiri C488 which is capable of transforming human T cells. *Virology* 314: 471–487.

Heck E, Friedrich U, Gack MU, et al. (2006) Growth transformation of human T-cells by herpesvirus saimiri requires multiple Tip–Lck interaction motifs. *Journal of Virology* 80: 9934–9942.

Li HW and Ding SW (2005) Antiviral silencing in animals. *FEBS Letters* 579: 5965–5973.

Pfeffer S, Zavolan M, Grasser FA, et al. (2004) Identification of virus-encoded microRNAs. *Science* 304: 734–736.

Searles RP, Bergquam EP, Axthelm MK, et al. (1999) Sequence and genomic analysis of a rhesus macaque rhadinovirus with similarity to

Kaposi's sarcoma-associated herpesvirus/human herpesvirus 8. *Journal of Virology* 73: 3040–3053.

Schäfer A, Lengenfelder D, Grillhösl C, *et al.* (2003) The latency-associated nuclear antigen homolog of herpesvirus saimiri inhibits lytic virus replication. *Journal of Virology* 77: 5911–5925.

Relevant Websites

http://en.wikipedia.org – Cebidae; Herpesviridae; Rhadinovirus.
http://www.ncbi.nlm.nih.gov – International Committee on Taxonomy of Viruses, ICTVdb.

Simian Immunodeficiency Virus: Animal Models of Disease

C J Miller and M Marthas, University of California, Davis, Davis, CA, USA

© 2008 Elsevier Ltd. All rights reserved.

Glossary

Dysplasia Abnormal growth of tissues, organs, or cells.

Endemic Prevalent in a particular group of animals or people.

Epitope The part of a macromolecule that is recognized by the immune system.

Hypergammaglobulinemia Increased blood levels of gamma globulin (IgG).

Hyperplasia An abnormal increase in the number of cells in an organ or a tissue with consequent enlargement.

MHC-I (major histocompatibility complex I) T-cell epitopes are presented on the surface of an antigen-presenting cell, where they are bound to MHC molecules. T-cell epitopes presented by MHC class I molecules are typically peptides between 8 and 11 amino acids in length, while MHC class II molecules present longer peptides, and nonclassical MHC molecules.

Taxonomy and Discovery

Simian immunodeficiency viruses (SIVs) are members of the genus *Lentivirus* and the family *Retroviridae*. The genus *Lentivirus* includes viruses that infect ungulates (maedi-visna virus of sheep; caprine arthritis-encephalitis virus, CAEV), horses (equine infectious anemia virus, EIAV), cows (bovine immunodeficiency virus, BIV), wild and domesticated cats (feline immunodeficiency virus, FIV), nonhuman primates (NHPs; SIV), and humans (human immunodeficiency virus, HIV). All retroviruses are spherical, 80–100 nm in diameter, and have a diploid, single-stranded RNA genome and viral enzymes inside a viral protein case, or core, which is enveloped by a host cell membrane studded with viral glycoproteins. The diploid genome of single-stranded RNA is linked noncovalently near the 5' end of the molecules. The 5' end of the viral RNA is capped and the 3' end is polyadenylated. DNA synthesis by the viral encoded reverse transcriptase (RT) is primed by host tRNA that is base-paired to the viral RNA. The double-stranded DNA provirus is integrated into the host chromosomes by a viral encoded integrase and the remaining events in transcription, translation, and assembly are host cell dependent. Depending on the cell type infected, particles assemble and bud through the plasma membrane or into membrane-lined intracytoplasmic vesicles. The morphological feature that distinguishes lentiviruses from other retroviruses is the cone or rod shape of the viral core protein in a mature virion.

SIV_{mac} was the first member of the group to be identified following isolation of a retrovirus from a captive rhesus macaque housed at a US primate center. Apparently, the virus was introduced into captive macaque populations in the US during experiments into the transmission of kuru and leprosy, at which time material derived from tissue of SIV-infected African monkeys was deliberately introduced into Asian macaques. SIV has been eliminated from captive populations of macaques in the US through a rigorous testing program. Fortunately, serology provides an effective and economical tool for screening for infection. Extended quarantine periods and regular serologic testing programs are rigorously applied to maintain the SIV-free status of captive macaque colonies.

Nomenclature and Classification of Primate Lentiviruses

Primate lentiviruses are classified as originating in NHPs (SIVs) or humans (HIVs). SIVs are given specific names based on the NHP species from which they are isolated, for example, SIV_{cpz} was isolated from chimpanzees and SIV_{deb} from De Brazza's monkeys.

Recently, a hierarchical nomenclature has been developed to describe the high level of genetic diversity found among the many HIV isolates. HIV is first divided into two broad types, designated HIV-1 and HIV-2. Both HIV types are subdivided into groups consisting of phylogenetically similar viruses that resulted from different nonhuman-primate–human SIV transmission events (**Figures 1** and **2**). Within a group, phylogenetic clusters of viruses are termed subtypes. There is a movement toward a unified, standard nomenclature to describe genetic diversity within and among SIVs in which a three-letter abbreviation of the vernacular name of the NHP species is used (**Figure 3**). For example, the SIV from chimpanzees is termed SIV_{cpz} and subspecies designation is provided a three-letter abbreviation – thus $SIV_{cpz.Ptt}$ is derived from *Pan troglodytes troglodytes*. SIVs of African green monkeys (SIV_{agm}) are divided into four subtypes, each named for the African green monkey subspecies from which it was isolated (**Figure 3**).

Genome Organization

Like all retroviruses, lentivirus genomes contain long terminal repeats (LTRs) at each end and genes encoding three virion structural components: core (*gag*), polymerase (*pol*), and envelope (*env*). In addition, all primate lentivirus genomes also contain five accessory genes: *vif*, *vpr*, *rev*, *tat*, and *nef*. The genomes of a subset of primate lentiviruses

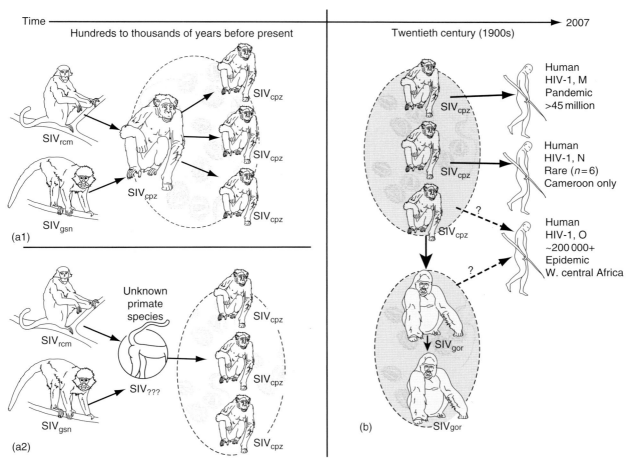

Figure 1 Origin of HIV-1 inferred from phylogenetic analyses of SIV and HIV-1 genomes. (a) The earliest event is transmission of SIVs from African monkeys (red-capped mangabey (RCM) infected with SIV (SIV_{rcm}) and greater spot nose monkey (GSN) infected with SIV (SIV_{gsn})) to chimpanzees followed by virus recombination that resulted in SIV_{cpz}. Two scenarios are consistent with the data: (a1) SIV_{rcm} and SIV_{gsn} initially infected a chimpanzee; or (a2) SIV_{rcm} and SIV_{gsn} first infected an unknown primate species where virus recombination occurred – in either case, a recombinant SIV was transmitted among chimpanzees producing SIV_{cpz}. (b) Three separate transmissions of SIVs to humans occurred from distinct populations of chimpanzee (SIV_{cpz}) or gorilla (SIV_{gor}) in the 1900s producing three HIV-1 groups, M (main), N, and O (outlier). Recent, but limited, phylogenetic data show that HIV-1 group O is most closely related to SIV_{gor}. The most likely scenario suggested by the data is that chimpanzees transmitted SIV_{cpz} to gorillas which resulted in SIV_{gor}; it is unknown whether HIV-1 group O was transmitted to humans from gorillas or from an as yet unidentified chimpanzee reservoir infected with SIV_{cpz} more closely related to HIV-1 O. Only HIV-1 groups M and O have established efficient human-to-human transmission. Group M HIV-1 is pandemic, infecting over 45 million persons; group O HIV-1 is epidemic having infected an estimated several thousands of persons in Africa. In contrast, only six individuals from Cameroon are known to be infected with HIV-1 group N. Dotted circles indicate distinct populations of a primate species; shading indicates SIV isolated from individuals of this species; solid arrows show inferred transmissions of SIV; dashed arrows indicate where SIV transmission is hypothesized, but there are insufficient data to confirm.

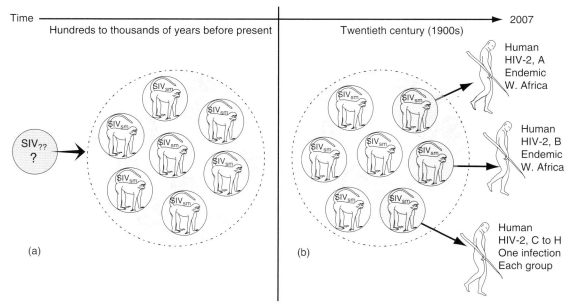

Figure 2 Origin of HIV-2 inferred from phylogenetic analyses of SIV and HIV-2 genomes. (a) SIV probably infected sooty mangabeys (SMs) thousands of years ago. Multiple separate transmissions of SIV_{sm} from distinct populations of SMs to humans occurred during the 1900s giving rise to eight HIV-2 groups, A–H (b). Only HIV-2 groups A and B have resulted in efficient human transmission and established epidemics; a single person is known to be infected with each of the HIV-2 groups C–H. Dotted circles indicate distinct populations of SMs; shading indicates SIV_{sm} isolated from individual SM in a population; arrows show inferred transmissions of SIVs.

have one of two unique genes: *vpr* or *vpu*. Three types of genomic structure are observed for primate lentiviruses (**Figure 4**). Group A includes SIVs only from African NHPs; these SIVs cause no disease in the host species from which the SIV was isolated: SIV_{agm}, SIV_{syk}, SIV_{lhoest}, SIV_{mnd-1}, SIV_{sun}, and SIV_{col} (A, **Figure 4**). SIVs in group A all have the basic primate lentivirus genome structure, which contains only five accessory genes. Group B includes HIV-2 and SIVs most closely related to HIV-2 (SIV_{sm}, SIV_{mac}, SIV_{rcm}, and SIV_{mnd-2}); each group B virus has a *vpx* gene which is absent in the Group A viruses (B, **Figure 4**). Group C viruses include HIV-1, SIV_{cpz}, and other SIVs with a *vpu* gene, but lacking *vpx* (SIV_{gsn}, SIV_{mon}, and SIV_{mus}; C, **Figure 4**).

As retroviruses, all lentiviruses also have two complete RNA genomes in a single virion. Therefore, when a host cell is infected with two or more genetically different primate lentiviruses, two different genomic viral RNAs can be packaged into the same virion, generating recombinant viral genomes in the next viral replication cycle. *vpx* shares sequence similarity with *vpr* and is thought to have originated by recombination among SIV genomes. Thus, viral recombination has been a dominant force in the evolution of primate lentiviruses as suggested by the diagrams of primate lentivirus genomes in **Figure 4**. As SIVs are isolated from more primate species and their full genome sequences compared, it was found that some primates harbored SIV with apparently 'mosaic' genomes which included portions of structural genes derived from SIVs of different primate species. Thus, recombination among SIVs has resulted in novel viruses; for example, the current SIV_{cpz} evolved from a recombinant between two SIVs – SIV_{rcm} from red-capped mangabeys and SIV_{gsn} from greater spot nose monkeys (**Figure 1**).

Evolution

All primate lentiviruses are more closely related to each other than to lentiviruses from nonprimates. This genetic similarity suggests primate lentiviruses co-evolved with their host species and that primates have not been infected by lentiviruses from other mammals, such as ungulates or felines. SIVs have been isolated from many wild African NHP species, but no SIVs have been found to infect any wild Asian or New World NHPs. This suggests that lentiviruses became established in primates sometime after the divergence of Old and New World primate species (~35–40 million years ago). Although they lack an endemic SIV, Asian macaques are susceptible to experimental infection with a variety of SIV isolates and develop acquired immune deficiency syndrome (AIDS), similar to HIV-infected humans. The estimated origin of the macaque genus is ~6 million years ago and emigration of macaques from Africa to Eurasia began ~5 million years ago. Thus, it is probable that lentiviruses were not widespread among African primates 5 million years ago.

Lentiviral replication generates high genetic diversity in two ways: mutation caused by the highly error-prone retroviral polymerase (RT) and recombination between

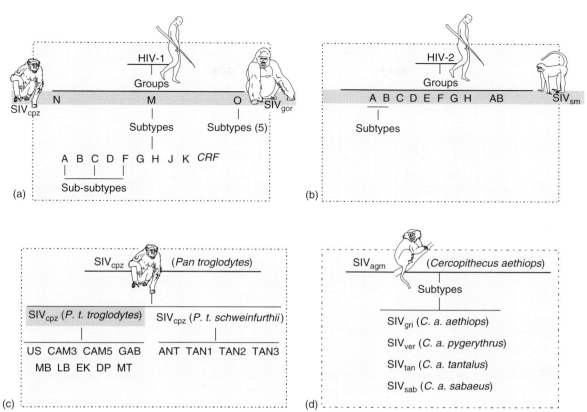

Figure 3 Classification and phylogenetic relationships of primate lentiviruses. (a) HIV-1 lineage. HIV-1 is divided into three groups: M (main), N (non-M/non-O or 'new'), and O (outlier); based on phylogenetic clustering with SIVs, each group represents a separate chimpanzee- or gorilla-to-human transmission event (see also **Figure 1**). Subtype diversification occurred in humans after each cross-species transmission; CRF indicates inter-subtype recombinants. Clusters of diversity within a subtype are called sub-subtypes. The shaded bar indicates the SIVs which gave rise to the HIV-1 groups. (b) HIV-2 lineage. HIV-2 is divided into eight groups (A–H); based on phylogenetic clustering with SIVs, each group represents a separate sooty mangabey-to-human transmission event (see also **Figure 2**); AB indicates recombinant virus. The shaded bar indicates SIV_{sm} giving rise to the HIV-2 groups. (c) SIV_{cpz} lineage. SIV_{cpz} is designated as one subtype although SIV_{cpz}'s have been isolated from two subspecies of chimpanzees, *Pan troglodytes troglodytes* and *Pan troglodytes schweinfurthii*; multiple SIV_{cpz}'s from each of these two subspecies have been isolated and sequenced (abbreviations are listed). SIV_{cpz} from *P. t. troglodytes* (shaded bar) is most closely related to HIV-1. (d) SIV_{agm} lineage. There are four subtypes of SIV_{agm}, one for each of the four subspecies of African green monkey, *Cercopithecus aethiops*: *C. a. aethiops* (grivet), *C. a. pygerythrus* (vervet), *C. a. tantalus* (tantalus), and *C. a. sabaeus* (sabaeus).

two different viral genomic RNA molecules packaged in a virion. The majority of African NHP species have genetically distinct SIVs. The isolation of more than one distinct SIV from a primate species provides evidence for regular interspecies SIV transmission (**Figures 1** and **3**). A notable example is isolation of two SIVs from mandrills (SIV_{mnd-1} and SIV_{mnd-2}; **Figure 4**) that are as genetically different from each other as are HIV-1 and HIV-2.

Thus, genetic diversity of primate lentiviruses has evolved in two ways: (1) within a single host species, and (2) by transmission of SIV from one primate host species to another (cross-species transmission). Cross-species SIV transmission has occurred in both wild and captive primates. The most likely modes of SIV transmission between wild primate species are thought to be fighting among different primates or hunting and eating of one primate species by another; for example, wild chimpanzees are known to kill and eat a variety of monkeys that share the same geographic range. When a cross-species SIV transmission event occurs in an individual already SIV infected, there is the opportunity for viral recombination and, thus, the generation of SIV variants with novel phenotypes, including increased pathogenicity. Such a dual SIV infection is proposed to have been the origin of SIV_{cpz} (**Figure 1**). Dual (or multiple) SIV infection of individuals can also occur when a primate species has genetically distinct SIV variants in one or different populations of animals. Thus, intraspecific SIV recombination increases the genetic diversity within an SIV lineage (**Figure 3**).

Virology

The life cycle of SIV is similar to all retroviruses; once uncoated from the capsid, the RNA genome is reverse-transcribed into a full-length DNA genome that

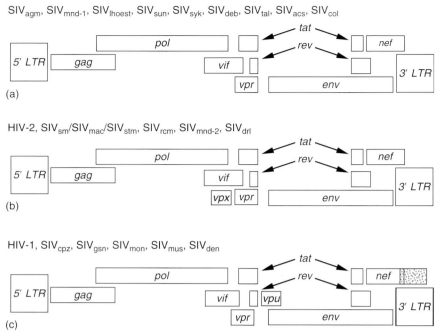

Figure 4 Genomic structure of primate lentiviruses. Viral genes are represented by rectangles with names inside and arrows show genes with two separate coding regions. Group A includes SIVs only from African NHPs (agm, African green monkey; mnd, mandrill; lhoest, l'Hoest monkey; sun, sun-tailed monkey; syk, Sykes monkey; deb, de Brazza's monkey; tal, talapoin; acs, *ascanius*; col, colobus). Group B includes HIV-2 and SIVs most closely related to HIV-2; each group B virus has a *vpx* gene (shaded box), that is absent in the Group A viruses (sm, sooty mangabey; mac, macaque; stm, stump-tailed macaque; rcm, red-capped mangabey; mnd, mandrill; drl, drill). Group C viruses include HIV-1, SIV$_{cpz}$, and other SIVs with a *vpu* gene, but lacking *vpx* (cpz, chimpanzee; gsn, greater spot nose monkey; mon, mona monkey; mus, mustached guenon; den, Dent's mona monkey). For HIV-1 and SIV$_{cpz}$ genomes only, the *nef* gene (stippled) does not overlap the *env* gene.

is transported to the nucleus and integrated into host chromosomes. Once integrated, the provirus produces a variety of RNA species that can be spliced to produce all the proteins required for virion assembly and egress. In T cells the immature virions bud from the plasma membrane into the extracellular space, but in macrophages virions often bud into intracellular vesicles that can eventually fuse with the plasma membrane of the cell releasing the progeny virions.

SIV can be propagated in many human T-cell and macrophage cell lines derived from tumors, mitogen-stimulated peripheral blood mononuclear cells (PBMCs), monocyte-derived macrophages, etc., although the range of permissive cell types varies from isolate to isolate. In PBMC cultures and many cell lines, viral infection results in obvious cytopathic effects including ballooning degeneration and syncytia formation due to fusion of cell membranes. Syncytium formation is mediated by the envelope glycoprotein and allows direct cell-to-cell spread of the infection.

The CD4 molecule is the primary cellular receptor for SIV. As with HIV, infection is primarily mediated by CD4 but interaction with CD4 alone is not sufficient to allow viral entry into the cell. A number of chemokine receptors are required co-receptors for SIV and HIV infection. The SIV envelope glycoprotein (gp120) binds to CD4 resulting in a conformational shift in the glycoprotein that exposes a co-receptor binding site. This secondary binding site interacts with a chemokine receptor on the cell surface, and fusion of cell and virion membranes results. Although various SIV strains can use a variety of chemokine receptors (CCR5, BOB, Bonzo) as a co-receptor *in vitro*, CCR5 is the most important co-receptor for SIV *in vivo* as it widely expressed by target cells (T cells, macrophages, dendritic cells (DCs)) in the body. Chimeric viruses have been constructed using SIV as a backbone for inserting HIV genes. In HIV infection, differential co-receptor usage by HIV variants has been implicated as a determinant in virulence and pathogenesis. Some of the SIV/SHIV (SHIV) viruses have been constructed using the *env* from HIV variants known to preferentially use CxCR4 or CCR5 as co-receptors. Although differential co-receptor usage can be demonstrated by the chimeric viruses *in vitro*, the extent to which altered replication capacity independent of co-receptor usage contributes to observed differences in pathogenesis has been difficult to determine.

Host cellular factors have recently been identified that confer resistance or susceptibility to productive SIV infection. In host cells resistant to another species' lentivirus, the host protein APOBEC3G is incorporated into the virion during particle formation. During reverse transcription, virion-incorporated APOBEC3G deaminates the minus strand of viral DNA and inactivates/degrades the viral genome. In permissive cells, the *vif* gene permits SIV to replicate by targeting APOBEC3G for degradation in the proteosome and preventing its incorporation into the virion. However, the activity of *vif* is species specific. Thus human APOBEC3G is inhibited by HIV-1 Vif but not by SIV_{agm} Vif, and AGM APOBEC3G is inhibited by SIV_{agm} Vif, but not by HIV-1 Vif. TRIM5alpha, a member of the poorly understood tripartite motif (TRIM) family of proteins, is another host-restriction factor. TRIM5alpha restricts HIV replication in rhesus macaques, by interacting with the virion capsid and blocking the uncoating of HIV-1 after entry but before reverse transcription.

Transmission and Dissemination

When experimentally inoculated onto intact mucosal surfaces, SIV rapidly infects local DCs, macrophages, and T cells. Although the number of SIV-infected cells is generally higher in the lamina propria adjacent to regions of epithelial damage, SIV-infected DCs and T cells can be found within the intact epithelium of the genital tract, indicating that the virus can cross the intact genital mucosa. Within 24 h of infection, SIV-infected cells can be detected in the lymph nodes that drain a mucosal inoculation site. Initially, there is limited viral replication in local tissues prior to widely disseminated infection in peripheral lymphoid tissues, but by 1 week, post inoculation (PI), viral replication is explosive with dramatic increases in viral RNA levels in all lymphoid tissues. At 10–14 days PI, the peak of viral replication and plasma vRNA levels occurs, and adaptive immune responses develop. By day 28–56 PI, plasma vRNA levels decline to moderate levels, where they remain for a variable period of time. Eventually viral replication increases, CD4+ T-cell levels decline further and AIDS develops. Examples of plasma viral levels in rhesus macaques inoculated with two pathogenic SIV variants are provided in **Figure 5**.

Disease Pathogenesis

While the endemic SIV infections of African monkeys and apes seem to be minimally pathogenic, SIV infections of Asian macaques cause simian AIDS. This disease results from the profound destruction of CD4+ T cells in lymphoid organs throughout the body. To an even greater degree than HIV infection of people, SIV replicates to very high levels in the infected host with as many as 10^9 vRNA copies per ml of plasma during peak infection. This high replicative capacity coupled with the error-prone viral RT produces viral quasi-species of tremendous genetic diversity in infected individuals. This viral quasi-species makes the virus very adaptable to changing conditions in the host as antiviral immune responses develop or successive waves of target cells are destroyed. As in HIV infection, the level of SIV RNA in plasma is an excellent indicator of clinical prognosis in Asian macaques.

The central feature of AIDS is the destruction of CD4+ T cells and the crippling effect this has on the host's ability to control opportunistic, or latent, infections. The majority of the CD4+ T cells in the body reside in the gastrointestinal (GI) tract and other mucosal sites, and these cells are activated memory and $CCR5^+$ T cells. Thus the GI tract is a major site for the destruction of CD4+ T cells. As viral replication occurs, the level of immune activation and lymphocyte proliferation dramatically

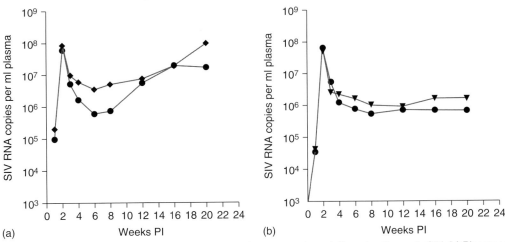

Figure 5 Plasma SIV RNA levels in rhesus macaques following intravenous inoculation of pathogenic SIV. (a) Plasma vRNA levels in two macaques inoculated with SIVmac251. (b) Plasma vRNA levels in 2 macaques inoculated with SIVmac239. Note in all four animals there was a peak in plasma vRNA at 2 weeks PI followed by a variable period of declining or stable plasma vRNA. While setpoint vRNA levels remained stable in both SIVmac239 inoculated animals over the period of observation, plasma vRNA had begun increasing by 24 weeks PI in both SIVmac251-infected animals, suggesting that these animals would soon develop clinical AIDS.

increases as the host attempts to mount an immune response. Many of the transcriptional signals involved in the host's innate and adaptive immune responses are controlled by nuclear factor-kappa B (NF-κB) transcriptional factors. NF-κB is also a key promoter for SIV replication as the viral LTR contains binding sites for NF-κB transcription factors. Thus the host immune response to SIV generates activated T cells with high levels of critical molecules (dNTPs, transcriptional factors) and, paradoxically, these are the cellular targets needed for optimal viral replication.

In addition to the effects of CD4+ T-cell depletion on the host defenses, SIV can directly produce disease. Thus SIV crosses the blood–brain barrier, infecting resident and transient cells of the macrophage lineage, which can result in viral encephalitis. In addition, the high levels of viral antigen and strong antibody responses can produce immune complex glomerulonephritis, while SIV infection of bone marrow macrophages is presumed to be the cause of the anemia and other hematologic abnormalities that SIV infection produces. The above features are also common in HIV infection, but SIV infection of macrophages can produce histiocytic inflammation with syncytial cell formation in the lung, lymphoid tissues, and GI tract.

The above discussion describes the pathogenesis of AIDS in SIV-infected Asian macaques, but in the African primates that are natural hosts of SIV there is little clinical effect of infection. A key difference lies in the response of natural hosts and Asian macaques to SIV infection. Thus SIV-infected African primates mount a complete adaptive immune response but there is little inflammation and no chronic immune activation associated with these immune responses despite continuous high-level viral replication. In marked contrast, SIV infection of macaques, as with HIV infection of humans, produces chronic immune activation and inflammation that is associated with depletion of central memory T-cell pools and loss of effector memory T cells. Thus despite levels of viral replication that are similar to Asian macaques, SIV infection rarely produces AIDS in natural hosts indicating that inflammation and immune activation are necessary for AIDS pathogenesis.

Specific strains of SIV have altered pathogenesis with a number of molecular clones and biologic isolates being attenuated for replication in Asian macaques. Although these variants do not produce disease in the typical time course of viruses with higher replicative capacity, some eventually produce disease in adults or infant macaques, while other attenuated SIV variants (SIVmac1A11) have failed to produce disease in infected macaques after more than 15 years of observation. In addition, some strains of SIV have specific tropism for monocyte/macrophages that are mediated by post-entry events. Infections with these macrophage-tropic SIV strains are more commonly associated with SIV-induced meningoencephalitis than other SIV strains.

Immune Response to SIV Infection

As SIV infection is lifelong, the host immune response cannot clear the infection but, in some cases, it can exert considerable control on the level to which SIV replicates. Before day 5 after mucosal SIV inoculation, there is little evidence of innate immune responses with only modest increases in type 1 interferon levels and interferon-stimulated gene levels at the site of inoculation. At days 6–7 PI, there is a dramatic and simultaneous increase in innate antiviral immune responses in all lymphoid tissues that coincides with the dramatic explosion of viral replication in these tissues. By days 10–14 PI, SIV-specific CD8+ T-cell responses are present in blood and in the mucosal sites of inoculation, with strong antiviral T-cell responses widespread by day 28 PI. This rapid increase in antiviral effector T cells occurs as viral replication and plasma vRNA levels decline from a peak at 14 days PI. The temporal relationship between the decline in plasma viremia and the appearance of antiviral T cells has been interpreted to be evidence that CD8+ T cells are critical in the control of HIV replication. Further evidence for the control of SIV replication by CD8+ lymphocytes has been obtained by using monoclonal antibodies to the α-chain of the CD8 complex to transiently deplete T cells and natural killer (NK) cells. In chronically infected animals, plasma vRNA levels dramatically increase during the period of CD8+ lymphocyte depletion and then rapidly fall as the lymphocyte population is replenished. Finally, as in HIV infection, detailed studies of well-characterized MHC-I-restricted T-cell epitopes in infected macaques have shown that as the host T-cell response targets a specific peptide sequence, viral variants with proteins that contain a variant of the targeted epitope become increasingly common. The effect of these mutated epitopes on viral replication is variable. In some cases, these variants are relatively unfit and the virus gains little advantage; however, in other cases, these immune escape variants are relatively fit and viral replication increases dramatically as the escape variants appear in the viral population. Finally, as in humans, specific MHC-I alleles of macaques (i.e., Mamu-A*01) are associated with particularly strong CD8+ T-cell responses, enhanced immune control of viral replication, and increased disease-free survival times after infection. Taken together, these observations argue that CD8+ T-cell responses can play a significant role in controlling viral replication. B-cell responses also have a role in controlling HIV and SIV replication. Thus it has been shown that passive transfer of high-titer SIV-specific gammaglobulin or neutralizing anti-HIV antibodies inhibits SIV and SHIV replication and retards the pace of disease progression. Further, if macaques are prevented from developing anti-SIV IgG antibodies by B-cell depletion prior to infection with highly pathogenic SIV, they rapidly develop uncontrolled viral replication, and progress to AIDS in a few months PI, while most SIV-infected macaques make strong antibody responses and develop AIDS at 10–24 months PI.

Clinical Features, Pathology, and Histopathology of SIV Infection in Asian Macaques

End-stage disease in SIV infection is indistinguishable from human AIDS and can be divided into four broad categories: (1) opportunistic infections, (2) SIV-mediated inflammatory diseases, (3) neoplastic diseases, and (4) diseases of unknown etiology. The clinical course of SIV varies with the strain involved but rhesus macaques infected with common pathogenic SIV strains develop AIDS within 6–24 months PI. In the first few weeks of infection, all animals initially develop lymphocytosis consisting largely of CD8+ T cells, which resolves in 4–6 weeks. Lymphadenopathy and splenomegaly are apparent by 2–4 weeks PI and these conditions remain manifest until a very late stage of the disease when lymphoid collapse can occur. As in HIV infection, hypergammaglobulinemia due to polyclonal B-cell activation is often a feature of the disease. Anemia can also be a feature of the clinical disease. Weight loss and diarrhea are very common in SIV-infected animals and can be the result of opportunistic infections (*Mycobacterium avium* complex, cytomegalovirus, adenovirus, *Cryptosporidium* sp., *Ameoba* sp., *Balantidium coli*) or an unknown etiology. Often these enteric conditions do not respond to antibiotic therapy. Lymphomas are the only neoplastic condition of significance in SIV-infected macaques. In fact, a B-cell lymphoma arising in a 19-year-old sooty mangabey infected with SIV may represent the best candidate for the case of an SIV-related fatal disease in a naturally infected African primate.

Lymphoid tissues, including the mucosal-associated lymphoid tissues of the GI, reproductive, and respiratory tracts, are the targets of SIV infection, and a range of histopathologic changes occur in these tissues from follicular hyperplasia to dysplasia, followed by follicular collapse and expansion of the paracortex, ultimately ending in lymphoid depletion, collapse, and fibrosis. These changes occur independent of any opportunistic infections. The same pattern of histologic changes occurs in lymphoid tissues of HIV-1-infected humans and these histologic changes accurately reflect the clinical stage of the infection in both SIV and HIV.

SIV Infections as Animal Models of AIDS

The HIV-1 pandemic continues unabated and developing effective vaccines and therapies is the greatest current public health need. Samples from HIV-infected individuals have provided key insights into AIDS pathogenesis. However, direct experimental testing of specific hypothesis arising from these studies cannot be undertaken in humans for ethical reasons and HIV-1 infection of chimpanzees does not produce AIDS. Thus, SIV infection of Asian macaques is the most widely accepted animal model of HIV pathogenesis. The biology of the macaque immune system, and the key organ systems (gut, lymphoid tissue, and reproductive tract) involved in AIDS pathogenesis and HIV transmission are very similar to humans, and SIV and HIV are closely related phylogenetically. Thus, SIV infection of macaques closely mimics the pathogenesis, virology, immunology, and pathology of HIV infection in the human. The model has been used to show that infection with a molecular clone of SIV is sufficient to cause AIDS, that the GI tract is a major site of CD4$^+$ T-cell depletion, and that some vaccine strategies elicit immune responses that can provide considerable control of viral replication. The macaque monkey model of AIDS has been used to define the molecular determinants of viral pathogenesis, the basis for non-pathogenic infections in natural hosts. These animal models have been particularly valuable for defining the mechanisms of HIV transmission and events in acute infection that are very difficult to study in humans.

Sexual HIV transmission can occur through oral, anal, or vaginal intercourse. Allowing SIV-discordant macaques to mate normally would seem to be most similar to sexual HIV infection. However, sexual SIV transmission occurs at a low or variable rate after natural mating, and, in addition, there is significant biting behavior during mating and thus significant potential for blood-borne SIV transmission. When SIV-discordant juvenile rhesus macaques were housed in groups, SIV transmission most commonly occurred when uninfected, dominant animals bit their SIV-infected, subdominant cagemates as part of social interactions. Thus, once deposited in the mouth of an uninfected macaque SIV-infected blood-transmitted infection therefore biting during mating would confound transmission studies.

A more controlled experimental approach to reliably transmit SIV across the genital mucosa is to intravaginally inoculate female macaques with a known quantity of well-characterized SIV stock. Most studies have used suspensions of cell-free SIV virions, but infection can be transmitted by intravaginal inoculation with SIV-infected cells. In order to reliably transmit SIV to monkeys by a single intravaginal inoculation, relatively high doses of cell-free virus (2–3 log 10 more virus) are used compared to intravenous inoculation. In addition, exogenous progestins have been used to ensure reliable intravaginal SIV transmission, as these hormones thin up the genital mucosa, lower the barrier to transmission, and enhance SIV transmission. Finally, intravaginally inoculating animals with relatively low doses of virus, approximately the same dose needed for IV transmission, repeatedly over the course for several months eventually produced infection in all exposed animals, although the number of exposures needed to acquire infection is variable; this strategy probably best models the level of virus exposure that occurs during HIV sexual transmission.

HIV transmission by sexual, intravenous, and perinatal routes is often associated with the acquisition of a limited distribution of genetic variant. In many instances, the transmitted variant represents a minor variant in the donor's virus population. Given the extent of genetic diversity between HIV isolates, these findings have been interpreted to suggest that HIV transmission may involve selective entry or selective amplification of specific viral variants. However, the inherent limitations of all studies using human samples include small sample sizes and uncertainty as to the genetic identity and the extent of genetic diversity of the virus population in the donor at the time of transmission. Thus, the mechanisms that underlie the sexual transmission of HIV variants are unclear, and are difficult to assess because of the difficulties in establishing the precise time of infection. The SIV macaque model for HIV transmission is particularly valuable for evaluating the role of viral selection during transmission. The model allows access to information that is usually unattainable in human studies, such as: the genotypic and phenotypic properties of the infecting virus; knowledge of the exact time of virus exposure; and the characteristics of viral variants in the infected host immediately after transmission. Both IV and IVAG SIV inoculation transmit genetically diverse populations of SIV env V1–V2 variants to macaques. However, compared to the complex SIV populations in the IV inoculated animals, most IVAG inoculated animals are infected with SIV populations that have relatively low genetic diversity in the env gene. The finding that genetically diverse SIV populations are transmitted to some IVAG inoculated monkeys is consistent with the observation that a more genetically diverse population of viral variants is sexually transmitted from HIV-infected men to women. The model has also been used to show that recombination can occur readily *in vivo* after mucosal SIV exposure and thus viral recombination contributes to the generation of viral genetic diversity and enhancement of viral fitness in the peracute stages of infection. The model has also been used to show that the mucosal barrier of the female genital tract greatly limits the infection of cervicovaginal tissues after intravaginal SIV inoculation, and thus the initial founder populations of infected cells are small. Despite limited foothold, SIV rapidly disseminates to distal sites, and continuous seeding from an infection in the genital tract is likely critical for the later establishment of a productive disseminated systemic infection.

Perinatal HIV transmission can occur at any time during gestation, delivery, or breast-feeding. Allowing female macaques infected with virulent SIV before or during pregnancy is most similar to perinatal HIV infection, because SIV/SHIV transmission can occur any time during gestation, delivery, or breast-feeding. However, mother–infant SIV transmission occurs at a low or variable rate and the timing of virus transmission is unknown. Alternatively, female macaques can be infected with SIV after delivery and the infant allowed to breast-feed normally. This model best mimics natural HIV breast milk transmission by eliminating fetal exposure to virus and transplacental transfer of maternal virus-specific antibodies to the infant. Thus, this approach controls more variables and transmission rates are high; however, a major limitation is that the time at which breast-feeding infants become SIV infected varies substantially (i.e., a few weeks to several months). Finally, direct oral inoculation of infant macaques with SIV or SHIV can be performed without infecting their dams. This approach controls most of the important variables related to the viral inoculum (dose, number, timing, and duration of virus exposure), maternal host immune response (level and quality of anti-HIV-specific maternal/passively transferred antibodies), and infant rearing (by uninfected dams or in a primate nursery). This system has been used to show that after oral inoculation of infant rhesus macaques with virulent SIVmac251, virus disseminates to distal lymphoid tissues faster than after oral inoculation of juvenile macaques or vaginal inoculation of adult macaques with the same virus.

In addition to helping define critical steps in AIDS pathogenesis, these SIV mucosal transmission models are ideal for testing vaccines and microbicide strategies designed to prevent HIV transmission. The demonstrated utility of the SIV model for vaccine testing contrasts sharply with the inability of models using macaques infected with CxCR4 SHIVs (SHIV 89.6P) to meaningfully segregate HIV-1 vaccine candidates by relative efficacy. Thus the SIVs will continue to be the most valuable model of HIV infection and a critical tool in AIDS research.

See also: Bovine and Feline Immunodeficiency Viruses; Human Immunodeficiency Viruses: Antiretroviral agents; Human Immunodeficiency Viruses: Molecular Biology; Human Immunodeficiency Viruses: Origin; Human Immunodeficiency Viruses: Pathogenesis; Simian Immunodeficiency Virus: General Features; Simian Immunodeficiency Virus: Natural Infection.

Further Reading

Butler IF, Pandrea I, Marx PA, and Apetrei C (2007) HIV genetic diversity: Biological and public health consequences. *Current HIV Research* 5: 23–45.

Gardner MB (2003) Simian AIDS: An historical perspective. *Journal of Medical Primatology* 32: 180–186.

Gordon S, Pandrea I, Dunham R, Apetrei C, and Silvestri G (2005) The call of the wild: What can be learned from studies of SIV infection of natural hosts? In: Leitner T, Foley B, Hahn B, et al. (eds.) *HIV Sequence Compendium 2005, LA-UR 06–0680*, pp 2–29. Los Alamos, NM: Los Alamos National Laboratory.

Greenier JL, Miller CJ, Lu D, et al. (2001) Route of simian immunodeficiency virus inoculation determines the complexity but not the identity of viral variant populations that infect rhesus macaques. *Journal of Virology* 75: 3753–3765.

Jayaraman P and Haigwood NL (2006) Animal models for perinatal transmission of HIV-1. *Frontiers in Biosciences* 11: 2828–2844.

Kestler H, Kodama T, Ringler D, et al. (1990) Induction of AIDS in rhesus monkeys by molecularly cloned simian immunodeficiency virus. *Science* 248: 1109–1112.

Kim EY, Busch M, Abel K, et al. (2005) Retroviral recombination *in vivo*: Viral replication patterns and genetic structure of simian immunodeficiency virus (SIV) populations in rhesus macaques after simultaneous or sequential intravaginal inoculation with SIVmac239Deltavpx/Deltavpr and SIVmac239Deltanef. *Journal of Virology* 79: 4886–4895.

Lifson JD and Martin MA (2002) One step forwards, one step back. *Nature* 415: 272–273.

Long EM, Martin HL Jr., Kreiss JK, et al. (2000) Gender differences in HIV-1 diversity at time of infection. *Nature Medicine* 6: 71–75.

Marthas ML and Miller CJ (2007) Developing a neonatal HIV vaccine: Insights from macaque models of pediatric HIV/AIDS. *Current Opinion in HIV and AIDS* 2(5): 367–374.

Miller CJ (1994) Mucosal transmission of SIV. In: Desrosiers RC and Letvin NL (eds.) *Current Topics in Microbiology and Immunology*, pp. 107–122. Berlin: Springer.

Miller CJ, Alexander NJ, Sutjipto S, et al. (1989) Genital mucosal transmission of simian immunodeficiency virus: Animal model for heterosexual transmission of human immunodeficiency virus. *Journal of Virology* 63: 4277–4284.

Miller CJ, Li Q, Abel K, et al. (2005) Propagation and dissemination of infection after vaginal transmission of simian immunodeficiency virus. *Journal of Virology* 79: 9217–9227.

Veazey RS and Lackner AA (2004) Getting to the guts of HIV pathogenesis. *Journal of Experimental Medicine* 200(6): 697–700.

Relevant Website

http://www.hiv.lanl.gov – HIV Sequence Database: Nomenclature Overview (modified 26 April 2007), HIV Databases by Los Alamos National Laboratory.

Simian Immunodeficiency Virus: General Features

M E Laird and R C Desrosiers, New England Primate Research Center, Southborough, MA, USA

© 2008 Elsevier Ltd. All rights reserved.

History

Simian immunodeficiency virus (SIV) was first isolated in 1984 from captive rhesus macaques (*Macaca mulatta*) at the New England Primate Research Center (NEPRC). This virus was originally called STLV-III because it displayed similar morphology, growth characteristics, and antigenic properties to the newly described immunosuppressive virus HTLV-III of humans. When HTLV-III was renamed human immunodeficiency virus (HIV), the name STLV-III was also changed to SIV. Retrospective studies have shown that SIV was introduced to the NEPRC when a group of rhesus macaques with immuosuppressive disease was delivered from another primate center 15 years prior to the initial SIV isolation. The original cohort of rhesus monkeys was most likely accidentally infected with SIV from wild-caught sooty mangabey monkeys at the same institution. SIV has been subsequently isolated from other captive macaque species (*M. fascicularis*, *M. nemestrina*, and *M. arctoides*) that were dying of immunosuppression-associated diseases, and from many species of feral asymptomatic African nonhuman primates (**Table 1**).

Taxonomy and Classification

SIVs belong to the genus *Lentivirus* of the family *Retroviridae*. Related lentiviruses have been isolated from sheep, goats, horses, cattle, cats, and humans. Based on host species and genetic analysis, 14 discrete evolutionary groupings of primate lentiviruses are now recognized (**Figure 1**). Even within a single grouping, discrete sub-groupings are defined based on host subspecies, geography, and genetic distance. Within a specific subgroup whose host range covers an extensive geographical area, discrete genetic sub-subgroups are further defined that correlate with monkey subspecies and precise natural geographic habitat.

The lentiviruses have a common morphogenesis and morphology that distinguish them from other retrovirus subgroups. Lentivirus particles are 80–100 nm in diameter and consist of an RNA genome and viral enzymes enclosed in viral protein core that is encased by a cell-derived membrane spiked with viral envelope glycoproteins. In lymphocytes, immature lentiviruses bud from the plasma membrane without a preformed nucleoid; mature particles contain a characteristic conical or rod-shaped nucleoid. Classification of lentiviruses by morphology alone is consistent with classification by phylogenetic analysis of polymerase (*pol*) gene sequences. The *pol* gene exhibits the greatest degree of sequence conservation and viruses classified as lentiviruses have *pol* gene sequences more closely related to one another than to other retroviruses.

Lentiviruses also share similarities in certain biological properties and genome organization. All lentiviruses have a propensity to replicate in macrophages and produce long-term, persistent infections in susceptible hosts.

Table 1 Detailed listing of primate lentiviruses[a]

Virus designation	Primate Lentivirus grouping	Species (common)	Species (formal)	Subspecies isolates
HIV-1	HIV-1/SIVcpz	Humans	Homo sapiens	
SIVcpz	HIV-1/SIVcpz	Chimpanzees	Pan troglodytes	P. t. troglodytes
				P. t. schweinfurthi
SIVsm	SIVmac/SIVsm/HIV-2	Sooty mangabeys	Cercocebus atys	
SIVmac	SIVmac/SIVsm/HIV-2	Macaques	Macaca mulatta	M. arctoides
				M. nemestrina
				M. fascicularis
HIV-2	SIVmac/SIVsm/HIV-2	Humans	Homo sapiens	
SIVagm	SIVagm	African green monkeys	Chlorocebus aethiops	C. a. grivet
				C. a. tantalus
				C. a. sabeus
				C. a. alboqularis
				C. a. nictitans
SIVsyk	SIVsyk	Sykes' monkeys	Cercopithecus mitis	
SIVgsn	SIVgsn/SIVmon/SIVmus	greater spot-nosed monkey	Cercopithicus mitis	
SIVmon	SIVgsn/SIVmon/SIVmus	mona monkey	Cercopithicus mona	
SIVmus	SIVgsn/SIVmon/SIVmus	mustached monkey	Cercopithicus cephus	
SIVlhoesti	SIVsun/SIVlhoesti	L'hoest monkey	Cercopithicus lhoesti	C. l. lhoesti
SIVsun	SIVsun/SIVlhoesti	Sun-tailed monkey	Cercopithicus lhoesti	C. l. solatus
SIVdeb	SIVdeb	DeBrazza monkey	Cercopithicus neglectus	
SIVden	SIVdeb	Dent's mona monkey	Cercopithicus mona denti	C. m. denti
SIVrcm	SIVrcm	Red-capped mangabey	Cercocebus torguatus	C. t. torguatus
SIVmnd	SIVmnd 1	Mandrill	Mandrillus sphinx	
SIVmnd	SIVmnd 2	Mandrill	Mandrillus sphinx	
SIVdrl	SIVmnd 2	Drill	Mandrillus leucophaeus	
SIVcol	SIVcol	Querza colobus	Colobus querza	
SIVolc	SIVolc	Olive colobus	ProColobus badius	
SIVwrc	SIVwrc	Western red colobus	Pilocolobus badius	
SIVtal	SIVtal	Angolia-talapoin monkey	Miopithicus talapoin	
SIVtal	SIVtal	Gabon talapoin monkey	Miopithicus ogouensis	

[a]Partial pol sequences have also been obtained from a black mangabey (*Loplcocebus aterrimus*) and from a Schmidt's guenon (*Cercopithicus ascanius schmidti*). In addition to the primate lentiviruses listed, serologic surveys for the detection of antibodies to SIV have suggested SIV infection of a variety of other species.

SIVs use CD4 as the first of two receptors used sequentially for viral entry into cells. The chronic disease induced by SIV includes immunodeficiency, undoubtedly because of the targeting of CD4+ lymphocytes for infection through the use of CD4 as the primary receptor. In addition to the *gag, pol,* and *env* genes that are found in all simpler retroviruses, lentiviruses have a number of auxiliary genes (**Figure 2** and **Tables 2** and **3**).

The SIVs are named according to the primate species of origin, for example, SIVmac from macaques or SIVsmm from sooty mangabey monkeys. Widespread availability of DNA sequencing has allowed an in-depth understanding of phylogenetic relationships among SIVs. However, to date, only 39 of the 69 recognized species of nonhuman primates that inhabit sub-Saharan Africa have been surveyed; additional distinct SIV groupings will likely be identified.

Geographic Distribution and Host Range

Many different species of African nonhuman primates are known to be infected with SIV in their natural habitats. However, few studies have investigated the distribution or extent of natural SIV infection. There has recently been an effort to overcome this shortfall of information by identifying and assessing the extent of SIV infection in wild primate populations. In 2002, Peeters *et al.*, collected and screened 788 blood samples from wild-caught monkeys for the prevalence of SIV infection in 13 different monkey species in Cameroon. The study examined the rates of SIV infection in monkeys hunted for bushmeat and those captured as pets as possible routes of zoonotic transmission. It was reported that 18.4% of bushmeat samples and 11.6% of pets tested positive for SIV infection. These results identified four species of monkeys not previously known to harbor SIV (*Cercocebus agilis, Lophocebus albigena, C. pogonias,* and *Papio anubis*) and likely underestimate the extent of SIV prevalence as not all native primate species were screened.

The origins of both HIV-2 and HIV-1 in humans are believed to have occurred through cross-species transmission events from SIV-infected simians relatively recently in history. SIVsmm is closely related to HIV-2 with the same genome organization and both viruses group together phylogenetically apart from the other 13 groups

Simian Immunodeficiency Virus: General Features 605

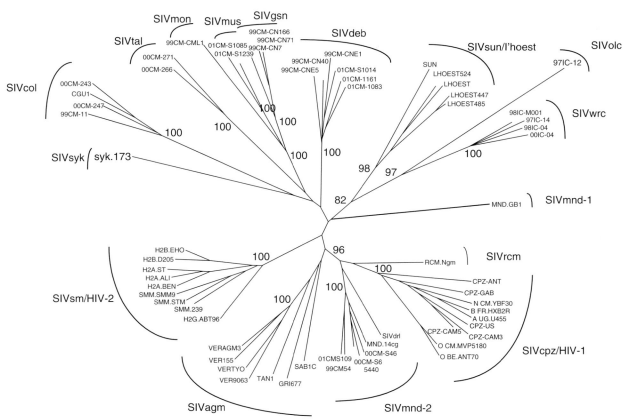

Figure 1 Phylogeny of primate lentiviruses. The 14 groupings of primate lentiviruses are shown. Please see **Table 1** for species abbreviations. Adapted from Courgnaud V, Formenty P, Koffi CA, et al. (2003) Partial molecular characterization of two simian immunodeficiency viruses (SIV) from African colobids: SIVwrc from Western Red Colobus (*Piliocolobus badius*) and SIVolc from Olive Colobus (*Procolobus verus*). *Journal of Virology* 77(1): 744–748, with permission from American Society for Microbiology.

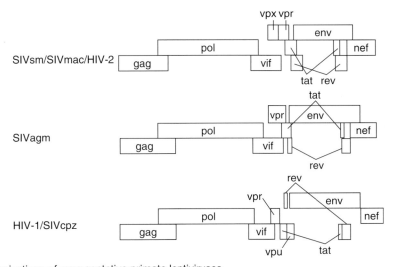

Figure 2 Genome organizations of representative primate lentiviruses.

of primate lentiviruses. The sooty mangabey monkey is native to the coastal forest regions of western Africa, where HIV-2 emerged and is now endemic in the human population. Thus, strong circumstantial evidence involving both viral sequences and geographic distribution link the monkey SIVsmm and HIV-2 in western Africa. In 2006, Keele and co-workers examined the wild chimpanzee (*Pan troglodytes troglodytes*) populations of Cameroon to determine the extent of natural SIVcpz infection and to investigate which infected populations may be responsible

Table 2 Presence of auxiliary genes in SIV species

	SIVsm/SIVmac/HIV-2	SIVagm	SIVsyk	SIVsun/SIVl'hoesti	HIV-1/SIVcpz	SIVgsn/SIVmon/SIVmus	SIVrcm
vif	+	+	+	+	+	+	+
vpu	−	−	−	−	+	+	−
vpr	+	+	+	+	+	+	+
vpx	+	−	−	−	−	−	+
tat	+	+	+	+	+	+	+
rev	+	+	+	+	+	+	+
nef	+	+	+	+	+	+	+

Different SIV species may vary with respect to the auxiliary genes that they carry. The presence or absence of these auxiliary genes do not always associate with phylogenetic clustering. For example, SIVmnd2 has a *vpx* gene, but SIVmnd1 does not. SIVden from a pet Dent's Mona monkey (*Cercopithecus mona denti*) has a *vpu* gene, although it clusters more closely to the SIV from DeBrazza monkeys, SIVdeb. SIVdeb has a *vpr* gene, but no *vpx* or *vpu* genes.

Table 3 Auxiliary gene function in SIV

Auxiliary gene product	Essential for replication?	Early gene product?	Function
tat	Yes	Yes	Potent activator of viral gene expression; enhances LTR-driven transcription
rev	Yes	Yes	Required for efficient transport of unspliced and singly spliced viral RNAs into the cytoplasm
nef	No	Yes	Functional activities include CD4 downregulation, MHC downregulation, infectivity enhancement and lymphocyte activation
vif	Yes/no[a]	No	Blocks restricting activity of innate cellular proteins, APOBEC-3G and APOBEC-3F
vpr	No	No	Involved in G_2/M phase cell-cycle arrest; mediates apoptosis of CD4+ T cells
vpx	No	No	Facilitates nuclear import of the preintegration complex in nondividing cells

[a]SIV strains containing a deletion in *vif* (SIVΔvif) can be grown in a *vif*-complementing cell line; when inoculated into animals no virus or PCR-amplifiable sequences could be recovered from PBMCs; however, monkeys still developed lowlevel antibody titers suggesting highly attenuated infection.

for the cross-species transmission events that initially introduced HIV-1 into the human population. By measuring the presence of anti-HIV cross-reactive antibodies and the ability to amplify viral gene sequences it was determined that SIV infection was widespread but uneven among the chimpanzee populations, with the prevalence of infection ranging from 23% to 35% in some isolated groups to a 4–5% infection rate in others, while still others had a complete absence of SIV infection. Sequence and phylogenetic analyses of the newly identified wild SIVcpz strains supported that distinct geographical chimpanzee groups acted as the sources of HIV-1 groups M and N in the human population.

Although SIVs naturally infect a variety of African nonhuman primates, a single example of natural infection of Asian Old World monkeys is yet to be reported. SIVsmm and SIVagm, when used to infect macaque monkeys (Asian Old World primates), can persist and cause an AIDS-like disease. Accidental introduction of SIVsmm into macaque monkeys in captive United States colonies occurred and was spread unknowingly into other macaques for more than a decade before it was identified and eliminated. At least one clear case of laboratory-acquired infection of a human with SIVmac has been documented.

Virus Propagation and Receptor Use

SIVs can be propagated in mitogen-stimulated primary peripheral blood mononuclear cells (PBMCs), in monocytes/macrophages from the primate host, and in many cultured cell lines including human tumor-derived CD4+ T-lymphocyte cell lines. The types of cells that can be infected by different strains of SIV correlate with the receptor(s) that are expressed on the cell surface. Some acutely pathogenic strains are unusual in their ability to replicate in lymphocytes of resting PBMC cultures without prior stimulation. In PBMC cultures and many cell lines, viral infection results in the fusion of cellular membranes producing large syncytial cells. Syncytium formation, which is mediated by *env*, allows the virus to spread directly from cell to cell in addition to direct infection. Several isolates also grow well in cultured macrophages derived from lung, blood, or bone marrow.

As with HIV-1, the SIVs use both CD4 and a chemokine receptor for viral entry. The SIVsmm/SIVmac/HIV-2 and SIVagm groups of viruses are known to use CCR5 as their principal co-receptor. However, a variety of other chemokine receptors, CCR2b, CCR3, STRL33 (Bonzo), GPR15 (Bob), and GPR1, also can be used as the co-receptor, depending on the individual virus isolate. A larger percentage of isolates from the SIVsmm/SIVmac/HIV-2 group of viruses show less dependence on CD4 for entry than do HIV-1 variants. Isolates of SIVsmm/SIVmac appear to use CXCR4 as the principal co-receptor much less frequently than HIV-1 isolates. SIVs from red-capped mangabeys (*Cercocebus torquatus torquatus*) predominantly use CCR2b as the principal co-receptor.

Genetics

SIV, like other retroviruses, replicates its genome through a proviral DNA intermediate. From the 5' cap to the 3' polyadenylation site, the SIV genome is approximately 9.6 kbp in length. The viral particle contains a diploid genome of single-stranded RNA that is linked noncovalently near the 5' end of the molecules. The 5' end of the viral genome is capped and the 3' end is polyadenylated. DNA synthesis by the viral-encoded reverse transcriptase is primed by host tRNA that is base-paired to viral RNA. The double-stranded proviral DNA is integrated into the host cell chromosome by a viral-encoded integrase and further replication events of transcription, translation, and particle assembly depend on cellular components. Particles then assemble at and bud through the plasma membrane. Because of this replication strategy, cloned DNA representing the entire proviral genome can yield infectious virus.

All retroviruses contain certain standard features in their genomic organization (**Figure 2**). Sequences regulating DNA synthesis, integration, transcription, and other functions are contained in the long terminal repeat (LTR) region at each end of the provirus. Open reading frames (ORFs) encoding the major structural and nonstructural proteins lie between the LTRs. Genes are encoded in any of three possible ORFs; overlaps between ORFs are common. All retroviruses contain three standard genes called *gag* (group-specific antigen), which encodes the core proteins; *pol* (polymerase), which encodes the viral reverse transcriptase, protease, and integrase; and *env* (envelope), which encodes the envelope glycoproteins.

Env is essential for virus replication. The envelope glycoproteins are responsible for binding the receptor and co-receptor on the cell surface and mediating viral entry. The *env* products are the main targets of antibodies that can neutralize infection. In addition, determinants of cell and tissue tropism often map to the *env* gene. Derivatives in which the SIVmac *env* has been replaced by envelope of HIV-1 are replication competent in macaque cells and are capable of infecting rhesus monkeys. These recombinant viruses are known as simian–human immunodeficiency viruses (SHIVs). Serial passage of several SHIV strains has resulted in second-generation SHIVs that are consistently pathogenic in macaques.

In addition to *gag*, *pol*, and *env*, all lentiviruses, including SIV, encode additional accessory genes not found in other simple retroviruses (**Tables 2** and **3**). Both SIV and HIV encode *tat* (transactivator protein), *rev* (regulator of gene expression), *vif* (viral infectivity factor), *nef* (originally termed negative factor), and *vpr* (viral protein 'r'). SIVagm and SIVsmm/HIV-2/SIVmac encode an additional gene, *vpx* (viral protein 'x'), thought to be a duplicated homolog of *vpr*. The ORF for *vpu*, found in HIV-1/SIVcpz and SIVgsn/SIVmon/SIVmus/SIVden, is not contained in HIV-2 or in other SIVs. These auxiliary genes likely contribute to the complex life cycle of lentiviruses, including persistent viral replication and immune evasion.

Some of the accessory proteins found in SIV can be deleted without abrogating the ability of the virus to replicate *in vivo* and *in vitro*, specifically *nef*, *vpr*, and *vpx*. However, the presence and the conservation of these genes in several different subgroups of lentiviruses suggest that they contribute to the virus' ability to replicate and persist *in vivo*. Cloned proviral DNAs containing deletions in these auxiliary ORFs have been used to study the contributions to replication and functional activity of the auxiliary genes in the context of experimental animal infection. Auxiliary gene functional data are summarized in **Table 3**.

Evolution

Comparison of genetic sequences among human and simian immunodeficiency suggests that there are at least 14 discrete groups of primate lentiviruses in existence: HIV-1/SIVcpz; SIVmnd-2; SIVagm; SIVsmm/HIV-2; SIVsyk; SIVcol; SIVtal; SIVmon/SIVmus/SIVgsn; SIVdeb/den; SIVsun/l'hoest; SIVolc; SIVwrc; SIVmnd-1; and SIVrcm (**Figure 1**). It is thought that HIV-1 and HIV-2 evolved from simian viruses that entered the human population through cross-species transmission events relatively recently in human history. Cross-species transmission among non-human primates occurring in nature may have generated further pathogenic variants; however, it is likely that most of the primate immunodeficiency viruses have long been present in the natural host but not recognized until recently. The SIVs and HIVs are more closely related to each other than to any other nonprimate retrovirus, suggesting that they are inherently primate viruses, not derived from nonprimate viruses that were introduced via other cross-species transmission events.

Serologic Relationships and Genetic Diversity

The *pol* gene generally contains the greatest degree of sequence conservation and therefore is most often used for the comparison of lentiviruses from different groups or subgroups to assess relatedness. Pol sequences from one subgroup of SIV (e.g., SIVsmm) will generally contain only a 55–60% amino acid identity when compared to another SIV subgroup (e.g., SIVagm). When different lentivirus groups are compared (e.g., EIAV with SIV) the amino acid identity in pol is often 35% or less, and the sequence homology found in other genes is even less. Antiserum to the Gag protein is generally cross-reactive to different strains within a group, whereas antiserum to the envelope is not and can be used to distinguish between isolates within a group.

Epidemiology

SIV has been found in many species of nonhuman primates throughout sub-Saharan Africa, but in most cases infection does not seem to cause an AIDS-like disease in the natural host. There are only a few examples of immunodeficiency in monkey species naturally infected with SIV. In contrast, though infection does not appear in nature, SIV infection of Asian macaques in captivity induces an AIDS-like disease similar to that observed in HIV-infected humans.

Transmission

There is little information regarding natural modes of SIV transmission. A study of wild grivet monkeys in Awash National Park in Ethiopia analyzed SIVagm serologic status as compared to age, sex, and risk. Infection was found overwhelmingly in females of reproductive age and was nearly absent among younger female animals. In the male population, infection was only observed in monkeys that were fully adult. These data support a predominantly sexual mode of transmission among the grivet population. SIV transmission through contact with infected blood from aggressive contacts (e.g., bite and scratch wounds) may also be a prominent mode by which SIV may be spread. In addition, maternal–infant transmission of SIV has been observed in captive animals.

Experimental infection of laboratory animals has most commonly been performed by direct needle inoculation. However, mucosal exposure is being used more frequently, especially in vaccine studies, as a model for the most common routes of HIV infection.

Features of Infection

While SIV infection of the natural host is usually not associated with any disease progression, SIV infection of macaques induces both acute and chronic disease symptoms that are similar to that which HIV-1 causes in human patients. SIV infection of rhesus macaques is generally thought to be the closest model of AIDS in humans.

The main sites of pathogenic SIV replication shortly after infection have been localized to the gastrointestinal (GI) tract, thymus, spleen, and other lymphoid tissues. SIV has been detected at early time points within periarteriolar lymphoid sheaths in the spleen, paracortex of lymph nodes, and medulla of the thymus. SIV infection of rhesus monkeys results in a dramatic and selective depletion of CD4+ T cells in the GI tract within days of infection, before depletion is evident in the peripheral lymphoid tissues. Coincident with the loss of CD4+ T cells in the GI tract is the productive infection of large quantities of mononuclear cells at this site. It is now clear that SIV replicates principally in the CD4+ CCR5+ cells of the memory T-cell phenotype. These cells predominate in the gut and other mucosal sites and are found at much lower levels in peripheral lymphoid tissues. Therefore, the GI tract appears to be a major site of SIV replication and CD4+ memory T-cell depletion early in the course of infection. Within the thymus, marked depletion of thymic progenitor cells has been observed 21 days post infection with pathogenic SIV; this cell depletion is followed by increased cell proliferation in the thymus and a marked increase in thymocyte progenitors. SIV can also be found in the central nervous system (CNS) at early time points following infection. The cells that are targeted in the brain, either early in infection or in late-stage SIV-induced encephalitis, are primarily cells of the monocyte/macrophage lineage.

Viral loads in infected animals decrease with the onset of immune responses, and this decrease correlates with CD8+ T-cell lymphocytosis and the rise of SIV-specific antibodies. Following immune system activation, animals enter an asymptomatic period of infection of variable duration. Viral replication persists during this period, inducing immune abnormalities, including gradual declines in CD4+ T-cell count, CD4/CD8 ratio, and the ability to respond to mitogens. Infected animals can also exhibit chronic diarrhea and wasting resulting in up to a 60% loss of the original body weight. Over the course of disease, dramatic changes, including hyperplasia and/or atrophy, take place in most of the lymphoid tissues. Terminal stages of SIV infection are characterized by a range of diseases that can be grouped into four broad categories: SIV-related inflammatory disease, opportunistic infections associated with SIV-induced immunosuppression, neoplastic diseases, and diseases of unknown pathogenesis. The tropism of SIV strains for monocytes/macrophages correlates with dramatic inflammatory and degenerative changes seen in the CNS, lung, digestive tract, and other organs that are separate from the pathogenesis of opportunistic infections. The characteristics and frequency of inflammatory lesions in SIV-infected macaques closely resembles those seen in HIV-infected patients. SIV-induced encephalitis is frequent, although the appearance of brain lesions is dependent on the infecting SIV strain. At necropsy, 30–50% of SIVmac-infected animals have characteristic multinucleate giant cell encephalitis that closely resembles that seen in HIV-associated encephalitis. The opportunistic infections seen in infected monkeys are also similar to those observed in HIV-infected individuals and include *Pneumocystis carinii*, *Mycobacterium avium*, *Crytposporidia* sp., *Toxoplasma gondii*, rhesus Epstein–Barr virus (rhesus lymphocrytovirus), cytomegalovirus, polyomavirus (SV40), and adenovirus. Neoplastic diseases during SIV infection are primarily limited to lymphomas, the frequency of which varies from study to study. Lymphoma induction in SIV-infected macaques has additionally been associated with the co-infection of rhesus Epstein–Barr virus. Diseases of unknown pathogenesis include generalized lymphoproliferative syndrome, arteritis, and arteriopathy.

Acutely pathogenic strains exist, such as SIVsmPBj14, which can be acutely lethal in infected macaques. SIVsm PBj14 infection causes death in infected monkeys within 14 days. These animals have very high viral loads, severe GI disease, cytokine disregulation, lymphoproliferative disease, and organ system failure. The increased pathogenicity of this virus has been attributed to the creation of an ITAM motif by a tyrosine at residue 17 of the nef protein. This change allows the virus to induce lymphocyte activation and replicate to high titers in PBMC cultures without prior stimulation. However, the disease induced by the majority of SIV infections is typically chronic and manifests itself over the course of 1–3 years post infection.

Several independent research groups have constructed recombinant forms of SIVmac with HIV-1 *env*, *tat*, *rev*, and, in some cases, *nef* and/or *vpu* genes renamed 'SHIV' (for simian–human immunodeficiency virus) that have been passaged in macaques to establish pathogenic virus strains. Although most of the HIV-1 envelopes that were used to construct these SHIVs were dual-tropic, in that they can use either CCR5 or CXCR4 as a co-receptor, the large majority of SHIV viruses appear to target primarily CXCR4-expressing cells when used to infect rhesus macaques. The pathogenic SHIVs consistently, rapidly, and irreversibly deplete CD4+ T cells from the periphery and can be acutely lethal.

Immune Response and Persistence

SIV-infected macaques typically produce high levels of antiviral antibodies and high-frequency cytotoxic T-lymphocyte (CTL) responses to the infecting virus. These immune responses persist for the lifetime of the infected host in both natural and experimental infection. SIV deletion mutants that are progressively more attenuated based on viral load measurements generate progressively weaker anti-SIV antibody responses. Anti-SIV CTL responses have been demonstrated as being major histocompatibility complex (MHC) restricted. Detailed investigation of CTL responses has been impeded by a lack of information regarding MHC types in different monkey species. However, a considerable amount of new information regarding MHC alleles and their cognate peptides is emerging for rhesus macaques.

The importance of CD8+ lymphocytes in limiting the extent of SIV or SHIV replication has been definitively shown using CD8+ T-cell depletion. Extensive depletion of CD8+ cells was accomplished by intravenous administration of large doses of anti-CD8 monoclonal antibodies. When CD8+ T cells were depleted during primary infection, viral replication continued unabated after the usual peak of viral loads, 10–14 days post infection; this was in stark contrast to undepleted animals, in which intact immune responses typically act to decrease viral loads after 14 days post infection. During chronic infection, elimination of CD8+ lymphocytes through depletion resulted in a rapid and dramatic increase in viremia, which was again suppressed upon removal of anti-CD8 antibody and the return of the SIV-specific CD8+ lymphocytes. Depletion of CD8+ T cells in animals infected with SHIV viruses has facilitated the appearance of the more highly pathogenic, passaged variants.

In macaques developing SIV-induced disease from wild-type strains of SIV, viral-specific, proliferative responses of CD4+ T cells are typically weak or absent all together. However, infection by attenuated SIV mutants containing a deletion of the *nef* gene produces strong, SIV-specific, CD4+ helper cell proliferative responses. This situation recalls that of HIV infection in humans, in which HIV-specific CD4+ proliferative responses in progressing patients are usually weak or absent, but are often very strong in nonprogressors that are able to control their infection. It seems that as CD4+ helper T cells try to respond to SIV at sites of infection, they arrive in the location where they are the ideal target cells for the invading virus. In pathogenic infections, the virus wins the battle between it and the responding CD4+ T cells.

All lentiviruses persist in the infected host through chronic active viral replication. Over the course of the months and years of chronic infection, macaques infected with SIV are producing and turning over millions of viral particles and infected cells every day. Although active replication persists throughout the course of infection, there are some cells that are most likely infected in a quiescent or latent fashion. The extent of chronic active replication may also differ depending on the infecting virus strain and the host. Consistent with prolonged antigen expression and chronic replication is the long-term persistence of a high level of circulating antibody and viral-specific CTLs. Nonpathogenic SIV derivatives also continue to replicate at low levels over long periods, as

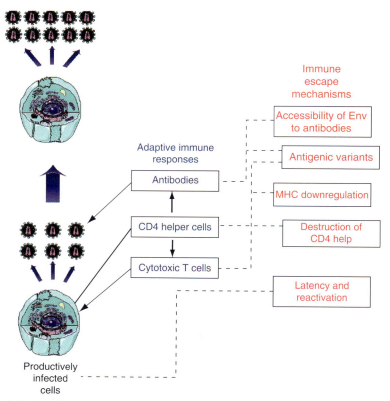

Figure 3 Simian immunodeficiency virus uses a variety of strategies to avoid recognition and clearance by both the humoral and cellular arms of the immune system.

demonstrated by accumulated sequence changes in these viral genomes and persistent antibody titers.

The dilemma of all lentiviruses is how to replicate persistently in the face of an apparently strong immune response. The levels of antiviral antibody and the frequency of CTLs in the infected individual have been measured and appear to be consistently high. Several strategies are used by SIV and other lentiviruses to allow persistent replication and evade the immune response. These are summarized in **Figure 3**.

Prevention and Control

Extensive testing and removal programs have essentially eliminated SIV from captive macaque colonies. However, continued vigilance is required to minimize the chance that breeding colonies may again become accidentally exposed to the virus. Animals can be easily and conveniently tested serologically for infection. Cases of SIV infection in humans are extremely rare and laboratory personnel that work with SIV follow the same precautions when working with SIV as working with HIV-1. Disposable gloves and surgical gowns are used, all work with live virus is performed in a biosafety hood, procedures creating aerosols are avoided, and any use of glass or needles in conjunction with live virus is minimized.

Future Perspectives

The development of a safe, effective, affordable vaccine for HIV/AIDS is one of the greatest challenges of our time. SIV will inevitably play an important role in instructing what is needed for protective immunity and how best to make a vaccine. Because SIVs are the closest known relatives of HIVs, the induction of AIDS in macaques by infectious molecular clones of SIV represents the best existing animal model for AIDS. SHIV infections are also extensively used in vaccine development because it allows for the analysis of HIV-1 envelope-containing vaccine products in an established system. Vaccine studies in animal models are vitally important as they provide useful information in several different ways. Head-to-head comparisons of different vaccine approaches can be performed to investigate which approach is more effective, at least within defined experimental conditions. Further, more in-depth analyses of specific vaccine approaches may also provide fundamental insights into what is needed to establish persistent, protective immunity to SIV and/or HIV infection. The worldwide crisis of HIV infection and AIDS has brought attention to the SIV system as a source of information that will shed light on the human condition. SIV as an animal model can contribute to further understanding of the most critical issues for future progress, including better understanding of pathogenesis, improvements in therapy and, most importantly, the development of a safe, effective, and affordable vaccine.

See also: Human Immunodeficiency Viruses: Antiretroviral agents; Human Immunodeficiency Viruses: Molecular Biology; Human Immunodeficiency Viruses: Pathogenesis.

Further Reading

Campbell RSF and Robinson WF (1998) The comparative pathology of lentiviruses. *Journal of Comparative Pathology* 119: 333–395.

Courgnaud V, Formenty P, Akoua-Koffi C, et al. (2003) Partial molecular characterization of two simian immunodeficiency viruses (SIV) from African colobids: SIVwrc from Western Red Colobus (*Piliocolobus badius*) and SIVolc from Olive Colobus (*Procolobus verus*). *Journal of Virology* 77: 744–748.

Daniel MD, Letvin NL, King NW, et al. (1985) Isolation of T-cell tropic HTLV-III-like retrovirus from macaques. *Science* 228: 1201–1204.

Desrosiers RC (2001) Nonhuman Lentiviruses. In: Knipe DM and Howley PM (eds.) *Fields Virology*, 4th edn., pp. 2095–2122. Philadelphia, PA: Lippincott Williams and Wilkins.

Desrosiers RC (2004) Prospects for an AIDS vaccine. *Nature Medicine* 5: 723–725.

Johnson WE and Desrosiers RC (2002) Viral persistence: HIV's strategies of immune system evasion. *Annual Review of Medicine* 53: 499–518.

Keele BF, Van Heuverswyn F, Li Y, et al. (2006) Chimpanzee reservoirs of pandemic and nonpandemic HIV-1. *Science* 313: 523–526.

Kestler H, Kodama T, Ringler D, et al. (1990) Induction of AIDS in rhesus monkeys by molecularly cloned simian immunodeficiency virus. *Science* 248: 1109–1112.

Koff WC, Johnson PR, Watkins DI, et al. (2006) HIV vaccine design: Insights from live attenuated SIV vaccines. *Nature Immunology* 7: 19–23.

Peeters M, Courgnaud V, Abela B, et al. (2002) Risk to human health from a plethora of simian immunodeficiency viruses in primate bushmeat. *Emerging Infectious Disease* 8: 451–457.

Veazey RS, DeMaria M, Chalifoux LV, et al. (1998) Gastrointestinal tract as a major site of CD4+ T cell depletion and viral replication in SIV infection. *Science* 280: 427–431.

Simian Immunodeficiency Virus: Natural Infection

I Pandrea, Tulane National Primate Research Center, Covington, LA, USA
G Silvestri, University of Pennsylvania, Philadelphia, PA, USA
C Apetrei, Tulane National Primate Research Center, Covington, LA, USA

© 2008 Elsevier Ltd. All rights reserved.

Glossary

APOBEC Human protein superfamily that interferes with the HIV/SIV replication. This family of proteins has cytidine deaminase activity and has been suggested to play an important role in innate antiviral immunity.

Catarrhines Relating to or being any of a division of primates (*Catarrhina*) comprising the Old World monkeys, great apes, and hominids that have nostrils close together and directed downward, 32 teeth, and a tail, when present, which is never prehensile.

Chemokine receptors Family of approximately 20 different G-protein-coupled receptors that have seven transmembrane segment polypeptides, and which cause cell activation. Each receptor subtype is capable of binding multiple chemokines within the same family.

Endemic (1) Natural to or characteristic of a specific place; native; indigenous. (2) A disease which persists in a given population or locality.

Giant cell disease Pathologic condition specific for nonhuman primates with severe immunodeficiency characterized by infiltration with syncytial cells in multiple tissues.

Guenon An Old World monkey native to sub-Saharan Africa; possesses a round head with beard, and 'whiskers' at side of face; slender, with long hind legs and tail; some species with colorful coats.

Sympatric Of two or more population or taxa, inhabiting the same geographic area.

Introduction

More than 40 different types of simian immunodeficiency viruses (SIVs) naturally infect different species of monkeys and apes. Two of these, SIVcpz, which naturally infects chimpanzees, and SIVsmm, which naturally infects sooty mangabeys (SMs), are the ancestors of HIV-1 and HIV-2, respectively. Inadvertent cross-species transmission of SIVsmm from naturally infected SMs to different species of macaques resulted in severe immunodeficiency and subsequent development of the animal models for AIDS. Unlike HIV/SIV infection of humans and macaques, which normally progresses to AIDS, natural SIV infection is generally nonpathogenic in African nonhuman primates (NHPs). The mechanisms behind this lack of disease progression are currently under investigation.

History

The history of SIVs began two decades prior to virus discovery, when two outbreaks of opportunistic infections and lymphoma occurred in rhesus macaques (RMs) (1968) and in stump-tailed macaques (1973) at the California National Primate Research Center (CNPRC). SIVmac was discovered in 1985, during another outbreak of lymphomas in RMs at the New England Primate Research Center from monkeys transferred from CNPRC. However, occurrence of SIV infection in macaques was perplexing since tests carried out on RMs in Asia failed to reveal any evidence of SIV circulation in the wild. In 1986, at the Tulane National Primate Research Center, attempts to transmit leprosy from SMs to RMs resulted in a new AIDS outbreak in RMs, establishing the link between pathogenic SIVs in macaques and an African monkey species. During the following 20 years, more than 40 SIVs were identified in different African NHP species.

SIVsmm is also the ancestor of HIV-2. At least eight cross-species transmissions in West Africa resulted in the emergence of the eight HIV-2 groups (AH). In 1989, the discovery of SIVcpz in chimpanzees from Gabon identified the ancestor of HIV-1. At least three cross-species transmissions were at the origin of HIV-1 groups M, N, and O. Groups M and N resulted from cross-species transmission of SIVcpz from *Pan troglodytes troglodytes* in West-Central Africa. Group O is more closely related to the recently discovered SIVgor from *Gorilla gorilla*.

The arguments to support cross-species transmission from NHPs as the origin of HIVs are: (1) similarities in viral genome organization; (2) phylogenetic relatedness; (3) prevalence in the natural host; (4) geographic coincidence; and (5) plausible routes of transmission. All these criteria are fulfilled by both SIVsmm/HIV-2 and SIVcpz+SIVgor/HIV-1. Therefore, the discovery of SIVs in African species identified the origin of HIVs. However, the events behind HIV emergence are still under debate. Some authors consider that human exposure to SIVs through bush meat consumption is the original source of AIDS ('cut-hunter theory'). Others believe that simian exposure is necessary but insufficient for HIV emergence as a human pathogen, which requires adaptation to the new human host. Reuse of needles and syringes or transfusions may have played a role in triggering SIV adaptation to humans. It is probable that deforestation, political unrest, increase in urbanization and travel in the second half of the twentieth century also acted as cofactors of HIV emergence. Altogether, the action of these factors explain why HIVs only emerged in the second half of the twentieth century while people in sub-Saharan Africa were exposed to SIVs for millennia.

Virology

Currently, there are 47 fully sequenced SIV genomes from 21 NHP species. Partial genomic sequences are available for 13 additional SIVs, and serological evidence only of SIV infection has been obtained for seven primates (**Table 1**). Asian species of Old World monkeys (colobine and macaques), as well as some African species (such as baboons) do not carry a species-specific SIV, suggesting that the last common ancestor of the catarrhines (Old World monkeys and apes) was not SIV-infected 25 million years ago and that SIV emerged after species radiation, from a nonprimate source.

Classification and Taxonomy

In most instances, the infected NHP species represents the reservoir of that virus type, which is designated by a three-letter abbreviation of the vernacular name of the host (**Table 1**). When related NHP subspecies are infected, the subspecies name is included in virus designation. Thus, for chimpanzee subspecies infected by SIVs, the two SIVcpz are identified as SIVcpz.Ptt (from *Pan troglodytes troglodytes*) and SIVcpz.Pts (from *P. t. schwenfurthii*). For individual isolates, nomenclature includes SIV type and the country of origin: SIVrcmGB1 is a red-capped magabey virus isolated from Gabon. The year of sample can also be included: SIVsmmSL92 is an SM virus isolated from Sierra Leone in 1992. This feature is useful in tracking the origin and evolution of viruses.

Phylogenetic Relationships

SIVs have a starburst phylogenetic pattern, suggesting the evolution from a single ancestor, and a high genetic divergence, forming six SIV lineages with genetic distances of up to 40% in Pol proteins (**Table 2** and **Figure 1**). Each SIV lineage is represented by two or more strains. The l'hoesti lineage is unique in being formed by SIVs circulating in distantly related species. The relationship between SIV lineages and newly characterized SIVs is complicated by sequence diversity and recombination that results in different clustering patterns when different genomic regions are analyzed.

To better understand SIV phylogenetic relationships, a brief presentation of primate species radiation follows. African NHPs belong to two different groups: Old World monkeys and anthropoid primates (apes). Two ape genera are endemic in Africa: *Pan* (formed by the four subspecies of chimpanzees and bonobo) and *Gorilla*. The Old World monkeys (family Cercopithecidae) are divided into two subfamilies (*Cercopithecinae* and *Colobinae*), separated 11 million years ago (**Figure 1**). *Cercopithecinae*

Table 1 African apes and monkeys infected with SIV

Species/subspecies	Virus type	Geographic location[a]	Seroprevalence	Pathogenicity	Cross-species transmission
Common chimp (Pan troglodytes troglodytes)	SIVcpz.Ptt	Central Africa (Cameroon, Gabon, Congo)	<10%	Not reported	Humans, HIV-1 groups M and N P. t. velerosus
Eastern chimp (Pan troglodytes schweinfurthii)	SIVcpz.Pts	East Africa (Tanzania, Democratic Republic of Congo-DRC)	<10%	Thrombocytopenia	Not reported
Pan troglodytes velerosus	SIVcpz.Ptt[b]	Zoo in Cameroon		Not reported	
Gorilla gorilla	SIVgor	West-Central Africa (Cameroon)	?	Not reported	Humans, HIV-1 group O
Sooty mangabey (Cercocebus atys)	SIVsmm	West Africa (Sierra Leone, Liberia, Ivory Coast)	20–58%	AIDS	Humans, HIV-2 Experimentally to M. mulatta (SIVmac), M. nemestrina (SIVptm), M. fascicularis and M. arctoides (SIVstm): AIDS. Accidentally to L. aterrimus (AIDS)
Red-capped mangabey (Cercocebus torquatus)	SIVrcm	West-Central Africa (Gabon, Cameroon, Nigeria)	10–20%	Not reported	Agile mangabey Experimentally, to Macaca mulatta and M. fascicularis (no AIDS)
Agile mangabey (Cercocebus agilis)	SIVagi	West-Central Africa (Cameroon)	0–10%	Not reported	Not reported
White-crowned mangabey (Cercocebus lunulatus)	SIVagm.ver[b]	Zoo in Tanzania		Not reported	
Gray-crested mangabey (Lophocebus albigena)	?	Central Africa	?	?	?
Black mangabey (Lophocebus aterrimus)	SIVbkm	Central Africa (DRC)	Not known	Not reported	Not reported
Mandrill (Mandrillus sphinx)	SIVmnd-1	Central Africa (Gabon)	50%	AIDS in captivity	Not reported
	SIVmnd-2	West-Central Africa (Cameroon, Gabon)	50%	AIDS in captivity	Experimentally, to M. mulatta (transient infection)
Drill (Mandrillus leucophaeus)	SIVdrl	West-Central Africa (Nigeria, Cameroon, Gabon, Bioko)	Not known	Not reported	Not reported
Yellow baboon (Papio cynocephalus)	SIVagm.ver[c]	Tanzania	Not known	Not reported	Not reported
Chacma baboon (Papio ursinus)	SIVagm.ver[c]	South Africa	Not known	Not reported	Not reported
Allen's monkey (Allenopithecus nigroviridis)	?	Central Africa	?	?	?
Talapoin (Miopithecus talapoin, M. ougouensis)	SIVtal	Central Africa (Gabon, Angola, Cameroon)	11%	Not reported	Transient infection in Rh upon experimental transmission

Continued

Table 1 Continued

Species/subspecies	Virus type	Geographic location[a]	Seroprevalence	Pathogenicity	Cross-species transmission
Patas (*Erythrocebus patas*)	SIVagm.ver[c]	West Africa	Not known	Not reported	Not reported
Grivet (*Chlorocebus aethiops*)	SIVagm.gri	East Africa	>50%	Not reported	Not reported
Vervet (*Chlorocebus pygerythrus*)	SIVagm.ver	East and South Africa	>50%	AIDS in a monkey co-infected with STLV	Naturally, to baboons in the wild and white-crowned mangabeys in captivity; Experimentally, to *M. nemestrina* (AIDS) and *M. mulatta* (transient infection)
Tantalus (*Chlorocebus tantalus*)	SIVagm.tan	Central Africa	>50%	Not reported	Not reported
Sabaeus (*Chlorocebus sabaeus*)	SIVagm.sab	West Africa	>60%	Not reported	Naturally transmitted to patas (No AIDS); experimentally transmitted to Rh (no AIDS)
Diana (*Cercopithecus diana*)	?	West-Central Africa	?	?	?
Greater spot-nosed monkey (*Cercopithecus nictitans*)	SIVgsn	Central Africa	4–20%	Not reported	Potential source virus for SIVcpz
Blue monkey (*Cercopithecus mitis*)	SIVblu	Central-East Africa	>60%	Not reported	Not reported
Syke's monkey (*Cercopithecus albogularis*)	SIVsyk	East Africa	30–60%	Not reported	Experimentally, to *M. mulatta* (transient infection)
Mona (*Cercopithecus mona*)	SIVmon	West-Central Africa (Cameroon, Nigeria)	Not known	Not reported	Not reported
Dent's mona (*Cercopithecus denti*)	SIVden	Central Africa	10%		
Crested mona (*Cercopithecus pogonias*)	?	West Africa	?	?	?
Campbell's mona (*Cercopithecus campbelli*)	?	West Africa	?	?	?
Lowe's mona (*Cercopithecus lowei*)	?	West Africa	?	?	?
Mustached monkey (*Cercopithecus cephus*)	SIVmus	Central Africa	3%	Not reported	Potential source virus for SIVcpz
Red-tailed monkey (*Cercopithecus ascanius*)	SIVasc/SIVschm	Central Africa (DRC)	Not known	Not reported	Not reported
Red-eared monkeys (*Cercopithecus erythrotis*)	SIVery	Central Africa (Bioko)	Not known	Not reported	Not reported

Continued

Table 1 Continued

Species/subspecies	Virus type	Geographic location[a]	Seroprevalence	Pathogenicity	Cross-species transmission
De Brazza's monkey (*Cercopithecus neglectus*)	SIVdeb	West-Central and Central Africa	40%	Not reported	Not reported
Owl-faced monkey (*Cercopithecus hamlyni*)	?	Central Africa	?	?	?
L'Hoest's monkey (*Cercopithecus lhoesti*)	SIVlhoest/SIVlho	East Africa	50%	Not reported	Experimentally, to *M. nemestrina* (AIDS)
Sun-tailed monkey (*Cercopithecus solatus*)	SIVsun	Central Africa	Not known	Not reported	Source virus for SIVmnd-1; Experimentally, to *M. nemestrina* (AIDS); Experimentally, to *M. fascicularis* (transient infection);
Preuss's monkey (*Cercopithecus preussi*)	SIVpre	Central Africa (Bioko)	Not known	Not reported	Not reported
Mantled colobus (*Colobus guereza*)	SIVcol	Central Africa	28%	Not reported	Not reported
Western Red colobus (*Piliocolobus badius*)	SIVwrc	West Africa	40%	Not reported	Not reported
Olive colobus (*Procolobus verus*)	SIVolc	West Africa	40%	Not reported	Not reported

[a]Countries listed correspond to reported evidences of SIV circulation in that NHP species and not to species distribution.
[b]Cross-species transmission in captivity.
[c]Cross-species transmission in the wild.

Table 2 SIV clusters based upon phylogenetic relationships

Cluster	Species	SIV strain	Comments
1	Arboreal guenons (*Cercopithecus*)	SIVsyk, SIVblu, SIVgsn, SIVdeb, SIVmon, SIVden, SIVmus, SIVasc, SIVtal, SIVery	Ancestral source of SIVcpz/SIVgor/HIV-1 (SIVgsn, SIVmon, SIVmus, and SIVden harbor a *vpu* gene); lineage formed by all arboreal guenons; partial sequences from SIVbkm from the black mangabey cluster in this lineage
2	Sooty mangabey	SIVsmm	Ancestral virus of SIVmac/HIV-2; SMs from Ivory Coast harbor SIVsmm strains related to the epidemic HIV-2 groups A and B; those from Sierra Leone are the sources of HIV-2 groups C-H
3	African Green Monkey	SIVagm (SIVagm.ver, SIVagm.tan, SIVrcm.gri, SIVagm.sab)	Four different SIV subtypes described for each species in the genus *Chlorocebus*, suggesting host-dependent evolution; SIVagm.sab is a recombinant between an SIVagm ancestor and a SIVrcm-like virus
4	L'Hoest supergroup, mandrill	SIVlhoest, SIVsun, SIVmnd-1, SIVpre	Host-dependent evolution for monkeys in the C.l'hoesti supergroup; cross-species transmission from solatus guenon to mandrills
5	Red-capped Mangabeys	SIVrcm, SIVagi	Originally considered recombinants, now appear to be 'pure' viruses; SIVagi is cross-species transmitted from RCMs
6	Mantled colobus	SIVcol	First virus isolated from *Colbinae*; other viruses from Western colobus species do not cluster with SIVcol

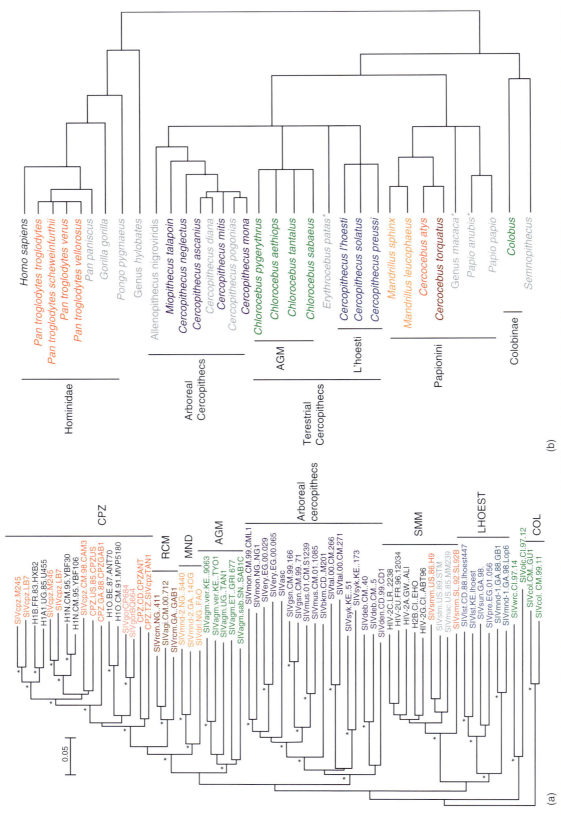

Figure 1 Comparison between SIV phylogeny (a) and primate phylogeny (b). Neighbor-joining tree constructed from available SIV sequences (a); primate phylogeny is a schematic using relationships cited in the text (b). While general alignment of host vs. virus can be observed, cross-species transmissions and viral recombination events make this correlation less than absolute. Asterisks indicate significant bootstrap values. Adapted from VandeWoude S and Apetrei C (2006) Going wild: Lessons from T-lymphotropic naturally occurring lentiviruses. *Reviews in Clinical Microbiology Reviews* 19: 728–762, with permission from American Society for Microbiology.

are divided into two tribes: Papionini (mangabeys (*Cercocebus* and *Lophocebus*), baboons (*Papio*), mandrills and drills (*Mandrillus*), gelada (*Theropithecus*), and the Asian genus, *Macaca*) and Cercopithecini (25 species in 3 arboreal genera: *Allenopithecus*, *Miopithecus*, and *Cercopithecus*, and three terrestrial genera: *Erythrocebus*, *Chlorocebus*, and *Cercopithecus lhoesti* supergroup).

The approximate equidistance among the major SIV lineages does not always match the relationships among their hosts (**Figure 1**). Thus, terrestrial monkeys form a single clade indicating that the evolutionary transition between arboreality and terrestriality has occurred only once among the extinct lineages. However, each of the terrestrial genera are infected with specific viral lineages. Arboreal guenons are infected with a cluster of viruses sharing biological properties and structural features. Papionini monkeys are infected with related viruses, though a higher proportion of recombinant viruses can be observed in these monkeys.

Genome Organization and Composition

SIVs have a complex genomic structure with three structural genes – *gag* (group antigen gene), *pol* (polymerase), and *env* (envelope) – and several accessory genes whose number varies in different SIVs. The accessory genes *vif* (virus infectivity factor), *tat* (transcriptional trans-activator), and *rev* are facilitators of viral transcription and activation; *tat* and *rev* each consists of two exons; *nef* induces CD4 and class I downregulation. Three accessory genes are specific for primate lentiviruses: *vpr*, *vpx*, and *vpu*. All primate lentiviruses harbor *vif*, *rev*, *tat*, *vpr*, and *nef*. The presence of *vpx* and *vpu* is variable and defines three patterns of genomic organization (**Figure 2**): (1) SIVsyk, SIVasc, SIVdeb, SIVblu, SIVtal, SIVagm, SIVmnd-1, SIVlhoest, SIVsun, and SIVcol contain no *vpx* or *vpu*; (2) Papionini viruses, SIVsmm, SIVmac, SIVstm, SIVrcm, SIVmnd-2, SIVdrl together with HIV-2 harbor a *vpx* gene, acquired following a nonhomologous recombination and duplication of the *vpr*; and (3) SIVcpz, SIVgor, SIVgsn, SIVmus, SIVmon, SIVden together with HIV-1 encode a *vpu* gene. *vpu* first appeared in cercopithecines, which appear to be the reservoir for viruses in the SIVcpz/HIV-1 lineage. SIVblu, SIVolc, SIVwrc, SIVbkm, SIVery, and SIVagi have not yet been completely sequenced; therefore, their classifications are pending.

SIV Recombination

Virus substrain recombination is a hallmark of SIVs and represents an important intrinsic mechanism other than rapidly accumulating point mutations for developing strains adapted to evade host defense mechanisms or for cross-species transmission. The most critical recombination of SIVs appears to be that involving SIVgsn/mon/mus and SIVrcm which resulted in the origin of the chimpanzee SIVcpz. Subsequent cross-species transmission from chimpanzees to humans is the root cause of the HIV/AIDS

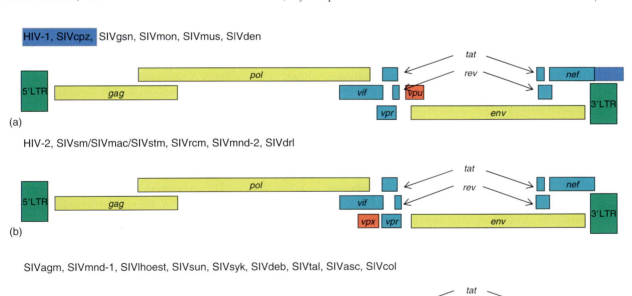

Figure 2 Genomic organization of the SIV strains belonging to different genomic types. SIV classification based on genomic structures is not superimposable on phylogenetic relationships. For references, see **Table 1**. Adapted from VandeWoude S and Apetrei C (2006) Going wild: Lessons from T-lymphotropic naturally occurring lentiviruses. *Reviews in Clinical Microbiology Reviews* 19: 728–762, with permission from American Society for Microbiology.

pandemic. Other recombinant SIVs include SIVagm.sab (containing SIVrcm-like fragments) and SIVmnd-2/SIVdrl (mosaic between SIVrcm and SIVmnd-1).

Host Range

Typically, SIVs are restricted to their host species (species-specific). However, cross-species transmissions can occur as rare events, the most notable of which were those of SIVrcm and SIVgsn/mon/mus to chimpanzees to generate SIVcpz and those of SIVcpz and SIVsmm to humans to generate HIV-1 and HIV-2, respectively. Macaque exposure to SIVsmm, SIVagm, SIVsun, and SIVlho resulted in persistent infection and induced AIDS. In the wild, SIVagm has been isolated from a yellow baboon (*Papio cynocephalus*), a chacma baboon (*P. ursinus*), and patas monkey (*Erythrocebus patas*). In captivity, in Kenya, SIVagm.ver was transmitted to a white-crowned mangabey (*Cercocebus lunulatus*). None of these recipient species have been reported to date to carry a specific SIV, which may explain the higher susceptibility to cross-species transmitted infection. It is not known if SIVagm is endemic or pathogenic in these species. SIVmnd-1 also resulted from cross-species transmission of SIVsun from sympatric *Cercopithecus lhoesti solatus*. The reasons for success of cross-species transmission infections, and the factors necessary for infection to result in pathogenicity, have not been clearly delineated.

Coreceptor Usage

Similar to HIV-1, SIVs use CD4 as the binding receptor, and chemokine co-receptors such as CCR5 and CXCR4. Most of the SIVs naturally infecting African NHPs use CCR5 as the main co-receptor. Different from HIV-1, for which a switch in viral tropism from R5 ('macrophage' tropic) to X4 ('lymphocyte' tropic) occurs with disease progression, no correlation between co-receptor usage and pathogenesis *in vivo* can be established for SIVs. SIVmnd-1, SIVagm.sab and some strains of SIVsmm use CXCR4, with no pathologic consequence. Experimental infection of sabeaus AGMs with SIVagm.sab (an X4/R5 virus) did not show a particular pattern of viral replication or disease progression. SIVrcm uses the CCR2b co-receptor for viral entry as a consequence of a 24-bp deletion in the CCR5 gene. As such, this example illustrates selected viral evolution, similar to CXCR4 infection of humans who possess the delta-32 mutation in the CCR5 gene.

Diagnosis

Antibody Detection

Serology is the gold standard for studying the prevalence of SIVs in NHPs. Commercial HIV-1/HIV-2 enzyme-linked immunosorbent assay (ELISA) and Western blot assays can be used for anti-SIV antibody screening in NHPs due to cross-reactivity with other lentiviral lineages. For a more sensitive detection of SIVs, two strategies are available: use of a highly sensitive line assay (INNO-LIA HIV, Innogenetics) as a screening test; more than 10 different new SIV types have been identified using this strategy. Alternatively, the use of SIV-specific synthetic peptides allows for increased sensitivity (Gp41/36 peptide) and specificity (V3 peptides); several SIVs have been discovered using this technique.

Propagation and Assay in Cell Culture

The efficiency of *in vitro* isolation of SIVs varies widely. The ability to replicate in human PBMC or T-cell lines has been documented for SIVcpz, SIVsm/SIVmac, SIVagm, SIVlhoest, SIVmnd-1, SIVrcm, SIVmnd-2, and SIVdrl and constitutes the major argument for the threat that these viruses may pose for humans (**Table 3**). SIVs' ability to infect human macrophages has also been reported for SIVsmm, SIVagm, and SIVmnd. SIVsun and SIVsyk cannot replicate in human peripheral blood mononuclear cells (PBMCs) or macrophages. SIVagm replication in human PBMCs is strain specific. SIVmnd-1 was reported to replicate in human PBMCs but not in macrophages. Some SIVs might require special culture conditions.

Epidemiology

Prevalence in the Wild

Due to their number, genetic diversity, and large distribution in sub-Saharan Africa, guenons (tribe Cercopithecini) are the largest reservoir species for SIV. SIV prevalence is high (50–60%) in some monkey species (AGMs, SMs, mandrills, l'Hoest and Syke's monkeys) and significantly lower (4–5%) in others (greater spotted nose monkeys, mustached monkeys, or agile mangabeys). Only two chimpanzee subspecies and one of gorilla were reported to carry SIVs at low prevalence levels. Geographical foci of SIVcpz infection were defined within the endemic area. The highest SIVcpz prevalence was observed in Cameroon (10%), in agreement with HIV-1 emergence in that area. The SIVgor strains also originated from Cameroon. No evidence of SIV infection was thus far reported for some African species, most notably baboons and some species of mangabeys.

Modes of Transmission

Epidemiologic patterns of SIV seroconversion in natural hosts showed the most efficient virus transmission during adult contact, similar to HIV-1, which is spread by sexual contact via primarily mucosal exposure. Horizontal

Table 3 Host range of *in vitro* replication SIV in different blood subsets and human T-cell lines

Growth support	Cell description	SIVcpz	SIVsm	SIVmac	SIVagm	SIVlhoest	SIVsun	SIVmnd-1	SIVsyk	SIVtal	SIVrcm	SIVmnd-2	SIVdrl
human PBMC[a]		+	+	+	±	+	−	+	−	−	+	+	+
human MDM[b]		−	+	+	±	+	−	−	−	−	±	−	+
macaque PBMC			±	+							−	+	−
chimpanzee PBMC		+	−			+		+					−
MT2	T-cell line		±		+	−	+			−		−	−
C8166	T-cell line		+	+	+	+	+		+	−	+	−	+
H9	Cloned from Hut78			+	−	−	+				−	−	−
MT4	T-cell line			+	+	+	+						+
U937	Promonocytic cell line			−	−	−	−		−			−	−
SupT1	T-cell line	+ +	+	−	+	+	+	+	+	−	+	+	+
PM1	T-cell line		±		−	−	−					−	−
Hut78	T-cell line		+	+	−	−	+					+	+
Molt 4 Clone 8	T-cell line	+ +	−	−	+	+	+		−		−	+	+ +
CEMss	T-cell line	+ +			−	−	−	−			−	−	−
CEMx174	T-cell–B-cell hybrid line		+	+	−	+	−		+ +	+		+	+

[a]PBMC-peripheral blood mononuclear cells.
[b]MDM-monocyte-derived macrophages.

transmission also occurs by biting or aggressive contact for dominance. While maternal to offspring transmission has been reported, it is relatively rare compared to horizontal transmissions.

Pathogenesis and Pathology

Pathogenicity of Natural SIV Infection

For 20 years, it was believed that natural SIV infections were nonpathogenic. This was a major paradox given the context of an active viral replication and high prevalence levels. However, occasionally, natural SIV infection of mandrills, AGMs, and SMs may eventually lead to the development of immunodeficiency. Cases of progression to AIDS in African NHP hosts are rare, possibly because host↔virus adaptation has occurred, resulting in a long-term persistent infection with an incubation period that exceeds the normal life span of the naturally infected animal.

AIDS was also reported to develop in African NHPs after infection with heterologous viruses (an SIVsmm-infected black mangabey, HIV-2-infected baboons, and a subset of HIV-1-infected chimpanzees). In these cases, disease progression occurred earlier than in naturally infected African NHP hosts, and the outcome of cross-species transmitted SIV infections varied widely, with some animals clearing the cross-species transmitted SIV, others being persistently infected (albeit without disease progression) and the rest progressing to AIDS.

Cell and Tissue Tropism

Upon infection, SIVs are disseminated to tissues by the blood. The target cells are CD4+ T lymphocytes and macrophages, with lymphocytes vastly predominating in terms of infected cells. The major sites of SIV replication are the gastrointestinal tract, lymph nodes (LNs), spleen, and other lymphoid tissues. Natural hosts for SIV infection (SMs, AGMs, mandrills, and chimpanzees) express lower levels of CCR5 on CD4+ T cells in blood and mucosal tissues, compared to immunodeficiency-susceptible hosts (macaques, baboons, and humans). Moreover, chimpanzees, which are more recent hosts of SIV, show an intermediate level of CD4+ CCR5+ T cells. As CCR5 is the main co-receptor for SIVs, African species with endemic naturally occurring SIVs may be less susceptible to pathogenic disease because they have fewer receptor-expressing targets for infection, leading to an evolutionary mechanism of 'passive co-existence' between SIV replication and natural host immune system function.

Virus Replication *In Vivo*

The lack of disease in African NHPs is not associated with effective host containment of viral replication, this is in contrast to HIV/SIV pathogenic infections for which levels of plasma viral loads (VLs) are the best predictor of the disease progression. Experimental SIV infections in natural hosts (SMs, AGMs, and mandrills) showed a consistent pattern of SIV replication with a peak of viremia (10^6–10^9 copies per ml of plasma) occurring around days 9–11 post infection, followed by a sharp decline (1–2 logs) and attainment of a stable level of VL (set point), which is maintained at high levels (10^5–10^6 copies per ml of plasma) during chronic infection (**Figure 3**). Experimental data were confirmed in naturally SIV-infected African NHPs, where chronic VLs were higher than in HIV-1 chronically infected asymptomatic patients and remained relatively constant over years. Some species-specific differences in viral replication between different African NHP species can occur without significant pathogenic consequences: SIV VLs are generally lower in AGMs than in other African NHP species. SIV proviral loads in the LNs are also 100-fold lower in AGMs than in naturally infected SMs or MNDs.

Immune Response and Persistence

Acute SIV infection in the natural host induces massive mucosal CD4+ T-cell depletion, of the same magnitude as in pathogenic HIV/SIV infection. However, during chronic infection, in spite of persistent high viral replication, there is a partial immune restoration of CD4+ T-cells in natural hosts, probably as a consequence of the preservation of a mucosal immunologic barrier and normal levels of immune activation. In marked contrast to HIV infection, in which a failure of the lymphoid regenerative capacity is an important factor in the pathogenesis of the immunodeficiency, the regenerative capacity of the CD4+ T-cell compartment is fully preserved in natural hosts and may play a key role in determining the lack of disease progression. Interleukin-7 (IL-7) has a critical role in preserving T-cell regeneration and in avoiding CD4+ T-cell depletion and disease progression in natural infections.

The high level of viral replication during chronic SIV infection in natural hosts is associated with low level of immunologic pressure, T-cell activation and proliferation, and apoptosis, resulting in limited bystander pathology. Thus, natural hosts of SIV are not confronted with the massive tissue destruction observed in HIV-1/SIVmac infection. This equilibrium is probably disrupted after cross-species transmission of viruses, when virus penetrates a new ecological niche, inducing a different immune response.

In natural SIV infections, *de novo* immune responses are muted compared to pathogenic HIV/SIVmac infections that induce robust neutralizing and cellular immune responses and continuous immune escape. Therefore, it is believed that natural SIV infections are characterized by tolerance to the virus or to specific antigens or epitopes,

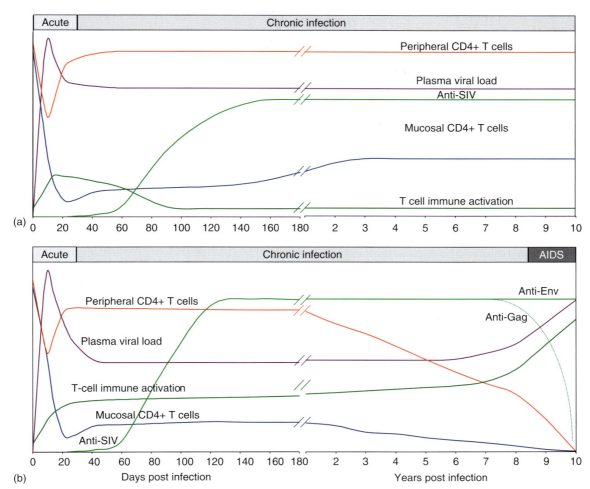

Figure 3 Pathogenesis of SIV infection in natural hosts (a) compared to the pathogenesis of HIV-1 infection (b). No significant difference can be observed between the two models during acute infection. During chronic SIV infection, natural hosts harbor higher viral loads, lower antibody titers, lower levels of T-cell immune activation. CD4+ T-cell levels are maintained at near pre-infection levels in natural hosts. Mucosal CD4+ T cells are partially restored during chronic infection, despite high levels of viral replication. Progression to AIDS is absent in most cases, being a rare outcome of SIV natural infection.

or by immune responses that differ qualitatively and quantitatively from those observed in pathogenic infections. Antibody responses are observed in natural SIV infection, but mainly directed to Env rather than Gag, as in SIVmac infection. The intensity of the antibody response is lower in natural hosts: for equivalent VL, antibody titers in SMs are about one log lower than in RMs.

Neutralizing antibodies are rarely detected in SIV-infected SMs, while SIVagm is susceptible to neutralization depending on the cell line used in the assay. In contrast to HIV-1, SIVagm infectivity is enhanced by the addition of soluble CD4 and this enhanced infectivity can be abrogated by SIVagm-specific antibodies. However, very high amounts of passively transferred specific immunoglobulins failed to prevent SIVagm infection suggesting that humoral immune response in AGMs is largely ineffective.

SIV-infected SMs and AGMs develop cytotoxic T-lymphocyte responses that are functional to some degree in controlling viral replication. Moreover, there is evidence of transient but massive expansion of CD8+ T cells in infected AGMs, SMs, and mandrills during acute infection, showing that the immune system of natural hosts is influenced by SIV infection. It is not yet clear if CD8+ T-cell expansion results from active stimulation of the specific immune response or from a nonspecific stimulation of the immune system in general.

SIV-specific T-cell responses can be detected in the majority of naturally SIV-infected NHPs. However, their magnitude is generally lower than in HIV-infected patients. In addition, no correlation was found between breadth or magnitude of SIV-specific T-cell responses and either VLs or CD4+ T-cell counts. Moreover, the magnitude of the SIV-specific cellular responses did not appear to determine the level of T-cell activation and proliferation in SMs and AGMs. Therefore, the presence of a strong and broadly reactive T-cell response to SIV antigens is not a requirement for the lack of disease progression in natural infections; conversely, the complete suppression of SIV-specific T-cell responses (i.e., immunologic tolerance and/or ignorance) is not required for the low levels of T-cell activation that are likely instrumental in avoiding AIDS.

Both acute and chronic natural SIV infections (SMs, AGMs, and mandrills) are associated with lower levels of T-cell activation, pro-inflammatory responses, immunopathology, and bystander apoptosis than pathogenic HIV/SIV infection. In addition, natural hosts of SIV maintain T-cell regenerative capacity, with normal bone marrow morphology and function, normal levels of T-cell receptor excision circle (TREC)-expressing T cells and preserved LN architecture. This downregulation of the immune response favors preservation of CD4+ T-cell homeostasis and is completely different from pathogenic HIV/SIV infections for which chronic immune activation and proliferation drive excessive activation-induced T-cell apoptosis, and ultimately result in the collapse of the immune system and progession to AIDS. Altogether, this is consistent with the hypothesis that chronic immune activation is a major determinant of disease progression during HIV infection.

In AGMs, low immune activation levels are due to a strong anti-inflammatory response (with induction of TGF-β1 and FOXP3, and a significant increase in IL-10 expression), which occurs early in the SIVagm infection. Together with an early increase in the levels of CD4+ CD25+ T cells, this results in the rapid establishment of an anti-inflammatory environment which may prevent damages to the mucosal immunologic barrier, microbial translocation, and thus aberrant chronic T-cell hyperactivation that is correlated with progression to AIDS during HIV-1/SIVmac infection.

Virulence

Viral factors may be related to a lack of virulence in natural SIV infections. In RMs, SIVmac *nef* gene deletion mutants were reported to replicate poorly *in vivo* and to be nonpathogenic, which corroborates the description of *nef* gene mutations of HIV-1-infected long-term progressors. Nef downregulates CD4, CD28, and the class I major histocompatibility complex, resulting in virus immune evasion. Nef may also enhance the responsiveness of T cells to activation, but this effect is not uniformly observed among SIVs. All SIVs from African monkeys have open reading frames corresponding to a functional Nef, and therefore the *nef* structure cannot account entirely for differences in pathogenicity. Nef proteins from the great majority of primate lentiviruses, including HIV-2, down-modulate T-cell receptor-CD3, which subsequently blocks T-cell activation. In contrast, Nef proteins derived from HIV-1 and closely related SIVs do not induce CD3 downregulation, which may have predisposed the simian precursor of HIV-1 to greater pathogenicity in humans. However, simian counterparts of HIV-1 (SIVcpz and SIVgsn/SIVmon/SIVmus), which do not induce CD3 downregulation, do not typically induce AIDS in their natural hosts. Further, SIVmac, which induces CD3 downregulation, is even more pathogenic in RMs than HIV-1 is in humans, suggesting that this accessory gene does not solely account for virulence.

Clinical and Pathologic Features

In contrast to macaques in which SIV infection constantly leads to an AIDS-like disease characterized by opportunistic infections and cancers, SIV infection in natural hosts generally does not show any clinical or pathological abnormalities. During progression to AIDS, SIV-infected RMs initially develop lymphadenopathy with confluent follicular hyperplasia, followed by LN atrophy due to lymphoid depletion and fibrosis. In contrast, SIV-infected natural hosts display normal LN morphology without evidence of either hyperplasia or depletion. No follicular trapping or CD8+ T-cell infiltration of the germinal centers and no replacement of the normal LN architecture with connective tissue is observed in chronic SIV infection of natural hosts. Also, no thymic disinvolution, nodular lymphocytic infiltrates, or giant cell disease are seen in SIV-infected natural hosts. The few AIDS cases reported in African NHP hosts presented with the entire range of diseases and pathologic lesions of AIDS.

Host Genetic Resistance

SIV species specificity has typically been ascribed to factors such as virus–host receptor compatibility and cellular machinery needed to direct viral replication. Specific host factors also prevent SIV cross-species infections *in vitro*.

Cytidine deaminase and Vif

Vif is involved in species specificity of SIVs. Its cellular target is a member of the cytidine deaminase APOBEC family. The cellular deaminase is incorporated into the virion during the reverse transcription to direct the deamination of cytidine to uridine on the minus strand of viral DNA. Deamination results in catastrophic G-to-A mutations followed by inactivation and/or degradation of the viral genome. At least two primate APOBEC family members (APOBEC3G and APOBEC3F) play a central role in antagonizing viral replication because they are expressed in natural targets of SIVs, including lymphocytes and macrophages. Lentiviruses are able to successfully infect and replicate in host target cells containing APOBEC when host-adapted Vif interferes with this mechanism. Vif activity is species specific: human APOBEC3G is inhibited by HIV-1 Vif but not by SIVagm Vif, whereas AGM APOBEC3G is inhibited by SIVagm Vif, but not by HIV-1 Vif. This specificity relies on a single amino acid change that can alter the ability of *vif* to interfere with APOBEC activity.

TRIM5-α

The cytoplasmic body component TRIM5-α, previously referred to as REF-1 and LV-1, restricts HIV-1 infection of monkey cells. TRIM5-α interferes with the viral uncoating step that is required to liberate viral nucleic acids into the cytoplasm upon viral binding and fusion with the target cell. Sensitivity to TRIM5-α restriction is dictated by a small region in the viral capsid gene, previously shown to be involved in cyclophilin A binding. Subtle amino acid differences in this region influence the strength of binding to TRIM5-α and, hence, relative sensitivity to its restriction. HIV-2, but not closely related SIVmac, is highly susceptible to RM TRIM5-α. Furthermore, HIV-2 is weakly restricted by human TRIM5-α, which may contribute to the lower pathogenic potential of the HIV-2 vs. HIV-1 in humans. Strategies to exploit TRIM5-α restriction for intervention of HIV-1 replication are under development.

See also: Host Resistance to Retroviruses; Human Immunodeficiency Viruses: Molecular Biology; Human Immunodeficiency Viruses: Origin; Human Immunodeficiency Viruses: Pathogenesis; Retroviruses: General Features; Simian Immunodeficiency Virus: Animal Models of Disease; Simian Immunodeficiency Virus: General Features.

Further Reading

Beer BE, Bailes E, Sharp PM, and Hirsch VM (1999) Diversity and evolution of primate lentiviruses. In: Kuiken CL, Foley B, Hahn B, et al. (eds.) *Human Retroviruses and AIDS 1999*, pp. 460–474. Los Alamos, NM: Theoretical Biology and Biophysics Group and Los Alamos National Laboratory.

Chakrabarti LA (2004) The paradox of simian immunodeficiency virus infection in sooty mangabeys: Active viral replication without disease progression. *Frontiers in Biosciences* 9: 521–539.

Drucker E, Alcabes PG, and Marx PA (2001) The injection century: Massive unsterile injections and the emergence of human pathogens. *Lancet* 358(9297): 1989–1992.

Gordon S, Pandrea I, Dunham R, Apetrei C, and Silvestri G (2005) The call of the wild: What can be learned from studies of SIV infection of natural hosts? In: Leitner T, Foley B, Hahn B, et al. (eds.) *HIV Sequence Compendium 2004*, pp. 2–29. Los Alamos, NM: Theoretical Biology and Biophysics Group and Los Alamos National Laboratory.

Hirsch VM (2004) What can natural infection of African monkeys with simian immunodeficiency virus tell us about the pathogenesis of AIDS? *AIDS Reviews* 6(1): 40–53.

Muller MC and Barre-Sinoussi F (2003) SIVagm: Genetic and biological features associated with replication. *Frontiers in Biosciences* 8: D1170–D1185.

Norley S and Kurth R (2004) The role of the immune response during SIVagm infection of the African green monkey natural host. *Frontiers in Biosciences* 9: 550–564.

Sharp PM, Shaw GM, and Hahn BH (2005) Simian immunodeficiency virus infection of chimpanzees. *Journal of Virology* 79(7): 3891–3902.

VandeWoude S and Apetrei C (2006) Going wild: Lessons from T-lymphotropic naturally occurring lentiviruses. *Clinical Microbiology Reviews* 19: 728–762.

Simian Retrovirus D

P A Marx, Tulane University, Covington, LA, USA

© 2008 Elsevier Ltd. All rights reserved.

Glossary

Codon The basic unit of the genetic code consisting of three consecutive nucleotides in DNA that specify the order of amino acids in a protein.

Monocistronic mRNA Messenger RNA that codes for a single protein.

Monoclonal antibody Antibody made *in vitro* that reacts with a single protein.

Pneumocystis carinii A fungus that causes pneumonia, especially in persons or animals with suppressed immune systems.

Polyprotein A precursor protein that is cut after its synthesis into smaller functional proteins.

Ribosomal frameshifting A change in the translational reading frame that allows the synthesis of different proteins from a single region of overlapping genes. For example, if the genetic code begins on the first of the three nucleotides of a codon, the out-of-frame code would begin on the second nucleotide. A second frameshift beginning on the third nucleotide is also possible. Frameshifting is an economical way to code for proteins.

Syncytium Cell-derived structure in culture or *in vivo* giant cell consisting of fused cells having more than one nucleus.

Virion Infectious viral particle containing the genome of the virus and a full complement of viral proteins required for growth and replication.

Western blot Nitrocellulose blotting paper containing viral polypeptides sorted by molecular size. The blot is soaked in diluted serum suspected of containing antibodies. A positive test requires binding to polypeptides from at least two different viral genes. More weight is often given to env polypeptide reactions. False results against a single *gag* coded protein commonly occur and are reported as indeterminant.

Zoonosis Microbial infection (virus, bacteria, or parasite) acquired directly through animal contact that progresses to a clinical disease.
Zoonotic infection Microbial infection (virus, bacteria, or parasite) acquired directly through animal contact that may or may not progress to disease.

Introduction

Simian type D retroviruses (SRVs) and Mason–Pfizer monkey viruses (MPMVs) are members of the family *Retroviridae* and comprise one species belonging to the genus *Betaretrovirus*. These viruses naturally infect members of the nonhuman primate genus *Macaca*, commonly known as macaque monkeys. SRV appears to be highly species specific, although the virus will replicate in the cells of other primate species, including tissue culture cells of human origin. Betaretroviruses are characterized by their simple genomic structure consisting of 4 genes, *gag, pro, pol*, and *env* (**Figure 1**).

MPMVs and SRVs are exogenous retroviruses, meaning that the virus is acquired from another infected macaque monkey. This transmission mechanism distinguishes SRVs from endogenous D retroviruses that are inherited and are a part of the genome of the host species. Endogenous D retroviruses occur in both New and Old World monkeys, including African baboons as well as Asian langur and South American squirrel monkeys.

Seven distinct isolates are known and each can be distinguished by serum neutralization tests or competitive binding tests between homologous p27 *gag* proteins. They are named SRV serotypes 1–7. Serotype 3 is also known as MPMV. A nomenclature has been devised that is similar to that used for influenza viruses and is an aid for tracking isolates and their origins. For example, D1/rhe/CA/84 represented the D retrovirus that was isolated at the California Primate Center Davis, CA from a rhesus macaque with an acquired immune deficiency syndrome (AIDS)-like disease and was reported in 1984. D2/Cel/OR/85 was a 1985 SRV-2 isolate from a Celebes black macaque with retroperitoneal fibromatosis (RF) at the Oregon Primate Center. MPMV is the prototype of this viral species and retains its original name for historical reasons.

History

In 1971, the first D retrovirus isolate was reported. The virus was recovered from a rhesus monkey with a

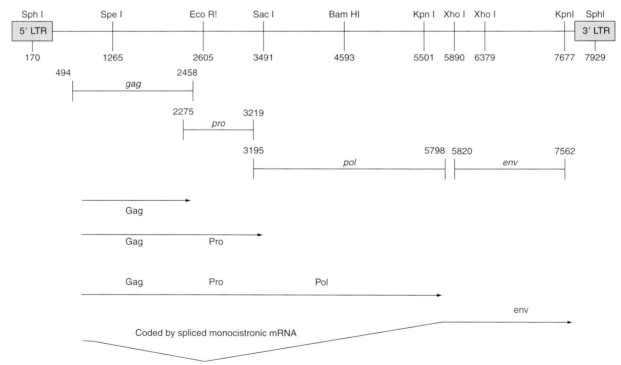

Figure 1 Map of the genome of primate D retroviruses. Top is the provirus that is 8105 nt and codes for four genes in three separate reading frames, *gag, pro, pol*, and *env*. The messenger RNAs consist of two forms, a genomic-length mRNA that is translated into the polyprotein Gag. After frameshifting(s), Gag-Pro and Gag-Pro-Pol precursors are made. Polyproteins are cleaved into smaller functional virion proteins. The second mRNA form is a spliced mRNA that encodes the Env proteins, SU (also named gp70) and TM (also named gp20). Reprinted from Marracci GH, Avery NA, Shiigi SM, *et al.* (1999) Molecular cloning and cell-specific growth characterization of polymorphic variants of type D serogroup 2 simian retroviruses. *Virology* 261: 43–58, with permission from Elsevier.

mammary carcinoma. The virus was named after the Mason Institute in Worcester, Massachusetts, where the macaque had been housed. The Pfizer Pharmaceutical Company had supported research using this female monkey, hence the name Mason–Pfizer monkey virus (MPMV). The impact of MPMV was greatly enhanced because the virus was isolated directly from cancerous breast tissue. Because of this association with neoplasia, MPMV was classified as an oncornavirus (Onco RNA for cancer-causing RNA virus). This classification persisted for many years.

The finding of MPMV in monkey breast cancer tissue prompted numerous reports in the 1970s of D retroviruses in association with human breast cancer. With the development of sensitive detection techniques along with complete sequencing of the MPMV and SRV genomes, it became apparent that nonhuman primate D retroviral sequences were not causally associated with human tumors. A 1975 report on immune impairment in neonatal monkeys infected with MPMV marked the beginning of our understanding of primate D retroviruses as a causative agent of simian AIDS. The modern era of type D retrovirus research deals with its immunosuppressive properties.

AIDS, the disease, was discovered in 1981 in persons displaying symptoms of a severe and fatal immunodeficiency. The first detailed report consisted of four men with *Pneumocystis carinii* pneumonia, evidence of cytomegalovirus infection (CMV), and Kaposi sarcoma, a skin cancer not commonly seen in young men. Early theories on the cause of AIDS implicated drug-induced impairment of the immune system. Drugs were known to suppress the immune system, but the idea that severe and fatal immune deficiency could be caused by a virus was not yet appreciated. A breeding group of rhesus monkeys that was housed outdoors at the California National Primate Research Center in Davis, CA, played a significant role in changing that perception. This breeding group had a higher mortality rate than monkeys in other outdoor groups only a short distance away. The affected animals had diarrhea, cytomegalovirus infections, and fibrosarcoma, a tumor that resembled Kaposi sarcoma upon microscopic inspection. A link was made between AIDS in monkeys and AIDS in humans. The study of this group of monkeys resulted in the isolation and characterization of a virus related to MPMV. This distinct D retrovirus was named simian retrovirus (SRV) and was used to prove the retroviral etiology of this fatal immune deficiency disease in monkeys. A similar virus was isolated from rhesus macaques at the New England Primate Research Center.

Replication

The synthesis of new progeny virus begins with the DNA provirus that is integrated into a host DNA chromosome (**Figure 1**). The provirus is a double-stranded (ds) DNA copy of the virus genome. The provirus is flanked by two repeated sequences called the long terminal repeat (LTR). The order of the coding and noncoding regions are the LTR, *gag* (for group antigen), *pol* (polymerase), *env* (envelope), and a second copy of the LTR. The viral genome is synthesized from the minus DNA strand of the provirus by host cell enzymes. The newly made viral genome is the same sense as messenger RNA (+-stranded) and is exported from the nucleus without being spliced. The cellular requirement for splicing is bypassed by the interaction of the constitutive transport element (CTE) with cellular proteins. This genomic RNA is translated in the cytoplasm into Gag and Pol viral proteins in a process that involves several steps.

The *gag* gene codes for 654 amino acids that are synthesized as a 78 kDa polyprotein precursor of smaller functional virion proteins. The *gag*-encoded polyprotein is cleaved by a protease enzyme, which in turn is coded by the *pro* gene. This Gag precursor is cleaved into six mature virion proteins, the p10 matrix protein (polypeptide 10 000 Da molecular weight (MW)), pp24 (phosphoprotein), p12 (assembly scaffold domain), p24 major capsid protein, p14 nucleocapsid protein, and p4 (chaperonin binding domain). The major capsid protein, p24 (in M-PMV, p27), is the most abundant protein in the mature virion (**Figures 2(b)** and **3**). The p10 protein is the matrix protein and resides just under the external envelope of the virion (**Figure 3**). The nucleocapsid protein p14 forms a tightly associated complex with the RNA genome. Unlike most other retroviruses, D-type retroviruses form immature capsids that are preassembled in the cytoplasm (**Figure 2** – see arrows). This is facilitated by a cytoplasmic targeting and retention signal in the matrix domain of Gag, which binds to the microtubule motor dynein and targets translating polysomes to the pericentriolar region of the cell where assembly occurs. The p12 region of Gag acts to facilitate this intracytoplasmic assembly process.

The protease (*pro*) gene overlaps the *gag* region by 61 codons (183 nt). The protease is an enzyme that cleaves protein and is highly conserved across the genus with 82.8% and 83.8% identity between SRV-1 and MPMV, respectively. The protease itself is translated from the genome-length mRNA (**Figure 1**) as two polyproteins (Gag-Pro) and Gag-Pro-Pol. Because the *pro* gene is out of frame with respect to *gag*, the polyprotein that contains the protease results from a ribosomal frameshifting mechanism during translation of genome-length mRNA. The final protease protein is produced by two autocatalytic cuts, one at the N-terminus producing a 17 kDa product and a second cut producing the final 13 kDa protease enzyme. The protease is incorporated into the immature capsids in the cytoplasm (**Figure 2** – see arrows). It functions to cleave the *gag*-coded polyprotein into smaller proteins that make up the internal capsid structure of

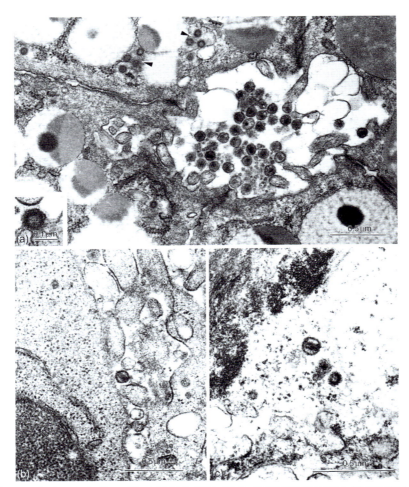

Figure 2 Electron microscopic demonstration of type D retroviral particles in (a) salivary gland, (b) germinal center of a lymph node, and (c) spleen of rhesus monkeys with AIDS. In the salivary gland (a), numerous mature extracellular particles in a small acinar lumen, as well as immature intracytoplasmic A particles (arrows) and a budding particle (inset) are seen. In the germinal center (b), a single mature virion particle adjacent to a lymphoid cell and numerous cellular processes possibly belonging to follicular dendritic cells are seen. (c) In the spleen, a single mature particle is seen in the extracellular space. Modified from Lackner AA, Rodriguez MH, Bush CE, et al. (1988) Distribution of a macaque immunosuppressive type D retrovirus in neural, lymphoid, and salivary tissues. *Journal of Virology* 62: 2134–2142, with permission from American Society for Microbiology.

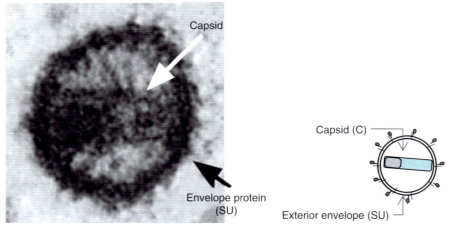

Figure 3 Mature D retrovirus virion showing the external envelope and the major capsid core. Modified from Marx PA, Maul DH, Osborn KG, et al. (1984) Simian AIDS: Isolation of a type D retrovirus and transmission of the disease. *Science* 223: 1083–1086, with permission from AAAS.

the virion. The three-dimensional (3-D) structure of the protease has been resolved and this 3-D structure is well conserved across the retrovirus family.

A second ribosomal frameshift within the *pro-pol* overlap region generates a larger polyprotein precursor containing the pol proteins (**Figure 1**). The *pol* gene encodes the reverse transcriptase (RT)-endonuclease and integrase proteins. The gene encoding the integrase of MPMV is located at the 3′ end of the *pol* open reading frame and the RT is encoded by the 5′ end of the gene. The RT, integrase, and endonuclease enzymes function after the progeny virus infects a new cell, and are incorporated into the immature capsids as Gag-Pro and Gag-Pro-Pol precursors. These polyproteins undergo post-translational cleavage during virion maturation to produce separate RT and endonuclease/integrase proteins. They are required to initiate the early steps of replication after the virus enters a host cell to begin a new cycle of replication.

The *env* gene is expressed from the 3′ end of the genome (**Figure 1**). *env* encodes two proteins, the external envelope spike protein (SU for surface) and the transmembrane protein (TM). The mRNA coding of the Env proteins is spliced in the nucleus to remove *gag* and *pol* coding regions. The mRNA is exported to the cytoplasm and is handled like a monocistronic mRNA. The *env* gene product is cleaved into the SU or gp70 (glycoprotein 70 000 Da) (**Figure 1**) and the TM (or gp20). Both the SU gp70 and TM gp20 proteins have sugar molecules bound to some of the amino acids that make up these two proteins. The SU glycoprotein can potentially contain 22 sugar residues. The TM is a membrane-spanning protein that anchors the complex in the virion membrane. An immunosuppressive motif is present within the TM protein and this protein motif may play a role in immune suppression. A monoclonal antibody is available against the TM-gp20 and this antibody is useful for diagnostic testing. The external glycoprotein is the least conserved among the macaque D retroviruses and is reflected in different serotypes. Neutralizing antibodies are defined by binding to the external glycoprotein. Therefore the major biological differences between SRV-1, SRV-2, MPMV, SRV-4, and SRV-5 are in fact differences in the amino acid sequence of Env.

Immature capsids (see arrows in **Figure 2**) are transported to the plasma membrane where they bud through the membrane (**Figure 2** inset) and acquire the Env proteins. The morphology of the completed D-type virion consists of an envelope and a cylindrical or rod-shaped core (**Figure 3**; a virion grown in tissue culture (**Figure 2**), and virions in the extracellular tissue space of an infected rhesus monkey).

The extracellular infectious virion initiates a new round of replication by attaching itself to a susceptible cell via a specific receptor on the surface membrane of the host cell. The receptor molecule for SRV and MPMV is an amino acid transport protein (ATBo) that is present in human tissues and cell lines, including white blood cells.

After entry in the cell, the RT synthesizes a minus DNA strand from the +-sense incoming SRV RNA genome. The RT next catalyzes the synthesis of a complementary plus sense DNA strand, making a dsDNA copy of the D retrovirus RNA genome that contains two LTRs. The LTRs flank the viral genes and, through their interaction with the viral integrase, ensure that the genome will integrate in the correct orientation for expression of the viral genes. Usually DNA codes for RNA, but the RT reverses the process by making DNA from an RNA template, hence its name reverse transcriptase. The integrase then enzymatically integrates the linear dsDNA copy of the original RNA genome into host cell DNA. This dsDNA copy of the virus genome is called a provirus. With transcription of the provirus to form full-length genomic mRNA and spliced env mRNA, the replication cycle is completed and begins a new round.

Natural Hosts

The natural host of SRV and MPMV is the Asian monkey genus *Macaca* or macaques. The known hosts are primarily common laboratory monkeys, such as the rhesus monkey (*M. mulatta*) of northern India, southern China, and geographical areas in between. SRV infection of rhesus macaques in China has been reported. Other macaque species housed at Primate Research Centers are also commonly infected. They include the cynomolgus macaque (*M. fascicularis*), also known as the long-tailed macaque and crab-eating macaque of Southeast (SE) Asia; *M. nemistrina*, the pig-tailed macaque of SE Asia; *M. nigra*, the Sulawesi or Celebes black macaque; and the bonnet macaque, *M. radiata*, of southern India. Feral cynomolgus macaques are infected, showing a natural Asian origin of this virus. Type D retroviruses are a significant problem for laboratory research, since infected animals may develop an AIDS-like disease. The most common SRV isolate in primate centers is SRV-2, followed by SRV-1, SRV-5, and MPMV. SRV-4 is only known from a single outbreak in cynomolgus macaques at a public health laboratory in Berkeley, California.

Transmission

SRV is present in blood, urine, saliva, lymphoid, and nonlymphoid tissues of infected macaques (**Figure 2**). With the possible exception of brain tissue, infectious SRV is found throughout the body. Inoculation of any of these fluids or tissues to rhesus monkeys will transmit the infection and disease. Although infection is easily transmitted under laboratory conditions, a classic experiment on SRV infections transmitted in outdoor enclosures strongly pointed to natural transmission by bite from infected saliva. In these experiments at the California Primate Center, it was first shown that uninfected monkeys must be in physical contact with infected monkeys for transmission to occur. The need for this contact was proven by keeping uninfected monkeys in the same enclosure with infected monkeys, but separated by two fences creating a 10-foot barrier. The barrier prevented contact between infected and uninfected groups, but allowed birds, rodents, and insects to move freely between the enclosures as well as allowing rainwater to wash back and forth. The result was that only monkeys in physical contact with infected monkeys became SRV infected even after 5 years of testing.

SRV transmission to healthy monkeys was linked to healthy carriers, in particular to female healthy carriers that occupied socially dominant roles in the enclosure. Healthy carriers were infected for as long as 10 years without developing the AIDS-like disease that is characteristic of SRV infections. Repeated testing of the saliva of healthy carriers showed >1 million infectious units of SRV per ml of saliva, hence the likely link between saliva and transmission. Healthy carriers were a major problem for laboratory colonies because they could spread the SRV immune deficiency and themselves remain undetected, compromising the usefulness of these research colonies. The problem is now largely controlled by screening for SRV infection by antibody enzyme-linked immunosorbent assay (ELISA) and polymerase chain reaction (PCR) testing. Infected animals are removed from contact with the rest of the laboratory colony. This test and removal program is a highly successful approach to developing specific pathogen-free colonies for research.

Broad *In Vitro* and *In Vivo* Cell Tropism

SRV has a broad cellular tropism and infects lymphoid, monocytes, and epithelial cells *in vitro*. SRV readily infects human T-cell lines, HuT-78, CEM-SS, MT-4, and SubT-1. The human B Raji cell line is also infected and SRV induces syncytia formation in this cell line. Adherent mononuclear cells from rhesus macaques were also infected *in vitro*. Peripheral blood mononuclear cells collected from infected animals and separated into CD4+ and CD8+ T-lymphocytes were also infected.

In vivo studies carried out on SRV-1-infected rhesus macaques showed infection of epithelial and lymphoid cells in the gut and elsewhere (**Figure 2**). The oral cavity and salivary glands have also been examined. Mucosal epithelia cells were heavily infected as early as 1 month post inoculation. Rarely, Langerhans cells were also shown to be infected using immunohistochemical techniques. Southern blot analyses showed salivary glands and lymphoid tissue to be more heavily infected than brain tissue. The infection of epithelia cells *in vivo* is striking in its widespread nature.

Pathogenesis

The pathogenesis of SRV, MPMV, and SRV-2 is well described. Inoculation of any of these three grown in rhesus monkey tissue culture will induce an AIDS-like disease in most of the infected macaques. Pathogenesis studies are best carried out using virus grown in cells of the original host. Cultivation of the virus in human cells, especially Raji B cells, may attenuate the virus. The clinical course for SRV-1 is typically one-third developing disease in less than 6 months, one-third developing disease in 6 months to 2 years, and a third recover but remain antibody positive for SRV. Recovered animals may develop AIDS after a long clinically latent normal period. The disease is very similar to AIDS in human beings and includes opportunistic infections such as generalized cytomegalovirus disease and its associated pneumonia, wasting, chronic diarrhea unresponsive to therapy, and severe anemia. Infecting with a molecular clone, ruling out adventitious agents, proved the pathogenesis of SRV-1.

The disease induced by SRV is AIDS, but SRV should not be confused with simian immunodeficiency virus (SIV), a lentivirus related to human immunodeficiency virus 2 (HIV-2) that also causes AIDS in rhesus macaques. Macaques are not the natural hosts of SIV, since SIV naturally occurs in only sub-Saharan African cercopithecine monkeys and apes. The species naturally infected with SRV are all Asian macaque species. These two infections can be easily distinguished by specific Western blot. AIDS caused by SRV and SIV are both severe immunodeficiency syndromes, but differ in that SIV-induced AIDS is frequently associated with pneumocystis pneumonia, atypical tuberculosis, and B-cell lymphomas. In contrast, Kaposi sarcoma-like neoplasias are seen exclusively in type D SRV-induced AIDS.

Simian RF

A tumor that occurs in the space behind the abdominal cavity (retroperitoneum) is associated with SRV-2 infections and the retroperitoneal fibromatosis herpesvirus (RFHV). RFHV is the macaque homolog of the human rhadinovirus that is associated with the Kaposi's sarcoma herpesvirus (KSHV). This monkey tumor is therefore an excellent model for Kaposi's sarcoma, one of the tumors that commonly occur in AIDS patients. DNA sequence data identified KSHV related herpes viruses in the RF tissue of pig-tailed and rhesus macaques. The basic fibroblast growth factor was found to be associated with RF tissues in SRV-2-infected macaques. The fibrosarcoma associated with SRV-1-induced AIDS is also a Kaposi-like tumor.

Vaccines

Two types of vaccines have been successfully tested in the SRV AIDS model. The first used a killed-virus formulation. The virus was inactivated with formalin and injected intramuscularly to induce antibody. The SRV challenge virus used to test the efficacy of the vaccine was prepared in isogenic rhesus monkey kidney cells, therefore ruling out induction of anticellular antibody as a protective mechanism. Protection was associated with neutralizing antibody. This was the first vaccine shown to be effective against a primate retrovirus. The second vaccine used a recombinant vaccinia virus vector. The vaccinia virus vector expressed the envelope proteins (gp70 and gp20) of SRV-1 and MPMV. Upon challenge with live SRV-1 by the intravenous route, both MPMV- and SRV-1-immunized animals were protected. The vaccine therefore conferred cross-protection for SRV types 1 and MPMV, demonstrating *in vivo* their close relationship. The neutralizing antibody did not cross-react with the more distant SRV-2.

Human Infections with SRV and Related Viruses

D-type retroviruses easily infect tissue culture cells derived from humans. Consequentially, several research groups have reported contamination of tissue cell lines. Therefore, there is a strong risk of contaminating cell cultures with this virus in the laboratories. Evidence for human infection that is based only on isolation of a D-type retrovirus from cultured human cells must be viewed with caution.

Human infections with SRV or MPMV have been reported in widely different diseases such as cancer and schizophrenia. Evidence of D-type virus has been reported in children with Burkett's lymphoma and in adult humans with breast cancer. However, direct evidence of SRV as an etiologic agent of human disease is thus far lacking.

An MPMV serological survey of European and African blood donors in Guinea-Bissau revealed 1 of 61 to be weakly positive for the MPMV p27 Gag polypeptide using Western blots. Squirrel monkey retrovirus-specific Western blots also revealed a few additional positive samples. Reaction in a blood donor consisting of only a single polypeptide is usually reported as an indeterminant result and is a false positive reaction. Nevertheless, D virus is widespread in nature and sequences related to the SRVs are found in the genomes of several primate species. Therefore, exposure to the virus is clearly possible. Nevertheless, extensive studies on over 1000 persons with various diseases have failed to find evidence of SRV infections in humans. Diseases tested included lymphoproliferative disease patients, HIV-1 infected persons, persons with unexplained low CD4 lymphocyte counts, blood donors, and intravenous drug users from the USA and Thailand. Serum samples were screened for antibodies against SRV by an ELISA, and reactive samples were re-tested by Western blot. None of the samples were seropositive.

The most convincing evidence for nonpathogenic SRV infections in humans comes from a serosurvey of workers occupationally exposed to macaque D-type retroviruses at primate centers in the USA. Occupational exposure could come from an accidental bite from infected monkeys during routine care such as cage cleaning or accidental exposure by a contaminated needle stick. Two of 231 persons tested were strongly positive by Western blot. Each person displayed antibody to more than one gene of SRV (**Figure 4**). Lanes 1 and 2 showing a reactive band at the 70, 31, 24, and 20 Da markers indicated reaction with the *pol*, *env*, and *gag* gene products. One individual had neutralizing antibody to SRV-2, providing convincing evidence of an

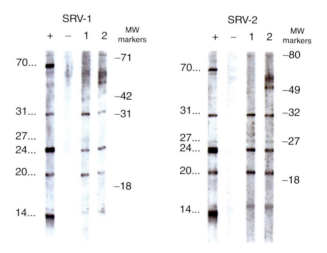

Figure 4 WB reactivity against SRV-1 and SRV-2 on initial screening in sera from two persons occupationally exposed to nonhuman primates. Lanes: (+)-lane is known SRV-positive serum from a rhesus monkey; (–)-lane is SRV-negative serum from a known negative rhesus monkey; lane 1, subject 1, lane 2, subject 2. MW, molecular weight (weights are in thousands of daltons). gp70 is the surface spike of SRV. Reproduced from Lerche NW, Switzer WM, Yee JL, et al. (2001) Evidence of infection with simian type D retrovirus in persons occupationally exposed to nonhuman primates. *Journal of Virology* 75(4): 1783–1789, with permission from American Society for Microbiology.

active SRV infection in the past. The most important tests to document active infections were negative. Repeated attempts to amplify the genomic DNA specific for SRV from specimens, as well as attempts to recover infectious virus, were not successful.

These findings are nevertheless important for understanding zoonotic infections and the related but more dangerous outcome of zoonosis. Zoonotic infections are fairly common under the right exposure conditions, but a zoonosis is a rare outcome. Even though SRV may infect human blood cells *in vivo* and grow temporary in the new host, the immune system of the non-natural human host is capable of eliminating the zoonotic infection. The past infection will be evident from trace amounts of specific antibody that was induced during the transient infection. Infection of humans with simian D and other simian retroviruses is well documented, but evidence of zoonosis has not been found.

See also: Mouse Mammary Tumor Virus; Replication of Viruses; Simian Immunodeficiency Virus: General Features.

Further Reading

Colcher D, Spiegelman S, and Schlom J (1974) Sequence homology between the RNA of Mason–Pfizer monkey virus and the RNA of human malignant breast tumors. *Proceedings of National Academy of Sciences, USA* 71(12): 4975–4979.

Daniel MD, King NW, Letvin NL, Hunt RD, Sehgal PK, and Desrosiers RC (1984) A new type D retrovirus isolated from macaques with an immunodeficiency syndrome. *Science* 223: 602–605.

Gardner MB and Marx PA (1985) Simian acquired immunodeficiency syndrome. In: Klein G (ed.) *Advances in Viral Oncology*, vol. 5, pp. 57–81. New York: Raven Press.

Heidecker G, Lerche NW, Lowenstine LJ, et al. (1987) Induction of simian acquired immune deficiency syndrome (SAIDS) with a molecular clone of a type D SAIDS retrovirus. *Journal of Virology* 61(10): 3066–3071.

Lackner AA, Rodriguez MH, Bush CE, et al. (1988) Distribution of a macaque immunosuppressive type D retrovirus in neural, lymphoid, and salivary tissues. *Journal of Virology* 62: 2134–2142.

Lerche NW, Marx PA, Osborn KG, et al. (1987) Natural history of endemic type D retrovirus infection and acquired immune deficiency syndrome in group-housed rhesus monkeys. *Journal of the National Cancer Institute* 79(4): 847–854.

Lerche NW, Switzer WM, Yee JL, et al. (2001) Evidence of infection with simian type D retrovirus in persons occupationally exposed to nonhuman primates. *Journal of Virology* 75(4): 1783–1789.

Linial M and Weiss R (2001) Other human and primate retroviruses. In: Knipe DM and Howley PM (eds.) *Fields Virology*, 4th edn., pp. 2123–2139. Philadelphia: Lippincott Williams and Wilkins.

Marracci GH, Avery N, Shiigi S, et al. (1999) Molecular cloning and cell-specific growth characterization of polymorphic variants of type D serogroup 2 simian retroviruses. *Virology* 261: 43–58.

Marx PA, Maul DH, Osborn KG, et al. (1984) Simian AIDS: Isolation of a type D retrovirus and transmission of the disease. *Science* 223: 1083–1086.

Marx PA, Pedersen NC, Lerche NW, et al. (1986) Prevention of simian acquired immune deficiency syndrome with a formalin-inactivated type D retrovirus vaccine. *Journal of Virology* 60(2): 431–435.

Rose TM, Strand KB, Schultz ER, et al. (1997) Identification of two homologs of the Kaposi's sarcoma-associated herpesvirus (human herpesvirus 8) in retroperitoneal fibromatosis of different macaque species. *Journal of Virology* 71(5): 4138–4144.

Sonigo P, Barker C, Hunter E, and Wain-Hobson S (1986) Nucleotide sequence of Mason–Pfizer monkey virus: An immunosuppressive D-type retrovirus. *Cell* 45(3): 375–385.

Stromberg K, Benveniste RE, Arthur LO, et al. (1984) Characterization of exogenous type D retrovirus from a fibroma of a macaque with simian AIDS and fibromatosis. *Science* 224(4646): 289–292.

Wolfheim JH (ed.) (1983) *Primates of the World*. Washington: University of Washington Press.

Simian Virus 40

A L McNees and J S Butel, Baylor College of Medicine, Houston, TX, USA

© 2008 Elsevier Ltd. All rights reserved.

History

Simian virus 40 (SV40) was discovered in 1960 as a contaminant of poliovaccines. Hundreds of millions of people worldwide were inadvertently exposed to infectious SV40 in the late 1950s and early 1960s when they were administered contaminated virus vaccines prepared in rhesus macaque kidney cells. SV40 had unknowingly contaminated batches of both the inactivated and live attenuated forms of the poliovaccine and preparations of some other viral vaccines. Although primary cultures of monkey cells were known to be commonly contaminated with indigenous viruses and safety testing was carried out, SV40 had escaped detection in part because it failed to induce cytopathic effects in rhesus cells. However, when it was inoculated into African green monkey kidney cells, a prominent cytoplasmic vacuolization developed. Originally christened as 'vacuolating virus', the name was later changed to SV40 to conform to a numerical system of designating simian virus isolates.

Concern about the vaccine contaminations heightened considerably when it was found in 1962 that SV40 was tumorigenic in newborn hamsters and could transform many types of cells in culture. Subsequently, manufacturers treated poliovirus vaccine seed stocks to remove infectious SV40, and screening methods were implemented to increase detection of infectious SV40 in vaccine lots.

Because of the potential risk to public health posed by the previous distribution of contaminated poliovaccines, SV40 became the focus of intensive investigation. For scientists, SV40 has turned out to be an invaluable tool for dissecting molecular details of eukaryotic cell processes. Numerous techniques now commonly used in molecular biology were pioneered in the SV40 system. It continues to serve as a model for basic studies of viral carcinogenesis.

Taxonomy and Classification

Viral species *Simian virus 40* is classified in the genus *Polyomavirus* in the family *Polyomaviridae* (**Table 1**). It was previously classified as a member of the *Papovaviridae* family. The International Committee on Taxonomy of Viruses has designated SV40 as the type species of the 13 members of the family *Polyomaviridae*, which includes murine polyomavirus, human polyomaviruses BK virus (BKV) and JC virus (JCV), as well as isolates from hamsters, rabbits, cattle, birds, and baboons. The human and animal polyomaviruses are antigenically distinct, and in most cases there is only one recognized serotype for each virus.

Properties of the Virion

SV40 particles are small and spherical, with a diameter of approximately 45 nm. Infectious virions have a sedimentation coefficient of 240S in sucrose and band at a density of $1.34\,\text{g}\,\text{ml}^{-1}$ in CsCl; empty capsids have a density of $1.29\,\text{g}\,\text{ml}^{-1}$. The molecular mass of the SV40 virion has been estimated at 270 kDa. The DNA content is 12.5% (w/w). The major capsid protein (VP1) accounts for 75% of the total virion protein. VP2 and VP3 are minor capsid proteins. Cellular histones (H2A, H2B, H3, H4) are used to condense the viral DNA for packaging and are present in the core of the particle. There is no lipid envelope. SV40 does not agglutinate erythrocytes.

SV40 particles exhibit icosahedral symmetry. The virion is composed of 72 pentameric capsomeres composed of the VP1 protein arranged on a $T=7d$ icosahedral surface lattice. This structure (hexavalent capsomeres having pentameric substructure) demands nonequivalent contacts between pentamers. This seems to be accomplished by the C-termini of the VP1 polypeptides, which extend as arms from one pentamer and fit into binding sites on adjacent pentamers, providing the necessary flexibility to build a capsid. The N-terminal arm of VP1 is completely internal in the virus particle. Minor capsid proteins VP2 and VP3 are predominantly internal as well and do not contribute to the basic structure of the virus outer shell.

The virus particles are very resistant to heat inactivation but are relatively labile when heated in the presence of divalent cations. Whereas SV40 is stable at $50\,°C$ for hours, incubation in the presence of $1\,M\,MgCl_2$ at $50\,°C$ for 1 h will effectively inactivate the virus if it is monodispersed. At a higher temperature ($60\,°C$), ~99% of infectious virus is inactivated within 30 min in the absence of divalent cations. Purified virions can be disrupted by strong alkaline conditions (pH 10.5), by lower pH (9.2) plus a reducing agent, or by detergent treatment. Virus particles are resistant to acid treatment (pH 3.0). Intact virus particles are not affected by nucleases, but in the presence of a reducing agent nuclease can enter the virion and cleave the viral DNA. SV40 is efficiently inactivated by UV light irradiation, following single-hit kinetics.

Properties of the Viral Genome

The SV40 genome is a circular, covalently closed, double-stranded (ds) DNA molecule of about 5 kbp (**Figure 1**). The native DNA assumes a superhelical configuration (form I) that sediments at 21S in a neutral sucrose gradient. A single-stranded (ss) nick generates relaxed circular dsDNA molecules (form II) that sediment at 16S, whereas a ds break produces linear dsDNA (form III, 14S). Alkaline denaturation of form I DNA produces dense cyclic coils that sediment at 53S. Form II DNA is converted to ss circular (18S) and ss linear (16S) molecules by denaturation. The supercoiled (form I) molecules can be separated from relaxed circular and linear forms by centrifugation of a DNA preparation in CsCl gradients with ethidium bromide. The form I molecules will band in a lower

Table 1 Properties of SV40

Classification	Family *Polyomaviridae*
Strain variation	Genetically stable; one serotype, multiple strains
Virion	Icosahedral, 45 nm in diameter, no envelope
Genome	Circular covalently closed dsDNA, 5200 bp
Proteins	Three structural proteins, VP1, VP2, VP3; cellular histones condense DNA in virion; nonstructural replication protein, T-antigen, is potent oncoprotein
Replication	In certain primate kidney cells; nucleus; stimulates cell DNA synthesis; long growth cycle
Natural host	Asian macaques, especially the rhesus monkey
Diseases	Asymptomatic persistent infections in natural hosts; tumors in experimentally infected rodents; associated with human tumors and with kidney disease
Historical note	Contaminant in early poliovaccines administered to millions of people

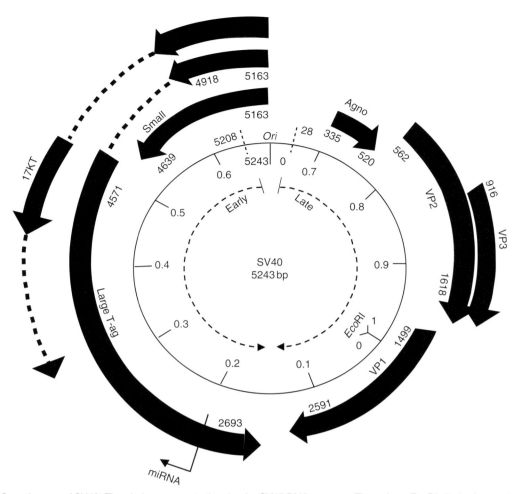

Figure 1 Genetic map of SV40. The circle represents the circular SV40 DNA genome. The unique EcoRI site is shown at map unit 0/1. Nucleotide numbers begin and end at the origin (Ori) of viral DNA replication (0/5243). Boxed arrows indicate the open reading frames that encode the viral proteins. Arrowheads point in the direction of transcription; the beginning and end of each open reading frame is indicated by nucleotide numbers. Note that T-ag is coded by two noncontiguous segments on the genome. The genome is divided into 'early' and 'late' regions that are expressed before and after the onset of viral DNA replication, respectively. Only the early region is expressed in transformed cells.

position in the gradient. The DNA forms also separate during electrophoresis in a neutral agarose gel; the supercoiled molecules migrate the fastest, the linear forms move at an intermediate speed, and the relaxed circles migrate the slowest.

The viral DNA both in virions and infected cells is associated with cellular histones H2A, H2B, H3, and H4. The histones are assembled in 24–26 nucleosomes on the viral DNA. The nucleosome structure and histone composition of the viral minichromosome mimic the chromatin structure of cellular DNA.

SV40 DNA was the first eukaryotic viral genome to be physically mapped by restriction endonuclease analysis (1971) and to be completely sequenced (1978). The DNA of reference strain 776 with a duplicated enhancer region contains 5243 bp for a calculated molecular weight of 3.5×10^6 Da. The genome is numbered in a clockwise direction from 1 to 5243, the central nucleotide of the unique BglI recognition site being assigned as 0/5243. Numbering continues through the late region in the 'sense orientation' and the early region in the 'antisense orientation'. The numbering system begins and ends (0/5243) in the middle of the functional origin of DNA replication. The unique EcoRI site at nucleotide 1782 was arbitrarily chosen as a point of reference and assigned a value of 0/1.0 on the circular map. Laboratory-adapted strains of SV40 contain a duplication or rearrangements of the 72 bp element in the enhancer region, whereas most natural isolates do not.

The SV40 genome has compact regulatory sequences and overlapping genes. The single origin of replication (core Ori = 64 bp in size) is embedded within a nontranslated regulatory region. These elements control transcription and replication and span about 400 bp. The coding regions are expressed early or late in infection, and represent the 'early' nonstructural genes and the 'late' structural genes, respectively, and are transcribed off opposite strands of the viral DNA.

There is a variable domain at the 3' end of the T-ag gene, encompassing about 270 bp. Nucleotide changes within this region can be used to distinguish strains of SV40. Although genetic variation among SV40 strains is minimal, three different genogroups have been distinguished. Sequence variations in this region are detected in human-tumor-associated SV40 sequences. Possible *in vivo* biological differences among SV40 strains are unknown but variations in oncogenic potential in rodent models have been observed.

Properties of Viral Proteins

SV40 encodes seven gene products: three 'early' nonstructural proteins (large T-antigen (T-ag), small t-antigen (t-ag), 17K T-ag (17KT)), three 'late' structural proteins (VP1, VP2, VP3), and a maturation protein (LP1 or agnoprotein).

The nonstructural proteins are expressed early in infection, before the onset of viral DNA synthesis. The coding regions of the two T-ags and 17KT overlap; alternative splicing of viral transcripts determines each protein sequence. Large T-ag of strain 776 (**Table 2**) contains 708 amino acids ($\sim$90 kDa), and small t-ag contains 174 residues ($\sim$20 kDa). The large and small T-ags share 82 N-terminal amino acids, whereas the remainder of each protein is unique. The T-ag/t-ag common exon contains a 'J-domain', believed to modulate hsc70 activity in the assembly and disassembly of multiprotein complexes.

Large T-ag is an essential replication protein required for initiation of viral DNA synthesis. It stimulates host cells to enter S-phase and undergo DNA synthesis and is the SV40 transforming protein. Large T-ag contains a nuclear transport signal (126-Pro-Lys-Lys-Lys-Arg-Lys-Val-132) that targets the protein into the nucleus. However, about 10% of the T-ag in the cell is found in the cytoplasm and the plasma membrane. The biology of small t-ag is enigmatic. It is a cytoplasmic protein that is not essential for viral replication in cultured cells. It associates with the regulatory and catalytic subunits (36 and 63 kDa) of protein phosphatase 2A and is believed to cause cellular growth stimulation. It is required for transformation of some human cells by SV40. Its role during natural infections by SV40 remains to be elucidated. The function of 17KT is unknown.

The functions of large T-ag in SV40 DNA replication are regulated by phosphorylation (**Figure 2**). The sites of phosphorylation are clustered near the ends of the molecule, one region lying between residues 106 and 124 and the other between residues 639 and 701. The majority of the phosphorylated residues are serines, although two threonine residues also become phosphorylated. Unlike many oncoproteins, T-ag is not phosphorylated at tyrosine residues.

T-ag is a DNA-binding protein that recognizes multiple copies of the sequence GAGGC in three T-ag-binding sites in the viral *Ori*. The minimal origin-specific DNA-binding domain of T-ag lies between residues 131 and 259. T-ag is predicted to have a zinc finger motif, typical of DNA-binding proteins, between amino acids 302 and 320. T-ag-specific ATPase and helicase activities are required in addition to DNA-binding activity in order for T-ag to function in initiation of DNA replication. The ATP-binding domain of T-ag is similar in structure to other ATP-binding proteins and is located between residues 418 and 627.

Large T-ag forms complexes with several cellular proteins. Such interactions are involved in T-ag functions in viral DNA replication, induction of cellular DNA synthesis, and cell-cycle progression. Target cellular proteins found in heterooligomeric structures with T-ag include transcriptional coactivators (CBP, p300, p400), tumor suppressor proteins (p53, pRb, p107, p130), DNA polymerase α, the molecular chaperone heat-shock protein hsc70, cell-cycle regulatory proteins cdc-2 and cyclin, and tubulin. The

Table 2 Properties and functions of SV40 T-ag

Structural properties	
Size[a]	708 amino acid residues, 82 N-terminal residues shared with t-ag, 81 632 Da, M_r 90 000–100 000 Da
Modifications	Phosphorylation, N-terminal acetylation, *O*-glycosylation, poly-ADP-ribosylation, palmitylation, adenylation
Supramolecular structure	Zinc finger, nuclear localization signal, J domain, monomers, dimers, higher homooligomers; heterooligomers with transcriptional coactivators (CBP, p300, p400); heterooligomers with DNA polymerase α; hsc70; cdc-2, cyclin, tubulin, Bub1, TEF-1, Cul7, Nbs1, Fbw7; tumor suppressor proteins (p53, pRb, p107, p130)
Subcellular distribution	Predominantly nuclear
Functions	
Replication and transactivation	Specific DNA binding (viral origin of replication), initiation of viral DNA replication; ATPase activity, helicase activity; autoregulation of viral early transcription, induction of viral late transcription
Host cell effects and transformation	Entry of cells into S phase and initiation of cellular DNA replication, complex formation with cellular proteins p53, pRb, p107, p130; adenovirus helper function; effect on host range, initiation and maintenance of cellular transformation, induction of immunity to SV40 tumor cells, target for cytotoxic T cells

[a]SV40 strain 776.

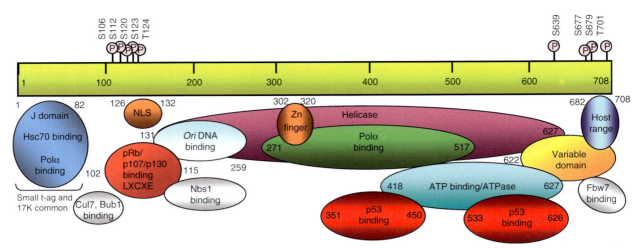

Figure 2 Functional domains of SV40 large T-ag. The numbers given are the amino acid residues using the numbering system for SV40-776. Regions are indicated as follows. Small t-ag common: region of large T-ag encoded in the first exon. The amino acid sequence in this region is common to both large T-ag and small t-ag. Polα binding: regions required for binding to polymerase α-primase. Hsc70 binding: region required for binding the heat shock protein hsc70. Cul7 binding: region required for binding of Cul7. pRb/p107/p130 binding: region required for binding of the Rb tumor suppressor protein, and the Rb-related proteins p107 and p130. NLS: contains the nuclear localization signal. Ori DNA binding: minimal region required for binding to SV40 Ori DNA. Nbs1 binding: region required for binding of Nbs1. Helicase: region required for full helicase activity. Zn finger: region which binds zinc ions. p53 binding: regions required for binding the p53 tumor suppressor protein. ATP binding/ATPase: region containing the ATP binding site and ATPase catalytic activity. Host range: region defined as containing the host range and Ad helper functions. Variable domain: region containing amino acid differences among viral strains. Fbw7 binding: region required for Fbw7 binding. The circles containing a P indicate sites of phosphorylation found on large T-ag expressed in mammalian cells. S indicates a serine and T indicates a threonine residue.

indicated cellular proteins are not all found in the same T-ag-associated complex; many subpopulations of T-ag exist in a cell.

The variable domain at the C-terminus of T-ag encompasses the host range–adenovirus helper function exhibited by SV40 (and mapped to T-ag) in some monkey kidney cell lines. The significance of the variable domain to natural infections by the virus is unknown.

The structural (capsid) proteins are expressed late in infection, after the onset of DNA replication. They are synthesized in much greater abundance than the early proteins. The major capsid protein, VP1, contains 362 amino acids (~45 kDa). The minor structural proteins are VP2 (352 residues, ~38 kDa) and VP3 (234 residues, ~27 kDa). The coding regions for VP2 and VP3 overlap, and they are translated in the same reading frame, so the sequence of VP3 is identical to the C-terminal two-thirds of VP2. VP3 is synthesized by independent initiation of translation via a leaky scanning mechanism, not proteolytic cleavage of VP2. The N-terminal portion of VP1 is derived from sequences that encode the C-termini of VP2 and VP3. However, VP1 is translated in a different reading frame from a different spliced transcript, so it shares no sequences with VP2 and VP3. VP1 is modified by phosphorylation and acetylation.

The late proteins are required only for the assembly of progeny virions during lytic infection. They are not involved in the early phases of viral replication. They are synthesized in the cytoplasm and move into the nucleus where particle morphogenesis occurs. The minor capsid proteins contain nuclear transport signals. The VP2/3 signal is Gly-Pro-Asn-Lys-Lys-Lys-Arg-Lys-Leu (VP2, residues 316–324; VP3, residues 198–206). For VP1, two clusters of basic residues within the N-terminal 19 residues are independently important for nuclear targeting. Mutations in VP1 affect capsid assembly and/or virion stability. Mutations in VP2 and VP3 affect the uncoating process when virions penetrate new host cells.

The agnoprotein LP1 is synthesized late in infection but is not found in virus particles. It is a small (62 residue, ~8 kDa) basic protein involved in particle assembly. It is believed that LP1 interacts with VP1 molecules to inhibit self-polymerization until they interact with viral minichromosomes in the nucleus to form virions.

Replication

Overview of SV40 Replication Cycle

The SV40 replication cycle is cleanly divided into early and late events, with the onset of viral DNA replication being the dividing landmark. SV40 virions attach to receptors on the cell surface, become internalized, and are transported to the cell nucleus where the viral DNA is uncoated. After uncoating, the half of the genome that contains the early region is transcribed ('early' mRNAs).

Viral early proteins (T-ags) are synthesized, cellular genes are expressed, and the cells enter S-phase. Viral DNA replication then begins. 'Late' mRNAs are transcribed from the other half of the viral genome (the opposite strand), and viral structural proteins are synthesized. Virus particles are assembled and SV40 is released from the cell surface in a manner dependent on intracellular vesicular transport. The SV40 multiplication cycle can take 24–72 h to complete. New particles are detected by 24 h. The time course of the virus growth cycle is dependent upon the virus strain and host tissue, the viral multiplicity of infection, and the growth state of the host cell at the time of infection.

Strategy of Replication of Nucleic Acid

SV40 DNA is replicated in the cell nucleus as a free unintegrated minichromosome. The only viral components required are the viral origin of replication on the DNA and the T-ag protein; all other factors are provided by the host cell replication machinery. T-ag is required for the initiation of DNA replication. The specific T-ag functions required are its DNA-binding ability and its ATPase/helicase activities. The relative simplicity of the SV40 system has allowed the development of cell-free replication systems and the identification of factors involved in mammalian DNA replication.

T-ag binds to the viral *Ori*, a 64 bp segment that contains binding site II for T-ag. In an ATP-dependent process, T-ag causes localized unwinding of the *Ori* region; cellular ss binding protein is required to stabilize the unwound single strands. The cellular DNA polymerase α-primase complex initiates DNA replication, and replication proceeds bidirectionally, with the two forks advancing at equal rates. Elongation involves DNA polymerase α, DNA polymerase δ, and proliferating cell nuclear antigen. Termination occurs 180° away from the viral *Ori*; topoisomerase II segregates the newly synthesized daughter molecules. Cellular histones are added to the new strands during the process of DNA replication.

Replication of SV40 DNA occurs in certain cell types of humans, monkeys, and possibly hamsters, and this permissiveness seems to depend in part on the nature of the DNA polymerase α-primase complex.

Characterization of Transcription

Transcription of the viral DNA is carried out by the cellular RNA polymerase II. In the noncoding region of SV40 DNA near the origin of replication are early and late promoter structures and enhancer elements. Early transcription begins at about nucleotide 5237, proceeds in the counterclockwise direction, and ends at the polyadenylation site at nucleotide 2694. The early promoter contains a TATA box about 30 bp upstream of the early RNA initiation site. (This start site is about 70 nucleotides upstream of the initiation codon shared by the early proteins.) There are three G + C-rich regions, the '21 bp repeats', located 40–103 nucleotides upstream, which are binding sites for the Sp1 cellular factor. Even farther upstream are the SV40 enhancer elements, the 72 bp elements, which contain binding sites for other cellular factors that regulate transcription. The primary early transcripts are differentially spliced to generate the mRNAs that code for the three T-ags.

There is no requirement for virus-encoded proteins, but early transcription is regulated by T-ag. T-ag regulates its own synthesis as it first binds to site I and then to sites II and III on the viral DNA. The presence of T-ag at site II blocks the binding of RNA polymerase.

Late transcription begins after viral DNA synthesis is underway. The abundance of late transcripts is much greater than the early transcripts because progeny DNA molecules are utilized as templates. A heterogeneous collection of late mRNAs is made, with late transcription beginning at multiple sites between nucleotides 120 and 482 and proceeding in the clockwise direction, ending at a polyadenylation site at nucleotide 2674. Both the 21 bp repeats and the 72 bp elements have positive effects on late transcription. The late transcripts are alternatively spliced into two size classes (19S, 16S). VP1 is synthesized from 16S RNA, and both VP2 and VP3 are translated from the 19S species. The agnoprotein is synthesized predominantly from the most abundant species of 16S RNA.

SV40 microRNAs (miRNAs) have been described that accumulate during the late phase of the infection cycle and are complimentary to early viral mRNAs. The SV40 miRNAs function to target the early mRNAs for cleavage, which effectively reduces the expression of T-ag.

Post-Translational Processing

No post-translational cleavages are involved in the production of SV40 proteins. As noted above, T-ag and VP1 are modified in various ways, including phosphorylation.

Uptake and Release of Virions

The attachment of SV40 particles to the cell surface is mediated by VP1. The major histocompatibility class I molecules and the ganglioside GM1 molecules function as receptors. Attached particles are internalized by caveolae-mediated endocytosis into vesicles that contain caveolin-1. These vesicles fuse with caveosomes and the virus particles are transferred to the endoplasmic reticulum, from which they exit and then enter the nucleus. Conformational changes are thought to occur that expose the nuclear localization signals on capsid proteins and allow the virions to squeeze through the nuclear pore complex. The capsid disassembles in the nucleus, releasing the viral DNA.

Maturation of progeny virions occurs in the nucleus, where the viral nucleic acid is replicated. Viral proteins are synthesized in the cytoplasm off viral transcripts exported from the nucleus, and the proteins are then transported back into the nucleus. The structural proteins condense around the viral minichromosomes. There is a packaging signal on SV40 DNA that includes the *Ori* and part of the enhancer element. During the maturation process, the agnoprotein is released and is not retained as a component of mature virions. There are size constraints for packaging DNA – molecules ranging from 3.5 to 5.7 kbp can be encapsidated into SV40 particles.

Some progeny virions are released from the surface of infected cells via a mechanism dependent on intracellular vesicular transport, but the majority stay associated with the cell until cell death. The release of virus from ruptured and fragmenting cells may also be a mechanism of virus exit from infected cells. SV40 infections are not lytic and host cells are killed as the result of a variety of effects, including the release of lysosomal enzymes into the cytoplasm and damage to the cell mitochondria. Late in infection, monkey kidney cells develop a characteristic cytopathic effect, cytoplasmic vacuolization. As many as 10^4 virus particles can be produced by an infected cell, although some cells produce fewer particles.

Geographic and Seasonal Distribution

The geographic distribution of SV40 can only be inferred, as no comprehensive surveys have been conducted. SV40 is found naturally in wild populations of certain Asian macaque species, and its geographic distribution in the wild presumably reflects its narrow host range. Infections in humans are more widespread geographically, possibly because contaminated poliovaccines were broadly distributed. Nothing is known about seasonal effects on natural infections by SV40.

Host Range and Virus Propagation

Polyomaviruses, in general, have a narrow host range, with each virus infecting only one or a few closely related species. Based on antibody surveys of wild populations of primates, the natural hosts for SV40 appear to be a few species of Asian macaque monkeys, especially the rhesus (*Macacca mulatta*). In captivity, several related species are easily infected, including the cynomolgus macaque (*M. fascicularis*) and the African green monkey, which belongs to the same family as macaques (Cercopithecidae). The virus grows poorly in more distantly related primates. SV40 can infect humans.

SV40 is propagated in tissue culture in established cell lines derived from kidneys of African green monkeys. Characteristic vacuolated cells appear in response to viral replication. The virus grows in rhesus kidney cell lines in which it establishes a persistent infection but produces no cytopathic effects.

SV40 typically does not cause tumors in its natural hosts. To demonstrate its tumorigenic potential, the virus must be inoculated into experimental animals (newborn hamsters are most susceptible). Many types of cells can be transformed in culture, including those of rodent, monkey, and human origin.

Genetics

SV40 is genetically stable, although rearrangements in the regulatory region can occur *in vivo*. Sequence variations exist at the $3'$ end of the T-ag gene among different isolates, which are stable on passage *in vivo* and *in vitro*. Adaptation of natural isolates to tissue culture often involves the selection of viruses with duplications or rearrangements in the viral regulatory region. The origins of several SV40 strains are listed in **Table 3**.

Serologic Relationships and Variability

Only one serotype of SV40 is known. The virus does not undergo noticeable antigenic variation. Perhaps restrictions imposed by the symmetry of the capsid permit only minimal deviation in amino acid sequence of the structural proteins, making most changes lethal for the virus.

There is a genus-specific antigenic determinant on the major capsid protein, VP1, that is shared by all animal and human polyomaviruses. It is expressed in infected cells and is internal in the virion. Antibodies are elicited against it by immunization with disrupted capsids or with purified VP1 protein. Antibodies against the shared determinant are not neutralizing, as the site is not exposed on the surface of virus particles. The structural proteins of SV40 and the two human polyomaviruses (BKV and JCV) are antigenically distinct, but display some cross-reactive determinants in enzyme-linked immunosorbent assay (ELISA) tests. The T-ags of SV40, BKV and JCV show antigenic cross-reactivity.

Epidemiology

Most adults of the Asian macaque species believed to be natural hosts for SV40 have neutralizing antibodies to the virus. Few of the juvenile animals of those species, in the wild, have antibodies. However, in captivity the young

Table 3 Origin of SV40 strains

Virus strain	Year isolated	Source
SV40-776[a]	1960	Adenovirus type 1 seed stock prepared in monkey kidney cells
VA45-54	1960	Uninoculated rhesus kidney cells
Baylor	1961	Type 2 Sabin oral poliovaccine prepared in 1956 in monkey kidney cells
A2895	c. 1961	Tumor from hamster injected with rhesus monkey kidney cells
777	1962	Inactivated poliovaccine
Rh911	1962	Uninoculated rhesus monkey kidney cells
N-128	1965	Uninoculated rhesus monkey kidney cells (Russia)
SVPML-1	1970	Cultured human brain cells from patient with progressive multifocal leukoencephalopathy
SVMEN[b]	1984	Human meningioma (cloned directly, Germany)
SVCPC[b]	1995	Human choroid plexus carcinoma
SV40-K661	1998	Brain from rhesus monkey coinfected with simian immunodeficiency virus (SIV)
SV40-H328[a]	1998	Brain from rhesus monkey coinfected with SIV
SV40-T302	1998	Brain from rhesus monkey coinfected with SIV
SV40-I508	1998	Brain from rhesus monkey coinfected with SIV

[a]SV40-776 and SV40-H328 are identical except for differences in the viral regulatory region.
[b]SVMEN and SVCPC are identical.
Data taken from Forsman ZH, Lednicky JA, Fox GE, et al. (2004) Phylogenetic analysis of polyomavirus simian virus 40 from monkeys and humans reveals genetic variation. *Journal of Virology* 78: 9306–9316.

animals are readily infected if they have contact with a virus-positive animal.

Serologic surveys have detected SV40 neutralizing antibodies in humans with prevalences ranging from 2% to 10%. However, whether SV40 infection typically induces a detectable and sustained humoral or cellular immune response in humans has not been determined. SV40 DNA has been detected in human tumors, but epidemiological studies based on SV40-reactive serum antibodies in ELISA tests do not support an association of SV40 exposure with human cancers. The results of such epidemiological studies should be interpreted with caution, however, due to an inability to differentiate with certainty those who were exposed to an SV40-contaminated vaccine and those who were not. The working stocks in use between 1961 and 1978 by a major eastern European manufacturer of poliovirus vaccines have been shown to contain infectious SV40. This finding raises questions as to whether all poliovirus vaccines used worldwide after 1963 were free from SV40 contamination.

Transmission and Tissue Tropism

SV40 establishes persistent infections in the kidneys, and possibly lymphocytes, of susceptible hosts. The level of persistent virus present may be very low. Modes of transmission are not known, but transmission probably occurs due to virus shed in the urine or stool. Experiments have established that susceptible animals can be infected by the oral, respiratory, or subcutaneous routes. Both viremia and viruria occur in infected animals. SV40 may cause neurologic disease in immunocompromised hosts.

The major known source of human exposure to SV40 was via the administration of contaminated viral vaccines before SV40 was recognized. Human exposure could also occur by contact with infected monkeys, a situation limited to small numbers of animal handlers. Transmission between human hosts is hypothesized to occur but has not been documented. It is presumed that patterns of tissue tropism and transmission similar to those described in monkeys would be observed in humans infected by SV40.

Pathogenicity and Pathology

SV40 infections in normal monkeys appear to be asymptomatic and harmless. However, SV40 has been associated with a fatal case of pulmonary and renal disease, as well as with cases of progressive multifocal leukoencephalopathy, in unhealthy rhesus monkeys. SV40 can cause widespread infections in monkeys suffering from simian acquired immune deficiency syndrome and has been found in a brain tumor; no tumors have been found in immunocompetent, natural hosts. Transgenic mice carrying wild-type SV40 DNA develop choroid plexus papillomas and die rapidly because of the physiological importance of the tumor site. When foreign tissue-specific regulatory sequences are substituted for the native promoter-enhancer of the virus, SV40 expression can be directed to other tissues in transgenic animals and lethal tumors usually appear. Intraperitoneal inoculation of SV40 into weanling hamsters produces mainly mesotheliomas, intravenous inoculation of SV40 leads to leukemia, reticulum cell sarcoma, and osteogenic sarcoma, and subcutaneous inoculation induces undifferentiated carcinomas or sarcomas. SV40 DNA has been detected in several types of human cancers, including brain tumors (especially those from children in the first decade of life), mesotheliomas, osteosarcomas, and non-Hodgkin's lymphomas. SV40 DNA is sometimes found in tumors arising in persons too young to have been exposed to the contaminated vaccines in use between 1955 and 1963. The role SV40 may have played in the induction of those tumors is under investigation.

Immune Response

SV40, like other members of the genus *Polyomavirus*, induces an asymptomatic, persistent infection in natural hosts. An antibody response to capsid antigen is elicited that can be detected in neutralization assays. It is well documented with the human viruses BKV and JCV that impaired cell-mediated immunity is associated with virus re-activation, showing that viral replication is under the influence of the immune system of the host; the same presumably applies to SV40.

Little is known about the immune response of humans to infection by SV40. Small numbers of individuals exposed to contaminated vaccines were analyzed for neutralizing antibody responses to SV40. Humoral responses were detected in some vaccinees and were variable and dependent on the size of inoculum and route of inoculation. Recent serological surveys have detected SV40 neutralizing antibody in 2–10% of persons not exposed to SV40-contaminated viral vaccines. Antibodies to SV40 were most often detected in people with some type of immune suppression. ELISA-based assays detected some cross-reactive antibodies against SV40, BKV, and JCV. SV40 T-ag specific cellular immune response has been detected in some patients with SV40 DNA-positive tumors.

Experimental studies have shown that animals with active infections by SV40 may produce humoral antibodies against the replication oncoprotein, T-ag. It should be noted that a T-ag antibody response could not be used to monitor SV40 infections in humans because of the cross-reactivity among the T-ags of SV40, BKV, and JCV.

In vitro and *in vivo* studies demonstrate that SV40 tumor-bearing rodent animals develop a strong immune response to T-ag. Both humoral and cell-mediated responses occur and are sufficient to prevent tumor growth in some cases. In studies using the Syrian golden hamster model, inoculated animals frequently produced virus-neutralizing antibody and T-ag antibody responses, and the titers of both tended to be higher in tumor-bearing animals. In murine models, cytotoxic T cells directed against T-ag determinants at the cell membrane limit tumor progression.

Interferon is induced only weakly by the polyomaviruses and is not thought to be an important component of the host response to SV40.

Prevention and Control

No control measures are available to prevent SV40 infection.

Future Perspectives

The reports of antibodies to SV40 in humans and the infrequent association of SV40 markers with human tumors suggest that SV40 may be present in the human population. It is important to determine the natural history of SV40 in humans, including modes of transmission and factors affecting susceptibility to infection. Because of its small genetic content and dependence on host cell functions, SV40 will continue to be a useful model system for discerning mechanisms of cellular processes, such as mammalian cell DNA replication, cell cycle progression, and growth control processes altered in neoplasia.

See also: Polyomaviruses of Humans; Polyomaviruses of Mice.

Further Reading

Ahuja D, Sáenz-Robles MT, and Pipas JM (2005) SV40 large T antigen targets multiple cellular pathways to elicit cellular transformation. *Oncogene* 24: 7729–7745.

Cole CN and Conzen SD (2001) *Polyomaviridae*: The viruses and their replication. In: Knipe DM, Howley PM, Griffin DE, et al. (eds.) *Fields Virology*, 4th edn., pp. 2141–2174. Philadelphia: Lippincott.

Cutrone R, Lednicky J, Dunn G, et al. (2005) Some oral polio vaccines were contaminated with infectious SV40 after 1961. *Cancer Research* 65: 10273–10279.

Dang-Tan T, Mahmud SM, Puntoni R, and Franco EL (2004) Polio vaccines, simian virus 40, and human cancer: The epidemiologic evidence for a causal association. *Oncogene* 23: 6535–6540.

Forsman ZH, Lednicky JA, Fox GE, et al. (2004) Phylogenetic analysis of polyomavirus simian virus 40 from monkeys and humans reveals genetic variation. *Journal of Virology* 78: 9306–9316.

Gazdar AF, Butel JS, and Carbone M (2002) SV40 and human tumours: Myth, association or causality? *Nature Reviews Cancer* 2: 957–964.

Hahn WC, Dessain SK, Brooks MW, et al. (2002) Enumeration of the simian virus 40 early region elements necessary for human cell transformation. *Molecular and Cellular Biology* 22: 2111–2123.

Liddington RC, Yan Y, Moulai J, et al. (1991) Structure of simian virus 40 at 3.8-Å resolution. *Nature* 354: 278–284.

Schell TD and Tevethia SS (2001) Control of advanced choroid plexus tumors in SV40 T antigen transgenic mice following priming of donor CD8$^+$ T lymphocytes by the endogenous tumor antigen. *Journal of Immunology* 167: 6947–6956.

Shah K and Nathanson N (1976) Human exposure to SV40: Review and comment. *American Journal of Epidemiology* 103: 1–12.

Stewart AR, Lednicky JA, and Butel JS (1998) Sequence analyses of human tumor-associated SV40 DNAs and SV40 viral isolates from monkeys and humans. *Journal of Neurovirology* 4: 182–193.

Vilchez RA and Butel JS (2004) Emergent human pathogen simian virus 40 and its role in cancer. *Clinical Microbiology Reviews* 17: 495–508.

Smallpox and Monkeypox Viruses

S Parker, Saint Louis University School of Medicine, St. Louis, MO, USA
D A Schultz, Johns Hopkins University School of Medicine, Baltimore, MD, USA
H Meyer, Bundeswehr Institute of Microbiology, Munich, Germany
R M Buller, Saint Louis University School of Medicine, St. Louis, MO, USA

© 2008 Elsevier Ltd. All rights reserved.

Glossary

Enanthem Eruptive lesion of mucous membranes.
Exanthem Eruptive lesion of skin.
Papule A small, solid, elevated lesion of the skin that is often inflammatory.
Pock A pustule on the body caused by an eruptive disease.
Pustule A small elevated pus-containing lesion of skin.
Vesicle A small cavity within the epidermis containing serous fluid.

Variola Virus

Smallpox is caused by variola major virus (VARV), which is a member of family *Poxviridae*, subfamily *Chordopoxviriniae*, genus *Orthopoxvirus*. VARV is a 200–250 nm brick-shaped enveloped virus with a double-stranded DNA genome of approximately 186 kbp. Compared with other orthopoxviruses, VARV exhibits gene conservation toward the center of the genome, with genetic variation increasing toward the termini. The genes in the terminal regions appear to encode virulence factors that differ among orthopoxviruses. Virus replication occurs in the cytoplasm of the host cell where intracellular mature virions (IMVs) and extracellular enveloped virions (EEVs) are produced. Humans are the only natural host of VARV, although monkeys can be infected when exposed to artificially high doses and baby mice and rabbits can briefly propagate the virus. For additional information regarding the replication and architecture of VARV, the reader is refered to the article on vaccinia virus in this encyclopedia. Vaccinia virus is an extensively studied orthopoxvirus that has many biological similarities to VARV.

History

Smallpox, so named to differentiate it from great-pox (*syphilis*), was described by Edward Jenner as "the most dreadful scourge of the human species." Although the exact number of deaths during Jenner's time is unknown, it is estimated to have been approximately 400 million in the twentieth century alone. Historically, smallpox has had a close association with humans. The origin of VARV remains unknown, but the dubious accolade probably goes to Egypt or India. Unmistakable descriptions of smallpox were documented in fourth-century China, seventh-century India, and tenth-century Mediterranean and southwestern Asia. Moreover, Egyptian mummies buried over 3000 years ago have skin lesions that are consistent with smallpox. Before the fifteenth century, smallpox was generally confined to the Eurasian landmass. However, European colonists introduced smallpox to the Americas, central and southern Africa, and Australia between the fifteenth and eighteenth centuries with devastating consequences, as indigenous populations were decimated with case–fatality rates approaching 90%. Smallpox enabled a handful of conquistadors such as Cortez and Pizarro to subjugate large parts of central and South America to Spanish rule, thereby permanently altering the future of these regions. This was not an isolated pattern.

By the end of the nineteenth century a milder and less lethal form of smallpox, named variola minor, became apparent. This virus was first documented in South Africa during 1904, but had been clinically apparent in the USA since 1896. Originally described as Amass (alastrim in South America), this virus eventually became recognized in Brazil during the 1960s and in Botswana, Ethiopia, and Somalia during the 1970s. The variola minor derivatives of variola major (classical smallpox) are believed to have originated in several places throughout the globe as the virus adapted to humans. The case–fatality rates for variola major were 16–30% and 1% for variola minor.

Clinical Features

Clinically, smallpox in an unvaccinated person has a 7–19 day incubation period from the time infection is established within the respiratory tract until the first symptoms of fever, malaise, headache, and backache occur, culminating in the start of the characteristic rash. The rash starts with papules, which sequentially transform into vesicles and then pustules; a majority of these lesions are located on the head and limbs (often confluent) compared to the trunk. The rash is typically centrifugal (head and limbs), but centripetal (trunk) rashes have been reported. Lesions are 0.5–1 cm in diameter and can spread over the entire body. Once pustules have dried, scabs will form which eventually desquamate during the following 2–3 week period. The resultant feature of these cutaneous lesions is the formation of the classic pock scar that is apparent on the skin of surviving patients.

Two clinical variations of smallpox have been identified. Flat-type smallpox is a rare form of the disease (about 6% in unvaccinated people), and is characterized by lesions that remain level with the skin. It was more frequently observed in children and usually resulted in death. Another variation of the disease is hemorrhagic smallpox (<2% in unvaccinated people), which occurred mainly in adults. Although this was a rare form of the disease, it also had a high mortality rate and is characterized by hemorrhages into the skin and/or mucous membranes early in the course of illness. Subconjuctival hemorrhages were most common as well as bleeding from the gums and other parts of the body.

Although smallpox is typically spread by respiratory droplets over a short distance, some examples of long-distance transmission are evident. One such case occurred in 1978 at the University of Birmingham, UK, where a woman who was vaccinated 12 years earlier died of smallpox. She is identified as the last human fatality of the disease. It is widely believed that VARV traveled up through an air duct that connected a smallpox virology laboratory to her work station. Other theories suggest that this woman was exposed by using the laboratory telephone or simply from laboratory personnel. Another case occurred in a hospital at Meschede, Germany in 1970. In this case, a recent returnee from Pakistan is believed to have initiated 19 other cases of smallpox on all levels of a large general hospital, despite being isolated for the 5 days of his stay. Factors that enabled VARV to travel long distances in the hospital were likely: a building design that facilitated strong rising air currents when it was heated, the patient's severe cough, and the humidity level in the hospital.

Variolation and Vaccination

Historically, it was understood that humans who survived an initial smallpox infection never developed the disease again. Furthermore, persons infected with VARV by cutaneous scratches suffered a less severe form of the disease. For these reasons, the practice of inoculating naive persons with pustular fluid collected from smallpox victims became a common practice; this type of inoculation was called variolation. Variolation usually induced a milder form of the disease, which was typified by a severe, local, cutaneous lesion at the site of variolation with smaller satellite cutaneous lesions; however, variolation sometimes led to generalized rashes with associated deaths.

Variolation was likely developed in both India and China and was subsequently introduced to Egypt and the rest of Africa in the seventeenth century and Europe and its colonies in the eighteenth century.

By the end of the eighteenth century, variolation had been widely accepted throughout the world as a means to prevent smallpox. The widespread use of variolation in Great Britain and her North American colonies widely reduced the impact of the virus in the upper classes but not in the general population as a whole. Despite its successes, the mortality rate and the frequent development of classical smallpox in many patients, with associated disease transmission, meant that variolation was less than ideal. Fortunately, a solution emerged from the common observation that milkmaids were rarely susceptible to smallpox. This lack of susceptibility was widely attributed to a zoonotic disease, cowpox. Based on these observations, Edward Jenner inoculated a boy with cowpox virus and observed the child's resistance to smallpox. Over time, more children were inoculated, exposed, and subsequently resisted smallpox. By the beginning of the nineteenth century this method of vaccination (*vacca*, Latin for cow) had become accepted, as it afforded the same level of protection as variolation without the associated risks of mortality and transmissibility. In some countries (France in particular), vaccination was substituted with orthopoxviruses causing horsepox (a method called equination). Moreover, in a recent study it was found that the causative agent of horsepox is closely related to the vaccinia virus strains derived from the historic smallpox vaccine, supporting the hypothesis that horsepox, or close relatives, replaced cowpox virus as the preferred virus used for worldwide vaccination at some point in the mid-nineteenth century. By the 1950s, endemic smallpox had been eliminated from most industrial nations.

All smallpox vaccines used during the eradication campaign were prepared from vaccinia virus; however, this vaccine was not without complications. In some cases, atypically severe lesions developed, coupled with severe symptoms and occasionally death. The most common complications of vaccination were noted in persons with eczema, where the eczematous region became rapidly inflamed and necrotic with frequent spread of the virus to healthy tissue. Immunocompromised individuals also presented with complications because the vaccination site failed to heal and secondary lesions appeared and spread; this complication was typically fatal.

Smallpox Eradication

In May 1959 the World Health Assembly tasked the World Health Organization (WHO) with the goal of eradicating smallpox globally. Although this was not the first eradication program, it was the only successful one.

Approximately 20 years later, in 1980, the WHO declared that smallpox had been eradicated from the world. Despite plans to destroy all stocks of the virus by the end of the twentieth century, none have been realized. Their preservation is due to the stated need (among others) to develop and evaluate antiviral agents, as VARV is a potential bioweapon. The virus which once dominated the health of mankind is categorized as a bio-safety level (BSL)-4 agent, and virus collections are still kept officially in two BSL-4 WHO reference laboratories in the USA and Russia (as of 2007). Any work with live VARV has to be approved by a WHO advisory committee. The genome sequences of 45 VARV strains are freely available online (see the 'Relevant website' section).

Monkeypox Virus

Monkeypox virus (MPXV) is also a member of the genus *Orthopoxvirus*. The MPXV virion is consistent in structure with other orthopoxviruses, that is, a 200–250 nm brick-shaped, enveloped virus with characteristic surface tubules and a dumbbell-shaped core (**Figure 1(b)**). Its genome is approximately 199 kbp of double-stranded DNA. Orthopoxviruses have host specificities ranging from narrow (e.g., VARV and ectromelia virus) to broad (e.g., cowpox and vaccinia viruses), and the capability of MPXV to infect rodents, nonhuman primates, and humans places it in the latter group. MPXV and VARV cause similar diseases in humans, although they are distinct viruses.

History

MPXV was first isolated in 1958 from the vesiculo-pustular lesions found on infected cynomolgus macaques imported to the State Serum Institute of Copenhagen, Denmark. During the next few years, similar outbreaks were reported in monkey colonies in the USA and in a zoo in Rotterdam, The Netherlands. In the latter case, the first animals affected were giant anteaters from South America, but the disease spread to various species of apes and monkeys. The viruses isolated from these animals were found to be similar to each other and to represent a species of orthopoxvirus that had not been described previously. Currently, MPXV is classified as a BSL-3 agent for animal studies.

Human Monkeypox

MPXV remained primarily of academic interest throughout most of the 1960s. Attitudes changed radically when it was realized that MPXV could infect humans in known smallpox-free locales. This gave rise to concerns that MPXV could fill the niche vacated by VARV. However, a WHO-driven campaign suggested that this was unlikely. It was generally assumed that MPXV infections in humans had been occurring before VARV was expunged, but that they were masked under the guise of smallpox.

The most severe human MPXV infections have been reported in the Congo Basin area of Africa, whereas attenuated human infections have generally occurred in West African countries. Human infections usually result from handling MPXV-infected animals (bush-meat); however, cases of human-to-human transmission have been reported. In 2003, an MPXV outbreak occurred when MPXV-infected West African rodents were imported into the USA and thereafter infected native prairie dogs destined for sale in the pet industry. Human infections were initiated by several routes, which appeared to affect the clinical manifestations of the disease. No fatalities occurred, but the virus was of the less aggressive West African type (see below). Nevertheless, this incident demonstrated the ease with which MPXV can penetrate the interspecies barrier.

Clinical Features

MPXV-infected humans develop a skin rash and follow a disease course similar to that observed in smallpox victims (see above). However, some differences exist between smallpox and human monkeypox. First, humans infected with MPXV frequently present with severely swollen lymph nodes (lymphadenopathy of the neck, inguinal and axillary regions), which is not clinically apparent with smallpox victims. Second, a hemorrhagic form of monkeypox has not been reported (however, it has been reported in some laboratory-housed African dormice infected with MPXV). Interestingly, humans infected with MPXV typically present with a rash similar to that observed in less severe cases of smallpox; to quantify this somewhat, approximately 58% of smallpox patients and 11% of human monkeypox (Congo Basin strain) cases had >100 pocks, respectively.

Epidemiology of Human Monkeypox

In humans, MPXV is a zoonotic infection that has limited capacity to transmit within the population. Between 1970 and 1979, only 47 cases of human monkeypox were reported in five African countries. Most (81%) of the cases were in the Democratic Republic of the Congo (DRC), and mathematical modeling experiments concluded that MPXV could not sustain itself in the unvaccinated human population without zoonotic amplification. Conversely, between 1970 and 2005, 2131 cases were reported in 12 countries. Most (94%) of the cases were in the DRC, and many were not confirmed in the laboratory. The majority of cases were reported in villages within, or close to, the tropical rainforest. African children were the most affected, with a 10% case–fatality rate. The reason for the upsurge, particularly in Africa, is unknown, but it has been suggested that a number of reported cases were

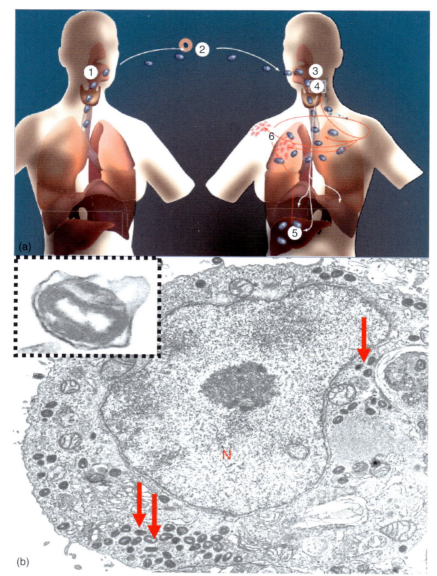

Figure 1 (a) Schematic of the natural life cycle of MPXV. (1) Release of the virus from the original host via the oropharyngeal mucosa. (2) Aerial dispersion of virus particles. (3) Seeding of the respiratory mucosa in a new host. (4) Initiation of replication and neutralization of the host's immune response. (5) Primary viremia and infection of internal organs and lymphatic system (white arrows). (6) Secondary viremia and development of exanthem and enanthem (red arrows). (b) A transmission electron micrograph (approximately ×10 000) of a BSC-1 cell infected with MPXV (red arrows). Virus replication occurs in the cytoplasm (not the nucleus, N). The inset image is an MPXV EEV particle (approximately ×165 000). Note the biconcave core and loose-fitting outer membrane. Adapted from Parker S, Nuara A, Buller RML, and Schultz DA (2007) Human monkeypox: An emerging zoonotic disease. *Future Microbiology* 2: 17–34, with permission from Future Medicine Ltd.

actually cases of chickenpox, caused by varicella-zoster virus (a herpesvirus). A likely contributing factor is the cessation of smallpox vaccination by the WHO *c.* 1980, because recent vaccination with vaccinia virus is 85% effective at protecting against severe MPXV-induced disease. The broad host range of MPXV is also likely to permit additional species to become reservoirs or incidental hosts, thus increasing the exposure risk for humans.

Studies between 1981 and 1986 revealed that most human monkeypox cases occurred as single sporadic infections after contact with animals. The first-generation secondary attack rate in nonvaccinated household contacts was approximately 9% (compared to 58% with smallpox). The attack rate decreased over the second and third generations, and fourth generation attacks were very rare. That said, it has recently been reported that the transmissibility of MPXV is increasing in human hosts. Genetic changes that improve transmissibility to similar levels as those seen in VARV would be required for MPXV to infect humans in an endemic fashion.

Genetics

Although MPXV and VARV present similar disease profiles in humans, neither of the viruses is believed to have given rise to the other. Rather, both are considered to have arisen from a progenitor virus(es) similar to the cowpox virus lineage. MPXV isolates from West Africa are less virulent and less transmissible in human populations than isolates from the Congo Basin. Consistent with other orthopoxviruses, MPXV strains exhibit gene conservation toward the center of the genome, with variation increasing in frequency toward the termini, which encode for specific virulence genes. Genomic differences between strains have been mapped with restriction fragment length polymorphism studies and DNA sequencing techniques. Sequence analyses of strains from West Africa and the Congo Basin have revealed that isolates are approximately 95% identical to each other and approximately 96% identical to VARV. This value increases to 99% when comparing isolates from only West African or only the Congo Basin regions, allowing a separation into two groups or clades. The genomic sequences of eight MPXV strains are available online (see the 'Relevant website' section).

Ecology

The broad host range of MPXV and seroprevalence studies suggest that several animal species, rather than a single species, may act as reservoirs for MPXV in nature. In the latter part of the twentieth century, several field studies were conducted in the lowland tropical forests of the Congo Basin and West Africa; these studies revealed that MPXV can infect many animal species, including squirrels (*Funisciurus* spp. and *Heliosciurus* spp.) and nonhuman primates (such as *Cercopithecus* spp.). Species that are seropositive for MPXV antibodies have some similar and some dissimilar traits in relation to diet and habitat preference; approximately 40% are arboreal, 40% are semiterrestrial, and 20% are terrestrial. Therefore, MPXV infects species that inhabit all levels of the lowland tropical rainforest in the Congo Basin and West Africa. For a thorough discussion of this topic, the reader is directed to the article by Parker *et al.* listed in the 'Further reading' section.

New Hosts and Geographic Expansion

The interaction of MPXV with reservoir and incidental hosts is still poorly understood, as is the potential for virus transmission to humans within and outside its geographical range. Until the outbreak in the USA in 2003, MPXV had remained fairly localized to a handful of countries in central and West Africa, with the majority of cases detected in the DRC. The USA outbreak added to the breadth of host species capable of supporting MPXV replication, and demonstrated the potential of the virus to expand its geographical range. No human infections were attributed to the shipment of animals that entered the USA from Africa; rather, most patients had direct contact with infected native prairie dogs that had been housed with the imported African rodents. Subsequent to the USA outbreak, prairie dogs that were experimentally infected with MPXV were found to have ulcerative lesions on their lips, tongues, and buccal mucosa. High titer MPXV could be cultured from the nasal discharge and oropharynx of these animals for up to 22 days post infection, indicating that transmission from prairie dogs was likely via the respiratory and mucocutaneous routes. The USA outbreak was caused by the less virulent West African strain, which probably made it easier to bring under control.

A more recent example (2005) of MPXV expanding its environs is found in 19 human monkeypox infections discovered in the previously MPXV-free country of Sudan. This outbreak occurred some 300 miles northeast of the edge of the tropical rainforest – the traditional home of the Congo Basin strain of MPXV. From experience with vaccinia virus, it would not be surprising if MPXV continues to adapt to new species. Such adaptation has already occurred with prairie dogs in the USA and is possibly the reason for the outbreak in Sudan.

Person-to-Person Transmission

MPXV and VARV transmission are likely similar. Human-to-human transmission can be separated into six steps, as demonstrated in **Figure 1**. Step 1: release of virions from lesions in the oropharyngeal mucosa and their aerosolization into the new host's breathing space. Step 2: virus particles, most likely in the EEV form, are transmitted by aerosols. Step 3: seeding of the new host's respiratory mucosa is initiated. Step 4: MPXV replication creates foci of infection and production of specific proteins to neutralize the immune response. Step 5: the primary viremia denotes successful virus replication and spread from the initial site(s) of infection to lymphoid tissues and internal organs. Step 6: the secondary viremia occurs when the virus moves from the infected lymphoid tissues and internal organs to the cornified and mucosal epithelium to cause the exanthem and enanthem, respectively. Transmissibility is dependent on the number of lesions in the host oropharyngeal mucosa, virus survivability in the face of the host immune response, and the ability of the virus to produce infectious virions for exhalation from the respiratory tract. As an explanation of the increased transmissibility of VARV over MPXV, it is likely that VARV produces more infectious virions in the respiratory mucosa than MPXV.

Treatment

Currently, the public health importance of human monkeypox is minor compared to that of VARV before the 1980s. However, MPXV is becoming a more common infection in central Africa, where there seems to be a rise in the number of transmission generations observed during outbreaks. MPXV could be controlled in the human population by vaccination against smallpox. However, considering the current poor transmissibility in human populations and zoonotic nature of MPXV, this would need to be weighed carefully against the adverse reactions expected from vaccination. As of 2007, two antiviral drugs with activity against orthopoxviruses are being evaluated in animal models.

Diagnostics

Historically, biological properties have been used to identify and differentiate orthopoxviruses. Growth characteristics in embryonated chick eggs were particularly useful during the smallpox eradication campaign. However, this approach is labor and time consuming and requires a high level of skill. Even today, electron microscopy is a first-line technique, but does not allow for differentiation between orthopoxvirus species. Real-time polymerase chain reaction (PCR) is now regarded as the technique of choice for species differentiation, and several protocols are available specifically to identify and differentiate VACV and MPXV from other poxviruses.

Future Perspectives

VARV has been exterminated from the world's human population and could only be reintroduced by artificial release of clandestinely stored stocks. Such an incident would have devastating ramifications. MPXV is of minor public health significance when compared to VARV, but human MPXV infections are increasing. Elimination of MPXV is not possible because, unlike VARV, the virus is likely to have several animal reservoirs. Lastly, the possibility that terrorist groups or rogue nations might bioengineer VARV or MPXV to enhance virulence and transmissibility is a potential threat. Since the molecular biology of both viruses (and other orthopoxviruses) is fairly well understood, genetic tinkering aimed at enhancing virulence is a possibility, although techniques to increase transmissibility are less well developed.

See also: Cowpox Virus; Mousepox and Rabbitpox Viruses; Poxviruses; Vaccinia Virus; Varicella-Zoster Virus: General Features.

Further Reading

Fenner F, Henderson DA, Arita I, Jezek Z, and Ladnyi ID (eds.) (1988) *Smallpox and Its Eradication*. Geneva, Switzerland: World Health Organization.

Jezek Z and Fenner F (1988) Human monkeypox. In: Melnick JL (ed.) *Monographs in Virology*, vol. 17. Basel, Switzerland: Karger.

Parker S, Nuara A, Buller RML, and Schultz DA (2007) Human monkeypox: An emerging zoonotic disease. *Future Microbiology* 2: 17–34.

Tulman ER, Delhon G, Afonso CL, *et al.* (2006) Genome of horsepox virus. *Journal of Virology* 80: 9244–9258.

Relevant Website

http://www.biovirus.org – Viral Bioinformatics Resource Center (VBRC).

Sobemovirus

M Meier, A Olspert, C Sarmiento, and E Truve, Tallinn University of Technology, Tallinn, Estonia

© 2008 Elsevier Ltd. All rights reserved.

Glossary

−1 Ribosomal frameshifting Event occurring during translation elongation when the ribosome shifts its frame for reading the mRNA exactly one position in the upstream direction.

Icosahedral particle Spherical viral particle that is a polyhedron having 20 faces.

Leaky scanning mechanism Mechanism during translation initiation for escaping the first start codon, which occurs when the first AUG resides in a very poor context and therefore only some ribosomes initiate translation at that point.

Polycistronic RNA Contains the genetic information to translate more than one protein.

Satellite RNA (satRNA) Subviral agent consisting of RNA that becomes packaged in protein shells made from coat protein of the helper virus and whose replication is dependent on that virus.

$T=3$ particle Icosahedral virus particle that contains three chemically identical coat protein monomers in the icosahedral asymmetric unit; $T=3$ particle contains 180 coat protein molecules.

VPg Protein that is attached to the 5′ end of viral genomic RNA.

Introduction

It was proposed in 1969 that single-component-RNA beetle-transmitted viruses be placed into a southern bean mosaic virus group. Since 1995, this group has been recognized by the International Committee on Taxonomy of Viruses (ICTV) as the genus *Sobemovirus* (sigla from *so*uthern *be*an *mo*saic *virus*) unassigned to any family. The establishment of this virus group was based on similarities in particle morphology, capsid stabilization, sedimentation coefficients, sizes of protein subunits and genomic RNA, features in mode of vector transmission, and distribution of the particles within the cell. In the *Eighth Report of the International Committee on Taxonomy of Viruses*, 13 virus species were accepted as definite species of the genus *Sobemovirus* and four tentative species were proposed (**Table 1**).

In addition, two viruses presently not recognized by the ICTV are closely related to the sobemoviruses. Nucleotide sequence comparison of the polymerase, VPg, and coat protein (CP) genes of papaya lethal yellowing virus (PLYV) shows high homology to sobemoviruses (about 41–51% with lucerne transient streak virus (LTSV), southern bean mosaic virus (SBMV), southern cowpea mosaic virus (SCPMV), and cocksfoot mottle virus (CfMV)). Snake melon asteroid mosaic virus (SMAMV) RNA-dependent RNA polymerase (RdRp) fragment possesses 71% amino acid sequence similarity to rice yellow mottle virus (RYMV) RdRp.

The host range of sobemoviruses is usually narrow; individual viruses can naturally infect plants from one family only. The exception is sowbane mosaic virus (SoMV) that infects plants from the families Chenopodiaceae, Vitaceae, and Rosaceae. Some sobemoviruses (SBMV, SCPMV, SoMV) are distributed throughout the world; others are limited to one continent or even to one country (**Table 1**).

Sobemoviruses are readily transmitted mechanically. RYMV, for example, is efficiently transferred from plant to plant by farming operations, donkeys, cows, grass rats, wind-mediated leaf contacts, soil, etc. In addition, sobemoviruses are transmitted by vectors. The most common vectors are different species of beetles that transmit sobemoviruses in a semipersistent manner. However, blueberry shoestring virus (BSSV) and SoMV are transmitted by aphids, SoMV also by leafminers and leafhoppers, and velvet tobacco mottle virus (VTMoV) by mirids. Several viruses in the genus are seed-transmissible (**Table 1**).

Genome Organization and Replication

The full-length genomic nucleotide sequences have been determined for members of nine sobemovirus species. Their genome sizes vary from 4.0 to 4.5 kb. The genomic (as well as subgenomic) RNA has a viral genome-linked protein (VPg) covalently bound to its 5′ end. The 3′ terminus of the genomic RNA is nonpolyadenylated.

All the sequenced sobemoviruses have a polycistronic positive-sense single-stranded RNA (ssRNA) genome that consists of four open reading frames (ORFs) (**Figure 1**). The genome is compact, as for most viruses all ORFs overlap. ORFs 1, 2a, and 2b are all translated from the genomic RNA. The initiation of translation from the genomic RNA is facilitated at least in case of CfMV by the translational enhancer in the 5′ untranslated region (UTR) of the genome. Translation of ORF1 and ORF2a occurs via a leaky scanning mechanism. ORF2b is expressed as a fusion protein with ORF2a through a −1 ribosomal frameshift mechanism. Previously, it was reported that some sobemoviruses express the RdRp from a single in-frame polyprotein, not via a −1 translational frameshift. Recently, it has been demonstrated, however, that the difference between the two kinds of genomic organization resulted from a single erroneous extra nucleotide in the genomic sequences of these sobemoviruses that RdRp was thought to be expressed without the −1 ribosomal frameshifting mechanism. Thus, all sequenced sobemovirus species possess similar genomic organization.

A genome 3′-proximal ORF is translated from the subgenomic RNA (sgRNA). The sgRNA has been detected in sobemovirus-infected tissues as well as in virus particles. In addition to the genomic and sgRNA, some sobemoviruses (LTSV, RYMV, subterranean clover mottle virus (SCMoV), Solanum nodiflorum mottle virus (SNMoV), VTMoV) encapsidate a circular viroid-like satellite RNA (satRNA). The sizes of these circular satRNAs range from 220 to 390 nt, the 220 nt RYMV satRNA being the smallest known naturally occurring circular RNAs.

The replication of sobemoviruses uses is poorly understood. The genomic RNA of incoming virus particles is probably uncoated by the co-translational disassembly mechanism that is followed by RNA replication. Little is known about the signals needed for the replication. The 5′ terminal nucleotides of the genomic RNA are ACAAAA for SCPMV, ACAA for RYMV, ACAAA for LTSV and ryegrass mottle virus (RGMoV), ACAAAA for SCMoV, CACAAAA for Sesbania mosaic virus (SeMV) and SBMV, and CAAAA for turnip rosette virus (TRoV). For all these viruses, the 5′ ACAAA or ACAAAA sequence is also present upstream from the CP translation initiation codon, indicating the possible 5′ terminus of sgRNA. This sequence is also characteristic of 5′ termini of polero-, diantho-, and barnaviruses. Due to the conservation of this sequence, it or its complementary sequence in (−)-strand has been predicted to function in viral RNA replication by promoting or enhancing the binding of viral RdRp. Different from other sobemoviruses, such motif is present neither at the 5′ end nor upstream from the CP translation initiation codon in CfMV genome. In addition, all sobemoviruses contain a polypurine tract 5′-aAGgAAA near the beginning of their genomic RNA. Nearly nothing is known about signals at the 3′ end of the genomic RNA essential for the initiation of the synthesis of the genomic

Table 1 Viruses of the genus *Sobemovirus* and their biological properties

Virus	Abbr.	Distribution	Natural host	Insect vector	Transmission Mechanical	Seed
Definitive species						
Blueberry shoestring virus	BSSV	USA (Maine, Michigan, New Jersey, Virginia), Canada (New Brunswick, Ontario, Quebec)	*Vaccinium corymbosum*, *V. angustifolium*	*Masonaphis pepperi* (aphid)	Yes	No
Cocksfoot mottle virus	CfMV	Europe (France, Germany, Norway, Russia, UK), New Zealand, Japan	*Dactylis glomerata*, *Triticum aestivum*	*Lema melanopus*, *L. lichensis* (beetles)	Yes	No
Lucerne transient streak virus	LTSV	Australia (Victoria, Tasmania), New Zealand, Canada	*Medicago sativa*	ND	Yes	No
Rice yellow mottle virus	RYMV	Africa (Benin, Burkina Faso, Cameroon, Chad, Côte d'Ivoire, Gambia, Ghana, Guinea, Guinea Bissau, Kenya, Liberia, Madagascar, Mali, Malawi, Mauritania, Mozambique, Niger, Nigeria, Rwanda, Senegal, Sierra Leone, Tanzania, Togo, Uganda)	*Oryza sativa*, *O. longistaminata*	*Chaetocnema pulla*, *Sesselia pusilla*, *Trichispa sericea* (beetles)	Yes	No
Ryegrass mottle virus	RGMoV	Japan, Germany	*Lolium multiflorum*, *Dactylis glomerata*	ND	Yes	ND
Sesbania mosaic virus	SeMV	India (Andra Pradesh)	*Sesbania grandiflora*	ND	Yes	ND
Solanum nodiflorum mottle virus	SNMoV	Australia (Queensland, New South Wales)	*Solanum nodiflorum*, *S. nitidibaccatum*, *S. nigrum*	*Epilachna sparsa*, *E. doryca australica*, *E. guttatopustulata*, *Psylliodes* sp. (beetles), *Cyrtopeltis nicotianae* (mirid)	Yes	No
Southern bean mosaic virus	SBMV	USA (Arkansas, California, Lousiana), North and South America (Brazil, Colombia, Mexico), Africa (Côte d'Ivoire, Morocco), Europe (France, Spain), Iran	*Phaseolus vulgaris*	*Ceratoma trifurcata*, *Epilachna variestis* (beetles)	Yes	Yes
Southern cowpea mosaic virus	SCPMV	USA (Wisconsin), Africa (Botswana, Ghana, Kenya, Nigeria, Senegal), Asia (India, Pakistan)	*Vigna unguiculata*	*Ootheca mutabilis* (beetle)	Yes	Yes
Sowbane mosaic virus	SoMV	USA, Canada, Central and South America, Europe (Bulgaria, Czech Republic/Slovakia, Croatia, France, Hungary, Italy, Moldova), Japan, Australia (Queensland, New South Wales, Victoria, Tasmania)	*Chenopodium* spp., *Atriplex subrecta*, *Spinacia oleracea*, *Vitis* sp., *Prunus domestica*, *Alisma plantago-aquatica*, *Danae racemosa*	*Myzus persicae* (aphid), *Liriomyza langei* (leafminer), *Circulifer tenellus* (leafhopper), *Halticus citri* (fleahopper)	Yes	Yes
Subterranean clover mottle virus	SCMoV	Australia (New South Wales, South Australia, Tasmania, Victoria, Western Australia)	*Trifolium subterraneum*	ND	Yes	Yes
Turnip rosette virus	TRoV	UK	*Brassica campestris*, *B. nigra*	*Phyllotreta nemorium* (beetle)	Yes	ND

Continued

Table 1 Continued

Virus	Abbr.	Distribution	Natural host	Insect vector	Transmission Mechanical	Seed
Velvet tobacco mottle virus	VTMoV	Australia (Northern Territory, Queensland, South Australia)	Nicotiana velutina	Cyrtopeltis nicotianae (mirid), Epilachna spp. (beetle)	Yes	No
Tentative species						
Cocksfoot mild mosaic virus	CMMV	Europe (Czech Republic/Slovakia, Denmark, France, Germany, Norway, UK), Canada (Ontario)	Phleum pratense, Dactylis glomerata, Agrostis stolonifera, Bromus mollis, Festuca pratensis, Poa trivialis, Triticum aestivum	Myzus persicae (aphid), Lema melanopus (beetle)	Yes	No
Cynosurus mottle virus	CnMoV	Europe (Germany, UK, Ireland), New Zealand	Cynosurus cristatus, Agrostis tenuis, A. stolonifera, Lolium perenne X L. multiflorum	Lema melanopus (beetle), Rhopalosiphum padi (aphid)	Yes	ND
Ginger chlorotic fleck virus	GCFV	India, Malaysia, Mauritius, Thailand	Zingiber officinale	ND	Yes	ND
Rottboellia yellow mottle virus	RoMoV	Nigeria	Rottboellia cochinchinensis	ND	Yes	ND

minus strand. A potential tRNA-like structure has been attributed to the 3′ end of some sobemoviruses, but no experimental data are available on that.

Mutation and recombination rates associated with the replication by sobemoviral RdRps are largely uncharacterized. No intra- or interspecies recombinant sobemoviruses have been described so far. Based on the phylogenetic analysis of RYMV sequences, it has been concluded that RYMV has evolved in the absence of recombination events. No recombinants were also detected between CfMV and RGMoV in doubly infected plants. However, several defective interfering (DI) RNA molecules have been cloned from CfMV-infected plants containing the 5′ end of the genomic RNA linked to 850–950 nt of the 3′ terminus. SatRNAs have several interesting interactions with the replication of the helper sobemoviruses. For example, LTSV supports the replication of satRNA of SNMoV but SNMoV does not replicate LTSV satRNA. At the same time, LTSV satRNA replication is supported by CfMV, SBMV, SoMV, and TRoV, whereas this support is host dependent.

Gene Products and Their Functions

P1

P1 is encoded by the 5′ terminal ORF of the viral genomic RNA and its translation occurs with poor efficiency, as the translation initiation context is suboptimal. The molecular masses of different P1s range between 11.7 and 24.3 kDa and, surprisingly, there is no similarity between the P1 amino acid sequences, making this region the most variable one in the genome of sobemoviruses.

The most-studied P1 proteins are the ones of RYMV, SCPMV, and CfMV. All these proteins are required for systemic infection and are dispensable for viral replication. The P1 of SCPMV has also been shown to be nonessential for viral assembly. In addition, RYMV P1 is described as an important pathogenicity determinant. Both CfMV P1 and RYMV P1 act as suppressors of RNA silencing in *Nicotiana benthamiana*, a nonhost species. P1 of CfMV binds ssRNA in a sequence-independent manner and it does not bind double-stranded small interfering RNAs (siRNAs).

The transient gene expression of ORF1 from CfMV and from SCMoV demonstrated that P1 – as a green fluorescent protein (GFP) fusion protein – is involved in movement. However, the cell-to-cell movement of P1, independent of other viral components, was very limited.

Little is known about the subcellular localization of P1. In the case of CfMV, the P1 is coupled with cellular membranes and/or heavily aggregated when overexpressed in insect cells.

It is worth mentioning that the 5′ terminal half of the genomes of sobemoviruses and poleroviruses is similar in organization. Moreover, poleroviral P0, encoded by the 5′ terminal ORF, shares many features with sobemoviral P1: it acts as suppressor of RNA silencing, it is the

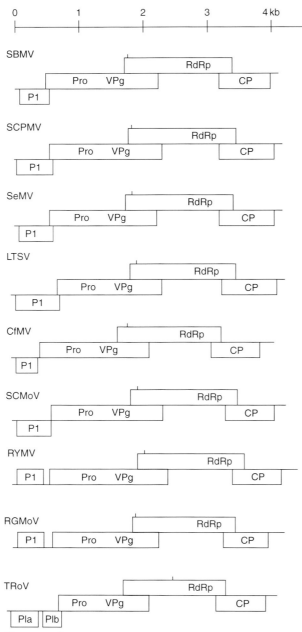

Figure 1 Genome organization of sobemoviruses.

most divergent protein of the viral genome, and it has a poor translation initiation context.

Polyprotein

Translation of the P2a or P2a2b polyprotein takes place from sobemoviral genomic RNA via a leaky scanning mechanism. The comparison of the sequences surrounding the initiation codons for ORF1 and ORF2a/ORF2a2b of sobemoviruses with the consensus sequences established for plant mRNAs shows that the sequence surrounding the ORF2a initiation codon is always in a more favorable context for translation by plant ribosomes.

In vitro translation of sobemoviral RNAs as identified a protein with a molecular mass of about 100–105 kDa. This protein is encoded by partially overlapping ORF2a and ORF2b due to the signal for −1 ribosomal frameshifting. The −1 ribosomal frameshift is needed to regulate the production of sobemoviral RdRp which is encoded by ORF2b. Although no sobemoviral RdRp has been molecularly characterized yet, the presence of the highly conserved GDD motif with its surrounding (characterized as SGSYCTSSTNX$_{19-35}$GDD) feature for RdRps of positive-strand ssRNA viruses has been identified by computer-based sequence analysis of sobemovirus genomes. In the case of CfMV, the consensus signal for −1 ribosomal frameshift event has been shown to consist of a slippery sequence UUUAAAC and a stem–loop structure located 7 nt downstream of it. The slippery sequence UUUAAAC, followed by a simple stem–loop structure, is absolutely conserved for all sequenced sobemoviruses. In a wheat germ extract, CfMV-derived −1 ribosomal frameshifting takes place with an efficiency of *c.* 10%. It has been proposed that the C-terminal processing products of ORF2a-encoded protein regulate the efficiency of the frameshifting.

ORF2a encodes the N-terminal part of the polyprotein that contains at least a viral serine protease and a VPg. The position of the VPg in between the viral protease and replicase (Pro-VPg-RdRp) is unique to sobemo-, polero-, enamo-, and barnaviruses. It is proposed that sobemoviral VPgs exist in a 'natively unfolded state' lacking both secondary and tertiary structures. The only conserved amino acid sequence element observed among sobemoviral VPgs is a WAD or WGD motif followed by a D- or E-rich region.

The proposed consensus amino acid sequence for the catalytic triad of the serine proteases of sobemo-, polero-, enamo-, and barnaviruses is $H(X_{32-35})[D/E](X_{61-62})$ TXXGXSG. The glycine and histidine residues downstream from the catalytic residues (H181, D216, S284 for SeMV) are suggested to be the site for substrate binding. The crystal structure of the protease domain has been determined for SeMV at 2.4 Å resolution. The structure exhibits the characteristic features of trypsin fold.

Mutations of active site residues of H181A, D216A, or S284A render the SeMV protease inactive. The SeMV serine protease domain lacking the N-terminal 70 amino acids is inactive *in trans*. According to *in silico* analysis, the N-termini of the sobemoviral polyprotein sequences (except LTSV that lacks the domain respective to N-termini of other sobemoviral polyproteins) show the presence of high-propensity membrane helices. Interestingly, the presence of the VPg domain at the C-terminus of the SeMV protease domain lacking the N-terminal 70 amino acids restores the protease activity both in *cis* and in *trans*. Furthermore, the substitution of conserved W43 in SeMV VPg to phenylalanine or the deletion of the entire VPg domain abolishes the proteolytic processing of the viral polyprotein.

The N-terminal sequencing of SBMV, CfMV, and SeMV VPgs attached to viral genomes as well as other approaches have indicated that the polyprotein is processed at E/T or E/N sites between the protease, VPg, and RdRp domains. Additionally, A/V cleavage site was identified at the N-terminus of SeMV polyprotein by mass spectrometric analysis. As the alanine and valine residues are small enough to be accommodated in the active site without steric hindrance, it is suggested that the specificity of sobemoviral proteases depends not only on the sequence but also on the conformation of the polypeptide.

Virion Structure and Coat Protein

The virions of members of the genus have an icosahedral capsid roughly the size of 30 nm, which is assembled according to $T=3$ symmetry. The capsid contains 180 molecules of a single ~30 kDa CP, which is translated from an sgRNA. The single-stranded genomic RNA and sgRNA together with VPg are packaged inside the virion. The three-dimensional structures of SCPMV, SeMV, RYMV, RGMoV, and CfMV virions have been determined utilizing X-ray crystallography. Each icosahedral unit of the virion comprises of three quasi-equivalent subunits A, B, and C, which have minor differences in conformation (**Figure 2(a)**). The A subunits group at fivefold axes while pairs of B and C subunits meet at threefold axes. In addition to protein–protein and protein–RNA interactions each icosahedral unit is stabilized by three calcium-binding sites located between subunits AB, BC, and CA.

The CP is divided into two domains: C-terminal S (shell) domain, which has an eight-strand jellyroll β-sandwich topology, common to nonenveloped icosahedral viruses, and N-terminal R (random) domain, which is disordered in subunits A and B, but is partially structured in subunit C. The S domain is the building block of the virion, whereas the R domain is involved in the regulation of the capsid structure. The primary sequences of CPs among the members of the genus are quite different. However, the three-dimensional structures are nearly identical, for example root mean square difference between RYMV and SCPMV is 1.4 Å. Regardless of that, two different structures of the R domain in C subunit have been found (**Figure 2(b)**). In SCPMV and SeMV the N-terminus of subunit C, makes a U-turn and extends toward the threefold axis nearest to C, where it makes a β-structure together with R domains from analogous C subunits. In RYMV and CfMV there is no U-turn, instead the N-terminal arm of subunit C extends toward subunit B and makes a similar structure at the distal threefold axes closest to subunit B. When the R domain of SCPMV or SeMV CP is removed, only $T=1$ particles are formed, indicating the importance of the region in $T=3$ particle formation.

Due to its complex nature, very little is known about the mechanism of capsid formation. Yet, there is evidence suggesting that the virion assembly could be nucleated by AB dimers at icosahedral fivefold axes since pseudo $T=2$ SeMV particles comprise of groups of A and B subunits. Studies with SCPMV, SeMV, and RYMV particles demonstrate that the stability of the virions depends greatly on pH and the availability of calcium ions. Upon alkaline pH or removal of the cations the virus particles swell and become less stable. Mutation analysis of SeMV CP calcium-binding sites demonstrates that cation-mediated interactions are mainly needed for particle stability. The R domain of all sobemovirus CPs is rich in basic amino acid residues and contains an arginine-rich motif (ARM). Studies with SCPMV and SeMV CP demonstrate that ARM is essential for RNA encapsidation but not for particle formation. However, the presence of RNA enhances the overall stability of capsids. The N-terminus of SCPMV CP withholds a potential to form an α-helix and can interact with membranes *in vitro*. It is proposed that the R domain of all sobemoviral CPs contains a nuclear localization signal.

The functions of sobemoviral CPs in viral life cycle besides capsid formation are not fully understood. Studies with full-length clones of SCPMV and RYMV display that the CP is dispensable for virus replication but systemic virus movement is completely abolished in the absence of CP. Sobemoviral CPs have also been reported to complement the long-distance movement of taxonomically distinct plant viruses.

Subcellular Localization, Short- and Long-Distance Movement

Sobemoviral particles are found mainly in mesophyll and vascular tissues, but also in epidermal, bundle sheath, and guard cells. The quantity of particles present is in correlation with the severity of symptoms. In vascular tissues there are reports of virus particles in both xylem and phloem. CfMV, SCPMV, and SBMV virus particles have been found in phloem companion cells, whereas RYMV particles have been detected predominantly in xylem. RYMV particles accumulated in xylem parenchyma cells and vessels; additionally association with intervascular pit membranes was observed. For RYMV, the common belief is that the virus is transported between xylem cells through pit membranes.

Subcellularly virus particles are found at least in cytoplasm, vacuoles, and nuclei. Virus particles in cytoplasm or vacuoles are known to form crystalline structures, sometimes particles are found in vesicles. No particles have been found in mitochondria and chloroplasts, but the latter are noted to form finger-like extrusions in infected cells. Studies with RYMV suggest that vacuoles of xylem parenchyma cells become the storage compartments for virions in late phase of infection. It is proposed that swollen and less compact virions coexist in

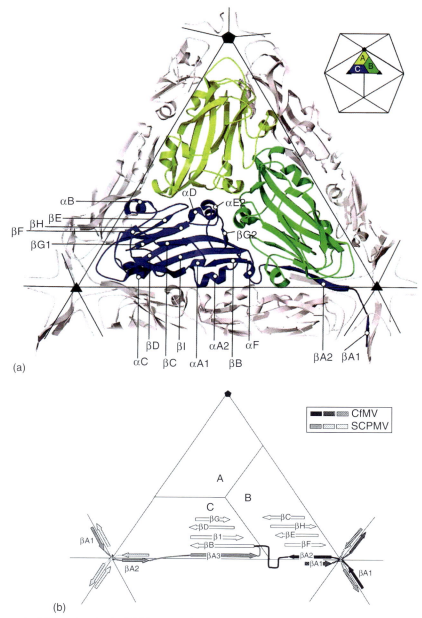

Figure 2 (a) Arrangement of CP molecules in CfMV capsid. (b) A schematic representation of the N-terminal arm in sobemoviruses. Reproduced from Tars K, Zeltins A, and Liljas L (2003) The three-dimensional structure of cocksfoot mottle virus at 2.7 Å resolution. *Virology* 310: 288–289, with permission from Elsevier.

the cytoplasm, whereas vacuoles with their lower pH and higher Ca^{2+} concentration contain compact virions.

Studies with SCPMV, SBMV, and RYMV emphasize that cell-to-cell and vascular movement of sobemoviruses are two distinct processes, whereas the long-distance movement is dependent on the correct capsid formation.

Pathology, Economic Importance, Resistance

Sobemoviral infections can cause a variety of disease symptoms: mild or severe chlorosis and mottling, stunting, necrotic lesions, vein clearing, sterility. Subcellularly, sobemoviruses form crystalline arrays and tubules in the cytoplasm, some of which are enveloped in endoplasmic reticulum-derived vesicles. It has been observed that nucleolus of the RYMV-infected cell enlarges. Probably the most dramatic change induced by RYMV occurs in the cell walls of parenchyma and mature xylem cells, where middle lamellae of the wall are disorganized. RGMoV has been reported to induce apoptotic cell death in oat leaves. The outcomes of these histopathological changes range from symptomless infections of sobemoviruses to severe diseases and death of plants. RYMV infection also causes important changes in the abundance of many host proteins.

For instance, the expression levels of several defense- and stress-related proteins like superoxide dismutase and different heat shock proteins increase several times.

Several sobemoviruses are economically important pathogens. RYMV causes one of the most damaging and rapidly spreading diseases of rice in Africa. Yield losses fluctuate between 10% and 100%, depending on plant age prior to infection, susceptibility of the rice variety, and environmental factors. PLYV, causing serious chlorosis, is responsible for an important disease of papaya in northeast Brazil. SCMoV decreases seed and herbage production in Australia. Over time, SCMoV-infected pastures become weedy and unproductive. SBMV infections in common bean leads to the mosaic and distortion of pods and reduced size and number of seeds.

Natural resistance to sobemoviruses has been detected at least for CfMV in cocksfoot, for CnMoV in *Cynosurus cristatus*, for RYMV in rice and *Oryza glaberrima*, for SBMV in beans, for SCMoV in subterranean clover, and for SCPMV in cowpea. The molecular mechanisms conferring resistance have only been described for RYMV in *Oryza* species. Namely, the recessive resistance gene *Rymv-1* encodes the eukaryotic translation initiation factor eIF(iso)4G whose interaction with viral VPg is responsible for the high-resistance trait. In parallel, a quantitative trait locus (QTL) is described conferring partial resistance against

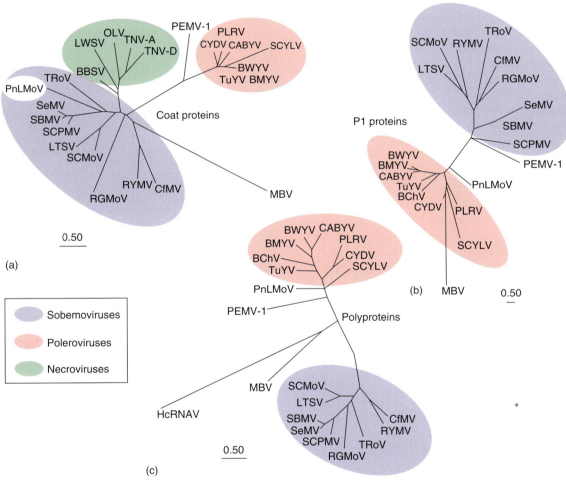

Figure 3 Unrooted phylogram of sobemoviral proteins and the respective proteins of related viruses using maximum-likelihood method. (a) Coat proteins, (b) P1 proteins, and (c) polyproteins. The bar shows the scale of branch length. The protein sequences were aligned with ClustalW and the phylogenetic trees were constructed using PHYLIP PROML 3.6.1 and visualized using Treetool. Viruses (with GenBank accession number) involved in the analyses presented in this figure were the following. Sobemoviruses: CfMV, cocksfoot mottle virus (DQ680848); LTSV, lucerne transient streak virus (U31286); RYMV, rice yellow mottle virus (AJ608206); RGMoV, ryegrass mottle virus (AB040446); SeMV, Sesbania mosaic virus (AY004291); SBMV, southern bean mosaic virus (AF0558871); SCPMV, southern cowpea mosaic virus (M23021); SCMoV, subterranean clover mottle virus (AF208001); TRoV, turnip rosette virus (AY177608). Poleroviruses: BChV, beet chlorosis virus (AF352024); BMYV, beet mild yellowing virus (X83110); BWYV, beet western yellows virus (AF473561); CYDV, cereal yellow dwarf virus (Y07496); CABYV, cucurbit aphid-borne yellows virus (X76931); PLRV, potato leafroll virus (D00530); ScYLV, sugarcane yellow leaf virus (AF157029); TuYV, turnip yellows virus (X13063). Enamovirus: PEMV-1, pea enation mosaic virus 1 (L04573). Barnavirus: MBV, mushroom baciliiform virus (U07551). Necroviruses: BBSV, beet black scorch virus (AF452884); LWSV, leek white stripe virus (X94560); OLV-1, olive latent virus 1 (X85989); TNV-A, tobacco necrosis virus A (M33002); TNV-D, tobacco necrosis virus D (D00942). Unclassified viruses: PnLMoV, poinsettia latent mottle virus (AJ867490); HcRNAV, Heterocapsa circularisquama RNA virus (AB218609).

several RYMV isolates, but the molecular mechanisms responsible for that trait are unknown. Pathogen-derived transgenic resistance against RYMV has been achieved by transforming plants with constructs expressing either RdRp or CP sequences of the virus.

Phylogenetic Relationships

Within the genus *Sobemovirus*, phylogenetic analysis of different proteins indicates that three species – SeMV, SBMV, and SCPMV – are very closely related to each other. Also LTSV and SCMoV as well as CfMV and RYMV cluster into two corresponding subgroups within the genus (**Figure 3**).

When the sobemoviral sequences are compared to all other sequences available, it is evident that the 5′ terminus of the sobemoviral genomes together with ORF1 are unrelated to any other known viral genera (**Figure 3**). The middle part of the genomes (encoding for the successive domains of Pro-VPg-RdRp), however, is similar to those of genera *Polerovirus* and *Enamovirus* from the family *Luteoviridae*. In contrast, the 3′ part of the sobemoviral genomes encoding for the CP is more closely related to CP genes of the genus *Necrovirus* from the family *Tombusviridae* (**Figure 3**). These similarities indicate that possibly early recombination events have played an important role during the evolution of sobemoviruses, luteoviruses, and tombusviruses. This possibility is further supported by the existence of poinsettia latent mottle virus (PnLMoV) whose sequence (AJ867490) shows a close relationship to poleroviruses within the first three quarters of its genome (encoding P1 and polyprotein), but rather to sobemoviruses in the last quarter (encoding CP gene) (**Figure 3**).

A virus belonging to the single species *Mushroom bacilliform virus* of the family *Barnaviridae* (genus *Barnavirus*) has a genomic organization similar to sobemoviruses (except that it lacks ORF1 of sobemoviruses) and its Pro-VPg-RdRp and CP genes are related to the same sequences of different sobemoviruses. In addition, putative protease and RdRp of a positive-sense ssRNA virus infecting a marine dinoflagellate *Heterocapsa circularisquama*, HcRNAV (Heterocapsa circularisquama RNA virus, AB218609), are similar to those of the sobemo- and poleroviruses as well as of PnLMoV and MBV.

See also: Barnaviruses; *Necrovirus*; Rice Yellow Mottle Virus.

Further Reading

Albar L, Bangratz-Reyser M, Hébrard E, Ndjiondjop M-N, Jones M, and Ghesquière A (2006) Mutations in the eIF(iso)4G translation initiation factor confer high resistance of rice to rice yellow mottle virus. *Plant Journal* 47: 417–426.

Aus dem Siepen M, Pohl JO, Koo BJ, Wege C, and Jeske H (2005) Poinsettia latent mottle virus is not a cryptic virus, but a natural polerovirus–sobemovirus hybrid. *Virology* 336: 240–250.

Gayathri P, Sateshkumar PS, Prasad K, Nair S, Savithri HS, and Murthy MRN (2006) Crystal structure of the serine protease domain of Sesbania mosaic virus polyprotein and mutational analysis of residues forming the S1-binding pocket. *Virology* 346: 440–451.

Hull R and Fargette D (2005) Genus *Sobemovirus*. In: Fauquet CM, Mayo MA, Maniloff J, Desselberger U, and Ball LA (eds.) *Virus Taxonomy: Eighth Report of the International Committee on Taxonomy of Viruses*, pp. 885–890. San Diego, CA: Elsevier Academic Press.

Kouassi NK, N'Guessan P, Albar L, Fauquet CM, and Brugidou C (2005) Distribution and characterization of rice yellow mottle virus: A threat to African farmers. *Plant Disease* 89: 124–133.

Mäkelainen K and Mäkinen K (2005) Factors affecting translation at the programmed −1 ribosomal frame-shifting site of cocksfoot mottle virus RNA *in vivo*. *Nucleic Acids Research* 33: 2239–2247.

Meier M and Truve E (2007) Sobemoviruses possess a common CfMV-like genomic organization. *Archives of Virology* 152: 635–640.

Qu C, Liljas L, and Opalka N (2000) 3D domain swapping modulates the stability of members of an icosahedral virus group. *Structure* 8: 1095–1103.

Tamm T and Truve E (2000) Sobemoviruses. *Journal of Virology* 74: 6231–6241.

Tars K, Zeltins A, and Liljas L (2003) The three-dimensional structure of cocksfoot mottle virus at 2.7 Å resolution. *Virology* 310: 288–289.

St. Louis Encephalitis

W K Reisen, University of California, Davis, CA, USA

© 2008 Elsevier Ltd. All rights reserved.

Glossary

Bridge vector Vector responsible for carrying virus from the primary cycle to tangential hosts such as humans.

Diapause Insect hibernation.

Gonotrophic cycle Recurrent cycle of blood feeding and egg laying by female mosquitoes.

Maintenance vector Vector responsible for transmission of virus among primary vertebrate host species.

> **Neuroinvasive** Ability of virus to invade the central nervous system.
> **Vector competence** Ability of an insect to become infected with and transmit a pathogen.
> **Viremia** Concentration of virus within peripheral blood.
> **Viremogenic** Ability to elicit an elevated viremia response.

History

St. Louis encephalitis virus (SLEV) probably has been present in the New World within its enzootic cycle for thousands of years. The arrival of European settlers in the 1600s and the extensive agricultural development that followed greatly altered the landscape by clearing and irrigating vast areas of North America and establishing extensive urban centers. These changes probably increased the abundance of peridomestic *Culex* mosquito species and avian hosts such as house finches and mourning doves, introduced new avian hosts such as house sparrows, intensified human–vector mosquito contact, and probably increased the incidence of human infection. However, diagnosis of diseases caused by arbovirus infections such as SLEV assuredly was confounded with other infections causing fever and central nervous system (CNS) disease during summer.

During the summer of 1933, a major encephalitis epidemic with more than 1000 clinical cases occurred in St. Louis, Missouri. These cases occurred during the middle of an exceptionally hot, dry summer and were concentrated within areas of the city adjacent to open storm water and sewage channels that produced a high abundance of *Culex* mosquitoes. A virus, later named St. Louis encephalitis virus, was isolated at autopsy from human brain specimens. Mouse protection assays using convalescent human sera demonstrated that SLEV differed from other viruses causing seasonal CNS disease, such as the equine encephalitides, poliomyelitis, and vesicular stomatitis. The epidemiological features of this epidemic included the late summer occurrence of cases (especially in persons over 50 years of age), exceptionally warm temperatures, and elevated *Culex* mosquito abundance associated with a poorly draining wastewater system. These features remain the hallmark of SLEV epidemics to date.

A multidisciplinary team of entomologists, vertebrate ecologists, epidemiologists, and microbiologists from the University of California subsequently investigated an SLEV epidemic in the Yakima Valley of Washington State during 1941 and 1942 and established the components of the summer transmission cycle, including wild birds as primary vertebrate hosts and *Culex* mosquitoes as vectors.

The isolations of SLEV from *Culex tarsalis* and *Culex pipiens* mosquitoes were among the first isolations of any virus from mosquitoes and stimulated the redirection of mosquito control in North America from *Anopheles* malaria vectors and pestiferous *Aedes* to *Culex* encephalitis vectors.

Understanding the basic transmission cycle, an appreciation of the wide range of clinical symptoms, and the development of laboratory diagnostic procedures provided an expanding view of the public health significance of SLEV, with epidemics or clusters of cases recognized annually throughout the US. Wide geographic distribution and consistent annual transmission since 1933 has resulted in >1000 deaths, >10 000 cases of severe illness, and >1 000 000 mild or subclinical infections. The largest documented SLEV epidemic occurred during 1975 in the Ohio River drainage, with >2000 human cases documented. Other substantial human epidemics involving hundreds of cases have occurred in Missouri (1933, 1937), Texas (1954, 1956, 1964, 1966), Mississippi (1975), and Florida (1977, 1990). Smaller outbreaks have been recognized in California (1952), New Jersey (1962), and several other states plus Ontario (1975), Canada. Cases reported annually to the Centers for Disease Control and Prevention (CDC) since 1964 are shown in **Figure 1**.

Distribution

SLEV is distributed from southern Canada south through Argentina and from the west to the east coasts of North America and into the Caribbean Islands. Historically, human cases have been detected in Ontario and Manitoba, Canada, all of the continental US (except the New England States and South Carolina, **Figure 2**), Mexico, Panama, Brazil, Argentina, and Trinidad. The low number of human cases in Canada probably reflects the warm temperature requirements for SLEV replication in the mosquito host, whereas the low numbers of cases from tropical America may reflect inadequate laboratory diagnosis, the circulation of attenuated virus strains, and/or enzootic cycles involving mosquitoes that feed infrequently on humans. Support for this geographic distribution comes from laboratory-confirmed human cases, SLEV isolations from birds, mammals, and mosquitoes, and serological surveys of mammal and avian populations.

Classification

Taxonomically, SLEV is classified within the Japanese encephalitis virus (JEV) complex in the genus *Flavivirus* of the family *Flaviviridae*. Related viruses within this group include Japanese encephalitis, Murray Valley encephalitis, West Nile, and Usutu. SLEV consists of a positive-sense, single-stranded RNA enclosed within a

654 St. Louis Encephalitis

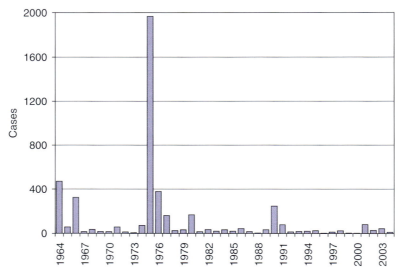

Figure 1 Number of clinical cases of St. Louis encephalitis reported to the US CDC, 1964–2004. Data provided by ArboNet, Center for Disease Control and Prevention, Ft. Collins, CO, USA.

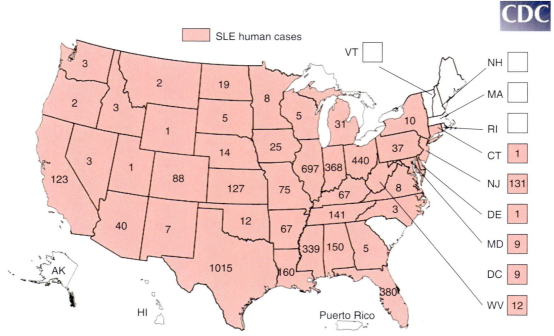

Figure 2 Distribution of human St. Louis encephalitis cases in the US, 1964–2004. Map provided by ArboNet, Center for Disease Control and Prevention, Ft. Collins, CO, USA.

capsid composed of a single polypeptide (C) and surrounded by an envelope containing one glycosylated (E) and one nonglycosylated (M) protein. Marked differences in the severity of SLEV epidemics stimulated interest in possible differences among isolates made over time and space. Detailed studies by the CDC during the 1980s clearly demonstrated geographic variation among 43 different SLEV isolates using oligonucleotide finger printing and virulence in model vertebrate hosts. These strains are grouped into six clusters: (1) east central and Atlantic USA, (2) Florida epidemic, (3) Florida enzootic, (4) eastern USA, (5) Central and South America with mixed virulence, and (6) South America with low virulence. Changes in virulence were attributed, in part, to differences in mosquito vector competence and were supported by the historical presence or absence of human cases. Subsequent genetic sequencing studies extended the understanding of SLEV genetics and provided further insight into patterns of geographical variation. Sequences of the envelope gene from SLEV strains isolated in California from 1952 to 1995 varied temporally and spatially, but indicated regional persistence in the Central Valley for at least 25 years as well as sporadic introduction and extinction. Studies in Texas using a single-strand conformation polymorphism technique showed that multiple SLEV strains circulate concurrently and remain highly focal, whereas other strains amplify and disseminate aggressively during some summers, but then disappear. Further analyses of sequences from 62 isolates made throughout the known geographical range of SLEV indicated that there have been seven lineages that overlapped somewhat with the six groups the CDC defined previously using oligonucleotide fingerprinting: (1) western USA, (2) central and eastern USA and three isolates from Mexico and Central America, (3) one mosquito isolate from Argentina, (4) five isolates from Panama mosquitoes, (5) South American strains plus an isolate from Trinidad, (6) one Panama isolate from a chicken, and (7) two isolates from Argentina rodents. Collectively, these data indicated that SLEV strains vary markedly in virulence and that the frequency and intensity of epidemics in the US may be related to genetic selection by different host systems. Interestingly, transmission within the Neotropics appears to have given rise and/or allowed the persistence of less-virulent strains that rarely amplify to produce epidemic-level transmission, a scenario duplicated by West Nile virus in the Americas.

Host Range

Arthropods

Although a wide variety of mosquitoes occasionally have been found infected in nature, three avian-feeding species within the genus *Culex* appear to be the most frequently infected and important arthropod hosts: *C. pipiens* (including the subspecies *C. pipiens quinquefasciatus* at southern latitudes, *C. p. pipiens* at northern latitudes, and intergrades) in urban and periurban environments throughout North and South America, *C. tarsalis* in irrigated agricultural settings in western North America including northern Mexico, and *Culex nigripalpus* in the southeastern US, the Carribean, and parts of the Neotropics. Although these species feed predominantly on birds, they also feed on mammals including humans, and therefore function as both maintenance and bridge vectors. Other *Culex* species such as *stigmatosoma* in the west, *restuans* and *salinarius* in the east, and perhaps species in the subgenus *Melanoconion* in the Neotropics also may be important in local transmission. Ticks have been found naturally infected, but their role in virus epidemiology most likely is minimal.

Wild Birds

The importance of avian host species appears to be related to vector *Culex* host-selection patterns as well as to avian susceptibility to the virus. Species can be separated into those frequently, sporadically, and never found infected in nature, and these groupings are related directly to their nocturnal roosting/nesting behavior and the questing behavior of *Culex* vectors. Wild birds do not develop apparent illness following experimental infection, but their viremia response varies markedly, depending upon virus strain, bird species, and bird age. Titers sufficient to infect mosquitoes typically are limited to 1–5 days post-infection. Based on serological surveys during or after epidemics, peridomestic passerifoms (including house finches, house sparrows, cardinals, and blue jays) and columbiforms (including mourning oves and rock doves or domestic pigeons) seem to be infected most frequently. In house sparrows, SLEV strains isolated from *C. pipiens* complex mosquitoes from the central and eastern USA produced elevated viremias, whereas strains isolated from *C. tarsalis* from the western USA were weakly viremogenic. Although host competence studies have been limited, the adults of few bird species seem to develop elevated viremias. However, nestling house finches, house sparrows, and mourning doves produce high viremias that readily infect mosquitoes. Therefore, the nesting period of multibrooded species may be critical for virus amplification. Regardless of their viremia response, most experimentally infected birds produce antibody and, although titers typically decay rapidly, these birds remain protected for life.

Humans

Humans are incidental hosts and do not produce viremias sufficient to infect mosquitoes. Like most arboviruses that cause CNS disease, infection with SLEV does not result in a clear clinical picture in humans and most infections remain unrecognized, unless associated with an epidemic. When presented with such diverse symptoms, few physicians initially suspect SLEV, even in endemic areas. Most SLEV infections, especially in young or middle age groups, fail to produce clinical disease, and infected individuals rarely experience more than a mild malaise of short duration with spontaneous recovery.

Domestic Animals

Although frequently antibody positive during serosurveys, SLEV infection does not produce elevated viremias or cause clinical illness in domestic animals, including equines, porcines, bovines, or felines. In a single experiment, dogs (purebred beagles) produced a low-level viremia, with only two of eight dogs developing clinical illness. Similar to wild birds, immature fowl <1 month old (including chickens and ducks) consistently developed sufficient viremia to infect mosquitoes, but did not develop clinical illness. Adult chickens (>22 weeks old) usually failed to develop a detectable viremia, and along with immature birds, developed long-lasting antibodies.

Wild Mammals

The response of wild mammals to natural or experimental infection varies. Serosurveys occasionally have shown higher SLEV prevalence in mammals than in birds, but these data could be confounded because mammalian hosts typically live longer than avian hosts and therefore have a longer history of exposure. Rodents in the genera *Ammospermophilus* and *Dipodomys* were susceptible to infection after subcutaneous (s.c.) inoculation, whereas *Spermophilus, Rattus, Sigmodon,* and *Peromyscus* were refractory. Similarly varied were lagomorphs: *Lepus* was susceptible, whereas four species within *Sylvilagus* ranged from refractory to susceptible. Raccoons and skunks were refractory, whereas opossums and woodchucks were susceptible. Like birds, susceptible mammals produced an immediate viremic response that generally persisted for <1 week, and all species produced detectable antibodies regardless of their viremia response. SLEV frequently has been isolated from bats (*Tadarida, Myotis,* etc.), and many populations exhibit a high prevalence of neutralizing antibody. Overall, the role of mammalian infection in SLEV epidemiology is complex and difficult to interpret. All reputed *Culex* vectors feed most frequently on avian hosts, occasionally on large mammals and lagomorphs, rarely on rodents, and almost never on bats.

Pathogenicity

In humans, clinical disease due to SLEV infection may be divided into three syndromes in increasing order of severity: (1) 'Febrile headache' with fever, headache possibly associated with nausea or vomiting, and no CNS illness; (2) 'Aseptic meningitis' with high fever and stiff neck; and (3) 'Encephalitis' (including meningoencephalitis and encephalomyelitis) with high fever, altered consciousness, and/or neurological dysfunction. The onset of illness may be sudden (<4 days after infection) and acute, leading rapidly to encephalitis, or insidious, progressing gradually through all three syndromes. Symptoms may resolve spontaneously during any stage of the illness, with full recovery. Acute illness may be followed by 'convalescent fatigue syndrome' in <50% of patients, with complaints of general weakness, depression, and the inability to concentrate that generally resolve within 3 years. Other sequelae include headache, disturbances in gait, and memory loss.

Pathogensis in SLEV follows a course similar to other flaviviruses in the JEV complex. The extent of illness usually is dependent upon viremia level and duration. Virus replication occurs within the lymphatic system soon after infection, and resulting viremias reflect the balance between virus production and release by the lymphatic system and clearance mediated by phagocytes of the liver and spleen. The probability of CNS involvement is directly correlated with the extent and duration of the viremia, although the mechanism of neuroinvasion remains unclear. Movement from peripheral to central nervous tissue most likely is by passive transport through neuron cytoplasm and then by transport across associated membranes after cell lysis. CNS pathology consists of necrosis of neurons and glia cells and inflammatory changes. Inflammatory changes typically are most important in slowly progressing or sublethal CNS disease and sequelae. Viral clearance is dependent upon a functional immune system and the rapid production of neutralizing antibody, which usually appears within 7 days after infection.

Epidemiology

Transmission of SLEV is complex and requires that the virus replicate in and avoid the immune responses of alternating insect and vertebrate hosts under temperatures ranging from below 0 °C in diapausing mosquitoes to more than 40 °C in febrile avian hosts. Annual transmission activity may be divided into overwintering, vernal and/or summer amplification, and autumnal subsidence periods.

Overwintering

Three possible mechanisms may explain the persistence of SLEV at temperate latitudes; however, few supportive field data are available.

Persistence in mosquito populations
Three mechanisms may explain SLEV overwintering within vector mosquito populations. First, low-level vertical passage of SLEV from infected females to F1 progeny has been demonstrated repeatedly in laboratory experiments. Although not detected for SLEV in nature, vertical transmission has been documented for other viruses in the JEV complex, including JEV and West Nile virus (WNV). Second, *C. p. pipiens* females destined for diapause have been shown to take small blood meals during late summer and early fall without ovarian development. Two isolations of SLEV made from diapausing *C. p. pipiens* females collected resting during winter in Maryland were considered to have been infected by this mechanism, although infection by vertical transmission also was possible. Third, *Culex p. quinquefasciatus* and *C. nigripalpus* do not enter reproductive diapause, remain reproductively active throughout winter at southern latitudes and, depending upon ambient temperature, could maintain SLEV by continued, infrequent transmission among resident birds. Experimentally infected, reproductively active *C. p. quinquefasciatus* females have been shown

to survive winter as gravid females and to then transmit SLEV to recipient birds throughout the following spring.

Persistence in vertebrate populations

SLEVs may also persist over winter within vertebrate host populations. Passeriform birds infrequently develop chronic infections that persist as long as a year following experimental infection. However, attempts to demonstrate natural relapse or to trigger relapse experimentally have not been successful. Flaviviruses, including SLEV, have also been isolated repeatedly from bats, and experimental infections in bats destined for hibernation have been maintained for 20 days at 10 °C. When returned to room temperature, SLEV was detected in the brown fat and at low levels in the blood. These data indicated that bats could function as an overwintering host. However, studies of mosquito host-selection patterns indicated that bats rarely, if ever, were fed upon by host-seeking mosquitoes.

Reintroduction of virus

An alternative hypothesis to local persistence involves annual or periodic reintroduction of virus into northern latitudes from southern refugia. Long-distance movement of SLEV has been indicated indirectly from genetic evidence as well as by the reappearance of SLEV after years of absence. Two possible hypotheses address reintroduction, but neither is well supported by field evidence. Many species of birds and some bats have long-distance annual migrations that could allow the transport of virus from foci active during winter in southern latitudes or south of the equator to receptive areas north of the equator during spring. These vertebrate migrations typically are very consistent in their summer and winter destinations, and this would allow the same or similar genetic strains to reappear each summer at the same locality. However, molecular genetic studies of North, Central, and South American isolates indicate that they are relatively distinct, thereby implying infrequent genetic exchange. In addition, migratory birds do not seem to be frequently involved in transmission because they infrequently are found positive for virus or antibody.

Amplification

Regardless of the persistence mechanism, summer enzootic amplification transmission in North America involves *Culex* mosquitoes and primarily birds in the orders Passeriformes and Columbiformes. Humans become infected tangentially to the primary cycle, do not develop viremias sufficient to infect mosquitoes, and are considered to be 'dead end' hosts (**Figure 3**). Transmission appears to be initiated after the *Culex* vectors resume blood-feeding and reproductive activity, and ambient temperatures warm sufficiently to allow the replication of virus in the mosquito host. Infection is acquired when a female *Culex* blood feeds on a viremic avian host. Virus imbibed within infectious blood meals taken in early in spring when ambient temperatures average <17 °C may lay dormant until warm conditions or changes in mosquito physiology stimulate replication. Under warm temperatures, virus replicates rapidly, disseminates within the mosquito during the ensuing extrinsic incubation period, and then may be transmitted by bite after the female oviposits and attempts to imbibe a subsequent blood

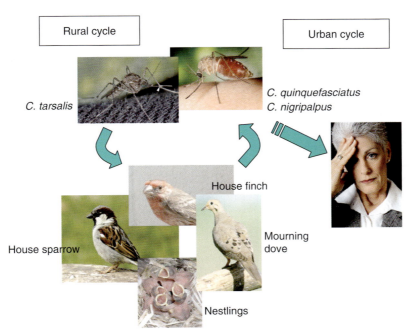

Figure 3 Amplification transmission cycle of St. Louis encephalitis virus in North America.

meal. The duration of the extrinsic incubation period is temperature dependent and requires >10 days and perhaps two mosquito gonotrophic cycles when temperatures average 22 °C. In contrast, the viremia response in susceptible avian hosts typically is of short duration, lasting 2–4 days.

Four distinct transmission cycles of SLEV are defined by differences in the biology of the primary vector mosquito species and their distribution, and include: (1) rural North America, west of the Mississippi River transmitted by *C. tarsalis*; (2) rural and urban central and eastern North America transmitted by members of the *C. pipiens* complex; (3) Florida, Caribbean, and parts of Central America transmitted by *C. nigripalpus*; and (4) urban and rural South America transmitted by *C. pipiens* complex and mosquitoes of other taxa.

Subsidence

Intensity of enzootic transmission and occurrence of new human cases always subsides rapidly during autumn. Cool evening temperatures slow the replication of SLEV within infected mosquito hosts, decreasing the efficiency of transmission and, concurrently, the combination of cool water temperature and shortening days during larval development initiates reproductive diapause (*C. tarsalis*, *C. p. pipiens*) or quiescence (*C. p. quinquefasciatus*, *C. nigripalpus*) in vector females emerging during fall. The fall mosquito population declines in abundance and divides into newly emerged females that do not routinely blood-feed and survive the winter, and remnants of the summer population that continue reproductive activity, but fail to survive winter. The critical day length that triggers the onset of diapause in *C. p. pipiens* may occur in late summer at northern latitudes, markedly shortening the SLEV transmission season. During warm days, however, females may become infected when taking partial blood meals from viremic birds, survive winter, and then transmit the virus after diapause is terminated by warm spring temperatures. *Culex p. quinquefasciatus* does not undergo diapause, so that reproductive activity may continue through winter, albeit at a rate slowed by winter temperatures. Populations exploiting underground storm water systems for resting or for larval development may be exposed to relatively warm temperatures throughout winter.

Risk Factors for Human Infection

Five factors have been associated with human risk of SLEV infection.

Residence

Clearly, place of residence markedly affects the risk of infection, with geographic regions in the southern USA having the greatest numbers of human cases and greatest incidence of disease (**Figure 2**). Based on experimental infection patterns in laboratory mice, virus strains from this geographical area also exhibit greater neurovirulence than strains from the western USA or South America. Because of mosquito abundance relative to humans and host-selection patterns, urban residents seem to be at greater risk for SLEV infection than rural residents. However, these conclusions may be confounded by protective immunity acquired early in life that may be greater among rural residents and by low apparent: inapparent case ratios that require a substantially large population to produce recognizable clusters of human cases.

Age

In the absence of acquired immunity, clinical illness and fatality rates, but not necessarily infection rates, increase dramatically with age. Infection seems to occur equally among different age classes as indicated by the increase in antibody as a function of age in endemic areas and by cohort seroconversion rates determined after epidemics in previously unexposed populations. For example, using data following the 1964 Houston (Texas) epidemic, seroprevalence rates remained similar among cohorts, whereas the case–incidence rates increased from 8.2 per 100 000 for the 0–9-year-old cohort to 13.5–27.6 for the 10–59-year-old cohorts and to 78.0 for the >60-year-old group; apparent to inapparent ratios decreased concomitantly from 1:806 to 1:490–1:239 and to 1:85, respectively. Case–fatality rates among 2288 cases reported to the CDC from 1971–83 increased from <6.7% for 0–64-year-old age classes to 9.5% for the 65–74-year-old class to 18% for the >75-year-old class.

Occupation

In the West, where SLE historically was a rural disease, infection risk was greatest among male agricultural workers who frequently lived in suboptimum housing and worked at night. However, infection patterns during recent urban outbreaks indicated that attack rates were highest among elderly women. These data indicated that there may be differences in risk related to vector species, with elderly women infected most readily during urban outbreaks associated with the *C. pipiens* complex, and men working outdoors at greatest risk during rural outbreaks associated with *C. tarsalis*.

Socioeconomic status

Historically, socioeconomic status has been related closely to the distribution of cases during urban epidemics. Homes and municipal drainage systems frequently were not well maintained in low-income neighborhoods, and this was related to the distribution of human cases, but not necessarily the occurrence of virus within the enzootic transmission cycle. TV and air conditioning ownership that brought people indoors during the evening *Culex* host-seeking period was found to reduce risk.

Weather

Climate variability affects temperature and precipitation patterns, mosquito abundance and survival, and therefore SLEV transmission. Annual temperature changes based on the El Niño/southern oscillation in the Pacific alter precipitation and temperature patterns over the Americas and cycle with varying intensity at 3–5 year intervals. These cycles alter storm tracks that affect mosquito and avian abundance, the intensity and frequency of rainfall events, and groundwater depth, all related to SLEV risk. Above-normal temperatures have been especially necessary for northern latitude SLEV epidemics, because elevated temperatures are required for effective SLEV replication within the mosquito host.

Prevention and Control

Effective vector control remains the only approach available to suppress summer virus amplification and prevent human infections. Best results are achieved using an integrated management approach that focuses on mosquito vector population suppression through habitat inspection and larviciding. Failure of larval management can be followed by emergency adult control focusing on reducing the force of transmission and preventing human infection. Protection of the human population by vaccination does not seem cost-effective or prudent, because there is no human-to-human transmission, few human infections produce disease, and infection rates remain relatively low, even during epidemics. However, if regional infection rates were to become high, thereby placing selected cohorts at high risk for disease, then selective vaccination may be warranted. There currently is no approved commercial vaccine for SLEV, although vaccination against other flaviviruses such as JEV may impart some protection. Control of avian hosts such as house sparrows and pigeons in urban situations could be done, but this approach is not generally acceptable to the public. Notification of the public of infection risk through the media and the wide scale use of personal protection through changes in behavior (staying indoors after sunset) and/or repellent application were credited with reducing the number of infections during the 1990 epidemic in Florida.

See also: Japanese Encephalitis Virus; Tick-Borne Encephalitis Viruses.

Further Reading

Day JF (2001) Predicting St. Louis encephalitis virus epidemics: Lessons from recent, and not so recent, outbreaks. *Annual Review of Entomology* 46: 111–138.

Kramer LD and Chandler LJ (2001) Phylogenetic analysis of the envelope gene of St. Louis encephalitis virus. *Archives of Virology* 146: 2341–2355.

Monath TP (1980) In: *St. Louis Encephalitis*, 680pp. Washington, DC: American Public Health Association.

Monath TP and Tsai TF (1987) St. Louis encephalitis: Lessons from the last decade. *American Journal of Tropical Medicine and Hygiene* 37: 40s–59s.

Reeves WC, Asman SM, Hardy JL, Milby MM, and Reisen WK (eds.) (1990) In: *Epidemiology and Control of Mosquito-Borne Arboviruses in California, 1943–1987*, 508pp. Sacramento, CA: California Mosquito and Vector Control Association.

Reisen WK (2003) Epidemiology of St. Louis encephalitis virus. In: Chambers TJ and Monath TP (eds.) *The Flaviviruses: Detection, Diagnosis and Vaccine Development*, pp. 139–183. San Diego, CA: Elsevier.

Sweetpotato Viruses

J Kreuze and S Fuentes, International Potato Center (CIP), Lima, Peru

© 2008 Elsevier Ltd. All rights reserved.

Glossary

Cultivar decline A vegetatively propagated cultivar suffering from reduced vigor as a result of a chronic (but often symptomless) disease.

Differential hosts Special species of plants varying in susceptibility to a given disease agent, such that their distinctive symptoms facilitate a presumptive identification of the causal agent.

Indexing Any procedure for demonstrating the presence of a pathogen(s) in susceptible plants. The virus indexing combines information on viruses with methodologies for their detection to assure effective safe movement of sweetpotato germplasm.

Introduction

Sweetpotato

Sweetpotato (*Ipomoea batatas* (L.) Lam.) is a dicotyledonous, perennial plant, producing edible tuberous roots. It belongs to the family Convolvulaceae, the morning glory.

Sweetpotato Viruses

This family contains about 55 genera. The genus *Ipomoea* is thought to contain over 500 species with ploidy levels ranging from $2x$ to $6x$. Sweetpotato is the only *Ipomoea* species of economic importance as a food crop, and has both $4x$ and $6x$ forms ($2n = 4x = 60$ or $2n = 6x = 90$). Thousands of cultivars of sweetpotato are grown throughout the tropics and subtropics. With an annual production of more than 133 million tons globally, sweetpotato currently ranks as the seventh most important food crop on a fresh-weight basis in the world, and fifth in developing countries after rice, wheat, maize, and cassava. The production is concentrated in East Asia, the Caribbean, and tropical Africa, with the bulk of the crop (88%) being grown in China (**Figure 1**). Sweetpotato performs well in relatively poor soils, with few inputs, and has a short growing period. Among the major starch staple crops, it has the largest rates of production per unit area per unit time: in some areas up to three harvests per year can be achieved. Sweetpotato roots are rich in vitamin C and essential mineral salts. Due to the high beta-carotene content of yellow and orange-fleshed storage roots, they are being promoted to alleviate vitamin A deficiency in East Africa and Eastern India.

Viruses of Sweetpotato

Until recently, viruses of sweetpotato have been relatively poorly studied as compared to viruses of other crops. Still, more than 20 different viruses have been described infecting sweetpotato worldwide, but only 15 of these are currently recognized by the International Committee on Taxonomy of Viruses (ICTV; **Table 1**). This number, however, will most likely increase by additional surveys (**Figure 2**) and by indexing germplasm collection (**Figure 3**).

Vegetative propagation, usually by taking cuttings from a previous crop, increases the risk of a buildup of viruses. The importance of virus diseases and their buildup in farmers' planting materials has been shown convincingly in China, where sweetpotato cultivars planted using pathogen-tested materials yielded 30–40% more, on average, than those grown from farm-derived planting materials. Next to weevils, virus diseases form the most important biotic production constraint in sweetpotato. Most sweetpotato-infecting viruses, however, show only mild or no symptoms when in single infection and the damages caused by sweetpotato viruses are mostly through synergistic mixed infections. Viruses of the families *Potyviridae* and *Geminiviridae* as well as sweetpotato chlorotic stunt virus (SPCSV) are particularly significant in relation to sweetpotato cultivar decline.

Due to low virus titers and absence of symptoms from single infections in sweetpotato by most viruses, grafting and in some cases sap transmission onto indicator plants is often required to increase virus concentration and detect viruses reliably (**Figure 3**). Commonly used indicator plants are *Ipomoea setosa*, *I. nil*, *I. purpurea*, *I. aquatica*, and in some cases *Nicotiana benthamiana* and *N. clevelandii*.

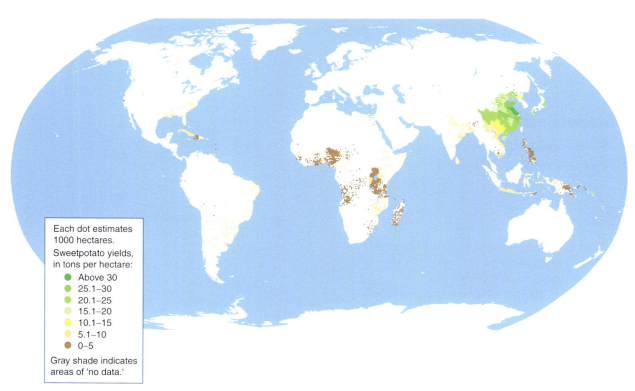

Figure 1 Sweetpotato cultivation: production areas and average yields.

Table 1 List of viruses that have been reported infecting sweetpotato

Genus/family	Species/virus	Reported distribution[a]	Vector	Detection methods[b]	Strains/ serotypes	Component of synergistic disease[c]	Sequence from GenBank[d]
Begomovirus	Ipomoea yellow vein virus (IYVV)	Spain, Italy	Whiteflies?	PCR	None reported		AJ132548
Begomovirus	Sweetpotato leaf curl Georgia virus (SPLCGV)	USA, Puerto Rico	Whiteflies	NA hybridization, PCR	None reported		AF326775
Begomovirus	Sweetpotato leaf curl virus (SPLCV)	Far East, USA, China, Taiwan, Japan, Korea, Europe, Africa?, Peru	Whiteflies	NA hybridization, PCR, qPCR	None reported	Unnamed (with SPFMV)	AF104036
Carlavirus	Sweetpotato chlorotic fleck virus (SPCFV)	Africa, South America, Asia, Cuba, Panama, China, Taiwan, Japan, Korea, New Zealand	Unknown	Serology	East Africa type, Asia type	Camote kulot, unnamed (with SPCSV)	NC_006550
Carlavirus?	C-6 virus	USA, Peru, Cuba, Dom. Rep., Indonesia, Philippines, P. Rico, Egypt, Uganda?, Kenya, South Africa, New Zealand	Unknown	Serology	None reported	Camote kulot, unnamed (with SPCSV)	None available
Caulimovirus?	Sweetpotato caulimo-like virus (SPCaLV)	South Pacific Region, Madeira, China, Egypt, Puerto Rico, Uganda, Kenya?, Nigeria	Unknown	Serology, PCR	None reported		None available
Crinivirus	Sweetpotato chlorotic stunt virus (SPCSV)	Worldwide	Whiteflies	Serology, NA hybridization, RT-PCR, qRT-PCR	EA, WA	SPVD, SPCD, Camote Kulot, SPSMD	RNA1: AJ428554, RNA2: AJ428555
Cucumovirus	Cucumber mosaic virus (CMV)	Israel, Egypt, Kenya, Uganda?, Japan, South Africa, New Zealand	Aphids	Serology	None reported	Unnamed (with SPCSV)	RNA1: D00356, RNA2: D00355, RNA3: D10538
Geminiviridae	Ipomoea crinkle leaf curl virus (ICLCV)	Israel	Whiteflies	NA hybridization	None reported		None available
Ilarvirus	Tobacco streak virus (TSV)	Guatemala	Thrips	Serology	None reported		RNA1: U80934, RNA2: U75538, RNA3: X00435
Ipomovirus	Sweetpotato mild mottle virus (SPMMV)	Africa, Indonesia, China, Philippines, Papua New Guinea, India, Egypt, New Zealand	Whiteflies?	Serology, RT-PCR	Different strains	Camote Kulot, SPSMD	Z73124
Ipomovirus	Sweetpotato yellow dwarf virus (SPYDV)	Taiwan, Far East	Whiteflies	Serology	None reported		None available
Luteoviridae	Sweetpotato leaf speckling virus (SPLSV)	Peru, Cuba	Aphids	NA hybridization, RT-PCR	None reported		DQ655700

Continued

662 Sweetpotato Viruses

Table 1 Continued

Genus/family	Species/virus	Reported distribution[a]	Vector	Detection methods[b]	Strains/serotypes	Component of synergistic disease[c]	Sequence from GenBank[d]
Nepovirus	Sweetpotato ringspot virus (SPRSV)	Papua New Guinea, Kenya?	Unknown	Serology	None reported		None available
Potyvirus	Sweetpotato feathery mottle virus (SPFMV)	Worldwide	Aphids	Serology, NA hybridization, RT-PCR, qRT-PCR	EA, RC, O, C	SPVD, SPCD, Camote Kulot	D86371
Potyvirus	Sweetpotato latent virus (SPLV)	Africa, Taiwan, China, Japan, India, Philippines, Indonesia, Egypt	Aphids	Serology, NA hybridization	None reported	Camote Kulot	X84011*, X84012*
Potyvirus	Sweetpotato mild speckling virus (SPMSV)	Argentina, Peru, Indonesia, Philippines, China, Egypt, Uganda?, Kenya?, South Africa, Nigeria, New Zealand	Aphids	Serology	None reported	SPCD, Camote Kulot	U61228*
Potyvirus	Sweetpotato vein mosaic virus (SPVMV)	Argentina	Aphids	No available	None reported		None available
Potyvirus	Sweet potato virus G (SPVG)	China, Japan, USA, Egypt, Ethiopia, Nigeria, Barbados, Peru	Aphids	Serology, RT-PCR, qRT-PCR	None reported		Z83314*, AJ515380*
Potyvirus	Sweetpotato virus 2	Taiwan, USA, China, South Africa, Portugal, Australia, Barbados	Aphids	Serology, RT-PCR, qRT-PCR	Geographical/genetical lineages	Unnamed (with SPCSV)	AY178992*, AY232437*
Tobamovirus	Tobacco mosaic virus (TMV)	USA	None	Serology	None reported		X02144, AJ132845

[a]?, signifies unconfirmed.
[b]NA, nucleic acid; RT-PCR, reverse transcription PCR; qPCR, quantative real-time PCR; qRT-PCR, quantative real-time reverse transcription PCR.
[c]SPVD, sweetpotato virus disease (Africa, Peru); SPCD, sweetpotato chlorotic dwarf (Argentina); SPSMD, sweetpotato severe mosaic disease (Africa); Camote kulot (Philippines).
[d]Complete genome sequence except those marked with (*).

Figure 2 Sweetpotato plants showing symptoms of vein clearing (a and b), leaf curling (c), mosaic (d), and chlorosis, stunting, and leaf deformation. Some times the vein clearing is surrounded by purple pigmentation.

Potyviruses

Several potyviruses infect sweetpotato and usually only cause transient, mild, or no symptoms when infecting sweetpotato by themselves. The most widespread of these, and the one studied in most detail is sweetpotato feathery mottle virus (SPFMV, genus *Potyvirus*, family *Potyviridae*) that occurs wherever sweetpotato is grown. In many cases infection of sweetpotato plants with SPFMV causes mild or no symptoms, although certain strains can cause qualitative damage due to internal cork or cracking of the tuberous roots. However, quantitative losses due to reduced plant vigor associated with chronic infection with SPFMV have also been experienced. Yet, it is as a component of complex virus diseases that SPFMV probably causes the greatest damage.

Sweetpotato Feathery Mottle Virus

SPFMV has flexuous filamentous particles measuring 830–850 nm in length. They contain a single-stranded, positive-sense RNA genome of about 10.6 kb, which is

Figure 3 *Ipomoea setosa* graft-inoculated with scions from sweetpotato plants collected in farmer's fields in Peru. Infected with sweetpotato feathery mottle virus (a–d) and sweetpotato virus G (e). Faint chlorotic spots (f) induced by an unknown virus. The causal virus tested negatively to available antisera in NCM-ELISA.

larger than the average size (9.7 kb) of a potyvirus genome. The coat protein (CP) of SPFMV is also exceptionally large (38 kDa) as compared to other potyviruses, which is largely due to the insertion of a contiguous sequence at the 5′-end of the CP cistron. SPFMV is transmitted by several aphid species (i.e., *Aphis gossypii, A. craccivora, Lipaphis erysimi, Myzus persicae*) in a nonpersistent manner. However, these aphids rarely colonize sweetpotato under field conditions and therefore itinerant alate aphids are probably the most efficient vectors of SPFMV. The experimental host range of SPFMV is narrow and mostly limited to plants of the family Convolvulaceae, and especially to the genus *Ipomoea*, although some strains have been reported to infect *N. benthamiana* and *Chenopodium* spp. Symptoms, host range, and serology have been used to group SPFMV isolates into two strains, the common strain (C) and the more severe russet crack (RC) strain. However, based on phylogenetic analysis of 3′ nt sequences of an extensive number of isolates it is now clear that SPFMV can be distinguished in four phylogenetic lineages RC, O (ordinary), EA (East Africa), and C (**Figure 4**). Based on molecular data the C strain is rather distantly related to the remaining strains and may in the future be classified as a separate virus.

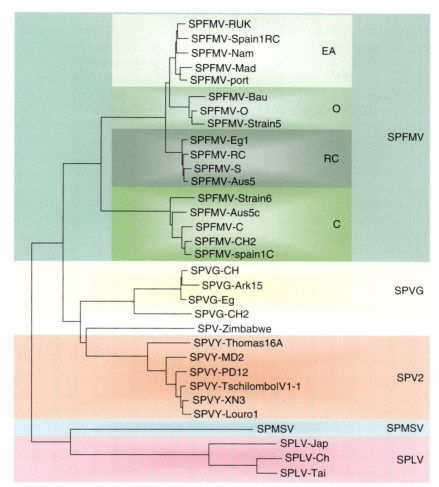

Figure 4 Phylogeny of representative isolates of sweetpotato-infecting potyviruses based on complete CP sequences. Clades corresponding to described virus species are shaded in different colors. The four recognized strain groups of SPFMV are also indicated and shaded in different colors of green. One sequence, originally described as SPFMV from sweetpotato in Zimbabwe, is distinct from all other named species.

Other Potyviruses

Several other potyviruses are known to infect sweetpotato, some of which are distributed widely. They have in common that they are mostly symptomless and in low titers when infecting sweetpotato by themselves, but can be distinguished by their symptoms induced in indicator plants such as *I. nil* and *I. setosa* (**Figure 3**), or by symptoms induced through synergism with SPCSV. With the exception of sweetpotato vein mosaic virus (SPVMV) sequence information is available for these viruses and their phylogenetic relationships are depicted in **Figure 4**.

Sweetpotato virus G (SPVG) was first reported from China, but is now known to occur also in the Americas and Africa. Another virus, originally identified as sweetpotato virus 2 in Taiwan, was recently further characterized simultaneously by two groups suggesting the species names sweetpotato virus Y and ipomoea vein mosaic virus, respectively, and has been reported from Taiwan, China, USA, South Africa, Australia, and Barbados. The name SPV2 is currently being considered for this virus by the ICTV. Sweetpotato latent virus (SPLV) was also first reported from China, but is now known to occur in most major sweetpotato-growing areas in Asia. The two remaining potyviruses reported were first described from Argentina and are SPVMV and sweetpotato mild speckling virus (SPMSV). Whereas SPMSV has also been detected in Peru and Indonesia, neither antibodies nor sequence information are available for SPVMV, and the virus has not been reported from elsewhere. The number of sweetpotato-infecting potyvirus species will certainly still increase through future surveys, for example, a potyvirus distinct from any of the above mentioned was isolated from infected sweetpotato in Zimbabwe and remains to be characterized in more detail (**Figure 4**).

Sweetpotato mild mottle virus (SPMMV, genus *Ipomovirus*) is transmitted by whiteflies in a nonpersistent manner. It should be noted, however, that since the initial report, it has not been possible to confirm independently the whitefly transmissibility of SPMMV. A study of the variability of SPMMV in Uganda using 3′ nt sequences showed the virus consisted of a population of distinct sequence variants (>85.9% nt identity in the CP), which however did not show any particular clustering. A distinct feature of SPMMV as compared to other sweetpotato-infecting viruses is its exceptionally broad host range including species in 14 families. SPMMV has serologically been detected throughout Africa, Indonesia, China, Philippines, Papua New Guinea, India, Egypt, and New Zealand. Another whitefly-transmitted potyvirus with properties distinct from SPMMV was described in Taiwan and named sweetpotato yellow dwarf virus (SPYDV). The relationship between these two viruses is however unclear as no sequence information is available for SPYDV.

Begomoviruses

Sweetpotato leaf curl diseases typical of geminivirus infection (**Figure 2(c)**) have been reported from Peru, Japan, Taiwan, Korea, China, Puerto Rico, Costa Rica, Spain, United States, Africa, and other countries. The host range of the *Ipomoea*-infecting begomoviruses (family *Geminiviridae*) is narrow and mostly restricted to species in the family Convolvulaceae (especially to the genus *Ipomoea*), but *N. benthamiana* (Solanaceae) can also be infected. They cause leaf curl on some hosts, and yellow vein or leaf distortion and chlorosis symptoms on others. Some sweetpotato cultivars are symptomless. As they induce mild, transient symptoms in the standard virus indicator, *I. setosa*, a universally reliable biological indicator is lacking. Some *Ipomoea* species (*I. aquatica*, *I. cordatotriloba*, *I. purpurea*) can however be used as differential hosts to distinguish *Ipomoea*-infecting begomoviruses.

Only three sweetpotato-infecting begomovirus species have been officially recognized by the ICTV (**Table 1**): *Sweetpotato leaf curl virus* (SPLCV), *Sweetpotato leaf curl Georgia virus* (SPLCGV), and *Ipomoea yellow vein virus* (IYVV). DNA B components have not been identified for any of these viruses indicating they are monopartite begomoviruses. Sequences of AC1 gene fragments from isolates obtained from infected *Ipomoea* spp. from USA, China, Taiwan, Korea, Puerto Rico, Spain, and Italy all fall into two groups corresponding to SPLCV/IYVV and SPLCGV. A third phylogenetically distinct group of begomoviruses was identified from infected sweetpotato in Spain, representing a putative fourth species. Similarly the sequence of a SPLCV isolate from China shows <83% identity to other published sequences indicating yet another virus species. Another virus, proposed to be named ipomoea crinkle leaf curl virus (ICLCV) and reported from Israel, differed in host range from SPLCV but its exact relationship to other identified viruses remains unclear.

From a taxonomic point of view, the *Ipomoea*-infecting begomoviruses are a curiosity; phylogenetic analysis of IYVV and several strains of SPLCV and SPLCGV revealed that these viruses form a separate unique cluster within the genus *Begomovirus*, dissimilar to both the New World and Old World begomoviruses. Additionally, the relatively poor transmission rates of the *Ipomoea*-infecting begomovirus by *Bemisia tabaci* may be a reflection of the low CP amino acid sequence identity (46%) between them and the other begomoviruses.

Sweetpotato Leaf Curl Virus

The virus has geminate particles of *c*. 18 × 30 nm with a genome of 2828 nt. Its genomic DNA and organization is similar to that of monopartite begomoviruses. SPLCV virus was first reported from Japan and Taiwan but now it is known to occur in several countries on different continents. SPLCV can cause up to 30% reductions in yield of storage roots. Various *Ipomoea* species are susceptible to SPLCV, such as *I. purpurea* causing leaf curl and stunt, *I. aquatica* causing yellow vein symptoms, *I. nil*, *I. setosa*, and *N. benthamiana* causing leaf curl symptoms. Co-infections of SPFMV and SPLCV in *I. setosa* and *I. nil* induce severe leaf distortion, general chlorosis, and stunting.

Ipomoea Yellow Vein Virus

IYVV was first found in Spain infecting *I. indica* plants showing yellow vein symptoms, but has since then been found infecting cultivated sweetpotato as well. Its properties, including typical geminivirus symptoms, detection of geminate particles by electron microscopy, and complete nucleotide sequence confirmed its begomovirus nature. Based on nucleotide sequence similarity of >89% over the entire genome, IYVV should be considered a strain of SPLCV, and will probably be revised accordingly in the future. However, contrary to SPLCV which is transmitted by *B. tabaci* biotype B, IYVV-[Spain] was not transmitted by biotypes Q, S, or B of *B. tabaci*.

Sweetpotato Leaf Curl Georgia Virus

SPLCGV, previously called ipomoea leaf curl virus, has a genome of 2773 nt with an organization typical of other Old World monopartite begomoviruses. The complete nucleotide sequence of SPLCGV is 82–85% identical to those of SPLCV and IYVV, which is below the species threshold of 89% nucleotide sequence identity for begomoviruses. Like SPLCV, SPLCGV is transmitted by *B. tabaci* biotype

B and induces similar symptoms in most *Ipomoea* species (various degrees of leaf curling). The means to distinguish SPLCGV from SPLCV are the use of *I. aquatica* and *I. cordatrotiloba* as differential hosts and a combination of polymerase chain reaction (PCR) and restriction enzyme digestion.

Sweetpotato Chlorotic Stunt Virus

Due to its ability to mediate severe synergistic viral diseases with several other sweetpotato-infecting viruses including potyviruses, cucumoviruses, and carlaviruses, SPCSV is probably the most devastating virus of sweetpotato worldwide. In single infection, this virus causes no symptoms at all or usually only mild stunting combined with slight yellowing or purpling of older leaves, symptoms which are easily confused with nutritional deficiencies.

SPCSV belongs to the genus *Crinivirus* of the family *Closteroviridae*. The particles of SPCSV are 850–950 nm in length and 12 nm in diameter. The size of the major coat protein is 33 kDa, which is similar to other criniviruses. SPCSV is transmitted by whiteflies (e.g., *B. tabaci*, *B. afer*, and *Trialeurodes abutilonea*) in a semipersistent, noncirculative manner, and is not mechanically transmissible. Similar to most other sweetpotato-infecting viruses, the host range of SPCSV is limited mainly to the family Convolvulaceae and the genus *Ipomoea*, although *Nicotiana* spp. and *Amaranthus palmeri* are reportedly susceptible. SPCSV has also been detected in the ornamental species Lisianthus (*Eustoma grandiflorum*). SPCSV can be serologically divided into two major serotypes. One of the serotypes (EA for East Africa) was first identified in East Africa, and also occurs in Peru, while the other serotype (WA for West Africa) was first identified in West Africa and occurs additionally in the Americas, and the Mediterranean, but not in East Africa. The two serotypes correlate to two genetically distantly related strain groups based on coat protein as well as Hsp70h gene similarities.

The genome of an SPCSV isolate from Uganda has been entirely sequenced and consists of two RNA segments of 9407 and 8223 nt. The SPCSV genome encodes two unique proteins, p22 and RNase3, not known to occur in any other RNA viruses. mRNAs corresponding to these proteins are transcribed early during infection and they are cooperatively able to suppress RNA silencing, a process that requires the RNA binding and double-stranded RNase activity of the RNase3 protein.

Other Viruses

Some viruses recognized by ICTV that affect other crops (i.e., cucumber mosaic virus, tobacco mosaic virus, and tobacco streak virus), are sporadically found infecting sweetpotato. Information on these viruses is extensively available. Other sweetpotato-specific viruses have not yet being assigned to a genus (i.e., sweetpotato leaf speckling virus – SPLSV) or recognized as a species (i.e., sweetpotato chlorotic fleck virus – SPCFV; C-6 virus; sweetpotato caulimo-like virus – SPCaLV) by the ICTV (**Table 1**).

Sweetpotato Leaf Speckling Virus

SPLSV has isometric particles *c*. 30 nm in diameter. The virus is transmitted by *Macrosiphon euphorbiae* in a persistent manner. It has a restricted geographical distribution. The sequence of the CP and 17K encoding region has been determined and is characteristic of luteo- and poleroviruses sequences, with highest similarity to potato leafroll virus (PLRV). Although these two viruses can be detected with heterologous CP probes, they are not serologically related.

Sweetpotato Chlorotic Fleck Virus

SPCFV has filamentous particles measuring about 800 nm in length consisting of its genome encapsidated by polypeptide subunits of 33.5 kDa. The virus appears to have a wide geographic distribution in sweetpotato crops in South America, Africa, and Asia. SPCFV has a narrow host range in the families Convolvulaceae and Chenopodiaceae, but some strains/isolates infect *N. occidentalis*. SPCFV is mostly symptomless in its natural host; hence, it was also referred to as sweetpotato symptomless virus in Japan.

Sequence analysis of the entire genome of SPCFV (9104 nt) provides unambiguous evidence for the assignment of SPCFV as a distinct species in the genus *Carlavirus*. The RNA of SPCFV is larger than that of the other carlaviruses due to its considerably larger replicase (238 vs. 200–223 kDa) and a long untranslated region of 236 nt between ORF4 and ORF5. Phylogenetic analysis of CP sequences suggests that there is some geographically associated variation among SPCFV isolates.

Complex Virus Diseases of Sweetpotato

Multiple virus infections are common in sweetpotato and synergistic interactions are often involved. The most common of these disease complexes, known as sweetpotato virus disease (SPVD), is caused by simultaneous infection with SPFMV and SPCSV (**Figures 5** and **6**). This disease is characterized by chlorosis, small, deformed leaves, and severe stunting, and can reduce yields of infected plants by up to 99%. Despite the apparent broad meaning of the name SPVD, the symptoms are so characteristic that the name has become restricted to the disease

Figure 5 Sweetpotato plants affected by viruses in Peru. Sweetpotato field with a large number of plants affected by sweetpotato virus disease complex (a) and a close up of an affected plant showing stunting, mosaic, and leaf deformation (b).

with these symptoms and caused by these viruses. SPVD is the most serious disease of sweetpotato in Africa and Peru, and may be the most important virus disease of sweetpotato globally.

Other viral disease complexes have also been described, which invariably seem to involve SPCSV. In Israel and Egypt cucumber mosaic virus (CMV, genus *Cucumovirus*, family *Bromoviridae*) is found infecting sweetpotato together with SPCSV and usually also SPFMV, producing symptoms similar to SPVD and causing up to 80% reduction in yield. It was shown that CMV could only infect sweetpotato if the plants were first infected with SPCSV. Interestingly, this seems not to be the case for CMV in Egypt, where it is found infecting sweetpotato with or without SPCSV. In Argentina, a disease locally known as chlorotic dwarf is caused by infection with SPCSV and SPFMV and/or SPMSV, and is the most important disease of sweetpotato in the country. Once again, the symptoms resemble those of SPVD and are most severe when all three viruses infect sweetpotato simultaneously. In the Philippines SPCSV together with several other viruses causes a disease locally known as Camote Kulot.

In all the mentioned disease complexes, infection with each virus separately causes only mild or no symptoms in sweetpotato. They are thus caused by a synergistic interaction between the viruses. As both SPFMV and SPCSV are involved in all these diseases, the variation in the strains of these viruses should be important factors determining the disease severity.

Experimentally SPCSV can induce synergism with all tested potyviruses (including SPMMV), CMV as well as carlaviruses, and is always associated with an increase in the titers of the co-infecting virus and reduced yield of storage roots. For instance, the dual infection of SPCSV and SPMMV has been named 'sweetpotato severe mosaic disease (SPSMD)'. Yet there are reports indicating that strains of SPFMV and SPV2 may differ in their ability to cause synergism with SPCSV. Triple infections involving SPCSV, SPFMV, and an additional virus are even more severe, leading to further increase of the SPFMV

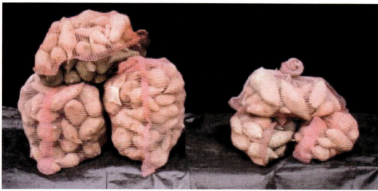

Figure 6 Effect of viruses on sweetpotato plants (top) and yield of storage roots (bottom). Observe yield reduction on sweetpotato variety Costanero (30 plants each treatment).

titers, whereas the titers of the third virus may either increase or decrease. It is thought the synergistic effects of SPCSV on other viruses may be due to interference with RNA silencing, because they are associated with substantially increased accumulation of co-infecting viruses.

See also: Carlavirus; Citrus Tristeza Virus; Ilarvirus; Luteoviruses; Plant Antiviral Defense: Gene Silencing Pathway; Plant Resistance to Viruses: Geminiviruses; Potyviruses; Viral Suppressors of Gene Silencing.

Further Reading

Ames T (ed.) (2002) *Acta Horticulturae. 583, ISHS: Proceeding of the First International Conference on Sweetpotato. Food and Health for the Future.* Brugge, Belgium: ISHS.

Clark CA and Moyer JW (1988) *Compendium of Sweetpotato Diseases*, 74pp. Minnesota: The American Phytopathological Society.

Kreuze J (2002) *Acta Universitatis agriculturae Sueciae. Agraria, vol. 335: Molecular studies on the Sweet Potato Virus Disease and Its Two Causal Agents.* Doctoral Dissertation, Dept. of Plant Biology, SLU (ISBN 91–576–6180–4; ISSN 1401–6249).

Loebenstein G, Fuentes S, Cohen J, and Salazar LF (2003) Sweet potato. In: Loebenstein G and Thottappilly G (eds.) *Virus and Virus-Like Diseases of Major Crops in Developing Countries*, pp. 223–248. Dordrecht, The Netherlands: Kluwer Academic Publishers.

Nakasawa Y and Ishiguro K (eds.) (2000) *Proceedings of International Workshop on Sweetpotato Cultivar Decline Study.* Kyushu National Agricultural Experimental Station (KNAES), 8–9 September 2000, Miyakonojo, Japan. http://www.knaes.affrc.go.jp/acre/Old/Workshop/WS2000/workshop/index_workshop.html (accessed July 2007).

Tairo F, Mukasa SB, Jones RAC, et al. (2005) Unravelling the genetic diversity of the three main viruses involved in sweet potato virus disease (SPVD), and its practical implications. *Molecular Plant Pathology* 6: 199–211.

Set
978-0-12-373935-3

Volume 4 of 5
978-0-12-374115-8